AF616340

Encyclopedia of Applied Physics

SPONSORS

AMERICAN INSTITUTE OF PHYSICS
DEUTSCHE PHYSIKALISCHE GESELLSCHAFT
JAPAN SOCIETY OF APPLIED PHYSICS
PHYSICAL SOCIETY OF JAPAN

DEUTSCHE
PHYSIKALISCHE
GESELLSCHAFT

JAPAN SOCIETY
OF APPLIED PHYSICS

PHYSICAL SOCIETY
OF JAPAN

ENCYCLOPEDIA OF APPLIED PHYSICS

VOLUME 13
Order-Disorder Transitions to
Physics and Engineering Organizations

Edited by
GEORGE L. TRIGG

Associate Editors
EDUARDO S. VERA
WALTER GREULICH

Managing Editor
EDMUND H. IMMERGUT

Assistant Managing Editor
CHRISTOPHER THOMAS MORAN

Editorial Assistant
HILARY GRANSON

George L. Trigg
275 Beaver Dam Road
Brookhaven, New York 11719

Walter Greulich
Ziegeleiweg 30
D-69488 Birkenau
Federal Republic of Germany

Eduardo S. Vera
Science and Technology Information Center (ICT)
University of Chile
Beaucheff 850
Santiago, Chile

Edmund H. Immergut
Christopher Thomas Moran
Hilary Granson
2 Sidney Place
Brooklyn, New York 11201

Library of Congress Cataloging-in-Publication Data

Encyclopedia of applied physics.
(Revised for Vol. 13)

"Sponsors, American Institute of Physics . . . [et al]."
Includes bibliographical references and index.
1. Physics—Encyclopedias. 2. Engineering—Encyclopedias. I. Trigg, George L. II. Vera, Eduardo S. III. Greulich, Walter. IV. American Institute of Physics.
QC5.E543 1991 530'.05 91-8738
ISBN 1-56081-072-6 (VCH Publishers)

British Library Cataloguing in Publication Data

Encyclopedia of applied physics.
Vol. 13
I. Trigg, George L. (George Lockwood) II. Vera, Eduardo S. III. Greulich, Walter
621
ISBN 1-56081-058-0 (set)
ISBN 1-56081-072-6 (vol. 13)

Printed in the United States of America.

ISBN 3-527-28135-5 (Volume 13) VCH Verlagsgesellschaft mbH
ISBN 3-527-26841-3 (set) VCH Verlagsgesellschaft mbH

Printing History:
10 9 8 7 6 5 4 3 2 1

Published jointly by:

VCH Publishers, Inc.
220 East 23rd Street
New York, NY 10010

VCH Verlagsgesellschaft mbH
P.O. Box 10 11 61
69451 Weinheim
Federal Republic of Germany

VCH Publishers (UK) Ltd.
8 Wellington Court
Cambridge CB1 1HZ
United Kingdom

ADVISORY BOARD

EDITORIAL CONSULTANTS

MAIN ENTRIES

08-D	Order-Disorder Transitions	1
05-A	Oscillators	17
02-A	Oscilloscopes and Oscillographs	37
17-D	Osmosis	59
16-C	Packaging Technology	73
20-D	Paleomagnetism	89
13-D	Paramagnetism	101
03-A	Particle Colliders	123
02-A	Particle Detectors, Tracking	141
12-D	Particle Impact Phenomena	175
03-E	Particle Physics	193
03-A	Particle Sources—Ion	213
03-D	Particles, Elementary	223
01-E	Patents and Intellectual Property	233
01-C	Perturbation Methods	245
18-B	Petroleum	269
11-D	Phase Equilibria	297
11-D	Phase Separation	323
11-D	Phase Transitions: Renormalization and Scaling	343
11-D	Phase Transitions, Structural	373
01-E	Philosophy of Physics	389
01-E	Philosophy of Technology	417
11-D	Phonons in Crystal Lattices	439
14-A	Photodetector Devices	459
12-D	Photoemission and Photoelectron Spectra	477
14-D	Photoluminescence	497
19-D	Photosynthesis	513
14-A	Photovoltaic Devices	533
01-F	Physics and Engineering Organizations	559
	Contents of Previous Volumes	589

The subject matter in the *Encyclopedia of Applied Physics* is presented in approximately 500 individual articles, arranged alphabetically. The topics can be classified into 20 sections, similar to the AIP Physics and Astronomy Classification Scheme (PACS):

01	General Aspects: Mathematical, Computational, and Information Techniques
02	Measurement Science, General Devices and/or Methods
03	Nuclear and Elementary Particle Physics
04	Atomic and Molecular Physics
05	Electricity and Magnetism
06	Optics (classical and quantum)
07	Acoustics
08	Thermodynamics and Properties of Gases
09	Fluids and Plasma Physics
10	Condensed Matter A: Structure and Mechanical Properties
11	Condensed Matter B: Thermal, Acoustic, and Quantum Properties
12	Condensed Matter C: Electronic Properties
13	Condensed Matter D: Magnetic Properties
14	Condensed Matter E: Dielectrical and Optical Properties
15	Condensed Matter F: Surfaces and Interfaces
16	Materials Science
17	Physical Chemistry
18	Energy Research and Environmental Physics
19	Biophysics and Medical Physics
20	Geophysics, Meteorology, Space Physics, and Aeronautics

Each article has been assigned a code number consisting of two digits which denotes the section, and a letter which gives the type of article. There are six types: A = Devices, Equipment; B = Materials; C = Methods, Processes; D = Phenomena, Effects; E = Scientific or Technological Fields; F = Institutions, Companies, Societies and other organizations.

CONTRIBUTORS

A. Acevski, Department of Physics, Bath University, Bath, BA2 7AY U.K.
Particle Detectors, Tracking

Michael Albert, Department of Physics and Astronomy, Vanderbilt University, Nashville, TN 37235-1807, U.S.A.
Particle Impact Phenomena

Nellie Brock, Tektronix, Inc., P.O. Box 500, M/S 39-353, Beaverton, OR 97077, U.S.A.
Oscilloscopes and Oscillographs

Ian G. Brown, Lawrence Berkeley Laboratory, Building 53-013, University of California, Berkeley, CA 94720, U.S.A.
Particle Sources—Ion

J. E. T. Channell, Department of Geology, 1112 Turlington Hall, University of Florida, Gainesville, FL 32611, U.S.A.
Paleomagnetism

Paul D. Coleman, Department of Electrical and Computer Engineering, University of Illinois, 1406 W. Green Street, Urbana, IL 61801, U.S.A.
Oscillators

A. G. Every, Physics Department, University of the Witwatersrand, PO WITS 2050, Johannesburg, South Africa
Phonons in Crystal Lattices

Benny Freeman, Department of Chemical Engineering, North Carolina State University, Raleigh, NC 27695-7905, U.S.A.
Osmosis

Govindjee, Department of Plant Biology, University of Illinois, 265 Morrill Hall, 505 S. Goodwin Avenue, Urbana, IL 61801, U.S.A.
Photosynthesis

Dan Green, Research Division, Fermilab, PO Box 500, Batavia, IL 60510, U.S.A.
Particle Physics

Claude Benoit à la Guillaume, Groupe de Physique des Solides, Tour 23, Université Paris 7, 2 Place Jussieu, F-75251 Paris Cedex 05, France
Photoluminescence

Donald L. Hartill, Cornell University Laboratory for Nuclear Studies, Ithaca 14850, NY, U.S.A.
Particle Colliders

F. J. Himpsel, IBM/Watson Research Center, PO Box 218, Yorktown Heights, New York, NY 10598, U.S.A.
Photoemission and Photoelectron Spectra

Stephen D. Holmes, Fermi National Accelerator Laboratory, Batavia, IL 60510, U.S.A.
Particle Colliders

Max Jammer, Department of Physics, Bar-Ilan University, Ramat-Gan 52900, Israel
Philosophy of Physics

H. William Koch, 12105 Ambassador Drive, Apt. 106, Colorado Springs, CO 80921, U.S.A.
Physics and Engineering Organizations

I. Lindau, Department of Synchrotron Radiation Research, Sölvegatan 14, S-223-62 Lund, Sweden
Photoemission and Photoelectron Spectra

Joseph J. Loferski, Division of Engineering, Brown University, Box D, Providence, RI 02912, U.S.A.
Photovoltaic Devices

Erwin Marquit, School of Physics and Astronomy, University of Minnesota, Minneapolis, MN 55455, U.S.A.
Philosophy of Technology

Glennys Mensing, Department of Physics and Astronomy, Vanderbilt University, Nashville, TN 37235-1807, U.S.A.
Particle Impact Phenomena

T. C. Meyer, PPE Division, CERN, CH-1211 Genèva 23, Switzerland
Particle Detectors, Tracking

Melvin Month, Accelerator Development Department, Building 902A, Brookhaven National Laboratory, Upton, NY 11973, U.S.A.
Particle Colliders

James Murdock, Department of Physics, Iowa State University, Ames, Iowa 50011, U.S.A.
Perturbation Methods

Robert B. Palmer, Brookhaven National Laboratory, Upton, NY 11973, U.S.A.
Particle Colliders

Arthur D. Pelton, École Polytechnique, Casa

Postale 6079, Succursale Centre-ville, Montreal, PQ, H3C 3A7, Canada
Phase Equilibria

Fred H. Poettmann, Petroleum Engineering Department, Colorado School of Mines, 1500 Illinois Street, Golden, CO 80401, U.S.A.
Petroleum

Maria Reichmanis, Registered Patent Attorney, 2300 Hickory Road, Aiken, SC 29803, U.S.A.
Patents and Intellectual Property

John J. Ritsko, IBM T.J. Watson Research Center, P.O. Box 218, Yorktown Heights, NY 10598, U.S.A.
Packaging Technology

Ronald D. Ruth, Stanford Linear Accelerator Center, Stanford, CA 94305, U.S.A.
Particle Colliders

E. K. H. Salje, Department of Earth Sciences, University of Cambridge, Downing Street, Cambridge CB2 3EQ, U.K.
Phase Transitions, Structural

J. M. Sanchez, Center for Materials Science and Engineering, ETC 8.180, The University of Texas, Austin, TX 78712, U.S.A.
Order-Disorder Transitions

Michael P. Schmidt, Physics Department, Yale University, PO Box 6666 Yale Station, New Haven, CT 06511, U.S.A.
Particles, Elementary

Franz Schwabl, Institut für Theoretische Physik, Technische Universitat München, Physik-Department, James-Franck-Strasse, D-85747 Garching, Germany
Phase Transitions: Renormalization and Scaling

Robert H. Siemann, Stanford Linear Accelerator Center, Stanford, CA 94305, U.S.A.
Particle Colliders

Susan R. Sloan, Hewett-Packard Company, 1412 Fountaingrove Parkway, M/S 1US-K, Santa Rosa, CA 94503, U.S.A.
Photodetector Devices

Uwe Claus Taüber, Lyman Laboratory of Physics, Harvard University, Cambridge, MA 02138, U.S.A.
Phase Transitions: Renormalization and Scaling

Robert S. Thompson, Petroleum Engineering Department, Colorado School of Mines, 1500 Illinois Street, Golden, CO 80401, U.S.A.
Petroleum

Norman Tolk, Department of Physics and Astronomy, Vanderbilt University, Nashville, TN 37235-1807, U.S.A.
Particle Impact Phenomena

John Whitmarsh, Department of Plant Biology, University of Illinois, 265 Morrill Hall, 505 S. Goodwin Avenue, Urbana, IL 61801, U.S.A.
Photosynthesis

F. Karl Willenbrock, 1740 New Hampshire Avenue NW, Unit E, Washington, DC 20009, U.S.A.
Physics and Engineering Organizations

Pierre Wiltzius, AT&T Bell Laboratories, Room 1D-428, 600 Mountain Avenue, Murray Hill, NJ 07974, U.S.A.
Phase Separation

Yako Yafet, 47 Curtiss Place, Maplewood, NJ 07040, U.S.A.
Paramagnetism

OPTOELECTRONIC DEVICES

See QUANTUM OPTOELECTRONIC DEVICES

ORDER-DISORDER TRANSITIONS

J. M. Sanchez, *Center for Materials Science and Engineering, University of Texas, Austin, Texas, U.S.A.*

	Introduction: General Aspects of Order and Disorder	1
1.	**Characterization of Ordering**	4
1.1	The Manifestations of Ordering: Examples	4
1.1.1	Ordering in Alloys: Intermetallic Compounds	4
1.1.2	Long-Period Structures	5
1.1.3	Ordering in Low-Dimensional Systems	6
1.2	Formal Description of Order	8
1.2.1	Occupation Numbers and Correlations in Crystals	8
1.2.2	Long- and Short-Range Order	11
1.2.3	Ordering at Finite Temperatures	12
2.	**Summary**	15
	Works Cited	16
	Further Reading	16

INTRODUCTION: GENERAL ASPECTS OF ORDER AND DISORDER

In nature, order is a recurrent theme. As such, we are conditioned to expect and recognize order in most physical systems. This experience has left us with an intuitive concept of order that, although it has served us well, requires some elaboration to be applicable in the realm of science. In most instances, order can be defined precisely in terms of mathematical concepts and be uniquely characterized using experimental probes, such as diffraction of x rays or electrons. Thus, the definition of order can be made void of subjective and/or semantic overtones. It is perhaps more difficult to rely on our intuition to define the concept of *disorder*, i.e., the *lack of order*, which appears, at first sight, to be a relative concept. However, we can find equally precise mathematical definitions of disorder. We will come back to these in the following sections so that we can expand upon the intriguing phenomenon of the transition from a disordered to an ordered state.

Undoubtedly the most apparent transition from a state of order to one of disorder is that of melting. In this case, an "ordered" crystalline state, which is stable at low temperatures, gives way to a "disordered" liquid state at high temperatures. Among several important features of this transition is the fact that the discrete translational symmetry characteristic of the crystalline state (see Crystalline State) is replaced by the continuous translational symmetry of the liquid. In the latter, all points in space are equivalent. The consequences of melting, in terms of physical properties, are generally quite apparent.

Other order-disorder transitions are less evident, although they also have profound effects on physical properties. In alloys, for example, and without ever abandoning the ubiquitous crystalline state, atoms of different kinds may be distributed more or less randomly (disordered state) or following a well-defined pattern in a unit cell (ordered state). Examples of these two states are schematically shown in Fig. 1 on the unit cell of a body-centered-cubic (bcc) crystalline lattice. Figure 1(a) shows a possible ordered state in which atoms of one kind (open circles) occupy all the cube corners, whereas the other type (full circles) occupy the cube centers. The crystal itself is constructed by repeating this unit which, in crystallographic terminology, is referred to as a simple cubic unit cell with two atoms per cell. We note in passing that this is the structure adopted by many ionic crystals, a typical case being cesium chloride (CsCl), and by a number of metallic systems,

3-527-28135-5/95/$5.00 + .50

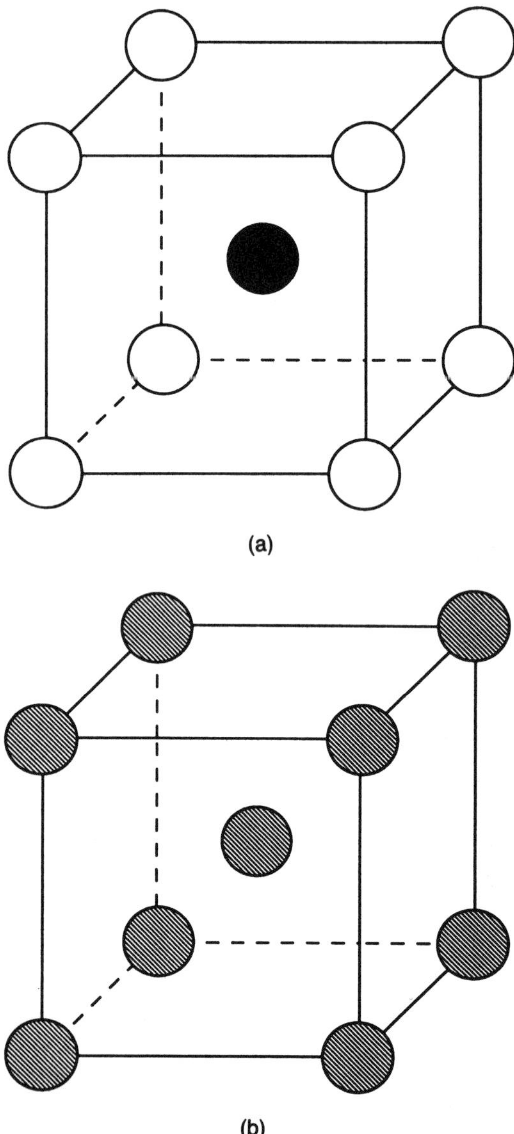

FIG. 1. Unit cells of (a) the primitive cubic CsCl and (b) the disordered bcc structures.

such as the low-temperature phase of copper-zinc (CuZn). Furthermore, since there are as many cube-corner as cube-center sites in the crystal, the numbers of Cs and Cl ions in CsCl, or of Cu and Zn atoms in CuZn, must be the same for perfectly stoichiometric compounds.

The representation of the disordered state is considerably more difficult since, strictly speaking, the unit cell is as large as the crystal itself and, thus, contains of the order of 10^{23} atoms. The disordered atomic arrangement is schematically shown in Fig. 1(b), in which an "average" atom (gray circle) is placed at each lattice site. As in the case of melting, the transition between the ordered and disordered states is also effected by temperature, and results in significant changes in the strength, brittleness, and electrical properties of the material.

Another easily recognizable order-disorder phenomenon is the spontaneous alignment of microscopic magnetic moments. Consider atoms having a well-defined magnetic moment and arranged on a crystalline lattice. At high temperatures, these atomic magnetic moments will be oriented in all possible directions such that the macroscopic, total magnetization of the crystal vanishes. Such a crystal is schematically shown in Fig. 2(a), which could be thought as a crude model of paramagnetic bcc iron. In many instances, including the ferrite phase of iron, the interactions between these atomic magnets are such that the energy of the crystal is lowered when the moments are aligned parallel to each other. Thus, the ground state, corresponding to the lowest energy state that can be achieved at low temperatures, is a ferromagnetic arrangement as shown in Fig. 2(b). In the ferromagnetic state, the macroscopic magnetization is nonzero and the material behaves as a permanent magnet (ferromagnet). Once again we find that macroscopic physical properties change significantly as the system undergoes an order-disorder transition.

There are certain common aspects to the three examples of order-disorder transitions discussed above. First and foremost is the fact that some physical property changes drastically in all cases. In melting the change is, primarily, in the ability of the material to withstand shear stresses; in the case of ordering in alloys, as in CuZn for example, it is in the strength or hardness of the material; and in the ferromagnetic transition is in the appearance of a permanent magnetic moment. These changes may reflect into either useful or detrimental properties. A material that is strong and ductile may become brittle upon ordering or, alternatively, some other desirable property, such as its electrical conductivity or high magnetic permeability, may be altered. These important practical aspects

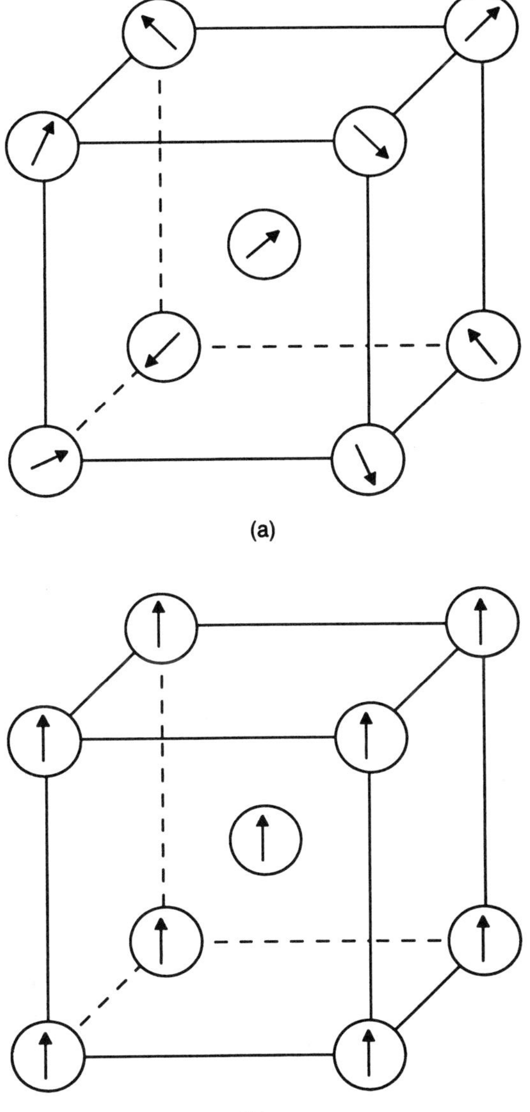

FIG. 2. Schematic representation of (a) paramagnetic and (b) ferromagnetic structures for the body-centered-cubic lattice.

have added to the interest in the study of order-disorder transitions in general terms as well as in specific systems.

A second aspect of commonality is that the transitions take place upon changes in some thermodynamic variable. In the cases discussed above, the thermodynamic variable is temperature although the transitions may also be effected by changes in pressure, in magnetic and/or electric fields, and, in the case of alloys, mixtures, or solutions, in the concentration of the different atomic or molecular components. Thus, the reasons for order-disorder transitions are to be found in the fundamental laws of thermodynamics and, more precisely, in the behavior of proper thermodynamic potentials or free energies.

A third aspect common to all order-disorder transitions is the existence of an *order parameter*. As a matter of convention, the order parameter is defined such that it vanishes in the disordered state whereas it takes finite values in the ordered state. The definition of the order parameter is not always intuitive, and it requires some characterization of the transition at hand before it can be identified. In the case of the paramegnetic-to-ferromagnetic transition, the order parameter may be defined as the macroscopic magnetization which, in turn, is the expectation value, or the average over the whole crystal, of the atomic magnetic moments. This average clearly vanishes in the paramagnetic state, where atomic magnetic moments point in all possible directions, and it is nonzero in the ferromagnetic state, where there is preferential orientation of these moments. In the case of ordering in alloys, as, for example, in the CuZn system, the order parameter is somewhat more difficult to visualize. However, a very precise definition can be given in terms of the probability of occupancy of cube-corner and cube-center sites by one type of atom, let us say Cu. Thus, the order parameter is given by the probability of a Cu atom occupying a cube-corner site *minus* the probability of a Cu atom occupying a cube-center site. Once again, the order parameter vanishes in the disordered state where Cu (and Zn) atoms occupy these two type of sites with equal probability and is nonzero in the order state. Unlike the case of the ferromagnet, where a nonzero order parameter can be put into evidence quite easily, the measurement of the order parameter in the case of the CuZn requires more sophisticated equipment such as an x-ray diffractometer.

A point of distinction between order-disorder transitions and general phase transitions is to be noted. Order-disorder transitions always involve a change in symmetry, which is not necessarily the case for phase transitions in general. For example, the translational symmetry is lost in melting, rotational symmetry is lost in the ferromagnetic state, and the *space group* of the crystal changes upon ordering in CuZn. In the last case, the crystal goes from a simple cubic structure in the ordered state to a bcc structure in the disordered state. This behavior is in contrast, for example, to the vapor-liquid phase transition at which the symmetry remains unchanged. The symmetry change characteristic of order-disorder transitions, generically known as *symmetry breaking*, usually provides powerful clues in the identification of the order parameter.

1. CHARACTERIZATION OF ORDERING

This section aims at providing a brief overview of the tools used in the description and characterization of ordering. The section begins with a few examples of ordered structures found in nature. These examples are chosen, primarily, because of their simplicity and as a way to introduce and highlight some of the relevant concepts. The elements needed for a more formal description of the ordered and disordered states are also given. The section concludes with a brief description of the thermodynamic principles involved in order-disorder transition, followed by examples of equilibrium phase diagrams.

1.1 The Manifestations of Ordering: Examples

1.1.1 Ordering in Alloys: Intermetallic Compounds Like most solids at low temperatures, alloys, which are mixtures of different atomic elements, adopt a crystalline structure. The phenomenon of ordering in these systems refers to the way in which the atomic elements are distributed in the crystal sites. An example of this type of ordering was discussed above and shown schematically in Fig. 1(b) for the case of the Cu_xZn_{1-x} alloys, where x stands for the concentration of Zn in the alloy. Clearly, the structure described by the unit cell of Fig 1(b) can be achieved only if the concentration x equals 0.5, i.e., at 1:1 stoichiometry. The ordered structure in this case is an example of an *intermetallic compound*. Compounds of other stoichiometries are always possible. In the case of Cu_xZn_{1-x}, two other low-temperature intermetallic compounds are observed in addition to CuZn, occurring at stoichiometries Cu_5Zn_6 ($x = \frac{5}{11}$) and $CuZn_4$ ($x = 0.2$). Because of their crystalline nature, intermetallic compounds necessarily belong to one of the 230 possible crystallographic space groups listed in the Henry and Lonsdale (1965) (see CRYSTALLINE STATE). Although the space-group notation provides an unambiguous way of describing the symmetry of the compounds, it is not particularly helpful in describing the atomic arrangements, i.e., the ordering of atoms in the unit cell. Thus, this space-group notation is not universally used to describe ordered structures, and a variety of other conventions abound. Among them is the use of a prototype structure to refer to all structures of a certain type. In this mode, one would refer to ordered CuZn as the "CsCl structure" [see Fig. 1(a)]. It is also common practice to use a symbol in order to identify a given ordered structure. For example, in this notation, the "CsCl" structure is known as the B2 structure, and the disordered body-centered-cubic CuZn is identified by the symbol A2. The most extensive compilation of compounds to date is to be found in Villars and Calvert (1985). Pearson introduced his own notation which is also commonly used in the literature. For example, the cubic Primitive structure of CuZn becomes the cP2 structure in Pearson's notation, where the 2 shows that two atoms are present in the unit cell.

Two closely related intermetallic compounds, which are manifestations of ordering in the face-centered-cubic lattice, are those known as the $L1_0$ and $L1_2$ structures (see Fig. 3). The $L1_2$ structure is simple (primitive) cubic (space group $Pm\bar{3}m$), whereas $L1_0$-type ordering, usually accompanied by a slight tetragonal distortion, results in a simple tetragonal structure (space group $P4/mmm$). The $L1_0$ and $L1_2$ structures are representative of the simplest type of ordering that can be realized in fcc lattices. As such, they are observed to occur in a large number of alloys. For example, in the Cu-Au system, the two possible $L1_2$ structures, namely Cu_3Au and $CuAu_3$, as well as the $L1_0$ CuAu are present. Furthermore, all three intermetallic compounds go through an order-disorder transition in the solid state, adopting an fcc structure at high temperatures before, eventually, melting. In view of this, the intermetallic compounds CuAu and Cu_3Au are commonly

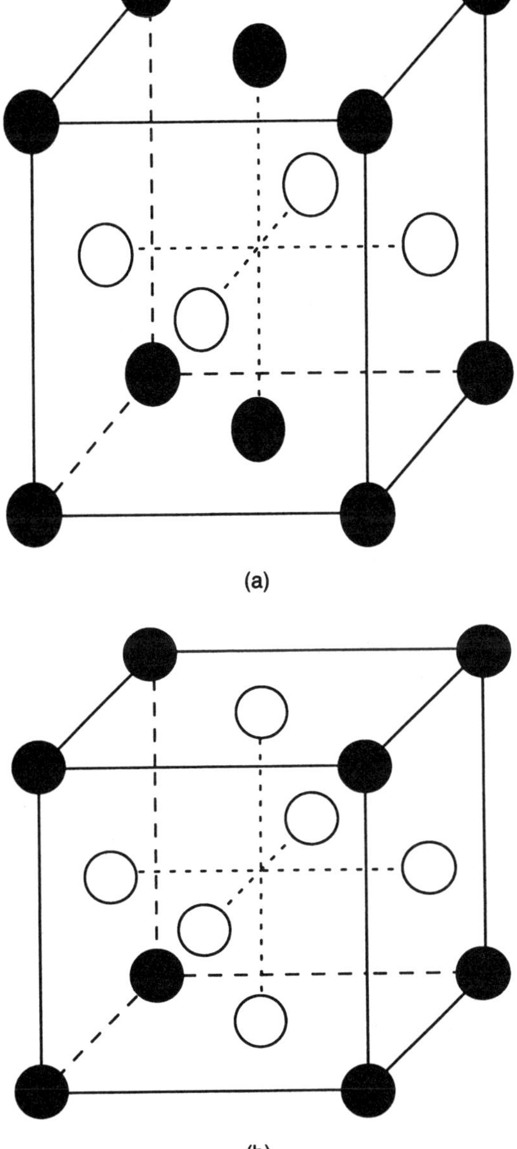

FIG. 3. Unit cells of (a) the body-centered tetragonal $L1_0$ and (b) the primitive cubic $L1_2$ structures.

used as the prototypes for the $L1_0$ and $L1_2$ structures, respectively.

1.1.2 Long-Period Structures The simplicity of the low-temperature ordered compounds present in the Cu-Au systems is in sharp contrast to the appearance, around 1:1 stoichiometry and in a narrow temperature range of approximately 25 K, of an orthorhombic ordered structure known as CuAuII. Although CuAuII is closely related to the tetragonal $L1_0$ structure, the long orthorhombic lattice constant is approximately ten times that of $L1_0$. Structures such as that of CuAuII are known as *long-period* structures. At present, a seemingly endless list of long-period structures has been reported, with those related to the $L1_2$ structure of Cu_3Au being particularly common. Examples include Ag_3Mg, Cu_3Pd, Au_3Zn, Al_3Ti, etc. A characteristic feature of these long-period structures is that they can be constructed by introducing a periodic array of so-called *antiphase boundaries* (APB). Antiphase boundaries are planar defects that occur when two ordered domains meet. For example, an APB results if two otherwise identical ordered domains are displaced relative to each other by a distance that is less than a period of the lattice. Figure 4 shows schematically such an APB in the $L1_2$ structure.

A rather common structure, realized, for example, in the intermetallic compounds Al_3Ti and Ni_3Mo, can be obtained by arranging APBs every other $L1_2$ unit cell. The unit cell of the resulting tetragonal structure, known as DO_{22}, is shown in Fig. 5. The structure itself is made of a periodic distribution of APBs that separate domains two unit cells thick. Another example is the structure with periodicity 4, sometimes referred as the DO_{23}, shown in Fig. 6. DO_{22} and DO_{23} belong to a relatively simple sequence of structures for which the periodicity of the APBs is $2n$, where n is an integer number. It is interesting to note that the $L1_2$ structure corresponds to an infinite periodicity n and, thus, the terminology long-period structures is rather subjective.

More complex sequences of APBs are also

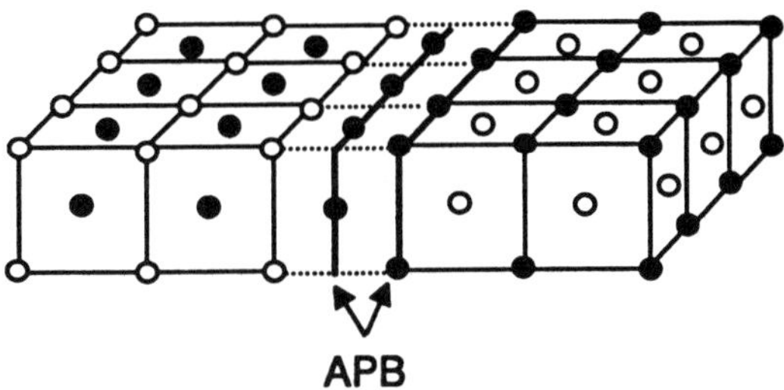

FIG. 4. Schematic representation of an antiphase boundary between two $L1_2$ ordered domains.

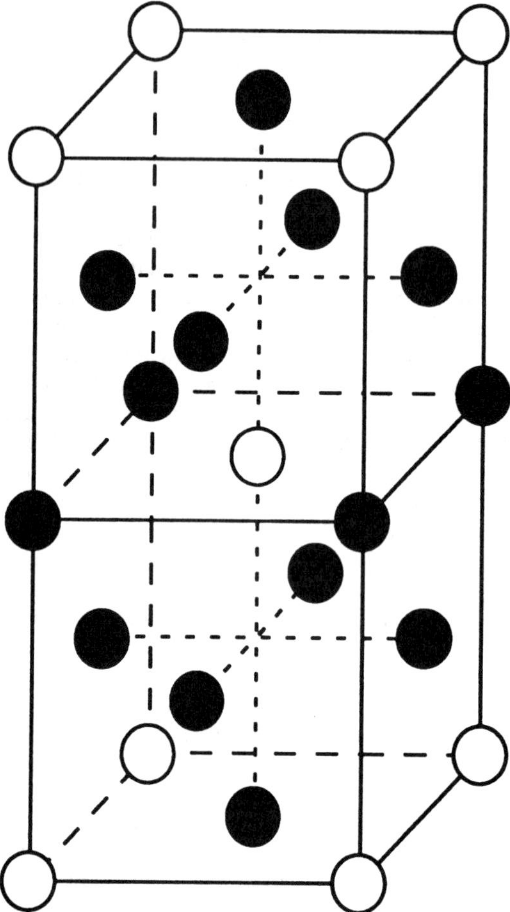

FIG. 5. Unit cell of the DO_{22} structure.

found in nature. To describe these structures a notation that indicates the sequence and the domain thickness in terms of the unit cell of the parent structure is commonly used. For example, the structure $\langle n_1 n_2 \ldots\ n_q \rangle$ corresponds to a sequence in which a domain of thickness n_1 is followed by an antiphase domain of thickness n_2, etc. As an example, the unit cell of the $\langle 211 \rangle$ ordered structure reported, for example, in the Pt_3V system is shown in Fig. 7.

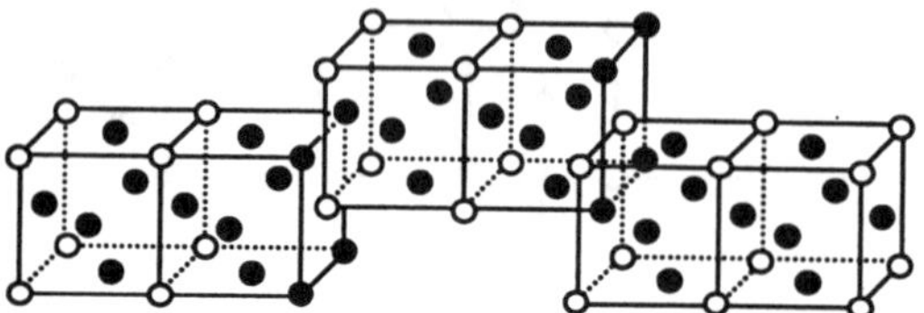

FIG. 6. Unit cell of a long-period structure generated by antiphase boundaries with periodicity 4.

1.1.3 Ordering in Low-Dimensional Systems When a physical phenomenon is strictly constrained to a surface, an interface, or an atomic plane in a crystal, the system may be construed as two dimensional. This reduced dimensionality has important consequences on general physical behavior and, in particular, on order-disorder transitions. The interest in the study of low-dimensionality systems is twofold. First, the physical understanding of phase transitions in two-dimensional systems is on very solid ground. For some simple models, such as the Ising model of two-dimensional ferromagnets, analytical results are known. Thus, these low-dimensional systems offer an excellent opportunity to test our most fundamental physical laws, models, and/or theories.

On the practical side, surfaces and interfaces are dominant in engineering systems and structures. For example, a modern microelectronics device is typically composed of several layers of different materials such as semiconducting silicon, noble metal gold, and transition metals—e.g., tungsten, titanium, and nickel—in close contact with each other. Surfaces also play a key role in friction, wear, and fracture of materials as well as in the performance of catalysts, adhesives, and composites and in biomaterials used for implants and prosthetics.

The order-disorder transition at a surface, for example, is generally found to differ considerably from that in the bulk. In many instances it is seen that, while the bulk undergoes a discontinuous or abrupt order-disorder transition, the surface orders continuously in the characteristic fashion of second-order transitions. This is the phenomenon of surface disorder or wetting, whereby a layer of the "disordered" or high-temperature phase develops at the surface as the transition temperature is reached from below. In the context of the solid-liquid transition, surface disorder is closely related to the phenomenon of premelting in which the surface region is seen to lose crystalline order at temperatures several degrees below the melting temperature of the bulk. The phenomenon of surface ordering has also been reported in ferromagnetic materials. In this case, the surface magnetization remains fi-

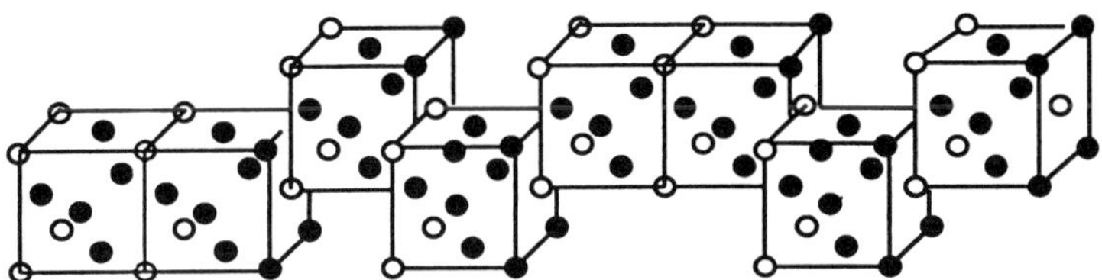

FIG. 7. Schematic representation of the ⟨211⟩ long-period structure.

nite at temperatures above the Curie temperature of the bulk.

The surface behavior in intermetallic compounds is governed by the interplay between at least one order parameter and the surface concentration. Furthermore, the symmetry breaking observed in the order-disorder transition is now constrained by the surface geometry and the crystallographic relation between the surface and the bulk. Consequently, surface-ordering phenomena and behavior in intermetallics are more complex than in the bulk. A case in point is the behavior of the (110) surface of Cu_3Au, for which two different surface-ordering transitions are reported to occur. This is in contrast to the single disorder-$L1_2$ transition taking place in the bulk at 663 K. Figure 8 depicts the atomic arrangements for each of the (110) surface phases, labeled 2x1 and 4x1, found in Cu_3Au. Note that perpendicular to the surface, the Cu_3Au crystal is made of alternating pure Cu and mixed (equiatomic) Cu-Au (100) planes. Thus, the 2x1 structure corresponds to a surface termination at a mixed (110) plane, whereas the formation of the 4x1 structure requires segregation of Cu to the surface.

Occasionally, a system is found where the order-disorder transitions can be described as a relatively simple two dimensional system. One such case is provided by the ordering of oxygen in the superconducting oxides, as exemplified by the celebrated $YBa_2Cu_3O_{6+x}$ system. The ordering of oxygen takes place, primarily, on one type of crystallographic plane. This ordering drives an important structural transition in the oxide from a high-temperature tetragonal structure to a low-temperature orthorhombic structure. The unit cells of both the tetragonal and orthorhombic structures are shown in Fig. 9 for the stoichiometric compound $YBa_2Cu_3O_7$. In the high-temperature tetragonal phase, shown in Fig. 9(a), the O1 and O5 oxygen sites are occupied with an equal probability of $\frac{1}{2}$, whereas the sites O2 and O3 are equivalent and occupied with probability 1. In the orthorhombic phase, Fig. 9(b), the O1 sites are now occupied with probability 1, leaving the O5 sites empty. In reference to the discussion of ordering in alloys, note that the $YBa_2Cu_3O_7$ system is

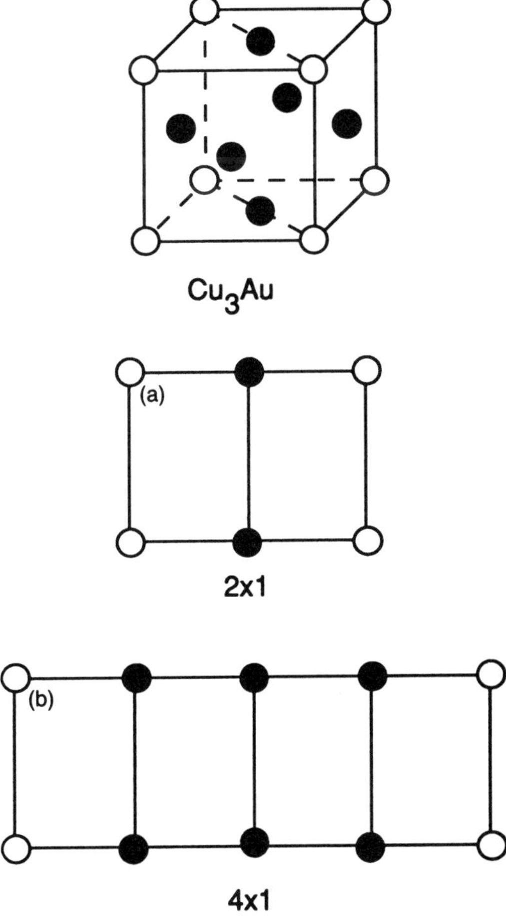

FIG. 8. Two-dimensional unit cells of ordered structures observed in the (110) surface of Cu_3Au. (a) The ordered structure with periodicity 2x1 appears at a temperature of 654 K, slightly lower than the bulk transition of 660 K due to surface segregation effects. (b) The transition to a 4x1 structure occurs at approximately 425 K.

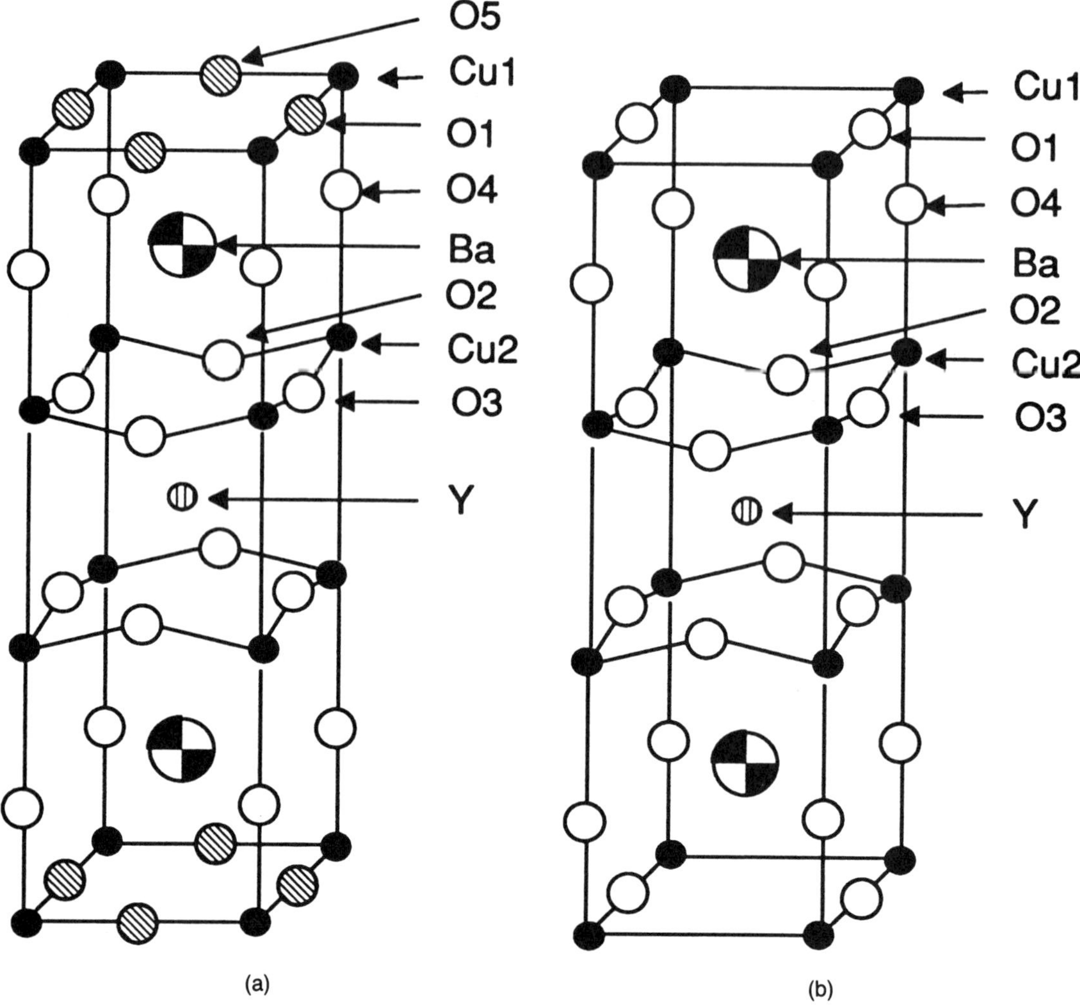

FIG. 9. The unit cells of the (a) high temperature tetragonal and (b) orthorhombic phases of $YBa_2Cu_3O_7$.

equivalent to that of a binary alloy in which the ordering species are vacant sites and oxygen ions. A number of possible ordered structures have been reported in $YBa_2Cu_3O_{6+x}$, a fact that reflects the relative complexity of the oxygen-oxygen interactions. However, two of the most prominent structures can be easily understood in terms of a simple model based on the repulsion between neighboring oxygen ions, coupled with an effective attraction between oxygen ions mediated by a Cu atom. The two-dimensional unit cells for these structures are shown in Fig. 10, where the repulsive interactions between nearest- (V_1) and next-nearest-neighbor (V_2) oxygen atoms and the attractive Cu-mediated interaction (V_3) are also shown.

The use of simple models base on short-ranged atomic interactions has traditionally played an important role in the classification and prediction of ordered structures in both two- and three-dimensional systems. For these models, the enumeration of the ordered states with the lowest energy, generically known as ground states, can be cast in the form of the classical linear programming problem of enumerating vertices of a convex polyhedron (see below).

1.2 Formal Description of Order

1.2.1 Occupation Numbers and Correlations in Crystals A useful concept in the description of order in crystalline systems is

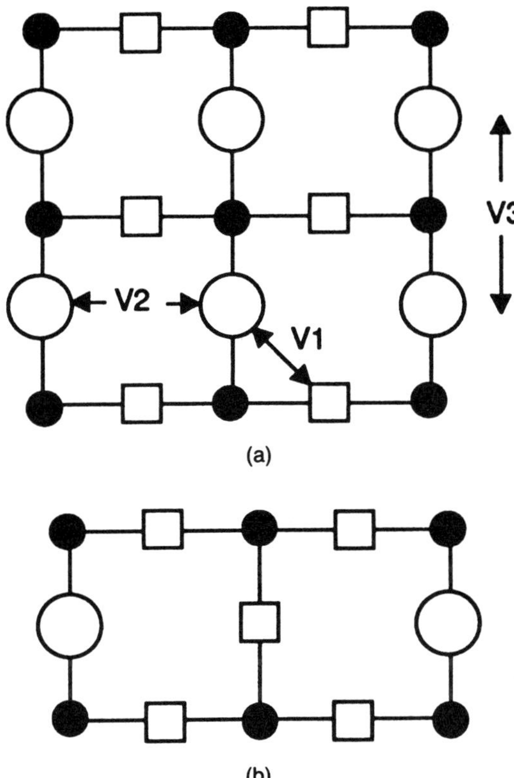

FIG. 10. Two-dimensional representation of oxygen ordering in the basal plane of the $YBa_2Cu_3O_{6+x}$ system corresponding to orthorhombic structures observed around (a) $x = 1$ and (b) $x = 0.5$.

that of an occupation number. In a binary alloy, for example, the occupation number σ_i takes two values, let us say +1 and −1. Each value of the occupation number indicates the atomic species located at the particular lattice site, which is labeled by the index i. The concept can be extended so that the occupation number refers, for example, to any number of atomic species, to the direction and magnitude of atomic magnetic moments, or, as in the case of the superconducting oxides discussed above, to vacant or occupied sites, A microscopic state of the system in which the identity of each atom at each crystal site is known can be fully specified by an N-dimensional vector $\boldsymbol{\sigma} = \{\sigma_1,\sigma_2,\ldots,\sigma_N\}$, where N stands for the total number of sites in the crystal. The total number of possible arrangements of A and B atoms, i.e., microstates, in a crystal with N lattice sites is 2^N.

A physical description of the system settles for a much less detailed knowledge in which only expectation values, or averages based on the equilibrium probability distribution, of certain key quantities are known. The calculation of such averages falls within the realm of statistical mechanics and thermodynamics and is essential for the study of the effect of temperature, pressure, and other thermodynamic variables on the order-disorder transition.

For a binary system, the average of the occupation number, denoted $\langle\sigma_i\rangle$, depends only on the concentration of each atomic species. If σ_i equals 1 for A atoms and −1 for B atoms, then $\langle\sigma_i\rangle = x_A - x_B$, where x_A and x_B are the concentrations of A and B atoms, respectively. The expectation value $\langle\sigma_i\rangle$ determines the composition of the binary alloy, but conveys no information on the state of order of the system. Some information regarding the atomic arrangement is provided by the pair correlation function $\langle\sigma_i\sigma_j\rangle$. These pair correlations depend strongly on the distance between sites i and j in the crystal. Note that for the perfectly ordered CuZn structure [Fig. 1(a)], the nearest-neighbor pair correlation equals −1, i.e., nearest-neighbor pairs are made of different atomic species, whereas it equals +1 for next-nearest neighbors in which case the two atoms are of the same type. On the other hand, for a random atomic distribution (totally disordered), the pair correlations vanish for 1:1 stoichiometry ($x_A = x_B = 0.5$). This stems from the fact that, for a random distribution, averages of products of the occupation numbers equal products of the averages; i.e., $\langle\sigma_i\sigma_j\rangle_{\text{rand}} = \langle\sigma_i\rangle\langle\sigma_j\rangle = (x_A - x_B)^2$.

The pair correlation functions contain enough information to distinguish between ordered and disordered states, although the characterization is not complete. For a complete description of the ordered state, all possible correlation functions are needed. The complete set of correlation functions includes averages of products of occupation numbers located on all possible sets of crystal sites. These sites can be visualized as forming well-defined geometrical figures, or clusters. In general, for a cluster α composed of crystal sites $\{i,j,\ldots,k\}$, i.e., $\alpha = \{i,j,\ldots,k\}$, the correlation function ξ_α is given by

$$\xi_\alpha = \langle\sigma_i\sigma_j\ldots\sigma_k\rangle.$$

As implied above, a full description of partially ordered and partially disordered states requires the characterization of an infinite number of correlation functions. Since, in practice, this cannot be achieved, approximations to the problem are called for. One approach consists of describing correlation functions up to a maximum cluster size. In this cluster approximation, the state of order is described by a finite set of correlation functions $\{\xi_\alpha\}$ corresponding to each of the subclusters of the maximum cluster. For example, a regular tetrahedron cluster made of nearest-neighbor sites is commonly used to describe ordering in the fcc lattice. At this level of approximation, four correlation functions are involved corresponding to the point, pair, triangle, and tetrahedron clusters. These clusters are shown schematically in Fig. 11. In general, the M-dimensional space spanned by the finite set of correlation functions ($M = 4$ for the tetrahedron approximation in the fcc lattice) is known as the configurational space of the system; i.e., any state of partial order corresponds to a point in this space. These M correlation functions are also subject to linear geometric constraints. For example, if the point correlation function $\xi_1 = \langle\sigma_1\rangle$ in a binary system is close to 1, implying that the system is made up almost entirely of A atoms, then the nearest-neighbor pair correlation function ξ_2 is also constrained to being close to 1. It can be shown that these constraints are linear, and that the space spanned by the correlation functions is a convex configurational polyhedron.

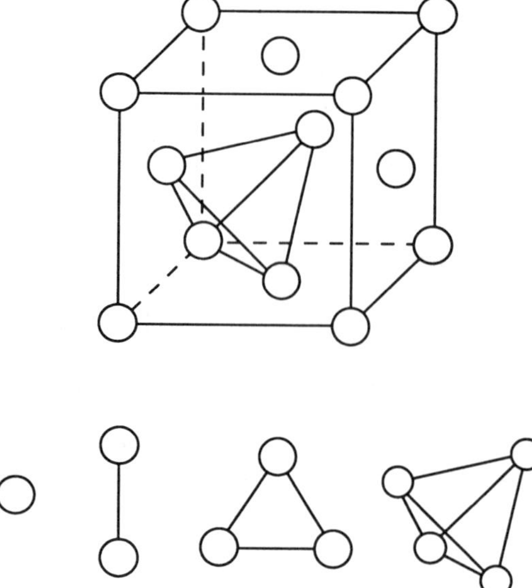

FIG. 11. Subclusters of the regular tetrahedron in the fcc lattice.

The cluster description of order is particularly useful for studying model systems with relatively short-range interactions. For these models, the expectation value of the energy is exactly given by a linear combination of correlation functions:

$$\langle E\rangle = \sum_{\beta\subseteq\alpha} V_\beta \xi_\beta,$$

where the sum is carried over all the subclusters β of the maximum cluster α, and where the coefficients V_β are known as *effective cluster interactions*. Since the energy is linear in the correlation functions and the configurational polyhedron is convex, it follows that the ground states are given by the vertices of the configurational polyhedron. Thus, a ground-state search is equivalent to finding all such vertices, itself a classical problem in linear programming. This procedure, although trivial in low dimensions (two or three), becomes rather cumbersome when the number of correlation functions is large, as would be the case for physical systems with long-range interactions.

The approximate description of order using finite clusters has been useful to describe and classify order structures in real systems. For example, a ground-state search for the fcc lattice with nearest- and next-nearest-neighbor effective interactions results in the structures shown in Fig. 12. In this simple model, the relevant parameter is the ratio R of first- to second-neighbor effective interactions. For example, for negative values of R, the model predicts the commonly observed A_3B ($Pm\bar{3}m$, Cu_3Au, $L1_2$) and AB ($P4/mmm$, CuAu, $L1_0$) structures. For $0 < R < 0.5$ the ground states are those labeled A_2B_2 ($I4_1/amd$), A_2B ($Immm$, Pt_2Mo), A_3B ($I4/mmm$, $TiAl_3$, DO_{22}) and A_5B. Finally, for R larger than 0.5, the ground states are indicated as AB ($R\bar{3}m$, CuPt, $L1_1$) and A_2B. The labeling scheme in Fig. 12 gives first the compound formula, which is followed, if available, by the space group, the prototype structure, and the corresponding structural symbol. This simple one-parameter model predicts a relatively large number of ordered structures, some of which are observed in nature. It should be noted that the number of possible ordered structures grows exponen-

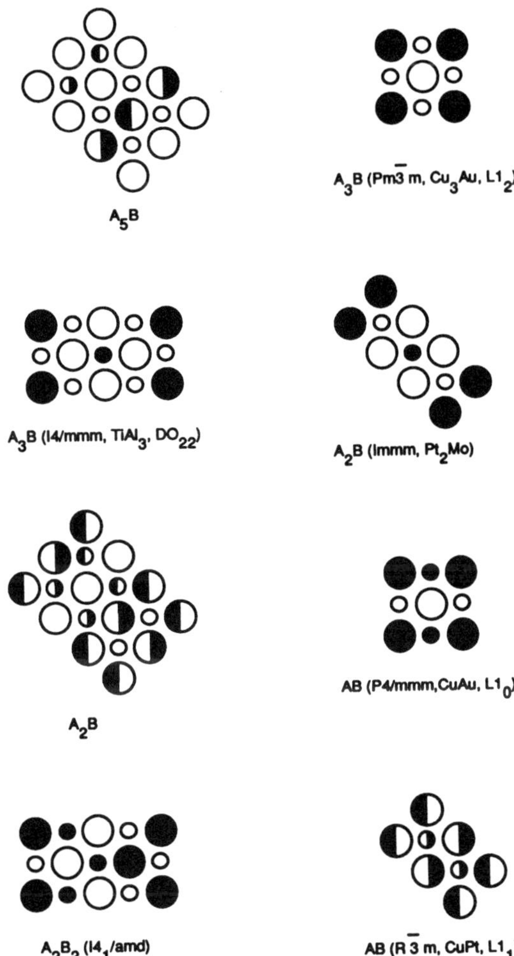

FIG. 12. Schematic representation of the unit cells for the ground states of the fcc lattice with nearest- and next-nearest-neighbor interactions. The two-dimensional drawings are projections of the unit cells onto the (001) crystallographic planes. Circles of different sizes corresponds to atoms located on consecutive (001) planes, with open and full circles representing *A* and *B* atoms, respectively. Half-open circles represent *A* and *B* atoms alternating along the direction perpendicular to the paper.

tially with the range of interactions, underscoring the richness of ordering phenomena in crystalline systems.

1.2.2 Long- and Short-Range Order The state of perfect order and that of total disorder are idealizations seldom realized in nature. The correlation functions introduced above provide a hierarchical framework to measure states of partial order (or disorder) that are relevant in the physical world.

In the disordered state, the correlation functions measure deviation from randomness and, as such, they provide an infinite set of what are generically known as *short-range order* (SRO) parameters. In practice, slight modifications of the correlation functions, the so-called Warren–Cowley SRO parameters, are used. These parameters have been adopted almost universally since they can be measured rather precisely using x-ray or neutron diffraction techniques. For a binary system, the Warren–Cowley SRO parameters are defined by

$$\alpha(r_{ij}) = \frac{\langle\sigma_i\sigma_j\rangle - \langle\sigma_i\rangle\langle\sigma_j\rangle}{(1 - \langle\sigma_i\rangle\langle\sigma_j\rangle)},$$

where r_{ij} stands for the distance between crystal sites i and j. These SRO parameters vanish for the random state, and are normalized in such a manner that they equal 1 for $r_{ij} = 0$, i.e., when the crystal sites are equal and, thus, $\sigma_i^2 = 1$.

If in the disordered state atoms of the same type have the tendency to cluster together, then the pair correlations for short distances will be positive and larger than the value corresponding to a random atomic distribution. Thus, in this case, the SRO parameters are positive. If the system has an ordering tendency, however, the SRO parameters will oscillate with distance, being negative (positive) for distances at which atoms of different (same) types are preferentially found. In the disordered state, the amplitude of these oscillations decreases with distance, vanishing for sufficiently large separation between the two crystal sites. That is to say, the atomic occupations of distant sites are uncorrelated.

The SRO parameters are functions of the thermodynamic state of the system and, therefore, change with temperature, pressure, and composition. At sufficiently high temperatures the system becomes increasingly disordered, and the Warren–Cowley parameters tend to zero. On the other hand, as the temperature is lowered, and, let us say, an order-disorder transition is approached, the oscillations of the SRO parameters in the disordered state persist over longer distances, decaying much more slowly with distance. At and below the order-disorder transition, the characteristic decay distance becomes infinite. Thus, *long-range order* (LRO) sets in and the atomic occupancy of distant sites is now strongly correlated.

The onset of LRO at the transition is described, appropriately enough, by one or more LRO parameters. As is the case with short-range order, there is significant latitude in the choice of LRO parameters. However, all practical definitions are related to the symmetry breaking that takes place at the transition. In particular, the change of symmetry taking place at the transition establishes uniquely the number of order parameters needed to describe the transition. It is common practice to group the different order parameters into the components of a vector, and to refer to the number of components as the dimension of the LRO parameter.

Returning to the example of CuZn, consider the transition from a disordered bcc structure [Fig. 1(b)] to the ordered $B2$ structure [Fig. 1(a)]. In addition to being endowed with the translational symmetry of the simple cubic lattice, the disordered system is invariant to a translation by a vector equal to half the cube diagonal, i.e., $(a/2)[1,1,1]$, where a is the cubic lattice constant. This symmetry element is lost in the ordered $B2$ structure, for which the cube-corner and cube-center sites are not equivalent. Thus, the expectation values of the occupation number at the cube corners, $\langle\sigma_0\rangle$, and at cube centers, $\langle\sigma_1\rangle$, are different. In this case, the transition is described by a unique LRO parameter, which can be defined as $\eta_1 = (\langle\sigma_1\rangle - \langle\sigma_0\rangle)/2$. This one-dimensional order parameter is nonzero in the ordered state, vanishing identically in the disordered state where $\langle\sigma_0\rangle = \langle\sigma_1\rangle$.

The concept of sublattice is often used to emphasize the nonequivalence of different sites in the unit cell of the ordered structure. A sublattice, which is not necessarily a Bravais lattice, connects all equivalent sites in a crystal. In the $B2$ structure, two interpenetrating simple cubic sublattices, displaced from each other by half the cube diagonal, are easily identified [see Fig. 1(a)]. The independent averaging of the occupation numbers on each sublattice leads naturally to a nonzero order parameter in the ordered state. This procedure is referred to as sublattice averaging and is commonly used in the thermodynamic description of the ordered state.

Sublattices and the associated LRO parameters may be defined for any ordered structure. Knowledge of the space group of the low-temperature structure uniquely determines the number of nonequivalent sites within the unit cell. These nonequivalent sites are known as Wyckoff positions and are listed in Henry and Lonsdale (1965), together with all symmetry elements, for all the 230 space groups. If the number of Wyckoff positions for a given structure is $m + 1$, the dimensionality of the LRO parameter is m, and the different components are given by

$$\eta_k = (\langle\sigma_k\rangle - \langle\sigma_0\rangle)/2 \quad (k = 1, \ldots, m).$$

For example, the number of nonequivalent sites in the primitive cubic $L1_2$ structure is 2, so that its LRO parameter is of dimension 1. On the other hand, the unit cell of the body-centered-tetragonal DO_{22} structure, with space group $I4/mmm$ (Fig. 5), contains three different Wyckoff positions, and the dimension of the LRO parameter is 2.

1.2.3 Ordering at Finite Temperatures

The transition between the ordered and disordered states is governed by the behavior of a thermodynamic potential or free energy. The relevant free energy is dictated by external constraints. For example, at a fixed temperature, the relevant thermodynamic potential is either the Helmholtz or the Gibbs free energy, depending on whether the volume or the external pressure is kept constant.

The fundamental thermodynamic principle involved in the study of phase stability and order-disorder transitions is that, under the proper set of conditions, the free energy is minimum for the equilibrium phase. Although the actual determination of the free energy, either experimentally or theoretically, may in itself be a rather involved task, much can be learned from general principles. An elegant and relatively simple theory of order-disorder transitions was advanced by Landau and Lifshitz (1969) on the basis of the expansion of the free energy in powers of the LRO parameter. Subsequent developments have established that the free energy is singular at the transition and, thus, such expansions are not generally valid. Nevertheless, the theory of Landau and Lifshitz is frequently used since it captures the most important aspects of order-disorder transitions.

For the case of a one-dimensional order parameter η, a fourth-order Landau and Lifshitz expansion is of the form

$$F(\eta,T,P) = F_0(T,P) + A(T,P)\eta^2 + B(T,P)\eta^3 + C(T,P)\eta^4,$$

where $F_0(T,P)$ is the free energy of the disordered state, i.e., for $\eta = 0$, and the coefficients $A(T,P)$, $B(T,P)$, and $C(T,P)$ are, in general, functions of temperature (T) and pressure (P). A fourth-order expansion of the free energy is usually sufficient to describe the thermodynamic behavior of the system provided that the coefficient $C(T,P)$ is positive. In the event that the fourth-order coefficient is negative, the free-energy expansion must be carried out to, at least, the next even power with a positive coefficient. In general, the calculation of the expansion coefficients is a difficult task that requires the adoption of a specific model for the atomic interactions. However, some general thermodynamic properties can be established using symmetry and group-theoretical arguments. In particular, Landau and Lifshitz studied the conditions under which the coefficient B in the free-energy expansion vanishes. Their analysis led to a set of celebrated rules for the occurrence of continuous phase transitions.

A vanishing third-order coefficient B, which is solely determined by the symmetry of the ordered phase, implies that the free energy is an even function of the LRO parameter, i.e., $F(\eta,T,P) = F(-\eta,T,P)$. Thus, assuming $C > 0$, a plot of the free energy vs the LRO parameter looks as shown schematically in Fig. 13(a). The several curves in the figure correspond to free energies at temperatures above, at, and below the order-disorder transition temperature T_c. At high temperatures, the minimum in the free energy occurs at $\eta = 0$ and the equilibrium state is disordered. At the transition temperature, the second-order coefficient $A(T,P)$ vanishes, becoming negative below T_c. Thus, below T_c, two minima in the free energy develop at finite values of η, indicating that the ordered state is the equilibrium thermodynamic phase. These two states, with equal free energies and ordered parameters of opposite signs, correspond to the same thermodynamic phase. This can be seen, for example, for the $B2$ structure shown in Fig. 1(b). In this case, a change in the sign of the order parameter corresponds to a simple exchange of cube-corner with cube-center sites. This operation leaves the $B2$ structure invariant and, therefore, symmetry imposes the requirement that the free energy be an even function of the LRO parameter; i.e., $F(\eta,T,P) = F(-\eta,T,P)$. In turn, this implies the vanishing of the coefficient B in the free-energy expansion.

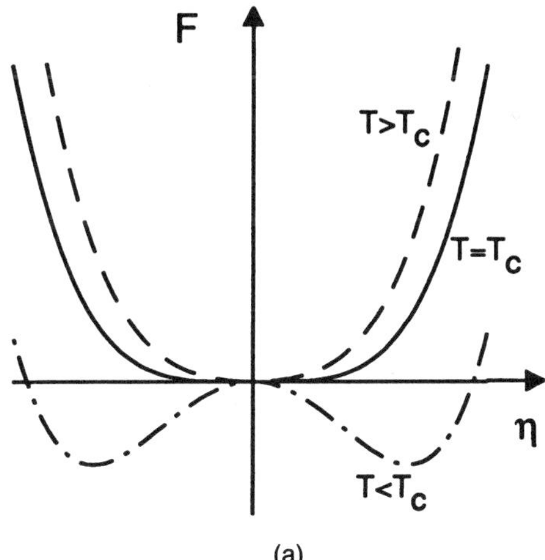

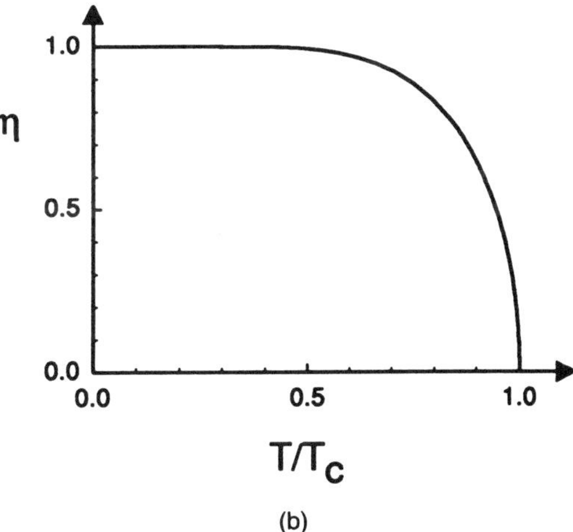

FIG. 13. Schematic (a) free-energy curves and (b) equilibrium long-range order parameter for systems displaying a second-order phase transition.

A plot of the value of η at which the minimum in the free energy takes place vs temperature is shown in Fig. 13(b). Note that the LRO parameter vanishes continuously as T_c is approached from below. This behavior of the LRO parameter is characteristic of *second-order* phase transitions.

The situation is significantly different when the third-order coefficient B in the free-energy expansion is nonzero, as shown in

Fig. 14. The free-energy curve may develop a local minimum at temperatures above the order-disorder transition. At the transition temperature, the LRO parameter jumps discontinuously from zero to a finite value corresponding to the ordered state. In this case, the order-disorder transition is said to be of *first order*.

For alloys, intermetallic compounds, and mixtures, the free energy is also a function of the concentrations of the different atomic or molecular components. For second-order transitions this concentration dependence of the free energy does not significantly alter the picture of order-disorder transitions given above. This is because at a second-order transition the compositions of the ordered and disordered phases are equal. For first-order transitions, however, an additional complication arises since the transition may now take place between phases of different average compositions. The situation is illustrated in Fig. 15, which shows schematic free-energy curves as a function of concentration for ordered and disordered phases in a binary alloy. The equilibrium condition of minimum free energy translates into a common-tangent rule that is also shown schematically in Fig. 15. The two points of tangency in Fig. 15 define three distinct regions: (I) a region where the disordered phase is stable; (II) a two-phase region where ordered and disordered phases coexist; and (III) a region where the ordered phase is stable. The information on the stability and coexistence between phases is commonly collected into graphs called *phase diagrams*. These phase diagrams are graphic representations of the temperature and composition ranges of stability and coexistence of all the thermodynamically stable phases in a given system. For example, the phase diagram of the Cu-Zn system is shown in Fig. 16, where the single-phase fields of the solid phases are indicated by Greek letters. The unlabeled regions in the figure correspond to two-phase fields, within which the adjacent phases coexist. The assignation of Greek letters to the stable phases is arbitrary and bears

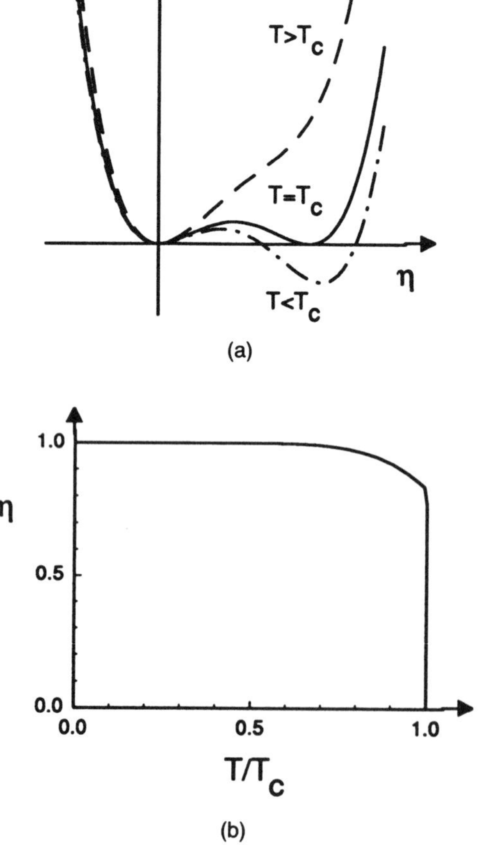

FIG. 14. Schematic (a) free-energy curves and (b) equilibrium long-range order parameter for systems displaying a first-order phase transition.

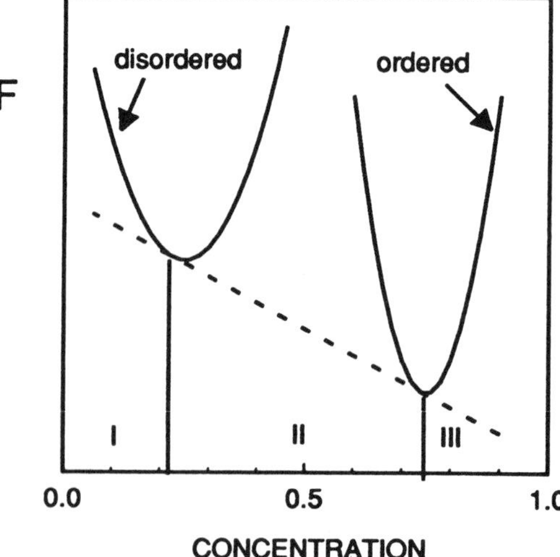

FIG. 15. Schematic free-energy curves vs composition for the ordered and disordered phases of a binary alloy that undergoes a first-order transition. The ordered and disordered phases are stable in regions I and III, respectively. For concentrations in region II both phases coexist.

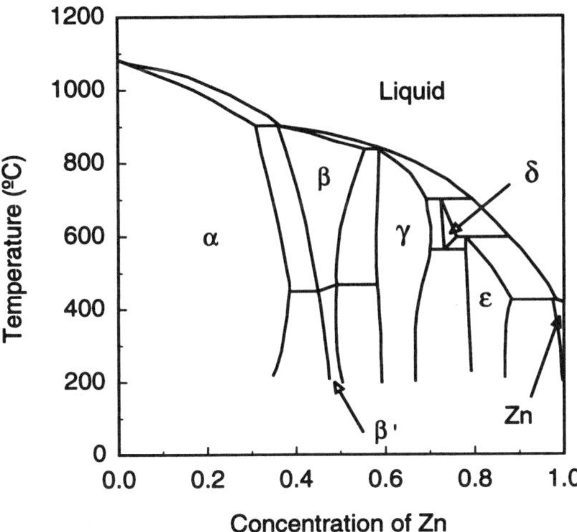

FIG. 16. Equilibrium phase diagram for the Cu-Zn system.

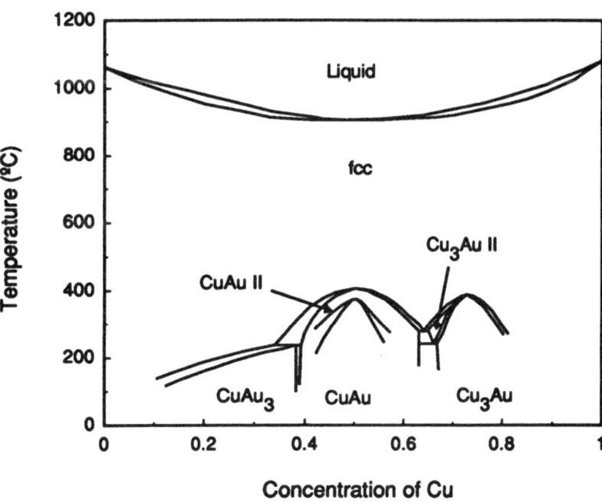

FIG. 17. Equilibrium phase diagram for the Cu-Au system.

no relation to the structure or to the symmetry of the phases. The phase indicated as α is a disordered fcc alloy. As the concentration of Zn increases beyond approximately 0.3, the α phase transforms, via first-order transitions, either to a bcc alloy (β) at high temperatures, or to an ordered *B*2 structure (β') at lower temperatures (Pearson's symbol *cP*2). The two intermetallic compounds labeled γ and ε have Pearson's symbols *tP*4 and *hP*2, respectively. The δ phase, occurring in a small concentration region at high temperatures, also has the *B*2 structure. At the Zn end of the phase diagram, the equilibrium phase is a hexagonal-closed-packed solid solution, with relatively low solubility of Cu.

Phase diagrams convey, in a concise and efficient manner, a great deal of information on the stability of different phases, on the occurrence of order-disorder transitions, and on the nature of these transitions. A well-known example is the CuAu phase diagram shown in Fig. 17. In this case, the system crystallizes from the liquid into a disordered fcc phase. As seen in the phase diagram, this disordered phase is stable over the entire composition range of the alloy. As the temperature is lowered, however, the phase diagram indicates that the disordered Cu-Au alloys undergo several ordering transitions. Around a Cu concentration of 0.25, the transition is from fcc solid solution to the $CuAu_3$ intermetallic compound that has the $L1_2$ structure. For larger concentrations of Cu the complexity and richness of the phase diagram increases: near a 1:1 stoichiometry the disordered solid solution first transforms into the CuAuII long-period structure, changing, at slightly lower temperatures, into the short-period CuAu ($L1_0$) phase. A similar sequence of transitions is found for higher Cu concentrations: the disordered phase also transforms to a long-period structure, labeled Cu_3AuII in Fig. 17, which, in turn, transforms to the Cu_3Au ($L1_0$) structure at lower temperatures. Finally, since all phases in Fig. 17 are separated by two-phase regions, it also follows from the phase diagram that all transitions are of first order.

2. SUMMARY

A few examples of well-known physical systems have been selected with the idea of exposing the fundamental and common aspects involved in the phenomenon of order-disorder transitions. Some well-known examples of ordering systems are followed through, starting with a description of their structure and symmetries and ending with a description of their equilibrium phase diagrams. The expectation is that readers will be able to transfer the knowledge gained here to other cases of interest since, quite clearly, a full treatment of the subject matter is beyond the scope of this article. There are excellent monographs, articles, and compilations that bring together the collective effort of over 50 years of intense research on order-disorder transitions. Some of these are listed below and recommended for further reading.

On a general note, it is perhaps worth while to reiterate two points that hopefully follow from the previous sections. First is the fact that the transition from order to disorder is accompanied by drastic changes in physical properties. Second, the diversity of ordered structures, either seen in nature or yet to be discovered, appear, if not endless, extremely rich. Consequently, understanding and controlling order-disorder transitions will continue to the be a key element in our efforts to tailor physical systems with desired properties.

Works Cited

Henry, N. F. M., Lonsdale, K. (Eds.) (1965), *International Tables for X-Ray Crystallography*, Birmingham, England: Kynoch Press.

Landau, L. D., Lifshitz, E. M. (1969), *Statistical Physics*, 2nd ed., Oxford: Pergamon Press.

Villars, P., Calvert, L. D. (1985), *Pearson's Handbook of Crystallographic Data for Intermetallic Phases*, Warrendale, OH: ASM International.

Further Reading

Krivoglaz, M. A., Smirnov, A. A. (1964), *The Theory of Order-Disorder in Alloys*, London: McDonald.

Ducastelle, F. (1991), "Order and Phase Stability in Alloys," in: F. R. de Boer, D. G. Pettifor (Eds.), *Cohesion and Structure*, Vol. 3, Amsterdam: North-Holland.

Kachaturyan, A. G. (1983), *The Theory of Structural Transformations in Solids*, New York: Wiley.

OSCILLATORS

PAUL D. COLEMAN, *Department of Electrical and Computer Engineering, University of Illinois, Urbana, Illinois, U.S.A.*

	Introduction	17
1.	**Oscillator Principles**	17
2.	**Low-Frequency Triode Transistor Oscillators**	19
2.1	Two-Port Devices	19
2.2	One-Port Devices	21
3.	**Microwave Conduction-Current Oscillators**	24
3.1	Reflex Klystrons	24
3.2	Magnetron	25
3.3	Backward-Wave Oscillators	27
3.4	Gyrotron	28
3.4.1	Gyrotron Principles	28
3.4.2	Mathematical Models of the Gyrotron	30
4.	**Novel Conduction-Current Oscillators**	30
4.1	High-Energy Free-Electron Oscillator	30
4.2	Coherent Cherenkov Oscillators	32
4.2.1	The Boundary-Value Problem	32
4.2.2	Self-Excited Cherenkov Oscillator	33
5.	**Relaxation Oscillators**	34
6.	**Polarization-Current (Laser) Oscillators**	34
	Works Cited	36

INTRODUCTION

The first generation of electromagnetic radiation was by H. Hertz in 1886, but the electronic age was born with the invention of the triode vacuum tube by L. de Forest in 1906. Vacuum-tube oscillators, in their many forms (triodes, klystrons, magnetrons, gyrotrons, backward-wave tubes, etc.), were the only sources of coherent electromagnetic radiation until the transistor was invented in 1947. In the last 40 years, the transistor and other solid-state devices have replaced the vacuum tube in all applications requiring power levels of less than a few watts at frequencies up to the order of 100 GHz. The vacuum tube cannot compete with the small size, cost, and heat dissipation of the transistor. However, for high-power applications, the vacuum tube is difficult to challenge, with relativistic free-electron oscillators, for example, being able to generate very large amounts of power from microwave to optical frequencies.

The development of vacuum-tube oscillators was driven by the challenge of extending their frequency of operation to higher and higher frequencies. The development of transistor and solid-state oscillators has been driven by circuit integration, cost, size, and heat dissipation. Both the vacuum tube and transistor are conduction-current devices and, hence, share basic physics problems and limitations.

In this article, the point of view taken is that only two types of amplifier oscillators exist: conduction current and polarization current. This point of view permits one to concentrate on the basic oscillator physics of each class rather than to regard various oscillators as fundamentally different.

1. OSCILLATOR PRINCIPLES

Sinusoidal oscillators in general consist of an amplifier with positive feedback and a resonator. Two classes of amplifiers exist: those based upon conduction currents and those based upon polarization currents. From a circuit's point of view, conduction-current devices can be two-port (three-terminal) or one-port (two-terminal), while polarization-current amplifiers are two-terminal devices.

The basic elements of a two-port oscillator are illustrated in Fig. 1. Power conservation for the steady state requires that

$$G\mathcal{P}_i = \mathcal{P}_i + \mathcal{P}_R + \mathcal{P}_F + \mathcal{P}_o, \qquad (1)$$

3-527-28135-5/95/$5.00 + .50

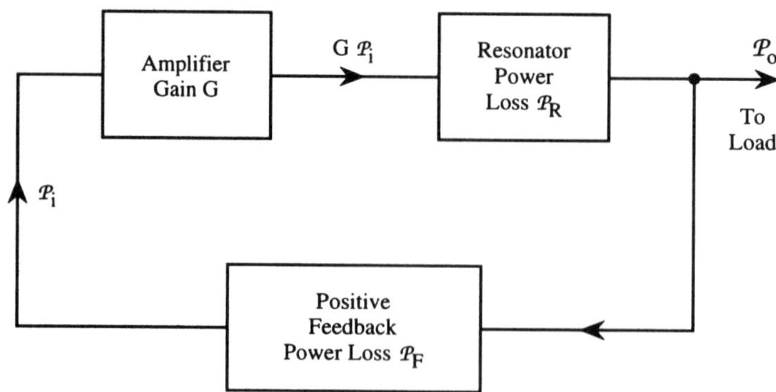

FIG. 1. Generic oscillator circuit.

where $\mathcal{P}_R$ and $\mathcal{P}_F$ are the power losses in the resonator and feedback system, $\mathcal{P}_o$ is the power output to the load, $\mathcal{P}_i$ the input power to the amplifier, and G the power gain. Gain nonlinearity must exist in the amplifier to limit $G\,\mathcal{P}_i$ to satisfy Eq. (1), although it could also be built into the feedback network. The oscillations build up from noise that always exists in a circuit at finite temperature, or from spontaneous radiation in the case of lasers. The two Barkhausen conditions for oscillation are as follows:

1. G must be greater than one; and
2. the phase shifts around the closed loop must be $2\pi n$ for the voltage and/or field to be single valued.

Maxwell's equations

$$\nabla \times \mathbf{E} + \mu_0 \frac{\partial \mathbf{H}}{\partial t} = -\mu_0 \frac{\partial \mathbf{M}}{\partial t}$$

and (2)

$$\nabla \times \mathbf{H} - \epsilon_0 \frac{\partial \mathbf{E}}{\partial t} = \mathbf{J} + \frac{\partial \mathbf{P}}{\partial t}$$

along with Newton's and Schrödinger's equations are the starting points in determining the sources for electromagnetic (EM) fields and conditions for gain. It is readily seen that only two sources for EM fields exist: conduction currents $\mathbf{J}$ and polarization terms $\partial\mathbf{P}/\partial t$ and $-\mu_0\partial\mathbf{M}/\partial t$. From the invention of the triode vacuum tube by de Forest (1907) to the invention of the laser in the 1950s, the only amplifiers of EM fields were conduction-current devices. Now that polarization-current ($\partial\mathbf{P}/\partial t$) laser devices have been invented, one may wonder if any new breakthroughs on oscillators in a fundamental sense will be made in the future.

Powers ($\mathcal{P}$) absorbed or generated by a $\mathbf{J}$ or $\partial\mathbf{P}/\partial t$ device are given by the expressions

$$\mathcal{P} = \int \mathbf{E} \cdot \mathbf{J}\, dv$$

or (3)

$$\mathcal{P} = \int \mathbf{E} \cdot \frac{\partial \mathbf{P}}{\partial t}\, dv,$$

where v is the volume.

Since the coherent field $\mathbf{E}$ of interest will vary as $e^{j\omega t}$, $\mathbf{J}$ and/or $\partial\mathbf{P}/\partial t$ must have a Fourier component at ω for the average of $\mathcal{P}$ over a cycle to be different from zero. In particular, $\mathbf{J}$ and/or $\partial\mathbf{P}/\partial t$ must be out of phase with $\mathbf{E}$ for power to be generated. This means that in $\mathbf{J}$ devices the electron beam cannot be a dc beam but must be "bunched" or modulated in some manner. For a $\partial\mathbf{P}/\partial t$ or laser device, inverted populations or nonequilibrium distribution of the laser system must be obtained by means of the appropriate excitation or pumping scheme.

Given a two-level system with an electric dipole transition, it can be shown (Pantell and Puthoff, 1969) that $\mathbf{P}$ is given by the equation

$$\frac{\partial^2 \mathbf{P}}{\partial t^2} + \frac{2}{T_2}\frac{\partial \mathbf{P}}{\partial t} + \Omega^2 \mathbf{P} = \frac{2\Omega}{3\hbar}|\mu_{12}|^2 \times (N_1 - N_2)\mathbf{E}, \quad (4)$$

where $\hbar\Omega$ is the energy difference of the two levels, μ_{12} is the dipole matrix element (a measure of the coupling of the levels), N is the population, T_2 is a relaxation time, and

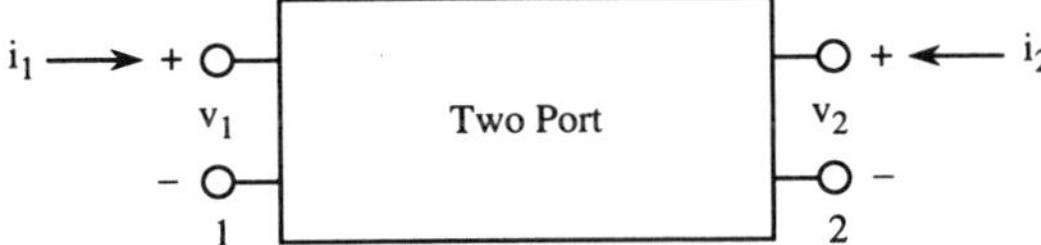

FIG. 2. Two-port circuit.

E is the field. Notice that for **E** sinusoidal, $\partial \mathbf{P}/\partial t$ will be sinusoidal, and that $\partial \mathbf{P}/\partial t$ will change signs when $N_2 > N_1$, the conditions for $\mathcal{P}_{AV}$ to be negative. From this point of view, there was no need to mention the words stimulated emission of radiation.

A principal practical motivation for the study of oscillators has been the quest for high oscillator frequencies, to provide more bandwidth and information capability. Conduction-current devices, with the exception of the free-electron laser, appear to have reached their practical frequency limit around 100 GHz. Lasers ($\partial \mathbf{P}/\partial t$ devices) can generate EM radiation almost across the spectrum from UHF to the ultraviolet.

In the next sections, various types of **J** and $\partial \mathbf{P}/\partial t$ devices will be examined from the point of view of the basic oscillator principles just described.

2. LOW-FREQUENCY TRIODE OR TRANSISTOR OSCILLATORS

2.1 Two-Port Devices

Low-frequency three-terminal, vacuum-tube and transistor oscillators employ lumped-circuit *LCR* circuit elements. These devices can be represented by the two-port network (Millman and Halhias, 1972) displayed in Fig. 2 with the hybrid equations

$$v_1 = h_{11}i_1 + h_{12}v_2, \tag{5}$$

$$i_2 = h_{21}i_1 + h_{22}v_2, \tag{6}$$

where the voltage and current variables represent small variations about the quiescent operating point. The h-parameter equivalent circuit, where terminals 1 and 2 are common, can be readily seen to be that given in Fig. 3. While the active elements on the circuits described here are semiconductor devices, earlier circuits used vacuum tubes in similar circuit configurations.

Given the h parameter of the amplifying two-port device, one needs to add an appropriate positive-feedback circuit to make an oscillator. The transfer gain A_f with feedback is given by the relationship

$$A_f = A/(1 + \beta A), \tag{7}$$

where β is a complex function of the frequency and A is zero-feedback gain. Oscillations can occur when $\beta A = -1$, the Nyquist criterion for instability.

The basic physics is concerned with processes within the "black box" two-port that yields the desirable h parameters of Eqs. (5) and (6) and how they might be changed. In both triode vacuum tubes and transistors, gain (or energy addition) is obtained by converting charge kinetic and/or potential energy into EM energy. In triodes and transistors, both dc ($\mathbf{E}_0$) and ac ($\mathbf{E}_1 \sin\omega t$) fields are present in the same region. Hence, the energy relations are from Newton's equation

$$m\mathbf{u}\cdot d\mathbf{u} = q\mathbf{E}_0\cdot d\mathbf{r} - q\mathbf{E}_1 \sin\omega t\cdot d\mathbf{r}. \tag{8}$$

The dc field is performing work on the charge q, which in turn is performing work on the ac field.

A key feature of a triode tube or bipolar junction transistor (BJT) transistor is the ability of the grid to control the flow of currents to the plate of the tube or the collector current to control the base current, respectively. By this means, charges are allowed into the interaction region of the device mainly

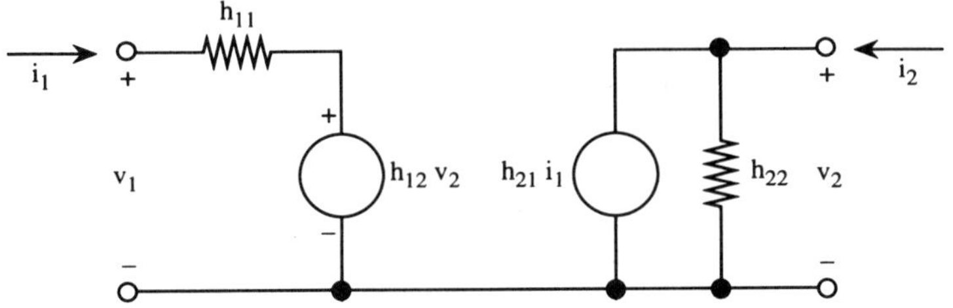

FIG. 3. Equivalent h-parameter circuit of a transistor.

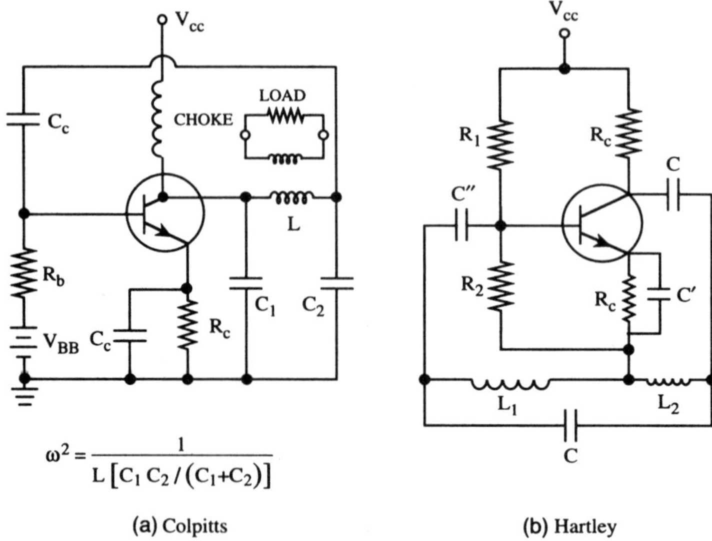

(a) Colpitts

(b) Hartley

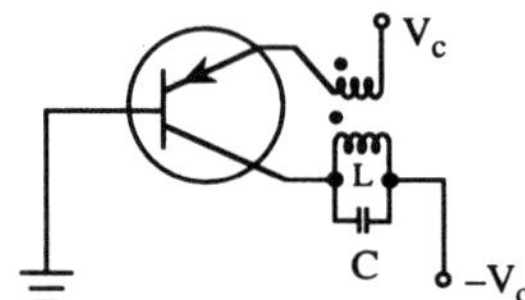

(c) Basic Transistor Oscillator Using L-C Circuit as the Resonator

FIG. 4. Basic Colpitts and Hartley oscillator configurations.

when they can deliver kinetic energy to the ac field.

A number of 180° feedback circuits have been developed to go with vacuum tubes and transistors to make oscillators, two of which have been given distinguishing names, Hartley and Colpitts. These low-frequency *L-C* oscillators (Hunter, 1956) are shown in Fig. 4, along with a simplified circuit to illustrate better the 180° feedback feature.

A phase shift of 180° in the positive-feedback circuit of an amplifier-oscillator can also be provided by a *RC* circuit as illustrated in Fig. 5. The loop gain condition yields the results

$$\omega_0 = [RC\sqrt{6 + 4R_C/R}]^{-1}$$

and (9)

$$\frac{R}{R_C} = \frac{h_{21} - 23}{58} > 0.$$

Piezoelectric quartz crystals with characteristic resonant frequencies can also be used in positive-feedback amplifier circuits to make an oscillator as shown in Fig. 6. Crystal oscillators can be made to be very frequency stable, having been used as frequency standards for many years.

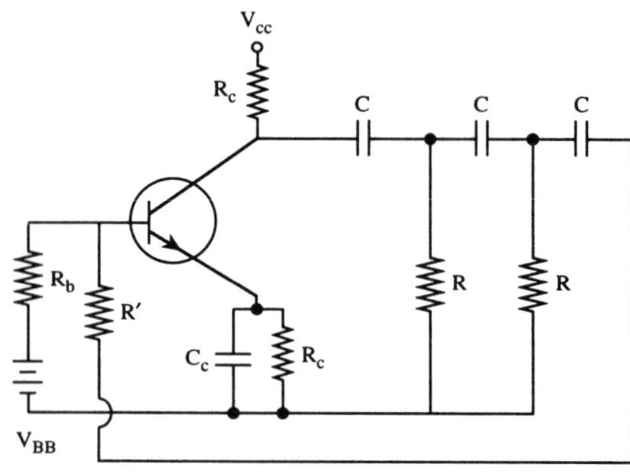

FIG. 5. Phase-shift oscillator circuit.

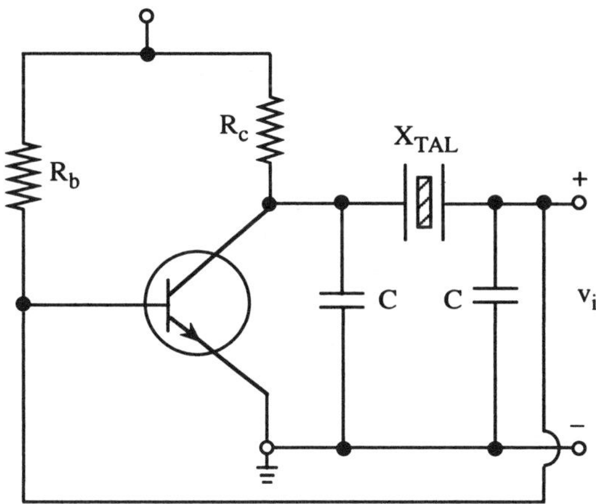

FIG. 6. Crystal oscillator circuit.

2.2 One-Port Devices

A number of two-terminal, one-port conduction-current devices have been discovered that display a negative ac resistance based upon a variety of physical processes. Their *I-V* characteristics are of the general form displayed in Fig. 7. Over the voltage range $V_1 < V < V_2$, assuming a linear relationship, it is seen that

$$I = a - V/r$$

and (10)

$$\frac{dV}{dI} = -r,$$

i.e., the device displays a negative differential resistance r.

To see, from a circuit point of view, how this diode amplifies, consider the simplified circuit of Fig. 8. The ac circuit equations are

$$e_s = i_D R + e_D = i_D(R - r)$$

with (11)

$$e = i_D R = e_s \frac{R}{R - r},$$

where e_s is the ac voltage signal and e the voltage output. Clearly, the output voltage e can exceed the input voltage e_s.

If the ac current and voltage amplitudes exceed the values $I_1 - I_0 = \Delta I$ and $V_2 - V_0 = \Delta V$, losses will occur to limit their values. The currents I_1 and I_2 and voltages V_1 and V_2 define the limits of the negative differential resistance region. Hence, the maximum ac power available from the diode is

$$\mathcal{P}_{AV} \leq \tfrac{1}{2}(\Delta I)(\Delta V). \tag{12}$$

The two-terminal devices of major interest are (Coleman, 1982; Liao, 1985) the Esaki tunnel diode, the Gunn diode, the IMPATT diode, and the resonant tunneling diode.

The Esaki diode consists of a heavily doped *p-n* junction where the Fermi levels are located within the bands themselves. As seen from Fig. 9, when a positive bias is applied electrons can tunnel from the filled electron states on the *n* side to the empty states on the *p* side at constant energy. With further

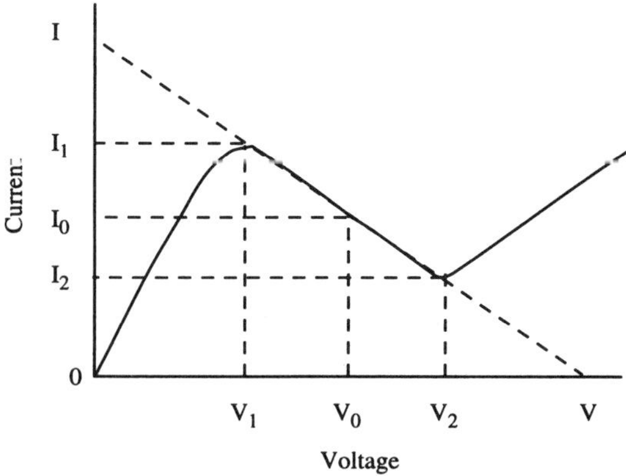

FIG. 7. Schematic *I–V* curves of a negative differential resistance device.

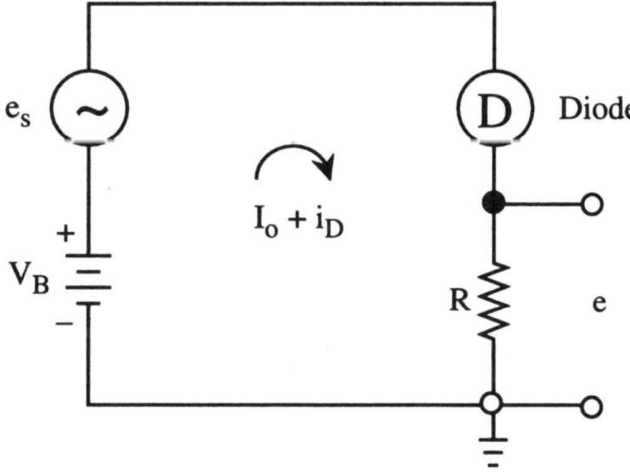

FIG. 8. Negative differential resistance diode amplifier circuit.

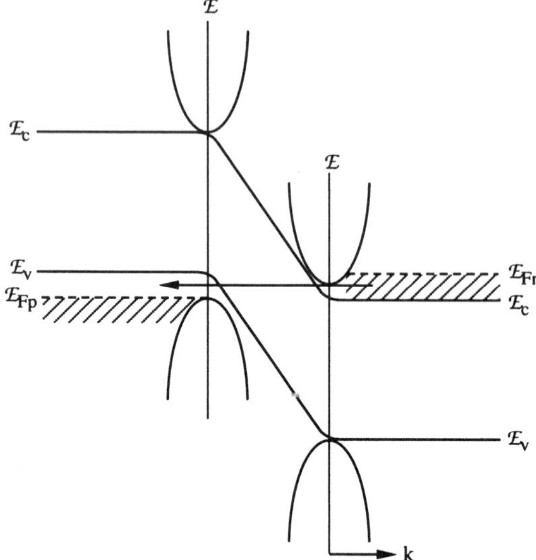

(a) Esaki Diode - Forward Bias

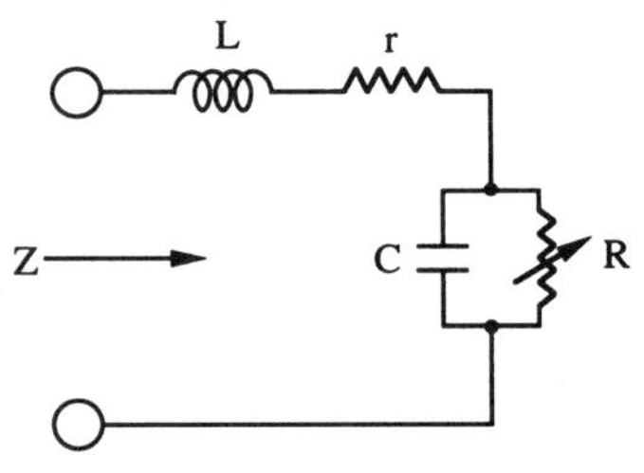

(b) Equivalent Circuit

FIG. 9. Esaki diode characteristics.

positive bias, the current decreases since there are fewer available states opposite the filled states. While the tunneling process is very fast, the negative resistance of the diode goes to zero at the frequency f_c given by the relationship

$$f_c = \frac{\sqrt{|R|/r - 1}}{2\pi C|R|}, \tag{13}$$

as can be determined from the equivalent circuit. With typical values for the circuit parameters, $f_c \sim$ 10–40 GHz.

The resonant tunnel diode (RTD) is a heterostructure with a double-barrier single-quantum well as illustrated in Fig. 10. Maximum current (resonant tunneling) occurs when the energy of the tunneling electrons is approximately in line with the quantized energy level for electrons in the quantum well. The equivalent circuit for the RTD is the same as that for the Esaki diode, but the cutoff frequency f_c can be made much greater, exceeding 100 GHz. Unfortunately, the product $(\Delta I) \times (\Delta V)$ of Eq. (12) is small so that only small amounts of power can be produced with this diode.

The Gunn diode is a bulk semiconductor electron-transfer device wherein electrons are transferred from the central valley of the material (i.e., GaAs) to an upper satellite valley where their velocity $(\hbar^{-1} d\mathcal{E}/dk)$ decreases when an electric field is applied to the semiconductor. This results in a negative differential (du/dE) velocity–field curve for the electric field greater than E_T as illustrated in Fig. 11. The usual semiconductor equations for the current J, electric field E, and electron density n (for one dimension),

$$J = qnu(E) - q\frac{\partial D(E)n}{\partial x} + \epsilon\frac{\partial E}{\partial t}, \tag{14}$$

$$\frac{\partial n}{\partial t} + \frac{1}{q}\frac{\partial J}{\partial x} = 0, \tag{15}$$

$$\frac{\partial E}{\partial x} = \frac{q(n - n_0)}{\epsilon}, \tag{16}$$

apply with the electron velocity u now a function of E. In general, these equations can only be solved by numerical methods. It is found that a number of modes of behavior exist (Copeland, 1967), viz., a transit-time dipole-layer mode, a limited space-charge accumulation mode, and a quenched dipole-layer mode.

Consider a dipole layer, illustrated in Fig. 12, moving from the negative to positive electrode of the semiconductor. The electric field inside the dipole is greater than the fields on either side. Thanks to the velocity property of the material, the electrons to the right of the dipole are moving more rapidly (running away from the dipole) leaving a depletion region while the electrons to the left of the dipole are catching up to form an accumulation region. The dc electron beam is being "bunched" to form the necessary ac current to drive the oscillator, with the frequency $f = u/L$. The dipole layer moves

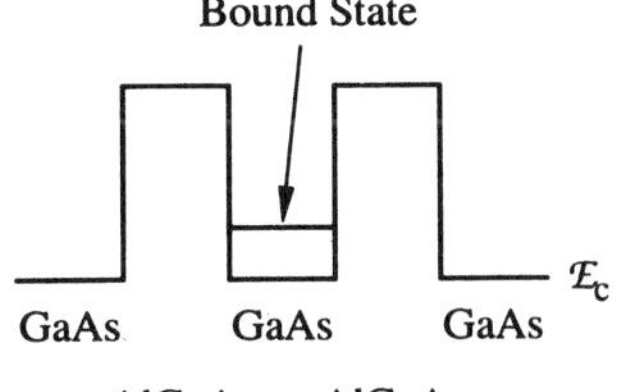

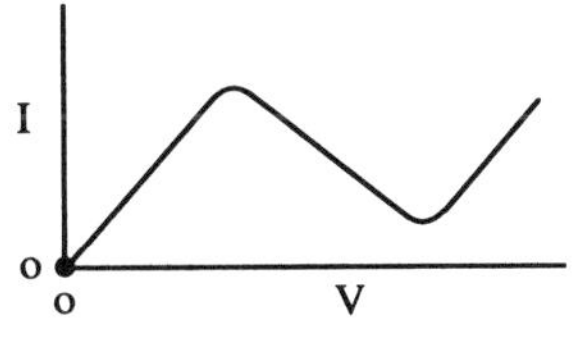

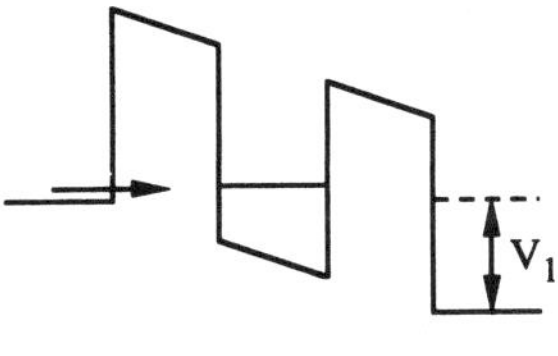

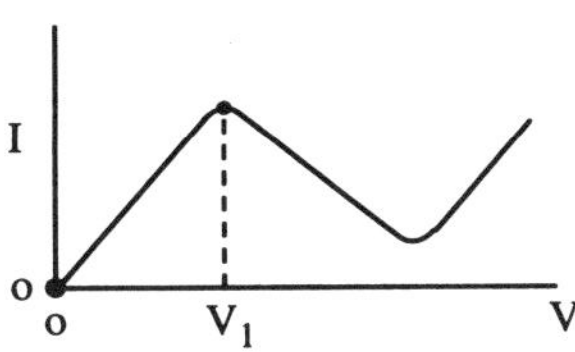

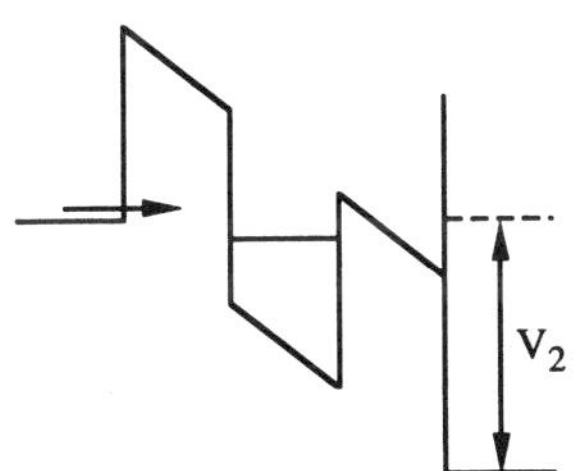

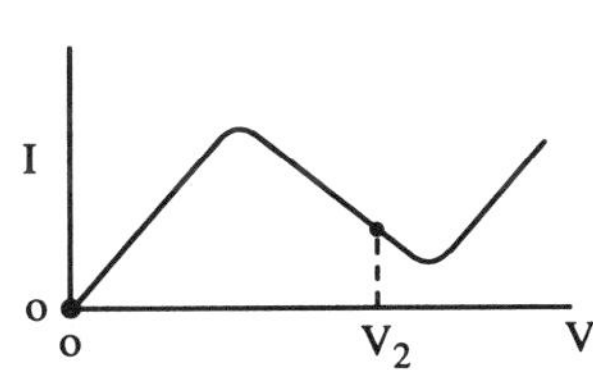

FIG. 10. Resonant tunneling diode characteristics.

through the crystal and disappears at the anode. The electric field then rises to form a new dipole layer to repeat the process. Gunn-diode oscillators can yield several watts of pulsed power at frequencies as great as 100 GHz.

An IMPATT diode is an n^+pip^+ semiconductor structure shown schematically in Fig. 13. It consists of a narrow p avalanche region

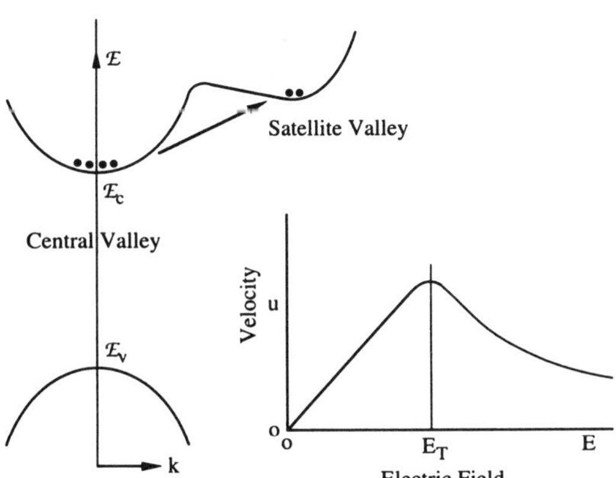

FIG. 11. Gunn diode characteristics.

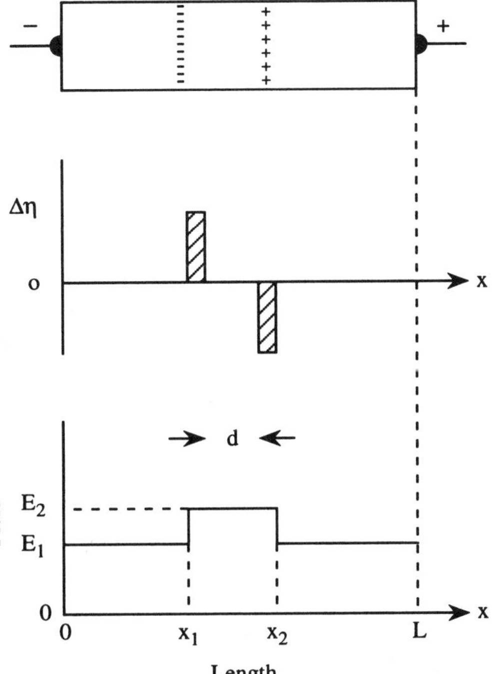

FIG. 12. Schematic of dipole drift in a Gunn diode.

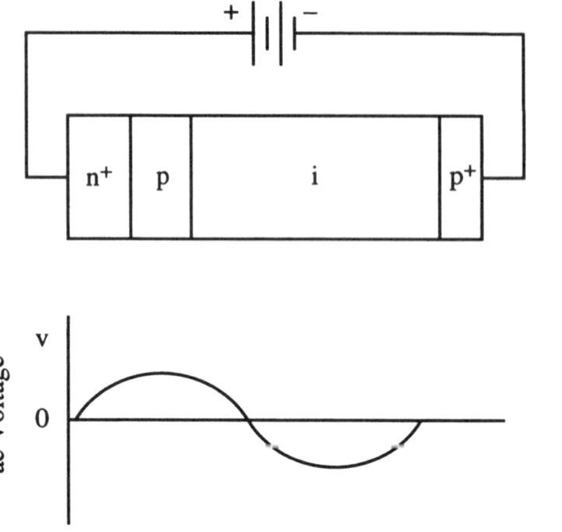

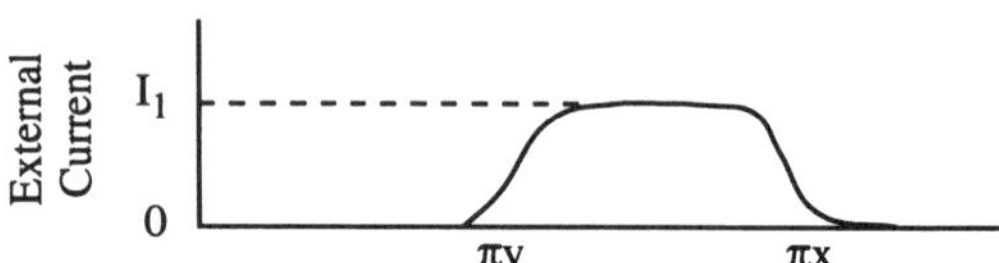

FIG. 13. Voltage and external current in an IMPATT diode.

where carrier multiplication by impact ionization occurs and a drift region where the carriers drift at saturated velocity. A phase-shift delay is caused by the avalanche process and the finite transit time of the carriers drifting through the drift region.

The average power for the wave forms shown is given by the expression

$$\mathcal{P}_{AV} = \frac{1}{2\pi}\int_{y\pi}^{x\pi} I_1 V_1 \sin\omega t(d\omega t) = \frac{V_1 I_0}{\pi}\left(\frac{\cos y\pi - \cos x\pi}{x - y}\right), \tag{17}$$

where $I_1 = 2I_0/(x - y)$ and I_0 is the dc current. Maximum power produced by the diode is seen to be

$$(\mathcal{P}_{AV})_{\max} = -\frac{V_1 I_0}{\pi}\left(\frac{1 + \cos(\pi x)}{x - 1}\right). \tag{18}$$

The avalanche process contributes a 90° inductive phase lag to the current generated in the avalanche region. This current is then injected into the drift region where an additional 90° phase in the external current occurs, thus resulting in the ac voltage and current being 180° out of phase. The diode will oscillate at a frequency and level where the diode plus external circuit admittance is zero.

3. MICROWAVE CONDUCTION-CURRENT OSCILLATORS

The oscillators discussed thus far were essentially lumped-circuit devices with *LCR* positive-feedback circuits, although Esaki and resonant tunneling diodes plus Gunn and IMPATT diodes could be embedded into transmission lines and microwave resonators. At microwave and millimeter-wave frequencies, propagation effects are present that require wave and field analysis to describe the amplification process. For tube or free-electron devices, methods need to be employed to cope with circuit reactance, circuit-tube loss, and electron transit-time limitations to obtain these higher frequencies.

3.1 Reflex Klystrons

The reflex klystron (see Fig. 14) consists of an electron-emitting cathode, a current-control grid, an anode-reentrant microwave resonator, and a repeller electrode. Electrons injected from the cathode are first velocity

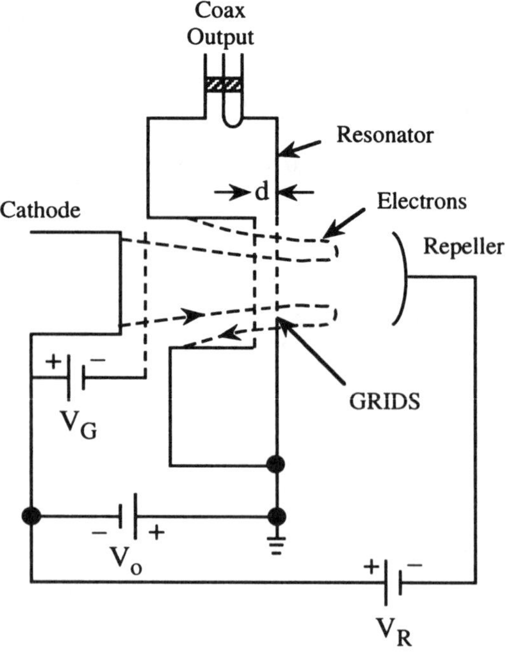

FIG. 14. Schematic diagram of a reflex klystron.

modulated by the cavity–gap field where the first electrons are decelerated and the later electrons accelerated to play a "handicap race" so as to yield a bunched ac electron beam. The electrons are then turned around by the repeller field and pass through the cavity–gap in the reverse direction 180° out of phase with the resonator field to give up a portion of their kinetic energy.

The determination of the ac currents in the bunched electron beam is obtained from Newton's equation as follows:

$$\int_{u_0}^{u} du = \frac{qE_1}{m} \int_{t_a - T/2}^{t_a + T/2} e^{j\omega t} dt, \tag{19}$$

where the transit time T across the resonator gap is given by $u_0 T \simeq d$ where d is the gap spacing, u_0 the dc beam velocity, and t_a the time the electrons reach the midpoint of the gap. Integration of Eq. (19) yields

$$\begin{aligned} u &= u_0\left[1 + \frac{2V_1}{V_0}\left(\frac{\sin\theta/2}{\theta/2}\right)e^{i\omega t_a}\right] \\ &= u_0[1 + 2\alpha A e^{i\omega t_a}], \end{aligned} \tag{20}$$

where $V_1 = E_1 d$, $mu_0^2 = 2qV_0$, $\theta = \omega T = \omega d/u_0$, $\alpha = V_1/V_0$, and $A = \sin\frac{1}{2}\theta/\frac{1}{2}\theta$. The electrons are returned to the resonator gap at time t_b given by

$$t_b = t_a + \frac{\ell}{u} \simeq t_a + \frac{\ell}{u_0}\left(1 - \frac{A\alpha}{2}\sin\omega t_a\right). \tag{21}$$

It is seen from Eqs. (20) and (21) that the transit time contained in the factor A is a key limitation in extending the frequency of a conduction-current device.

Finally, from conservation of electrons, the return current is

$$I_b = I_o\left|\frac{dt_a}{dt_o}\right| = I_0\left|\frac{1}{1 - k\cos\omega t_a}\right|, \tag{22}$$

where $k = (A\alpha/2)(\omega d/u_0) = (A\alpha/2)(\omega T_0)$.

The current I_b is a periodic function, which can be analyzed into its Fourier components to yield for the fundamental

$$I_{b_1} = j2I_0 J_1(k)\, e^{-j\omega t_o}, \tag{23}$$

where $J_1(k)$ is a first-order Bessel function, and the power generated,

$$\mathcal{P}_{AV} = V_1 I_0 J_1(k) \sin\omega t_o. \tag{24}$$

A reflex klystron can be frequency tuned mechanically by varying the gap spacing d, and electronically of the order of 1% by varying the repeller voltage. Reflex klystrons can be made to operate up to the order of 100 GHz. Power levels of the order of 10 MW can be obtained in the microwave range. For many years, they have been very useful oscillators for laboratory measurements, local oscillators in radars, linear accelerators, and missiles. Harmonic signals obtained from reflex klystrons driving Schottky diodes provided the first coherent millimeter–submillimeter radiation for the microwave spectroscopy of molecules. Microwave spectroscopy made key contributions to the invention of the NH_3 laser.

3.2 Magnetron

The magnetron (Coleman, 1982; Liao, 1985) is a crossed-field ($\mathbf{E}\cdot\mathbf{B} = 0$), termed M-type electron tube that was invented by A. W. Hull in 1921, although it was not developed into a high-power generator until World War II when the need for radar existed. Its high efficiency of 50%–80% makes it useful for many commercial uses such as in a microwave oven. It can produce pulsed powers as large as 10 MW in the microwave range, and it can be made to operate at frequencies approaching 100 GHz.

There are a number of different magnetron structures: a split-anode, a cylindrical traveling-wave slot, a cylindrical traveling-wave hole and slot, a rising sun, and a linear slotted magnetron, three of which are shown in Fig. 15. The electrons are tightly bunched when they are emitted from the large, high-current cathode, which leads to the high efficiency of the tube.

The fields in the interaction space are not simple and the charge motion, including space charge, is very complicated so that an analytical solution for a magnetron oscillator is not possible. Some of the features of a magnetron can be seen by considering a linear magnetron configuration.

Consider first the motion of electrons in crossed dc ($\mathbf{E}\cdot\mathbf{B} = 0$) fields where $\mathbf{E} = \hat{\mathbf{y}}\mathbf{E}$ and $\mathbf{B} = \hat{\mathbf{z}}\mathbf{B}$. Then Newton's equations are

$$\ddot{x} = -\omega_c \dot{y}$$

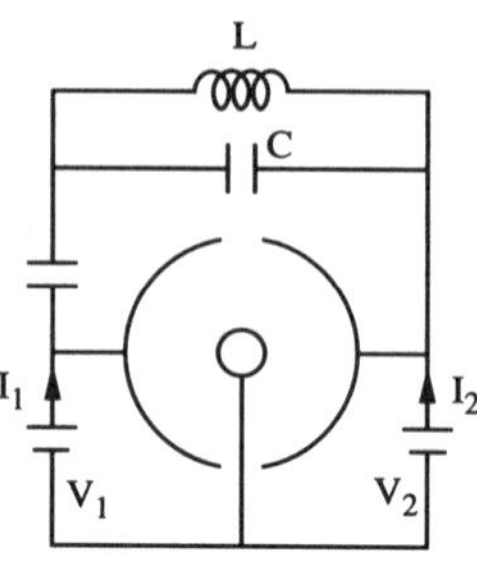

(a) Circuit Configuration

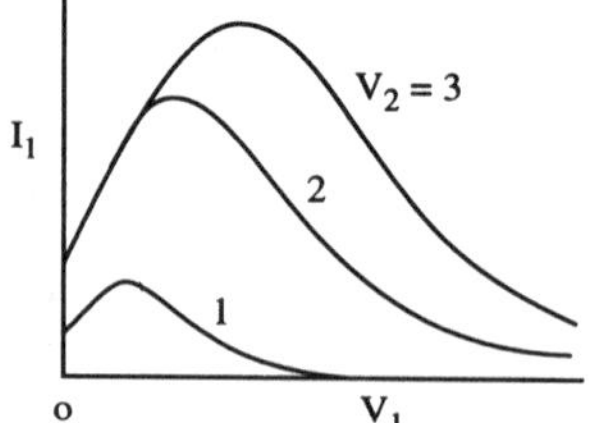

(b) Current-Voltage Characteristics

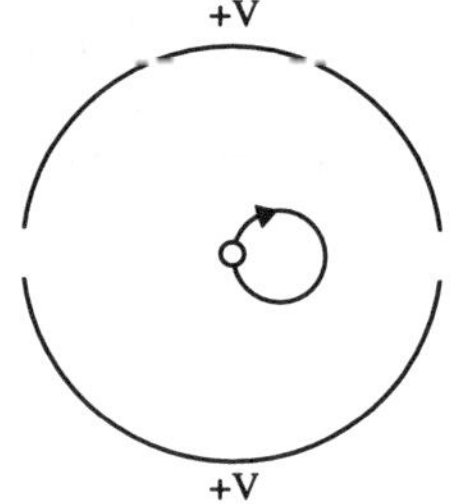

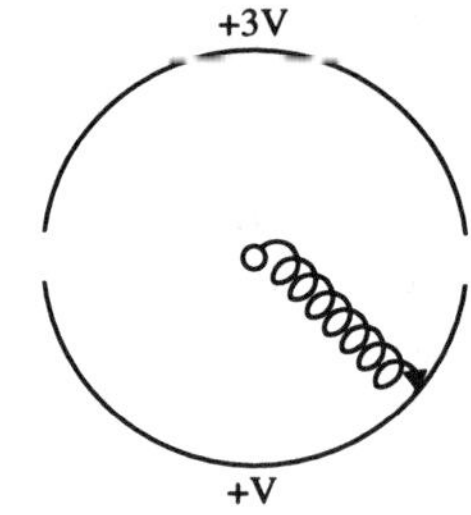

(c) Electron Paths in Split and DE Magnetron

(d) Lines of Force in a Cylindrical Magnetron

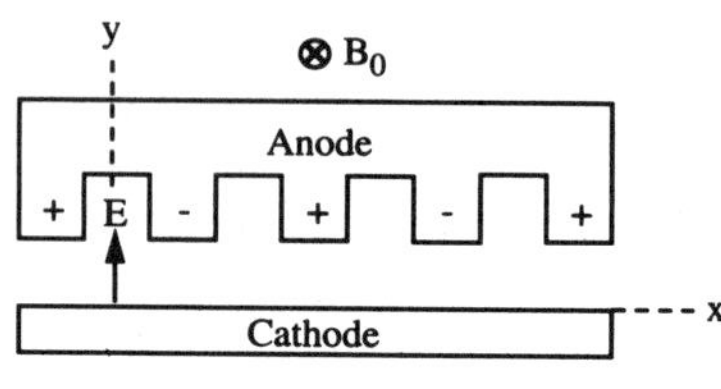

(e) Linear Magnetron Configuration

FIG. 15. Various magnetron configurations.

and

$$\ddot{y} = -\, eE/m + \omega_c \dot{x}, \tag{25}$$

where $\omega_c = eB/m$, with the solution for $\dot{x}$ being

$$\dot{x} = -\omega_c R \sin(\omega_c t - \alpha) + E/B \tag{26}$$

and

$$y = y_0 + R \sin(\omega_c t - \alpha). \tag{27}$$

Equation (26) suggests that to realize an oscillator a travelling-wave field moving with the drift velocity E/B should be used. Since E/B will be much less than the velocity of light c, a heavily loaded slotted circuit, shown in Fig. 15(e), will be needed to reduce the phase velocity to E/B.

Adding rf fields to Eq. (20) results in a problem too difficult to solve analytically. Hence, a word description of the results will be given. Electrons with certain phases drift toward the cathode while other electrons in other phases drift toward the anode and do work in the rf field. Here the kinetic energy stays approximately constant (E/B ~ constant), but the full potential energy is delivered to the rf field. Only a small amount of energy is lost to the cathode.

For a cylindrical magnetron, Newton's equations become in polar coordinates (r,θ)

$$\ddot{r} - r\dot{\theta}^2 = -\omega_c r\dot{\theta} + \frac{e}{m}\frac{dV}{dr}, \tag{28}$$

where $E_r = -dV/dr$, and

$$\frac{1}{r}\frac{d}{dt}(r^2\dot{\theta}) = \omega_c\dot{r}, \tag{29}$$

where $\omega_c = eB/m$. Given that the radial and angular velocity are zero at the cathode where $r = r_c$, Eq. (29) yields

$$\dot{\theta} = \tfrac{1}{2}\omega_c[1 - (r_c/r)^2]. \tag{30}$$

Now the energy equation is

$$\tfrac{1}{2}m(\dot{r}^2 + r^2\dot{\theta}^2) = eV. \tag{31}$$

Eliminating $\dot{\theta}$ between Eqs. (30) and (31) and using the condition $\dot{r} = 0$ for $r = r_p$ yields the cutoff voltage

$$V_c = \frac{er_p^2}{8m}B\left[1 - \left(\frac{r_c}{r_p}\right)^2\right]^2, \tag{32}$$

the voltage below which no electrons will reach the plate. Oscillations are usually observed in the vicinity of cutoff.

The slot magnetron structure of Fig. 15(d) is a resonator with certain allowed frequencies, which can be found by requiring the phase shift around the circumference to be $2\pi m$ to bring the field back to its initial value. Hence, given the phase factor β_0 of the rotating electromagnetic wave, it is seen that

$$\beta_0 NL = 2\pi m, \tag{33}$$

where N is the number of slots, L the periodic distance, and m an integer.

The mode for $m = N/2$ where N is even is termed the π mode of the resonator. Other modes are double in that they allow traveling waves to propagate in either of two possible directions around the magnetron. The large currents in a magnetron require that space charge be included in the analysis of charge bunching, again a problem that cannot be solved analytically. Numerical calculations yield the results that the electrons bunch into a configuration like the spokes of a wheel (see Fig. 16) that rotate in synchronism with the tangential electric field.

Magnetrons are used in a pulsed mode with pulse lengths typically 1–10 μs at pulse repetition frequencies of the order of 100–400 Hz.

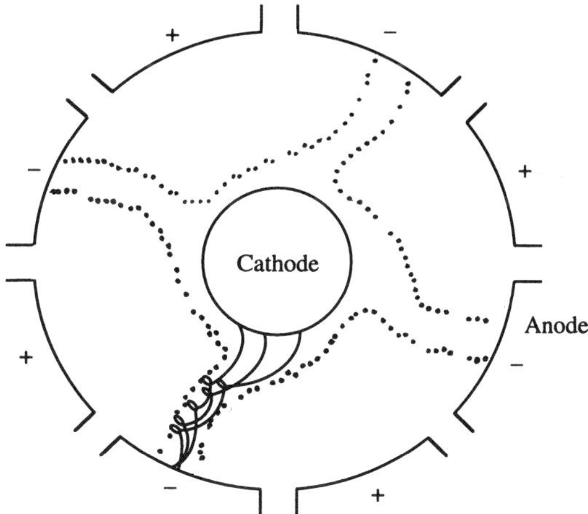

FIG. 16. Schematic of electron trajectories in a cylindrical magnetron.

For microwave oven applications, they can be operated for many minutes at power levels from 400 to 800 W.

3.3 Backward-Wave Oscillators

In order for a linear electron beam to interact with an electromagnetic traveling wave, be bunched, and deliver energy to the field, the electron velocity u_0 must be slightly greater than the phase velocity v_p of the wave. Since the normal transmission line or waveguide has a phase velocity $v_p \geq c$ where c is the velocity of light, some means must be found to reduce the phase velocity by a substantial amount if relativistic beam voltages are not used (Karp, 1955; Kompfner, 1947). Phase-velocity reduction of a waveguide can be made by making the structure periodic as illustrated in Fig. 17. What is also needed is a transverse-magnetic (TM) mode of propagation, where an axial field E_z exists on the axis and the transverse field is zero on the axis.

Periodic loading of the waveguide changes the ω vs β characteristics as shown in Figs. 17(c) and 17(d) so that the loaded guide now takes on filter stop-band properties. The calculation of β as a function of ω is a difficult task, but some general results can readily be obtained. Since the field is spatially periodic, one can expand the axial electric field in a

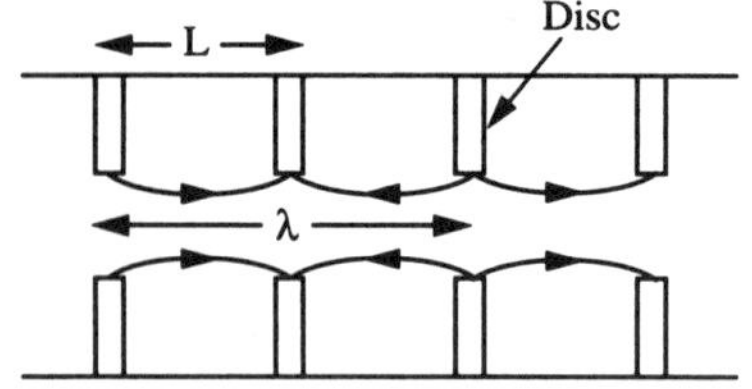

(a) Unloaded Guide

(b) Periodically Loaded Guide

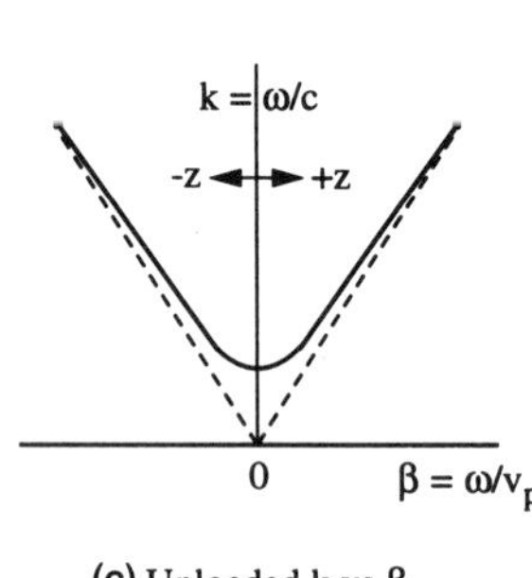

(c) Unloaded k vs β

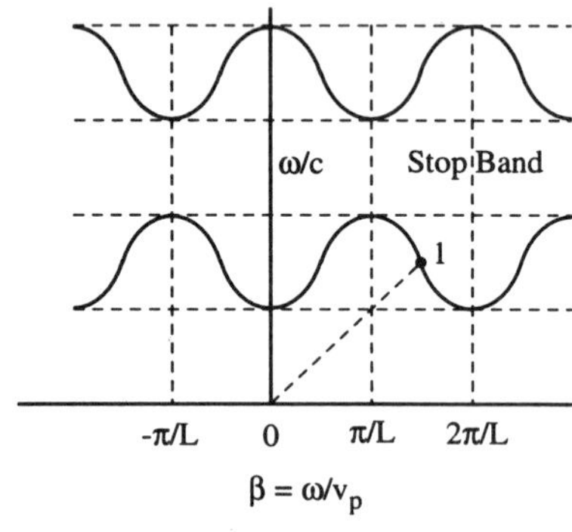

(d) Loaded k vs β

FIG. 17. Characteristics of unloaded and periodically loaded waveguides.

Fourier series:

$$E_z = \sum_n E_n \exp\{j[\omega t - (\beta_0 + 2\pi n/L)z]\}$$
$$= \sum_n E_n \exp[j(\omega t - \beta_n z)], \quad (34)$$

where

$$\beta_n = \beta_0 + 2\pi p/L = \omega/v_{pn}, \quad (35)$$

i.e., the phase velocity will get smaller as n increases.

It will also be observed from Fig. 17(d) that for $n = 1$ (the first spatial harmonic), the group velocity v_{gr} is negative:

$$\frac{1}{v_{gr}} = \frac{d\omega}{d\beta_n} = \frac{d\omega}{d\beta_0} < 0$$

for operation at point 1 on the ω vs β curve.

The gain analysis of a backward-wave oscillator (BWO) is the same as a traveling-wave-tube amplifier (TWT). However, since the group velocity of the spatial harmonic field being used is negative, the power output is taken at the cathode end of the tube as illustrated in Fig. 18. The feedback is internal in the structure. If the beam velocity ($u_0 \sim v_p$) is changed, the frequency of the BWO is changed as point 1 moves along the ω vs β curve.

Backwave oscillators (also termed O-Carcinotrons), as developed by French scientists (Convert *et al.*, 1959), have reached 1 THz in frequency. These nonrelativistic beam tubes are strong competitors as continuous sources of millimeter and submillimeter waves, having achieved 1 W cw at 300 GHz with a beam voltage of the order of 10 kV.

3.4 Gyrotron

3.4.1 Gyrotron Principles The requirement of a synchronization of the electron velocity u_o and phase velocity v_p in a TWT or

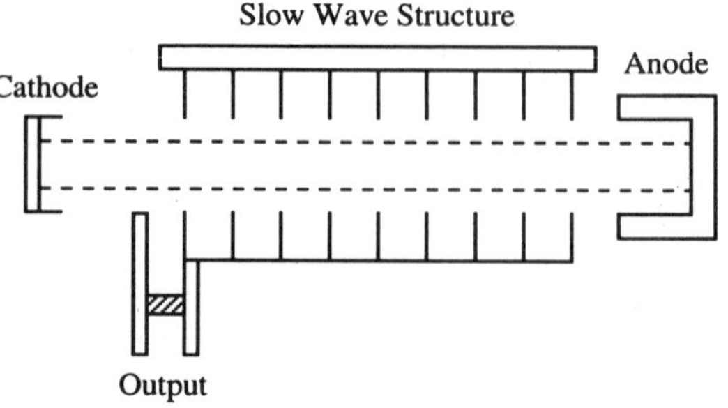

FIG. 18. Backward-wave oscillator schematic.

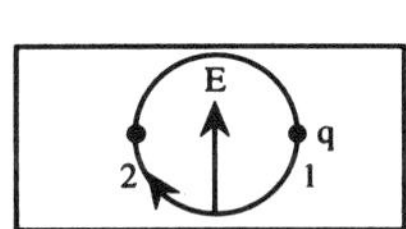

(a) End View of Charge Motion

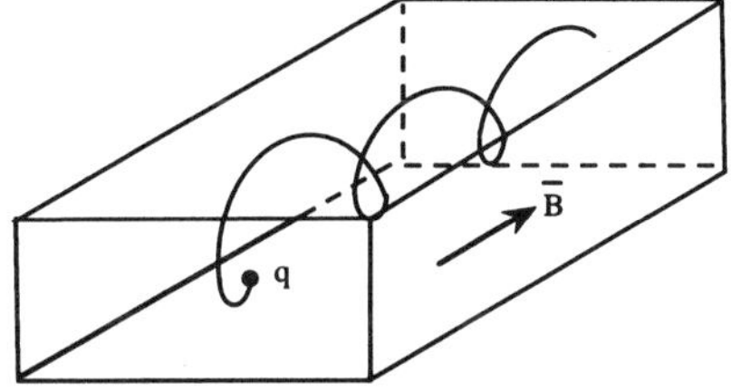

(b) Charge Motion in Waveguide

FIG. 19. Quasi-velocity matching.

BWO meant that the waveguiding structure of the tube had to be periodically loaded to achieve the condition $u_o \simeq v_p$. Loading the waveguide reduced its cross-sectional area. As the frequency of the tube was extended, the periodically loaded guide became uncomfortably small and the electron beam had to approach the loading elements more closely to maintain efficiency. What was necessary to solve this problem was to remove the periodicity from the waveguide and introduce the periodicity on the electron beam (Sprangle and Drobot, 1977). A quasisynchronism between the wave and beam is achieved by allowing the wave to skip past the beam, with the interaction taking place with different "parts" of the wave in consecutive cycles.

For example, consider a spiral electron beam in a field $\mathbf{B} = \hat{\mathbf{z}}B$ in a TE-mode waveguide as illustrated in Fig. 19(b). At point 1 of Fig. 19(a), the charge is doing work on the field. When the charge rotates with angular velocity $\omega_c = qB/m$ to position 2, the field $\mathbf{E}$ needs to reverse directions to continue the interchange. Hence, it can be seen that

$$t_{12} = \frac{\pi}{\omega_c} = \frac{d}{u_z} = \frac{d + \lambda q/2}{v_p} = \frac{d + \pi v_p/\omega}{v_p}; \qquad (36)$$

solving for ω, one obtains a Doppler frequency relation

$$\omega = \frac{\omega_c}{1 - u_z/v_p} \qquad (37)$$

for the quasisynchronous condition, where d is the distance traveled in the z direction with velocity u_z.

The electrons follow a helical trajectory around the lines of the axial dc magnetic field inside a TE-mode waveguide ($v_p > c$) as illustrated in the gyrotron schematic of Fig. 20.

In the case of a linear, nonperiodic beam, the electron bunching is longitudinal. For the gyrotron with its periodic electron beam, the electron bunching is transverse or azimuthal phase bunching as indicated in Fig. 21. This bunching occurs because the cyclotron frequency qB/m depends on the relativistic mass $m = \gamma m_o$. It is not possible to solve the relativistic trajectory equations in closed form. However, the "handicap race" can readily be seen. The electron at position 1 is losing kinetic energy to the field; hence, it rotates faster with a smaller radius trying to catch up with electron at position 2. The electron at position 3 is gaining kinetic energy from the field and rotates more slowly with a larger radius, allowing electron 2 to catch up in position. At a later period of time, electrons 1, 2, and 3 will be lined up in azimuth.

The gyrotron shown in Fig. 20 uses a hollow electron beam and a TE_{01}-mode waveguide to obtain an E_ϑ field so that there are many orbits of the electrons as displayed in Fig. 21(b). Also, the electron beam must have most of its energy transverse to the axis of the tube (typically by a factor of 1.5–2).

In 1982, Temkin *et al.* (1982) built a gyrotron at 140 GHz with a pulsed power of 100

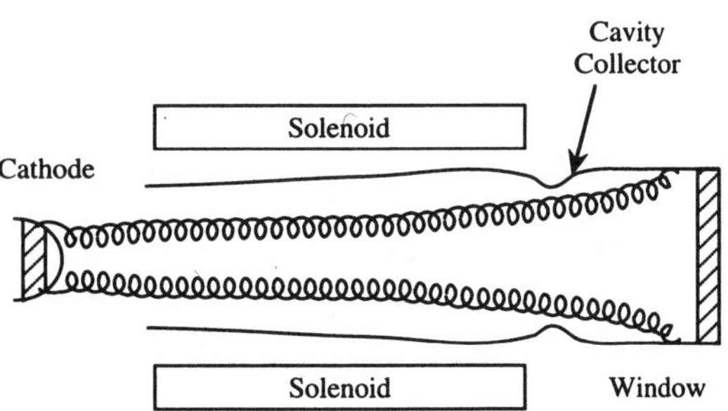

FIG. 20. Gyrotron circuit configuration.

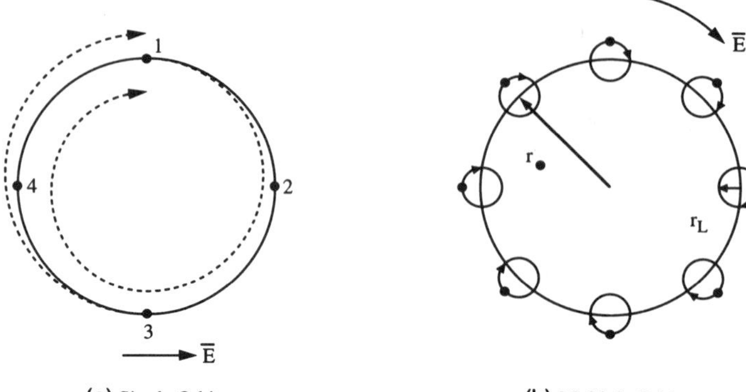

FIG. 21. Transverse electron bunching in a gyrotron.

kW and efficiency of 17% using a TE_{031}-mode resonator, 80-kV beam, and 5.4-T magnetic field. At lower frequencies (~30 GHz), scientists of the former Soviet Union achieved efficiencies of over 40%. For high-power pulsed operation, the gyrotron is in a class by itself.

3.4.2 Mathematical Models of the Gyrotron In the treatment of oscillators in this article, it was implied that conduction-current (**J**) devices could be analyzed by solving a Newton problem for the current and Maxwell's equations for the gain and fields. Of course, any classical problem can be quantized, even an *LC* circuit, but quantum effects would be indeed difficult to observe at 100 MHz. On the other hand, a polarization-current ($\partial\mathbf{P}/\partial t$) device requires a quantum treatment since the polarization **P** cannot be calculated classically. In this article, a $\partial\mathbf{P}/\partial t$ device has been called a laser and a **J** device a free-electron oscillator irrespective of how it is analyzed. The mathematical model cannot change the physics of the problem.

From this perspective, the gyrotron has had an interesting history. In the initial work to determine the operating principles of the gyrotron, a quantum-mechanical approach was used and the phase "electron cyclotron maser" was coined. It was implied that a gyrotron was a $\partial\mathbf{P}/\partial t$ device with observable quantum effects. While this is an interesting point of view, a quantum-mechanical model contributes little to the detailed analysis of the gyrotron interaction process.

4. NOVEL CONDUCTION-CURRENT OSCILLATORS

4.1 High-Energy Free-Electron Oscillator

In this review, oscillators have been classified as (1) conduction- or (2) polarization-current devices. The high-energy or relativistic free-electron oscillator has been termed a laser in the literature in spite of its being a conduction-current oscillator. Basically, the oscillator consists of an accelerator to produce a pulsed electron beam, a "wiggler" magnet to force the electrons to oscillate, and a Fabry–Pérot mirror system, as shown in Fig. 22. Because of relativistic effects, the radiation is strongly forward directed and Doppler shifted.

The wavelength λ_d of the radiation can be approximately calculated as seen in Fig. 22(b), by observing that while the radiation emitted by an electron at position 1 reaches position 3, the electron will have moved from position 1 to 2. Both the electron and wave will have undergone one cycle.

The "wiggler" magnet concept first appeared in the dissertation of P.D. Coleman (1951). Phillips (1960) developed the periodic magnetic field idea with his invention of the Ubitron, and in 1970 Madey (1971) extended the concepts to the optical range.

If the Fabry–Pérot mirror were not present, the electron emission would be strictly incoherent. With an optical field present to provide feedback, the electrons can be accelerated and decelerated slightly so that the faster electrons (more mass) can catch up with the slower electrons (less mass) to form

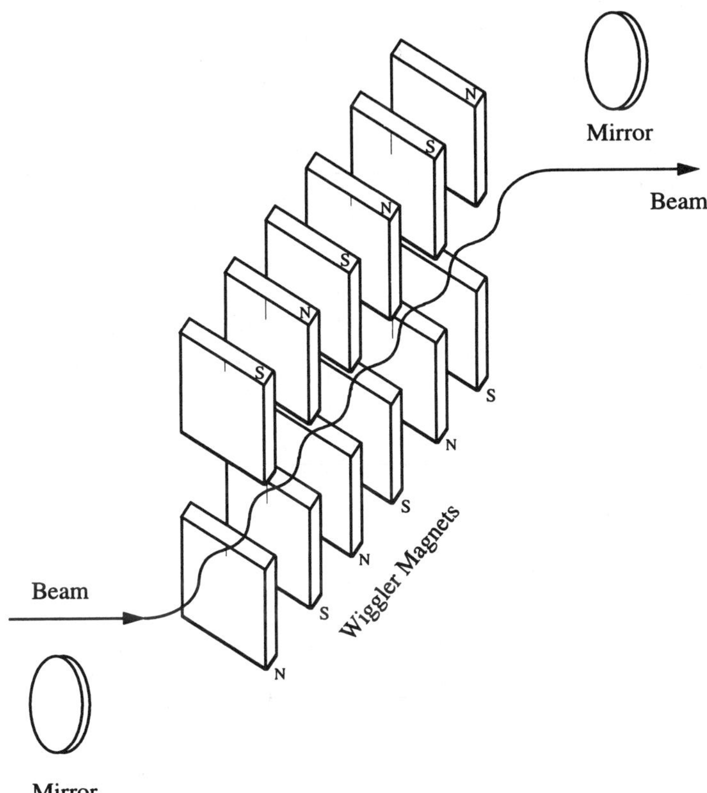

(a) Free Electron Oscillator Schematic

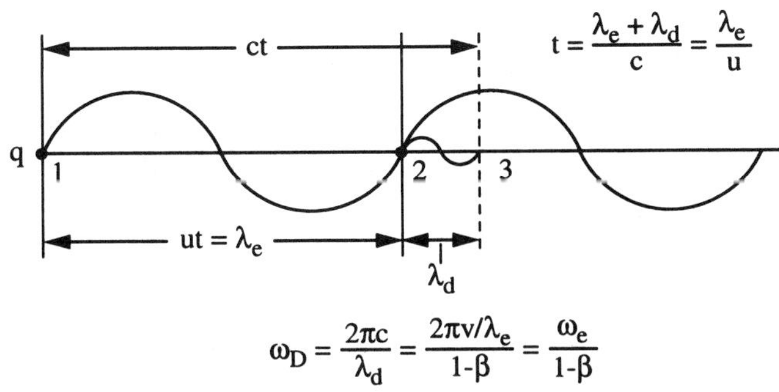

(b) Doppler Frequency Shift

FIG. 22. Schematic drawing of a free electron oscillator.

bunches spaced at optical wavelengths (Parker *et al.*, 1982).

The relativistic equation of motion is

$$\frac{d}{dt}(m_o\gamma\mathbf{u}) = -e[\mathbf{E} + \mathbf{u} \times \mathbf{B}] + \mathbf{F}, \tag{38}$$

where $\gamma = (1 - \beta^2)^{-1/2}$; $\beta = u/c$, $\mathbf{F}$ is the self-force, and $\mathbf{B} \simeq \mathbf{B}_0 \sin(2\pi x/\lambda_e)$. If the self-force is neglected, then

$$m_o c^2 \frac{d\gamma}{dt} = -e\mathbf{E} \cdot \mathbf{u} = \mathbf{u} \cdot \frac{d}{dt}(m_o\gamma\mathbf{u}), \tag{39}$$

where $\mathbf{E}$ is the transverse electromagnetic (TEM) optical field in the resonator.

While the electron velocity is modulated, the greater change is in the modulation of the electron mass $m_o\gamma$. The electrons and wave are in resonance on the forward pass and out of resonance on the backward pass. It is a complex problem to calculate the velocity $\mathbf{u}$ and acceleration $\dot{\mathbf{u}} = d\mathbf{u}/dt$ of the charge e because of the relativistic equations.

The power radiated can be estimated from Larmor's formula (Stratton, 1941)

$$\begin{aligned} \mathcal{P} &= \frac{Q^2\dot{u}^2}{6\pi\epsilon c^3} \frac{1 - \beta^2 \,|(\mathbf{u}/u) \times (\dot{\mathbf{u}}/\dot{u})|^2}{(1 - \beta^2)^3} \\ &\to \frac{Q^2\dot{u}^2}{6\pi\epsilon c^3(1 - \beta^2)^2}, \end{aligned} \tag{40}$$

where if the electron bunch is small compared to the radiation wavelength $Q^2 = (ne)^2$, when n is the number of electrons in the bunch.

Pulse power levels (Szöke, 1981; Granatstein and Roberson, 1985; Brau and Newnam, 1987; Süsskind, 1988) in a free-electron oscillator can be enormous compared to other types of conduction-current devices. Roberson *et al.* (1987) have reported a peak laser power output of 6×10^8 W at a wavelength of 514 nm with a free-electron oscillator driven with a 1.3-GHz linear accelerator having the pulsed current output shown in Fig. 23. The efficiency of this oscillator is seen to be

$$\begin{aligned} \frac{\mathcal{P}_{\text{LASER}}}{\mathcal{P}_{\text{BEAM}}} &= \frac{6 \times 10^8}{100(120 \times 10^6)} \\ &= \frac{0.6\ \text{GW}}{12\ \text{GW}} = 0.05. \end{aligned} \tag{41}$$

Power levels in the optical resonator can readily reach levels that damage the metal mirror of the Fabry–Pérot system.

It is tempting to use Larmor's formula [Eq. (40)] to estimate the peak power. The peak acceleration $\dot{u}$ can be approximated from the relation

$$\dot{u} \simeq c(qB/m). \tag{42}$$

Then

$$\mathcal{P} = \frac{(nq)^2(qB/m_o)^2}{6\pi\epsilon c(1 - \beta^2)^2} \simeq 10^{-12}n^2\ \text{W}. \tag{43}$$

The number of electrons in the 100-A, 20-ps pulse is 1.25×10^{10}; hence, $\mathcal{P} \simeq 0.16$ GW at the peak magnetic field.

4.2 Coherent Cherenkov Oscillators

4.2.1 The Boundary-Value Problem The generation of electromagnetic radiation (oscillators) can be based upon a number of physical phenomena. In the case of classical EM theory described by Maxwell's equations, it is the solutions to the equations, not the equations themselves, that provide a

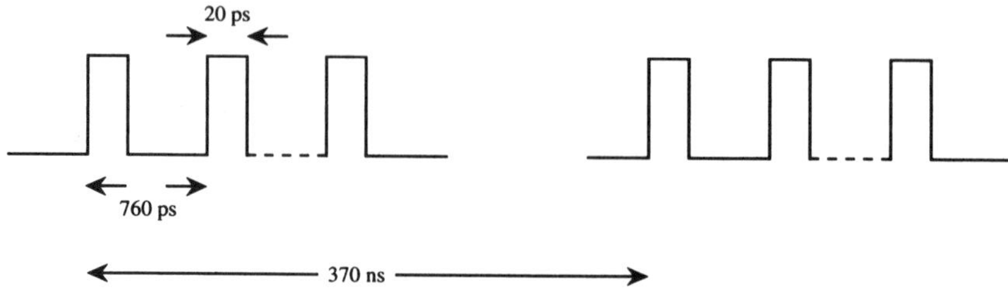

FIG. 23. Current pulses from accelerator driving the free electron oscillator.

mathematical description of the physical phenomena. From this point of view, the last fundamental phenomenon derived from Maxwell's equations was the Cherenkov effect, viz., a charge moving through a polarizable medium with a *constant* velocity faster than the medium's phase velocity will radiate. The Cherenkov effect can be studied by considering the excitation of the elementary dipoles by the passing charge or as a straightforward Maxwell boundary-value problem. It is not necessary for the charge to pass through the medium, just close to it.

While the Cherenkov effect was studied extensively by P. A. Cherenkov from 1924 to 1938 and a theory was constructed by Frank and Tamm (1937), coherent Cherenkov radiation in the microwave range was not demonstrated until 1960 by Coleman and Enderby (1960).

Consider the boundary-value problem of a bunched (ac) electron beam passing through a cylindrical hole in a dielectric as illustrated in Fig. 24. The equation for the vector potential $\mathbf{A} = \hat{\mathbf{z}}A$ is

$$\nabla^2\mathbf{A} - \mu\epsilon\frac{\partial^2\mathbf{A}}{\partial t^2} = -\mu\mathbf{J}_n e^{jn\omega_0(t-z/u)} \tag{44}$$

for the nth harmonic current density $\mathbf{J}_n$.

In region II outside of the beam, Eq. (44) becomes

$$\frac{1}{r}\frac{d}{dr}\left(r\frac{dR}{dr}\right) + \left(\frac{n\omega_0}{v}\right)^2[\mu\epsilon u^2 - 1]R = 0, \tag{45}$$

where $A_z = R(r)e^{jn\omega_0(t-z/u)}$. If

$$\mu\epsilon u^2 = (u/v_p)^2 > 1 \tag{46}$$

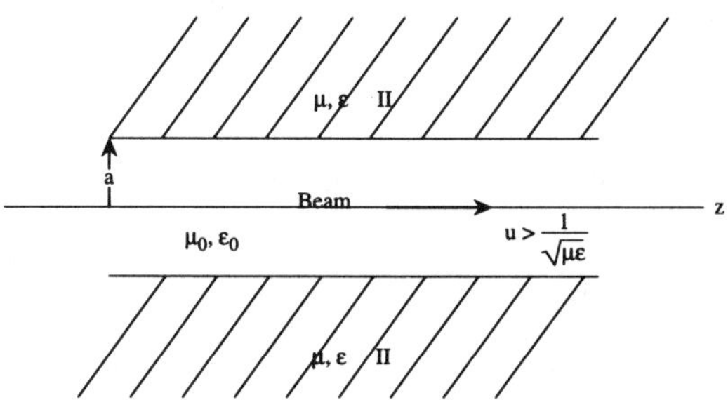

FIG. 24. Bunched cylindrical electron beam passing through hole in a dielectric.

(Cherenkov condition), then

$$R(r) = H_0^{(2)}(br)$$

(Hankel function—outgoing waves), where

$$b = (n\omega_0/v)[\mu\epsilon u^2 - 1]^{1/2}$$

For the driven oscillator system considered, coherent radiation will be produced at all harmonic frequencies $n\omega_0$ for which the Cherenkov condition is satisfied.

4.2.2 Self-Excited Cherenkov Oscillator
A self-excited Cherenkov oscillator can be made by passing an electron beam through a dielectrically lined double-slab waveguide or a dielectrically lined cylindrical waveguide as illustrated in Fig. 25. Here the Cherenkov effect takes the form of a traveling-wave amplifier with internal feedback wherein the beam velocity u_0 is greater than the wave phase velocity.

It is of interest to briefly consider the physics of electronic motion in a traveling-wave field (Hutter, 1960). Here Newton's equation for one dimension takes the form

$$m_o\frac{du}{dt} = m_o\left(\frac{\partial u}{\partial t} + u\frac{\partial u}{\partial z}\right) = qE_1e^{j\omega t}e^{-\Gamma z}, \tag{47}$$

with

$$u = u_0 + u_1e^{j\omega t}e^{-\Gamma z},$$

where u_0 is a constant velocity and u_1 is a small amplitude. Solving Eq. (47) one obtains

$$u_1 = \frac{qE_1/m}{j\omega - \Gamma u_0}. \tag{48}$$

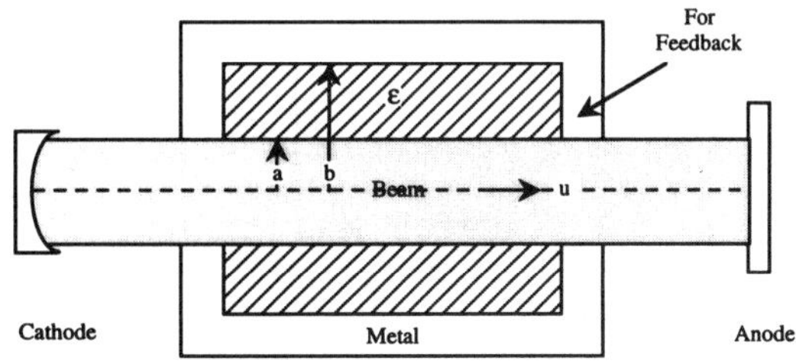

FIG. 25. Configuration of a Cherenkov oscillator.

Next let the charge density ρ and current J be

$$\rho = \rho_0 + \rho_1 e^{j\omega t} e^{-\Gamma z}, \tag{49}$$

$$J = \rho u = J_0 + J_1 e^{j\omega t} e^{-\Gamma z}, \tag{50}$$

with

$$\frac{\partial J}{\partial z} + \frac{\partial \rho}{\partial t} = 0. \tag{51}$$

Solving for J_1 yields the result

$$J_1 = \frac{j\omega}{u_0} \frac{J_0 E_1 e/m}{(j\omega - \Gamma u_0)^2}. \tag{52}$$

If the propagation constant Γ is purely imaginary, J_1 and E_1 are out of phase, no electron bunches are formed, and no work is done. However, if Γ is complex, work can be done, leading to attenuated or amplified waves. The value of Γ can be found by using J_1 in Maxwell's equations, applying boundary conditions for the problem, and obtaining a self-consistent solution.

Garate (1987) has analyzed a dielectrically lined double-slab waveguide driven by a relativistic electron-beam wave, using the procedure outlined, and obtained the data shown in Fig. 26. Von Laven *et al.* (1982) using a 1–40-A, 100–25-keV beam with a pulse duration of 4 μs, obtained 30–100 kW of pulsed power in the 30–50-GHz range using the structure shown in Fig. 25.

The use of microwave-type transmission lines and resonators in a Cherenkov oscillator will limit its frequency most likely below 300 GHz. The beauty of a Cherenkov oscillator is its simple beam-coupling structure. No high-technology, submicron fabrication techniques are needed to make the oscillator.

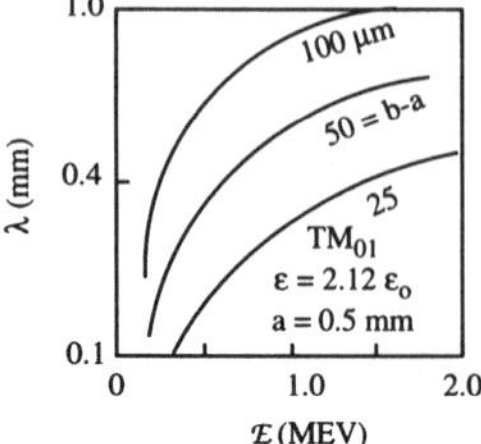

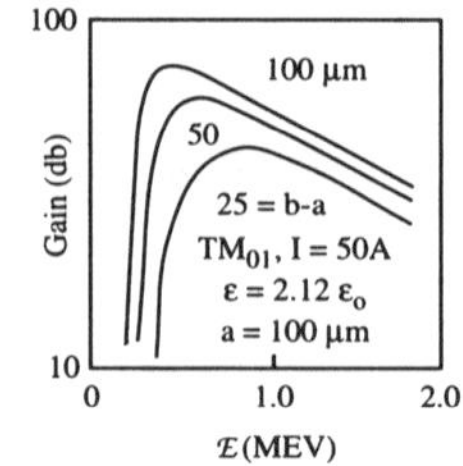

FIG. 26. Wavelength and gain vs electron energy of a Cherenkov oscillator.

5. RELAXATION OSCILLATORS

While the generation of sine waves is important, especially at high frequencies, there are many uses for other wave forms (pulses, rectangles, triangles, trapezoids, etc.)—generators, for example, in timing applications (Chance *et al.*, 1949). The relaxation oscillator that will be considered here is the astable plate-coupled multivibrator, which can provide nearly rectangular pulses of a specific duration. This multivibrator circuit has two triode tubes with *RC* coupling from the plate of each tube to the grid of the other tube as shown in Fig. 27.

The cycle of operation consists of alternate periods when one tube and then the other is cut off with rapid switching from one condition to the other. Suppose the plate current i_{b_2} increases, thereby lowering the voltage e_{b_2}. This in turn causes a sequence of actions with e_{c_1} decreasing, decreasing i_{b_1}, increasing e_{b_1}, increasing e_{c_2}, further increasing i_{b_2}. As the charge on capacitor C_1 decreases, e_{c_1} rises toward zero and, as soon as the cutoff voltage e_{c_0} is reached, turns tube 1 from off to on. Thus, the current is suddenly transferred from the plate circuit of one tube to that of the other. The various voltage wave forms are also displayed in Fig. 27.

The periods τ_1 and τ_2 depend on the *RC* time constant and the ratio of the voltages involved. For example,

$$\tau_2 = R_{g_2} C_2 \ln(e_2/e_{c_0}). \tag{53}$$

Switching times are much smaller than 1 μs, and the periods τ_1 and τ_2 can be changed by varying R_{g_1} and R_{g_2}.

6. POLARIZATION-CURRENT (LASER) OSCILLATORS

From a Maxwell-equation, mathematical point of view, conduction-current (**J**) and polarization-current ($\partial \mathbf{P}/\partial t$) oscillators represent different driving terms for the EM fields. Given **J** or $\partial \mathbf{P}/\partial t$, the mathematics of solving for the fields is the same. In this case, **J** or $\partial \mathbf{P}/\partial t$ must be out of phase more than 90° with

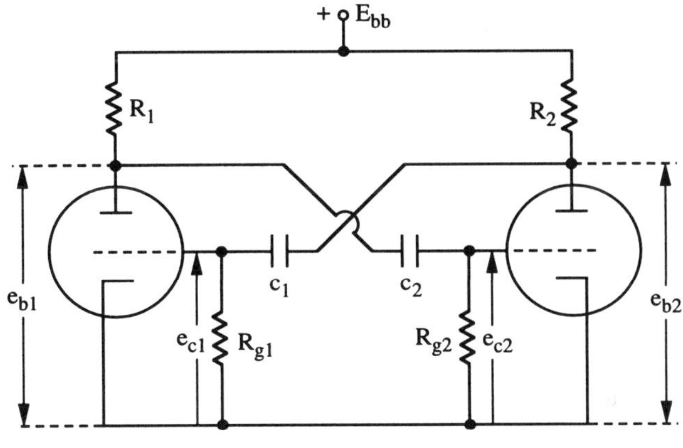

(a) Circuit Diagram

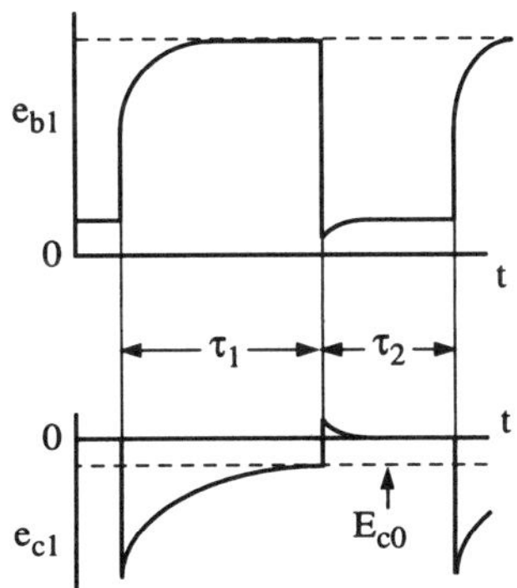

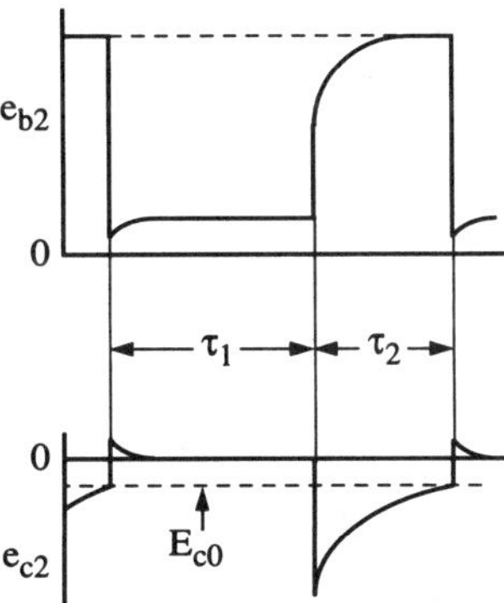

(b) Waveforms

FIG. 27. Astable multivibrator oscillator.

the field **E** to deliver energy. For a **J** device, this means generating an ac or bunched electron beam, and for a $\partial \mathbf{P}/\partial t$ device it means creating a non-thermodynamic-equilibrium distribution in a quantum system. In some cases, both can use the same feedback or resonator configuration, as, for example, the free-electron oscillator, which uses a Fabry–Pérot resonator. One key difference in the two devices is that the laser oscillator does not have a transit-time limitation on its frequency range.

However, a basic physics change had to be made in the analysis of **J** and $\partial \mathbf{P}/\partial t$ oscillators in that quantum mechanics has to replace Newtonian mechanics to calculate $\partial \mathbf{P}/\partial t$. This is not surprising since the laser is a quantum system. The equation for **P** for a two-level system, assuming an electric dipole transition, has been given in Eq. (4). It is to be observed that this second-order linear equation is of the same mathematical form as an *LRC* circuit driven by a voltage source. Since sinusoidal oscillations are being considered, this too is not unexpected.

The sign of $\partial \mathbf{P}/\partial t$ can be changed by making $N_2 > N_1$, i.e., a population inversion or change in the distribution function away from its thermodynamic equilibrium value. Equation (4) allows one to calculate the electric susceptibility χ_e since

$$\mathbf{P} = \epsilon_0 \chi_e \mathbf{E} = \epsilon_0(\chi_e' - j\chi_e'')\mathbf{E}, \tag{54}$$

and the loss α (or gain) of the EM wave traversing the medium,

$$\alpha = \omega \chi_e''/v_g, \tag{55}$$

where v_g is the group velocity.

Various types of lasers are described in other articles (see LASER PHYSICS; LASERS,

FREE-ELECTRON; LASERS, GAS; LASERS, SEMICONDUCTOR; LASERS, SOLID STATE; ORGANIC DYE LASERS; TUNABLE DYE LASERS) and hence are discussed only to make comparisons with **J** devices. Conduction-current oscillators, as has been seen, have the challenges of transit time, circuit sizes, and losses, as they are extended to higher frequencies. Lasers have the challenges of devising specific excitation schemes to create highly non-thermodynamic-equilibrium distributions in the face of fast relaxation and damping processes.

In this author's opinion, while technological advances will be made on **J** devices in the future, finding new physics phenomena to exploit is becoming more difficult. No basic new **J** device has appeared for many years. A possible new frontier exists for $\partial\mathbf{P}/\partial t$ laser devices for nuclear energy-level systems.

Works Cited

Brau, C. A., Newnam, B. E. (Eds.), (1987), Special Issue on Free Electron Lasers, *IEEE J. Quantum Electron.* **QE-23**, 1468–1665.

Chance, B., Hughes, V. W., MacNichol, E. F., Jr., Sayer, D., Williams, F. C. (1949), *Waveforms*, MIT Radiation Laboratory Series, Vol. 19, New York: McGraw-Hill.

Coleman, J. T. (1982), *Microwave Devices*, Reston, VA: Reston Publishing Company.

Coleman, P. D. (1951), Ph.D. Dissertation, Massachusettes Institute of Technology.

Coleman, P. D., Enderby, C. E. (1960), "Megavolt Electronics Cerenkov Coupler for the Production of Millimeter and Submillimeter Waves," *J. Appl. Phys.* **31**, 1695–1696.

Convert, G., Yeou, T., Pasty, A. (1959), "Millimeter Wave O-Carcinotron," *Microwave Res. Inst. Symp. Ser. Polytech. Inst. Brooklyn* **9**, 313–339.

Copeland, J. A. (1967), "Characterization of Bulk Negative Resistance Diode Behavior," *IEEE Trans. Electron Devices* **ED-14**, 461–463.

de Forest, L. (1907), U.S. Patent 841387.

Frank, I. M., Tamm, I. (1937), "Coherent Visible Radiation of Fast Electrons Passing Through Media," *C. R. Acad. Sci. USSR* **14**, 109–114.

Garate, E. P. (1987), "High-Gain Cerenkov Free-Electron Laser at Far Infrared Wavelengths, *J. Quantum Electron.* **23**, 1627–1632.

Graratstein, V. L., Roberson, C. W. (Eds.) (1985), Special Issue on Free Electron Lasers, *J. Quantum Electron.* **QE-21**, 804–1119.

Hunter, L. P. (1956), *Handbook of Semiconductor Electronics*, New York: McGraw-Hill.

Hutter, R. G. E. (1960), *Beam and Wave Electronics in Microwave Tubes*, New York: D. Van Nostrand Co.

Karp, A. (1955), "TWT Experiments at Millimeter Wavelengths," *Proc. IRE* **43**, 41–46.

Kompfner, R. (1947), "The TWT as an Amplifier at Microwaves," *Proc. IRE* **35**, 124–127.

Liao, S. Y. (1985), *Microwave Devices and Circuits*, Englewood Cliffs, NJ: Prentice-Hall.

Madey, J. M. J. (1971), "Stimulated Emission of Bremsstrahlung in a Periodic Magnetic Field," *J. Appl. Phys.* **42**, 1906–1913.

Millman, J., Halhias, C. C. (1972), *Integrated Electronics*, New York: McGraw-Hill.

Pantell, R. H., Puthoff, H. E. (1969), *Fundamentals of Quantum Electronics*, New York: Wiley.

Parker, R. K., Jackson, R. H., Gold, S. H., Freund, H. P., Granatstein, V. L., Efthimion, P. E., Herdon, M., Kinkead, A. K. (1982), "Axial Magnetic Field Effects in a Collective Interaction Free-Electron Laser and Millimeter Wavelengths," *Phys. Rev. Lett.* **48**, 238–242.

Phillips, R. M. (1960), "The Ubitron," *IRE Trans. Electron Devices* **ED-7**, 231–241.

Robenson, K. E., Quimby, D. C., Slater, J. M., (1987), "The Tapered Hybrid Undulator of the Free-Electron Laser Oscillator Experiment," *J. Quantum Electron.* **9**, 1497–1513.

Sprangle, P., Drobot, A. T. (1977), "The Linear and Self-Consistent Nonlinear Theory of the Electron Cyclotron Maser," *IEEE Trans. Microwave Theor. Techniques* **MTT-25**, 528–544.

Süsskind, C. (Ed.) (1988), Special Issue Commemorating the Centennial of Heinrich Hertz, *IEEE Trans. Microwave Theor. Techniques* **MTT-36**, 802–858.

Szöke, A. (Ed.) (1981), Special Issue on Free Electron Lasers, *J. Quantum Electron.* **QE-17**, 1326–1514.

Stratton, J. (1941), *Electromagnetic Theory*, New York: McGraw-Hill.

Temkin, R. J., Kreischer, K. E., Mulligan, N. J., MacCabe, S., Fetterman, H. R. (1982), "A 100 kW, 140 GHz Pulsed Gyrotron," *Int. J. Infrared Millimeter Waves* **3**, 427–438.

Von Laven, S., Branscum, J., Golub, J., Layman, R., Walsh, J. (1982), "High Power Cerenkov Maser Oscillators," *Appl. Phys. Lett.* **41**, 408–410.

OSCILLOSCOPES, ANALOG AND DIGITAL

NELLIE BROCK, *Tektronix, Inc., Beaverton, Oregon, U.S.A.*

	Introduction	37
1.	**Basic Oscilloscope Concepts**	38
1.1	Analog Oscilloscopes	39
1.2	Digital Storage Oscilloscopes	41
1.3	General Wave-Form Capture Requirements	41
1.3.1	Bandwidth and Rise Time	41
1.3.2	Probing Effects	42
1.3.3	Transient vs Repetitive	43
2.	**Resolution and Accuracy**	43
2.1	Analog Display Resolution	44
2.2	Digital Storage Resolution	44
2.3	Accuracy	44
2.4	Digital Time-Base Accuracy	44
3.	**Memory and Record Length**	45
4.	**Analog-to-Digital Conversion**	46
4.1	Digitizing Rate	46
4.2	Real Time and Equivalent Time	46
4.3	Converter Types	47
4.3.1	Successive Approximation	47
4.3.2	Flash Conversion	48
4.3.3	Charge-Coupled Device (CCD)	48
4.3.4	Scan Conversion	48
4.4	Vertical Quantizing	48
4.5	Effective Bits	49
5.	**Display Techniques**	49
5.1	Dot Displays	49
5.2	Interpolated Displays	50
5.3	Peak Detection and Enveloping	50
5.4	Pre- and Post-Trigger Displays	51
6.	**Useful Storage Bandwidth and Rise Time**	52
6.1	Useful Storage Bandwidth	53
6.2	Useful Storage Rise Time	53
7.	**Signal Processing**	53
7.1	Save-on-Delta	53
7.2	Averaging	53
7.3	Integration and Differentiation	54
7.4	Fast Fourier Transform	54
7.5	Convolution and Correlation	54
8.	**Measurement Automation**	54
8.1	Interfacing Techniques	54
8.2	Software Considerations	55
9.	**Conclusion**	55
	Glossary	56
	Works Cited	57
	Further Reading	57

INTRODUCTION

Oscillographs are a means of graphically studying oscillatory or cyclical phenomena—hence the term "oscillo" + "graph." With this fundamental definition, probably the earliest use of oscillography was the sundial. This simple device portrays the movement of the sun across the sky as a shadow cast across a scale to indicate the time of day.

A more sophisticated and widely familiar example is the seismograph used to record earth tremors and quakes. Here, a graphical record of seismic activity is penned onto a paper chart for detailed study. The process uses a slowly rotating drum onto which the blank graph paper is attached. The drum's rotation is controlled to correspond to a time scale. A seismic sensor detects earth tremors and converts them to a corresponding electrical signal which is used to drive a pen mechanically back and forth across the moving chart paper. The result is a graph of seismic amplitude vs time.

Historically, numerous general-purpose and highly specialized oscillographic devices have been developed to further the recording and study of physical phenomena. Prior to the advent of electronic devices, however, oscillography was generally confined to chart recorder and photographic methods.

It was not until the age of radio that an unavoidable need emerged for observing electronic wave forms. Engineers needed to

3-527-28135-5/95/$5.00 + .50

see into the radio circuits they were developing in order to measure wave-form distortion and other amplifier effects. Using cathode-ray tubes (CRT), which had previously been developed in the study of x rays, engineers built their own rudimentary oscilloscopes based on designs passed along through symposia and journals. These first instruments were referred to as oscillographs, but gained the name oscilloscope by the 1940s to distinguish them from the broader category of oscillographs. The term uses the root word scope, from the Greek "skopos," meaning an instrument for viewing, as in telescope.

The basic function of an oscilloscope is to provide an *x-y* graphical view of physical phenomena. In the most common applications today, the display is voltage amplitude vertically on the *y* axis vs time horizontally on the *x* axis. This is because an oscilloscope is an electronic instrument designed for acquisition and display of electrical signals. However, it is of the utmost importance to realize that the displayed electrical signal, or wave form, can represent virtually any physical phenomenon.

What the oscilloscope display represents is determined by the application and by the type of probe or sensor used at the oscilloscope input. For example, optical events can be captured and displayed via an optical-to-electrical converter. Similarly, a microphone can provide an audio signal input, an accelerometer can provide a motion signal, and a thermocouple can provide a temperature signal input.

Thus, by judicious choice of sensors or probes, an oscilloscope is potentially able to provide monitoring, measurement, and analysis of almost any physical phenomenon or event. The phrase "potentially able to" is used here because actual display, measurement, or analysis depends on the type of oscilloscope selected and its various capabilities.

The most basic selection issue, of course, is adequate frequency bandwidth of the oscilloscope for the event signals being investigated. For example, a 20-MHz oscilloscope is inadequate for acquiring and displaying 100-MHz signals. Even with adequate bandwidth, however, there can be certain classes of signals that still cannot be displayed adequately on certain types of oscilloscopes. For example, a fast transient signal often will not create a visible trace on the display of a conventional analog oscilloscope, but will be clearly visible on the display of a real-time digital oscilloscope or transient digitizer. On the other hand, extremely fast transients may only be brought to light by the use of fast analog techniques and special CRT photography.

The point here is that careful and informed oscilloscope selection is paramount in successful display, measurement, and analysis of physical phenomena. This requires an understanding of basic oscilloscope concepts, the different oscilloscope technologies available, and the key advantages of each. This article provides the basic information necessary for informed oscilloscope selection.

1. BASIC OSCILLOSCOPE CONCEPTS

The most basic elements of any oscilloscope are acquisition, processing, and display. These elements may be implemented in a wide variety of manners with either analog or digital technologies, or a combination of both.

Acquisition begins, fundamentally, as an analog process. It includes, first of all, probing or sensing the phenomenon to be displayed or measured. It also includes triggering on a particular event or aspect of the phenomenon and generally includes some preconditioning (amplification or attenuation) of the input signal.

Processing, in general, is anything that is done to or in conjunction with the input signal to provide a useful display of the signal. This almost always includes use of a time base to set a calibrated display sweep or to control a sampling interval. It may also include other functions, such as subtracting a ground reference or even performing a fast Fourier transform on the signal.

Display is the method of formatting and presenting the acquired signal for observation. This may include scale factors, measurement cursors, and even measurement results. Most often, the primary display is the oscilloscope screen. However, displays may include separate numeric readouts as well as results sent to a printer.

There are two classes of oscilloscopes available for signal acquisition, processing, and display. These are generally referred to as the *analog oscilloscope* and the *digital storage oscilloscope.* Figure 1 shows a basic block diagram of each oscilloscope class for purposes of comparison in the following discussion.

1.1 Analog Oscilloscopes

Beginning with the input signal from a probe or sensor, the first operation of an analog oscilloscope is input signal conditioning. This consists of either attenuation or amplification to provide an amplitude-calibrated signal within the vertical display range of the oscilloscope. This input signal drives the display trace vertically, similar to the y-axis pen drive on a plotter or chart recorder.

The input signal amplitude or y-axis drive provides one dimension of the display. The second dimension, the x-axis drive, is provided by a second signal. When the oscilloscope is operated in what is called an x-y mode, the x-axis drive signal is a second input signal. This produces an x-y display of one signal against another, such as in creating Lissajous patterns. However, typical oscilloscope usage, which is the focus here, is to plot or display signal amplitude (y) against time on the x axis. This is done by using a ramp signal generated within the oscilloscope to provide the x-axis or horizontal drive. The ramp signal used is often referred to as the horizontal sweep signal. The duration of this ramp and how it is triggered are key factors in providing the desired time-calibrated display of the signal being input to the oscilloscope.

As shown in Fig. 1(a), triggering information is obtained at the signal input stage. The setup of the triggering and horizontal systems also determines how the signal is displayed. This is illustrated in Fig. 2.

The trigger system settings determine the point on the input signal that causes triggering to occur. Triggering starts sweep generation by the horizontal system. The sweep signal is a linear ramp that drives the display trace left to right across the display at a constant sweep speed. The duration of this ramp, and thus the horizontal sweep speed, are determined by the oscilloscope's time-base setting. This provides horizontal time calibration of the display.

In single-shot triggering, the trigger is manually armed by pressing a trigger arm button. One single-sweep display occurs when the input signal crosses a preset trigger level. There are no further displays after that until the trigger is manually rearmed and another triggering event occurs. The single-shot mode is intended for use in acquiring single events such as a laser pulse, a sonar ping, or a mechanical impact signal.

A repetitive triggering mode is used for acquiring repetitive signals such as sinusoids. This is the triggering mode shown in

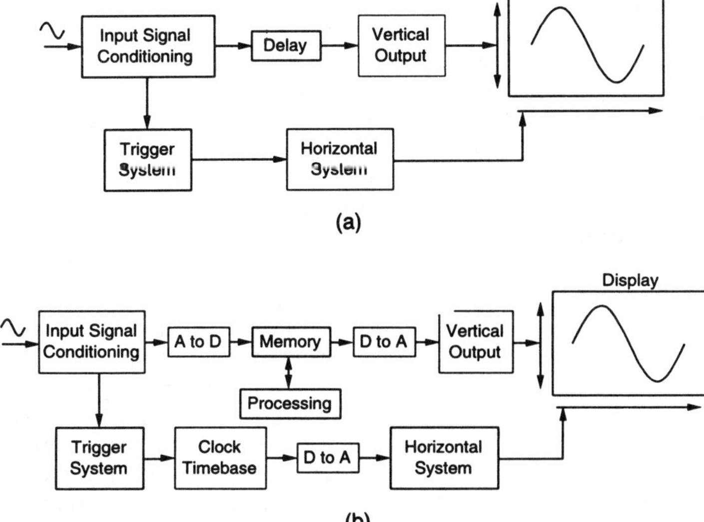

FIG. 1. Block diagram of (a) an analog oscilloscope and (b) a digital storage oscilloscope (DSO). The DSO has the same analog oscilloscope front-end requirements for signal conditioning and triggering.

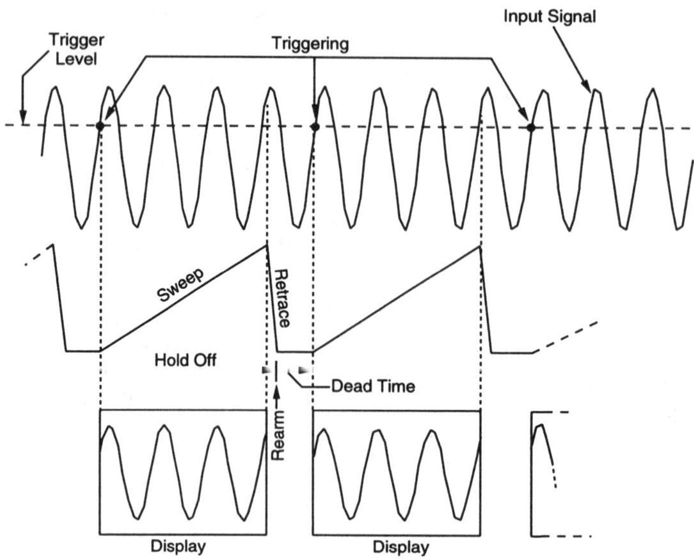

FIG. 2. A trigger level and triggering conditions are set up to determine the beginning point of waveform acquisition and display. Triggering also starts the sweep ramp for driving the display trace horizontally.

Fig. 2. Note that a sweep, and thus a display, are generated when the input signal reaches the trigger level setting. Once the sweep starts, further triggering is held off until sweep retrace is completed. At this point, the triggering circuitry is automatically rearmed. The next sweep (and display) then occurs when the input signal again crosses the trigger level.

The display on an analog oscilloscope is created by driving an electron beam across the phosphor-coated screen of a cathode ray tube (CRT). The input signal drives the CRT's vertical deflection plates to move the beam vertically, and the triggered sweep ramp (or other signal) drives the CRT's horizontal deflection plates to move the beam horizontally. The beam electrons strike and excite the CRT phosphor, causing it to emit light along the path traced by the electron beam.

In the case of single-shot triggering, the phosphor is excited by a single pass of the electron beam. However, the faster the beam is traveling (higher sweep speeds), the less the excitation and the dimmer the trace. The limit where portions of the displayed trace are no longer visible defines the CRT's writing speed. Thus, for viewing single-shot events on an analog oscilloscope, it is important to have a high–writing-speed CRT.

For repetitively triggered events, such as illustrated in Fig. 2, CRT writing speed is less of a concern. This is because the repetitively triggered sweeping of the beam regularly re-excites the phosphor in the trace path. As a result, repetitively triggered wave forms provide bright traces even when acquired at extremely fast sweep speeds.

It is important to note that the wave-form display process is analog and occurs in real time. That is, the actual analog signal being acquired is used directly to create the display trace. Thus, changes in the input signal are immediately reflected on the oscilloscope display in real time, as the signal changes actually occur. This is a key advantage when using the oscilloscope as a monitoring aid in making amplitude or timing adjustments on signals.

Another advantage is that momentary wave form disturbances, such as amplitude noise or time jitter, are also easily distinguishable. Such momentary or random signal disturbances cause corresponding deviations from the predominant trace path. Over multiple triggered traces, the result is less intense traces surrounding the more intense trace of the statistically dominant trace path. This intensity variation provides a subjective indication of signal noise statistics.

The real-time aspect of analog oscilloscopes can be a disadvantage, too. Since the display is a real-time display, transitory signal changes may fade from view before they can be studied. In the case of single pulses or other transient signals, the trace is a single brief flash across the screen. If the CRT's writing speed is not sufficient, there may be no trace at all. At the other extreme, a very low-frequency wave form may appear only as a bright dot moving slowly across the screen.

Many of these real-time display problems have been addressed through variations in CRT technologies. High-persistence phosphors have been used, for example, to slow trace fading in single-shot or low-frequency applications. Trace viewing time has been extended even more by use of storage meshes behind the CRT's phosphor. Another alternative is the use of CRT photography to capture transitory real-time displays for more detailed study.

1.2 Digital Storage Oscilloscopes

The digital storage oscilloscope (DSO) was developed to answer some of the shortcomings of analog oscilloscopes. Primarily, the DSO digitizes the input signal and stores it in digital memory. The key advantage is that the stored wave form can be easily reconstructed on the display at any time, even long after the actual analog signal is gone.

For transient events, the DSO essentially freezes the event on screen for detailed study. By contrast, single-shot capture of transient events with an analog oscilloscope provides only a brief trace flash on the display. Any detailed study of such analog wave-form displays must usually be done via a photograph of the display.

For any signal—single-shot or repetitive—digital storage offers the advantages of detailed analysis. For example, the mean value, root mean square value, or any other signal parameter can easily be computed from the stored values of the digitized signal. With appropriate computational support, advanced analysis techniques can also be applied. This can include performing fast Fourier transforms, convolution, correlation, or any other analytic process that can be programmed.

While the digital storage and processing of a DSO offer numerous advantages, there can also be some disadvantages. Understanding both the advantages and disadvantages relative to analog oscilloscopes is an important part of oscilloscope selection. However, before making a digital vs analog selection, it is important to consider the general wave-form capture requirements that apply to both.

1.3 General Wave-Form Capture Requirements

All oscilloscopes, regardless of their type, have an analog signal input system. This is necessary since the real world is analog. Even the digital signals used in digital systems are, in actuality, analog signals; they are time-continuous voltages that are simply level switched to represent digital data.

Consequently, whether considering an analog oscilloscope or a DSO, the first issue is the analog signal input requirements. This includes not only the oscilloscope's input signal conditioning but also the effects of probing or sensing.

1.3.1 Bandwidth and Rise Time Oscilloscopes are most commonly specified in terms of frequency bandwidth—e.g., a 200-MHz oscilloscope, a 200-MHz DSO. The bandwidth specification refers to the maximum-frequency sinusoid that can be acquired and displayed without significant signal attenuation. In other words, a 200-MHz oscilloscope can be expected to display any sinusoid up to 200 MHz without significant signal attenuation.

In the case of nonsinusoidal signals, the bandwidth specification must be interpreted in terms of the highest frequency component of significance in the signal being acquired. For example, repetitive nonsinusoidal wave forms, such as a square-wave signal, are made up of the fundamental frequency and some combination of harmonics. In general, such signals must be preserved out to the fifth harmonic for reasonably accurate observations and measurements. Thus, acquisition of a 40-MHz square wave would typically require a 200-MHz (5×40 MHz) oscilloscope to preserve signal amplitude and shape characteristics.

In many applications, the key consideration is measurement of a signal's rise time or some other timing aspect. For such applications, the oscilloscope's input bandwidth specification needs to be converted to a rise-time specification. According to Tektronix (1993a), the expression for bandwidth/rise-time conversion is

$$T_r = 0.35/f_{BW}, \qquad (1)$$

where T_r is rise time and f_{BW} is frequency bandwidth. Thus, the rise-time specification for a 200-MHz oscilloscope is 0.35/(200 MHz) = 1.75 ns. This expression can be derived from the Fourier transform relation between a Gaussian pulse's rise time and bandwidth. It provides a good approximation for most systems having a constant delay vs frequency.

An oscilloscope's rise time should be faster than the signal rise time being measured. How much faster depends on the measurement accuracy desired. The graph in Figure 3 provides a means of determining the necessary rise time for a desired accuracy. For example, if the oscilloscope's rise time is 3 times the rise time being measured (3:1 rise-time ratio), a 5% measurement error can be expected.

1.3.2 Probing Effects The type of probe or sensor used with the oscilloscope must also be considered as part of the overall measurement system. This is because the probe's impedance causes circuit-loading effects.

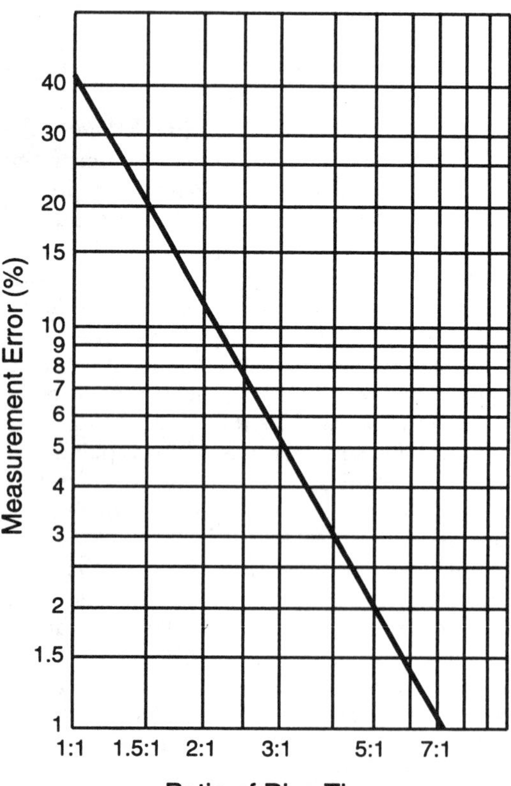

FIG. 3. Chart for determining rise-time measurement error. The amount of error depends on the ratio of the oscilloscope's rise time to the rise time being measured. A 3:1 ratio corresponds to 5% error.

For signal frequencies below 5 kHz, resistance is the dominant circuit-loading element. To minimize resistive loading, the probe used should have a resistance that is at least two orders of magnitude greater than the impedance of the circuit or sensor output being measured.

As frequencies progress above 5 kHz, probe inductance and capacitance have increasing loading effects. There are several steps that can be taken to reduce this reactive loading. To minimize inductive effects, the probe should be securely grounded with as short a ground lead as possible. To reduce capacitive effects from the probe, an attenuator probe (e.g., 10× attenuation) can be used. (Essentially, added resistive attenuation at or near the probe tip reduces the shunting effects of stray capacitance in the probe body and probe cable to the scope input.)

Additionally, the probe used should be adjusted to compensate for variations in the oscilloscope's input capacitance. Probe compensation is a simple procedure that is described in most oscilloscope manuals. Yet it is also one of the most often overlooked procedures, the result of which can be serious measurement errors, as reported by Tektronix (1993b).

Probes, like oscilloscopes, have rise-time and bandwidth limitations. As a result, probe selection must also be made to ensure adequate bandwidth or rise time for the signals to be measured. However, the probe's bandwidth or rise time cannot be considered alone. This is because the probe is an integral part of the oscilloscope's or DSO's input system.

Thus, the overall input system rise time must be considered in terms of the following relationship:

$$T_{r(\text{system})} = \sqrt{T^2_{r(\text{scope})} + T^2_{r(\text{probe})}}. \tag{2}$$

The system rise time computed by Eq. (2) can be re-expressed as a system bandwidth using Eq. (1).

The key point to be realized from Eq. (2) is that inappropriate probe selection can significantly degrade the input bandwidth of an oscilloscope or DSO. To avoid this, manufacturers of high-quality instruments generally optimize their instruments for specific probe models. As reported by Tektronix (1993b), oscilloscope manufacturers often specify oscilloscope bandwidth at the probe tip when specific probe models are used.

1.3.3 Transient vs Repetitive Even with sufficient bandwidth or rise time, it is still possible that the instrument will not be able to provide an adequate display of the acquired signal. The classic example of this is in acquiring transient wave forms such as static discharges, mechanical impacts, and other one-time events.

When acquired with an analog oscilloscope, the transient signal causes a single-sweep trace across the display screen. For slow transients, the trace speed is slow enough that ample energy is imparted to the screen phosphor for a brief, but visible, trace in normal room light. On faster transients, the faster trace speed imparts less energy to the screen phosphor and the trace is dimmer.

As transient speed (and thus single-sweep trace speed) increases, a point is reached where the visual writing speed of the analog oscilloscope's display is exceeded. When this occurs, the trace is no longer visible in normal room light. However, the trace may still provide enough phosphor excitation to be visible through use of a viewing hood or via single-shot display photography. When the transient and its trace speed exceed the display's photographic writing speed, the trace is too dim for capture even with specialized photographic techniques.

In general, the photographic writing speed limit of an analog oscilloscope is significantly less than the bandwidth and horizontal sweep-speed limits. This means that the transient wave-form viewing capabilities will be substantially less than the repetitive wave-form viewing capabilities. The exception to this is microchannel plate (MCP) display technology. MCP oscilloscopes provide trace intensity multiplication of about 1000 times, as reported by Gustafson *et al.* (1985). This MCP intensity amplification is usually sufficient for single-sweep viewing up to the oscilloscope's rise-time specification at bandwidths of as much as 1 GHz.

With DSOs, display writing speed is not an issue. This is because the display trace does not have to be a real-time representation of the acquired signal. Instead, the signal is digitized and stored in memory, and then a display is generated from the stored wave-form data. As a result, both transient and repetitive signals can be displayed with equally clear and bright displays. Moreover, because the display is reconstructed from memory, transients are no longer a brief flash on the display, but are frozen on-screen for easy viewing.

The DSO's real-time sample rate is the limitation for capturing and displaying transient signals. Many DSOs use equivalent-time sampling for capture of repetitive wave forms up to the DSO's bandwidth limit. However, equivalent-time sampling, as will be discussed later, is a method of using lower real-time sample rates to create, effectively, high-speed sampling on repetitive wave forms. For single-sweep transient capture, the limit is always the actual real-time sampling rate.

Certain types of DSOs use real-time sampling exclusively to allow capture of transients up to the DSO's bandwidth and rise-time limit. Such DSOs are generally designed specifically for transient signal capture and are often referred to as transient digitizers.

The limitation on transient capture imposed by real-time sampling is essentially the time resolution provided by the sample rate. Similarly, the maximum equivalent-time sample rate imposes the upper limit on repetitive wave-form capture with a DSO. By comparison, analog oscilloscope limits for transient wave form capture are the display's writing speed and the bandwidth/rise-time specification for repetitive wave forms.

There is one other special limitation that needs to be considered regarding repetitive wave forms. This is the limitation of viewing very slow wave forms on analog scopes. Sub-hertz wave forms must be acquired on the lowest sweep-speed settings. As a result, the display on an analog oscilloscope will be just a bright dot moving across the screen. For such cases, there are two solutions for viewing a full trace of the slowly changing signal. One solution is to use an analog display storage oscilloscope or other high-persistence display technology. The other solution is to use a DSO to digitize the signal and re-create a full trace from memory.

2. RESOLUTION AND ACCURACY

Resolution and accuracy are sometimes confused in discussing oscilloscopes. In many cases they are wrongly taken to be synonymous.

Accuracy is the degree of conformity to an accepted standard. Resolution is the ability to distinguish differences. Thus, measurements between different things cannot be more accurate than the ability to resolve the differences. Consequently, to take full advantage of accuracy, there must be at least enough resolution to match accuracy and preferably more.

2.1 Analog Display Resolution

Analog oscilloscopes display wave forms against a calibrated display graticule. Generally, this graticule consists of eight major vertical divisions by ten major horizontal divisions. Each division is one centimeter square.

Typical spot sizes for the beam trace representing an acquired signal will range from 1/25 to 1/50 of a division, as reported by Tektronix (1992). The spot size can resolve half of a trace width. Therefore, the vertical display resolution is one part in 400 to one part in 800, while horizontal resolution is one part in 500 to one part in 1000. This is adequate for most on-screen measurement needs. Also, many analog oscilloscopes now have measurement cursors, or crosshairs, that can be positioned on the display to simplify waveform time and voltage measurements. It should be remembered, however, that such analog cursors are actually measuring screen position. Therefore, the cursors must be carefully positioned on the wave-form trace for accurate wave-form measurements.

2.2 Digital Storage Resolution

Digital storage oscilloscopes operate by sampling the input signal and digitizing the sample values. Thus, the time resolution of the stored signal data corresponds to the sampling interval (Δt), which is the reciprocal of the acquisition sampling rate.

DSO resolution may also be expressed in terms of record length. When a signal is captured with a DSO, a block of samples is taken for digitizing and storage. The number of stored samples for any given signal is referred to as the record length. Typical DSO record lengths are 512 and 1024 samples, although longer records are often available. A record length of 1024 samples implies a horizontal resolution of one part in 1024.

Vertical resolution corresponds to the digitizing resolution, which is one part in 2^B, where B is the number of digitizing bits. An 8-bit digitizer, for example, provides an amplitude resolution of one part in $2^8 = 256$; a 10-bit digitizer provides one part in $2^{10} = 1024$.

2.3 Accuracy

Oscilloscopes and DSOs have a variety of accuracy specifications. The most important of these include the amplitude accuracy of the vertical system and the time-base accuracy of the horizontal system.

Amplitude accuracy is simply the degree of accuracy to which a signal amplitude can be measured using the oscilloscope or DSO. In other words, how accurately can the amplitude of a 1-V, peak-to-peak sinusoid be measured. A fairly typical accuracy is 5%, which corresponds to one part in 20. This is well within the basic vertical resolution requirement cited above in 2.1 and 2.2 (resolution is greater than accuracy).

Time-base accuracy refers to ramp or clocking accuracy in reflecting a calibrated time horizontally across the display. In analog oscilloscopes, this is most directly related to the linearity of the ramp signal that drives the horizontal sweep. This is critical since the ramp horizontally drives the display, and any ramp nonlinearity translates to an effective speeding or slowing of the represented time. High-performance oscilloscopes address this by keeping sweep nonlinearities to 3% or less.

In DSOs, a ramp is often used to drive the horizontal sweep of the display system. The accuracy of this ramp is also important for measurements made directly from the display. However, since the display is reconstructed from digital memory, rather than real-time acquisition, it is not necessary to use as wide a range of ramp speeds as in analog oscilloscopes. Instead, DSO timing accuracy is based on the clocking accuracy of the digital time base used to drive sampling.

2.4 Digital Time-Base Accuracy

Digital time bases are generally driven by crystal-controlled clocks. This provides greater long-term accuracy and stability than the real-time methods of analog oscilloscopes.

In a DSO, the digital time base clocks the occurrence of signal sampling. The stability and accuracy of the clock determine the accuracy of where samples fall relative to each other on the signal. This is particularly important for equivalent-time sampling techniques since a complete sample record is built up over many signal repetitions. The longer it takes to build such a record, the more dominant clock instability and drift become in placing samples relative to each other and to the signal.

It is important to note that the sample interval (Δt) is assigned to the stored signal record as a single number. This sample interval is the reciprocal of the sample (or clock) rate. Thus, instabilities in actual clock rate result in a deviation of sample placement relative to sample interval. As indicated in Fig. 4, not only does this create a timing error, but it also results in a sample amplitude error for changing signal levels.

3. MEMORY AND RECORD LENGTH

DSOs, by definition, require memory for storage of the digitized wave form. However, it is common today for analog oscilloscopes to have digital memory, too. In analog oscilloscopes, this memory is used for storing front-panel setups for easy recall. Most DSOs offer similar setup storage and recall features. The advantage is that, once a complex measurement setup has been made, it can be stored in memory. Then, at the press of a few buttons, the stored settings can be recalled from memory for automatic setup of the oscilloscope or DSO.

A typical memory specification is the number of front-panel setups that can be stored in the DSO or oscilloscope. Another memory specification is the number of wave forms that can be stored in memory. The latter specification is exclusive to DSOs since analog oscilloscopes do not allow wave-form storage.

The number of wave forms that can be stored by a DSO is a function of available memory and the wave-form record length. As pointed out in previous discussion, record length is the number of stored samples used to represent the digitized wave form. The longer the record length (e.g., 1024 samples vs 512 samples), the greater the memory requirement for each stored wave form.

In some DSOs, record length is selectable. Thus, selecting a shorter record length allows more wave forms to be stored. On the other hand, a longer record length provides greater time resolution, but uses more wave-form memory, allowing fewer wave forms to be stored in a given memory size.

Also, memory can be partitioned differently in various DSOs. In some cases, there is fixed partitioning, allowing a fixed number of wave forms and front-panel setups to be stored. In other cases flexible partitioning allows trade-offs between the number of wave forms and setups stored.

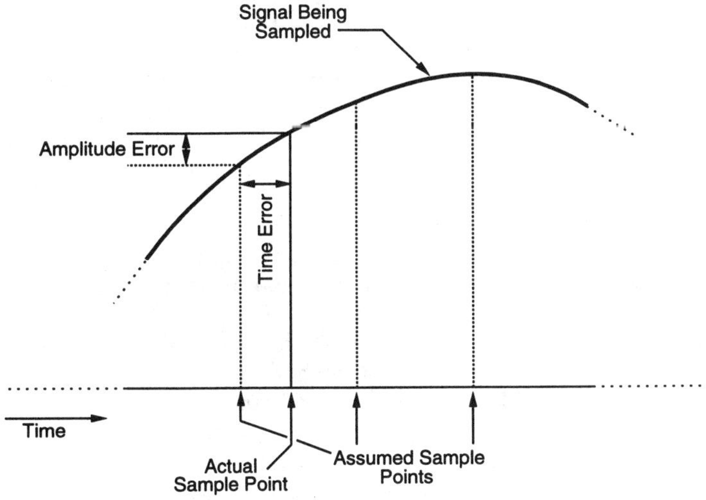

FIG. 4. Sample clocking inaccuracy results in two types of error. One is error in time placement of the sample. This results in a sample amplitude error that varies on the signal slope being sampled.

The key point is that, while memory and wave-form record length are important, the way memory is used is equally important. Applications requiring storage of numerous wave forms and few setups need memory partitioning in favor of wave forms. Applications requiring numerous front-panel setups generally benefit from memory partitioning in favor of setup storage.

4. ANALOG-TO-DIGITAL CONVERSION

Sampling and digitizing are the basic steps in analog-to-digital conversion (ADC) of wave forms. Up to this point, ADC has been discussed in a general sense. In reality, there are numerous ADC methods available. Each method has different advantages and limitations for wave-form acquisition and storage.

4.1 Digitizing Rate

The term "digitizing rate" is sometimes used in place of "sampling rate." To be correct, sampling rate is the rate at which amplitude samples are taken, and digitizing rate is the speed of converting an amplitude sample to a digital value. Thus digitizing and storage of an amplitude sample value must be completed before the next sample arrives for digitizing.

In terms of wave-form representation, the digitizing or sampling rate must be fast enough to define the wave form adequately. In the strictest sense, ADC must, at minimum, meet the Nyquist sampling rule. The Nyquist rule states that there must be two samples per cycle for the highest frequency of significance in the wave form being sampled.

For example, consider the case where the highest frequency of significance in a 20-MHz wave form is the fifth harmonic. Then, according to the Nyquist rule, the wave form must be sampled at a rate that is twice the frequency of the fifth harmonic. This would be a sampling rate of 200 MHz for the fifth harmonic of a 20-MHz wave form.

For most purposes, it is easier to think of sampling rate in terms of the DSO's bandwidth. At a minimum, the Nyquist sampling limit should be twice the DSO's bandwidth. For a 200-MHz DSO, this corresponds to a 400-MHz sampling rate at minimum. However, that also means that a 400-MHz sinusoid would be acquired with only two samples per cycle. While two samples per cycle may be adequate for mathematical description of the sinusoid, it is inadequate for graphical display of a recognizable sinusoid. Because of this, many DSOs use oversampling to provide more samples per cycle at the bandwidth limit. The effects of various numbers of samples per cycle on graphical representation are shown in Fig. 5.

4.2 Real Time and Equivalent Time

Real-time ADC means that all wave-form amplitude samples are taken, digitized, and stored in real time, as the wave form occurs. This is the ADC method necessary for capturing single-shot events. Samples are taken one after the other, in sequence, as the event transpires.

The difficulty with real-time ADC is the rates needed for adequate oversampling of the DSO's bandwidth. For example, with a 200-MHz DSO, oversampling at just 5 times the bandwidth means an ADC rate of 1 GHz. While 1-GHz and higher real-time ADC rates are certainly achievable, the technology is not inexpensive. Moreover, such high rates are not necessary when dealing with repetitive wave forms.

Repetitive wave forms can be repetitively acquired in a triggered mode (see Fig. 2). Since the wave-form repetitions are acquired over and over, a few samples can be taken at different points on each triggered acquisition. Over many triggered acquisitions, enough samples can be accumulated for a complete record of the wave form. This concept is illustrated further in Fig. 6. Essentially, the resulting composite of wave form samples is an equivalent-time representation of what would be achieved with a very fast real-time sampling rate.

Equivalent-time DSOs can capture repetitive wave forms with significant oversampling, even at their bandwidth limit. This is done economically by using low real-time sample rates and the correspondingly less-expensive ADC technology. For example, a 1-MHz real-time ADC can be used to achieve equivalent-time rates of hundreds of megahertz and even gigahertz for capturing repetitive wave forms. This assumes, of course,

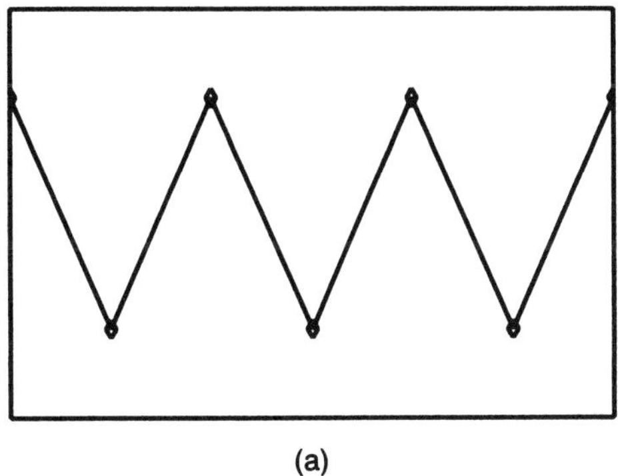

(a)

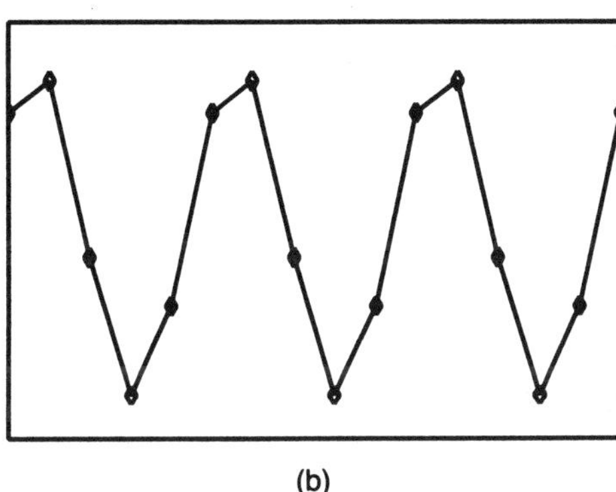

(b)

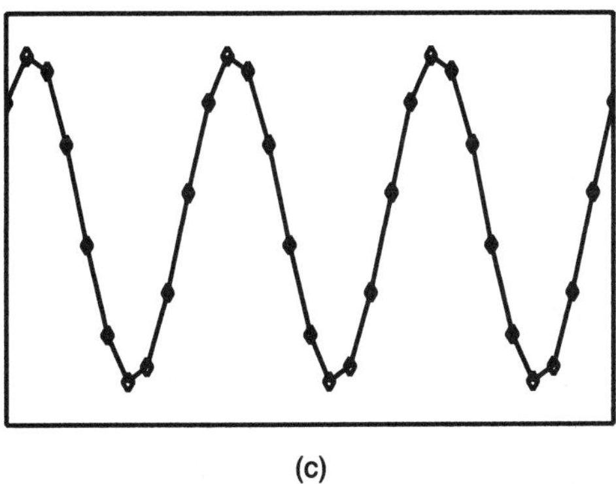

(c)

FIG. 5. The effects of various numbers of samples per cycle in accurate portrayal of a sinusoid—(a) two samples per cycle, (b) six samples per cycle, (c) 12 samples per cycle.

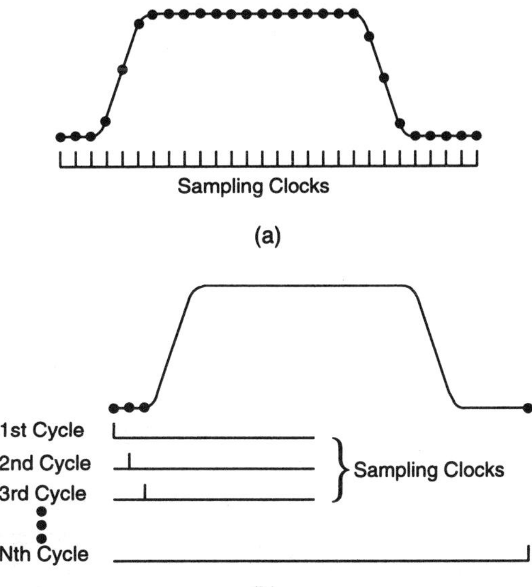

FIG. 6. Comparison of sampling methods. (a) In real-time sampling, samples are taken in sequence as the wave form actually occurs. (b) In equivalent-time sampling, samples are accumulated from numerous triggered acquisition cycles. For example, a 512-point record may take 512 acquisitions for complete sampling.

adequate analog bandwidth in the oscilloscope circuitry preceding the ADC.

There are some important cautions, however, when considering equivalent-time DSOs. Some DSOs with high equivalent-time capabilities gain their economic edge by using the lesser technology of very low real-time digitizing rates. This results in a very low single-shot capture capability for transients since single-shot capability is always limited to the real-time sample rate. Also, with a low real-time rate, full equivalent-time sampling can take longer because more repetitive acquisitions are necessary to build up a full wave-form sample record. Finally, any wave-form jitter or amplitude variations will have a degrading effect on the accuracy and appearance since the final record is a composite of many acquisitions over time. However, degradations from random effects can be reduced through the process of signal averaging.

4.3 Converter Types

A variety of converter types have been developed to meet various speed, resolution, and economic requirements. The major categories of converter types used in DSOs are successive approximation, flash conversion, charge coupled devices (CCDs), and scan conversion.

4.3.1 Successive Approximation These converters compare the input signal sample level to the output of a digital-to-analog converter (D to A). A bit-by-bit conversion is made starting with the most significant bit (MSB) and ending with the least significant bit (LSB).

Since each analog sample is digitized bit by bit, there is a trade-off between conversion speed and digital resolution (e.g., 8-bit vs 10-bit). For higher resolutions, longer conversion times must be accepted. To overcome the conversion-time trade-off, equivalent-time sampling methods are almost always used with successive approximation. However, the real-time conversion rate is still an important consideration for overall wave-form ac quisition time and for any single-shot capture needs.

4.3.2 Flash Conversion These converters use a binary string of resistive dividers and comparators. The number of divider resistors and comparators is equal to the binary resolution of the converter. For example, an 8-bit converter would use an eight-level string of dividers and comparators.

The analog sample is applied to the top of the divider string, and the comparators output the corresponding binary levels in parallel. Since the conversion to all bits takes place in parallel, the method is quite fast, hence the name "flash" conversion.

However, there are several trade-offs to consider in flash conversion. One trade-off is that for higher resolution, more resistive elements and comparators must be strung together. This translates to greater converter complexity and expense. Also, the resistive divider string must be carefully designed to compensate for changing values at higher frequencies. Designing a high-performance flash converter is not a trivial task. Still, flash conversion offers an economical method for fast equivalent-time DSOs with real-time rates that are substantially higher than those using successive approximation.

4.3.3 Charge Coupled Device (CCD) A CCD is actually a very fast analog sampler and not a digitizer. Essentially, the CCD consists of a set of input cells, a memory section, and a set of output cells.

The key is that the CCD takes in analog samples at a high rate and transfers them to the memory section. The analog samples are held in memory and can then be read out and digitized at a slower rate, by successive approximation for example.

The advantage of the CCD technique is that high real-time sample rates can be achieved at modest cost. As reported by Tektronix (1992), a 100-MHz sample rate, 1-K record length, and 8-bit vertical resolution can be achieved at a modest cost compared to other methods used to achieve the same performance level.

4.3.4 Scan Conversion Scan converters typically provide the highest available real-time performance. As a result, they are only used in the class of DSOs referred to as transient digitizers.

A scan converter is basically a double-ended cathode-ray tube (CRT). One end provides a writing beam similar to that in a conventional analog oscilloscope CRT. However, instead of writing the wave-form trace onto the phosphor of a viewing screen, the trace is written onto a semiconductor target in real time. The target holds the trace, usually in the form of charge in a diode array. The reading beam at the other end of the CRT then scans the target and reads out the stored trace. This reading process can be done at a slow rate, allowing digitizing to be done by one of the more economical digitizing processes.

During scan conversion (reading of the target) no further wave-form information can be captured. However, this usually does not present a problem in transient wave-form capture since transients are infrequent occurrences.

4.4 Vertical Quantizing

Vertical quantizing is often referred to as quantizing noise. It is a function of digitizing, and thus occurs regardless of the method of digitizing.

Quantizing occurs when the analog value of a sample does not correspond exactly to one of the available digital levels. Since the digitizer cannot represent the analog level exactly, it must default to the closest digital level above or below.

This quantizing process has the effect of adding noise to the captured signal, hence the term "quantizing noise." The amplitude of this quantizing noise (or error) depends on the vertical resolution of the digitizer. An 8-bit digitizer, because of its lower resolution, causes higher quantizing noise amplitude than a 10-bit digitizer.

Quantizing noise is unavoidable. However, under certain circumstances, it can be significantly reduced, as reported by Ramirez (1985). This reduction occurs when repetitive wave forms are signal averaged by a processor having greater resolution than the digitizer. The result of such averaging can be that 8-bit digitizing can be cleaned up to be equivalent to 10-bit or even 12-bit digitizing.

4.5 Effective Bits

Digitizer resolution is generally stated as static resolution. That is, an 8-bit digitizer provides 8 bits of resolution in digitizing a dc voltage level. It also will generally provide 8-bit resolution for slowly changing wave forms. However, jitter, sampling uncertainty, and other effects begin to have a greater influence as wave-form frequency increases. These effects manifest themselves as increased noise on the digitized signal.

Since increased noise from sampling uncertainty is digitizing noise, it can, in effect, be categorized as quantizing noise. Thus, a higher quantizing noise translates to a lower digitizing resolution. This means that an 8-bit digitizer can actually appear to be of much lesser resolution when digitizing higher frequency signals. For example, at 75% of its sample rate limit, an 8-bit digitizer may appear to effectively be a 5-bit digitizer.

As mentioned previously, random noise effects can be overcome by signal averaging when repetitive wave forms are being acquired. Signal averaging cannot be done on single-shot or transient waveform acquisition. As a result, most transient digitizers have effective-bits specifications for different points across their bandwidth. In general, the higher the effective-bits specification, the higher the transient digitizer's quality.

5. DISPLAY TECHNIQUES

Digitized wave forms are stored in memory as an array of discrete amplitude values. These discrete values are then processed to present a visual display representing the original analog wave form. How this display is created is a matter of preference as well as a matter of how various types of wave forms should be represented for accurate interpretation.

5.1 Dot Displays

Dot displays are the simplest of all displays. These consist of simply displaying each sample value from memory as a dot. The vertical position of the dot represents amplitude and the horizontal position represents time.

In truth, the dot display is the most accurate representation of the acquired wave form. This is because the dots are direct plots of the acquired samples, and the acquired samples are the only recorded wave-form data. What happens between samples is speculation because no wave-form information was acquired between samples.

In practice, dot displays can be horribly confusing. This is shown in Fig. 7, where several sinusoids of different frequencies are represented by dot displays. The problem is that, when the dots are moved closer together horizontally, the eye is tricked into picking out different wave-form patterns. Generally, the eye picks out lower-frequency

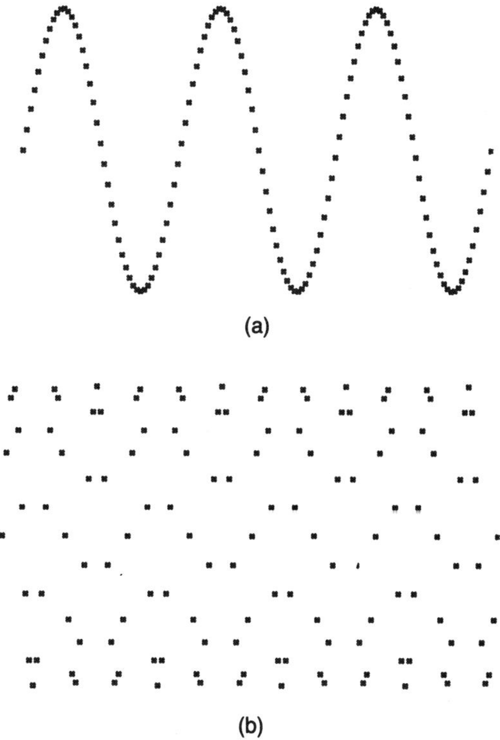

FIG. 7. The effects of perceptual aliasing. (a) A few sinusoid cycles are clearly distinguishable with a dot display. (b) With many sinusoid cycles, the dot order becomes visually confusing and the eye tends to pick out lower-frequency aliases of the actual wave form.

wave forms than actually represented by the dots. This effect is referred to as perceptual aliasing.

5.2 Interpolated Displays

To avoid perceptual aliasing, most DSOs provide interpolated displays. Most often the interpolation is linear. That is, a dot display is transformed by connecting the dots in time order with straight lines.

The effect of straight line interpolation is shown in Fig. 8. Note that, in comparison to the dot displays of Fig. 7, interpolated displays eliminate any confusion about the actual wave form path.

Linear interpolation provides a very close approximation of the actual wave shape for wave forms composed primarily of linear segments—e.g., square waves, ramps, triangular waves. For wave forms that follow nonlinear forms, such as sinusoids, other forms of interpolation may be preferred. Not surprisingly, sine interpolation is the most common form used in DSOs after linear interpolation. This is because sine interpolation provides a closer approximation for smooth wave forms.

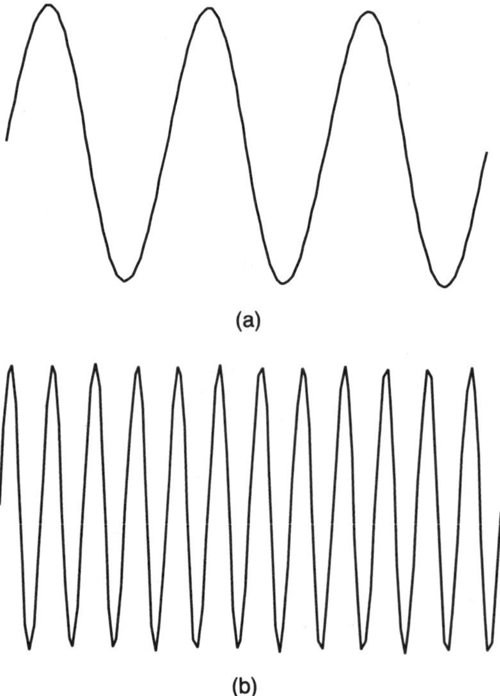

FIG. 8. Linear interpolation or vector display. When a dot display is straight-line connected in dot order, wave forms are easier to distinguish. Compare (b) here to (b) in Fig. 7.

5.3 Peak Detection and Enveloping

Peak detection and enveloping are both special methods of handling samples to reveal special wave-form situations. Peak detection, in general, ensures capture of wave-form peaks or spikes that many times are not captured with ordinary sample processing. Enveloping is somewhat different and allows an envelope of maximum wave-form excursions to be captured.

With ordinary sample processing, the wave-form record is filled with samples taken at equally spaced points in time (Δt). The time spacing, or Δt, corresponds to the display-time window (total horizontal sweep time) divided by the displayed record length. For example, say that the DSO's sweep-speed setting is 0.01 s/division. For a 10-division horizontal display, this corresponds to a display-time window of 0.1 s. Thus, with a displayed record of 1024 sample points, the sample spacing (Δt) is 0.1 s/1023 = 0.97752 μs.

The problem with ordinary sampling is that short-duration details, such as a noise spike, can easily fall between samples and be missed. The likelihood of this happening increases with the greater sample spacing that accompanies acquisition of slower wave forms.

Peak detection minimizes the loss of small details, such as spikes, that could fall between samples. The method involves always sampling at full speed regardless of the sweep-speed (time base) setting. With a 100-MHz real-time sample rate, this corresponds to a fixed sample interval (Δt) of 1 ns. However, on longer time windows this results in far more samples than there is record length storage. A 0.1-s window with a Δt of 1 ns corresponds to 100,000,000 samples. For a stored record length of 1024 samples, only selected samples are stored and the excess samples are discarded.

For peak detection, the maximum-amplitude sample in any stored sample interval is retained for the stored wave-form record. This is illustrated in Fig. 9. The result is that for any sweep speed, no matter how slow, there is a very high probability of capturing wave-

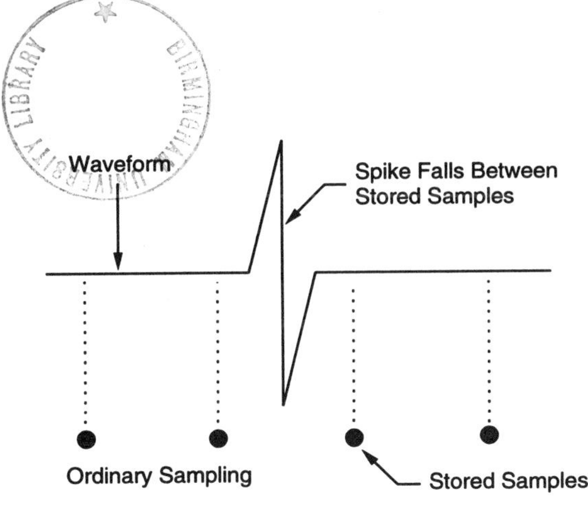

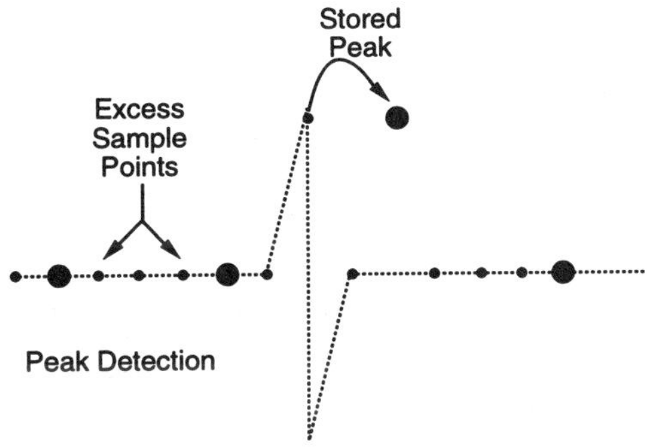

FIG. 9. Peak detection. With ordinary sampling, narrow wave-form spikes may occur between samples and may not be captured. Peak detection uses oversampling and picks the maximum valued sample for storage. This guarantees at least one sample on narrow spikes.

form peaks (or spikes) as narrow as the smallest sample interval. In the case of 100-MHz maximum sampling, this is referred to as 1-ns peak detection since peaks to 1 ns wide can be detected.

The key benefit of peak detection is that there is always an indication that signal spikes exist, even when operating at the slowest sweep speed. With this knowledge, a higher sweep speed can be used for more detailed capture and examination of the spikes. Without the initial indication provided by peak detection, signal spike problems could easily be missed.

Enveloping, another very useful process, can be implemented with or without peak-detection sampling. When implemented in conjunction with peak detection, enveloping is far more sensitive to minor wave-form changes.

Basically, enveloping involves the use of two stored wave-form records. One is used for maximum sample values and the other for minimum sample values. Over multiple wave-form acquisitions, an envelope display is created of maximum and minimum excursions of the wave form. Such an envelope display is very useful for detecting and studying jitter, drift, or other wave-form variations over time.

5.4 Pre- and Post-Trigger Displays

The majority of the displayed trace on an analog oscilloscope is a post-trigger display. It is the trace of the wave form occurring in time immediately after the trigger point (refer to Fig. 2).

In some oscilloscopes, a trigger hold-off feature allows the display to be delayed to a selected time after the trigger point. This is useful for viewing wave-form activity that occurs some time after a selected triggering point or triggering event. For example, the firing pulse for a laser could be used as a trigger, and trigger hold-off could be used to study laser activity some time after the firing pulse.

Often, it is important to see wave-form activity prior to the trigger point. This is referred to as pretrigger acquisition. Pretrigger acquisition is provided in some analog oscilloscopes by placing a fixed delay line in the signal path after the trigger-system pick-off point [refer to Fig. 1(a)]. This allows the displayed signal trace to include a portion of the signal occurring prior to the trigger point. The duration of this pretrigger portion of the display is equal to the time delay of the inserted delay line.

The pretrigger portion of the display shows wave-form activity prior to the trigger point. This is quite useful, for example, in capturing the starting point of pulses or transient wave forms. However, in analog oscilloscopes, delay-line pretriggering may not be sufficient for many needs. This will be especially true when the trigger level must be set high in order to avoid false triggers on noise or other aberrations (see Fig. 10).

DSOs get around the analog limits to pretrigger length in a fairly simple manner. This involves continuously sampling the input signal and shifting the samples through the record in memory. The segment of samples that is locked into memory as a stored wave-form record is determined by triggering on

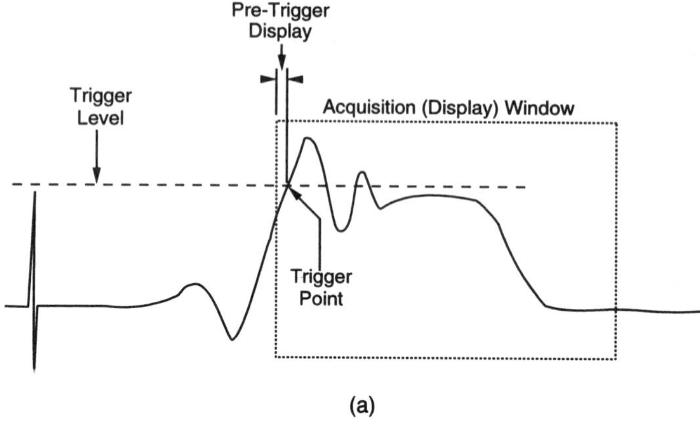

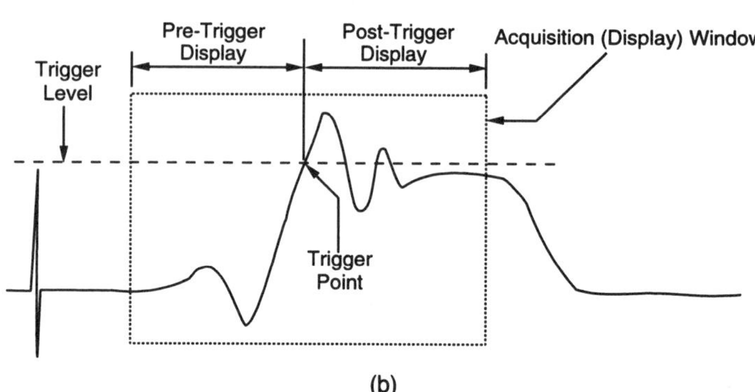

FIG. 10. Pre- and post-triggering. (a) Analog oscilloscopes use a fixed delay line to provide a small amount of pretrigger display. (b) Digital storage oscilloscopes can provide varying amounts of pre- and post-triggered wave-form information. This allows triggering above noise while still displaying the entire leading edge of a pulse.

the wave form and the amount of time that has been selected for pre- or post-triggering.

For example, let us say the DSO has been set up to provide half the display as pretrigger information and the other half as post-trigger information. The DSO is continuously acquiring samples and shifting them through memory while waiting for the triggering event. At the moment triggering occurs, samples existing in memory represent a record of 100% pretriggering information. Since the example setup calls for a record of 50% pretrigger and 50% post-trigger samples, the DSO continues acquiring samples after triggering until half a record length of post-trigger samples is in memory. At this point, the samples are frozen in memory as a wave-form record of half pretrigger and half post-trigger information, as shown in Fig. 10.

There are several key benefits to the adjustable pre- and post-triggering offered by DSOs. The first benefit is that trigger levels can be set high on the wave form to avoid false triggers, and the full wave form can still be viewed by using pretriggering to include the wave-form starting point. Another key benefit or use is the ability to look well ahead of the triggering point or event to study activity leading up to the triggering event. An example of the latter use would be study of charge buildup, corona, or other effects preceding high-voltage arcing.

6. USEFUL STORAGE BANDWIDTH AND RISE TIME

Previous discussion mentioned bandwidth and rise time as key specifications in selection and analog oscilloscope. The same analog specifications also apply to the signal input circuitry of a DSO. This is necessary for accurate presentation of the analog signal to the sampling and digitizing circuitry. Additionally, there are various other factors that affect the usefulness of the digitized signal display. Dot displays and perceptual aliasing are two of the factors already discussed.

Rather than trying to interpret the interactions of analog bandwidth and rise time and other various digitizing effects, combined figures of merit have been established for overall DSO evaluation. These figures of merit are referred to as useful storage bandwidth and useful storage rise time.

6.1 Useful Storage Bandwidth

In general, useful storage bandwidth (USB) is defined as follows for dot displays:

$$f_{USB} = (\text{maximum digitizing rate})/25. \quad (3)$$

The value of 25 in the above relationship is the number of samples per cycle generally required for visual recognition of wave shapes on a dot display. For linearly interpolated display (vector connection of dots), the divisor value can be changed to 10.

6.2 Useful Storage Rise Time

With a DSO, the displayed rise time depends upon where samples fall on the actual rising wave-form transition. With a vector-interpolated display, the maximum rise-time errors on the DSO display closely follow the form of an analog oscilloscope's errors when the analog system has a rise time of 1.6 sample intervals. As a result, useful storage rise time (T_r) can be defined as follows:

$$T_r = (\text{minimum sample interval}) \times 1.6. \quad (4)$$

The above figure of merit applies to the worst-case error for rise-time measurement on a DSO vector display.

7. SIGNAL PROCESSING

One of the more powerful possibilities of a DSO is the ability to provide a wide variety of wave-form processing and analysis capabilities. Since the stored wave forms are essentially a series of numbers, virtually any kind of numerical analysis can be performed. Whether or not such analysis is available as a built-in feature depends on the particular DSO.

Most DSOs do provide measurement cursors for reading out the amplitude and time values of stored wave forms. Additionally, most DSOs provide the simpler built-in analysis functions such as wave-form maximum, minimum, average, and rms (root mean square) values.

High-end DSOs generally provide all of the simpler built-in functions as well as one or more advanced built-in functions. Some of the advanced functions that might be provided are briefly discussed below.

7.1 Save-on-Delta

Save-on-Delta is a function that relies on first storing an envelope of expected wave-form limits. Depending on the DSO implementation, the upper and lower limit traces can be generated manually or by downloading data to the DSO.

Once the upper and lower limit data, or envelope, is in place, Save-on-Delta acquisition of an actual wave form can begin. During acquisition, each sample of the wave form is compared to the upper and lower limit envelope. If the acquired samples do not violate the limits envelope, the wave form is not saved. If any sample does violate the limits envelope, the wave form is saved.

This acquisition mode is very useful for catching intermittent or infrequent wave-form conditions. A DSO can be left unattended in Save-on-Delta mode for as long as needed to capture limit-violation wave forms.

7.2 Averaging

Averaging can be applied to repetitively acquired wave forms to provide improvements in signal-to-noise ratio (SNR). Whenever possible, averaging should be applied to improve measurement accuracy.

The usual form of averaging is to select a number of acquisitions to be averaged—for example, 64 averages. Averaging then takes place by acquiring the first wave form. A second wave form is then acquired and automatically added to the previous wave form. This acquisition and summing of amplitude samples continues until the specified number of wave forms has been acquired and summed. The summed wave forms are then divided by the specified averaging number.

The result of averaging is that randomly occurring events (noise) on repetitively acquired wave forms become reduced, as reported by Ramirez (1985). When averaging is done at computation resolutions exceeding the digitizer's amplitude resolution, the effects of quantizing noise can also be reduced. This can result in a higher effective measurement resolution on the averaged wave form.

7.3 Integration and Differentiation

Stored wave forms can be integrated or differentiated using various numerical analysis routines. Typically integration and differentiation are done in a point-by-point (two point) process.

Integration is simply the process of calculating the rectangular area represented by a pair of points and summing those areas. The result of such integration provides a point-by-point plot of energy contained under the wave-form curve.

Differentiation is simply the point-by-point slope of a wave form. This is useful for defining break points on wave forms. A particularly useful application, as reported by Ramirez (1985), is in acquiring the impulse response of electronic circuits. Since a step voltage is easier to generate and control as a stimulus signal, a circuit step response is easier to acquire in most cases. When acquired with a DSO, the step response can be differentiated to provide an impulse response display.

7.4 Fast Fourier Transform

The classical Fourier transform is an analog method of transforming a time-domain signal to the frequency domain. This allows detailed analysis of a signal's frequency content. The discrete Fourier transform (DFT) is the classical numerical analysis approach to obtaining Fourier transform data. The fast Fourier transform (FFT) is simply a computationally efficient method of computing the DFT.

Some DSOs contain a built-in FFT algorithm. This allows acquired wave forms to be transformed from time-series displays to frequency-spectrum displays.

7.5 Convolution and Correlation

Convolution and correlation functions are very easy to compute in the frequency domain. For those DSOs having an FFT algorithm, the additional steps of providing convolution and correlations are often taken.

Convolution is a key computation used in transfer-function analysis. Correlation has many uses in signal identification and matching.

8. MEASUREMENT AUTOMATION

Since most oscilloscopes today are digitally controlled, they offer the opportunity for measurement automation. One form of automation has already been mentioned. That form is push-button storage and recall of instrument front-panel setups. Another form of automation just discussed is the signal processing capabilities that can be provided as built-in functions of DSOs.

Both forms of automation can be taken another step further by interfacing the instrument to a computer. When this is done, the automation resources are vastly expanded through software programming, disk storage, and setup and wave-form information transfers between the computer and instrument.

8.1 Interfacing Techniques

The basis for most stand-alone instrument interfacing today is the General Purpose Interface Bus (GPIB) as defined by the IEEE-488 specification. A single GPIB loop can contain up to 16 devices. One of the devices must be the GPIB controller (a personal computer for example), which leaves a potential of 15 devices for GPIB control. These devices can be DSOs, programmable analog oscilloscopes, arbitrary wave-form generators, power supplies, or any other type of GPIB instrumentation.

Another popular form of instrument interfacing is through an RS-232 port. This has become popular because most personal computers come standard with one or more RS-232 ports. The actual interfacing is still usually via the GPIB standard, but it is accomplished through RS-232/GPIB translation in the instrument's interfacing hardware and firmware.

RS-232 interfacing has another important advantage in that it is compatible with telephone modems. This allows an instrument to be remotely located and accessed through telephone lines and modem connection. If an RS-232/GPIB controller is used at the remote site, an entire GPIB instrument system can be remotely accessed and controlled over a single phone line.

8.2 Software Considerations

There are two key considerations in looking at software for automation applications. One consideration is the instrument interfacing and control software. The other is the wave-form handling and analysis software.

Most GPIB instrument manufacturers provide software drivers for their specific instrument models. These drivers are usually created so that they can be linked as modules into standard software platforms, such as BASIC or UNIX. A more important consideration, however, is command and syntax consistency across various instruments. If the manufacturer does not follow some form of codes and formats standard in creating drivers for various instruments, control of multiple instruments can become extremely complex.

After instrument control and wave-form transfer to the computer, there is the consideration of wave-form processing. This can be done by writing programs from the ground up using BASIC, C, or any other popular software platform. However, a tremendous amount of work can be saved by using any of several available wave-form processing packages. These packages provide specialized wave-form processing capabilities and often include many advanced analysis features such as FFT, convolution, correlation, modal analysis, and so forth. Essentially, once the acquired wave forms have been transferred to a computer, the analysis possibilities are endless.

9. CONCLUSION

Virtually any wave-form capture and analysis need can be met by one of the analog oscilloscopes or DSOs on the market today. The primary issue is matching application needs to a specific instrument type. This includes the following major steps:

1. Defining analog bandwidth and rise-time requirements.
2. Determining whether analog display or stored wave-form display is best for the application. Generally, digital storage is preferred for very slow wave forms, transient wave forms, or any wave form where substantial measurement or analysis is necessary. Analog oscillscopes are preferred for real-time wave-form viewing needs, such as in making circuit adjustments, and for multi-GHz transient recording.
3. Determining digital resolution in terms of both amplitude (digitizing resolution) and record length (generally longer records are needed for fast-rise, slow-decay wave forms).
4. Specifying sampling technique and rates—real time for single-shot transients and equivalent time for repetitive waveforms.
5. Wave-form storage requirements. Built-in instrument storage may be adequate for applications focused on a few wave forms, but large quantities of wave-form data usually require external storage (e.g., interfacing to a computer).
6. Considering the benefits that might be derived from the automated instrument control and advanced analysis techniques available through computer interfacing.

Careful consideration of the above issues in terms of specific application requirements will lead to a set of specific wave-form acquisition needs. This set of specific needs provides a clear definition of the specific instrument requirements for the application. This matching the tool to the job—rather than bending the job to fit an existing tool—leads to greater success in quickly obtaining useful measurement and analysis results.

It is important, however, to realize that both analog and digitizing oscilloscope technologies are marching to a quick pace and often leapfrog each other. This makes it risky to formulate any set of rules that dictates exclusive use of one technology over another for any given application need. Thus, evaluation of measurement needs against available oscilloscope technologies is an ongoing process.

As far as future trends are concerned, it is almost always safe to say that acquisition bandwidths and speeds will continue to increase. Also, with the digital age in full bloom, both analog and digitizing oscilloscopes are offering an expanding array of automation and measurement processing features. As a sidelight to this, industry observers speculate from time to time that oscilloscopes will eventually be compressed to a card that plugs into a personal computer. Current industry trends, however, suggest that the opposite is actually happening. More and more specialized computing power is being compressed into oscilloscopes. Whatever course future implementations take, the key oscillographic concepts will remain the same—acquisition, processing, and display of wave forms.

GLOSSARY

Aliasing: A condition where one frequency takes on the appearance of another frequency. In sampling, a minimum of two samples per cycle is required to avoid aliasing. In dot displays, perceptual aliasing can be prevented by connecting the dots in their correct time order.

Bandwidth: The continuous band of frequencies passed or transmitted by a device without undue attenuation. The bandwidth or pass-band edges are often defined as the points where amplitude becomes attenuated by 3 dB (half power).

Cathode-Ray Tube (CRT): A vacuum-tube display device consisting of an electron beam, an electronic drive and deflection system, and a phosphor-coated display screen. The deflection system traces the electron beam across the phosphor coating on the back of the screen causing luminescence of the phosphor along the beam path for a visible display of the trace on the front of the screen.

Equivalent-Time Bandwidth: The highest attainable bandwidth using equivalent-time sampling (one-half the equivalent-time sample rate).

Equivalent-Time Sampling: Any wave-form sampling method where a full wave-form record is built up over multiple, repetitive acquisitions of a wave form. Equivalent-time sampling allows capture of high-speed repetitive wave forms while using lower speed sampling and analog-to-digital conversion. Equivalent-time sampling cannot be applied to the capture of single-shot, transient wave forms.

Fast Fourier Transform (FFT): An efficient computer algorithm for computing the discrete Fourier transform of sampled wave-form data. Time-domain samples are converted to frequency-domain data or vice versa.

Persistence: In reference to CRT phosphors, the property of persisting in light emission after excitation has ceased. Oscilloscope traces fade more slowly with high-persistence phosphors than with low-persistence phosphors.

Probe: Device used to sense or pick up signals and convey them to an instrument for observation or measurement. Voltage probes are commonly used with oscilloscopes, but current and temperature probes are generally available, too. In many cases, the probe is used to pick up the output of another sensor such as a strain gauge or accelerometer.

Probe Compensation: Adjustment of probe capacitance so that the probe attenuates all input signals equally across a range of frequencies.

Real-Time Bandwidth: The highest attainable bandwidth using real-time sampling (one-half the real-time sample rate).

Real-Time Sampling: Any wave-form sampling method where samples are taken sequentially in real time as the wave form actually occurs. Real-time sampling is required for the capture of single-shot, transient events and is equally applicable to the capture of any other class of wave form (repetitive, random, etc.).

Record Length: The number of samples acquired and stored as a digitized representation of a wave form.

Rise Time: The time required for the leading edge of a pulse to rise from 10 to 90% of its final value.

Sweep Speed: The speed at which the CRT display's electron beam is driven horizontally across the screen. Sweep speeds are usually in terms of seconds per display division, providing a time-calibrated display.

Time Base: In oscilloscopes, an electronic sweep or clocking system used to provide time-calibrated acquisition and display of wave-form information.

Triggering: The process where selected wave-form conditions cause acquisition of the wave form to begin.

Trigger Hold Off: A period of time when triggering is prevented. This is usually done to prevent a second trigger during the horizonal sweep period. Trigger hold off may also be used to provide a delayed sweep, allowing acquisition to occur at a selected period of time after the triggering event.

Trigger Point: The point on a wave form that causes triggering of wave-form acquisition.

Works Cited

Gustafson, M., Johnson, L., Caryl, M., Gifford, T., White, D. (1985), "State-of-the-Art Trends in Oscilloscopes," *Test & Measurement World*, April, 105–113.

Ramirez, R. W. (1985), *The FFT Fundamentals and Concepts*, Englewood Cliffs, NJ: Prentice-Hall, pp. 101, 151.

Tektronix (1992), *An Introduction to Digital Storage*, Report No. 41W-6051–4, Beaverton, OR: Tektronix, Inc.

Tektronix (1993a), *The Effects of Bandwidth on Measurement Accuracy*, Report No. 85W-8907–0, Beaverton, OR: Tektronix, Inc.

Tektronix (1993b), *ABC's of Probes*, Report No. 60W-6053–4, Beaverton, OR: Tektronix, Inc.

Further Reading

IEEE (1994), *IEEE Standard for Digitizing Waveform Instruments*, IEEE Standard 1057-1994, New York: IEEE.

Baumgart, C., Dunham, M., Moses, J. (1991), "Curve Fitting and Error Modeling for the Digitization Process Near the Nyquist Rate," *IEEE Trans. Instrum. Meas.* **IM-40**, 553–557.

Peetz, B. (1983), "Dynamic Testing of Waveform Recorders," *IEEE Trans. Instrum. Meas.* **IM-32**, 12–17.

Ramirez, R. W. (1985), *The FFT: Fundamentals and Concepts*, Englewood Cliffs, NJ: Prentice-Hall, Inc.

Tektronix (1992), *An Introduction to Digital Storage*, Report No. 41W-6051-4, Beaverton, OR: Tektronix, Inc.

Textronix (1992), *Real-Time Digital Signal Processing Techniques Address High-Resolution Measurement Requirements*, Technical Brief No. 3GW-8222-1, Beaverton, OR: Tektronix, Inc.

Tektronix (1992), *The XYZs of Using a Scope*, Report No. 03W-8605-0, Beaverton, OR: Tektronix, Inc.

Tektronix (1992), *Real-Time Versus Equivalent-Time Sampling*, Technical Brief No. 55W-8628-0, Beaverton, OR: Tektronix, Inc.

Tektronix (1993), *ABC's of Probes*, Report No. 60W-6053-4, Beaverton, OR: Tektronix, Inc.

Tektronix (1993), *Sampling Oscilloscope Techniques*, Technique Primer No 47W-7209, Beaverton, OR: Tektronix, Inc.

Tektronix (1993), *The Effect of Probe Input Capacitance on Measurement Accuracy*, Technical Brief No. 60W-8910-0, Beaverton, OR: Tektronix, Inc.

Tektronix (1993), *The Effects of Bandwidth on Measurement Accuracy*, Technical Brief No. 85W-8907-0, Beaverton, OR: Tektronix, Inc.

OSMOSIS

Benny Freeman, *Department of Chemical Engineering, North Carolina State University, Raleigh, North Carolina, U.S.A.*

	Introduction	59
1.	**Molecular and Thermodynamic Description of Osmosis**	60
1.1	Molecular Basis of Osmosis	60
1.2	Thermodynamic Analysis of Osmosis	61
1.2.1	Osmotic Pressure of an Ideal, Nonionic Solution	61
1.2.2	The Influence of Solute Molecular Weight on Colligative Properties	62
1.2.3	Osmotic Pressure of a Solution Containing Several Solutes	63
1.2.4	Osmotic Pressure of a Dilute Ideal Electrolyte Solution	63
1.2.5	Osmotic Pressure of Dilute Ideal Protein Solutions	64
1.2.5.1	Donnan Equilibrium	65
1.2.6	Effect of Solution Nonideality on Osmotic Pressure	66
2.	**Experimental Methodology for the Determination of Osmotic Pressure**	68
3.	**Applications**	69
3.1	Reverse Osmosis	69
3.2	Dialysis	69
3.3	Electrodialysis	70
	Glossary	70
	Works Cited	71
	Further Reading	71

INTRODUCTION

Osmosis [from the Greek *ōsmos*, the action of pushing, derived from *ōthein*, to push (Morris, 1981)] is the phenomenon resulting from the tendency of a pure solvent to diffuse into a solution separated from the solvent by a membrane that is permeable only to the solvent (Atkins, 1990). Many important examples of osmosis are found in nature. Osmosis drives the transport of oxygen and carbon dioxide across capillary walls separating blood from air in the lungs. Fresh fruit will shrivel when placed in contact with strong sugar solutions, and dried fruit can swell in contact with water as a result of osmosis (Marick, 1969). The delicate balance of chemical substances inside living cells is maintained, in part, by osmosis. Furthermore, osmosis has been exploited scientifically as a sensitive probe of molecular weight of natural and synthetic macromolecules (Tanford, 1961). Reverse osmosis is an important industrial process for making drinking water from saltwater. A feed stream such as brackish water or seawater, is contacted with a membrane that is permeable to water but impermeable to microorganisms, colloids, dissolved salts, and organic species. The feed stream is pressurized to force permeation of pure, drinking-quality water across the membrane (Bhattacharyya and Williams, 1992). In the 1940s, the concept of artificial kidney dialysis was demonstrated and has become widely used to prolong the lives of people suffering from renal failure (Kessler and Klein, 1992). This technology is based on the osmotic transport of metabolic waste products from blood into a saline solution. The blood and saline solution are separated by a membrane that is permeable to the waste products.

Osmosis can be observed by separating a solution of sugar water from pure water by a semipermeable membrane made of a material such as cellulose, which permits small water molecules to permeate through the membrane but does not allow permeation of the larger solute molecules (in this case sugar) (Atkins, 1990). At the start of the experiment, the osmometer shown in Fig. 1(a), containing a sugar-water solution, is dipped into a bath of pure water to a depth such that the sugar solution in the tube is at the same level as

3-527-28135-5/95/$5.00 + .50

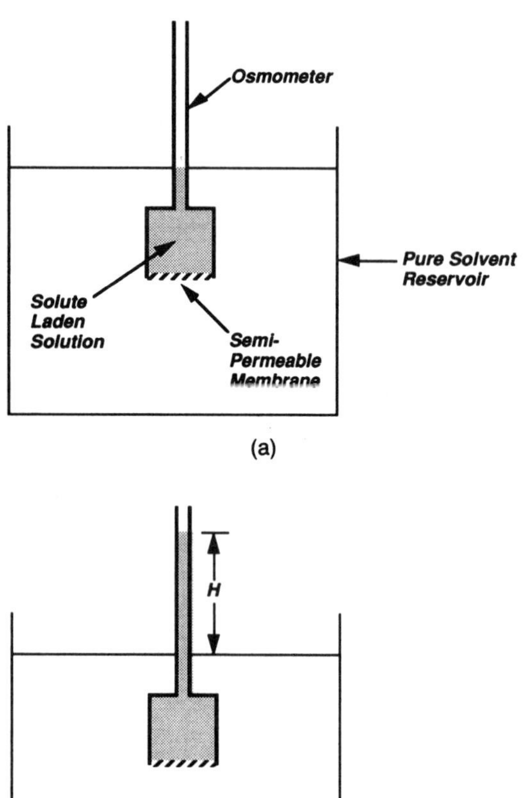

FIG. 1. Schematic of experiment to detect osmotic pressure of sugar water solution.

the pure water. As time proceeds, pure water diffuses across the semipermeable membrane into the solute-containing solution. As the pure water enters the sugar-water solution, it raises the level of the solution in the tube, and the pressure of the solution at the membrane face increases. Figure 1(b) illustrates the system at equilibrium, when there is no net transfer of solvent across the membrane, and the pressure of the sugar-water solution is higher than the pressure of the pure water in contact with the membrane. The value of the pressure difference across the membrane is $(\rho - \rho_{air})gH$, where ρ is the solution density, ρ_{air} is the density of air in the tubes above the solvent and solution, g is the average gravitational acceleration, and H is the difference between the level of sugar water in the capillary above the sugar-water reservoir and the level of pure water. This pressure difference is the osmotic pressure π.

Osmotic pressure is a thermodynamic property of a solution related to the solvent chemical potential in the solution (Denbigh, 1981). The experiment described in Fig. 1 is, therefore, an operational method for determining osmotic pressure, but the definition of osmotic pressure does not rely on the presence of a membrane or a bath of pure solvent. Like freezing point or boiling point temperatures, it is an intrinsic property of the solution. In this regard, the osmotic pressure does depend on solute concentration and nature since it is the presence of the solute that depresses the solvent chemical potential below its value in pure solvent. This chemical potential decrease, in turn, leads to experimentally observable manifestations of osmotic pressure (such as the increase in the height of solution in the capillary column described above). To provide some perspective of the magnitude of the osmotic effect, Table 1 presents values of osmotic pressure for several systems of interest. In this table, the increase in solution osmotic pressure with increasing solute concentration is observed.

1. MOLECULAR AND THERMODYNAMIC DESCRIPTION OF OSMOSIS

1.1. Molecular Basis of Osmosis

The presence of a solute in a solution can depress the solvent chemical potential by diluting the solvent (i.e., the solvent mole fraction is less in the solution than in pure solvent) (Denbigh, 1981). When a reservoir of solution is separated from pure solvent by a semipermeable membrane, the solvent chemical potential difference between the two solutions is a driving force for molecular diffusion of solvent from the solvent reservoir across the membrane into the solution. As solvent diffuses across the membrane, a pressure difference between the two reservoirs is established. The magnitude of this pressure difference increases until the solvent chemical potential in the solution is exactly equal to the chemical potential of the pure solvent. The equality of solvent chemical potential on both sides of the membrane is the criterion

Table 1. Typical values of osmotic pressure at 25 °C (Riley, 1990).

Solute	Concentration (mg/L)	Concentration (mol/L)	Osmotic pressure (bar)
NaCl	2 000	0.034 2	1.57
NaCl (isotonic solution)[a]	9 050	0.155	7.85
NaCl	35 000	0.60	27.4
Brackish water	2 000–5 000	0.034–0.086	1.0–2.7
Seawater	32 000		23.4
Sucrose	1 000	0.002 92	0.072
Dextrose	1 000	0.005 5	0.14

[a]Isotonic with mammalian plasma (Woodbury 1965).

for thermodynamic equilibrium (Denbigh, 1981). At equilibrium, there is solvent exchange between the two reservoirs due to molecular diffusion, but there is no net transport of solvent from one side of the membrane to the other. The pressure difference across the membrane at equilibrium is defined as the osmotic pressure.

1.2. Thermodynamic Analysis of Osmosis

The chemical potential μ_0 of a solvent in a solution at temperature T and pressure P relative to the chemical potential of pure solvent at the same temperature and pressure may be expressed as (Denbigh, 1981)

$$\mu_0 = \mu_0^0 + RT \ln a_0, \tag{1}$$

where μ_0^0 is the chemical potential of the pure solvent at temperature T and pressure P, R is the gas constant, and a_0 is the solvent activity. If the solution pressure is increased from P to $P + \pi$, the solvent chemical potential becomes

$$\mu_0 = \mu_0^0 + RT \ln a_0 + \int_P^{P+\pi} \bar{v}_0 dP, \tag{2}$$

where $\bar{v}_0$ is the partial molar volume of solvent in the solution (Prausnitz, 1969). Solvent in a solution will be in equilibrium with pure solvent when

$$\ln a_0 = -\frac{1}{RT}\int_P^{P+\pi} \bar{v}_0 dP. \tag{3}$$

The osmotic pressure π is, therefore, the pressure that must be applied to the solution to raise the solvent chemical potential in the solution to its pure-solvent value. The solvent partial molar volume is, to a very good approximation, independent of pressure (Denbigh, 1981), and therefore the osmotic pressure may be written as

$$\pi = -(RT/\bar{v}_0) \ln a_0. \tag{4}$$

1.2.1 Osmotic Pressure of an Ideal, Nonionic Solution

The thermodynamic activity of solvent in an ideal solution is related to the solvent concentration as

$$a_0 = x_0, \tag{5}$$

where x_0 is the mole fraction of solvent (Denbigh, 1981). The reference state for chemical potential chosen for Eq. (1) is pure solvent and, accordingly, the solvent activity is 1.0 when the mole fraction of solvent is 1.0. As the solvent is diluted by solute, the solvent mole fraction and, in turn, activity decrease, which leads to a decrease in the solvent chemical potential according to Eq. (1). The solvent partial molar volume is equal to the pure-solvent molar volume in an ideal solution (Prausnitz, 1969). For an ideal solution, therefore, Eq. (4) becomes

$$\ln x_0 = -\pi\hat{v}_0/RT, \tag{6}$$

where $\hat{v}_0$ is the pure solvent molar volume. For dilute solutions containing a single solute,

$$\ln x_0 = -\pi\hat{v}_0/RT = \ln(1 - x_1) \approx -x_1, \tag{7}$$

where x_1 is the solute mole fraction. The solute mole fraction is the number of moles of solute, m_1, divided by the total number of moles of solvent and solute in the solution,

m_{tot}. In dilute solution, the total volume of solution, V, is not appreciably different from the product of m_{tot} and $\hat{v}_0$. In this case, Eq. (7) reduces to

$$\pi V = m_1 RT. \tag{8}$$

Equation (8) is the van't Hoff equation for the osmotic pressure of a dilute ideal solution (Atkins, 1990). The similarity of this result with the ideal gas law illustrates the analogy between an ideal solution and an ideal gas (Atkins, 1990). Experimental determinations of the osmotic pressure of dilute solutions, which led to the empirical formulation of Eq. (8) well before its derivation from thermodynamic principles, inspired van't Hoff to develop the theory of ideal solutions based on the analogy with ideal gases (Mason, 1991). From Eq. (8), the osmotic pressure in an ideal solution is directly proportional to the number of solute molecules in the solution but does not depend upon the chemical nature of the solute.

Osmometry, the technique of using osmosis to determine the molecular weight of macromolecular species, has been widely practiced. To illustrate the sensitivity of osmotic pressure to solute molecular weight, Eq. (8) may be rewritten as follows:

$$\pi / c_1 = RT/M_1, \tag{9}$$

where c_1 is the solute concentration (mass of solute/volume of solution) and M_1 is the solute molecular weight (g/mol). Experimentally, osmotic pressure is determined as a function of solute concentration at fixed temperature because, in real solutions, π/c_1 may not be independent of c. By plotting the ratio π/c_1 vs c_1 and extrapolating to infinite dilution (i.e., $c_1 \to 0$), RT/M_1 and, in turn, solute molecular weight may be determined. An example of using this technique to determine the molecular weight of a polymeric solute is presented later in this article.

1.2.2 The Influence of Solute Molecular Weight on Colligative Properties Solution properties such as the osmotic pressure of an ideal solution, which depend on the number, but not the chemical nature, of solute molecules in solution, are called colligative properties (Atkins, 1990). Other colligative properties, such as boiling point elevation, freezing point depression, and vapor pressure lowering, may be measured to determine solute molecular weight (Billmeyer, 1984). However, solution osmotic pressure is often the most practical colligative property to measure for high–molecular-weight solutes such as polymers. For example, Table 2 presents the coefficients characterizing the sensitivity of colligative properties to solute concentration when a high–molecular-weight solute is dissolved in benzene. For a solute molecular weight of 10 000 g/mol and a solute concentration of 0.1 g/cm^3, for example, the osmotic pressure would be 2.8 mm of benzene,

Table 2. Effect of solute molecular weight on colligative properties of a benzene solution. The pure-solvent normal freezing point is T_f. The freezing point depression is ΔT_f. The pure-solvent density and molar enthalpy of fusion are given by ρ and ΔH_f, respectively. The pure-solvent normal boiling point and enthalpy of vaporization are T_b and ΔH_v, respectively. The osmotic pressure coefficient and vapor pressure depression coefficient were estimated at 20 °C. The solute is assumed to be involatile. Moreover, it is presumed to be insoluble in solid benzene since these ideal freezing-point depression calculations are based on the assumption that the liquid containing the solute is in equilibrium with pure frozen solvent.

M_1 (g/mol)	Osmotic pressure coefficient (mm benzene cc/g)	Cryoscopic coefficient (K cc/g)	Ebullioscopic coefficient (K cc/g)	Vapor pressure depression coefficient (mm Hg cc/g)
	$\frac{\pi}{c_1} = RT\frac{1}{M_1}$	$\frac{\Delta T_f}{c_1} = \frac{RT_f^2}{\rho \Delta H_f}\frac{1}{M_1}$	$\frac{\Delta T_b}{c_1} = \frac{RT_b^2}{\rho \Delta H_v}\frac{1}{M_1}$	$\frac{\Delta p_d}{c_1} = p_0^* \hat{v}_0 \frac{1}{M_1}$
1 000	280	5.8	3.2	6.7
10 000	28	0.58	0.32	0.67
100 000	2.8	0.058	0.032	0.07
500 000	0.56	0.012	0.006 5	0.013
1 000 000	0.28	0.005 8	0.003 2	0.006 7

which is easily measurable using an apparatus similar to the one presented in Fig. 1. In contrast, the freezing point depression of benzene would be only 0.058 °C, the boiling point elevation of benzene would be even less, 0.032 °C, and the vapor pressure depression would be 0.067 mm Hg, a difference which would be difficult to detect experimentally.

1.2.3 Osmotic Pressure of a Solution Containing Several Solutes If there are N chemically distinct solutes in an ideal solution, the solvent activity in the solution is still given by Eq. (5). However, Eq. (7) must be replaced by

$$\ln x_0 = -\frac{\pi \bar{v}_0}{RT} = \ln\left(1 - \sum_{i=1}^{N} x_i\right) \approx -\sum_{i=1}^{N} x_i, \tag{10}$$

where x_i is the mole fraction of component i. If the solution is dilute, the total number of moles in the solution is approximately equal to the number of moles of solvent and, consequently, Eq. (10) becomes

$$\pi = RT \sum_{i=1}^{N} \frac{c_i}{M_i}, \tag{11}$$

where c_i and M_i are the concentration and molecular weight of solute species i. Using for the total solute concentration the definition

$$c = \sum_{i=1}^{N} c_i, \tag{12}$$

Eq. (11) becomes

$$\frac{\pi}{c} = \frac{RT}{c} \sum_{i=1}^{N} \frac{c_i}{M_i} = \frac{RT}{\bar{M}_n} \tag{13}$$

where $\bar{M}_n$ is the so-called number average molecular weight, which is defined as the mass of solute divided by the number of moles of solute molecules (Billmeyer, 1984). Osmotic pressure and, in fact, all colligative properties provide a measure of the number average molecular weight of solutes (such as polymers). Other relationships describing ideal-solution colligative properties may be generalized to solutions containing species of different molecular weight by replacing the solute concentration c_1 and solute molecular weight M_1 in the equations in Table 2 by the total solute concentration c and number average solute molecular weight $\bar{M}_n$, respectively.

1.2.4 Osmotic Pressure of a Dilute Ideal Electrolyte Solution The solute in many solutions of practical interest is an electrolyte (e.g., salt in saltwater or proteins in cells). The dissociation of the electrolyte must be accounted for in determining the osmotic pressure of the solution, since the number of individual solute molecules will change depending on the degree of dissociation of the electrolyte. This section presents the analytical fundamentals of osmotic pressure of electrolyte solutions.

If an electrolyte undergoes dissociation according to

$$M_{\nu_+}A_{\nu_-} \leftrightarrow \nu_+ M_{z_+} + \nu_- A_{\nu_-}, \tag{14}$$

where n_+ and n_- are the number of cations and anions, respectively, formed by dissociating one molecule of the parent electrolyte $M_{\nu_+}A_{\nu_-}$ into n_+ cations M_{ν_+} of charge z_+ and n_- anions A_{ν_-} of charge z_-, then the mole fraction of solvent in the solution is given by

$$x_0 = \frac{1}{1 + (M_0/1000)(n_+ + n_- + n_u)}. \tag{15}$$

The molality of dissolved cations (i.e., the number of moles of dissolved cations per kilogram of solvent) is n_+, n_- is the dissolved anion molality, and n_u is the molality of dissolved electrolyte that is undissociated. M_0 is the molecular weight of solvent in units of g/mole. Many electrolytes are dissolved in water, and Eq. (15) ignores the influence of hydronium and hydroxyl ions, which are always present in such solutions. If the electrolyte concentration is low (below approximately 0.01 molal) and the pH is markedly acidic or basic (i.e., $p\text{H} < 4$ or $p\text{H} > 10$), then Eq. (15) must be rewritten as follows to include the number of hydroxyl and hydronium ions in the solution (Tanford, 1961):

$$x_0 = \frac{1}{1 + (M_0/1000)(n_+ + n_- + n_u + n_{\text{OH}^-} + n_{\text{H}^+})}. \tag{16}$$

The remainder of this discussion is limited to cases where the concentrations of hydroxyl and hydronium ions are low relative to the concentration of electrolyte components. The following discussion must be modified accordingly if this assumption is not valid.

If n moles of parent electrolyte are added to 1 kg of solvent, the electrolyte concentrations are related by the following equation:

$$n = n_u + n_+/\nu_+ = n_u + n_-/\nu_-. \tag{17}$$

This relationship insures that the solution is electrically neutral (i.e., that the number of positive charges in the solution is exactly equal to the number of negative charges in the solution) (Atkins, 1990). The dissociation of the electrolyte is characterized by an apparent equilibrium dissociation constant, K, which is defined as (Denbigh, 1981)

$$K \equiv \frac{n_+^{\nu_+} n_-^{\nu_-}}{n_u} = \nu_+^{\nu_+} \nu_-^{\nu_-} \frac{\alpha^{\nu_+ + \nu_-}}{1 - \alpha} n^{\nu_+ + \nu_- - 1}, \tag{18}$$

where α is the fraction of parent electrolyte which dissociates ($\alpha \equiv 1 - n_u/n$). Values of K are tabulated for many electrolytes in standard references (e.g., Weast, 1979). As the tendency for the solute to undergo dissociation increases, the K value increases and vice versa. In aqueous solutions, the dissociation constant varies over an enormous range. For example, the dissociation constant of $(HPO_2)^{2-}$ to H^+ and $(PO_4)^{3-}$ is 2.2×10^{-13}, while K for hydriodic acid (HI) is approximately 10^{11} (Atkins, 1990).

The solvent activity of an ideal solution containing an electrolyte is equal to the solvent mole fraction, as before, and the osmotic pressure of an electrolyte solution is obtained by combining Eqs. (10) and (15) to give

$$\begin{aligned}\pi &= \frac{RT}{\hat{v}_0} \ln \left(1 + \frac{M_0}{1000} (n_+ + n_- + n_u) \right) \\ &\approx \frac{RT}{\hat{v}_0} \frac{M_0}{1000} (n_+ + n_- + n_u) \\ &= RT \frac{\rho_0}{1000} n[1 + \alpha(\nu_+ + \nu_- - 1)],\end{aligned} \tag{19}$$

where ρ_0 is the pure solvent density.

1.2.5 Osmotic Pressure of Dilute Ideal Protein Solutions Osmometry is a useful tool for characterizing the molecular weight of proteins such as hemoglobin, bovine serum albunim, and ovalbumin (Tanford, 1961). Proteins are macromolecules containing both weakly acidic and weakly basic structural elements. At low pH, they may carry a positive charge as high as +10 to +20 (measured in units of the charge of a proton) per 10 000 Da [1 Da (dalton) = 1/16 of the mass of an oxygen atom] (Tanford, 1961). As pH increases, weakly acidic groups on the protein chains acquire a negative charge, and the net charge of the protein molecule decreases. At very high pH, a protein molecule can carry a net negative charge of 10 to 20 per 10 000 Da (Tanford, 1961). The pH at which the protein has a net charge of zero is called the isoelectric point.

If protein molecules carrying a net charge of Z per molecule (measured in units of the charge of a proton) are dissolved at a concentration of n_p moles of protein per kilogram of solvent in a solution containing n moles of a fully dissociated electrolyte MA ($n_+ = n_- = 1$) per kilogram of solvent, the concentration of protein molecules will be related to the cation and anion concentration in the solution by (Tanford, 1961)

$$Zn_p = n_- - n_+ + (n_{OH^-} - n_{H^+}). \tag{20}$$

Equation (20) represents the condition for electrical neutrality. In many practical situations, the terms in parenthesis, which represent the effect of water ionization on electrical neutrality, may be neglected since hydroxyl and hydronium ion concentrations are often orders of magnitude lower than cation and anion concentrations. Ionization of water is therefore neglected in the following discussion. The protein molality of the solution would be n_p, the molality of cations would be n, and the molality of anions would be $n + Zn_p$.

The chemical potential of a fully dissociated electrolyte in solution is customarily written as the algebraic sum of the anion and cation chemical potentials. This convention is used since the individual anion and cation chemical potentials cannot be measured separately as a consequence of the requirement that the bulk solution remain electrically neutral (Denbigh, 1981). In a solution containing a polymer electrolyte (such as the protein solution described above), the chemical potential of a polymer molecule would represent, therefore, the chemical potential of the macromolecular ion plus the chemical potential of $|Z|$ small ions of opposite charge (Tanford, 1961). In order to define a chemical potential for macromolecular ions that more nearly reflects the properties of only the macromolecular component, Scatchard has suggested that the polymer electrolyte solution be viewed as comprised of the following three neutral pseudocomponents (Tanford, 1961):

- Component 0, solvent;
- Component 1, n_p moles of protein ions + $Zn_p/2$ moles of anions (A^-) − $Zn_p/2$ moles of cations (M^+);
- Component 2, neutral salt (MA).

Using these definitions, the chemical potential of component 1 will be equal to the algebraic sum of the chemical potential of the protein ions plus $Z/2$ times the difference between the chemical potentials of the anions and cations. For small ions, the difference in chemical potentials between cations and anions is believed to be very small (Tanford, 1961). Therefore, the chemical potential of component 1 in the mixture would be representative of the chemical potential of only the macromolecular ions.

Using Scatchard's convention, the concentrations of macromolecular ions, cations, and anions are given by (Tanford, 1961)

$$\begin{aligned} n_p &= n_1 \\ n_+ &= n_2 - Zn_1/2 \\ n_- &= n_2 + Zn_1/2. \end{aligned} \tag{21}$$

In these expressions, n_2 is the molality of neutral salt MA, and n_p is the polymer molality. The concentrations of anions and cations, expressed as molalities, are given by n_- and n_+, respectively. The polymer is added to the solution as $M_{|Z|}P$ if the polymer is negatively charged or as PA_Z if the polymer is positively charged. The total number of ions per kilogram of solvent is $n_p + 2n_2$. The solution can be formed experimentally by adding n_p moles of polymer and $n_2 - Zn_p/2$ moles of neutral salt to 1 kg of solvent. Note that to use this convention of component definition, the concentrations must be such that the number of moles of neutral salt added, $n_2 - Zn_p/2$, is positive.

The osmotic pressure of such a solution is given by an extension of Eq. (19) (Tanford, 1961):

$$\begin{aligned} \pi &= \frac{RT}{\hat{v}_0}\ln\left(1 + \frac{M_0}{1000}(n_+ + n_- + n_p)\right) \\ &\approx \frac{RT}{\hat{v}_0}\frac{M_0}{1000}(n_+ + n_- + n_p). \end{aligned} \tag{22}$$

1.2.5.1 Donnan Equilibrium. Living cells contain charged protein molecules and salt ions in an aqueous solution called plasma. The cells, which are bathed in a salt-containing, aqueous interstitial fluid, are impermeable to protein but permeable to the small salt ions. The presence of the protein inside the cell results in an unequal partitioning of the small ions between the inside and outside of the cell. This effect, called Donnan equilibrium, plays a central role in cell metabolic functions related to, for example, the propagation of nerve impulses (Woodbury, 1965). It is also of great importance in maintaining an osmotic pressure balance critical to the operation of living cells such as red blood cells, which contain proteins (hemoglobin) and are permeable to water and small ions such as Na^+, K^+, Cl^-, etc. (Tanner, 1959).

To develop a quantitative understanding of the Donnan effect, consider two solutions in contact via a membrane which is permeable to small ions and solvent but not to macromolecules. The solution on side I contains n_p^{I} moles of macromolecular ions, such as protein molecules, and $n^{\mathrm{I}} - Zn_p^{\mathrm{I}}/2$ moles of neutral salt per kilogram of solvent. The solution on side II of the membrane contains n^{II} moles of neutral salt per kilogram of solvent. The small cations and anions (always in equal amounts) can migrate across the membrane. At equilibrium, the chemical potential or, equivalently, activity of the neutral salt must be the same on both sides of the membrane. The activity may be expressed as the product of the constituent ion activities (Denbigh, 1981):

$$a = a_+a_-. \tag{23}$$

The condition of thermodynamic equilibrium requires that

$$a_+^{\mathrm{I}}a_-^{\mathrm{I}} = a_+^{\mathrm{II}}a_-^{\mathrm{II}}. \tag{24}$$

In a dilute solution, ion activity is taken to be equal to ion molality (e.g., $a_+^{\mathrm{I}} = n_+^{\mathrm{I}}$). Thus, the concentrations of small-molecule cations and anions on each side of the membrane are related by (Tanford, 1961)

$$n_+^{\mathrm{I}} n_-^{\mathrm{I}} = n_+^{\mathrm{II}} n_-^{\mathrm{II}}. \quad (25)$$

When combined with the electroneutrality condition expressed by Eq. (20), the concentration of ions may be written as

$$(n_+^{\mathrm{II}})^2 = n_+^{\mathrm{I}}(n_+^{\mathrm{I}} + Zn_p)$$
$$(n_-^{\mathrm{II}})^2 = n_-^{\mathrm{I}}(n_-^{\mathrm{I}} - Zn_p). \quad (26)$$

From Eq. (26), the cation concentration will be higher on side II of the membrane than on side I and the anion concentration will be lower on side II if the macromolecular ions are positively charged (i.e., $Z > 0$) and vice versa if $Z < 0$. The difference in osmotic pressure between the solution containing the macromolecular ions and the solution without macroions is (Tanford, 1961)

$$\Delta\pi \equiv \pi^{\mathrm{II}} - \pi^{\mathrm{I}} = -\frac{RT}{\hat{v}_0}\ln\left[\frac{1 + (M_0/1000)(n_+^{\mathrm{I}} + n_-^{\mathrm{I}} + n_p)}{1 + (M_0/1000)(n_+^{\mathrm{II}} + n_-^{\mathrm{II}})}\right]. \quad (27)$$

1.2.6 Effect of Solution Nonideality on Osmotic Pressure Solvent activity in a nonideal solution is not equal to the mole fraction of solvent as shown in Eq. (5). Rather, the activity is written as the product of a convenient concentration scale (i.e., mole fraction, molality, volume fraction, etc.) and an activity coefficient γ_0, which characterizes the departure of the system from ideality (Prausnitz, 1969).

Many theoretical and empirical models are available for activity coefficients (Reid *et al.*, 1977). For uncharged polymer solutions, a widely used model is due to Flory and Huggins. In this model, the volume fraction of solvent in the solution is the concentration scale. The Flory–Huggins model yields the following expression for the solvent activity (Prausnitz, 1969):

$$a_0 = \phi_0 \exp\left[\left(1 - \frac{1}{r}\right)(1 - \phi_0) + \chi(1 - \phi_0)^2\right], \quad (28)$$

where ϕ_0 is the solvent volume fraction; χ and r are model parameters. The parameter χ characterizes the total intermolecular interaction energy between the molecules in the solution relative to the total intermolecular interaction energy for the pure components (Freeman, 1992). Good solvents have χ values less than 0.5; poor solvents have χ values greater than 0.5. A solvent for which χ is equal to 0.5 is called a theta solvent for the polymer (Billmeyer, 1984). The ratio of the polymer molecular volume to the solvent molecular volume is characterized by r, which is often set equal to the ratio of molar volumes of polymer and solvent. For high–molecular-weight polymers, $1/r$ is often negligibly small (Prausnitz, 1969).

Using the Flory–Huggins model, the following expression may be obtained for the osmotic pressure of a dilute, high–molecular-weight, polydisperse polymer solution (Young and Lovell, 1991):

$$\frac{\pi}{c} = \frac{RT}{\bar{M}_n}\left[1 + \frac{\bar{M}_n}{\hat{V}_0\rho_p^2}\left(\tfrac{1}{2} - \chi\right)c\right], \quad (29)$$

where ρ_p is the polymer density and, as in Eq. (12), c is the total polymer concentration. As indicated by Eq. (29), the ideal solution expression for the osmotic pressure [i.e., Eq. (13)] is recovered in the limit as polymer concentration goes to zero or as the Flory–Huggins χ parameter goes to $\frac{1}{2}$.

To characterize deviations from ideality that are not accurately described by the Flory–Huggins theory, Eq. (29) is often written as a power series (or virial series) in concentration (Tanford, 1961):

$$\frac{\pi}{c} = \frac{RT}{\bar{M}_n}[1 + Bc + Cc^2 + \ldots], \quad (30)$$

where the coefficients B and C are the second and third virial coefficients, respectively. The second virial coefficient can be written in term of the Flory–Huggins interaction parameter as (Young and Lovell, 1991)

$$B = \bar{M}_n\left(\tfrac{1}{2} - \chi\right)/(\hat{V}_0\rho_p^2). \quad (31)$$

Experimental osmotic pressure data for the polymer nitrocellulose in several solvents are presented in Fig. 2. The data presented in Fig. 2 are well represented by truncating Eq. (30) after the term containing the second virial coefficient. In thermodynamically good solvents, such as acetone, the second virial coef-

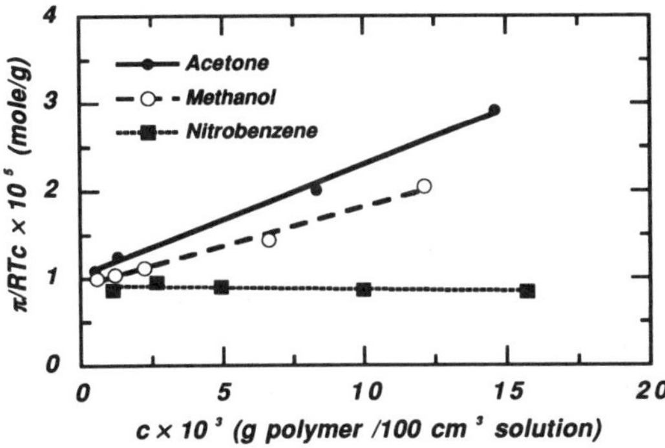

FIG. 2. Influence of solvent on results of osmotic pressure experiment to determine number average molecular weight of nitrocellulose.

ficient is large and positive, and the ratio of osmotic pressure to polymer concentration increases with increasing polymer concentration. The osmotic pressure in this nonideal solution depends, therefore, on both the number of solute molecules and the chemical nature of the solute (or, more precisely, on the interactions between the solute and solvent). In a thermodynamically poorer solvent, such as methanol, the sensitivity of π/c to polymer concentration is reduced. In a very poor solvent, such as nitrobenzene, the ratio π/c is almost insensitive to polymer concentration, indicating a χ value of approximately 0.5 or, equivalently, a vanishing second virial coefficient. When π/c is independent of polymer concentration, the solution is ideal, and the osmotic pressure depends only on the number of solute molecules, not on their chemical nature.

In systems containing electrolytes, activity coefficient models are available in the literature to account for nonideal solution behavior (Atkins, 1990). As described in Eq. (23), the activity for an electrolyte solution is typically written as the product of the constituent ion activities as (Tanford, 1961)

$$a = a_+a_- = n_+n_-\gamma_\pm^2, \tag{32}$$

where $\gamma_\pm$ is the mean activity coefficient for the cations and anions. A widely used model for the activity coefficient is given by the extended Debye–Hückel law (Atkins, 1990):

$$\log(\gamma_\pm) = -\frac{A|z_+z_-|}{1 + BI^{1/2}} I^{1/2}, \tag{33}$$

where A and B are constants, and I is the ionic strength of the solution. The value of A is 0.509 $(\text{kg/mol})^{1/2}$ for an aqueous solution at 25 °C, and, in general, A is a function of the temperature and dielectric constant of the solution (Tanford, 1961). B is typically treated as an adjustable constant. The ionic strength of the solution is given by

$$I \equiv \tfrac{1}{2} \sum_i z_i^2 n_i, \tag{34}$$

where the summation extends over all different ions in the solution, and z_i and n_i are the charge and molality, respectively, of ion i (Tanford, 1961). This model can account for activity coefficients at concentrations as high as 0.1 mol/kg. There is no available theory for solutions which are substantially more concentrated (Atkins, 1990).

An example of osmosis in biological systems is presented in Fig. 3, which illustrates the dependence of hemoglobin osmotic pressure on hemoglobin concentration in water. As shown by the dashed line in Fig. 3, the osmotic pressure is not a linear function of hemoglobin concentration above approximately 7 g/(100 cm^3), indicating that hemoglobin solutions are nonideal above this concentration.

As a practical application of the material presented in Sec. 1.2.5, it is interesting to note that at the concentration of hemoglobin in the blood [33 g/(100 cm^3)], the osmotic pressure inside red blood cells due to hemoglobin is approximately 0.5 bar (380 mm Hg) (Lightfoot, 1974). The osmotic pressure of the serum proteins in contact with red blood cells is, on

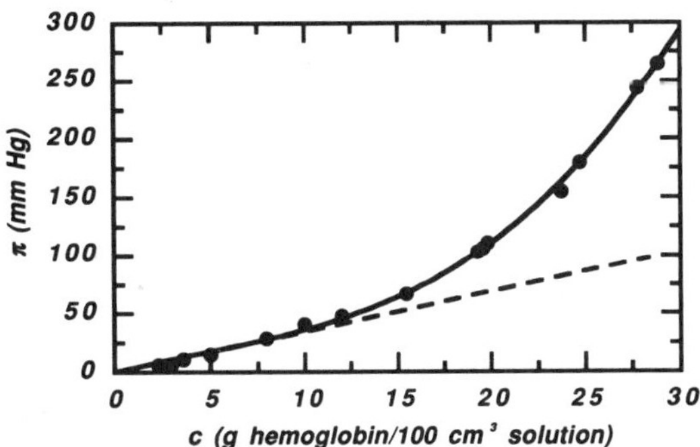

FIG. 3. Osmotic pressure of a hemoglobin solution at 0 °C. The dashed line indicates ideal solution behavior. (Lightfoot, 1974.)

average, 25 mm Hg (Lightfoot, 1974). The cell wall cannot support this large difference in osmotic pressure arising from the substantial difference in intracellular and intercellular protein concentration. Therefore, an opposing imbalance in electrolyte concentration is maintained which compensates for the unequal protein concentrations inside and outside the cell wall (Lightfoot, 1974).

2. EXPERIMENTAL METHODOLOGY FOR THE DETERMINATION OF OSMOTIC PRESSURE

Osmotic pressure may be determined directly using two broad classes of devices: static osmometers and dynamic osmometers (Young and Lovell, 1991). In both types of osmometers, the osmotic pressure is determined as a function of concentration [solute concentration is commonly less than 20 g/L (Young and Lovell, 1991)] at fixed temperature. The extrapolation of the ratio of π/c to zero concentration gives $RT/\bar{M}_n$, and the solute number average molecular weight may be computed directly.

In a static osmometer, the osmotic pressure is determined by measuring the equilibrium pressure difference across a membrane separating two solutions containing different concentrations of solute. A schematic of a static osmometer is presented in Fig. 4. Older static osmometers (e.g., Pinner–Stabin osmometers) have large cell volumes (3–20 cm^3) on either side of the membrane (Young and Lovell, 1991). Small-bore capillaries are used for the tubes above the cell. The use of small capillaries decreases the amount of solvent which must permeate through the membrane before the attainment of equilibrium. Normally, the amount of solvent added to the solution during the equilibration time does not significantly change the solute concentration (Young and Lovell, 1991). Modern static osmometers (e.g., Knauer and Wescan osmometers) have small cell volumes (on the order of 0.3–2 cm^3) and large membrane areas to minimize equilibration times (typically 30–60 minutes) (Young and Lovell, 1991). Pressure differences between the two sides of the membrane are typically monitored indirectly using pressure transducers (Billmeyer, 1984). Osmotic pressure is sensitive to temperature, and so temperature is typically controlled to ±0.01 °C (Young and Lovell, 1991).

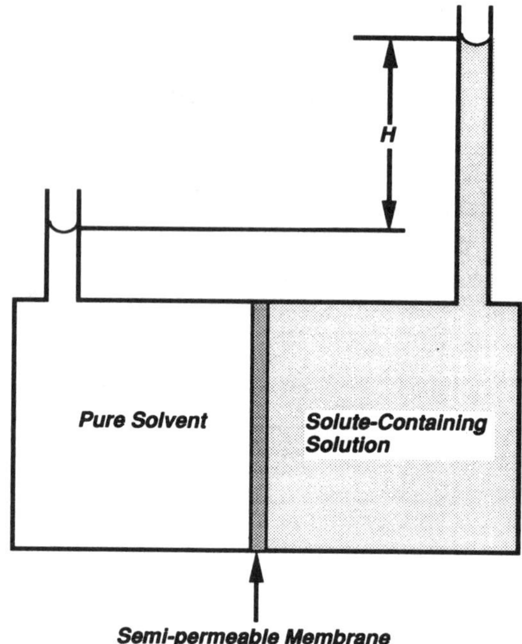

FIG. 4. Schematic of osmometer.

In a dynamic osmometer, the flow of solvent across the membrane is monitored, and a hydrostatic pressure just sufficient to prevent solvent flow through the membrane is applied to the solution (Young and Lovell, 1991). This hydrostatic pressure is the osmotic pressure of the solution.

Number average molecular weights between approximately 5×10^3 and 5×10^5 Da can often be determined to an accuracy of approximately 1% (Rodriguez, 1989). Real membrane materials are not completely impermeable to solutes. Most osmometer membrane materials are based on cellulose, and at molecular weights below 5×10^3 Da, many solutes can permeate through the membrane and will not, therefore, contribute to the observed osmotic pressure of the solution (Rodriguez, 1989; Young and Lovell, 1991). For this reason, membrane osmometry can seriously overestimate the number average molecular weight of samples containing a large number fraction of low–molecular-weight species. Above a molecular weight of approximately 5×10^5 Da, the osmotic pressure of the solution is very small, and the precision of the technique is limited, therefore, by the precision with which small pressure differences (between the two chambers connected by the membrane) can be determined (Billmeyer, 1984).

3. APPLICATIONS

3.1. Reverse Osmosis

The role of osmosis in industrial separation technologies such as reverse osmosis, dialysis, and electrodialysis are described in several review articles (Bhattacharyya and Williams, 1992; Eykamp and Steen, 1987; Kessler and Klein, 1992; Klein *et al.*, 1987; Kulkarni *et al.*; 1992, Strathmann, 1992). In reverse osmosis (also called hyperfiltration), two solutions of different solute concentrations are separated by a semipermeable membrane. The objective of the separation is to drive solvent (e.g., water) from a region of high solute concentration (e.g., seawater) across the membrane to a region of low solute concentration (e.g., fresh water). In order to transport solvent from the high–solute-concentration side to the low–solute-concentration side of the membrane, the solvent chemical potential on the high-concentration side must be raised to a value that is above the solvent chemical potential on the low–solute-concentration side. A hydrostatic pressure applied to the high–solute-concentration side of the membrane is typically used to increase the solvent chemical potential on the high–solute-concentration side of the membrane and drive solvent transport.

At steady state, the flux of solvent across the membrane is given by (Bhattacharyya and Williams, 1992; Kessler and Klein, 1992)

$$J_0 = A(\Delta p - \sigma \Delta \pi), \qquad (35)$$

where σ is the Staverman reflection coefficient (defined below), J_0 is the solvent flux (volume of solvent permeating membrane per unit membrane area per unit time), A is the solvent permeability constant, Δp is the impressed hydrostatic pressure difference across the membrane, and $\Delta \pi$ is the osmotic pressure difference across the membrane (Bhattacharyya and Williams, 1992). Theoretical models are available to predict the dependence of the solvent permeability constant on material parameters (such as solvent diffusion coefficient and solubility in the membrane) and membrane thickness (Bhattacharyya and Williams, 1992).

At a solvent flux equal to zero, the Staverman reflection coefficient σ is the ratio of the applied pressure difference Δp to the osmotic pressure difference $\Delta \pi$ across the membrane. If σ is equal to 1, then the hydrostatic pressure difference that must be maintained across the membrane to prevent solvent flow due to osmosis is equal to the osmotic pressure difference between the two solutions. In this limit, the membrane is completely impermeable to solute. If the membrane is permeable to solute, the pressure difference needed to maintain zero solvent flux is less than the osmotic pressure difference predicted based on bulk solute concentrations and σ is, accordingly, less than 1 (Bhattacharyya and Williams, 1992).

The importance of osmosis in membrane separations such as reverse osmosis can be understood from Eq. (35). The osmotic pressure difference $\Delta \pi$ must be overcome by an applied hydrostatic pressure difference across a membrane to transport solvent through the membrane and, for example, desalinate water. If the osmotic pressure difference across the membrane is large, the applied hydrostatic pressure difference will be large which, in turn, increases the cost of the technique. From Table 1, the osmotic pressure of seawater is approximately 20 times that of brackish water. Consequently, reverse osmosis, which is very competitive economically with other technologies (such as multistage flash distillation, multiple effect evaporation, ion exchange, and electrodialysis) for brackish water desalination, is much less competitive for seawater desalination (Riley, 1990).

The desalination of water is the most widely known application of reverse osmosis. However, in addition to desalination of water, reverse osmosis is applied in the food industry to concentrate products such as milk, in the paper industry for the concentration of lignosulfonates, and in the chemical process industries to concentrate and recover a variety of industrial solutes (Eykamp and Steen, 1987).

3.2. Dialysis

Dialysis (*q.v.*) is a separation process in which a feed solution is passed across a semipermeable membrane that is permeable to solvent and small solutes in the feed but impermeable to large macromolecules in the feed (Klein *et al.*, 1987). Unlike reverse osmosis, dialysis membranes are designed to encourage the passage of small solute molecules while retaining only the species of highest molecular weight in the mixture. An important application of dialysis is hemodialysis. In this process, blood containing small–molecular-weight metabolic waste products is passed across a membrane that is permeable to the low–molecular-weight wastes but impermeable to higher–molecular-weight blood proteins (Klein *et al.*, 1987). This application has sparked the development of dialysis for use in industrial separations such as the desalting of cheese whey solids and the reduction of alcohol content in beer (Klein *et al.*, 1987). Dialysis based upon the Donnan effect described earlier is being tested in applications such as the softening of feedwater to reverse-osmosis membrane units and the removal of toxic or otherwise harmful cations from industrial wastewater streams (Klein *et al.*, 1987).

3.3. Electrodialysis

Electrodialysis involves the use of electrically charged membranes. An electric field is applied across the membrane to separate ionic solutes from an aqueous feed stream (Strathmann, 1992). The membranes used in such devices are ion selective (that is, they allow positively charged ions to permeate while excluding negatively charged ions or vice versa). The principal commercial applications of electrodialysis are desalination of water, the production of table salt (Strathmann, 1992), desalination of food such as cheese whey and sugar, and the production of chlorine in so-called "chlor-alkali cells" (Lonsdale, 1982).

GLOSSARY

Chemical Potential: An intensive thermodynamic function that is defined, mathematically, as the derivative of the extensive thermodynamic functions of internal energy, enthalpy, Gibbs free energy, and Hemholtz free energy with respect to the amount of one component in a mixture. More importantly for osmosis, the condition of chemical equilibrium between two or more phases is established when each substance has an equal value of its chemical potential in all phases into which this substance may partition. Difference in the chemical potential of a substance that may partition between two phases is the origin or driving force for the diffusion of that substance from a region of high chemical potential to a region of lower chemical potential.

Dalton (Da): A standard unit of molecular weight. One dalton is 1/16 the mass of an oxygen atom.

Dialysis: A separation process in which a feed solution is passed across a semipermeable membrane that is permeable to solvent and small solutes in the feed but impermeable to large macromolecules in the feed. Small solute molecules are separated from the feed solution in this process.

Electrodialysis: A separation process involving the use of electrically charged membranes and the application of an electric field across the membrane to separate ionic solutes from an aqueous feed stream.

Isotonic Solution: A solution that has the same osmotic pressure as a reference solution. The reference solution in biomedical applications is typically mammalian plasma.

Molality: The molal concentration of a solute. A one molal solution contains one mole of solute molecules per kilogram of solvent.

Osmotic Pressure: Operationally, the excess pressure that must be placed on a solution separated by a semipermeable membrane from a reservoir of pure solvent to prevent any net mass transport of solvent through the membrane. It is a thermodynamic property of the solution and is a measure of the depression in the solvent chemical potential due to the presence of solute molecules.

Osmometry: The analytical technique for measuring the osmotic pressure of a solution.

Reverse Osmosis: (Also called hyperfiltration) a separation process in which a solution at high pressure is passed across a semipermeable membrane that is permeable to solvent but impermeable to solute. The result is a concentration of the solute in the feed stream to the unit and the production of a stream of pure solvent that has permeated through the membrane.

Semipermeable Membrane: A membrane that is permeable to solvent molecules but impermeable to solute molecules.

Works Cited

Atkins, P. W. (1990), *Physical Chemistry*, San Francisco: W. H. Freeman.

Bhattacharyya, D., Williams, M. E. (1992), in: W. S. W. Ho, K. K. Sirkar (Eds.), *Membrane Handbook*, New York: Van Nostrand Reinhold, p. 263.

Billmeyer, F. W., Jr. (1984), *Textbook of Polymer Science*, New York: Wiley.

Denbigh, K. (1981), *The Principles of Chemical Equilibrium, 4th ed.*, Cambridge, U.K.: Cambridge Univ. Press.

Eykamp, W., Steen, J. (1987), in: R. W. Rousseau (Ed.), *Handbook of Separation Process Technology*, New York: Wiley, p. 826.

Freeman, B. D. (1992), in: S. L. Aggarwal, S. Russo (Eds.), *Comprehensive Polymer Science*, Oxford: Pergamon, p. 167.

Kessler, S. B., Klein, E. (1992), in: W. S. W. Ho, K. K. Sirkar (Eds.), *Membrane Handbook*, New York: Van Nostrand Reinhold, p. 161.

Klein, E., Ward, R. A., Lacey, R. E. (1987), in: R. W. Rousseau (Ed.), *Handbook of Separation Process Technology*, New York: Wiley, p. 954.

Kulkarni, S. S., Funk, E. W., Li, N. N. (1992), in W. S. W. Ho, K. K. Sirkar (Eds.), *Membrane Handbook*, New York: Van Nostrand Reinhold, p. 392.

Lightfoot, E. N. (1974), *Transport Phenomena with Living Systems*, New York: Wiley.

Lonsdale, H. K. (1982), "The Growth of Membrane Technology," *J. Membrane Sci.* **10**, 81–181.

Marick, L. (1969), in: W. H. Nault (Ed.), *The World Book Encyclopedia*, Chicago: Field Enterprises Education, p. 657.

Mason, E. A. (1991), "From pig bladders and cracked jars to polysulfones: an historical perspective on membrane transport," *J. Membrane Sci.* **60**, 125–145.

Morris, E. (1981), *The American Heritage Dictionary of the English Language*, Boston: Houghton Mifflin.

Prausnitz, J. M. (1969), *Molecular Thermodynamics of Fluid-Phase Equilibria*, Englewood Cliffs, NJ: Prentice-Hall.

Reid, R. C., Prausnitz, J. M., Sherwood, T. K. (1977), *The Properties of Gases and Liquids*, New York: McGraw-Hill.

Riley, R. L. (1990), in: R. W. Baker, W. Eykamp, R. L. Riley (Ed.), *Membrane Separations Systems: A Research Needs Assessment*, Washington, DC: Department of Energy, Chap. 5.

Rodriguez, F. (1989), *Principles of Polymer Systems*, New York: Hemisphere Publishing.

Strathmann, H. (1992), in: W. S. W. Ho, K. K. Sirkar (Eds.), *Membrane Handbook*, New York: Van Nostrand Reinhold, p. 217.

Tanford, C. (1961), *Physical Chemistry of Macromolecules*, New York: Wiley.

Tanner, R. (1959), in: J. L. Oncley (Ed.), *Biophysical Science—a Study Program*, New York: Wiley, p. 342.

Weast, R. C. (Ed.) (1979), *CRC Handbook of Chemistry and Physics*, Boca Raton, FL: CRC Press.

Woodbury, J. W. (1965), in: T. C. Ruch, H. D. Patton (Eds.), *Physiology and Biophysics*, Philadelphia: W. B. Saunders, p. 26.

Young, R. J., Lovell, P. A. (1991), *Introduction to Polymers*, London: Chapman & Hall.

Further Reading

Bhattacharyya, D., Williams, M. E. (1992), in: W. S. W. Ho, K. K. Sirkar (Eds.), *Membrane Handbook*, New York: Van Nostrand Reinhold, p. 263.

Billmeyer, F. W., Jr. (1984), *Textbook of Polymer Science*, New York: Wiley.

Denbigh, K. (1981), *The Principles of Chemical Equilibrium, 4th ed.*, Cambridge, U.K.: Cambridge Univ. Press.

Eykamp, W., Steen, J. (1987), in: R. W. Rousseau (Ed.), *Handbook of Separation Process Technology*, New York: Wiley, p. 826.

Kessler, S. B., Klein, E. (1992), in: W. S. W. Ho, K. K. Sirkar (Eds.), *Membrane Handbook*, New York: Van Nostrand Reinhold, p. 161.

Klein, E., Ward, R. A., Lacey, R. E. (1987), in: R. W. Rousseau (Ed.), *Handbook of Separation Process Technology*, New York: Wiley, p. 954.

Kulkarni, S. S., Funk, E. W., Li, N. N. (1992), in: W. S. W. Ho, K. K. Sirkar (Eds.), *Membrane Handbook*, New York: Van Nostrand Reinhold, p. 392.

Tanford, C. (1961), *Physical Chemistry of Macromolecules*, New York: Wiley.

Young, R. J., Lovell, P. A. (1991), *Introduction to Polymers*, London: Chapman & Hall.

PACKAGING TECHNOLOGY

JOHN J. RITSKO, *IBM T.J. Watson Research Center, Yorktown Heights, New York, U.S.A.*

	Introduction	73
1.	**General Principles**	74
1.1	Interconnection Hierarchy	74
1.2	Electrical Design	74
1.3	Materials Characteristics	75
2.	**Chip-to-Package Interconnections**	77
2.1	Wirebonding	77
2.2	Tape Automated Bonding (TAB)	77
2.3	Solder-Bump Interconnections	77
3.	**Molded Plastic Packaging**	78
3.1	Lead-Frame Materials	78
3.2	Encapsulation Materials	79
4.	**Printed-Circuit Cards and Boards**	79
4.1	Structures	79
4.2	Materials and Processes	79
5.	**Ceramic Packaging**	80
5.1	Metal-on-Ceramic Packages	82
5.2	Multilayer Ceramic Substrates	82
5.3	Ceramic Materials	82
6.	**Thin-Film Wiring**	83
6.1	Materials	83
6.2	Sequential Patterning	83
7.	**Package-to-Package Interconnections**	84
7.1	Pins	84
7.2	Surface-Mount Technology	84
7.3	Connectors	85
7.4	Cables	85
8.	**Heat Transfer**	85
8.1	Thermal Engineering	86
8.2	Cooling Structures	86
9.	**Future Challenges, Emerging Technologies**	86
9.1	Optoelectronics	86
9.2	Electromagnetic Management	88
9.3	Low-Power, Low-Cost, Light-Weight Systems	88
	Works Cited	88

INTRODUCTION

The term "packaging" is used to describe the methods and structures by which electronic components are interconnected to make a functional system. The interconnection package must perform several key roles. It must provide mechanical support for the components as well as protection from damage due to handling. Sensitive active components such as transistors and integrated circuits must be protected from the external environment. The package must distribute power to active devices and remove the heat generated by them as well as providing for all required signal interconnections between components. Packaging technology is the set of designs, materials, and processes of which the interconnection packages are made.

In the earliest electronic systems, individual components, such as resistors, capacitors, tubes, and laser semiconductor devices, were interconnected by soldering their terminals or plugging them into sockets with connections made by discrete wires. As the complexity of electronic systems increased, denser and denser interconnection packages were required to reduce cost and save space as well as to increase the overall system performance. Printed circuit cards were introduced to interconnect discrete components efficiently, and integrated circuits allowed many active devices to be placed and interconnected on a single silicon chip. Nevertheless, the increasing complexity of electronic systems required ever more sophisticated packaging technology. Today, powerful computers utilize hundreds of thousands or even millions of circuits, and many systems are designed in such a way that their performance depends as much on packaging technology as on chip technology.

3-527-28135-5/95/$5.00 + .50

Modern packaging technology is a complex field concerning the fabrication of many different mechanical, electrical, and thermal structures using many different metals, polymers, glasses, and ceramics, which are formed in a wide variety of processes and which are optimized for both cost and performance. This article describes the principles and materials considerations of importance to packaging. It explains how integrated circuit chips are connected to the interconnection package and the various types of simple and complex packages in use today. Since only a general description can be provided here and the details are very important in the engineering of different packaging technologies, the reader is encouraged to refer to several recent books on packaging cited as references.

1. GENERAL PRINCIPLES

1.1 Interconnection Hierarchy

Complex systems consist of many levels of interconnections, as shown for a typical system in Fig. 1. The first-level package encapsulates the chip for protection from handling damage and the environment and provides metallic leads for power input as well as signal connections between chips. The simplest and lowest-cost package is a plastic or ceramic single-chip module (SCM). The chip-to-chip interconnections are then provided by the second level of packaging or multichip module (MCM), which may be a laminated plastic-based card or board (MCM-L) or a ceramic-based substrate (MCM-C), or a hybrid containing sequentially built thin-film wiring (MCM-D). For the highest-performance electrical interconnections between chips, bare chips are mounted on a multichip module that is then encapsulated as a whole, thus combining the first and second packaging levels.

1.2 Electrical Design

Signal interconnection between chip and package and between different packaging levels requires both making the number of required interconnections and providing the optimal electrical performance. The number of terminals required by a component is re-

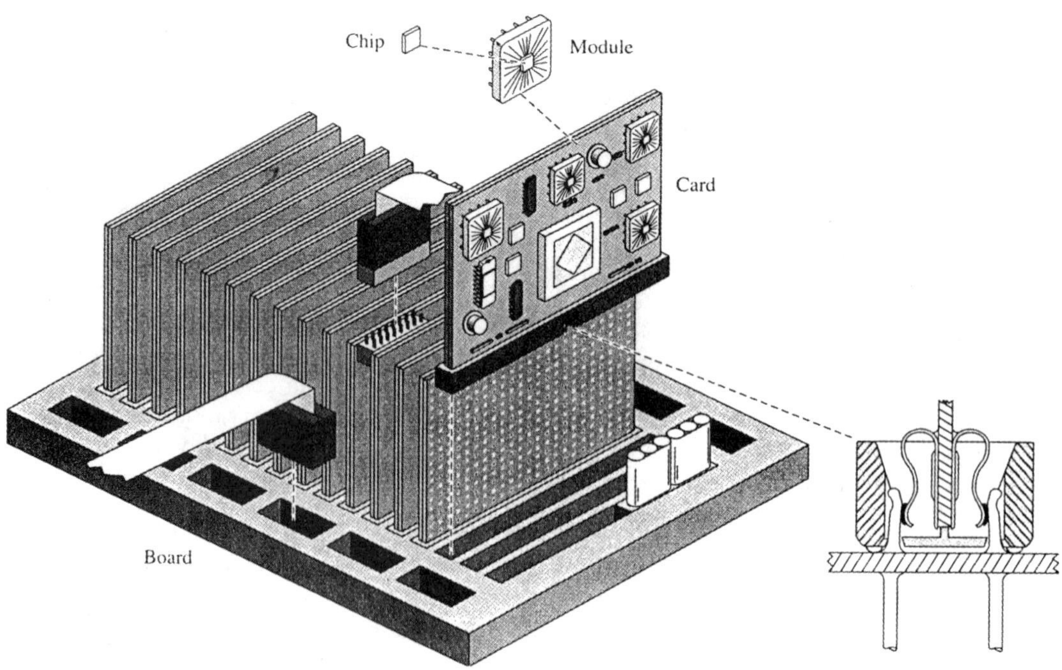

FIG. 1. Chips are mounted in single- or multiple-chip modules. The modules are soldered or pin attached to cards, which are plugged into connectors in a board. Cables connect signals between boards. (From Tummala and Rymaszewski, 1989, reproduced by permission.)

lated to the communication needed between this component and the rest of the system. A useful design guideline was given in an empirical rule devised by E. Rent in 1960 (Radke, 1969):

$$I = bC^p,$$

where I is the number of input and output terminals (I/O), C is the number of circuits on a package, b is a constant equal to approximately 2, and p is approximately $\frac{2}{3}$. In recent years, the degree of functionality contained in a package component has increased, requiring fewer interconnections to the next level than predicted by the original Rent's rule, but the above formula with lower values of p is still useful (Seraphim *et al.*, 1989). Bipolar chips containing 5000–20 000 circuits require 200–400 signal I/O, and large multichip modules containing millions of circuits require thousands of interconnections. The larger the number of I/O terminals on a given package the greater will be the amount of interconnection wiring required on the next level of packaging.

The delay caused by signal propagation between chips often limits overall system performance. The resistance R, capacitance C, and inductance L of the package wiring are key to improved performance. Proper electrical design minimizes both signal delay and noise. For short nets the RC time constant determines the package delay. Over longer distances signals travel in transmission lines in cards, boards, ceramic modules, or thin-film wiring with typical dimensions shown in Fig. 2. These signals travel with the velocity $V = c/\sqrt{\epsilon}$, where c is the speed of light and ϵ is the insulator dielectric constant. The time delay T in traveling a distance l in the package is then

$$T = (l/c)\sqrt{\epsilon}.$$

This delay is minimized by shortening the distance signals must travel by using finer, denser wiring and by reducing the dielectric constant of the insulator. The impedance of the transmission line, Z_0, is $\sqrt{L/C}$ where L is the inductance and C the capacitance per unit length. Z_0 is typically 40–80 Ω. In addition, the transmission-line designs shown in Fig. 2 contain parallel planes of wires, and the dimensions are chosen to reduce cross-coupling noise between lines to typically <10% of the signal amplitude by minimizing mutual inductance and interline capacitance.

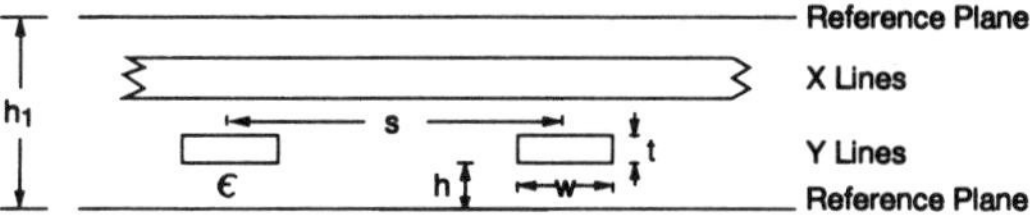

	Card/board	Ceramic module	Thin-film wiring
W	150–100	80	10–25
t	30–50	25	3–15
h	270–430	100	5–15
h_1	500–1200	300	20–36
s	500–400	400	25–75
ϵ	4.4–4.0	5–10	3–3.5
Z_0 (Ω)	80	50	40–50

FIG. 2. Cross section of transmission lines with typical dimensions (in micrometers). The term ϵ is the insulator dielectric constant and Z_0 the characteristic impedance.

Power must be distributed to the chips on an MCM by low-resistance power planes to minimize voltage drops across the package. When many drivers, N, are switched simultaneously on the chip, a noise voltage ΔV is induced on the chip reference voltage, related to the rate of driver current switching, di/dt, by the effective package inductance L_{eff}:

$$\Delta V = NL_{\text{eff}} \frac{di}{dt}.$$

The package must provide low enough L_{eff} to keep ΔV below typically 10–20% of the supply voltage. The low inductance is achieved by making the connection between the chip and power distribution planes as short as possible.

1.3 Materials Characteristics

Package structures are made from a wide variety of materials (Harper, 1991; Minges, 1989). The characteristics of some typical conductors and insulators are given in Tables 1 and 2. The most important properties are the electrical conductivity of metals, the dielectric constant of insulators, the thermal conductivity of the heat dissipation structures, and the relative thermal expansion coefficient of dissimilar materials that are bonded or joined together.

Table 1. Properties of insulator materials.

Insulator	Dielectric constant	Thermal expansion coefficient (10^{-7}/°C)	Thermal conductivity W/m°K
SiC	42	37	270
Si	11.7	27	170
Alumina	9.4	66	20
AlN	8.8	33	230
BeO	6.8	68	240
BN	6.5	37	600
Mullite	6.2	45	10
Diamond (CVD)	5.7	23	1500
Glass-ceramics	4–8	30–50	5
Epoxy-glass (FR-4)	4.7	110–150	0.2
Borosilicate glass	4.0	30	5
Epoxy-Kevlar (60%)	3.6	60	0.2
Polyimides	3–3.5	70–500	0.2
Benzocyclobutene	2.6	350–600	0.2
Polytetrafluoroethylene	2.2	200	0.1

Since the resistance of a conducting element increases with length but is inversely proportional to cross-sectional area, different types of interconnection make trade-offs between dimensions and resistivity to optimize performance. Copper is the conductor of choice for all interchip wiring because of its low resistivity and its superior chemical stability to silver; however, because of relative ease of fabrication higher-resistivity metals are often used, particularly in ceramic packages. Joining metals such as solders or brazes have high resistivity but are used for short, wide-area connections. Kovar is used for pins because of its mechanical properties, while the relatively large size of pins allows the use of a high-resistivity metal.

Since the performance of the package improves with lower dielectric constant and shorter wiring lengths, new materials or processes have been introduced to improve performance. Low-ϵ glass-ceramics have been developed to replace the more common alumina ceramic. Thin-film copper wiring (which produces the highest wiring density) is made with polymer insulators.

The thermal conductivity of plastics is relatively poor and heat is conducted away from high-power chips by ceramic or metal heat spreaders to air- or water-cooled heat sinks.

Electronic packages are comprised of dissimilar materials bonded or joined together, which expand at different rates on heating. The thermal cycles that occur during fabri-

Table 2. Properties of conductor materials.

Conductor	Electrical resistivity (10^{-6} Ωcm)	Thermal expansion coefficient (10^{-7}/°C)	Thermal conductivity W/m K
Alloy 42 (42% Ni, 58% Fe)	70	43	16
Kovar	50	53	17
Invar	46	15	11
Silver–palladium	20	140	150
Chromium	20	63	66
Pb 95%, Sn 5%	19	290	63
Pb 40%, Sn 60%	17	245	51
Au 80%, Sn 20%	16	159	57
Palladium	10.8	110	70
Platinum	10.6	90	71
Nickel	6.8	133	92
Molybdenum	5.2	50	146
Tungsten	5.5	45	200
Aluminum	4.3	230	240
Gold	2.2	142	297
Copper	1.7	170	393
Silver	1.6	197	418

cation or when systems are turned on and off result in stresses and/or strains which are proportional to the difference in thermal expansion coefficients of the different materials and the change in temperature. These stresses can result in fatigue failure such as delamination at interfaces of a multilayer structure, solder cracking, and metal fatigue. The design of a package for maximum reliability requires the choice of materials with as closely matched thermal expansion coefficients as possible given the fabrication and electrical design requirements.

2. CHIP-TO-PACKAGE INTERCONNECTIONS

2.1 Wirebonding

Wirebonding is the most common chip connection technology. Wires are directly and individually bonded to the input/output pads on the perimeter of the chip and then bonded to the package. The bonding process is ultrasonic, thermocompression, or thermosonic bonding applied by mechanical tools. The wires are aluminum (with 1% Si) or aluminum alloys or gold and are 25–38 μm in diameter for signal and as much as 125–150 μm in diameter for power connections if required. Chip pads can be placed on a 100–150-μm pitch, thus allowing several hundred I/O from large chips. Wirebonding is most commonly used with molded plastic SCMs (described below).

2.2 Tape Automated Bonding (TAB)

Tape automated bonding or TAB was developed as a highly automated technique for high-volume packaging of individual chips with relatively few connections, although it has evolved to accommodate over 300 I/O terminals. The TAB process begins with a continuous tape up to 70 mm or more wide as shown in Fig. 3(a). The tape is processed in a reel-to-reel method and contains sprocket holes for movement and alignment as shown in the figure. The tape is made of stable plastics such as polyimides, polyesters, etc., and contains copper conductors patterned using photolithographic techniques and metal etching processes. In order to isolate electrically the metal patterns associated with a given chip, mechanical punching is used. The plastic tape is either etched or prepunched to form a window in which the metal leads extend beyond the tape so that they can be bonded to chips containing metal bumps on their I/O pads. The bonding of all leads is done simultaneously in a thermocompression bonding process in an inner-lead bonding tool shown in Fig. 3(b). In this tool the chip is placed with its aluminum or gold bumps facing up. The TAB tape is placed over the chip with the inner lead bonds aligned to the metal bumps. A heated metal tip or thermode then uses compression to form the bonds between leads and bumps. The chips are retained on the tape for subsequent testing, encapsulation (if desired), and burn-in followed by excising of the individual chip-lead packages from the tape. TAB is a chip connection technology that can be used as part of a molded plastic- or ceramic-based single-chip module, but it can also function as a first-level package if the chip is encapsulated in plastic on the tape and the outer TAB leads are bonded directly to the second-level package.

2.3 Solder-Bump Interconnections

Solder-bump chip connections, also known as controlled-collapse chip connection (C4) or flip-chip bonding, involves solder bumps deposited by metal evaporation in a dense area array on wettable metal terminals on the chip and a matching footprint of solder-wettable terminals on the interconnection package. The upside down chip is aligned to the substrate and all joints are made simultaneously by reflowing the solder. The high surface tension of the solder causes self-alignment. For a given chip size this technology allows the greatest number of interconnections. The small bump size results in very low inductance. In a typical application, 100-μm-diam bumps are placed on a 225-μm grid allowing space for 784 connections on a 6.5-mm square chip. An example of this technology is shown in Fig. 4. Smaller solder bumps and tighter grids resulting in thousands of connections have been demonstrated (Tummala and Rymaszewski, 1989).

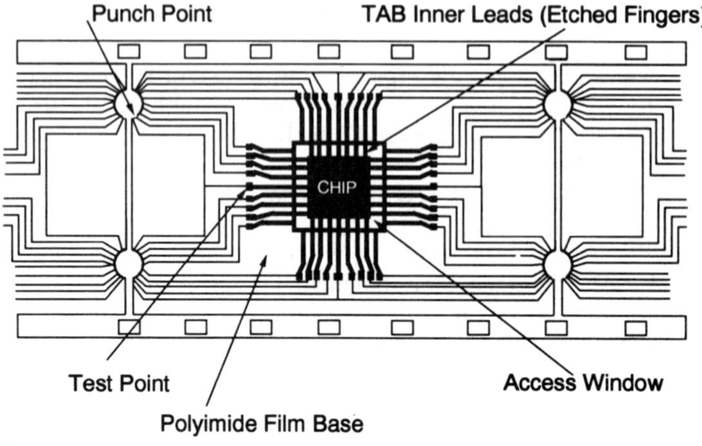

FIG. 3. (a) Tape automated bonding (TAB) assembly tape (adapted from Sinnadurai, 1985). (b) Inner lead bonding process (adapted from Tummala and Rymaszewski, 1989).

3. MOLDED PLASTIC PACKAGING

Transistors encapsulated in plastic appeared in the late 1950s, and because of the low cost and effectiveness of the technology, molded plastic packages are the most widely used single-chip modules today. They are called by many names such as dual in-line package (DIP), and variations like SIP and ZIP, the small outline package (SOP and SOJ), the plastic leaded chip carrier (PLCC), and plastic quad flat pack (PQFP), as well as others. Some of these are illustrated in Figure 5 where a cutaway view shows their basic construction. The lead frame is the supporting structure to which the chip is physically attached. Electrical connections are made by wirebonding and then the whole assembly is covered with a thermosetting molding compound (Goosey, 1985) in a transfer molding process.

3.1 Lead-Frame Materials

The most widely used metal for lead frames is Alloy 42 (42% Ni, 58% Fe), which has a thermal expansion coefficient closely matched to the Si chip, although its low thermal conductivity presents a problem for high-power chips since the lead frame is the main heat conduit away from the chip. To overcome the thermal limitation a variety of other lead-frame alloys are used (Tummala and Rymaszewski, 1989). The lead frames are 0.15–0.25-mm thick and are chemically etched or mechanically stamped from rolled strip stock. Chips are bonded to the lead frames by silicon/gold eutectic bonding or by use of silver-filled epoxy or polyimide adhesive paste. After encapsulation the leads are formed to special shapes for soldering to cards. In Fig. 5 the DIP leads are wider at the top than at the bottom, creating an integral standoff when

(a)

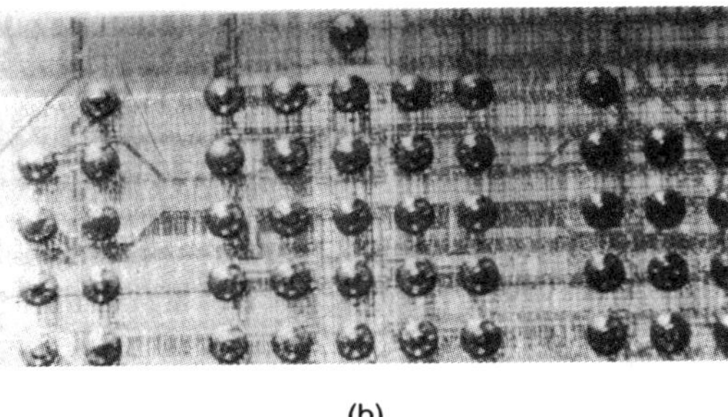

(b)

FIG. 4. (a) A 5000-circuit chip with 354 solder bumps in an area array; (b) a close-up view of solder bumps (courtesy of IBM Corp).

the leads are inserted into holes in cards. The PLCC leads shown in Fig. 5 have a "J bend" and the SOIC leads are "gull wing" shaped for surface-mount soldering.

3.2 Encapsulation Materials

Molding compounds are mixtures of materials optimized for a specific application. The most commonly used resin is cresol novolac epoxy, which has relatively low shrinkage on curing, low moisture absorption, chemical inertness, good adhesive properties, and a glass transition temperature of 150 °C. Other resins have glass transition temperatures between 125 and 200 °C. For applications requiring high heat resistance, low dielectric properties, and low-temperature properties silicone materials can be used. In addition to the resin, curing accelerators and curing agents, flame retardants, colorants, and inert fillers make up the molding compound. The filler comprises about 70% by weight and is needed to control shrinkage, improve thermal conductivity, and enhance thermal shock resistance. Powdered fused silica (SiO_2) is the primary filler used today. Nevertheless, the thermal expansion coefficients of molding compounds are typically (120–200) $\times 10^{-7}$/°C. Care must be taken in the transfer molding process and the physical design of the chip, lead frame, and wire bonds to avoid stress-induced failures (Manzione, 1990).

4. PRINTED-CIRCUIT CARDS AND BOARDS

4.1 Structures

Printed-circuit technology is the lowest-cost method for interconnecting a large number of diverse components. The basic structure is shown in Fig. 6. The simplest cards are single sided and have wiring on only one surface. More complex cards are double sided and have wiring and components on both outside surfaces. The most complex cards have multilayer wiring and can support components on one or both outside surfaces. In cards and boards, plated copper wiring is connected between layers by plated through holes in a metal/polymer laminate. The plated through holes (PTH) are also used to mount chip carriers, passive devices, and connectors. The voltage reference planes distribute current, provide shielding, and control the impedance of the signal transmission lines. This basic structure is used to make a range of second-level packages, from inexpensive small cards containing one or two planes with signal linewidths of 1.0 mm and PTH diameter of 0.84 mm to high-cost, high-performance boards 600 × 700 mm in size and containing as many as 50 planes with signal linewidths of 0.05 mm and PTH diameter of 0.35 mm. Fig. 7 is a picture of a large card used in workstations.

4.2 Materials and Processes

The most common insulator is FR-4 (fire-retardant, epoxy-impregnated glass cloth), although polyimide-glass laminates are also used, and a variety of polymers and fillers are used for specific applications requiring low thermal expansion coefficient, low dielectric

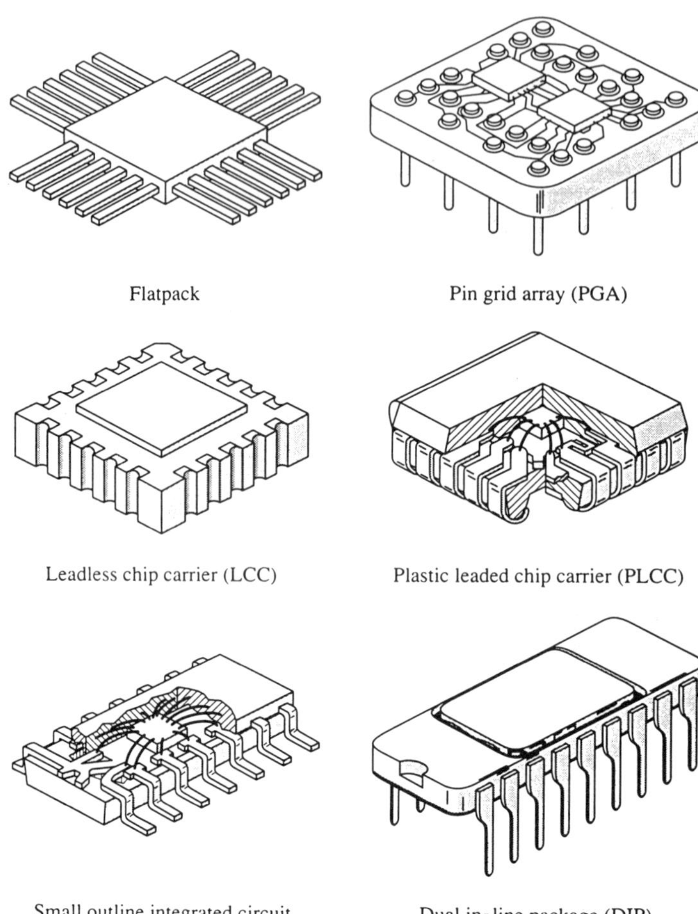

FIG. 5. First-level packages (from Seraphim *et al.*, 1989, reproduced by permission) are commonly attached by solder to second-level wiring structures. In the LCC, the solder bridges into the groves shown.

constant, or high thermal stability (Seraphim *et al.*, 1989). High thermal stability is achieved with a number of polyimide materials. Low–dielectric-constant boards with very low dissipation at high frequencies are made of polytetrafluoroethylene (PTFE) (Goosey, 1985).

The fabrication process involves laminating copper on both sides of the insulating material (FR-4 or equivalent), drilling holes, and joining the two sides by electroless copper plating of the hole walls and surfaces. The copper wiring is then created by either a subtractive or an additive process. Both use photolithographic techniques to define a wiring pattern. In the subtractive process, the unwanted copper is etched away after through-hole plating. In the additive process, as the through holes are plated, copper is added in a pattern to the surface and finally coated with tin before the photoresist is removed and the base copper etched away. To construct a multilayer structure, partially cured "prepreg" (epoxy-glass laminate alone) is placed between each of the metallized double-sided laminates or cores to form a stack. Copper surface planes may be added and the entire stack laminated and treated as a core. Holes are drilled where desired to intersect internal copper pads joined to lines or voltage planes and the whole unit is plated again (Seraphim *et al.*, 1989).

5. CERAMIC PACKAGING

Ceramics are used to fabricate many of the chip carriers shown in Fig. 5. There are ceramic dual in-line packages (CERDIPS) and leadless chip carriers. While the shapes are similar to molded plastic packages, ceramic chip carriers are more rugged and have the

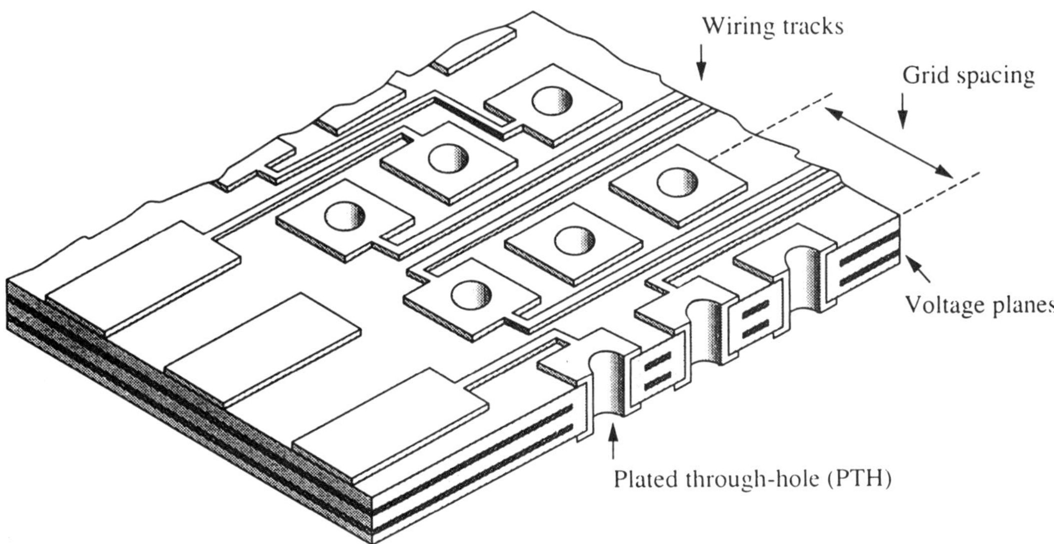

FIG. 6. Printed circuit card with two signal layers. There are two wiring tracks within the grid spacing (two lines per channel). (From Seraphim *et al.*, 1989, reproduced by permission.)

advantage that they can be made hermetic, i.e., the interior is totally isolated from the ambient atmosphere. Hermeticity is not achievable with plastic packages because of the gas permeability of plastics. Ceramic packages are sealed with metal or ceramic lids. Moreover, ceramics have thermal expansion coefficients well matched to semiconductors, thus reducing bonding stress, and they have relatively high thermal conductiv-

FIG. 7. This printed circuit card or "planar" is 250 × 420 mm^2 and interconnects the components of a workstation computer, including surface-mounted single-chip modules with heat sinks and connector blocks for additional small cards (courtesy of IBM Corp).

ity. In addition, the strength and rigidity of ceramics easily accommodate a large-area array of pins for dense interconnection to higher package levels.

5.1 Metal-on-Ceramic Packages

With the exception of the pin-grid array (PGA), the chip carriers in Fig. 5 have wire-bond connections between chip and package I/O leads. In the ceramic PGA there is a single layer of metal on the ceramic surface to provide wiring between the components mounted on the PGA and make connections to the pins. This wiring is made by either thick- or thin-film processing. In the thick-film process, a screen or metal stencil is prepared with openings defining the circuit pattern. The screen is placed over the ceramic substrate. An ink paste containing metal particles is made to fill the openings in the stencil by use of a compliant blade or "squeegee" as in common silk-screen printing processes. The stencil is removed and the substrate is fired at high temperature, sintering the metal particles in the circuit pattern. In the thin-film process, metal is sputter deposited on the ceramic and patterned by photolithography and metal etching.

5.2 Multilayer Ceramic Substrates

Multichip ceramic substrates with multiple wiring layers are made by cosintering a ceramic and metal system. A powdered ceramic is mixed with solvents and organic binder polymer and tape cast and dried into a "green sheet," which is cut to size. Holes are punched for via interconnections and a thick-film metallization process used to fill the holes and define the circuit lines. After inspection the good sheets are stacked and laminated and then sintered at high temperature to form a solid body with metal conductors and ceramic insulation. Surface metals are overplated as required and pins brazed to bottom pads. Chips are mounted directly on the module, and it is hermetically sealed. An example is pictured in Fig. 8. The most complex ceramic MCMs today contain more than 60 layers of low-dielectric glass-ceramic with several hundred meters of cosintered copper wiring and are used to interconnect over a hundred chips.

5.3 Ceramic Materials

Alumina (Al_2O_3) is the most common ceramic in use. Thick-film processes typically employ 96% Al_2O_3, 3–3.5% SiO_2, and a small

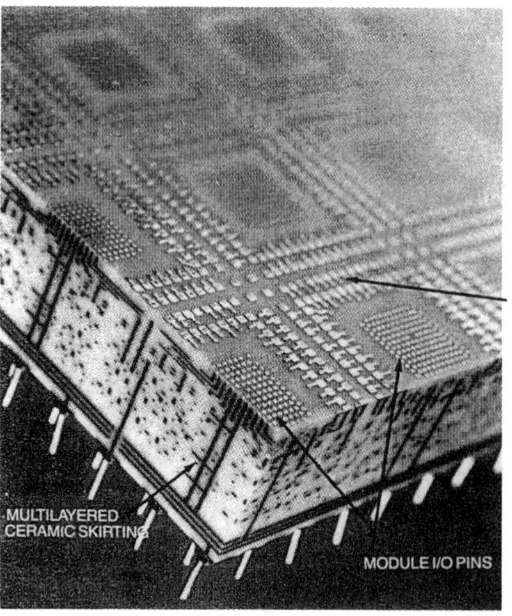

FIG. 8. Cross section of multilayer ceramic showing dense wiring (courtesy of IBM Corp., from Tummala and Rymaszewski, 1989, reproduced by permission).

amount of MgO and CaO. Other ceramics such as BeO, SiC, and AlN are used where higher thermal conductivity is desired, and glass-ceramics or mixtures of glass and ceramic that are compatible with copper metallization are used when a low dielectric constant is required for electrical performance. Since metals must be cosintered with the ceramic, the choice of metallization for a given ceramic is limited since the sintering temperature of the ceramic must be somewhat lower than the melting point of the metal. (Tummala and Rymaszewski, 1989).

6. THIN-FILM WIRING

Multilayer thin-film wiring is applied to the surface of a printed circuit card, ceramic package, or passive mechanical support where high performance, high density, and extreme compactness of interconnection wiring are required. The dimensions of transmission lines given in Fig. 2 indicate that roughly ten times as many wiring tracks can be made in a wiring layer using thin-film wiring as in thick-film card or ceramic wiring. Thus, roughly ten thick-film layers are equivalent to one thin-film layer in wirability, and thin-film layers are each only one tenth as thick.

6.1 Materials

The high electrical performance of thin-film wiring is due to the use of low–dielectric-constant insulators such as spray- or sputter-deposited borosilicate glass or quartz or, more commonly, polymers such as polyimides. Polyimides are chosen for their ability to form high-quality, defect-free thin films by spin or spray application and for high thermal stability, low moisture uptake, and good solvent resistance. Copper is the metal of choice, although aluminum and others are used. Since copper does not adhere well to polymers and can react with uncured material, very thin layers (≲1000 Å) of chromium, titanium, or other materials are placed between the copper conductor and the polymer to act as adhesion promoters and passivants.

6.2 Sequential Patterning

The very fine dimensions of thin-film wiring require sequential fabrication techniques; an example is given in Fig. 9. There

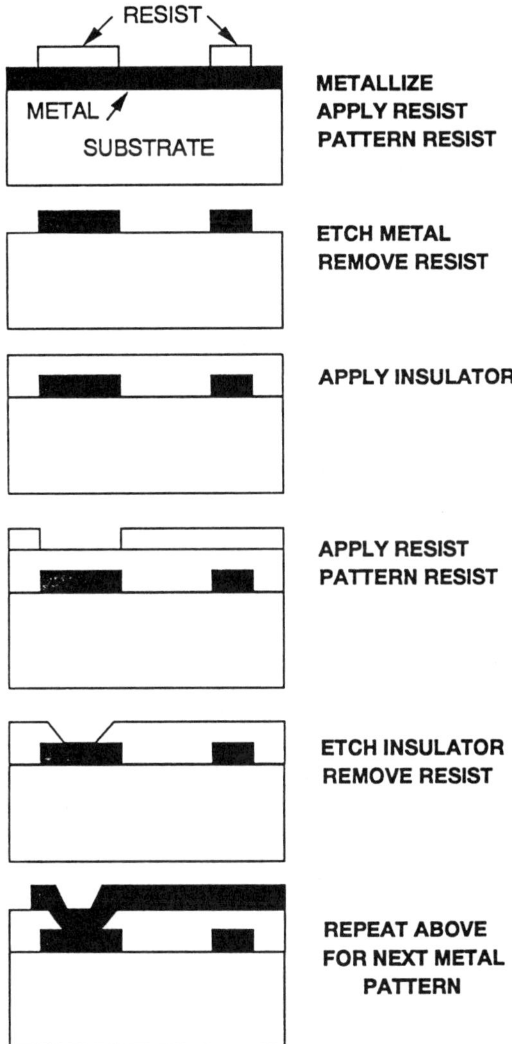

FIG. 9. Typical thin-film wiring fabrication process. In this figure, the sequential process flow is from top to bottom.

are a variety of methods for metal and polymer deposition and patterning, and many different processes are in use. Metals may be deposited by evaporation, sputtering, electroless plating, or electroplating. They may be patterned using photolithographic techniques by wet etching as used in printed circuit cards, by plating in a resist pattern, or by lift-off processes used in making on-chip wiring. Polymer insulators may be patterned by wet chemical etching, reactive ion etching, or laser ablation, or they may be intrinsically photosensitive and directly patterned optically like a photoresist.

The choice of a particular patterning process for metal or polymer is dependent on cost and is often dictated by the aspect ratio (the ratio of height to width) of the most demanding feature required by the electrical design. Processes that are capable of forming high–aspect-ratio structures such as electroplating of a metal in a patterned photoresist or reactive ion etching of a polymer are generally more expensive than processes that can only form low–aspect-ratio structures such as wet etching of metals or polymers.

7. PACKAGE-TO-PACKAGE INTERCONNECTIONS

7.1 Pins

The most common interconnection method between first- and second-level packages is to solder the leads or pins of one component into the plated through holes of a card or board. Typically tin-lead eutectic solder (63% Sn, 37% Pb), which is liquid above 183 °C, is used in the wave soldering process shown schematically in Fig. 10. After flux is applied, the populated card is preheated and passed over a fountain of flowing solder where capillary forces cause the solder to fill the plated through holes.

In pin-grid array (PGA) and multilayer ceramic packages, pins are attached to the package before mounting on the card or board. In the simplest packages, pins are swaged into the package while more complex structures require pins that are brazed to pads on the bottom of the package. For multilayer ceramic packages, Ni/Au-plated Kovar pins are attached with eutectic Au-Sn (20% Au, 80% Sn) braze, which melts at 280 °C. Higher melting point (779 °C) but less expensive eutectic copper-silver braze or other materials are also used.

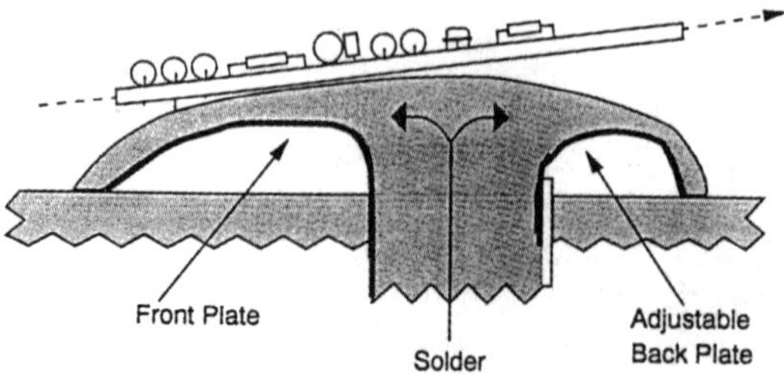

FIG. 10. Wave soldering (adapted from Seraphim *et al.*, 1989).

7.2 Surface-Mount Technology (SMT)

Relative to pin-in-hole mounting, surface-mount technology reduces package size and cost and improves board utilization. It does this by joining leads from components to the board surface on a finer pitch than the 2.54-mm hole spacing currently in use, thus allowing the whole board cross-sectional area to be used for wiring unobstructed by holes. Currently SMT connections are made with a pitch as fine as 0.3 mm, allowing several hundred perimeter bonding leads for a single chip module. Surface-mounted packages are either leadless or leaded chip carriers, some of which are shown in Fig. 5. For these packages a very short surface solder joint provides both electrical and mechanical connection between package and board. In leadless chip carriers, the solder must accommodate the shear strain caused by the relative displacement of package and board during thermal excursions. In leaded chip carriers, the strain is accommodated by elastic deformation of the leads. Since solder fatigue failure is a concern for SMT components, the contact metallurgy and intermetallic compound formation as well as the solder time and temperature cycle must be carefully controlled. Typically, eutectic Sn-Pb solder is applied by dipping or wave soldering, paste screening, or stenciling, plating, or application of preforms. The major joining techniques are wave soldering, vapor-phase reflow, infrared or hot-gas reflow, laser soldering, or hot-knife soldering. In vapor-phase reflow, the populated board with applied solder is placed in the saturated vapor of a boiling fluorocarbon. The vapor condenses on the board raising the temperature of all components to the fluorocarbon boiling point (typically 215 °C) thus reflowing the solder and joining all components simultaneously at uniform temperature.

While the vast majority of surface-mounted components are attached by solder, the use of conductive polymer adhesives is growing. Metal-particle–filled polymers such as epoxies form acceptable conductors in some applications. The principal metals used are silver, copper, or gold although silver/palladium alloys have been utilized (Goosey, 1985).

Highly loaded polymer films are isotropic conductors. Less highly loaded polymer films can be insulating until compressed between opposing metal pads. These materials form anisotropic "z-axis" conductive films and are finding increasing applications.

7.3 Connectors

A connector is a device to separate and reconnect electronic circuits repeatedly, and they come in many sizes and shapes. All involve the application of force by springs or other elements and a remakable low-resistance electrical contact surface. An example of a spring-loaded connector for a package with pins is shown in Fig. 11. The cross section of a card-edge connector is shown in Fig. 1, where spring-loaded contacts join to pads along both edges of a printed circuit card. Metal spring materials are most frequently beryllium copper or phosphor bronze, and interface contact metals are gold, palladium, or tin. The insulation that supports the metal contacts is most often an injection-molded plastic housing. Very high-density connectors are made from an area array of metal contact elements separated by a plastic insulator. The individual contact elements may be solid metal structures or shaped metal springs in a rigid plastic housing. The contact elements may also be metal structures in a compressible plastic medium that provides the contact force.

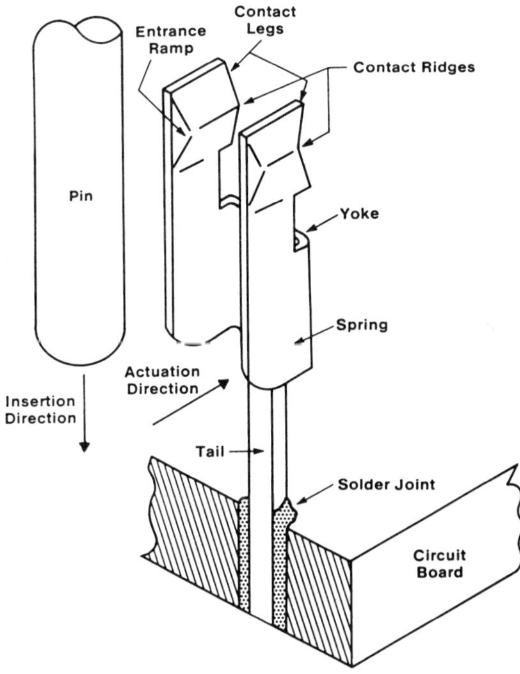

FIG. 11. Mechanical pin connector (from Tummala and Rymaszewski, 1989, reproduced by permission).

7.4 Cables

Cables are used to carry power, sense very low voltages or current, or transmit data over relatively long distances. Signal cables are flexible transmission lines containing a signal wire and a ground return conductor (a wire or shield, braid or continuous) separated by a dielectric and covered with a protective polymer jacket. Typical transmission lines are twin-lead, tri-lead, or coaxial. Since data are transmitted over long distances, the speed of transmission and the attenuation in the cable are important. For high speed, low–dielectric-constant polymers are used, and air is introduced to lower ϵ further. For example, expanded polytetrafluoroethylene has a dielectric constant of 1.3 so that signals travel at 88% of the speed of light. At high frequencies current flows in the skin depth, δ, on the outer surface of the inner conductor and the inner surface of the outer conductor. The skin depth δ is proportional to $1/\sqrt{f}$, where f is the frequency. Hence, the signal is attenuated by the skin-effect resistance and insulator dielectric loss. For low attenuation at high frequencies conductors are coated with silver or copper.

Electrical cables are used to interconnect packaged components within a system or for communication between systems. For short distances electrical cables can be used for high-speed data transmission, but for the highest data transmission rates or high-speed transmission over long distances, optical fiber cables are used. These cables consist of a graded-index SiO_2-based glass fiber in a protective plastic sheath. Optical fiber cables can carry data with minimal losses over several kilometers at rates in excess of Gbits/s.

8. HEAT TRANSFER

In order to maintain active device characteristics and ensure long lifetimes, semiconductor junctions must be maintained typically at 85 °C or less. The function of the package is to cool the chips by transferring the heat generated to an air or water reservoir or heat sink.

8.1 Thermal Engineering

The cooling structure of the package is characterized by the thermal resistance R, in °C/W, that it presents to a device generating heat Q (in watts, W), resulting in a temperature difference ΔT (in °C) between the device and the cooling fluid (a gas or liquid):

$$R = \Delta T/Q.$$

In low-power applications, R may be ~100 °C/W, but for densely packaged high-power chips, R must be engineered to be ≤10 °C/W. Heat is usually transferred to the cooling fluid by convection, which may be natural or forced (by a fan or pump) or due to boiling. The ability of the fluid to accept heat from area A (in m^2) is characterized by the heat transfer coefficient h (in W/m^2 °K):

$$Q = hA\Delta T.$$

Ranges of values for h are given in Table 3, but since h encompasses all the complex effects of convection, including fluid motion, fluid properties, and system geometry, the exact value for a particular cooling structure must be modeled and ultimately determined empirically.

8.2 Cooling Structures

Low-power electronic systems use natural-convection air cooling, but most computers use forced-air cooling, relying on circulating fans and finned metal heat spreaders as required to cool components. The multichip module packages require the most sophisticated cooling structures because they achieve high system performance by closely interconnecting many powerful chips. Figure 12 shows a few of the wide variety of structures in use. Depending on the thermal resistance required between the device and the cooling fluid, the heat may be removed from the chips through the interconnection package or from the back of the chip through various thermal structures such as solder, adhesives, paste, ceramic structures, metal plates, or metal pistons.

9. FUTURE CHALLENGES, EMERGING TECHNOLOGIES

Progress in package technology is characterized by steady evolutionary improvement of existing technology as well as revolutionary changes associated with new materials. For example, the outer lead bond pitch of surface-mount packages has continued to shrink over a 20-year period, and wiring density of cards and ceramic modules is increasing. There is ongoing work to reduce the dielectric constant of processable ceramics and polymers by introducing new materials. Other new materials such as synthetic diamond for thermal management could open new dimensions in packaging. But the key challenges to package technology will be, as always, to package new and ever more powerful systems at the lowest possible cost.

9.1 Optoelectronics

Optoelectronics devices and fiber-optic links are increasingly being used for high-bandwidth communications between electronic systems. While semiconductor devices used as transmitters (such as lasers and light-emitting diodes) and receivers can be interconnected and powered using the technologies described above, special package designs are required to couple the optical fiber to the semiconductor transmitter or receiver. For example, to avoid signal losses the alignment tolerance for single-mode fibers is about 0.5 μm. The coupling of fiber to device can

Table 3. Heat transfer coefficients (W/m^2 °K).

	Air (or N_2)	Fluorocarbon liquid	Water
Free convection	3–12	100–300	—
Forced convection	10–100	200–2000	3000–7000
Boiling	10 000	2000–6000	3000

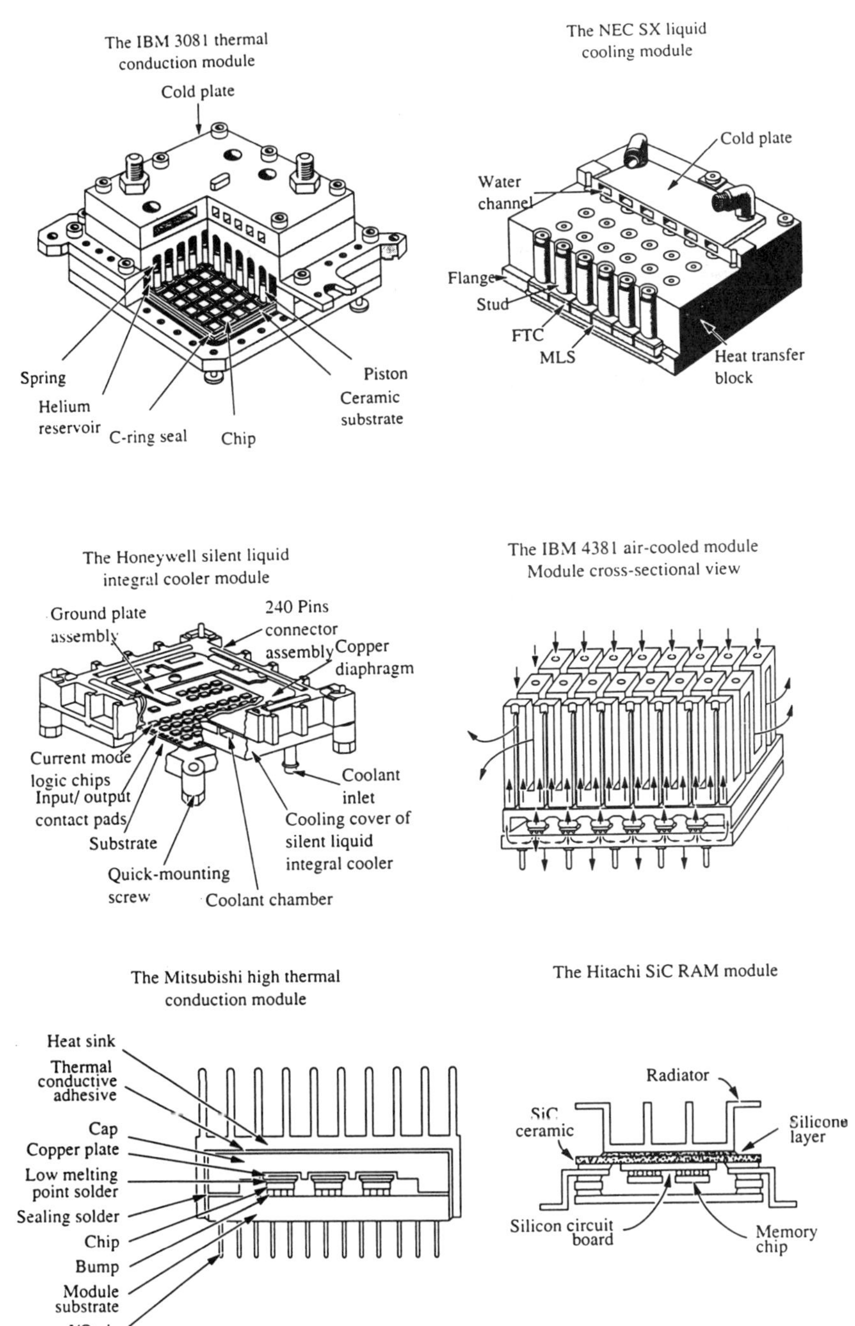

FIG. 12. Schematic diagrams of multichip-module cooling structures (from Seraphim *et al.*, 1989, reproduced by permission).

be accomplished by precision mechanical mounting (with or without lenses), an adherent butt joint, or thin-film optical waveguides (Tummala and Rymaszewski, 1989). Future systems may require dense arrays of fibers coupled to small high-speed semiconductor switches that include optical transmitters and receivers.

9.2 Electromagnetic Management

Future packaging technology must successfully handle processor chips running at frequencies of hundreds of megahertz and the electromagnetic emissions and interference associated with these systems. In addition, as wireless communication becomes dominant for portable systems, the fabrication of high-frequency transmitters and receivers of electromagnetic waves will become important. Electromagnetic field modeling, component placement, and shielding will become key aspects of package technology. The need for low-cost, light-weight, moldable shielding materials could open up a range of applications for conducting polymer materials.

9.3 Low-Power, Low-Cost, Light-Weight Systems

As semiconductor technology reduces device size and power and careful design puts more and more function on a single chip, potentially very low-cost, reasonably powerful systems will be possible. Package engineers will be increasingly challenged to build thin, light-weight, but very robust structures that provide signal interconnection, power distribution, and cooling for these future portable systems.

Works Cited

Goosey, M. T. (1985), *Plastics for Electronics*, London: Elsevier Applied Science Publishers.

Harper, C. A. (Ed.) (1991), *Electronic Packaging and Interconnection Handbook*, New York: McGraw-Hill, Inc.

Manzione, L. T. (1990), *Plastic Packaging of Microelectronics Devices*, New York: Van Nostrand Reinhold.

Minges, M. L. (Tech. Chairman) (1989), *Electronic Materials Handbook*, Vol. 1, *Packaging*, Materials Park, OH: ASM International.

Radke, C. (1969), in: N. Garaffa, Jr. (Chmn.), *Proceedings of the 6th Annual SHARE–ACM–IEEE Design Automation Workshop*, New York: IEEE, pp. 257–267.

Seraphim, D. P., Lasky, R. C., Li, C. Y. (1989), *Principles of Electronic Packaging*, New York: McGraw-Hill Inc.

Sinnadurai, F. N. (1985), *Handbook of Microelectronics Packaging and Interconnection Technologies*, Ayr, Scotland: Electrochemical Publications Ltd.

Tummala, R. R., Rymaszewski, E. J. (1989), *Microelectronics Packaging Handbook*, New York: Van Nostrand Reinhold.

PALEOMAGNETISM

J. E. T. CHANNELL, *Department of Geology, University of Florida, Gainesville, Florida, U.S.A.*

	Introduction	89
1.	**Relevant Characteristics of the Geomagnetic Field**	89
2.	**Origin of Magnetization in Rocks and Sediments**	90
2.1	Magnetic Minerals	90
2.2	Natural Magnetization Processes in Rocks and Sediments	91
3.	**Laboratory and Analytical Techniques**	91
3.1	Demagnetization Techniques	91
3.2	Analysis of Demagnetization Data	92
3.3	Statistics	93
3.4	Field Tests for the Age of Magnetization	93
3.5	Identification of Carriers of Natural Remanence	94
4.	**Applications of Paleomagnetic Studies**	95
4.1	Paleomagnetic Poles and Continental Drift	95
4.2	Tectonic Rotations in Mountain Belts	95
4.3	Magnetic Polarity Stratigraphy	95
4.4	Secular Variation Magnetic Stratigraphy in Lake Sediments	96
4.5	Rock Magnetic Stratigraphy	96
4.6	Insights into Geomagnetic Field Behavior from Secular Variation, Excursions, Polarity Transitions, Paleointensities, and Polarity Bias	96
	Glossary	97
	Works Cited	98
	Further Reading	98

INTRODUCTION

Paleomagnetism is the study of the magnetization of rocks and sediments, and provides a record of the direction and intensity of the ancient geomagnetic field.

Paleomagnetism was a cornerstone of the "revolution" in the earth sciences that began in the 1950s and led to the development of the theory of plate tectonics. A series of landmark papers published in the 1950s on paleomagnetic data from Britain and North America provided the first land-based quantification of continental drift (e.g., Runcorn, 1956). Forty years on, paleomagnetism remains the primary means of reconstructing the relative positions of rock bodies in the geologic past (on scales from individual thrust sheets to entire continents). In addition, paleomagnetism has led to an improved understanding of the mechanisms for generation of the geomagnetic field, for secular variations, and for polarity reversals. The record of polarity reversals in rocks and sediments provides a unique means of global correlation, and has become the thread on which we hang our geologic time scales.

1. RELEVANT CHARACTERISTICS OF THE GEOMAGNETIC FIELD

About 90% of the geomagnetic field at the Earth's surface can be modeled by a dipole close to the center of the Earth. The nondipole components of the geomagnetic field contribute to the short-term (secular) variations of the field, and are responsible for the deviation of the north geomagnetic pole from the north pole (situated at the rotation axis of the Earth). It is generally accepted, on the basis of records for the last 10^5 years, that secular variation averages out over $\sim 10^4$ years. The time-averaged geomagnetic field is thought to approximate a geocentric axial

3-527-28135-5/95/$5.00 + .50

dipole field. This is of paramount importance to paleomagnetic studies because it allows us to predict the location of the ancient north pole relative to an observation point where the time-averaged field direction is recorded by ancient rocks or sediments.

The dipole component of the geomagnetic field has changed polarity numerous times in the geologic past. The present state of "normal" polarity has existed for about the last 780 000 years. Our template for the sequence of geomagnetic polarity reversals since about 150 Ma B.P. (before present) is based on magnetic anomalies from the oceans, which are produced by normally and reversely magnetized ocean crust formed at the mid-oceanic ridges. The so-called "seafloor spreading" process often occurred at rather constant rates over long periods of geologic time and spreading was usually symmetric either side of the ridge. For these reasons, the record of geomagnetic reversals in the ocean floor is often a faithful record of the sequence of polarity reversals through time. Statistical analysis of the reversal record since about 150 Ma B.P. indicates that there are too few short polarity intervals for a Poisson process; however, this may be attributed to the inadequacy of the record. It is generally considered that the lack of memory characteristic of the Poisson process is also a characteristic of polarity reversal history. Because of the dipole nature of the main field and the "geologically instantaneous" nature of the reversal process (the process takes about 3000 years), the record of individual reversals in rocks and sediments provides a unique global time marker, a unique means of global correlation.

2. ORIGIN OF MAGNETIZATION IN ROCKS AND SEDIMENTS

2.1 Magnetic Minerals

Iron oxide and iron sulfide minerals are the most common carriers of remanent magnetization in rocks and sediments. Titanomagnetites [$x\mathrm{Fe_2TiO_4} \cdot (1-x)\mathrm{Fe_3O_4}$] are present in many types of igneous and metamorphic rocks, and are therefore a detrital constituent in many sedimentary rocks. The titanomagnetites are ferrimagnetic at room temperature for values of x less than about 0.8. Many sediments, particularly deep-water pelagic limestones, contain magnetite (Fe_3O_4) or low-titanium titanomagnetite that may be biogenic in origin, being derived from marine organisms, such as bacteria and some molluscs, which incorporate magnetite in their metabolism (see BIOMAGNETISM AND GEOMAGNETIC FIELD DETECTION BY ORGANISMS). Biogenic magnetite usually lies in the single-domain (SD) grain-size range, which is the optimal grain size for retention of magnetic remanence over geologically significant time intervals.

Titanohematites [$x\mathrm{FeTiO_3} \cdot (1-x)\mathrm{Fe_2O_3}$] are antiferromagnetic for $0 < x < 0.5$. The weak net magnetic moment in hematite (α-Fe_2O_3) and titanohematites is due to a slight canting of the spin moments. Titanohematites are constituents in some igneous and metamorphic rocks. Hematite is a common constituent of sedimentary rocks and is an important carrier of magnetic remanence in clastic and chemical sediments. Hematite may occur in sediments as detrital "specularite" or as an authigenic pigment that grows during diagenesis.

Maghemite (γ-Fe_2O_3) has the spinel structure of magnetite and the chemical composition of hematite. Maghemite is ferrimagnetic and has similar magnetic properties to magnetite. Maghemite is common as a low-temperature oxidation product of magnetite in both igneous and sedimentary rocks. It reverts to hematite at about 350 °C.

Goethite (α-FeOOH) is the most common iron oxyhydroxide mineral and is antiferromagnetic, the net moment being due to canting of spin moments. Goethite is an important product in low-temperature oxidation, particularly of iron sulfides, and is a constituent of manganese nodules and the Pacific red clays where it may precipitate directly from seawater.

Pyrrhotite (FeS_{1+x} where $x = 0$–0.14) is the most commonly occurring iron sulfide mineral that is ferrimagnetic. The ferrimagnetism is restricted to a limited compositional range ($0.1 < x < 0.14$). Pyrrhotite occurs in basic igneous rocks and can grow during sediment diagenesis in certain reducing environments. The most common iron sulfide mineral (pyrite, FeS_2) is paramagnetic, and does not therefore carry magnetic remanence.

2.2 Natural Magnetization Processes in Rocks and Sediments

Thermal remanent magnetization (TRM) is the magnetization acquired by a magnetic mineral grain during cooling through its blocking temperature, above which the magnetic ordering of a grain is randomized by thermal energy. The blocking temperature is a function of grain size and composition and lies below the Curie (Néel) temperature. Igneous rocks cooling from magma generally acquire a TRM in the direction of the ambient geomagnetic field.

Chemical remanent magnetization (CRM) is produced by growth of a magnetic mineral grain through its blocking volume, below the blocking temperature of the grain. The blocking volume is the critical volume for a particular grain shape and composition, above which the grain can carry magnetic remanence. Magnetic mineral grains acquire a CRM as they grow through their blocking volume in the ambient geomagnetic field, and the direction of the CRM is usually maintained during further grain growth. CRM is an important mode of remanence acquisition during sediment diagenesis.

Detrital or depositional remanent magnetization (DRM) is an important magnetization process in sediments. Detrital or biogenic magnetic mineral grains with preexisting TRM or CRM can align themselves with the ambient field during deposition. The alignment occurs in the upper few centimeters of the soft sediment, and is therefore often referred to as postdepositional remanent magnetization (pDRM). The lock-in depth of DRM has been estimated experimentally and theoretically to be of the order of 10 cm, but will depend on grain size, porosity, sedimentation rate, and bioturbation depth.

Viscous remanent magnetization (VRM) is the remanence produced by a weak external field (such as the geomagnetic field) over a long period of time. VRM is an important source of secondary magnetization in rocks and sediments. The intensity of VRM is proportional to the logarithm of time that the sample experiences the weak field. The proportionality constant (viscosity coefficient) increases with temperature. Many rocks and sediments carry a VRM acquired during the present Brunhes interval of "normal" polarity of the geomagnetic field.

3. LABORATORY AND ANALYTICAL TECHNIQUES

The introduction of cryogenic magnetometers in the mid 1970s resulted in a significant reduction in magnetometer noise level, and opened up paleomagnetic research to a wider range of rocks and sediments (see Collinson, 1983). Modern cryogenic magnetometers can measure sample magnetization intensities as low as $\sim 10^{-5}$ A/m with sufficient precision for precise determination of paleomagnetic direction.

Rocks and sediments can become magnetized at any time during their history, and hence the natural remanent magnetizations (NRM) of rock and sediment samples are often the resultant of more than one magnetization component. TRM in igneous rocks and DRM in sedimentary rocks can usually be associated with the original cooling of the igneous rock or original deposition of the sedimentary rock and, in such cases, are referred to as "primary" magnetizations. Subsequent chemical alteration or heating of the rock can yield a CRM or TRM that would then be a "secondary" magnetization. VRM often has recent origin, having formed since the last geomagnetic polarity reversal. Laboratory techniques in paleomagnetism are designed to resolve the various magnetization components that constitute the total magnetization of the rock or sediment. The objective is usually to resolve the "primary" magnetization component, that associated with the formation of the rock or sediment. The age of the "primary" magnetization component can then be derived from the paleontological/radiometric age of the rock or sediment.

3.1 Demagnetization Techniques

The external field strength required to alter the magnetic remanence of a grain will depend on the coercivity of the individual grain, which is controlled by composition and grain size/shape. The dependence of coercivity on composition and grain size is utilized in *alternating-field* (AF) *demagnetization* to resolve the magnetization components carried by different grain populations. If the NRM of the sample is a composite of individual magnetization components that have distinct coercivity spectra, the AF method is suitable for resolving the individual components. The normal procedure is to demagnetize the sample progressively by a stepwise increase in peak alternating field, measuring the remaining magnetic remanence after each step. The range of peak alternating fields is typically 5–100 mT. For each demagnetization step, the sample is placed in a magnetically shielded coil and the alternating field is smoothly ramped down to zero in the absence of background field. Grain remanences with coercivity lower than the peak alternating field will be reoriented as the peak field is ramped down, and their contributions to the remaining remanence will cancel out. The net effect is to demagnetize these grains. As the peak alternating field is progressively raised, a larger proportion of the total magnetic grain population becomes demagnetized and ceases to contribute to the remanence of the sample. For a full description of the design of alternating-field demagnetizers see Collinson (1983).

The time required for magnetic remanence of a grain to decay in zero external field (the relaxation time) is a function of ambient temperature, grain size, and composition. The blocking temperature is the temperature at which the relaxation time falls within the time range of laboratory experiments ($\sim 10^3$–10^4 s). As the relationship between relaxation time and temperature is exponential, the exact length of the experiment does not, within reason, affect the observed blocking temperature. Two populations of magnetic grains with different origins, and therefore different compositions and/or grain sizes, might be expected to have distinct blocking-temperature spectra. This principle is utilized in *thermal demagnetization* to separate magnetization components carried by different magnetic grain populations. Normal practice is to heat the sample in progressive temperature increments (until the magnetization falls below the magnetometer noise level), allowing the sample to cool in field-free space after each heating increment. The remanence of all grains with blocking temperature less than the maximum temperature will become randomized as the sample cools in field-free space. The magnetization measured after each heating step can be used to determine the directions of magnetization components with distinct blocking-temperature spectra. For a description of the design of thermal demagnetizers, see Collinson (1983).

Leaching with hydrochloric acid can provide a means of selective destruction of magnetic grains. Such *chemical demagnetization* is applicable to siliceous rocks, particularly red sandstones, which contain a secondary magnetization due to authigenic growth of fine-grained pigmentary hematite. The dissolution rate of coarse-grained hematite of detrital origin is lower than that of the finer-grained pigmentary hematite of diagenetic origin, because of the difference in the ratio of surface area to grain volume. Leaching in 3–10 N hydrochloric acid, and remanence measurement at certain time intervals during the leaching, provides a means of resolving magnetization components with distinct "leaching spectra."

3.2 Analysis of Demagnetization Data

The most widely used method for analyzing changes in magnetization direction and intensity during demagnetization is the orthogonal projection, also referred to as the Zijderveld diagram (see Zijderveld, 1967; Dunlop, 1979; Collinson, 1983). This is a simple, but extremely useful, means of displaying the change in magnitude and direction of the magnetization vector during demagnetization. Straight-line segments on these plots indicate that the magnetization vector being removed has constant direction. This direction can be computed, and defined as a component of the magnetization. It is important to note that straight-line segments on orthogonal projections may not always represent single components. They may also represent composites of two or more components with identical demagnetization (coercivity, blocking, or leaching) spectra. The magnetization components cannot be resolved if the overlap in demagnetization spectra is too great. However, if the unblocking temperature spectra overlap, this does not necessarily mean that the coercivity spectra will overlap. The choice of demagnetization technique will be critical to whether the components of magnetization can be resolved.

The points on the orthogonal projection that define straight-line segments are commonly picked by eye. The component direction associated with a particular straight-line segment can then be computed by a least-squares technique (Kirschvink, 1980). This method gives a maximum angular deviation

(MAD) value, which provides an estimate of the goodness of fit of the straight line to the points on the projection.

In the special case when three magnetization components with overlapping demagnetization spectra are present, the Hoffman–Day method (Hoffman and Day, 1978) may help to resolve the component directions. The method involves plotting subtracted vectors on an equal-area stereographic projection, at each stage in the progressive demagnetization process. In a demagnetization interval characterized by two overlapping demagnetization spectra, the subtracted vectors will plot on a great circle between the two component directions. If three overlapping demagnetization spectra are present, the intersection of the two great circles can give an estimate of the direction associated with the component with intermediate blocking temperature, even though this component direction is not directly resolved by the demagnetization process. Orthogonal projection in this case would not resolve the intermediate component.

If two magnetization components with partially overlapping demagnetization spectra are present in the sample, the resultant vector will trace a great circle in stereographic projection as the overlapping interval is demagnetized (Halls, 1978). If one of the two components has dispersed directions, the great circles from individual samples will converge to give the direction of the undispersed component. The method is only applicable in the special case where one of the components is dispersed, as there will be no appreciable convergence of the great circles if both components have single directions. The dispersion of one of the components can occur in a number of situations. For example, the primary component in conglomerate clasts will be dispersed relative to the secondary component that postdates conglomerate deposition. In this case, the secondary component could be resolved by remagnetization circle analysis.

3.3 Statistics

When the the magnetization components for individual samples have been determined, the sample magnetization components are combined to produce a site mean direction. The site mean directions may then be combined to compute an overall mean direction.

The probability density function generally used in paleomagnetism to determine the mean and dispersion estimates is referred to as the Fisher distribution (Fisher, 1953). This distribution assumes circular symmetry about the mean direction. Each observation is considered as an independent unit vector, and the mean direction is determined by summing the Cartesian coordinates of each of the vectors. A resultant vector **R** of the distribution can be calculated as the vector sum of the vectors. The concentration parameter k for N vectors is calculated by

$$k = (N - 1)/(N - R).$$

For $k > 10$, the radius of the 95% circle of confidence, in which the true mean should lie at the 95% confidence level, can be calculated by

$$\alpha_{95} \sim 140^\circ/(kN)^{1/2}.$$

The use of Fisher statistics implies that the population of vectors has a Fisherian distribution, a condition that is often not satisfied. Although the Fisherian distribution is almost universally used as the basis for statistical treatment in paleomagnetism, the Bingham distribution is preferred by some workers as it is better suited to the bimodal (normal and reversed polarity) and azimuthal asymmetry which is inherent in paleomagnetic data (Onstott, 1980). When using Fisher statistics, each vector should be equally independent and it is therefore important to maintain the sample/site hierarchy.

3.4 Field Tests for the Age of Magnetization

As rocks and sediments can become magnetized at any time in their history, field tests that constrain the age of the magnetization are critical to any paleomagnetic study. The *fold test* involves the determination of dispersion of paleomagnetic vectors during unfolding of the structure, and therefore allows determination of the age of magnetization relative to the age of folding. The age of folding can usually be determined by stratigraphic constraints. The fold test is best suited to individual folds, although it can be adapted for regional use on tilted beds that are not part of a single structure. The test involves analysis of the dispersion of paleomagnetic vectors during progressive unfolding of the structure, by rotating the limbs of the structure to the horizontal about the strikes of the individual limbs. In cases where the structure has plunge, the plunge should be rotated to the horizontal prior to the rotation of the limbs about the strike. McElhinny (1964) used the ratio of the concentration parameter (k) before and after tilt correction to determine whether the fold test was positive (i.e., whether the magnetization was prefolding). Critical values of k_2/k_1 (k_1 and k_2 are values of k before and after tilt correction of the structure, respectively) are listed by McElhinny (1964) for vector populations of 2–20. In the last ten years, it has become increasingly apparent that magnetization components are sometimes coeval with deformation/folding. Statistical criteria for analysis of the dispersion of paleomagnetic vectors that are more rigorous and better suited to the progressive unfolding of structures have recently been introduced (e.g., McFadden, 1990).

The second category of field test is the *conglomerate test*, which can be applied to conglomerates, breccias, or agglomerates. If the magnetization component directions from the clasts are randomized, the magnetization component predates the formation of the clasts. The same magnetization component recognized in an adjacent *in situ* outcrop of the same lithology can then be considered to have age very similar to the stratigraphic age of outcrop. The test for randomness (Watson, 1956) can be applied to the vectors derived from the clasts.

The third category of field test is the *baked contact test* which can be applied to igneous intrusions and their thermally metamorphosed contact with the country rock (Wilson, 1962). If the igneous intrusion and the baked country rock carry the same magnetization component direction, and the country rock outside the aureole carries a different (preceding) magnetization, then this is considered to constitute a test for a "primary" TRM in the igneous body.

3.5 Identification of Carriers of Natural Remanence

Magnetic mineral phases tend to have characteristic modes of occurrence, and therefore, the identification of magnetic remanence carriers provides clues to the age of magnetization components. Recent advances in instrumentation have allowed complete magnetic hysteresis loops to be measured even on weakly magnetized sediments. For example, the alternating-gradient force magnetometer (μMAG) can be used to determine values of saturation magnetization (M_s), saturation remanence (M_{rs}), and coercive force (H_c) even on weakly magnetized limestones, giving good constraints on the magnetic mineral content of the sample. High-temperature experiments designed to determine Curie (Néel) temperatures, and low-temperature experiments designed to monitor transitions in magnetic behavior (Verwey and Morin transitions), are important means of determining the magnetic mineral content of rocks and sediments. However, in order to determine the carriers of natural remanent magnetization (NRM), the characteristics of the NRM itself must be determined as fully as possible.

As described above, progressive demagnetization is the standard procedure for determining the component content of the NRM. Thermal and alternating-field demagnetization characteristics indicate the blocking temperature and coercivity spectra, respectively, of NRM components.

Blocking temperature is a function of composition and grain size of the magnetic carrier. For single-domain (SD) grains of constant composition, the blocking temperature will be a function of grain size. Because of the large grain size of hematite at the SD/MD (single-domain to multidomain) threshold (~2 mm), the SD range for hematite covers most of the remanence-carrying grain sizes encountered in sediments. Therefore, in the case of hematite, blocking temperature is a useful guide to grain size. For titanomagnetites [$x\mathrm{Fe_2TiO_4}\cdot(1 - x)\mathrm{Fe_3O_4}$], the blocking temperature is a function of composition and grain size and shape. As x decreases, the Curie point increases almost linearly, and as blocking temperature cannot exceed the Curie point, the available blocking temperature range will increase as x decreases. For SD titanomagnetite grains of uniform shape and composition, the blocking temperature will be a function of grain volume/shape. The blocking temperature range in itself is not generally diagnostic of magnetic mineralogy, other than the fact that the maximum blocking temperature cannot exceed the Curie (Néel) point of the mineral composition. In the case of minerals with relatively low Curie (Néel) points, such as geothite (~120 °C), the blocking-temperature range is rather diagnostic because it is so restricted. Mineralogical changes as a result of heating during thermal demagnetization can obscure the blocking-temperature spectrum. For example, maghemite alters to hematite at about 350 °C, and iron sulfides will oxidize in air to form magnetite at about 300 °C.

The coercivities of remanence of hematite and geothite are typically about two and three orders of magnitude greater, respectively, than that of SD magnetite. For this reason, hematite and goethite magnetizations are very resistant to alternating-field (AF) demagnetization. For titanomagnetites, coercive force increases with x and with decreasing grain size. The grain-size effect will dominate in natural samples with fairly uniform composition.

4. APPLICATIONS OF PALEOMAGNETIC STUDIES

4.1 Paleomagnetic Poles and Continental Drift

The focus of the first modern paleomagnetic studies, carried out in the 1950s, was the documentation of continental drift. Deviating paleomagnetic data from European and North American rocks of the same age were shown to reflect the development of the North Atlantic Ocean since the formation of the rocks. As a result of the dipolar nature of the main time-averaged geomagnetic field, the paleolatitude of the rock body can be determined from the paleomagnetic inclination and the coordinates of the sampling site. It is customary to express a paleomagnetic direction recorded at a particular location as a virtual geomagnetic pole (VGP), determined assuming an axial geocentric dipole. Such paleomagnetic poles are a standard means of expressing paleomagnetic data, and the time-space distribution of paleomagnetic poles for a particular crustal block or continent is known as an apparent polar wander path (APWP). The APWP for a particular continent is an expression of the movement of the crustal block relative to the rotation axis of the Earth. True polar wander, the movement of the rotation axis relative to the entire lithosphere, is generally considered to be insignificant compared to apparent polar wander. Apparent polar wander paths are a principal means of reconstructing the relative position of crustal blocks in the geologic past and are continually updated by addition of new data. For a recent compilation of the APWPs for the major continents and the constraints that these data place on paleogeographic reconstructions, see Van der Voo (1993).

4.2 Tectonic Rotations in Mountain Belts

Paleomagnetic data have proved useful in documenting rotations of small-scale crustal blocks during deformation in mountain belts. One of the aims of geological studies in fold and thrust belts is the "palinspastic" reconstruction of the deformed belt by field mapping and seismic reflection studies of the subsurface structure. Such reconstructions require knowledge of the relative rotation of thrust sheets in the deformed belt. Paleomagnetic component directions that predate the deformation can be used to monitor these rotations. In zones of oblique convergence, large-scale vertical-axis rotations of thrust sheets relative to the adjacent stable continent have been recorded (e.g., Channell *et al.*, 1990). Clockwise and anticlockwise rotations are usually associated with dextral and sinistral shear, respectively, although the fault patterns that accommodate the observed rotations have not been well documented.

4.3 Magnetic Polarity Stratigraphy

Because of the dipole nature of the geomagnetic field, the recording of ancient polarity reversals in rocks and sediments provides a unique means for global correlation. No other type of event (isotopic, paleontologic, lithologic, or geochemical) recorded in rocks or sediments provides the same potential for precise global correlation. As our knowledge of the history of geomagnetic polarity reversal has improved, the geomagnetic polarity time scale (GPTS) has become central to our determination of geologic time. The template for the GPTS for the last 150 Ma is based on oceanic magnetic anomalies, which are produced by alternating zones of normal and reversed magnetized ocean crust formed at the mid-oceanic ridges. One of the roles of magnetic polarity stratigraphy is to correlate the GPTS to paleontologic events, radiometric ages, and isotopic events. Almost any stratigraphic study benefits from magnetostratigraphic data on account of the potential for precise global correlation, and for correlation to the GPTS and current geologic timescales (see King and Channell, 1991).

4.4 Secular Variation Magnetic Stratigraphy in Lake Sediments

"Magnetic stratigraphy" usually refers to magnetic polarity stratigraphy. Short-term "secular" variations of the geomagnetic field are generally considered to average out over intervals of about 10^4 years. Directional changes associated with secular variation provide a potential means of correlation in recent sediments. Whereas conventional polarity stratigraphy depends on the recognition of directional changes of remanent magnetization of about 180°, secular variations rarely exceed a few tens of degrees. Hence, sampling and measurement techniques in secular variation studies are particularly critical. The technique has been largely restricted to lake sediments because of their high sedimentation rates and relative lack of bioturbation. Most marine sediments are too bioturbated and have sedimentation rates that are too low for a high-fidelity record of secular variation. As secular variation is largely due to the nondipole components of the geomagnetic field, secular variations are not globally synchronous but can be used for local/regional correlation.

4.5 Rock Magnetic Stratigraphy

Rock magnetic stratigraphy involves the utilization of nondirectional magnetic properties of sediments as a means of correlation. The first rock magnetic parameter to be widely used for correlation was magnetic susceptibility. Magnetic susceptibility essentially monitors the concentration of ferrimagnetic minerals. The parameter provides a rapid means of sediment correlation, and is now a standard technique for correlation of lake and deep-sea sediment cores. The field of rock magnetic stratigraphy has broadened considerably in the last ten years with the use of magnetic parameters that are sensitive not only to magnetic mineral concentration but also to grain size and mineralogy. These techniques have been applied to the study of environmental changes in lake catchments by monitoring changes in magnetic mineralogy of the sediments (Thompson and Oldfield, 1986). Similar methods have been applied to marine sediments to study changes in sediment provenance and clastic/biogenic fluxes, which can often be related to paleoclimatic change (see King and Channell, 1991).

4.6 Insights into Geomagnetic Field Behavior from Secular Variation, Excursions, Polarity Transitions, Paleointensities, and Polarity Bias

Globally distributed secular-variation records provide a means of determining the history of drift of centers of secular variation, and hence establish the longevity of the westward drift observed in historical observatory records. The time-space configuration of the nondipole field revealed by detailed secular-variation-records will provide insights into geomagnetic field-generation mechanisms.

Brief departures of the geomagnetic field from the full polarity state (geomagnetic excursions) appear to have been recorded in the present normal-polarity interval (the Brunhes chron) at numerous locations, but many recorded excursions have not been documented at multiple sites and therefore cannot be firmly established as geomagnetic phenomena. The short duration ($\sim 10^3$–10^4 years) of geomagnetic excursions leads to difficulty in recording these phenomena in rock or sediment sections, as the rock/sediment record is usually episodic on these time scales. In addition, age control is often insufficient to establish unequivocal correlation for Brunhes excursions recorded at different locations. The best-documented excursion is from sediments at Mono Lake (California) where the excursion (dated at 27–29 ka B.P.) is followed by four successive recurrences of the excursion wave form (Lund *et al.*, 1988). An additional excursion at about 120 ka B.P. (Blake event) has been recorded at several locations in the Mediterranean region and in Atlantic cores. These and other records imply that the geomagnetic field exhibits a complete spectrum of behavior from high-amplitude secular variation to full polarity reversals. Brief polarity reversals that never stabilize, where the dipole field immediately reverts back to its initial polarity, have also been recorded. Detailed high-fidelity records of the process of polarity reversal, recorded in sediments with high accumulation rates, have tended to indicate that the dipole configuration of the field is maintained during the reversal process (Clement, 1992), and that the paleomagnetic pole follows preferred longitudinal paths during the reversal process.

Absolute geomagnetic paleointensity determinations in volcanic rocks generally utilize a method based on the so-called Thellier technique (Thellier and Thellier, 1959). The majority of the determinations are for the last few thousand years, and display considerable scatter due to short-term (secular) variation in geomagnetic intensity, nonrigorous experimental techniques, and use of unsuitable archaeological and volcanic materials. As volcanic eruption is discontinuous, the volcanic record of paleointensity is discontinuous. Each determination requires independent age control, and the uncertainties in these ages inhibit correlation to other paleointensity determinations.

More continuous paleointensity records for the last 100–140 ka have been obtained from deep-sea sediment cores (see Tauxe, 1993). These relative paleointensity records are obtained by normalizing the remanence (NRM) intensity using anhysteretic remanence (ARM), isothermal remanence (IRM), or susceptibility to compensate for variations in concentration of magnetic minerals. Good correlation of paleointensity records from widely spaced locations indicates that these data record the time-averaged dipole moment of the geomagnetic field, effectively averaging out secular variation. The scope of sediment paleointensity studies has been expanded back to 3.9 Ma B.P. by recent records from the equatorial Pacific (Valet and Meynadier, 1993). Several intriguing features appear in these records. Paleointensity lows in the Brunhes chron appear to correlate to short "polarity events" or excursions. Reversals tend to coincide with lows in the virtual axial dipole moments (VADMs), with abrupt increases immediately after reversal. The longer the duration of the polarity chron, the greater the abrupt increase in VADM at its base; and each polarity chron appears to be characterized by sawtooth decay of VADM leading up to the next reversal. This has led to the suggestion that the reversal frequency, and the occurrence of "polarity events" or excursions, are ultimately controlled by the intensity of the geomagnetic dipole moment, which also controls the stability of a particular polarity state.

As our record of polarity reversal history improves and is pushed back in time, we will be able to ascertain whether there is any long-term pattern or polarity bias in the record that may signify changes in convection in the Earth's core. Our present record is fairly complete for the last 150 Ma; however, this is a small fraction of the entire history of polarity reversal, which goes back at least 3000 Ma.

GLOSSARY

Anhysteretic Remanent Magnetization (ARM): The magnetic remanence produced in the laboratory by smooth decay of a strong alternating field in the presence of a weak direct field.

Apparent Polar Wander Path: The time-space distribution of paleomagnetic poles for a particular crustal block.

Blocking Temperature: The critical temperature above which the magnetic moment of a mineral grain becomes randomized by thermal energy. The relaxation time of the grain magnetization may increase from a few seconds to millions of years as the grain passes through its blocking temperature.

Blocking Volume: The threshold volume above which a grain can carry remanence at room temperature. Below this volume, the grain is superparamagnetic and the blocking temperature is below room temperature.

Coercivity of Remanence: The back field required to reduce a remanent magnetization to zero, after saturation.

Curie (Néel) Temperature: Temperature above which saturation magnetization is zero. It is a characteristic of a particular magnetic mineral.

Demagnetization: The process of progressive demagnetization of a magnetic remanence (using alternating fields, thermal, or chemical techniques). Progressive demagnetization is used as a means of resolving the component content of the remanence. The process is successful where the components have nonuniform responses to one of the demagnetization techniques.

Isothermal Remanent Magnetization (IRM): The magnetic remanence produced in the laboratory by temporary application of a strong direct (dc) field.

Magnetic Domains: Uniformily magnetized regions in a magnetic grain.

Magnetic Susceptibility: Measure of degree of magnetization of a substance in a weak direct field, equal to M/H (ratio of induced magnetization to applied field).

Multidomain Grain: A grain which contains more than one magnetic domain.

Natural Remanent Magnetization (NRM): The naturally occurring remanent magnetization of a sediment, rock, or mineral.

Normal Polarity: Polarity of the geomagnetic dipole field as of today.

Paleomagnetic Pole or Virtual Geomagnetic Pole (VGP): The pole calculated from a magnetization direction recorded at a particular location assuming that the geomagnetic field responsible for the magnetization direction was a geocentric dipole.

Reversed Polarity: Polarity of the geomagnetic dipole field opposite to that of today.

Secular Variations: Geologically short-term geomagnetic field variations, which are generally believed to average out in $\sim 10^5$ years.

Virtual Axial Dipole Moment (VADM): The magnetic moment of an axial geocentric dipole calculated from paleointensities observed at a particular location.

Works Cited

Channell, J. E. T., Oldow, J. S., Catalano, R., D'Argenio, B. (1990), "Paleomagnetically Determined Rotations in the Western Sicilian Fold and Thrust Belt," *Tectonics* **9**, 641–660.

Clement, B. M. (1992), "Evidence for Dipolar Fields during the Cobb Mountain Geomagnetic Polarity Reversals," *Nature* **358**, 405–407.

Collinson, D. W. (1983), *Methods in Rock Magnetism and Palaeomagnetism*, London: Chapman and Hall, 503 pp.

Dunlop, D. J. (1979), "On the Use of Zijderveld Vector Diagrams in Multicomponent Paleomagnetic Samples," *Phys. Earth Planet. Sci. Lett.* **20**, 12–24.

Fisher, R. A. (1953), "Dispersion on a Sphere," *Proc. R. Soc. London A* **217**, 295–305.

Halls, H. C. (1978), "The Use of Convergent Remagnetization Circles in Palaeomagnetism," *Phys. Earth Planet. Interiors* **16**, 1–11.

Hoffman, K. A., Day, R. (1978), "Separation of Multicomponent NRM: A General Method," *Earth Planet. Sci. Lett.* **40**, 433–438.

King, J. W., Channell, J. E. T. (1991), "Sedimentary Magnetism, Environmental Magnetism, and Magnetostratigraphy," *Rev. Geophys.* Suppl. (April), 358–370.

Kirschvink, J. L. (1980), "The Least-Squares Line and Plane and the Analysis of Paleomagnetic Data," *Geophys. J. R. Astr. Soc.* **62**, 699–718.

Lund, S., Liddicoat, J. C., Lajoie, K. R., Henyey, T. L., Robinson, S. W. (1988), "Paleomagnetic Evidence for Long Term (10^4 years) Memory and Periodic Behavior in the Earth's Core Dynamo Process," *Geophys. Res. Lett.* **15**, 1101–1104.

McElhinny, M. W. (1964), "Statistical Significance of the Fold Test in Paleomagnetism," *Geophys. J. R. Astron. Soc.* **8**, 338–340.

McFadden, P. L. (1990), "A New Fold Test for Paleomagnetic Studies," *Geophys. J. Int.* **103**, 163–169.

Onstott, T. C. (1980), "Application of the Bingham Distribution Function in Paleomagnetic Studies," *J. Geophys. Res.* **85**, 1500–1510.

Runcorn, S. K. (1956), "Paleomagnetic Comparisons Between Europe and North America," *Geol. Assoc. Can. Proc.* **9**, 77–85.

Tauxe, L. (1993), "Sedimentary Records of Relative Paleointensity of the Geomagnetic Field: Theory and Practice," *Rev. Geophys.* **31**, 319–354.

Thellier, E., Thellier, O. (1959), "Sur l'Intensité du Champ Magnétique Terrestre dans le Passé Historique et Géologique," *Ann. Geophys.* **15**, 285–376.

Thompson, R., Oldfield, F. (1986), *Environmental Magnetism*, London: Allen and Unwin, 227 pp.

Valet, J.-P., Meynadier, L. (1993), "Geomagnetic Field Intensity and Reversals during the Past Four Million Years," *Nature* **366**, 234–238.

Van der Voo, R. (1993), *Paleomagnetism of the Atlantic, Tethys and Iapetus Oceans*, Cambridge, U.K.: Cambridge University Press, 411 pp.

Watson, G. S. (1956), "A Test for Randomness of Directions," *Mon. Not. R. Astron. Soc.* **7**, 160–161.

Wilson, R. L. (1962), "The Paleomagnetism of Baked Contact Rocks and Reversals of the Earth's Magnetic Field," *Geophys. J. R. Astron. Soc.* **7**, 194–202.

Zijderveld, J. D. A. (1967), "A.C. Demagnetization of Rocks, Analysis of Results," in: D. W. Collinson, K. M. Creer, S. K. Runcorn (Eds.), *Methods in Paleomagnetism*, Amsterdam: Elsevier, 254–286.

Further Reading

Butler, R. F. (1992), *Paleomagnetism*, Boston: Blackwell Scientific Publications, 319 pp. A recent text dealing with basic principles of paleomagnetism.

Collinson, D. W. (1983), *Methods in Rock Magnetism and Palaeomagnetism*, London: Chapman and

Hall, 503 pp. A useful text, particularly for laboratory methods and instrumentation.

O'Reilly, W. (1984), *Rock and Mineral Magnetism*, Glasgow, Scotland: Blackie, 220 pp. Basic principles of rock magnetism.

Tarling, D. H. (1983), *Palaeomagnetism, Principles and Applications in Geology, Geophysics and Archaeology*, London: Chapman and Hall. A general text on paleomagnetism.

Thompson, R., Oldfield, F. (1986). *Environmental Magnetism*, London: Allen and Unwin, 227 pp. A text dealing with environmental magnetism, the use of magnetic properties of rocks and sediments as a monitor of environmental change, also some discussion of rock magnetic stratigraphy and secular variation stratigraphy.

Van der Voo, R. (1993), *Paleomagnetism of the Atlantic, Tethys and Iapetus Oceans*, Cambridge, U.K.: Cambridge University Press, 411 pp. A recent text dealing with the application of paleomagnetism to paleogeography and tectonics, including a useful database of paleomagnetic poles.

PARAMAGNETISM

YAKO YAFET, *AT&T Bell Laboratories (retired), Murray Hill, New Jersey, U.S.A.*

	Introduction	101
1.	**The Magnetic Moment**	102
1.1	Orbital Moment	102
1.2	Some Definitions and Units	103
1.3	Spin Moment	103
1.4	The Moment in a Magnetic Field	104
2.	**Atomic Paramagnetism**	104
2.1	Atomic States	104
2.2	Hund's Rules	105
2.3	Spin–Orbit Interaction	105
2.4	Landé g Factor	106
2.5	Magnetic Susceptibility	107
2.6	Saturation	108
3.	**Paramagnetism in Solids**	109
4.	**Ionic Solids**	110
4.1	Quenching of the Orbital Moment	111
4.2	Kramers's Theorem	112
4.3	Jahn–Teller Effect	112
4.4	g and g'	115
5.	**Conduction-Electron Paramagnetism**	116
5.1	Pauli Susceptibility	116
5.2	Measurement of χ_P	117
5.3	Localized Moments and Kondo Effect	117
5.4	Nonlocal Susceptibility and RKKY Interaction	118
5.5	Spin-Density Waves	118
5.6	Large g Factors in Semiconductors	118
6.	**Applications**	118
6.1	Magnetic Cooling	118
6.2	The Maser	119
6.3	Tunable Spin-Flip Raman Laser	119
7.	**Summary**	119
8.	**Acknowledgments**	120
	Glossary	120
	Works Cited	121
	Further Reading	121

INTRODUCTION

Paramagnetism may be defined as the study of the properties of matter that are related to the existence of free magnetic moments. A primary manifestation of paramagnetism is a positive magnetic susceptibility, but there are many phenomena that are connected with the existence of free moments and we shall examine some of them in this article.

Michael Faraday was the first person to study systematically the magnetic properties of matter. This work was published in three volumes in London from 1839 to 1855 under the title *Experimental Researches in Electricity*, and it provides a fascinating glimpse into Faraday's great mind. Having subjected all kinds of substances (including rarefied gases and such common items as dry mutton and fresh beef) to the action of a powerful electromagnet (Faraday, 1846a), he concluded, "All matter appears to be subject to the magnetic force as universally as it is to the gravitating, the electric and [other] forces." (Faraday, 1846b). He observed that *all* substances respond to a magnetic field in one of two ways: They orient their long dimension either across the "lines of force," or alongside them. He called the first kind "diamagnetic" (see DIAMAGNETISM) and kept the term "magnetic" for the second. "Magnetic" was then a general term for substances that are attracted to the lodestone and to magnetic metals such as iron. Diamagnetic materials are repelled by a magnet. The word "paramagnetic" was coined in 1850 from the Greek prefix "para," which means alongside (as in parallel), to describe paramagnetic behavior. "Ferromagnetic" was used by W. Thomson (Lord Kelvin), also in 1850, in the expression "feebly ferromagnetic," which might have meant "paramagnetic." A clear distinction between "ferromagnetic" and "paramag-

3-527-28135-5/95/$5.00 + .50

netic" as they are used today was made by P. Curie in 1895.

The magnetic properties of matter are due primarily to the electrons. Nuclear magnetism is interesting in its own right, but it is of order 10^{-6} smaller than electronic magnetism and will not be considered here. Diamagnetism is a universal property of matter that follows from Lenz's law (see ELECTRODYNAMICS, CLASSICAL): All electrons contribute a diamagnetic moment that, omitting considerations of anisotropy, points in the direction opposite to that of the magnetic field. Paramagnetism on the other hand occurs only in materials that have spontaneous, randomly oriented moments in their ground state (with the exception of Van Vleck paramagnetism, which is discussed in Sec. 2.5). These moments tend to align themselves parallel to an applied field, but thermal agitation, or in the case of metals the exclusion principle, opposes that tendency and the net result is partial alignment along the field. This is paramagnetic behavior, which again may be subject to anisotropies.

Static paramagnetism, as all equilibrium magnetic phenomena, is quantum mechanical in origin. This follows from the Bohr–Van Leeuwen theorem (Van Vleck, 1932), which shows, in essence, that in classical systems all dependence of the Hamiltonian function on an applied magnetic field can be eliminated without affecting the values of thermodynamic averages. Our treatment will therefore be quantum mechanical, but nevertheless thinking along classical lines is still useful in providing physical insight. It is interesting that 30 years before quantum mechanics, Langevin (1905), by postulating the existence of magnetic moments, was able to derive the temperature dependence of paramagnetic susceptibilities that had been observed ten years earlier by P. Curie (Curie, 1905).

In this article, the paramagnetic properties of free atoms, of ionic salts containing metallic elements having partially filled shells, and of metals are considered. Some applications of paramagnetic substances, such as magnetic cooling, the maser, and the tunable spin-flip Raman laser, are also mentioned.

The basic concept in the theory of magnetism is that of the magnetic moment. In the next section, we define this and show that it is given by the derivative of the Hamiltonian with respect to the applied magnetic field. This property is used in Sec. 2.5 to calculate the thermal average of the moment per unit volume, and the susceptibility.

1. THE MAGNETIC MOMENT

The magnetic moment $\boldsymbol{\mu}$ of an electron is comprised of two parts, the orbital moment $\boldsymbol{\mu}_l$ and the spin moment $\boldsymbol{\mu}_s$. These two quantities will be defined, and the total moment $\boldsymbol{\mu} = \boldsymbol{\mu}_l + \boldsymbol{\mu}_s$ will be shown to satisfy the relation

$$\mu_\alpha = -\frac{\partial \mathcal{H}}{\partial H_\alpha}, \tag{1}$$

where $\mathcal{H} = T + V$ is the Hamiltonian function, the sum of the kinetic and potential energies, and α denotes a Cartesian component of the applied field $\mathbf{H}$ and of $\boldsymbol{\mu}$. Unless otherwise specified, the magnetic field will be taken in the z direction. The moments of individual electrons i are additive so that in many-electron systems, Eq. (1) is valid for the total moment, $\boldsymbol{\mu} = \Sigma_i \boldsymbol{\mu}_i$. To prove Eq. (1) in a one-electron system, the dependences of $\mathcal{H}$ on the orbital and spin moments will be considered separately.

1.1 Orbital Moment

The definition of the magnetic moment of a moving electron of charge $-e$ (note that e is positive) with respect to a point of origin is

$$\boldsymbol{\mu}_l = -(e/2c)\,\mathbf{r} \times \dot{\mathbf{r}}, \tag{2}$$

where c is the velocity of light $\mathbf{r}$ the position, and $\dot{\mathbf{r}}$ the velocity. The effect of an applied field $\mathbf{H}$ on the electron's motion is obtained by adding a vector-potential term (Van Vleck, 1932, Ch. I) to the momentum operator $\mathbf{p}$ in the Hamiltonian, thus:

$$\mathcal{H} \rightarrow (1/2m)[\mathbf{p} + e\mathbf{A}/c]^2 + V, \tag{3}$$

where $\mathbf{H}$ is derived from the vector potential $\mathbf{A}$,

$$\mathbf{H} = \nabla \times \mathbf{A}. \tag{4}$$

It is easily shown that the velocity is given by

$$\dot{\mathbf{r}} = (1/m)[\mathbf{p} + e\mathbf{A}/c], \tag{5}$$

and substituting for $\dot{\mathbf{r}}$ in Eq. (2) gives

$$\boldsymbol{\mu}_l = -(e/2mc)\,\mathbf{r} \times [\mathbf{p} + e\mathbf{A}/c]. \tag{6}$$

In atomic systems, it is convenient to use the symmetric gauge

$$\mathbf{A} = \tfrac{1}{2}\mathbf{H} \times \mathbf{r}, \tag{7}$$

and substitution of Eq. (7) into Eq. (3) followed by differentiation with respect to H gives, for a field in the z direction,

$$-\frac{\partial \mathcal{H}}{\partial H} = -\frac{e}{2mc}\left[(xp_y - yp_x) + \frac{eH}{2c}(x^2 + y^2)\right]. \tag{8}$$

The right-hand (r.h.) side of Eq. (8) is identical with the z component of $\boldsymbol{\mu}_l$ given by Eq. (6), and this proves Eq. (1) as far as the dependence on the orbital moment is concerned. The second term on the r.h. side of Eq. (8), proportional to H, is the induced diamagnetic moment, which always contributes a positive term to the energy. This term does not contribute to the paramagnetism of localized electrons and it will be neglected here. The first term, independent of H, is proportional to the z component of the orbital angular momentum. For non-s states, it represents a spontaneous, paramagnetic moment that, if the potential V is spherically symmetric, is free to take different orientations in space and hence to lower the magnetic energy.*

*This clear separation into a paramagnetic, H-independent moment and a diamagnetic moment linear in H is meaningful only when the electron's wave function is localized in space, as it is in atoms and molecules. The wave functions of electrons in energy bands (Bloch functions) extend over the whole volume of the sample, and the coordinates x and y can take on arbitrarily large values. The contributions of the two terms to the energy are then comparable, giving rise to the orbital diamagnetism of Bloch electrons, and it is simpler to deal directly with the complete Hamiltonian [Eq. (3)].

1.2 Some Definitions and Units

Before introducing the spin moment, it is convenient to give some definitions and establish a notation. We define the dimensionless angular momentum $\mathbf{l} = \mathbf{r} \times \mathbf{p}/\hbar$, where $\hbar$ is Planck's constant divided by 2π. The eigenvalues of the components l_α are integers ranging from $-l$ to $+l$ where l is the orbital quantum number. In the absence of the diamagnetic term, the orbital moment reduces to

$$\boldsymbol{\mu}_l = -(e\hbar/2mc)\,\mathbf{l}. \tag{9}$$

As $\mathbf{l}$ is dimensionless, it is natural to take as unit of magnetic moment the Bohr magneton $\beta = e\hbar/2mc$, which has the value 0.927×10^{-20} in centimeter-gram-second (cgs) units. The magnetomechanical ratio γ is the algebraic ratio of the magnetic moment to the angular momentum. For orbital motion, Eq. (9) gives the value

$$\gamma_{\text{orb}} = -e/2mc. \tag{10}$$

The magnetomechanical factor g' is defined as the absolute value of the ratio of magnetic moment in units of the Bohr magneton to the angular momentum in units of $\hbar$. For orbital motion, $g'_{\text{orb}} = 1$.

1.3 Spin Moment

The existence of an intrinsic electron spin angular momentum was postulated by Uhlenbeck and Goudsmit (1925) in order to explain the doubling of the optical spectral lines in the alkali atoms. The components of the spin vector $\hbar\mathbf{s}$ obey the angular momentum commutation rules, and in order to give a multiplicity of 2 to the optical spectra, the components of $\mathbf{s}$ must have magnitude $\frac{1}{2}$. To be consistent with the anomalous Zeeman effect (see below), the magnetomechanical ratio of the spin moment must have the value $2\gamma_{\text{orb}}$ with γ_{orb} given by Eq. (10), and hence the magnetomechanical factor of the spin moment must be twice that of the orbital moment, or $g'_s = 2$ (the more precise value, including radiative corrections is $g'_s = 2.0023$). With $g_s \equiv g'_s \cong 2$, the spin magnetic moment becomes $\boldsymbol{\mu}_s = -2\beta\mathbf{s}$ and the interaction of the total magnetic moment with an external field $\mathbf{H}$, called the Zeeman interaction, becomes finally

$$\mathcal{H}_Z = -(\boldsymbol{\mu}_l + \boldsymbol{\mu}_s)\cdot\mathbf{H} = \beta(\mathbf{l} + 2\mathbf{s})\cdot\mathbf{H}. \tag{11}$$

Differentiation with respect to H gives Eq. (1).

1.4 The Moment in a Magnetic Field

In his investigations of the influence of a magnetic field on light propagating through glass, Faraday (1846c) had discovered that the field produces a rotation of the plane of polarization of plane-polarized light (the Faraday effect). Convinced that light and magnetism were related, he set out to find the effect of a magnetic field on the spectrum of a light source but was unable to reach a conclusion (as were other workers who tried in the following 50 years), primarily because the effect is very small and requires sensitive means of detection. Zeeman (1896) succeeded in showing that a magnetic field produces a definite broadening of the spectrum but he was not able to resolve the line splitting into what is now the familiar "Zeeman effect." From the different polarization properties of the center and edges of the broadened spectrum, he concluded that the line was split into three components. This observation was the first indication of the space quantization of angular momentum, even though it was still explainable by Lorentz's classical theory of a harmonically bound electron. Had Zeeman been able to resolve the broadened line, he would have found that it was not split into three (the normal Zeeman effect), but into a larger number of components, and this was no longer explicable by Lorentz's theory. The anomalous Zeeman effect, as this is called, depends in detail on the initial and final states of the emitting atom, and requires for its explanation that the spin moment have a magnetomechanical factor $g_s' \equiv g_s = 2$, and be coupled (by the spin–orbit interaction) to the orbital moment.

The Zeeman splittings determine the magnitude of the induced paramagnetic moment and of the susceptibility. Experimentally, the latter is determined by measuring the force exerted on the induced moment by an inhomogeneous field. In the Gouy method, this is done as follows. The specimen, in the form of a long thin rod of cross section A, is suspended between the vertical pole faces of an electromagnet with the field H at the center of the pole faces pointing horizontally. The specimen is midway between the poles but vertically off the magnet axis so that it is partly in the fringing field region where H has a gradient in the vertical direction x. If H is not too large ($H \lesssim 1$ T), the induced moment is proportional to H and the magnetization, which is the moment per unit volume M, can be written $M = \chi H$ where χ is independent of H. The force df_x exerted on the volume element $dv = Adx$ is

$$df_x = dm\,\frac{dH}{dx}. \tag{12}$$

where $dm = \chi H A dx$ is the moment induced in the volume element. Substituting for dm and integrating Eq. (12) gives

$$f_x = \tfrac{1}{2}\chi A(H_{\max}^2 - H_{\min}^2), \tag{13}$$

where $H_{\max}$ and $H_{\min}$ are the maximum and minimum values, respectively, of the field inside the specimen. Knowledge of A, $H_{\max}$, and $H_{\min}$ combined with a measurement of f_x determines the susceptibility.

The calculation of the paramagnetic susceptibility of free atoms is described in the next section. Knowledge of atomic paramagnetism provides the basis for analyzing the paramagnetic properties of solids.

2. ATOMIC PARAMAGNETISM

At a temperature T, the states that contribute to the paramagnetism are the ground state and those states that are within a thermal energy kT above it, k being Boltzmann's constant. A brief discussion of the structure of the atomic states is thus in order.

2.1 Atomic States

The electronic wave functions of many-electron atoms cannot be known exactly, and in order to classify and calculate them one makes use of a scheme based on perturbation theory and known as the central-field approximation (see Electronic Structure of Atoms and Molecules). This is as follows.

The total atomic Hamiltonian is split into three parts in order of decreasing magnitude,

$$\mathcal{H} = \mathcal{H}_1 + \mathcal{H}_2 + \mathcal{H}_{so}. \tag{14}$$

The term $\mathcal{H}_1$, the major part, is the sum over all electrons of the kinetic energy and of a suitably chosen spherical potential. $\mathcal{H}_1$ can be solved exactly in terms of one-electron orbitals (nl) of energy ϵ_{nl} where n and l are the principal and orbital quantum numbers (q.n.) respectively. The electrons are assigned the lowest energy levels in accordance with the Pauli principle, and they successively fill the (nl) shells, which can accommodate $2(2l + 1)$ electrons each. Filled shells have zero orbital and spin momenta and hence zero magnetic moment. Only atoms with partly filled shells are paramagnetic. The set of q.n.s (nl) that specify an eigenstate of $\mathcal{H}_1$ is called a configuration, and for paramagnetism it is sufficient to specify only the number of electrons in open shells. It is customary to use spectroscopic notation for the l values; thus l = 0, 1, 2, 3, . . . is written s, p, d, f, . . ., etc. For example singly ionized iron, which has six electrons in the $3d$ and one in the $4s$ shells, is written $3d^6 4s^1$. Being the sum of one-electron energies ϵ_{nl} that are independent of "orientation" (i.e., azimuthal q.n.), the energy of a configuration that represents a partially filled shell is highly degenerate.

The effect of $\mathcal{H}_2$, which includes the electron–electron interaction minus an average one-electron potential that is included in $\mathcal{H}_1$, is to remove partially the orientational degeneracy of a configuration, thus splitting the configuration into "terms." Because $\mathcal{H}_2 < \mathcal{H}_1$, the mixing of configurations usually can be neglected. The total orbital ($\mathbf{L} = \Sigma_i \mathbf{l}_i$) and spin ($\mathbf{S} = \Sigma_i \mathbf{s}_i$) angular momenta, summed over all the electrons, commute with $\mathcal{H}_1$ and $\mathcal{H}_2$ and hence can be used to label the term states, each of which is associated with particular L and S values and has degeneracy $(2L + 1)(2S + 1)$. This classification scheme, which is based on the smallness of the spin–orbit interaction $\mathcal{H}_{so}$ [the last member of Eq. (14)], is known as Russell–Saunders coupling. Lastly, when $\mathcal{H}_{so}$ is taken into account, the degeneracy within each term is partially removed: Since $\mathcal{H}_{so}$, as well as $\mathcal{H}_1$ and $\mathcal{H}_2$, commutes with the total angular momentum $\mathbf{J} = \mathbf{L} + \mathbf{S}$, $\mathbf{J}$ is a good q.n. and the main effect of $\mathcal{H}_{so}$ is to split each term into multiplets (LSJ) with J running from $|L - S|$ to $L + S$. Each multiplet still has degeneracy $2J + 1$, and this is ultimately removed by an applied magnetic field that splits the multiplet into $2J + 1$ equidistant Zeeman levels. It is this splitting that causes the anomalous Zeeman effect. Because $\mathcal{H}_{so}$ does not commute with $\mathbf{L}$ or $\mathbf{S}$, the q.n.s L and S are now only approximate, and a multiplet (LSJ) may contain small admixtures of other multiplets ($L'S'J$) with $L' = L \pm 1$ and correspondingly $S' = S \mp 1$.

Multiplets having $J \neq 0$ possess a spontaneous moment while multiplets having $J = 0$ give rise only to an induced paramagnetic moment. The next two subsections describe how the lowest-energy term and the lowest multiplet belonging to that term are found.

2.2 Hund's Rules

Hund noted the empirical fact that the lowest-energy term of a configuration satisfies the following two conditions:

1. It has the largest S; and
2. if several terms have that same S value, it has the largest L among them.

Physically, these rules follow from the repulsive character of the electron–electron interaction: Concerning **1**, the Pauli principle keeps electrons of parallel spins apart from one another and hence lowers their mutual interaction energy. The larger the number of electrons with parallel spins, the larger the energy reduction. And concerning **2**, a large value of L allows the electrons to distribute themselves among a large number of orientations and thus to lower their energy even further.

Once the lowest term of a configuration is identified through Hund's rules, the ground state, which is the lowest multiplet level of that term, is determined by the spin–orbit interaction.

2.3 Spin–Orbit Interaction

An electron moving in an electric field $\mathbf{E}$ sees a magnetic field $\mathbf{H}_{orb} = \mathbf{v} \times \mathbf{E}/c$ proportional to the velocity $\mathbf{v}$. In the spherical potential of the atom, $\mathbf{E}$ is radial; hence $\mathbf{H}_{orb}$ is parallel to the orbital moment and there results a coupling between the spin and orbital moment of each electron i, given by

$$h_{so}(i) = \zeta_{nl} \mathbf{l}_i \cdot \mathbf{s}_i, \tag{15}$$

where ζ_{nl} is the spin–orbit parameter of the

shell (nl) obtained by integration over the square of the radial part of the corresponding wave function. The total spin–orbit interaction is $\mathcal{H}_{so} = \Sigma_i h_{so}(i)$.

By making use of the angular momentum commutation relations, it is shown that $\mathcal{H}_{so}$ commutes with **J** and that as a result of the Wigner–Eckart theorem it is equivalent, within the manifold of states belonging to the term (LS), to the operator

$$\mathcal{H}_{so} \rightarrow \Lambda \mathbf{L} \cdot \mathbf{S}. \tag{16}$$

The constant Λ is given by

$$\Lambda = \pm \zeta_{nl}/2S, \tag{17}$$

where the + and − signs apply respectively to configurations with less than half-filled and more than half-filled shells. In addition to the matrix elements of $\Lambda \mathbf{L} \cdot \mathbf{S}$, $\mathcal{H}_{so}$ has matrix elements between terms that differ in their (L,S) values by $\Delta L = \pm 1$, $\Delta S = \mp 1$, but this will not concern us here as their effect is usually small.

From Eq. (16), the multiplet energies with respect to the unsplit term energy are found to be

$$E_J = \tfrac{1}{2}\Lambda[J(J+1) - L(L+1) - S(S+1)]. \tag{18}$$

The energy difference between neighboring multiplets is thus proportional to the J value of the multiplet with the largest J,

$$\Delta E_J = E_{J+1} - E_J = \Lambda(J + l), \tag{19}$$

a relation that is known as the Landé interval rule.

It follows from Eqs. (17) and (18) that the ground multiplet state of a partly filled shell is $J = |L - S|$ or $J = L + S$ according to whether the shell is respectively less than or more than half-filled. Figure 1 shows the Hund's-rule ground term and the lowest multiplet for the different configurations d^n of a d shell. A similar figure, with maximum values $L = 6$ and $S = \frac{7}{2}$, would represent the $4f^n$ configurations of the rare earths (no confusion should arise between the superscript n and the principal quantum number n).

In the particular case of a shell that is half-filled, the ground state has $L = 0$, $S = J = (2l + 1)/2$, and the magnetic moment is entirely

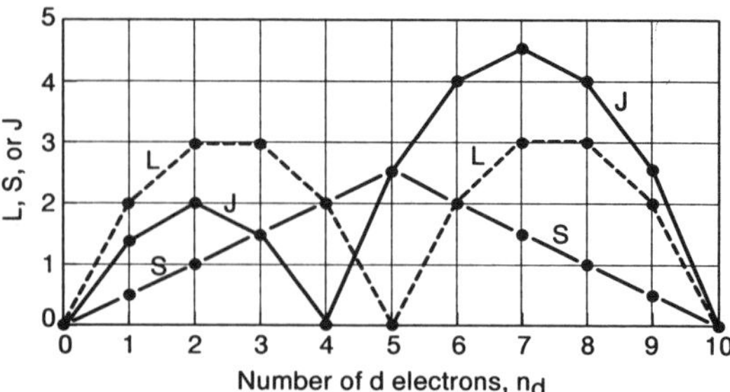

FIG. 1. Values of L, S, and J for the ground state of iron-group ions as the d shell is filled. Because of the change in sign of the spin-orbit constant Λ at $n_d = 5$, the values of J at a given L are higher for $n_d > 5$ than they are for $n_d < 5$. Only the values at integral n_d have physical meaning; the straight lines are a guide to the eye.

due to the spin with $g = 2$. A shell that is one electron short of being half-filled has a ground state $L = l$, $S = l$, and $J = 0$.

2.4 Landé *g* Factor

A magnetic field lifts the $(2J + 1)$-fold degeneracy of the multiplet states $|JJ_z\rangle$. Unlike the operator J_z, the magnetic moment $\beta(L_z + 2S_z)$ is not a constant of the motion, and in free ions it has matrix elements between multiplet states differing by $\Delta J = \pm 1$. However, the Zeeman interaction $\mathcal{H}_Z = \beta(\mathbf{L}_z + 2\mathbf{S}_z) \cdot \mathbf{H}$ is in practice much smaller than the spin–orbit interaction, and these off-diagonal matrix elements are usually unimportant. The Zeeman energies are then given by the expectation values of the magnetic moment over the $|JJ_z\rangle$ states,

$$\mathcal{H}_Z(JJ_z) = \langle JJ_z|\beta\mathbf{H} \cdot (\mathbf{L}_z + 2\mathbf{S}_z)|JJ_z\rangle \equiv g_J\beta H J_z, \tag{20}$$

where g_J is defined as the Landé g factor. The energy separation between consecutive levels is called the Zeeman splitting, equal in free atoms to $g_J\boldsymbol{\beta}H$. The fact that these expectation values are proportional to J_z follows from the Wigner–Eckart theorem, which in systems having spherical symmetry requires the matrix elements of any vector operator (such as $\boldsymbol{\mu}$) over the states $|JJ_z\rangle$ to be proportional to those of the angular momentum itself. The coefficient of proportionality here is g_J.

The physical meaning of g_J can be seen in a semiclassical picture: The orbital and spin momenta, being strongly coupled by $\mathcal{H}_{so}$, precess around one another and around the direction of their resultant **J**. The magnetic moment precesses rapidly around **J** in the same manner, and since its components normal to **J** average to zero, it is only the component of $\boldsymbol{\mu}$ along **J** that couples to an external field. The term g_J is then obtained from the expectation value of this component in the direction of **J**,

$$g_J = \frac{\langle \boldsymbol{\mu} \cdot \mathbf{J} \rangle}{J(J+1)} \frac{1}{\beta}. \tag{21}$$

According to Eq. (21), g_J is the magnetomechanical ratio g' (as defined in Sec. 2.1) for the states of the (LSJ) multiplet. Explicit calculation of $\langle \boldsymbol{\mu} \cdot \mathbf{J} \rangle$ gives

$$g_J = \tfrac{3}{2} + \frac{S(S+1) - L(L+1)}{2J(J+1)}. \tag{22}$$

The proportionality factor in the Zeeman energies, Eq. (20), is traditionally called the spectroscopic splitting factor and denoted by g. From Eq. (20), g too is equal to g_J. The equality $g = g'$ is particular to spherical systems. For ions in solids, it will be seen in Sec. 4.4 that the two are in general different.

2.5 Magnetic Susceptibility

The magnetic susceptibility is defined as the derivative of the magnetization with respect to the applied magnetic field. At sufficiently low fields, the thermal average of the induced moment is linear in the field and the atomic susceptibility is a constant that can be calculated by expanding the energies of the low-lying states, n, to second order in the field. From Eq. (11),

$$E_n(H) = E_n^{(0)} - (\mu_z)_{nn}H + \sum_{n' \neq n} \frac{|(\mu_z)_{n'n}|^2}{E_n^{(0)} - E_{n'}^{(0)}} H^2, \tag{23}$$

where $E_n^{(0)}$ is the unperturbed energy and $\mu_z = -\beta(l_z + 2s_z)$. From Eq. (1) and the definition of the thermodynamic free energy, it follows that the paramagnetic moment is given by the negative of the derivative of the free energy with respect to H. The moments of atoms obey Boltzmann statistics, and hence for an assembly of N atoms per unit volume the magnetization $M(H,T)$ is given by

$$M(H,T) = -N \frac{\partial F(H,T)}{\partial H}, \tag{24}$$

where $F(H,T)$, the free energy per atom, is defined through

$$\exp\left[-\frac{F}{kT}\right] = \sum_n \exp\left[-\frac{E_n H}{kT}\right]. \tag{25}$$

The low-field ($\beta H \ll kT$) susceptibility follows from these relations and is given by

$$\begin{aligned} \chi(T) &= -N\left[\frac{\partial^2 F}{\partial H^2}\right]_{H\to 0} \\ &= N\left[\sum_n \left[\frac{(\mu_z)_{nn}^2}{kT} + 2\sum_{n'} \frac{|(\mu_z)_{n'n}|^2}{E_{n'}^{(0)} - E_n^{(0)}}\right] \right. \\ &\quad \left. \cdot \exp\left(-\frac{E_n^{(0)}}{kT}\right)\right]\left[\sum_n \exp\left(-\frac{E_n^{(0)}}{kT}\right)\right]^{-1}, \end{aligned} \tag{26}$$

where the states n and n' belong to the low-lying multiplets of the lowest term (LS).

The two terms in the numerator of Eq. (26) have different physical meanings: The first one is the contribution of a spontaneous moment that can orient itself in the external field. It exists only when n belongs to a multiplet having $J \neq 0$. The second one is a temperature-independent contribution of an induced moment known as Van Vleck paramagnetism (Van Vleck, 1932, Ch. 7). At low temperature, this term is important only when the ground multiplet has $J = 0$, but at temperatures comparable to the multiplet splitting* it makes a significant contribution even if $J \neq 0$. This is the reason for the discrepancies between the experimental and theoretical values of the effective magneton number in the cases of samarium and europium noted below (Table 1). In the induced moment, selection rules restrict the difference in the J values of the states n and n' to $\Delta J = \pm 1$.

A detailed account of the temperature variation of $\chi(T)$ is given by Van Vleck (1932, Ch. 9), and may be summarized as follows:

*Various units of energy are used in atomic physics, among them the kelvin (K), the wave number (cm^{-1}), and the electron volt (eV), which is equal to 1.6×10^{-19} J. In terms of the other units, 1 eV = 8065 cm^{-1} = 11 600 K.

Table 1. Effective magneton numbers p for trivalent rare-earth ions (near room temperature) [from Kittel (1986), p. 405].

Ion	Configuration	Basic level	p(calc.) $= g[J(J+1)]^{1/2}$	p(exp.), approximate
Ce^{3+}	$4f^1 5s^2 p^6$	$^2F_{5/2}$	2.54	2.4
Pr^{3+}	$4f^2 5s^2 p^6$	$^3H_4$	3.58	3.5
Nd^{3+}	$4f^3 5s^2 p^6$	$^4I_{9/2}$	3.62	3.5
Pm^{3+}	$4f^4 5s^2 p^6$	$^5I_4$	2.68	...
Sm^{3+}	$4f^5 5s^2 p^6$	$^6H_{5/2}$	0.84	1.5
Eu^{3+}	$4f^6 5s^2 p^6$	$^7F_0$	0	3.4
Gd^{3+}	$4f^7 5s^2 p^6$	$^8S_{7/2}$	7.94	8.0
Tb^{3+}	$4f^8 5s^2 p^6$	$^7F_6$	9.72	9.5
Dy^{3+}	$4f^9 5s^2 p^6$	$^6H_{15/2}$	10.63	10.6
Ho^{3+}	$4f^{10} 5s^2 p^6$	$^5I_8$	10.60	10.4
Er^{3+}	$4f^{11} 5s^2 p^6$	$^4I_{15/2}$	9.59	9.5
Tm^{3+}	$4f^{12} 5s^2 p^6$	$^3H_6$	7.57	7.3
Yb^{3+}	$4f^{13} 5s^2 p^6$	$^2F_{7/2}$	4.54	4.5

1. At low temperatures, $kT \ll \Delta E_J$ where ΔE_J is the energy separation of the ground multiplet J from the next higher one, only the states of the ground multiplet are populated. Usually $J \neq 0$, in which case the second term of Eq. (26), which reduces to a constant $N\alpha$, is negligible. The matrix elements $(\mu_z)_{nn}$ are equal to $-g_J J_z \beta H$, and summation over J_z gives the well-known Curie susceptibility,

$$\chi = N g_J^2 J(J+1)\,\beta^2/3kT. \tag{27}$$

The factor $p = g_J\sqrt{J(J+1)}$ is called the effective magneton number and provides a convenient way to tabulate the experimental data. Table I shows the values of p for trivalent rare-earth (4f) ions, deduced from room-temperature measurements, and for comparison the Hund's-rule values p(calc.). With the exception of Sm and Eu, the agreement between the two is very good even though the measurements were not made on free ions but on solid hydrated sulfates. As will be seen in Sec. 4.1, the reason for the good agreement is that rare-earth ions are little affected by their environment.

2. The discrepancy for Eu^{3+} arises because the Hund's-rule multiplet has $J = 0$ (the f shell of Eu^{3+} is one electron short of being half-filled). However the next level, $J = 1$, is separated in energy only by $\Delta E_J = 330$ K, which is comparable to room temperature. It turns out that the overall multiplet width is $21\Delta E_J$ and to obtain the susceptibility accurately, the Van Vleck part of Eq. (26) must be summed over all the multiplets. The result, which obviously depends on T, agrees with the measured value of p. The discrepancy in the case of Sm^{3+} has a similar origin: Even though the ground multiplet has $J = 5/2$, higher multiplets must be taken into account. These two cases illustrate the behavior of $\chi(T)$ when T is comparable with ΔE_J.

3. Finally when T is large compared to the overall width of the ground term multiplets, the expression of $\chi(T)$ become simple again. The Van Vleck terms cancel out in Eq. (26) since the Boltzmann factors effectively can be taken equal to unity, and L and S are decoupled and free to assume random orientations with respect to one other. The susceptibility is then the sum of independent orbital and spin contributions,

$$\chi = \frac{N\beta^2}{3kT}[L(L+1) + 4S(S+1)]. \tag{28}$$

2.6 Saturation

When the magnetic field is so large and the temperature so small that the ground-state Zeeman splitting is comparable to the temperature ($g_J\beta H \gtrsim kT$), the low-field expansion is not useful and the free energy must be calculated exactly. Replacing g_J by g in order to lighten the notation, Eq. (25) becomes

$$\exp\left(-\frac{F}{kT}\right) = \sum_{J_z=-J}^{+J} \exp\left(\frac{gJ_z\beta H}{kT}\right) \tag{29}$$

and the magnetization follows as

$$M(H,T) = NgJ\beta B_J(gJ\beta H/kT), \tag{30}$$

where in terms of $x = gJ\beta H/kT$, the function

$$B_J(x) = \frac{2J+1}{2J}\coth\frac{2J+1}{2J}x - \frac{1}{2J}\coth\frac{x}{2J} \tag{31}$$

is the Brillouin function, which approaches unity as $x \to \infty$. Thus, the magnetization saturates when the Zeeman energy is large compared to kT.

The classical limit of the Brillouin function is the Langevin function $L(x) = \coth x - 1/x$, which is obtained by letting the value of J approach infinity while $gJ\beta$ is kept constant and equal to the value of the classical moment μ. For large J, $B_J(x)$ approaches $L(x)$.

Figure 2 shows measurements of the magnetization in the saturation region, $x \gg 1$, for three ions having J values of $\frac{7}{2}$, $\frac{5}{2}$, and $\frac{3}{2}$ respectively. Gd^{3+} and Fe^{3+} have L = 0 in the ground state, and the moment is that of the spin alone, with $g = 2$. The ion Cr^{3+} has the configuration d^3 and as seen in Fig. 1, the free ion has $L = 3$, $S = \frac{3}{2}$, and $J = \frac{3}{2}$, resulting in a g value of $\frac{2}{5}$. However, the orbital moment is quenched in the solid (Sec. 4.1), and the magnetization is due entirely to the spin. The theoretical curve for Cr^{3+} in Fig. 2 is calculated with $g = 2$. Had $g = \frac{2}{5}$ been used instead, the curve would have strongly disagreed with the experimental points, thus providing confirmation of the quenching of the orbital moment.

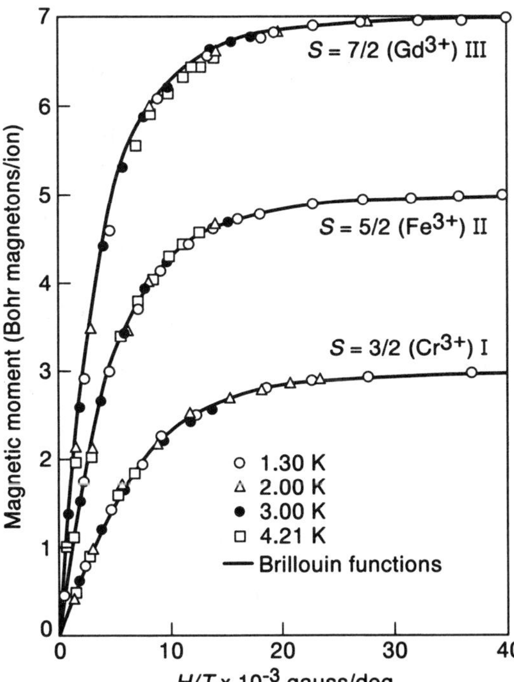

FIG. 2. Plot of magnetic moment vs H/T for (I) potassium chromium alum, (II) iron ammonium alum, and (III) gadolinium sulfate hexahydrate, taken at temperatures $T = 1.3$, 2.0, 3.0, and 4.21 K. The abscissa is proportional to the argument x of the Brillouin function in Eq. (31). Near-complete saturation is achieved at $T = 1.3$ K and $H = 5$ T (after Henry, 1952).

3. PARAMAGNETISM IN SOLIDS

The type of solid that forms when atoms come together depends on the valence electrons. Elements with open s or p shells form either metals, semiconductors (covalent solids), or molecular solids. In the latter two cases, strong bonds form between neighboring atoms, the electron spins pair off, the orbital moments are quenched as described below, and the solid is diamagnetic. Among paramagnetic solids are metals and salts of metallic ions having partially filled shells. The conduction electrons in metals and in doped semiconductors show a weak paramagnetism (Pauli paramagnetism) that is due predominantly to the spin. In addition, they have an orbital diamagnetism with susceptibility χ_L named after Landau (see DIAMAGNETISM) that in the free-electron model only is equal to minus one-third of the Pauli susceptibility. In real metals, χ_L depends on the band structure and the total susceptibility, including the filled-core diamagnetism and the Van Vleck paramagnetism of any partially filled bands, can be either positive or negative. Pauli paramagnetism is discussed in Sec. 5.

Strong, atomiclike paramagnetism occurs in the salts of the elements that have partially filled d shells (of the $3d$, $4d$, and $5d$ series) or f shells (the $4f$ rare earths and the $5f$ actinides). Most intensively studied have been the iron group ($3d$ series) and the $4f$ group, and these are the subject of this section. At high temperatures, the susceptibility obeys a Curie–Weiss law

$$\chi = N\,C/(T - \theta), \tag{32}$$

where the Curie constant C [see Eq. (27)] depends on the value of the effective Bohr magneton of the ion in the solid environment. In the rare-earth ions, according to Table 1, C is given by the full atomic moment, but Table 2 shows that in the ions of the iron group, C is given essentially by the spin-only moment, $2\beta S$: The orbital momentum is quenched (Sec. 4.1). The constant θ accounts for the exchange interactions between moments. In the mean-field approximation, these are represented by a molecular field proportional to the magnetization, $H_{mol} = \lambda M$, and the value that results for θ is $\theta = \lambda C$.

4. IONIC SOLIDS

The study of paramagnetic ions in solids is of interest because it reveals the changes that they undergo in a crystalline environment. These changes in turn provide information about the electronic state of the whole solid. Aside from its own interest, such a study provides opportunities for designing materials with properties suitable for applications, such as maser or laser materials. The susceptibility is a macroscopic, global quantity, which cannot by itself give much information about these effects. In the decades following World War II, the paramagnetism of solids was extensively studied by electron spin resonance (see ELECTRON PARAMAGNETIC RESONANCE), which measures directly the Zeeman splitting of the paramagnetic ground state. If the ions were free, this would of course be given by the Landé g factor, but in a solid, even the smallest interaction of the ion with the environment is often sufficient to change its Zeeman splittings radically.

The theoretical treatment of these interactions is known as "crystal-field" or "ligand-field" theory, and it is the subject of several excellent books. Discussions of the "spin Hamiltonian" and of the effective spin concept may be found in these. Only a brief account that emphasizes the physics is given here.

The ionic species under investigation is usually at the center of a cage of "ligands," which are negative ions such as F^- or O^{2-} ions belonging to a molecule such as H_2O. The ligands are often at the corners of an octahedron or a cube, and hence the symmetry of the central ion is cubic. Usually a small trigonal or tetragonal distortion is also present. Symmetry is very important in determining the g value in a spin-resonance [electron paramagnetic resonance (EPR)] experiment. The central ion together with the ligands forms a "complex" or complex ion that is a molecular entity that maintains its identity in a water solution. The effect of the charged ligands on the paramagnetic ion is represented by a local electric field (the crystal field) that has the site symmetry of the ion.

The notion of an electric field should not be taken too literally for it serves mostly as a convenient way to parametrize the interaction of the ion with the environment, including for instance the effect of the covalency with the ligands. The crystal-field interaction is written $\mathcal{H}_{cf}$, and it is added to the Hamiltonian [Eq. (14)]. The manner in and degree to which the states of the free ion are modified depend on the magnitude of $\mathcal{H}_{cf}$ relative to $\mathcal{H}_2$ and $\mathcal{H}_{so}$, and three cases can be distinguished: $\mathcal{H}_{so} > \mathcal{H}_{cf}$; $\mathcal{H}_2 > \mathcal{H}_{cf} > \mathcal{H}_{so}$; and $\mathcal{H}_{cf} > \mathcal{H}_2$.

Table 2. Effective magneton numbers for iron-group ions [from Kittel (1986), p. 406].

Ion	Configuration	Basic level	p(calc.) $= g[J(J+1)]^{1/2}$	p(calc.) $= 2[S(S+1)]^{1/2}$	p(exp)[a]
Ti^{3+}, V^{4+}	$3d^1$	$^2D_{3/2}$	1.55	1.73	1.8
V^{3+}	$3d^2$	$^3F_2$	1.63	2.83	2.8
Cr^{3+}, V^{2+}	$3d^3$	$^4F_{3/2}$	0.77	3.87	3.8
Mn^{3+}, Cr^{2+}	$3d^4$	$^5D_0$	0	4.90	4.9
Fe^{3+}, Mn^{2+}	$3d^5$	$^6S_{5/2}$	5.92	5.92	5.9
Fe^{2+}	$3d^6$	$^5D_4$	6.70	4.90	5.4
Co^{2+}	$3d^7$	$^4F_{9/2}$	6.63	3.87	4.8
Ni^{2+}	$3d^8$	$^3F_4$	5.59	2.83	3.2
Cu^{2+}	$3d^9$	$^2D_{5/2}$	3.55	1.73	1.9

[a]Representative values.

1. $\mathcal{H}_{so} > \mathcal{H}_{cf}$ is the weak-field case, which occurs when the ligands have a minimal effect. In the rare-earth ions, the $4f$ shells lie inside the filled $5s$ and $5p$ shells, and as a result the inner $4f$ electrons are kept away from the ligands. As seen in Table 3, $\mathcal{H}_{cf}$ is small, of the order of 100 cm^{-1}, and its effect is limited to splitting the degenerate states of each multiplet and mixing their wave functions according to the local symmetry.
2. $\mathcal{H}_2 > \mathcal{H}_{cf} > \mathcal{H}_{so}$ is the intermediate crystal-field case, which occurs in many salts of the $3d$ group. The spatial extent of the $3d$ orbitals is comparable to that of the $4f$ in the rare earths, but there are no outer $5s$ or $5p$ shells to screen them from the ligands. The ground state is still given by Hund's rule, but the orbital levels are split by the crystal field and the orbital moment is partly or totally quenched. The term J is not a quantum number.
3. $\mathcal{H}_{cf} > \mathcal{H}_2$ is the strong crystal-field case, which occurs in some salts of the transition groups. The perturbation by the environment is so strong that the atomic one-electron orbitals (nl) have their $(2l + 1)$-fold degeneracy lifted, and the many-electron states must be expressed in terms of these split states. Hund's rule does not necessarily hold, and in fact, in octahedral symmetry the configurations d^4–d^7 do not obey it. Such ions are said to be in the low-spin state, $S = 1$ for d^4, $S = \frac{1}{2}$ for d^5, etc. When the ligands are so close that their wave functions overlap substantially those of the paramagnetic ion, covalent bonding takes place and the magnetic electrons become delocalized into molecular orbitals that are shared with the ligands. The orbital moments and the spin–orbit coupling are then affected through "orbital reduction" factors as the d orbitals now constitute only part of the wave function. Covalency strongly enhances the crystal-field splittings, which are as high as 30 000 cm^{-1} in some cyanides.

4.1 Quenching of the Orbital Moment

The $(2L + 1)$-fold degeneracy of the orbital states $|LL_z\rangle$ of a free ion reflects the freedom of the orbital momentum to assume any orientation (classically speaking) in its spherically symmetric environment. In a crystal the internal electric field has lower symmetry and is not radial. The orbit of an electron moving classically in such a field will deviate from being planar and precess in space in some complicated way. The angular momentum will fluctuate in time and either will average to zero or, if the symmetry is high, will have a nonzero average value only for certain orbits. This picture corresponds quantum mechanically to a mixing of the states having different values of L_z. The mixing of states strongly affects the expectation values of the Zeeman energies, and hence of the spectroscopic splitting factor. If the orbital degeneracy is completely lifted, the wave function must be real and the expectation value of the orbital momentum, which is a purely imaginary operator, must vanish. The orbital momentum is then completely quenched. The foregoing discussion has completely ignored the spin–orbit coupling, which must now be taken into account.

The spin–orbit coupling can be viewed as an internal magnetic field generated by the spin moment and acting on the orbital motion. Through the interaction $\zeta \mathbf{s} \cdot \mathbf{l}$, this field is coupled to the orbit of the electron *around* the nucleus, and in this way it opposes the quenching of $\mathbf{l}$. Whether quenching occurs or not depends therefore on the relative magnitudes of $\mathcal{H}_{cf}$ and $\mathcal{H}_{so}$, which are measured by the ratio of the crystal-field splitting Δ of the ground multiplet to the spin–orbit parameter ζ_{4f} or ζ_{3d}.

In the rare earth ions, Table 3 shows that $\Delta \ll \zeta_{4f}$. The orbital moment is not quenched, but the crystal field mixes the $|JJ_z\rangle$ states of the ground multiplet and splits the multiplet energy. The angular momentum J remains as an approximate quantum number for clas-

Table 3. Ranges of the energies that determine the nature of the ground electronic state of a paramagnetic rare-earth ($4f$) ion, or an iron-group ($3d$) ion in a solid environment. E_c is the ground-term splitting, ζ_{nl} is the spin–orbit parameter, and Δ is the ground-state crystal-field splitting.

	E_c (cm^{-1})	ζ_{nl} (cm^{-1})	Δ (cm^{-1})
$4f$ ions	$(5\text{–}30) \times 10^3$	600–3000	~100
$3d$ ions	$(10\text{–}40) \times 10^3$	60–800	$(10\text{–}30) \times 10^3$

sifying the levels, and the ratio of magnetic to mechanical moment, g', is still given by the Landé g_J factor. However, the spectroscopic splitting factor is no longer equal to g_J but depends on the admixture coefficients of the $|JJ_z\rangle$ states in the crystal-field states. At temperatures T smaller than Δ, the susceptibility deviates from the Curie value both because of the splitting of the states and because of the difference of g from g_J. For T large compared to the overall Δ, the susceptibility should return to the free-ion value with $p = g_J\sqrt{J(J+1)}$ as the effect of the splittings averages out. This is borne out by the experimental results in Table 1, which show that the room-temperature susceptibility of rare-earth ions is not affected by their environment.

The ions of the iron group are usually at the opposite extreme, $\Delta \gg \zeta_{3d}$. The orbital momentum is then quenched and neither J nor g_J retains any validity. There is only a slight deviation of the g factor from $g_s = 2$ that is due to an admixture of orbital moment by the spin–orbit interaction. Table 2 shows that with the exception of Co^{2+}, the measured values of p are close to the spin-only value.

Although affecting the Curie susceptibility only slightly through the deviation of g from 2, the orbital moment contributes now a Van Vleck susceptibility χ_{VV}, which arises from the splitting of the d states by the crystal field and the fact that the orbital moment has matrix elements between the occupied and unoccupied crystal-field states. Being proportional to the product of the number of occupied and number of unoccupied states, χ_{VV} tends to be largest for half-occupied shells. In ionic salts, χ_{VV} is expected to be only a very small fraction of the Curie susceptibility, but in the transition metals the d electrons become delocalized into moderately broad energy bands and the spin susceptibility χ_s is reduced by a factor of order 10^2 compared to the Curie susceptibility (see Sec. 5). The term χ_{VV} becomes then comparable with χ_s. Indirect experimental evidence and an estimate of χ_{VV} came first in Knight-shift and susceptibility measurements in the intermetallics V_3Ga and V_3Si. More direct evidence was obtained later by a measurement of the g' factor in the transition metals themselves (see Sec. 4.4).

4.2 Kramers's Theorem

There is a limit to the removal of degeneracies by a crystalline field. A free ion has degeneracy $2J + 1$, which is even if the number of electrons is odd and, vice-versa, odd if the number of electrons is even. The ion may be a complex ion. Kramers's theorem states that in the presence of an arbitrary electric field and in zero magnetic field, all states of a system having an *odd* number of electrons must still have even degeneracy. The theorem is a consequence of the time-reversal invariance of the Hamiltonian including the spin–orbit interaction. The theorem is proved by showing that when the number of electrons is odd, the operation of time reversal applied to an eigenstate produces another eigenstate, which is orthogonal to the first and hence must be independent of it. The number of states that are degenerate with each other must therefore be even.

Ions with an *even* number of electrons, by contrast, may have their *spin* degeneracy lifted by an electric field of low symmetry. This occurs for instance in the eight-electron Ni^{2+} ion, which has an orbitally nondegenerate ground state with a spin of 1. The threefold spin degeneracy is completely lifted by a rhombic crystalline field acting in conjunction with the spin-orbit coupling (Griffiths and Owen, 1952).

In solids, the ground states of $3d$ and $4f$ ions with an odd number of electrons are usually Kramers doublets.

4.3 Jahn–Teller Effect

In the previous parts of this section, it has been tacitly assumed that paramagnetic ions situated at sites of high symmetry may have a ground state that is orbitally degenerate. By contrast, if the site is without any symmetry, then barring accidental degeneracy, all the states must be orbitally nondegenerate. Jahn and Teller (1937) asked the question, is the stability of a high-symmetry environment compatible with an orbitally degenerate electronic ground state? The Jahn–Teller (J–T) theorem states that excepting the case when the paramagnetic ion is part of a linear molecule (which does not concern us here), stability and orbital degeneracy are not compatible. When the spin is included, the theorem holds for the degeneracy including spin, except for the Kramers degeneracy, which cannot be removed by purely electrical forces. We shall not consider the spin any further here except to say that sometimes the spin–orbit coupling may stabilize the symmetric configuration. A simple system where the J–T theorem applies is illustrated in the following example, taken from their paper.

A planar molecule consists of a *diamagnetic* ion A located at point P and surrounded by four identical ligands B_1, B_2, B_3, B_4 situated symmetrically at a distance a [configuration C_0 in Fig. 3(a)]. This configuration is stable. Ion A is now replaced by ion M, which in the symmetric configuration has an orbitally degenerate ground state: For instance, M has one p electron outside of closed shells, and the orbitals p_x and p_y with wave functions proportional to x and y, respectively, are degenerate as a result of the fourfold local symmetry. The J–T theorem asserts that the configuration C_0 is now unstable with respect to the distortions C_1 and C_2 of Fig. 3(b). The physical reason for the instability is simply that the two degenerate states point in mutually orthogonal directions and a distortion of the symmetric configuration such as C_1 or C_2 will lower the energy of one of the states while raising that of the other, as shown in Fig. 3(c). An elementary calculation that assigns one electron charge, $-e$, to the ligands shows that if ions B_1 and B_3 are moved outward by a distance d while B_2 and B_4 are moved inward by the same distance (configuration C_1), the additional potential seen by an electron situated at the point (x,y,z) with respect to P is, in first order in d.

$$\Delta V(x,y) = -\tfrac{9}{2}(e^2/a^4)(x^2 - y^2)d. \qquad (33)$$

Averaging Eq. (33) over the wave functions of p_x and p_y shows that the energy of p_x is lowered by

$$\Delta E(p_x) = -Vd \qquad (34)$$

where the quantity V is positive, while the energy of p_y is raised by the same amount. The important fact is that Eq. (34) is linear in the nuclear displacement d. A nuclear configuration cannot be stable if the electronic energy of a neighboring configuration depends linearly on the nuclear displacement since the elastic energy $U = \frac{1}{2}Kd^2$, which provides the restoring force for displacements away from a position of stable equilibrium, is quadratic in the displacement. Minimizing the sum of the energy shift [Eq. (34)] and U, and setting $K = M\omega^2$ where M is the nuclear mass and ω the vibration frequency, the equilibrium value of d is found to be $d_{eq} = V/M\omega^2$, and the net lowering of the energy, or stabilization energy, is obtained as the negative of

$$E_{J-T} = V^2/2M\omega^2. \qquad (35)$$

Note that the J–T theorem does not say how the system (the molecule) will respond to the instability, but only that the symmetric configuration is unstable. The simplest answer that comes to mind is that the nuclear configuration will distort in the manner suggested above until d assumes the value d_{eq} and the energy is lowered to its minimum, $-E_{J-T}$. Such a response of the system is called the "static J–T effect" as it assumes implicitly that the nuclei have infinite mass: Once at their position of minimum energy, they stay there forever. The consequence of such a response in an ionic solid would be the removal of all orbital degeneracies, and the

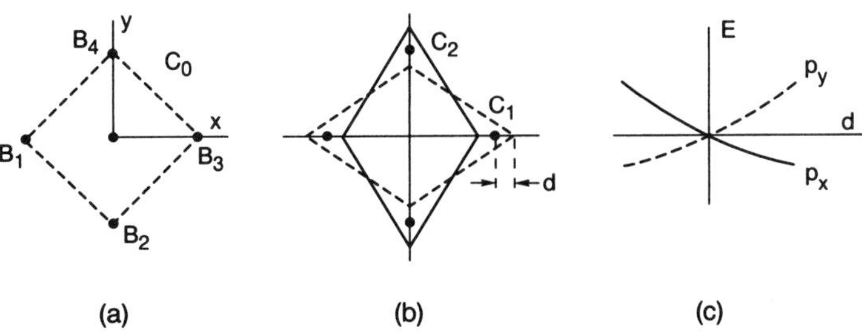

FIG. 3. The Jahn–Teller effect of a *p*-state ion in a symmetric planar molecule: (a) In the undistorted configuration C_0, the ion has fourfold symmetry and the states p_x and p_y are degenerate. The configuration C_0 is unstable against distortion. (b) Two equivalent distortions, C_1 and C_2. The symmetry of the central ion is lowered and the degeneracy is removed. (c) For small distortions, the splitting of the energies is linear in the amplitude *d* of the distortion.

lowering of the site symmetry should be observable as an anisotropic line (or g value) in an EPR experiment. In fact, the first observation of the J–T effect in a solid, made by EPR at room temperature (Bleaney and Ingram, 1950), showed a g value that was isotropic and different from $g_s = 2$.

The experiment was done on the d^9, Cu^{2+} ion at the center of an octahedral complex where, in the absence of any distortion, it would have had the twofold orbitally degenerate Γ_3 level as its ground state. A distorted octahedron, elongated along one of the [100] cubic axes, should have shown a resonance characteristic of tetragonal symmetry with $g_\parallel \neq g_\perp$. The observed isotropic g value was inconsistent with this, but it was also inconsistent with an undistorted configuration that, because of the twofold degeneracy, should have shown two resonance lines, split by the spin–orbit interaction, and having cubic but not spherical symmetry with $g_{111} \neq g_{100}$. The observed spectrum was explained by Abragam and Pryce (1950) as a manifestation of the "dynamic J–T effect," this terminology having been introduced later. The labeling of the effect as "dynamic" here means that because of their finite mass, the nuclei are able to move over an energy barrier from one distorted configuration to another under the action of the thermal vibrations of the surrounding lattice. At room temperature, these thermal excitations are sufficient to cause the activated transitions to occur at a rate that is large compared to the resonance frequency, and the three resonance lines that would have resulted from tetragonal distortions along any one of the three equivalent [100] cubic axes are motionally narrowed to a single isotropic line. Two years later (Bleaney and Bowers, 1952), the same system at 20 K showed in fact the three anisotropic lines of the "static J–T effect," indicating that at 20 K the transition rate between configurations was smaller than the EPR frequency. The same kind of distortion may thus appear as dynamic or static depending on the time resolution of the measuring apparatus as compared with the transition rate of the distortion between the different configurations.

The condition for the static J–T effect to occur is that the stabilization energy be large compared to the zero-point energy of the nuclei, $E_{J-T} \gg \hbar\omega$. This is referred to as the "strong J–T" condition. When this is not the case and $E_{J-T} \leq \hbar\omega$ ("weak J–T effect"), it is necessary to include the kinetic energy of the nuclear vibrations from the start in solving the problem, i.e., to consider the coupled electronic-vibrational system as a single dynamical system. The Hamiltonian of this system is called the vibronic Hamiltonian, and its states the vibronic states. The full dynamical nature of the J–T effect becomes apparent in the properties of these vibronic states.

It is clear that the vibronic Hamiltonian, inclusive of the J–T coupling, is invariant under the symmetry operations of the undistorted complex since these operations merely replace one distorted configuration by another. This can be seen in Fig. 3(b), where a rotation of a distorted configuration, accompanied by a rotation of the two electronic states, leaves things unchanged. It follows from this invariance that the J–T coupling *cannot lift* a degeneracy that is due to high symmetry, but only *replace* a purely electronic degeneracy by a vibronic degeneracy of the same character, i.e., the vibronic and the original electronic states must transform in the same way under the symmetry operations of the undistorted system. Physically, the electronic states and the nuclear vibrations have become strongly correlated.

Having the same twofold Γ_3 degeneracy as the unperturbed electronic states, the vibronic states should show an EPR spectrum having the same features as those of the electronic doublet described above. Such a spectrum was observed at very low temperatures, independently by Coffman (1965) and by Höchli and Estle (1967), on systems similar to those investigated by Bleaney and Ingram but having a weaker J–T coupling, $E_{J-T} \leq \hbar\omega$. Two anisotropic lines with $g_{111} \neq g_{100}$, characteristic of a Γ_3 level, were observed, but the anisotropy was smaller than that predicted by the known value of the spin–orbit constant. This difference was experimental proof of the presence of a vibronic doublet because such orbital reduction factors are predicted as a consequence of the smaller than unity overlap between the vibrational states associated with different electronic states. From the degree of the reduction, an estimate of the Jahn–Teller coupling could be obtained.

In principle, the static susceptibility should be affected too since the orbital reduction factors cause a change in the g values.

The experimental manifestations of the

dynamic J–T effect are diverse and fascinating. The interested reader (see Further Reading) may consult an excellent account by Sturge that includes physical discussions of the theory and the experimental results. More formal theoretical treatments can be found in the treatise by Abragam and Bleaney and in the review article of a key contributor to the theory, F. Ham. This section owes much to these sources.

4.4 *g* and *g′*

The magnetomechanical ratio in the z direction of a paramagnetic ion in a solid is defined as the ratio of the expectation values:

$$g' = -\frac{\langle \mu_z \rangle/\beta}{\langle J_z \rangle/\hbar}, \tag{36}$$

where μ_z and J_z are the z components of the magnetic moment and angular momentum, respectively. The negative sign reflects the sign of the electron charge so that g' as defined by Eq. (36) is positive. The spectroscopic splitting factor is a measure of the Zeeman splitting in a spin-resonance experiment, and for an effective spin of $\frac{1}{2}$ is defined by

$$g\beta H = \hbar\omega, \tag{37}$$

where ω is positive. For a free spin, $g = g_s = 2.0023$. The sign of g, which can be positive or negative, reflects the sense of precession of the moment around **H** in the solid (Blume *et al.*, 1969).

Early measurements of g and g' in the ferromagnetic 3*d* metals showed that they are not the same. Although both were found to have values close to $g_s = 2$, thus reflecting strong orbital quenching, they differed from 2 in opposite directions with $g > 2$ and $g' < 2$ (Kittel, 1949). The measured g values were close to those deduced from magnetic susceptibility measurements on the salts of the metals.

The first measurement of g' in a paramagnetic salt was made by Sucksmith (1930) on Dy_2O_3. Within experimental error, the result was $g' = g_j = \frac{4}{3}$, implying negligible orbital quenching in agreement with our remarks in Sec. 4.1.

The fact that $g \neq g'$ shows that the torque equation

$$\frac{d\langle \mathbf{J} \rangle}{dt} = \langle \boldsymbol{\mu} \rangle \times \boldsymbol{H} \tag{38}$$

does not hold in solids because it would imply $g = g'$. Thus from Eq. (38), $d\langle \mathbf{J} \rangle/dt$ is proportional to $\omega\langle \mathbf{J} \rangle$ and hence to $g\langle \mathbf{J} \rangle$, while $\langle \boldsymbol{\mu} \rangle$, from Eq. (36) is proportional to $g'\langle \mathbf{J} \rangle$.

The reason Eq. (38) does not hold is that through the interaction with the crystalline field, an exchange of angular momentum between the lattice and the moment is taking place as the latter precesses around the magnetic field (in the case of strong orbital quenching, this exchange is mediated by the spin–orbit interaction). Hence, part of the torque $\langle \boldsymbol{\mu} \rangle \times \mathbf{H}$ that is transferred by the microwave field to the angular momentum of the electron escapes into the lattice (or alternatively an additional torque is exerted by the lattice), and Eq. (38) does not hold. The situation is very similar (Kittel, 1949) to that of an electron in a lattice that is acted on by an external electric field: In free space, an electron of mass m acted on by a force **F** has an acceleration $\langle \dot{\mathbf{v}} \rangle = \mathbf{F}/m$, while in a crystal lattice, as a result of momentum transfer between the electron and the lattice, the acceleration is given by $\langle \dot{\mathbf{v}} \rangle = \mathbf{F}/m^*$, where m^* is the effective mass, here assumed isotropic. In this analogy, the g factor corresponds to m^* and g' corresponds to m.

From Eqs. (36) and (37), g' and g can be expressed as

$$g' = \frac{\langle L_z \rangle + 2\langle S_z \rangle}{\langle L_z \rangle + \langle S_z \rangle}, \tag{39}$$

$$g = \frac{\langle L_z \rangle + 2\langle S_z \rangle}{1/2}, \tag{39'}$$

where $\langle L_z \rangle$ and $\langle S_z \rangle$ are in units of $\hbar$.

In the case of 3*d* ions, strong orbital quenching gives $\langle S_z \rangle = \frac{1}{2}$ and $\langle L_z \rangle \ll \langle S_z \rangle$. Both g and g' are then close to 2, and Eqs. (39) and (39′) show that they satisfy the relation

$$g' = g/(g - 1). \tag{40}$$

This is the Kittel–Van Vleck relation, and it is in fair agreement with experiment. Equation (40) has been generalized by Blume *et al.*

(1969) to cases where the g factor can be large and of either sign as for instance in the cases of some rare-earth ions and of the current carriers in semiconductors with very small band gaps.

As mentioned in Sec. 4.1, the $3d$ metals are expected to have a Van Vleck orbital susceptibility χ_{VV} (for which $g' = 1$) that is of the same order of magnitude as the spin susceptibility χ_s for which $g' = 2$. The actual value of g' can be measured in a gyromagnetic experiment and the relative magnitudes of χ_{VV} and χ_s can thereby be inferred. Such experiments, which are very delicate in paramagnetic metals as a result of the smallness of the χ's, were performed by Hughenin *et al.* (1971) on several transition metals. In V and Cr, the measured g' is close to 1.2, indicating the dominance of the Van Vleck term in these metals, while in Pd and Pt, which are at the end of the $4d$ and $5d$ series, the values of g' are about 1.7. The spin susceptibility is strongly exchange-enhanced in Pd and Pt and χ_{VV} is smaller than in V or Cr because the d shells are nearly full. As a result, the spin susceptibility is dominant.

5. CONDUCTION-ELECTRON PARAMAGNETISM

In metallic crystals, the valence states of neighboring atoms overlap strongly and the wave functions become delocalized. Ignoring at first the difficult problem of exchange and correlation, the delocalized, or conduction, electrons are assumed to move independently in an average potential that has the periodicity of the lattice. The valence states are broadened into energy bands of "Bloch states" $\Psi_{\mathbf{k}}$, which are modulated plane waves that depend on the wave vector $\mathbf{k}$, and the conduction electrons occupy these states subject to the Pauli principle that allows a maximum of two electrons of opposite spins to occupy each Bloch state. At $T = 0$, these are filled up to a maximum energy E_F, the Fermi energy, which ranges in different metals from about 3 to 15 eV, a value considerably larger than the room-temperature energy of $kT = 0.025$ eV. (No confusion between Boltzmann's constant k and $\mathbf{k}$ should arise.)

5.1 Pauli Susceptibility

One of the early successes of the theory of metals was the explanation by Pauli (1927) of their temperature-independent paramagnetism. The Langevin theory predicts a susceptibility $\mu^2/3kT$ per atom, which at the typical atomic densities of metals results in a volume susceptibility at room temperature of order 10^{-3}–10^{-4}. Instead, the observed susceptibilities were temperature independent and of order 10^{-6}. The explanation, well known now, is that the Pauli principle prevents the electrons that are well below the Fermi surface from reorienting their spin. Only those electrons having energies within a range kT of the Fermi surface are free to do so, contributing each a susceptibility $\mu^2/3kT$. For N conduction electrons per unit volume, the paramagnetic susceptibility becomes

$$\chi_P^{(0)} \sim N\frac{kT}{E_F}\frac{\mu^2}{3kT} \sim N\frac{\mu^2}{3E_F}. \tag{41}$$

This is independent of T and smaller than the Curie value by a factor 10^2, as experimentally observed. For a free-electron gas, the actual calculated value is $\chi_P^{(0)} = 3N\beta^2/2E_F$, and for a general band structure having a density of states of one spin direction $n(E)$ per unit volume and unit energy range, the calculated value at zero temperature is

$$\chi_P^{(0)} = \tfrac{1}{2}g^2\beta^2 n(E_F), \tag{42}$$

where $g = 2$ for a free spin and $n(E)$ is evaluated at the Fermi energy. The orbital moment contributes to the Pauli susceptibility only through the spin–orbit coupling $\mathcal{H}_{so}$. With the exception of points in $\mathbf{k}$ space at which Bloch states may be degenerate, the expectation value of $\mathcal{H}_{so}$ over a Bloch state vanishes, and only admixtures of states from other bands by $\mathcal{H}_{so}$ may give an orbital contribution to the g factor. These are usually small and shift the g value by a quantity Δg that ranges from 10^{-3} in Na to 0.2 in the heavy metals (Yafet, 1963). The reason for the smallness is that the interband energies at constant $\mathbf{k}$ are usually large compared to $\mathcal{H}_{so}$. The g values at the Fermi surface, $g_{\mathbf{k}}$, depend on $\mathbf{k}$, but in a spin-resonance experiment the EPR line is narrowed by motional and exchange averaging to a width that is below the inhomogeneous width associated with the distribution of g values.

When one goes beyond the independent-electron approximation and includes the Coulomb interactions, the susceptibility can be calculated only approximately. Just as exchange interactions in free atoms are responsible for Hund's rule of maximum spin, here exchange interactions (somewhat tempered by correlations) between the conduction electrons tend to maximize the total number of electron pairs with parallel spins when a magnetic field is present. As a result, the susceptibility is enhanced over the noninteracting value $\chi_P^{(0)}$, and in the simplest approximation, the interacting susceptibility χ_P is given by

$$\chi_P = \chi_P^{(0)}/(1 - \nu). \tag{43}$$

The constant ν is the product of an effective exchange energy per electron and the density of states at the Fermi level normalized to one electron. If ν reaches the value 1, the system becomes ferromagnetically unstable. Paramagnetic metals have $\nu < 1$.

Since $\chi_P^{(0)}$ can be obtained from the calculated band structure, it is of interest to measure χ_P directly in order to compare the actual enhancement factor with the theoretical one.

5.2 Measurement of χ_P

The contribution of χ_P to the total χ was isolated by Schumacher and Slichter (1956) through the use of magnetic resonance. The spin paramagnetism χ_P is related to the area under the resonance absorption signal plotted as a function of the frequency. Precise measurements of the absolute value of such areas are very difficult to make. The problem was circumvented by observing the resonances of both electrons and nuclei on the same sample, changing only the magnetic field and leaving the frequency and all other parameters unchanged. The ratio of the areas under the two absorption curves, which involves only a relative measurement that can be done accurately, gives the ratio between the electron and the nuclear susceptibilities, and since the latter is simply the nuclear Curie susceptibility, χ_P may be deduced.

5.3 Localized Moments and Kondo Effect

When dissolved in nonmagnetic metals (Cu, Zn, Au), many elements of the 3d transition series show a Curie susceptibility, which indicates that they have a free localized moment. This is not totally unexpected as Hund's rule indicates that if the d electrons remain on the atom (as they must in order to keep it neutral) then they might well possess a moment. Friedel (1958) and Anderson (1961) developed the idea of the virtual bound state according to which the d electrons can transfer back and forth between the localized state and the conduction states of the host metal. This transfer broadens the localized level, and the higher the rate at which it occurs (which depends on the density of states), the larger the broadening. If the level becomes too broad, the two spin directions become equally populated in order to minimize the total band energy and the impurity atom becomes nonmagnetic. If the level remains fairly narrow, then Hund's rule wins, the energy of the unoccupied spin state rises above the Fermi level, and the impurity atom acquires a local moment.

The moment is coupled by exchange to the conduction electrons, and the exchange integral can be shown to be negative, which favors an antiparallel arrangement of the local spin with respect to the spins of the conduction electrons. Anomalous low-temperature behavior in the resistivity of normal metals with small amounts of magnetic impurities, and its interpretation by Kondo, led to the discovery that this exchange coupling causes the formation of a collective (with the conduction electrons) bound state that is a spin singlet. This Kondo state, as it is known, is at an energy below the Fermi level so that at zero temperature it is fully occupied. Being a singlet, it possesses no magnetic moment, and thus the local moment is compensated collectively by the spin moments of the conduction electrons. The binding energy depends exponentially on the reciprocal of the exchange integral and may thus vary over a broad range. The magnetic susceptibility of the impurity ion, which is Curie-type, $\chi = C/T$ at high temperature, saturates to a constant value $\chi = C/T_K$ as T approaches zero. The term T_K is a measure of the singlet binding energy.

5.4 Nonlocal Susceptibility and RKKY Interaction

It might be expected that if it were possible to apply a local magnetic field extending only within the space occupied by one atom in a metal, the polarization of the conduction electrons would extend up to some distance away because of the delocalized nature of their orbitals. Such a magnetic field is effectively realized by the exchange coupling between a local moment and the conduction electrons. The response of the electron gas to such a field is conveniently analyzed in terms of the wave-vector–dependent susceptibility: The local field is decomposed into Fourier components $H_{\mathbf{q}} \exp(i\mathbf{q}\cdot\mathbf{r})$. Each $H_{\mathbf{q}}$ induces a magnetic-moment density of the same wave vector, proportional to $M_{\mathbf{q}}^{(0)} = \chi_q^{(0)}$ where for convenience the noninteracting susceptibility is used. The dependence of $\chi_q^{(0)}$ on $\mathbf{q}$ is primarily determined by the shape of the Fermi surface. At wave vectors $\mathbf{q} = \mathbf{q}_c$ that span *locally parallel* areas of the Fermi surface (FS), $\chi_{q_c}^{(0)}$ has singularities that depend in detail on the local shape of the FS (Roth, 1966).

These singularities determine the dependence of the magnetization $\mathbf{M}(\mathbf{r})$ at large distances r. It is found that

1. $\mathbf{M}(\mathbf{r})$ is oscillatory in space, with a wave vector related to the spanning vector appropriate to the direction of $\mathbf{r}$; and
2. the amplitude of the oscillations decreases with distance as $1/r^3$ provided the FS has no unusual features at the spanning points.

For a gas of free electrons, the FS is a sphere and the oscillations are isotropic with a wave vector that is twice the Fermi wave vector, $2k_F$ (Ruderman and Kittel, 1954).

The long-range polarization $\mathbf{M}(\mathbf{r})$ causes an exchange interaction to occur between any two localized spin moments, such as those of the nuclear spins of the constituent atoms, or those of any localized electronic moments. This interaction is named the Ruderman–Kittel–Kasuya–Yosida (RKKY) interaction. It is responsible for the magnetic helical orderings of the core 4f moments in the heavy rare-earth metals (see Further Reading).

5.5 Spin-Density Waves

The interacting wave-vector–dependent susceptibilities have an enhancement factor similar to that of the Pauli susceptibility in Eq. (43), i.e.,

$$\chi_q = \chi_{\mathbf{q}}^{(0)}/(1 - \Omega v_c \chi_{\mathbf{q}}^{(0)}). \qquad (44)$$

If it should happen that the denominator vanishes at some $\mathbf{q}_n$, then χ_{q_n} diverges, which means that the electron gas acquires a spontaneous long-range magnetization having spatial dependence $\exp(i\mathbf{q}_n \cdot \mathbf{r})$ as the spins order themselves in an antiferromagnetic helix. Such a configuration occurs as the ground state in metallic Cr below the Néel temperature $T_N = 311$ K (see Further Reading).

5.6 Large g Factors in Semiconductors

In the semiconductor InSb, the energy gap between conduction and valence band occurs at the center of the zone, $\mathbf{k} = 0$, and the magnitude of the gap is small compared to the spin–orbit splitting of the valence band. The magnitude of the g factor of the electrons at the bottom of the conduction band was predicted (Roth *et al.*, 1959) to be very large, $g \approx -50$, even though the wave function at this value of $\mathbf{k}$ is practically of pure s atomic character. Such large g factors were shown by Herring (1963) to arise from interatomic orbital currents circulating over large numbers of atomic sites by virtue of the smallness of the effective mass. Even larger g factors have been observed in the semimetal Bi.

6. APPLICATIONS

6.1 Magnetic Cooling

An early application of paramagnetism was the attainment of very low temperatures by the method of adiabatic demagnetization, proposed independently by Debye (1926) and Giauque (1927). This method makes use of the fact that a paramagnetic salt at a low temperature T has only two kinds of degrees of freedom, namely those of the vibrational motion of the nuclei that constitute the lattice, and those of the orientational dependence of the electronic magnetic moments of the ions (the nuclear moments are left out here). In the absence of a magnetic field, the entropy of the disordered, noninteracting moments is very large as each moment contributes its full orientational entropy $k \ln(2J + 1)$. A magnetic field H such that $g\beta H/kT \gg 1$ lowers this entropy dramatically as the moments become oriented. By contrast, the entropy of the lattice is very small at small T, only of order $k(T/\Theta)^3$ per degree of freedom, where the Debye temperature Θ is of order 400 K. The principle of the method is as follows. Starting from an initial $T_i \sim 1$ K that is reached by pumping on liquid He, the salt is magnetized isothermally so that its temperature remains at T_i. The initial field H_i is of order 1–5 T. The thermal contact with the outside is then broken, and the magnetic field is adiabatically and reversibly reduced to a very low value, $H_f \sim 10^{-3}H_i$.

From Eq. (29), the free energy and hence the entropy of the magnetic moment (or "spin") system depends only on the ratio $g\beta H/k_BT$. Hence, at constant entropy, the final temperature of the spin system is $T_f^{(s)} = (H_f/H_i)T_i^{(s)}$. The final temperature of the salt is attained after the lattice comes into thermal equilibrium with the spins by transfer of entropy between them. But since the entropy of the lattice is so much smaller than that of the spins, this transfer is negligible and the final temperature is simply $T_f \approx (H_f/H_i)T_i$. The limit of attainable values of T_f is determined by the temperature T_c at which the weak interactions (dipolar and exchange) between the spins lead to the formation of a cooperative magnetic state. Hence the choice of a suitable salt is guided by a compromise between minimizing these interactions and maintaining a sufficient density of magnetic ions. This is done by choosing suitable hydrated salts, and temperatures of order 10^{-3} K are thus reached. Lower temperatures, of order 10^{-6} K, are obtained by demagnetizing the nuclei of Cu.

6.2 The Maser

The maser (acronym for Microwave Amplification by Stimulated Emission of Radiation) is basically a microwave amplifier with a very low noise figure. It can be implemented by different physical systems, among them a paramagnetic salt having three low-lying spin levels A, B, C in order of increasing energy (Bloembergen, 1956). With the frequency differences denoted by ω_{AB}, ω_{BC}, and ω_{AC}, a microwave field of frequency ω_{AC} (the "pump" frequency) is used to induce transitions A → C, thereby equalizing the populations of these two levels. The signal that one wishes to amplify has frequency ω_{BC}. Application of this signal to the device induces a net rate of transitions C → B as long as the population of level C is kept higher than that of level B by the pump. The emitted radiation is in phase with and proportional to the amplitude of the applied signal, and hence the device acts as an amplifier. If the material is chosen such that the relaxation rates between the levels are favorable, the device can also operate as an oscillator. The noise figure is low because the maser must of necessity operate at a low temperature. The maser is thus particularly useful for amplifying weak signals that come from outer space and have themselves a low noise figure.

6.3 Tunable Spin-Flip Raman Laser

An infrared laser in the frequency range just below the energy gap ($E_G = 0.23$ eV) that separates the conduction and valence bands in the semiconductor InSb has been realized by inducing electronic Raman transitions in the conduction band (Patel and Shaw, 1971). A large magnetic field is used to quantize the valence- and conduction-band states into sharp Landau levels, and transitions between the two spin states of the lowest Landau level in the conduction band are induced by scattering incident light at a frequency ω_0 slightly below the gap energy.

The scattering cross section for spin reversal is large for two reasons:

1. The spin–orbit coupling is large compared to the band gap, which results in the Raman matrix element for spin reversal being of the same order as or larger than the matrix element for orbital transitions, which is proportional to the magnetic field (Yafet, 1966); and
2. the photon energy $h\omega_0$ is chosen to be nearly in resonance with the gap energy, which enhances the cross section further.

The laser operates at the frequency of the scattered light, $\omega_1 = \omega_0 - |g|\beta H/\hbar$, and since $|g|$ is of the order of 50, the frequency ω_1 can be tuned in a limited range by varying H. This flexibility allows the laser to be used as an analytical tool.

7. SUMMARY

A material is paramagnetic if it contains magnetic moments that are free to orient in an external magnetic field. Electron paramagnetism has been an invaluable probe, mainly through the use of paramagnetic resonance, for exploring the magnetic and electronic structure of solids. By this method, detailed studies of the low-lying states of paramagnetic ions in solids and of the relaxation processes of the magnetic energy have been made; subtle phenomena such as the Jahn–Teller effect have been elucidated.

The thermodynamic properties of spin systems are special in that the total number of states, and hence the entropy, are finite. In addition, the entropy can easily be manipulated by an external magnetic field. This flexibility has been used to reach very low temperatures, and also to achieve very high temperatures (or population inversion) by selective absorption of radiation. Spin systems are even used as model systems in the teaching of thermodynamics.

8. ACKNOWLEDGMENTS

The author is indebted to Prof. Sir Roger Elliott for making available his unpublished lecture notes on magnetism, and to Prof. Michael Sturge for a critical reading of the manuscript.

GLOSSARY

Boltzmann Factor: The population factor of a state in an ensemble that represents thermal equilibrium.

Complex: A compound ion that consists of a central metal ion surrounded by a cage of ligands and is sufficiently stable to retain its identity in a water solution.

Crystal Field: Electric field set up by the ligands that reflects the symmetry of the complex.

Debye Temperature: The temperature above which all acoustic modes of the lattice vibrations are appreciably excited and below which modes begin to freeze out.

Effective Mass: The apparent inertial mass, m^*, of conduction electrons in metals and semiconductors. Under the effect of an external force F, the acceleration of a wave packet in a spherical band is $\dot{v} = F/m^*$.

Effective Spin: If a paramagnetic ion in a solid has a level with degeneracy $2S' + 1$, the effective spin of that level is S'. The effective spin is defined in a more general way in the spin-Hamiltonian formalism.

Energy Band: A set of itinerant one-electron states in metals or semiconductors whose energies vary continuously within an energy range as the wave vector of the states ranges over the reduced Brillouin zone.

EPR: Electron paramagnetic resonance.

Fermi Surface: The surface of constant energy in **k** space that corresponds to the Fermi energy.

***g* Factor:** The proportionality coefficient that measures the splitting of a degenerate level by a magnetic field. In free atoms, g is equal to the Landé g factor.

Knight Shift: The shift of the nuclear resonance frequency in a metal, produced by the polarization of the conduction electrons in the applied magnetic field.

Ligand: A molecule or ion attached to the central ion of a complex (or complex ion).

Line of Force: A line formed by connecting neighboring points in space that are situated along the direction of the local magnetic field with respect to one another.

Magnetic Susceptibility: A measure of the response of a substance to an applied magnetic field. More precisely, the derivative of the magnetization with respect to the applied field.

Magnetization: The magnetic moment per unit volume of a piece of matter.

Paramagnet: A substance that contains elementary magnetic moments that tend to orient themselves along an applied field but that do not exhibit a macroscopic moment or any kind of magnetic ordering in the absence of a field.

Polarization of Light: The specification of the direction of the electric field of a light wave. In plane-polarized light, this direction stays the same at all points of space, at all times, while in circularly polarized light the direction of the field at each point in space rotates uniformly with time.

Van Vleck Susceptibility: The paramagnetic susceptibility that arises from off-diagonal matrix elements of the orbital magnetic moment.

Vibronic State: In a system that undergoes the Jahn–Teller effect, a state in which the electronic state of the paramagnetic ion and the state of the surrounding nuclei are strongly coupled. The total wave function is no longer a simple product of an electronic function and a function of the nuclear displacements, but a sum of such products.

Zeeman Energy: The energy of a spontaneous magnetic moment in a magnetic field.

Zeeman Splitting: The energy splitting of a multiplet level or of any degeneracy connected with the existence of a spontaneous moment by an external magnetic field.

Works Cited

Abragam, A., Pryce, M. H. L. (1950), *Proc. Phys. Soc. A* **63**, 409–411.

Anderson, P. W. (1961), *Phys. Rev.* **124**, 41–53.

Bleaney, B., Bowers, K. D. (1952), *Proc. Phys. Soc. A* **65**, 667–668.

Bleaney, B., Ingram, D. J. E. (1950), *Proc. Phys. Soc. A* **63**, 408–409.

Bloembergen, N. (1956), *Phys. Rev.* **104**, 324–327.

Blume, M., Geschwind, S., Yafet, Y. (1969), *Phys. Rev.* **181**, 478–487.

Coffman, R. E. (1965), *Phys. Lett.* **19**, 475–476.

Curie, P. (1905), *Ann. Chim. Phys. 7ème Ser.* **5**, 289.

Debye, P. (1926), *Ann. Phys.* **81**, 1154.

Faraday, M. (1846a), *Philos. Trans. R. Soc.* **136**, 21–40, articles 2279, 2280. Also in (1855) *Experimental Researches in Electricity*, Vol. 3, Series 20, London: Taylor & Francis.

Faraday, M. (1846b), *Philos. Trans. R. Soc.* **136**, 41–62, article 2420. Also in (1855) *Experimental Researches in Electricity*, Vol. 3, Series 21, London: Taylor & Francis.

Faraday, M. (1846c), *Philos. Trans. R. Soc.* **136**, 1–20, article 2152. Also in (1855) *Experimental Researches in Electricity*, Vol. 3, Series 19, London: Taylor & Francis.

Friedel, J. (1958), *Suppl. Nuovo Cimento* **7**, 287–311.

Giauque, W. F. (1927), *J. Am. Chem. Soc.* **49**, 1864–1870.

Griffiths, J. H. E., Owen, J. (1952), *Proc. R. Soc. A* **213**, 459–473.

Henry, W. E. (1952), *Phys. Rev.* **88**, 559–562.

Herring, C. (1963), in Yafet (1963), pp. 93–96, Appendix B.

Höchli, V. T., Estle, T. L. (1967), *Phys. Rev. Lett.* **18**, 128–130.

Hughenin, R., Pells, G. P., Baldock, D. N. (1971), *J. Phys. F: Met. Phys.* **1**, 281–289.

Jahn, H. A., Teller, E. (1937), *Proc. R. Soc. A* **161**, 220–235.

Kittel, C. (1949), *Phys. Rev.* **76**, 743–748.

Langevin, P. (1905), *J. Phys.* **24**, 678; *Ann. Chim. Phys. 8ème Ser.* **5**, 70.

Patel, C. K. N., Shaw, E. (1971), *Phys. Rev. B* 1279–1295.

Pauli, W. (1927), *Z. Phys.* **41**, 81.

Roth, L. M., Lax, B., Zwerdling, S. (1959), *Phys. Rev.* **114**, 90–104.

Roth, L. M. (1966), *Phys. Rev.* **149**, 519–525.

Ruderman, M. A., Kittel, C. (1954), *Phys. Rev.* **96**, 99–102.

Schumacher, R. T., Slichter, C.P. (1956), *Phys. Rev.* **101**, 58–65.

Sucksmith, W. (1930), *Proc. R. Soc. A* **128**, 276–293.

Thomson, W. (1850), *Philos. Mag.* **37**, 250.

Uhlenbeck, G. E., Goudsmit, S., (1925), *Naturwissenschaft 13*, 953–954; (1926) *Nature* **117**, 264–265.

Van Vleck, J. H. (1932), *The Theory of Electric and Magnetic Susceptibilities*, London: Oxford University.

Yafet, Y. (1963), in: F. Seitz, D. Turnbull (Eds.), *Solid State Physics*, Vol. 14, New York: Academic, pp. 1–98.

Yafet, Y. (1966), *Phys. Rev.* **152**, 858–862.

Zeeman, P. (1896), *Proc. R. Acad. Amsterdam* **5**, 181; also (1897) *Philos. Mag.* **43**, 226–239.

Further Reading

Historical

Vershuur, G. L. (1993), *Hidden Attraction (The History and Mystery of Magnetism)*, New York: Oxford University Press.

General

Ashcroft, N. W., Mermin, N. D. (1976), *Solid State Physics*, New York: Saunders College, Chaps. 31 and 32.

Kittel, C. (1986), *Introduction to Solid State Physics*, 6th ed., New York: Wiley, Chaps. 14 and 16.

White, R. M. (1983), *Quantum Theory of Magnetism*, 2nd ed., Berlin: Springer, Chaps. 1–4 and 7.

Paramagnetism of Ions in Solids

Abragam, A., Bleaney, B. (1970), *Electron Paramagnetic Resonance of Transition Ions*, Oxford: Clarendon Press.

Ham, F. S. (1972), "Jahn-Teller Effect in Electron Paramagnetic Resonance Spectra," in: S. Geschwind (Ed.), *Electron Paramagnetic Resonance*, New York: Plenum, pp. 1–119.

Martin, D. H. (1967), *Magnetism in Solids*, Cambridge, MA: MIT Press.

Owen, J., Thornley, J. H. M. (1966), "Covalent Bonding and Magnetic Properties of Transition Metal Ions," *Rep. Prog. Phys.* **29**, 675–730.

Sturge, M. (1967), "The Jahn-Teller Effect in Solids," in: F. Seitz, D. Turnbull (Eds.), *Solid State Physics*, Vol. 20, New York: Academic, pp. 92–211.

Taylor, K. N. R., Darby, M. I. (1972), *Physics of Rare Earth Solids*, London: Chapman and Hall.

Paramagnetism of Metals

Fawcett, E. (1988), "Spin-Density-Wave Antiferromagnetism in Chromium," *Rev. Mod. Phys.* **60**, 209–283.

Freeman, A. J. (1972), "Energy Band Structure, Indirect Exchange Interactions and Magnetic Ordering," in: R. J. Elliott (Ed.), *Magnetic Properties of Rare Earth Metals*, London: Plenum, pp. 245–333.

Heeger, A. J. (1969), "Localized Moments and Non-moments in Metals: the Kondo Effect," in: F. Seitz, D. Turnbull (Eds.), *Solid State Physics*, Vol. 23, New York: Academic, pp. 283–411.

Applications

Hudson, R. P. (1972), *Principles and Application of Magnetic Cooling*, Amsterdam: North-Holland.

Schulz-Du Bois, E. O. (1960), "The Three-Level Solid State Maser," in: K. Mendelsohn K (Ed.), *Progress in Cryogenics*, Vol. 2, New York: Academic, pp. 175–231.

PARTICLE ACCELERATORS

See ACCELERATORS, LINEAR; ACCELERATORS, POTENTIAL DROP; BETATRONS; CYCLOTRONS; SYNCHROTRONS

PARTICLE BEAMS

See INTENSE PARTICLE BEAMS

PARTICLE COLLIDERS

DONALD L. HARTILL, *Cornell University Laboratory for Nuclear Studies, Ithaca, New York, U.S.A.*

STEPHEN D. HOLMES, *Fermi National Accelerator Laboratory, Batavia, Illinois, U.S.A.*

MELVIN MONTH AND ROBERT B. PALMER, *Brookhaven National Laboratory, Upton, New York, U.S.A.*

RONALD D. RUTH AND ROBERT H. SIEMANN, *Stanford Linear Accelerator Center, Stanford, California, U.S.A.*

Introduction 123
1. **Collider Basics** 124
1.1 Parameters 124
1.2 Energy and Luminosity Limits 125
2. **Hadron Colliders** 125
3. **Electron–Positron Ring Colliders** 128
3.1 Operating Colliders and Future Plans 128
3.2 ϕ Factories 128
3.3 τ–Charm Factories 128
3.4 CESR Luminosity Upgrade .. 129
3.5 *B* Factories 130
3.6 LEP 200 130
4. **Linear Colliders** 131
5. **New Technologies for Colliders** 134
5.1 General Issues 134
5.2 Very High-Field Bending Magnets 134
5.3 Advanced Acceleration Techniques 135
5.3.1 Conventional Acceleration at Shorter Wavelengths 135
5.3.2 Exotic Acceleration 135
5.3.2.1 Inverse Cherenkov Effect 135
5.3.2.2 Single-Pulse Acceleration 135
5.3.2.3 Plasma Acceleration 135
5.3.2.4 Inverse Free-Electron Laser (IFEL) 136
5.4 Gamma–Gamma and Electron–Gamma Colliders .. 136
5.5 Muon–Muon Collider 136
Glossary 137
Further Reading 139

INTRODUCTION

Since the invention of the cyclotron in the early 1930s, the particle accelerator and storage ring have been the instruments of choice for scientists engaged in particle physics research. Historic advances in particle physics have been, and are expected to continue to be, dependent upon advances in accelerator facilities and the underlying technologies upon which they are based. Over the past 60 years, these technologies have provided nearly a millionfold increase in beam energy, accompanied by facilities whose size is measured in kilometers rather than meters, and whose costs are now measured in hundreds of millions, or even billions, of dollars.

Accelerators in operation today generally fall into two classes: hadron accelerators and electron accelerators. The majority of contemporary high-energy physics experiments utilize colliders, which are accelerators with opposing beams brought into collision. This mode of operation provides the most efficient utilization of beam energy for producing high-mass particles. The highest energies are attained in hadron colliders, whereas electron–positron colliders are used in applications in which a premium is placed on having all available energy carried by the interacting constituents.

The highest-energy accelerator facility in the world today is the Tevatron at Fermilab. This facility is the first high-energy accelerator ever constructed utilizing superconducting magnets and is the only accelerator of any

3-527-28135-5/95/$5.00 + .50

type at which the top quark can be produced and its properties measured. Construction and operation of the Tevatron have made possible a further increase in energy at the Large Hadron Collider now under construction at CERN.

Electron colliders come in two types and operate in two distinct energy regimes. Colliders in the low-energy regime are based exclusively on circular accelerators, and typical operations are tuned to a specific energy associated with a state of matter such as a quark–antiquark bound state. Higher-energy electron–positron collisions must rely on linear rather than circular colliders because of limitations imposed by synchrotron radiation. The first linear collider, the Stanford Linear Collider located at Stanford University, is a prototype of many aspects of a future high-energy linear collider.

Technologies currently or nearly in hand are sufficient to allow exploration of physics at the TeV scale, but moving beyond that within the confines of physical size and cost will require either the development of new technologies and paradigms such as plasma accelerators, muon colliders, or photon–photon colliders, or significant cost reduction in existing technologies. Future colliders will be based on the developments in these directions.

1. COLLIDER BASICS

1.1 Parameters

The two key parameters describing an accelerator are the center-of-mass energy $E_{\text{c.m.}}$ and the luminosity L, which is defined as the proportionality between the cross section σ and the event rate, R:

$$R = L\sigma.$$

The center-of-mass energy of a *fixed-target experiment*, where a beam of particles of mass m and total energy E strikes a stationary target of mass M, is

$$E_{\text{c.m.}} = \sqrt{2EM + m^2 + M^2},$$

and the luminosity is given by

$$L = n_b n_t,$$

where n_b is the beam flux and n_t is the target density. For some typical numbers, consider Experiment 772 performed at the Tevatron (McGaughey *et al.*, 1994). There the beam and the target were protons, the beam flux during part of the experiment was $n_b = 10^{11}\ \text{s}^{-1}$, and the 0.51-m-long liquid hydrogen target has a density $n_t = 2.2 \times 10^{24}\ \text{cm}^{-2}$. This results in $E_{\text{c.m.}} = 27.4$ GeV and $L = 2.2 \times 10^{35}\ \text{cm}^{-2}\ \text{s}^{-1}$. The center-of-mass energy is relatively low, and the luminosity is high. This is a common feature of most fixed-target experiments. The center-of-mass energy scales as $E^{1/2}$, and increasing it substantially requires colliding beams—at the cost of lower luminosity because the target density has to be reduced by replacing the target by another beam.

With a beam of energy E_1 colliding head-on with a beam of energy E_2, the center-of-mass energy, neglecting rest masses, is

$$E_{\text{c.m.}} = \sqrt{4E_1E_2},$$

which is linear in the beam energy when $E_1 = E_2$. If n_1 particles in one beam collide with n_2 particles in the other beam with a collision frequency f_c, the luminosity is

$$L = \frac{n_1 n_2 f_c}{2\pi\sqrt{(\sigma_{x1}^2 + \sigma_{x2}^2)(\sigma_{y1}^2 + \sigma_{y2}^2)}},$$

where the denominator is the effective area of overlap of the two beams. The σ's are the rms sizes in the two transverse distributions, which are assumed Gaussian. The usual case is to have $\sigma_{x1} = \sigma_{x2} = \sigma_x$ and $\sigma_{y1} = \sigma_{y2} = \sigma_y$; then

$$L = (1/4\pi)\, n_1 n_2 f_c/\sigma_x\sigma_y.$$

Beam size at the collision point depends on beam phase-space density and accelerator focusing magnets. The rms emittance ϵ is the transverse phase-space area that contains 46.6% of the beam. (The conventional phase-space coordinates in accelerator physics are position and angle.) It depends on synchrotron radiation in electron storage rings and is determined by the beam energy and the focusing and bending magnets in that case. In hadron and linear colliders, the emittance is determined by the particle source, and it decreases with energy as

$$\epsilon = \epsilon_n/\gamma$$

in the absence of effects such as beam-generated electromagnetic fields that can increase emittance. In this equation, ϵ_n is the normalized emittance and is determined by the particle source, and γ is the energy in units of rest energy.

The focusing produced by the accelerator quadrupole magnets is described by a Hill's equation for the beam envelope. The amplitude or β function is related to the solution of this equation. The beam envelope is proportional to $\beta^{1/2}$, and regions of small β correspond to small beam size and large angular divergence. The horizontal and vertical β functions are minimized at the interaction point for high luminosity, and are denoted as β_x^* and β_y^* for the horizontal and vertical, respectively. The rms beam sizes at the collision point are given by

$$\sigma_x = \sqrt{\beta_x^* \epsilon_x} \quad \text{and} \quad \sigma_y = \sqrt{\beta_y^* \epsilon_y}.$$

This follows from the definitions of the rms emittances.

High luminosity requires that the beams be "bunched," confined longitudinally to short bunches with length

$$\sigma_L \leq \min(\beta_y^*, \beta_x^*).$$

When there are k bunches around the circumference of a circular collider or k bunches per rf pulse for a linear collider, the luminosity can be rewritten in terms of the total number of particles in each beam, N_1 and N_2, and the revolution frequency (or rf pulse frequency) f_0 as

$$L = (1/4\pi)\, N_1 N_2 f_0 / k\sigma_x\sigma_y.$$

If there is a limit on the total number of particles—for example from the particle source—it pays to keep the number of bunches small. However, instabilities and beam–beam effects tend to place restrictions on the number of particles per bunch, and when those restrictions are important, the only way to increase the luminosity is to increase the number of bunches and the total number of particles.

1.2 Energy and Luminosity Limits

The constraints limiting energy and luminosity can be technological or they can arise from the beam dynamics. Among the most important ones are the following:

1. Magnetic field strengths. These determine the minimum bending radius and maximum focusing possible in hadron colliders and thereby the maximum energy possible for a given sized collider.
2. Synchrotron radiation. This limits circular electron–positron colliders to $E_{\text{c.m.}} \leq$ 200 GeV and requires that linear colliders be used for higher energies.
3. High-efficiency rf production. Achievable accelerating gradient is not the energy limit for linear electron–positron colliders. Instead that limit comes from the ability to accelerate particles efficiently.
4. Single-beam current limits. These arise from the interaction of a beam with self-generated fields, called "wake fields." These lead to instabilities in circular accelerators and to emittance increase in linear accelerators. They can limit the number of particles accelerated per bunch or the total number of particles.
5. Beam–beam interaction. The primary interaction between the beams is through the electromagnetic fields they produce. These fields produce multiturn resonant effects in circular colliders, and backgrounds and energy spread in linear colliders.

The sections that follow will discuss specific colliders and how the interplay of these effects determines energy and luminosity.

2. HADRON COLLIDERS

Hadron colliders provide the highest-energy particle collisions of any of the collider types operational in the world today. The first hadron collider was the Intersecting Storage Rings (ISR), a facility that commenced operations in 1971 at the European Center for Nuclear Research (CERN) in Geneva, Switzerland. Countercirculating proton beams were brought into collision at energies as high as 31 GeV per beam. The operating luminosity grew substantially over the lifetime of the facility, peaking at about 1.4×10^{32} cm^{-2} s^{-1} when operations ceased in 1983. Two hadron collider facilities have been constructed and operated following the ISR. The first was the CERN proton–antiproton collider. This facility operated with protons and antiprotons circulating in opposite directions in a single storage ring, with collisions observed at two locations. Operations commenced in 1981 and were curtailed in 1991. The beam energy in the collider was 315 GeV, and the facility ultimately operated with a luminosity of 6×10^{30} cm^{-2} s^{-1}. Following construction of the CERN collider, a proton–antiproton collider, known as the Tevatron, commenced operations in 1986 at the Fermi National Accelerator Laboratory in Batavia, Illinois, U.S.A. (Fig. 1). This facility, the first based on superconducting magnets and the highest-energy collider of any type in the world today, continues to operate with a beam energy of 900 GeV and at a luminosity in excess of 1×10^{31} cm^{-2} s^{-1}. While the basic design of these accelerators has largely changed only in scale, advances in the underlying technology have allowed a nearly 30-fold increase in the center-of-mass energy achieved in these facilities over the last two decades. The advent of superconducting magnets and the invention of stochastic cooling have spurred the latest advances.

High beam energy and luminosity are the primary goals in the design and construction of a hadron collider. For all hadron colliders constructed to date, energy has been limited by the magnitude of the magnetic field that can be created in the dipole magnets used to bring the beams around in a circle. Hadron colliders are very different from electron colliders in this regard. The maximum achievable magnetic fields in accelerator magnets today are roughly 20 kG with copper conductor operating at room temperature, and 88 kG with superconductors operating in the range 2 to 4 K. For a specified magnetic field B, the radius of a circular accelerator required to achieve a beam energy E is given by

$$R = (44 \text{ m kG GeV}^{-1})E/B.$$

Here the factor of 44 assumes 75% utilization of the entire circumference of the accelerator for dipole magnets. Since for hadron accelerators built to date synchrotron radiation is a negligible effect, acceleration and beam storage can typically be accomplished with an rf system operating at a few MV. The sole limitation on energy is given by the product of the magnetic field one can produce and the size of the accelerator one is willing to build.

Because of the higher achievable fields, superconducting magnets are the technology of choice in hadron accelerators today. The Tevatron, the first superconducting high-energy accelerator, was commissioned in 1983. Operation as a proton–antiproton collider was initiated in 1986. Today this facility operates at a beam energy of 900 GeV based on dipole magnetic fields of 40 kG. The field is pro-

FIG. 1. The interior of the Main Ring enclosure at Fermi National Accelerator Laboratory in Batavia, Illinois, U.S.A. The Main Ring, a 150-GeV accelerator constructed in the early 1970s utilizing room-temperature copper-conductor magnets, is shown on top. Underneath lies the 900-GeV Tevatron superconducting accelerator in which countercirculating proton and antiproton beams are stored. Each of the two accelerators is approximately 6280 m in circumference and lies 7 m below the land surface.

duced by a coil supporting a current distribution approximating a $\cos\theta$ (in azimuth) distribution, leading to a nearly pure dipole field. The coil is made from NbTi embedded in a copper matrix. The NbTi is superconducting and supports a significant (1800 A/mm^2) current density at 4.6 K. The magnet aperture is about 75 mm, and each magnet is 6 m in length. A total of 774 dipoles and 216 quadrupoles are used in the Tevatron.

Rapid advances in superconductor technology and in cryostat design allowed the development of superconducting magnets for the Superconducting Super Collider (SSC) project. Magnets with an operating field of 66 kG and with a greatly reduced heat leak compared with the Tevatron were developed for that machine. The plans for the SSC envisioned the construction in Waxahachie, Texas, U.S.A., of an accelerator 87 km in circumference, providing collisions of 20-TeV proton beams with a luminosity of 1×10^{33} cm^{-2} s^{-1}. Construction on this project, which began in 1990, was canceled by the U.S. Congress in 1993. The current state of the art in superconducting magnet design is represented by the magnet designed for the proposed Large Hadron Collider (LHC) project at CERN. The magnet, a two-in-one design operating at 87 kG, represents a significant extension of the SSC design. The LHC magnet is felt to be near the limit of the technology represented by the NbTi/$\cos\theta$ designs.

The luminosity achievable in hadron colliders is given by the interplay of a variety of effects. These include antiproton production (for antiproton–proton colliders), beam size at the collision point, and the instantaneous interaction rate tolerable by the experimenters performing physics research. The CERN proton–antiproton collider and the Fermilab Tevatron both produce(d) collisions between protons and antiprotons. The advantage of doing this is that the antiprotons and protons can circulate in opposite directions in a common accelerator (remember $\mathbf{F} = q\mathbf{v} \times \mathbf{B}$), effecting considerable cost savings by eliminating the need to build two accelerators. The downside is that copious sources of antiprotons do not exist. Facilities called antiproton sources have been constructed both at CERN and at Fermilab to produce antiprotons for use in the collider facilities. Antiprotons are produced by directing high-energy protons onto a metal target. Some fraction of the interactions produces antiprotons, which are then collected in a storage ring. The antiprotons collected are quite diffuse, and many hours of accumulation are required to build up a large enough quantity to meet the needs of the collider. The entire accumulation scheme depends on a technology called stochastic cooling. This technique, pioneered by CERN in the late 1970s, increases the antiproton beam densities during accumulation by measuring the position deviation from nominal of individual antiprotons with a device known as a pickup, and transmitting a signal to a device referred to as a kicker, which applies the appropriate correction. The performance is directly related to the time resolution, or bandwidth, of the system. In the implementation at Fermilab, 2.5×10^{12} protons, at an energy of 120 GeV, are directed onto a nickel target every 2.4 s. Approximately 3×10^7 antiprotons are produced in these collisions and are collected and cooled at 8.9 GeV by stochastic cooling systems with bandwidths up to 4 GHz. Approximately 5×10^{10} antiprotons can be produced and collected every hour in this facility.

The luminosity in both proton–proton and proton–antiproton colliders is also highly dependent on the beam size at the collision point. The beam size depends on both the beam emittance and the focusing properties of the storage ring. Beam emittance is limited by a variety of collective effects in which macroscopic electromagnetic fields produced by the beam influence the motion of particles either within the beam itself or within its colliding partner beam. Included are space charge, intrabeam scattering, and beam–beam effects, as well as a whole series of collective instabilities related to fields produced by interaction of the beam with the conducting walls of the accelerator itself. In the Fermilab facility, beams containing six bunches, each with up to 2×10^{11} protons/bunch and 5×10^{10} antiprotons/bunch, with beam sizes of approximately 40 μm at the interaction point, are routinely achieved.

The future of hadron colliders currently points in a variety of directions. The Main Injector upgrade, currently under construction at Fermilab, is expected to yield a luminosity of 1×10^{32} cm^{-2} s^{-1} in the Tevatron proton–antiproton collider by the end of the 1990s. Further improvements in either energy or luminosity are under study. A Relativistic Heavy

Ion Collider (RHIC) is currently under construction at Brookhaven National Laboratory in Upton, New York, U.S.A. This facility, consisting of two superconducting accelerator rings of circumference 3.6 km, will accelerate and store fully ionized gold atoms to 100 GeV/nucleon. The expected luminosity is 2×10^{26} cm^{-2} s^{-1}. Operations are expected to commence in 1999. The LHC complex, as currently proposed for construction in Geneva, Switzerland, is very similar to the SSC except for the smaller scale. The design energy is 14 TeV in the center-of-mass system with a luminosity of 1×10^{34} cm^{-2} s^{-1}. The plan is to construct the LHC in the existing 27-km LEP tunnel. If this project proceeds as currently proposed, it is expected that operations will commence around the year 2005. A hybrid electron–proton collider, HERA, is currently in operation in Hamburg, Germany. This facility produces collisions between 800-GeV protons and 30-GeV electrons. The technology for construction of a hadron collider at 30 TeV, or greater, per beam is likely to exist early in the next decade. Whether such a facility will ever be constructed will be left to a future generation of scientists and politicians.

3. ELECTRON–POSITRON RING COLLIDERS

3.1 Operating Colliders and Future Plans

Currently, five e^+e^- circular colliders are in operation (VEPP 2M at the Budker Institute in Novosibirsk, Russia; BEPC at the Institute for High Energy Physics in Beijing, China; CESR at Cornell University in Ithaca, New York, U.S.A.; TRISTAN at KEK in Tsukuba, Japan; and LEP at CERN in Geneva, Switzerland). The total center-of-mass energy for the beams in these colliders ranges from the ϕ mass (1.020 GeV) to the Z^0 mass (91.2 GeV). For circular colliders, the emphasis is on luminosity and on providing a very clean background environment for the detectors. Upgrade plans for VEPP 2M, BEPC, and CESR are to increase their luminosities, and the operating energy of LEP will be increased to above W^+W^- threshold by the installation of approximately 200 multicell superconducting rf accelerating cavities. The luminosity of LEP will also be increased by a more aggressive pursuit of multibunch operation with pretzels (see Glossary), with initial operation planned for 1996.

Future plans are for the completion of the construction of PEP II, a B factory at SLAC in Stanford, California, U.S.A., and KEK B, a B factory at KEK. Both machines are being constructed in existing tunnels at these laboratories and will operate with asymmetric beam energies to provide a significant Lorentz boost for the center of mass in the laboratory frame. This makes it possible to observe the time evolution of the B^0 decays from the $\Upsilon(4S)$ and thereby the violation of CP invariance for the first time outside the K system. At Frascati in Italy, a phi (ϕ) factory, DaΦne, is under construction with initial operation expected in early 1996. At Cornell, CESR's luminosity is being increased to the region of 10^{33} cm^{-2} s^{-1} to provide a large data sample at the $\Upsilon(4S)$ for the study of rare B decays in an upgraded CLEO detector. The remainder of this short section will be a small expansion of these remarks on each of the colliders.

3.2 ϕ Factories

VEPP 2M at the Budker Institute in Novosibirsk has been operating for the past several years with a peak luminosity of 5×10^{30} cm^{-2} s^{-1} and currently has the highest luminosity of any collider in this energy region. An upgrade in luminosity is planned for the end of 1995 by installing a solenoidal final focusing system. This will permit operation with round beams and has the potential of increasing the luminosity by a factor of 20. It will be the first extended operation of round beams in a circular e^+e^- collider.

DaΦne is being constructed at Frascati in buildings previously occupied by ADONE, one of the first successful modern e^+e^-colliders, which was retired in April 1992. The main features of DaΦne are listed in Table 1.

3.3 τ–Charm Factories

This term refers to a high-luminosity e^+e^- collider operating at a center-of-mass energy of 3 to 4 GeV. The collider at the High Energy Institute in Beijing, BEPC, continues to be the premier facility in this energy region. An upgrade of the luminosity of the collider is planned, and an improvement program for the detector is currently under way. The pro-

Table 1. DaΦne main features.

Luminosity	1.3×10^{32} cm^{-2} s^{-1} (initial, with ×3 increase in 3 years)
Number of interaction regions	Two (KLOE, CP-invariance violation; FINUDA, hypernuclei)
Circumference	97.69 m
Number of bunches	30 (initial), 120 (final)
Number of particles per bunch	8.9×10^{10}
Rf frequency	368.25 MHz
Feedback	Similar to PEP II
Initial operation	Early 1996

jected peak luminosity performance is 3×10^{31} cm^{-2} s^{-1}, which will be an increase of approximately a factor of 5 over present peak performance.

A project to construct a high-luminosity collider in Spain by a consortium of European and U.S. physicists is currently on indefinite hold and is likely to remain so. As a result, BEPC will continue to be the facility for this energy region for the foreseeable future.

3.4 CESR Luminosity Upgrade

CESR has achieved the highest luminosity of any collider. The layout of the machine is typical of circular colliders and provides a good example of these machines. Figure 2 shows the tunnel at Cornell containing the 10-GeV synchrotron that provides on-energy particle injection into the storage-ring CESR.

The phased upgrade of CESR to peak luminosities of 10^{33} cm^{-2} s^{-1} is making good progress. Operation with ±2-mrad horizontal crossing angle for the two beams is now routine. With head on-collisions, the best peak luminosity was 3×10^{32} cm^{-2} s^{-1}. Installation of large-aperture final-focus quadrupoles and beam tubes along with the new silicon strip vertex detector in CLEO (a particle detector system) in early 1995 will make operation much easier and allow for bunch train operation—multiple closely spaced bunches to enhance beam stability during collisions—with low backgrounds. Initial test runs with two bunches per train (position of original single bunch) spaced by 28 ns yielded peak luminosities of 2.4×10^{32} cm^{-2} sec^{-1} with acceptable detector backgrounds. The total beam current in these tests was comparable to the normal operating currents. For the last month of 4*S* running in 1994, two bunches per train was the normal operating mode, and the monthly integrated luminosity was the second highest recorded at CESR.

The key to the success of the upgrade plan is the lowering of the higher-order mode impedance of the ring. An initial beam test in CESR in August 1994 of a 500-MHz single-cell superconducting rf cavity with all higher-order modes well damped went well. The

FIG. 2. Photograph of CESR tunnel with Prof. B. D. McDaniel, former laboratory director. CESR is on the right and the 10-GeV synchrotron is on the left.

cavity delivered 155 kW of power to the beam and achieved an accelerating gradient of 6 MV/m. A current of 220 mA was stored with the cavity powering the beam in conjunction with the conventional rf system. With CESR operating at 4.5 GeV, the single-cell cavity was able to store beam with the conventional rf system detuned. Operation of superconducting cavities in the harsh environment of a high-current storage ring should be similar to that of conventional cavities on the basis of this very successful test.

The elements of the upgrade appear to be in hand, and what remains to be achieved is successful integration of these elements into a smoothly working system. A peak luminosity of 10^{33} cm^{-2} s^{-1} by 1998 seems possible.

3.5 *B* Factories

The luminosity frontier will be the main thrust of these new colliders operating with a center-of-mass energy at the $\Upsilon(4S)$. The goal is to produce as many *B* mesons as possible in the clean environment of an e^+e^- collider. These colliders have the added advantage of operating with asymmetric beam energies to boost the *B*'s from the 4*S* decay so that the time dependence of the *B* decays can be determined to study CP-invariance violation. The clean environment will enable very sensitive tests of rare *B* decays. The two projects represent the future of circular colliders and promise a rich physics return.

PEP II at SLAC in the U.S. is well under way with completion of the machine scheduled for fall 1998. The Japanese project at KEK, KEK B, is receiving significant funding this fiscal year with an agreed-upon plan that calls for commissioning of the storage rings in late 1998. Both machines use as much existing infrastructure as possible to minimize the cost of the facilities. Because of the large currents and large numbers of bunches, both designs work hard to minimize the higher-order mode impedance of the beam environment and feature state-of-the-art feedback systems. A short summary of the main machine parameters is given in Table 2. For details, the reader is referred to the latest versions of the design reports. Achieving these parameters will be a challenge.

3.6 LEP 200

Adding 192 superconducting cavities to LEP to achieve a center-of-mass energy above the W^+W^- threshold will approach the upper limit of what is sensible for the maximum energy of a circular collider using electrons and positrons. Above this energy, the tradeoff between circumference and total length and voltage gain of the rf system forces a very large cost for the storage ring. Linear colliders are the sensible approach above LEP 200 energies.

The present plan for installation of the superconducting cavities in LEP calls for operation at energies above the W^+W^- threshold in 1996. Because the cross section for *W*-pair production is significantly smaller than the cross section at the Z^0, a luminosity upgrade is also planned. This will make use of pretzel orbits to increase the number of bunches from 4 to 8. The integrated luminosity goal is to accumulate 500 pb^{-1} (1 pb^{-1} = operation of the collider at a luminosity of 10^{32} cm^{-2} s^{-1} for 10^4 s) by the end of 1999. A short summary of the main machine parameters is given in Table 3. The latest accelerator conference reports should be consulted for details.

Table 2. Some parameters of PEP II and KEK B.

	PEP II		KEK B	
Parameter	e^+	e^-	e^+	e^-
Beam energy (GeV)	3.1	9.0	3.5	8
Circumference (m)	2199		3016	
Current (A)	2.14	1.48	2.6	1.1
Number of bunches	1658	1658	5120	5120
Bunch length (cm)	1.0	1.0	0.32	0.4
β_V/β_H (Intersection point) (cm)	1.5/37.5	3.0/75	1.0/33.0	1.0/33.0
Crossing angle (mrad)	0		±10	
Luminosity (10^{33} cm^{-2} s^{-1})	3.0		10.0	

Table 3. Summary of machine parameters for LEP 200.

Luminosity	7×10^{31} cm^{-2} s^{-1}
Circumference	26.7 km
Bend radius	3.28 km
Maximum beam energy	90 GeV
Number of superconducting rf cavities	192
Rf voltage	1900 MV
Rf power	15.6 MW
Bunch current	1 mA
Number of bunches	8

4. LINEAR COLLIDERS

There is now broad agreement in the high-energy physics community that to continue exploring the energy frontier in e^+e^- interactions, circular colliders have to be abandoned in favor of linear colliders. This realization has led to active research throughout the world aimed at the next generation of linear colliders. The past few years have seen great strides in the understanding of both the accelerator physics and the technology of linear colliders. At this point, the design of such a Next Linear Collider, or NLC, can be discussed in fair detail.

The first linear collider, the SLAC Linear Collider (SLC), is now in operation, colliding highly polarized electrons with positrons at a beam energy of about 47 GeV. The SLC, shown in Figs. 3 and 4, is a hybrid design in that the electrons and positrons are accelerated in the same linac and are bent around arcs to collide inside the SLC Large Detector (SLD). Although the SLC has this folded design, the basic approach is common to future linear collider designs: The electrons and positrons are accelerated in a linear accelerator, focused to a small spot to collide, and then discarded. The SLC is now running at an energy tuned to produce the Z^0 boson. By taking advantage of the polarized electrons at the SLC, the present running seeks to measure the "mixing angle" between the electric and weak forces to very high precision. The

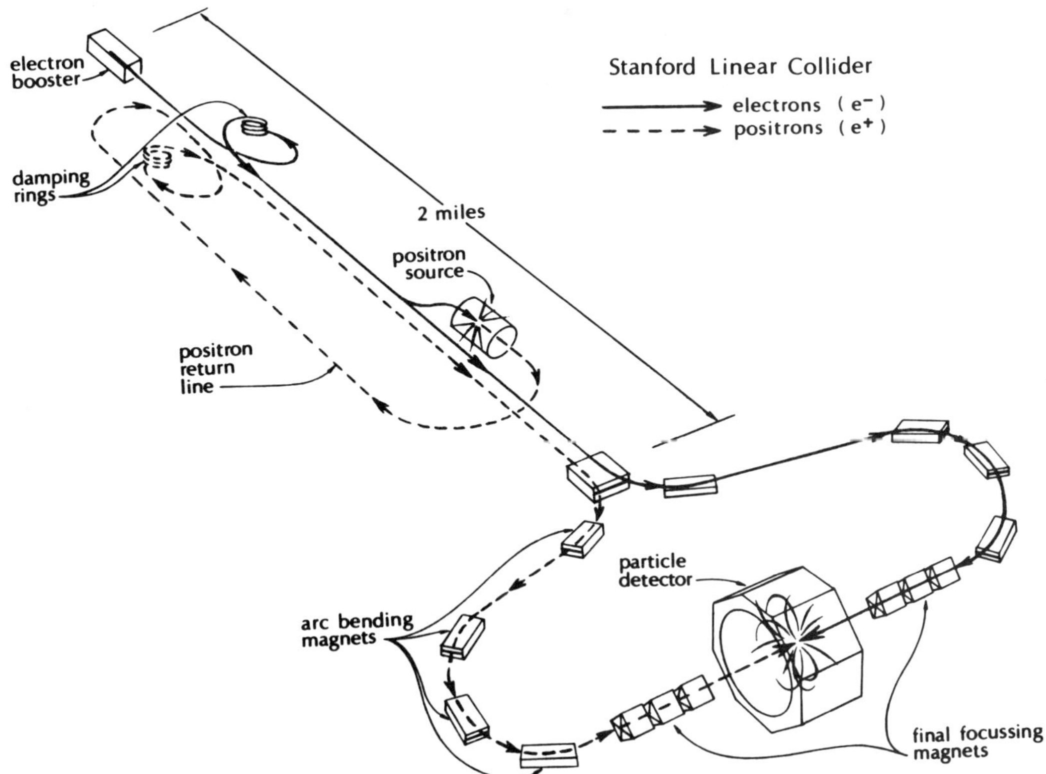

FIG. 3. The layout of the Stanford Linear Collider (SLC).

FIG. 4. The Stanford Linear Accelerator Center (SLAC).

luminosity of the SLC is currently about 5×10^{29} cm^{-2} s^{-1}, but the ultimate goal is to exceed 10^{30}.

A broad consensus has emerged over the past several years that the energy of the NLC should be 0.5 TeV (total electron plus positron energy), upgradable to at least 1.0 TeV and perhaps to 1.5 TeV. One reason for this choice of energy range is the great potential of such a collider for significant high-energy physics research. Another is that this energy range is a natural next step; it is a factor of 5 to 10 beyond that of the present SLC. In order to obtain a sufficient event rate to perform detailed measurements, the luminosity of the collider should increase with the square of its energy. For a NLC in the TeV energy range, a luminosity of 10^{33} to 10^{34} cm^{-2} s^{-1} is required.

The energy can be increased by a factor of 5 to 10 in two ways: by increasing the collider length to 10 to 20 times that of the SLC (3 km), or by raising the accelerating field to 5 to 10 times the SLC gradient (20 MV/m). Several approaches under investigation around the world span this range of possibilities. The TESLA approach uses superconducting technology with proposed acceleration gradients comparable to those of the present SLC. Other approaches use conventional copper accelerating structures that range in frequency from the SLC frequency of 2.8 GHz up to 33 GHz. For example, KEK and SLAC are pursuing 11.4-GHz acceleration with gradients ranging from 50 to 100 MV/m, up to five times the SLC gradient. The ultimate choice of technology and the tradeoff between length and accelerating field are governed by the overall cost and upgradability of the design.

The choice of luminosity range also greatly influences the design of the linear collider. In principle, one could increase the luminosity simply by raising the repetition rate of the accelerator, but the wall-plug power increases in direct proportion. In a reasonable design, the wall-plug power should be in the range 100 to 200 MW. Given this constraint, a good way to increase the luminosity is to shrink the beam size at the interaction point (IP). In addition, the beam cross section must be kept flat at the IP in order to minimize the amount of "beamstrahlung" radiation emitted as energetic electrons or positrons interact with the electromagnetic field of the opposing bunch.

The luminosity can be further enhanced by accelerating several bunches on each machine cycle. A single bunch of particles can, in practice, extract only a few percent of the energy available in the accelerating structure. With additional bunches, we get both greater luminosity and higher efficiency of energy transfer to the beam. The number of particles in each bunch, another factor that directly affects the luminosity, is limited by the rf energy that can be stored in the accelerating structure and by the amount of beamstrahlung radiation that can be tolerated. The obvious solution is to generate trains of successive bunches, each with a fairly moderate number of particles.

Given these goals and constraints, we can now sketch a rough design of a linear collider able to achieve both the desired energy and luminosity. A possible layout is shown in Fig. 5. There are two complete linear accelerators, one for electrons and the other for positrons. Each linac is supplied with particle beams by a damping ring followed by a preacceleration section consisting of two bunch compressors and a low-energy linac. After being accelerated through the main linacs and focused to a small spot in the final focus system, the beams collide at a small crossing angle inside a large particle detector like the Stanford Linear Detector (SLD) at SLAC.

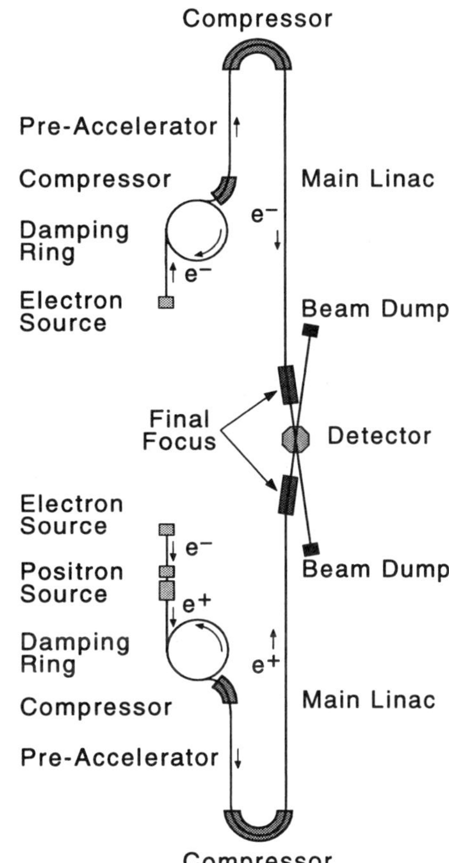

FIG. 5. A possible layout of a future linear collider.

The basic features of the NLC operation can be illustrated by following some electron bunches through the collider. A sequence of several bunches is created at the source and accelerated up to about 2 GeV in a preaccelerator. This batch of bunches is then injected into a damping ring that serves to reduce the transverse and longitudinal phase space occupied by the electrons in each bunch. At the proper moment, these bunches are extracted from the ring and then compressed along their direction of motion by a bunch compressor, after which they are accelerated up to about 10 GeV and compressed a second time just prior to injection into the main, high-gradient linac. The entire batch is carefully steered and focused as the electrons are accelerated up to full energy in the linac. Precision magnets in the final focus system squeeze the bunches down by about a factor of 300 just before they collide at the IP with similar bunches of positrons. Except for the fact that they were created differently, from the shower of particles that occurs when a bunch of electrons hits a metal target, these high-energy positron bunches have followed a similar evolution. After the beams collide, their debris is channeled out of the detector area and into shielded dumps.

In order to test the enabling technologies for the NLC, several test facilities are under construction worldwide: The TESLA collaboration is constructing a 500-MV linac to test the superconducting approach. CERN is testing various aspects of acceleration at 33 GHz using the two-beam approach to transform energy from a low-energy, high-current beam to a high-energy, low-current beam. At KEK, a damping ring is being constructed together with an injector linac. At SLAC, the Final Focus Test Beam was constructed by an international collaboration; this facility has recently achieved a 70-nm vertical spot size (designs range from 3 to 100 nm). A 500-MV test accelerator for high-gradient acceleration at 11.4 GHz is being constructed at SLAC. The SLC, the first linear collider, continues to teach the lessons that can be learned only in a full-scale operating accelerator with the demands of low background and reliability required by a high-energy experimental program.

In order to facilitate communication and cooperation in the above process, a Linear Collider World Collaboration has recently been established. This collaboration will review and monitor progress of the technical approaches. As the designs and test facilities are completed in the last half of the 1990s, it is anticipated that a technology will be selected and that a linear collider will be constructed by an international coalition.

5. NEW TECHNOLOGIES FOR COLLIDERS

5.1 General Issues

Cost played a pivotal role in the demise of the Superconducting Super Collider (SSC) and is a major issue in the Large Hadron Collider (LHC), both proton–proton machines. Cost is also likely to be a major issue in the Next electron–positron Linear Collider (NLC) discussed above. It would be very desirable to develop technologies that could reduce these costs. It may be imperative to do so if yet higher-energy machines are to be built.

As the energy scale of the phenomenon being studied rises, the size, and thus cross section, of the phenomenon falls. For a given rate of events of interest, the luminosity must then be increased, to compensate for the falling cross sections. No technical schemes for producing luminosities much beyond those of the LHC and NLC are in hand.

Thus new colliders encounter both economic and technical challenges. This section explores the possibility of radical solutions: advanced acceleration techniques, advanced radio-frequency power sources, gamma–gamma colliders, and muon–muon colliders. None of the techniques discussed below are clearly applicable to the problem, but they need to be, and are being, studied.

Proton and proton–antiproton colliders can be circular, and their size, for a given bending magnetic field, rises linearly with their energy. The LHC will be 27 km in circumference. The SSC was to have been 85 km in circumference. Yet larger machines do not seem economically practical. In Sec. 5.2, we discuss the possibility of higher bending magnetic fields.

The effective energy (i.e., the energy seen by an experimenter) of an electron collider is at least a factor of 10 higher than the effective energy of a proton collider with the same beam energy. One would thus expect an electron machine, for the same effective energy, to be smaller. But in order to avoid synchrotron radiation, high-energy electron colliders must be straight, and their length, with current accelerating gradients, is long. In Secs. 5.3 and 5.4, we discuss technology for raising such gradients, and thus reducing the length.

In Sec. 5.5, we discuss the possibility of obtaining gamma–gamma collisions using a linear electron collider and powerful lasers. In Sec. 5.6, we introduce the idea of a muon–muon collider that would have the effective energy advantage of the electron machine, but need not be straight, and could thus be smaller.

5.2 Very High-Field Bending Magnets

The size of a proton collider, for a given energy, is inversely proportional to the magnetic field in its bending magnets. The higher the field, the smaller the ring. The highest-energy operating collider, the Tevatron at Fermilab, has bending fields of about 4.5 T. The Superconducting Super Collider (SSC) was to have used magnets with 6.7 T, and the Large Hadron Collider (LHC) now plans to use magnets with about 8.5 T. Studies of bending magnets with fields up to 16 T are under way, and small-bore hybrid (superconducting and copper) solenoids have been built with fields over 40 T. An SSC-scale machine, with such fields, would have a circumference of only about 10 km instead of 85 km. But the cost of such very high-field magnets would be excessive, and would almost certainly negate the cost savings from the smaller machine size. It is not clear at what field the cost would be at a minimum.

Early studies for the SSC indicated a cost minimum at a field close to 6.7 T. This cost-optimized field may have risen since then, but not by much. Even with an ideal infinite-current–carrying high-temperature superconductor, the cost-minimum field will not go very high. The cost will become dominated by the constraints of the enormous forces (proportional to the fields squared), and the copper conductors needed to protect the magnet from self-destruction in the event of a quench (when the superconductor gets too warm and becomes resistive).

5.3 Advanced Acceleration Techniques

5.3.1 Conventional Acceleration at Shorter Wavelengths. The acceleration gradient (energy gain per unit of length) in conventional structures is limited by breakdown (sparking) and effects of field emission (electrons emitted from surfaces due to high electric fields). Both are reduced, almost linearly, as the wavelength of the fields is reduced. An *S*-band 1-TeV collider (wavelength 10 cm) would be some 80 km long. Next Linear Collider (NLC) technology (wavelength 2.6 cm) would reduce this length to about 25 km. With 1-cm radiation, the length could be reduced to 10 km, and at laser frequencies it might be only 3 km.

Shorter wavelengths also reduce the stored energy and allow higher repetition rates. But at shorter wavelengths (1 cm and below), there is a lack of suitable power sources of radio-frequency power. The most commonly used source is a klystron. Since the energy and current density of a klystron's drive beam are limited, its power falls as the square of the wavelength. The power can be maintained by raising the voltage or using a device whose beam dimensions are less constrained by the wavelength.

The drive-beam energy can be raised by using induction units or superconducting cavities. If this beam, after giving up some of its energy, is reaccelerated and used again, then the machine is referred to as a "two-beam accelerator" (TBA). Alternatively, the constraint on the beam dimension can be relieved by the use of larger beams in the rf structures. Devices that would allow this include ring and sheet-beam klystrons, gyroklystrons, and free-electron lasers, and all of these are being studied. But the best solution to the problem is not yet clear. At very short wavelengths, lasers are available with adequate power, but with somewhat limited efficiency, and high cost.

Even if a suitable power source were available, problems would arise in designing a linear accelerator using very short wavelengths. The cavities would become smaller and harder to machine. An extreme example would be a laser-driven open structure that might be generated lithographically. But in such a small structure, wake fields, which tend to blow up the beam and thus reduce luminosity, are larger. The problem could be alleviated by higher-precision construction and alignment, and by designing special cavities, but this may not be enough.

5.3.2 Exotic Acceleration

5.3.2.1 Inverse Cherenkov Effect. The need for a mechanical cavity, and the problems in its construction for short wavelengths, can be overcome by accelerating in a gas with finite refractive index. The mechanism is the opposite of the radiation (Cherenkov) that is emitted when a particle passes through such a gas. Demonstrations so far have been at relatively low gradients, but high gradients are certainly possible. Unfortunately, Coulomb scattering in the gas increases the beam size (emittance), which probably makes the mechanism unsuitable for linear colliders.

5.3.2.2 Single-Pulse Acceleration. In conventional radio-frequency acceleration, the electrical fields are alternating, and continue for many thousands of cycles. Higher accelerating gradients can be obtained without breakdown if the electromagnetic power is provided as a single one-sided pulse. Such a pulse could be generated by electrical switching, or from a drive beam consisting of a single intense bunch. But such acceleration appears expensive, and in such nonresonant systems it may be hard to obtain the degree of field symmetry needed to preserve emittance.

5.3.2.3 Plasma Acceleration. Even higher gradients can be obtained in plasma waves, where no further breakdown is possible. The plasma waves can be excited by beating two laser pulses of slightly different frequency, by one or a train of short laser pulses, or by one or a train of intense particle bunches. Gradients as high as 3 GeV/m have already been achieved. A 1-TeV collider with this gradient would be less than a kilometer long. But the demonstrations so far have been only over short distances, and only of a small number of particles. It remains to be seen whether such mechanisms can be used to achieve high energies at the required high efficiencies while preserving the required small beam sizes (emittances).

5.3.2.4 Inverse Free-Electron Laser (IFEL). There can be no breakdown at a focus of intense electromagnetic fields in a vacuum, but such fields do not accelerate an unperturbed particle. If the particle is wiggled by an external magnetic field, then acceleration is possible. There is no obvious limit to the gradients attainable, and the beam size (emittance) would not be blown up during such acceleration. But, unfortunately, loss of energy from direct synchrotron radiation sets an effective limit on the maximum energy achievable. For electrons, this limit is at only about 50 GeV, i.e., at an energy already obtained. Once again, the technology seems unlikely to be applicable to high-energy colliders.

5.4 Gamma–Gamma and Electron–Gamma Colliders

Such colliders have been considered as add-ons to electron–positron linear colliders. The motivation has been to exploit the different initial states available for detailed physical studies.

The electron–positron collider, for this application, would be operated with electrons on both sides. Just prior to the intersection point, the electron beam would be intercepted by a powerful pulse of laser light. A substantial fraction of the electrons would backscatter on the laser beam, converting the electron beam into a gamma (i.e., very high-energy light) beam aimed at the intersection point. The remaining unscattered electrons could be swept aside by a magnetic field or a plasma lens, leaving a pure gamma beam. For gamma–gamma interactions, this process would be duplicated on the other side. For gamma–electron collisions, the other side would be left unmodified.

Since the gamma beams do not produce synchrotron radiation (beamstrahlung) when they pass through the interaction region, many constraints that apply to electron–positron colliders are removed. In addition, a positron source is no longer needed, and less emittance damping is required. It might be hoped that these conditions would allow higher luminosity and efficiency in a specially designed collider. One might even hope that this would allow the design of higher-energy gamma–gamma colliders than are practical with electron–positron collisions. But this is yet to be convincingly demonstrated, and seems unlikely.

5.5 Muon–Muon Collider

In general, the physics that could be studied with a muon–muon collider would be the same as that with an electron–positron machine. Muons are just heavy electrons. But muons, because they are heavy (about 200 times heavier than an electron), produce much less synchrotron radiation. There are two advantages:

1. They give little beamstrahlung (synchrotron radiation as the bunches collide). As a result, far more muons per bunch can be used, yielding greater luminosity.
2. They can be accelerated and stored in circular rings, like protons, at energies at which an electron–positron collider has to be linear.

The effective energy in the collision of point particles like muons (or electrons or gammas) is at least ten times greater than that in the collision of the more diffuse protons. Thus a circular muon machine would be about a factor of 10 smaller than its proton equivalent. A 4-TeV muon collider, with 8-km circumference, would have an effective energy similar to that of the 85-km-circumference Superconducting Super Collider (SSC). The muon machine would also be smaller than a conventional electron linear collider of the same energy: A 4-TeV electron collider would be 50 km long, even assuming an accelerating gradient double that of the Next Linear Collider (NLC). But the smaller size of the muon machine does not guarantee that it would be cheaper. And it is not obvious that the needed luminosity can be obtained in the muon machine.

There are problems in generating enough muons, and in obtaining small enough beams (damping the muon emittance). One can certainly not make as many muons, with as small a beam size, as with electrons. But there is a compensating factor. In an electron linear collider, the particles are used once and discarded. In a muon machine, they can be circulated a thousand or more times before they decay. On paper, it is thus possible to generate comparable, or even higher, luminosities in a muon–muon machine. But muon colliders have not been studied in enough detail to show whether this is really true.

There is another problem with a muon–muon collider. The muons will, between interactions, be circulating in a storage ring. While circulating, they will be continuously decaying into electrons. These electrons will have lower energy than the muons, and will not remain trapped in the ring. They will spiral in, and collide with the walls, and make a large amount of undesirable radiation as they do so. Despite the best efforts to shield it, some of this radiation is bound to pass through the detectors being used to study the muon–muon interactions. Such radiation will certainly complicate, and possibly destroy, the ability of the physics users to see and understand the interactions of interest to them. The question is being studied, but we do not yet know whether this is an insuperable problem.

GLOSSARY

Acceleration Gradient: The increase in energy of a particle, per unit of length, in an accelerating field.

Antiproton Source: A facility designed for the collection and storage of antiprotons, usually generated through collisions of a high-energy proton beam with a metal target.

Beamstrahlung: A process in colliders producing radiation when a beam of electrons interacts with the electromagnetic field of the oncoming beam.

β Function: The function characterizing the focusing system in an accelerator.

Bunch Compressor: A device that compresses the length of the bunch while increasing its energy spread.

Cavity: A conducting box or other structure in which electromagnetic fields are constrained.

Charmed Quark: The fourth member of the family of six known quarks.

Collider: A facility in which high-energy elementary particles are brought into collision.

CP Invariance: Combined charge-conjugation invariance and parity symmetry. At present, the only observed violation is in kaon decay.

Damping Ring: An electron storage ring designed to shrink the beam size rapidly prior to acceleration in a linear collider.

Emittance: The phase-space area occupied by a particle beam.

Field Emission: The spontaneous emission of electrons from a metallic surface in the presence of a large electric field.

Final-Focus Test Beam: A test final-focus system for future linear colliders.

Free-Electron Laser: A source of electromagnetic radiation similar to a klystron, but in which the bunching and energy extraction are done in an undulating magnetic field, instead of in a cavity.

Gamma: A commonly used term for a single photon of electromagnetic energy.

Gyroklystron: A source of radio-frequency radiation similar to a klystron, but in which the bunching and energy extraction are done in, and from, the circumferential motion of a beam with large angular momentum.

Higher Modes: Electromagnetic radiation contained in a cavity at frequencies higher than those for which it was primarily designed.

Induction Units: Acceleration devices that employ a toroid of magnetic material that rings the beam. A change in the flux in the toroid induces acceleration in particles that pass through it.

IP: Interaction point, the point at which two beams collide in a particle collider.

Kicker: An electrical device that can deflect a collection of particles. Usually used in collaboration with a pickup.

Klystron: A source of radio-frequency power that uses an applied voltage to accelerate an electron beam, bunches that beam by modulating its energy and allowing the modulated beam to drift, and extracts power by decelerating the beam in one or more cavities.

Linac: Linear accelerator.

Linear Collider: A colliding-beam accelerator in which bending at high energy is avoided and the collided particles are not recirculated.

Luminosity: A measure of the rate of interactions produced in a collider, for a given reaction cross section, the cross section being a property of that reaction.

Phase Space: The multidimensional space describing particle motion in terms of both real space and momentum.

Pickup: An electrical device that can sense the position of a particle or collection of particles.

Plasma Lens: A length of plasma through which a beam is passed in order that the beam be focused by the induced self-fields.

Polarized Beam: A beam of particles in which the magnetic moments of the particles are predominantly aligned.

Pretzel: A description of the appearance of two counter-rotating electron–positron beams when they are separated by electrostatic kickers.

Quark: One of the fundamental building blocks of nuclear matter. Quarks exist in three families, each containing two partners: (up, down), (charmed, strange), and (top, bottom).

Stochastic Cooling: A technique, based on pickups and kickers, for increasing the density of a diffuse beam.

Superconductor: A material that loses electrical resistance when cooled to a low temperature.

Synchrotron Radiation: The radiation emitted when a charged particle is deflected by a magnetic field.

τ-Lepton: A charged lepton with mass about 1.8 GeV. It is the third elementary particle of the group consisting of electron, muon, and tau.

Y(4S): A bound state of bottom quark–antiquark pair with mass about 10.6 GeV.

***W* and *Z* Bosons:** Particles that together with the photon mediate the electro-weak force. Their discovery confirmed the unification of electromagnetic and weak forces.

Wake Field: The electromagnetic field left by a beam of particles that can influence trailing particles in the beam. It is a strong function of the vacuum chamber enclosing the beam.

HIGH-ENERGY–ACCELERATOR LABORATORIES

CERN: Center for Particle Physics Research in Geneva, Switzerland.

DESY: A particle physics laboratory in Hamburg, Germany, which in the early 1990s commissioned the first hybrid electron–proton collider.

Fermilab: Fermi National Accelerator Laboratory, a national laboratory for particle physics in Batavia, Illinois.

KEK: National Laboratory for particle physics in Tsukuba, Japan.

SLAC: Stanford Linear Accelerator Center in Palo Alto, California.

COLLIDER FACILITIES, PROJECTS, AND DETECTORS

Accelerator Facilities

ADONE: An e^+e^- ring collider at Frascati, retired in 1992.

BEPC: An e^+e^- ring collider of center-of-mass energy more than 2 GeV at the High Energy Institute in Beijing, China.

CESR: An e^+e^- ring collider at the Laboratory for Nuclear Studies, Cornell University.

HERA (Hadron Elektron Ring Anlage): An electron–proton collider operating at the DESY Laboratory in Hamburg, Germany.

ISR (Intersecting Storage Rings): A proton–proton collider that operated at the CERN Laboratory in Geneva, Switzerland through 1983.

LEP (Large Electron Positron Collider): An electron–positron collider currently operating at the CERN Laboratory in Geneva, Switzerland. At more than 50 GeV per beam, it is the highest-energy electron–positron circular collider in the world, capable of producing the *Z* boson.

SLC (SLAC Linear Collider): A modified Linear Collider—the first of its kind.

Tevatron: A proton–antiproton collider facility currently operating at the Fermi National Accelerator Laboratory in Batavia, Illinois, the highest-energy hadron collider (proton–antiproton) in the world today.

TRISTAN: An e^+e^- ring collider of center-of-mass energy more than 30 GeV at KEK.

VEPP IIM: An e^+e^- ring collider at the Budker Institute in Novosibirsk.

A Few e^+e^- Detectors

CLEO: A particle detector used in the CESR Collider.

SLD: Particle detector at SLC interaction region currently taking data.

Accelerator Projects

***B* Factory:** An e^+e^- collider designed for the copious production of *B* mesons. Two are in design stage: PEP II, a *B* factory being designed at SLAC; KEK B, a *B* factory being designed at KEK.

DaΦne: An e^+e^- ring collider (ϕ-meson factory) being constructed at Frascati, Italy.

LHC (Large Hadron Collider): A proton–proton collider currently proposed for construction at the CERN Laboratory in Geneva, Switzerland.

Main Injector: A facility aimed at improving performance of the Tevatron collider, currently under construction at the Fermi National Accelerator Laboratory in Batavia, Illinois, U.S.A.

NLC (Next Linear Collider): A project being considered by SLAC. If proposed, would be the first full-fledged linear collider. Also, a generic term used for future linear colliders.

RHIC (Relativistic Heavy Ion Collider): A heavy-ion collider facility currently under construction at Brookhaven National Laboratory in Upton, New York, U.S.A.

SSC (Superconducting Super Collider): A proton–proton collider upon which construction was initiated, but not completed, in Waxahachie, Texas, U.S.A.

TESLA: An international collaboration, including DESY, Cornell, and Fermilab, for the design of a TeV Linear Collider using superconducting technology.

τ–Charm Factory: A high-luminosity e^+e^- collider operating at a center-of-mass energy of 3 to 4 GeV.

Further Reading

Bryant, P., Johnsen, K. (1993), *The Principles of Circular Accelerators and Storage Rings*, Cambridge, UK: Cambridge Univ. Press.

Edwards, D. A., Syphers, M. J. (1993), *An Introduction to the Physics of High Energy Accelerators*, New York: Wiley.

Edwards, Helen (1985), "The Tevatron Energy Doubler: A Superconducting Accelerator," *Annu. Rev. Nucl. Particle Sci.* **35**, 605–660.

IEEE (1989), *Proceedings of the 1989 IEEE Particle Accelerator Conference: Engineering and Technology, March 20–23, 1989, Chicago, Illinois*, New York: IEEE.

IEEE (1991), *Proceedings of the 1991 IEEE Particle Accelerator Conference: Engineering and Technology, May 6–9, 1991, San Francisco, California*, New York: IEEE.

IEEE (1993), *Proceedings of the 1993 IEEE Particle Accelerator Conference: Engineering and Technology, May 17–20, 1993, Washington, DC*, New York: IEEE.

Lederman, L. M. (1991), "The Tevatron," *Sci. Am.* **264** (3), 48–55.

Lindstrom, E. R., Taylor, L. S. (Eds.) (1987), *Proceedings of the 1987 IEEE Particle Accelerator Conference: Engineering and Technology, March 16–19, 1987, Washington, DC*, New York: IEEE.

Rees, John (1989), "Stanford Linear Collider," *Sci. Am.* **261** (4), 58–65.

Wurtele, Jonathan (1994), "Advanced Accelerator Concepts," *Phys. Today* **47** (7), 33–40.

PARTICLE DETECTORS

See DETECTORS, PARTICLE, CALORIMETRIC; DETECTORS, SCINTILLATION; DETECTORS, SEMICONDUCTOR; PARTICLE DETECTORS, TRACKING

PARTICLE DETECTORS, TRACKING

A. Acevski, *Department of Physics, Bath University, Bath, United Kingdom*

T. C. Meyer, *CERN, Geneva, Switzerland*

	Introduction	141
1.	**Particle Detectors**	142
1.1	Gaseous Detectors in Proportional Mode	143
1.1.1	The Ionization Chamber	143
1.1.2	The Proportional Wire Chamber	144
1.1.3	The Drift Chamber	145
1.1.4	The Parallel-Electrode (Plate) Chamber	146
1.1.5	The Multistep Chamber	146
1.1.6	The Proportional Gas Chamber with Stimulated Scintillation	147
1.1.7	The Time Expansion Chamber	147
1.1.8	The Blade Chamber	148
1.2	Proportional Detectors for High-Rate Applications	149
1.2.1	The Induction Drift Chamber	149
1.2.2	The Microstrip Gas Chamber (MSGC)	149
1.2.3	The Straw-Tube Chamber	152
1.3	Gaseous Detectors in Nonproportional Mode	153
1.3.1	Limited Streamer Mode	153
1.3.2	The Gaseous Pixel Chamber	154
1.3.3	The Geiger–Müller Counter	154
1.3.4	The Spark Chamber	155
1.3.5	The Resistive Plate Chamber	155
1.4	Solid-State Detectors	156
1.4.1	The Silicon Strip Detector	156
1.4.2	The Silicon Drift Chamber	157
1.4.3	The Pixel Detector	157
1.4.4	The Gallium-Arsenide (GaAs) Detector	158
1.4.5	The Diamond Detector	158
1.5	Scintillating Fiber Detectors	158
1.5.1	The Organic Scintillating Fiber	159
1.5.2	The Inorganic or Glass Scintillating Fiber	160
1.5.3	The Liquid Scintillating Capillary Fiber	160
1.6	Visual Tracking Detectors	160
1.6.1	The Cloud Chamber	160
1.6.2	The Bubble Chamber	160
2.	**Tracking Systems**	161
2.1	The Imaging Drift Chamber	161
2.2	The Time Projection Chamber	161
2.3	The Vertex Detector	162
2.4	The Multidrift Tube Detector	162
2.5	The Honeycomb Strip Detector	164
2.6	The Wall-less Drift Chamber	165
3.	**Applications of Particle Detectors in the Future**	166
4.	**Summary**	168
	Glossary	171
	Works Cited	172
	Further Reading	174

INTRODUCTION

Particles can be emitted in a variety of physical processes. They differ by their energy, the way they interact with other particles, and their mass. In the majority of cases, they are detectable by their electromagnetic interactions, directly if they are charged, and by their decay products if they are uncharged (the latter case applies to the conversion of photons into an electron or electron-positron pair). In a few cases, detectors are directly sensitive to the energy deposited in a medium—for example, the bubble chamber and the superconductive bolometer.

3-527-28135-5/95/$5.00 + .50

Charged particles are detected when crossing solid, liquid, or gaseous media by the ionization they produce or by the production of photons as in scintillation, Cherenkov light, and transition radiation. This can be exploited in electronic detectors to obtain information on the particle's energy, velocity, and charge. Normally, in the field of high-energy physics, a combination of detectors must be used to identify particles, since each detector can only measure a single parameter.

The types of detector that the physicist has at his disposal are ionization chambers, wire chambers (in proportional and nonproportional mode), drift chambers, multistep avalanche chambers, scintillation counters, liquid detectors, solid-state detectors, Cherenkov counters, and transition-radiation detectors. While most detectors deliver a signal for a specific particle interaction, it is sometimes necessary to have a full image of particle trajectories in various media. For this, tracking devices play a major role in today's nuclear and subnuclear physics and have been the backbone of large detectors in the past. The development of particle detectors started with the earliest cloud chambers and progressed to nuclear photographic emulsions and then bubble chambers, spark chambers, and streamer chambers, and finally reached the combinations of multiwire and drift chambers used today.

When consideration is made of the ever-increasing complexity of particle collisions being examined today, particle tracking becomes an essential part of the field of nuclear physics, having applications in astrophysics, nuclear imaging in the medical profession, biological analysis, and even archeology.

1. PARTICLE DETECTORS

Particle detectors (with the exception of photographic imaging detectors) deliver an electrical signal after they interact with charged particles. Stable neutral particles such as photons and neutrons are only detectable via an intermediate conversion step, in which their kinetic energy is used to produce secondary charged particles in a nuclear (neutron-proton recoil) or electromagnetic (photon–electron/positron conversion) interaction. These detectors are used to count the number of particles emitted by unstable nuclei or produced in nuclear collisions, to measure their energies or trajectories, and to evaluate their masses. The type of detector depends on the intrinsic properties of the particles to be detected, but generally, it is the electromagnetic interaction of the particle with its surroundings that is observed; typically the interaction involves the production of electron-ion pairs in a medium. The result may be current pulses in wire chambers, electron-hole production in solid-state devices, photons in a scintillating material, Cherenkov light, or transition radiation.

The electric fields in gaseous detectors are controlled by electrode configurations and are not always of uniform intensity. The nature of the gas filling is strongly dependent on the applications foreseen for a detector. Further, the nature of the interaction or the radiation to be detected may be a factor leading to the choice of some components. For instance, in order to detect slow neutrons with high efficiency, BF_3 or ^3He is used. For soft x rays, xenon or krypton fillings are favorable. For charged particles, the criterion is sometimes the maximum permissible gain in order to simplify electronics. Because of its very low cost, argon is often used as the major component. The additional gases (methane, alcohol, CO_2, etc.) have various functions: quenching of the metastable states of the noble gases that are responsible for UV photon emission, reduction of the resolution time by increasing the drift velocity, enhancement of the visible photon emission when the optical detection of the avalanches is required, reducing the aging rate due to polymerization of the vapors, and more. The average energy loss of a charged particle in a gas varies between 20 and 35 eV per liberated ion. In large systems, the gas filling is flushed continuously and released into the atmosphere.

Photosensitive gases, used as an additive in a main gas carrier, have opened a new field of applications for gaseous detectors. Triethylamine (TEA), e.g., has an ionization potential of 6.5 eV. It has a peak quantum efficiency for UV photons larger than the best photocathodes used in photomultipliers, and it permits the detection and imaging of photons emitted by Cherenkov radiation or by scintillators like BF_3, with spatial accuracies close to 1 mm.

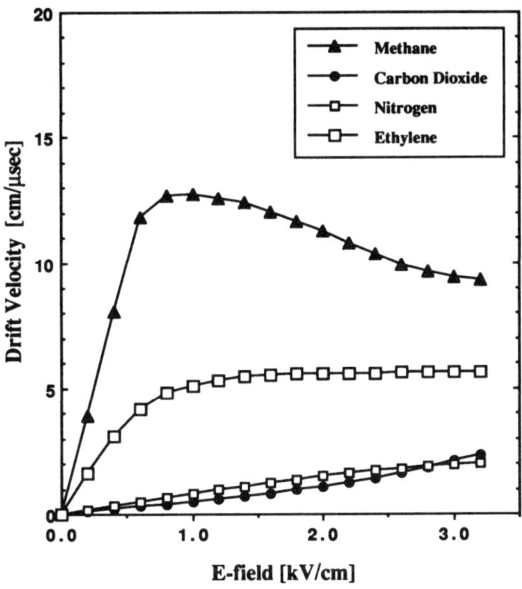

FIG. 1. Electron drift velocity for various gases.

1.1 Gaseous Detectors in Proportional Mode

When a charged particle crosses a gaseous medium, it can collide with the atoms of that gas, and if it can transfer more than the electron binding energy, then electrons can be liberated from the medium. These can either stay free or be rapidly recaptured by parent ions or atoms with (unwanted) electron affinity, such as oxygen.

The introduction of an electric field in the detection region leads to many other effects that need to be taken into account in the design of detectors or the selection of detector types. This is determined by the nature of the gas and the field intensity. The presence of a field leads to two possibilities:

1. If the field is strong enough, electrons will be accelerated enough between collisions to knock another electron out of the next atom they hit, ultimately causing an electron avalanche.
2. If the field is not sufficiently strong, electrons freed from the material will only collide with atoms and drift along field lines until they reach the field anode, where they are captured.

The average velocity of these drifting electrons is determined by the properties of the gas and the strength of the electric field; Fig. 1 shows electron drift velocity vs electric field for four different gases (Sauli, 1977). This finite velocity results in a time delay between emission and detection of a given electron, which is used to determine the point at which the electron was liberated (see Sec. 1.1.3 on drift chambers). For further discussion, see Palladino and Sadoulet (1974) and Sauli (1977).

1.1.1 The Ionization Chamber In an ionization chamber, a weak homogeneous electric field is used to collect the electrons at the anode, without their undergoing charge multiplication (Fig. 2). The amount of charge collected gives a measure of the average radiation intensity and the energy loss (dE/dx) that the particles have suffered in traversing

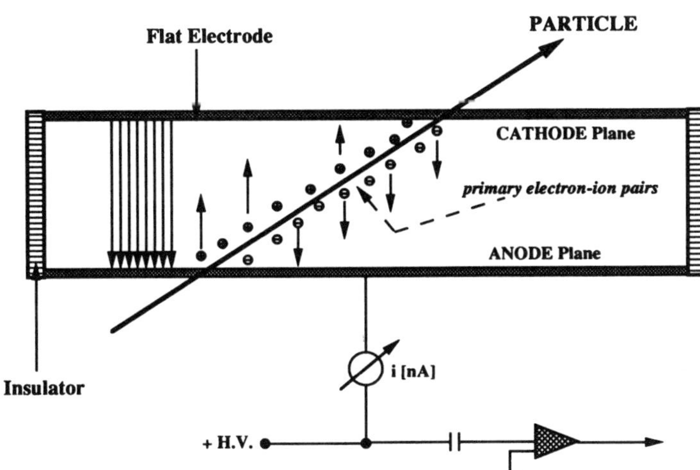

FIG. 2. The ionization chamber.

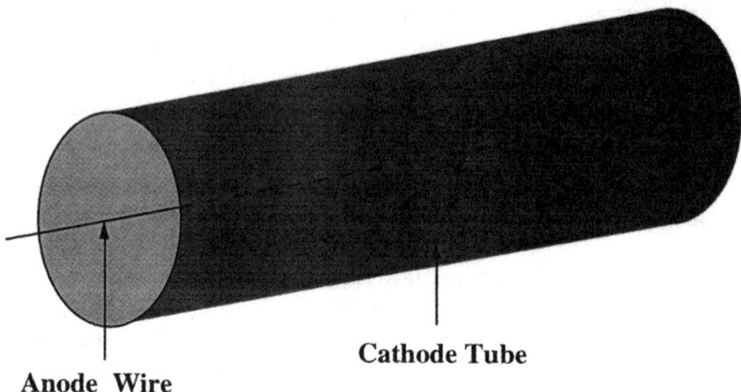

FIG. 3. The proportional wire tube.

the gas gap of the chamber. This instrument is widely used in dosimetry and calorimetry [where a liquid is often used, such as liquid argon (LAr), xenon (LXe), or tetramethyl pentane (TMP) (Ankoviak *et al.*, 1989), instead of a gas]. As a result of not relying on gas gain, it has a very stable and uniform response, and it can measure average fluxes with an accuracy of 10^{-4}.

Ionization chambers can detect a single particle if that particle has enough energy and gas volume to liberate many electrons. Depending on the quality of the amplifier used, the noise equivalent can be between about 100 and 1000 electrons, so that at least this many must be produced for a detectable signal to be generated by the chamber.

Generally, ionization chambers have little or no spatial granularity (except in calorimetry) and are rarely used in tracking. They often serve as charge-integrating rate meters, such as accelerator beam monitors.

1.1.2 The Proportional Wire Chamber The simplest proportional chambers (Fig. 3) consist of a thin wire held taut between two points, along the axis of a hollow cylinder. The wire acts as the anode, and the cylinder, which is the cathode, is filled with gas that can be maintained at a pressure of between 100 Pa and 10 MPa, although it generally is kept at atmospheric pressure for mechanical reasons. The parameters of size and wire potential are chosen to give the correct signal gain (and hence electron multiplication factor) close to the wire. The detector is strictly operated in a regime where the response of the detection cell is proportional to the primary charge produced within the sensitive volume.

The simple cylindrical wire counter is still used for limited applications where the detection of soft x rays (in the range of 1–50 keV) is required. It delivers pulses with an energy resolution of 13% [full width at half maximum (FWHM)] at 6 keV, varying like $E^{-1/2}$. This is much poorer than what can now be achieved with solid-state detectors (Sec. 1.4), but it is much less expensive for a given volume.

More elaborate detectors—for example, the multiwire proportional chamber (MWPC) (Charpak *et al.*, 1979)—have been developed from these simple chambers and are in use at CERN and other research facilities because of their versatility, good resolution, and large surface coverage (Fig. 4). With the possibility of very accurate localization of ionizing events, they allow precise imaging for high-energy collisions.

In multiwire chambers, there are two electric field regions: one far from the wire, where the field is uniform, and the other close

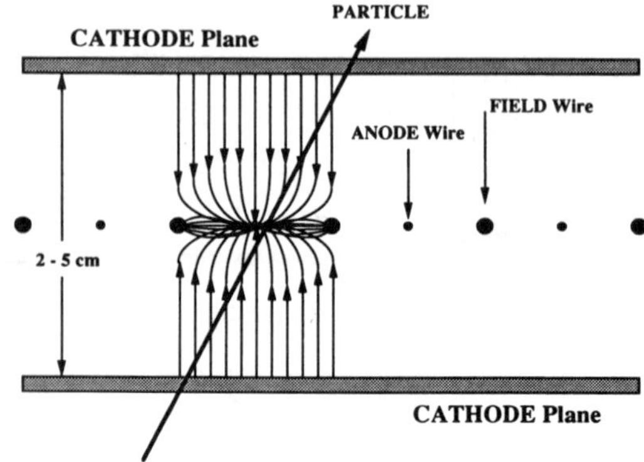

FIG. 4. The multiwire proportional chamber (MWPC).

to the wire, where the field is radial. In the latter region, electrons experience inelastic collisions leading to an avalanche effect, which gives the detector its signal gain; i.e., one electron outside this region leads to many electrons hitting the wire.

Typically, with mean free paths for the electrons of about 1 μm and proportional gains of $\leq 10^5$, the multiplication process occurs over a distance of about 15 μm. The electrons from the avalanche are rapidly collected by the wire, whereas the positive ions must drift across the full potential gap of the chamber before hitting the cathode. The electrons generated in the avalanche result in a sharp current pulse followed by a steady decay because of the induced current from the receding positive ions. It is important to notice that the collection time of electrons induces a very small pulse on the wire since the induced charge is proportional to the potential drop experienced by the moving charge and, for electrons, is only a minute fraction of the total applied potential difference. In contrast, the heavy positive ions have to cross the entire anode-cathode distance, and it is their motion that induces most of the observed pulse. The motion is fast close to the wire, because of the intense electric field, and slow after a small distance of a few tens or hundreds of microns.

The resolution time of a proportional counter, i.e., the accuracy with which one can ascertain the initial time of the ionization process, is equal to the fluctuations in the time required for an electron liberated in the gas to reach the multiplication region. It may be of the order of 0.5 μs for a counter of 10 mm diam. This resolution time is much larger than that reached by scintillation counters.

In planar geometries sense (anode) and field (shaping) wires are interspaced at regular distances (order mm) between an upper and lower cathode plane (Fig. 4). This arrangement builds individual and isolated detection segments with a sense wire at the center. The cell size, also known as the wire pitch, determines the potential accuracy of the detector. In such planar geometries an ionizing event can be measured with an accuracy of 10^{-2} times the wire spacing. This is done by using the values of the current pulses measured from neighboring field wires to construct a Gaussian envelope, whose peak lies at the position of the ionizing event. To obtain information *along* the wire in the same chamber, two methods are most commonly employed in MWPCs:

1. Cathode segmentation: If the cathode surface of the detector is divided into strips or pads along the wire, the charges induced on individual cathode segments can be measured as functions of distance from the electron-wire impact. The mutual sharing results in a Gaussian-shaped distribution whose peak position describes the impact point with very high precision (order tens of microns). This method, however, requires adequate granularity and yields a large number of electronic channels. This property is essential for the imaging of soft radiations, such as x rays or neutrons, when the two coordinates have to be determined from a single totally absorbing interaction.
2. Charge division: If a resistive alloy (several kΩ/m) is used for the sense wire, signals detected at *both* ends of the wire will suffer different attenuation depending on the distance traveled. The charge asymmetry can then be used to determine the avalanche impact point on the wire. The precision of this method depends crucially on the gain (signal/noise ratio) and is typically 1% of the total wire length.

Multiwire proportional chambers have shaped more than two decades of particle tracking. No other detector has ever acquired the same versatility, experience, and confidence in operational safety, even at present system standards. In maintaining the basic operating principles they have gradually paved the way to more elaborate and sophisticated devices that will be discussed in the following sections.

For further discussion see Charpak *et al.* (1979) and Walenta (1983).

1.1.3 The Drift Chamber The drift chamber is an extension of the proportional wire chamber in which the position of the ionizing event normal to the plane is found by examining the drift time of the electrons. This is done by using an external triggering device to start a clock and exploiting the fact that the electrons have constant drift velocity until they enter the nonuniform field region close to the sense wires. The signal, which is proportional to the ionization of the track segment as projected on the receiving sense wire cell, stops the clock.

Most drift chambers operate with the drift (E) field oriented transverse or parallel to the direction of a magnetic bending field (B). Therefore, nonparallel E and B fields strongly influence the velocity and direction of drifting electrons, usually leading to loss of efficiency and resolution (Schmidt, 1980; Atac *et al.*, 1986). The time resolution of such a chamber depends also on how much the electrons diffuse away from the ionization cluster while they undergo drifting, which again depends on the magnetic field and combined electric and magnetic effects. However, choosing suitable gases (fast or slow) in combination with high drift fields can restore chamber performance.

A typical drift-chamber cell is shown in Fig. 5. Using inexpensive electronics, which permits reaching an accuracy of the order of 1 ns in the measurement of the drift time (corresponding to order 100 μm for the position of the initial electron), it is easy to reach the same spatial resolution as the wire planes achieve in the proportional chamber. The drift distance can also be large (of the order of 10 cm), which reduces the need for large numbers of sense wires, helping to cut the cost of such detectors.

For further discussion see Breskin *et al.* (1975) and Sauli (1977).

1.1.4 The Parallel-Electrode (Plate) Chamber Parallel-electrode or plate chambers produce uniform electric fields in which nonetheless electron multiplication can take place, but it is more difficult to achieve high gains with this type of detector without the strong radial fields created by wires. The gain of such a detector is proportional to an exponential function of the distance from the ionization point to the anode. The result is that this type of detector has been preferred for particles that do not require high gains (i.e., strongly ionizing particles). However, for average gas gains of 5×10^4 over a 1-mm gap, an efficiency of 90% for minimum ionizing particles can be maintained. Moreover, these chambers have fast response and excellent time resolution of a few ns, since all electrons from an avalanche will arrive at the anode at the same time, eliminating the time-dispersion characteristics of the wire chamber, in which the charge multiplication is confined to a small region near the wire (Fig. 6). As a result of this, the parallel-plate detector can sustain very high rates of charged particles (up to 10^7 cm^{-2} s^{-1}) and, because of its highly polished conducting surface, will provide radiation resistance 100 times larger than conventional wire chambers. This, combined with geometrical flexibility and low production costs, makes the device suitable for operation at high luminosity and large backgrounds, such as the forward regions of detectors at hadron colliders.

1.1.5 The Multistep Chamber Parallel-plate counters have experienced a renewal of interest for two reasons, one being the introduction of multistep structures (Fig. 7), the other being the fact that this structure is most favorable for the imaging of avalanches from light emitted by some vapor additives in the counter gas.

The multistep chamber allows the use of more than one stage of gas amplification: A wire grid is used as the anode for the first

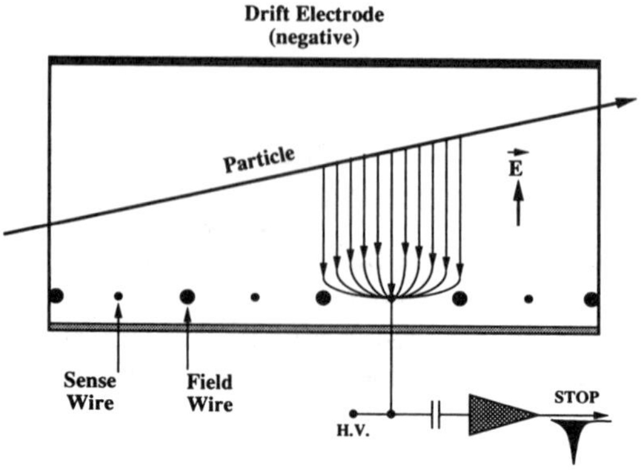

FIG. 5. The drift chamber.

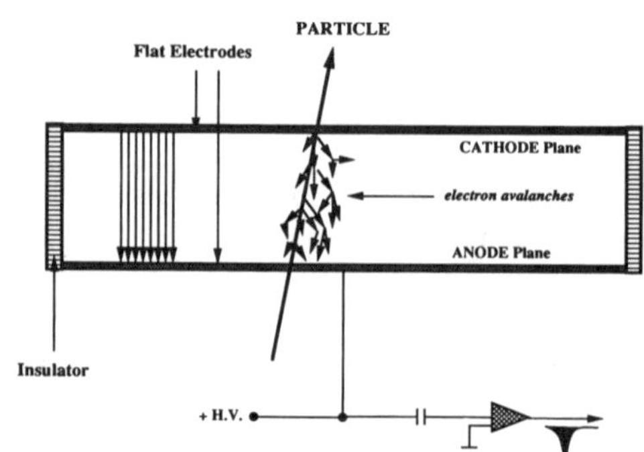

FIG. 6. The parallel-plate chamber.

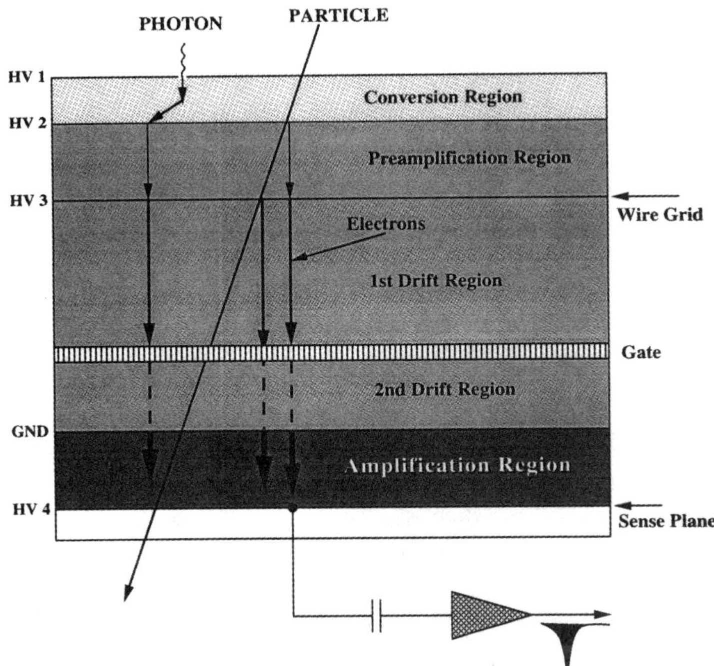

FIG. 7. The multistep chamber.

stage of amplification, where the electrons from the avalanche are extracted and transferred to a second parallel-plate structure where they start a second round of avalanches. This technique can achieve gains of 10^6 or even 10^7, with fewer secondary effects such as sparking at the anode, which produces undesirable ultraviolet radiation or positive ions. This high level of gain easily allows the detection of a single electron, making this type of detector ideal for measuring weakly ionizing particles and imaging vacuum ultraviolet photons emitted by Cherenkov radiators.

For further discussion, see Breskin *et al.* (1980), Bouclier *et al.* (1983), and Charpak and Sauli (1984).

1.1.6 The Proportional Gas Chamber with Stimulated Scintillation This device is used primarily for the measurement of a particle's energy. It converts the kinetic energy of the particle into potential energy by freeing electrons (in an ionization process) after which they are accelerated and collide with other atoms. Electrons have a certain mean free path in a gas, and in this type of chamber the electric field is adjusted so that electrons do not have quite enough energy after traveling the equivalent of one mean free path to ionize a gas atom. Instead, they will simply excite the bound electrons that will then radiate this extra energy in the form of photons, visible or otherwise. With pure noble gases it is possible to obtain emission of photons in the vacuum ultraviolet region with an abundance of about 1 photon per 10 V of potential excursion of a drifting electron. These can then be detected by photomultiplier cells or other photosensitive devices, giving pulses proportional to the number of electrons released in the original ionization process. This detector has far better energy resolution than those relying on electron multiplication processes, which are inherently prone to fluctuations.

Typical resolutions (FWHM) are 50 eV for photons of 150 eV, 400 eV for 6-keV, and 2 keV for 50-keV photons. At higher energies the solid-state detectors have the considerable advantage of greater efficiency with better resolution as well (Sec. 1.4), but are far more expensive for large surfaces.

For further discussion see Suzuki *et al.* (1988).

1.1.7 The Time Expansion Chamber The internal structure of the time expansion chamber (TEC) is very similar to that of the multiwire proportional chamber, but the principle of operation is slightly different

(Fig. 8). The electrons drift much more slowly in the drifting region than in the multiplication region. Usually, these two regions are separated by another wire grid held at ground. Using this technique, it is possible to improve the resolution dramatically, since there is a much longer time interval between successive ionized clusters arriving. The mean drift-time interval from subsequent single-ionization clusters of a relativistic particle in the TEC can be made large compared to the width of the anode signal to allow the recording of the clusters separately. In other detectors, these usually appear as one long ionized track and cannot be resolved because they arrive too quickly. Tracks that are parallel to the wire plane are resolved by reducing the wire spacing. The result is that one can distinguish individual ionized clusters as opposed to track segments made up of many clusters. Basic ionization phenomena and drift properties can be measured at the single-electron level.

The problems associated with this type of chamber arise from the relatively long drift time of some of the clusters, which will tend to fade out through diffusion and electron capture by electronegative impurities at the ppm level, e.g., oxygen. However, in TEC chambers a "cool" gas is used giving a low drift velocity ($v_d \sim E/p$) and resulting in low diffusion of $a/x^{1/2} \sim 100\ \mu\text{m/cm}^{1/2}$. The effect of diffusion can be further reduced if, instead of measuring the arrival of the first electron (leading-edge discrimination), the centroid of the nearly Gaussian electron cloud of n electrons can be measured. In that case one ideally obtains $\sigma = \sigma_0/n^{1/2}$, where σ_0 is the single-electron diffusion. A point of concern is also the transition from the low-field to the high-field region. The resulting deformation of the arriving electron cloud by the converging field lines in the amplification region can normally be sharpened by altering the choice of gas, chamber geometry, and applied voltages so as to define a narrow amplification region.

For further discussion see Walenta (1979) and Anderhub *et al.* (1985).

1.1.8 The Blade Chamber This chamber, developed in the framework of the Lepton Asymmetry Analyzer (LAA) project at CERN, is still in the development stage but is based on the use of thin, sharp-edged blades instead of wires to create the high electric field regions for charge multiplication. The shape of the blades means that the high electric field is concentrated at the edge of the blade that points toward the grounded chamber roof, as

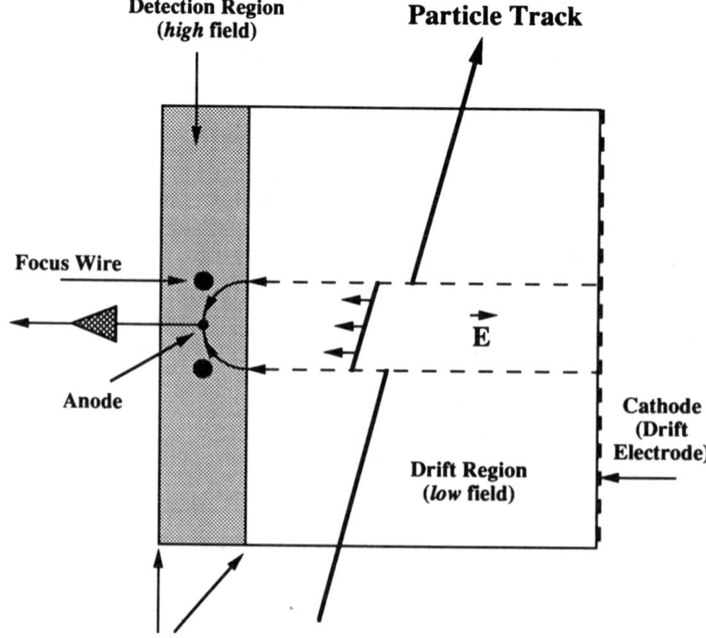

FIG. 8. The time-expansion chamber.

in Fig. 9. Under suitable conditions, the gain of the structure can be pushed to the limited streamer regime (see Sec. 1.3.1). The advantages of such a chamber are the robust nature of the blade as opposed to a wire and the fact that the blade can be bent to accommodate an irregular chamber geometry, without any support problems. Authors have demonstrated good efficiency for minimum-ionizing particle detection, and a spatial resolution of about 250 μm. In addition, a measurement of the charge asymmetry on the two walls of the cell resolves the left-right ambiguity. See Ambrosi *et al.* (1990a,b).

In more recent tests, chambers were constructed with a modular design of grounded walls running parallel to each blade, with an insulated roof, as shown in Fig. 10. This allows additional pickup strips to be mounted on the roof, perpendicular to the blade, in order to determine the remaining position coordinate of an ionized cluster.

1.2 Proportional Detectors for High-Rate Applications

For proportional gas detectors to cope with very high particle rates (i.e., >10 MHz/cm^2) the detector cell area must be sufficiently small to avoid signal pileup (Walenta, 1983). This effect is further enhanced by long drift times of electrons and the very low mobility of positive ions generated in the charge avalanche. Unfortunately, in attempting to achieve very short drift times the drift volume of gaseous chambers cannot be made arbitrarily thin, because the ionization of gas atoms by charged particles proceeds via a nonuniform clustering process. This limits the efficiency and operation of any gas-filled detector. On the other hand, positive ions copiously produced at the anode migrate to the cathodes at a speed more than a thousand times slower than electrons. Their presence in the drift space reduces the effective fields. The effect of this is twofold: a change of drift velocity for electrons and a reduction of the amplification field near the anode.

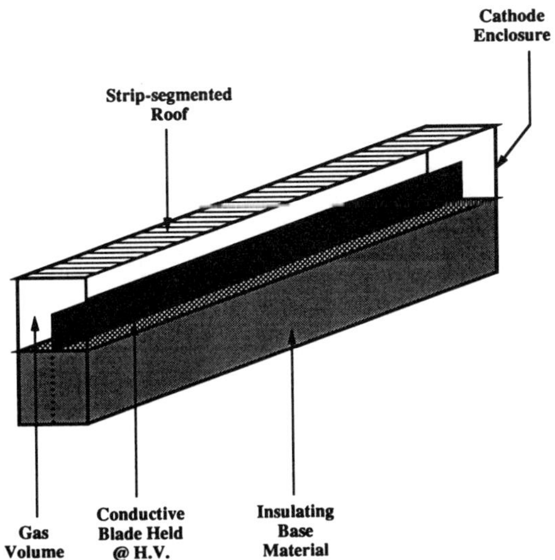

FIG. 9. One cell of a blade chamber.

The best way to overcome these drawbacks is to reduce the detector cell size to submillimeter dimensions. Fast gases with electron drift velocities of more than 100 μm/ns can help for certain (nonmagnetic) applications as well. However, these gases operated in a magnetic field face the problem of large drift angles for electrons subject to the Lorentz force (Atac *et al.*, 1986; Schmidt, 1980).

1.2.1 The Induction Drift Chamber This device, in close resemblance to the classical MWPC, uses common field and sense wires arranged in a plane having, however, a reduced cell width of 600 μm. It has an upper and lower drift volume, each being 3 mm thick, yielding better efficiency from double the number of electrons per track (Fig. 11). For best results with this chamber, particle tracks should be nearly perpendicular to the detection plane. If the weighted-charge method on field wires is used (see Sec. 1.1.2), this detector can achieve a resolution of better than 30 μm for particles at normal incidence. Again, charged-particle rates in excess of 10^7 (events/cm^{2}s^{-1}) can be tolerated without loss of efficiency. It should be noted that this chamber is still in the research-and-development stage and prototypes up to only 10 × 10 cm^2 in size have been constructed. The application of such a chamber in a real detector would require much larger areas to be covered. See Roderburg *et al.* (1986).

1.2.2 The Microstrip Gas Chamber (MSGC) Microstrip chambers are, as the name implies, made of very thin, narrow, conducting strips that lie on a thin and flat glass or ceramic substrate similar to printed circuit boards. For more "flexible" applica-

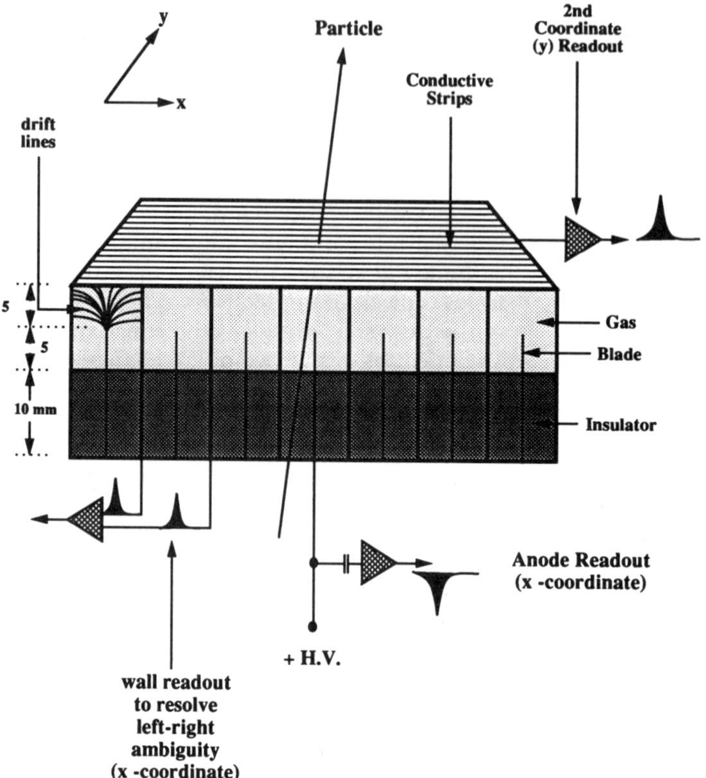

FIG. 10. The blade chamber.

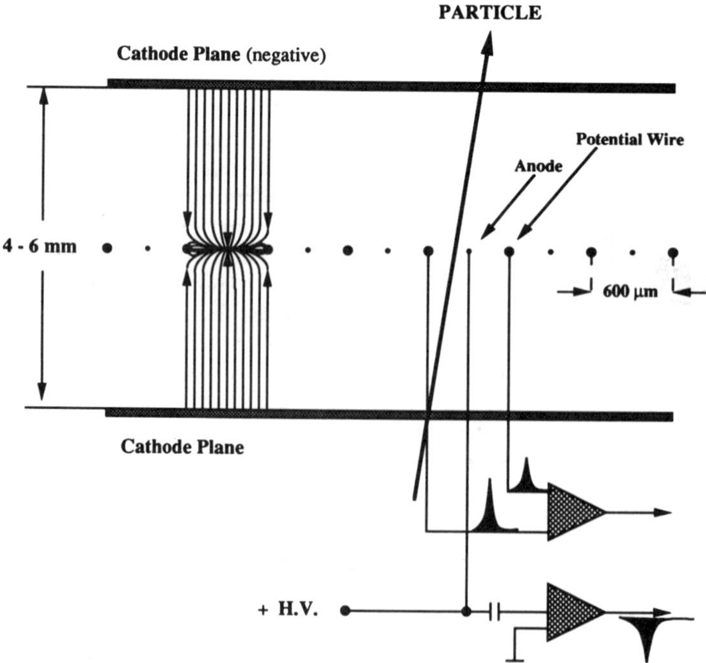

FIG. 11. The induction drift chamber.

tions plastic foils made of Kapton or Tedlar with similar strip designs have been shown to work satisfactorily in nonplanar, say cylindrical, detector geometries (Bouclier *et al.*, 1992a,b). These strips, whose pitch is ≤200 μm, replace the anode and cathode wires in the traditional multiwire proportional chamber, which normally uses much larger wire separation. Similar to the induction drift chamber, the drift region of this detector lies between these strips and the cathode plane (2–5 mm above) as shown in Fig. 12. The strips are formed by electron-beam or photo lithography, which can give a spatial accuracy of 0.1 μm for the position of the strips. Despite a modest potential difference of 200–500 V between the anode and cathode strips, a very high anode field of 100 kV/cm is produced (Florent *et al.*, 1992). Apart from the close separation of electrodes, this must be attributed to the edges of the strips and the high dielectric constant of the insulating substrate. Thus, almost the entire gain is established between the anode and cathode strips with almost no effect from the upper cathode plane. The result is a very homogeneous proportional gain ($\leq 10^4$) giving rise to excellent energy resolution (12% FWHM for 6-keV photons), and a high rate capability [$\geq 10^7$ events/(cm^2 s^{-1})] due to the short drift path of the positive ions.

The small detection cell lends itself to another important feature of this chamber: spatial resolution. In the classical readout mode, only the anode strips are read out, resulting in a modest space resolution comparable to the cell size divided by $\sqrt{12}$ (≈60 μm at 200-μm cell size). A more elaborate method makes use of the induced signal on the field or cathode strips next to the signal-producing anode strip (refer to Sec. 1.1.2). Using carefully calibrated analog electronics, the relative charge sharing between the two cathode strips allows a weighted interpolation inside the cell width with a precision of better than 30 μm. It should be noted that the resolution of this detector is spoiled by angular tracks, where the charge is deposited over several detection cells (Angelini *et al.*, 1992).

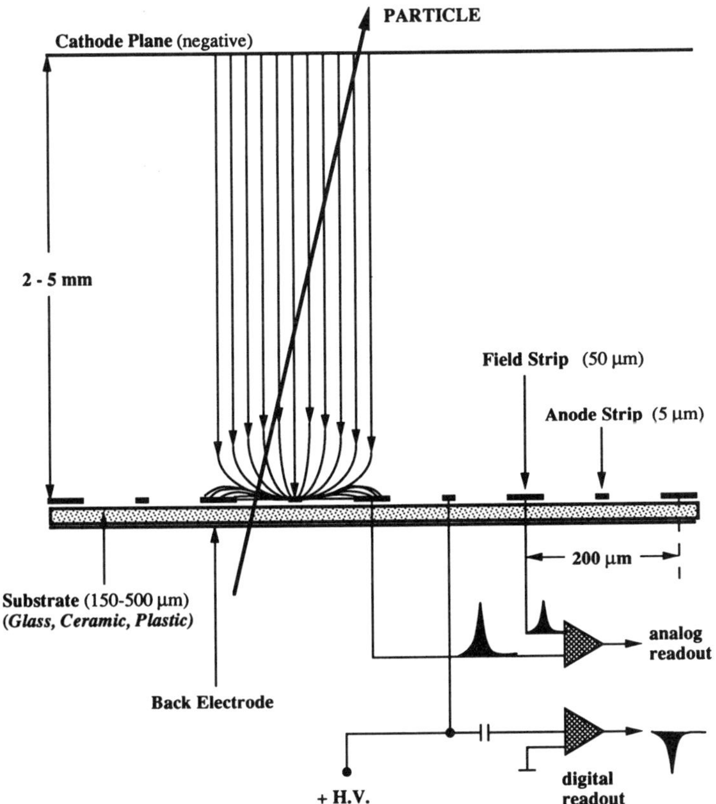

FIG. 12. The microstrip gas chamber.

To exploit the full measuring potential of MSGCs the back electrode of the substrate (see Fig. 12) can also be segmented into strips or pads of any suitable shape, which, in analogy to silicon double-sided detectors (refer to Sec. 1.4.1), can be used for the readout of the second hit coordinate. The capacitive coupling of the anode signal into the back plane is achieved by placing the back electrodes into close proximity (separated only by an insulating SiO_2 film of ~1 μm) to the front (anode-cathode) plane (Angelini *et al.*, 1993). In these so-called double-sided MSGCs the backplane potential not only ensures an optimum field configuration for avoiding charge buildup (as for single-sided MSGCs), but plays an essential role in the gain mechanism because of its close distance from the anode. Double-sided MSGCs are presently considered state-of-the-art detectors for running in future supercollider experiments.

For operation in high-rate environments, gases with high cluster statistics and drift velocity must be chosen. Two gases that simultaneously satisfy these constraints are dimethyl ether (DME)-CO_2 (60:40) and Xe-DME-CO_2 (30:30:40). Efficiencies are maintained at nearly 100%, even for high discriminator thresholds ($\geq 5 \times$ noise level). Under operating fields close to 10 kV/cm, these gases reach drift speeds of 70 μm/ns.

Early test samples operated at low rates have shown significant gain drops (>30%) within a few hours (Fig. 13), most likely due to surface charging as a consequence of the high surface and bulk resistivity of the detector substrate (Bouclier *et al.*, 1992b). The applied potentials on the substrate favor the depletion of charge carriers within the bulk, thus further increasing the resistivity of the material. By choosing substrates of low bulk resistivity (i.e., 10^{10} Ω cm) this effect can be avoided or substantially reduced.

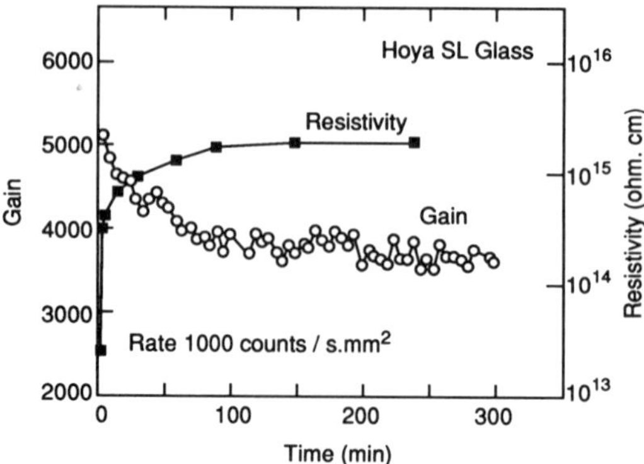

FIG. 13. Gain drop in a microstrip gas chamber after voltage turnon and simultaneous rise in bulk resistivity for a substrate made of HOYA SL glass.

MSGCs exposed to very high rates of ionizing radiation experience flux-dependent dynamic gain modifications. As in the case of spontaneous charge buildup, these can be avoided by a proper choice of a low-resistivity support and suitably chosen bias voltages. Despite gain changes, the detectors can be safely operated at fluxes up to 10^8 cm^{-2} s^{-1} (Bouclier *et al.*, 1993). More serious is the phenomenon of aging in MSGCs as observed by Bouclier *et al.* (1992a) leading to an unrecoverable loss of performance due to permanent damage of the electrode area. The mechanisms seem to be comparable to those observed in ordinary proportional chambers. Therefore, similar precautions in the selection of material and counting gases with very low impurity concentrations must be taken.

For further discussion see Oed (1988) and Angelini *et al.* (1991).

1.2.3 The Straw-Tube Chamber As the name implies, straw-tube detectors are long ($\geq$1 m) cylindrical chambers with a central wire for charge collection (Fig. 3). Resembling closely the classical Geiger–Müller tube in basic geometry, they differ by size and operational mode. The radial size of a few millimeters (diameters mostly between 4 and 6 mm) is chosen to employ the detector in high-rate environments following a philosophy similar to the multidrift tubes (see Sec. 2.4). Correct choice of gas and electric field make it work as a proportional counter that is fast, simple, and relatively radiation hard while not sacrificing the good spatial resolution ($\leq$100 μm) derived from electron drift time. Straws can equally well be used to form detector modules or layers in a geometry dictated by other experimental components (Fig. 14). Typical problems encountered in long (>1 m) straws are the precise positioning of the sense wires and gravitational sag. In some applications (SDC, 1990a) straws are 3 m long to span the available tracking volume. In these cases it will be necessary to provide centering for the sense wires at about 1-m intervals to ensure electrostatic stability. This, however, creates dead zones in addition to those

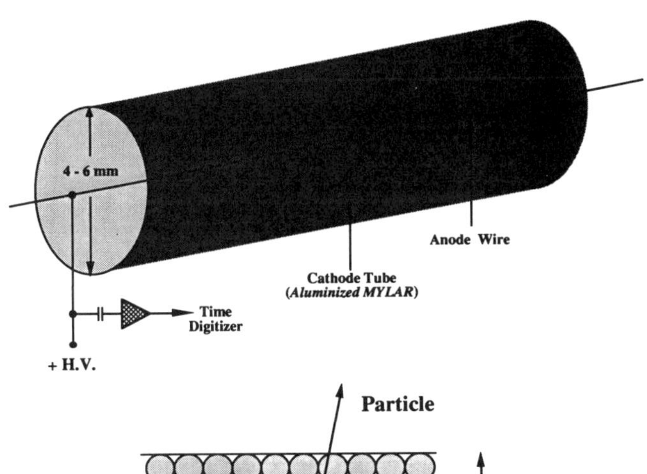

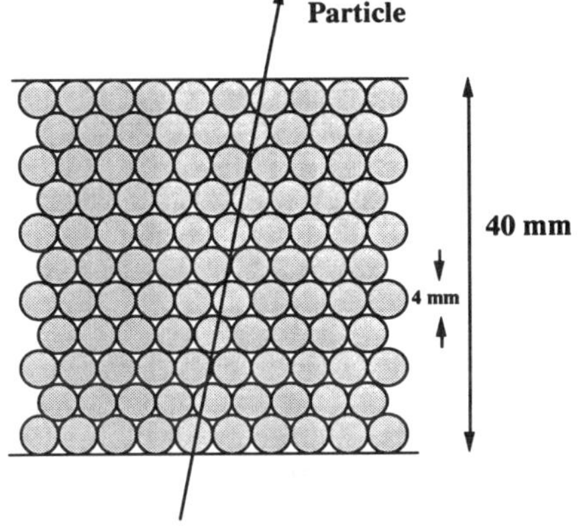

FIG. 14. A single straw tube and a straw-tube detector.

at the end pieces of the tube. Long straws also suffer from signal attenuation due to the resistivity of the wire and cathode surface. The attenuation length (gain drop to e^{-1}) was found to be ~5 m (SDC, 1990b).

Straw-tube chambers have been widely used as vertex detectors (Frautschi *et al.*, 1993) and have been considered as first-choice tracking detectors for the new generation of hadron colliders (SSC and LHC). In comparison to the previously discussed microcell proportional chambers, straw tubes are still large-surface devices that benefit from high drift velocities and simple construction only. Recent studies on their performance in high-rate experiments have shown, however, that their use is limited to accelerator luminosities not exceeding 10^{33} cm^{-2} s^{-1} (Corden, 1990).

1.3 Gaseous Detectors in Nonproportional Mode

If, in a gaseous detector, the electric field is above a certain critical value, then the ultraviolet photons emitted in an avalanche will have sufficient energy to generate further ionization elsewhere in the chamber. This is determined by the choice of wire diameter and the gas composition in the chamber. If the appropriate gases (i.e., with a reduced concentration of organic quencher) are used that do not impede the propagation of the photons, then secondary avalanches can occur far from the initial one.

1.3.1 Limited Streamer Mode In this mode of operation, avalanches occur at the wires but are not confined and extend *radially* out from them to form "streamers." However, the electric field is not high enough for a complete, ionized channel to be formed between anode and cathode, which would lead to an unwanted spark. Depending on the geometry of the chamber, the streamer can finish close to one face or two streamers may be formed on neighboring wires that do not quite meet each other. This method of detection leads to "fast" pulses, also 100 times larger than in the proportional mode of operation. However, the locally high concentration of positive space charge causes the effective field in the avalanche region of the wire to drop so low that during the migration of

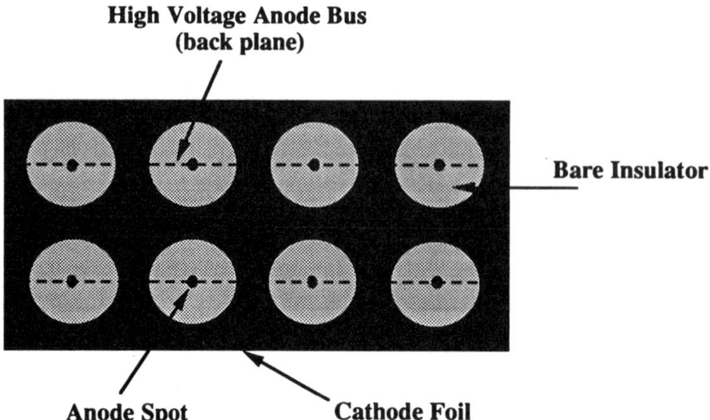

FIG. 15. The gaseous pixel detector (top view).

the positive ions toward the cathode, charge multiplication cannot occur. These are the typical dead regions formed in limited streamer mode detectors. See Brehin *et al.* (1975); Sauli (1977).

1.3.2 The Gaseous Pixel Chamber The pixel chamber is based on etched, printed circuit boards instead of wires, using a small-scale modular approach, making it useful for moderate rate applications where detector occupancy would not allow the use of large wire chambers. The accuracy of measurement is determined by the film printing and etching process used on the board; this means that relatively cheap chambers can be constructed with high levels of accuracy on a large scale. The chamber is constructed from a thin metal foil, used as a cathode plane, with small holes for the anodes, overlaid on a circuit board as shown in Fig. 15.

A second foil is employed 8 mm above the anode-cathode plane, on which are mounted cathode strips for field shaping, as shown in Fig. 16. These strips are linked together in one direction in the detector plane and the anode points are linked in the other direction in order to give a two-dimensional picture of each hit. The third spatial coordinate is obtained from the electron cluster drift time as determined by the choice of chamber gas, as in other types of chambers.

For further discussion, see Antreasyan *et al.* (1990).

1.3.3 The Geiger—Müller Counter When one is using ("counting") gases with essentially no photon-absorbing additives, elec-

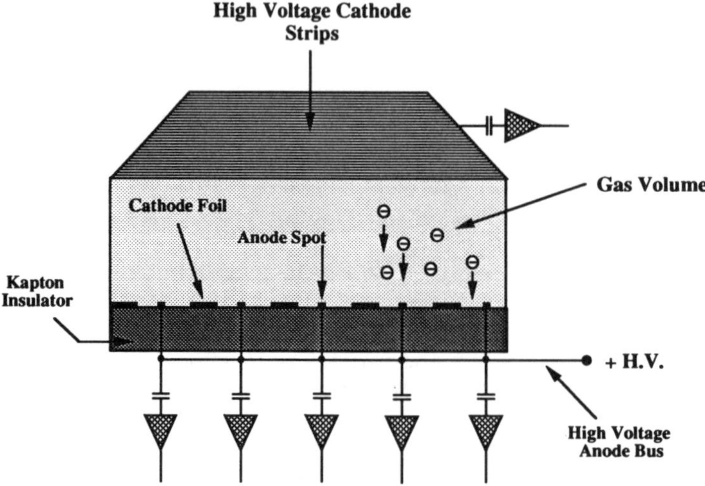

FIG. 16. The gaseous pixel chamber.

tron avalanches will copiously produce photons that are not suppressed and consequently will lead to more charge multiplication along the wire (in contrast to radial streamers). This process will continue until the entire region is saturated with ions, causing the electric field near the wire to drop sufficiently to prevent further avalanches. The result is a charge at the output of the detector of almost constant value for every particle entering the detector. This tends to make counting particles easy, and this type of chamber was used extensively in the past for such measurements. The disadvantage of this counter is the long busy time experienced (1 ms approximately), which is why it has been superseded, for the most part, by solid-state devices. The Geiger–Müller counter can also be used in proportional mode. In this mode, the dead time is the same but is limited to a region of about 0.1 mm length instead of extending over the full length of the wire. See Charpak and Sauli (1971).

1.3.4 The Spark Chamber This chamber is constructed in a parallel-plate configuration, which produces a uniform field throughout the gas volume. A spark is caused as follows: An ionizing particle passes through the chamber and ionizes a track; at some stage during this process, the particle causes a trigger to be set that rapidly raises the voltage across the chamber to a breakdown level, suitable for discharges to occur anywhere in the chamber. At this point, a complete ionized channel is formed from anode to cathode leading to a rapid (ns) discharge, or spark. This results in a large number of positive ions in the detector, which are removed by the application of a clearing field while the chamber is waiting to be triggered again. The spark itself can be detected by charge measurement or sonic methods (Meyer, 1977); often, small microphone pickups are used to detect and localize the sonic pulse from the spark (Fig. 17). Spark chambers are extremely robust and easy to handle. Because of the high spark energy, simple and low-cost electronics can be used. They are, however, limited for use at very low-rate particle detection.

1.3.5 The Resistive Plate Chamber The operating principle of the resistive plate chamber differs greatly from that of the classical spark chamber although it is one. Using

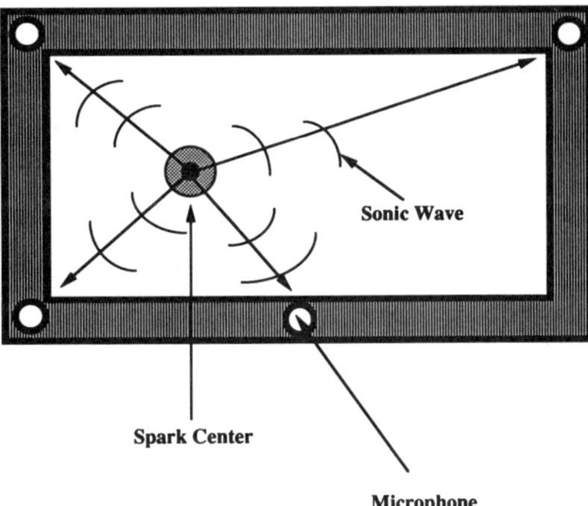

FIG. 17. The sonic spark chamber. One microphone is placed "off-corner" to avoid ambiguities.

flat electrodes with high resistivity allows one to operate this type of spark detector in a continuous, i.e., nontriggered, mode. The violent discharge that is produced by an ionizing particle does not propagate nor continue for long because of the instant *local* breakdown of the electric field at the point of impact. The resistivity of the electrode material prevents an immediate charge rebuildup within an area of typically 0.1 cm^2 around the spark center. Furthermore, gases with very high photon-absorption cross section (quenching) must be used to avoid secondary sparks induced by UV photons generated in the first spark.

The immediate advantages of the resistive plate chamber lie in the ease of construction (costs) and excellent time resolution (<1 ns), pulse rise time (3–4 ns), and pulse duration (about 10 ns). The generated pulses are so large that they induce sufficient charge (300–500 mV) on so-called pickup electrodes exterior to the high field electrodes. These pickup electrodes can be arranged in x and y giving the detector spatial granularity and modularity (Fig. 18). This permits use of low-cost amplifiers for large-scale operation as, e.g., outer muon detectors in colliding-beam experiments at the SSC and LHC (CMS, 1992; L3P, 1992).

It must be noted, however, that in spite of the fast response of the chamber the rate capability is largely limited by the field restoring time. Resistive plate detectors therefore operate only at rates not exceeding 100 Hz/cm^2.

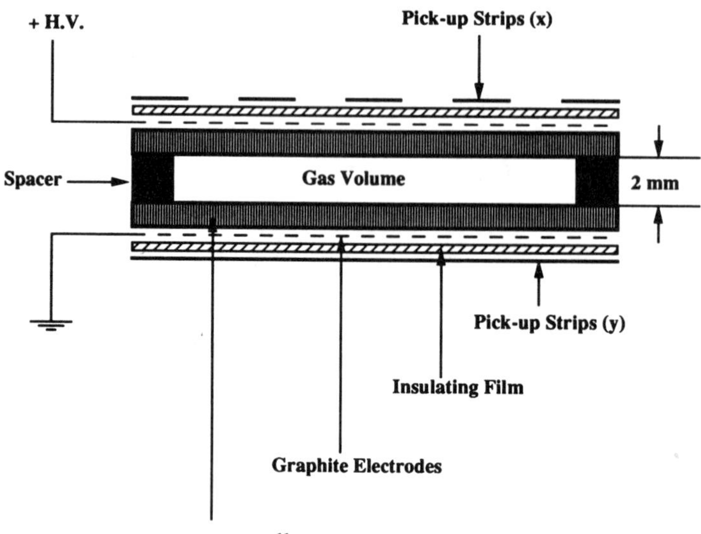

FIG. 18. The resistive plate chamber.

For further discussion see Cardarelli and Santonico (1981) and Cardarelli *et al.* (1988).

1.4 Solid-State Detectors

Solid-state or semiconductor detectors are based on a reverse-biased *n-p* semiconductor junction in which a carrier-free zone, the depletion zone, is created. In this way, they are used as ionization chambers. Charged particles interacting with the material in this area will liberate electron-hole pairs at the expense of only 3.6 eV (in Si) energy loss per pair, against 30 eV for a gas and 1000 eV for the combination of a scintillator and a photomultiplier. Therefore, compared to gaseous and liquid ionization detectors, larger primary signals and high charge statistics are expected, yielding exceptional performance in energy and position resolution. These detectors have revolutionized fields where the detection of γ rays played an essential role, like the study of nuclear structure. These detectors are also used in many fields where one wants to identify isotopes by their characteristic x- or γ-ray emission lines.

Silicon and germanium are the most widely used. The latter is mostly used for keV and MeV photon detection, but it has to be operated at the temperature of liquid nitrogen. The energy resolution varies with the energy of the photons. At 6, 60, and 661 keV it reaches 0.15, 3.3, and 4.3 keV (FWHM), respectively. This is ten times better than what can be obtained with the best scintillator.

Silicon can be used at room temperature. It is commonly used in low-energy nuclear physics for the energy loss of particles and their identification (Comiso *et al.*, 1975).

For high-energy physics applications, these detectors can be made very compact, with excellent granularity and precision. They are therefore first-choice technologies for today's vertex detection and tracking, predominantly in congested high-rate environments.

1.4.1 The Silicon Strip Detector Silicon has become the most widely used semiconductor material in electronics application today. Its features, quality, and production costs have also made it a nearly ideal candidate for a solid-state detector. The basic detector size is determined by the area available on a 100–150-mm- (4–6-in-) diam wafer. An area less than this is available for the active detector region, which will be subdivided into strips or pads. A convenient choice is dimensions of ~70 × 56 mm^2 with an active area of 64 × 51.2 mm^2. The most common design is that of a strip detector where thin metal strips are deposited on the *p* or junction side of the silicon wafer (single-sided readout) (Bencze *et al.*, 1992). Perpendicular strips or suitable pad structures on the back plane (*n* side) of the wafer provide two-dimensional track information (double-sided readout) (Sadrozinski, 1989; O'Shaughnessy *et al.*, 1990;

Brenner *et al.*, 1992). Diffusion during charge collection allows interpolation from strip pulse heights. Reading, e.g., every strip (20 μm wide) results in a single-track resolution of $\sigma < 3$ μm, and a double-track resolution of $\sigma < 50$ μm (Sadrozinski and Seiden, 1989).

The band-gap energy of silicon is only 1.3 eV. Thus, in a substrate of 300 μm thickness, a minimum-ionizing particle generates a signal of ~25 000 electrons, which can easily be interfaced and processed with integrated circuits, sometimes even mounted on the same substrate. Signal collection time is limited by the mobility of the holes, which is roughly one-third of the electron mobility (0.15 m^2/V s). Most detectors have collected the full charge within tens of ns. Interstrip capacitance and strip resistance of 6-cm-long strips have little influence on signal quality and signal propagation and are found to be adequate even for operation at highest particle rates expected at hadron colliders. In SSC and LHC experiments, however, budget restrictions may impose substantially larger (factor 2) strip sizes in view of the total detector area to be covered. Under these circumstances degradation in signal speed and signal dispersion are a matter of prime concern for very high-speed operation. Furthermore, the effects of radiation on the detector performance (e.g., *n*- to *p*-type inversion, bulk damage, surface charging) are presently being studied and may ultimately limit the use of silicon in high-radiation environments (Wunstorf *et al.*, 1989; RD-20, 1992; Barberis *et al.*, 1992).

1.4.2 The Silicon Drift Chamber By making use of the uniform and constant drift velocity of charge in silicon, a simple and low-cost detector can be built from this material providing two-dimensional readout for tracking. This detector employs anode pads for signal readout and junction strips, spaced 150 μm, for creating a uniform drift field of 150 V/cm transverse to the bias field (Fig. 19). The drift velocity within the device is approximately 2 μm/ns.

One track coordinate is derived from the measurement of the drift time, the other from anode pad interpolation. Both result in about 20-μm resolution over 4 mm of drift distance. Because of the weak drift field, very pure and uniform doping of the *n*-type Si crystal is needed. Furthermore, this method will result in inferior double-track separation as compared with classical silicon strip detectors with similar strip spacing.

1.4.3 The Pixel Detector Under extreme conditions, such as in high-luminosity hadron colliders, particle densities become so large that individual tracks cannot be resolved, even using the most granular strip detectors. Particle multiplicities in high-energy jets with opening angles of order 20 mrad are likely to be 30–70. A device that lends itself to the precise detection and recognition of dense tracks in a crowded environment is the pixel detector (Campbell *et al.*, 1990; Heijne *et al.*, 1990; Anghinolfi *et al.*, 1992). The sensitive cell size of this detector is reduced to pads or pixels of 100 × 100 μm^2 or less,

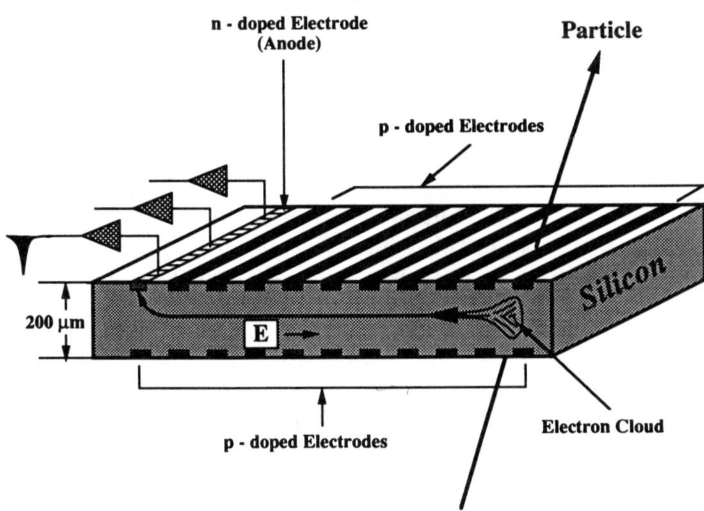

FIG. 19. The silicon drift chamber.

operated similarly to the silicon strip detector. Silicon pixel detectors with true two-dimensional direct readout [as opposed to charge-coupled-device- (CCD) shifted readout] promise to provide the resolving power and precision in three dimensions. Being still a rather novel technique, this detector will ultimately combine detection and readout within the same substrate. In this so-called smart pixel detector, empty channels (i.e., those that have received no hits) can be suppressed locally (zero skipping), thereby significantly reducing the data flow.

1.4.4 The Gallium-Arsenide (GaAs) Detector Solid-state detectors are often subject to high radiation fluences because they are located close to the beam-interaction vertex (Groom *et al.*, 1989; Stevenson, 1992). For physics reasons they must be placed close to the decay origin of short-lived particles. Costs in terms of number of channels have become a limiting factor in the design and placement of these detectors as well. Therefore, silicon detectors are particularly prone to radiation damage by charged particles and neutrons, resulting, e.g., in high leakage currents from bulk defects in the crystal (Wunstorf *et al.*, 1989) (see also Sec. 1.4.1).

GaAs as a semiconducting material operated in a similar fashion as silicon can tolerate higher radiation levels than silicon before undergoing gradual destruction (Smith, 1990). The signal produced by GaAs detectors is, however, inferior to that from silicon by approximately a factor of 2. Since research in this field is still new, further improvements with GaAs detectors can be expected.

1.4.5 The Diamond Detector For reasons of even higher radiation tolerances, diamond crystals have become an attractive alternative to silicon and GaAs devices. For a long time, natural diamond crystals have been found to function as homogeneous conduction counters offering a few unique features for application as tracking detectors. Their major advantages over comparable detection techniques are survivability after radiation exposure to tens of Mrad and more than 10^{14} neutrons/cm^2 year, very fast charge collection ($\sim$1 ns) due to roughly equal electron-hole mobility (0.18/0.12 m^2/V s), and lower strip capacitance due to different dielectric properties of the crystal. The high resistivity of the crystal lattice ($>10^{12}$ Ω cm) as compared with silicon (10^5) reduces leakage currents to negligible levels at high bias voltages allowing saturation velocity of the charge carriers. Unfortunately, high bulk resistivity also leads to polarization effects that, from accumulation of charges, render the detector inoperable when exposed to high rates. A band-gap energy more than five times as high as in silicon requires higher energy losses per ionizing particle to yield comparable signals. Thus, pulse heights of diamond detectors, at present, are roughly 3–4 times lower than silicon signals.

To form a full-scale tracking detector, some of the inherent drawbacks in this technology have yet to be solved. Natural diamond, although produced in quantities of several tons (i.e., millions of carats) per year (Kozlov *et al.*, 1975), is still unaffordable, where sensitive detector areas in the order of several m^2 are required. Furthermore, even rather pure crystals contain traces of impurities (these create the different colors) such as nitrogen that trap free carriers and thereby reduce the lifetime of drifting electrons and holes. A commercial process called chemical vapor deposition (CVD) produces low-cost, thin (order 100 μm) sheets of artificial diamond that is chemically purer than natural crystals but suffers from lattice defects that act as trapping centers for drifting charges (DIAMAS, 1990–1991).

1.5 Scintillating Fiber Detectors

One of the first detectors in particle physics is the by now classical scintillation counter combined with a photomultiplier tube. Still used today for fast triggering and low-cost-beam hodoscopes, it benefits from the very fast (<ns) scintillation process induced by ionizing radiation in various transparent media. Optical readout via rugged photomultiplier tubes has been simple and fast, as well, and has progressed via wavelength-shifting techniques into many fields of physics, such as tracking and calorimetry. Scintillators most commonly used were plastics, inorganic crystals [e.g., NaI(Tl)], and liquids. They did not play an important role in modern tracking detectors until a finer-grained form of scintillator became available; the scintillating fiber.

1.5.1 The Organic Scintillating Fiber Organic fibers are the most commonly used fibers today, commercially and in physics applications. They draw from a widespread knowledge and experience in mass production and are mostly used in telecommunications. As such, the fiber was *a priori* not usable for particle physics, for it did not scintillate but served as a light guide only.

Therefore, scintillating fibers consist of an organic core (in most cases polystyrene with index of refraction $n_1 = 1.58$) doped with a scintillator (see FIBER OPTICS). The core is surrounded by a cladding with lower refractive index n_2, such that scintillation light is transmitted to the fiber ends by total reflections on the core-cladding interface (Fig. 20). Reflection losses, which depend on the quality of this interface zone, are expressed in terms of reflection length, i.e., the fiber length at which the light intensity drops to $1/e$ of its initial value. Similarly, absorption phenomena also limit the length of scintillating fibers. Consequently, both reflection and absorption length make up the attenuation length, a characteristic parameter for optical fibers. Thin (50 μm) fibers, e.g., can be efficiently operated up to 1–2 m in length (Baton *et al.*, 1992).

On the other hand, spatial and double-track resolution are basically defined by the fiber diameter, which, for this application, typically ranges between 20 and 1000 μm. For example, fibers with 30-μm diam provide 35-μm positional precision (limited by photon yield) and 80-μm double-track separation (Leutz, 1989).

Fibers are usually read out in bundles. Hence, the precision and uniformity in bundling is a crucial factor in the overall accuracy of track reconstruction (D'Ambrosio *et al.*, 1990).

High-speed readout devices are necessary to preserve the fast response of the fiber to ionizing particles. They must also provide the spatial granularity (pixel size) for individual fibers within the bundle. With only few (order 1) photons arriving at the fiber end, these photosensitive detectors are required to have high quantum efficiency. Size, power consumption, and experimental circumstances, such as operation in a magnetic field, determine the choice of readout element. Most widely in use are, e.g., charged-coupled devices (CCD cameras), position-sensitive photomultipliers, and avalanche photodiodes. Another device, the optoelectronic delay tube (Gys *et al.*, 1990), can store single fiber hits in bunches of tens of ns for several μs until a (late) trigger selects the associated bunch for event reconstruction. This method of data "pipe-lining" is crucial at high-rate accelerators such as the SSC and LHC.

Another method for tracking could be provided by scintillating fibers coupled to visible-light photon counters (VLPCs). VLPCs are silicon-based, solid-state photomultipliers, with their visible-light response (around 550 nm) optimized in both quantum efficiency and speed. The quantum efficiency has been mea-

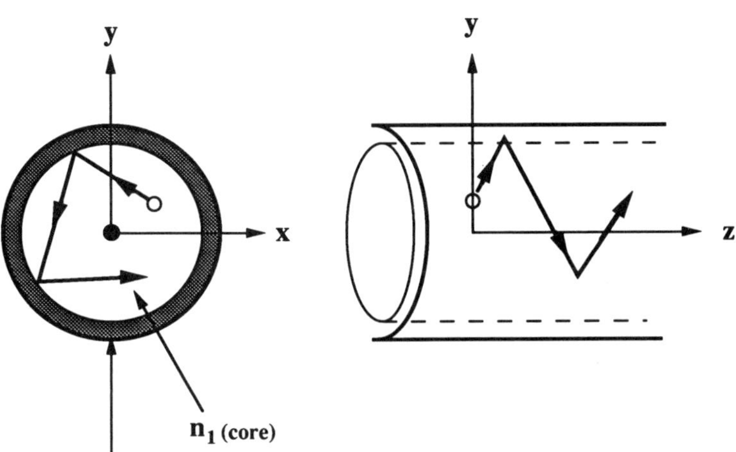

FIG. 20. Light propagation in a scintillating fiber.

sured as high as 85%, and the device has an intrinsic rise time of a few ns.

VLPCs need to be cryogenically cooled. The optimum temperature is near 6.5 K. The activation energy of the electrons from the impurity band to the conduction band of the extrinsic semiconductor, Si-As, is about 0.05 eV. Therefore, the thermal-electron counting rate would be intolerably high at temperatures above 10 K. The scintillation light is brought into the cryostat via long clear fibers, while the preamplifiers remain outside at room temperature. The device can be operated with an average gain of 3×10^4, with a gain dispersion of less than 30% (SDC, 1990a,b; Atac *et al.*, 1992a,b; Rockwell International, 1992).

1.5.2 The Inorganic or Glass Scintillating Fiber A serious limitation to organic scintillation fibers is radiation damage by charged and neutral particles as described in Sec. 1.4.4. Radiation damage manifests itself in a gradual drop of attenuation length, i.e., loss of photons (Majewski *et al.*, 1989; Fabian *et al.*, 1990). Following exposure to several Mrad of radiation, the optical transmission characteristics of the fiber can deteriorate by more than a factor of 2. Cerium-doped glass fibers, e.g., have been found to be more radiation resistant than any other fiber material (Fisher, 1988). Their predominant drawback, however, is a much lower attenuation length to begin with than their organic counterparts. Production costs have inhibited widespread use of glass fibers in particle physics, as well, while encouraging progress toward higher radiation tolerances for organic scintillators is made.

1.5.3 The Liquid Scintillating Capillary Fiber Another promising development in the field of scintillating fibers is the glass capillary filled with liquid scintillator (Artamonov *et al.*, 1991). Capillaries of 50 μm diam with 5-μm-thick walls have been successfully tested, and 20-μm diameters are technically feasible. The refractive index of the liquid can be made so high that the capillary serves as an efficient optical light guide with surface reflection losses at the core-cladding interface of less than one part in 10^5. In addition it incorporates the high photon yield obtained in scintillating plastic fibers. The scintillating light pulse is fast ($\leq$5 ns), and, in the context of radiation damage, the scintillator is replaceable making it suitable where heavy radiation damage of conventional organic scintillators is expected.

For use in particle tracking, coherent capillary bundles of appropriate size must be produced. Using a multiple drawing procedure, hexagonal bundles of about 5 mm diam can be obtained. These bundles consist of roughly 30 000 single capillaries with an inner diameter of 20 μm.

1.6 Visual Tracking Detectors

1.6.1 The Cloud Chamber This is the simplest and most straightforward form of tracking technique, using visual means to observe the tracks of charged particles. The basic construction is simple: a closed chamber containing a supersaturated vapor and a window to view the interior (see Henderson, 1970). When ionized clouds form inside the chamber, small droplets of liquid condense out of the gas at these points and can be seen. The chamber can be used only for the study of reactions that are relatively abundant, because a large chamber of, say, 0.5 m^3 volume can be triggered only once every few minutes. This type of detector has been superseded by the more sophisticated and accurate bubble chamber.

1.6.2 The Bubble Chamber Discovered by D. Glaser in 1956, the bubble chamber has for almost two decades been the major tool in particle physics. It relies on the properties of a superheated liquid for its operation, although it is basically similar to the cloud chamber (see Henderson, 1970). The liquid in question is held well above its boiling point, and under pressure to prevent expansion. To bring the chamber into operation, at the point when the trigger is set, the pressure is released slightly and the liquid is free to vaporize. Bubbles form preferentially along the paths of charged particles in the chamber and can be recorded visually using photographic emulsion or CCD imaging systems. The further growth of bubbles is stopped by recompression. The repetition cycle is typically of the order of seconds. With separated beams of charged particles, this instrument has permitted the gathering of information on resonant states and new particles, which has been the basis for our present concept of particle structure. The intrinsic accuracy in the measurement of bubble position is of the order of 100 μm. A great variety of liquids has been used, from liquid hydrogen to xenon.

Extensive data analysis and processing are then necessary in order to obtain three-dimensional track information (a common factor in all types of tracking systems). This chamber is rarely used nowadays, with physicists choosing fast electronic detectors which are more reliable and easier to control.

A brief summary of the different detector technologies is given in Table 1.

2. TRACKING SYSTEMS

A complete tracking system is composed of various tracking techniques and different types of chamber, combined in such a way as to be complementary in operation. The ultimate aim is to completely pin down a particle in space and time as it passes through the detector. Following are some of the tracking systems in operation today, complete with details of position, momentum, and time resolution, where they are of importance.

2.1 The Imaging Drift Chamber

This is based on the classical drift chamber using large (order tens of cm) drift distances. The principal difference is the use of multiimaging techniques as opposed to single-hit electronics, which means that multitrack events can be recorded in full, even if they land on the same sense wire. This is done by using flash analog-to-digital converters (FADCs) and fast sampling rates to record digitally the charge distribution and history of the event in the entire drift region. From the recorded pulse height and time information, the location of tracks in space can be ascertained. The reconstructed pulse height is a measure of energy loss (dE/dx) of the particle within the detection cell. This can be used to identify particles within a certain energy regime.

As an example of an imaging drift chamber, we show the UA1 central detector (Beingessner *et al.*, 1987) in Fig. 21. This detector, 6 m long and 2 m in diameter, was part of a colliding-beam experiment at the CERN proton-antiproton accelerator (SppS), which provides center-of-mass energies of about 700 GeV. In this chamber, large 18-cm-wide drift gaps (Fig. 22) allow the detection of multiple particle hits, generated by collimated jets, up to a spatial resolution of ≤ 200 μm. Thus, complex event structures such as those in Fig. 23 can be resolved to very high precision without ambiguities.

2.2 The Time Projection Chamber

The time projection chamber (TPC) is a more sophisticated version of the imaging drift chamber. One of its most important features is that the electric and magnetic fields are parallel (Clark *et al.*, 1976). Under the combined action of the fields, it is possible to drift electrons over large distances (up to several meters), while retaining their precise location in space and time (Fig. 24) (ALEPH, 1990). Thus the detector essentially consists of a large gaseous drift volume with a homogeneous drift field similar to the classical parallel-plate configuration. One end generates the electric field necessary to transport electrons, while the other, at low potential, receives and measures the arriving charge. Similar to a photographic plate, this so-called end cap of the detector records the arrival of

Table 1. Summary and special features of the different detector technologies.

Detector technology	Major application	Pros	Cons
Gaseous detectors	Large volume and high-rate tracking, muon triggering	Rugged, cheap, relatively radiation hard, tailored to experimental needs	Sensitive to magnetic field, limited precision
Solid-state detectors	Vertex detection, high-rate tracking	High precision and good energy resolution	Radiation sensitive (Si), expensive
Scintillating fiber detectors	Vertex detection, high-rate tracking and triggering	Fast, insensitive to magnetic field	Low light yield, critical readout
Visual tracking detectors	Vertex measurement (outdated)	High track precision (holographic)	Very slow, dangerous

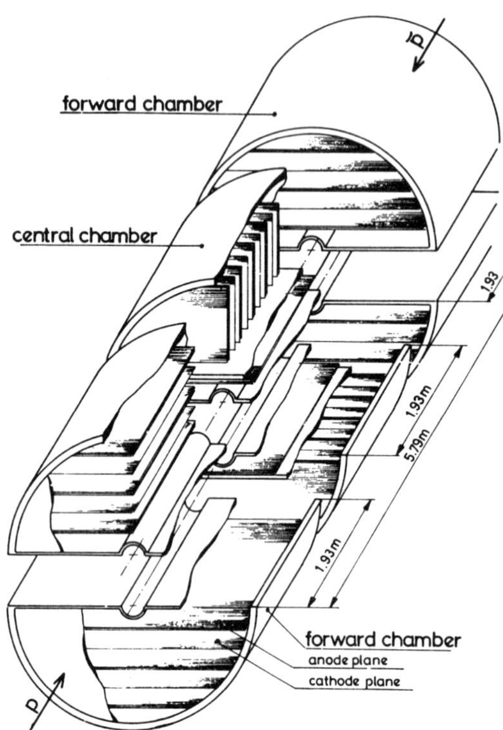

FIG. 21. Exploded view of the UA1 central detector. The detector is made of six modules: four forward and two central chambers. Wire planes are adapted to the event signatures in *p-p* collisions, i.e., vertical drifting for forward particles (jets) and horizontal drifting for central tracks. This assures a maximum of measured track samples per detector volume.

information (tracks) in short time slices, such that the full history of an event can be reproduced.

The actual detector elements are proportional chambers shaped into sectors so as to adapt to the overall geometry of the TPC (Fig. 25). They have fine-grained cathode segmentation (pad structure) for high-resolution tracking. Analogous to photographic techniques, these sectors can even be equipped with electronic shutters (known as gates) to prevent unwanted charge from entering them. Given the size of such a detector (typically 25 m^3), impressive resolution in all three space coordinates can be achieved.

A good example of such a detector is the ALEPH TPC at CERN. It has a diameter of 3.6 m and a length of 4.4 m. The two end caps are segmented into 18 sectors each, which come in three different shapes, designed to make optimum use of the space and to lose a minimum of data along the sector boundaries (Fig. 26). This detector has a spatial accuracy in the event projection on the end plates $(x - y) \leq 200\ \mu m$. This is kept low by having the magnetic and electric fields parallel, so that electrons do not diffuse laterally but are forced to orbit about the center of the ionized cluster. The z-coordinate precision is determined by the longitudinal diffusion of electrons along their drift path and is of the order of 1 mm (Atwood *et al.*, 1991).

These values for the resolution also depend on the angle at which tracks hit the anode wires, since, in the avalanche region, the magnetic field is no longer parallel to the electric field and there is a smearing effect on each track segment along the wire, caused by the electromagnetic interaction with the electrons.

This detector concept is specially suited to colliding-beam experiments with sufficiently long ($\approx 10\ \mu s$) time intervals between successive beam interactions. Figure 27 illustrates a typical two-jet event from the decay of a Z^0 boson into a quark–antiquark pair as produced at the 100-GeV LEP electron-positron collider (ALEPH, 1991).

2.3 The Vertex Detector

This type of detector, as the name implies, is usually the innermost one, nearest to the interaction vertex. Many types of detection system can be used, although, because of the high particle density, those with the highest resolution are preferred such as solid-state strip or pixel detectors, microstrip gas and induction drift detectors, and small-diameter fiber detectors.

The purpose of such a detector is to measure the starting point of an event with great accuracy (DELPHI, 1986; Bouclier *et al.*, 1986) and, in addition, to track the decay origin of any short-lived particles that may be created. Typically, a resolution of $\leq 50\ \mu m$ is required of such a detector, measuring particles with lifetimes of around 10^{-13} s.

2.4 The Multidrift Tube Detector

This type of chamber is based on wire amplification in hexagonal drift cells, as shown in Fig. 28. Each anode wire is surrounded by six cathode wires constituting one drift cell, with many (typically around 70) similar cells

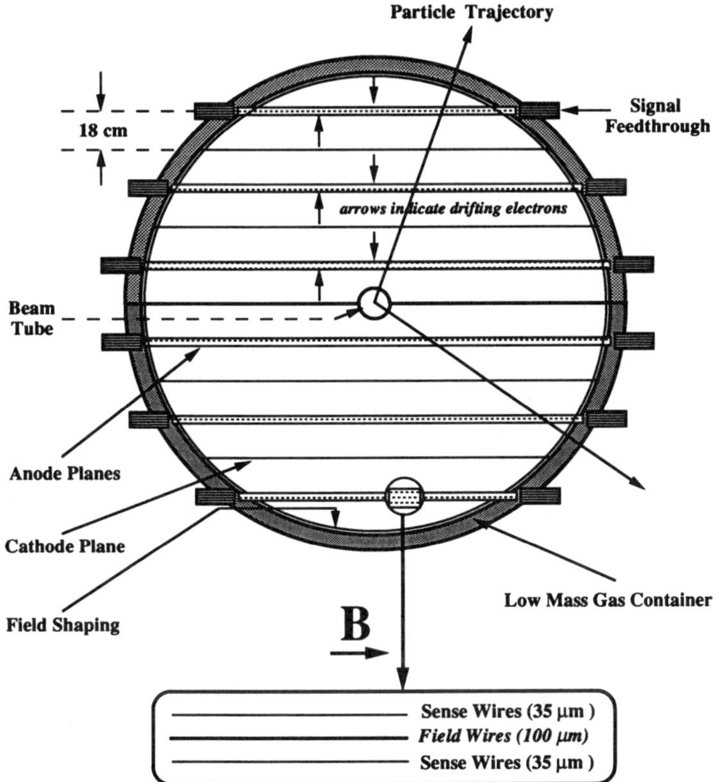

FIG. 22. Cross section of the UA1 central detector (forward section).

arranged hexagonally in the detection region, surrounded and supported by a carbon fiber tube. See Bouclier *et al.* (1986, 1988, 1989); Sauli (1981).

The design of this chamber was aimed at tracking under extreme rate conditions, where the high track density and radiation levels would render other detectors inoperative. With this aim, each individual detection cell is made small so that it can give fast signal extraction and recovery. Designed for very high-rate applications such as vertex detec-

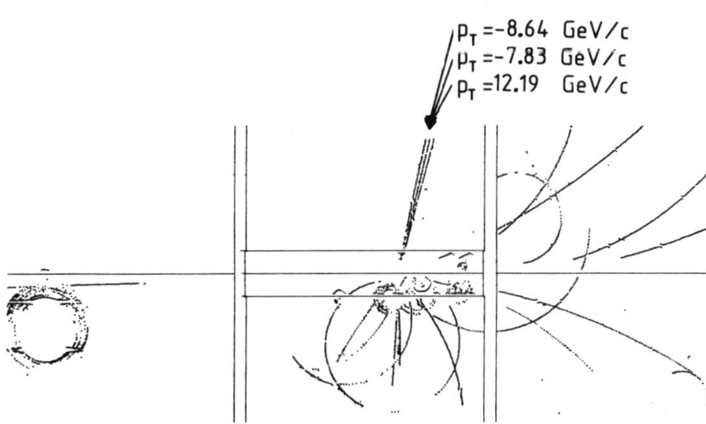

FIG. 23. Particle tracks of a *p-p* collision in the UA1 central detector. The event shown is the decay *W*-ν + τ-three prong jet.

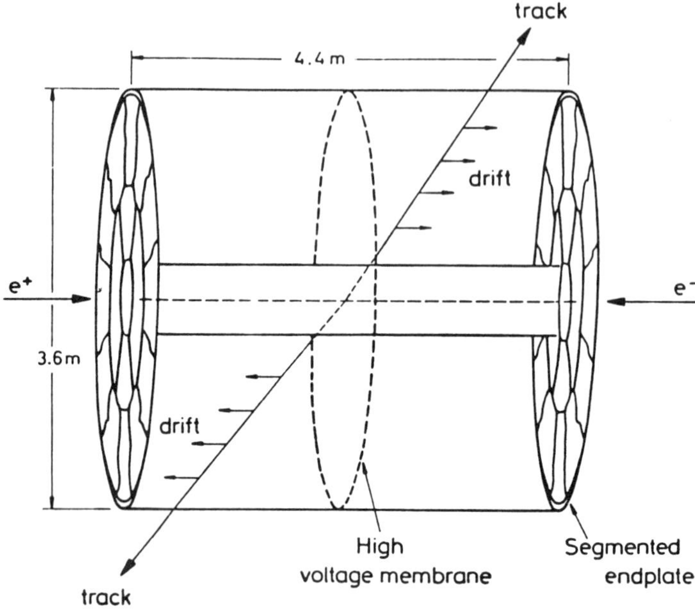

FIG. 24. Principle of the time projection chamber (TPC). Long-distance electron drift without lateral diffusion is assured by the parallel arrangement of electric and magnetic field.

tion or tracking in future proton-proton colliders, this detector achieves good spatial resolution (about 60 μm transverse to the wire, and $\approx$1% of the wire length in the coordinate along the wire) while avoiding high channel occupancy. Furthermore, it achieves an impressive radiation hardness using dimethyl ether as the drift gas, giving 1 C/cm wire lifetime (charge in terms of coulombs absorbed per cm of wire is the standard measure of radiation hardness in proportional chambers).

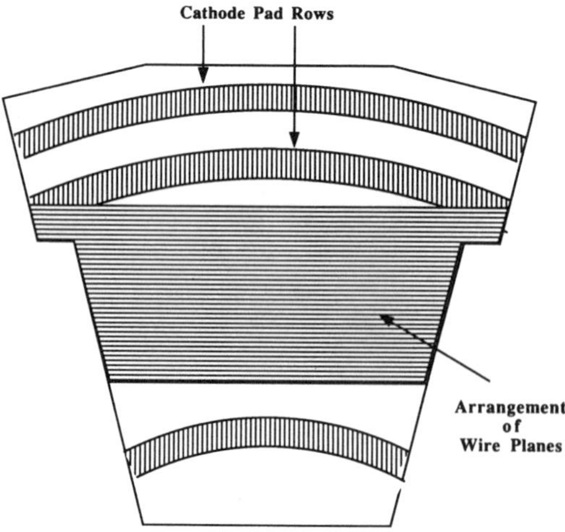

FIG. 25. Schematic of an end-cap sector of the ALEPH TPC. Sectors of a TPC are standard multiwire proportional chambers, specially shaped in order to avoid long cracks for particle tracks projected onto the end plates. Cathode pad shape and their circular arrangement are optimized for high spatial and double-track resolution.

2.5 The Honeycomb Strip Detector

Muon detection in high-energy physics experiments is usually accomplished by using thick absorbers as particle filters followed by multiwire proportional or planar drift chambers. The dimensions of these absorbers (typically iron) are order of meters to prevent "punch through" of hadrons. Therefore, if for muon detection good coverage in solid angle is required, muon detectors tend to be large and rather costly.

A new mass-production technique of a position-sensitive detector has been introduced recently to overcome cost without sacrificing in precision and reliability (Van der Graaf, 1991). The honeycomb strip chamber consists of a stack of folded aluminized Mylar foils forming a hexagonal, honeycomblike structure (Fig. 29). The nature of this structure gives the detector the necessary rigidity while being light and easy to handle at large sizes. Each hexagonal cell (outer diameter about 12 mm) is equipped with a central sense wire and, hence, serves as a traditional proportional tube. Further spatial granularity is

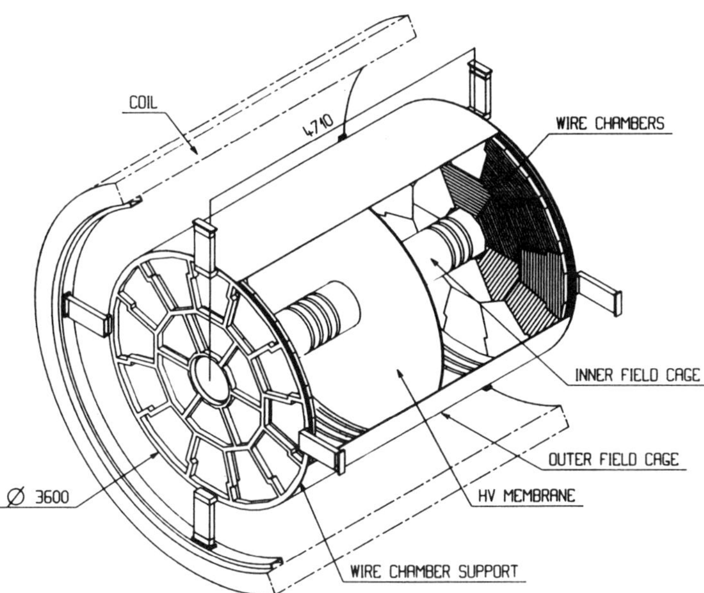

FIG. 26. Three-dimensional view of the ALEPH TPC.

provided by segmentation of the cathode wall into annular strips (etched on the Mylar sheets prior to folding) along the wire (width: 4 mm) (Fig. 30).

For a particle traversing the cell, the distance from the wire can be determined from the drift time, while the position along the wire can be obtained from the induced charge on the strips. If the relative charge sharing of three adjacent strips nearest to the track is measured, excellent space resolution along the wire can be achieved. This center-of-gravity method is intrinsically more accurate than the time measurement (Fig. 30). In practice, however, secondary effects such as inclined tracks and **E** × **B** forces "smear" the charge trajectory out and thereby deteriorate the intrinsic cell resolution. For strictly perpendicular tracks and no magnetic field present, the spatial resolution of the honeycomb strip chamber can be as high as 60 μm. Double tracks can be resolved by the strips if track separation is greater than 15 mm. However, if the separation is as little as 1 mm the presence of a double track causes a broadening of the induced charge distribution. The charge profile can also be used to discard tracks accompanied by electromagnetic showers originating from muon bremsstrahlung or hard scattering. Double-track resolution in the other coordinate is determined by the wire spacing (12.7 mm).

For large-scale applications, such as full-coverage muon detection, neighboring strips but of different hexagons are ganged together in order to reduce the number of readout channels to a tolerable level. Since this method increases confusion due to overlapping tracks it can be applied to low-rate environments only, where the strip occupancy is low enough for acceptable pattern recognition.

2.6 The Wall-less Drift Chamber

This novel concept for LHC muon chambers (Faissner *et al.*, 1992) emphasizes a combination of good spatial and timing accuracy with a high degree of robustness. Each chamber (Fig. 31) is a rigid monolith made up of 10 layers of fine-grained drift chambers for position measurement, which in turn are composed of multicells, each housing 16 individual drift cells, 14 mm wide. Individual drift cells are separated by two cathode wires or a thin (~0.05 mm) foil, whereas the cell covers are shaped to create a sensitive volume of well-defined, limited range to avoid long signal pulses (from low field and long path field lines) and also to prevent distortions of the field lines near the anode (Fig. 32). The multicell is fabricated from two identical extrusion Al (or plastic) profiles, about 23 cm wide and 1 cm thick. Since no cathode readout is foreseen the chambers are

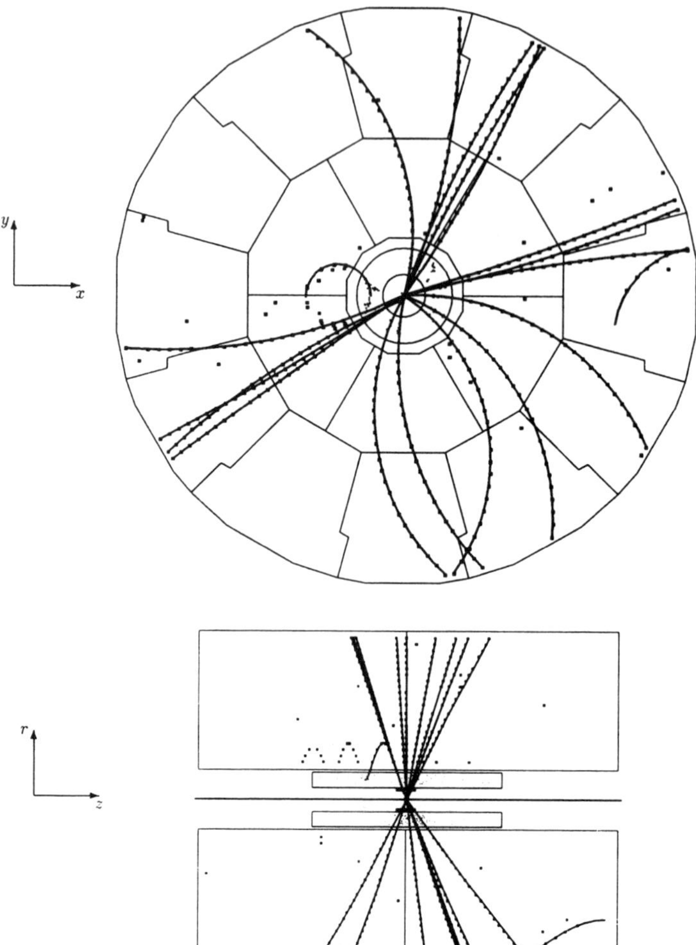

FIG. 27. Decay of a Z^0 into hadrons (jets) seen by the ALEPH TPC. Each point on a track corresponds to a pad-row hit (*x*-*y* plane).

arranged in a 90° *x*-*y* fashion to measure both track coordinates.

Prototypes of this chamber achieved a single-cell resolution of 250 μm with a simple nonflammable gas such as Ar-CO_2.

3. APPLICATIONS OF PARTICLE DETECTORS IN THE FUTURE

Particle tracking is, in its true sense, a visualization of complex event structures in nuclear and subnuclear reactions. To disentangle and measure individual tracks with increasingly higher precision has been the aim of every tracking detector in the past and continues to be so at present. As a result of continuous improvement in the hardware technologies combined with powerful pattern recognition algorithms, tracking systems have acquired state-of-the-art performance in terms of accuracy and speed. Nearly reaching the grain size of particle tracks as "seen" in holographic bubble chambers (order microns), today's electronic detectors measure, filter, and analyze track data within microseconds.

Yet progress in particle physics and its discovery potential come along with steadily decreasing cross sections for the production of new physics phenomena. In the new generation of future accelerators only one out of 10^{12} events will carry signatures and evidence for new physics. The new proton supercolliders, now in preparation in the United States (SSC) and Europe (LHC), provide tremendous luminosities in order to overcome the low production rates. Hence, these ma-

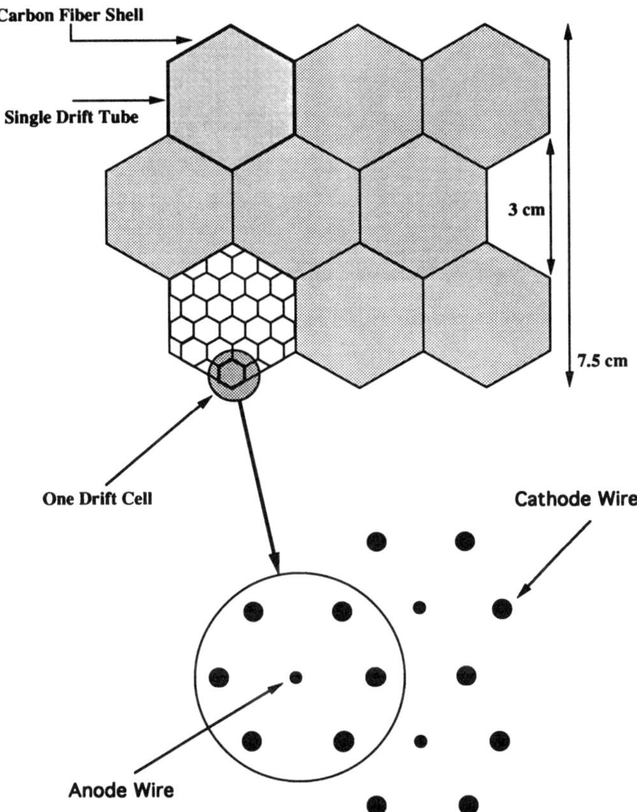

FIG. 28. Multidrift tube arrangement.

chines set the scale and define new challenges for future tracking detectors (Figs. 33 and 34) (SDC, 1990a,b; CMS, 1992; L3P, 1992).

The enormous center-of-mass energy of up to 40 TeV (SSC) releases hundreds of charged particles and numerous collimated, pencil-like jets in every beam interaction (Fig. 35). Because of the bunch structure of the proton beams these crossings occur every 16 ns (SSC) and 25 ns (LHC) in order to operate at maximum luminosity. (Notice that present accelerators, such as LEP or Tevatron, run in μs or longer intervals.)

The tasks of tracking, though unchanged,

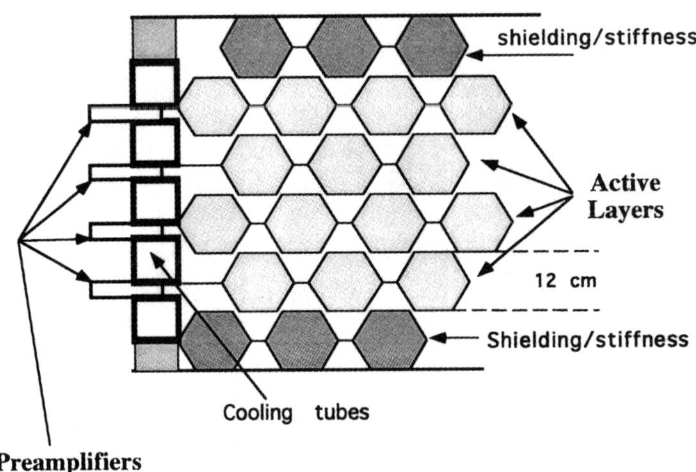

FIG. 29. An array of honeycomb drift chambers. In this layout the upper and lower chambers are used for mechanical reasons only. Note that a true honeycomb structure, i.e., one without voids, cannot be used because of the capacitive coupling between neighboring cells, causing cross-talk.

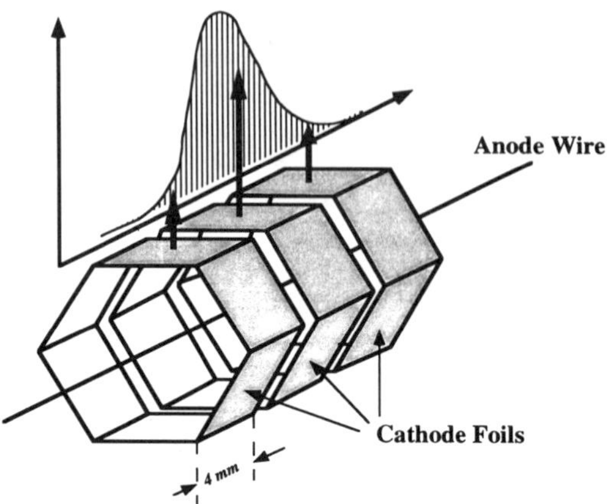

FIG. 30. Schematic of one honeycomb cell with segmented cathode. The heavy arrows indicate the pulse heights induced on the three segments, and the superimposed curve gives the precise position of the hit on the wire.

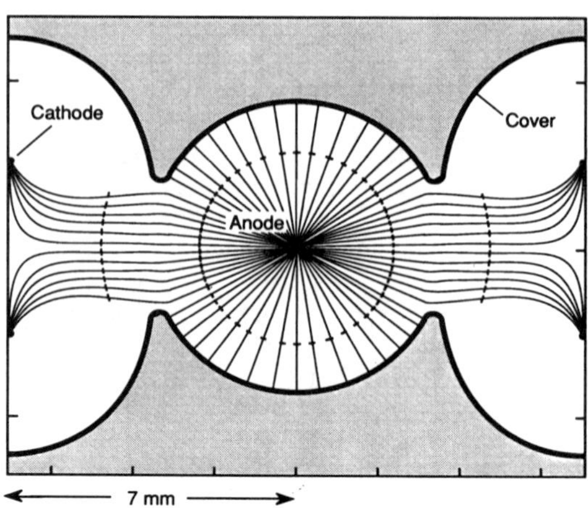

FIG. 32. Field lines in one cell of a wall-less drift chamber. Only field lines reaching the anode are displayed. The ticks on the field lines mark points of equal drift time.

must be seen in the light of this new environment never experienced before. (Notice that similar, though still orders of magnitude less severe, scenarios will be expected at the new electron-proton collider facility HERA at Deutsches Elektronen-Synchrotron, Hamburg, in operation since 1992.) Only the fastest and most highly segmented detectors, i.e., those with small (10 mm^2) cells and several ns response, avoid or reduce pileup effects due to overlapping tracks. Furthermore, the long-term effects of radiation damage limit the use of any detector under these circumstances.

Still, the role of tracking has become even more important than ever before, for only this technique provides the redundancy in rejecting backgrounds and recognizing interesting event signatures in a congested environment. It is by playing this role that tracking in the new generation of detectors provides the credibility in the rare physics we are searching for.

4. SUMMARY

Proportional gas chambers have been highly successful as tracking and energy-measuring devices in a wide range of applications. They have played the role of a true workhorse in the heart of many experiments

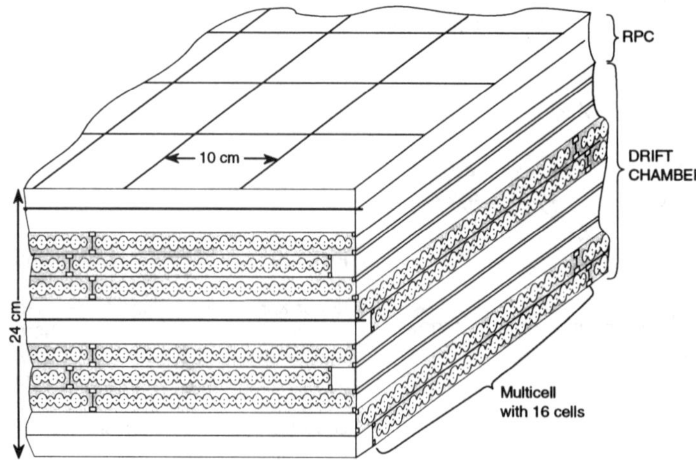

FIG. 31. A stack of wall-less drift chambers (WLDC). The upper layers are resistive plate chambers (RPCs) for triggering purposes only.

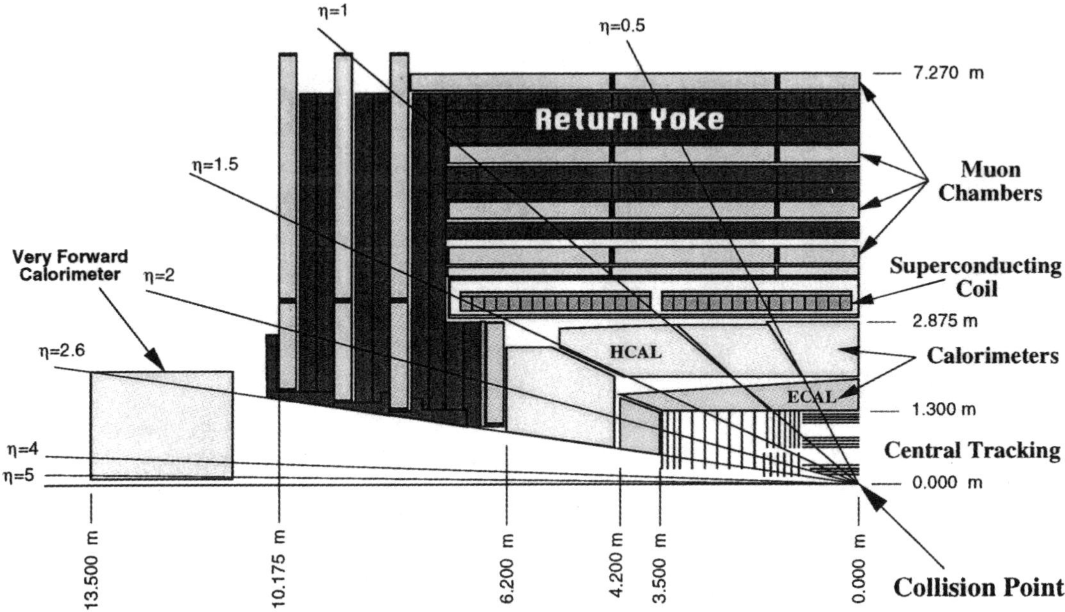

FIG. 33. The CMS (Compact Muon Solenoid) detector at LHC. One-quarter of the detector along the beam line is shown.

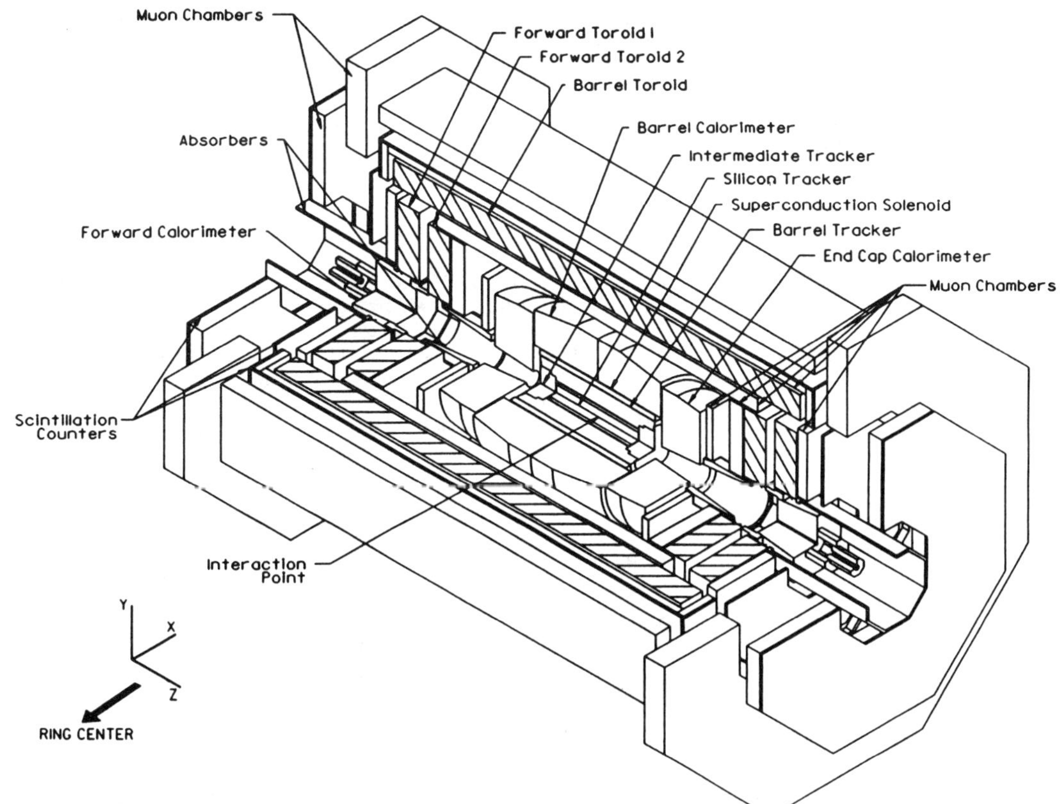

FIG. 34. The SDC (Solenoidal Detector Collaboration) detector at SSC.

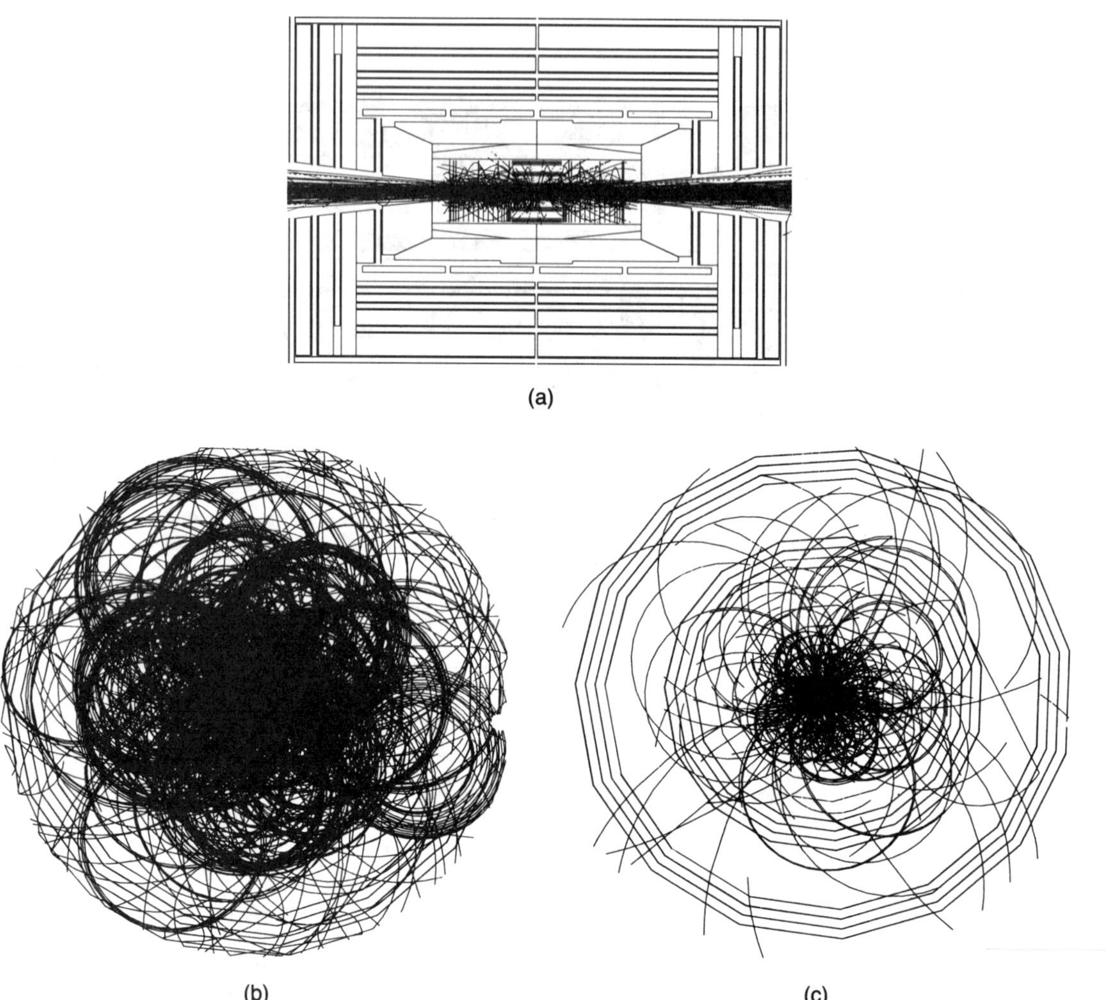

FIG. 35. A simulated high-multiplicity event at LHC as seen by the full CMS Detector (r-z view). About several hundred tracks are shown in the central tracking cavity surrounded by calorimeters. The high solenoid field confines low-momentum tracks to the inner and forward part of the tracker. (b) A high-multiplicity interaction as seen in the bending (r-ϕ) plane, with all tracks projected over the total length (z) of the detector, (c) The same event as seen by a high-resolution detector of CMS segmented into 10-cm cells along the beam axis (z). Outer tracks can be unambiguously measured by microstrip gas chambers (of 200-μm pitch), while the inner core is resolved with finer-grained (50-μm pitch) silicon microstrip detectors.

in the past and still do in almost all detectors at accelerators and nuclear laboratories worldwide. Many of these experiments will continue to run into the next century until a new generation of particle accelerators comes into operation. The challenge for all detection techniques to be used in the future will be to be able to cope with high levels of radiation without sustaining damage and to remain efficient in the process. Yet the classical principle of electron drift and thin-wire amplification is still and will continue to be competitive with semiconductor and scintillating-fiber technologies. Because of its enormous versatility and ruggedness, the proportional gas counter may even outperform the more recently developed techniques in tracking. With the development of the microstrip gas chamber, it has been shown that this type of detector can reach the performance of silicon microstrip detectors at low radiation risk and cost.

GLOSSARY

Affinity: Ability to bind charge carriers to neutral molecules. Oxygen and Freon, e.g., are known to capture free electrons in drift gases. These charges are lost in the final signal arriving at the anode.

ALEPH: A large (about 4000 tons) 4π solenoidal detector for the LEP electron-positron storage ring facility, in operation since 1989.

Bunch: In most particle accelerators and storing ring facilities particles are accelerated in a noncontinuous way. For optimum efficiency and luminosity, particles are injected and accelerated in discrete packages or bunches that, depending on their number, are separated in time from several msec to tens of ns. The spatial dimensions of a bunch are typically a few μm laterally and of order cm longitudinally (i.e., along the direction of motion).

Calorimeter: In particle physics, a detector to measure the total energy of neutral (photons) and charged particles by total absorption in a dense medium such as iron, lead, uranium, or others. One distinguishes between two classes of calorimetry: electromagnetic calorimetry for the energy measurement of photons and electrons (positrons), and hadronic calorimetry for the interaction of charged hadrons (primarily pions).

CERN: Centre Européenne pour la Recherche Nucléaire: The world's largest laboratory in high-energy particle physics, based in Geneva, Switzerland. CERN, financed by 17 European member states, at present, operates the largest accelerator for particle research (LEP) having a circumference of 27 km.

Charged-Coupled Device (CCD): Light-sensitive solid-state device that allows the storage of charge (from electrical signals) in so-called time buckets for a certain time prior to readout.

Chemical Vapor Deposition (CVD): A catalytic process to produce from carbon-based raw materials thin sheets (up to 100 μm) of diamond.

Cherenkov Light: Electromagnetic radiation emitted by a particle moving (with constant velocity) through a material medium at a velocity greater than the phase velocity of light in that medium. This phenomenon is exploited for the identification of charged particles passing through matter.

DESY: (Deutsches Elektronen-Synchrotron): A German-based research facility for high-energy particle physics similar to CERN.

Electron-Beam Lithography: A process known in semiconductor technology to produce by means of photographic masks thin metal layers (order μm) on solid-state or insulating substrates. The short wavelength of the electron beam allows accuracies below 1 μm.

Flash Analog-to-Digital Converter (FADC): A device that samples and digitizes the time response and pulse height of analog signals (coming from a detector) in time slices comparable to the speed of the signal itself. The reconstruction of the time buckets allows, similar to the CCD, a recovery of the original signals and their timing history.

FWHM: Full width at half maximum.

Gaussian Curve, Gauss Function: The "normal," bell-shaped, distribution, which is symmetric about its mean value.

Hadron Collider: An accelerator primarily for protons (antiprotons) and heavy ions. These machines are in contrast to electron (positron) facilities, which are lower in energy because of energy losses from synchrotron radiation.

HERA (Hadron-Electron Ring Accelerator): This facility for colliding electrons of 30 GeV on protons of 820 GeV is now operational at DESY, Germany.

Hodoscope: An arrangement of detectors, usually an array of scintillators with photomultipliers, used to count and measure the position and trajectory of incoming particles to an experiment.

Holographic Bubble Chamber: Photographic method of recording particle tracks in a bubble chamber using interferometry. This improves the spatial resolution (bubble size) significantly.

Jet: In most cases, the fragmentation products of a quark decay into hadrons. Depending on the initial energy of the parent particle, jets appear as collimated particle tracks with high multiplicity. Their high density of tracks makes them particularly difficult to measure in common detectors because of insufficient separation of tracks.

LAA Project: The Lepton Asymmetry Analyzer project, which aims at developing new techniques and methods for the detection of particles and the readout and processing of detector signals.

LHC (Large Hadron Collider): This storage ring accelerator will make use of the present underground tunnel built for LEP and ultimately provide center-of-mass energies for protons of 2 × 8 TeV (1 TeV = 1000 GeV).

Luminosity: In analogy to optical terms, luminosity describes the probability of accelerated particles to interact with each other in head-on collisions. The luminosity depends on the particle density (usually confined in a bunch) *and* the interaction cross section. Since rare phenomena have very low production cross sections, accelerators and colliders are designed for maximum luminosity.

Minimum-Ionizing Particle: Charged particles whose energy loss (depending on their velocity) in a medium is near or inside the minimum of the Bethe–Bloch function. Most charged particles produced in a high-energy physics interaction have a velocity range where the ionization from energy losses (dE/dx) falls into this minimum. This makes them difficult to detect, predominantly in gas-filled detectors, where only about 100 electrons/cm are liberated by these particles.

Occupancy: Term describing the "busy time" of a single detector channel due to ionizing radiation, noise, and pickup, thus creating true plus uncorrelated hits. It is commonly expressed in percent and depends primarily on the incoming flux of particles and the intrinsic speed of the detector itself; i.e., gaseous detectors with long signal-processing time (e.g., drift time, pulse shaping) suffer from higher occupancies than fast devices such as scintillators. High occupancies (i.e., >10%) result in substantial detector dead time and lead to confusion in the pattern recognition of genuine tracks. To reduce detector occupancy in high-background environments to tolerable levels, small cell sizes must be chosen.

Pattern Recognition: Sophisticated software algorithms to disentangle complex event structures and select physically interesting tracks from confusing backgrounds.

Photomultiplier: A device consisting of a photosensitive surface inside a vacuum tube where electrons are liberated from the cathode. Each electron is multiplied in a succession of dynodes by secondary emission under the influence of the electric field. Photomultipliers are essentially sensitive to single photons and provide very fast signals in the ≤ns range.

Pitch: The distance between adjacent cells in detectors with repetitive cell structures. Also known as cell width.

Scintillation Light: Light produced by quantum effects in atoms excited by charged particles traversing the medium.

SSC (Superconducting Super-Collider): A colliding-beam facility, similar to the LHC in Europe, proposed for construction near Dallas, Texas, with center-of-mass energies of 2 × 20 TeV for protons.

Strongly Ionizing Particle: Typically low- (MeV) and medium-energy (order 100 MeV) charged particles whose energy loss in matter is large to produce a high rate of electron–positive-ion pairs. These are in contrast to minimum-ionizing particles (order GeV) which are more common in high-energy physics.

Tevatron: Proton-antiproton collider with 2-TeV center-of-mass energy located at Fermi National Laboratory, Batavia, Illinois.

Vertex: The origin of a particle interaction. This interaction can be the collision of two accelerated beam particles or the location of a parent particle prior to its decay. In the latter case, the position of the decaying particle is reconstructed from the tracks of its decay products.

Works Cited

ALEPH (1990), *Nucl. Instrum. Methods A* **294**, 127–178.

ALEPH (1991), *Phys. Lett. B* **273**, 181–192.

Ambrosi, G., *et al.* (1990a), *Nucl. Instrum. Methods A* **289**, 351–355.

Ambrosi, G., *et al.* (1990b), "New Results on the Blade Chamber," CERN–European Committee on Future Accelerators Proceedings, Vol. 3, CERN Report No. 90-10, ECFA Report No. 90-133, pp. 489–497.

Anderhub, H., *et al.* (1985), *Nucl. Instrum. Methods A* **235**, 267–278.

Angelini, F., *et al.* (1991), *Nucl. Phys. B Proc. Suppl.* **23A**, 254–260.

Angelini, F., *et al.* (1992), *Nucl. Instrum. Methods A* **323**, 229–235.

Anghinolfi, F., *et al.* (1992), *IEEE Trans. Nucl. Sci.* **NS-39**, 654–661.

Ankoviak, K., *et al.* (1989), *Nucl. Instrum. Methods A* **279**, 83–90.

Antreasyan, D., *et al.* (1990), "A Device for Particle Detection at Future Hadron Colliders: The

Gaseous Pixel Chamber," CERN–European Committee on Future Accelerators Proceedings, Vol. 3, CERN Report No. 90-10, ECFA Report No. 90-133, pp. 230–233.

Artamonov, A., *et al.* (1991), *Nucl. Instrum. Methods A* **300**, 53–62.

Atac, M., *et al.* (1986), *Nucl. Instrum. Methods A* **249**, 265–276.

Atac, M., *et al.* (1992a), *Nucl. Instrum. Methods A* **314**, 56–62.

Atac, M., *et al.* (1992b), *Nucl. Instrum. Methods A* **320**, 155–160.

Atwood, W. B., *et al.* (1991), *Nucl. Instrum. Methods A* **306**, 446–458.

Barberis, E., *et al.* (1992), paper presented at the 6th European symposium on Semiconductor Detectors, Report No. UCR SSC 92-01.

Baton, J. P., *et al.* (1992), DAPNIA Report No. 92-02, CEN-Saclay, Gif-sur-Yvette, France.

Beingessner, S. P., *et al.* (1987), *Nucl. Instrum. Methods A* **257**, 552–555.

Bencze, G., *et al.* (1992), "Silicon Tracker for LHC," CERN Report No. CMS TN/92-10.

Bouclier, R., *et al.* (1983), *Nucl. Instrum. Methods* **205**, 403–423.

Bouclier, R., *et al.* (1986), *Nucl. Instrum. Methods A* **252**, 373–378.

Bouclier, R., *et al.* (1988), *Nucl. Instrum. Methods A* **265**, 78–84.

Bouclier, R., *et al.* (1989), *Nucl. Instrum. Methods A* **283**, 509–514.

Bouclier, R., *et al.* (1992a), *Nucl. Instrum. Methods A* **323**, 240–246.

Bouclier, R., *et al.* (1992b), *IEEE Trans. Nucl. Sci.* **NS-39**, 650–653.

Bouclier, R., *et al.* (1993), "Performance of Gas Microstrip Chambers on Glass Substrates with Electronic Conductivity," CERN Report No. CERN-PPE/93-04.

Brehin, S., *et al.* (1975), *Nucl. Instrum. Methods* **123**, 225–229.

Brenner, R., *et al.* (1992), *Nucl. Instrum. Methods A* **315**, 502–506.

Breskin, A., *et al.* (1975), *Nucl. Instrum. Methods* **124**, 189–214.

Breskin, A., *et al.* (1980), *Nucl. Instrum. Methods* **178**, 11–25.

Campbell, M., *et al.* (1990), *Nucl. Instrum. Methods A* **290**, 149–157.

Cardarelli, R. R., Santonico, R. (1981), *Nucl. Instrum. Methods* **1887**, 377–380.

Cardarelli, R. R., *et al.* (1988), *Nucl. Instrum. Methods A* **263**, 20–25.

Charpak, G., Sauli, F. (1971), *Nucl. Instrum. Methods* **96**, 363–367.

Charpak, G., Sauli, F. (1984), *Nucl. Instrum. Methods* **225**, 627–635.

Charpak, G., *et al.* (1979), *Nucl. Instrum. Methods* **162**, 405–428.

Clark, A. R., *et al.* (1976), "Proposal for a PEP Facility Based on the TPC," SLAC Report No. PEP-4.

CMS (1992), "The Compact Muon Solenoid," letter of intent, CERN Report No. CERN/LHCC 92-3.

Comiso, J., *et al.* (1975), *Phys. Rev. Lett.* **35**, 13–15.

Corden, M. (1990), "Properties of a Straw Tube/Scintillating Fiber Hybrid Central Tracking Chamber," Florida State University Report No. FSU-SCRI-90C-178.

D'Ambrosio, C., *et al.* (1990), "Present Status and Future Program of Scintillating Fibers for Central Tracking," CERN-European Committee on Future Accelerators Proceedings, Vol. 3, CERN Report No. 90-10, ECFA Report No. 90-133, pp. 255–260.

Della Negra, M. (1992), "CMS—A Compact Solenoid Detector for LHC," CERN, Proceedings of the General Meeting on LHC Physics and Detectors, Evian, March 1992, pp. 165–217.

DELPHI (1986), "DELPHI Microvertex Detector," Report No. DELPHI 86-86 GEN-52.

DIAMAS (1990–1991), "A Diamond Detector for the SSC," expression of interest submitted to the Superconducting Super Collider, Dallas, TX.

Fabian, H., *et al.* (1990), "Producing Radiation Hard All Silica Fibers—State of the Art and Future Aspects," CERN–European Committee on Future Accelerators Proceedings, Vol. 3, CERN Report No. 90-10, ECFA Report No. 90-133, pp. 736–742.

Faissner, H., *et al.* (1992), CERN Report No. CMS TN/92-02.

Fisher, C. (1988), *Nucl. Instrum. Methods A* **263**, 159–162.

Florent, J. J., *et al.* (1992), "The Electrostatic Field in Microstrip Chambers and Its Influence on Detector Performance," CERN Report No. CERN-PPE-92-078.

Frautschi, M., *et al.* (1993), "The AMY Inner Tracking Chamber," University of California, Davis, Report (unpublished).

Groom, D. E., *et al.* (1989), *Nucl. Instrum. Methods A* **279**, 1–6.

Gys, T., *et al.* (1990), "Opto-Electronic Delay Tubes," CERN–European Committee on Future Accelerators Proceedings, Vol. 3, CERN Report No. 90-10, ECFA Report No. 90-133, pp. 261–264.

Heijne, E., *et al.* (1990), "Development of Silicon Pixel Detectors for the LHC," CERN–European Committee on Future Accelerators Proceedings, Vol. 3, CERN Report No. 90-10, ECFA Report No. 90-133, pp. 237–240.

Henderson, C. (1970), *Cloud and Bubble Chambers*, London: Methuen.

Kozlov, S. F., *et al.* (1975), *IEEE Trans. Nucl. Sci.* **NS-22**, 160–170.

Leutz, H. (1989), "Central Tracking of Particles with Scintillating Fibers," CERN–European Committee on Future Accelerators Proceedings, Vol. 1,

CERN Report No. 89-10, ECFA Report No. 89-124, pp. 176–181.

L3P (1992), "Lepton and Photon Precision Physics," letter of intent, CERN Report No. CERN/LHCC 92-5.

Majewski, S., *et al.* (1989), "Comparative Study of the Radiation Resistance of Selected Plastic Scintillating Fibers," CERN–European Committee on Future Accelerators Proceedings, Vol. 1, CERN Report No. 89-10, ECFA Report No. 89-124, pp. 239–249.

Meyer, T. C. (1977), "A Search for Particle Unstable ^{4}He," Ph.D. thesis, Physics Department, University of Virginia.

Oed, A. (1988), *Nucl. Instrum. Methods A* **263**, 351–359.

O'Shaughnessy, K., *et al.* (1990), "Conceptual Design for a Silicon Tracker at the SSC," CERN–European Committee on Future Accelerators Proceedings, Vol. 3, CERN Report No. 90-10, ECFA Report No. 90-133, pp. 234–236.

Palladino, V., Sadoulet, B. (1974), "Application of the Classical Theory of Electrons in Gases to Multiwire Proportional and Drift Chambers," Lawrence Berkeley Laboratory Reports No. LBL-3013, No. UC-37, and No. TID-4500-R62.

RD-20 (1992), "Status Report," Report No. CERN/DRDC/92-28.

Rockwell International (1992), "VLPC Cryogenic Conceptional Design," Report No. C91-1000.1/801.

Roderburg, E., *et al.* (1986), *Nucl. Instrum. Methods A* **252**, 285–291.

Sadrozinski, H. F.-W. (1989), "Tracking with Silicon Detectors at the SSC," CERN–European Committee on Future Accelerators Proceedings, Vol. 1, CERN Report No. 89-10, ECFA Report No. 89-124, pp. 297–300.

Sadrozinski, H. F.-W., Seiden, A. (1989), "Resolution of a Silicon Tracking Device for the SSC," Santa Cruz Institute for Particle Physics Report No. SCIPP 89/50.

Sauli, F. (1977), "Principles of Operation of Multiwire Proportional and Drift Chambers," CERN Report No. 77-09.

Sauli, F. (1989), *Proceedings of the 1988 DPF Suumer Study*, Singapore: World Scientific, pp. 774–778.

Schmidt, B. (1980), "Elektronendrift in Zählgasen unter dem Einfluß elektrischer und magnetischer Felder," Diplomarbeit, Physikalisches Institut der Universität Heidelberg.

SDC (1990a), "The Solenoidal Detector Collaboration for the SSC," proposal.

SDC (1990b), "Hybrid Central Tracking Chamber Collaboration," summary report, Part 1, August 1990, submitted to Physics Research Division, Superconducting Super Collider Laboratory, Dallas, TX.

Smith, K. M. (1990), "Gallium Arsenide Detectors," CERN–European Committee on Future Accelerators Proceedings, Vol. 3, CERN Report No. 90-10, ECFA Report No. 90-133, pp. 244–247.

Stevenson, G. R. (1992), "On the Prediction of Radiation Levels in LHC Experiments," paper presented at the 3rd International Conference on Advanced Technology and Particle Physics, Como, Italy, CERN Report No. CERN/TIS/RP/92-09/CF.

Suzuki, M., *et al.* (1988), *Nucl. Instrum. Methods A* **263**, 237–242.

Thun, R. (1988), *Nucl. Instrum. Methods A* **273**, 157–160.

Van der Graaf, H., *et al.* (1991), *Nucl. Instrum. Methods A* **307**, 220–230.

Walenta, A. H. (1979), *IEEE Trans. Nucl. Sci.* **NS-26**, 73–80.

Walenta, A. H. (1983), *Nucl. Instrum. Methods* **217**, 65–76.

Wunstorf, R., *et al.* (1989), "Radiation Damage in Silicon Detectors," CERN–European Committee on Future Accelerators Proceedings, Vol. 2, CERN Report No. 89-10, ECFA Report No. 89-124, pp. 321–323.

Zhou, B., *et al.* (1989), *Nucl. Instrum. Methods A* **287**, 439–446.

Further Reading

Agoritsas, V., *et al.* (1991), "Ultrafast Readout of Scintillating Fibers Using Upgraded Position Sensitive Photomultipliers," CERN Report No. CERN/DRDC 91-8.

D'Ambrosio, C., *et al.* (1991), "Scintillating Fibers for Central Tracking of Charged Particles," CERN Report No. CERN/LAA-SF 91-2, Rev.

D'Auria, S., *et al.* (1990), "Results of Radiation Hardness Tests of GaAs Solid State Detectors," CERN–European Committee on Future Accelerators Proceedings, Vol. 3, CERN Report No. 90-10, ECFA Report No. 90-133, pp. 725–731.

Hall, G. (1990), "Radiation Resistance of Semiconductor Detectors and Associated Electronics," CERN–European Committee on Future Accelerators Proceedings, Vol. 3, CERN Report No. 90-10, ECFA Report No. 90-133, pp. 693–705.

Jackson, J. D. (1975), *Classical Electrodynamics*, 2nd ed., New York: Wiley.

Lindström, G., *et al.* (1990), "Radiation Damage of Silicon Detectors by Monoenergetic Neutrons and Electrons," CERN–European Committee on Future Accelerators Proceedings, Vol. 3, CERN Report No. 90-10, ECFA Report No. 90-133, pp. 706–720.

PARTICLE IMPACT PHENOMENA

MICHAEL ALBERT, GLENNYS MENSING, AND NORMAN TOLK, *Department of Physics and Astronomy, Vanderbilt University, Nashville, Tennessee, U.S.A.*

	Introduction	175
1.	**Scattering from a Coulomb Potential**	177
2.	**Electron Impact**	178
2.1	Interaction Mechanisms	178
2.1.1	Collisions with Electrons in the Solid	178
2.1.2	Scattering from the Lattice	179
2.1.3	Bremsstrahlung	180
2.2	Phenomena	180
2.2.1	Plasmons	180
2.2.2	Radiative Transitions	180
2.2.3	Auger Processes	181
2.2.4	Desorption Induced by Electronic Transitions (DIET)	181
2.2.5	Secondary Electron Emission	181
2.3	Applications	182
2.3.1	Scattered-Beam Electrons	182
2.3.2	Electrons from the Target	182
2.3.3	Fluorescence Measurements	182
3.	**Ion Impact**	182
3.1	Interaction Mechanisms	182
3.1.1	Elastic Collisions with Target Atoms	182
3.1.2	Collisions with Electrons in the Solid	183
3.2	Phenomena	184
3.2.1	Depth Dependence of Backscattering	184
3.2.2	Sputtering	184
3.2.3	Channeling	184
3.2.4	Electronic Excitations	185
3.3	Applications	185
3.3.1	Measuring Beam Particles	185
3.3.2	Recoil Target Atoms	185
3.3.3	Fluorescence Measurements	185
4.	**Photon Impact**	185
4.1	Interaction Mechanisms	185
4.1.1	Electronic Absorption	185
4.1.2	Nuclear Scattering	186
4.2	Phenomena	186
4.2.1	Photoelectric Effect	186
4.2.2	Compton Scattering	186
4.2.3	Raman Scattering	186
4.2.4	Two-Phonon Absorption	187
4.2.5	Resonant Bond Breaking	187
4.2.6	Laser Ablation	187
4.3	Applications	187
4.3.1	Scattered or Absorbed Incident Light	187
4.3.2	Electrons from the Target	187
4.3.3	Measuring Scattered Light	188
5.	**Neutron Impact**	188
5.1	Interaction Mechanisms	188
5.1.1	Scattering from Target Nuclei	188
5.1.2	Elastic Scattering from a Lattice	188
5.1.3	Neutron Activation	188
5.2	Neutron-Bombardment Phenomena	189
5.2.1	Phonon Emission and Absorption	189
5.2.2	Decay-Product Energy Loss	189
5.3	Applications	189
5.3.1	Elastic Scattering	189
5.3.2	Inelastic Scattering	189
5.3.3	Reaction Products	189
6.	**Conclusion**	190
	Glossary	190
	Works Cited	191
	Further Reading	192

INTRODUCTION

In the last century, there has been a revolution in our understanding of the microscopic nature of matter, and much of this new insight was gained by observing the impact of beams of particles and photons on surfaces. Rutherford made his historic discovery of the nucleus by studying the scattering of a beam of α particles as they passed through

3-527-28135-5/95/$5.00 + .50

thin foils. Davisson and Germer confirmed the existence of de Broglie matter waves by observing the diffraction of a beam of electrons by crystalline targets. In fact, many of the most important experiments of the twentieth century relied on impact phenomena (see PHYSICS AND TECHNOLOGY, HISTORY OF).

While high-energy particle beams continue to be the vanguard of research into the fundamental nature of matter, the techniques that played major roles in the development of modern physics have now become indispensable tools for materials analysis in nuclear, atomic, medical, and condensed-matter physics, as well as physical chemistry and materials science (Kittel, 1986). Much of the new technology that defines the modern world resulted from progress in these fields, and in turn, this progress depends upon a basic understanding of impact phenomena. In fact, impact phenomena play a prominent role throughout this Encyclopedia. This discussion, however, focuses on the rich variety of phenomena associated with beam interactions with condensed matter.

Analysis techniques that employ impact phenomena have in turn relied heavily upon progress made in related fields (Feldman and Mayer, 1986). Useful particle beams can only be produced in vacuum, and surfaces are difficult to characterize in the presence of air at normal temperatures and pressures. Therefore, vacuum technology has been particularly important to the development of impact techniques (see VACUUM TECHNIQUES AND INSTRUMENTATION). Also, the emerging laser technologies and synchrotron-radiation sources are revolutionizing the way materials are characterized (see LASER PHYSICS). Of course, progress in particle accelerator designs in both the high- and low-energy regimes has been crucial. Finally, the development of a tremendous variety of detectors and measurement devices has greatly expanded the range and sensitivity of experiments.

The nature of the interaction between a particle or photon beam and a target is highly dependent on the wavelength of the incident species. For a particle beam, the wavelength is given by the de Broglie relation

$$\lambda = h/\sqrt{2mE}, \tag{1}$$

where h is Planck's constant, m is the mass of the particle, and E is the energy (Shankar, 1980). The wavelength of a photon is just

$$\lambda = hc/E, \tag{2}$$

where c is the speed of light. If the wavelength is much larger than the interatomic distance, then the incident particle wave function interacts with several target atoms at the same time. Consequently, in the infrared, visible, and ultraviolet atomic resolution is not possible. For photons, near-field techniques can circumvent this restriction somewhat (see OPTICAL MICROSCOPY). If the wavelength is the same order of magnitude as the interatomic distance in the target material, as it is in the x-ray region of the electromagnetic spectrum, then interference effects occur, allowing diffraction experiments, which can reveal periodic atomic structure. If the wavelength is much smaller, collisions can be characterized as simple two-particle interactions, and true atomic resolution is possible. Table 1 shows for several beam species the energy required to obtain wavelengths of 1 Å, which is about the size of an atom, and 5 Å, which is typical of interatomic distances in crystals.

As the table indicates, with photons only x rays have short enough wavelength for diffraction experiments. Electrons, on the other hand, have a very convenient mass so that diffraction and atomic resolution are possible with modest energies easily achieved in the laboratory. In contrast, nucleons have such large mass that interference effects are difficult to produce. Nevertheless, neutrons at thermal energies can be collimated into beams and used for diffraction experiments. For example, the energy for a 1-Å neutron given in the table would correspond to a temperature

Table 1. The energy, in eV, required to achieve wavelengths of 1 and 5 Å for various species.

	Photon	Electron	Nucleon	Helium	Argon
1 Å	12 400	150	0.0819	~0.02	~0.002
5 Å	2 480	6.02	0.00328	~0.0008	~0.00008

of 650 °C. This illustrates that each species has inherent difficulties and advantages.

Another central issue is the degree to which the probe itself modifies the target. Almost all experiments damage the target on a microscopic level, and high-energy or high-flux beams can do macroscopic damage. With the recent developments in constructing and characterizing nanostructures, minimizing even microscopic damage from analysis techniques has become crucial. Unfortunately, a useful description of damage requires a very thorough description of the extremely complicated dynamics and is beyond the scope of this article (see RADIATION DAMAGE IN CRYSTALS).

In this article, Sec. 1 describes the basic Coulomb interaction, which is the basis for most simple models of particle impact and is therefore applicable to several of the subsequent sections. The remaining sections are organized by the identity of the impacting beam. Section 2 treats electron bombardment. Section 3 deals with ion–solid interactions. Section 4 discusses photon interactions, and Sec. 5 presents neutron interactions. Although the article is organized in this manner, many of the concepts of impact are applicable to more than one beam species. Within each section, the basic mechanisms of interaction are described, the observed phenomena are indicated, and finally, the most common applications are outlined.

1. SCATTERING FROM A COULOMB POTENTIAL

Many of the impact phenomena treated in this article begin with two-particle Coulomb scattering (Fetter and Walecka, 1980). Since this model is central to the discussion, a basic description is included. Charged particles traveling through a target material will feel a Coulomb repulsion or attraction to the electrons and nuclei in the material (see ELECTROSTATICS). Because materials are typically macroscopically neutral, it is assumed that long-range interactions are screened, and only individual classical two-particle collisions are significant. Because of the limited validity of this assumption, some of the later treatments will include correction factors that will not be derived.

Low-mass particles scattering from heavy target particles will transfer very little energy. Large scattering angles are possible, and scattering behavior can be determined from the relation of impact parameter to scattering angle. The impact parameter b is defined as the perpendicular distance from the target particle to the initial line of flight. Figure 1 indicates the relation of impact parameter to scattering angle. The differential cross section $d\sigma$ can be defined by

$$d\sigma = 2\pi b db. \tag{3}$$

From this model, it is seen that the differential cross section per unit solid angle is

$$\frac{d\sigma}{d\Omega} = \frac{b}{\sin\theta}\left|\frac{db}{d\theta}\right|. \tag{4}$$

Integrating this over all solid angles yields the total cross section. The scattering problem is now reduced to determining $db/d\theta$, which can be inferred from the equations of motion in a Coulomb potential. In the center-of-mass frame, a particle with mass m and charge ze scattering from a particle of mass M and charge Ze will satisfy the equation

$$\mu\ddot{\mathbf{r}} = zZe^2\mathbf{r}/r^3, \tag{5}$$

where μ is the reduced mass, $\mu = mM/(M + m)$, and $\mathbf{r}$ is the vector from the scattering center to the incident particle. The resulting orbits lead to the following relation between b and θ:

$$b = \frac{|zZ|e^2}{\mu v^2}\cot\frac{\theta}{2}, \tag{6}$$

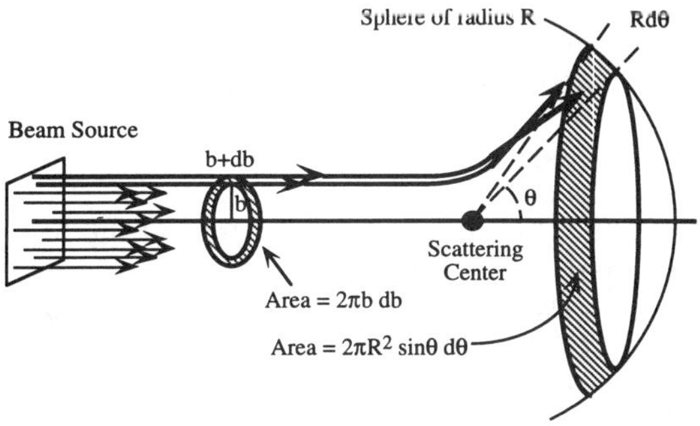

FIG. 1. The relation of differential cross section to impact parameter.

where v is the velocity of the incident particle. Note that the sign of the charges is indistinguishable. With reference to Fig. 1, an attractive potential would draw the particle down rather than push it up, but the scattering angle would be the same. Using this relation, the differential cross section may be written

$$\frac{d\sigma}{d\Omega} = \left(\frac{zZe^2}{4E\sin^2(\theta/2)}\right)^2, \tag{7}$$

where E is the energy of the incident particle in the center-of-mass frame. If the scattering center is much more massive than the incident particle, the center-of-mass energy is approximately equal to the laboratory energy. Thus, this derivation is extremely useful for describing the scattering of light particles from heavy particles.

On the other hand, a heavy particle scattering from a light particle will change direction very little. In this case, the interesting parameter is the amount of energy transferred. Assume that the heavy particle does not deflect at all and that the interaction is sudden so that the target particle cannot move very much. The net impulse in the direction of travel is zero as a result of the symmetry of the incoming and outgoing paths. However, integrating the Coulomb force, the momentum transfer in the perpendicular direction Δp is

$$\Delta p = 2zZe^2/bv. \tag{8}$$

This then gives the transferred energy directly:

$$\Delta E = \frac{\Delta p^2}{2m} = \frac{mz^2Z^2e^4}{Mb^2E}, \tag{9}$$

where the velocity of the incident particle has been written in terms of its kinetic energy:

$$E = \tfrac{1}{2}mv^2 \Rightarrow v^2 = 2E/m. \tag{10}$$

Equation (9) may be inverted to find the impact parameter in terms of the transferred energy. Using this for the impact parameter in Eq. (3), the differential cross section may be written in terms of the energy transfer as

$$d\sigma = \frac{\pi m z^2 Z^2 e^4}{ME\Delta E^2}\, d\Delta E. \tag{11}$$

2. ELECTRON IMPACT

2.1 Interaction Mechanisms

2.1.1 Collisions with Electrons in the Solid The interaction of a beam of electrons impinging on a solid or liquid target can be modeled to a good approximation by classical two-particle scattering from target electrons. Although the masses of the incident particle and target particle are identical, the impulse approximation applies if the energy of the incoming electron is sufficiently high. Here, the change in momentum Δp and thus the change in energy ΔE as a function of impact parameter b reduce to

$$\Delta p = \frac{2e^2}{bv} \Rightarrow \Delta E = \frac{e^4}{Eb^2}, \tag{12}$$

where e is the electron charge, v is the velocity of the incident electron, and E is the energy of the incident electron. Thus the differential cross section $d\sigma$ is

$$\begin{aligned} d\sigma(E) &= -2\pi b\,db \\ &= (\pi e^4/E\Delta E^2)d\Delta E. \end{aligned} \tag{13}$$

Integrating this gives the cross section for a particular transition with energy E_t:

$$\sigma = \pi\frac{e^4}{E}\left(\frac{1}{E_t} - \frac{1}{E}\right). \tag{14}$$

If the beam energy is much greater than the transition energy, the cross section simplifies to

$$\sigma \cong \pi e^4/EE_t. \tag{15}$$

The possible electronic transitions are determined by the electronic structure of the target. The transition corresponding to E_t could be from a bound state, valence-band state, or conduction-band state to any higher energy state including free-particle states which, if the surface is nearby, may result in escaping electrons. For example, a valence-band electron could be promoted to the conduction band resulting in an electron–hole pair, or a bound core electron could be ejected from the solid.

Assuming one transition dominates the absorption, the rate of energy loss as a function of distance traveled in the target may also be estimated by integrating energy times the differential cross section. This yields

$$-\frac{dE}{dx} = \frac{2\pi e^4 n}{E} \ln\left(\frac{E}{E_t}\right), \tag{16}$$

where n is the density of electrons in the initial state. Since energy can only be transferred in discrete amounts, the mean free path can be estimated from the rate of energy loss. For example, the dominant interaction in a metal is plasmon excitation, and so the mean free path λ can be estimated from the plasmon energy $\hbar\omega$:

$$\frac{1}{\lambda} = -\frac{m\omega e^2}{2\hbar E} \ln \frac{4E}{\hbar\omega}, \tag{17}$$

where m is the mass of the electron. For a 1-keV electron incident on a typical metal, this initial penetration depth is a few nanometers. This formula applies equally well to the escape depth of ejected electrons in the solid.

Note that the cross section and thus the mean free path depend on the energy of the incident particle. Since the particle loses energy after each collision, λ must be integrated to find the average range of travel of the incident electron in the target. This yields an equation for the path length (range) R of the form

$$R = (K/\rho)\, E^{\gamma}, \tag{18}$$

where ρ is the density of the material and K and γ are constants. Since this derivation is approximate, the form of the electron range as shown in Eq. (18) is assumed, but the parameters K and γ are fitted empirically. The parameter K is material independent, and for R in microns, ρ in g/cm^3, and E in eV the value for K is 0.064. The parameter γ depends on the target material and is typically between 1.2 and 1.7. While this estimates the length of the electron's trajectory, each interaction results in a random scattering angle so that the trajectory is in general far from straight.

2.1.2 Scattering from the Lattice Because of the wave nature of the incident electron, it may also scatter from the atoms in the target without exciting an electronic state. Quantum mechanically, the wave function interacts with many atoms at the same time (Wichmann, 1971). The scattered wave function interacts with itself, resulting in quantum interference effects, and scattering is enhanced in those directions for which the interference is constructive. Since the mass of the target is so much greater than the mass of the electron, the momentum transfer is negligible, and the electron will lose almost no energy. In order for diffraction to occur, the wavelength of the incident particle must be near the size of the structure to be investigated. At 150 eV, an electron will have a wavelength of about 1 Å, and so electrons are very well suited to diffraction studies of crystals.

If the atoms in the target are regularly spaced in a lattice, the interference will also have a regular pattern. This can easily be demonstrated by considering Bragg reflection from adjacent planes of atoms (Fig. 2). A beam reflecting from parallel planes separated by a distance d will have a path-length difference of $2d \sin\theta$ where θ is the angle of incidence. If these paths are such that constructive interference occurs, the path-length difference must equal an integer number of wavelengths:

$$n\lambda = 2d \sin\theta \text{ (Bragg condition).} \tag{19}$$

There are many ways in a three-dimensional lattice to choose parallel planes of atoms, and each of these facets imposes another Bragg condition. Constructive interference will be greatest in scattering directions that best match all of these Bragg conditions, so that scattering in those directions is highly

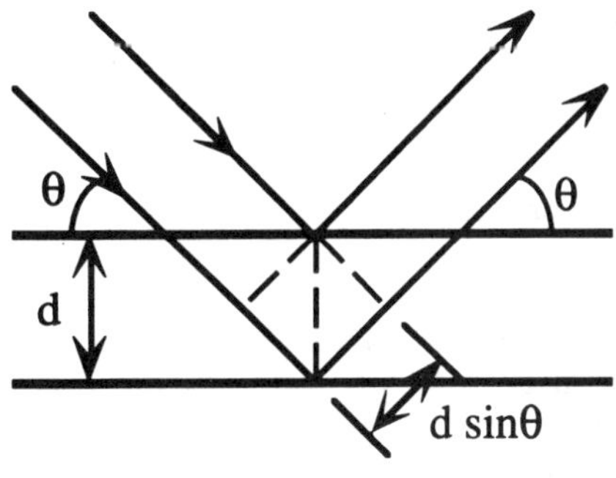

FIG. 2. Bragg reflection.

probable. On a detector screen, the set of favorable scattering directions forms a distinct pattern of points referred to as a diffraction pattern. The Bragg condition along with some simple geometry indicates that the distance between adjacent points in the diffraction pattern is inversely proportional to the distance between planes in the lattice (for a more detailed discussion of diffraction, see X-RAY DIFFRACTION; NEUTRON DIFFRACTION; ELECTRON DIFFRACTION).

2.1.3 Bremsstrahlung In scattering from the target atoms, the incident electron is decelerated, and so must emit radiation. While this is true for any charged particle, it is particularly important for electrons whose light mass readily allows them to experience large decelerations. A continuous spectrum of radiation will be produced with intensity decreasing with increasing energy up to the energy of the incident electron. The radiation spectrum can be calculated from classical electromagnetic theory. For nuclear scattering of nonrelativistic electrons, the spectrum is found to be

$$I = \frac{16Z^2e^6}{3m^2v^2c^3} \ln \left(\frac{2 - \hbar\omega/E + 2\sqrt{1 - \hbar\omega/E}}{\hbar\omega/E} \right), \tag{20}$$

where I is the spectral intensity and Z is the atomic number of the target species. Note that the effect is greatest for high–atomic-number targets. The ratio of bremsstrahlung energy E_B to other losses is approximately

$$\frac{dE_B}{dE_{\text{other}}} \approx \frac{4Z}{3\pi 137} \left(\frac{v}{c} \right)^2. \tag{21}$$

For an energy of 100 keV and a target with $Z \approx 30$, roughly 4% of the beam power goes into bremsstrahlung.

2.2 Phenomena

2.2.1 Plasmons In metals, plasmon excitation will dominate the interaction. A plasmon is a density wave in the quasifree gas of conduction electrons. It is important to note that electrons interact and produce correlated behavior in insulators as well as in metals; however, the following discussion applies only to metals. Consider a volume of negative charge embedded in a stationary positive charge of similar volume. Displacing the negative charge creates a dipole field that in turn produces a linear restoring force. The plasmon is thus a harmonic oscillation with the frequency

$$\omega = \sqrt{4\pi ne^2/m}, \tag{22}$$

where n is the density of electrons. The energy of these oscillations is $\hbar\omega$. For typical metals, this falls in the range of 10 to 15 eV. Because of the surface boundary conditions, there is a separate surface-plasmon mode that has a characteristic frequency of $\omega/\sqrt{2}$. For very small particles, boundary conditions complicate the plasmon model, and the plasmon frequency will depend on the size and shape of the particle. (see COLLECTIVE PHENOMENA IN SOLIDS)

2.2.2 Radiative Transitions Electrons in the solid excited by impact may relax by emitting a photon. Also, if a core target electron is ejected, other electrons in the solid either from the same atom or neighbors may fall into the vacancy, emitting a photon. For radiative transitions, the energy of the photon will equal the difference in energy of the initial and final states of the radiating electron.

The easiest of these transitions to understand conceptually and interpret experimentally is a transition of a bound electron to fill a vacancy in the same atom. From perturbation theory, assuming only dipole interactions, the rate R for a spontaneous transition from a filled state to an empty state is found to be

$$R = \frac{4e^2}{3\hbar^4c^3} \Delta E^3 \, |\langle f|\mathbf{r}|i\rangle|^2, \tag{23}$$

where ΔE is the transition energy and the labels i and f refer to the initial and final electron states respectively. The rate depends heavily on the transition energy, but many transitions are disallowed by the dipole matrix element. Thus the transition energy dependence is most apparent when comparing similar transitions in atoms with differing atomic number. The dipole matrix element between the initial and final state indicates that transitions between highly overlapping states are preferred. Since the change in angular momentum of the absorber (or emitter) must equal the angular momentum of the photon, the matrix element automatically imposes dipole selection rules. Allowed dipole transitions have $\Delta l = \pm 1$ and $\Delta j = \pm 1$ or 0, where l and j are the angular-momentum quantum numbers. However, other multipole transitions may occur with different restrictions.

2.2.3 Auger Processes An electron making a transition to a core vacancy may also release the excess energy by exciting another electron. This is called an Auger process (see AUGER SPECTROSCOPY), and the resulting excited electron is called an Auger electron (Fig. 3). The electron that falls into the vacancy can most easily transfer energy to an electron in the same subshell. One falls into the vacancy, and the other carries the excess energy to a higher state, possibly the vacuum state, leaving two new vacancies in the higher shell or band. The transition rate also depends on the overlap of the initial and final states of the electron that fills the vacancy. Unlike radiative transitions, a vacancy may be filled by an electron from a higher orbital substate of the same primary shell as the vacancy. Since the initial and final states for such an intrashell transition are very similar, these transitions have higher rates than might be expected from the previous argument. These intrashell transitions are called Coster–Kronig transitions.

To first order, the energy of an escaping Auger electron can be estimated from the binding energies of the initial and final states. If a vacancy in state a is filled by an electron originally in state b, resulting in an electron in state c being ejected, the energy of the liberated electron is

$$E = E_a - E_b - E_c, \tag{24}$$

where these are the normal binding energies of the states. However, there is a significant correction due to the vacancies, which alter the normal binding energies. A fairly good estimate can be constructed from the binding energies of similar states in an atom with one more proton:

$$E = E_a^Z - E_b^Z - E_c^Z - \tfrac{1}{2}(E_c^{Z+1} - E_c^Z + E_b^{Z+1} - E_b^Z). \tag{25}$$

2.2.4 Desorption Induced by Electronic Transitions (DIET) When a *localized* excited state is created in a solid, it dramatically changes the local electronic environment. If the state is sufficiently long lived, the electromagnetic forces can alter the position of an atom or molecule in the lattice, resulting in defect formation or even desorption (Tolk *et al.*, 1983). Consider an atom adsorbed on a surface. An electron in some bonding orbital holds the atom to the surface. If that electron is excited to a less-bonding or even an antibonding state, the bond may break, resulting in desorption. Somewhat more spectacular is the highly repulsive state that results when the electron of one atom makes a transition into the core vacancy of a neighboring atom. Several other mechanisms exist by which the electronic energy can be channeled into kinetic energy of the atoms or molecules in the lattice, resulting in damage.

2.2.5 Secondary Electron Emission In addition to directly ejected electrons or Auger electrons, a target will also emit a broad band of electrons with energies less than 100 eV that are relatively insensitive to the energy of the incident beam. These low-energy electrons are called secondary electrons because they clearly do not result from a single-step process or even a two-step process like Auger electrons. Although secondary electrons arise from several sources, an argument based on plasmons provides a clue to

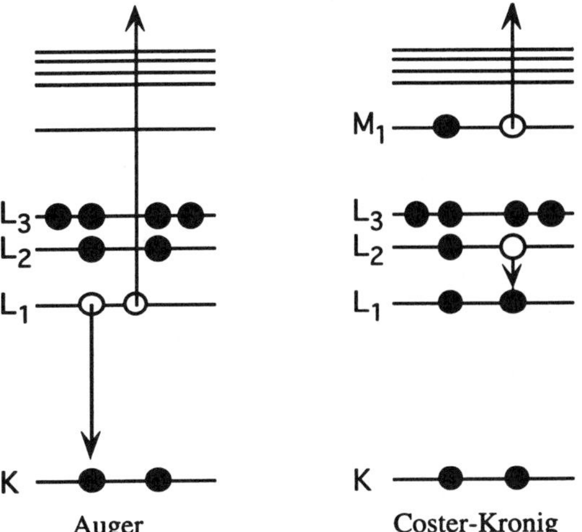

FIG. 3. Auger processes.

their origin. One or more plasmons may couple to a single electron, thus giving that electron enough energy to escape the surface. Since plasmons have fixed energies, the shape of the energy spectrum of emitted electrons would not depend on how the plasmons were created. This and other similar processes result in an energy spectrum of which only the amplitude depends on the incident energy.

2.3 Applications

2.3.1 Scattered-Beam Electrons Since the probability for electron scattering depends on the density of electrons in the material, the scattering of electrons for a fixed target thickness will vary with different materials. This effect for example provides the contrast for transmission electron microscopy (TEM) (see ELECTRON MICROSCOPY).

The incident electron can only lose energy in discrete amounts determined by the allowed transitions in the target material. Thus one can learn a great deal about the electronic structure of the target by measuring the energy of beam electrons that pass through the target. This is the basis of electron energy-loss spectroscopy (EELS). If the same measurement is made for the reflected portion of a low-energy beam, the measurement is surface sensitive. This is the principle behind high-resolution electron-energy-loss spectroscopy (HREELS).

Since elastically scattered electrons may interfere, the structure of the lattice can be determined from the spatial distribution of scattered electrons. This is the basis of electron diffraction. For example, the transmitted-beam electrons in a transmission electron microscope demonstrate a diffraction pattern. Low-energy electrons reflected at normal incidence from a crystalline surface will also show a diffraction pattern, which is primarily surface sensitive. This is called low-energy electron diffraction (LEED). Because the component of energy into the surface is small, a high-energy electron beam reflected at grazing incidence will also produce a surface-sensitive diffraction pattern. This technique, called reflection high-energy electron diffraction (RHEED), provides a more convenient geometry than LEED in that the surface is more accessible for simultaneous experiments (see ELECTRON DIFFRACTION).

2.3.2 Electrons from the Target In addition to scattered-beam electrons, Auger electrons may also escape from the target. Measuring the energy of electrons reveals peaks at energies corresponding to the Auger transitions, which are characteristic of the target atoms. Thus Auger-electron spectroscopy (AES) can detect target species down to the 0.1% level. Elaborate techniques are usually required to separate the Auger peaks from the background of scattered beam electrons.

2.3.3 Fluorescence Measurements To avoid the background problem of Auger-electron spectroscopy, the characteristic photons produced by similar radiative transitions may be detected. Since the size of the target volume probed by the electron beam can be reduced to about a micron, this technique is referred to as electron microprobe analysis (EMA). Characteristic photons are typically in the hard ultraviolet or x-ray region and thus have their own detection challenges.

3. ION IMPACT

3.1 Interaction Mechanisms

3.1.1 Elastic Collisions with Target Atoms Particles such as ions or molecules striking a surface will interact primarily by elastic collisions with target atoms. By conservation of energy and momentum, it is easily shown that the fraction of initial energy retained by the incident particle is

$$\frac{E_1}{E_0} = \left[\frac{(M_2^2 - M_1^2 \sin^2\theta)^{1/2} + M_1 \cos\theta}{M_1 + M_2}\right]^2, \qquad (26)$$

where M_1 is the mass of the incident particle, M_2 is the mass of the target particle, and θ is the scattering angle. This ratio is referred to as the kinematic factor and is often denoted by K. The kinematic factor as a function of angle for several choices of the ratio M_2/M_1 is plotted in Fig. 4. The fraction of energy transferred to the target particle is similarly given by

$$\frac{E_2}{E_0} = \frac{4M_1M_2}{(M_1 + M_2)^2}\cos^2\phi, \qquad (27)$$

where ϕ is the recoil angle. For particular

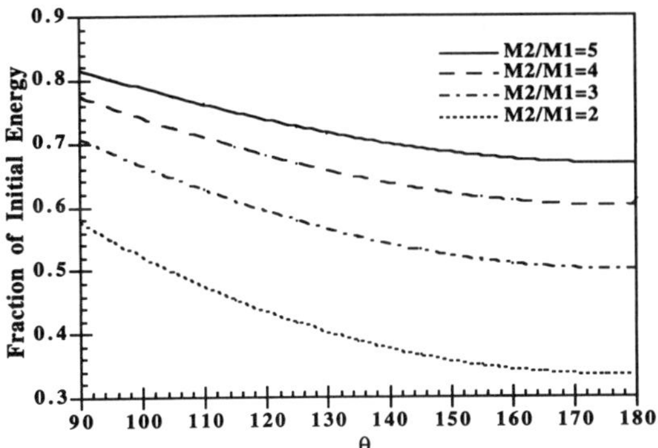

FIG. 4. The kinematic factor as a function of scattering angle for several choices of the ratio M_2/M_1.

scattering angles, these relations become quite simple. For example, in the case of backscattering, $\theta = \pi$ and $\phi = 0$ so that

$$\frac{E_1}{E_0} = \left(\frac{M_2 - M_1}{M_1 + M_2}\right)^2 \quad \text{and} \quad \frac{E_2}{E_0} = \frac{4M_1M_2}{(M_1 + M_2)^2}. \tag{28}$$

Electron screening of target nuclei complicates heavy particle scattering. Nevertheless, if the energy of the incident particle is high enough to penetrate well within the electron cloud, the simple scattering model will be approximately correct. This condition will be met if

$$E_0 > zZe^2/a_0, \tag{29}$$

where z is the atomic number of the incident atom, Z is the atomic number of the target species, and a_0 is the Bohr radius. In any case, screening will produce some deviation from the simple model, and in general a correction factor must be included. Ignoring electron screening, the cross section is as given above,

$$\sigma(\theta) = \left(\frac{zZe^2}{4E\sin^2(\theta/2)}\right)^2. \tag{30}$$

3.1.2 Collisions with Electrons in the Solid An incident heavy particle will also collide with electrons in the solid. However, as a result of momentum conservation, a heavy particle can transfer only a small fraction of its momentum to an electron. Thus as a high-energy particle traverses a target, it will slowly lose energy through numerous electron collisions in an almost continuous fashion. These collisions may also be modeled with the impulse approximation. The derivation of the rate of energy loss is completely analogous with that for electron–electron scattering. Thus the rate of energy loss as the particle traverses the target is

$$-\frac{dE}{dx} = \frac{2\pi z^2 e^4 n}{E}\frac{M}{m_e}\ln\frac{4Em_e}{IM}, \tag{31}$$

where M is the mass of the incident particle, z is its atomic number, I is the average excitation energy of electrons in the solid, and n is the density of electrons in the solid. I is complicated, but, for heavier elements, it is approximately equal to $10Z$. This is usually converted to a "stopping power" per atom (Fig. 5). The stopping power ϵ per atom is defined as

$$\epsilon = \frac{dE/dx}{N}, \tag{32}$$

where N is the density of the target atoms. The stopping power can then be used to determine dE/dx for compounds. According to Bragg's rule, for a compound A_nB_m, the stopping power is

$$\epsilon_{A_nB_m} = n\epsilon_A + m\epsilon_B, \tag{33}$$

where n and m are the relative atomic abundances of the respective constituents. Thus multiplying by the molecular density, the rate of energy loss in a compound target may be determined.

Although electronic stopping is the most important contribution to the energy loss for

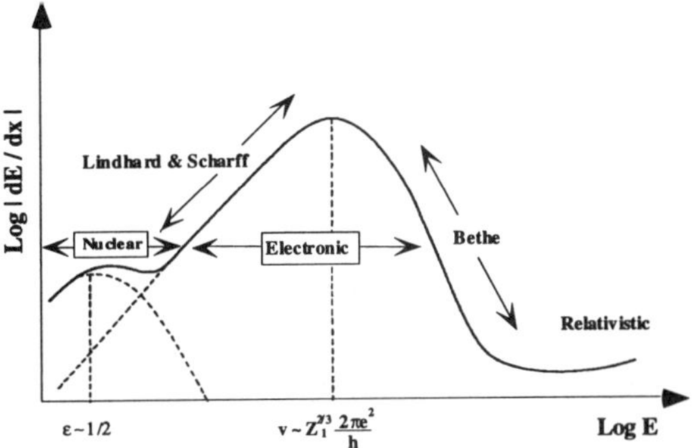

FIG. 5. Stopping power.

a large range of energies, other factors such as scattering from the atomic cores are also important. The convoluted path resulting from numerous scattering events further complicates this picture. To compute the total depth beneath the surface attained by an incident particle, all of these factors must be considered. Table 2 indicates the median penetration depth for various species and energies of particles incident on amorphous alumina.

3.2 Phenomena

3.2.1 Depth Dependence of Backscattering Since a heavy particle will lose energy to electrons as it penetrates a target, the electron scattering produces a depth dependence in the energy of atomic scattering. For example, consider a light ion with energy E scattering from an atom at a depth d within the target. To find the total energy lost to electron scattering, dE/dx should be integrated to the penetration depth. However, for modest depths, the fractional energy loss will be small, and the total energy loss is approximately

$$\Delta E \cong \frac{dE}{dx} d, \tag{34}$$

where dE/dx is evaluated at the incident energy E. The particle with this reduced energy now backscatters from a target atom, and the energy is reduced by the kinematic factor above. The particle then traverses the target again before escaping. Since the particle may have been scattered at an angle to the surface normal, the distance traveled on the way out is $d/\cos\theta$. The energy loss is approximated as above with dE/dx now evaluated at the appropriate backscattered energy. Thus the total energy lost to the electrons both in entering and in exiting the sample is

$$\Delta E = d\left(K\frac{dE}{dx}\bigg|_{\text{in}} + \frac{1}{|\cos\theta|}\frac{dE}{dx}\bigg|_{\text{out}}\right), \tag{35}$$

where K is the kinematic factor and "in" and "out" indicate at which energy dE/dx should be evaluated. Note that the total energy lost to electron scattering is directly proportional to the penetration depth.

3.2.2 Sputtering Atomic collisions impart significant energy to the target atoms, and in general they will in turn collide with other target atoms. The resulting collision cascade can give some atoms sufficient momentum to escape from the target, and so the ion beam erodes the target. This process is called sputtering (*q.v.*) (Behrisch, 1981). The sputtering yield is defined as the number of target atoms emitted per incident particle. This depends not only on the nature of the target material but also on the beam species and energy, and the possible values vary tremendously. Although calculation of sputtering yields is understandably daunting, theory has produced good results. Nevertheless, for practical applications empirical values are used.

3.2.3 Channeling If the target is crystalline, an interesting phenomenon known as channeling may occur (Fig. 6). If the incident particles are aligned along one of the highly oriented directions of the crystal, they cannot collide solidly with atoms in the bulk of the target. Essentially, the atoms on the surface shadow the rest of the atoms. However, the effect is greater than that simple explanation indicates. Consider the regular lattice to be parallel planes of atoms. If the incident particle is moving nearly tangential to these planes, it will merely glance off the planes at a shallow angle. This costs very little energy, and the incident particle will move down this

Table 2. Penetration depth, in $\mu g/cm^2$, for various incident species and energies, derived from implantation of amorphous alumina (Jespergård and Davies, 1967).

Energy (eV)	Na	K	Kr	Xe
10	4.5	2.6	2.5	2.3
100	35	20	14	11
1000	500	260	170	100

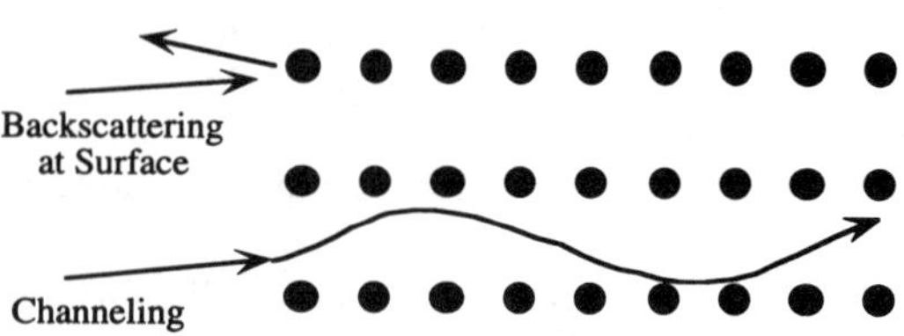

FIG. 6. Channeling.

channel, making a series of shallow collisions with the "walls." Thus a particle incident in a channeling direction may penetrate very deeply into the target. The ultimate depth achieved is primarily determined by electron stopping. Furthermore, backscattering will only occur at the surface.

3.2.4 Electronic Excitations In colliding with the electrons, a heavy particle excites electronic states in the target just as in the case of electron bombardment. Therefore, many of the same phenomena are seen in each case. Electrons will be ejected, electron–hole pairs will be created, plasmons will be stimulated, secondary electrons will be produced, and most of the other previously discussed electronic excitations will occur. Of particular interest are the radiative transitions to fill a core hole created by the incident particle. The origins of the core hole are irrelevant, and the process is exactly as described in the electron section.

3.3 Applications

3.3.1 Measuring Beam Particles The energy of a light ion that has been scattered is uniquely determined by the species from which it scattered. If the energy of scattered-beam particles is measured, the composition of the target can be determined. This is the basis of Rutherford backscattering spectrometry (RBS). For simplicity, the measurement is usually made as close as possible to the backscattering direction. Since the energy has a simple dependence on depth due to electron scattering, depth profiling is also possible. Finally, since subsurface atoms cannot readily backscatter when the beam is aligned along a channeling direction, surface sensitivity of RBS can be enhanced by channeling. For example, RBS can easily detect whether a surface is reconstructed, relaxed, or covered with an adsorbate.

To enhance surface sensitivity, an ion beam with energy in the range of one to a few keV may be used in conjunction with an electrostatic energy analyzer. Since particles that penetrate the surface are almost always neutralized, only particles scattered from the surface will be detected by the electrostatic energy analyzer. The low-energy beam insures that most of the ions will be scattered from the surface. This is the basis of low-energy ion scattering (LEIS).

3.3.2 Recoil Target Atoms Backscattering is inefficient for lighter target materials and kinematically impossible if the target material is lighter than the beam species. Rather than measuring backscattered beam particles, measuring the energy of forward recoiling target atoms gives the same kinematic information. This is the basis of forward recoil spectrometry (FRS). The beam is usually incident at a shallow angle to the target surface, and the recoiling atoms are detected at some small angle from the surface in the forward direction.

3.3.3 Fluorescence Measurements Heavy particles may also produce core-hole vacancies in target atoms. The mechanisms for filling the vacancy are exactly the same as described above. In particular, the target may be characterized by measuring the radiation from such transitions. This is the basis of proton (or particle) induced x-ray emission (PIXE). Because the larger mass of the incident particle results in far less bremsstrahlung, PIXE has less background noise than electron microprobe analysis.

4. PHOTON IMPACT

4.1 Interaction Mechanisms

4.1.1 Electronic Absorption A photon incident on the surface of a solid or liquid will excite higher electronic states in the target. If the photon is modeled as an oscillating perturbation $H'e^{-\omega t}$, from first-order perturbation theory the probability of stimulating a transition is given by Fermi's "golden rule,"

$$P_{i\to f} = t\,\frac{2\pi}{\hbar}\,|\langle f|H'|i\rangle|^2\,\delta(E_f - E_i - \hbar\omega), \qquad (36)$$

where t is the length of time that the perturbation is applied, and the labels i and f indicate initial and final states of the *unperturbed* system. The terms E_i and E_f are the energies of the unperturbed initial and final states. Note that the transition energy must be equal to the energy of the photon. This seems to imply that the whole photon must be absorbed, and scattering will not occur.

However, the excited state may relax by emitting another photon with either equal or lower energy, resulting in an effective scattering. Also, according to the uncertainty principle noneigenstates can exist for a short time, so photons may be scattered even when no resonant transition exists. Nevertheless, the probability of interaction is much greater at resonance, i.e., when there exists a transition with a difference in energy equal to the energy of the photon. Therefore, light will be strongly absorbed at frequencies that correspond to allowed electronic transitions. The matrix element indicates the strength of coupling of the photon to the electronic states. In general, the coupling is proportional to the dipole moment. However, as in the case of atomic hydrogen, time dependence allows some coupling where no dipole moment exists. Finally, because a photon has angular momentum, only transitions with $\Delta l = 1$ are allowed with high probability in the dipole approximation.

4.1.2 Nuclear Scattering An incident photon may also scatter from the target atom without leaving the atom in an excited state. The large mass of the nuclei allows momentum to be conserved without significantly changing the energy of the photon. Since the energy does not change, the photon maintains its initial wavelength, and interference may occur. As with electrons, the resulting diffraction pattern is indicative of the long-range order of the target. The diffraction relations are exactly the same as discussed previously for particle interference.

4.2. Phenomena

4.2.1 Photoelectric Effect If the incident photon has greater energy than the binding energy of electrons in the target, electrons may be liberated. The maximum energy of the ejected electron will be

$$E = E_b - \hbar\omega, \tag{37}$$

where E_b is the binding energy of the electron. Since the final state is a free particle, selection rules can always be satisfied, and all electrons with binding energy less than or equal to the photon energy may be liberated. Applying Fermi's "golden rule" to hydrogenic states yields the cross section for a photon of energy $\hbar\omega$ liberating an electron with binding energy $E_b \ll \hbar\omega$:

$$\sigma = \frac{128\ \pi e^2 \hbar E_b^{5/2}}{3mc(\hbar\omega)^{7/2}} = \frac{7.45\ \text{eV Å}^2}{\hbar\omega}\left(\frac{E_b}{\hbar\omega}\right)^{5/2}. \tag{38}$$

4.2.2 Compton Scattering For photon scattering, if the energy transferred to the electron is much greater than its binding energy, the scattering will be nearly identical to scattering from a free electron. The resulting scattering behavior can be deduced simply from energy and momentum conservation. If a photon initially with energy $\hbar\omega$ scatters at an angle θ from an electron initially at rest, conservation of energy requires

$$E_p + E_e = E_p' + E_e' \Rightarrow \hbar\omega + mc^2 = \hbar\omega' + E', \tag{39}$$

where the subscripts refer to photon and electron, and unprimed and primed stands for before and after the collision, respectively. The term E' is the subsequent relativistic energy of the electron. Momentum conservation requires

$$\mathbf{p}_p = \mathbf{p}_p' + \mathbf{p}_e' \Rightarrow (\mathbf{p}_p - \mathbf{p}_p')^2 = \mathbf{p}_e'^2, \tag{40}$$

which may be written

$$\frac{\hbar^2\omega^2}{c^2} + \frac{\hbar^2\omega'^2}{c^2} - \frac{2\hbar^2\omega\omega'}{c^2}\cos\theta = p_e'^2. \tag{41}$$

Combining Eqs. (38) and (41) yields

$$\omega' = \frac{\omega}{1 + (\hbar\omega/mc^2)(1 - \cos\theta)} \quad \text{or}$$

$$\lambda' = \lambda + \frac{2\pi\hbar}{mc}(1 - \cos\theta). \tag{42}$$

Note that the wavelength of the scattered photon depends on the scattering angle. Also, the wavelength shift is independent of the initial wavelength. The shift only becomes significant for fairly short initial wavelengths, and hard x rays must be used to see the effect.

4.2.3 Raman Scattering For a photon to couple to one of the collective nonlocalized modes of a lattice, momentum must be conserved. A photon traversing a material can emit or absorb an optical phonon, but in or-

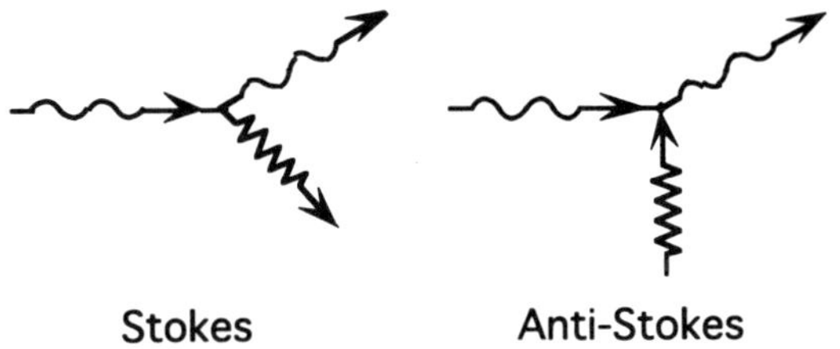

FIG. 7. Raman processes.

der to conserve momentum and energy, the frequency and direction of the photon must change. If a phonon is created, the interaction is called a Stokes process. If a phonon is absorbed it is an anti-Stokes process (Fig. 7). Conservation of energy requires $\omega = \omega' \pm \Omega$ where ω is the frequency of the incident photon, ω' is the frequency of the scattered photon, and Ω is the frequency of the phonon that is created or absorbed by the photon. To conserve momentum, $\mathbf{k} = \mathbf{k}' \pm \mathbf{K}$ where these are the wave vectors for the incident photon, scattered photon, and phonon respectively (see RAMAN SCATTERING).

4.2.4 Two-Phonon Absorption Similarly, a photon cannot be completely absorbed into a single phonon because a phonon with the same energy will not have the same wave vector. In order to conserve momentum, the photon must create two phonons (Fig. 8).

4.2.5 Resonant Bond Breaking Desorption induced by electronic transitions (DIET, Sec. 2.2.4) is possible with photons as well (Tolk *et al.*, 1983). If the incident photon has the appropriate resonant energy to excite an electron initially in a bonding state to an antibonding state, the bond may be broken.

4.2.6 Laser Ablation An incident photon may liberate electrons and desorb ionized target atoms. With a high-intensity laser

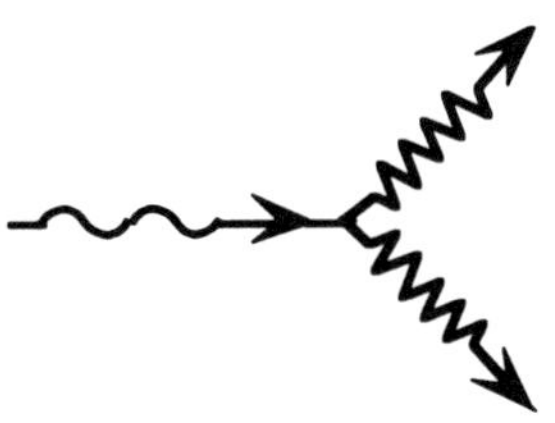

FIG. 8. Two-phonon absorption.

beam, electrons and ions may be created in sufficient quantity to make a dense plasma near the surface of the target. Once the plasma exists, the laser may couple directly to the plasma, and the superheated plasma then ablates the target by collisions with plasma particles.

4.3 Applications

4.3.1 Scattered or Absorbed Incident Light Photons with energies resonant with electronic transitions in the target will be strongly absorbed. Thus if the transmission of light as a function of frequency is measured, the electronic structure of the target may be indirectly mapped. This is the basis of absorption spectroscopy. Because the resonances correspond to the difference in energy of two electronic states rather than the absolute energy of a single electronic state, the information is somewhat convoluted. However, absorption spectra for most materials are well known, and so information about the target can be obtained without a detailed knowledge of the electronic structure. Also, if the target is not transparent, the same information can be inferred from reflection (see SPECTROSCOPY OF SOLIDS).

4.3.2 Electrons from the Target In order to map the electronic states directly, the energy of photoelectrons can be measured. Since the energy of a photoelectron is just the energy of the photon minus the electron binding energy, the spectrum of photoelectron energies is a direct measure of the electronic states of the target. This is the basis of x-ray photoelectron spectroscopy (XPS). However, this will only indicate filled states. In order to map the normally empty excited states, the excited states can be populated prior to measurement in a pump–probe experiment. For example, the density of states of the conduction band in a semiconductor can be mapped in this manner.

In semiconductors, some of the electronic structure of the target can be directly mapped without reverting to x rays as in XPS. If a target is biased with a small voltage and the current flow is measured, the current will be enhanced when photons promote electrons to the conduction band. If this photocurrent is measured as a function of photon energy, the derivative of the resulting spectrum is a direct map of the density of states below the band gap. This is the basis of internal photoemission spectroscopy (IPE). The simplicity of the measurement and the lower photon energy make this method preferable to XPS (see PHOTOEMISSION AND PHOTOELECTRON SPECTRA).

The energy spectrum of photoelectrons will also have some additional structure. This is due to a characteristic energy loss as an electron is liberated from an inner shell, coupled with an interference effect as the photoelectron interacts with target atoms as it escapes the target. This is the basis of extended x-ray absorption fine-structure (EXAFS) measurements. The advantage of EXAFS is the extremely localized sensitivity due to the short range of electrons in solids. Thus the primary peak indicates the species of the scattering atom from the characteristic energy loss, and the fine structure indicates the lattice positions of its neighbors.

4.3.3 Measuring Scattered Light Since the phonon-dispersion relation is a unique function of the target material, Raman scattering will also be unique to the target material. The shift in energy of a scattered photon is a direct measure of the energy of the optical phonon with wave vector nearly equal to zero. This is the basis of Raman spectroscopy. If the incident beam is focused to a small spot, the identity of crystallites in a heterogeneous target may be spatially mapped as well (see RAMAN SCATTERING).

Exactly as in the electron case, the diffraction pattern of x rays scattered from a crystalline target is a measure of the long-range order of the target. This is the basis of x-ray diffraction (see X-RAY DIFFRACTION).

5. NEUTRON IMPACT

5.1 Interaction Mechanisms

5.1.1 Scattering from Target Nuclei Unlike ions or electrons, Coulomb scattering does not apply to neutrons because of their charge neutrality. A complete discussion of the neutron-scattering process is not possible here, but a few points should be made. A neutron can only scatter from the atomic nuclei in the target, which means neutron scattering is insensitive to the electronic structure of the target material. Since the cross section for neutron scattering is relatively small, neutrons usually penetrate well into the target, and neutron bombardment consequently is not surface sensitive. In addition, the scattering amplitude depends on the relative orientations of the neutron spin and nuclear spin so that a polarized neutron beam is sensitive to magnetization. In many cases, neutron beams may have very low energies of less than 0.1 eV per particle. With other beam particles, interactions could be treated as simple two-particle collisions. For a low-energy neutron beam, this is not the case. In fact, interactions with phonons dominate.

5.1.2 Elastic Scattering from a Lattice A low-energy neutron may also scatter from the nuclei in the target as a whole. As in the case of electrons, this elastic scattering results from the wave nature of the neutron, and as before, quantum interference effects result. For constructive interference, the Bragg condition must be satisfied, and specific scattering angles are preferred, resulting in a diffraction pattern. Note that the Bragg condition depends on the de Broglie wavelength, which in turn depends upon the energy. Thus, while electrons are much preferred for diffraction studies, neutron diffraction is still very important as a means of monochromatizing a neutron beam.

5.1.3 Neutron Activation For many target atoms, energetic neutron bombardment may produce a nuclear reaction, resulting in an unstable isotope (see NUCLEAR REACTIONS). The unstable isotope will then decay producing radiation. Since the target is now said to be "radioactive" or "activated," this process is called neutron activation. If N_x is the number of atoms available to be activated in the target and N_y is the number of activated atoms, then the growth of the number of activated nuclei is given by

$$\frac{dN_y}{dt} = N_x\sigma\phi - N_y\lambda_y, \tag{43}$$

where σ is the reaction cross section, ϕ is the flux of neutrons, and λ_y is the decay constant of the unstable nuclei. The first term in this equation is the rate of activation while the second term is the rate of decay of already activated atoms. Integrating this gives the number of activated atoms as a function of the time of bombardment t:

$$N_y = \frac{N_x \sigma \phi}{\lambda_y}(1 - e^{-\lambda_y t}). \tag{44}$$

The activity A_t, which is the number of decays per second, is just the number of activated atoms times the decay constant:

$$A_t = N_x \sigma \phi (1 - e^{-\lambda_y t}), \tag{45}$$

It is apparent that after a long bombardment time, the activity saturates. In fact, the activity reaches 50% of saturation in one half-life and 75% in two half-lives. This gives experimental control over other undesired processes. Processes with much longer half-lives than the desired reaction can be avoided by stopping activation after two half-lives. Processes with much shorter half-lives than the desired reaction can be eliminated by waiting a full half-life after bombardment ceases before measuring the activity.

5.2 Neutron-Bombardment Phenomena

5.2.1 Phonon Emission and Absorption Excluding nuclear processes, a low-energy neutron will interact with target nuclei primarily by inducing normal-mode vibrations or phonons. In a given direction in a given lattice, only particular phonon energies and momenta are allowed. Since momentum must be conserved, neutrons will scatter most readily in those directions for which the change in momentum of the neutron corresponds to an allowed phonon in the lattice. In addition, during absorption or emission of a phonon, the lattice as a whole may give or take specific quantities of momentum as indicated by the reciprocal lattice vectors $\mathbf{G}$. Thus, if $\mathbf{q}$ is the wave vector of an allowed phonon and $\mathbf{k}$ is the propagation vector of the neutron, then the change in the propagation vector of the neutron $\Delta\mathbf{k}$ satisfies the following condition:

$$\Delta\mathbf{k} = \pm\mathbf{q} + \mathbf{G}. \tag{46}$$

The plus-minus sign indicates that the neutron may either absorb or emit a phonon. The above condition specifically applies to the bulk phonon structure since neutrons penetrate deeply into the target, and neutron bombardment is not sensitive to surface phonons.

5.2.2 Decay-Product Energy Loss If the decay product of an activated nucleus is a charged particle, then it will lose energy as it escapes from the target through scattering from electrons in the target material. This is exactly the same as the electronic energy loss discussed in the ion section. Assuming that the fractional energy loss is small and that the target material is fairly uniform, the energy loss is found to be proportional to the distance traveled through the material. Decay-product energy loss differs from the Rutherford backscattering case in one very important way: The charged particle only has to travel through the target material once. Typically, these decay products have energies of a few MeV and can escape from depths up to a few microns beneath the surface.

5.3 Applications

5.3.1 Elastic Scattering Diffraction of neutrons by a crystal depends on the energy of the neutrons. In order to produce low-energy neutron beams with well-defined energies, a beam of thermal neutrons is scattered from a crystal and collimated. The diffraction angle determines the energy and can be varied by rotating the crystal and apertures. Neutron monochromators operate according to these principles.

5.3.2 Inelastic Scattering Measuring the scattering angle and energy of neutrons scattered from a crystal maps the phonon structure of the lattice. The change in momentum of the neutron corresponds to the momentum of the phonon, and the change in energy of the neutron is the energy of the phonon. These kinds of measurements are routinely carried out with a triple-axis neutron spectrometer.

5.3.3 Reaction Products If the cross section for activation and the lifetime of the unstable isotope are known, then the concentration of a particular species in a target can be calculated directly from the measured activity. This is the basis of neutron activation analysis. Some care is necessary to reduce the background of other activation processes.

If the reaction product is a charged particle, then a depth profile of the activated species can be obtained by measuring the energy of the escaping particles. This is the basis of neutron activation depth profiling.

6. CONCLUSION

Particle impact onto condensed matter results in a great variety of phenomena. In this article, the most important effects have been presented, but these are by no means all of the possible results of an impact event. More importantly, the reader should understand that any given impact experiment will produce many of the phenomena described above. Experiments are usually designed to emphasize one particular aspect of the impact event. However, while an electron energy-loss experiment may emphasize plasmon excitation, the experimenter must be aware that the electron beam may also be diffracting, undergoing collision cascades, ejecting core electrons, creating electron–hole pairs, or producing lattice defects. It is this great variety of effects that makes directed particle or photon beams powerful probes for measuring the properties of materials.

GLOSSARY

AES (Auger-Electron Spectroscopy): An experimental technique for measuring the atomic species in the region very near the surface of a solid by measuring the energies of Auger electrons.

Anti-Stokes Process: A Raman scattering process in which a photon is scattered by absorbing a phonon.

Auger Electron: An electron excited and possibly even liberated by the energy released when another electron falls into a vacant low-energy state.

Auger Process: When an electron falls into a lower-energy vacancy, energy is liberated. If that energy is transferred to another electron, resulting in that electron being excited to a higher state or the vacuum state, the process is called an Auger process, and the excited electron is called an Auger electron.

Bragg Reflection: The reflection of light or particles from parallel planes that have spacing such that self interference determines the allowed scattering angles.

Bremsstrahlung: German for "braking radiation." Bremsstrahlung is the broad spectrum of electromagnetic radiation produced when charged particles are decelerated by impact with a target.

Channeling: Phenomenon in which particles incident on a crystal in a high-symmetry direction reach unusually great depths by traveling through the spaces between the atoms.

Coster–Kronig Transitions: A particular Auger transition in which the vacancy is filled by an electron from the same subshell.

Coulomb Scattering: A model for scattering of charged particles in which the Coulomb repulsion or attraction due to the electric charge is assumed to dominate.

Depth Profiling: A technique in Rutherford backscattering experiments that exploits the nearly uniform deceleration of a charged particle in a solid to map the distribution of atomic species as a function of depth into the target.

DIET (Desorption Induced by Electronic Transitions): The ejection of an atom or molecule from a surface by a repulsive Coulomb force resulting from an electronic excitation.

Diffraction Pattern: The interference pattern produced at a detector or phosphor screen by particles or light elastically scattered from a lattice.

Dipole Approximation: Approximation of the interaction of light with electronic states that only considers the dipole term in the expansion of the electronic state.

EELS (Electron Energy-Loss Spectroscopy): An experimental technique for determining the electronic structure of a material by measuring the change in energy of a beam of electrons as they pass through a material.

EMA (Electron Microprobe Analysis): An experimental technique in which a tightly focused electron beam produces vacancies in the target atoms, and the radiation produced when other electrons fill the vacancies is spectroscopically analyzed.

EXAFS (Extended X-Ray Absorption Fine Structure): An experimental technique for measuring the lattice structure in the neighborhood of a particular impurity by carefully measuring the energy of photoelectrons produced by x-rays.

Fermi's "Golden Rule": An approximation of the probability of a photon producing a particular electronic transition calculated by treating the photon as a short-lived oscillatory perturbation and considering only dipole interactions.

FRS (Forward Recoil Spectroscopy): An experimental technique for measuring light atomic species in a thin film by measuring the energy of particles ejected in the forward direction by a high-energy ion beam.

HREELS (High-Resolution Electron Energy-Loss Spectroscopy): An experimental technique for measuring the electronic structure of the surface of a solid by measuring the change in energy of a low-energy beam of electrons reflected from a surface.

Impact Parameter: The perpendicular distance from the target particle to the initial line of flight of the incident particle.

Impulse Approximation: An approximation based on the assumption that the interaction between a fast-moving beam particle and a stationary target particle happens so quickly that neither particle has time to change its initial motion much while the interaction is occurring.

IPE (Internal Photoemission): An experimental method for mapping the density of states of the valence band of a semiconductor by measuring the photocurrent produced by a tunable light source.

Kinematic Factor: In an elastic two-particle collision, the fraction of energy retained by the incident particle.

LEED (Low-Energy Electron Diffraction): An experimental technique for measuring the near-surface lattice structure of a crystal by diffracting a low-energy beam of electrons from the surface.

LEIS (Low-Energy Ion Spectroscopy): An experimental technique for measuring the atomic species near the surface of a solid by reflecting a low-energy beam of ions from the surface and measuring the ion energy electrostatically.

Mean Free Path: The average distance traveled between interactions.

Neutron Activation Analysis: An experimental technique for measuring the concentration of atomic species in a sample by neutron bombardment activation.

Phonon: A collective vibration of a lattice that carries momentum and energy and is therefore treated as virtual particle.

PIXE (Proton- or Particle-Induced X-Ray Emission): An experimental technique for measuring the atomic species in a target, in which particle bombardment produces vacancies in the target atoms, and the radiation produced when other electrons fill the vacancies is spectroscopically analyzed.

Plasmon: A collective oscillation of the electrons of a material, which carries momentum and energy and is therefore treated as a virtual particle.

Quantum Interference: Interference effects that result from the quantum wave nature of particles.

Raman Scattering: Photon scattering from the lattice either producing or absorbing a phonon.

Reconstruction: The rearrangement of the lattice structure at the surface of a crystal into a symmetry that differs from the bulk solid.

RBS (Rutherford Backscattering): An experimental technique for determining the atomic species in a target by measuring the energy of reflected ions.

RHEED (Reflection High-Energy Electron Diffraction): A convenient technique for determining the surface structure of a sample by diffracting a grazing-incidence high-energy beam of electrons from the surface.

Scattering Angle: The angle between the initial line of flight continued through the scattering center and the subsequent line of flight.

Spectral Intensity: The radiative power per unit frequency.

Sputtering: The ejection of target atoms from a surface by collision cascade.

Sputtering Yield: The number of target atoms sputtered from the surface per incident particle.

Stokes Process: A Raman process in which a photon is scattered from the lattice by emitting a phonon.

Stopping Power: The distance rate of energy loss normalized to the number of atoms.

TEM (Transmission Electron Microscopy): The classic electron microscopy in which a high-energy beam of electrons is passed through a very thin sample and the beam is then imaged on a phosphor screen.

XPS (X-Ray Photoemission Spectroscopy): An experimental technique for determining the electronic density of states of a sample by measuring the energy of photoelectrons produced by X rays.

Works Cited

Behrisch, R. (Ed.) (1981), *Sputtering by Particle Bombardment I*, New York: Springer-Verlag.

Feldman, L., Mayer, J. (1986), *Fundamentals of*

Surface and Thin Film Analysis, New York: North-Holland.

Fetter, A., Walecka, J. (1980), *Theoretical Mechanics of Particles and Continua*, New York: McGraw-Hill.

Jespergård, P., Davies, J. A. (1967), *Can. J. Phys.* **45**, 2983–2994.

Kittel, C. (1986), *Introduction to Solid State Physics*, 6th ed., New York: Wiley.

Shankar, R. (1980), *Principles of Quantum Mechanics*, New York: Plenum.

Tolk, N. *et al.* (1983), *Desorption Induced by Electronic Transitions*, New York: Springer-Verlag.

Wichmann, E. (1971), *Quantum Physics*, New York: McGraw-Hill.

Further Reading

General:

Woodruff, D. P., Delchar, T. A. (1994), *Modern Techniques of Surface Science*, 2nd ed., New York: Cambridge Univ. Press.

Prutton, M. (1994), *Introduction to Surface Physics*, Oxford: Oxford Science Publications.

Electron Impact:

Bundle, C. R., Baker, A. D. (Eds.) (1981), *Electron Spectroscopy: Theory, Techniques and Applications*, New York: Academic Press.

Ibach, H. (Ed.) (1977), *Topics in Current Physics*, Vol. 4, Berlin: Springer-Verlag.

Ion Impact:

Tolk, N. H., Traum, M. M., Tully, J. C., Madey, T. E. (Eds.) (1983), *Desorption Induced by Electronic Transitions, DIET I*, New York: Springer-Verlag.

Photon Impact:

Cohen-Tannoudji, C., DupontRoc, J., Grynberg, G. (1992), *Atom-Photon Interactions*, New York: Wiley.

Neutron Impact:

Izyumov, Y. A., Chernoplekov, N. A. (1992), *Neutron Spectroscopy*, New York: Consultants Bureau.

Skold, K., Price, D. L. (1986), *Neutron Scattering*, Part A, Vol. 23, Boston: Academic.

PARTICLE PHYSICS

DAN GREEN, *Fermi National Accelerator Laboratory, Batavia, Illinois, U.S.A.*

	Introduction	193
1.	**The Standard Model—Synthesis and Overview**	194
1.1	The SM Constituents	194
1.2	SM Forces	195
1.3	Frontier Issues in Particle Physics	196
1.4	QCD Calculational Implications	198
2.	**Accelerators and Other Facilities**	198
2.1	Accelerator-Based Particle Physics	198
2.1.1	Fixed-Target Experiments	199
2.1.2	e^+e^- Collider Experiments	199
2.1.3	*ep* Collider Experiments	200
2.1.4	*pp* and $p\bar{p}$ Collider Experiments	200
2.1.5	Applications	200
2.2	Nonaccelerator Particle Physics	200
3.	**Mapping the SM Constituents onto the Detectors**	201
3.1	Confinement and Jets—Quarks and Gluons	201
3.2	Missing Energy and Neutrinos	201
3.3	Charged Leptons and EM Interactions	201
3.4	*W*, *Z*, and τ Decays	202
3.5	Displaced Secondary Vertices	202
4.	**A Model $pp/\bar{p}p$ Collider Detector**	202
4.1	"Soft" Backgrounds	202
4.2	Magnetic Analysis	203
4.3	Si Inner Tracker	204
4.4	Outer Tracker	204
4.5	Calorimetry	204
4.5.1	EM Segment	204
4.5.2	Hadronic Segment	204
4.6	Muon Detection	204
4.7	Trigger and DAQ	205
4.8	Analysis	205
5.	**Other Experimental Techniques**	205
5.1	Ionization Detectors	205
5.2	Photomultiplier Tubes, Scintillator	205
5.3	Slow-*n* Detection	206
5.4	Cherenkov Detectors	206
5.5	Transition-Radiation Detectors	206
6.	**Future Developments**	206
6.1	The Energy Frontier at the LHC	206
6.2	Precision Studies at New Facilities	207
6.3	The Far Future	207
	Glossary	209
	Works Cited	210
	Further Reading	210

INTRODUCTION

The goal of particle physics is to discover the most elementary particles and to study the most basic interactions among them. Over the last 30 years, remarkable strides have been made, culminating in what is called the standard model (SM) of particle physics. This model, which exhibits a deep simplicity and beauty, along with the classical theory of gravity (general relativity, GR), together explain all phenomena that have been observed to date with a compelling economy and power. The SM is valid at all distances down to $\sim 10^{-16}$cm, or by uncertainty-principle arguments up to energies ~ 200 GeV $= 2 \times 10^{11}$ eV. The development of the SM should be ranked as one of the great scientific triumphs of the 20th century.

The SM is both a stunning success and the goad to further progress in the field. A brief description of the SM is given below, followed by a discussion of the present frontier issues in the field. The experimental program

3-527-28135-5/95/$5.00 + .50

that has been mounted to confront those issues is then described with a subsequent explanation of the experimental tools presently in active use in particle physics. Finally, new facilities are mentioned as are the questions that they are designed to explore.

1. THE STANDARD MODEL—SYNTHESIS AND OVERVIEW

1.1 The SM Constituents

We now believe that all matter is ultimately made up of pointlike elementary particles. The great complexity and diversity exhibited by the world are thought to proceed from the interaction of only a few basic entities. At normal energy scales, the world can be described as composed of neutrons and protons, the electron, and the photon, which is the quantum of the familiar electromagnetic (EM) interaction between charged particles. The SM asserts that the neutron and proton are themselves composite bound states of pointlike u and d (up and down) quarks. That belief is partially due to the fact that the strong interactions and other properties of the proton and neutron are very similar.

At higher energies (shorter distances), the strong interactions that are responsible for binding the nucleus together must be considered. An analogy is perhaps useful. In atomic physics, the EM interaction (photon quanta) between charged protons and electrons binds atoms together. A residual van der Waals force exists between neutral atoms, which in turn binds atoms into molecules. Similarly, quarks interact strongly by means of a quantum number called color through an intermediate field with quanta called gluons. The bound-state "neutral" colorless composites—for example, protons and neutrons—interact to form bound nuclei. The fundamental theory is called quantum chromodynamics (QCD), which is the theory of the interaction of colored pointlike quarks via massless gluons (see STRONG INTERACTIONS).

Gluons and quarks are believed to be permanently confined by the strong forces that they experience. However, their existence has been inferred from the observation of pointlike behavior in experiments scattering electrons off protons. These experiments are the direct analog of the famous Rutherford scattering experiment by which the nucleus within the atom was inferred by scattering off atoms (Friedman and Kendall, 1972). Even if atoms could not be ionized, and the nuclei and electrons separately examined, the Rutherford scattering experiment would allow one to infer the existence of the nucleus. Similarly, quarks and gluons are confined, but their existence is confirmed in scattering experiments. The existence of the quark was also inferred by the observed regularities and multiplet structure exhibited by the many hadrons, or strongly interacting colorless quark composites (Gell-Mann, 1964; Zweig, 1964).

The discovery of radioactive decay required the existence of a new force, the weak interaction (*q.v.*). The weakness of this interaction can be seen by comparing the lifetime of a typical strongly interacting hadron resonance to the neutron lifetime. The lifetime ratio is $\sim 10^{26}$, which would seem to imply that the weak force is unrelated to the EM force. The decay products of weak decays were found to include neutral massless neutrinos. For example, the neutron is unstable, decaying into a proton, an electron, and a neutrino. One of the profound insights that led to the formulation of the SM was that the weak and electromagnetic interactions are unified (Glashow, 1961; Weinberg, 1967; Salam, 1968) despite the apparent great disparity in strengths. The Fermi coupling constant describing the strength of weak interactions, G_F, is describable at a deeper level as due to a dimensionless coupling constant g_W, just as is the EM coupling constant e. In fact, the weak and EM interactions have approximately the same basic strength. The ratio is measured by the Weinberg angle, θ_W (Prescott *et al.*, 1978):

$$\begin{aligned} G_F &\sim g_w^2/M_w^2, \\ e &\equiv g_W \sin\theta_W, \\ M_W &\sim e/\sqrt{G_F} \sim 90\ \text{GeV}. \end{aligned} \quad (1)$$

Therefore, electromagnetism and the weak force are unified into the electroweak (EW) theory (see UNIFIED FIELD THEORIES). This great triumph of the SM is akin to the prior unification of electricity and magnetism by Maxwell. It culminated in the discovery of the charged and neutral quanta of the weak force, the W and Z gauge bosons. Note that the EW theory predicted the W and Z masses [see Eq.

(1)] prior to their discovery (Arinson *et al.*, 1983; Banner *et al.*, 1983). The EW unification is only evident at the very high energies probed by accelerators. At lower energies, the unified relationship is masked by the large mass of the W and Z, and the weak and electromagnetic forces appear to have vastly different interaction strengths. It seems that Nature becomes simpler and more symmetric at higher energies. The EM and weak potentials differ only slightly in coupling strength, α and a_W, but profoundly in the range of the interaction:

$$V_{EM} = \alpha/r,$$
$$V_W = (\alpha_W/r)e^{-r/\lambda_W},$$
$$\lambda_W = \hbar/M_W c. \qquad (2)$$

At low energies, much less than M_W, the exponential factor causes V_W to be $\ll V_{EM}$, while at high energies, $V_W \cong V_{EM}$.

The constituents of the SM are given in Table 1. The quanta of the two basic interactions, QCD and EW, are the gluons, the photon, and the massive W and Z weak bosons. There are three dimensionless coupling constants, describing the strength of the strong, EM, and weak forces. All three basic interactions are mediated by spin-1 quanta. (See PARTICLES, ELEMENTARY.)

The matter with which these bosons interact has been discussed already. Not shown in Table 1 are the antiparticle partners that exist for each quark and lepton. For example, the positron is paired to the electron. The leptons have only EW interactions, while the quarks also possess strong (QCD) interactions. Both quarks and leptons are spin-$\frac{1}{2}$ fermions. The first-generation quarks and leptons (e,ν_e,u,d) describe our world at low energies, as mentioned before. What has been discovered by experiment is that this basic pattern is repeated in "generations," whose members differ from those of the first largely only in mass. For example, the muon appears to interact exactly as if it were a heavy electron. Why the pattern recurs is unknown. How many generations are there? Recently, both cosmology (primordial helium abundance) and high-energy physics (Z decay width) have converged on the conclusion that there are three and only three generations (Burkhardt and Steinberger, 1991; Steigman *et al.*, 1977). Thus, the "periodic table" of particle physics, as given in Table 1, is complete except for the Higgs boson, which is discussed in Sec. 6. In fact, the discovery of the top quark, t, has very recently completed the entries in Table 1 (CDF Collaboration, 1995; D0 Collaboration, 1995). Indirect inferences of the top mass had previously been available (Burkhardt and Steinberger, 1991).

1.2 SM Forces

The unfolding of the SM has also resulted in a much deeper understanding of the basic forces. It is remarkable that all forces now thought to be fundamental appear to be derivable from a gauge principle. The similarity of all the fundamental forces holds out the hope that they can one day be unified into a

Table 1. The SM constituents.

	Generations			Charge[a] Q	
Matter (spin 1/2)[b]	$\begin{pmatrix} e \\ \nu_e \end{pmatrix}$	$\begin{pmatrix} \mu \\ \nu_\mu \end{pmatrix}$	$\begin{pmatrix} \tau \\ \nu_\tau \end{pmatrix}$	$\begin{pmatrix} 1 \\ 0 \end{pmatrix}$	Leptons
	$\begin{pmatrix} u \\ d \end{pmatrix}$	$\begin{pmatrix} c \\ s \end{pmatrix}$	$\begin{pmatrix} t \\ b \end{pmatrix}$	$\begin{pmatrix} 2/3 \\ -1/3 \end{pmatrix}$	Quarks
	Quanta	Force	Coupling	No. of quanta	
Interactions (spin 1)	g	QCD	$\alpha_s = g_s^2$	8	Gluons
	γ	EW	$\alpha = e^2$	1	Photon
	W,Z	EW	$\alpha_w = g_w^2$	2,1	Weak bosons
Mass (spin 0)	H^0	EW	Yukawa	1	Higgs boson

[a]Units are electron charge, e.
[b]Units are $\hbar$.

grand unified theory (GUT) (see UNIFIED FIELD THEORIES). The gauge principle is one of the great organizing insights brought about by the development of the SM. Just what is the gauge principle, expressed nonmathematically?

As a familiar example, consider classical relativity. If the global space-time coordinate transformations of special relativity are required to hold at arbitrary local points, then a new field, the gravitational field, is required, and the interaction of that field with matter is largely specified (Utiyama, 1956; see also GRAVITATION AND GENERAL RELATIVITY). Clearly, local gauge invariance is a powerful and compact principle.

In quantum mechanics, the overall phase of the wave function Ψ is irrelevant since only $|\Psi|^2$ is an observable. Requiring that physics be unchanged under an arbitrary local choice of overall phase implies the existence of the EM field A_μ and specifies the interaction of that field with the basic charged fermions (Weyl, 1929). The gauge replacement of the derivative is familiar from both classical Hamiltonian mechanics and introductory quantum mechanics. It indicates that gauge invariance requires a new field A_μ:

$$\begin{aligned} P_\mu &\to P_\mu - eA_\mu, \\ \partial_\mu &\to \partial_\mu - ieA_\mu. \end{aligned} \tag{3a}$$

Making this replacement in the free Lagrangian specifies the interactions of the gauge field with the fermions:

$$\begin{aligned} &\bar{\Psi}(\gamma^\mu \partial_\mu - m)\Psi \to \\ &H_{\text{int}} = -ie(\bar{\Psi}\gamma_\mu \Psi)A^\mu \\ &\quad = j_\mu A^\mu, \quad j_\mu = \text{fermion current}. \end{aligned} \tag{3b}$$

This elegant and simple gauge principle then defines the theory of electromagnetism (see QUANTUM ELECTRODYNAMICS). Similar considerations allow one to write the gauge theories of QCD and EW physics. In these cases, the quanta themselves couple together, in contrast to the photon which is uncharged. This type of behavior is familiar from the prototypical gauge theory, GR. The gravitational field possesses energy, and since all energy has mass ($E = mc^2$), the field itself gravitates. Thus, GR is a nonlinear theory, as is QCD, for example, since the gluons themselves carry color. Nonlinear theories are difficult to treat mathematically but imply a very rich phenomenology—perhaps as rich as the world itself.

1.3 Frontier Issues in Particle Physics

The SM has been spectacularly successful as a synthesis of our understanding of all fundamental phenomena. However, it is thought to be incomplete. For example, why are there three generations of quarks and leptons? The SM gives us no answer to this question. Why is the top quark so heavy (or why are the other fermions so light)? A basic understanding of the mass spectrum of the SM constituents eludes us. The masses of the particles given in Table 1 are plotted in Fig. 1. The charged-lepton masses for the three generations increase smoothly. In the SM, neutrinos are assumed to be massless, and experimental upper limits are indicated in Fig. 1. Note that the photon and gluon are required to be massless by the gauge principle, but that there is no compelling reason for neutrinos to be rigorously massless. Indeed, there are some indications, such as the deficit of observed solar neutrinos (Davis *et al.*, 1989), that neutrinos may possess a mass. The search for neutrino masses is a high priority in particle physics. If nonzero masses are found, then a candidate for the nonluminous matter, dark matter, inferred from galactic rotation curves exists (Fich and Tremaine, 1991).

The quark masses are also shown in Fig. 1. There are intriguing regularities within and between the three generations. Note that the SM is mute about any relationship between leptons and quarks, yet the masses of quarks and leptons seem to exhibit a pattern. The situation is akin to that in which atomic spectroscopists found themselves before the advent of quantum mechanics. Clearly, there were patterns in the spectra, but the underlying dynamics was not available. For example, if the generations are simply excited states of composites of more basic particles, then why does the spectrum terminate at three? Speculation abounds, but particle physics is a data-driven field, and new information is needed in order to make progress.

The W and Z are believed to gain their mass by interacting with the hypothesized Higgs boson, H^0. This interaction allows for the evasion of the rule that gauge bosons are massless, e.g., the photon and gluon. The sit-

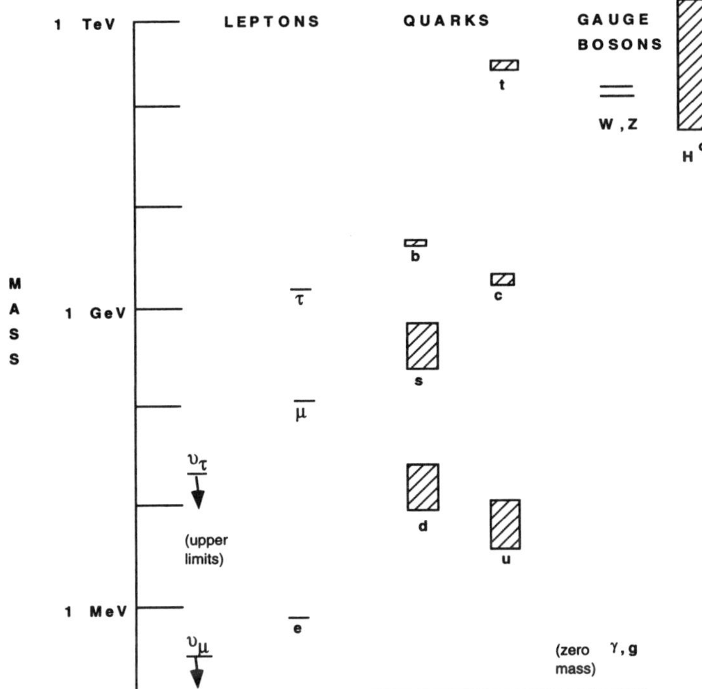

FIG. 1. Masses of the SM constituents; arrows indicate limits.

uation is similar to that of a massless photon propagating in a superconducter (Meissner effect) wherein the photon acquires a mass by interacting with the medium. The Higgs ground-state field is thought to be the analog of that medium and to pervade all of space-time. This background field is held responsible not only for the generation of the *W* and *Z* mass, but for the generation of the masses of all quarks and leptons. A more basic understanding of the Higgs field and its coupling to quarks and leptons should shed light on the regularities of the masses displayed in Fig. 1. Note that the Higgs boson has not yet been discovered.

The generations shown in Table 1 refer to quantum-number labels of states produced in strong interactions where quark–antiquark pairs are produced. As stated above, decays are due to the weak interactions, whose conservation laws are different from those obeyed by the strong interactions. Thus, the weak eigenstates are not necessarily the same as the strong-interaction eigenstates listed in Table 1.

In general, there is a unitary mixing matrix that connects the strong and weak eigenstates: the Cabibbo–Kobayashi–Maskawa (CKM) matrix. The SM does not give us any guidance about this matrix; its elements are determined purely experimentally (Cabibbo, 1963) so that, for example, $|V_{du}|^2$ and $|V_{su}|^2$ represent respectively the relative strengths of $d \to u$ and $s \to u$ weak transitions. Note that, should neutrinos be found to have a mass, there will be another 3×3 unitary matrix for neutrinos, analogous to the CKM matrix for quarks, that also needs to be determined experimentally.

A parametrization of the CKM matrix is shown in Fig. 2. Note that it is almost diagonal, $\theta \sim 0.2$ and $A \sim 1$. Again the data exhibit fascinating regularities. For example, the magnitude of the off-diagonal elements follows a pattern. The origin of these patterns is, as yet, beyond our understanding. For example, why is the weak decay of the *b* quark, controlled in part by V_{bc}, so much slower than that of the *c* quark, controlled by V_{cs}? Note also that the matrix is complex, indicating that invariance under the operation of charge conjugation followed by parity inversion (*CP*) is violated in the SM. This fact has been known for 30 years (Fitch, 1981; Cronin, 1981), but a deeper explanation of the phenomenon still eludes us. The SM allows for *CP* nonconservation, yet does not really explain it. Moreover, *CP* nonconservation has profound

$$
V(\text{CKM}) = \begin{bmatrix} 1 & \theta & A\theta^3\rho e^{i\delta} \\ -\theta & 1 & A\theta^2 \\ A\theta^3(1-\rho e^{-i\delta}) & -A\theta^2 & 1 \end{bmatrix} \begin{matrix} u \\ c \\ t \end{matrix}
$$
$$
\qquad\qquad d \qquad\qquad s \qquad\qquad b
$$

FIG. 2. A representation of the CKM matrix in terms of four parameters.

cosmological implications. It is required in order that the presently observed preponderance of matter over antimatter in the universe and the matter-to-entropy ratio appear gracefully (Sakharov, 1967).

Dramatic progress in physics in the past has occurred when a unification of forces has taken place, such as the recent electroweak unification. It is possible that the similarities in mass between quarks and leptons indicate that they are related in some more basic manner. In that case, the heavier quarks would eventually decay into light leptons and, for example, the proton would not be absolutely stable. Present experiments set a lower limit of $\sim 10^{31}$ years on the proton lifetime. Note that some violation of baryon-number conservation is in fact also needed along with *CP* nonconservation if the observed excess of matter in the universe is to be explained. Although the idea of unification of QCD and EW forces is tantalizing, no truly compelling evidence for such unification yet exists.

1.4 QCD Calculational Implications

As formulated, QCD has no free parameters except the coupling constant. Thus, for example, one should be able to calculate the bound-state hadron mass spectrum from first principles. However, the fact that the quarks and gluons are permanently confined speaks to the strength of the forces involved. Hence perturbation theory, a familiar tool used throughout all areas of physics, is not generally applicable.

At present, the mass spectrum and quantum numbers of the hadrons are found by experiment. Similarly, the distribution of quarks and gluons bound in the proton is determined by scattering experiments. Recently, using strong-coupling techniques adopted from solid-state physics, a cutoff approximation to the full QCD theory has been formulated. QCD is formulated on a mesh with finite spacing (limited momentum) in a "lattice gauge theory." The numerical problems are still formidable and massively parallel computing is needed to complete the calculations, but preliminary indications are heartening.

Given that we possess the complete theory of QCD, it is appropriate to begin to explore its full nonlinear richness. For example, it is an open question whether there exist phase transitions in the quark-gluon system. One hopes that new phenomena that arise because of the strong coupling in QCD will become accessible. There is further discussion of this topic in Sec. 6.

2. ACCELERATORS AND OTHER FACILITIES

After this very brief explanation of the state of the art, we turn to a description of the facilities available to confront the frontier issues. Experimental particle physics is the subject of worldwide study, which takes place in a variety of facilities. The primary focus for experimental particle physics has traditionally been accelerators of ever-increasing energy. The increase in energy is driven by the desire to produce heavier particles together with the Einstein relation between mass and energy, $E = mc^2$. This growth in energy will, naturally, have some intrinsic limits. Hence, other avenues for particle physics may gain in importance in the future.

The program described below constitutes a broad-based attack on the validity of the SM. It is focused on the outstanding problems alluded to in Sec. 1.3, which constitute the frontier issues in particle physics.

2.1 Accelerator-Based Particle Physics

The locations and energies of the accelerator facilities operating at the highest energies are shown in Table 2. There are two

Table 2. Operational experimental facilities at accelerators.

Type	Location	Name energy	CM energy $\sqrt{s}$ (GeV)	Experiments
Fixed target	KEK, Japan	12 GeV	5.0	[Spectroscopy (QCD)
	BNL, USA	AGS		ν beams (ν mass, mixing)
		33 GeV	8.0	K beams (CP nonconservation,
	SLAC, USA	50 GeV	9.8	rare decays)
	Serpukhov, Russia	76 GeV	12.0	e,γ beams
	CERN, Switzerland	SPS	29.2	c,b production (CKM elements)
		450 GeV		e,μ,ν beams (structure
	Fermilab, USA	Tevatron	38.8	functions)]
		800 GeV		
e^+e^- collider	Beijing, China	BEPS	4	$c\bar{c}$ spectroscopy
	Cornell, USA	CLEO	16	$b\bar{b}$ spectroscopy
	KEK, Japan	Tristan	60	Jet phenomena
	CERN, Switzerland	LEP	100	Z properties
	SLAC, USA	SLC	100	Z properties
ep collider	DESY, Germany	HERA	314	Proton structure function
				c,b production
$\bar{p}p$, pp collider	CERN, Switzerland	S $\bar{p}pS$	620	[QCD jets
	Fermilab, USA	TeV collider	1800	$W/Z/\gamma$ properties
				b spectroscopy
				t search
				New phenomena search]

[ ] Indicates a set of common experimental topics studied at all corresponding facilities.

broad categories of particle physics experiments: fixed-target experiments where a beam of accelerated particles strikes a stationary target, and collider experiments, wherein beams of particles collide head-on (see PARTICLE COLLIDERS). The former type of experiment is unrivaled in the study of rare interactions, while the latter is the technique of choice in exploring high mass scales since all the beam energy in such a collision is available for particle production. This distinction appears in Table 2 where beam energy is distinct from center-of-momentum (cm) energy, which is roughly the energy available to create new particles.

2.1.1 Fixed-Target Experiments Fixed-target experiments are currently in progress at several facilities in Russia, Europe, Japan, and the USA. Secondary beams of unstable particles of different types are the exclusive purview of fixed-target experiments. Examples include photon, muon, pion, kaon, and neutrino (γ,μ,π,K,ν) beams.

One component of the program is QCD studies. The interacting hadrons contain within them quarks and gluons. Hadron interactions are therefore a source of underlying quark and gluon interactions.

Searches for neutrino mass are an active part of research using neutrino beams. Searches for the origin of CP nonconservation are also carried out. Thus far, CP nonconservation has only been observed in rare kaon decays. Second- and third-generation (c,b) heavy quarks also possess a rich spectroscopy (CKM matrix element determination), and in addition their production dynamics provides interesting QCD tests.

Experiments with incident electrons, muons, and neutrinos (e,μ,ν) scattering off protons also provide sources of valuable data that measure the distribution of quarks and gluons within the proton. Lepton beams are chosen because the results are most easily interpreted, since leptons are pointlike and do not have QCD interactions, only EW. Since these distributions, or structure functions, are, as yet, not calculable from first principles, the experiments provide essential data.

2.1.2 e^+e^- Collider Experiments Heavy quarks form quark–antiquark bound states, "onia," in analogy to the electron-positron bound states of positronium. These states occur at high enough mass scales that QCD perturbation theory may be expected to apply. The QCD force becomes weaker at high mass, as will be discussed in Sec. 6. The explication of the spectroscopy of the bound-state systems is a vital test of QCD, just as positronium spectroscopy was a crucial test for QED.

In addition, at c.m. energies slightly above the masses of the bound states, almost pure sources of quark–antiquark pairs become available, when strong decay into heavy-meson pairs becomes energetically possible. Thus, these experiments are also major contributors to the knowledge of the spectroscopy of mesons containing heavy quarks.

At elevated c.m. energies, the EW bosons are resonantly formed, and thus e^+e^- collisions are a clean source of Z bosons. Very precise EW tests are performed at these Z "factories," which yield the most accurate determinations of the EW parameters. In fact, the level of precision of these data now allows a comparison with EW theory in higher orders and a subsequent indirect determination of the top-quark mass. The EW theory is being tested beyond leading order in the perturbation expansion similarly to the precise tests of QED made 40 years ago.

2.1.3 *ep* Collider Experiments At present, the distribution of quarks and gluons in the proton is determined experimentally. The data come from lepton-proton scattering with fixed proton targets, and neutrino, electron, and muon incident beams. The range of such experiments is limited by the available c.m. energy. Recently, an *ep* colliding-beam accelerator has become operational. The goal of experiments at this facility is to extend the range of measurements of the proton constituents and, of course, to search for new and unexpected phenomena.

2.1.4 *pp* and *p$\bar{p}$* Collider Experiments The proton is a composite object and can be thought of as consisting of valence *uud* quarks, quark–antiquark pairs, and gluons of different momenta as measured in lepton-proton scattering experiments. Therefore, pp and $p\bar{p}$ colliders are capable of simultaneous studies of many aspects of current investigations in particle physics. Examples include QCD studies in jet phenomena, $W/Z/\gamma$ production and decay properties, heavy-flavor (b) production dynamics and decay spectroscopy, searches for the top (t) quark, and searches for new phenomena outside the SM.

2.1.5 Applications Accelerators have also been a fruitful source of applications to areas outside particle physics. The large-scale use of superconductivity was pioneered in accelerators for high-energy physics. Large-scale high-vacuum and distributed-control systems are another outgrowth of accelerator "technology transfer." Electron accelerators are used as synchrotron light sources (see SYNCHROTRON RADIATION). These new facilities are heavily subscribed to by materials and solid-state physicists at many locations throughout the world.

In addition, there are numerous examples of medical applications. The detectors used in particle physics are widely used in medicine. High-field magnets similar to those built for accelerators are used in magnetic resonance imaging. Finally, accelerators themselves are used in treatment. Examples include electron, neutron, and proton irradiation. For example, proton beams of variable energy allow precise doses to be placed deep in the body with minimal damage to intervening tissues. Electron linear accelerators are, however, simpler and therefore the most widespread therapeutic devices.

2.2 Nonaccelerator Particle Physics

Cosmic rays extend up to energies that are still inaccessible at the accelerator facilities given in Table 2. Therefore, it is natural to search for new phenomena at the highest cosmic-ray energies. In addition, stimulated by unification speculations, several experiments have been set up to search for proton decay, all without success to date. However, they have set useful limits, and since they are located deep underground in order to minimize cosmic-ray backgrounds, they also function as active neutrino "observatories." Detectors of this type—for example, at Kamioka in Japan, Gran Sasso in Italy, and IMB and Soudan in the USA—detected the neutrino pulse from supernova SN1987A. The utility of these "observatories" indicates the increasing interconnection between particle physics, astrophysics, and cosmology.

The Kamioka detector has also been able

to detect neutrinos from the sun, following the pioneering work at the Homestake mine in the USA. Other specialized detectors that measure the solar neutrino flux have recently come on line at Baksan in Russia and at Gran Sasso (GALLEX) in Italy. There are tantalizing indications that the number of neutrinos arriving from the sun is not in accord with the Standard Solar Model prediction (Davis *et al.*, 1989). These experiments may signal a nonzero neutrino mass.

If the neutrino mass is nonzero, then different lepton generations will mix. Neutrino beams prepared to be composed of only one neutrino generation at an accelerator can be directed toward distant underground observatories where the appearance of a different generation of neutrino can be searched for. Such long-baseline neutrino mixing experiments would improve the determination of neutrino masses and mixing parameters. A finite neutrino mass would clearly imply the existence of new physics beyond the SM.

3. MAPPING THE SM CONSTITUENTS ONTO THE DETECTORS

Having described the SM constituents and the experimental program, one can now briefly describe the specifics of experimental particle physics. Detectors are used to measure the vector position and momentum of all the constituents of the SM emanating from a collision and to distinguish between them (particle identification, "id"). In principle, a perfect detector would measure and identify all the particles in both the initial and final states of a collision. The design of detectors is dictated by the characteristic interactions that the SM constituents participate in. The different categories are shown in Table 3 and are explained below.

3.1 Confinement and Jets—Quarks and Gluons

Quarks and gluons carry the color quantum number and interact strongly. It has been found that quarks and gluons cannot be detected as distinct free states. Rather, they appear in detectors as collimated "jets" of colorless hadrons moving near to the direction and with roughly the momentum of the parent q or g. These jets are the hadronic remnants of the quark or gluon. Finely segmented energy-measuring detectors (calorimeters) are used to measure the positions and energies of these ensembles of particles. Thus, one may infer the properties of the q or g parent.

Table 3. Detection/identification methods.

Signature	Detector	Particle
Jet of hadrons	Calorimeter	$u, c, t \rightarrow Wb$ d, s, b g
Missing Energy	Calorimeter	ν_e, ν_μ, ν_τ
Electromagnetic shower	Calorimeter	$e, \gamma, W \rightarrow e\nu$
Only ionization interactions	Muon absorber	$\mu, \tau \rightarrow \mu\nu\bar{\nu}$ $Z \rightarrow \mu\mu$
Decay with $c\tau \gtrsim 100\ \mu$m	Si tracking	s, c, b, τ

3.2 Missing Energy and Neutrinos

Leptons, by definition, interact only via the EW interaction. In particular, neutrinos are uncharged, and therefore only interact weakly. Therefore, neutrinos pass right through a calorimeter system that absorbs and detects all strongly and electromagnetically interacting particles, without interacting with the detecting medium.

The presence of neutrinos is then signaled by the existence of the "missing energy" that they carry off undetected. Since the initial-state energy is well defined (for example, in a symmetric $p\bar{p}$ collider it is twice the beam energy), the energy taken away by noninteracting neutrinos may be inferred using calorimeters with complete angular coverage.

3.3 Charged Leptons and EM Interactions

Charged leptons interact electromagnetically. Extremely relativistic photons and electrons interact with a calorimeter medium by pair production and bremsstrahlung respectively. The resulting cascade is a characteristic signature. Therefore, a calorimeter can be used to perform e/γ particle id by observation of the characteristic electromagnetic shower.

The muon is unstable, with a lifetime characterized by the product $c\tau \sim 658$ m. Therefore, it will typically pass through a detector before it decays. Muons are "heavy electrons"—second-generation leptons identical to electrons save for mass. The large mass means that with the identical forces experienced by e and μ, the acceleration of a muon is much reduced with respect to an electron. Since radiation is caused by acceleration, the muon cascade occurs only at much higher energies than the e cascade. For example, muons at energies below about 300 GeV only ionize the detection medium, thereby leaving a minimum-ionizing particle, or MIP, signal in particle detectors. This signature is the basis for muon identification, since all other charged particles traversing thick material deposit more energy.

3.4 *W*, *Z*, and τ Decays

The W and Z gauge bosons decay rapidly into lighter final states containing either lepton or quark pairs. The cleanest experimental signatures are the decays into final-state leptons, $W \rightarrow l + \nu$ and $Z \rightarrow l^+ + l^-$ where l denotes charged lepton. Final states containing quarks, such as $W^+ \rightarrow u + \bar{d} \rightarrow \text{Jet}_1 + \text{Jet}_2$, are subject to large backgrounds.

The τ lepton often decays into lighter final-state leptons—for example, $\tau \rightarrow \mu + \bar{\nu}_\mu + \nu_\tau$. Since the τ has a measured lifetime of $c\tau \sim 89\ \mu\text{m}$, the experimental signature is the existence of both a secondary decay vertex (see Sec. 3.5) and a primary production vertex.

3.5 Displaced Secondary Vertices

All the elementary constituents of matter except electrons, protons (a *uud* bound state), and massless neutrinos are unstable. The leptons decay leptonically with substantial branching fractions—for example, $\mu \rightarrow e + \bar{\nu}_e + \nu_\mu$. The leptonic decay rate scales as the available energy to the fifth power. Therefore, the light muon can be considered stable during its detection, while the heavy τ must be detected by measuring its decay products.

The separation of light (u,d) from heavy (s,c,b) quark states can be accomplished by the observation of a secondary decay vertex. The quarks can decay semileptonically—for example, $s \rightarrow u + l + \bar{\nu}_l$. Hadrons containing s quarks have lifetimes $c\tau$ of the order of centimeters, while c and b quark decays lead to hadron lifetimes, $c\tau \sim 200\ \mu\text{m}$.

The top quark appears to be unusually massive, as seen in Fig. 1. Assuming that it is simply a heavy sequential quark, the favored decay mode is $t \rightarrow W^+ + b$. The large mass and consequent short lifetime imply that top id is necessarily performed by observation of the final-state decay products.

4. A MODEL $pp/\bar{p}p$ COLLIDER DETECTOR

The characteristic interactions of the SM constituents are used in their detection as outlined above. In this section, a typical detector that is used in experimental particle physics is described. Fixed-target detectors are often special-purpose instruments optimized to explore a specific type of physics—for example, neutrino interactions. Collider detectors tend to be general-purpose instruments designed to study a large variety of reactions simultaneously as mentioned above. For this reason, a collider detector can be discussed "generically." Generally, the backgrounds and trigger strategies are more challenging at a hadron collider than at an e^+e^- collider, and for this reason the former is described. A schematic of a typical detector and event is shown in Fig. 3.

4.1 "Soft" Backgrounds

At a c.m. energy of 1800 GeV, the pp inelastic cross section is $\sim 50\ \text{mb} = 5 \times 10^{-26}\ \text{cm}^2$. The corresponding inelastic interaction rate at the Fermilab Tevatron is presently 500 kHz. The overwhelming majority of these interactions are soft, meaning that they consist of produced hadrons with limited transverse momentum, $\langle p_t \rangle \sim 0.5$ GeV, where p_t is the component of secondary particle momentum transverse to the incident beam direction. The secondary particles are distributed in longitudinal momentum, $p_\parallel$, according to single-particle phase space:

$$\text{single-particle phase space} = d^4p\delta(p^2 - m^2) = d^3p/E \sim dydp_t^2, \tag{4}$$

where m is the particle mass, E the particle energy, and y the rapidity, which is defined by $\tanh y = p_\parallel/E$. The nonrelativistic limit for one-particle phase space is d^3p/m, which is familiar from classical statistical mechanics.

The collisions of greatest interest occur at

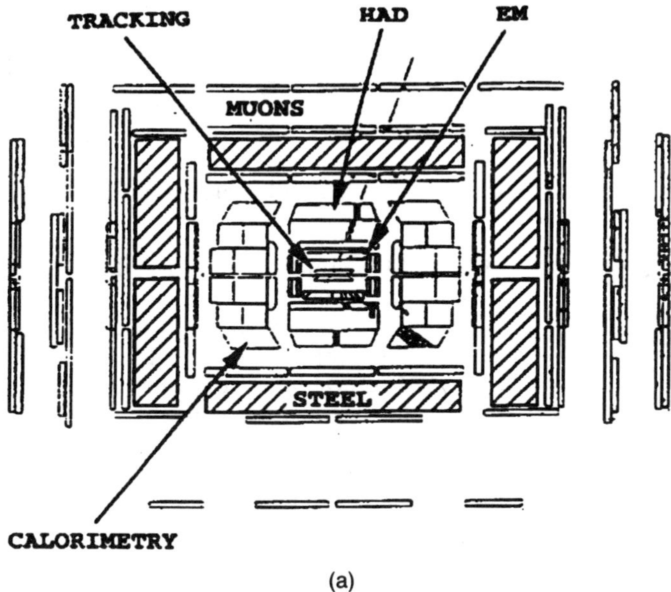

(a)

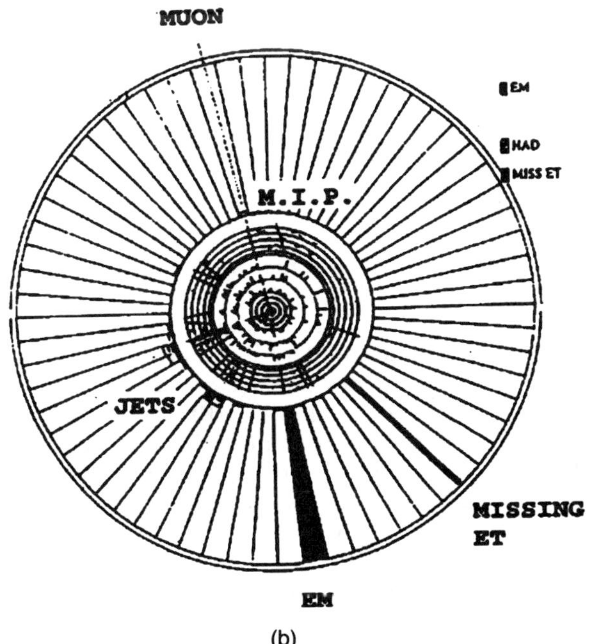

(b)

FIG. 3. A typical $p\bar{p}$ collider detector showing tracking, calorimetry, muon electromagnets, and muon detectors. (a) A muon track is evident in the calorimeter and muon system. (b) Charged tracks and EM and HAD energy deposits are shown, indicating the existence of two jets, an electron, a muon, and a neutrino in the interaction.

large transverse momentum, since they probe small distances. Therefore, the detection, triggering, and data-acquisition (DAQ) strategy adopted for these detectors puts a strong emphasis on collecting the rare "hard" collisions, those containing high-p_t secondary particles.

4.2 Magnetic Analysis

Typically, the detector contains a large electromagnet, which provides a volume filled with a magnetic field. A uniform magnetic field causes all charged particles to move on helical paths, with radii of curvature determined by the particle momentum and charge. Thus, determination of the particle trajectories in the field volume determines the charge, vector momentum, and position of all the secondary charged particles produced in the collision.

4.3 Si Inner Tracker

Since b and c quarks and the τ leptons have lifetimes of order 200 μm, an inner solid-state detector (see DETECTORS, SEMICONDUCTOR) consisting of strips of Si with a strip spacing of ~50 μm is often employed as the first detector element encountered by a secondary particle after it exits the vacuum pipe containing the incident beams. The Si detector strip spacing is well matched to the lifetimes of the particles. The Si detector also serves as the initial element in a tracking system that measures the particle trajectory in the magnetic field.

4.4 Outer Tracker

The Si elements are fairly expensive. For that reason, most tracking systems employ a different detector technology at larger radii. Gas-filled wire chambers (see PARTICLE DETECTORS, TRACKING) are the standard. The size of these outer trackers is typically 1–2 m in radius.

A perfect tracking system in principle allows the experimenter to perform particle identification, or id, on the c and b quarks and the τ lepton. The e and μ charged leptons are also measured and, by extension, the leptons from τ, W, and Z decays. All charged-hadron trajectories are also determined. A view of an event as seen in the tracking system is shown in Fig. 3(b).

4.5 Calorimetry

The next system encountered by the outgoing particles is usually the calorimeter (see DETECTORS, PARTICLE, CALORIMETRIC). Within the calorimeter, particles cascade in a dense medium and deposit their energy. The calorimeter is segmented transversely so that this system measures, destructively, the position and energy of all particles, both charged and neutral, that possess strong or EM interactions. The calorimetric medium can be either completely active (crystals) or interspersed heavy absorber plates with active detection layers (sampling calorimetry). Particle id is achieved by using the transverse and longitudinal segmentation of the calorimetry to distinguish between the differences in cascade development of the different incident particles—for example, electrons and hadrons.

4.5.1 EM Segment The first longitudinal segment encountered typically contains Pb or some other element with high atomic number that induces e and photon cascades via the EM interaction. Since hadrons interact differently, the EM segment is used to identify e and γ. The tracking system is then used for e–γ separation since e will leave an ionizing track while the uncharged photon will not.

The depth of the EM segment is sufficient to contain the EM cascade fully. It is typically ~20 radiation lengths or ~12 cm of Pb. By comparison, the ratio of radiation length to hadron interaction length is ~30 in Pb, so that most hadrons will not even begin to produce a cascade in the EM segment. This difference in interaction-length scales, and a similar difference in the transverse size of the EM and hadronic cascades, is the basis of clean e/γ particle id with respect to hadrons.

4.5.2 Hadronic Segment The EM segment is followed by a hadronic (HAD) segment. Its task is to absorb the energy of all the hadrons. The scale for hadronic interactions defines the size of the calorimeter system. For example, ~10 nuclear absorption lengths (or 167 cm of Fe) are needed to absorb most of the hadronic energy. Any particle that passes through the calorimeter unscathed is a candidate for identification as a muon.

Muons will only deposit ionization energy in the calorimeter. Thus, a particle that acts like a minimum-ionizing particle in the calorimeter can be provisionally identified as a muon.

4.6 Muon Detection

Detectors external to the calorimetry are often employed to provide independent muon identification and a redundant measurement of the muon position and momentum. This redundancy is similar to that provided by the two measurements of an electron: by tracking in a magnetic field and by the measured energy deposited in the EM segment of the calorimetry.

Redundant muon information is supplied by a second magnetic analysis of the muon after most of the secondary particles in the interaction are removed by absorption in the calorimeter. This analysis may occur either in air or in a solid steel electromagnet. Wire chambers surrounding this additional magnetic volume typically provide the trajectory.

4.7 Trigger and DAQ

Since the inelastic cross section is large and the cross sections for the processes of interest are often small, the trigger for readout must be very selective. The ratio of all events to the events of primary interest may be as small as 10^{-6}. With such a small signal-to-noise value, redundant measurement and particle id are absolutely necessary. The situation for e^+e^- interactions is somewhat more relaxed.

The detection systems described above are designed to supply information very rapidly. That information is used in a hierarchy of triggers that select the events of interest (see DATA ACQUISITION SYSTEMS). The trigger hierarchy is designed to reduce the total inelastic rate (500 kHz at the Fermilab Tevatron) to a rate logged to permanent medium (e.g., magnetic tape) of a few hertz. Clearly, the trigger system cannot accomplish this prodigy (reduction by a factor 5×10^5) in one step. Often, several successively slower but more incisive trigger algorithms are used.

4.8 Analysis

Often, crude analysis of complete events consisting of the full set of information supplied by the detector is part of the highest level of triggering. Thus, triggering and data acquisition merge smoothly into off-line analysis of the events. Typically, a complete event is assembled from the individual detector systems by the DAQ system and loaded into a high-level workstation where the highest-level triggering is performed. Off-line analysis of the events stored on a permanent medium can be more leisurely and more exacting.

5. OTHER EXPERIMENTAL TECHNIQUES

Detectors are designed to address specific physics processes. Some detector systems are of a more specialized nature than the generic detectors for collider experiments that were discussed in the previous two sections.

All detectors of necessity must utilize the interactions of the particles of interest with a detecting medium. In the previous sections, ionization in the tracking system, EM/hadronic cascades in the calorimetric media, and ionization energy loss in the muon system were discussed. Much of the discussion in this section deals with particle id in other detection systems or using other measurements.

5.1 Ionization Detectors

Ionization refers to the interaction of incident charged particles with the atomic electrons in a medium, which leads to ionization of the atoms. The energy loss by ionization varies with incoming particle velocity v as $1/v^2$ for slow incident particles. At high velocity, the ionization loss approaches a minimum; such a track is referred to as that of a MIP (minimum ionizing particle).

Tracking detectors use the detected ionization to determine a point on the charged-particle trajectory, as already mentioned. If, in addition, the value of the deposited energy is measured several times, the velocity is determined. With the momentum known from the curvature of the path in the magnetic field, the mass of the particle may thus be inferred. Therefore, ionization detectors can be used for slow-particle id if many redundant samples of the ionization are made.

5.2 Photomultiplier Tubes, Scintillator

Scintillating plastics coupled to photomultiplier tubes (PMT) can be used to measure the particle time of flight between two sets of counters to subnanosecond (10^{-9} s) accuracy (see DETECTORS, SCINTILLATION). The measurement of time over a fixed distance is a velocity measurement. Thus, scintillation-counter arrays are also useful in particle id for slow ionizing particles.

5.3 Slow-*n* Detection

Neutrons are uncharged and hence do not ionize. Nevertheless, they can be detected easily by measuring the proton recoils from elastic *np* interactions. The recoil protons take up a significant fraction of the initial neutron energy, $\frac{1}{2}$ on average. Thick organic scintillation counters can both supply the proton targets and detect the recoil protons via their large ionization losses in the plastic.

5.4 Cherenkov Detectors

A charged particle with a velocity v that exceeds the velocity of light in a medium of index of refraction n, $v > c/n$, emits a Mach cone of radiation. This is called the Cherenkov effect. Cherenkov counters consist, typically, of a gas-filled vessel outfitted with mirrors to collect the emitted light, and PMT to convert the light into a fast electrical signal.

The angle of the emitted light, θ_C, is determined by $\beta = v/c$ and n: $\cos\theta_C = 1/\beta n$. Clearly, a measurement of the angle of the emitted light is a measure of the particle velocity. Such counters are called ring- (from the Mach cone image) imaging Cherenkov counters (RICH).

5.5 Transition-Radiation Detectors

A transition-radiation detector (TRD) may be thought of as utilizing the diffraction pattern arising from the radiation of a Cherenkov detector of finite length. The radiation is confined to very forward angles. In addition, it is in the x-ray frequency range, with an intensity that is proportional to $\gamma = 1/\sqrt{1-\beta^2}$. Therefore, the TRD is a device that is particularly useful for particle id with very relativistic particles. The TRD is a detector complementary to those, like the Cherenkov detector, which lose discrimination power as $\beta \rightarrow 1$ ($v \rightarrow c$).

6. FUTURE DEVELOPMENTS

Particle physics is a dynamic and vital field with an extremely broad-based experimental program, as has been indicated in the last four sections. In general, the program seeks to probe the SM either with increasing precision or with higher energies where phenomena appear to become simpler. To date, the SM has withstood this assault and proven itself capable of very precise and testable predictions. In order to make decisive progress, new facilities are now under construction in order to press forward on these two complementary fronts.

6.1 The Energy Frontier at the LHC

As mentioned in the first section, the breaking of the symmetry between electromagnetism and the weak interactions creates the mass of the W and Z bosons and maintains the masslessness of the photon. In fact, the coupling of the Higgs field to matter is thought to produce the masses of all the elementary particles. Unfortunately, the true origin of these couplings and their regularities is not presently understood.

The SM does predict the existence of a neutral scalar Higgs particle, but does not constrain the mass (see Fig. 1). The lower limit shown in Fig. 1 is that imposed by searches performed in experiments at LEP. One can estimate an upper limit to the Higgs mass by asking that the EW interactions do not become strong. For example, the S-wave amplitude A_0 for the EW process $e^+ + e^- \rightarrow W^+W^-$ at a c.m. energy squared of s is

$$A_0(ee \rightarrow WW) = (\alpha_w/4\pi)\,(s/M_w^2). \tag{5a}$$

Since A_0 grows with energy, we expect that new phenomena will occur in order to restore partial-wave unitarity. This restoration guarantees that new physics will be seen if the H^0 mass range shown in Fig. 1 can be explored:

$$A_0(ee \rightarrow WW) < 1$$
$$\Rightarrow M_H < \sqrt{\frac{4\pi}{\alpha_w}}\,M_w \sim 1600\ \text{GeV}. \tag{5b}$$

The discovery of the Higgs boson or some other new EW phenomenon is crucial if we are either to progress to a deeper understanding of the SM or to pass beyond the SM. The full exploration of the TeV (1000 GeV) mass scale is the province of the next machine at the energy frontier, the Large Hadron Collider (LHC). New facilities are shown in Table 4. The LHC is the single new facility

Table 4. New facilities.

Type	Location	CM energy $\sqrt{s}$ (GeV)	Experiments
Nucleus-nucleus collider	BNL, USA RHIC	400	Quark-gluon plasma
e^+e^- collider	CERN, Switzerland LEP II	200	$WW\gamma$ coupling
	SLAC, USA SLAC B Factory	10	CP nonconservation
	KEK, Japan	10	b spectroscopy
Fixed target or $\bar{p}p$ collider	Fermilab, USA Main Injector	39 (17) 2 000	K beams γ beams, ν beams t decays b spectroscopy, CP nonconservation
pp collider	CERN, Switzerland Large Hadron Collider (LHC)	14 000	Higgs search t factory b spectroscopy, CP nonservation

approved to operate at the EW-breaking mass scale.

The LEP II project at CERN will approach the EW-breaking energy scale, but the discovery of fundamentally new phenomena is not assured. The goal of LEP II is to produce W pairs, thus opening up the direct study of the self-coupling between gauge bosons, $e^+e^- \rightarrow \gamma \rightarrow W^+W^-$. As noted earlier, the nonlinear interaction among gauge bosons is a basic feature of the fundamental forces.

6.2 Precision Studies at New Facilities

The other frontier is at the higher-precision experiments. At Fermilab, the Main Injector project will allow for much higher reaction rates, which will extend the measurements of rare processes in the fixed-target program (see Table 2). In the collider program, the higher rates should allow the top-quark studies to progress from simple discovery to measurement of some of the basic properties of the top quark.

At a fundamental level, one asks why the universe is composed of matter and not antimatter. The violation of CP invariance is a necessary ingredient (Sakharov, 1967, Primakoff and Rosen, 1981). CP nonconservation can be studied experimentally by measuring the complex elements of the CKM matrix. These experiments require an exquisite precision. At present, e^+e^- colliders built specifically for this purpose, "B factories," are being constructed at the Stanford Linear Accelerator Center (SLAC) and the Japanese National Laboratory for High Energy Physics (KEK). The smallness of the effects to be studied may require the higher reaction rates available at hadron colliders. The LHC and the upgraded Tevatron collider are both being considered as possible sites for experiments to study CP nonconservation in b decays.

It has previously been noted that QCD is a nonlinear and strongly coupled theory. The full implication of the strong coupling for the dynamics is still a computational challenge. On the experimental side, at Brookhaven National Laboratory (BNL) the Relativistic Heavy Ion Collider (RHIC) machine is being built to collide heavy ions. These collisions are believed to be capable of generating a state of very high density and temperature. Phase transitions may occur wherein a plasma of quarks and gluons is expected to be formed.

6.3 The Far Future

New fundamental particles and/or interactions are expected to appear at the EW-breaking mass scale of ~1 TeV or below. It is impossible to predict the shape of this new physics. One possibility is that the objects now thought to be elementary are really composites. At each point in the past—atoms to nuclei to nucleons to quarks—this hypothesis has proved to be true. It remains clear that particle physics will maintain its vitality by again pushing to higher energies and more precise experiments. New facilities beyond those now under construction might include a TeV-scale e^+e^- linear collider or a multi-TeV pp collider, or both. The choice of the appropriate tools clearly must await the insights provided by the new data produced at the facilities presently under construction.

Many of the questions to be answered are contained within the success of the SM. A partial list might include these: What is the origin of EW symmetry breaking? Why are there three generations? What explains the pattern of the fermion masses? What is the origin of the CKM matrix? What is the origin of *CP* nonconservation? Is there a neutrino mixing matrix? Are the QCD and EW forces unified at some energy? What is the connection of gravity to the other fundamental gauge interactions? What is the correct quantum theory of gravity?

There is a mass scale where the gravitational interaction measured by Newton's constant, becomes strong; its value, known as the Planck mass, is 1.2×10^{19} GeV. At that mass scale, a quantum theory of gravity is required. Such theories based on pointlike fundamental particles appear to suffer from insuperable technical difficulties. Extended theories in which the basic objects are two-dimensional strings avoid these difficulties and, in addition, appear potentially to contain within them the standard model. Thus, in principle one can construct a "Theory of Everything" (ToE), which unifies the standard model containing the strong and electroweak forces with the gravitational interaction (see UNIFIED FIELD THEORIES). Unfortunately, as yet testable predictions are scant.

There is a second very large mass scale, which is the grand unification scale. Its value requires some explanation. It is well known (vacuum polarization) that the EM coupling constant is not a constant but increases with decreasing distance of observation because of the screening effect of the virtual cloud of e^+e^- pairs present in the vacuum. For example, $\alpha(M = 0) \sim 1/137$, while $\alpha(M = M_Z) \sim 1/128$. We have asserted that the QCD quanta are colored gluons while the EM quanta are uncharged photons. This implies that QCD is weak at large mass and strong at small mass; i.e., $\alpha_s(M = 0) \rightarrow \infty$, which leads to confinement. Therefore, since $\alpha_s(M = M_Z) \sim 1/10$, as the mass increases further α_s will drop and α will rise to meet it at some mass.

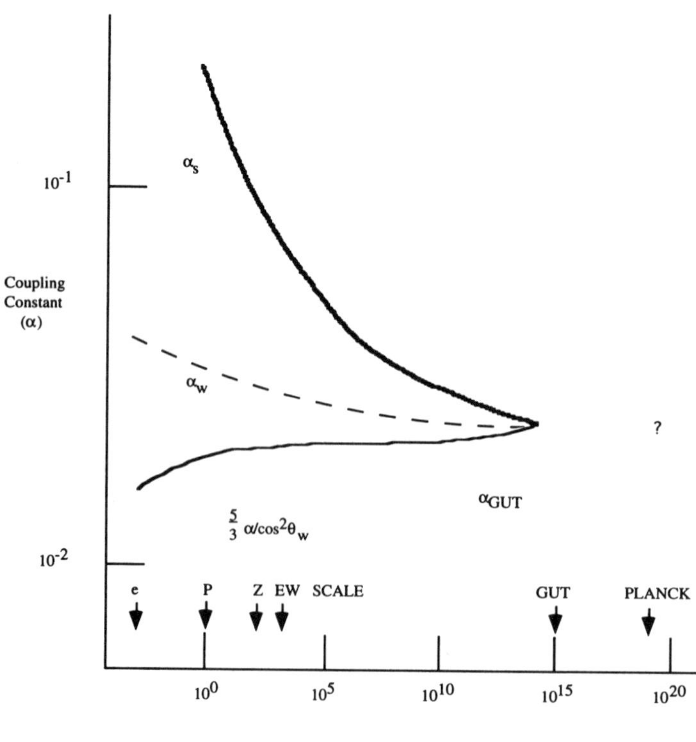

FIG. 4. Plot of the energy dependence of the coupling constants for QCD (α_s), weak (α_w), and EM (α) interactions.

Since the W and Z self-couple similarly to the gluons, α_w will also drop. It is remarkable that within errors α_s, α_w, and α all roughly meet at a mass scale $\sim 10^{15\pm1}$ GeV. At this scale, QCD and EW forces are expected to be fully unified (Pati and Salam, 1973; Georgi and Glashow, 1974). The energy dependence of the coupling constants is shown in Fig. 4. Note that they are well measured only for energies $<M_Z$. Extrapolation by 13 orders of magnitude to the GUT scale leads to a potential unification of all forces, if no new physics intervenes. In point of fact, precision measurements of the coupling constants may decisively allow the conclusion that new physics does intervene. This is an example of the use of precision studies to probe presently inaccessible mass scales.

Both the Planck and unification masses are many orders of magnitude above the capabilities of any feasible accelerator. The intellectual challenge of these high mass scales is to design incisive experiments at available energies. This task will require considerable ingenuity. Note that all of 20th-century particle physics has been an exploration, which on Fig. 4 occupies only $\frac{1}{2}$ of the logarithmic energy range between the EW and GUT mass scales.

It is humbling to note that we do not even appear to understand the nature of the matter that constitutes the bulk of the universe, the dark matter. Clearly, much remains to be understood, and the intellectual vitality of particle physics is assured for the foreseeable future.

GLOSSARY

Baryon: Half-integral–spin hadron made up of qqq valence quarks.

Boson: Integral-spin particle.

Calorimeter: A device that provides a signal proportional to the energy deposited by a particle incident upon it.

Color: The quantum number carried by quarks and gluons; the "charge" of QCD.

Confinement: The hypothesis that colored quarks and gluons cannot exist as free particles.

***CP* Symmetry:** A symmetry of interactions under the combined parity and charge-conjugation operation.

Fermion: Half-integral–spin particle.

Flavor: Approximately conserved quantum number of quarks. Pair production of strange, charmed, bottom, and top ($s\bar{s}$,$c\bar{c}$,$b\bar{b}$,$t\bar{t}$) quarks is required. Different generations of quarks have different flavor labels.

g_s: Strong coupling constant.

g_w: Weak coupling constant; analogous to e in QED.

Gauge Symmetry: A symmetry arising from requiring that physics be independent of the local phase convention chosen at arbitrary space-time points.

Generations: The replication of the basic quark/lepton four elementary fermions, u,d,e,ν_e, at higher masses.

Gluon: The massless quantum of QCD.

Hadron: A strongly interacting, colorless composite of quarks.

Higgs Boson: A hypothesized particle whose nonzero vacuum field gives rise to the masses of all fermions and bosons.

Interaction Length: The characteristic length of material within which a hadron interacts by a strong interaction with a nucleus.

Jet: An ensemble of hadrons that are the final-state remnants of a quark or gluon parent that is permanently confined. This ensemble approximately retains the direction and momentum of the parent.

Kaon: A meson composed of strange (s) and u or d quarks; e.g., $K^+ = (u\bar{s})$.

Lepton: Fermion without color (QCD interactions).

Lifetime: A characteristic time for the exponential decay of unstable particles.

Mach Cone: The characteristic shock-wave pattern at fixed angle that appears when an object exceeds the speed of propagation of a wave in the medium.

Meson: Integral-spin hadron made up of $q\bar{q}$ valence quarks.

Neutrino: Uncharged lepton, assumed to be massless.

"onium": Bound state of heavy quark and antiquark; e.g., $c\bar{c}$ is charmonium. Analog of positronium = e^+e^-.

Photon: The massless quantum of QED.

Pion: The lightest meson composed of u and d quarks, e.g., $\pi^+ = (u\bar{d})$.

Planck Mass: The mass scale where gravity becomes strong: $M_{\rm PL} = \sqrt{\hbar c/G_{\rm N}}$ where $G_{\rm N}$ is Newton's constant.

QCD: Quantum theory of colored quarks and gluons.

QED: Quantum theory of charged fermions and the photon.

Quark: Fermion with color (QCD interactions).

Radiation Length: The characteristic length of material within which an electron interacts by emission of radiation caused by acceleration by a nucleus.

Rapidity: A variable, conventionally denoted by y, that defines longitudinal velocity and is invariant under Lorentz transformations: $\tanh y = (p_\parallel/E)$, where E is energy and $p_\parallel$ is momentum along a defined axis. Nonrelativistically $y \rightarrow \beta_\parallel = v_\parallel/c$.

Spin: The intrinsic, quantized angular momentum possessed by SM constituents. Units are $\hbar$.

Structure Function: Experimentally determined distribution of momenta of constituents within a composite bound state (e.g., proton).

Uncertainty Principle: The inverse relationship between the possible simultaneous precision of knowledge of momentum and position, $\Delta x \Delta p_x = \hbar$.

Units: The convention $\hbar = c = 1$ is adopted. The dimensions of energy, momentum, 1/length, and 1/time are then all expressible in electron volts (eV). Conversion using $\hbar c = 2 \times 10^{-14}$ GeV cm, $\hbar = 0.66 \times 10^{-25}$ GeV s is straightforward.

Vacuum Polarization: In QED, the observed charge of an electron is modified by the sea of e^+e^- pairs virtually present in the vacuum. Since the neutral sea reduces the effective charge, and since the sea extends only over distances $\sim\hbar/m_e c$, the effective charge increases with decreasing distance.

Wire Chamber: A detector with an array of wires maintained at high electric field that localizes, amplifies, and detects the ionization deposited by charged particles traversing a volume of gas.

W, Z: The massive quanta of the weak force.

Yukawa Coupling: The coupling of fermions to massive quanta analogous to EM coupling. Massive quanta give rise to short-ranged forces.

Works Cited

Arnison, G., *et al.* (1983), *Phys. Lett.* **122B**, 103–116.

Banner, M., *et al.* (1983), *Phys. Lett.* **122B**, 322–328.

Burkhardt, H., Steinberger, J. (1991) *Annu. Rev. Nucl. Part. Sci.* **41**, 55–96.

Cabibbo, N. (1963), *Phys. Rev. Lett.* **10**, 531–533.

CDF Collaboration (1995), *Phys. Rev. Lett.* **74**, 2626–2631.

Cronin, J. W. (1981), *Rev. Mod. Phys.* **53**, 373–383.

D0 Collaboration (1995), *Phys. Rev. Lett.* **74**, 2632–2637.

Davis, R., *et al.* (1989), in: J. Schnepps, T. Kafka, W. A. Mann (Eds.), *Proceedings of the 13th International Conference on Neutrino Physics and Astrophysics*, Singapore: World Scientific.

Fich, M., Tremaine, S. (1991), *Annu. Rev. Astron. Astrophys.* **29**, 409–445.

Fitch, V. L. (1981), *Rev. Mod. Phys.* **53**, 367–371.

Friedman, J. I., Kendall, H. W. (1972), *Annu. Rev. Nucl. Sci.* **22**, 203–254.

Gell-Mann, M. (1964), *Phys. Lett.* **8**, 214–215.

Georgi, H., Glashow, S. L. (1974), *Phys. Rev. Lett.* **32**, 438–441.

Glashow, S. L. (1961), *Nucl. Phys.* **22**, 579.

Pati, J. C., Salam, A. (1973), *Phys. Rev. D* **8**, 1240–1251.

Prescott, C., *et al.* (1978), *Phys. Lett. B* **77**, 347–352.

Primakoff, H., Rosen, S. P. (1981), *Annu. Rev. Nucl. Part. Sci.* **31**, 145–192.

Sakharov, A. D. (1967), *Zh. Eksp. Teor. Fiz. Pis'ma Red.* **5**, 31–35 (*JETP Lett.* **5**, 24–30).

Salam, A. (1968), in: N. Svartholm (Ed.), *Elementary Particle Theory*, Nobel Symposium No. 8, Stockholm: Almqvist and Wiksells.

Steigman, G., Schramm, D. N., Gunn, J. (1977), *Phys. Lett. B* **66**, 202–204.

Utiyama, R. (1956), *Phys. Rev.* **101**, 1597–1607.

Weinberg, S. (1967), *Phys. Rev. Lett.* **19**, 1264–1266.

Weyl, H. (1929), *Z. Phys.* **56**, 330–352.

Zweig, G. (1964), CERN Report 8182/TH.401.

Further Reading

The Standard Model—Synthesis and Overview

Aitchison, I. J. R., Hey A. J. G. (1989), *Gauge Theories in Particle Physics*, 2nd ed., Philadelphia: Adam Hilger.

Misner, C., Thorne, K., Wheeler, J., (1973), *Gravitation*, San Francisco: W.H. Freeman and Co.

Okun, L. B., (1984), *Particle Physics, The Quest for the Substance of Substance*, Chur, U.K.: Harwood.

Perkins, D. (1987), *Introduction to High Energy Physics*, Redwood City, CA: Addison-Wesley.

Quigg, C. (1983), *Gauge Theories of the Strong, Weak, and Electromagnetic Interactions*, Reading, MA: Benjamin Co.

Accelerators and Other Facilities

Particle Data Group (1992), *Phys. Rev.* **45** (11), Pt. II, SI.1–SIX.20.

Mapping the SM Constituents onto the Detectors

Barger, V. D., Phillips, R. J. N. (1987), *Collider Physics*, Redwood City, CA: Addison-Wesley.

Eichten, E., Hinchliffe, I., Lane, K., Quigg, C. (1984), "Supercollider Physics," *Rev. Mod. Phys.* **56**, 579–707.

Shapiro, M., Siegrist, J. (1991), "Hadron Collider Physics," *Annu. Rev. Nucl. Part. Sci.* **41**, 97–132.

A Model $pp/p\bar{p}$ Detector

Ferbel, T. (1987), *Experimental Techniques in High Energy Physics*, Redwood City, CA: Addison-Wesley.

Kleinknecht, K. (1985), *Detectors for Particle Radiation*, Cambridge, U.K.: Cambridge Univ. Press.

Other Experimental Techniques

Jackson, J. D. (1962), *Classical Electrodynamics*, New York: Wiley.

Future Developments

Berkelman, K., Stone, S. (1991), "Decays of B Mesons," *Annu. Rev. Nucl. Part. Sci.* **41**, 1–28.

Gunion, J., Haber, H., Kane, G., Dawson, S. (1990), *The Higgs Hunter's Guide*, Redwood City, CA: Addison-Wesley.

Ross, G. G. (1987), *Grand Unified Theories*, Cambridge, U.K.: Cambridge Univ. Press.

PARTICLE SOURCES AND TARGETS

See PARTICLE SOURCES—ION; PHYSICS AND TECHNOLOGY OF ION AND ELECTRON SOURCES; SOURCES, PARTICLE—ELECTRON; TARGETS FOR PARTICLE ACCELERATORS

PARTICLE SOURCES—ION

IAN G. BROWN, *Lawrence Berkeley Laboratory, University of California, Berkeley, California, U.S.A.*

	Introduction	213
1.	**Plasma Formation**	213
2.	**Beam Formation and Related Issues**	214
3.	**Some Representative Ion Sources**	217
4.	**Applications**	219
5.	**Summary**	220
	Glossary	220
	Works Cited	221
	Further Reading	221

INTRODUCTION

Ions are formed from neutral atoms by one kind or another of ionization mechanism. The various ways in which the ionization can be done include electron-impact ionization, photoionization, field ionization, surface ionization, and others. When incorporated within an ion source, often the ionization method lends its name to the ion source; thus for example one can have microwave ion sources, field-emission ion sources, vacuum-arc ion sources, and so on. Most ion sources are plasma based. Plasma might be formed from a gas (gaseous plasma) as in microwave, Penning, beam-plasma, and duoplasmatron ion sources, or from a metal (metal plasma) as in vacuum-arc and some laser ion sources. In some ion sources, the ions are formed without going through the plasma state, as in the field-emission or liquid-metal ion source. Because of the predominance of ion sources in which the ions are formed via the plasma state, here we restrict consideration to this kind of source.

Plasma-based ion sources are comprised of two basic parts—a means of plasma production (the "plasma source"), and a means of beam formation (the "extractor"). A simplified schematic of an elementary generic ion source is shown in Fig. 1. The requirements of the plasma source and the extractor may be either trivial or very complicated depending on the required ion-source performance and ion-beam characteristics. For example, the production of a low-energy, low-current (say a few μA at a few keV), singly charged argon-ion beam with no particular demands on beam geometry or divergence is rather easily accomplished; but the generation of a beam of very highly charged ions of low divergence at high energy and beam current is quite a different story.

Note the terminology used. An ion source is a device that produces an ion beam, and it (usually) contains a plasma source. Even though the plasma indeed contains ions and so the plasma source is in a sense a source of ions too, it is more appropriate to refer to the device that produces an extracted beam of more or less energetic ions, i.e., the ion-beam generator, as an ion source, and to use the term plasma source for the generator of charge-neutral plasma.

Here we briefly review the general characteristics of plasma-based ion sources and examine a few specific kinds of sources that have become well known and that are used extensively, including both plasma-based sources as well as one non-plasma-based source, the liquid-metal ion source.

1. PLASMA FORMATION

The plasma formation process is the heart of the ion source. The properties of the ion source and of the extracted ion beam are determined in large part by the plasma source, and the name given to the ion source is often that of the kind of plasma generator used. Some of the most basic ion-beam characteristics, such as ion species, beam current, beam purity, and ion charge state are determined by the equivalent parameters of the plasma

3-527-28135-5/95/$5.00 + .50

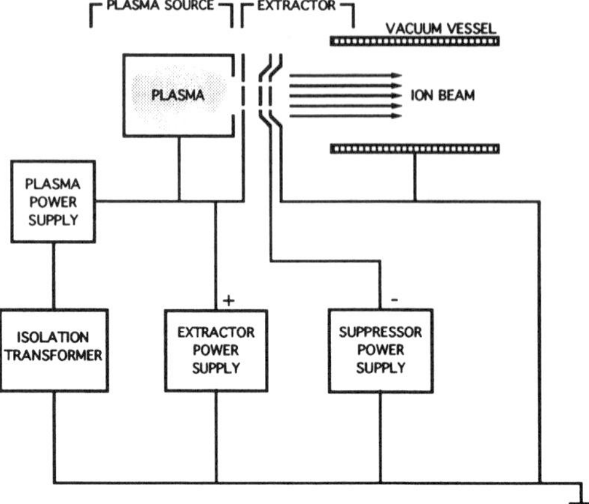

FIG. 1. Simplified schematic of a "generic" ion source.

source—plasma ion species, plasma density, plasma composition, and plasma ion charge-state distribution. For example, in an ECR ion source, the plasma is formed in an ECR (electron cyclotron resonance) plasma discharge and the ion-beam characteristics reflect the properties of the ions that are produced in an ECR plasma device; most notably, they are highly charged. If a high-current ion source is required, the plasma source should be of high density and high ion-production rate. As for all plasma sources, the means by which the plasma is confined is vitally important, and this is no less so when embodied within an ion source. Magnetic field configurations of various kinds have been employed using both electromagnetic field coils and high-field, rare-earth permanent magnets. The "magnetic bucket" geometries that have been used so successfully for the production of large, uniform plasmas have also been used in ion sources for the production of large-area ("broad") ion beams. Such ion sources are often referred to as bucket sources.

Whereas a plasma source as commonly used in plasma physics research is basically at ground potential, in an ion source the plasma and its electrical system are necessarily elevated to extractor potential (see Fig. 1), which might be many tens of kilovolts. This requirement adds complexity and expense to the ion-source configuration. Simple methods of plasma preparation have an advantage in this respect. For high-power dc ion sources, the requirements of the floating power supplies and their control and protection electronics can be a major part of the overall system hardware and cost.

2. BEAM FORMATION AND RELATED ISSUES

The plasma in an ion source is more or less confined within the region in which it is generated except for the place where the beam is to be formed, where the beam-formation electrodes are located. The plasma ion and electron components are separated at the beam-formation electrodes, the ions are accelerated to the required beam energy, and the ion beam is formed. Note that the beam is not really "extracted" from the plasma. What takes place is that the plasma ions drift across a Child–Langmuir sheath at a rate determined by the plasma density and the electron temperature, and not at all by the applied "extraction voltage"; it is only consequent to the ion–electron separation at the sheath that the ions experience the full applied voltage and are thereby accelerated to the requisite beam energy. Thus, the terms "ion extractor," "beam extraction electrodes," etc., are in fact misleading in terms of forming a good conceptual model of the ion source. Nevertheless the terms are used widely, and with the qualifications presented we use this terminology here too.

The ion-beam current that can be extracted from an ion source is given by the Child–Langmuir equation for ion current flow under space-charge–limited conditions (Child, 1911; Langmuir and Compton, 1931; Green, 1974, 1976),

$$I = \tfrac{4}{9}\varepsilon_0 S \left(\frac{2q_i}{M_i}\right)^{1/2} \frac{V^{3/2}}{d^2} \tag{1}$$

$$= 1.72 S \left(\frac{Z}{A}\right)^{1/2} \frac{V^{3/2}}{d^2}, \tag{2}$$

where ε_0 is the permittivity of free space, S the open extractor area, $q_i = eZ$ the ion charge, e the electronic charge, Z the ion charge state, $M_i = Am_{\mathrm{amu}}$ the ion mass, A the atomic weight (amu), V the extractor voltage, and d the extractor gap (separation between the main beam-forming electrodes), and in the second expression I is in mA, V in kV, S in cm^2, and d in cm. This equation is strictly valid only under special conditions, the most important of which are

1. the assumption of space-charge–limited ion flow, and
2. plane-parallel geometry throughout.

Although condition **1** is often well satisfied, condition **2** is often only poorly met. But nevertheless, the Child–Langmuir equation in its simple form usually provides a good approximation to the ion current that can be expected.

The beam-formation electrodes, or extractor grids, are in the form of aligned apertures in metallic electrodes across which the ion acceleration voltage, or extraction voltage, is applied. A pair of single, circular holes might constitute the extractor, as for example is commonly used in ECR ion sources, or an array of a large number, perhaps thousands, of small holes each producing its own "beamlet," as for example in some rf ion sources, "Kaufman" ion sources, and sources used for space thrusters, or an array of parallel slits as for example in some of the broad-beam ion sources developed for the formation of large, high-power neutral beams used in the controlled thermonuclear fusion research program. A photograph of the large and high-precision extractor used for the 100-keV, 75-A neutral beam source developed at the Lawrence Berkeley Laboratory (Kunkel, 1990) is shown in Fig. 2.

To maximize the beam current produced by the source, the extractor gap needs to be as small as it can be for a given extraction

FIG. 2. Beam formation electrodes from a 10 x 40-cm^2 long-pulse ion source showing the multiple-slot, rail-type plasma grid facing upward.

voltage while not allowing breakdown between the grids under a wide range of plasma conditions. Designing for these conditions is where the art and skill come in. The Child–Langmuir equation provides a good estimate of the beam current that can be expected from an ion source when the plasma density and the extraction voltage are optimized with respect to the extraction geometry (also called the perveance match condition). When the mechanical geometry and the plasma and operational parameters are matched to each other in this way, the beam divergence is minimum, determined now by the ion temperature and not the extraction optics, and the beam current density is highest. Calculation of beam formation is a complicated process, and a number of computer codes have been written to predict the extracted ion trajectories (Spädtke, 1989). An example of such a calculation is shown in Fig. 3 (taken from Spädtke, *et. al.*, 1994).

In the simplest scenario, the ion-beam extractor can be made from just a pair of grids. These are often called the plasma grid and the ground grid, where the plasma grid is in contact with the plasma and the ground grid is held at ground potential. The extraction voltage, typically some tens of kilovolts, is put across the extractor, which means that the whole plasma formation system is now elevated to high positive voltage—ions created in the plasma are accelerated across this high extraction voltage; this is where the beam energy comes from.

It is usual that the ion beam is injected into and transported through a vacuum system in which ionization of the residual gas by the ion beam can occur. The cold electrons so created serve an important purpose, especially for high-current ion beams: they provide neutralization of the positive space-charge potential of the beam due to the ions themselves. In this way, space-charge blowup of the beam is minimized. The tendency of an ion beam to space-charge blowup increases with ion mass and beam current density. An expression has been given by Holmes (1989) for the current density for which the beam space charge of an incompletely neutralized beam would cause a 10% increase in beam diameter, which for the nonrelativistic limit can be written as

$$j_0 = \frac{3.24 \times 10^{-7}\, U^{3/2} Z^{1/2}}{A^{1/2} z^2 (1-h)}, \qquad (3)$$

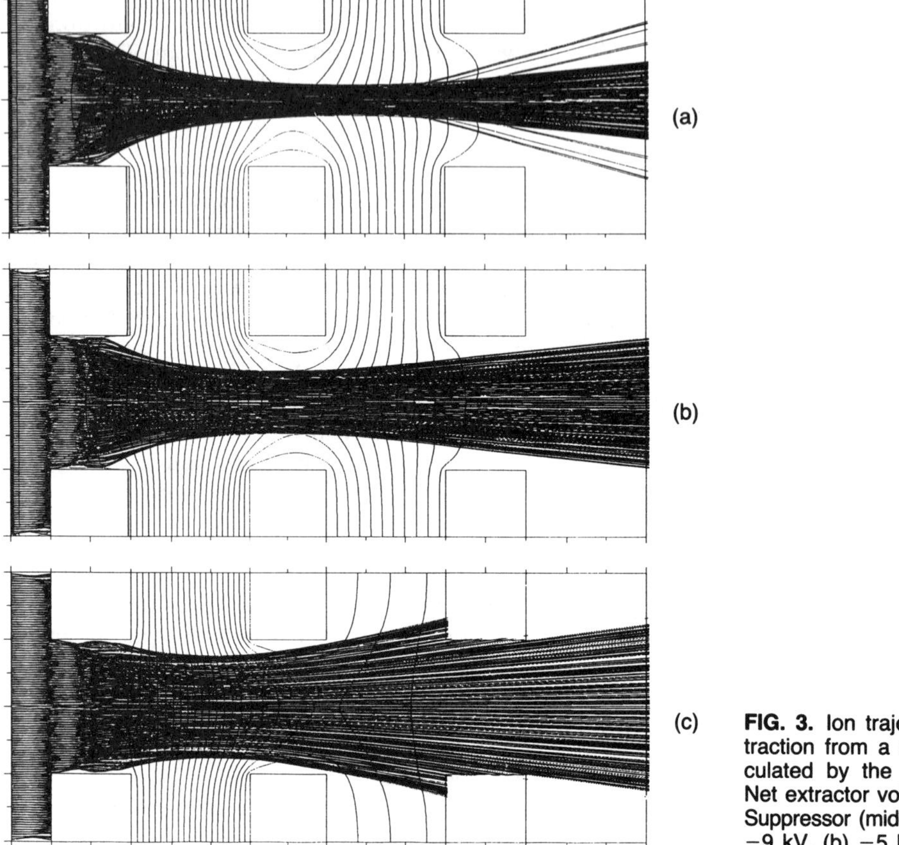

FIG. 3. Ion trajectory plots for extraction from a single hole as calculated by the AXCEL-INP code. Net extractor voltage 5 kV; Ti ions. Suppressor (middle) grid voltage (a) −9 kV, (b) −5 kV, (c) −1 kV.

where j_0 is the beam current density in A/m^2, U is the beam energy expressed in volts, Z is the ion charge state, A is the ion mass in amu, z is the beam propagation distance in meters, and h is the fraction of the beam space charge that is neutralized.

The space-charge–neutralizing electrons form a background sea for the energetic ion beam. The cold electrons can stream backward to the ion-source extractor, where they see an accelerating field in the direction opposite to the extracted ions, i.e., back into the plasma chamber. This electron backflow can sometimes cause damage to the source because of its power deposition, and in any case it shows up as an added electrical load—the measured "extractor current" (power supply drain) is composed not only of the forward ion current but also the backward electron current, and the electron current can easily be several times greater than the true ion current. For these reasons, it is common to make the extractor with a third electrode between the plasma grid and the ground grid, usually called the "suppressor grid," since its function is to suppress the electron backflow. It is biased negatively, say at about 10% or so of the full extraction voltage. The extraction shown in Fig. 3 contains three grids—plasma, suppressor, and ground grids. Sometimes, when design of the beam-ion current flow is particularly critical, yet more grids are added so as to be able to fine-tune the shape of the electric field within the extractor. A cutaway view of an extractor consisting of four rail grids is shown in Fig. 4 (from the Lawrence Berkeley Laboratory neutral-beam development program). In low-current ion sources the suppressor electrode is often omitted; and in ion sources where the extraction is across a magnetic field, the magnetic field serves to suppress electron backflow and so an additional electrostatic suppression is not needed.

FIG. 4. Cut-away view of a four-grid multiple-slot extractor.

The electron backflow can also give rise to quite substantial x-ray generation when the electron stream is stopped on electrodes within the ion source. This can pose a significant personnel safety hazard. Proper operation of the suppressor can vastly reduce the x-ray intensity. In any case, whenever an ion source is operated at voltages near or above about 20 kV, it is important to monitor carefully the x-ray intensity in the vicinity. Lead shielding is commonly used to reduce the intensity to acceptably low levels.

An ion beam may be focused to submicron dimensions or it may be of very wide area. Ion sources producing small beams often do so by forming the ions at a tiny point, as for example in the liquid-metal ion source where ions are produced by field emission from the pointlike tip of the "Taylor cone" formed electrostatically in the liquid metal. Production of large-area beams is done by ion sources in which the plasma is produced or expanded to large area before being presented to the extractor. The mechanical design of broad-beam extractors is not straightforward, especially when the electrodes can be heated by the beam and suffer thermal expansion while good alignment between the grids must be maintained. The 50-cm extractor from a broad-beam vacuum-arc ion source is shown in Fig. 5. This source produced a pulsed titanium ion beam of energy 100 keV and pulsed current about 10 A (Brown, 1994).

3. SOME REPRESENTATIVE ION SOURCES

Ion sources can be classified in a great many different ways, corresponding to a wide range of source and beam characteristics, parameters, and applications. One can categorize according to beam current intensity, ion charge state, or ion species, or according to whether the source is short pulse, long pulse, or dc, etc. Here we briefly describe several different kinds of ion sources that are broadly used in a number of important fields. These sources are for the most part "mature" in that within their field of use they have been studied and developed by a relatively large number of workers at numerous laboratories around the world. They convey some of the important principles involved in the physics and technology of ion sources. These sources as well as a wide range of other kinds of ion sources also are well described in the Proceedings of the International Conferences on Ion Sources (see Further Reading).

FIG. 5. 50-cm-diam multiple-aperture extractor from a vacuum-arc ion source.

The ECR (electron cyclotron resonance) ion source (Jongen and Lyneis, 1989) has achieved a high degree of prominence because of the very high charge states of the ions produced—species such as fully stripped argon, Ar^{18+}, for example, have been produced (Kutner, 1994). These sources are widely used for accelerator (cyclotron) injection and for atomic physics experiments. They are an outgrowth of the hot-electron plasma source research carried out as part of the world controlled-fusion research effort in the 1960s and 1970s. The early pioneering work to develop these kinds of plasma sources into usable ion sources was led by Geller and his group at Grenoble (Geller, 1976). In the ECR source, a hot-electron plasma is created by the resonant coupling of high-power microwaves to the plasma electrons trapped within a minimum-B stabilized magnetic mirror. The electrons can be heated to very high energies, up to hundreds of keV, and it is these hot electrons that provide the ionization energy to strip the plasma ions, also trapped within the magnetic mirror confinement geometry. A small fraction of ions is axially lost from the mirror trap, and it is this plasma loss flux from which the ion beam is then extracted, providing a beam of very highly stripped ions. The electron cyclotron frequency is given by $\omega_c = eB/m$, where B is the magnetic field strength and m is the electron mass, or $f_{ce} = 2.8B$ GHz, where now the field-strength B is in kG; for typical laboratory magnetic fields of, say, 3–10 kG, this means that the resonant frequency is in the low microwave band, around 10 GHz. A schematic of a typical ECR ion source is shown in Fig. 6. The ECR source is a dc source, producing stable beams for long periods of time (up to weeks). Beams of virtually all of the elements of the periodic table can be produced, including the solid elements either by using an oven or by plasma sputtering from an inserted target. Beam currents are typically in the 1- to 100-eμA (electrical microampere) range for a given charge-state component, dropping to fractional microamperes for the highest charge states.

Freeman sources are used extensively in the semiconductor ion implantation field. Their ion beams are particularly quiescent and stable, and they work well with B, P, and As, the ion species most commonly used for semiconductor doping. The source was initially developed at the Harwell Laboratory, U.K., for electromagnetic isotope separation application (Freeman, 1963), and was subsequently adapted for ion implantation when the need arose in the late 1960s. In the Freeman source, electron emission from a thick (several millimeter diameter) filament is used to control the ionization of material admitted to an arc chamber and the plasma so formed is magnetically confined by the field that is the sum of an externally applied homogeneous field parallel to the filament and that due to the filament itself. It is usual for the arc chamber to be fed by vapor from an oven, providing the feed material containing the wanted ion species. Ion extraction is through a long slit, typically several millimeters wide by several centimeters long, in the same direction as the filament and the external field. The extracted beam is thus of rectangular cross section. Typical ion beam current is up to the 10-mA range.

The EBIS, or electron beam ion source, holds the record for the highest charge-state production. These sources were developed initially by Donets and co-workers at Dubna (Donets, 1990). In the EBIS, a high-energy electron beam is injected into a strong, uniform magnetic field so as to compress the beam to high current density. The beam is surrounded by a number of cylindrical "drift tubes" that are electrically biased from an external source. The drift-tube structure is cryogenically cooled to liquid helium temperature so as to keep the background gas pressure very low, in the 10^{-12}-Torr region or less. Neutral gas is injected into the electron beam where a fraction is ionized and trapped

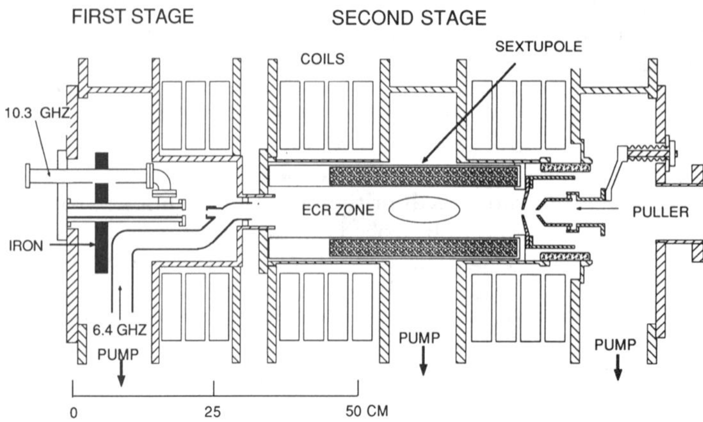

FIG. 6. Simplified schematic of a typical ECR ion source.

within the beam, where the ions are stripped to progressively higher charge states by successive electron impact. After maximal ion charge state is achieved, the drift-tube potentials are switched so as to allow the confined ions to be axially extracted from the device. Fully stripped xenon ions, Xe^{54+}, have been detected (Kutner, 1994).

4. APPLICATIONS

Historically, the first major application of ion sources was for injection into particle accelerators, initially for the relatively simple high-voltage accelerators and later for the growing spectrum of linear and circular machines. To match the capabilities of the accelerators the ion-source parameters required were in general quite modest, for example calling for beam currents only in the microampere range. As the diversity and performance of particle accelerators expanded, so also have the ion-source requirements grown, and sources used in this field today include Penning sources for linear accelerators, ECR sources for cyclotrons, EBIS and laser sources for synchrotrons, and many others. The injection requirements of modern accelerators can be quite severe, calling for very stable and reliable ion beams, often of high charge state and high current. The state of the art of ion sources as relevant to the field of particle accelerators is well presented at regular meetings of the community and can be found in the conference proceedings (see Further Reading).

Electromagnetic isotope separation (the deflection and collection of an ion beam in a magnetic field), in small quantities as an analytical tool and in larger quantities for their production for medical and industrial applications, is a field that also made early use of developments in ion source technology. By far the largest project of this kind was that undertaken in the United States as part of the Manhattan Project in World War II for the separation of uranium isotopes; thereafter these facilities, called calutrons, were used for industrial isotope production. The ion sources first employed were essentially vapor sources with electron-beam ionization. A drawback of the method is the very large investment of electric power required for other than minute quantities. For atomic mass analysis as a research tool, however, the technique is well developed, and the kinds of ion sources employed can be of very low current. The conference series, Electromagnetic Isotope Separators and Techniques Related to their Application, reports well on ion-source developments in this field (see Further Reading).

The field of ion implantation has grown steadily from its introduction in the early 1960s to its wide acceptance as a mature industrial technology for semiconductor material fabrication today. Ion implanters are manufactured and sold by a number of companies throughout the world, and this industrial research community is a well established one. The kind of ion sources used in "mainstream" implanters are the Freeman source and the Bernas source, where the qualities of beam quiescence and stability are paramount; other kinds of ion sources, such as microwave-driven sources, have found niche application. As described above, the beam species of particular application are B, P, and As, and a great deal of development of these sources has been done. The semiconductor ion-implantation community holds a regular biannual conference at which ion-source technology is a significant component; the proceedings of these meetings provide a wealth of detail about ion-source developments in this field (see Further Reading).

Ion implantation is a surface-modification technique with scientific and technological application wider than for semiconductor materials processing alone, and there are active research communities in nonsemiconductor areas such as, for example, metallurgical, ceramic, polymer, superconductor, and electro-optics materials synthesis and modification. For these applications the characteristics of the ion beams required vary widely and so also the kinds of ion sources needed. The field remains largely a research field, having not yet made the transition to widely used industrial application, and the diversity of the experimental configurations investigated and of the ion sources used is great. Relevant major conference series that address ion sources for these applications are those on Ion Beam Modification of Materials and the Surface Modification of Metals by Ion Beams (see Further Reading).

Applications such as ion-beam lithography, micromachining, and materials surface characterization using ion beams call for low-current, finely focused beams, and field-emission liquid-metal ion sources are often used for this purpose. Ion-beam focal spots as small as a few hundred angstroms have been produced.

The quest for propulsion of space vehicles and satellites by ion thrusters has motivated research internationally since the beginning of the space era. Though the size of this effort has decreased with time, work still continues, with emphasis on guidance and steering. A number of different kinds of ion sources have been developed and investigated for this potential application, including rf-driven sources, broad-beam sources, and liquid-metal sources. The requirements of the ion source needed are clear; not surprisingly, a major hurdle is in obtaining the specified ion-beam performance in an adequately small package with adequately high efficiency.

5. SUMMARY

The ion source is the beginning of the ion process, whatever the application is to be. The ion beam is created at the source, transported over distances from a meter or so up to many hundreds of meters, and delivered to the intended device or use. The physics and technology of beam transport is a sibling field to that of ion sources, which attains increasing importance as the required beam transport efficiency and the beam current density increase. The concern of space-charge blowup of the beam due to uncompensated electrostatic repulsive forces between ions is often critical, and as available ion-source beam currents steadily increase, this is becoming more commonly the limiting feature in some ion-beam facilities. In the absence of externally applied electric or magnetic fields, the beam can acquire a high degree of space-charge neutralization from the residual gas in the vacuum system, but the fields that are inherent to beam focusing and bending elements, such as dipole and quadrupole magnets or electrostatic lenses, disrupt the low-energy neutralizing electrons and beam neutralization can be diminished or lost. This concern is more or less limited to the high-current, low-energy regions that can exist in the beam transport line immediately after the ion source. It is usual that as the beam is transported further downstream toward the application, the current becomes lower and often some postacceleration (ion acceleration after initial ion extraction) adds energy to the ion beam; consequently beam space-charge blowup is less of a concern further away from the ion source. Often the ion source provides much more beam, perhaps several orders of magnitude more, than is finally available for the application. Beam loss is a major concern in many cases.

Understanding of the physical principles of ion-source design and operation has increased greatly in recent years, and the technology of their construction and application has similarly flourished. Fields of use both in fundamental science and in applied technology call for increasingly formidable ion-beam parameters and sophisticated ion-source performance, and the field has responded by generating a wide array of ion devices. Whereas in the past it was the case that the field was segmented into distinct and largely noncommunicating application areas, there is now a thriving international community of ion-source researchers with good cross-specialty communication links. The result is a healthy and fertile research area and a steady growth in ion-source technology.

GLOSSARY

ECR: Electron cyclotron resonance. Condition when an applied rf field equals the gyro frequency of electrons in the confining magnetic field. Usually in the microwave band.

Extractor: The electrode structure that forms an energetic, directed ion beam from a plasma.

Magnetic Bucket: Magnetic field geometry formed when an array of permanent bar magnets is distributed around the plasma confinement region.

Magnetic Mirror: Magnetic field geometry formed between two spaced field coils, with maxima in the plane of the coils and a central minimum.

Sheath: Boundary layer between plasma and electrode, where charge neutrality is not preserved.

Taylor Cone: Conical surface shape assumed by a liquid subjected to an electric field.

Works Cited

Brown, I. G. (1994), *Rev. Sci. Instrum.* **65**, 3061.

Child, C. D. (1911), *Phys. Rev. (Series 1)* **32**, 492.

Donets, E. D. (1990), *Rev. Sci. Instrum.* **61**, 225.

Freeman, J. H. (1963), *Nucl. Instrum. Methods* **22**, 30.

Geller, R. (1976), *IEEE Trans. Nucl. Sci.* **NS-23**, 904.

Green, T. S. (1974), *Rep. Progr. Phys.* **37**, 1257.

Green, T. S. (1976), *IEEE Trans. Nucl. Sci.* **NS-23**, 918.

Holmes, A. J. T. (1989), in: I. G. Brown (Ed.), *The Physics and Technology of Ion Sources*, New York: Wiley, p. 53.

Jongen, Y., Lyneis, C. M. (1989), in: I. G. Brown (Ed.), *The Physics and Technology of Ion Sources*, New York: Wiley, p. 207.

Kunkel, W. B. (1990), *Rev. Sci. Instrum.* **61**, 354.

Kutner, V. B. (1994), *Rev. Sci. Instrum.* **65**, 1039.

Langmuir, I., Compton, K. T. (1931), *Rev. Mod. Phys.* **3**, 251.

Spädtke, P. (1989), in: I. G. Brown (Ed.), *The Physics and Technology of Ion Sources*, New York: Wiley, p. 107. See also the codes AXCEL-INP and KOBRA-INP: INP, Junkernstr. 99, Wiesbaden D-65205, Germany.

Spädtke, P. *et. al.* (1994), *Rev. Sci. Instrum.* **65**, 1441.

Further Reading

Brewer, G. R. (1970), *Ion Propulsion Technology and Applications*, New York: Gordon and Breach.

Brown, I. G. (Ed.) (1989), *The Physics and Technology of Ion Sources*, New York: Wiley.

Forrester, A. T. (1988), *Large Ion Beams*, New York: Wiley.

Valyi, L. (1977), *Atom and Ion Sources*, New York: Wiley.

Wilson, R. G., Brewer, G. R. (1973), *Ion Beams with Applications to Ion Implantation*, New York: Wiley.

Conference Proceedings

The Proceedings of the biannual International Conferences on Ion Sources (ICIS) are excellent sources of information on ion sources of all kinds. See, for instance: *Rev. Sci. Instrum.* **61** (No. 1, Pt. 2), January 1990, (ICIS '89); *Rev. Sci. Instrum.* **63** (No. 4, Pt. 2), April 1992 (ICIS '91); *Rev. Sci. Instrum.* **65** (No. 4, Pt. 2), April 1994 (ICIS'93).

For ion sources of relevance to the field of particle accelerators, see the proceedings of the biannual IEEE Particle Accelerator Conferences (published by the IEEE as Conference Proceedings), and of the annual Conferences on the Application of Accelerators in Research and Industry (published in *Nuclear Instruments and Methods*).

Progress in ion sources as used for semiconductor ion implantation is reported at the International Conferences on Ion Implantation Technology, held every 2 years, with the proceedings published in *Nuclear Instruments and Methods*. The International Conference on Ion Beam Modification of Materials is focused more on nonsemiconductor implantation; these proceedings are also published in *Nuclear Instruments and Methods*. Another relevant conference series is the biannual Surface Modification of Metals by Ion Beams, proceedings published in *Surface and Coatings Technology*.

The proceedings of the conference series on Electromagnetic Isotope Separators and Techniques Related to their Application are archived in *Nuclear Instruments and Methods*, where ion-source and ion-beam developments and methods in the field are well covered.

PARTICLES, ELEMENTARY

MICHAEL P. SCHMIDT, *Physics Department, Yale University, New Haven, Connecticut, U.S.A.*

	Introduction	223
1.	**Building Blocks of the Standard Model**	224
1.1	Quarks and Leptons	224
1.2	Bound States of Quarks: Baryons and Mesons	225
2.	**The Forces**	226
2.1	The Electromagnetic Force	226
2.2	The Strong Force: Color	227
2.3	The Weak Force: Radioactive Decay	228
2.3.1	Flavor Conservation	229
2.3.2	Conservation of Baryon and Lepton Number	230
2.3.3	Parity and Charge Conjugation	230
2.3.4	Weak Neutral-Current Interactions and Electroweak Unification	231
3.	**Beyond the Standard Model**	231
4.	**Summary**	232
	Works Cited	232
	Further Reading	232

INTRODUCTION

The term "elementary particles" refers to the entities that are considered to be the fundamental constituents of matter. The definition of elementary depends on the distance scale relevant to the physics of interest. The behavior of bulk matter, under normal terrestial conditions, is largely determined by the properties of atoms, in the form of the chemical elements (the elementary particles of chemistry) and combinations of atoms into molecules. Atoms, with a typical size of 10^{-8} cm, have structure and are comprised of electrons orbiting a nucleus, with a typical size of 10^{-12} cm. Atomic nuclei are known to have structure as well and are built up out of protons and neutrons that are an order of magnitude smaller in size. Finally, protons and neutrons are believed to be made of *quarks*, which, along with electrons, appear to be elementary particles, in the sense that they reveal no internal structure down to distances as small as 10^{-16} cm.

Equally important to an understanding of the structure of matter are the forces of nature. Four forces have been identified, each with its domain of major importance. *Gravity* is the most familiar force, being responsible for the formation of stars and planets and the motion of celestial bodies. The *electromagnetic* force is also familiar and among other things leads to the binding of electrons to nuclei and the structure of atoms. On the subnuclear scale, less familiar forces play a fundamental role. The *strong* nuclear force "glues" quarks together into nucleons (protons and neutrons) and the nucleons together into nuclei. The *weak* nuclear force is responsible for the radioactive transmutation of nucleons and nuclei. At the subnuclear scale, it is also important to understand the *quanta*, also elementary particles, that mediate the fundamental forces. The most familiar example is the photon, the quantum of electromagnetic energy and mediator of the electromagnetic interaction.

Remarkable achievements occurring in the study of subnuclear particle physics over the last several decades have led to the development of what is now known as the Standard Model of elementary particle physics. This model is extremely successful in describing and predicting the relationships and interactions among the observed elementary particles and forces. Most notably, the Standard Model includes the unification of the electromagnetic and weak interactions, a discovery akin to that of Maxwell's unification of electricity and magnetism. The development and verification of the Standard Model is a result of the interplay between experimental and theoretical advances.

On the experimental side, the develop-

3-527-28135-5/95/$5.00 + .50

ment of high-energy accelerators as well as the sophisticated detectors and techniques used to study these particles has been essential. Progress in the field has largely been set by the development of accelerators capable of producing increasingly energetic and intense beams of particles (electrons, protons, etc.). Very high-energy particles when used as projectiles can serve as probes of subnuclear structures by virtue of their small Compton wavelength, $\lambda = h/mc$ (where h is Planck's constant and m is the particle mass); they can also give rise to very energetic collisions in which the energy released in the collision can materialize as massive particles in accordance with Einstein's equation, $E = Mc^2$, relating mass (M), energy (E), and the speed of light (c). Intense beams of particles are needed for precision measurements and the study of rare phenomena.

On the theoretical side, the development of a relativistic quantum field theory for electrodynamics was crucial. Its success in providing a means to calculate detailed properties of atomic structure led theorists to conclude correctly that there was hope in attempting to formulate the other subnuclear interactions in a similar fashion. The exploitation of symmetries and conservation laws also proved to be most valuable. The plethora of new particles discovered with the advent of particle accelerators was initially bewildering. Fortunately, regularities were uncovered in the systematics of the production and decays of these particles, and these regularities were successfully interpreted as resulting from symmetries (sometimes hidden or partially broken) of the physics underlying the properties of the particles and their interactions.

The development of the Standard Model represents a very important step in achieving a comprehensive understanding of subnuclear physics, but it is not yet the ultimate theory of elementary particles. Continued research will be required in order to answer fundamental questions left open by the Standard Model, such as the origin of the masses of the elementary particles. It is also hoped and suspected that further unification of the fundamental forces should be possible, including gravity.

The field of subnuclear physics, which grew out of the study of nuclear physics and cosmic rays, continues to evolve. For example, the field of particle astrophysics is growing in importance, building on the interplay between particle physics, the physics of the early universe, and cosmology. Collisions occurring at high-energy accelerators can create conditions not experienced since shortly after the Big Bang. In addition, the number and type and properties of the species of particles existing in the early moments of the universe affect its evolution.

1. BUILDING BLOCKS OF THE STANDARD MODEL

The elementary particles of the Standard Model can be divided into two broad classes: the particles that carry the properties (e.g., electric charge) that give rise to the forces of interaction between the particles and the particles that serve to communicate these forces between the particles. The quarks and leptons, discussed in the present section, are in the first class. The gauge bosons that mediate the forces are discussed in the following section.

1.1 Quarks and Leptons

The building blocks of the Standard Model are the *quarks* and *leptons*. Quarks are constituents of protons, neutrons, and related particles (previously considered elementary) known as *hadrons* that feel the strong nuclear force. The leptons form a separate group of particles that are not subject to the strong interactions. Examples of leptons are the electron and the neutrino. The neutrino, being electrically neutral, participates only in the weak interactions. The electron as well as the quarks experiences both the electromagnetic and weak interactions.

The quarks come in six *flavors* named up, down, charm, strange, top, and bottom. Leptons come in three pairs, each pair consisting of a charged lepton and an associated neutrino, and they are named after the charged lepton of the pair: electron, muon, and tau. The electric charges carried by the quarks and leptons are given in Table 1 (in units of the electron's charge). Electric charge is one of several *quantum numbers* associated with the elementary particles. These quantum num-

Table 1. The quarks and leptons[a]

Particle	Symbol	Spin	Baryon number	Charge	Mass (GeV)
Up quark	u	1/2	1/3	+2/3	$\approx 5 \times 10^{-3}$
Down quark	d	1/2	1/3	−1/3	$\approx 10 \times 10^{-3}$
Charm quark	c	1/2	1/3	+2/3	≈1.3
Strange quark	s	1/2	1/3	−1/3	≈0.2
Top quark	t	1/2	1/3	+2/3	≈170
Bottom quark	b	1/2	1/3	−1/3	≈4.3
Electron neutrino	ν_e	1/2	0	0	$<5.1 \times 10^{-9}$
Electron	e	1/2	0	−1	0.511×10^{-3}
Muon neutrino	ν_μ	1/2	0	0	$<0.27 \times 10^{-3}$
Muon	μ	1/2	0	−1	0.106
Tau neutrino	ν_τ	1/2	0	0	<0.031
Tau (lepton)	τ	1/2	0	−1	1.78

[a]Extracted from Particle Data Group (1994).

bers are found to be conserved quantities in the interactions of elementary particles. The determination of the conserved quantities has played an important role in establishing the relationships among the subnuclear particles. The significance of other quantum numbers such as lepton number and baryon number will be discussed later.

Each quark and lepton also has a corresponding *antiparticle* that carries equal but opposite quantum numbers; the antielectron, also known as the *positron*, carries +1 unit of electric charge. Particles and antiparticles have equal masses. In Table 1, the particle masses are expressed in terms of an energy (millions of electron volts) using the relation $E = mc^2$. Sufficiently energetic collisions of elementary particles can create conditions in the laboratory in which particle–antiparticle pairs are created. When this occurs through the electromagnetic or strong interaction, the various quantum numbers, including quark or lepton flavor, associated with the particles must be conserved. For example, a positron must be created in association with an electron so that no net electric charge is created from energy alone. However, as discussed later, the weak interactions do not conserve quark flavor.

All of the quarks and leptons are *fermions*, obey Fermi–Dirac statistics, and carry an intrinsic angular momentum, or spin, of 1/2 (in units of $\hbar$, Planck's constant divided by 2π). Angular momentum as well as energy and linear momentum are conserved in the interactions of elementary particles.

1.2 Bound States of Quarks: Baryons and Mesons

Spin plays an important role in determining the properties of the hadrons that are formed out of combinations of quarks and antiquarks by the strong interaction. As discussed more fully below, the strong interaction binds together groups of three of quarks (denoted as qqq) and quark–antiquark pairs (denoted by $q\bar{q}$). In principle, any flavor can be chosen for each of the quarks. The three-quark bound states qqq are known as *baryons*; they have baryon number +1 (from the three quarks, each contributing +1/3). The rules of addition for spin angular momentum result in baryons having odd half-integral spin (1/2, 3/2, etc.) and therefore being fermions. The proton and neutron are the most familiar baryons—the proton consisting of two up quarks and a down quark ($p = uud$) and the neutron consisting of two down quarks and an up quarks ($n = udd$). The electric charge of the proton and neutron follows from the algebraic sum of the electric charges of the constituent quarks. A partial list of observed baryons is given in Table 2, including their quark content, spin, and mass.

The quark–antiquark bound states, known as *mesons*, have baryon number equal to zero. They also have integral spin (0, 1, etc.) and so are *bosons* and obey Bose–Einstein statistics. Again, in principle, any flavor of quark can be chosen in forming the mesons. Some examples of mesons readily produced in high-energy particle collisions include the pi me-

Table 2. Some representative baryons and mesons[a]

Particle	Symbol	Quark content	Spin	Charge	Mass (GeV)
Proton	p	uud	1/2	+1	0.938
Neutron	n	udd	1/2	0	0.940
Lambda	Λ^0	uds	1/2	0	1.12
Sigma	$\Sigma^+, \Sigma^0, \Sigma^-$	uus, uds, dds	1/2	+1, 0, −1	1.19
Delta	$\Delta^{++}, \Delta^+, \Delta^0, \Delta^-$	uuu, uud, udd, ddd	3/2	+2, +1, 0, −1	1.23
Omega	Ω^-	sss	3/2	−1	1.67
Lambda (c)	Λ_c^+	udc	1/2	+1	2.29
Pions	π^+, π^0, π^-	$u\bar{d}, u\bar{u} + d\bar{d}, d\bar{u}$	0	+1, 0, −1	0.137
Kaons	$K^+, K^0, \bar{K}^0, K^-$	$u\bar{s}, d\bar{s}, s\bar{d}, s\bar{u}$	0	+1, 0, 0, −1	0.495
Rho	ρ^+, ρ^0, ρ^-	$u\bar{d}, u\bar{u} + d\bar{d}, d\bar{u}$	1	+1, 0, −1	0.770
Phi	ϕ^0	$s\bar{s}$	1	0	1.02
D mesons	$D^+, D^0, \bar{D}^0, D^-$	$c\bar{d}, c\bar{u}, u\bar{c}, d\bar{c}$	0	+1, 0, 0, −1	1.86
	D_s^+, D_s^-	$c\bar{s}, s\bar{c}$	0	+1, −1	1.97
J (Psi)	J/Ψ	$c\bar{c}$	1	0	3.1
B mesons	$B^+, B^0, \bar{B}^0, B^-$	$u\bar{b}, d\bar{b}, b\bar{d}, b\bar{u}$	0	+1, 0, 0, −1	5.28
	$B_s^0, \bar{B}_s^0$	$s\bar{b}, b\bar{s}$	0	+1, −1	5.37
Upsilon	Υ	$b\bar{b}$	1	0	9.46

[a]Extracted from Particle Data Group (1994).

sons, also known as pions. The pions come in three charged varieties: $\pi^+ = u\bar{d}$, $\pi^- = \bar{u}d$, and π^0, a linear combination of $u\bar{u}$ and $d\bar{d}$. It is noted that the π^- is the antiparticle of the π^+, and the π^0 is its own antiparticle. The K mesons, or kaons, are also relatively easily produced, being the least massive particles containing the strange quark: $K^+ = \bar{s}u$, $K^- = s\bar{u}$, $K^0 = \bar{s}d$, and $\bar{K}^0 = s\bar{d}$. Table 2 also gives a partial list of some of the observed mesons and their quark contents, spins, and masses.

2. THE FORCES

The interactions among elementary particles are the result of several forces. Three of these forces—the strong, electromagnetic, and weak—have been understood in terms of relativistic quantum field theories and in particular as *gauge* field theories. Each of the forces is described in terms of the exchange of quanta of the gauge field between the interacting particles. The strength of the gravitational force is sufficiently small that gravity can safely be ignored in studying the interactions of small numbers of elementary particles.

2.1 The Electromagnetic Force

The electromagnetic force is the most familiar of the forces relevant to the study of elementary particles. It is the attractive *Coulomb* force between the negatively charged electron and the positively charged proton that gives rise to the bound state known as the hydrogen atom. Ignoring relativistic effects, the potential energy of the bound state can be expressed as $V = e^2/r$, where e is the magnitude of the electron (or proton) charge and r is the distance separating the proton and electron.

The infinite range of the electromagnetic force is in keeping with the masslessness of the photon, the quantum of the electromagnetic field. The electromagnetic field of a charged particle is taken to be the result of the radiative emission and reabsorption of *virtual* photons by the charged particle. Virtual particles are those that do not satisfy the equation $E^2 - p^2c^2 = m^2c^4$; that is, their existence violates energy conservation for a short time interval, as allowed by the uncertainty principle of quantum mechanics: $\Delta E \Delta t \leq \hbar$.

In first approximation, the Coulomb force responsible for the binding of the hydrogen atom is due to the emission of a photon by the proton and subsequent absorption by the electron. The strength of the interaction due to the exchange of one photon is $\alpha = e^2/\hbar c = 1/137$ when expressed in dimensionless form. Here, α is associated with the quantum mechanical probability of one photon being exchanged between two particles with charge e. The full theory, known as quantum electrodynamics (*q.v.*) or QED, is used to determine the contributions to the force arising from the exchange of multiple photons. Perturbative methods can be employed and lead to converging results as a result of the smallness of α.

Even as it is recognized that the definition of what is an elementary particle depends on the distance scale in question, it is also recognized that the definition of the strength of an interaction depends on the distance scale being probed. The strength of the electromagnetic interaction, α, depends on the electric charge e. However, the magnitude of the effective charge can be shown to increase as the distance associated with the measurement of the charge decreases. This is due to the screening of the electron's charge at longer distances by the polarization of the vacuum surrounding the electron. The electron emits virtual photons to give rise to the electromagnetic field, and these photons can themselves create virtual electron–positron pairs. The electron–positron pairs are affected by the presence of the central electron charge and become oriented so as to have the positrons closer to the central electron. The screening effect is measurable, though small, changing α by about 10% for a change in the distance scale by a factor of roughly 100 000.

Quantum electrodynamics (QED) serves as the paradigm in the construction of theories of the strong and weak interactions. The gauge bosons that are involved in these theories are listed in Table 3. The significant differences observed between the interactions result in large part from the different symmetries inherent to the interactions. The properties of quantum electrodynamics result from a local gauge symmetry based on the unitary group U(1). This symmetry requires the invariance of the physics under a redefinition of the quantum mechanical wave functions of particles carrying electric charge by a space-time–dependent complex phase. The U(1) gauge invariance of QED is responsible for the conservation of electric charge.

2.2 The Strong Force: Color

The strong force, also called the nuclear force, is formulated in terms of a gauge field theory known as quantum chromodynamics (QCD) (see STRONG INTERACTIONS). The gauge symmetry for QCD is based on the special unitary group SU(3). In QED, there is one kind of charge, the electric charge, acting as the source for the electromagnetic field. In QCD, there are three kinds of charge, called color; each flavor of quark (u,d,c,s,t,b) can possess one of the three color charges, and the leptons carry no color charge. The three color charges can be named R, G, and B in analogy with the primary colors red, green, and blue. The properties of the color force are such that only color-charge neutral (i.e., colorless) bound states are observed in nature. This has the unusual consequence that free, isolated colored quarks are not observable in nature, a fact that so far has been borne out experimentally. The only color-neutral objects are the baryons and mesons of which more than a hundred different types have been observed.

The baryons are colorless bound states in which each of the three quarks carries a different color charge to obtain equal contributions of R, G, and B color charge. The color charges are independent of the quark flavor and, for example, the lambda baryon, $\Lambda = uds$ is a linear combination of $u_R d_G s_B$, $u_G d_B s_R$, $u_B d_R s_G$, etc. (where the subscript indicates the color charge of the quark). A linear combination of colors is required to assure that the net color charge of the particle is invariant under rotations in the abstract space associated with the color SU(3) symmetry group.

Antiquarks carry "anti"-color charges $\bar{R}$, $\bar{G}$, $\bar{B}$ (or, to follow the visual color theory

Table 3. The gauge bosons[a]

Particle	Symbol	Spin	Force mediated	Electric charge	Color charge	Mass. (GeV)
Photon	γ	1	Electromagnetic	0	0	0
Gluon	g	1	Strong	0	r, g, b	0
W boson	W^+, W^-	1	Charged weak	±1	0	80.2
Z boson	Z^0	1	Neutral weak	0	0	91.2

[a]Extracted from Particle Data Group (1994).

analogy, the complementary colors cyan, magenta, and yellow), just as the positron carries an electric charge opposite to that of the electron. The mesons are colorless objects by virtue of being comprised of a quark and antiquark having "opposite" colors. The positive pi meson is a linear combination of $u_R\bar{d}_R$, $u_G\bar{d}_G$, and $u_B\bar{d}_B$.

The analogue of the photon in QED for QCD is the gluon. The gluon, like the photon, is massless and has spin 1, and color charge is a conserved quantity. However, because of the *non-Abelian* property of the QCD gauge group, SU(3), there are in fact eight distinct types of gluons, and they all possess net color charge: $R\bar{G}$, $B\bar{R}$, etc. Photons do not carry electric charge and can only interact with each other via charged particles. Gluons, on the other hand, can interact directly with each other via their color charges, and this plays a crucial role in determining the properties of the strong force.

In quantum chromodynamics, the chromodynamic fields of quarks are created by the emission and reabsorption of virtual colored gluons. The vacuum can be polarized, as in the case for QED. However, for QCD, the polarization of the vacuum leads to an antiscreening of color charge, a result due to the fact that gluons carry color charge. The antiscreening implies that the strength of the color force decreases as the distance associated with the measurement decreases, and conversely, the force increases with increasing distance. The former result has been proven (by Gross and Wilczek and Politzer; see Wilczek, 1982) and is known as the *asymptotic freedom* property of QCD. The latter result is conjectured and is known as the *confinement* property of QCD. The phenomenological result (consistent with experimental observations) is that quarks are confined in protons and other hadrons at a distance scale of about 1 fm (10^{-13} cm) where the chromodynamic interaction coupling is strong ($\alpha_s > 1$). However, inside the hadrons, the quarks behave relatively freely.

Unfortunately, perturbative methods cannot be used to determine the masses of hadrons because α_s effective at a distance scale of 1 fm is not small. Nonperturbative techniques, such as lattice gauge theory, show promise for addressing these problems. It is expected and has been observed that the masses of hadrons containing the more massive quarks (c,b,t) are largely due to the mass of the massive quark.

The strong force that binds protons and neutrons into nuclei (and overcomes the large repulsive electric force existing between protons inside a nucleus) is then related to the structure of the nucleons (and other hadrons), as determined by the more fundamental chromodynamic interaction of quarks and gluons. This is similar to the relation of the forces at play between neutral atoms and molecules and the more fundamental electromagnetic interaction that establishes atomic structure. The nuclear binding force is strong but has a short range; the nucleons must be close enough to detect the distribution of color charge within each other.

Before the advent of the quark model of hadrons, it was suggested that the nuclear binding force could be due to the exchange of mesons between nucleons (protons and neutrons). The finite range of the force could be explained in terms of a potential of the form $V = (\alpha_s^2 \hbar c/r)e^{-mc^2 r/\hbar c}$, now known as the Yukawa potential, where m is the mass of the exchanged meson. The Compton wavelength, $\lambda = h/mc$, of pi mesons agrees with the observed effective range of the nuclear force between nucleons.

2.3 The Weak Force: Radioactive Decay

Only a few elementary particles are stable; the rest of the particles rapidly decay into other less massive particles. The study of decays has been an essential aspect of elementary particle research and has provided clues to understanding the relationships and interplay among the forces affecting elementary particles. The exponential decay lifetimes observed in the decays of elementary particles can be understood in terms of the forces responsible for the decay. A typical strong-interaction decay time is 10^{-24} seconds. Electromagnetic decays occur over time intervals in the range of 10^{-16}–10^{-20} seconds. Weak-interaction decay times are found in the range of 10^{-13}–10^{-6} seconds. The special properties of the weak interaction were initially revealed in the study of particle decays. The weak interaction was in fact first recognized as being the agent responsible for the radioactive β decay of nuclei. As an example, in the decay $n \rightarrow pe^- \bar{\nu}_e$, a neutron can be transformed by the weak interaction into the less massive proton and an electron (the β ray) plus an anti–electron neutrino.

As in the case of the electromagnetic and strong interactions, the weak interaction is understood in terms of a gauge field theory. The weak interaction is governed by the group SU(2) and mediated by the exchange of three types of spin-1 gauge bosons between the interacting fermions. Two of the gauge bosons, known as the W^+ and W^-, carry electric charge and mediate the *charged-current* weak interaction responsible for nuclear radioactive decay. The third boson, known as the Z^0, has zero electric charge and is responsible for the weak *neutral-current* interactions. Typical decays are given in Table 4.

2.3.1 Flavor Conservation The electromagnetic and strong interactions conserve quark and lepton flavors. The photon can couple with any charged fermion and conserves the fermion flavor: An electron can radiate a photon, or a very energetic photon can materialize as an electron–positron pair or any other particle–antiparticle pair of quarks or charged leptons. Electric charge as well as quark and lepton flavors must be conserved: A muon cannot just radiate a photon and as a result be transformed into an electron. The gluon can couple to any quark, as they carry color charge (leptons do not): A blue up quark can radiate a blue–anti-red gluon and be transformed into a red up quark, or a very energetic green–antiblue gluon can materialize as a particle–antiparticle pair of quarks [e.g., charm (green) plus anticharm (antiblue)]. Color and quark flavor must be conserved: A bottom quark cannot just radiate a gluon and be transformed as a result into a strange quark.

The "weak" charges carried by the quarks and leptons are such that the charged-current weak interactions lead to specific sets of transmutations that do not conserve quark flavor. The weak interactions view the quarks and leptons as coming in doublets:

$$\begin{pmatrix} u \\ d \end{pmatrix}, \begin{pmatrix} c \\ s \end{pmatrix}, \begin{pmatrix} t \\ b \end{pmatrix}; \begin{pmatrix} \nu_e \\ e \end{pmatrix}, \begin{pmatrix} \nu_\mu \\ \mu \end{pmatrix}, \begin{pmatrix} \nu_\tau \\ \tau \end{pmatrix}.$$

Referring to Table 1, it is noted that the elements of each doublet differ in electric charge by one unit. The W bosons carry one unit of electric charge and couple in an equivalent way to each of the weak fermion doublets and thereby give rise to transmutations between the fermions of a weak doublet. For example, a top quark (charge $+2/3$) can decay into a W^+ boson and a bottom quark (charge $-1/3$): $t \to bW^+$; the W^+ can then subsequently decay into a positron and an electron neutrino: $W^+ \to e^+\nu_e$. The process of neutron β decay, $n \to pe^-\bar{\nu}_e$, can be described in terms of the decay of one down quark in the neutron into an up quark and a virtual W^- that subsequently materializes as an electron plus an

Table 4. Some representative particle decays[a]

Particle	Decay	Interaction	Lifetime (s)
Photon	—	—	Stable
Electron	—	—	Stable
Neutrinos	—	—	Stable
Muon	$\mu^- \to e^-\bar{\nu}_e\nu_\mu$	Weak	2.2×10^{-6}
Tau	$\tau^- \to \mu^-\bar{\nu}_\mu\nu_\tau$	Weak	3×10^{-13}
Pion	$\pi^- \to \mu^-\bar{\nu}_\mu$	Weak	2.6×10^{-8}
Pion	$\pi^0 \to \gamma\gamma$	Electromagnetic	8.4×10^{-17}
Kaon	$K^+ \to \pi^+\pi^0$	Weak	1.2×10^{-8}
Rho	$\rho^0 \to \pi^+\pi^-$	Strong	4×10^{-24}
Neutron	$n \to pe^-\nu_e$	Weak	887
Lambda	$\Lambda^0 \to p\pi^-$	Weak	2.6×10^{-10}
Sigma	$\Sigma^0 \to \Lambda^0\gamma$	Electromagnetic	7.4×10^{-20}
Delta	$\Delta^+ + \to p\pi^+$	Strong	5×10^{-24}
Omega	$\Omega^- \to \Lambda^0 K^-$	Weak	8.2×10^{-11}
D meson	$D^0 \to K^-\mu^+\nu_\mu$	Weak	4.2×10^{-13}
Lambda − c	$\Lambda_c^+ \to pK^-\pi^+$	Weak	2×10^{-13}
B meson	$B^- \to D^0 e^-\bar{\nu}_e$	Weak	1.5×10^{-12}

[a]Extracted from Particle Data Group (1994).

electron antineutrino: $n \rightarrow (ud)d \rightarrow (ud)u + W^- \rightarrow pe^-\bar{\nu}_e$.

The weak interaction, like the strong interaction, and in contrast to the electromagnetic interaction, is very short ranged. The effective strength of the weak interaction is smaller than the electromagnetic interaction by a factor of about 10 000. Both of these properties result from the fact that the *W* (and *Z*) are very massive particles. The intrinsic strength of the weak interaction is in fact comparable to the electromagnetic interaction after one accounts for the suppressed probability of creating a virtual *W* (or *Z*) compared to that for creating a virtual photon.

2.3.2 Conservation of Baryon and Lepton Numbers The weak interactions (as well as the electromagnetic and strong interactions) conserve both the baryon and lepton numbers. In fact, separate lepton numbers are conserved, one for each of the three flavors of leptons: electron, muon, and tau. The electron and its companion neutrino (ν_e) each carry one unit of electron lepton number; the positron and electron antineutrino carry minus one unit while the muon and tau and their companion neutrinos carry zero units of electron lepton number. This pattern continues in the obvious way for muon and tau lepton numbers. The origin of baryon and lepton number conservation is not understood. This is in contrast to the case for electric charge conservation that follows from the local U(1) gauge invariance of electromagnetism.

2.3.3 Parity and Charge Conjugation Another essential difference of the weak interaction is that it violates the discrete symmetries of *parity* (*P*) and *charge conjugation* (*C*). Invariance under parity requires the physics of a system to be unchanged for an inversion of all coordinates through the origin. Invariance under charge conjugation requires the physics to be unchanged if all chargelike quantum numbers are reversed (i.e., particles and antiparticles interchanged). The violation of both symmetries by the charged-current weak interaction is manifest in the properties of the apparently massless neutrinos that partake only in the weak interaction.

Neutrinos are found to be left-handed objects, in the sense that their spin angular momentum is always directed in opposition (or antiparallel) to their line of motion. Under the assumption that neutrinos are massless and therefore move at the speed of light, no object can overtake their motion and thereby change their observed handedness. Antineutrinos are right-handed with their spin angular momentum always directed along their line of motion. A parity-reversed universe (left-handed spins interchanged with right-handed spins) or a charge-conjugated universe (lepton number interchanged with antilepton number) would contain right-handed neutrinos and left-handed antineutrinos—particles that are not observed. It should be noted, however, that a combined operation of charge conjugation and parity inversion (*CP*) transforms the observed left-handed neutrinos and right-handed antineutrinos into themselves.

The quarks and the charged leptons, having nonzero masses, occur in both left- and right-handed varieties. However, only the left-handed variety are subject to the weak force; the right-handed fermions do not carry the weak charge. The charged-current weak interactions are then completely left-handed while the electromagnetic and strong interactions are completely ambidextrous.

The weak transitions among quarks reveal another intriguing aspect of the charged-current weak interaction. For reasons that are not yet understood, the mass *eigenstates* and the weak-charge *eigenstates* of quarks are not identical. This leads to weak transitions among charge $+2/3$ quarks of one weak doublet with charge $-1/3$ quarks of another weak doublet. For example, the charm quark is observed to decay mainly to the strange quark, but it is also found less frequently to decay (about one chance in 20) into the down quark. The effect of the mismatch between the mass and weak-charge eigenstates of quarks can be described in terms of a unitary transformation acting on the charge $-1/3$ quark fields. This transformation can be parametrized as a complex rotation with three Euler-type angles and one complex phase. The values of these angles and the phase are fundamental parameters of the Standard Model. The angles govern the relative frequency of all of the possible transitions between the charge $+2/3$ and charge $-1/3$ quarks, and the complex phase is thought to be connected with a small violation of *CP* invariance, so far detected only in certain decays of neutral *K* mesons. It is believed that *CP* noninvariance is required to explain the gross asymmetry represented by the preponderance of matter over antimatter observed in the universe.

2.3.4 Weak Neutral-Current Interactions and Electroweak Unification The remaining agent of subnuclear interactions, the Z^0, is responsible for the weak neutral-current interactions. The Z^0 is massive enough that after being produced in high-energy collisions, it will decay into particle–antiparticle pairs of any flavor of lepton or quark (except for the very massive top quark.) Precision measurements of Z^0 decays have been used to rule out the existence of any additional flavors of massless neutrinos. These measurements have also been used to determine how the strength of the weak neutral-current interaction differs for various classes of fundamental fermions: the neutrinos, the charged leptons, and the charge $+2/3$ and charge $-1/3$ quarks. In addition, it has been found to be neither left-handed nor ambidextrous. These properties of the neutral-current interactions have a common cause: the unification of the electromagnetic and weak interactions, as proposed by Glashow, Salam, and Weinberg.

In the Standard Model of the electroweak interactions, the theory has an underlying symmetry based on the direct product group SU(2) × U(1). This symmetry requires that the electroweak gauge bosons (two electrically charged and two electrically neutral) and the fermions (quarks and leptons) are all massless—in contrast to what is observed in the real world. The symmetry is broken, however, by the solution nature has chosen from the equations governing the theory, and this gives rise to masses for the quarks and leptons as well as for the W and Z. The photon remains massless as a result of the fact that the U(1) gauge symmetry of electromagnetism remains unbroken. The breakdown of the initial gauge symmetry is often compared with the case of the spontaneous breakdown of rotational symmetry that occurs in a ferromagnet when the long-range correlation among the spins (and magnetic moments) results in the alignment of the spins along a random direction.

The spontaneous breakdown of the SU(2) × U(1) gauge symmetry in the Standard Model occurs in such a manner that the two electrically neutral gauge bosons associated with the original unbroken symmetry become mixed, leaving two linear combinations, one of which is the photon and the other the Z^0. The "peculiar" couplings of the Z^0 to the quarks and leptons thus reflect the specific "charges" associated with the unbroken gauge symmetries that are carried by the right- and left-handed fermions. In this way, three otherwise phenomenologically distinct interactions—electromagnetism and the charged- and neutral-current weak interactions—are unified in the sense of being inextricably intertwined. The masses of the W and the Z as well as the weak and electric charges are no longer independent but related in a predictable way that has been very precisely confirmed by experiment.

The price of this unification is the introduction of a new force, and the precise understanding of its origin is not yet clear. In the simplest version of the Standard Model, a set of four spin-0 particles (*scalar* fields) are introduced, and it is the pattern of the scalar boson interactions with the fermions, gauge bosons, and themselves that gives rise to the unification. The scalar fields carry the weak and electric charges; two are electrically neutral, and the other two carry $+1$ and -1 unit of electric charge. Under the spontaneous breakdown of the electroweak symmetry, the two charged scalar fields and one of the neutral scalar fields become absorbed in the definition of the massive W and Z and the final neutral scalar emerges as a detectable elementary particle known as the Higgs particle. The mass of the Higgs particle is not known, but it is expected to be less than 1 TeV. Discovering the Higgs particle, or whatever agent is in fact responsible for the electroweak-symmetry breakdown, is the next crucial step to make in order to reach a more complete understanding of the nature of matter.

3. BEYOND THE STANDARD MODEL

The development of the Standard Model of elementary particle physics is a great achievement. The quark model of hadrons and the unification of the electromagnetic and weak interactions successfully provide a coherent and simplifying framework in which to view the complex spectrum of subnuclear particles and their interactions. The predictive power of the Standard Model has been and continues to be thoroughly tested; so far, no discrepancy has emerged. However, several questions remain unanswered by the Standard Model. These questions include the following:

1. The origin of mass: What is the physics that causes the top quark to be so heavy or the electron to be so light? Are the neutrinos in fact massless?
2. What is the physical significance of there being three types of each class of fundamental fermions: neutrinos, charged leptons, and the charge +2/3 and −1/3 quarks? Why not four types?
3. What is the underlying rationale for the assignment of quantum numbers (for example electric charge) to the fundamental fermions?
4. What physics determines the magnitude of the Euler angles describing the weak quark transitions?
5. What is the origin of *CP*-invariance violation?

Some of these questions are undoubtedly related, and some extensions of the Standard Model have been contemplated that partially address these questions. It is possible that nature will reveal additional symmetries that will lead to a resolution of these issues. It is useful to mention briefly two examples. The first is supersymmetry: a postulated symmetry of nature between fermions and bosons, so that every fermion with a specific set of quantum numbers has a boson partner. While the symmetry has so far not been observed in nature (the supersymmetric partners have not been produced and identified for any of the fundamental fermions), it imposes constraints that make it very attractive in addressing the question of the origin of mass. A second class of conjectures has to do with further unification of the fundamental forces (see UNIFIED FIELD THEORIES). The possibility of unifying the symmetries governing the strong and electroweak interactions, SU(3) and SU(2) × U(1), into a larger group, such as SU(5), seems very attractive in that it can provide a rationalization for the assignment of quantum numbers to the fundamental fermions. The implication of this type of unification is the presence of additional interactions, most notably interactions that can give rise to proton decay ($p \rightarrow \pi^0 e^+$), which violates baryon (and lepton) number conservation and has so far not been observed in spite of very sensitive searches. Finally, there are even grander schemes involving an extension of quantum field theory (of point fields) to a supersymmetric quantum field theory of strings (*superstrings*), which appears to be the most promising path toward including gravity in the unified theory of the fundamental forces.

4. SUMMARY

The Standard Model in a sense plays a role in particle physics that the periodic table of the elements plays in chemistry. The rules of chemistry are established primarily by the electronic structure of atoms. It is pressing on and trying to understand the relative abundance of specific nuclear isotopes (atoms that differ only by the number of neutrons in the nucleus) that leads to the study of nuclear physics, rich in its own way. As research continues in an effort to understand the questions raised by the Standard Model, we can expect new insights into the properties of matter to emerge, whether it be in the form of massive new particles being produced at high-energy accelerators, unexpected effects discovered in precision experiments or searches for rare processes, or from astrophysical considerations and observations.

Works Cited

Particle Data Group (1994), "Review of Particle Properties," *Phys. Rev. D* **50**, 1173–1826.

Wilczek, F. (1982), *Annu. Rev. Nucl. Part. Phys.* **32**, 177–209.

Further Reading

Frauenfelder, H., Henley, E. (1991), *Subatomic Physics*, 2nd ed., Englewood Cliffs, NJ: Prentice Hall.

Pais, A. (1986), *Inward Bound*, New York: Oxford University Press.

Pickering, A. (1984), *Constructing Quarks*, Chicago: University of Chicago Press.

Williams, W. S. C. (1991), *Nuclear and Particle Physics*, New York: Oxford University Press.

PATENTS AND INTELLECTUAL PROPERTY

MARIA REICHMANIS, *Registered Patent Attorney, Aiken, South Carolina, U.S.A.*

	Introduction	233
1.	**Types of Intellectual Property: Patents, Trademarks, Copyrights, and Trade Secrets**	233
2.	**Patents**	235
2.1	The Patent Application	235
2.1.1	The Specification, Claims, and Drawings	236
2.1.2	Examination	237
2.2	Enforcing Patent Rights	237
2.2.1	Assignment vs License	237
2.2.2	Infringement	238
2.3	International Filing	238
2.3.1	International Filing and the Patent Cooperation Treaty	239
3.	**Trademarks, Service Marks, Certification Marks, and Collective Marks**	239
3.1	Enforcement of Trademark Rights	239
4.	**Copyrights**	240
4.1	Copyrightable Works	240
4.2	Exclusive Rights	240
4.2.1	Fair Use	240
4.3	Ownership	240
4.4	Notice, Registration, and Deposit	241
4.5	Enforcement of Copyrights	241
5.	**Trade Secrets**	241
6.	**Appendix**	242
	Glossary	242
	Works Cited	243
	Further Reading	243

INTRODUCTION

"Property" has been defined as the exclusive right to possess, enjoy, and dispose of a thing (Nolan and Connolly, 1979). Thus, property in the legal sense is not so much the thing itself, but rather the aggregate of rights in the thing. Property can be real (land; and whatever is erected on, growing upon, or affixed to the land) or personal (money; goods; stocks and bonds). It can be tangible (a house or an automobile) or intangible (contract rights). Intellectual property—patents, trademarks, copyrights, and trade secrets—is personal property.

The owner of intellectual property has the same rights with respect to the property as the owner of a house has with respect to her property: subject to applicable laws, she may prevent another from using, selling, or otherwise exercising any form of control over the house. However, she may choose to rent the house for a period of time, give it away, or sell it. The owner of intellectual property, like the owner of a house, may assign all her rights in the property to another, or grant a license giving the right to use the property under specified conditions.

1. TYPES OF INTELLECTUAL PROPERTY: PATENTS, TRADEMARKS, COPYRIGHTS, AND TRADE SECRETS

The four generally recognized forms of intellectual property are listed in Table 1. Each protects different aspects of an idea and confers different rights upon its owner.

A patent is a grant from a government that conveys to an inventor the exclusive right to make, use, and sell the patented invention for a specified term, usually a term of years. United States law provides for three types of patents: utility patents, for novel, useful, and nonobvious machines, processes, compositions of matter, and articles of manufacture; plant patents, for new and distinct, asexually reproduced varieties of plant; and design patents, for novel and nonobvious ornamental designs for articles of manufacture.

A trademark or service mark is any word,

3-527-28135-5/95/$5.00 + .50

Table 1. Patent, trademark, copyright, and trade secret protection.

	Patent	Trade secret	Copyright	Trademark, service mark
Subject matter	Process, machine, article of manufacture, composition of matter	Something that gives its owner an advantage over competitors that do not know it	Literary works; musical and dramatic works; pantomimes and choreographic works; pictorial, graphic, and sculptural works; motion pictures and other audiovisual works; sound recordings	A word, name, symbol, device, or combination thereof that identifies the product or service
Requirements for protection	Novel, useful, nonobvious	Keep it secret	Original; fixed in a tangible medium of expression	Distinctive; used as a mark
What is protected?	The substance of the idea	The substance or the form of the idea, or both	The expression of the idea	The appearance of the mark
How is it protected?	Prevent others from making, using, or selling it	Prevent others from using it, if obtained unlawfully	Prevent others from making copies or derivative works; prevent others from distributing, performing, or displaying the copyrighted work	Prevent others from using later-registered, confusingly similar marks
When is it protected?	Upon the grant of a patent	Upon creation	Upon fixation in a tangible medium of expression	Upon first use
For how long is it protected?	17 years (utility patent) 17 years (plant patent) 14 years (design patent)	As long as it is a secret	Life of author plus 50 years; if a "work for hire," 100 years after creation or 75 years after publication	As long as it is used

name, symbol, device, or any combination thereof, adopted and used to identify goods or services. Trademarks are used to identify the source of goods; service marks are used to identify the source of services. The owner of a trademark or service mark can prevent others from using it or confusingly similar marks, but trademark or service-mark protection alone does not give him the right to prevent others from providing the same product or service *without* the mark.

Copyright law deals with information and ideas reduced to tangible form. Unlike patents, which protect the substance of the idea, copyrights protect the expression of the idea.

A trade secret is any confidential information that gives its owner an advantage over competitors that do not know it. The secret may be a machine or a chemical process, an abstract idea such as a concept for a new consumer product, or a compilation of data such as a list of customers. Matters of common knowledge, or matters of general knowledge in a particular industry, cannot be protected as trade secrets.

Patent law protects the substance of an idea in order to encourage investment in new technology and invention, whereas trademark law is concerned with the protection of symbols that identify products and services. Copyright law protects the original expression of an idea, and trade secret law protects confidential information. A single product may be the subject of more than one form of intellectual property. For example, the design of an automobile body may be copyrightable or protectible by a design patent; the engine, brakes, seat belts, and other parts may be protected by utility patents; some of the parts may be made by secret processes; and the model name may be a trademark (McCarthy, 1990).

2. PATENTS

The United States Patent and Trademark Office (PTO) issues three types of patents: utility patents, plant patents, and design patents.

Utility patents are issued for novel, useful, and nonobvious processes, machines, articles of manufacture, or compositions of matter. A "nonobvious" invention may be the result of inspiration, routine study and experimentation, or pure chance—there is no single test of obviousness. The owner of the patent need not exploit the invention himself. However, he can *prevent others* from making, using, or selling the invention during the term of the patent regardless of whether or not he, himself, does so.

Fundamental laws of nature ($F = ma; E = mc^2$) and mathematical algorithms ($\sin^2\theta + \cos^2\theta = 1$) are not patentable, under the rationale that the result would be a monopoly on a scientific truth. This prohibition includes computer programs that are pure algorithms, although such programs may be the subject of copyright protection. However, a process that uses a mathematical algorithm may be patentable if the process itself, not merely the algorithm, is new and useful. Other nonpatentable categories include naturally occurring substances, methods of doing business, forms, and abstract ideas.

Even though naturally occurring substances are not patentable in themselves, highly purified products and extracts thereof may be patentable. In addition, they may form part of a patentable product or be used in a patentable process. Drugs, vitamins, food additives, and other products that are derived from natural products are patentable; processes for making such products may also be patentable. Genetically engineered life forms are patentable if they meet the statutory requirements of novelty, usefulness, and nonobviousness (Miller and Davis, 1983).

Plant patents are issued for new and distinct asexually reproduced plants, including mutants, hybrids, cultivated sports, and newly found seedlings other than tuber-propagated plants or plants found in an uncultivated state (sexually reproduced plants are protectible under the Plant Variety Protection Act). To be patentable, a plant must be nonobvious as well as new and distinct, but it need not be useful. Patent protection applies only to the plants themselves, not to flowers, fruits, bulbs, or seeds *per se*.

Design patents protect new, original, and ornamental designs for articles of manufacture, that is, the *appearance* of a product rather than its function. A design patent can protect surface ornamentation applied to a product or the overall shape of the product, but not a shape that is dictated by the function of the product. For example, a simple bed frame with four legs and a platform for supporting a mattress would not be a proper subject for design patent protection, since a bed needs a mattress support in order to *be* a bed. However, the shape and size of the legs, the headboard, and the footboard may be patentable, since these aspects of the design are not dictated by their utilitarian purpose (the requirement that the frame be capable of supporting a mattress). All of these features could easily be changed or even eliminated without changing the *function* of the product.

An invention may be the subject of both design and utility patent protection: a utility patent, for its function, and a design patent, for its ornamental appearance. When there is a choice between trademark registration of a nonfunctional design and a design patent, trademark protection may be preferred because of the potentially unlimited term (i.e., as long as the mark is used by the trademark owner).

2.1 The Patent Application

In most countries, patents are obtained by filing an application with the appropriate government agency. In the United States, a patent is obtained by filing an application with the Commissioner of the Patent and Trademark Office (PTO). The PTO examines the application to determine whether the three tests for patentability—novelty, utility, and nonobviousness—are met. If the invention in question satisfies these tests and all relevant fees are paid, a patent issues in due course.

In the United States, a patent application must be filed by the inventor or on behalf of the inventor; in some countries, an application may be filed by the owner of the invention—for example, the inventor's employer. A patent application includes a verbal description of the invention (the "specification"), drawings (where needed to understand the invention), and "one or more claims particularly pointing out and distinctly claiming the subject matter which the applicant regards as his invention." The application process is governed by a set of regulations developed by the PTO (see U.S. Patent and Trademark Office, 1992). Some applicants prepare their own patent applications; however, most work with a registered patent attorney or agent.

2.1.1 The Specification, Claims, and Drawings The specification includes the title of the invention, cross-references to related applications (if any), a discussion of the background of the invention explaining the context of the problem that is being addressed, a brief summary of the invention, a description of the drawings, a detailed description of the invention, the claims, an abstract, and the drawings, and an oath or declaration signed by the inventor(s). When included in a patent application, drawings must show every element of the claims. "Drawings" may include not only line drawings, but graphs, flow charts, and photographs where appropriate. The invention should be described as clearly as possible so that it is understandable by the person "of ordinary skill in the art to which it pertains." Either the specification, the claims, or both jointly must constitute an "enabling disclosure," that is, provide an explanation that is sufficient for others to practice the invention.

The claims provide a concise, formal definition of the invention. The claims define the metes and bounds of the invention, much as a deed defines the boundaries of a parcel of land. A claim contains three parts: the preamble, the transition, and the body. The preamble generically defines the invention, for example, "An apparatus for doing X." The transition relates the preamble to the body, typically with the words "comprising" ("including parts A, B, and C, and maybe D, E, . . . "), "consisting" ("including only A, B, and C"), or "consisting essentially of" ("including A, B, and C, and nothing else that matters"). The body describes elements A, B, and C that make up apparatus X. Because the claims describe the invention in words, they are inherently less precise than the description that appears in a deed to real estate. Any given invention can be described by many different combinations of words, ranging from broad ("a chair, comprising means for supporting a person in a sitting position") to narrow ("a chair, consisting of a round wooden seat, and three equally spaced cylindrical legs attached to the bottom of the seat").

It is sometimes helpful to view an invention as a territory analogous to real estate. Thus, a new, useful, and nonobvious widget is located in a region unoccupied by previously known, prior-art widgets A, B, and C (Fig. 1). A broad claim for the new widget encompasses as much previously unclaimed territory as possible, up to (but not crossing) the boundaries of the territories occupied by A, B, and C (Fig. 2). A narrow claim covers much less territory (Fig. 3). A patent application for the new widget might contain a series of claims, ranging from the broadest possible, covering many different ways of practicing the invention, to the narrowest that will protect the actual manufactured product. A single, broad claim would protect the invention, but narrower claims could protect

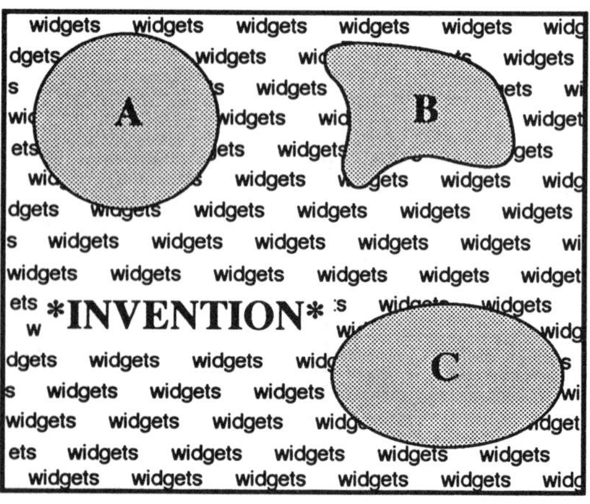

FIG. 1. The field of widgets. Previously known, prior-art widgets (A, B, C) and the invention occupy separate territories. The invention is outside the territories of widgets A, B, and C.

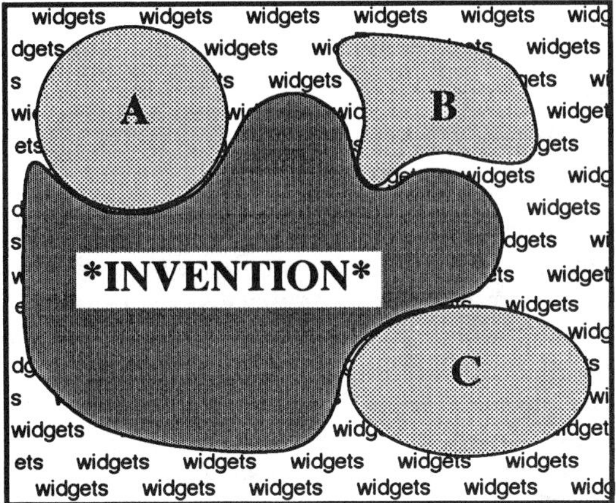

FIG. 2. Broad claim. In addition to the best mode, the claimed territory includes other modes of making the new widget, up to the boundaries of prior art widgets A, B, C.

the actual product if the broader claims are later found invalid.

2.1.2 Examination After a patent application is filed, the PTO conducts a search of prior United States patents, foreign patents, and publications (including the scientific literature) to determine whether the invention is novel and nonobvious. Frequently, an application is rejected and the applicant applies for reexamination. He may submit arguments in response to the examiner's stated reasons for rejection, with or without amending the application to address the reasons for the rejection. This process of examination, rejection, and amendment, known as "prosecution," may be repeated until eventual acceptance or a "final rejection" of the application. Procedures are available for appealing a rejection based on the substance of the application; rejections based on matters of form and procedure must be reviewed by petition to the Commissioner of Patents and Trademarks.

Occasionally an application claims the same invention as is claimed in another pending application or an issued patent. In that case, the PTO may conduct an interference proceeding to determine who is the first inventor and entitled to a U.S. patent on the invention.

If and when the PTO decides that a patent is to be granted, the applicant is sent a Notice of Allowance. If the issue fee is paid, the patent is granted. Thereafter, maintenance fees are due in the 4th, 8th, and 12th years to maintain the patent. If the fees are not paid, the rights lapse and the patented invention enters the public domain.

2.2 Enforcing Patent Rights

An issued patent is intellectual property having all the attributes of other forms of property. Most importantly, the owner has the right to exclude others from making, using, or selling the patented invention for a specified term (at present, 17 years in the United States; to compensate for delays in marketing caused by the regulatory process, the term may be extended for inventions that are subject to review by the Food and Drug Administration). At the end of that term, the invention enters the public domain.

2.2.1 Assignment vs License Although patent rights come into being only when a patent issues, patentable inventions may be treated as property even before issuance. For example, patent applications and notices of allowance may be viewed as property for income tax purposes. An inventor may assign rights to an invention at any time; in general, the further along the inventor has taken the application process, the more valuable his or her property rights in the invention.

The patentee may assign rights in a patent application or a patent, that is, sell all rights

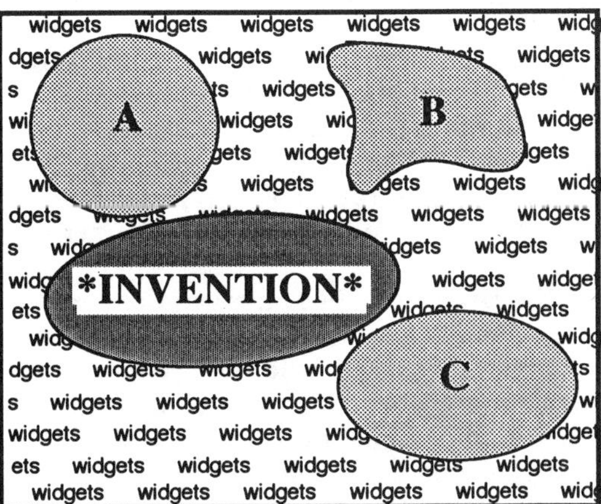

FIG. 3. Narrow claim. The claimed territory includes only the best mode of making the new widget; it does not reach to the boundaries of prior-art widgets A, B, C.

that he or she has in the patent. When he does so, he conveys the entire right to exclude others from making, using, or selling the invention. An assignment is equivalent to a sale of real estate. To pass legal title, an assignment must be in writing. While not essential, recording an assignment in the PTO establishes title against subsequent assignees who take from the same assignor.

A patentee may license the right to make, use, and sell the invention. A license may be exclusive or nonexclusive, for all or part of the claimed invention, for the remaining term of the patent or for a lesser term. It may be limited to a certain geographic area ("only in North Carolina and South Carolina"), provide for a fixed payment or royalties based on sales, or indeed any terms that suit the business needs of the parties.

2.2.2 Infringement Patent rights come into being only when the patent issues; others are free to make, use, and sell the invention before that time. The words "patent pending" may offer some psychological protection, since competitors do not know when the patent will issue, or what features of the product the inventor believes are patentable (U.S. patent applications are not available to the public), but afford no legal protection.

Anyone who makes, uses, or sells a patented invention during the term of the patent is a direct infringer. Anyone who actively encourages another to do so is an indirect infringer. Anyone who knowingly sells an item for which the only use is in connection with a patented invention is a contributory infringer.

Determining whether a patent has been infringed involves a thorough analysis of the claims and all the documents relating to the patent, including the original application, the prior art cited and discussed during prosecution of the application, written exchanges between the patent attorney or agent and the examiner, and the record of interference proceedings (if any). The test is whether the allegedly infringing product or process does the same work, in substantially the same way, and accomplishes substantially the same result as the patented invention.

Remedies for patent infringement include injunctions to prevent future infringement, damages to compensate the patent owner, attorneys' fees when appropriate, and costs. Damages may be trebled in the case of willful infringement (Miller and Davis, 1983).

2.3 International Filing

Patent laws vary from country to country, but certain features are common to all. Patents are granted for a specified period of time, and for novel and nonobvious inventions. In some countries, the invention must be "applicable to industry" (roughly equivalent to the "useful" requirement of U.S. law); some require that a patent, once granted, be exploited (i.e., the owner *must* make, use, or sell the invention in order to keep his rights in force).

Depending on the law of the country involved, applications may be filed by the actual inventor, in the name of the inventor, or by the assignee. Different types of patent protection may include patents of improvement, granted for an improvement of another patented invention; dependent patents, for inventions which cannot be practiced without infringing an earlier patent; and a type of patent known as an importation, revalidation, confirmation, or introduction patent, that gives the owner of a foreign patent local protection for a term limited to the unexpired term of the underlying patent. Inventor's certificates constitute recognition of inventorship and a grant of the rights and privileges provided by the national law; however, the exclusive rights are transferred to the State. Precautional patents provide that no patent relating to the subject matter of the patent will be granted without notice to the owner, who is given the opportunity to lodge objections within a stipulated term. Petty patents (for example, the German Gebrauchsmuster and the Japanese utility model) protect the shape or configuration of an object insofar as it serves some useful purpose (Baxter and Sinnott, 1992).

The United States is almost alone in having a "first to invent" patent system; most countries have a "first to file" system, where the first inventor to file a patent application is entitled to a patent, regardless of who actually invented first. In most countries, including the United States, the government has the right to make secret any patent application covering an invention which is of importance to the national defense. Once a secrecy order is given, the application proceeds through the normal examination stages, but is not published, nor is a patent granted, until the order is removed.

2.3.1 International Filing and the Patent Cooperation Treaty Before implementation of the Patent Cooperation Treaty (PCT) in 1978, patent applications were filed in each country where protection was desired. Although it does not provide for an "international patent" as such, the PCT has simplified the procedures that nationals or residents of member countries must follow when filing patent applications in other member countries. The PCT provides for centralized filing and searching of an international application (the "international stage"), followed by filing an application in each country where patent protection is desired (the "national stage"). Substantive examination of the application is carried out by each respective national patent office. The PCT preserves the right of priority of the Paris Union Convention and the Inter-American Convention, that is, an applicant who has filed in any member country may thereafter file a patent application for the same invention in any other member country within 12 months, with the effect that these later-filed applications will be treated as if they were filed on the first filing date.

In addition to the PCT route for obtaining international patent rights, the European Patent Convention (EPC) provides for centralized filing and examination leading to the grant of a "regional" patent, which matures into an individual patent for each of the member countries designated by the applicant. An applicant may, of course, file applications only in the desired countries.

Obtaining patent protection in African countries can be complicated by the variety of different patent systems involved. The African Intellectual Property Organization (OAPI) is responsible for implementing and applying common administrative procedures for the protection of industrial property (Baxter and Sinnott, 1992).

3. TRADEMARKS, SERVICE MARKS, CERTIFICATION MARKS, AND COLLECTIVE MARKS

Trademarks and service marks are believed to have originated with the medieval guilds, whose members affixed the mark of their guild to the goods they sold to identify it as the product of a particular craftsman or group of craftsmen. Like guild marks, modern trademarks serve two purposes: they identify the source of the goods, and they provide a marketing advantage for the owner of the mark (Miller and Davis, 1983).

A trademark is a symbol that allows a purchaser to identify goods or services that have been satisfactory in the past and those that have not. Service marks are identical to trademarks, except that they indicate the source of services, not goods. Trademarks and service marks encourage the production of quality products, and reduce the customer's costs of acquiring information about products and services. Most countries afford some form of trademark and service mark protection as an identification of the source and quality of goods and services (McCarthy, 1990). PTO trademark registration procedures are described in U.S. Patent and Trademark Office (1993).

A certification mark is any word, name, symbol, device, or combination thereof that is used to indicate that goods or services meet certain quality or regional origin standards. For example, the term "Roquefort" identifies a particular type of cheese that originates in a province of France; the "Underwriters Laboratories" and "Good Housekeeping" seals indicate that particular products meet the standards set by the owner of these marks. The mark must be one that buyers are likely to recognize as a symbol of guarantee or certification. Most titles or degrees such as Professor, Professional Engineer, MD, CPA, etc. are not used to certify the services performed by an individual, but rather to convey personal information about the individual performing those services.

Collective marks are used by members of a cooperative, an association, or a collective group or organization. While the organization itself neither sells goods nor performs services, its members do. Use of the mark allows members of the group to identify themselves as such (McCarthy, 1990).

3.1. Enforcement of Trademark Rights

Federal remedies for trademark infringement include injunctions, an accounting for profits, damages (including treble damages where appropriate), attorney's fees, and costs.

4. COPYRIGHTS

Copyright protection is available for original works of authorship. "Original" does not mean "new" or "novel"; it means only that the author did not copy the work from someone else. Consider *Romeo and Juliet*: an author could copyright a story about star-crossed lovers as long as it was not copied from Shakespeare.

4.1 Copyrightable Works

Copyrightable works include all original works of authorship "fixed in any tangible medium of expression, now known or later developed." Copyrightable works include literary works, musical and dramatic works, pantomimes and choreographic works, pictorial, graphic, and sculptural works, motion pictures and other audiovisual works, and sound recordings. Copyright protection does not extend to the idea embodied in the work, just the expression. Architectural plans, technical drawings, diagrams, and models are copyrightable (Nimmer and Nimmer, 1992).

Purely utilitarian works, including forms, contest rules, tests, and similar items are not protected by copyright. Since others are free to use the *idea* embodied in the work, there is very little protection available to the original author of, for example, an accounting system, a layout for a laboratory notebook or a method of doing business. Even though purely utilitarian designs are not copyrightable, fabric designs, jewelry designs, etc. may be copyrightable to the extent that the design is separable from its utilitarian aspect.

To the extent that a work is not "fixed," it is not copyrightable. For example, a dance routine is copyrightable only to the extent that it is filmed or written down in choreographic notation. Typefaces are usually considered to be noncopyrightable, perhaps because the subject matter—the alphabet—is limited and finite. However, the names of typefaces are subject to trademark protection.

Computer programs and databases are copyrightable to the extent that they incorporate the programmer's original expression of ideas, as distinguished from the ideas themselves (whether or not the idea is original). Microcode or firmware, which converts source code into object code, is itself copyrightable as a computer program (Nimmer and Nimmer, 1992). To the extent that a computer program lacks minimal originality or constitutes the only way of accomplishing a particular result, it is not copyrightable. Such a program embodies the idea itself rather than the expression of an idea.

4.2 Exclusive Rights

The owner of a copyright has the exclusive rights to do and to authorize the following: to reproduce the copyrighted work; to prepare derivative works based upon the work; to distribute copies or phonorecords of the work by sale, rental, lease, or lending; to perform the work publicly; and to display the work publicly. The right of reproduction allows the owner to exclude all others from reproducing the work in the form of a copy or phonorecord. Two exceptions allow radio, television, and cable television companies to record works temporarily for later transmission, and the owners of computer programs to make copies to utilize the program effectively.

The right of making derivative works allows the owner to prevent others from creating works based on the copyrighted work, such as translations, arrangements, dramatizations, films, abridgements, etc.

The right of distribution applies only to the first sale of a copy. Once a sale has occurred, the new owner can proceed to sell the work to another, but not copy the work.

4.2.1 Fair Use "Fair use" of a copyrighted work is not infringement. The concept of fair use balances the purpose and character of the use, the nature of the work, the amount and substantiality of the portion used, and the effect of the potential market for the work. One factor is whether the use is commercial in nature, or for nonprofit educational purposes. Educational purposes are traditionally a basis in favor of finding that the use is "fair." Under the current statute, limited photocopying by libraries and archives, and for classroom use, is allowed.

4.3 Ownership

A work may have a single author or joint authors. It may be a collection or a compilation, or a work for hire created for publication by another, either as an employee or an independent contractor. The ownership and duration of the copyright may hinge upon the classification of the work.

Ownership vests initially in the author or authors, or, in the case of a work made for hire, in the employer or other person for whom the work was prepared. Ownership may be transferred in whole or in part, may be bequeathed by will, or may pass as personal property. The individual rights may be transferred separately; thus, the right to make a movie and the right to translate a novel may be separated and transferred to different persons. Coauthors of copyrighted works are much as co-owners of other types of personal property; each has the right to use the property for his own purposes, but neither has the right to exclude the other from using it.

In the case of a single author, the copyright endures for the life of the author plus 50 years (70 years in most Berne Convention countries); for coauthors, for the life of the last surviving author plus 50 years. For works for hire, the term is 75 years from the date of first publication, or 100 years from the date of creation, whichever comes first.

The present U.S. copyright statute allows authors who have sold or otherwise transferred their rights to terminate the transfer by serving notice that they intend to do so during a five-year period beginning at the end of 35 years from the date of execution of the grant. This provision protects authors who have sold their works, but discover later in their careers that the works have become more valuable. If the author is dead, his or her spouse and children may exercise the right.

4.4 Notice, Registration, and Deposit

Under present U.S. law, an author has a copyright in a work as soon as it is fixed in a tangible medium of expression; a notice on published copies is no longer necessary to preserve the copyright. A notice of copyright consists of the symbol ©, the word "Copyright," or the abbreviation "Copr.," followed by the year of first publication and the name of the owner. For phonorecords, notice consists of the letter "P" in a circle, the year of first publication, and the name of the copyright owner.

Omission of notice does not invalidate the copyright, if

1. registration is obtained within five years, and
2. good faith efforts are made to affix notice to all remaining copies.

An innocent infringer incurs no liability for damages for infringing acts committed before receiving actual notice.

There is no need for examination: since others are free to create whatever they want as long as they do not copy the work, there is no need to insure that the work meets any particular standards. However, registration is a condition to filing a lawsuit for infringement.

Deposit of two copies of the work with the Library of Congress is required within three months of publication; however, deposit is not a condition of copyright protection. Registration requires deposit of at least one copy of a work. United States copyright registration procedures are described in U.S. Copyright Office (1984).

4.5 Enforcement of Copyrights

The owner of the copyright is entitled to sue for infringement. Available remedies include injunctions to prevent or restrain infringement, court-ordered licensing fees, and actual damages plus the defendant's profits. Costs and attorney's fees may also be recovered by the prevailing party. The defendant may establish his expenses to reduce the gross profit to a net amount. The plaintiff may choose to ask for statutory damages (a minimum of $250 and a maximum of $10,000) instead of actual damages, and a court may award as much as $50,000 statutory damages for willful infringement.

5. TRADE SECRETS

A trade secret has been described as any commercially valuable information that gives its owner an advantage over competitors that do not know it. A trade secret may be a machine or a process, an abstract idea such as a concept for a new product, or a compilation of data such as a list of customers. Trade secrets are confidential; however, not all confidential business information is a trade secret. Matters of common knowledge, or matters of general knowledge in a particular industry or trade, cannot be protected as trade secrets (McManis, 1988).

In some cases, an inventor may have the option of obtaining patent protection for an invention or relying on trade secret protection. A patent protects the invention for a period of time after it is disclosed; however, trade secret law protects it for as long as it remains secret.

Courts protect trade secrets under several theories. Some consider trade secrets to be a form of intellectual property, analogous to patents, copyrights, and trademarks. Others emphasize the breach of a confidential relationship, impropriety in ascertaining a secret, and use of a trade secret or idea known to have been improperly disclosed by another. For example, industrial espionage and bribery of a competitor's employees are an improper means of obtaining information, whereas "reverse engineering," discovery by one's own independent efforts or by inspection of a publicly distributed product, is usually considered to be proper.

6. APPENDIX

Professional societies involved in intellectual property law and related areas include the Section of Patent, Trademark and Copyright Law of the American Bar Association (ABA), the American Intellectual Property Law Association (AIPLA), the International Trademark Association (INTA, formerly the United States Trademark Association), and the World Intellectual Property Organization (WIPO). These groups sponsor numerous meetings, workshops, and courses throughout the year. Courses on various aspects of intellectual property are also offered by the Patent Resources Institute, Inc. (Charlottesville, VA), the Practicing Law Institute (New York, NY), and several universities.

Treatises, introductory texts, and practice handbooks dealing with intellectual property law, both U.S. and international, are available from Matthew Bender & Co., Inc. (New York, NY), Clark, Boardman, Callaghan (Deerfield, IL), and West Publishing Co. (St. Paul, MN), among others. The *Copyright Law Journal* and the *Trademark Reporter* deal with current developments in the areas of copyright and trademark law, respectively. The *United States Patents Quarterly* and the *Patent, Trademark & Copyright Journal*, both published by the Bureau of National Affairs, Inc. (Washington, DC), provide comprehensive coverage of U.S. court cases. Texts of newly issued patents are published in the *Official Gazette*. Information on court decisions, texts of recently issued patents, periodicals, news reports, etc. is also available on computer databases, including LEXIS (Mead Data Central) and WESTLAW (West Publishing Co.).

United States statutes dealing with intellectual property are found in Titles 15 (commerce and trade), 17 (copyright), and 35 (patents) of the United States Code (U.S.C.). Applicable regulations may be found in the Code of Federal Regulations (C.F.R.). Both are available online (LEXIS, WESTLAW). Trademarks, trade secrets, and unfair trade practices may also be the subject of state law.

The procedural aspects of the U.S. patent application process, federal trademark registration, and copyright registration are detailed in the *Manual of Patent Examining Procedure* (*MPEP*), *Trademark Manual of Examining Procedure* (*TMEP*), and *Compendium of Copyright Office Practices*, available from the Superintendent of Documents, U.S. Government Printing Office, Washington, DC 20402.

GLOSSARY

Assignment: A conveyance of all rights in property.

Certification Mark: A word, name, symbol, device, or any combination thereof, used to indicate that goods or services meet certain standards of quality or regional origin. The mark must be such that buyers are likely to recognize it as a symbol of guarantee or certification.

Claim: A numbered, single-sentence paragraph that provides a concise, formal definition of an invention; part of a patent application.

Collective Mark: A word, name, symbol, or device used by members of a cooperative, an association, a collective group, or an organization to identify themselves as such.

Copyright: A statutory right granted to the author of a literary or artistic work.

Design Patent: A patent granted for a new ornamental design.

EPO: European Patent Convention.

Infringement: A violation of the rights secured by a patent, trademark, or copyright; unauthorized use of the subject matter of a patent, trademark, or copyright.

Injunction: An equitable remedy granted by a court, forbidding a party to do some act, or restraining him in continuing to perform the act.

Intellectual Property: Includes patents, trademarks, copyrights, and trade secrets.

License: A conveyance of the right to make, use, or sell a patented article for a limited period or in a limited territory.

Notice of Allowance: A notice issued by the PTO, informing the applicant that a patent is to be granted.

OAPI: African Intellectual Property Organization.

Patent: A grant from a government that conveys to the patentee the exclusive right to make, use, and sell the invention for a specified term.

PCT: Patent Cooperation Treaty.

Plant Patent: A patent granted for a new, distinct, asexually reproduced variety of plant.

Property: The exclusive right to possess, enjoy, and dispose of a thing.

PTO: Patent and Trademark Office.

Service Mark: A word, name, symbol, device, or any combination thereof, adopted and used to identify services.

Specification: A detailed verbal description of an invention; part of a patent application.

Trade Secret: Any commercially valuable information that gives its owner an advantage over competitors that do not know it.

Trademark: A word, name, symbol, device, or any combination thereof, adopted and used to identify goods.

Utility Patent: A patent granted for a machine, process, composition of matter, or article of manufacture.

Works Cited

Baxter, J. W., Sinnott, J. P. (1992), *World Patent Law and Practice*, 3rd ed., New York: Matthew Bender & Co., Inc.

McCarthy, T. J. (1990), *McCarthy on Trademarks and Unfair Competition*, 3rd ed., Deerfield, IL: Clark, Boardman, Callaghan.

McManis, C. R. (1988), *Unfair Trade Practices*, St. Paul, MN: West Publishing Co.

Miller, A. R., Davis, M. H. (1983), *Intellectual Property—Patents, Trademarks and Copyright*, St. Paul, MN: West Publishing Co.

Nimmer, M. B., Nimmer, D. (1992), *Nimmer on Copyright. A Treatise on the Law of Literary, Musical and Artistic Property, and the Protection of Ideas*, New York: Matthew Bender, Inc.

Nolan, J. R, Connolly, M. J. (1979), *Black's Law Dictionary*, 5th ed., St. Paul, MN: West Publishing Co.

U.S. Copyright Office (1984), *Compendium of Copyright Office Practices*. Available from the Superintendent of Documents, U.S. Government Printing Office, Washington, DC 20402.

U.S. Patent and Trademark Office (1992), *Manual of Patent Examining Procedure (MPEP)*. Available from the Superintendent of Documents, U.S. Government Printing Office, Washington, DC 20402.

U.S. Patent and Trademark Office (1993), *Trademark Manual of Examining Procedure (TMEP)*. Available from the Superintendent of Documents, U.S. Government Printing Office, Washington, DC 20402.

Further Reading

Becker, S. A. (1992), *Patent Applications Handbook*, Deerfield, IL: Clark, Boardman, Callaghan.

Boorstyn, N. (1981), *Copyright Law*, Deerfield, IL: Clark, Boardman, Callaghan.

Cooper, I. P. (1982, revised annually), *Biotechnology and the Law*, Deerfield, IL: Clark, Boardman, Callaghan.

Epstein, M. A., Laurie, R. S., Elder, L. E. (1992), *International Intellectual Property: The European Community and Eastern Europe*, Englewood Cliffs, NJ: Prentice Hall.

Epstein, M. A. (1992), *Modern Intellectual Property*, Englewood Cliffs, NJ: Prentice Hall.

McManis, C. R. (1988), *Unfair Trade Practices*, St. Paul, MN: West Publishing Co.

Miller, A. R., Davis, M. H. (1983), *Intellectual Property—Patents, Trademarks and Copyright*, St. Paul, MN: West Publishing Co.

Nimmer, M. B., Geller, P. E. (1988, revised annually), *International Copyright Law and Practice*, New York: Matthew Bender, Inc.

Scott, M. D. (1992), *Scott on Computer Law*, Englewood Cliffs, NJ: Prentice Hall.

PERTURBATION METHODS

JAMES MURDOCK, *Department of Mathematics, Iowa State University, Ames, Iowa, U.S.A.*

	Introduction	245
1.	**Basic Concepts**	246
1.1	Perturbation Methods vs Numerical Methods	246
1.2	Perturbation Parameters	247
1.3	Perturbation Series	248
1.4	Uniformity	249
2.	**Nonlinear Oscillations and Dynamical Systems**	250
2.1	Simple Nonlinear Oscillators and Lindstedt's Method	251
2.2	Averaging Method for Single-Frequency Systems	253
2.3	Multiple-Scale Method for Oscillatory Problems	254
2.4	Multiple-Frequency Systems and Hamiltonian Systems	256
2.5	Perturbation of Stable Manifolds; Melnikov Functions	257
3.	**Initial and Boundary Layers**	258
3.1	Multiple-Scale Method for Initial-Layer Problems	258
3.2	Matching for Initial-Layer Problems	259
3.3	Boundary-Layer Problems	261
3.4	WKB Method	261
3.5	Fluid Flow	262
4.	**Perturbations of Spectra**	263
	Glossary	264
	Works Cited	266
	Further Reading	266

INTRODUCTION

In the most general sense, perturbation theory is concerned with the following question: a situation is given that admits of a mathematical description; one asks how the mathematical description changes when the situation is varied slightly, or "perturbed." Broadly speaking, there are two possibilities, either an abrupt qualitative change in the nature of the situation, or continuation of the situation with only small quantitative changes. Certain problems of the "abrupt" type, such as boundary-layer problems, and all problems of the "continuation" type, are usually classified together as *perturbation theory*. Other problems of the "abrupt" type are classified under the heading of *bifurcation theory* (see CHAOTIC PHENOMENA) and treated as a separate topic, although there are many points of contact between the two and it is not possible to discuss perturbation theory without occasional mention of bifurcations. The specific subject matter of the present article will be the following two topics.

1. A system of ordinary or partial differential equations is given, together with initial or boundary conditions. The system contains a small parameter, and is explicitly solvable when the parameter is zero. One asks how to construct approximate solutions (in explicit analytic form) when the parameter is small but nonzero; one asks for error estimates for these approximate solutions, and whether the approximate solutions exhibit the same qualitative behavior as the unknown exact solutions.
2. A matrix or linear transformation is given, depending on a small parameter. The eigenvalues and eigenvectors (or the Jordan normal form) are known when the parameter is zero, and one asks for approximate calculations of the eigenvalues or normal form when the parameter is small but nonzero.

The origins of perturbation theory lie in three classical problems, which continue to dominate the development of the theory to this day. These three root problems are the problem of planetary motion, the problem of viscous fluid flow past a wall, and the problem of spectra in quantum mechanics. The present article is structured in the same

3-527-28135-5/95/$5.00 + .50

threefold way: after an initial section presenting basic concepts common to the three areas, we take up in turn oscillatory problems and dynamical systems (Sec. 2), initial- and boundary-layer problems (Sec. 3), and spectra (Sec. 4). In an article of this length it is not possible to explain each method sufficiently to enable the reader to use it for new problems. The aim rather is to give an idea what each method is capable of, and to give references to more complete treatments. To conclude this introduction, we briefly describe the classical problem of each of the three basic types.

Isaac Newton showed that the inverse-square law of gravitational force implies that a single planet will move around the sun in an ellipse, satisfying Kepler's laws of planetary motion. The same law of gravity, however, implies that the several planets will exert attractive forces on each other, which will "perturb" their orbits. Newton did not completely compute the effect of these perturbations, and was of the opinion that God's intervention would be necessary from time to time to make small adjustments in the running of the solar system. Laplace did compute the perturbations, to a certain degree of approximation, and found that they were periodic; the solar system corrected itself. (Laplace is said to have told Napoleon that he did not need the hypothesis of God in his theory; the remark was probably directed at this specific point, rather than being a proclamation of atheism.) Laplace's techniques were laborious, and can be much simplified today by Hamiltonian mechanics; furthermore, he did not actually prove that the solar system is "stable" since his method, if carried to higher orders, does not necessarily converge. However, a great many of the ideas of modern perturbation theory originated here: variation of parameters, averaging, multiple scales, and the problems that in the twentieth century led to the Kolmogorov-Arnol'd-Moser theorem, the Nekhoroshev theorem, and other topics in dynamical systems theory.

In theory, a fluid that is viscous (even to the smallest degree) will adhere to the walls of any container (such as a pipe) in which it is flowing. However, at any reasonable distance from the walls, the fluid flows almost as if the wall were not there. In order to resolve this apparent paradox, L. Prandtl introduced the idea of a "boundary layer," a thin layer of fluid against the wall in which the fluid passes from rest to rapid motion. Here the "unperturbed" problem is the inviscid flow (which does not adhere to the wall), the "perturbation" is the small viscosity, and the effect of the perturbation is not a small correction of the motion but a quite drastic correction confined to a small region (the boundary layer). This example leads to the ideas of stretched or scaled coordinates, inner and outer solutions, and matching.

According to quantum mechanics, all observable quantities are the eigenvalues of operators on a Hilbert space. In simple problems, the Hilbert space will be finite dimensional and the operators are representable as matrices; in other cases they are partial differential operators. In modeling an atom, for instance, the eigenvalues will be related to the frequencies of light (the spectrum) emitted by the atom when electrons jump from one shell to another. These frequencies can be perturbed if, for instance, the atom is placed in a weak magnetic field. Mathematically, the resulting problem is to determine the changes in the "spectrum" (the set of eigenvalues) of a matrix or other operator when the operator is slightly perturbed. One striking difference between this problem and the first two is that quantum mechanics is entirely a linear theory, whereas in both the dynamical systems problems and the boundary-layer problems, nonlinearities can play a crucial role.

1. BASIC CONCEPTS

1.1 Perturbation Methods vs Numerical Methods

When one is confronted with a system of differential equations that is not explicitly solvable, an approximate solution method is required. The most useful approximate methods are numerical methods (implemented on a digital computer) and perturbation methods. Perturbation methods are usable only if the system is "close" in some sense to an explicitly solvable system—that is, if the system would become solvable if certain small changes were made, such as deleting small terms or averaging some term over a rapidly rotating angle. In such a case, perturbation theory takes the solution of the simplified problem as a "zero-th order approximation" that can be improved. The result obtained is an approximate solution in the form of a formula. Numerical methods operate without the restriction that the problem be nearly solvable, but give a solution in the form of numerical tables or graphs. The advantage of a formula is that one can see by inspection the manner in which each variable affects the solution. However, since both types of solution are approximate, it is often best to verify a perturbation solution by comparing it with numerical ones (or directly with experimental data), especially in cases where a mathematical proof of the validity of the perturbation solution is missing.

1.2 Perturbation Parameters

Mathematical models of physical phenomena typically contain several variables, which are divided into "coordinates" (of space or time) and "parameters." A spring/mass system with a cubic nonlinearity, for instance, will contain position and time coordinates as well as parameters for the mass and for the coefficients of the linear and cubic terms in the restoring force. The first step in preparing such a system for perturbation analysis is to nondimensionalize these variables. The second step is to look for solvable special cases that can serve as starting points for approximating the solution of nearby cases. Most often, these solvable cases will be obtained by setting some of the parameters equal to zero. For instance, the forced and damped nonlinear oscillator, given by

$$\ddot{y} + C\dot{y} + k^2y + Ay^3 = B\cos\omega t, \tag{1}$$

becomes easily solvable if $A = B = C = 0$; it is still solvable if only $A = 0$, but not quite as simply. It is therefore plausible to look for approximate solutions by perturbation theory if A, B, and C are small, and it may also be possible if only A is small.

Suppose that we choose to investigate the case when A, B, and C are small. Ideally, we could treat these as three small independent parameters, and considerable work is now being devoted to the investigation of such "multiparameter" perturbation methods, especially in the context of bifurcation theory (see Golubitsky and Schaeffer, 1985). However, most classical perturbation methods are developed only for single-parameter problems. Therefore, it is necessary to *make a choice* as to how to reduce Eq. (1) to a single-parameter problem. The simplest way is to write $A = a\epsilon$, $B = b\epsilon$, and $C = c\epsilon$, obtaining

$$\ddot{y} + \epsilon c\dot{y} + k^2y + \epsilon ay^3 = \epsilon b\cos\omega t. \tag{2}$$

We appear to have added a parameter instead of reducing the number, but now a, b, and c are regarded as constants, whereas ϵ, which is called the *perturbation parameter*, is taken as a small, but variable, quantity; typically we expect the perturbation solution to have the form of a power series in ϵ. We have in effect chosen to investigate a particular *path* leading from the origin in the space of variables A, B, and C. It is at once clear that other paths are possible; for instance

$$\ddot{y} + \epsilon^2c\dot{y} + k^2y + \epsilon ay^3 = \epsilon b\cos\omega t. \tag{3}$$

One might choose Eq. (3) over Eq. (2) if the goal is to investigate systems in which the damping is *extremely* small, small even compared to the cubic term and the forcing. But it is not clear, without experience, what the best formulation will be for a given problem. As an example of the role of experience, one might expect (knowing something about resonance in the linear case) that the results of studying Eq. (2) will be different if ω is close to k than if it is far away. But how do you express mathematically the idea that "ω is close to k"? Recalling that the only parameter available to express "smallness" is ϵ, the best answer turns out to be

$$\omega^2 = k^2 + \epsilon d, \tag{4}$$

where d is another constant. Substituting Eq. (4) into Eq. (2) leads to the "correct" formulation of the near-resonance problem. One can see that the setting up of perturbation problems is sometimes an art rather than a science. In the present article, we will for the most part assume that a perturbation problem has been chosen. Mathematical analysis of the problem may then suggest the use of *stretched* or otherwise *rescaled* variables, which in fact amount to a modification of the initial perturbation problem.

In recent years, a further consideration has come to the fore regarding the choice of parameters in a mathematical model. Physical considerations may have led to a model such as Eq. (1) above, and yet we know that this model is not exactly correct; there may, for instance, be very small nonlinearities other than the cubic term in the restoring force, or nonlinearities in the damping, or additional harmonics in the forcing. How many such effects should be included in the model? In the past, it was simply a matter of trial and error. But in certain cases, there now exists a mathematical theory that is able to determine just which additional small terms might make a *qualitative* (rather than just a tiny quantitative) difference in the solution. In these cases, it is sometimes best to *add all such significant small terms to the equation before attempting the solution, even if there is no evident physical reason for doing so*. The process of adding these additional terms is called finding the *universal unfolding* of the system. The advantage is that the universal unfolding will account for all possible qualitative behaviors that may be observed as a result of unknown perturbations. For instance, in the past, many bifurcations were not observed to occur exactly as predicted. They were called *imperfect bifurcations*, and each situation required that the specific perturbation responsible for the imperfection be discovered and incorporated into the model. Now it is often possible to determine all possible imperfections in advance by examining the universal unfolding of the original model. See Golubitsky and Schaeffer (1985).

1.3 Perturbation Series

Let us suppose that a perturbation problem has been posed, and let the exact solution (which we wish to approximate) be denoted $u(x,\epsilon)$. Here ϵ is the (scalar) perturbation parameter, x is a vector consisting of all other variables in the problem including coordinates and other parameters, and u is the quantity being solved for. [In the case of Eq. (2) above, $u = y$ and $x = (t,a,b,c,\omega)$, or, if Eq. (4) is used, then $x = (t,a,b,c,d)$.] The simplest form in which to seek an approximation is that of a (truncated) *power series* in ϵ:

$$u(x,\epsilon) \cong u_0(x) + \epsilon u_1(x) + \cdots + \epsilon^k u_k(x). \tag{5}$$

There are times when this is insufficient and we require a *Poincaré series*,

$$u(x,\epsilon) \cong \delta_0(\epsilon)u_0(x) + \delta_1(\epsilon)u_1(x) + \cdots + \delta_k(\epsilon)u_k(x), \tag{6}$$

where each δ_i is a monotone function of ϵ defined for $\epsilon > 0$ satisfying

$$\lim_{\epsilon\to 0} \frac{\delta_{i+1}(\epsilon)}{\delta_i(\epsilon)} = 0; \tag{7}$$

such functions δ_i are called *gauges*. [Of course, a Poincaré series reduces to a power series if $\delta_i(\epsilon) = \epsilon^i$.] Finally, there are times when not even a Poincaré series is sufficient and a *generalized series* is needed:

$$u(x,\epsilon) \cong \delta_0(\epsilon)u_0(x,\epsilon) + \delta_1(\epsilon)u_1(x,\epsilon) + \cdots + \delta_k(\epsilon)u_k(x,\epsilon). \tag{8}$$

With such a series it might appear that we could delete the gauges, or rather assimilate them into the u_i, since these are now allowed to depend upon ϵ; but the intention is that the dependence of u_i on ϵ should not affect its order of magnitude. An example is the following two-term generalized series in which the vector x consists only of the time t:

$$u(t,\epsilon) \cong \sin(1 + \epsilon)t + \epsilon \cos(1 + \epsilon)t. \tag{9}$$

Here, $u_0(t,\epsilon) = \sin(1 + \epsilon)t$ and $u_1(t,\epsilon) = \cos(1 + \epsilon)t$; the dependence of these coefficients upon ϵ modifies their period but not their amplitude, and the second term still has the order of magnitude of its gauge ϵ.

Notice that we have written only truncated series in the previous paragraph. While most perturbation methods allow in principle for the computation of infinite series, these series very often do not converge, and in practice it is impossible to calculate more than a few terms. The type of accuracy that we hope for in a perturbation solution is not convergence (improvement in accuracy as the number of terms increases), but rather *asymptotic validity*, which means improvement in accuracy as ϵ approaches zero. To explain this concept, let us consider a generalized series

$$u(\epsilon) \cong \delta_0(\epsilon)u_0(\epsilon) + \delta_1(\epsilon)u_1(\epsilon) + \cdots + \delta_k(\epsilon)u_k(\epsilon), \tag{10}$$

which contains no variables other than ϵ. We will say that this series is an *asymptotic approximation* to $u(\epsilon)$ if

$$u(\epsilon) = \delta_0(\epsilon)u_0(\epsilon) + \delta_1(\epsilon)u_1(\epsilon) + \cdots + \delta_k(\epsilon)u_k(\epsilon) + R(\epsilon), \quad (11)$$

where the remainder or error $R(\epsilon)$ satisfies a bound of the form

$$|R(\epsilon)| \leq c\delta_{k+1}(\epsilon) \quad (12)$$

for some constant $c > 0$ and some gauge δ_{k+1} that approaches zero more rapidly than δ_k as $\epsilon \to 0$. [If the series is vector-valued, then $|R(\epsilon)|$ denotes a vector norm.] Equation (12) is abbreviated with the "big-oh" notation

$$R(\epsilon) = O(\delta_{k+1}(\epsilon)). \quad (13)$$

The series (10) is called *asymptotically valid*, or *an asymptotic series*, if it is an asymptotic approximation (in the above sense) and if in addition every truncation of series (10) is also an asymptotic approximation, with the error being "big-oh" of the first omitted gauge. The case in which $u(x,\epsilon)$ depends upon the variable x in addition to ϵ is discussed in the following section on uniformity.

As a technical matter, any bound such as Eq. (12) is not intended to hold for all ϵ, but only for ϵ in some interval $0 \leq \epsilon \leq \epsilon_0$. A perturbation solution is never expected to be valid for large values of the perturbation parameter, and the meaning of "large" is relative. In this article, we will not continue to mention ϵ_0, but it is always lurking in the background.

To conclude this discussion of types of series in perturbation theory, it should be mentioned that Fourier series arise frequently in dealing with oscillatory problems, but a Fourier series is never a perturbation series. Rather, if a perturbation series such as $u(t,\epsilon) \cong u_0(t) + \epsilon u_1(t)$ depends periodically on time t, then the coefficients $u_k(t)$ may be expressed as Fourier series in t.

1.4 Uniformity

In the previous section, we have defined asymptotic validity of a perturbation series for a function $u(\epsilon)$ depending only on ϵ. This is adequate for a problem such as finding a root of a polynomial (supposing that the polynomial contains a perturbation parameter ϵ), because the root is a single number. But for most perturbation problems (such as differential equations), the solution is a function of space and/or time coordinates, and possibly various parameters, in addition to ϵ. For such problems, the previous definition of asymptotic validity is insufficient.

Let us return to the generalized series (8), and denote the error by $R(x,\epsilon)$. Now we may require that *for each fixed* x this error is of order $\delta_{k+1}(\epsilon)$:

$$|R(x,\epsilon)| \leq c(x)\delta_{k+1}(\epsilon). \quad (14)$$

Notice that the "constant" c here may change when we move to a new point x. In this case, we say that the error is *pointwise* of order δ_{k+1}.

Alternatively, we can choose a domain D of the variable x (remember that x may have several components, so D may be a subset of a vector space) and require that the error be *uniformly* of order δ_{k+1} for all x in D:

$$|R(x,\epsilon)| \leq c\delta_{k+1}(\epsilon). \quad (15)$$

Here, the constant c is truly constant. In this case, we say $R(x,\epsilon) = O(\delta_{k+1}(\epsilon))$ *uniformly in* x *for* x *in* D. If every truncation of a perturbation series is uniformly of the order of the first omitted gauge, we say that the series is *uniformly valid*. (In the last sentence we neglected to say "in D," but it is important to remember that such an expression has no meaning unless the domain D is understood.) Obviously, uniform asymptotic validity is stronger than pointwise validity, and it is safe to say that every method used in perturbation theory has been introduced in order to gain uniform validity for a problem for which previous methods only gave pointwise validity.

Now the definition of uniform validity calls for an estimate of the error of an approximation, and such an estimate is a difficult thing to come by. It would be convenient if there were an easier test for uniformity. In actuality there is not. However, it is possible to obtain a simple *necessary* (but not sufficient) condition for a series to be uniformly valid. Namely, it is not difficult to show that if a series (8) is uniformly valid in a domain D, then each coefficient $u_k(x)$ with $k \geq 1$ is bounded on D; that is, there exist constants

c_k such that

$$|u_k(x,\epsilon)| \le c_k \tag{16}$$

for x in D. If this is true, we say that the series (8) is *uniformly ordered*. We have already encountered the concept of a uniformly ordered series when discussing series (9) above. A uniformly ordered series is one in which each term after the first is of no greater order than is indicated by its gauge. It is easy to inspect a perturbation series, once it has been constructed, and determine whether it is uniformly ordered. If not, the series is called *disordered*, and it cannot be uniformly valid. On the other hand, if it is uniformly ordered, it does not follow that it is uniformly valid, since one has done nothing toward estimating the error. Almost all textbooks are misleading on this point, since they almost invariably claim to be showing the uniform validity of a series when in fact they are only showing that it is uniformly ordered. However, if a perturbation series is constructed on the basis of good intuitive insight into a problem, and if it is uniformly ordered, then it frequently turns out to be uniformly valid as well. (A very elementary example in which this is *not* the case is given in Sec. 2.1 below.)

With regard to uniform ordering, Poincaré series occupy a special place. Recall that a series is a Poincaré series if its coefficients do not depend on ϵ; see series (6). In this case, if the domain D is compact and the coefficients $u_k(x)$ are continuous, then the coefficients are automatically bounded and the series is uniformly ordered (but still not automatically uniformly valid). However, even a Poincaré series may fail to be uniformly ordered if D is not compact or if D is allowed to depend on ϵ. We have not considered this latter possibility until now, but in fact, in many problems one must consider domains $D(\epsilon)$ that depend upon ϵ. An important example is a boundary layer, a thin domain near the boundary of some larger region, whose thickness depends on ϵ. For such domains, the definitions (15) and (16) of uniform validity and uniform ordering are the same, except that they are to hold for all x in $D(\epsilon)$; that is, for each value of ϵ, Eqs. (15) or (16) hold for a different range of x.

It is now possible to explain one of the principal divisions in the subject of perturbation theory, the division of perturbation problems into *regular* and *singular*. A perturbation problem is regular if there exists a Poincaré series that is uniformly valid on the intended domain; it is singular if it is necessary to use a generalized series in order to obtain a uniformly valid solution on the intended domain. This is the only correct definition. Many textbooks give partial definitions such as "a perturbation problem for a differential equation is singular if ϵ multiplies the highest derivative." Such a definition, which refers only to the differential equation without stating an intended domain, cannot be correct. The presence of an ϵ multiplying the highest derivative does, of course, affect the domain on which a problem can be regular; we will see below that a problem such as $\ddot{u} + u + \epsilon u^3 = 0$ is regular on any fixed interval $[0,T]$ but singular on an expanding interval $[0,1/\epsilon]$, while a problem such as $\epsilon\ddot{u} + (t^2 + 1)\dot{u} + u = 0$ is regular only on a shrinking interval $[0,\epsilon]$ and singular on a fixed interval.

2. NONLINEAR OSCILLATIONS AND DYNAMICAL SYSTEMS

In this section we discuss the major perturbation methods used in solving nonlinear ordinary differential equations (dynamical systems) that reduce, when the small parameter vanishes, either to a linear oscillator, to an integrable Hamiltonian system, or to a system with a computable homoclinic orbit. These methods include Lindstedt's method (or the method of strained coordinates), the methods of multiple scales and averaging, and Melnikov functions (or integrals). It is easiest to approach these methods by way of specific examples. We begin each example by describing the qualitative behavior of the solutions, which can be determined by other methods from dynamical systems theory. It will be seen that a knowledge of the qualitative behavior provides the easiest way to select a suitable perturbation method. However, in facing a new problem, the qualitative behavior may be unknown. One may then choose either to study the problem by geometrical methods (dynamical systems theory) first, or to use perturbation methods in an exploratory fashion, hoping to discover the correct methods and the qualitative behavior simultaneously. It is important to bear in mind that very frequently, especially in nonlinear problems, a perturbation solution may appear correct but may predict qualitative features (such as periodicity, stability, or presence of chaos) erroneously. Many such faulty results appear in the literature, and more are published every year. As a general rule, qualitative results obtained from perturbation solutions (or any other approximate solutions) should be taken as conjectural, even if supported by numerical evidence, until they have been proved by rigorous mathematical methods.

2.1 Simple Nonlinear Oscillators and Lindstedt's Method

As a first example, we consider the "hard" nonlinear spring or unforced Duffing equation, given by

$$\ddot{u} + u + \epsilon u^3 = 0 \tag{17}$$

for $\epsilon > 0$. This can be expressed as a first-order system in the $(u,\dot{u})$ phase plane in the form $\dot{u} = v$, $\dot{v} = -u - \epsilon u^3$. In the phase plane, the orbits are closed curves surrounding the rest point at the origin; this may be seen from the conservation of energy. (The "soft" spring with $\epsilon < 0$ behaves differently.) Since the solutions of Eq. (17) with any initial conditions $u(0) = a$, $\dot{u}(0) = b$ are smooth functions of ϵ, they may be expanded as Taylor series having the form (if we retain two terms)

$$u(t,\epsilon) \cong u_0(t) + \epsilon u_1(t). \tag{18}$$

The coefficients may be determined by substituting series (18) into Eq. (17), expanding the u^3 term, dropping powers of ϵ higher than the first, and setting each order in ϵ separately equal to zero. This gives two *linear* equations, which can be solved (recursively) for u_0 and u_1. The result, for $b = 0$ (and it is enough to consider this case because every solution is at rest momentarily when its amplitude is at its maximum), is

$$u(t,\epsilon) \cong a \cos t - \epsilon \frac{a^3}{32}(\cos t + 12t \sin t - \cos 3t). \tag{19}$$

Upon examining Eq. (19) for uniform ordering, we discover that all functions of t appearing there are bounded for all t except for $12t \sin t$, which becomes unbounded as t approaches infinity. This is an example of a so-called "secular" term, one which grows over the "ages" (*saecula* in Latin). We conclude from this that Eq. (19) is uniformly ordered for t in any finite interval $D = [0,T]$ but not for $D = [0,\infty)$. According to the general principles discussed in section 1, this shows that Eq. (19) is not uniformly valid on $[0,\infty)$, and it *leads us to suspect, but does not prove*, that Eq. (16) is uniformly valid on $[0,T]$. In the present case, this conjecture is correct. If the intended domain D for the solution of Eq. (17) is a finite interval, then we have obtained a uniform approximation in the form of a Poincaré series, and the problem is a regular one. If a solution valid for a longer time is desired, the problem will prove to be singular.

In an effort to extend the validity of the solution, we recall that the actual solutions of Eq. (17) are periodic, whereas Eq. (19) is not. The problem is that the period of the exact solution depends upon ϵ, and there is no way that a Poincaré series can have such a period since the coefficients are not allowed to depend on ϵ. To remedy this, we seek to approximate the (unknown) frequency of the solution in the form

$$\nu(\epsilon) \cong \tilde{\nu}(\epsilon) = \nu_0 + \epsilon\nu_1 + \epsilon^2\nu_2, \tag{20}$$

with the solution itself being represented as

$$u(t,\epsilon) \cong u_0(\tilde{\nu}(\epsilon)t) + \epsilon u_1(\tilde{\nu}(\epsilon)t), \tag{21}$$

which is now a generalized series. Notice that we have carried series (20) to one more order than series (21). Now we substitute series (20) and (21) into Eq. (17) and attempt to determine ν_0, u_0, ν_1, u_1, and ν_2 recursively, in that order, in such a way that each u_i is periodic. The mechanics of doing this is explained in the next paragraph; the result is

$$u(t,\epsilon) \cong a \cos t^+ + \epsilon\left(-\frac{1}{32}a^3 \cos t^+ + \frac{1}{32}a^3 \cos 3t^+\right), \tag{22}$$

where

$$t^+ = \tilde{\nu}(\epsilon)t = \left(1 + \epsilon\frac{3}{8}a^2 - \epsilon^2\frac{21}{256}a^4\right)t. \tag{23}$$

Examining series (22), we see that it is uniformly ordered for all time, since the coefficients are bounded (there are no secular terms). One might, therefore, conjecture that the solution is uniformly valid for all time, *but this would be incorrect*! (This example is an excellent warning as to the need for proofs of validity in perturbation theory.) The difficulty is that t^+ uses the approximate frequency $\tilde{\nu}(\epsilon)$ in place of the exact frequency $\nu(\epsilon)$; there is no escape from this, as the exact frequency remains unknown. Therefore, series (22) gradually gets out of phase with the exact solution. The reason for taking series (20) to one higher order than series (21) is to minimize this effect. It can be shown that series (22) is uniformly valid on the *expanding interval* $D(\epsilon) = [0,1/\epsilon]$. (This is our first example of a domain that depends on ϵ, as discussed in Sec. 1.4.) If the intended domain is such an expanding interval, then series (22) provides a uniformly valid generalized series, and the problem is seen to be singular. (If the intended domain is all t, the problem is simply impossible to approximate asymptotically.)

In order to complete the example in the last paragraph, we must indicate how to obtain series (22) and (23). The easiest way is to substitute $\tau = \nu(\epsilon)t$ into Eq. (17) to obtain

$$\nu(\epsilon)^2 \frac{d^2u}{d\tau^2} + u + \epsilon u^3 = 0. \tag{24}$$

Then substitute series (20) and (21) into series (24), expand, and set the coefficient of each power of ϵ equal to zero as usual. It is easy to find that $\nu_0 = 1$ and $u_0 = a \cos\tau$ (which in the end becomes $a \cos t^+$ because we do not know the exact frequency ν). The crucial step arises when examining the equation for u_1, which is (writing $' = d/d\tau$)

$$\begin{aligned} u_1'' + u_1 &= -a^3 \cos^3\tau + 2\nu_1 a \cos\tau \\ &= (-\tfrac{3}{4} a^3 + 2\nu_1 a) \cos\tau \\ &\quad - \tfrac{1}{4} a^3 \cos 3\tau. \end{aligned} \tag{25}$$

From the Fourier series expansion [the second line of series (25)], we see that the term in $\cos\tau$ will be resonant with the free frequency, and hence produce unbounded (secular) terms in u_1, unless the coefficient of $\cos\tau$ vanishes. In this way, we conclude that

$$\nu_1 = \tfrac{3}{8} a^2, \tag{26}$$

and, after deleting the $\cos(\tau)$ term from series (25), we solve it for u_1. This procedure is repeated at each subsequent stage.

The previous example is typical of *unforced conservative* oscillators, in which every solution (at least in a certain region) is periodic. There are two additional classes of oscillators that must be mentioned, although we cannot give them as much space as they deserve: self-excited oscillators and forced oscillators.

The standard example of a self-excited oscillator is the *Van der Pol equation*,

$$\ddot{u} + \epsilon(u^2 - 1)\dot{u} + u = 0. \tag{27}$$

Instead of a region in the phase plane filled with periodic solutions, this equation has a single periodic solution (limit cycle) for $\epsilon > 0$. The Lindstedt method, described above, can be used to approximate the periodic solution, but must be modified slightly: the initial condition can no longer be assigned arbitrarily, because to do so will in general yield a nonperiodic solution for which the Lindstedt method fails. (These solutions can be found by averaging or multiple scales; see below.) Suppose that the limit cycle crosses the positive x axis (in the phase plane) at $(a(\epsilon),0)$ and has frequency $\nu(\epsilon)$. Then, the solution is sought in the form of series (20) and (21) together with an additional expansion $a(\epsilon) = a_0 + \epsilon a_1 + \ldots$; the coefficients u_i, ν_i, and a_i are determined recursively, choosing ν_i and a_i so that no secular terms arise in u_{i+1}. This example shows the effect of the dynamics of a system on the correct formulation of a perturbation problem.

The general nearly linear, periodically forced oscillator can be written

$$\ddot{u} + u = \epsilon f(u,\dot{u},\omega t), \tag{28}$$

where $f(u,v,\theta)$ is 2π-periodic in θ; thus, the period of the forcing is $2\pi/\omega$. The dynamics of Eq. (28) can be complicated, and we limit ourselves to one type of periodic solution, the *harmonic* oscillations, which are *entrained* by the forcing so that they have the same frequency ω; these harmonic solutions occur for ϵ small and ω close to 1 (the frequency of the solutions when $\epsilon = 0$). Since the problem contains two parameters (ϵ and ω) and we are limited to one-parameter methods, it is necessary to express the statement "ω is close to 1" in terms of the perturbation parameter ϵ. [A study using two independent parameters would uncover the phenomenon of "resonance horns" or "resonance tongues." See Murdock (1991), Sec. 4.5.] It turns out that an efficient way to do so is to write

$$\omega^2 = 1 + \epsilon\beta, \tag{29}$$

where β is a new parameter that is considered fixed (not small). With Eq. (29), Eq. (28) can have one or more isolated periodic solutions with unknown initial conditions $(a(\epsilon),b(\epsilon))$. (We can no longer assume $b = 0$.) On the other hand, the frequency of the solution this time is not unknown but equals ω. So the Lindstedt method undergoes another modification dictated by the dynamics: u, a, and b are expanded in ϵ and solved recursively, choosing the coefficients of a and b to eliminate secular terms from u. In contrast with the previous cases, there is no accumulating phase error since the frequency is known, and the perturbation approximations are uniformly valid for all time.

2.2 Averaging Method for Single-Frequency Systems

All of the systems discussed in the previous section, and a great many others, can be expressed in the form

$$\dot{x} = \epsilon f(x,t,\epsilon), \tag{30}$$

where $x = (x_1,\ldots,x_n) \in \mathbb{R}^n$ and where f is 2π-periodic in t. We will refer to Eq. (30) as *periodic standard form*. (A more general form is discussed in Sec. 2.4.) Once a system has been expressed in periodic standard form, its solutions may be approximated by the method of averaging. This method not only locates the periodic solutions (and proves their existence as well), but also determines their stability or instability and approximates the transient (nonperiodic) solutions. The method of averaging has been rediscovered many times and exists (with slight variations) under a variety of names, including the following: method of Van der Pol; method of Krylov-Bogoliubov-Mitropolski (KBM method); method of slowly varying amplitude and phase; stroboscopic method; Struble's method; Von Zeipel's method (in the Hamiltonian case); method of Lie series or Lie transforms. Some of the differences in these "methods" pertain to how the original system is put into periodic standard form, and others to details about how the near-identity transformations (described below) are handled.

To illustrate how a system may be put into periodic standard form, consider the Van der Pol Eq. (27). This may be written as a system

$$\begin{aligned} \dot{u} &= v \\ \dot{v} &= -u + \epsilon(1 - u^2)v. \end{aligned} \tag{31}$$

Then, rotating polar coordinates (r,φ) may be introduced by setting $u = r\cos(\varphi - t)$, $v = r\sin(\varphi - t)$. The result is

$$\begin{aligned} \dot{r} &= \epsilon[1 - r^2\cos^2(\varphi - t)]\, r\sin^2(\varphi - t) \\ \dot{\varphi} &= \epsilon[1 - r^2\cos^2(\varphi - t)]\sin(\varphi - t)\cos(\varphi - t), \end{aligned} \tag{32}$$

which is in periodic standard form with $x = (r,\varphi)$. The same result may be achieved by seeking a solution of Eq. (27) by variation of parameters, in the form $u = r\cos(\varphi - t)$, where r and φ are variables, and imposing the requirement that $\dot{u} = r\sin(\varphi - t)$; the motivation for these choices is that with r and φ constant, these solve Eq. (27) for $\epsilon = 0$. However, it is crucial to understand that the transformation to periodic standard form is merely a change of coordinates, not an assumption about the nature of the solutions.

In its crudest form, the method of averaging simply consists in replacing Eq. (30) by

$$\dot{z} = \epsilon\bar{f}(z), \tag{33}$$

where

$$\bar{f}(z) = \frac{1}{2\pi}\int_0^{2\pi} f(z,t,0)\,dt. \tag{34}$$

System (33) is easier to solve than Eq. (30), because it is autonomous. The form of system (33) can be motivated by the fact that for small ϵ, x in Eq. (30) is slowly varying and therefore nearly constant over one period; therefore, to a first approximation, we might hold x constant while integrating over one period in Eq. (30) to find the "average" influence due to f. But this sort of motivation gives no idea how to estimate the error or to extend the method to higher-order approximations. A much better procedure is to return to Eq. (30) and perform a near-identity change of variables of the form

$$x = y + \epsilon u_1(y,t,\epsilon), \tag{35}$$

where u is a periodic function of t, which is to be determined so that the transformed equation has the form

$$\dot{y} = \epsilon g_1(y) + \epsilon^2 \hat{g}(y,t,\epsilon), \tag{36}$$

where g_1 is independent of t. It turns out that such a transformation is possible only if we take $g_1 = \bar{f}$; by doing so, series (33) can be obtained from Eq. (36) simply by deleting the ϵ^2 term. When it is formulated in this way, the entire method of averaging is seen to consist of nothing but coordinate changes [first into periodic standard form, then into the form of Eq. (36)], followed by truncation; it is only the truncation that introduces error, and this error is easy to estimate. It is also clear how to proceed to higher orders; simply replace Eq. (35) by

$$x = y + \epsilon u_1(y,t,\epsilon) + \cdots + \epsilon^k u_k(y,t,\epsilon) \tag{37}$$

and Eq. (36) by

$$\dot{y} = \epsilon g_l(y) + \cdots + \epsilon^k g_k(y) + \epsilon^{k+1}\hat{g}(y,t,\epsilon); \tag{38}$$

the averaged equations are obtained by deleting $\hat{g}$. It is, of course, necessary to determine the u_i and g_i recursively in such a way that the u_i are periodic and the g_i are independent of t. This is where the technical details of various versions of averaging come into play. The most recent and most effective innovation here is the use of Lie series as a bookkeeping device for handling near-identity transformations. See Nayfeh (1973).

The final conclusion of the method of averaging is that if Eq. (38) is truncated to order ϵ^k and solved, and the solutions are put back into the transformation (37), the resulting approximate solutions of Eq. (30) will differ from the exact solutions (with the same initial condition) by $O(\epsilon^k)$ during a time interval of length $O(1/\epsilon)$. Under special conditions it is possible to obtain results on still longer intervals of time.

2.3 Multiple-Scale Method for Oscillatory Problems

The earliest perturbation problem, that of planetary motion, illustrates the appeal of the idea of multiple scales. A single planet under the influence of Newtonian gravitation would travel around the sun in an elliptic orbit characterized by certain quantities called the "Keplerian elements" (the eccentricity, major axis, and certain angles giving the position of the ellipse in space). Since the actual (perturbed) motion of the planets fits this same pattern for long periods of time, it is natural to describe the perturbed motion as "elliptical motion with slowly varying Keplerian elements." A simpler example would be a decaying oscillation of the form $e^{-\epsilon t}\sin t$, which could be described as a periodic motion with slowly decaying amplitude. Solutions of nonlinear oscillations obtained by the method of averaging frequently have this form, in which time appears both as t and as ϵt, the latter representing slow variation; sometimes other combinations such as $\epsilon^2 t$ appear. This leads to the question of whether it is possible to arrive at such solutions more directly, by postulating the necessary time scales in advance. The "method of multiple scales" is the result of such an approach, and is perhaps the most flexible general method in perturbation theory, since it is applicable both to oscillatory problems (such as those covered by averaging) and to boundary-layer problems (discussed below). However, this very flexibility is also its drawback, because the "method" exists in an immense variety of *ad hoc* formulations adapted to particular problems. There are two-scale methods using fast time t and slow time ϵt; two-scale methods using strained fast time $(\nu_0 + \epsilon\nu_1 + \epsilon^2\nu_2 + \ldots)t$ (similar to the strained time in the Lindstedt method) and slow time ϵt; multiple-scale methods using $t, \epsilon t, \epsilon^2 t, \ldots, \epsilon^n t$; and methods using scales that are nonlinear functions of t. See Nayfeh (1973) for examples of many of these variations.

All of these methods have the following points in common. The time scales to be used must be selected in advance, prior to attempting a solution; for instance, t, τ, σ with $\tau = \epsilon t$, $\sigma = \epsilon^2 t$. An approximate solution is sought as a series taken to a certain number of terms, such as

$$x_0(t,\tau,\sigma) + \epsilon x_1(t,\tau,\sigma) + \epsilon^2 x_2(t,\tau,\sigma). \tag{39}$$

In substituting series (39) into the differential equations to be solved, the definitions of the scales (such as $\tau = \epsilon t$) are used, so that ordinary derivatives with respect to t are replaced by combinations of partial derivatives with respect to the different scales; thus

$$\frac{d}{dt} = \frac{\partial}{\partial t} + \epsilon \frac{\partial}{\partial \tau} + \epsilon^2 \frac{\partial}{\partial \sigma}. \tag{40}$$

From this point on (until the very end, when τ and σ are again replaced by their definitions), the separate time scales are treated as independent variables. This has the effect of changing ordinary differential equations into partial differential equations that are highly underdetermined, so that various free choices are possible in expressing the solution. The point of the method is now to make these choices skillfully so that the final series (39) is uniformly ordered (and, it is hoped, uniformly valid) on the desired domain.

As an illustration, we return to the Van der Pol Eq. (27) with initial conditions $u(0) = a$, $\dot{u}(0) = 0$. Choosing time scales t and $\tau = \epsilon t$, and writing the solution as $u \cong u_0(t,\tau) + \epsilon u_1(t,\tau)$, one finds recursively that (with subscripts denoting partial derivatives) $u_{0tt} + u_0 = 0$ and $u_{1tt} + u_1 = -2u_{0t\tau} - u_0^2 u_{0t}$. The first equation gives $u_0(t,\tau) = A(\tau)\cos t + B(\tau)\sin t$, a modulated oscillation with slowly varying coefficients. The solution remains underdetermined, since there is nothing here to fix $A(\tau)$ and $B(\tau)$. The solution for u_0 is now substituted into the right-hand side of the differential equation for u_1, and $A(\tau)$ and $B(\tau)$ are chosen to eliminate resonant terms so that the solution for u_1 will remain bounded. (This is similar to the way the undetermined quantities are fixed in the Lindstedt method.) The result (see Fig. 1) is

$$u(t) \cong u_0(t,\epsilon t) = \frac{2a}{\sqrt{a^2 + (4 - a^2)e^{-\epsilon t}}}\cos t. \tag{41}$$

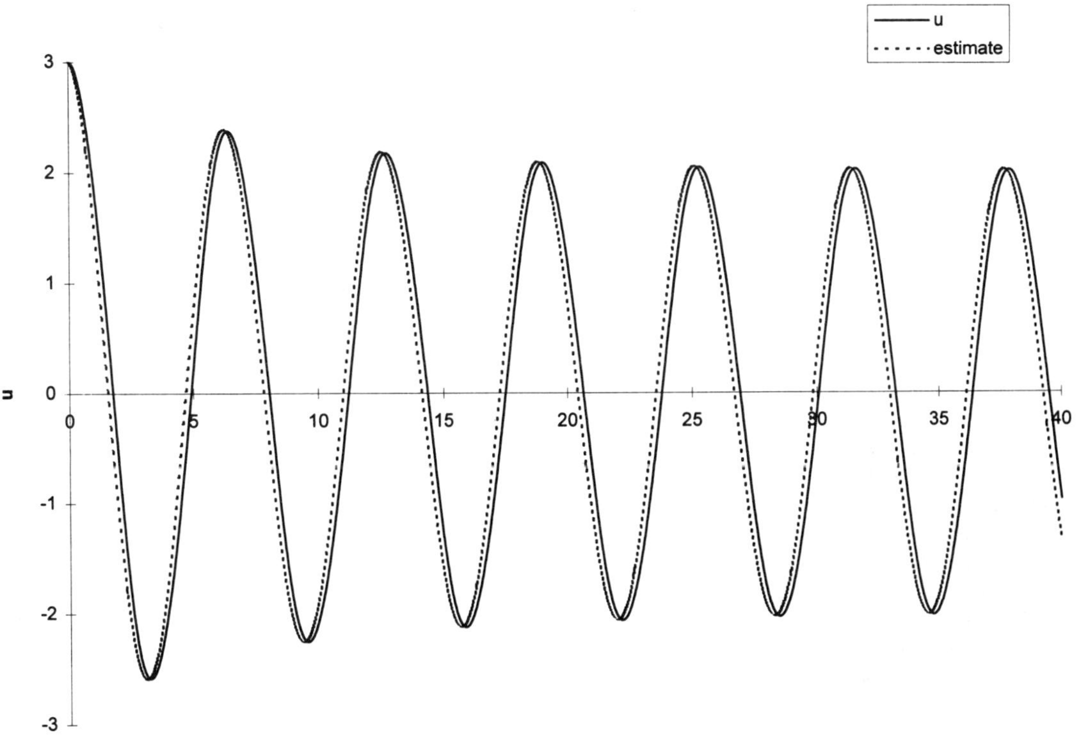

FIG. 1. Numerical (solid) and first-order asymptotic (dotted) solutions for the Van der Pol equation with initial conditions (3,0) and $\epsilon = 0.1$.

This is the same result (to first order) as would be found by applying averaging to system (32), but it has been found without any preliminary coordinate transformations. On the other hand, the possibility of constructing the solution depended on the correct initial guess as to the time scales to be used; the method of averaging generates the needed time scales automatically. The solution (41) exhibits oscillations tending toward a limit cycle that is a simple harmonic motion of amplitude 2. This is qualitatively correct, but the motion is not simple harmonic; carrying the solution to higher orders will introduce corrections.

2.4 Multiple-Frequency Systems and Hamiltonian Systems

Oscillatory problems that cannot be put into periodic standard form can often be put into the following *angular standard form*:

$$\dot{r} = \epsilon f(r,\theta,\epsilon)$$
$$\dot{\theta} = \Omega(r) + \epsilon g(r,\theta,\epsilon), \tag{42}$$

where $r = (r_1,\ldots,r_n)$ is a vector of amplitudes and $\theta = (\theta_1,\ldots,\theta_m)$ is a vector of angles (by which we mean that f and g are periodic in each θ_i with period 2π). This form includes the periodic standard form, by taking $m = 1$ and $\dot{\theta} = 1$. The "naive" method of averaging for system (42) would be to replace f and g by their averages over θ, for instance

$$\bar{f}(r) = \frac{1}{(2\pi)^m}\int_0^{2\pi}\cdots\int_0^{2\pi} f(r,\theta,0)d\theta_1\cdots d\theta_m. \tag{43}$$

In an attempt to justify this process, and to extend the method to higher orders, one can follow the general method of averaging by attempting to make a near-identity change of variables $(r,\theta) \to (\rho,\varphi)$, which will render the system independent of θ up through a given order k in ϵ. However, one encounters at once the famous difficulty of "small divisors," which make the existence of such a transformation very problematic. The nature of the difficulty is as follows. If f is expanded in a convergent multiple Fourier series

$$f(r,\theta,0) = \sum_\nu a_\nu(r)e^{i(\nu_1\theta_1+\cdots+\nu_m\theta_m)}, \tag{44}$$

then the transformation to averaged form necessarily involves the series

$$\sum_{\nu\neq 0}\frac{a_\nu(r)}{i[\nu_1\Omega_1(r)+\cdots+\nu_m\Omega_m(r)]}e^{i(\nu_1\theta_1+\cdots+\nu_m\theta_m)}, \tag{45}$$

which may not converge because the denominators $i(\nu_1\Omega_1+\cdots+\nu_m\Omega_m)$ may be small (or even zero), causing the coefficients to become large. It is of no use to say at this point that "perturbation theory is not concerned with the convergence of series," since the series in question are not being used for approximation, but to prove the existence of a transformation needed in order to justify a method.

Some preliminary progress can be made by considering the case in which the series (44), and hence series (45), are finite. In this case, convergence difficulties cannot arise, but there is still the difficulty that for some r one or more of the denominators of series (45) may become zero. Since $\Omega_i(r)$ are the frequencies of the free oscillations ($\epsilon = 0$) of system (42), we see that averaging must fail when there exists a *resonance relationship* among these free frequencies (more precisely, when there exists a resonance relationship defined by an integer vector ν for which $a_\nu \neq 0$). In general, for each ν there will be a manifold of r, called a *resonance surface*, for which the resonance relation defined by ν holds. On (or near) any such surface it is not permissible to average over all angles θ, although it may be possible to average over a subset of these angles or over certain integral linear combinations of them.

Results beyond these have been obtained in the important special case of Hamiltonian systems; the *Kolmogorov-Arnol'd-Moser* (or *KAM*) *theorem* and the *Nekhoroshev theorem* are the high point of modern perturbation theory and together give the definitive answer to the problem of the stability (in the sense of Laplace) of the (idealized Newtonian) solar system, with which this article began. Consider a system defined by a Hamiltonian function of the form

$$H(r,\theta,\epsilon) = H_0(r) + \epsilon H_1(r,\theta) + \epsilon^2 H_2(r,\theta) + \cdots, \tag{46}$$

where r and θ are as before except that $m = n$. Written in the form of system (42), this system is

$$\dot{r} = \epsilon \frac{\partial H_1}{\partial \theta} + \cdots$$
$$\dot{\theta} = -\frac{\partial H_0}{\partial r} - \epsilon \frac{\partial H_1}{\partial r} + \cdots . \quad (47)$$

Since H_1 is assumed to be periodic in the components of θ, it may be expanded in a multiple Fourier series like series (44); differentiating with respect to any component of θ then eliminates the constant term [$a_0(r)$, which is the average]. It follows that $\partial H_1/\partial \theta$ has zero average, so that the (naive) first-order averaged equation for r becomes

$$\dot{r} \cong 0. \quad (48)$$

This suggests that the motion is oscillatory with nearly constant amplitudes; if this is true, the solar system (and all other systems having the same general form) will be stable (in the sense of Laplace). Of course, the argument we have given does not prove the result, unless the small-divisor problem can be overcome in this situation. This is exactly what the KAM and Nekhoroshev theorems accomplish. The KAM theorem states that (if a certain determinant does not vanish) the great majority of initial conditions will lead to motion on an invariant torus close to a torus r = constant. The Nekhoroshev theorem states that even for those initial conditions that do not lie on invariant tori, the drift in r (called *Arnol'd diffusion*) takes place exponentially slowly (as $\epsilon \to 0$). (Notice that n-dimensional tori in $2n$-dimensional space do not have an inside, so the presence of many such invariant tori does not prevent other solutions from slowly drifting off to infinity.) For details see Lochak and Meunier (1988).

In all applications of averaging and related methods to Hamiltonian systems, it is necessary to have a means of handling near-identity transformations that preserve the Hamiltonian form of the equations; that is, one needs to construct near-identity transformations that are *canonical*. Classically, such transformations can be constructed from their *generating functions* (in the sense of Hamilton-Jacobi theory); averaging procedures carried out in this way are called *von Zeipel's method*. At the present time, this approach can be regarded as obsolete. It has been replaced by the method of *Lie transforms*, in which near-identity canonical transformations are generated as the flows of Hamiltonian systems in which ϵ takes the place of time. (The Lie method is not limited to Hamiltonian systems, but is particularly useful in this context.) Algorithmic procedures for handling near-identity transformations in this way have been developed, and they are considerably simpler than using generating functions. See Nayfeh (1973), Sec. 5.7.

2.5 Perturbation of Stable Manifolds; Melnikov Functions

The problems addressed in the previous sections are the classical initial-value problems of nonlinear oscillations and celestial mechanics. With the steadily increasing importance of nonlinear phenomena such as chaos and strange attractors, finding solutions of specific initial-value problems often becomes less important than finding families (manifolds) of solutions characterized by their qualitative behavior. Many of these problems are accessible by means of perturbation theory. We will briefly describe one example. If a dynamical system has a rest point of saddle type, there will exist a *stable manifold* and an *unstable manifold* of the saddle point; the former consists of all points that approach the saddle point as $t \to \infty$, the latter of points approaching the saddle as $t \to -\infty$. In some cases, the stable and unstable manifold will coincide; that is, points that approach the saddle in the distant future also emerged from it in the distant past. (The simplest case occurs in the plane when the stable and unstable manifolds form a figure-eight pattern with the saddle at the crossing point.) If such a system is perturbed, it is important to decide whether the stable and unstable manifolds separate, or continue to intersect; and if they intersect, whether they are transverse. (The latter case leads to chaotic motion.) The criterion that in many cases decides between these alternatives is called the *Melnikov function*; if this function has simple zeroes, the manifolds will intersect transversely and there will be a chaotic region. The Melnikov function is an integral over the homoclinic orbit of the normal component of the perturbation; the form of the integral is determined by applying regular perturbation methods to the solutions in the stable and unstable manifolds and measuring the distance between the approximate solutions. For details see Wiggins (1990).

3. INITIAL AND BOUNDARY LAYERS

3.1 Multiple-Scale Method for Initial-Layer Problems

The oscillatory problems considered in Secs. 2.1–2.4 are regular perturbation problems when considered on a fixed interval of time, but become singular when considered on an expanding interval such as $0 \leq t \leq 1/\epsilon$. We now turn to problems that are singular even on a fixed interval. It is not easy to solve these problems even numerically, because for sufficiently small ϵ they are what numerical analysts call "stiff." The simplest class of such problems comprises initial-value problems of the form

$$\begin{aligned} &\epsilon \ddot{u} + b(t)\dot{u} + c(t)u = 0 \\ &u(0) = \alpha \\ &\dot{u}(0) = \beta/\epsilon + \gamma. \end{aligned} \tag{49}$$

One may think, for instance, of an object of small mass ϵ subjected to a time-dependent restoring force and friction; at time zero the position and velocity have just reached α and γ when the object is subjected to an impulse imparting momentum β, increasing the velocity by an amount β/ϵ, which is large since ϵ is small. We will use this example to explain two methods, which are applicable to many problems in which a small parameter multiplies the highest derivative.

In approaching any perturbation problem, one first tries to understand the case $\epsilon = 0$, but here it does not make sense to set $\epsilon = 0$. On one hand, the differential equation drops from second order to first, and can no longer accept two initial conditions; on the other hand, the second initial condition becomes infinite. Progress can be made, however, by introducing the "stretched" time variable

$$\tau = t/\epsilon \tag{50}$$

Upon substituting Eq. (50) into Eqs. (49) and writing $' = d/d\tau$, we obtain

$$\begin{aligned} u' + b(\epsilon\tau)u' + \epsilon c(\epsilon\tau)u &= 0 \\ u(0) &= \alpha \\ u'(0) &= \beta + \epsilon\gamma. \end{aligned} \tag{51}$$

This problem is regular (for a fixed interval of τ) and can be solved readily. For a first approximation, it suffices to set $\epsilon = 0$ in Eqs. (51) to obtain $u'' + b_0u' = 0$ with $b_0 = b(0)$; the solution is

$$u_0^i = -\frac{\beta}{b_0} e^{-b_0\tau} + \alpha + \frac{\beta}{b_0}, \tag{52}$$

called the *first-order inner solution*. [Higher-order approximations can be found by substituting a perturbation series $u_0^i + \epsilon u_1^i + \cdots$ into Eqs. (51).] The name "inner solution" comes from the fact that Eq. (52) is only uniformly valid on an interval such as $0 \leq \tau \leq 1$, which translates into $0 \leq t \leq \epsilon$ in the original time variable; this is a narrow "inner region" close to the initial conditions. It is necessary somehow to extend this to a solution valid for a fixed interval of t. This is, of course, equivalent to an expanding interval of τ, and one might attempt to solve Eqs. (51) on such an expanding interval by previously discussed methods. The equation cannot be put in a form suitable for averaging. However, the method of multiple scales is flexible enough to be adapted to this situation. One takes as time scales τ and t, and seeks a solution in the form

$$\begin{aligned} u \cong &\{u_0^i(\tau) + u_0^{\text{cor}}(t)\} \\ &+ \epsilon\{u_1^i(\tau) + u_1^{\text{cor}}(t)\} + \cdots. \end{aligned} \tag{53}$$

[We could have taken $u_0(\tau,t) + \epsilon u_1(\tau,t) + \ldots$, but the solution turns out to be the sum of the previously calculated u^i and a "correction" u^{cor}, and so it is convenient to postulate this form initially.] One can differentiate Eq. (53) with respect to τ using Eq. (50) and substitute it into Eqs. (51), or equivalently differentiate with respect to t and substitute into Eqs. (49). Assuming that $u^{\text{cor}}(0) = 0$ (since the inner part u^i should suffice initially), one finds that u^i must satisfy Eqs. (51) as expected, and that u^{cor} satisfies a first-order differential equation together with the assumed initial condition $u^{\text{cor}}(0) = 0$; thus u^{cor} is fully determined. At the first order, u_0^{cor} in fact satisfies the differential equation obtained from Eqs. (49) by setting $\epsilon = 0$; this is the very equation that we initially discarded as unlikely to be meaningful. Upon solving this equation (with zero initial conditions) and adding the result to u_0^i, we obtain the *com-*

posite solution

$$u_0^c = u_0^i + u_0^{\text{cor}} = -\frac{\beta}{b_0} e^{-b_0 t/\epsilon} + \left(\alpha + \frac{\beta}{b_0}\right) \exp\left[-\int_0^t \frac{c(s)}{b(s)} ds\right]. \tag{54}$$

This solution is uniformly valid on any fixed interval of t. See Fig. 2.

3.2 Matching for Initial-Layer Problems

Although the method of multiple scales is successful for problems of this type, it is not used as widely as the *method of matched asymptotic expansions*, probably because multiple scales requires that the choices of gauges and scales be made in advance, whereas matching allows for the discovery of the correct gauges and scales as one proceeds. [Recall that gauges are the functions $\delta_i(\epsilon)$, usually just powers ϵ^i, that multiply successive terms of a perturbation series; scales are the stretched time or space variables used.] To apply the matching method to Eqs. (49), begin with the first-order inner solution (52), which is valid near the origin. Assume that at some distance from the origin, a good first approximation should be given by setting $\epsilon = 0$ in Eqs. (49) and discarding the initial conditions (which we have already seen do not make sense with $\epsilon = 0$); the result is

$$u_0^o = A \exp\left[-\int_0^t \frac{c(s)}{b(s)} ds\right], \tag{55}$$

called the *first-order outer solution*. Since we have discarded the initial conditions, the quantity A in Eq. (55) remains undetermined at this point. At this point, one compares the inner solution (52) with the outer solution (55) in an effort to determine the correct value of A so that these solutions "match." In the present instance, the inner solution decays rapidly (assuming $b_0 > 0$) to $\alpha + \beta/b_0$, while the outer solution has A as its initial value (at $t = 0$). One might try to determine where the "initial layer" ends, and choose A so that u_0^i and u_0^o agree at that point; but in fact it is sufficient to set $A = \alpha + \beta/b_0$ on the assumption that the inner solution reaches this value at a point close enough to $t = 0$ to allow taking it as the initial condition for the outer solution. Finally, we note that adding the inner and outer solutions would duplicate the quantity $\alpha + \beta/b_0$ with which one

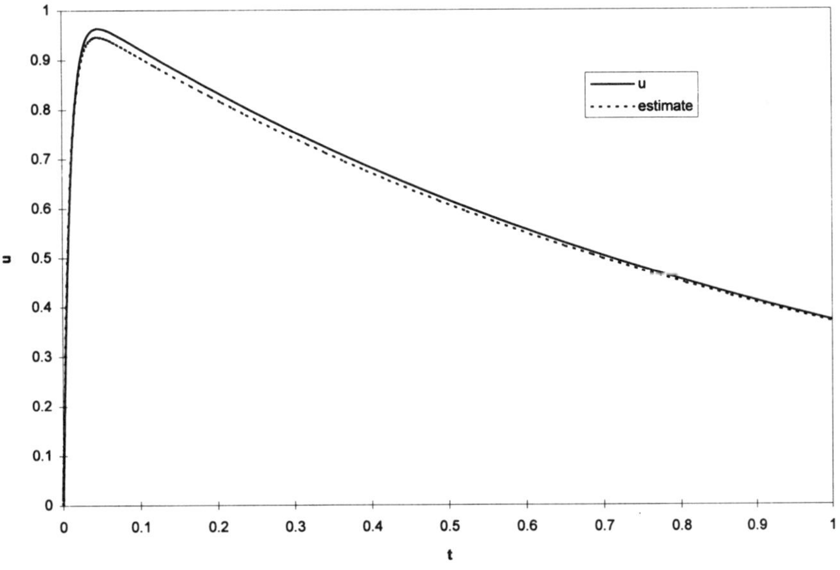

FIG. 2. Numerical (solid) and first-order asymptotic (dotted) solutions to Eqs. (49) with $\alpha = \gamma = 0$, $\beta = 1$, and $\epsilon = 0.1$.

ends and the other begins, so we subtract this "common part" u^{io} of the inner and outer solutions to obtain the composite solution

$$u^c = u^i + u^o - u^{io}, \tag{56}$$

which is equal to the result (54) obtained by multiple scales.

In the last paragraph, we have cobbled together the inner and outer solution in a very *ad hoc* manner. In fact, several systematic procedures exist for carrying out the matching of u^i and u^o to any order and extracting the common part u^{io}. The most common procedure consists of what are sometimes called the *Van Dyke matching rules*, details of which are given below. Although this procedure is simple to use, it does not always lead to correct results, in particular, when it is necessary to use logarithmic gauges. The other methods, *matching in an intermediate variable* and *matching in an overlap domain*, are too lengthy to explain here (see Lagerstrom, 1988), but give better results in difficult cases. None of these methods has a rigorous justification *as a method*, although it is often possible to justify the results for a particular problem or class of problems. Occasionally one encounters problems in which the inner and outer solutions cannot be matched. These cases sometimes require a "triple deck," that is, a third- (or even fourth-)layer time scale. In other cases there does not exist a computable asymptotic approximation to the exact solution.

To explain the Van Dyke matching rules, we first assume that the inner and outer solutions $u^i(\tau,\epsilon)$ and $u^o(t,\epsilon)$ have been computed to some order ϵ^k. In the problem we have been studying, the outer solution contains undetermined constants whose values must be determined, and the inner solution contains none, but in more general problems to be considered below there may be undetermined constants in both. It is important to understand that the inner and outer solutions are naturally computed in such a way that u^i is "expanded in powers of ϵ with τ fixed" while u^o is "expanded in powers of ϵ with t fixed." We are about to re-expand each solution with the opposite variable fixed. *The first step* is to express u^i in the outer variable t by setting $\tau = \epsilon t$. The resulting function of t and ϵ is then expanded in powers of ϵ to order ϵ^k holding t fixed. This new expansion is called u^{io}, the *outer expansion of the inner solution*. Notice that in computing u^{io} we retain only the terms of degree $\leq k$, so that in effect part of u^i is discarded because it moves up to order higher than k; the meaning of this is that the discarded terms of u^i are insignificant, at the desired order of approximation, in the outer region. *The second step* is to express u^o in the inner variable τ by setting $t = \tau/\epsilon$, and expand the resulting function of τ and ϵ in powers of ϵ to order k holding τ constant. The result, called u^{oi} or the *inner expansion of the outer solution*, contains those parts of u^o that are significant in the inner region (to order k), arranged according to their significance in the inner region. *The third step* is to set $u^{io} = u^{oi}$ and use this equation to determine the unknown constants. The rationale for this is that if the domains of validity of the inner and outer regions overlap, then, since the inner solution is valid in the overlap, but the overlap belongs to the outer region, u^{io}, which is the inner solution stripped of the part which is insignificant in the outer region, should be valid there; similarly, since the outer solution is valid in the overlap, but the overlap belongs to the inner region, u^{oi} should be valid there. Now, in setting $u^{io} = u^{oi}$ it is not possible to carry out the necessary computations unless both are expressed in the same variable, and so it is necessary to choose either t or τ and express both sides in that variable before attempting to determine the unknown constants. *The fourth step* is to compute the composite solution $u^c = u^i + u^o - u^{io}$. At this stage, u^{io} (which is equal to u^{oi}) is known as the *common part* of u^i and u^o; it is subtracted because otherwise it would be represented twice in the solution.

As a final example of a problem solvable by the methods of this section, we mention systems of the form

$$\begin{aligned} \dot{x} &= f(x,y) \\ \epsilon\dot{y} &= g(x,y), \end{aligned} \tag{57}$$

with x and y real. For $\epsilon = 0$, the motion is confined to the set $g(x,y) = 0$, called the *slow curve*. For $\epsilon \neq 0$, the entire plane is accessible, but (assuming $\partial g/\partial y < 0$, in which case we say the slow curve is *stable*) any point moves rapidly toward the slow curve and then slowly along it. These two stages of the motion can be approximated separately as inner and outer solutions and then matched. An interesting case arises when the slow curve is S-shaped, with the upper and lower branches stable and the middle section (the doubled-over part) unstable. A point can move along a stable branch until it reaches a vertical tangent point, then "fall off" and move rapidly to the other stable branch, then move along that branch to the other vertical tangent point and "fall off" the other way, leading to a cyclic motion called a *relaxation oscillation*. In a further, very unusual scenario, the point may actually turn the corner at the vertical tangent point and follow the unstable branch for some distance before "falling." This rather recently discovered phenomenon is called a *canard*. The explanation of canards is that several time scales become involved; the solution is actually "falling" away from the unstable branch all the time, but doing so at a rate that is slow even compared to the already slow motion along the branch. For relaxation oscillations and canards see Grasman (1987).

3.3 Boundary-Layer Problems

Problems in which a small parameter multiplies the highest derivative are encountered among boundary-value problems at least as frequently as among initial-value problems. Since the basic ideas have been covered in the previous sections, it is only necessary to point out the differences that arise in the boundary-value case. Either the multiple-scale or matching method may be used; we will use matching. The method is illustrated here with an ordinary differential equation; a partial differential equation will be treated in Sec. 3.5 below.

Consider the problem

$$\begin{aligned} &\epsilon y'' + b(x)y' + c(x)y = 0, \\ &y(0) = \alpha, \\ &y(1) = \beta, \end{aligned} \tag{58}$$

on the interval $0 \leq x \leq 1$. The differential equation here is the same as Eqs. (49), only the independent variable is a space variable rather than time in view of the usual applications. If $b(x)$ is positive throughout $0 < x < 1$, there will be a boundary layer at the left end point $x = 0$; if negative, the boundary layer will be at the right end point; and if $b(x)$ changes sign, there may be internal transition layers as well. We consider the former case. To the first order, the outer solution y^o will satisfy the first-order equation $b(x)y' + c(x)y = 0$ obtained by setting $\epsilon = 0$ in Eqs. (58); it will also satisfy the right-hand boundary condition $y(1) = \beta$. Therefore, the outer solution is completely determined. The first-order inner solution y^i will satisfy the equation $d^2y/d\xi^2 + b_0 y = 0$ with $b_0 = b(0)$, obtained by substituting the stretched variable $\xi = x/\epsilon$ into Eqs. (58) and setting $\epsilon = 0$; it will also satisfy the left-hand boundary condition $y(0) = \alpha$. Since this is a second-order equation with only one boundary condition, it will contain an undetermined constant, which must be identified by matching the inner and outer solutions. The differential equations satisfied by the inner and outer solutions are the same as in the case of Eqs. (49), the only difference being that this time the constant that must be fixed by matching belongs to the inner solution rather than the outer.

3.4 WKB Method

There are a great variety of problems that are more degenerate than the one we have just discussed, which can exhibit a wide range of exotic behaviors. These include internal layers, in which a stretched variable such as $\xi = (x - x_0)/\epsilon$ must be introduced around a point x_0 in the interior of the domain; triple decks, in which two stretched variables such as x/ϵ and x/ϵ^2 must be introduced at one end: and problems in which the order of the differential equation drops by more than one. The simplest example of the latter type is

$$\epsilon^2 y'' + a(x)y = 0. \tag{59}$$

This problem is usually addressed by a technique called the *WKB* or *WKBJ method*. This method is rather different in spirit from the others we have discussed, because it depends heavily on the linearity of the perturbed problem. Rather than pose initial- or boundary-value problems, one finds approximations for two linearly independent solutions of the linear Eq. (59) on the whole real line. The general solution then consists of the linear combinations of these two. If $a(x) = k^2(x) > 0$, these approximate solutions are

$$y^{(1)} \cong \frac{1}{\sqrt{k(x)}} \cos\left(\frac{1}{\epsilon} \int k(x)dx\right) \tag{60a}$$

and

$$y^{(2)} \cong \frac{1}{\sqrt{k(x)}} \sin\left(\frac{1}{\epsilon} \int k(x)dx\right). \tag{60b}$$

If $a(x) = -k^2(x) < 0$, the two solutions are

$$y^{(1),(2)} \cong \frac{1}{\sqrt{k(x)}} \exp\left(\frac{1}{\epsilon} \int \pm k(x)dx\right). \tag{61}$$

If $a(x)$ changes sign, one has a difficult situation called a *turning-point problem*. This can be addressed in various ways by matching solutions of the form of Eqs. (60) and (61) or by using Airy functions. The latter are solutions of the differential equation $y'' + xy = 0$, which is the simplest problem with a turning point at the origin. These Airy functions can be considered as known (they can be expressed using Bessel functions of order $\frac{1}{3}$), and solutions to more general turning-point problems can be expressed in terms of them. For an introduction to turning-point problems, see Lakin and Sanchez (1970), Chap. 2, and for theory see Wasow (1976), Chap. 8.

3.5 Fluid Flow

We conclude this section with a brief discussion of the problem of fluid flow past a flat plate, because of its historical importance (see the Introduction) and because it illustrates several aspects of perturbation theory that we have avoided so far: the use of perturbation theory for partial differential equations, and the need to combine undetermined scales with undetermined gauges. The classic reference for this material is Van Dyke (1975). Consider a plane fluid flow in the upper half plane, with a "flat plate" occupying the interval $0 \leq x \leq 1$ on the x axis; that is, the fluid will adhere to this interval, but not to the rest of the x axis. The stream function $\psi(x,y)$ of such a fluid will satisfy

$$\begin{aligned} &\epsilon(\psi_{xxxx} + 2\psi_{xxyy} + \psi_{yyyy}) \\ &\quad - \psi_y(\psi_{xxx} + \psi_{xyy}) + \psi_x(\psi_{xxy} + \psi_{yyy}) = 0, \end{aligned} \tag{62}$$

with $\psi(x,0) = 0$ for $-\infty < x < \infty$, $\psi_y(x,0) = 0$ for $0 \leq x \leq 1$, and $\psi(x,y) \to y$ as $x^2 + y^2 \to \infty$. The latter condition describes the flow away from the plate, and this in fact gives the leading-order outer solution as

$$\psi^o(x,y) = y. \tag{63}$$

To find an inner solution, we stretch y by an undetermined scale factor,

$$\eta = y/\mu(\epsilon), \tag{64}$$

and expand the inner solution using undetermined gauges, giving (to first order)

$$\psi^i = \delta(\epsilon)\Psi(x,\eta). \tag{65}$$

Substituting this into Eq. (62) and discarding terms that are clearly of higher order yields

$$\frac{\epsilon}{\mu} \Psi_{\eta\eta\eta\eta} + \delta[\Psi_x\Psi_{\eta\eta\eta} - \Psi_\eta\Psi_{x\eta\eta}] = 0. \tag{66}$$

The relative significance of ϵ/μ and δ has not yet been determined, but if either of them were dominant, the other term would drop out of Eq. (66) to first order, and the resulting solution would be too simple to capture the behavior of the problem. So we must set

$$\epsilon/\mu = \delta \tag{67}$$

and conclude that the first-order inner solution satisfies

$$\Psi_{\eta\eta\eta\eta} + \Psi_x\Psi_{\eta\eta\eta} - \Psi_\eta\Psi_{x\eta\eta} = 0. \tag{68}$$

It is not possible to solve Eq. (68) in closed form, but it is possible to express the solution as

$$\Psi(x,\eta) = \sqrt{2x}f(\eta/\sqrt{2x}), \tag{69}$$

where f is the solution of the ordinary differential equation $f''' + ff'' = 0$ with $f(0) = 0$, $f'(0) = 0$, and $f'(\infty) = 1$. In attempting to match the inner and outer solutions, it is discovered that this is only possible if $\delta = \mu$. Together with Eq. (67), this finally fixes the undetermined scales and gauges as

$$\delta = \mu = \sqrt{\epsilon}. \tag{70}$$

Upon attempting to continue the solution to higher orders, obstacles are encountered that can only be overcome by introducing triple decks and other innovations.

4. PERTURBATIONS OF SPECTRA

In this section we address the following question: if

$$A(\epsilon) = A_0 + \epsilon A_1 + \cdots \tag{71}$$

is a matrix or linear operator depending on a small parameter, and the spectrum of A_0 is known, can we determine the spectrum of $A(\epsilon)$ for small ϵ? For the case of a matrix, the spectrum is simply the set of eigenvalues (values of λ for which $Av = \lambda v$ for some column vector v called an eigenvector). More generally, the spectrum is defined as the set of λ for which $A - \lambda I$ is not invertible; for linear transformations on infinite-dimensional spaces (such as Hilbert or Banach spaces) this need not imply the existence of an eigenvector. Our attention here will be focused on the matrix case, but many of the procedures (excluding those which involve the determinant or the Jordan normal form) are applicable as well to any operators whose spectrum consists of eigenvalues. The classical reference for the general infinite-dimensional case is Kato (1966).

The most direct approach (although in the end we do not recommend it) to finding the eigenvalues of Eq. (71) in the matrix case would be to examine the *characteristic equation*

$$P(\lambda,\epsilon) = \det[A(\epsilon) - \lambda I] = 0, \tag{72}$$

the roots of which are the eigenvalues. There are standard perturbation methods for finding the roots of polynomials (see Murdock, 1991, Chap. 1), the simplest of which is to substitute

$$\lambda(\epsilon) = \lambda_0 + \epsilon\lambda_1 + \cdots \tag{73}$$

into Eq. (72) and solve recursively for λ_i. This method works if λ_0 is a simple root of $P(\lambda,0) = 0$; that is, it will work if A_0 has distinct eigenvalues. If there are repeated eigenvalues, then in general it is necessary to replace Eq. (73) with a *fractional power series* involving gauges $\delta_i(\epsilon) = \epsilon^{i/q}$ for some integer q that is most readily determined by using Newton's diagram. Although these are the best available perturbation methods for finding roots of general polynomials, they have two drawbacks in the case of eigenvalues: if the matrices are large, it is difficult to compute the characteristic polynomial; and, more importantly, these methods do not take account of the special features of eigenvalue problems. For instance, if $A(\epsilon)$ is a symmetric matrix, then its eigenvalues will be real, and fractional powers will not be required (even if A_0 has repeated eigenvalues). But the fact that A is symmetric is lost in passing to the characteristic polynomial, and one cannot take advantage of these facts.

For these reasons, it is best to seek not only the eigenvalues, but also at the same time the eigenvectors that go with them. The general procedure (which must be refined in particular situations) is to seek solutions of

$$A(\epsilon)v(\epsilon) = \lambda(\epsilon)v(\epsilon) \tag{74}$$

in the form

$$\begin{aligned} \lambda(\epsilon) &= \lambda_0 + \epsilon\lambda_1 + \epsilon^2\lambda_2 + \cdots \\ v(\epsilon) &= v_0 + \epsilon v_1 + \epsilon^2 v_2 + \cdots. \end{aligned} \tag{75}$$

In the first two orders, the resulting equations are

$$\begin{aligned} (A_0 - \lambda_0 I)v_0 &= 0 \\ (A_0 - \lambda_0 I)v_1 &= (\lambda_1 I - A_1)v_0. \end{aligned} \tag{76}$$

We now discuss how to solve Eqs. (76) under various circumstances.

The simplest case occurs if A_0 is real and symmetric (or complex and Hermitian), and also has distinct eigenvalues. In this case, the first equation of Eqs. (76) can be solved simply by choosing an eigenvector v_0 for each eigenvalue λ_0 of A_0. It is convenient to normalize v_0 to have a length of 1, that is, $(v_0,v_0) = 1$ where $(\cdot,\cdot)$ is the inner (or "dot") product. Now we fix a choice of λ_0 and v_0 and insert these into the second equation of Eqs. (76). The next step is to choose λ_1 so that the right-hand side lies in the image (or range) of $A_0 - \lambda_0 I$; once this is accomplished, it is possible to solve for v_1. To determine λ_1, we rely upon special properties of the eigenvectors of a symmetric matrix; namely, they are orthogonal (with respect to the inner product). Thus, there exists an orthogonal basis of eigenvectors in which A_0 is diagonal; examining $A_0 - \lambda_0 I$ in this basis, we see that its kernel (or null space) is spanned by v_0 and its image is spanned by the rest of the eigenvectors. Therefore, the image is perpendicular to the kernel. It follows that $(\lambda_1 I - A_1)v_0$ will lie in the image of $A_0 - \lambda_0 I$ if and only if its orthogonal projection onto v_0 is zero, that is, if and only if $(\lambda_1 v_0 - A_1 v_0, v_0) = 0$, or, using $(v_0,v_0) = 1$,

$$\lambda_1 = (A_1 v_0, v_0). \tag{77}$$

It is not necessary to find v_1 unless it is desired to go on to the next stage and find λ_2. (There is a close similarity between these steps and those of the Lindstedt method, Sec. 2.1, in which each term in the frequency expansion is determined to make the next equation solvable.)

If A_0 has distinct eigenvalues but is not symmetric, most of the last paragraph still applies, but the eigenvectors of A_0 need not be orthogonal. The vector space still decomposes as a direct sum of the image and kernel of $A_0 - \lambda_0 I$, but the inner product can no longer be used to effect the decomposition; λ_1 can still be determined but cannot be written in the form of Eq. (77).

If A_0 does not have distinct eigenvalues, the situation can become quite complicated. First, suppose $A(\epsilon)$ is symmetric for all ϵ, so that all A_i are symmetric. In this case, every eigenvalue has a "full set" of eigenvectors (as many linearly independent eigenvectors as its algebraic multiplicity). However, suppose that A_0 has an eigenvalue λ_0 of multiplicity two, with eigenvectors w and z. It is likely that for $\epsilon \neq 0$, the eigenvalue λ_0 splits into two distinct eigenvalues having separate eigenvectors. In this case, it is not possible to choose an arbitrary eigenvector v_0 from the plane of z and w to use in the second equation of Eqs. (76); only the limiting positions (as $\epsilon \to 0$) of the two eigenvectors for $\epsilon \neq 0$ are suitable candidates for v_0. Since these are unknown in advance, one must put $v_0 = az + bw$ (for unknown real a and b) into Eqs. (76), then find two choices of a, b, and λ_1 that make the second equation solvable. It also may happen that the degeneracy cannot be resolved at this stage but must be carried forward to higher stages before the eigenvalues split; or, of course, they may never split.

If $A(\epsilon)$ is not symmetric, the possibilities become even worse. The example

$$A(\epsilon) = \begin{bmatrix} 1 & \epsilon \\ 0 & 1 \end{bmatrix} \tag{78}$$

shows that a full set of eigenvectors may exist when $\epsilon = 0$ but not for $\epsilon \neq 0$; the contrary case (diagonalizable for $\epsilon \neq 0$ but not for $\epsilon = 0$) is exhibited by

$$A(\epsilon) = \begin{bmatrix} 1 & 1 \\ 0 & 1+\epsilon \end{bmatrix}.$$

The fact is that the Jordan normal form of $A(\epsilon)$ is not in general a continuous function of ϵ. But there is a normal form that is continuous, which is due to Arnol'd. Given A_0, one can determine its *universal unfolding*, which is a matrix $A(\mu_1,\ldots,\mu_k)$, depending on a certain number k of parameters (k is called the *codimension* of A_0); the universal unfolding is the simplest matrix giving all possibilities (up to similarity) for matrices near A_0. Then $\mu_1, \ldots, \mu_k$ can be determined as functions of ϵ so that the particular given unfolding $A(\epsilon)$ of A_0 is exhibited (up to similarity) within the universal unfolding. From this, all questions about the spectrum, eigenvectors, and generalized eigenvectors of $A(\epsilon)$ can in principle be answered. The theory is described in Arnol'd (1988), Secs. 29 and 30.

GLOSSARY

Asymptotic Approximation: An approximate solution to a perturbation problem that increases in accuracy at a known rate (the order, measured by a gauge function) as the perturbation parameter approaches zero. Usually the accuracy is expected to be uniform. See **Uniform**.

Asymptotic Order: An estimate of the rate at which a quantity decreases as a perturbation parameter approaches zero. A term is of order ϵ^2 if its absolute value is less than a constant times ϵ^2.

Asymptotic Series: A finite or infinite series, successive partial sums of which are asymptotic approximations of some function to successively higher order. Usually the accuracy is expected to be uniform. See **Poincaré Series, Generalized Series, Perturbation Series**, and **Uniform**.

Averaging: A method of constructing asymptotic approximations to oscillatory problems. In the simplest case, it involves replacing periodic functions by their averages to simplify the equations to be solved.

Bifurcation: Any change in the number or qualitative character (such as stability) of the solutions to an equation as a parameter is varied.

Boundary Layer: A transition layer located near the boundary of a region where boundary values are imposed. See **Transition Layer**.

Composite Solution: A solution uniformly valid on a certain domain, created by matching an inner solution and an outer solution each valid on part of the domain. See **Matching, Inner Solution**, and **Outer Solution**.

Gauge: A monotonic function of a perturbation parameter used to express the asymptotic order of a term in a perturbation series or of an error bound. See **Asymptotic Order**.

Generalized Series: A perturbation series of the form $\Sigma\delta_i(\epsilon)u_i(x,\epsilon)$ in which the perturbation parameter appears both in the gauges and in the coefficients. See **Poincaré Series**.

Hamiltonian System: A system of differential equations having the form $\dot{q}_i = \partial H(p,q)/\partial p_i$, $\dot{p}_i = -\partial H(p,q)/\partial q_i$ for $i = 1, \ldots, n$, where H is a given function of p and q called the Hamiltonian or the energy.

Harmonic Oscillation: An oscillation of a periodically forced system having the same frequency as the forcing. (Not to be confused with a *simple* harmonic oscillation, which is sinusoidal.)

Initial Layer: A transition layer located near the point at which the initial data for an initial-value problem are prescribed. See **Transition Layer**.

Inner Solution: An approximate solution uniformly valid within a transition layer.

Intermediate Variable: A scaled variable with a scaling intermediate between two other scaled variables (usually an inner and an outer variable). For instance, $\sqrt{\epsilon}t$ is intermediate between t and ϵt.

Lie Series: In perturbation theory, a means of representing a near-identity transformation by means of a generating function. If $W(x,\epsilon)$ is the generating function, then the solution of $dx/d\epsilon = W(x,\epsilon)$, with $x(0) = y$ for small ϵ, is a near-identity transformation of the form $x = y + \epsilon u_1(y) + \cdots$.

Lindstedt Method: A method of approximating periodic solutions whose frequency varies with the perturbation parameter. It consists in introducing a scaled time in terms of which the frequency does not vary, and then expanding the solution and the frequency separately in the small parameter.

Matching: Any of several methods for choosing the arbitrary constants in an inner and an outer solution so that they both approximate the same exact solution.

Multiple Scales: The simultaneous use of two or more variables having the same physical significance (for instance, time or distance) but proceeding at different "speeds" (in terms of the small parameter), for instance "normal time" t and "slow time" $\tau = \epsilon t$. The variables are treated as if they were independent during part of the discussion, but at the end are reduced to a single variable again.

Outer Solution: An approximate solution uniformly valid in a region away from a transition layer.

Overlap Domain: A region in which both an inner and an outer approximation are valid. For instance, if there is an initial layer at $t = 0$, the inner solution, which is usually considered to be valid on $0 \le t \le \epsilon$, may actually be valid (with a weaker error estimate) on the longer interval $0 \le t \le \sqrt{\epsilon}$. Then, if the outer solution is valid on $\epsilon \le t \le 1$, there is an overlap domain of $\epsilon \le t \le \sqrt{\epsilon}$.

Perturbation Parameter: A parameter, usually denoted ϵ, occurring in a mathematical problem, such that the problem has a known solution when $\epsilon = 0$ and an approximate solution is sought when ϵ is small but nonzero.

Perturbation Series: A finite or infinite series obtained as a formal approximate solution to a perturbation problem, usually a power series in the perturbation parameter or some generalization of a power series. The term "perturbation series" does not imply that the solution has been proved to be asymptotically valid. (See **Asymptotic Series**.)

Poincaré Series: A perturbation series of the form $\Sigma\delta_i(\epsilon)u_i(x)$ in which the perturbation parameter appears only in the gauges. See **Generalized Series**.

Regular Perturbation Problem: A perturbation problem having an approximate solution in the form of a Poincaré series that is uniformly valid on the entire intended domain. See **Singular Perturbation Problem**.

Relaxation Oscillation: A self-sustained oscillation characterized by a slow buildup of tension (in a spring, for instance) followed by a rapid release or relaxation. The rapid phase is an example of a transition layer.

Rescaled Coordinate: A coordinate that has been obtained from an original variable by a transformation depending on the perturbation parameter, usually multiplication by a scaling factor. For instance, time t may be rescaled to give a "slow time" ϵt (see **Multiple Scales**) or a "strained time" $(\omega_0 + \epsilon\omega_1 + \cdots)t$ (see **Lindstedt Method**).

Resonance: In linear problems, an equality of two frequencies. In nonlinear problems, any integer relationship holding between two or more frequencies, of the form $n_1\omega_1 + \cdots + n_k\omega_k = 0$, especially one involving small integers or one that produces zero denominators in a Fourier series.

Self-Excited Oscillation: An oscillation about an unstable equilibrium, which occurs because of the instability and has its own natural frequency, rather than an oscillation in response to an external periodic forcing.

Singular Perturbation Problem: A perturbation problem that does not have an approximate solution in the form of a Poincaré series uniformly valid on the entire intended domain, although it may have such solutions valid over part of the domain. For singular problems one seeks a solution in the form of a generalized series. See **Regular Perturbation Problem**.

Stable Manifold: Let x_0 be a saddle point of a system $\dot{x} = f(x)$ of differential equations; that is, $f(x_0) = 0$ and the matrix of partial derivatives of f at x_0 has no eigenvalues on the imaginary axis. Then, the stable manifold of x_0 is the set of points which, if taken as initial conditions, lead to solutions that approach x_0 as $t \to \infty$; the unstable manifold is the set of points that approach x_0 as $t \to -\infty$.

Stretched Variable: See **Rescaled Variable**.

Transition Layer: A small region in which the solution of a differential equation changes rapidly and in which some approximate solution (outer solution) that is valid elsewhere fails.

Triple Deck: A problem that exhibits a transition layer within a transition layer and that therefore requires the matching of three approximate solutions rather than only two.

Unfolding: A family of perturbations of a given system obtained by adding several small parameters. An unfolding is universal if (roughly) it exhibits all possible qualitative behaviors for perturbations of the given system using the least possible number of parameters.

Unstable Manifold: See **Stable Manifold.**

Works Cited

Arnol'd, V. I. (1988), *Geometrical Methods in the Theory of Ordinary Differential Equations*, 2nd ed., New York: Springer-Verlag.

Golubitsky, M., Schaeffer, D. G. (1985), *Singularities and Groups in Bifurcation Theory*, Vol. 1, New York: Springer-Verlag.

Grasman, J. (1987), *Asymptotic Methods for Relaxation Oscillations and Applications*, New York: Springer-Verlag.

Lochak, P., Meunier, C. (1988), *Multiphase Averaging for Classical Systems*, New York: Springer-Verlag.

Kato, T. (1966), *Perturbation Theory for Linear Operators*, New York: Springer-Verlag.

Lagerstrom, P. A. (1988), *Matched Asymptotic Expansions*, New York: Springer-Verlag.

Lakin, W. D., Sanchez, D. A. (1970), *Topics in Ordinary Differential Equations*, New York: Dover.

Murdock, J. A. (1991), *Perturbations*, New York: Wiley.

Nayfeh, A. (1973), *Perturbation Methods*, New York: Wiley.

Van Dyke, M. (1975), *Perturbation Methods in Fluid Mechanics, Annotated Edition*, Stanford, CA: Parabolic Press.

Wasow, W. (1976), *Asymptotic Expansions for Ordinary Differential Equations*, Huntington, NY: Robert E. Krieger Pub. Co.

Wiggins, S. (1990), *Introduction to Applied Nonlinear Dynamical Systems and Chaos*, New York: Springer-Verlag.

Further Reading

Bush, A. W. (1992), *Perturbation Methods for Engineers and Scientists*, Boca Raton: CRC Press.

Hinch, E. J. (1991), *Perturbation Methods*, Cambridge: Cambridge University Press.

Kevorkian, J., Cole, J. D. (1981; corrected second printing 1985), *Perturbation Methods in Applied Mathematics*, New York: Springer-Verlag.

Lagerstrom, P. A. (1988), *Matched Asymptotic Expansions*, New York: Springer-Verlag.

Murdock, J. A. (1991), *Perturbations*, New York: Wiley.

Nayfeh, A. (1973), *Perturbation Methods*, New York: Wiley.

Nayfeh, A. (1981), *Introduction to Perturbation Techniques*, New York: Wiley.

O'Malley, R. E. (1991), *Singular Perturbation Methods for Ordinary Differential Equations*, New York: Springer-Verlag.

Sanders, J. A., Verhulst, F. (1985), *Averaging Methods in Nonlinear Dynamical Systems*, New York: Springer-Verlag.

Smith, D. R. (1985), *Singular-Perturbation Theory*, Cambridge: Cambridge University Press.

PETROLEUM

FRED H. POETTMANN AND ROBERT S. THOMPSON, *Department of Petroleum Engineering, Colorado School of Mines, Golden, Colorado, U.S.A.*

	Introduction	269
1.	**Definition of Petroleum**	270
2.	**Physical Properties of Petroleum**	272
2.1	Gas	273
2.1.1	Dry Gases	273
2.1.2	Wet Gases	273
2.1.3	Retrograde Gas Condensates	274
2.2	Oils	274
2.2.1	Volatile Oils	274
2.2.2	Black Oils	275
3.	**Geology and Occurrence of Petroleum**	275
3.1	Reservoir Traps	275
3.2	Reservoir Rock Characteristics	275
3.2.1	Porosity	275
3.2.2	Permeability	276
3.2.3	Factors Influencing Rock Properties	276
4.	**Drilling and Completion Operations**	277
4.1	Directional and Horizontal Wells	277
4.2	Casing and Wellhead Design	278
4.3	Completion Types	279
4.4	Improving Well Performance	280
5.	**Petroleum Recovery Mechanisms**	280
5.1	Primary Recovery Mechanisms	281
5.1.1	Solution-Gas Drive	281
5.1.2	Gas-Cap Drive	282
5.1.3	Water-Drive Reservoirs	282
5.1.4	Gravity Drainage	282
5.1.5	Combination Drives	282
5.2	Secondary Recovery	282
5.2.1	Water Flooding	282
5.2.2	Gas Injection, Pressure Maintenance	283
5.3	Improved Oil Recovery Mechanisms	284
5.3.1	Polymer Flooding	284
5.3.2	Miscible Flooding	284
5.3.3	Thermal Methods	285
5.3.3.1	Steam Injection	285
5.3.3.2	*In Situ* Combustion	286
5.3.4	Micellar-Polymer Flooding	286
5.3.5	Alkaline-Surfactant-Polymer Flooding	287
6.	**Production Operations**	287
6.1	Natural Flow	287
6.2	Artificial Lift	287
6.3	Field Processing	289
6.3.1	Natural Gas Processing at the Wellhead	289
6.3.2	Wet Gas, Gas Condensates, and Volatile Oil Processing at the Wellhead	289
6.3.3	Black Oils Processing at the Wellhead	289
7.	**Petroleum Reserves and Economics**	289
7.1	Definitions	291
7.2	Classification of Reserves	293
7.3	Petroleum Resource Assessment	293
7.4	Petroleum Economics	293
	Glossary	293
	Works Cited	294
	Further Reading	295

INTRODUCTION

The word "petroleum" is derived from the Latin—"petra" meaning rock and "oleum" meaning oil, "rock oil." Crude oil, natural gas, retrograde gas condensate, and volatile oils are all varieties of petroleum and are referred to as petroleum.

The petroleum industry as we know it today got its start in the mid nineteenth century. The very rapid growth of the industry came about with the development of the in-

3-527-28135-5/95/$5.00 + .50

ternal combustion engine in the beginning of the twentieth century. However, archaeological evidence indicates that ancient man made use of oil and gas seeps. Asphalt, a weathered oil, was used to cement stones, to waterproof cisterns, and to caulk boats. Chinese references over 3000 years old indicate that the Chinese used petroleum in their lamps. The Chinese while looking for brine to make salt developed a drilling technique similar to modern cable tool rigs. The ancient Chinese also transported natural gas through bamboo tubing for the purposes of heating and light.

In more modern times, the exploration for and the production of petroleum were accelerated by the demand for low-cost energy necessary to satisfy our consumer-oriented society. Not just gasoline, but heating oils, diesel fuels, lubricants, and the raw materials for petrochemical manufacture are produced from petroleum. Fuel oil and natural gas have replaced coal for home heating purposes.

The various categories of petroleum can best be characterized by their temperature–pressure phase diagrams along with the temperature at which the fluid exists in the reservoir. This identifies the petroleum as being a dry natural gas, a wet natural gas, a retrograde gas condensate, a volatile oil, or a black oil.

Petroleum occurs in underground reservoirs consisting of porous and permeable rock into which the petroleum migrated from a source bed. In order to produce this petroleum, wells are drilled to provide a conduit between the reservoir and the surface. The wells serve as a pressure sink into which the petroleum flows. The initial or primary recovery of the petroleum is due to the energy initially present in the reservoir fluids. Gas in solution in the petroleum or hydrodynamic pressure provides this energy. Once the primary energy has been dissipated, a large fraction of oil may still remain in the reservoir as a result of the capillary nature of the reservoir rock. At this stage of production, water or gas may be injected into the reservoir to provide the energy to displace additional oil. This is an immiscible displacement and is referred to as secondary recovery. Improved or enhanced oil recovery results from the injection of special gases at miscibility pressures or of liquid chemicals, or the use of thermal energy. The injected fluids interact with the rock–oil system to create conditions favorable to oil recovery.

When the natural energy of the petroleum in the reservoir is no longer adequate to cause the petroleum to flow to the surface, artificial means are used to lift the petroleum to the surface. This can be done in a number of different ways, depending on the circumstance. Sucker-rod pumps and gas lift are the most common methods used.

Once the petroleum reaches the surface, it is processed before entering a pipeline. The purpose of surface processing of the petroleum is to provide the purchaser with a water-free, hydrocarbon-liquid–free natural gas or, in the case of a condensate or crude oil, to provide a water-free, emulsion-free, and solid-free liquid ready for processing at a refinery or gasoline plant.

The lead time required from project inception to the first production of petroleum may be anywhere from 5 to 10 years and require billions of U.S. dollars of investment. Reservoirs with large initial petroleum fluids in place are needed. The ultimate economics will depend on the rate of petroleum fluids recovery and the recovery efficiency of the petroleum from the reservoir.

1. DEFINITION OF PETROLEUM

Petroleum can be divided into five classifications: dry natural gas, wet natural gas, retrograde gas condensate, volatile oils, and black oils. Although this classification appears to be quite specific, there is a continuity of the phase state between that of a dry gas and a black oil. A dry natural gas consists primarily of methane with perhaps traces of ethane, propane, and heavier components. A wet natural gas also consists primarily of methane; however, it contains a higher percentage of ethane, propane, and butane with traces of pentane through hexane and heavier components. At surface conditions of temperature and pressure, a dry gas remains in a single phase, whereas a wet gas produces condensate (thus wet) consisting of propane through hexane and heavier components. In the reservoir, both dry and wet gases remain in a single gaseous phase.

Retrograde gas condensates most often are found as a single gas phase in the reservoir.

As the reservoir fluid is produced and the pressure in the reservoir drops below its retrograde dew point, liquid condensate precipitates within the pore space of the reservoir rock. Unless the condensate is very rich in intermediate hydrocarbons (i.e., ethane through hexanes), the liquid saturation seldom reaches the critical value where the condensate starts to flow.

Volatile oils are liquids in the reservoir. However, the reservoir temperature is very close to the critical temperature of the volatile oil, and once the reservoir pressure drops below the bubble-point pressure a large gas saturation forms very rapidly in the reservoir. Without doing a phase-behavior study, it is difficult in the field to distinguish a rich retrograde gas condensate from a volatile oil.

Black oils are found as liquid in the reservoir. The composition of the black oil is high in the C_{7+} fraction. The C_{7+} fraction characterizes the black oil. Black oils or crude oils are further broken down as being paraffinic, aromatic, naphthenic, and asphaltic. This means that in the case of a paraffinic crude the C_{7+} fraction consists primarily of paraffinic hydrocarbons. Penn Grade crudes are paraffinic. High–molecular-weight paraffinic hydrocarbons make excellent lubricating-oil stock. A barrel of Penn Grade crude oil may consist of as much as 60% lubricating-oil stock compared with an asphaltic crude, which may have as low as 10% lubricating-oil stock. If the crude is classified as aromatic, naphthenic, or asphaltic, it means that the C_{7+} fraction contains a predominance of aromatic, naphthenic, or asphaltic compounds.

Figure 1 is a plot of the C_{7+} concentration in petroleum against the initial gas/oil ratio (GOR) of the petroleum reservoir fluid (McCain, 1993). The gas/oil ratio is the number of standard cubic meters of gas that is liberated from a reservoir fluid per cubic meter of liquid or stock-tank oil at the surface. A GOR of infinity would be a dry gas whereas a GOR of less than 360 m^3/m^3 would be a black oil. As can be seen from the plot, there is a continuity between the gaseous and liquid states of matter. As the GOR increases, we approach volatile oil and retrograde gas condensates. In fact, these two classifications of reservoir fluid overlap on the plot. As the reservoir fluid or well-stream effluent becomes leaner in the C_{7+} fraction, the GOR increases. Wet gas can have a GOR as low as 9000 m^3/m^3 of oil. As in the case of volatile oils and retrograde gas condensates, retrograde gas condensate and wet gases can overlap. In summary, black oils are repre-

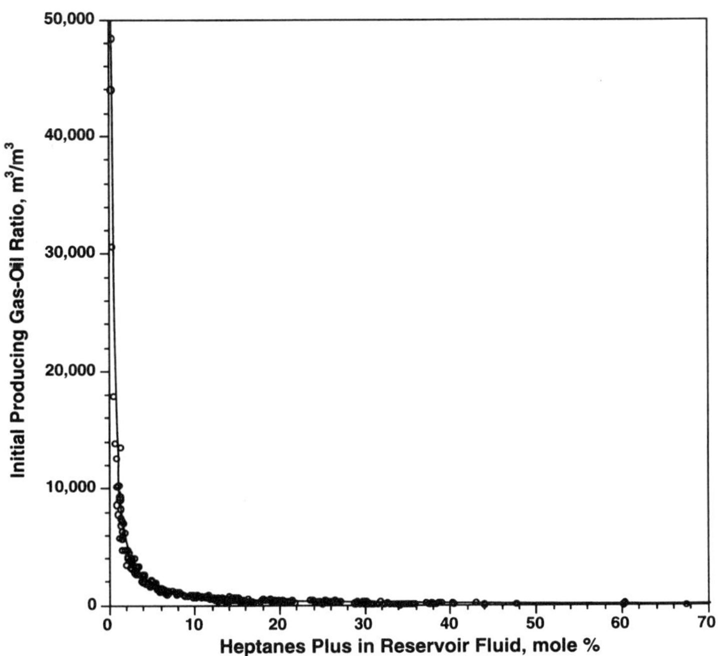

FIG. 1. Relationship between initial producing gas/oil ratio and heptanes-plus fraction in reservoir fluid (McCain, 1993).

Table 1. Classification of petroleum.

Classification	GOR (m^3/m^3)	CH_4 (%)	C_{7+} (%)
Dry gas	Infinity	>90	...
Wet gas	>9000	80–95	Trace
Retrograde gas condensates	530–9000	60–90	<12.5
Volatile oil	360–590	50–70	12.5–20
Black oil	<360	<45	>20

sented in the lower right of the figure and dry gases in the upper left. All other fluids are found in between with no sharp demarcation between these characterizations.

Table 1 is another way of illustrating this point. It shows the composition of petroleum under this classification system. Petroleum may also contain small amounts of nonhydrocarbons such as nitrogen, carbon dioxide, and hydrogen sulfide. Generally speaking, the concentration of nitrogen will range from 0 to 15 mole %, carbon dioxide from 0 to 4 mole %, and hydrogen sulfide from 0 to 6 mole % in petroleum. However, there are always exceptions. There are gas reservoirs that yield very high concentrations of both carbon dioxide and hydrogen sulfide. Table 2 illustrates some typical compositions of reservoir fluids.

2. PHYSICAL PROPERTIES OF PETROLEUM

The physical properties of petroleum reservoir fluids determine the calculation procedures that are used to predict the petroleum in place, the production performance, and the way the reservoir fluids are processed at the surface.

The five categories of petroleum reservoir fluids can best be described by their phase diagrams. The phase state of a reservoir fluid along with a knowledge of the rock and fluid properties is necessary to make reservoir performance calculations. The phase diagrams for these reservoir fluids are plots of pressure vs temperature that show the condition under which the fluid will exist as a gas, liquid, or both. Similarly, as shown in Fig. 1, phase diagrams also show a continuum between a dry gas and a black oil. The range of critical temperatures from that for a dry gas to that of a black oil can vary from

Table 2. Typical composition of reservoir fluids.

	Initial GOR (m^3/m^3)				
	Infinity Dry gas	27 660 Wet gas	1140 Gas cond	347 Volatile oil	145 Black oil
Component	Mole %				
N_2	...	0.30	0.88	1.55	...
CO_2	0.30	2.78	0.53	0.29	0.30
H_2S	...	...	9.42	...	...
C_1	98.50	94.41	67.98	51.97	41.30
C_2	0.87	1.44	6.21	11.72	4.93
C_3	0.17	0.34	2.37	9.23	3.85
C_4's	0.06	0.17	2.07	4.13	3.67
C_5's	0.02	0.08	1.21	1.82	2.36
C_6's	0.08	0.04	1.47	1.59	3.01
C_{7+}	...	0.44	7.88	17.70	40.58
	100.00	100.00	100.00	100.00	100.00
MWC_{7+} [a]	...	160	135	185	228
SGC_{7+} [b]	...	0.8190	0.7025	0.8231	0.8833
Reservoir T (°C)	...	126	120	173	76
Saturation pressure (kPa)	...	39 655	30 448	28 103	19 959

[a]Molecular weight of the C_{7+} fraction.
[b]Specific gravity of the C_{7+} fraction.

as low as −82.6 °C, the critical temperature of methane, to well over +600 °C for a black oil. The reservoir temperatures will range from as low as 32 °C to as high as 170 °C. Thus, in the case of a dry natural gas the reservoir temperature is considerably above the critical temperature of the natural gas, and in the case of a black oil the reservoir temperature is considerably below the critical temperatures of the black oil. In the case of a retrograde gas condensate, the reservoir temperature lies between the critical temperature and the cricondentherm or point of maximum temperature of the phase diagram.

The phase diagrams of natural gas will cover a much smaller temperature range because it consists of only very low–molecular-weight paraffin hydrocarbons when compared with a black oil that has in the mixture very large nonvolatile molecules. In going from a dry gas through wet gas, retrograde gas condensate, and volatile oil to black oil, the temperature range of the phase diagram increases dramatically.

2.1 Gas

2.1.1 Dry Gases Figure 2 shows the phase diagram of a dry gas. Since dry gas is primarily methane, the phase envelope is very small when compared with that of a black oil. The critical temperature of a dry gas will be in the range −82.6 to −20 °C, which is well below the reservoir temperature, line 1–2. The reservoir temperature is also well above the maximum temperature of the phase envelope. The fluid is a single-phase dry gas in the reservoir. Surface conditions are represented by point 3, also a single gaseous phase. As the reservoir is depleted, the pressure in the reservoir will drop from point 1 to point 2, the abandonment pressure.

The viscosity of the gas in the reservoir is low and its ability to flow will be primarily limited by the permeability of the rock matrix. Typically, the viscosity of dry gases at reservoir conditions will range between (0.013 and 0.030) × 10^{-3} Pa s.

2.1.2 Wet Gases The phase diagram of a wet gas is shown in Fig. 3. Here the reservoir temperature line and the pressure depletion path 1–2 in the reservoir lie much closer to the two-phase region of the phase diagram but still do not enter the two-phase region. The reservoir fluid remains in a single phase in the reservoir throughout the life of the field.

However, the surface condition of temperature and pressure lies within the two-phase region, point 3, and a low–specific-gravity liquid will condense at the surface. The viscosity of the wet gas at reservoir conditions of temperature and pressure will be in the range of (0.013 to 0.040) × 10^{-3} Pa s. The

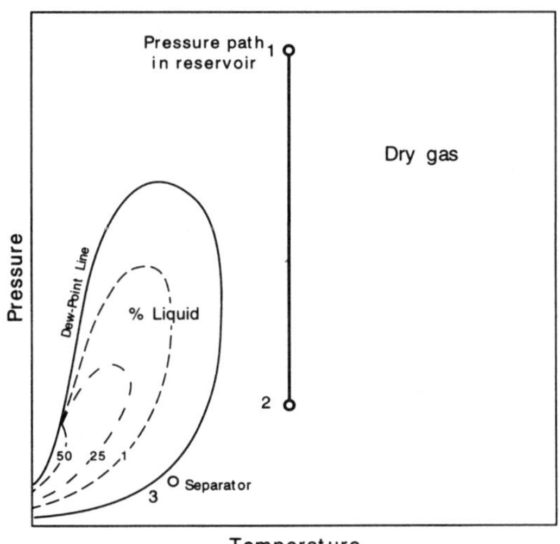

FIG. 2. Pressure–temperature diagram for dry gas reservoir.

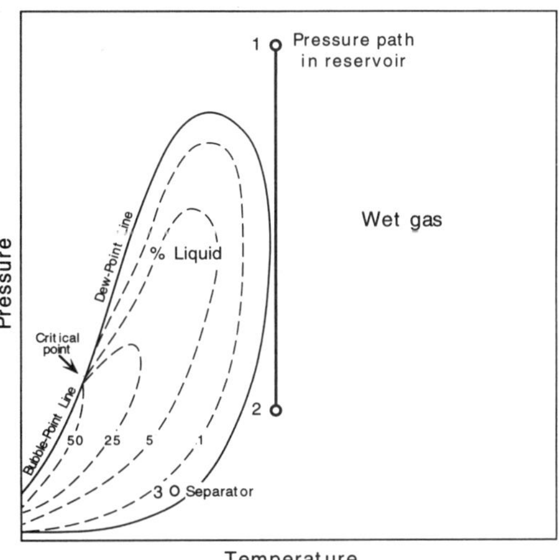

FIG. 3. Pressure–temperature diagram for wet gas reservoir.

stock-tank liquid or condensate produced from a wet gas is usually light yellow to water white and has a very low specific gravity in the range of 0.7389 to 0.8251. The specific gravity is calculated by comparing the density of the condensate with the density of water when both are at 15.56 °C (15.56/15.56 °C). During the 1920s and early 1930s, the condensate from a wet gas was used as gasoline to power oil-field vehicles.

2.1.3 Retrograde Gas Condensates In the case of a retrograde gas condensate, the reservoir temperature lies between the critical point and the point of maximum temperature, Fig. 4. At the original reservoir pressure, the gas exists as a single phase in the reservoir, point 1. As the fluid is produced, the reservoir pressure drops until the retrograde dew-point pressure is reached, point 2. At this point, liquid starts to condense in the reservoir. Generally, the liquid does not reach a saturation at which it flows. Upon further reduction in pressure, some of the liquid begins to revaporize; however, most of it is lost in the reservoir unless some enhanced recovery process is employed.

Point 3 represents the reservoir abandonment pressure. Point 4 represents the surface conditions of the separator. Retrograde gas condensate reservoirs generally range from 20 000 to 55 000 kPa and from 93 °C to as high as 170 °C. Fluid viscosity at reservoir conditions is in the range of (0.010 to 0.060) $\times$ 10^{-3} Pa s. The specific gravity of the stock-tank condensate is in the same range as the condensate of a wet gas, i.e., 0.7389 to 0.8251 at 15.56/15.56 °C. However, the condensate can range in color from water white through lightly colored to brown, orange, or green.

2.2. Oils

2.2.1 Volatile Oils In the case of a volatile oil, the reservoir temperature lies just below the critical temperature of the volatile oil. The critical temperature of a volatile oil is considerably lower than that of a black oil. When the reservoir pressure drops below the bubble-point pressure of the oil, a large gas saturation immediately forms in the reservoir. A pressure drop of (5–15) $\times$ 10^2 kPa may result in as much as a 50% gas saturation in the reservoir. The gas released in the reservoir is a retrograde gas and, consequently, a large part of the stock-tank liquid is produced from this retrograde gas. The stock-tank liquid has a specific gravity that is less than 0.8251 at 15.56/15.56 °C. The stock-tank liquid is usually orange to brown in color.

Figure 5 is the phase diagram for a volatile oil. Line 1–2–3 is at the reservoir temperature. Since the temperature is below the critical temperature, a volatile oil exists as a liquid in the reservoir, point 1. Point 2 is the

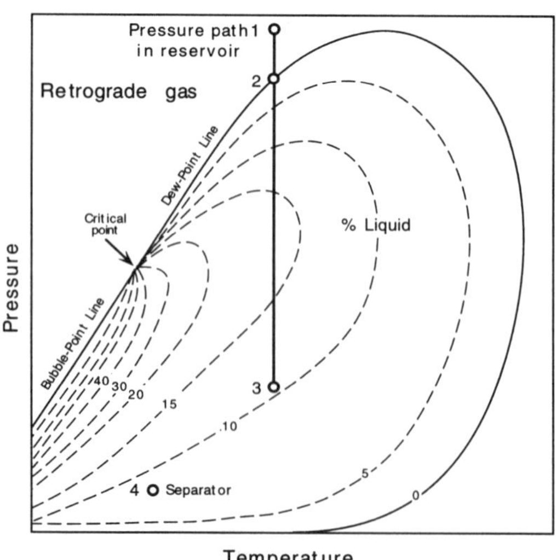

FIG. 4. Pressure–temperature diagram for retrograde gas reservoir.

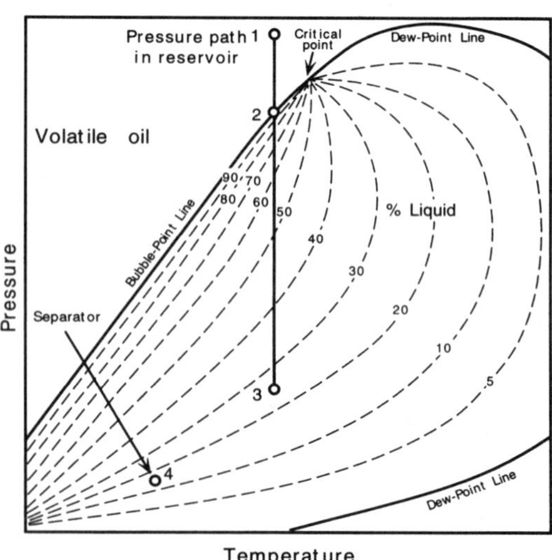

FIG. 5. Pressure–temperature diagram for volatile oil reservoir.

bubble point of the oil at reservoir temperature. Point 3 represents the reservoir abandonment pressure. Point 4 represents the separator conditions at the surface. At point 1, the reservoir fluid is undersaturated, that is, no gas evolves as the pressure drops until the bubble-point pressure is reached.

2.2.2 Black Oils The phase diagram of a black oil is shown in Fig. 6. The path 1–2–3 at constant reservoir temperature represents the reduction in pressure that occurs during the production of the reservoir. Point 4 represents the surface separator conditions. When the pressure is above point 2, the reservoir fluid is undersaturated, that is, no gas comes out of the solution as the pressure drops. At the bubble-point pressure, point 2, gas will start to come out of the solution and form a free gas phase in the reservoir. As the pressure drops, more gas comes out of the solution and flows to the surface separator with the oil. The specific gravity of the stock-tank oil is usually greater than 0.8251 at 15.56/15.56 °C. Stock-tank oil viscosity can range from 1.0×10^{-3} Pa s to very heavy viscous crudes having viscosities of many Pa s. The heavier crudes are usually void of the intermediate light hydrocarbons C_4 through C_{15}. The color of the stock-tank liquids usually range from light to dark brown. Pennsylvania Grade crudes, in which the heavy ends are highly paraffinic, are dark fluorescent green in color.

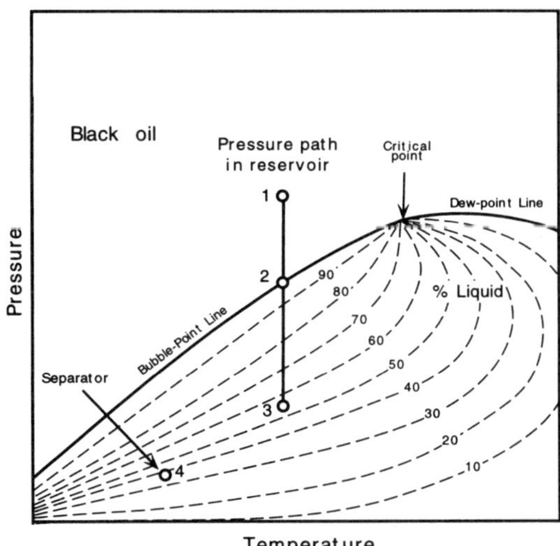

FIG. 6. Pressure–temperature diagram for black oil reservoir.

3. GEOLOGY AND OCCURRENCE OF PETROLEUM

Petroleum occurs in sediments deposited over geologic time. It is generally thought that petroleum originated in source beds consisting of mud and silt and was formed through the decomposition of plant and animal life. As the sediments were buried, the temperature and pressure increased, causing the organic matter to decompose thermally to form petroleum. The petroleum is expelled from the source rocks and migrates into layers of porous and permeable reservoir sands and carbonates. These sedimentary reservoir rocks are capped by an impermeable bed that "traps" the hydrocarbon and prevents further migration of the petroleum. The portion of the trap in which the petroleum occurs is called a petroleum reservoir.

3.1. Reservoir Traps

Reservoirs are generally classified according to the geological features under which they are formed. Cross sections of the different types of reservoir traps are presented in Fig. 7. Generally, traps are classified as structural, stratigraphic, or a combination structural/stratigraphic trap. In Fig. 7, example combination traps are referred to as strati-structural traps. Structural traps are a result of the deformation of the earth. Stratigraphic traps are caused by a change in the permeability and porosity in the reservoir bed.

3.2 Reservoir Rock Characteristics

Generally speaking, reservoir rocks are either sandstones or limestones; however, some reservoirs have been discovered in fractured shale. A knowledge of a reservoir's physical characteristics is of utmost importance in the management of the reservoir for maximum recovery of the petroleum it holds.

3.2.1 Porosity Porosity is the fraction of the total bulk volume of the rock not occupied by rock. Total porosity includes all the pore space in the rock, whereas effective porosity includes only the void space that is interconnected and through which fluids can flow. The pore space in a reservoir is filled with water and petroleum, that is oil and gas. Porosity is measured as a percentage of total

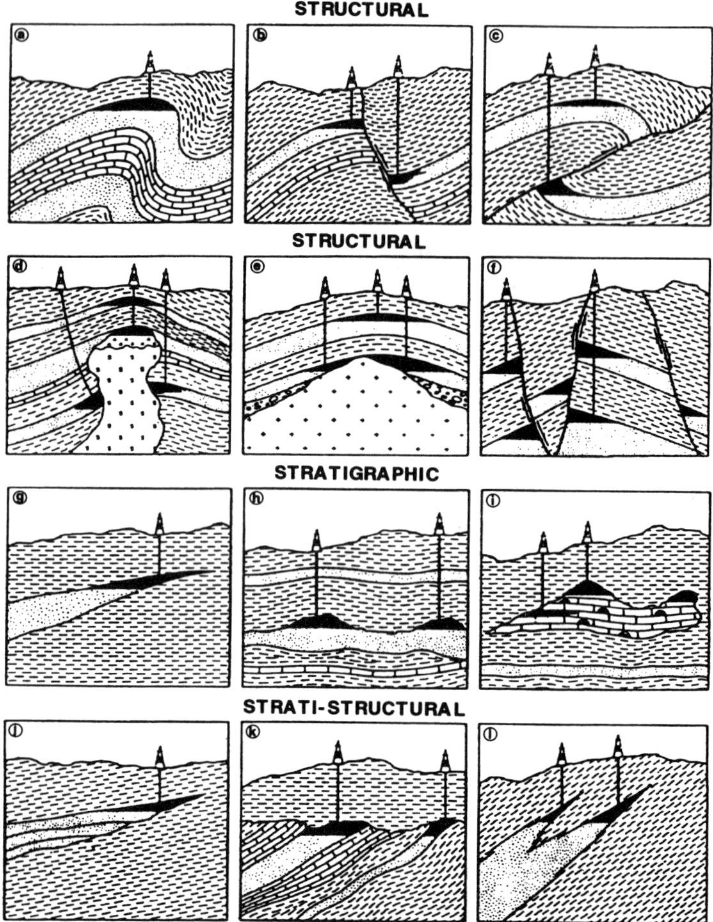

FIG. 7. Reservoir traps (courtesy Colorado School of Mines).

rock bulk volume, e.g., 20%–30% porosity. Porosity can vary from a few percent to as high as 35%–40%. Fluid content (saturation) of the pore space is measured as a percentage of the total pore space, e.g., 60% oil saturation and 40% water saturation.

3.2.2 Permeability Permeability is a measure of the ease with which a fluid can flow through the interconnected pore space of a rock. The unit of permeability is the square micrometer (μm^2), or 10^{-12} m^2. A permeability of 1 μm^2 will permit the flow of 1 m^3/s of fluid of 1 Pa s viscosity through an area of 1 m^2 under a pressure gradient of 10^{12} Pa/m. Or, stated another way, a permeability of 1 m^2 will permit the flow of 1 m^3/s of 1 Pa s viscosity through an area of 1 m^2 under a pressure gradient of 1 Pa/m.

Usually, there is more than one fluid flowing through the reservoir rock. Under these circumstances, we consider the relative permeability of the fluid, that is, the ease with which one fluid (of the two or more fluids present in the rock) will flow through the connecting pore spaces in the presence of the other fluids as compared with the ease with which the fluid will flow if it alone were present in the rock. Relative permeability then is the ratio of the permeability of a fluid flowing in the presence of other fluids to the permeability of the fluid if it alone were present in the rock.

3.2.3 Factors Influencing Rock Properties Sedimentary rocks are classified as sandstones, shales, limestones, and dolomites. The size and shape of the sedimentary materials as they were laid down, the presence of clay, the types of clay, and the presence of other minerals and the manner in which the sedimentary materials are cemented will affect the porosity and permeability of the reservoir rock.

4. DRILLING AND COMPLETION OPERATIONS

Drilling operations are conducted to provide a conduit between the reservoir rock and the surface for the production of oil and gas. Rotary drilling rigs are commonly used to drill wells. Drilling mud is pumped down the drill pipe and is returned to the surface in the annular space between the drill string and the earth, or a string of pipe called casing. The drilling mud must be capable of lifting the cuttings to the surface and controlling subsurface pressures encountered during the drilling process while minimizing damage to the reservoir rock.

The team responsible for the well gathers information during the drilling process to help evaluate the economic merit of the well. Drill cuttings are examined and identified. The drill cuttings may indicate zones of possible hydrocarbon accumulation. These zones are often tested with a tool run in the well on the drill string. This test, called a drill-stem test, gives an estimate of formation pressures, produced fluids (oil, gas, water), and potential flow rates for the well. Sometimes a larger sample of the formation is needed to estimate formation properties such as porosity and permeability. In these cases, a core is cut. A core is a solid, 7- to 10-cm-diameter cylindrical piece of the formation that may be up to 10 m or more in length. The core helps in characterizing the formation. The reservoir characterization helps the development team with reservoir management decisions such as well placement, completion methods, and secondary and enhanced recovery methods.

Well logs are another source of information for the evaluation team. Tools are run into a well on a cable to evaluate a section of the well bore before casing is set opposite this section. Logs run in this manner are called conventional open-hole wireline logs. These logs are run to estimate the porosity of the rock and the water saturation. The estimate of the water saturation permits an estimate of the hydrocarbon content of the formation. The porosity and water saturation of the rock give the evaluation team an estimate of the amount of hydrocarbon that may be present in the formation.

In some cases, it is desirable to have real-time porosity and saturation data rather than waiting for conventional wireline open-hole logs. In these cases, logging tools are run behind the drill bit as the well is drilled. These tools are called MWD (measurement while drilling) tools. When MWD tools are run, conventional wireline logs are still run.

The decision to abandon the well as a dry hole or to complete the well as a producer is made after all the available data are examined: the well logs, cores, and drill-stem tests. Petroleum must be present in an economic quantity as well as producible at economic rates before the decision is made to complete the well.

4.1 Directional and Horizontal Wells

Wells should be placed in the reservoir to maximize the economic recovery of oil and gas. In deep water, offshore fields can often only be economically developed from a single platform with multiple wells because of high platform costs. For this reason, offshore fields require the use of *directional* (or deviated) wells to develop the field efficiently and economically. In this case, the wells are drilled from an offshore structure and "steered" to the desired bottom-hole location selected by the development team of geologists, geophysicists, and petroleum engineers.

Another example of where directional wells are encountered is the Long Beach Unit near Los Angeles, California. Four man-made islands offshore act as "platforms" where over 1200 directional wells have been drilled to develop the Wilmington Field. A portion of the field lies beneath the city of Long Beach, California. The drilling rigs are soundproof and look like apartment buildings. Care is taken to survey each well bore so that the many well-bore paths do not accidently intersect.

Directional wells are drilled onshore whenever the bottom-hole objective cannot be reached by a vertical well.

When the well bore that penetrates the reservoir is horizontal, the well is referred to as a *horizontal* well. Horizontal wells are becoming an important method to improve the recovery from some reservoirs. Advances in reservoir characterization have identified reservoirs where horizontal placement of the well bore will enhance the drainage geometry and the ultimate recovery of oil from the well. Reservoirs having thin beds, vertical fracture systems, and compartmentalization are good candidates for horizontal drilling. Horizontal wells are more expensive to drill than a conventional or vertical well. Accordingly, the productivity from a horizontal well must be greater than a conventional well to justify the incremental cost of drilling. The incremental costs to drill a horizontal well may be as much as 30%–100% over the cost to drill a conventional vertical well. However, when the technology is appropriate, the productivity may be 2–6 times greater than a conventional completion (Fisher, 1988).

4.2 Casing and Wellhead Design

Pipe is placed in a well to provide a safe passage for the produced fluids and to protect formations (fresh water, for example) from contamination by the produced fluid; see Fig. 8. This pipe is referred to as casing and is cemented in the well for these reasons. Conductor pipe, set at 15–30 m, is cemented in the well, and provides a base for the surface wellhead equipment and an annular space for the drilling mud to be circulated to the surface.

After the conductor pipe is set, surface casing is set approximately 300 m below ground level to protect fresh-water sands. The surface casing is usually cemented completely from the surface to the bottom of the surface casing.

In deep wells, several strings of casing may be cemented in the well. The final casing that is set is referred to as the production casing or string. This casing is not set until after well logs are run and the decision is made to complete and produce the well. The production casing string and any intermediate strings may not be cemented all the way to the surface. Production casing is set to control well pressures and to protect formations during production. Figure 8 shows a production liner that has been set instead of running production casing from the surface to the producing formation. This is done to reduce costs.

When the reservoir fluid reaches the surface, it flows through the wellhead; see Fig. 9. Wellheads can vary greatly in design depending on the fluid pressure. The first element of a wellhead is the casing head, which helps support the weight of the casing or outer pipe. If there is more than one casing string, several casing heads may be used. Next is the tubing head, which is used to support the tubing string. Next comes the Christmas Tree; see Fig. 9. The Christmas Tree controls the

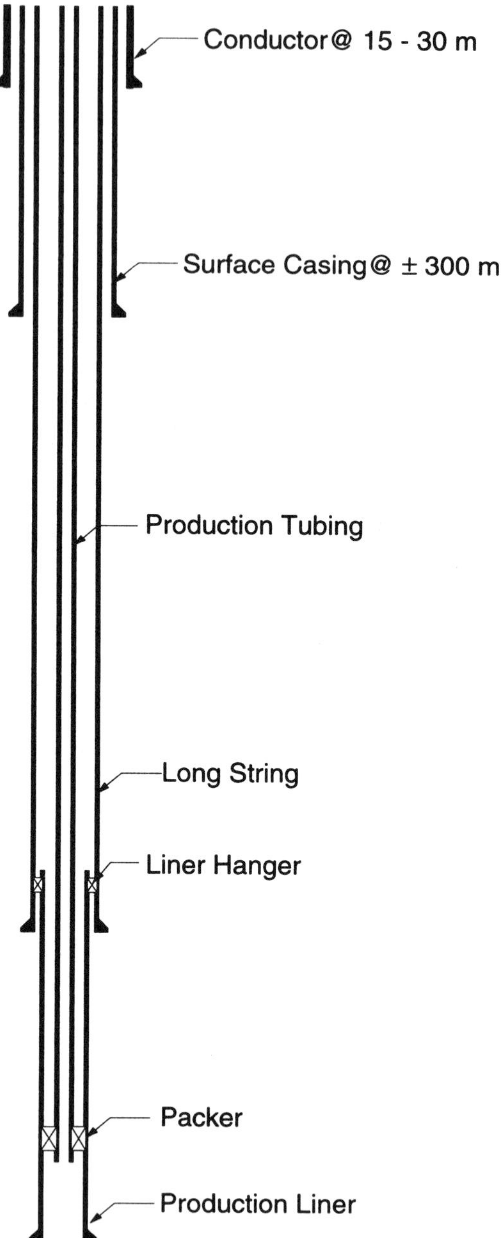

FIG. 8. Well completion design.

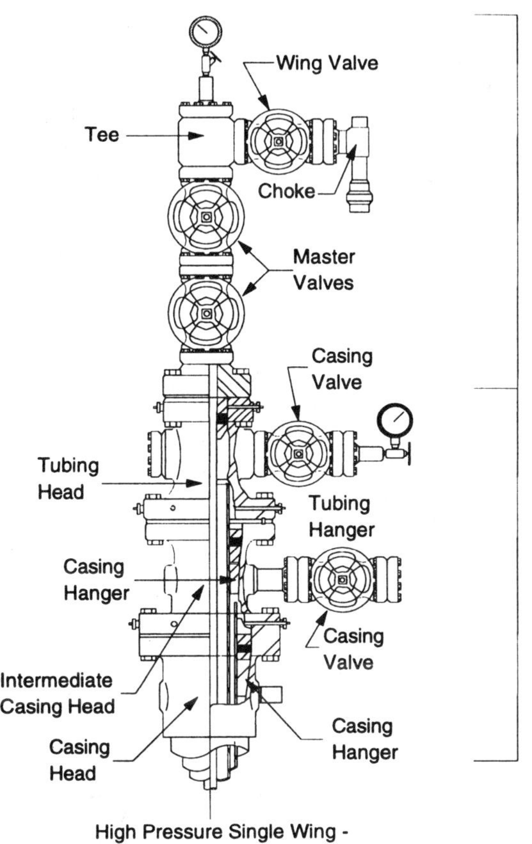

FIG. 9. Well-head equipment and Christmas Tree.

flow of gas and oil from the well and is installed on top of the tubing head. The Christmas Tree design can vary depending on the fluid pressure. Pressure gauges are installed on the Christmas Tree. Valves on the "Tree" control the flow from the well. A choke is also part of the Christmas Tree. The choke is an orifice in the line that controls the rate of flow of the well fluid. A choke can be fixed or adjustable.

4.3 Completion Types

Once the decision is made to complete the well, the type of completion must be considered. Completion refers to the manner in which the well bore opposite the producing formation is designed or equipped. Some of the advantages of the various completion methods will be presented.

Perforated production casing is a common method of completing a well in areas where the producing formation is consolidated. Casing is set through the production interval and perforated with a perforating gun; see Fig. 10. The holes in the pipe can be made with bullets or with a shaped charge. The jet from a shaped charge has a velocity of approximately 9500 m/s and can penetrate through the pipe and cement and more than a meter into the formation. The holes are approximately 1.3 cm in diameter and provide the conduit between the formation and the well bore. Production tubing is run from the surface to the perforated interval. The production tubing is anchored in the well by a tool called a packer. The packer also isolates the annulus between the production tubing and the production casing; see Fig. 10.

In some cases, the last string of casing does not run from the surface to the bottom of the well. This last casing string is referred to as a *production liner*; see Fig. 8. A production liner is a piece of pipe that is hung from the next to last casing string. The liner is cemented in the well and perforated. The pri-

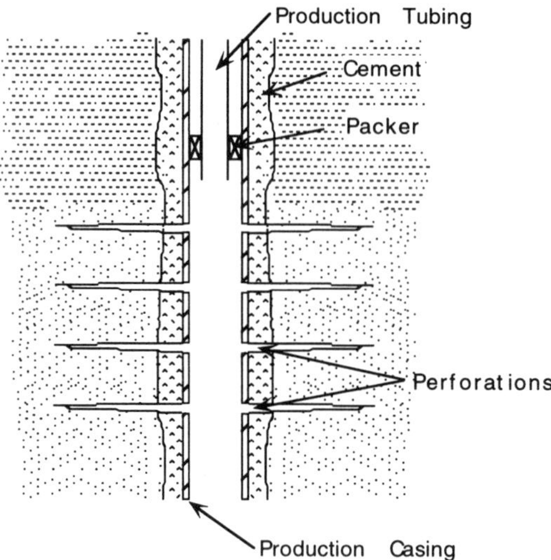

FIG. 10. Perforated casing completion.

mary advantage is a savings in the cost of casing since the last string does not need to be run from the surface to the bottom of the well. The disadvantage is a possible leak at the liner hanger.

In areas where the producing formation is unconsolidated and sand production is a problem, a *gravel-pack* completion is used to hold the sand in place and minimize the resistance to reservoir flow. In a gravel-pack completion, gravel is placed against the formation to hold the formation sand in place followed by a screen to keep the gravel in place. This method of completion is common in the Gulf Coast. New methods of completing unconsolidated formations are being tested today (1995).

A final completion method is an *open-hole* completion where the last string of casing stops above the productive zone. The well is drilled out below the casing into the zone of interest. In an open-hole completion, the restrictions to flow into the well bore are minimal compared with a perforated completion. This type of completion will only work in consolidated formations where the formation will not collapse.

4.4 Improving Well Performance

The value of the oil and gas is a function of how many barrels can be produced and the rate of production (deliverability). There are many formations where economic flow rates are not possible because the natural formation permeability is extremely low. *Hydraulic* fracturing is used to improve the formation flow rates. In these cases, a fluid is pumped down the well at high pressures and rates until the formation fractures. Once the formation has begun to fracture, a proppant (e.g., sand or bauxite beads) is pumped into the fractures. The pressure is then decreased, permitting the fractures to close on the proppant. The proppant-filled fractures provide a high-permeability conduit to the well bore for the produced fluids.

Acidizing a well may also improve the flow performance. Acidizing is the process of pumping an acid into the formation to improve the flow rate of the fluids from the formation into the well bore. Acidizing can be used to remove fine silica materials deposited during the drilling process. This fine silica material if *not* removed will restrict the flow of reservoir fluids. An acid can also be used to react with the formation rock and dissolve some of the rock. The formation may or may not be fractured during an acid job. Limestone and dolomite formations are good candidates for an acid treatment. An acid job in this case may include pumping 15% HCl (hydrochloric acid) that will react with the carbonate rock. There are acids that will react with quartz sandstone formations.

5. PETROLEUM RECOVERY MECHANISMS

Petroleum recovery can generally be classified into three steps: primary recovery, secondary recovery, and improved or enhanced recovery methods.

Primary recovery is that recovery in which the natural energy present initially in the reservoir is the main source of energy causing the displacement of petroleum to the producing well. The petroleum in the reservoir upon discovery is generally under pressure because of gas in solution in the petroleum liquid or hydrodynamic pressure or both. In the case of a gas reservoir, the gas itself is under pressure. When a well penetrates the reservoir, it provides the pressure sink to which the reservoir fluid flows. Because of the capillary nature of the reservoir rock and the flow of both oil and gas in the rock, oil recovery can vary from as low as 5% to as high as 50% of the liquid petroleum initially in place. On the average, primary recovery generally varies around 5%–30% of the oil initially in place.

In the case of a dry gas or wet gas reservoir, the production of the gas depletes the reservoir energy to the point where the production rate is no longer economical and the well is abandoned. In this case, the recovery factor varies from 60% to 90% of the gas initially in place.

In the case of a retrograde gas condensate reservoir, liquid condensate will drop out in the pore spaces of the reservoir rock as the pressure in the reservoir drops. Liquid dropout will reach a maximum and then as the pressure further declines some of the liquid condensate will revaporize. In the case of a rich retrograde gas condensate, the loss of liquid in the reservoir can be quite large when the abandonment pressure is reached. To prevent this liquid dropout, dry gas is quite often recycled into the reservoir to maintain the pressure and thus prevent the liquid from condensing.

In the case of a volatile oil reservoir, a small reduction in reservoir pressure below the bubble-point pressure of the volatile oil creates a large gas saturation. As a result, considerable liquid saturation can remain in the reservoir on abandonment by simple pressure depletion. Similarly, reinjection of dry gas into the reservoir to maintain pressure can reduce the loss of hydrocarbon.

After the primary energy in the reservoir has been dissipated, as much as 70%–95% of the oil may be left in the reservoir. The injection of water or gas into the reservoir to provide the energy to displace the oil to the producing well is referred to as *secondary recovery*. The displacement of the oil by gas or water is an immiscible fluid process, and because of the capillary effects, some of the oil is not displaced and a residual saturation of oil remains in the reservoir.

Improved or *enhanced oil recovery* results primarily from the injection of special gases at miscibility pressures or of liquid chemicals, or the use of thermal energy. The injected fluids interact with the rock–oil system to create conditions favorable to oil recovery by

1. lowering the interfacial tension between the oil and displacing fluids, thus reducing or eliminating capillary effects,
2. swelling the oil,
3. reducing the oil viscosity,
4. modifying the rock wettability, and/or
5. creating favorable phase behavior.

The viability of any improved oil recovery process depends on two factors, the unit displacement efficiency and the volumetric displacement efficiency. Unit displacement efficiency is the effectiveness of the displacing fluid displacing the oil in the *rock volume contacted*, whereas volumetric displacement efficiency refers to the fraction of total reservoir volume contacted by the displacing fluid.

Unit displacement efficiency is determined by the interfacial forces between the displacing fluids and displaced fluids. If the interfacial tension between the displacing fluid and displaced fluid is high, the capillary forces will be high and oil retention will be large. On the other hand, as the interfacial tension between the fluids is lowered the capillary forces will be lower. At an interfacial tension of zero, the fluids become miscible and the capillary forces are completely eliminated. At this point, the unit displacement efficiency is 100%.

The volumetric displacement efficiency is a function of mobility of the displacing and displaced fluids. The lower the viscosity of a fluid, the more mobile the fluid. A fluid of low mobility displacing a fluid of high mobility displaces the fluid in a pistonlike manner, whereas a fluid of high mobility displacing a fluid of low mobility will cause fingering and bypassing of the low-mobility fluid resulting in a lower volumetric displacement efficiency.

Improved oil recovery processes attempt to maximize both the unit displacement efficiency and the volumetric displacement efficiency. It is not always possible to do both at the same time. Application of improved oil recovery processes usually entails substantial risk because of their technological difficulties and front-loaded financial requirements.

5.1 Primary Recovery Mechanisms

5.1.1 Solution-Gas Drive In a solution-gas drive reservoir, the reservoir energy is derived from the light hydrocarbons, primarily methane, that are in solution in the oil in the reservoir. As the oil is produced and the reservoir pressure drops, light hydrocarbons come out of the solution and form a gas phase. The remaining gas in the solution plus the gas phase drives the oil to the well bore as the pressure declines. Since the gas is more mobile than the oil, the producing gas/oil ratio increases as the pressure decreases until a pressure is reached where the gas/oil ratio peaks and then starts declining. Up to this point, the reservoir energy is sufficient for the oil to flow up the tubing of the well. However, as the pressure decreases a pressure will eventually be reached where the well is put on artificial lift. Solution-gas drive reservoirs leave large quantities of oil in the reservoir. Recoveries are of the order of 10%–15% of the original oil in place.

5.1.2 Gas-Cap Drive Oil reservoirs can exist with a cap of gas over the oil. This gas cap is a source of energy along with the gas in solution in the oil to provide energy to move the oil to the well bore and lift it to the surface. As oil production proceeds, the gas cap expands into the reservoir originally containing the oil. The larger the gas cap, the smaller the rate of pressure drop as the oil is produced. This has the advantage of keeping gas in solution in the oil. As the gas cap expands and encroaches on a producing well, the gas/oil ratio may suddenly increase. Oil recovery will normally be in the range 20%–40% of oil originally in place.

5.1.3 Water-Drive Reservoirs In an active water-drive reservoir, the reservoir pressure declines relatively slowly as the oil is withdrawn from the reservoir and is replaced by water from an aquifer underlying the reservoir. The gas/oil ratio remains relatively constant because the pressure decreases very slowly, which keeps the gas from coming out of solution and forming a gas saturation. Water production from a well will increase gradually as the water encroaches into the reservoir. When water production gets too high, the well is abandoned since it becomes uneconomic to operate. Oil recovery from a water-drive reservoir will range from 30% to 75% of the oil originally in place, but is highly dependent on the depositional character of the reservoir as well as the drive mechanism.

5.1.4 Gravity Drainage In the case of gravity drainage, the fluids in a reservoir segregate in the reservoir according to their densities. Gravity drainage can play an effective role in the recovery of oil from a reservoir. By proper control of production rate, wells that are drilled into the bottom of the reservoir structure produce oil by gravity drainage of the oil into the well bore. Gravity drainage can be a very effective production mechanism, and recoveries can reach as high as 80%–90% of the oil originally in place.

5.1.5 Combination Drives Most often a combination of producing mechanisms exists in a reservoir under primary production. A common combination is one in which both water and gas-cap gas displace oil toward a producing well with perhaps some assistance from gas in solution in the oil. The degree to which each mechanism participates in the primary production of the oil is specific to the particular reservoir and the environment in which the reservoir exists. Since the water, oil, and gas are immiscible with one another, the capillarity of the reservoir rock traps large amounts of oil that are left behind once the primary production phase is completed. This oil provides the challenge to improved or enhanced oil recovery processes.

5.2 Secondary Recovery

5.2.1 Water Flooding The process of injecting water into a reservoir to displace oil was called a secondary recovery process because the process produced a second yield of oil after the field was depleted by primary production. Water flooding began accidentally in 1865 in the Pithold City area of western Pennsylvania. In the late 1800s, wells were being abandoned in the Bradford Field of Pennsylvania. In those days, there were no rules and regulations governing the abandonment of wells. Operators walked away from the wells, sometimes pulling casing, without plugging the wells. Fresh water from shallower horizons would enter the producing formation and stimulate production. These operations were named "Dump Floods." When operators realized that water was stimulating production, they deliberately started injecting water. By 1907, water injection had a significant impact on production in the Bradford Field. It was not until the 1930s that water flooding operations were reported in the midcontinent, i.e., Kansas, Oklahoma, and Texas. However, it was not until the late 1940s and early 1950s that water flooding operations became common practice. About 50% of the current oil production in the lower 48 states is attributed to water flooding.

Water flooding is usually conducted in the form of a pattern flood under different spacing conditions. Patterns can vary among line drives, 5-spot, 7-spot, and 9-spot: see Fig. 11. In some cases, a peripheral pattern is used, in which wells around the periphery of the field or lease are used for injection purposes. The type of spacing and pattern used depends on the location of existing wells, geometry of the reservoir, cost of drilling new wells, and the nature of the offset operations.

5.2.2 Gas Injection, Pressure Maintenance The injection of air or natural gas into a reservoir to stimulate oil recovery occurred well before the advent of water flooding. In the early 1900s, most secondary recovery projects involved gas injection. In gas-injection pressure maintenance operations, the gas can be directly injected into the oil-bearing formation. This is usually done through a pattern injection, that is, 5-spot or line-drive pattern, in order to obtain uniform distribution of the injected gas. Reservoir configuration must also be taken into consideration. In oil reservoirs having a gas cap, gas is often injected into the gas cap using wells in the crest of the structure. Such formations having structural relief and high permeabilities also benefit from gravity drainage. The number of wells required will depend on well injectivity and aerial coverage.

In the case of a retrograde gas condensate reservoir, economics may dictate the injection of a dry gas to maintain the reservoir pressure at or above the dew-point pressure to prevent the dropout of retrograde liquid condensate and thus the loss of liquid in the reservoir. If the retrograde gas condensate reservoir has been produced by pressure depletion, cycling dry gas through the reservoir will recover some of the liquid condensed in the reservoir by vaporization into the gas phase. Whether to cycle or not to cycle depends on the economics and the amount of liquid lost in the reservoir. If dry natural gas is unavailable, an inert gas such as a stack gas or nitrogen has been used. Nitrogen has

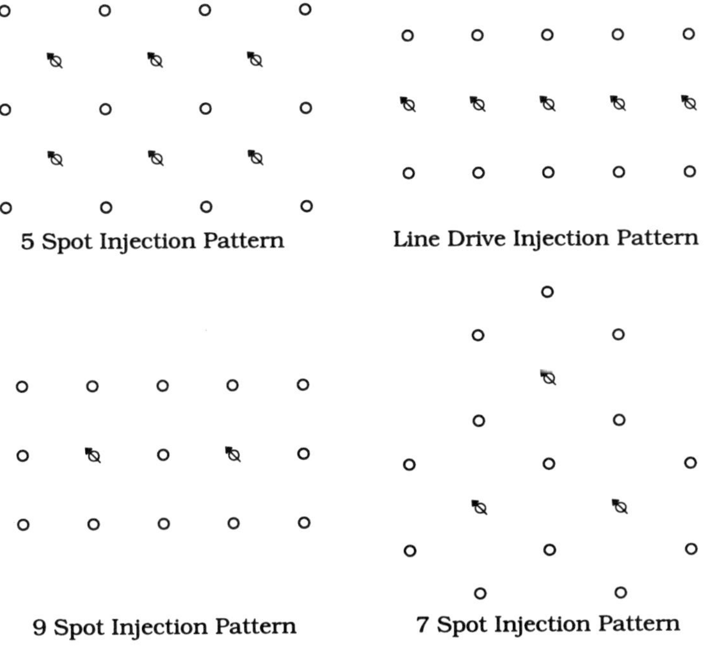

FIG. 11. Water flood patterns.

also been mixed with dry natural gas if not enough dry natural gas was available to maintain the desired pressure.

5.3 Improved Oil Recovery Mechanisms

Improved oil recovery results primarily from the injection of miscible gases and liquid chemicals or the use of thermal energy. Gases employed are hydrocarbon gases, carbon dioxide, or nitrogen. Liquid chemicals employed are polymer solutions, surfactant solutions, or hydrocarbon solvents. Thermal methods use steam or *in situ* combustion.

The lead time for the implementation of an improved oil recovery process from the time of first inception to the time of the first incremental volume of oil produced is anywhere from two to ten years depending on the complexity of the process and the reservoir to which it is to be applied.

The chemical requirements for improved oil recovery are also quite large, as shown in Table 3.

5.3.1 Polymer Flooding Polymer flooding is often referred to as improved water flooding. A high–molecular-weight polymer is added to water to lower the mobility of the water. Mobility of a fluid is defined as the permeability of the rock to the flowing fluid divided by the viscosity of the flowing fluid. The mobility of the water can be reduced either by increasing the viscosity of the water by adding polymer to the water or by reducing the permeability of the water to the rock by polymer adsorption and retention on the porous matrix of the rock. Improved volumetric sweep efficiency results. Polymer concentrations vary from 250 to 2000 ppm.

Table 3. Chemical requirements for improved oil recovery processes.

Process	Kg chemical/m^3 of oil recovered
Liquid chemicals	
Polymer flooding	
Polymer	5.5–8.5
Micellar polymer flooding	
Polymer	3.0–4.2
Surfactant sulfonates	20.0–48.0
Co-surfactant alcohols	2.8–28
Gases	
Carbon dioxide miscible flooding	1400–2700 standard m^3 CO_2/m^3 of oil recovery
Thermal	
Steam flooding	1–3 m^3 of water as steam/m^3 of oil recovered
In situ combustion	1400–5300 standard m^3 air/m^3 of oil recovered

There are a number of general requirements for the polymers used in polymer flooding. The polymer in solution should

1. reduce mobility of the water solution;
2. propagate through the reservoir rock without plugging;
3. experience minimum loss on the reservoir rock;
4. have adequate mechanical, chemical, thermal, and biochemical stability; and
5. be economical.

Polymer flooding should be initiated as early as possible during the course of a water flood while the oil saturation is still high. The higher the producing water/oil ratio at the start of the polymer flood, the higher the risk of failure. By initiating a polymer flood early, oil production is accelerated such that the cumulative water/oil ratio at the economic limit is much lower than if initiated later in the life of the flood.

There are two types of polymers used in polymer flooding, partially hydrolyzed polyacrylamides (PHPA) and biopolymers, the latter also known as polysaccharides and xanthan gums. PHPA is a synthetic polymer and is the most commonly used polymer for polymer flooding. Biopolymers are produced by fermentation. Biopolymers control mobility by viscosity increase only, whereas PHPA controls mobility by permeability reduction and viscosity increase.

The amount of polymer solution injection into a reservoir varies widely. Generally speaking, successful slug sizes are 15% of pore volume or greater. Incremental oil recovery from polymer floods over that of conventional water flooding may only be of the order of 2%–3% of the oil in place at the end of the water flood. Even so the process is still economic.

5.3.2 Miscible Flooding In miscible flooding, a fluid is injected into a reservoir that at reservoir pressures either is first-contact miscible with the reservoir crude oil or develops miscibility with the reservoir crude oil through a multicontact mixing process. The approach is applicable to medium and light crude oils. The fluids injected can be light hydrocarbon gases, nitrogen, flue gases, and carbon dioxide. Carbon dioxide is by far the most commonly used of the gases because of its unique properties. For a gas to become miscible with a crude through a multicontact mixing process, there is a minimum pressure requirement. This pressure is referred to as the minimum miscibility pressure, MMP. When carbon dioxide is injected into a reservoir at or above the MMP, light to intermediate components in the oil are transferred to the carbon dioxide, and when the enrichment of light to intermediate components in the carbon dioxide becomes large enough the gas at the leading edge becomes miscible with the oil and displaces the oil miscibly. As pointed out previously, the miscible displacement approaches 100% unit displacement efficiency. The MMP can be lowered by enriching the gas with intermediates. If, for example, the existing reservoir pressure is below the MMP for pure carbon dioxide, enrichment of the carbon dioxide with intermediate hydrocarbon will lower the MMP to the point that it will be at or below the reservoir pressure. In this case, there is a transfer of the intermediates into the crude oil, enriching the oil to a point where the injected rich gas becomes miscible with the crude. Although miscible displacement with a gas achieves high unit displacement efficiency, the mobility of the gas is higher than that of the crude oil and results in low volumetric displacement efficiency, i.e., fingering and bypassing of oil. To overcome this problem, alternate injection of water with carbon dioxide to improve the carbon dioxide mobility has been practiced with limited success.

Operating problems with carbon dioxide miscible flooding can be severe if the proper precautions are not taken. Corrosion, scaling, and asphalt precipitation are some of the problems encountered.

Carbon dioxide forms a highly nonideal solution with crude oil, resulting in swelling of the crude and in the lowering of the crude viscosity. Both of these effects enhance oil recovery.

The magnitude of additional oil recovery by CO_2 miscible flooding after water flooding is generally in the range of 8%–15% of the oil in place at the start of the flood; however, there are exceptions.

Large volumes of carbon dioxide are required for oil-field flooding. Gross carbon dioxide/oil ratios have varied from as low as 900 to as high as 4600 standard m^3 of carbon dioxide per m^3 of stock-tank oil recovered. Carbon dioxide cost will determine the cutoff. Sources of carbon dioxide can be varied: that occurring naturally in reservoirs, byproduct of ammonia plants, power-plant stack gas, and manufactured carbon dioxide. Reservoirs that contain carbon dioxide are presently being used to supply carbon dioxide for oil-field flooding, primarily because they can supply the volumes and rates at reasonable costs. Produced gases containing high concentrations of carbon dioxide are often recycled.

Of all the enhanced oil recovery processes, carbon dioxide miscible flooding is one of the most frequently applied.

5.3.3 Thermal Methods Thermal recovery methods can be classified into two broad categories: that in which heat is injected from an external source and that in which heat is generated *in situ*. Of the various possibilities, only steam injection and forward or concurrent combustion have proven viable commercially.

5.3.3.1 Steam Injection. Steam injection can be broken down into two different processes, steam flooding or continuous steam injection and cyclic steam injection or stimulation. About 72 187 m^3 of oil per day were produced in the United States in 1992 through steam injection (Moritis, 1994). The great majority of this oil is produced in California.

When steam is used, thermal expansion, viscosity reduction, and steam distillation are the principal mechanisms by which oil is recovered. Steam distillation is an important mechanism. The lighter the crude oil, the more important the distillation effect becomes. One characteristic that makes steam so useful as an injection fluid is that it carries a large amount of heat per unit mass compared with other fluids.

Steam flooding is generally applied to heavier crude oils ranging in specific gravity from 0.9000 to 1.000 and viscosities from 0.10

Pa s to several thousand Pa s. The viscosities of heavy crude oils are very sensitive to temperature. In continuous steam flooding, the steam moves out into the reservoir until it condenses, heating the rock and oil to steam temperature. The condensed water is produced, and an expanding steam zone efficiently displaces the crude oil. Generally, the equivalent of about one-third of the additional oil recovered is consumed to generate the required steam. However, because of environmental drawbacks, natural gas is now the preferred fuel to generate the steam.

Cyclic steam injection is a single-well operation where steam is injected into a producing well for a limited time. The well is then shut in for a short period of time before it is put on production. The well is produced anywhere from a few months to a year before the cycle is repeated. Each cycle produces a decreasing amount of oil. Sometimes only a few cycles are employed, and other times as many as 15 cycles are employed before the production rate becomes so low that it is not economic. Steam cycling works best for thick sands, viscous oils, and small well spacings. Cyclic steam recoveries are generally 10%–25% of the oil in place compared with 50%–60% by steam flooding.

5.3.3.2 In Situ Combustion The *in situ* generation of heat involves injection of air or an oxygen–nitrogen mixture into a reservoir and burning part of the crude oil in the form of coke deposited on the sand grains just ahead of the moving combustion front. The combustion zone moves in the same direction as the flow of fluids. A clean sand is left behind the combustion front. The temperature in the combustion zone ranges from 315 to 650 °C. The combustion-zone thickness is perhaps 10–20 cm. The *in situ* combustion process is ideal in that an undesirable fraction of the oil (heavy ends) is consumed as fuel in the form of coke, a clean sand is left behind, and the heat generated is used as efficiently as possible. Oil recoveries for successful projects can range from 60%–90% of the oil in place at the time the project was initiated.

Basically, three physical processes are taking place simultaneously, but not independently of one another. These processes are chemical reaction, heat transfer, and fluid flow. *In situ* combustion is initiated by simultaneously injecting air- or oxygen-enriched gas into a well and heating the well bore opposite the formation to the ignition temperature. Occasionally, in the case of a highly reactive crude oil, spontaneous ignition will take place. Well-bore heaters are usually electrical or gas burners.

When applicable, the *in situ* combustion process is conducted in sandstone reservoirs. *In situ* combustion is generally not applicable to limestone or dolomite reservoirs, although there have been a few exceptions.

A variation of *in situ* combustion is wet combustion, also known as COFCAW (Combination of Forward Combustion and Waterflooding). Water is injected either with the air or intermittently. The process takes advantage of the latent heat of vaporization of water to recover heat energy stored behind the combustion front. However, one needs to be careful, as injecting too much water can extinguish the combustion front.

The Tulare formation of the South Belridge Field has undergone both *in situ* combustion and steam drive, and as a consequence an excellent comparison can be made between these two processes in this reservoir (Gates *et al.*, 1978; Gates *et al.*, 1975). *In situ* combustion was much more energy efficient than that of steam, using only about one-quarter of the surface energy compared with steam. From an environmental standpoint, steam emits about 2.5 times as much flue gas as *in situ* combustion. Well failures were as bad for steam as for *in situ* combustion. The question then arises as to why *in situ* combustion was not used more in this formation. The answer lies in the fact that steam was probably applied first and was an established improved oil recovery process in this field.

5.3.4 Micellar-Polymer Flooding Micellar-polymer flooding is a fluid-injection process wherein a phase-stable, optically clear (transparent or translucent) solution of oil, water, and one or more surfactants, along with electrolytes or salts, is injected into a formation and is displaced by a solution of polymer. The polymer solution is in turn displaced by water. Oil and water are displaced ahead of the micellar solution, and a stabilized oil and water bank develops. The micellar solution (microemulsion) is the key to the process. The micellar solution displaces the oil and water either miscibly or at very low interfacial tensions and thus has a high unit displacement efficiency. In addition, the mobility of the micellar solution can be controlled, resulting in high volumetric displacement efficiencies. The process is applicable over a wide range of reservoir conditions. Micellar-polymer flooding is usually applicable if water flooding is applicable because of the mobility relationships. The polymers used are usually partially hydrolyzed polyacrylamide. The surfactants used are petroleum sulfonates and alcohols. The alcohol is referred to as a co-surfactant.

Because of the cost of the micellar solution, only a limited amount is injected. Oil recovery by micellar-polymer flooding can be over 60% of the oil in place at the time the project was initiated.

The current price ($18–$20 per barrel) of crude oil makes micellar-polymer flooding uneconomic. Micellar-polymer flooding is front-loaded financially; that is to say, it takes a large initial investment to initiate a project and the maximum financial exposure is quite high. Unless the price of crude oil goes up to the $35–$45 per barrel range in 1994 dollars, micellar-polymer flooding will not be a factor in improved oil recovery.

5.3.5 Alkaline-Surfactant-Polymer Flooding A relatively new flooding process has been developed in recent years and shows considerable promise as a result of a number of field pilot operations. Many oils are reactive with alkali and produce surfactants that lower the interfacial tension between the oil and brine. However, the optimum alkali concentration to produce this reduction in interfacial tension is quite low. By adding a low concentration of a surfactant to the system, a synergism results wherein the interfacial tension is considerably lower than alkali or surfactant acting by themselves. Polymer is also added to the system to insure the best possible mobility control. Alkalis used range among caustic soda, soda ash, sodium orthosilicates, and trisodium phosphate. The polymer generally used is partially hydrolyzed polyacrylamide. Surfactants used are usually high–molecular-weight organic sulfonates. It has been shown in a limited number of field tests that alkali surfactant polymer (ASP) flooding can recover just as much oil as micellar polymer flooding but at considerably less cost per barrel of oil recovered. In 1994 dollars, the costs range from $2 to $3 per barrel of incremental oil recovered (Surkalo *et al.*, 1993). The design of ASP flooding systems is quite involved, requiring detailed knowledge of reservoir geology, and fluid and rock chemistry.

6. PRODUCTION OPERATIONS

Artificial lift is installed to increase production from naturally flowing wells and to produce from wells when the natural energy in the reservoir is no longer adequate to lift the oil and gas from the well. The following sections discuss natural flow and the four major methods of artificial lift used in the oil and gas industry: gas lift, sucker-rod pumps, electrical submersible pumps, and hydraulic pumps.

6.1 Natural Flow

The ability of a well to flow naturally is a function of the formation pressures and deliverability, well-head pressures, tubing size, and depth. Except in the case of a very strong aquifer that is providing pressure support as fluids are produced from the reservoir, the pressure of the reservoir will decline with production. As the pressure declines, the energy is no longer sufficient to lift the fluids from the well bore. At this time, an external means of lifting the fluids from the reservoir is considered. The goal is to match the ability of the formation to deliver fluids at a specified well-bore pressure with a method of artificial lift that will add sufficient energy to lift the fluids to the surface. Wells may be placed on artificial lift before natural flow ceases. Economics will govern when it is viable to place the wells on artificial lift.

6.2 Artificial Lift

The selection of an artificial lift method depends on several factors. Capital costs, operating costs, reliability, and salvage value are major considerations when comparing the various alternatives for artificial lift. Flexibility to handle changing production conditions over the life of the well should also be considered. A "cradle to grave" economic analysis should be performed when selecting artificial lift methods.

Approximately 10% of the wells on artificial lift in the United States are *gas lifted* (Clegg *et al.*, 1993). Gas lift is a common method of artificial lift offshore. In gas lift, an external source of gas is injected as deep as possible into the production tubing to augment the formation gas: see Fig. 12. In this sense, gas lift is an extension of natural flow. The injection gas lightens the fluid column in the production tubing and lowers the flowing bottom-hole pressure at the formation. The effect is to increase the formation flow and extend "natural flow." In cases where flow rates cannot be established by the continuous injection of an external source of gas, the well may be produced by the intermittent injection of high volumes of gas. In these cases, the well is alternately produced and shut in. During the shut-in periods, formation fluids accumulate in the well bore. High-pressure gas is then injected into the tubing below the slug of formation fluid to displace the formation fluids to the surface.

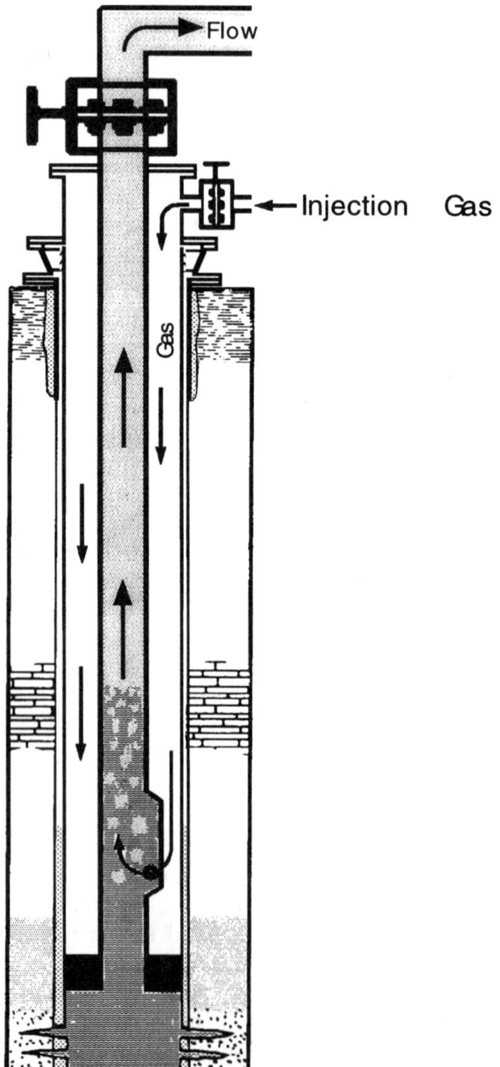

FIG. 12. Continuous gas lift.

The *sucker-rod pump* is the most common method of artificial lift: see Fig. 13. Approximately 85% of the wells on artificial lift in the United States are lifted with sucker-rod pumps. Sucker-rod pumps operate on the principle of positive displacement. The reciprocating motion of the down-hole pump is generated by a beam pumping unit at the surface and transmitted to the down-hole pump through steel rods called sucker rods. Part of the rod string may be made of fiberglass rods.

Sucker-rod pumps are easy to operate, and this is one reason for their popularity. Where space and weight are limiting factors, sucker-rod pumps are not a good choice. Sucker-rod pumps are rarely a good choice for offshore.

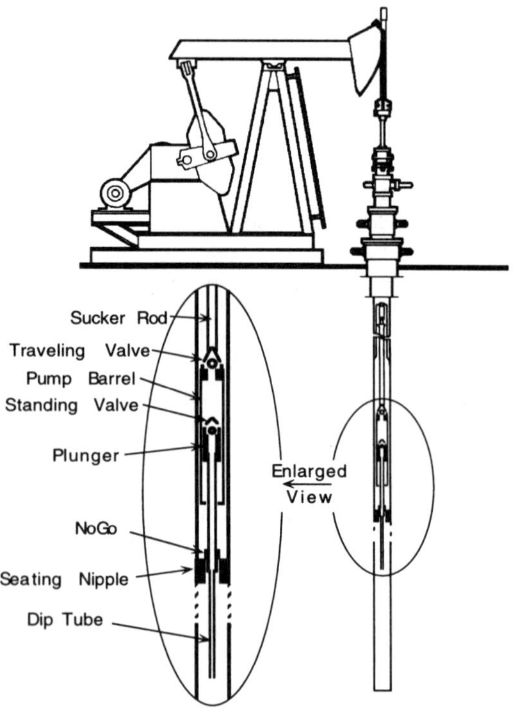

FIG. 13. Sucker-rod pump.

Depth of lift, especially for high-volume wells, is also a limiting factor for sucker-rod pumps.

Electrical submersible pumps (ESPs) are centrifugal pumps installed in the well on the end of the tubing and submerged in the well fluid. The down-hole components consist of an electric motor, motor protector, gas separator, pump, and power cable. The power cable, which is strapped to the tubing, transmits power from the surface to the down-hole motor.

ESPs are best suited for high-volume applications and are a common method of artificial lift where the reservoir is waterflooded. ESPs make up approximately 4% of wells on artificial lift in the United States (Clegg *et al.*, 1993).

There are two basic types of *hydraulic pumps* used in the petroleum industry, the positive displacement pump and the jet pump. In both cases, the pump is installed down hole to add energy to the produced fluid. Hydraulic pumps make up less than 2% of the wells on artificial lift in the United States (Clegg *et al.*, 1993).

6.3 Field Processing

When a reservoir fluid reaches the surface and flows through the wellhead, the reservoir fluid is further processed before entering a pipeline. In the case of a dry gas, the surface processing facilities can be quite simple, and in the case of a gas condensate or black oil they can be quite extensive. The objective of surface processing before the fluid enters the pipeline is to provide the purchaser with a water-free, hydrocarbon-liquid–free natural gas, or a clean condensate or crude oil.

6.3.1 Natural Gas Processing at the Wellhead Natural gas wells can produce both reservoir water and water that is in the form of vapor in the gas. Upon reaching the surface, water vapor condenses when the temperature is lowered. The presence of free water can cause corrosion and form solid hydrates of natural gas that can plug lines and valves. If the water problem is not too severe, the gas may be heated in an indirect heater to prevent water condensation and keep the temperature of the gas above the hydrate-forming conditions. Water produced from the formation is trapped in a surface separator. Gas sold to a pipeline must meet certain specifications in regard to its water content. To meet this specification, the gas is dehydrated, that is, the water is removed in equipment called a dehydrator. There are different types of dehydration equipment. One common unit uses glycol to remove the water from the gas.

6.3.2 Wet Gas, Gas Condensates, and Volatile Oil Processing at the Wellhead In the case of a wet gas, gas condensate, or volatile oil, the reservoir fluid must be separated at the surface into a liquid condensate and a gas stream. The reservoir fluid enters a separator that allows the gas and condensate to separate from one another. If free water is produced, three phases will exist in the separator. Quite often two separators are used and the pressure controlled on the separators to maximize the amount of liquid that enters the stock tank, which is at atmospheric pressure. The gas phase is then treated in a manner similar to that of a dry natural gas well.

6.3.3 Black Oils Processing at the Wellhead When black oils are produced from a reservoir, they may include solids such as sand and scale along with water. This is generally referred to as basic sediments and water. The produced water may occur in two forms, free water or an oil-in-water emulsion. The amount and degree of emulsification depend on the nature and amount of surfactants that are present in the crude oil. Some oil-field emulsions are difficult to break, and heat and chemicals must be used to separate the water from the oil. The well-stream effluent first flows into a separator where the gas, oil and/or emulsion, and free water are separated. The gas exits from the top of the separator, the oil and/or emulsion from the middle, and the free water from the bottom. The separators come in different designs, both horizontal and vertical.

The emulsion layer goes to a heater-treater where both heat and chemicals are applied to the emulsion. The emulsion breaks and water settles to the bottom of the vessel and clean oil floats to the top from where it is then piped to a stock tank. The stock-tank oil is then ready for the pipeline.

7. PETROLEUM RESERVES AND ECONOMICS

Industry and government agencies are interested in estimates of the future availability of petroleum. Unfortunately, consistent

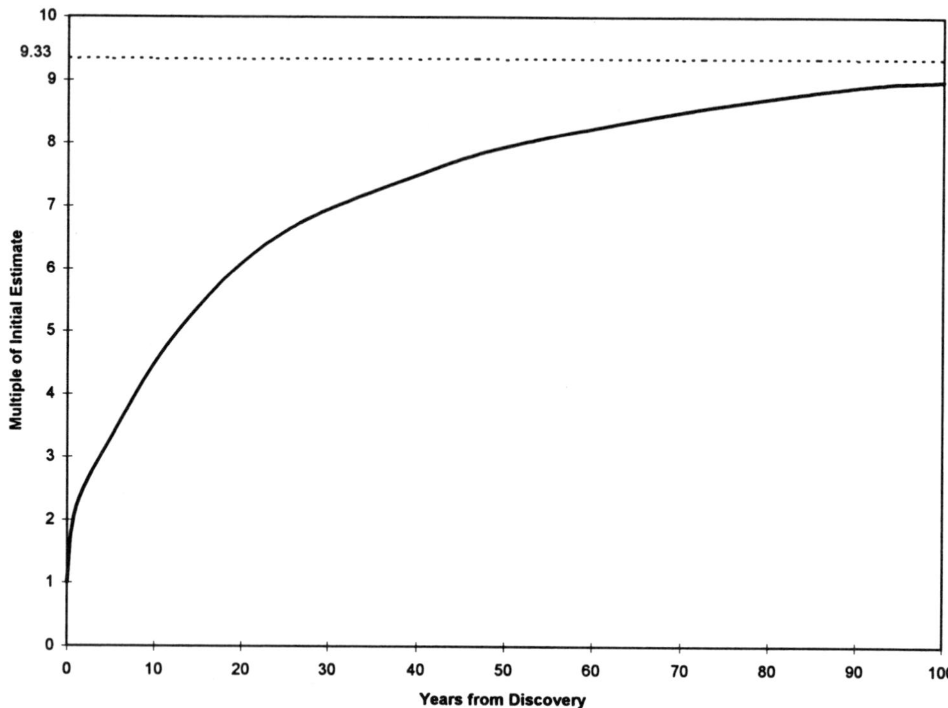

FIG. 14. Reserve appreciation factors for oil (Energy Information Administration, 1990).

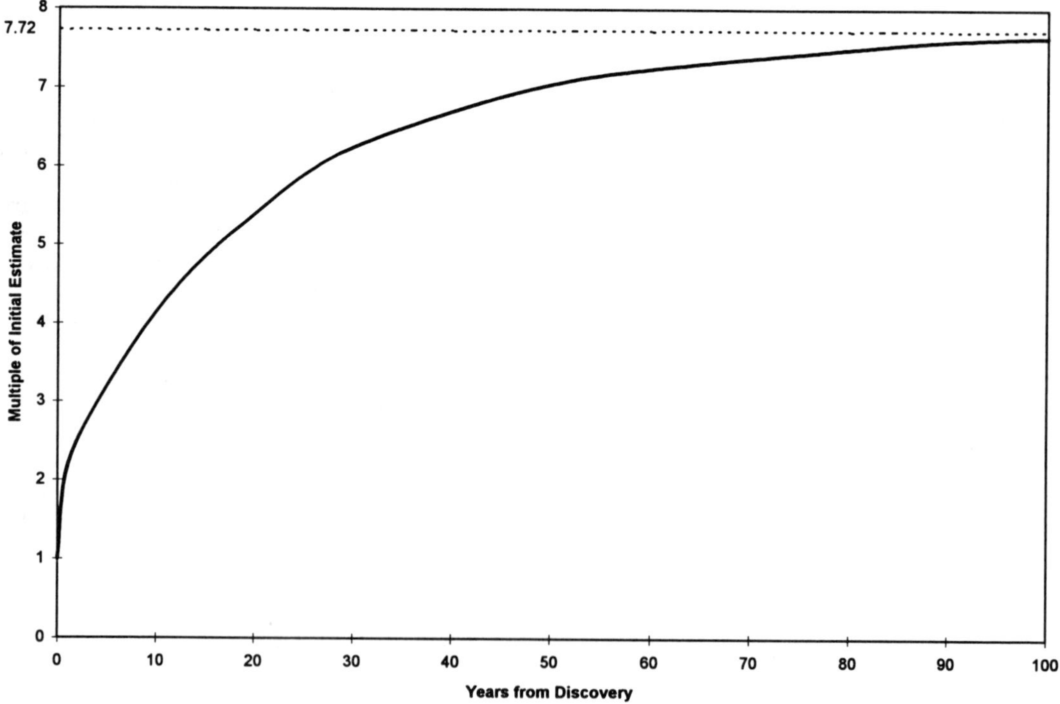

FIG. 15. Reserve appreciation factors for gas (Energy Information Administration, 1990).

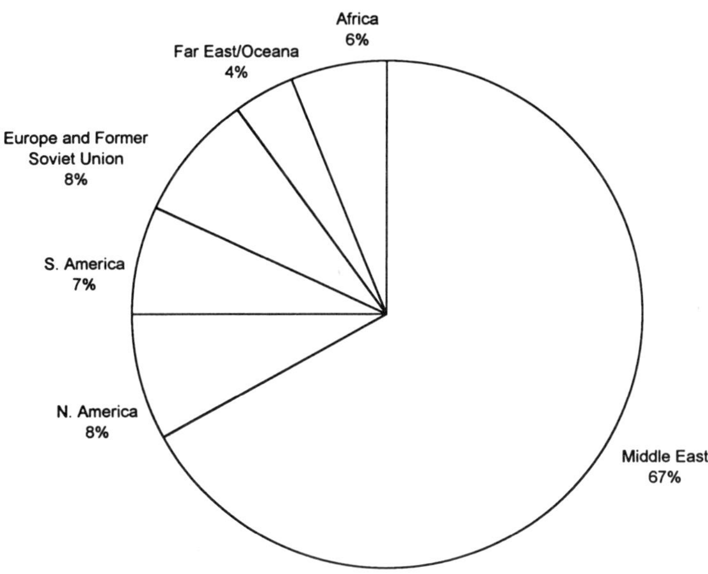

FIG. 16. Total world oil reserves.

terminology is not in use worldwide to describe future estimates of petroleum resources.

7.1 Definitions

The distinction between *resource base, resource*, and *reserves* is helpful in understanding resource assessment studies. The petroleum resource base is all the oil and gas in the earth regardless of the economic or technical feasibility of extraction. The term petroleum resource is used to denote the quantity of petroleum that is recoverable under *specified* economic and technological conditions (Lovejoy, 1965).

Petroleum reserves are the oil and gas that *remain* to be recovered under *current* economic conditions using *current* technology. This definition implies certain conditions:

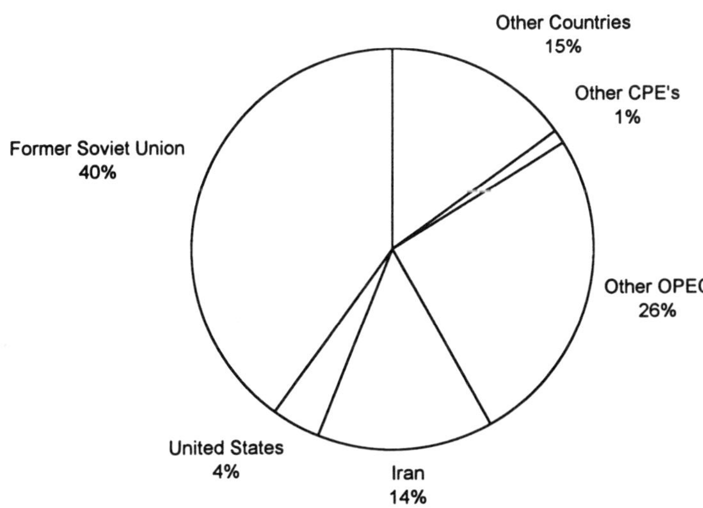

FIG. 17. Total world gas reserves.

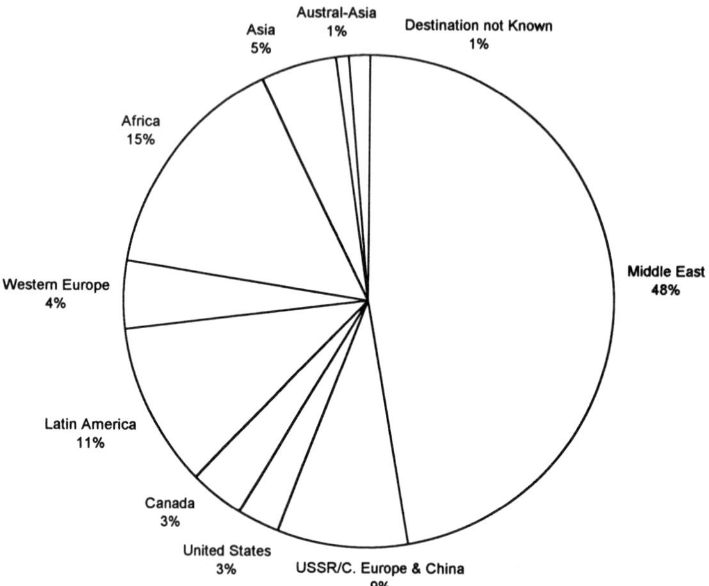

FIG. 18. World oil movements—exports.

First, in order to be called reserves, oil and gas must be physically producible. Second, it must be economic to produce the oil and gas. Third, since the resource has not been produced and is inaccessible, reserves cannot be measured but must be estimated. Finally, since reserves are remaining, there must be a time line associated with the estimate.

Hydrocarbons that have been produced up to a given date are referred to as cumulative recovery or production. The remaining reserves on this date are referred to as esti-

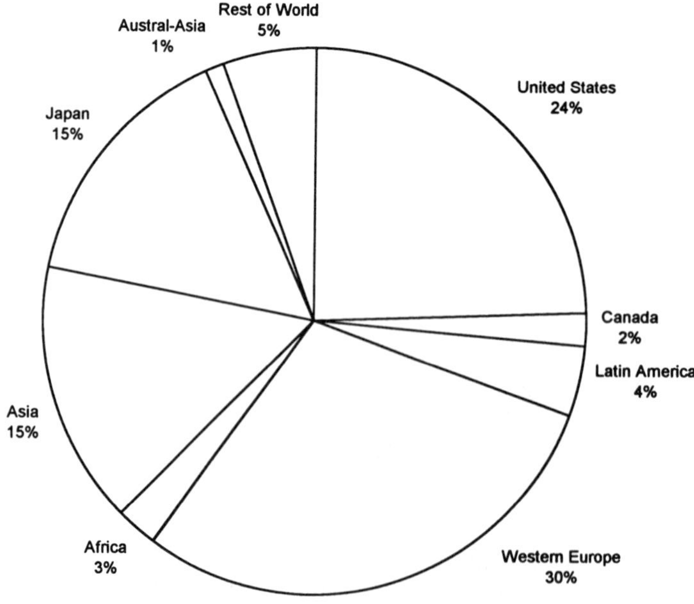

FIG. 19. World oil movements—imports.

mated remaining reserves (ERR). On a given date, the estimated ultimate recovery (EUR) would be equal to the cumulative recovery plus the estimated remaining reserves (Thompson, 1985).

7.2 Classification of Reserves

Classification systems commonly group reserves by the degree of uncertainty associated with the estimate. Three categories are proved, probable, and possible reserves. Proved reserves are those quantities that the estimator is *reasonably certain* are recoverable from known reservoirs under current economic and operating conditions. If the evidence does not support the proved reserves category, the reserves may be classified as probable. In cases where the evidence does not support either the proved or probable categories, the reserves may be classified as possible reserves.

7.3 Petroleum Resource Assessment

Since reserves are estimated, not measured, quantities, at the time of discovery there may be considerable uncertainty in the estimated ultimate recovery for a field. Historically, on an aggregate basis, estimates for the ultimate recovery for fields increase (or appreciate) from the initial estimate. The Energy Information Administration (EIA) has developed a database that contains the year of discovery, the initial estimate for ultimate recovery, and subsequent revisions in reserves on a field basis for the United States (Energy Information Administration, 1990). This information has been used to develop appreciation factors (also referred to as growth factors) that can be multiplied times the initial estimate for ultimate recovery to obtain a better estimate of the final ultimate recovery. The appreciation factors for oil and gas are shown in Figs. 14 and 15. The appreciation factors calculated from this database represent reserves additions from field extensions, new reservoirs, infill drilling, improved technology, enhanced recovery techniques, and longer productive life from an increase in oil and gas prices.

An interesting approach for estimating reserves growth for gas reservoirs has been developed by the Bureau of Economic Geology (Fisher, 1988). This method relates the depositional environment and the reservoir compartmentalization to the amount of reserves growth that may be expected from infill drilling gas reservoirs.

7.4 Petroleum Economics

On a macroeconomic scale, the changes taking place in Eastern Europe and the former Soviet Union and the worldwide distribution and movement of petroleum are examples of the global nature of the petroleum industry. The distribution of world oil and natural gas reserves is shown in Figs. 16 and 17 (Anon., 1991). The movement of oil, imports and exports, around the world is shown in Figs. 18 and 19 respectively (American Petroleum Institute, 1994).

On a microeconomic scale, actual or perceived supply disruptions are reminders of the capital-intensive and risky nature of the petroleum business and the long lead time from project inception to first production. Major projects, for example, often require 5 to 10 years and billions of United States dollars to complete.

The risk is often put into three categories: economic risk, political risk, and technological risk (Garb, 1988). The technical uncertainty includes the estimates of hydrocarbon volumes and the recovery rates. The economic risk includes prices and capital costs for drilling and development. Political risk includes local and national taxes, environmental regulations, and international instability. The petroleum industry is placing more emphasis on decision methods that help quantify the risks, including sensitivity analysis, portfolio analysis, and Monte Carlo simulation.

GLOSSARY

Christmas Tree: The control valves, pressure gauges, and chokes assembled at the top of the well to control the flow of oil and gas after the well has been drilled and completed.

Drilling Mud: The liquid that circulates through the well bore during rotary drilling. It brings cuttings to the surface, cools the bit and drill stem, controls subsurface pressure, and deposits a mud cake on the wall of the borehole to prevent fluid loss to the formation.

Enhanced Oil Recovery (EOR): The injection of special gases at miscibility pressures or of liquid chemicals or the use of thermal energy into a reservoir to interact with the rock–oil system to create conditions favorable to oil recovery. Factors such as lowering interfacial tension, lowering viscosity, changing rock wettability, and creating favorable phase behavior are EOR techniques.

Gas-Oil Ratio (GOR): The ratio of the volume of gas produced with a unit volume of oil measured at standard temperature and pressure.

Permeability: A measure of the ease with which a fluid can flow through the interconnected pore network of a porous rock.

Porosity: The fraction of the total bulk volume of the rock not occupied by rock solids.

Primary Recovery: Oil production from a reservoir in which the natural energy present initially in the reservoir fluids is the main source of energy causing the displacement of petroleum to a producing well.

Reservoir: A subsurface, porous, and permeable rock formation in which oil and/or natural gas have accumulated.

Retrograde Condensation: The condensation of a light hydrocarbon liquid that occurs when the pressure on a gas containing intermediate to heavy hydrocarbons is reduced at constant reservoir temperature.

Saturation Pressure: When a liquid is at its bubble-point pressure or a gas is at its dew-point pressure, the fluids are said to be at their saturation pressure.

Secondary Recovery: After the primary energy in a reservoir has been dissipated, the injection of water or gas into the reservoir to provide energy to displace the oil immiscibly to the producing well is referred to as secondary recovery.

Specific Gravity: For a liquid, the ratio of the weight of a unit volume of the liquid to that of a unit volume of water both measured at 15.56 °C and at atmospheric pressure. For a gas, the ratio of the weight of a unit volume of the gas to that of a unit volume of air under ideal gas conditions. This is equivalent to the ratio of the molal average molecular weight of the gas to the molal average molecular weight of air (28.96).

Well Completion: The activity and procedure necessary to prepare a well for production of oil and gas. The procedure depends on the characteristics of the producing formation. Examples of completion types include open hole, sand exclusion (gravel pack), and perforated casing.

Well Log: A systematic recording of data as a function of well depth such as a driller's log, mud log, electrical well log, or radioactivity log in order to obtain properties of the formations.

Well Stimulation: The treatment of a producing formation to obtain increased production of oil or gas, e.g., hydraulic fracturing, sand control, and acidizing.

Works Cited

American Petroleum Institute (1994), *Basic Petroleum Data Book*, Washington, DC: American Petroleum Institute, Vol. XIV, No. 3, Sec. IX, Tbl. 10j.

Anonymous (1991), "World Wide Production Report," *Oil Gas J.* (December 30, 1991), 43–88.

Clegg, J. D., Bucaram, S. M., Hein, N. W., Jr. (1993), Recommendations and Comparisons for Selecting Artificial Lift Methods, *J. Pet. Tech.* **45**, 1128–1131, 1163–1167.

Energy Information Administration (1990), *The Domestic Oil and Gas Recoverable Resource Base: Supporting Analysis for the National Energy Strategy*, Washington, DC: Energy Information Administration, pp. 15–21.

Fisher, W. L. (1988), *An Assessment of the Natural Gas Resources Base of the United States*, Austin, TX: Bureau of Economic Geology, p. 66.

Garb, F. A. (1988), "Assessing Risk in Estimating Hydrocarbon Reserves and in Evaluating Hydrocarbon-Producing Properties," *J. Pet. Tech.* **40**, 765–778.

Gates, C. F., Jung, K. D., Sierface, R. A. (1978), "Insitu Combustion in the Tulare Formation, South Belridge Field, Kern County, California," *J. Pet. Tech.* **30**, 798–806.

Gates, C. F., Brewer, S. W. (1975), "Steam Injection into the D and F Zone, Tulare Formation, South Belridge Field, Kern County, California," *J. Pet. Tech.* **27**, 343–348.

Lovejoy, W. F., Homan, P. T. (1965), *Methods of Estimating Reserves of Crude Oil, Natural Gas, and Natural Gas Liquids*, Baltimore, MA: Johns Hopkins, pp. 1–5.

McCain, W. D., Jr. (1993), "Chemical Composition Determines the Behavior of Reservoir Fluids," *Pet. Eng. Int.* **65** (20), 18–25.

Moritis, G. (1994), "EOR Dips in U.S. But Remains a Significant Factor," *Oil Gas J.* **92** (39), 51–79.

Surkalo, H., Pitts, M. J., Wyatt, K. (1993), "Design and Field-Implementation of Alkaline-Surfactant-Polymer Chemical Enhanced Oil Recovery Systems," in *Proceedings of the 7th European Symposium on Improved Oil Recovery (October 27–29, 1993), Moscow, Russia,* Vol. 2, Moscow: Russian Interbranch Scientific and Technological Complex "Oil Recovery," pp. 308–316.

Thompson, R. S., Wright, J. D. (1985), *Oil Property Evaluation*, Golden, CO: Thompson-Wright Associates, pp. 4-1–4-6.

Further Reading

Allen, T. O., Roberts, A. P. (1978), *Production Operations*, Vols. 1 and 2, Tulsa, OK: Oil & Gas Consultants International Inc.

Anderson, R. O. (1984), *Fundamentals of the Petroleum Industry*, Norman, OK: Univ. of Oklahoma Press.

Archer, J. S., Wall, C. G. (1986), *Petroleum Engineering—Principles and Practices*, Norwell, MA: Graham and Trotman.

Arnold, K., Stewart, M. (1986), *Surface Production Operations*, Vol. 1, Houston: Gulf Publishing Company.

Arnold, K., Stewart, M. (1989), *Surface Production Operations*, Vol. 2, Houston: Gulf Publishing Company.

Baker, R. (1985), *A Primer of Offshore Operations*, Austin, TX: Petroleum Extension Service, The University of Texas at Austin.

Bradley, Howard B. (Ed.) (1987), *Petroleum Engineering Handbook*, Dallas: Society of Petroleum Engineers.

Brock, J. (1986), *Applied Open-Hole Log Analysis*, Houston: Gulf Publishing Company.

Clark, N. J. (1969), *Elements of Petroleum Reservoirs*, Henry L. Doherty Series, Dallas: Society of Petroleum Engineers.

Craft, B. C., Hawkins, M. (1991), *Applied Petroleum Reservoir Engineering*, 2nd ed., Englewood Cliffs, NJ: Prentice-Hall.

Gerding, Mildred (Ed.) (1986), *Fundamentals of Petroleum*, 3rd ed., Austin, TX: Petroleum Extension Service, The University of Texas at Austin.

Golan, M., Whitson, C. H. (1986), *Well Performance*, Dordrecht, the Netherlands: D. Reidel Publishing Company.

Gray, F. (1986), *Petroleum Production for the Non Technical Person*, Tulsa, OK: PennWell Publishing Co.

Katz, D. L., Cornell, D., Kobayashi, R., Poettmann, F. H., Vary, J. A., Elenbaas, J. R., Weinaug, C. F. (1959), *Handbook of Natural Gas Engineering*, New York: McGraw-Hill.

McCain, W. D., Jr. (1990), *The Properties of Petroleum Fluids*, 2nd ed., Tulsa, OK: PennWell Publishing Company.

Morton-Thompson, Diana, Woods, Arnold M. (Eds.) (1992), *Development Geology Reference Manual*, Tulsa, OK: The American Association of Petroleum Geologists.

North, F. K. (1985), *Petroleum Geology*, Boston: Univin Hyman.

Poettmann, Fred H. (Ed.) (1983), *Improved Oil Recovery*, Oklahoma City, OK: Interstate Oil and Gas Compact Commission.

Skinner, D. R. (1982), *Introduction to Petroleum Production*, Vols. 1, 2, and 3, Houston: Gulf Publishing Company.

PHASE EQUILIBRIA

ARTHUR D. PELTON, *Centre de Recherche en Calcul Thermochimique, Ecole Polytechnique, Montréal, Québec, Canada*

	Introduction	297
1.	**Phase Diagrams of Single-Component Systems**	298
2.	**Binary Phase Diagrams—Descriptive**	299
2.1	Systems with Complete Miscibility in All Phases	299
2.1.1	Minima and Maxima in Two-Phase Regions	301
2.2	Eutectics	302
2.3	Miscibility Gaps	303
2.4	Polymorphism	304
2.5	Intermediate Compounds	305
2.5.1	Congruently Melting Compounds	305
2.5.2	Incongruently Melting Compounds	305
2.5.3	Nonstoichiometric Compounds	305
2.6	Topology of Binary Phase Diagrams	306
2.7	Phase Diagrams of Systems with Liquid-Crystalline Phases	307
3.	**The Gibbs Phase Rule**	308
4.	**Ternary Phase Diagrams**	309
4.1	Ternary Composition Triangle	309
4.2	Ternary Space Model	309
4.3	Polythermal Projections of Liquidus Surfaces	311
4.4	Ternary Isothermal-Isobaric Sections	311
5.	**Other Kinds of Phase Diagrams**	312
6.	**Thermodynamic Origin of Phase Diagrams**	314
6.1	Models of Solutions—Regular Solution Theory	315
6.2	Thermodynamic-Phase Diagram Optimization	317
7.	**Glass Formation**	318
8.	**Order-Disorder Transitions**	319
9.	**Interfacial Effects**	319
	Glossary	319
	Works Cited	320
	Further Reading	321

INTRODUCTION

In general, a chemical system will consist of several phases, solid, liquid, and gaseous, of different compositions. The study of phase equilibria and phase transformations is central to materials science, chemical engineering, geology, and many other disciplines.

A phase diagram is a graphical representation of the conditions of temperature, pressure, composition, etc. for which equilibrium among the phases of a system is established. A system at equilibrium is one that has achieved a state of minimum free energy and is unchanging even over an infinite amount of time. Although departures from equilibrium will occur in any real system, a knowledge of the equilibrium state under a given set of conditions is the starting point for the understanding of most processes.

We restrict ourselves to phase diagrams with pressure, temperature, composition, and, in a few examples, partial pressures as the axes. However, other variables, such as electrical, magnetic, and gravitational fields, can also affect phase equilibria, and phase diagrams with these as axes could also be drawn.

In Secs. 1 and 2, we describe phase diagrams of one-component and two-component systems without rigorously justifying the rules governing their construction. This justification follows, however, in Sec. 3 on the Gibbs phase rule. The thermodynamic origin of phase diagrams is discussed in more detail in Sec. 6. Some simple examples of the relation between the phase diagram of a system and the structure and bonding of the phases at the atomic level are also presented in Sec. 6.

3-527-28135-5/95/$5.00 + .50

1. PHASE DIAGRAMS OF SINGLE-COMPONENT SYSTEMS

A single-component system consists of only one chemical substance which does not dissociate. The phases present may be solid, liquid, or gaseous. The most common type of phase diagram of a one-component system is a *P-T* diagram in which the axes are temperature and pressure as shown for the system H_2O in Fig. 1. Volume is not an independent variable since, for any phase in a single-component system, the volume is uniquely given, as a function of *P* and *T*, by the equation of state. Figure 1 can be considered to be an orthogonal projection, upon the *P-T* plane, of the nodal lines of the *P-V-T* surface.

In Fig. 1(a), regions in which the entire system is solid, liquid, or gaseous are labeled. The coexistence of two phases occurs along the lines of the diagram. For example, ice transforms to liquid water at a constant temperature of 0 °C at a pressure of 101.3 kPa (1.0 atm), and the liquid and gas coexist in equilibrium at this pressure at 100 °C.

If a previously evacuated container is partially filled with water, sealed, and heated, the pressure in the container will follow the two-phase boundary between the liquid and vapor regions of Fig. 1 as long as the liquid and vapor phases are both present. This line thus gives the *vapor pressure* of H_2O as a function of *T*. Solid, liquid, and vapor can only coexist at one point, the *triple point*, which, for H_2O, occurs at 0.0098 °C and 0.611 kPa. At pressures and temperatures below this point, ice sublimes directly to vapor.

At low pressures, ice has a hexagonal crystal structure. This form of ice is known as ice-I. At very high pressures, other *polymorphs* (i.e., other crystal structures) of ice, known as ice-II, ice-III, etc., are observed. The phase diagram at high pressures is shown in Fig. 1(b). Again, the diagram consists of single-phase areas, two-phase lines, and three-phase triple points.

The (liquid + vapor) two-phase line in Fig. 1(a) terminates at a *critical point*. For H_2O, the *critical temperature*, T_c, and *critical pressure*, P_c, are 374 °C and 22.3 MPa. At this point, all the properties of the two phases become identical. Above the critical point there is no longer a distinction between liquid and vapor. As discussed above, if liquid and vapor are heated in a sealed container, the pressure follows the two-phase line. As the critical point is approached, the densities of the two phases approach each other, becoming identical at the critical point. At the molecular level, the behavior of a system very near its critical point is complex. For a discussion, see CRITICAL PHENOMENA.

The slopes of the two-phase lines in Fig. 1 are given by the *Clapeyron equation*,

$$dP/dT = \Delta H/T\Delta V, \qquad (1)$$

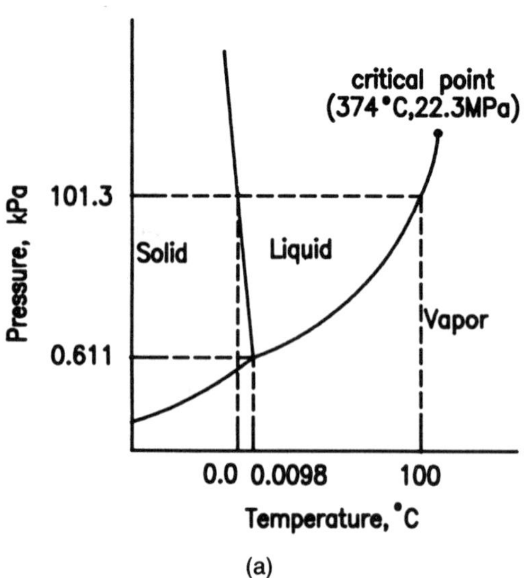

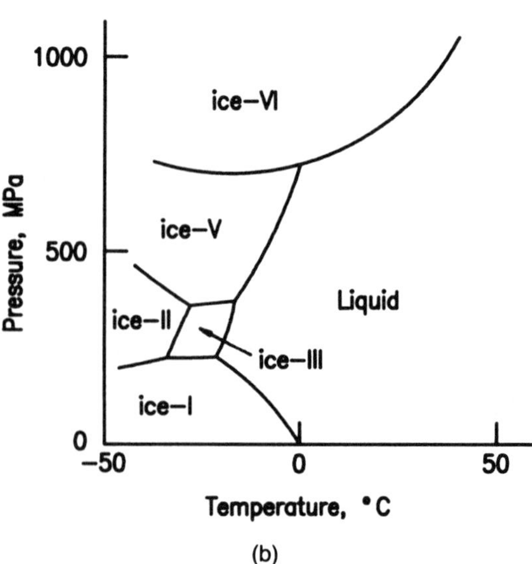

FIG. 1. *P-T* phase diagram of water at (a) low and (b) high pressures.

where ΔH is the heat (enthalpy) change for the transformation of one mole of substance from one phase to the other, ΔV is the corresponding molar volume change, and T is the temperature in kelvins. For the (liquid + vapor) line, ΔH is the heat of vaporization ($\Delta H > 0$) and ΔV is the volume change accompanying the vaporization of one mole of liquid. Hence, ΔH is very large and so, from Eq. (1), the slope of the line is small. For the (solid + liquid) boundary, however, ΔV of transformation from solid to liquid is small. Consequently, this line is very steep. It is interesting to note that H_2O is unusual in that $dP/dT < 0$ for the (solid + liquid) boundary. This is due to a negative ΔV of melting. (Ice floats on water.) For most substances, ΔV of melting is positive, and the (solid + liquid) boundary has a positive slope.

2. BINARY PHASE DIAGRAMS—DESCRIPTIVE

A binary system is one with two components. The precise definition of "component" is given in Sec. 3. However, the components of a system are usually evident. In this section we discuss binary temperature-composition (*T-X*) and pressure-composition (*P-X*) phase diagrams.

2.1 Systems with Complete Miscibility in All Phases

A *T-X* phase diagram at constant pressure $P = 101.3$ kPa for the methanol-water system is shown in Fig. 2. The abscissa is the composition, expressed as mole fraction of water, X_{H_2O}. Note that $X_{CH_3OH} = 1 - X_{H_2O}$. Phase diagrams are also often drawn with the composition axis as weight percent.

At a temperature and overall composition given by a point in the region labeled "vapor" in Fig. 2, the system, at equilibrium, will be entirely in the vapor phase. In the region labeled "liquid," the system, at equilibrium, will consist entirely of one miscible liquid phase. For example, it can be seen from Fig. 2 that pure water transforms from a liquid to a gas at 100.0 °C at $P = 101.3$ kPa (1.0 atmospheres), while the "normal boiling point" of methanol is 64.6 °C.

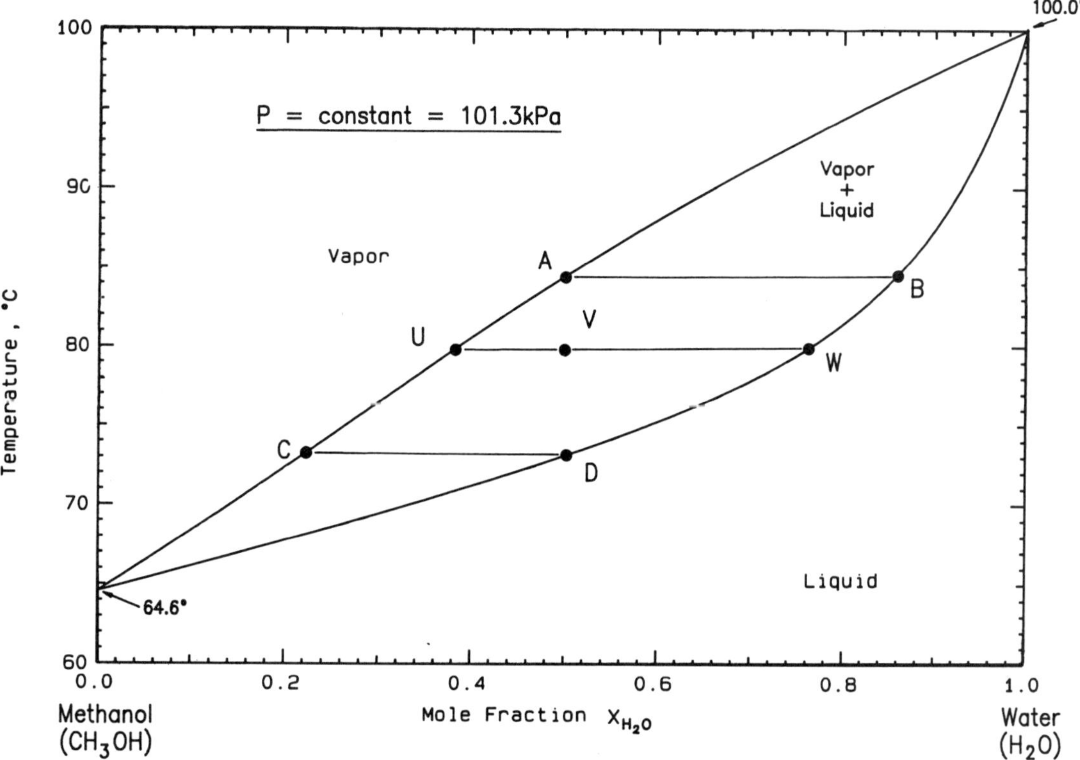

FIG. 2. *T-X* phase diagram at constant $P = 101.3$ kPa of the methanol-water system.

Suppose that a system at an overall composition of $X_{H_2O} = 0.5$ is held at 80 °C in a container at $P = 101.3$ kPa (i.e., at point V in Fig. 2). At equilibrium, two phases will be observed, a vapor phase of composition U ($X_{H_2O} = 0.385$) and a liquid solution of composition $W(X_{H_2O} = 0.76)$. The line UW is called a *tie-line* or *conode*.

If we were to add some water to the system to bring the overall composition to, say, $X_{H_2O} = 0.65$, keeping T constant at 80 °C, we would observe that the compositions of the vapor and liquid phases remain unchanged at points U and W, although the relative amount of the liquid phase would increase. From a simple mass balance, one can derive the *lever rule* for binary systems: (moles of vapor)/(moles of liquid) = VW/UV, where VW and UV are the lengths of the line segments.

Hence, at 80 °C and 101.3 kPa for a system with an overall composition $X_{H_2O} = 0.50$, the amounts of vapor and liquid at equilibrium are in the molar ratio (0.76 − 0.5)/(0.5 − 0.385) = 2.26. When the composition axis of a phase diagram is in weight percent, then the lever rule gives the weight ratio of the two phases.

Suppose that a gas with composition $X_{H_2O} = 0.5$ is slowly cooled at a constant pressure of 101.3 kPa. When the temperature has decreased to 84.5 °C the system is at point A on the upper phase boundary in Fig. 2. At this temperature (the "dew point") the first liquid condensate appears with a composition at point $B(X_{H_2O} = 0.86)$ on the lower phase boundary. As the temperature is decreased further, liquid continues to condense with the compositions of the equilibrium vapor and liquid phases at any temperature always lying on the phase boundaries. Condensation is complete at 73 °C, the last vapor to condense having composition $X_{H_2O} = 0.22$ at point C.

At pressures above (or below) 101.3 kPa, the T-X diagram in Fig. 2 will be shifted to higher (or lower) temperatures. It is often useful to plot a P-X phase diagram at constant T as illustrated in Fig. 3 for the methanol-water system. The interpretation of this diagram is analogous to that of Fig. 2. The tie-line UVW at 101.3 kPa at 80 °C in Fig. 3

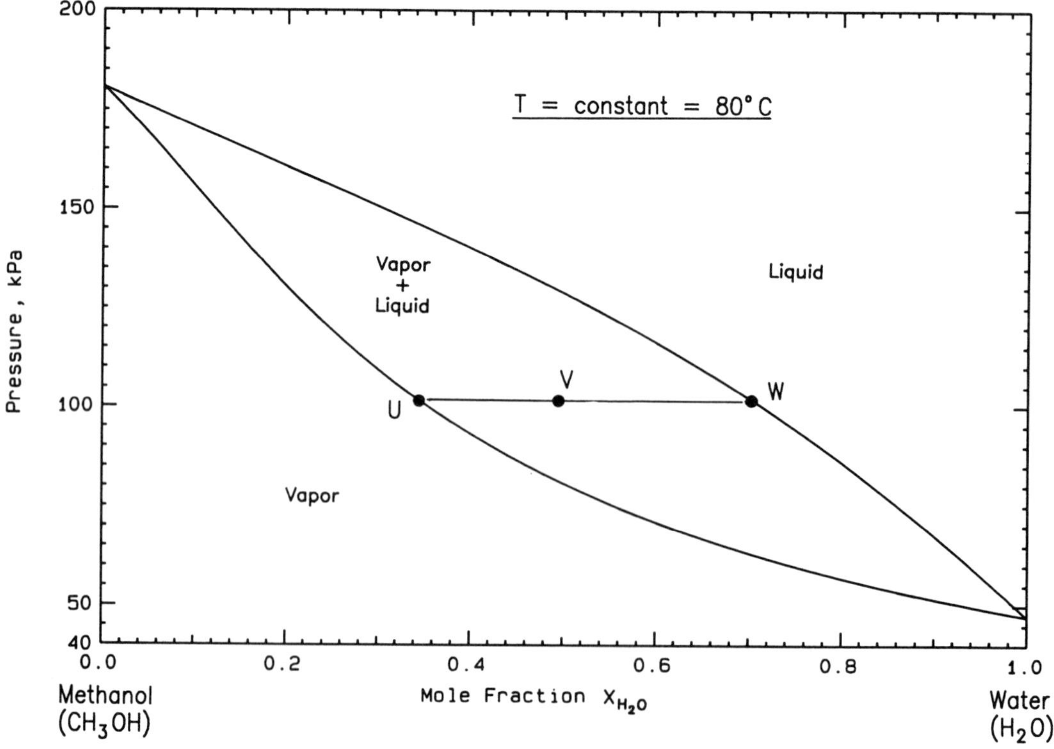

FIG. 3. *P*-*X* phase diagram at constant P = 80 °C of the methanol-water system.

is the same as the tie-line *UVW* at the same temperature and pressure in Fig. 2. The lever rule applies to *P-X* diagrams just as to *T-X* diagrams. At temperatures above (or below) T = 80 °C, the *P-X* diagram in Fig. 3 will be shifted to higher (or lower) pressures.

A *T-X* diagram at constant pressure of 101.3 kPa for the CaO-MnO system is shown in Fig. 4. The interpretation of this diagram is similar to that of Fig. 2. Pure CaO and MnO melt at 2572 °C and 1842 °C, respectively. CaO and MnO are completely miscible in the solid as well as the liquid state. The upper phase boundary, which gives the composition of the liquid phase in equilibrium with the solid phase, is called the *liquidus* line, while the lower phase boundary, the *solidus* line, gives the composition of the solid in equilibrium with the liquid.

Suppose that a liquid CaO-MnO solution with composition X_{MnO} = 0.60 is cooled very slowly from an initial temperature of about 2500 °C. When the temperature has decreased to the liquidus temperature 2270 °C (point *B*) the first solid appears, with a composition at point *A* (X_{MnO} = 0.28). As the temperature is decreased further, solid continues to precipitate, with the compositions of the two phases at any temperature being given by the liquidus and solidus compositions at that temperature and with their relative proportions being given by the lever rule. Solidification is complete at 2030 °C, the last liquid to solidify having composition X_{MnO} = 0.60 (point *C*).

The process just described is known as equilibrium cooling. At any temperature during equilibrium cooling the solid phase has uniform (homogeneous) composition. In the preceding example, the composition of the solid phase during cooling varies along the line *APC*. Hence, in order for the solid to have a uniform composition at any temperature, diffusion of CaO from the center to the surface of the growing grains must occur. Since solid-state diffusion is a relatively slow process, equilibrium cooling conditions are only approached if the temperature is decreased very slowly. If a sample of composition X_{MnO} = 0.60 is cooled very rapidly from the liquid, concentration gradients will be observed in the solid grains, with the concentration of MnO increasing toward the surface from a minimum of X_{MnO} = 0.28 (point *A*) at the center. Furthermore, in this case solidification will not be complete at 2030 °C, since at 2030 °C the average concentration of MnO in the solid particles will now be less than X_{MnO} = 0.60.

Since only condensed (i.e., nongaseous) phases are involved in Fig. 4, the phase diagram is very insensitive to pressure. Typically, phase boundaries involving only condensed phases are shifted by the order of one degree per 10 MPa (100 atm) change in pressure.

2.1.1 Minima and Maxima in Two-Phase Regions A two-phase region may exhibit a minimum or a maximum. For example, in the chloroform-methanol system shown in Fig. 5, a minimum occurs at 35 mol % methanol at 53.4 °C. A liquid of this composition, when heated at constant pressure of 101.3 kPa, will

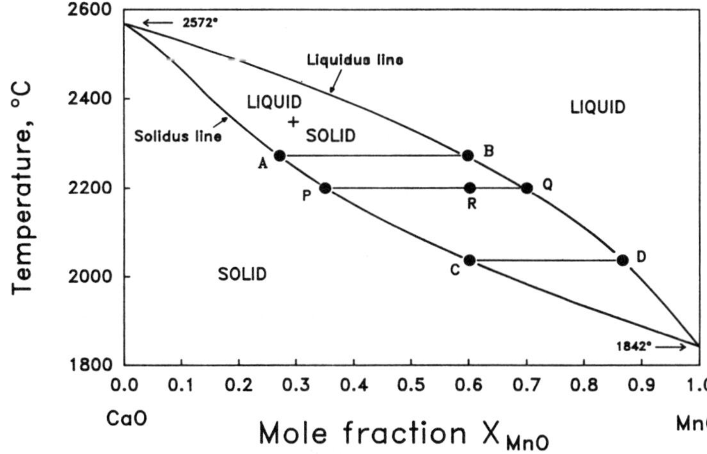

FIG. 4. Phase diagram of the CaO-MnO system at P = 101.3 kPa (after Schenck *et al.*, 1964, and Wu *et al.*, 1993).

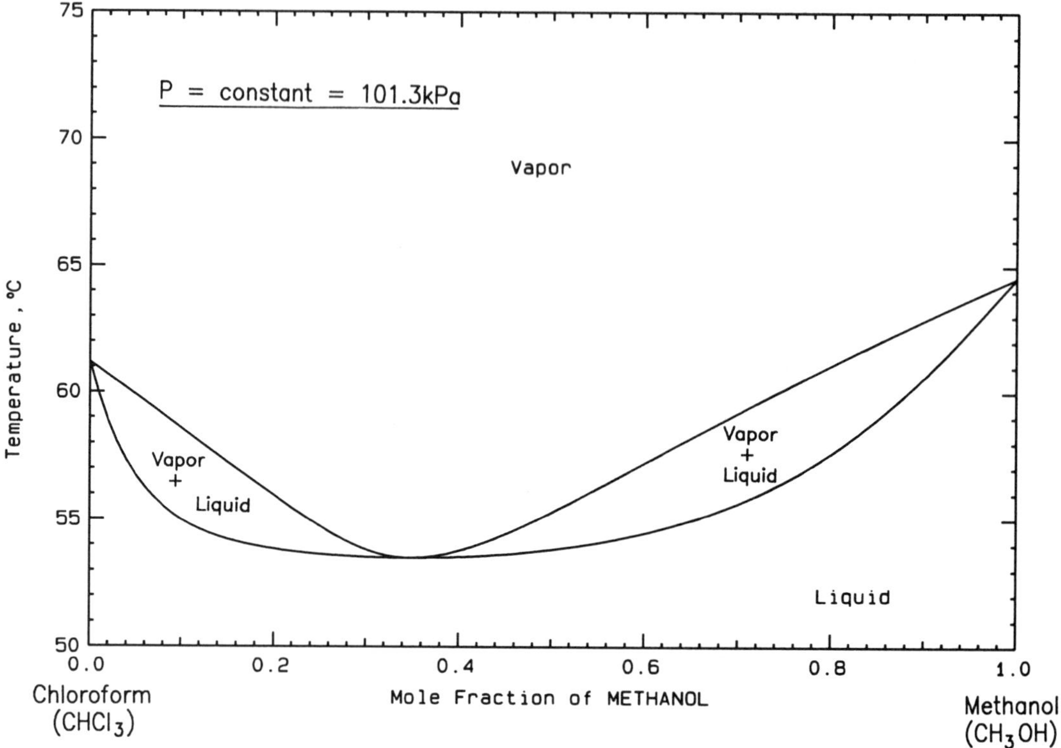

FIG. 5. *T-X* phase diagram at P = 101.3 kPa of the chloroform-methanol system (reprinted from Pelton, 1991).

transform completely to a vapor of the same composition at a constant temperature of 53.4 °C.

An extremum in a (liquid + vapor) two-phase region is called an *azeotrope*. Azeotropes are important in the design of distillation processes. In distillation, the aim is to separate the components of a mixture by virtue of the difference in composition between the liquid and vapor phases. At an azeotrope, however, both phases have the same composition.

Such extrema are not limited to (liquid + vapor) regions. Examples of extrema in (solid + liquid) regions are shown in Figs. 7, 9, and 19. Maxima in (solid + liquid) regions are usually associated with the existence of an intermediate compound as discussed in Sec. 2.5.3. At all such extrema, both phase boundaries must pass through their minimum or maximum at the same point.

2.2 Eutectics

The phase diagram of the Ag-Cu system is shown in Fig. 6. In this system, solid Ag and Cu have limited mutual solubility. At all compositions in the region labeled "solid S_1" a single-phase homogeneous solid solution of Cu in Ag will be observed. The *solvus line* shown in Fig. 6 indicates the solubility limit of Cu in Ag. Similarly, the solvus line on the Cu-rich side indicates the solubility limit of Ag in Cu. At overall compositions lying in the region labeled "2 solids," the system at equi-

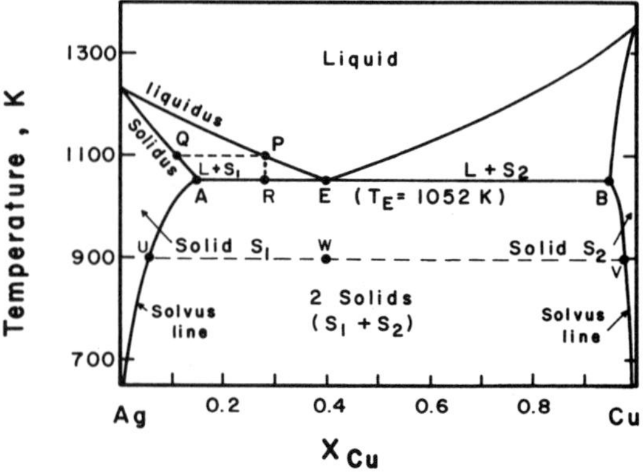

FIG. 6. Phase diagram at P = 101.3 kPa of the Ag-Cu system (after Hultgren *et al.*, 1973).

librium will consist of two phases: Ag saturated with Cu and Cu saturated with Ag. For example, at 900 °C, at an overall composition of 40 mol % Cu (point W in Fig. 6), the equilibrium microstructure will consist of two phases with compositions 5 mol % Cu (point U) and 97 mol % Cu (point V). The relative amounts of the two phases are given by the lever rule: (moles of S_1)/(moles of S_2) = $WV/UW = (0.97 - 0.4)/(0.4 - 0.05) = 1.63$.

The temperature T_E is called the *eutectic temperature*. At T_E, a liquid of composition E, the *eutectic composition*, is simultaneously in equilibrium with solid S_1 and solid S_2 of compositions A and B, respectively. Suppose a liquid Ag-Cu solution of composition $X_{Cu} = 0.29$ is cooled from the liquid state very slowly under equilibrium conditions. At 1100 °C, when the liquid is at point P, the first solid appears with composition Q. As T decreases further, solidification continues with the liquid composition following the liquidus curve from P to E and the solid composition following the solidus curve from Q to A. The relative amounts of the two phases at any T are given by the lever rule. At T_E, the overall system composition and temperature are given by point R. At a temperature just above T_E, two phases are observed, a solid and a liquid of compositions A and E, respectively, while at a temperature just below T_E, according to the diagram, two solids with compositions A and B will be observed. Therefore, at T_E, during cooling, the following *binary eutectic reaction* occurs:

$$\text{liquid} \rightarrow \text{solid}_1 + \text{solid}_2. \qquad (2)$$

Under equilibrium conditions, the temperature will remain constant at $T = T_E$ until all the liquid has solidified, and during the reaction the compositions of the three phases will remain fixed at points A, B, and E. For this reason, the eutectic reaction is called an *invariant* reaction.

After the eutectic solidification is completed, if the microstructure is examined at a temperature slightly below T_E, one will observe two types of grains. There will be grains of pure S_1 solid solution of composition A that solidified at temperatures above T_E. This is called the *primary* solid or *pro-eutectic* constituent. There will also be grains that formed during the eutectic reaction. These *eutectic* grains will have an overall composition at point E, and will consist of an intimate mixture of the phases S_1 and S_2 of compositions A and B. At high magnification it will be seen that this intimate mixture may consist of alternating lamellae of S_1 and S_2 or of globules or needles of one phase in the other, or of another morphology depending upon the system and upon kinetic factors. Although the pro-eutectic and eutectic *constituents* are not phases (the eutectic grains are two-phase mixtures), the lever rule may nevertheless be applied at a temperature just below T_E as follows: (moles of pro-eutectic constituent)/(moles of eutectic constituent) = RE/AR. If the composition axis is in weight fraction, then this ratio gives the weight ratio of the constituents.

Upon heating an alloy with overall composition between points A and B ($0.15 < X_{Cu} < 0.96$), the eutectic reaction (2) will occur in the reverse direction at T_E. Hence, 1052 K is the temperature of first appearance of a liquid phase for such alloys. For alloys with compositions outside this range the temperature of first appearance of a liquid phase, under equilibrium conditions, will be the solidus temperature. However, equilibrium conditions may not prevail. For example, an alloy containing 10 mol % Cu, which consists of two solid phases at lower temperatures, will consist of only one solid solution phase when at equilibrium above the solvus temperature of ~1020 °C. Hence, if the alloy is first held somewhat above 1020 °C until equilibrium is attained, before further heating, then the first liquid will appear at the solidus temperature of ~1110 °C. However, if the alloy is heated very quickly from room temperature to T_E, there will not be enough time for the S_2 phase to dissolve by diffusion in the S_1 phase. In this case, the alloy will arrive at T_E as a two-phase mixture, and a liquid phase will appear at T_E.

For further discussion of equilibrium and nonequilibrium heat treatment of alloys and other materials and the resultant microstructures, see MATERIALS TREATMENT and THERMAL ANNEALING.

2.3 Miscibility Gaps

The CaO-SiO_2 system is of great importance in the fields of ceramics, refractories, and metallurgy as well as in geology. The phase diagram of this system in Fig. 7 ex-

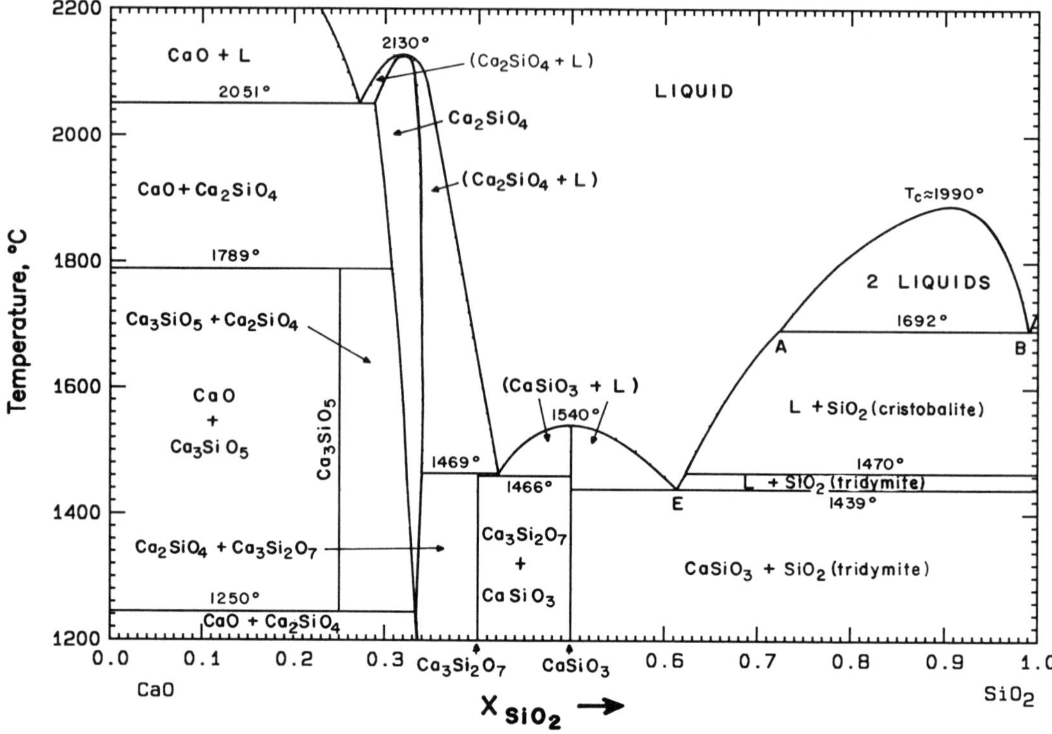

FIG. 7. CaO-SiO_2 phase diagram at P = 101.3 kPa (after Eriksson *et al.*, 1994).

hibits many features that are discussed in this section as well as in Secs. 2.4 and 2.5.

A *liquid-liquid miscibility gap* with a *consolute* or *critical point* at T_c = 1990 °C and $X_{c(SiO_2)} \approx 0.91$ is seen in Fig. 7.

Suppose that a liquid of composition $X_{SiO_2} = 0.80$ is cooled slowly from high temperatures. At ~1820 °C the miscibility gap boundary is crossed and a second liquid layer appears with an initial composition of $X_{SiO_2} \approx 0.966$. As the temperature is lowered further, the composition of each liquid phase follows its respective phase boundary until, at 1692 °C, the SiO_2-rich liquid has a composition of $X_{SiO_2} = 0.99$ (point B) and in the CaO-rich liquid $X_{SiO_2} = 0.725$ (point A).

At 1692 °C the following *invariant monotectic reaction* then occurs upon cooling:

$$\text{liquid } B \rightarrow \text{liquid } A + SiO_2(\text{solid}). \quad (3)$$

The temperature remains constant at 1692 °C and the compositions of the phases remain constant until all of liquid B is consumed. Cooling then continues with precipitation of solid SiO_2, with the equilibrium liquid composition following the liquidus from point A to the eutectic E at 1439 °C.

The critical point in Fig. 7 is similar to the critical point in Fig. 1 in that the properties of the two phases become identical at the critical point and above the critical point there is only one miscible liquid phase. For further discussion, see CRITICAL PHENOMENA.

2.4 Polymorphism

Solid compounds can often exist in two or more crystal structures depending upon the temperature and pressure. The example of water ice has already been given in Sec. 1. As another example, pure SiO_2 transforms from the tetragonal tridymite crystal structure to the cubic cristobalite structure at 1470 °C. The transformation involves an invariant reaction, which appears on the phase diagram in Fig. 7. As yet another example, the polymorphism of Fe is seen on the phase diagram in Fig. 9, as is discussed in Sec. 2.6.

In Fig. 7, SiO_2 is shown as a *stoichiometric* solid. That is, the solubility of CaO in solid SiO_2 is negligible, such that the liquid on the liquidus AE is in equilibrium with pure solid

SiO_2. In reality, no solid is ever exactly stoichiometric (that is, no solubility is ever identically zero), so that there is actually a solidus line and solvus line in Fig. 7 just as in Fig. 6. However, these lines are indistinguishable from the vertical axis at $X_{SiO_2} = 1$.

2.5 Intermediate Compounds

2.5.1 Congruently Melting Compounds In the CaO-SiO_2 system in Fig. 7, there is a compound $CaSiO_3$ that melts at a constant temperature of 1540 °C to form a liquid of the same composition. $CaSiO_3$ is thus called a *congruently melting* compound. There is a eutectic on either side of this compound in Fig. 7. In the eutectic reaction at 1439 °C, for example, liquid of composition E solidifies at constant T to give $CaSiO_3$ and SiO_2. In the eutectic reaction at 1466 °C, liquid at the eutectic composition at $X_{SiO_2} = 0.425$ solidifies to give $CaSiO_3$ and $Ca_3Si_2O_7$.

2.5.2 Incongruently Melting Compounds The compound $Ca_3Si_2O_7$ in Fig. 7 is said to be an *incongruently melting compound.* When heated, $Ca_3Si_2O_7$ decomposes at a constant temperature of 1469 °C to form solid Ca_2SiO_4 plus a liquid of composition $X_{SiO_2} = 0.42$. This is the temperature of first appearance of a liquid. The material will not be completely liquid, however, until it is heated to the liquidus temperature of ~1630 °C. When a liquid containing 40 mol % SiO_2 is cooled at equilibrium, these events occur in reverse. The first solid, Ca_2SiO_4, appears at the liquidus temperature of ~1630 °C. At the *peritectic temperature* of 1469 °C, the following *peritectic reaction* occurs:

$$\text{liquid} + \text{solid}_1 \rightarrow \text{solid}_2, \tag{4}$$

where solid_1 and solid_2 are Ca_2SiO_4 and $Ca_3Si_2O_7$, respectively.

Like the eutectic reaction, this is an invariant reaction, which, under equilibrium conditions, occurs at constant temperature and a fixed compositions of all three phases. However, upon cooling, peritectic reactions generally occur with formation of the product solid ($Ca_3Si_2O_7$) on the surface of the reactant solid (Ca_2SiO_4). This coating can prevent contact between the reactant solid and liquid. Further reaction may thus be greatly retarded so that equilibrium conditions can only be achieved by extremely slow cooling. In the present example, if a liquid containing 40 mol % SiO_2 were cooled rapidly, some unreacted Ca_2SiO_4 would remain at room temperature.

An incongruently melting compound is always associated with a peritectic invariant in a phase diagram.

2.5.3 Nonstoichiometric Compounds In the phase diagram of the CaO-SiO_2 system, Fig. 7, the phase labeled Ca_2SiO_4 is a *nonstoichiometric compound* with nominal composition Ca_2SiO_4 ($X_{SiO_2} = 0.333$). At 2000 °C, for example, the width of the single-phase region extends from $X_{SiO_2} = 0.28$ to $X_{SiO_2} = 0.335$. This is the *range of stoichiometry* over which the compound can exist as a single homogeneous phase. Generally, as in Fig. 7, the range of stoichiometry becomes narrower at lower temperatures.

The nonstoichiometry of Ca_2SiO_4 can also be regarded as a solid solubility of CaO in Ca_2SiO_4 on one side and of SiO_2 in Ca_2SiO_4 on the other (the solubility of SiO_2 being less than that of CaO).

All compounds exhibit some finite range of stoichiometry. Although the compounds $CaSiO_3$ and $Ca_3Si_2O_7$ in Fig. 7 are shown as stoichiometric "line compounds," in reality they have finite, but very narrow, ranges of stoichiometry. That is, a "stoichiometric" compound like $CaSiO_3$ is simply a limiting case of nonstoichiometric compound.

Ca_2SiO_4 is a congruently melting compound with a congruent melting point of 2130 °C. It can be seen that this congruent melting point can also be regarded as the point at which the solidus and liquidus defining the two-phase (Ca_2SiO_4 + L) region pass simultaneously through a maximum (see Sec. 2.1.1). In this regard, it can be appreciated that the congruent melting point of the 'stoichiometric' compound $CaSiO_3$ is simply a limiting case of this construction as the range of stoichiometry becomes negligibly small.

As an example of the practical consequences of nonstoichiometry, consider a ceramic with $X_{SiO_2} = 0.31$ in the single-phase region at 2000 °C. If this ceramic is cooled to, say, 1200 °C in the two-phase region and then held at this temperature, fine precipitates of CaO will form. If, on the other hand, it is rapidly quenched to room temperature, there will be insufficient time for precipitation to occur by solid-state diffusion, and the single-phase metastable Ca_2SiO_4 structure will be frozen in.

Generally, nonstoichiometry in a compound results from the presence of lattice *point defects*. Let A_mB_n be the ideal stoichiometry of a compound as given by its crystal structure. An excess of B, $A_{m-x}B_{n+x}$ with $x > 0$, can result from defects such as interstitial B atoms, B atoms occupying lattice sites normally occupied by A atoms, vacant A sites, etc. An excess of A can result from similar defects. The energy of forming such defects will determine the number of defects at equilibrium. If this energy is very high, few defects will be present and the range of stoichiometry will be narrow. A relatively low energy of formation, on the other hand, will result in a large number of defects and a wide range of stoichiometry. In Fig. 7, it can be seen that the energy of forming defects that permit an excess of SiO_2 in Ca_2SiO_4 is larger than the energy of forming defects permitting an excess of CaO.

2.6 Topology of Binary Phase Diagrams

From the preceding examples it can be seen that all binary *T-X* and *P-X* diagrams obey the same topological rules of construction. Scanning horizontally across the diagrams, at constant T and P, one encounters single-phase regions alternating with two-phase regions. Two-phase regions can pass through maxima or minima as discussed in Sec. 2.1.1, or they can terminate in a critical point as discussed in Sec. 2.3.

There are only two general types of three-phase *invariants* in binary phase diagrams. These are the *eutectic-type* and *peritectic-type invariants* as illustrated in Fig. 8. Let the three phases concerned be called α, β, and γ with β as the central phase as in Fig. 8. The α, β and γ can be solid, liquid, or gaseous phases. At a eutectic-type invariant, the following invariant reaction occurs isothermally as the system is cooled:

$$\beta \rightarrow \alpha + \gamma, \tag{5}$$

while, at a peritectic-type invariant, the invariant reaction upon cooling is

$$\alpha + \gamma \rightarrow \beta. \tag{6}$$

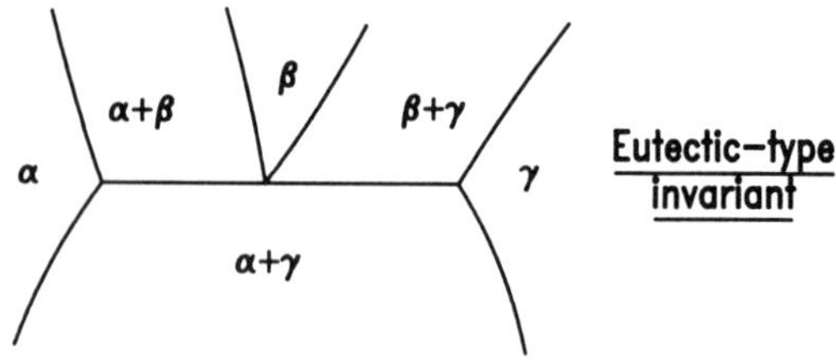

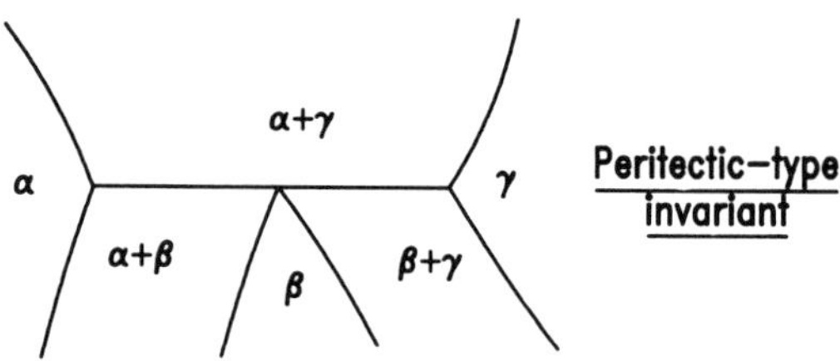

FIG. 8. The two types of invariants in binary *T-X* and *P-X* phase diagrams.

For example, at a *eutectic*, α = solid$_1$, β = liquid, γ = solid$_2$. The eutectic reaction is $l \rightarrow s_1, + s_2$. A *monotectic*, as seen in Fig. 7 at 1692 °C is also a eutectic-type invariant with α = liquid$_1$, β = liquid$_2$, γ = solid. The monotectic reaction is $l_2 \rightarrow l_1 + s$.

Other examples can be seen in the phase diagram of the Fe-Mo system in Fig. 9. *Peritectics*, in which α = liquid β = solid$_1$, γ = solid$_2$, are seen at 1540 °C, 1488 °C, and 1450 °C. The peritectic reaction is $l + s_2 \rightarrow s_l$. At 1235 °C and 1200 °C in Fig. 9 are examples of *eutectoids* (which are eutectic-type invariants) involving three solid phases. With α = solid$_1$, β = solid$_2$, and γ = solid$_3$, the eutectoid reaction is $s_2 \rightarrow s_1 + s_3$. Examples of *peritectoids* are seen at 1370 °C and 950 °C. These are peritectic-type invariants. The peritectoid reaction is $s_1 + s_3 \rightarrow s_2$. In Fig. 7 the invariants at 1250 °C and 1789 °C are, respectively, limiting cases of a eutectoid and a peritectoid where the range of homogeneity of the phase Ca_3SiO_5 is negligibly small.

Many other eutectic-type and peritectic-type invariants are observed such as *syntectics* ($l_1 + l_2 \rightarrow s_1$) and *catatectics* ($s_1 \rightarrow l + s_2$).

In Secs. 2.2 and 2.5.2, the microstructures resulting from heat treatments of systems with eutectics and peritectics were briefly described. Similar considerations apply in the case of eutectoids and peritectoids. For ex-

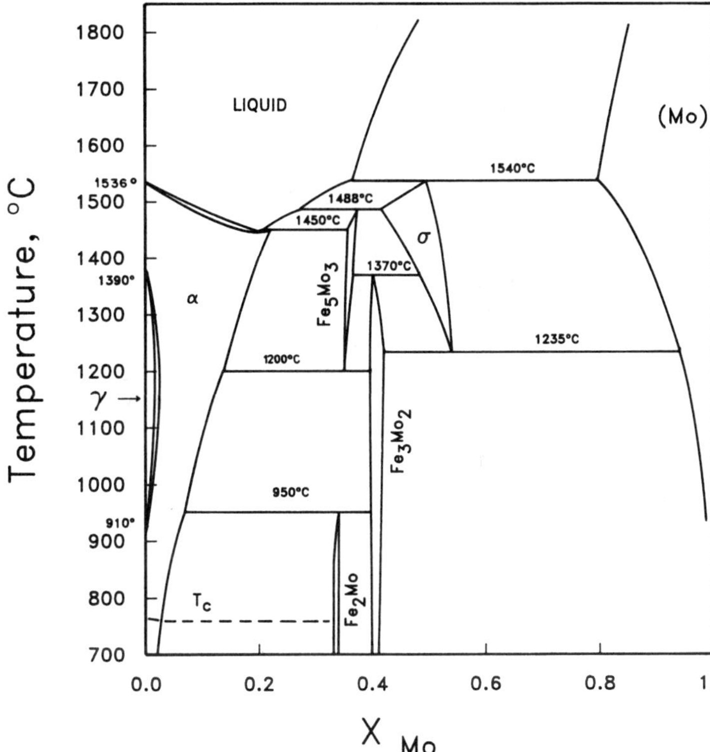

FIG. 9. Fe-Mo phase diagram at $P = 101$ kPa (Kubaschewski, 1982; reprinted from Pelton, 1991). The α and γ phases have the body-centered-cubic and face-centered-cubic structures, respectively. The σ phase is an intermediate solid phase.

ample, the best-known example of a eutectoid microstructure is that of pearlitic steel. For more information, see MATERIALS TREATMENT; METALLURGY, PHYSICAL; and STEEL AND IRON.

A final point of interest in Fig. 9 is the two-phase ($\alpha + \gamma$) γ *loop* between 910 °C and 1390 °C. Pure Fe adopts the face centered-cubic (fcc) γ crystal structure between these two temperatures, but exists as the body-centered-cubic (bcc) α phase at higher and lower temperatures. Mo, however, is much more soluble in the bcc than in the fcc structure. Hence, small additions of Mo stabilize the bcc structure, thereby producing the γ loop.

2.7 Phase Diagrams of Systems with Liquid-Crystalline Phases

In Sec. 2.4, solid polymorphism was discussed. In certain systems, liquid polymorphism is also observed. For example, the phase diagram of the system EEBC-TBBA is shown in Fig. 10.

All the phases shown with the exception of solid$_1$ and solid$_2$ are liquid phases. The high-temperature "isotropic" phase is a "normal" disordered liquid. In all the other phases there exists, on the average, an orientational order of the molecules. This orientational order may be distinguished from the positional order found in solid crystals. Generally, the molecules of liquids that form liquid crystals are elongated and rigid. If, on the average, the molecular axes are aligned, a *nematic* liquid-crystalline phase results. In other liquid-crystalline phases, the molecules tend to be both aligned and arranged in layers. Such phases are termed *smectic*. Various smectic phases can exist, differing, for example, in the angle between the molecular axes and the planes of the layers. In all respects, the phase diagram in Fig. 10 obeys the topological rules of binary *T-X* diagrams discussed in Sec. 2.6. For example, pure EEBC and TBBA transform from the smectic *A* to the nematic liquid crystal structure at 158.3 °C and 199.0 °C, respectively. At intermediate compositions there is a two-phase (smectic *A* + nematic) region, with a minimum at 156.2 °C. Several eutectic-like and peritectic-like invariants are

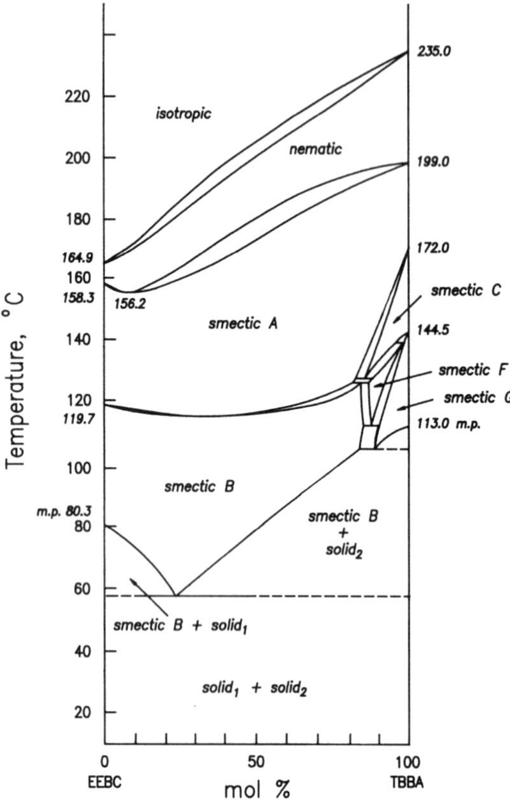

FIG. 10. Equilibrium *T-X* diagram at $P = 101.3$ kPa of the EEBC-TBBA system, which exhibits liquid-crystalline phases (after Richter *et al.*, 1981). (EEBC = ethyl 4-[4-ethoxy benzylidene-amino]-cinnamate: TBBA = terephthalylidene-*bis*-[4-*n*-butyl-aniline].)

seen in Fig. 10. For further discussion see LIQUID CRYSTALS, STRUCTURE OF.

3. THE GIBBS PHASE RULE

In Secs. 1 and 2 the topology of phase diagrams was described with no justification given as to why these topological rules should be obeyed. The present section addresses this question, although space restrictions do not permit a completely rigorous development to be made. For a more complete treatment see THERMODYNAMICS, EQUILIBRIUM.

First, the term *component* should be defined more precisely. At most, the number of components in a system is equal to the number of chemical elements. However, there may be chemical constraints such that certain elements always appear in fixed ratios. If these constraints apply *in every phase* that could possibly appear over the ranges of interest of T and P, then there will be a consequent reduction of the number of components. For example, in the methanol-water system, Figs. 1 and 2, the elements are always combined as CH_3OH and H_2O in every phase. Hence, this is a two-component system. A system of water and sodium chloride near ambient conditions is a two-component system since the number of Na^+ ions must equal the number of Cl^- ions in every phase. A system prepared by mixing the salts NaCl, KCl, NaF, and KF is a three-component system because the number of cations must always equal the number of anions. (A solution of any desired composition could always be formed by mixing only three of the four salts.)

Next we must determine the number of variables required to define the equilibrium state of a system. We choose these as temperature T, total pressure P, and the compositions of the phases at equilibrium. In the general case, other variables such as electrical, magnetic, and gravitational fields can also influence phase equilibria. However, we do not consider these here. It must be stressed that the overall (global) composition of the system is NOT a variable in the present context. We are interested in knowing which phases are present and their compositions, but not their relative amounts. For example, in a two-phase region in a binary system, as long as the overall composition lies on the tie-line between the compositions of the two phases, they will both be present and their compositions will not change.

Now, in order to specify the composition of a phase in a system with C components, $C - 1$ composition variables are required, since the sum of the mole fractions is unity. Hence, if P phases are present at equilibrium. $P(C - 1)$ composition variables are needed. Along with temperature and pressure, the number of variables is thus $P(C - 1) + 2$.

When we say that two phases are at equilibrium we mean that there is no net transfer with time of any component between the phases. Now, there must be some solution property that determines the direction of spontaneous transfer of a component between two phases in contact. This property is called the *chemical potential of component i*, and is given the symbol μ_i (where the components are numbered 1, 2, 3, . . ., C). Just as electrical charge is transferred spontaneously from high to low electrical potential, so component i will migrate spontaneously from a phase of higher chemical potential to a phase of lower chemical potential.

Therefore, if two phases, α and β are in equilibrium, then $\mu_i^\alpha = \mu_i^\beta$ for all components $1 \leq i \leq C$. If P phases are present at equilibrium, then there are $C(P - 1)$ such equations. However, μ_i, being a property of a phase, is a function of T, pressure, and the composition of the phase. That is, each of the $C(P - 1)$ equations is an equation relating those variables which specify the equilibrium state of the system.

The difference between the number of variables, $P(C - 1) + 2$, and the number of equations relating the variables, $C(P - 1)$, is called the *variance*, or the *number of degrees of freedom*, F, of the system. This then is the Gibbs phase rule:

$$F = C - P + 2. \tag{7}$$

That is, F is the number of variables which can and must be specified in order to specify completely the state of the system.

In a one-component system as in Fig. 1, $C = 1$. Hence, the phase rule reduces to $F = 3 - P$. The only variables being considered are temperature and pressure. In a single-phase area, $F = 3 - 1 = 2$. That is, T and pressure can be specified independently. These regions are thus called *bivariant*. When two phases are at equilibrium, $F = 3 - 2 = 1$. If, say, T is specified, then the pressure is fixed. For example, if liquid and vapor are in equilibrium at a given T, then the pressure must be the equilibrium vapor pressure. Such two-phase regions are thus called *univariant* and are represented by lines on Fig. 1. When three phases are in equilibrium, $F = 0$, the system is *invariant*, and the state is represented by a triple point in Fig. 1.

In a two-component system, $C = 2$ and $F = 4 - P$. When we consider a T-X phase diagram at a constant pressure, however, we have already eliminated one degree of freedom. Hence, for an isobaric T-X diagram, $F = 3 - P$. The variables of interest are now T and the compositions of the phases. In a single-phase area, $F = 3 - 1 = 2$. Hence, T and composition can be independently specified, and the system is bivariant. When two phases are in equilibrium, $F = 1$. If, say, T is specified, then the compositions of both phases are fixed by the ends of the tie-line. When three phases are at equilibrium, the system is invariant. That is, the compositions of all three phases as well as the temperature are fixed. This gives rise to eutectic-type and peritectic-type invariants.

It can thus be seen how the topology of phase diagrams is determined by the phase rule.

4. TERNARY PHASE DIAGRAMS

Only a very brief introduction to isobaric ternary T-X diagrams can be given here. For further discussion see Pelton (1983, 1991) and Prince (1966).

4.1. Ternary Composition Triangle

In a ternary system with components A, B, C, the sum of the mole fractions is unity: $X_A + X_B + X_C = 1$. Hence, there are two independent composition variables. A representation of composition, symmetrical with respect to all three components, may be obtained with the equilateral "composition triangle" as shown in Fig. 11(a) for the LiF-NaF-KF system. Compositions at the corners of the triangle correspond to the pure components. Along the edges of the triangle are found compositions corresponding to the three binary subsystems LiF-NaF, LiF-KF, and NaF-KF. Lines of constant mole fraction X_{LiF} are parallel to the NaF-KF edge, while lines of constant X_{NaF} and X_{KF} are parallel to the LiF-KF and LiF-NaF edges, respectively. For example, at point a in Fig. 11(a), $X_{\mathrm{LiF}} = 0.06$, $X_{\mathrm{NaF}} = 0.23$, and $X_{\mathrm{KF}} = 0.71$.

Similar equilateral composition triangles can be drawn with coordinates in terms of weight percent of the three components.

4.2 Ternary Space Model

A ternary temperature-composition "phase diagram" at constant total pressure may be plotted as a *three-dimensional space model* within a right triangular prism with the equilateral composition triangle as base and temperature as vertical axis. Such a space model for the simple eutectic ternary system LiF-NaF-KF is illustrated in Fig. 11(b). On the three vertical faces of the prism are found the

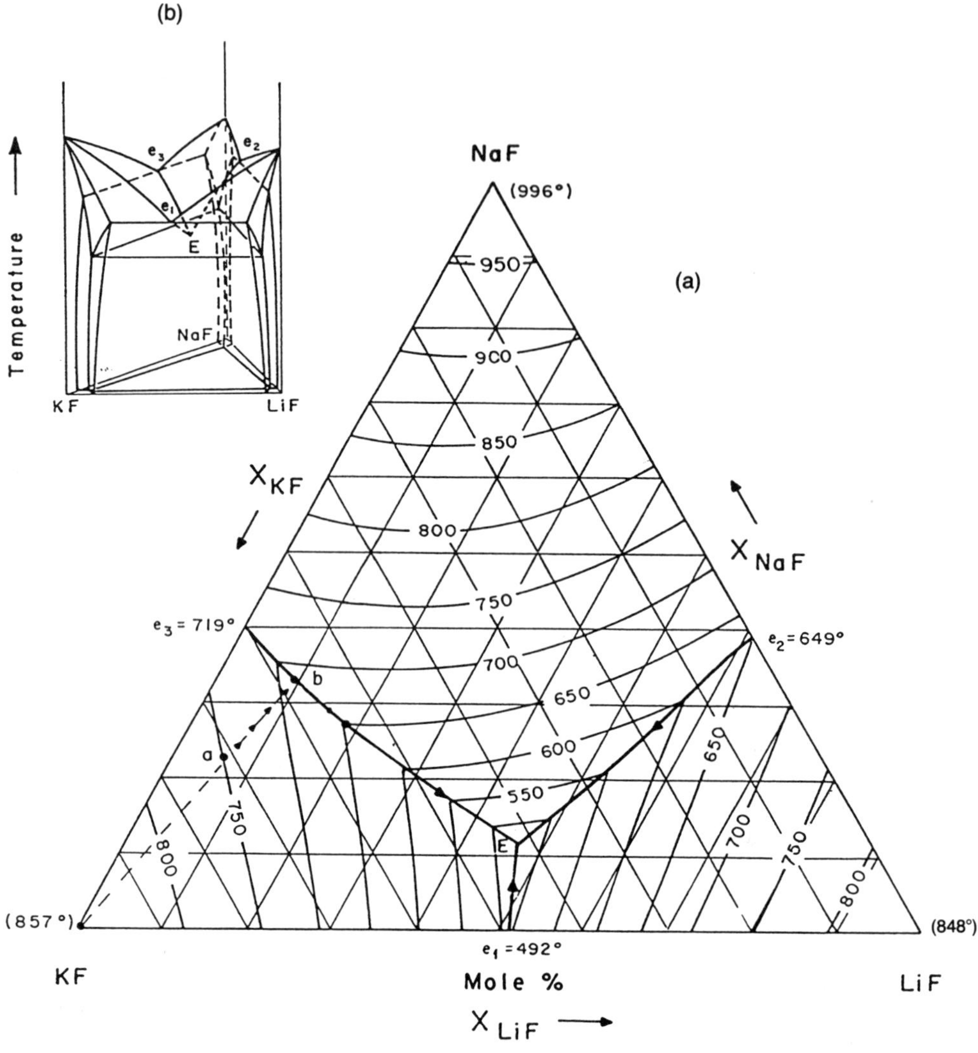

FIG. 11. (a) Polythermal projection of the liquidus surface of the LiF-NaF-KF system. Temperature in °C (Levin *et al.*, 1964.) (b) Perspective view of ternary space model of a simple eutectic system. ϵ_1, ϵ_2, and ϵ_3 are the binary eutectics and *E* is the ternary eutectic. The base of the prism is the equilateral composition triangle. (Reprinted from Pelton, 1983.)

phase diagrams of the three binary subsystems, LiF-NaF, LiF-KF, and NaF-KF, which in this example are all simple eutectic binary systems. The binary eutectic points are at e_1, e_2, and e_3. Within the prism are seen three *liquidus surfaces* descending from the melting points of pure LiF, NaF, and KF. Compositions on these surfaces correspond to compositions of liquids in equilibrium with LiF-, NaF-, and KF-rich solid phases.

The NaF and KF liquidus surfaces in Fig. 1(b) intersect along the line e_3E. Liquids with compositions along this univariant "valley" are therefore in equilibrium with NaF-rich and KF-rich solid phases simultaneously. The three univariant lines meet at the *ternary eutectic point E* at which the temperature and the compositions of all four phases in equilibrium are fixed.

In a 3-component system at constant pressure, the phase rule, Eq. (7), reduces to $F = 4 - P$. Therefore, for a liquid in equilibrium with one solid, $F = 2$. Hence, liquiduses are bivariant surfaces. The composition of a liquid in equilibrium with two solids is represented by a univariant line, and liquid can exist in equilibrium with three solids at a ternary eutectic invariant point.

4.3 Polythermal Projections of Liquidus Surfaces

A two-dimensional representation of the ternary liquidus surface may be obtained as an orthogonal projection upon the base composition triangle. Such a *polythermal projection* of the liquidus of the LiF-NaF-KF system is shown in Fig. 11(a). This is a simple eutectic ternary system. The constant-temperature lines in Fig. 11(a) are called *liquidus isotherms*. The univariant valleys are shown as the heavier lines. By convention, the arrows indicate the directions of decreasing temperature along these lines.

Let us consider the sequence of events that occur during the equilibrium cooling of a liquid of overall composition *a* in Fig. 11(a). Point *a* lies within the *field of primary crystallization* of KF. That is, it lies within the composition region in Fig. 11(a) in which KF-rich solid will be the first solid to precipitate upon cooling. As the liquid is cooled, the KF liquidus surface is reached at $T = 750$ °C. A solid KF-rich phase begins to precipitate at this temperature. Now, in this particular system, NaF and LiF are nearly insoluble in solid KF so that the solid phase is virtually pure KF. [Note that this fact cannot be deduced from Fig. 11(a) alone]. Therefore, as solidification proceeds, the liquid becomes depleted in KF, but the ratio X_{NaF}/X_{LiF} in the liquid remains constant. Hence, the composition path followed by the liquid (its *crystallization path*) is a straight line passing through point *a* and projecting to the KF corner of the triangle. This crystallization path is shown in Fig. 11(a) as the line *ab*. (Note that in the general case, in which a solid solution rather than a pure component or stoichiometric compound is precipitating, the crystallization path will not be a straight line.)

When the composition of the liquid has reached point *b* in Fig. 11(a) at $T \approx 680$ °C, the primary crystallization of KF is complete. Upon further cooling, the liquid composition follows the univariant valley from *b* to *E* while LF- and NaF-rich solids coprecipitate as a binary eutectic mixture. When the liquidus composition attains the ternary eutectic composition *E* at $T = 463$ °C, the invariant *ternary eutectic reaction* occurs:

$$\text{liquid} \rightarrow s_1 + s_2 + s_3, \tag{8}$$

where s_1, s_2, and s_3 are the three solid phases and where the compositions of all four phases, as well as *T*, remain fixed until all liquid is solidified.

As another more complex example, the polythermal liquidus projection is shown in Fig. 12 for the CaO-Al_2O_3-SiO_2 system. Liquidus surfaces, with isotherms, are seen for several primary crystalline phases. The compositions of the solid compounds are shown in Fig. 12 by points. For example, $CaSiO_3$ (pseudowollastonite) is a *binary compound* (since its composition is in the CaO-SiO_2 binary system), which melts at 1540 °C. For any liquid with composition in the field labeled "pseudowollastonite," $CaSiO_3$ will be the first solid to precipitate upon cooling. Anorthite, $CaAl_2Si_2O_8$, is a *ternary compound* which melts congruently at 1555 °C.

The liquidus surfaces for the various solid phases in Fig. 12 intersect along univariant lines. The arrows shown on these lines indicate the direction followed by the liquid composition during cooling in equilibrium with both solid phases. The univariant lines meet at invariant points. If the arrows on all three lines converge at such a point, then the point is a *ternary eutectic point*. If one or more arrows lead away from the point, then the point is a *ternary peritectic point*.

4.4 Ternary Isothermal-Isobaric Sections

If the solid phases in equilibrium with the liquid are stoichiometric, or nearly stoichiometric compounds, then a polythermal liquidus projection as in Figs. 11(a) and 12 gives complete information on the ternary space model. However, when solid solutions are present, the liquidus projection does not give information on the compositions of the solid phases. This information can be conveyed by a series of *isothermal sections*, which are constant-temperature slices through the space model.

An isothermal section at 25 °C of the Al-Zn-Mg system is shown in Fig. 13. Such isothermal sections follow strict topological rules. With three components, at constant *T* and pressure, the phase rule, Eq. (7), reduces to $F = 3 - P$. Single-phase regions are thus

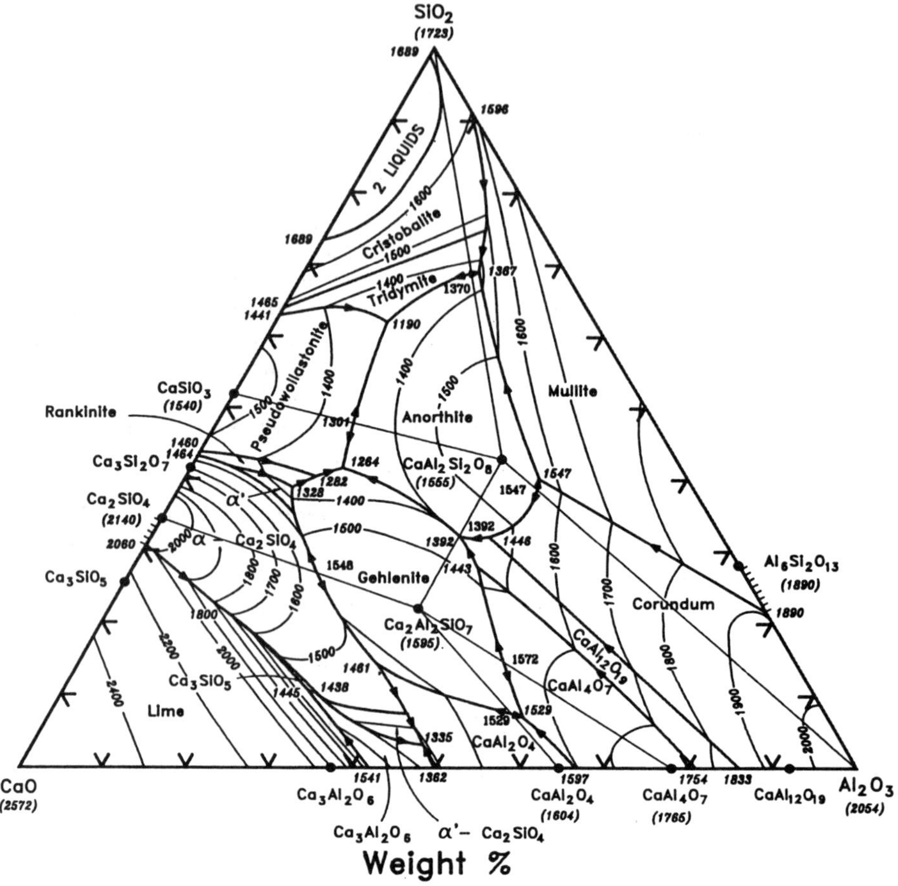

FIG. 12. Polythermal projection of the liquidus surface of the CaO-Al_2O_3-SiO_2 system. Temperatures in °C. (Eriksson and Pelton, 1993.)

bivariant ($F = 2$). These are shown as the black regions in Fig. 13. When two phases are in equilibrium, $F = 1$, and a tie-line results. These tie-lines together form two-phase regions. For example, in Fig. 13, a set of tie-lines joins the δ-phase and τ-phase regions. For an overall composition on any one of these tie-lines a δ-phase alloy with composition at one end of the tie-line is in equilibrium with a τ-phase alloy whose composition is at the other end. When three phases are at equilibrium, $F = 0$ and the compositions of all three phases are fixed. Three-phase regions are represented by *tie-triangles* as shown in Fig. 13. For an overall composition inside such a triangle, three phases are present at equilibrium, and their compositions are given by the corners of the triangle.

5. OTHER KINDS OF PHASE DIAGRAMS

The types of phase diagrams discussed in the preceding sections, and exemplified in Figs. 1 through 13, are the most commonly encountered. However, there are many other useful types of phase diagrams.

Vertical sections, known as *isopleths*, through a ternary space model like Fig. 11(b), are often a convenient method of representing ternary equilibria. In systems of four or more components, many techniques of presenting two-dimensional sections or projections have been proposed. For a discussion of these, see Prince (1963), Palatnik and Landau (1964), Prince (1966), and Hillert (1985).

Many types of phase diagrams result when the chemical potentials of components are

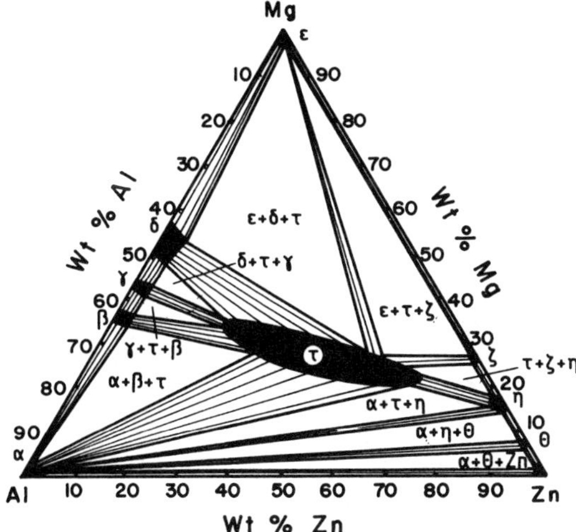

FIG. 13. Ternary isothermal section of the Al-Zn-Mg system at 25 °C (after Koster and Dullenkopf, 1936, reprinted from Pelton, 1983).

used as axis variables or are held constant. The chemical potential of a gas is easily controlled by fixing its partial pressure. Hence, the most useful and common phase diagrams of this class are those with partial pressures of gases as variables.

For example, Fig. 14 shows a *predominance diagram* of the Cu-S-O system at 1000 K. The axes are the equilibrium SO_2 and O_2 partial pressures. The diagram is divided into areas or domains of stability of the various solid Cu-containing compounds. The conditions for coexistence of two and three solids are indicated respectively by the lines and triple points on the diagram. Such diagrams have many applications in hot corrosion, roasting, chemical vapor deposition, etc. It can be seen that these diagrams obey the same topological rules of construction as *P-T* diagrams of one-component systems as in Fig. 1.

Another type of predominance diagram, which again obeys the same rules of construction, is shown in Fig. 15. This diagram indicates which solid Fe-containing phases are stable as a function of equilibrium oxygen partial pressure and temperature when the partial pressure of SO_2 is held constant.

Predominance diagrams, their thermodynamic calculation, and their extension to systems of several components have been discussed by Bale *et al.* (1986) and Bale (1990).

In Fig. 16 is shown a *chemical-potential–composition phase diagram* for the Fe-Cr-O system at 1300 °C. This is an isothermal plot of the equilibrium oxygen pressure vs the molar metal ratio, $n_{Cr}/(n_{Fe} + n_{Cr})$, where n_i = number of moles of i. The diagram consists of single-phase areas, two-phase regions with

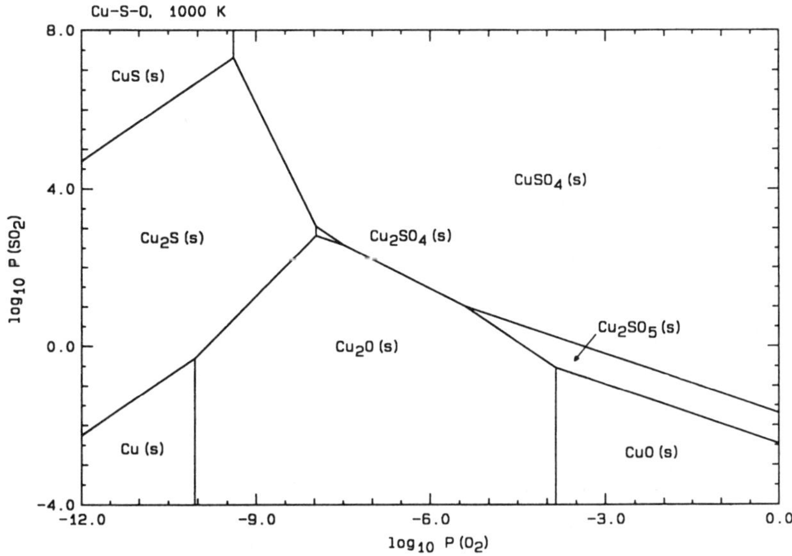

FIG. 14. Predominance diagram, $\log p_{SO_2}$ vs $\log p_{O_2}$ (atm) at 1000 K for the Cu-S-O system (Bale *et al.*, 1986).

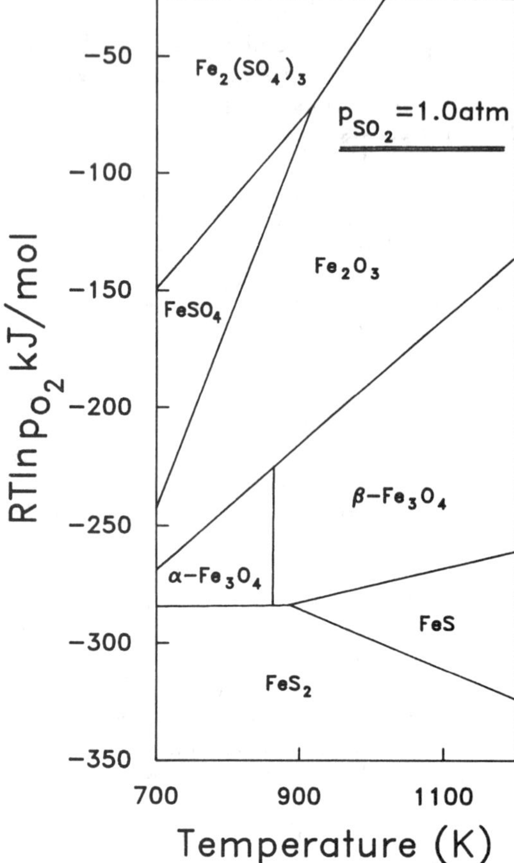

FIG. 15. Predominance diagram, $RT \ln p_{O_2}$ versus T at p_{SO_2} = 1.0 atm for the Fe-S-O system (reprinted from Pelton, 1991).

horizontal tie-lines (to which the lever rule applies), and three-phase eutectic-type and peritectic-type invariants. That is, topologically the diagram obeys the same rules of construction as binary T-X or P-X diagrams.

Many other phase diagrams can be drawn which have the same topologies as either one-component P-T diagrams, binary T-X and P-X diagrams, or ternary isothermal sections. Such phase diagrams and their classification have been discussed by Pelton and Schmalzried (1973).

6. THERMODYNAMIC ORIGIN OF PHASE DIAGRAMS

In this section we briefly discuss the thermodynamic origin of phase diagrams. This subject is treated in more detail by Pelton (1991). A basic knowledge of solution thermodynamics is required for this section. The principles are covered in THERMODYNAMICS, EQUILIBRIUM and by Pelton (1991).

Every system is characterized by a Gibbs energy function. The equilibrium state of the system is the state with minimum Gibbs energy. The isobaric T-X phase diagram of the Ge-Si system is shown in the lowest panel of Fig. 17. In the upper three panels of Fig. 17 the molar Gibbs energies of the solid and liquid phases, g^s and g^l, at three temperatures are shown to scale. As illustrated in the top panel, g^s varies with composition between the standard molar Gibbs energies of pure solid Ge and pure solid Si, $g_{Ge}^{0(s)}$ and $g_{Si}^{0(s)}$, while g^l varies between the standard molar Gibbs energies of the pure liquid components, $g_{Ge}^{0(l)}$ and $g_{Si}^{0(l)}$.

The quantities Δg_m^s and Δg_m^l are the molar Gibbs energies of mixing of the solutions. Each of these is equal to an ideal mixing term, $RT(X_1 \ln X_1 + X_1 \ln X_2)$, plus an excess term, g^E:

$$\Delta g_m = RT(X_1 \ln X_1 + X_2 \ln X_2) + g^E, \tag{9}$$

where X_1 and X_2 are the component mole fractions. The ideal mixing term is negative and is calculated statistically from the entropy of a random distribution of Ge and Si atoms over the lattice positions (see STATISTICAL MECHANICS, CLASSICAL).

As shown in the top panel of Fig. 17, the difference between $g_{Si}^{0(l)}$ and $g_{Si}^{0(s)}$ is equal to the standard molar Gibbs energy of fusion (melting) of pure Si, $\Delta g_{f(Si)}^0 = g_{Si}^{0(l)} - g_{Si}^{0(s)}$. Similarly for Ge, $\Delta g_{f(Ge)}^0 = g_{Ge}^{0(l)} - g_{Ge}^{0(s)}$. The Gibbs energy of fusion of a pure component may be written as

$$\Delta g_f^0 = \Delta h_f^0 - T\Delta s_f^0, \tag{10}$$

where Δh_f^0 and Δs_f^0 are the standard molar enthalpy and entropy of fusion.

Since to a first approximation, Δh_f^0 and Δs_f^0 are independent of T, Δg_f^0 is approximately a linear function of T. If $T > T_f^0$, then Δg_f^0 is negative. If $T < T_f^0$, then Δg_f^0 is positive. Hence, as seen in Fig. 17, as T decreases, the g^s curve descends relative to g^l. At 1500 °C, $g^l < g^s$ at all compositions. Therefore, by the principle that a system always seeks the state of minimum Gibbs energy at constant T and

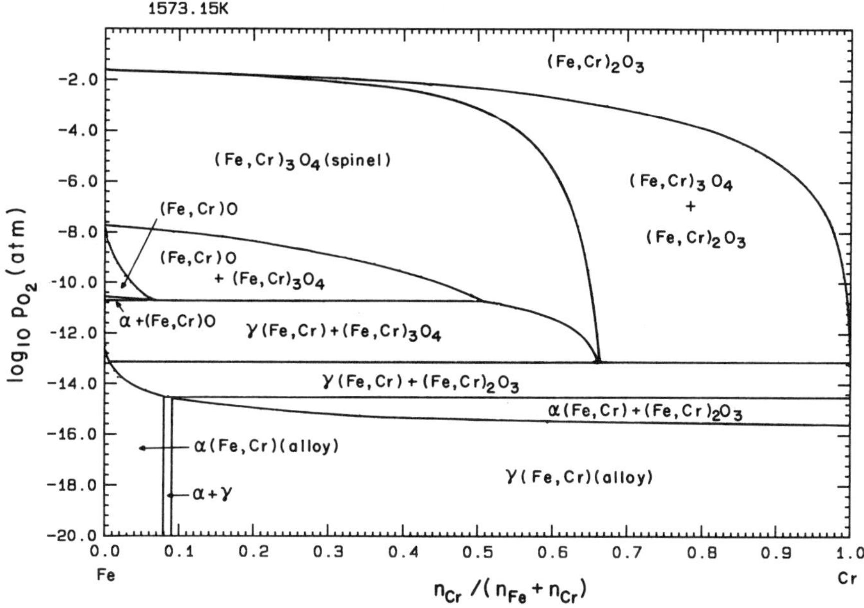

FIG. 16. Phase diagram of $\log p_{O_2}$ vs molar metal ratio for the Fe-Cr-O system at 1300 °C (Pelton and Schmalzried, 1973).

P, the liquid phase is stable at all compositions at 1500 °C.

At 1300 °C, the curves of g^s and g^l cross. The line P_1Q_1, which is the common tangent to the two curves, divides the composition range into three sections. For compositions between pure Ge and P_1, a single-phase liquid is the state of minimum Gibbs energy. For compositions between Q_1 and pure Si, a single-phase solid solution is the stable state. Between P_1 and Q_1, a total Gibbs energy lying on the tangent line P_1Q_1 may be realized if the system adopts a state consisting of two phases with compositions at P_1 and Q_1 and with relative proportions given by the lever rule. Since the tangent line P_1Q_1 lies below both g^s and g^l, this two-phase state is more stable than either phase alone. Furthermore, no other line joining any point on the g^l curve to any point on the g^s curve lies below the line P_1Q_1. Hence, this line represents the true equilibrium state of the system, and the compositions P_1 and Q_1 are the liquidus and solidus compositions at 1300 °C.

As T is decreased to 1100 °C, the points of common tangency are displaced to higher concentrations of Ge. For $T < 937$ °C, $g^s < g^l$ at all compositions. In this way, the lens-shaped Ge-Si phase diagram is generated.

If both Δg_m^l and Δg_m^s are ideal [i.e., if the g^E term in Eq. (9) is zero for both phases], then a lens-shaped two-phase region always results. However, in most systems, even approximate ideal behavior is the exception. Curves of g^s and g^l for a hypothetical system A-B are shown schematically in Fig. 18 at a constant temperature below the melting points of pure A and B such that the solid state is the stable state for both pure components. However, in this system $g^{E(l)} < g^{E(s)}$, so that g^s presents a flatter curve than does g^l and there exists a central composition region in which $g^l < g^s$. Hence, there are two common tangent lines, P_1Q_1 and P_2Q_2. Such a situation gives rise to a phase diagram with a minimum in the two-phase (solid + liquid) region similar to the diagram of Fig. 5.

Other forms of the Gibbs energy curves can give rise to phase diagrams with regions of immiscibility, eutectic-type and peritectic-type invariants, etc. The treatment can be extended to multicomponent systems.

6.1 Models of Solutions—Regular Solution Theory

The phase diagram of a system is determined by the thermodynamic properties of the phases which, in turn, depend upon their structure and bonding. To illustrate this re-

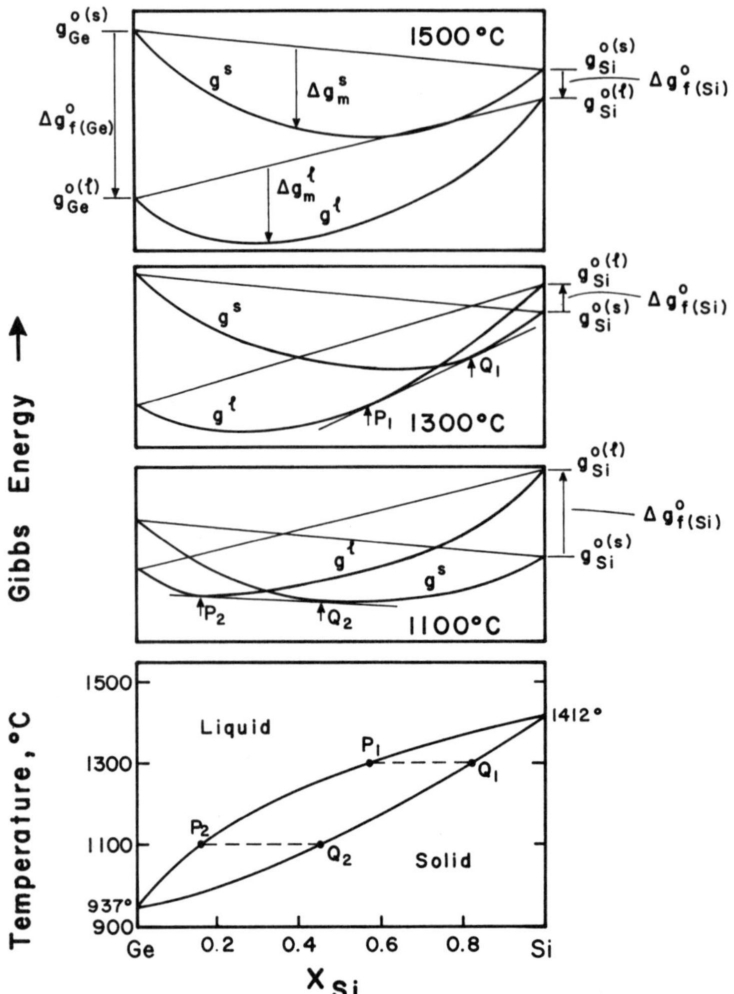

FIG. 17. The Ge-Si phase diagram (after Hansen, 1958) and Gibbs energy-composition curves at three temperatures illustrating the common tangent construction (reprinted from Pelton, 1983).

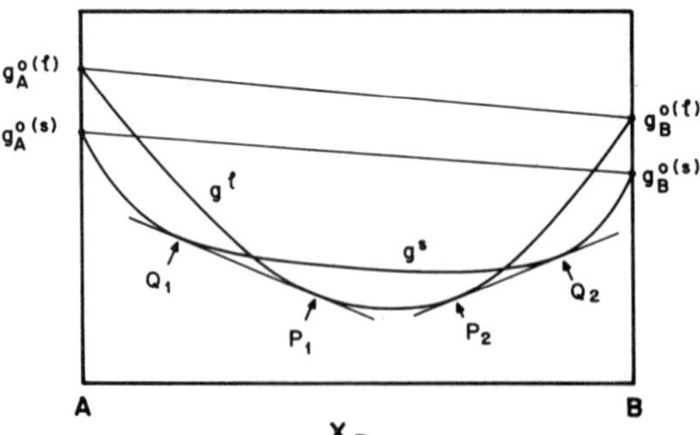

FIG. 18. Isothermal Gibbs energy- composition curves for solid and liquid phases in a system *A-B* in which $g^{E(l)} < g^{E(s)}$. A phase diagram with a minimum in the (solid + liquid) two-phase region results.

lationship, we shall briefly discuss a very simple solution model, that of a "regular" solution.

Consider a binary solution formed of atoms or molecules of the components *A* and *B* that mix substitutionally and nearly randomly. The entropy of mixing will then be nearly ideal and will give rise to the ideal mixing term in Eq. (9).

We now assume that the bond energies ϵ_{AA}, ϵ_{BB}, and ϵ_{AB} of nearest-neighbor pairs of atoms or molecules are independent of temperature and composition and that the average nearest-neighbor coordination number, *Z*, is also constant. Finally, we assume that the excess Gibbs energy of mixing, g^E, results mainly from the change in the total energy of nearest-neighbor pair bonds.

In one mole of solution there are $N^0Z/2$ nearest-neighbor pair bonds, where N^0 is Avogadro's number. Since the distribution is assumed to be random, the probability that a given bond is an A-A bond is equal to X_A^2. The probabilities of B-B and A-B bonds are, respectively, X_B^2 and $2X_AX_B$. The molar excess Gibbs energy of mixing is then equal to the sum of the energies of the nearest-neighbor bonds in one mole of solution, less the energy of the A-A bonds in X_A moles of pure A and the energy of the B-B bonds in X_B moles of pure B:

$$\begin{aligned} g^E &= (N^0Z/2)(X_A^2\epsilon_{AA} + X_B^2\epsilon_{BB} + 2X_AX_B\epsilon_{AB}) \\ &\quad - (N^0Z/2)(X_A\epsilon_{AA}) - (N^0Z/2)(X_B\epsilon_{BB}) \\ &= (N^0Z)[\epsilon_{AB} - (\epsilon_{AA} + \epsilon_{BB})/2]X_AX_B \\ &= \omega X_AX_B, \end{aligned} \tag{11}$$

where ω is a constant, and where the relation $X_A + X_B = 1$ has been used in the simplification of Eq. (11).

If A-B bonds are stronger than AA and BB bonds, then $\epsilon_{AB} < (\epsilon_{AA} + \epsilon_{BB})/2$ and $\omega < 0$, and $g^E < 0$. That is, the solution is rendered more stable. If the A-B bonds are relatively weak, then the solution is rendered less stable since $\omega > 0$ and $g^E > 0$.

In Fig. 19, several phase diagrams are shown, calculated for a hypothetical system A-B containing a solid and a liquid phase with melting points of $T^0_{f(A)} = 800$ K and $T^0_{f(B)} = 1200$ K and with entropies of fusion of both A and B set to 10 J/mol K, which is a typical value for metals. The solid and liquid phases are both regular, with temperature-independent excess Gibbs energies:

$$g^{E(s)} = \omega^s X_AX_B \quad \text{and} \quad g^{E(l)} = \omega^l X_AX_B.$$

The parameters ω^s and ω^l have been varied systematically to generate the various panels of Fig. 19.

In panel (n) both phases are ideal. Panels (l)–(r) exhibit minima or maxima depending upon the sign and magnitude of $g^{E(l)} - g^{E(s)}$. In panel (h) the liquid is ideal, but positive deviations in the solid give rise to a solid-solid miscibility gap. That is, A-B bonds are energetically so unfavorable that $g^{E(s)}$ becomes so positive that the solid solution splits into an A-rich phase and a B-rich phase.

On passing from panel (h) to panel (c), an increase in $g^{E(s)}$ results in a widening of the miscibility gap so that the solubility of A in solid B and of B in solid A decreases. Panels (a)–(c) illustrate that negative deviations in the liquid cause a relative stabilization of the liquid with resultant lowering of the eutectic temperature.

If the negative deviations in the liquid phase are sufficiently large, then the liquid solution will adopt a structure in which the number of A-B nearest-neighbor pairs in greater than the value $2X_AX_B$ for random mixing. Such *short-range ordering* is not predicted by the simple regular solution model, but can be taken into account by more sophisticated models such as the quasichemical model (Pelton and Blander, 1986).

6.2 Thermodynamic-Phase Diagram Optimization

In recent years, the development of solution models, numerical methods, and computer software has permitted a quantitative application of thermodynamics to phase-diagram analysis. For a great many systems it is now possible to perform a simultaneous critical evaluation of available phase-diagram measurements and of available thermodynamic data (calorimetric data, measurements of chemical activities, etc.) with a view to obtaining optimized equations for the Gibbs energies of each phase that best represent all the data.

The phase diagram can be calculated from these thermodynamic equations, and so one set of self-consistent equations can describe all the thermodynamic properties and the phase diagram. This technique of analysis greatly reduces the amount of experimental information needed to characterize a system fully. All data can be tested for internal consistency. The data can be interpolated and extrapolated more accurately, and metastable phase boundaries can be calculated. All the thermodynamic properties and the phase diagram can be represented and stored by means of a small set of coefficients. Finally and most importantly, it is often possible to estimate the thermodynamic properties and phase diagrams of ternary and higher-order systems from the assessed parameters for their binary subsystems. The *Calphad Journal* (Pergamon Press) is an in-

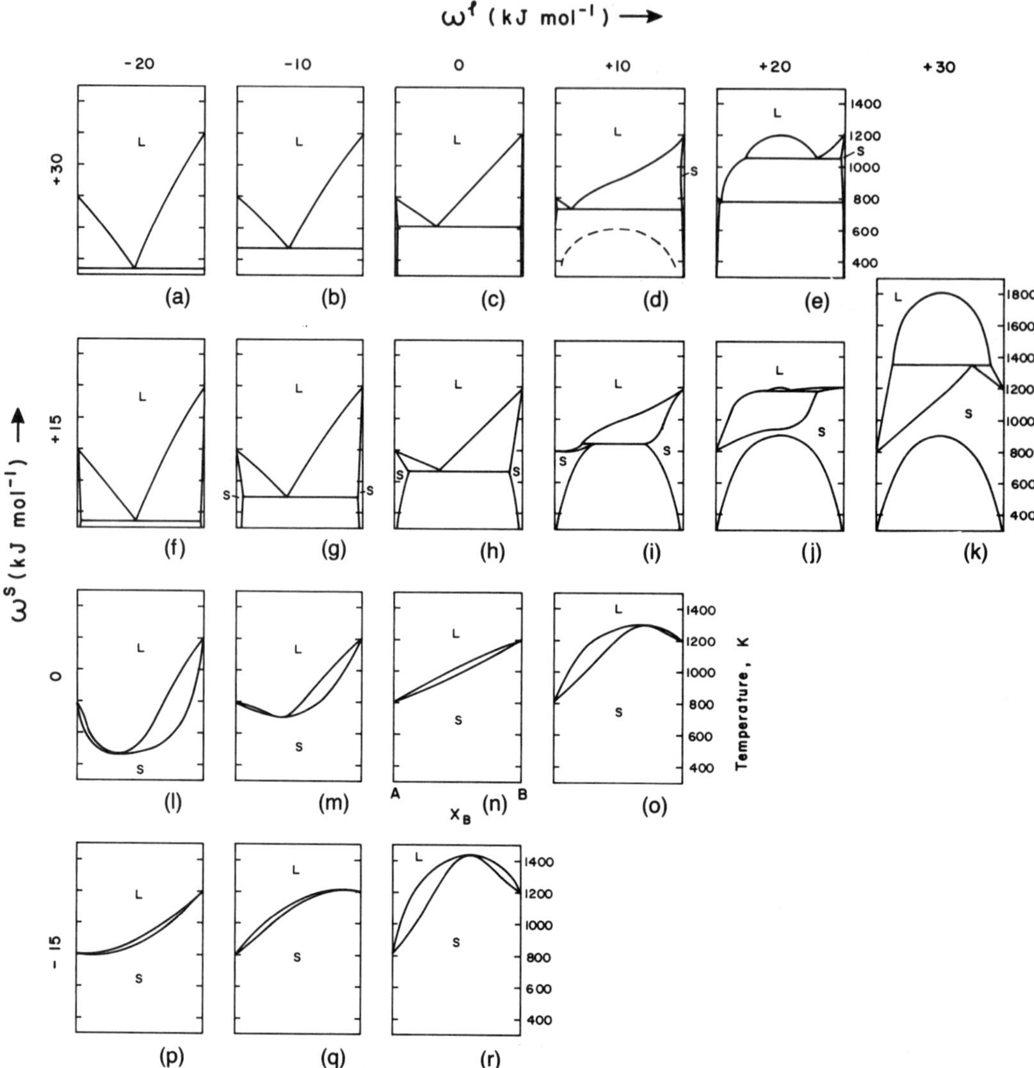

FIG. 19. Topological changes in the phase diagram for a system *A-B* with regular solid and liquid phases, brought about by systematic changes in the regular solution parameters ω^s and ω^l. Melting points of pure *A* and *B* are 800 and 1200 K. Entropies of fusion of both *A* and *B* are 10.0 J/mol K (Pelton and Thompson, 1975). Dotted curve in panel (d) is metastable liquid-liquid miscibility gap. (Reprinted from Pelton, 1983.)

ternational journal devoted to publishing this type of work.

7. GLASS FORMATION

In certain liquids, such as molten SiO_2, the nature of the chemical bonds is such that solidification to form an ordered crystal structure upon cooling the melt is a relatively slow process. Consequently, if such melts are cooled quickly, the disordered liquid structure will be frozen in, and the result at room temperature will be an *amorphous, glassy,* or *vitreous* phase, which can be considered to be a supercooled liquid. Glasses, being nonequilibrium phases, do not appear on equilibrium phase diagrams. Nevertheless, the equilibrium phase diagram is very useful in understanding the microstructure of materials with a glassy phase. For example, suppose that a $CaO\text{-}SiO_2$ melt of overall composition $X_{SiO_2} = 0.53$ (see Fig. 7) is annealed at 1500 °C and then quenched. The equilibrium phases at 1500 °C will be $CaSiO_3$ and liquid of composition $X_{SiO_2} = 0.57$ with the relative amounts given by the lever rule: (moles $CaSiO_3$)/(moles liquid) = (0.57 − 0.53)/(0.53 − 0.50) = 1.33. Upon rapid quenching, the entire liquid phase becomes a glass, and at room temperature we have (moles $CaSiO_3$)/(moles glass) = 1.33.

Most liquids, even metals, will form an amorphous phase if the quenching rate is sufficiently rapid. Generally, when a liquid is quenched to form a glass, the physical properties, such as viscosity, change from values typical of a liquid to values typical of a solid over a relatively narrow range of temperature about an average or *glass transition temperature* that depends upon the cooling rate. Glass formation is discussed in AMORPHOUS MATERIALS, STRUCTURE OF, and GLASSES.

8. ORDER-DISORDER TRANSITIONS

In solid solutions, long-range ordering can occur. For example, in the Ag-Mg system, a transformation occurs at approximately 392 °C at the composition Ag_3Mg from a low-temperature form with long-range ordering (superlattice formation) to a high-temperature form in which Ag and Mg are randomly distributed over the lattice sites. Unlike first-order phase transformations such as solidification, or the transformation between two different crystal structures, such *order-disorder transformation* can occur by atomic rearrangement over distances of the order of atomic dimensions.

A type of order-disorder transformation of importance in ferrous metallurgy is the magnetic transformation. Below its *Curie temperature* of 769 °C, Fe is ferromagnetic. Above this temperature it is not. The transformation involves a change in ordering of the atomic spins and is not first order. Additions of alloying elements will change the temperature of transformation as shown in Fig. 9.

For further discussion see ORDER-DISORDER TRANSITIONS and Pitsch and Inden (1991). Magnetic transformations are discussed by Miodownik (1982) and Inden (1982).

9. INTERFACIAL EFFECTS

In Sec. 6 it was shown that equilibrium between phases results from a minimization of the total Gibbs energy of the system. If a homogeneous solution is slowly cooled, a second phase will begin to precipitate at a temperature below which the formation of a second phase lowers the total Gibbs energy. In Sec. 6, however, it was assumed that the Gibbs energy associated with the formation of the interface between the phases is negligible. This is usually a good assumption unless the precipitating particles are very small. In the case of very small precipitates, however, the interfacial area is so large that the interfacial Gibbs energy cannot be ignored. An interface is a region in which the continuity of the structure and the bonding is interrupted. The Gibbs energy associated with the formation of an interface is thus positive. If the precipitates are very small, then, the precipitation temperature will be lowered below the value shown in the equilibrium phase diagram.

A further positive contribution to the Gibbs energy of forming a second phase arises if the interface is *coherent*, that is if there is a matching of the crystal structures across the interface. Since the lattice parameters of the two phases are not identical, such coherency gives rise to a positive strain energy which will also influence the phase equilibrium.

The existence of a strain energy in the region near the interface can result in concentration gradients in this region, since the formation of such gradients will partially decrease the strain energy. An applied stress on the sample can also affect the phase equilibria, as well as the kinetics of precipitation, because of interaction between the applied and coherency stress fields. For discussions of these topics, the reader is referred to Wagner and Kampmann (1991) and Purdy (1991).

GLOSSARY

Azeotrope: An extremum in a liquid + vapor two-phase region.

Binary System: A system of two components.

Bivariant: Having a variance of two.

Component: One of the chemical species required to define the composition of a system.

Congruent Melting: Melting to form a liquid of the same composition as the solid.

Conode: See **Tie-Line**.

Consolute Point: The summit of a miscibility gap. See also **Critical Point**.

Critical Point: A point on a phase diagram at which the properties of two phases become identical and a two-phase region ends.

Critical Pressure: The pressure at a critical point.

Critical Temperature: The temperature at a critical point.

Crystallization Path: The locus followed by the liquid composition during equilibrium cooling.

Equilibrium: A state of minimum free energy. A system at equilibrium is unchanging even over an infinite amount of time.

Eutectic Composition: The (constant) composition of the liquid during a eutectic reaction.

Eutectic Reaction: An invariant reaction during which, upon cooling, a liquid solidifies at constant temperature to give two solids (in a binary system), three solids (in a ternary system), etc.

Eutectic Temperature: The (constant) temperature at which a eutectic reaction occurs.

Eutectoid Reaction: An invariant reaction during which, upon cooling, a solid dissociates, at constant temperature, to two (or more) different solids.

Incongruent Melting: Melting (in a peritectic reaction) to form a liquid plus a solid of another composition.

Invariant: Having a variance of zero.

Isopleth: A constant-composition section through a multicomponent phase diagram.

Lever Rule: The rule by which the relative amounts of two phases existing at equilibrium can be calculated from a phase diagram.

Liquidus: The surface on a phase diagram giving the composition of the liquid phase in equilibrium with a solid phase.

Miscibility Gap: A region on a phase diagram in which two liquid phases (or two solid phases of the same crystal structure) are in equilibrium.

Mole Fraction: A composition variable. The mole fraction X_i of component i is equal to $n_i/\Sigma n_j$ where Σn_j is the sum over all the components.

Monotectic Reaction: An invariant reaction during which, upon cooling, a liquid reacts at constant temperature to give another liquid and one (or more) solids.

Nonstoichiometric Compound: A compound that exists as a single phase over a range of composition.

Peritectic Reaction: An invariant reaction during which, upon cooling, a liquid and one (or more) solids react, at constant temperature, to give one (or more) other solids. See also **Incongruent Melting**.

Phase: A region of a system of uniform physical and chemical properties.

Primary Solid: The first solid phase to crystallize upon equilibrium cooling of a single-phase liquid. Also called *pro-eutectic solid*.

Solidus: The surface on a phase diagram giving the composition of a solid phase in equilibrium with a liquid phase.

Solvus: The surface on a phase diagram giving the composition of a solid phase in equilibrium with another solid phase.

Space Model: The three-dimensional temperature-composition phase diagram of a ternary system.

Stoichiometric Compound: A compound of fixed composition.

Ternary System: A system of three components.

Tie-Line: A straight line on a phase diagram whose ends give the compositions of two phases at equilibrium. Also called *Conode*.

Tie-Triangle: A triangle on a phase diagram whose corners give the compositions of three phases at equilibrium.

Triple Point: At invariant point on a phase diagram at which three phases are at equilibrium.

Univariant: Having a variance of one.

Variance: The number of variables (temperature, pressure, compositions of the phases) which can and must be specified in order to completely specify the state of a system.

Works Cited

Bale, C. W., Pelton, A. D., Thompson, W. T. (1986), *Can. Metall. Quart.* **25**, 107–112.

Bale, C. W. (1990), *Can. Metall. Quart.* **29**, 263–278.

Eriksson, G., Pelton, A. D. (1993), *Metall. Trans. B*, **24B**, 807–816.

Eriksson, G., Wu, P., Blander, M., Pelton, A. D. (1994), *Can. Metall. Quart.* **33**, 13–22.

Hansen, M. (1958), *Constitution of Binary Alloys*, 2nd ed., New York: McGraw-Hill.

Hillert, M. (1985), *Int. Metall. Rev.* **30**, 45.

Hultgren, R., Desai, P. D., Hawkins, D. T., Gleiser, M., Kelly, K. K., Wagman, D. D. (1973), *Selected Values of the Thermodynamic Properties of the Elements and Binary Alloys*, Metals Park, OH: Am. Soc. of Metals.

Inden, G. (1982), *Bull. Alloy Phase Diag.* **2**, 412.

Köster, W., Dullenkopf, W. (1936), *Z. Metallk.* **28**, 309.

Kubaschewski, O. (1982), *Iron-Binary Phase Diagrams*, New York: Springer Verlag.

Levin, E. M., Robbins, C. R., McMurdie, H. F. (1964), *Phase Diagrams for Ceramists*, and (1969) Supplement; Levin, E. M., McMurdie, H. F. (1975) Supplement; Roth, R. S., Negas, T., Cook, L. P. (1981, 1983) Supplements; Roth, R. S., Denis, J. R., McMurdie, H. F. (1987) Supplement; Cook, L. P., McMurdie, H. F. (1989) Supplement; Mysen, B. O. (1990) Supplement; Stringfellow, G. B. (1992) Supplement, Columbus, OH: Am. Ceramic Soc.

Miodownik, A. P. (1982), *Bull. Alloy Phase Diag.* **2**, 406.

Palatnik, L. S., Landau, A. I. (1964), *Phase Equilibria in Multicomponent Systems*, New York: Holt, Rinehart and Wilson.

Pelton, A. D. (1983), in: R. W. Cahn, P. Haasen, (Eds.), *Physical Metallurgy*, Amsterdam: Elsevier, Chap. 7 (revised edition to appear in 1995).

Pelton, A. D., Blander, M. (1986), *Metall. Trans. B*, **17B**, 805–815.

Pelton, A. D., Schmalzried, H. (1973), *Metall. Trans.* **4**, 1395–1404.

Pelton, A. D., Thompson, W. T. (1975), *Prog. Solid State Chem.* **10**, Part 3, 119.

Pelton, A. D. (1991), in: R. W. Cahn, P. Haasen, E. J. Kramer (Eds.), *Materials Science and Technology*, Vol. 5, New York: VCH, Chap. 1.

Pitsch, W., Inden G. (1991), in: R. W. Cahn, P. Haasen, E. J. Kramer (Eds.), *Materials Science and Technology*, Vol. 5, New York: VCH, Chap. 9.

Prince, A. (1963), *Metall. Rev.* **8**, 213.

Prince, A. (1966), *Alloy Phase Equilibria*, Amsterdam: Elsevier.

Purdy, G. R. (1991), in: R. W. Cahn, P. Haasen, E. J. Kramer (Eds.), *Materials Science and Technology*, Vol. 5, New York: VCH, Chap. 5.

Richter, L., Demus, D., Sackmann, H. (1981), *Mol. Cryst. Liq. Cryst.* **71**, 269–291.

Schenck, H., Frohberg, M. G., Nunninghof, R. (1964), *Arch. Eisenhütten* **35**, 269–277.

Wagner, R., Kampmann, R. (1991), in: R. W. Cahn, P. Haasen, E. J. Kramer (Eds.), *Materials Science and Technology*, Vol. 5, New York: VCH, Chap. 4.

Wu, P., Eriksson, G., Pelton, A. D. (1993), *J. Am. Ceram. Soc.* **76**, 2065–2075.

Further Reading

Phase Diagram Compilations

Elliott, R. P. (1965), *Constitution of Binary Alloys*, First Supplement, New York: McGraw Hill. Second volume in the series started by Hansen (1958).

Hansen, M. (1958), *Constitution of Binary Alloys*, 2nd ed., New York, McGraw Hill. Classic compilation in the field of binary alloy phase diagrams.

Journal of Phase Equilibria (formerly *Bulletin of Alloy Phase Diagrams*) (1977–1994), Materials Park, OH: ASM International. Since 1979, the American Society for Metals, in collaboration with the National Institute of Science and Technology and with national organizations in several countries, has undertaken a project whose goal is a complete critical evaluation of all binary and ternary alloy phase diagrams. All available literature on phase equilibria, crystal structures, and often on thermodynamic properties, is cited and critically evaluated in great detail. The evaluations are peer reviewed and the majority are published in this journal.

Kubaschewski, O. (1982), *Iron-Binary Phase Diagrams*, New York: Springer Verlag. A critical compilation.

Levin, E. M., Robbins, C. R., McMurdie, H. F. (1964), *Phase Diagrams for Ceramists* (1969) Supplement; Levin, E. M., McMurdie, H. F. (1975) Supplement; Roth, R. S., Negas, T., Cook, L. P. (1981, 1983) Supplements; Roth, R. S., Denis, J. R., McMurdie, H. F. (1987) Supplement; Cook, L. P., McMurdie, H. F. (1989) Supplement; Mysen, B. O. (1990) Supplement; Stringfellow, G. B. (1992) Supplement, Columbus, OH: Am. Ceramic Soc. Phase Diagrams of over 8000 binary, ternary, and multicomponent ceramic systems, including oxides, halides, carbonates, sulfates, etc. Earlier volumes are noncritical compilations. Recent volumes include critical commentaries.

Massalski, T. (1986–1990), *Binary Alloy Phase Diagrams*, Metals Park, OH: Am. Soc. of Metals. Condensed versions of approximately 5000 of the critical evaluations published in the *Journal of Phase Equilibria*.

Moffatt, W. B. (1979–1993), *Handbook of Binary Phase Diagrams*, Schenectady, NY: General Electric Corp. A noncritical compilation of binary alloy phase diagrams.

Shunk, F. A. (1969), *Constitution of Binary Alloys*, Second Supplement, New York: McGraw Hill. Third volume in the series started by Hansen.

Wisniak, J. (1981). *Phase Diagrams: A Literature Source Book*, Vols. 1 and 2, New York: Elsevier. An extensive bibliography of binary and multicomponent phase diagrams of all types of systems: metallic, ceramic, aqueous, organic, etc.

General Reading

Alper, A. M. (Ed.) (1970–1978), *Phase Diagrams—Materials Science and Technology*, Vols. 1–5, New York: Academic. Many aspects of phase diagrams in materials science are discussed.

Bergeron, C. J., Risbud, S. H. (1984), *Introduction to Phase Equilibria in Ceramics*, Columbus, OH: American Ceramic Society. An introduction to phase diagrams with particular attention to applications in ceramic systems.

CALPHAD Journal (1977–1994), New York, Pergamon. A journal devoted to the thermodynamic calculation of phase diagrams.

Pelton, A. D. (1983), in R. W. Cahn, P. Haasen, (Eds.), *Physical Metallurgy*, Amsterdam: Elsevier, Chap. 7. Revised addition to appear in 1995. Relationships between thermodynamics and phase diagrams are discussed as are techniques of measuring phase diagrams.

Pelton, A. D. (1991), in: R. W. Cahn, P. Haasen, E. J. Kramer (Eds.), *Materials Science and Technology*, Vol. 5, New York: VCH, Chap. 1. Relationships between thermodynamics and phase diagrams are discussed.

Prince, A. (1966), *Alloy Phase Equilibria*, Amsterdam: Elsevier. Contains a detailed treatment of the geometry of multicomponent phase diagrams.

Raynor, G. V. (1970), in: R. W. Cahn (Ed.), *Physical Metallurgy*, 2nd ed., Amsterdam: North Holland, Chap. 7. Contains a discussion of techniques of measuring phase diagrams.

Rhines, F. N. (1956), *Phase Diagrams in Metallurgy*, New York: McGraw Hill. A classical discussion of phase diagrams in metallurgy.

PHASE SEPARATION

PIERRE WILTZIUS, *AT&T Bell Laboratories, Murray Hill, New Jersey, U.S.A.*

	Introduction	323
1.	**Phase Diagrams**	324
1.1	One-Component Systems	324
1.2	Multicomponent Systems	325
1.3	Thermodynamic Principles	326
2.	**Nucleation and Metastable States**	327
2.1	Free Growth	329
2.2	The Lifshitz–Slyozov–Ostwald Evaporation–Condensation Mechanism	330
2.3	Advanced Issues	332
3.	**Spinodal Decomposition and Unstable States**	333
3.1	The Early Regime	335
3.2	The Intermediate Regime	337
3.3	The Late-Stage Regime	338
4.	**Conclusions**	340
	Glossary	340
	Works Cited	341
	Further Reading	341

INTRODUCTION

Among the most fundamental notions about the physical world is the existence of matter in different phases. H_2O molecules, for example, can exist as either gas (water vapor), liquid (water), or crystals (ice, snow). The processes by which matter transforms from one phase to another are at the basis of numerous phenomena that either occur naturally or play a technologically important role. Of equal fundamental importance is the notion of phase coexistence (Stanley, 1971; Gunton *et al.*, 1983). The temperature of a block of ice rises as it is heated up until it reaches the melting temperature. Further addition of heat energy does not raise the temperature but is used to transform the ice crystal into water at constant temperature. Only after all the ice is melted does additional added heat raise the temperature further. The coexistence of water and ice is an example of phase coexistence for which numerous examples can be found in everyday life. In addition to the melting and freezing cycle, the evaporation and condensation cycle is commonly observed.

A further kind of phase coexistence can be found when the same chemical adopts different crystalline or amorphous structures, with sometimes very different properties. A well-known example is the element carbon, which can exist as amorphous carbon, diamond, graphite, or the recently discovered fullerenes. These different phases can coexist as evidenced by diamond crystals embedded naturally in carbon.

So far, only single-component systems have been considered, and their phase behavior is described in the first section below. If one considers two-component or multicomponent systems, the phase diagrams become rapidly more complex, and the possible combinations of coexisting phases grow rapidly in number. One example that is treated below is mixtures of two fluids. Crudely, fluid mixtures can be categorized as either miscible, like water and alcohol, or immiscible, like water and oil. There are well-known examples of binary fluids (say *A* and *B*), however, which at high temperatures are miscible and at low temperatures separate into two phases that coexist. One of the two phases is rich in fluid *A* and the other in fluid *B*. The above situation describes a system with an upper critical temperature, but many examples are known where the reverse occurs, i.e., at low temperatures a single phase exists and as the temperature is raised the physical system undergoes phase separation. This is described by a lower critical temperature. The phase separation can take place in a variety of binary mixtures ranging from liquids, polymers (de Gennes, 1979), and complex fluids (Safran and Clark, 1987) to glasses (Cahn and Charles, 1965) and metal alloys (Komura and Furukawa, 1988). In such sys-

3-527-28135-5/95/$5.00 + .50

tems, the phase-separation process can be induced by changing thermodynamic variables, such as temperature, pressure, or volume rapidly. How the phase-separation process occurs and how it might be controlled can affect the material properties considerably and be thus of great technological importance. Examples of phase diagrams of multicomponent systems are shown in Sec. 1.2.

A system that has been rapidly quenched from a one-phase equilibrium state to a point in the phase diagram where two phases coexist in a new equilibrium is at first in a state far from equilibrium. The processes by which the system evolves into its thermodynamic state of coexisting phases are called phase separation (Gibbs, 1948). The two main mechanisms are nucleation and growth, and spinodal decomposition. The path between the initial one-phase equilibrium and the final state of coexistence determines which one of these two mechanisms takes place. Under most circumstances, this path goes through a first-order phase-transition point, and nucleation and growth is observed. Such a phase transition is called first order or discontinuous, because even for an infinitesimally small change in parameter space, the two coexisting phases have a finite difference in composition. There are singular points in thermodynamic phase space, however, where a system can enter the two-phase region continuously, and if the final state is close enough to such a critical point the difference in composition between the two phases can become vanishingly small. Here the phase-separation process is called second order.

1. PHASE DIAGRAMS

1.1 One-Component Systems

The phases of a system with a single component can be described by an equation of state that relates thermodynamic variables, such as temperature, volume, and pressure. For a one-component system, three variables uniquely define all the possible phases. The equation of state defines a surface in the three-dimensional parameter space on which each point defines an equilibrium state. In Fig. 1, projections of the pressure/volume/temperature surface are shown schematically for what could be for instance a gas like argon. The projection onto the temperature–pressure plane has three well-known regions separated by lines. The region of the solid is separated from the liquid by the fusion curve and from the gas by the sublimation curve. These curves represent states where two or even three adjoining phases coexist.

Liquid and gas are separated by the vapor-pressure curve. This line terminates at a critical point, which is of great interest. In its

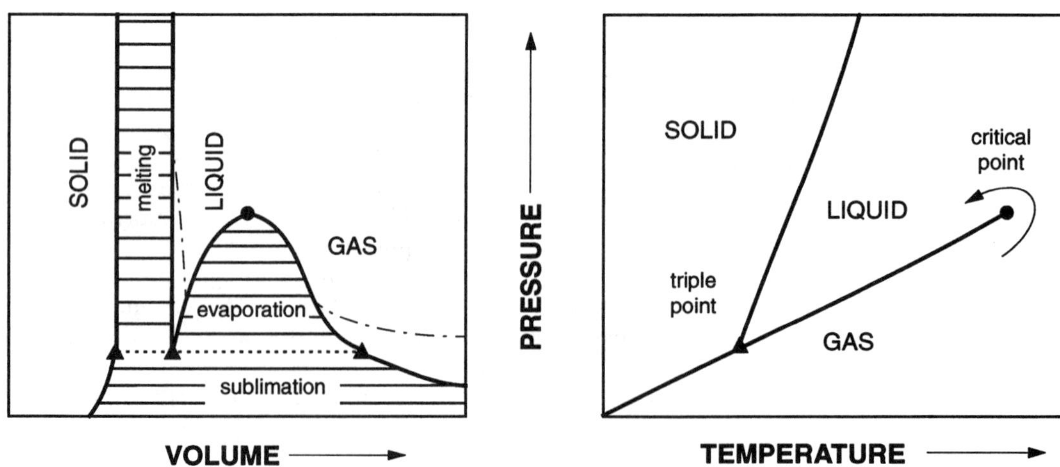

FIG. 1. Projections of the *PVT* surface of a one-component system that can exist in three different states, solid, liquid, and gas. The solid curves represent coexisting phases. The tricritical point is a point where all the phases coexist, and at the critical point two coexisting phases (gas and liquid) become identical.

vicinity, the gas and the liquid phases can be arbitrarily similar, and one can change the state of the system continuously from the gas to the liquid without crossing a phase-transition line, as indicated by the arrow. All the points on the various curves are first-order transition points. The phases coexisting on these lines are clearly distinguishable. The critical point is an exception. On approaching this point along the vapor-pressure curve, the gas and the liquid phases become less and less distinguishable and become ultimately identical. This point is also called a second-order phase-transition point. The search for a similar end point on the fusion curve has been in vain so far, and theoretical arguments have been made precluding its existence.

A further point of interest in this projection is the intersection of the three curves, which is a triple point. At this well-defined temperature and pressure, three phases of the substance coexist.

Another instructive projection is that onto the pressure/volume plane. Here the vapor-pressure curve (the gas-liquid coexistence curve) has the shape of a hump with the critical point at the top. As one reduces the volume of the system at constant temperature, the gas is compressed, and the pressure increases until the gas-liquid curve is reached, as indicated by the dashed-dotted line. Further volume reduction does not increase the pressure. Liquid droplets start to condense, and further reduction of the volume transforms more of the gas phase into the liquid phase. The horizontal lines under the coexistence curve are called tie-lines. After all the gas has been compressed to liquid, the pressure inside the liquid will increase rapidly until the coexistence between liquid and solid is reached.

The phase diagram in Fig. 1 is for illustrative purposes only. Real phase diagrams, even for one-component systems, can be much more complex. The phase diagram for helium has additional phases, such as the superfluid phase and associated with it a special point called λ point where the normal fluid, the superfluid, and the gas coexist. The phase diagram of water and ice is also of great complexity. For this single-component system, eight different crystalline structures and eight different triple points are known. Various ice phases can coexist with each other or with water. Interestingly, only one phase of liquid water is known.

The concepts described above for very familiar states of matter can be found in a multitude of dissimilar physical systems. Changing the temperature of certain materials can change them from being nonmagnetic to magnetic, conducting to superconducting, metallic to insulating. The same liquid-crystalline material can exist in a number of different structures depending on temperature.

1.2 Multicomponent Systems

An example of a two-component system is shown in Fig. 2. It is the schematic phase diagram of a mixture of a polymer and a liquid crystal (Doane *et al.*, 1986; Smith, 1993). At high temperatures, the two mix at all proportions. The pure liquid-crystalline material has a transition between an isotropic and a nematic state at a temperature T_{NI}. In the nematic phase, the liquid-crystalline molecules have orientational order but lack the translational order found in crystalline structures. At temperatures below T_{NI}, there

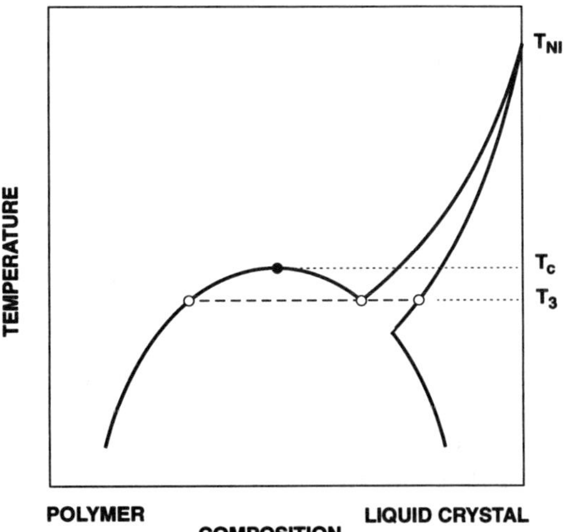

FIG. 2. Phase diagram for a mixture of a polymer and a liquid crystal in the temperature-composition space. T_{NI} is the nematic-to-isotropic transition temperature of the pure liquid-crystalline material. T_c is a critical point where the polymer-rich phase and the liquid-crystalline–rich phase become identical. T_3 is the temperature where three phases coexist, a polymer-rich phase and two liquid-crystalline–rich phases, one where the liquid crystal is isotropic and one where it is nematic.

are several regions of coexistence between two phases. Depending on composition and temperature, the two phases can be both isotropic or one can be isotropic and the other nematic. At one particular temperature T_3, three phases coexist. One is rich in polymer, and the two others are rich in liquid crystal. In the two latter phases, the liquid-crystalline material is in the isotropic or nematic phase, respectively. In addition, this phase diagram shows a critical point at T_c.

Figure 2 displays only a portion of the full phase diagram. At very high temperatures, there is a vapor phase, and at low temperatures the polymer can crystallize and the liquid crystal can assume a number of different structures with increasing degree of crystalline order. In addition, pressure and volume have to be considered as possible degrees of freedom for a full description in this case.

1.3 Thermodynamic Principles

Phase diagrams are based on thermodynamic principles and describe the conditions under which various phases of matter can coexist. The processes by which matter changes from a single-phase to a two-phase state upon sudden changes of parameters are called phase separation.

Of particular interest for the description of two fundamentally different mechanisms by which phase separation can occur are two-component fluid systems. These systems have been studied extensively and lend themselves readily for comparison between experiment and theory. The free energy of a mixture of two fluids, A and B, consists in its simplest form of two terms, an entropic term that favors disorder or mixing of the molecules A and B, and an enthalpic term that favors the proximity of similar molecules. The expression for the free energy of a mixture of two polymers has been provided by Flory and Huggins (Flory, 1953):

$$F = \frac{\phi \ln(\phi)}{N_A} + \frac{(1-\phi)\ln(1-\phi)}{N_B} + \chi\phi(1-\phi), \tag{1}$$

where ϕ is the volume fraction of polymer A and $1 - \phi$ the volume fraction of polymer B. The terms N_A and N_B are the polymerization indices or number of monomers in chains A and B, and χ is the Flory interaction parameter. It has the following temperature dependence:

$$\chi(T) = a/T + b. \tag{2}$$

The free energy is plotted schematically in Fig. 3 as a function of composition for two different temperatures corresponding to two different values of χ. At temperatures above a certain critical temperature T_c, the free energy has a single minimum and the two components A and B mix at all compositions. Separating mixture 1 into two mixtures at slightly higher and lower concentrations would increase the total free energy, and the two phases could achieve a state of lower energy by remixing. This is the case for the upper curve in Fig. 3. An example of this is a mixture of water and ethanol at room temperature, for instance. As one lowers the temperature of the A/B mixture below T_c, however, the free energy has two minima and the state of the mixture depends on its global composition. Three thermodynamically different regions can be distinguished. The phase behavior of the mixture depends on the sign

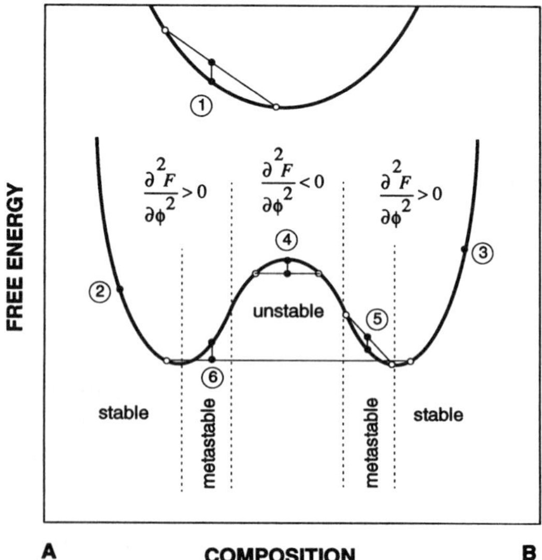

FIG. 3. Free energy as a function of composition for a two-component mixture. The upper curve is a high-temperature single phase, whereas the lower curve is at low temperatures where two phases coexist. Three regions (stable, unstable, and metastable) as well as various hypothetical samples (1–6) and their stability behavior are described in the text.

of the second derivative of the free energy. Very dilute mixtures, indicated by 2 and 3, behave like the high-temperature mixtures described above. They are miscible and form a single stable phase. Mixtures with compositions in the unstable region, where the second derivative of the free energy is negative, spontaneously separate into two coexisting phases. A mixture prepared at these compositions, as indicated by 4 in Fig. 3, can have a lower global free energy by forming domains of different compositions. One of the phases is rich in component *A* and the other in component *B*. Between the unstable and the stable regions is a metastable region. Here the second derivative of the free energy is positive and small separations in composition do not lower the overall free energy, as shown for mixture 5. Formation of a domain on the other side of the maximum in the free-energy curve, however, leads to a globally lower free energy. This is shown schematically for mixture 6. This region of the phase diagram is thus metastable. It takes a triggering event such as nucleation to induce phase separation.

The transition from a single-minimum free-energy curve to one with two minima is common to phase transitions. Even though such free-energy curves help us to understand the cause of the phase transition, they are not very convenient if one is interested in the phase behavior of a given system. A complete description of the temperature dependence would require many of these curves. The important points are the inflection points, defined by the condition

$$\frac{\partial^2 F}{\partial \phi^2} = 0, \tag{3}$$

and the points separating the metastable from the stable compositions, given by

$$\left.\frac{\partial F}{\partial \phi}\right|_{\phi_1} = \left.\frac{\partial F}{\partial \phi}\right|_{\phi_2}. \tag{4}$$

A plot of these points as a function of temperature is shown in Fig. 4. It is important to note that both Figs. 3 and 4 are for model systems, with symmetry about the 50% composition line. For most real systems, the curves are asymmetric, with the maximum coexistence temperature occurring at a composition different from 50%. In Fig. 4, the three regions in the temperature-composition diagram are clearly visible. They are separated by two lines, the coexistence curve, also called binodal curve, and the spinodal curve, or stability limit. Crossing these lines induces phase separation in a sample that was originally in a single-phase state at high temperatures. The process by which the separation occurs is fundamentally different depending on whether the quench point in the temperature/composition is in the metastable region (nucleation and growth) or the unstable region (spinodal decomposition). The volumes of the coexisting phases will not necessarily be identical. An illustration of this is given in Fig. 5. Here three samples were prepared at compositions corresponding to points 1, 2, and 3 in Fig. 4. In the one-phase region, $T > T_c$, no meniscus is observed within the sample. After a quench into the unstable region (sample 1) or the metastable region (samples 2 and 3), phase separation starts. Immediately after the mixtures are quenched, they are in a highly dispersed, nonequilibrium state that has a milky appearance (Fig. 5, sample 1, $T < T_c$, immediately after quench). Under the influence of gravity, the coexisting phases are allowed to settle. In sample 1, the *A*-rich and *B*-rich phases occupy equal volumes whereas for the off-critical samples the minority phases occupy a smaller volume. The three samples in Fig. 5 have reached their new equilibrium. Both processes are explained in detail in Secs. 2 and 3. Samples at different compositions will separate into phases with compositions at the same coexistence-curve values, if quenched to the same temperature.

2. NUCLEATION AND METASTABLE STATES

A system quenched into the region between the coexistence curve and the spinodal curve is thermodynamically metastable. For short enough times, it will not demix, but remain as a one-phase system. Very small fluctuations in concentration, such as those due to thermal activity, will decay as if the mixture were in thermodynamic equilibrium. In such a supersaturated mixture, a nucleation process is needed to initiate formation of a region having a composition of the coexisting minority phase. This nucleation process can

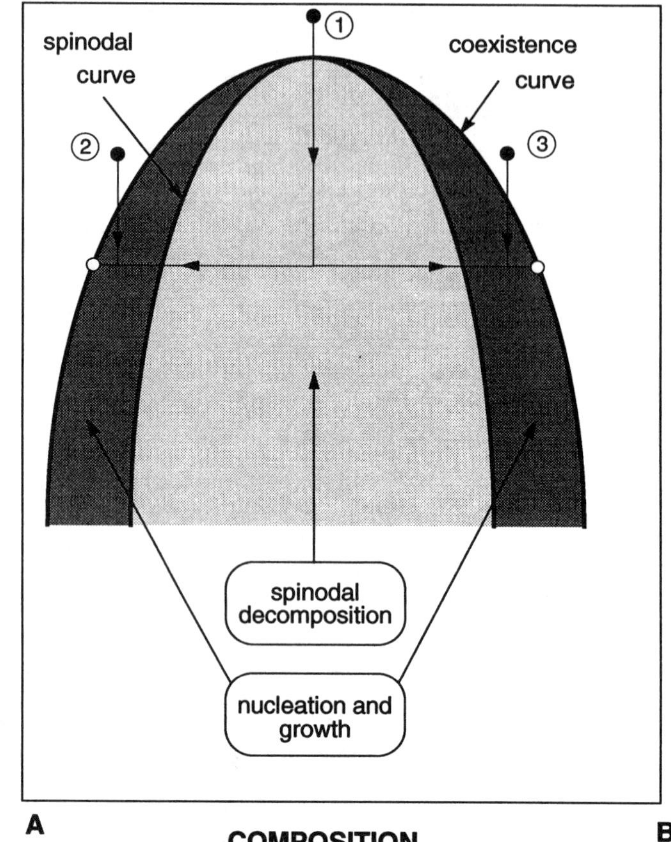

FIG. 4. Coexistence and spinodal curves for a two-component mixture in the temperature-composition space. The phase-separation process of sample 1 is called spinodal decomposition, whereas samples 2 and 3 undergo nucleation and growth.

be either heterogeneous or homogeneous. A heterogeneous process is induced by a third component, such as a contaminant or dirt in the bulk mixture, or a surface. This third component can locally decrease the free energy required for molecules of the minority phase to aggregate and create a domain of composition different from that of the initial mixture. During homogeneous nucleation, thermal fluctuations alone can cause the minority phase to nucleate without the presence of a third component. Far away from the spinodal curve, heterogeneous nucleation will dominate, whereas in the vicinity of the spinodal, the system is close enough to being thermodynamically unstable that thermal fluctuations can cause nucleation.

Once a droplet of the minority phase has been formed, it can evolve in two different ways. It can either grow by accreting more material from its supersaturated environment, or it can shrink and redissolve. The radius of the nucleated droplet determines which route it is likely to take.

Gibbs (1948) was among the first to recognize that a balance of competing effects leads to a critical droplet radius. Deep quenches or large supersaturation of the mixture favor the growth of a droplet. The deeper the quench, the more energy that can be gained by growth of the minority phase. Surface tension (interfacial free energy) is comparatively more costly for small droplets and opposes their existence. For every value of the supersaturation, there is a critical radius R_c. Droplets with a radius greater than R_c grow, and those with a smaller radius shrink. The critical radius depends on temperature, composition, and time elapsed since the system was quenched (see below).

Figure 6 shows a series of photomicrographs of quenched lead (Pb) and tin (Sn) mixtures (Hardy and Voorhees, 1988). The Sn-rich domains were held at 185 °C for varying periods of time. As time elapsed, the average particle size increased and the number of particles decreased, i.e., particle coarsening occurs.

$T > T_c$

$T < T_c$
right after
quench

$T < T_c$
very late
stages

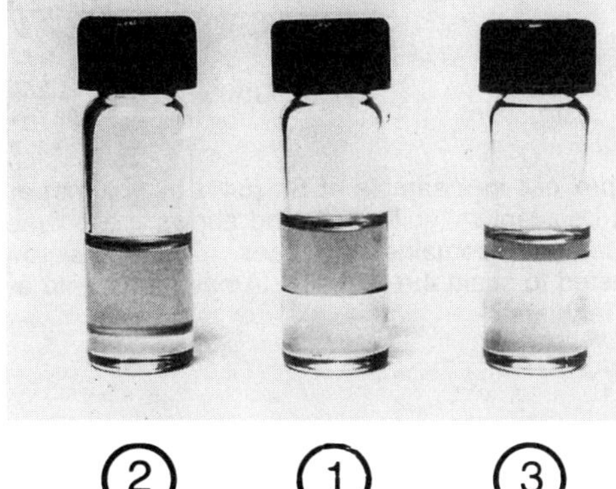

(2) (1) (3)

FIG. 5. Mixtures of two liquids at various compositions and temperatures. The mixture in the top vial is at a temperature above the critical temperature T_c and is in a single phase. Samples 1, 2, and 3 are prepared at compositions corresponding to the samples schematically depicted in Fig. 4. Immediately after being quenched to a lower temperature corresponding to a point in the two-phase region of the phase diagram, all three samples have a cloudy appearance. Many small domains of the two phases coexist, and with time these domains coalesce and coarsen. Finally, after the heavier phase has settled under gravity a meniscus appears separating the two coexisting phases. Note that for sample 1 the meniscus is in the middle (equal volumes of the two coexisting phases), whereas in samples 2 and 3 the volumes are unequal. These mixtures correspond to fluid *A* in Fig. 4 having a lower density than fluid *B*.

2.1 Free Growth

Droplets in a supersaturated environment are depicted schematically in Fig. 7. Here the *B* material is the minority component. Immediately after having been nucleated, each droplet is surrounded by the same composition ϕ_0, which is related to the supersaturation Δ by

$$\Delta = (\phi_0 - \phi_A)/(\phi_B - \phi_A). \qquad (5)$$

As long as the degree of supersaturation does not change, there will be a concentration gradient, called the depletion layer, favoring diffusion of material to droplet, where it accumulates. This leads to growth of the droplets (first and second frames in Fig. 7). At the same time as the droplets grow, their depletion layers also grow since the total amount of materials *A* and *B* must be conserved. If the depletion layers of individual droplets do not overlap, so-called free growth is observed, with no interactions between droplets. Under these conditions, the radius of any droplet that has been nucleated with a radius greater than R_c will grow asymp-

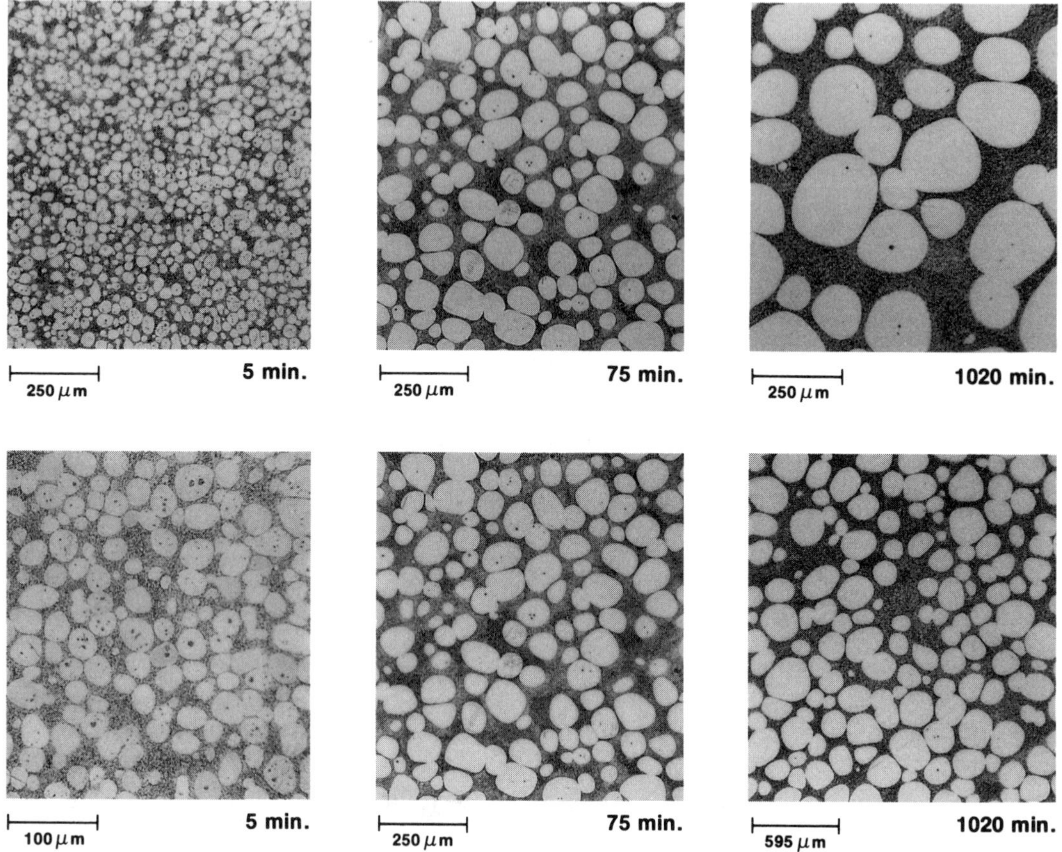

FIG. 6. Micrographs of the coarsening process in a sample of Sn (64% by volume) and Pb over a 17-h period at 185°C. The top row of pictures is at constant magnification and shows the coarsening process, where the average domain size grows and the number of domains decreases. In the lower row, the same micrographs are shown but with a magnification adjusted to scale the domains to have the same average size. Interestingly, the pictures show the same qualitative features.

totically as the square root of time. An example of this is shown in Fig. 8. Here the radii of droplets nucleated in a mixture of two polymers are shown as a function of time. The mixture had been quenched from a temperature in the one-phase region to five different temperatures in the metastable region, and the sizes of the droplets were measured by light scattering. The temperature differences ΔT cited are the depths below the coexistence curve. All the quenches were into the metastable region, since the spinodal curve lies about 1.2° below the coexistence curve for the composition used in the experiment. The radii are given in units of the critical radius, and the time is also in dimensionless units. Note that even though the raw data do not fall on a single curve, this scaled representation of the data supports the idea that a generalized dimensionless description of the growth phenomenon can be found. The slope of the experimental data is in agreement with the above-mentioned square-root growth law.

2.2 The Lifshitz–Slyozov–Ostwald Evaporation–Condensation Mechanism

The free-growth regime lasts only as long as the growing droplets are surrounded by a constant supersaturation layer. As more and more material diffuses through the depletion layers and accretes onto the droplets, the size of the depletion layers grows (see Fig. 7). Eventually, the depletion layers of neighboring droplets overlap, and competition for material from the supersaturated medium sets

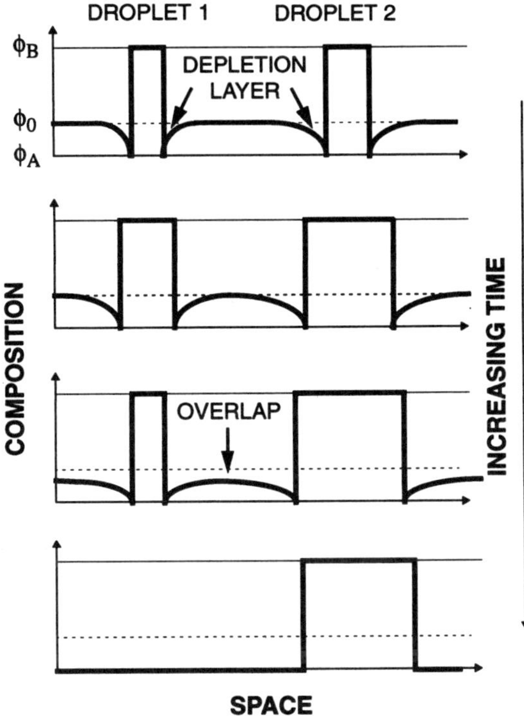

FIG. 7. Growth of droplets as a function of time in a supersaturated medium in the nucleation and growth regime. The average composition of the sample is ϕ_0, and the compositions of the coexisting phases are ϕ_A and ϕ_B, respectively. Droplet 2 is growing as a function of time, whereas droplet 1 is first growing and then shrinking.

in, leading to a decrease of the level of supersaturation (third panel of Fig. 7). This, in turn, increases the droplet critical radius R_c. Small droplets stop growing and can even become subcritical in size, leading to their evaporation as indicated for droplet 1 in Fig. 7. This process is called the evaporation–condensation mechanism. This leads to the late-stage regime of coarsening shown in the fourth panel of Fig. 7, where the supersaturation has vanished. Diffusion due to concentration gradients ceases, but processes reducing the surface free energy can lead to further changes in the droplet size distribution. In one such mechanism, diffusion of droplets and their collisions allows them to coalesce, thus reducing the total surface free energy. Coarsening processes that lead to reduction of the interfacial area are also known as Ostwald ripening.

An important feature during all phases of coarsening is that diffusion is normal, i.e., in the direction of a concentration gradient. This is fundamentally different from the case for spinodal decomposition, which is discussed in Sec. 3.

Interestingly, the late-stage coarsening laws do not depend on whether the coarsening occurs via evaporation, diffusion, and condensation of material molecule by molecule (as might happen in a metal alloy) or by global diffusion and collision of macroscopic droplets (as can happen in fluids). A significant

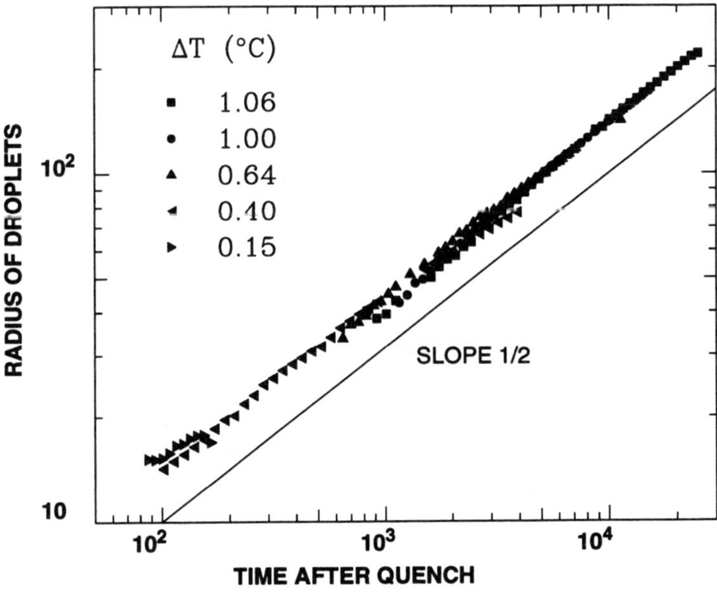

FIG. 8. The average radius of droplets of the minority phase in a phase-separating polymer mixture as a function of time for several quench depths in the metastable region of the phase diagram. A line of slope $\frac{1}{2}$ is shown for comparison.

theoretical body based both on analytical work and on numerical simulations predicts that such a system asymptotically evolves toward a polydisperse distribution of droplets. The breadth of the distribution is about 50%, and the average droplet size grows as the $\frac{1}{3}$ power of time. A full pictorial description of the various phases of the coarsening in the metastable region is shown in Fig. 9 [for details see Cumming *et al.* (1992)]. Here the evolution of the radii of individual droplets that have been "nucleated" at early times has been calculated using the differential equation proposed by Lifshitz and Slyozov (1961):

$$\frac{dR}{dt} = \frac{D}{R}\left[\Delta - \frac{\alpha}{R}\right]. \tag{6}$$

Here D is a diffusion coefficient, and α is the thickness of the interface between the droplet and the surrounding medium. Note that the absence of new droplets after time $t = 6$ is arbitrarily assumed. This case might describe a situation in which droplets are nucleated around nucleation centers of a third material (heterogeneous nucleation). Four regimes are observed. The nucleation process is assumed to be heterogeneous, since no new droplets are nucleated after a reduced time of 6. Only droplets with a radius greater than R_c (assumed to be unity here) are nucleated. After nucleation, the droplets grow as the square root of time during the free-growth regime. An interesting phenomenon is observed in this regime. An originally polydisperse distribution of droplets becomes more monodisperse. This monodispersity disappears, however, after the transition regime to the late-stage Lifshitz–Slyozov–Ostwald coarsening. The duration of the free-growth regime depends on the degree of the original supersaturation. Free growth lasts longer for deep quenches.

During the last phase, the smallest droplets become subcritical and evaporate rapidly, returning material to the majority phase, which is the source of growth of the largest droplets. This ongoing process keeps the size distribution quite broad.

2.3 Advanced Issues

The theoretical framework evoked above has been known to fail to predict correctly the quantitative features of particle size distributions. Figure 10 (Marder, 1987) shows a compilation of a number of experimentally obtained particle size distributions. It is found that the distributions are broader than the distribution predicted by Lifshitz and Slyozov (dashed line).

This failure has been attributed to the neglect of spatial distributions and interactions of the droplets that can play an important role

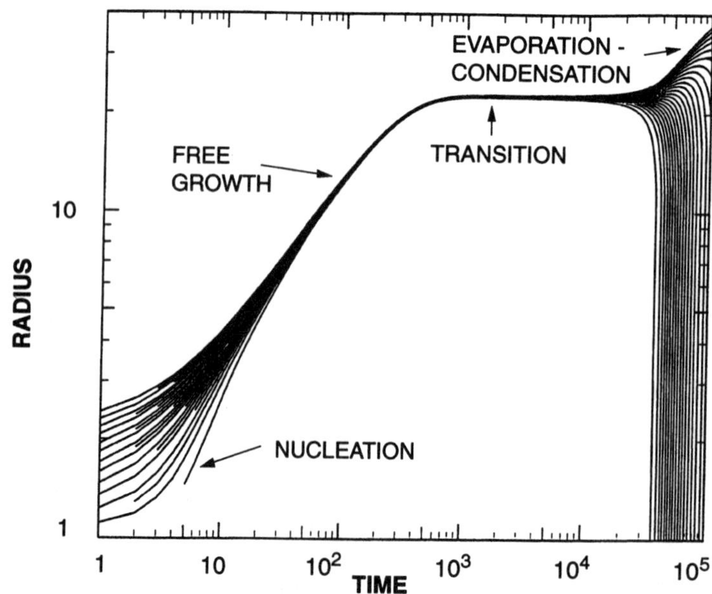

FIG. 9. Model curves obtained from numerical integration of a differential equation [Eq. (6)] proposed by Lifshitz and Slyozov. The various stages of the phase-separation process are shown. The nucleation process was halted artificially after times $t = 6$, which mimics a fast heterogeneous nucleation process at early times. During the free-growth period, the slope of the curves approaches $\frac{1}{2}$, whereas at late times an asymptotic slope of $\frac{1}{3}$ is found.

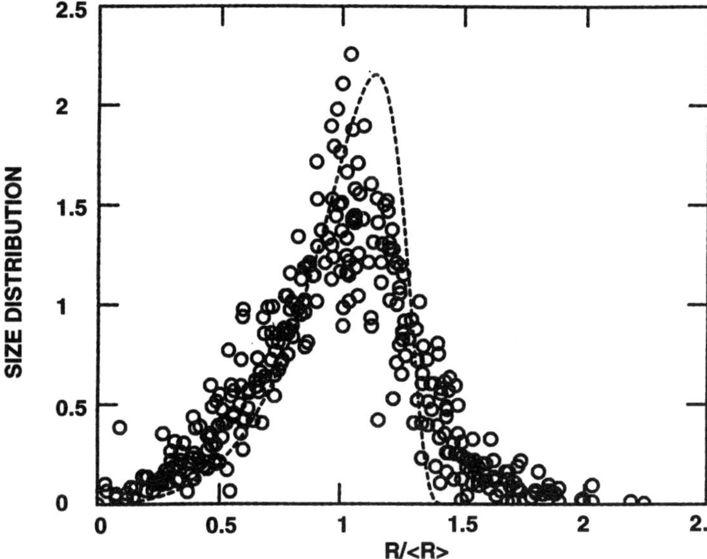

FIG. 10. Experimentally observed size distributions in a variety of phase-separating systems. For comparison, the distribution predicted by Lifshitz and Slyozov for the late-stage evaporation–condensation regime in Fig. 9 is shown (dashed line). [Adapted from Marder (1987).]

during the coarsening mechanism (Voorhees and Glicksman, 1984; Voorhees, 1992). The simplified theory predicts that all particles of a given size grow at the same rate, since only the global, mean supersaturation counts. This mean-field picture may be a good approximation for dilute systems, but in denser systems it is clear that a particle of a given radius will have a different evaporation rate depending on whether it is next to a large or a small sphere. These effects of spatial distributions around given particles are difficult to treat in analytical theories. They are currently investigated by means of large-scale numerical simulations.

3. SPINODAL DECOMPOSITION AND UNSTABLE STATES

Spinodal decomposition is the phase-separation process that occurs when a binary system is quenched from the one-phase region into the thermodynamically unstable two-phase region (see Fig. 4). A convenient way of observing this experimentally is to prepare a mixture at the critical composition where the spinodal curve and the coexistence curve coincide. For off-critical quenches, any trajectory at constant composition from the one-phase region into the two-phase region is always through a region of metastability, where the above-described nucleation and growth mechanisms can occur. In an off-critical quench, phase separation could begin locally, leading to composition inhomogeneities before spinodal decomposition begins.

Thermodynamic considerations dictate that a mixture of molecules A and B that is quenched rapidly into the unstable region separates into two different phases with compositions prescribed by the coexistence curve. Since the interfacial energy between these two phases should tend toward minimal value, the ultimate equilibrium state of a phase-separated fluid mixture in a test tube will have the denser phase at the bottom and the lighter phase on top of it with a meniscus separating the two. The scenario by which this final state is reached can be divided into at least three stages, as shown in Fig. 11. During the early, linear-stage regime, composition fluctuations with long enough wavelength will be amplified spontaneously. During this phase, the wavelength of the fluctuations, $L_m(0)$, is time independent, as explained below. During a second, or intermediate, regime, both the amplitude and the wavelength of the composition fluctuations grow. Finally, domains form with constant compositions given by the coexistence-curve values, ϕ_A and ϕ_B. Thermodynamically, the mixture is at its equilibrium point. Interfacial tension, however, drives further coarsening during the late-stage regime. These three regimes are discussed below in detail.

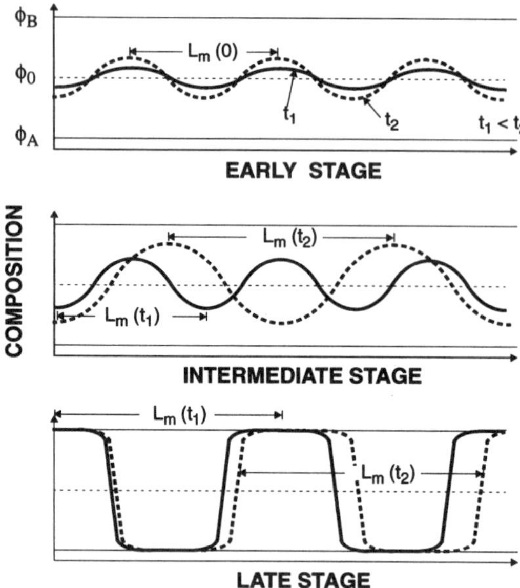

FIG. 11. Schematic description of the various stages of spinodal decomposition. $L_m(0)$ is the maximally unstable wavelength of a composition fluctuation.

The morphologies observed during the two phase-separation processes (nucleation and growth and spinodal decomposition) are quite different. Figure 12 shows photomicrographs of two different polymer mixtures undergoing phase separation. One of them (b) is prepared at an off-critical composition and has been quenched into the metastable nucleation and growth region of the phase diagram. The other (a) is at the critical composition, and it undergoes phase separation in the unstable, spinodal decomposition regime. The droplet morphology in the nucleation and growth region is obvious. Since in this case the sample was thick compared with the focal depth of the microscope, only the spheres close to the focal plane appear bright with a sharp interface. Droplets above or below the focal plane are out of focus and appear gray.

A spinodally decomposing sample (a) appears very different. This micrograph was taken during the intermediate stages where boundaries are not yet defined and only concentration gradients are present. There are

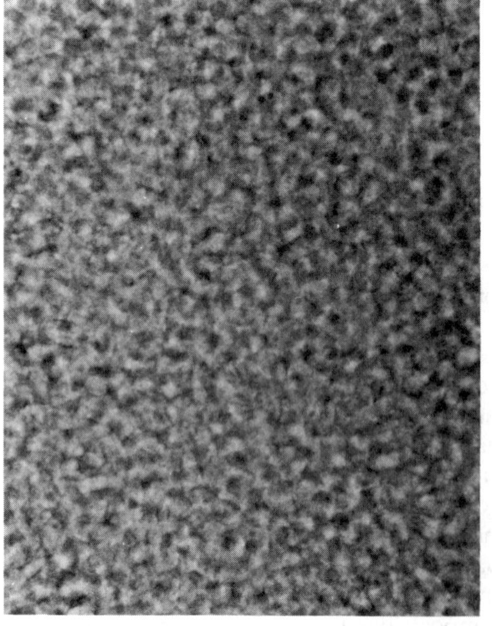

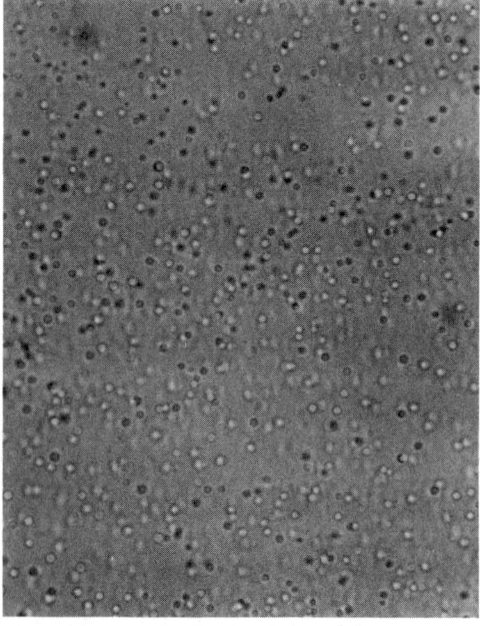

FIG. 12. Photomicrographs of two different polymer mixtures undergoing phase separation. The mixture shown in (a) was prepared at the critical composition and undergoes spinodal decomposition. The mixture shown in (b) was prepared at an off-critical composition and was quenched into the metastable region. Droplets of the minority phase nucleate and grow as a function of time.

no sharp interfaces to focus on, which explains the blurry image. More importantly, a characteristic length scale is apparent. This is even more evident in a Fourier transform, which has a ring pattern, indicative of a dominant length scale in the system. This dominant length is directly related to the maximally unstable wavelength q_m described below.

Since optical images of spinodally decomposing systems are not very easy to interpret, experimental work is typically done using light-scattering techniques (Goldburg, 1983; Hashimoto, 1988; Bates and Wiltzius, 1989; Cumming *et al.*, 1992). Figure 13 shows the time evolution of light scattered by a polymer blend undergoing phase separation. At time = 0, the sample is quenched from the one-phase region into the two-phase region. The 1-mm-thick sample is perfectly clear and does not scatter any light. The central spot is due to the laser beam directly passing through the sample without scattering. After 2 days, a ring of scattered light becomes visible. This ring is caused by fluctuations in the index of refraction (i.e., fluctuations in the concentration in the sample as schematically shown in Fig. 11). The most dominant fluctuations are those with a wavelength $L = \lambda/[2n \sin(\theta/2)]$. Here λ is the wavelength of the light, n is the index of refraction of the material, and θ is the scattering angle of the intensity maximum. As can be seen in Fig. 13, the ring becomes smaller and brighter in intensity as a function of time. This behavior reflects both an increase in the amplitude of the concentration fluctuations and a coarsening of the domains.

3.1 The Early Regime

Immediately after a quench, the thermodynamically favored state of a mixture is phase separation into two coexisting phases. This must occur, however, by transport of molecules A and B to the appropriate regions in space. From a thermodynamic point of view, a very finely divided arrangement of coexisting domains is the same as two macroscopic domains next to each other. The former arrangement of domains, however, has a huge interfacial area separating the two phases and a large energy cost associated with it. John Cahn (Cahn and Hilliard, 1958; Cahn, 1965), in seminal contributions, formalized these concepts and combined the thermodynamic free energy [Eq. (1)] with an interfacial energy into a functional form,

$$\frac{\Delta F}{k_B T} = \int \left(\frac{\phi \ln(\phi)}{N_A} + \frac{(1-\phi)\ln(1-\phi)}{N_B} + \chi(T)\phi(1-\phi) + r_p^2[\nabla\phi]^2 \right) dr, \tag{7}$$

where r_p is the range of interactions in the

0 HOURS

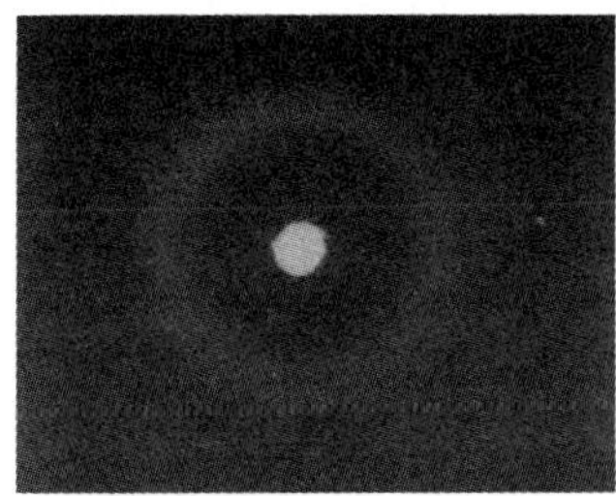
48 HOURS

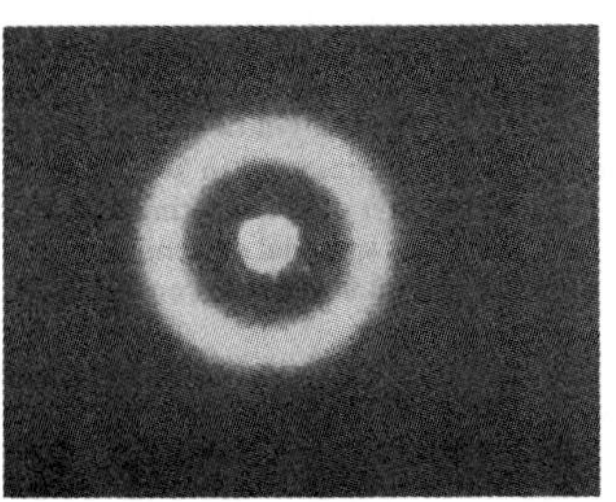
80 HOURS

240 HOURS

FIG. 13. Light-scattering patterns on a screen photographed at various times after a polymer mixture had been quenched to a temperature in the unstable two-phase region. (From Bates and Wiltzius, 1989.)

system and for polymer blends is of the size of the polymer chains. The phase separation in fluid system has to obey both a continuity equation,

$$\frac{\partial \phi}{\partial t} + \nabla J = 0, \qquad (8)$$

for the mass flux density J, and a transport equation,

$$J = -D\nabla \mu, \qquad (9)$$

relating J to the chemical potential μ via a mass diffusion coefficient D. Combining Eqs. (7)–(9) in reciprocal q space leads to the fundamental differential equation describing the phase-separation process in the unstable region,

$$\frac{\partial}{\partial t}\phi(q,t) = -Dq^2\mu(q,t). \qquad (10)$$

Unfortunately, this equation is highly nonlinear and analytically not soluble. Cahn's insight led him to develop a linear stability analysis of the differential equation. Rewriting the composition as

$$\phi(r,t) = \phi_0 + \delta\phi(r,t) \qquad (11)$$

with $\delta\phi(r,t) \ll \phi_0$ and keeping only terms linear in $\delta\phi$, one finds a simpler differential equations for small fluctuations in composition:

$$\frac{\partial}{\partial t}\delta\phi = R(q)\delta\phi, \qquad (12)$$

which has the solution

$$\delta\phi(q) = e^{R(q)t}. \qquad (13)$$

The amplification factor $R(q)$ in the exponential has the following form:

$$R(q) = -Dq^2\left\{\frac{\partial^2 F}{\partial \phi^2} + 2q^2 r_p^2\right\}. \qquad (14)$$

In the one-phase region (see Fig. 3), the second derivative of the free energy is always positive, and thus $R(q) < 0$ for all values of q. This means that spontaneous composition fluctuations of arbitrary wavelength decay exponentially, in agreement with our intuitive understanding of a system in thermodynamic equilibrium. In the two-phase region, however, $\partial^2 F/\partial\phi^2$ is negative for mixtures with compositions in the unstable region. For small enough values of q, the right-hand side of Eq. (14) is thus positive, leading to exponential growth of spontaneous fluctuations of those wavelengths as described by Eq. (13). Concentration fluctuations with a wavelength longer than a critical wavelength are unstable and grow as a function of time. Very short-wavelength fluctuations cannot grow, because they are too costly in terms of gradient or interfacial energy. Very long-wavelength fluctuations grow very slowly, because mass has to be transported over long distances. This makes it intuitively clear that there is an intermediate wavelength that grows at a maximal speed, the maximally unstable wavelength $L_m = 2\pi/q_m$. This wavelength depends on materials parameters and on the quench depth. For polymer blends, it can be expressed as follows (Binder, 1983; de Gennes, 1979):

$$q_m^2(t = 0) = \frac{3}{2R_G^2}\left(\frac{\chi - \chi_c}{\chi_c}\right), \qquad (15)$$

where R_G is the radius of gyration, or size, of the polymer chains and χ_c is the value of the Flory parameter χ at the critical temperature. During the early stages, where the linear analysis applies, L_m is also independent of time. Part of this scenario is schematically depicted in the first panel of Fig. 11. One spontaneous spatial composition fluctuation with wavelength $L_m(0)$ grows as a function of time. Fluctuations with longer wavelength would also grow as a function of time, albeit more slowly, whereas very short-wavelength fluctuations would decay the same way as they would in the one-phase–equilibrium region of the phase diagram.

It is important to realize that this process is happening by "uphill" diffusion, i.e., against the concentration gradient. Uphill diffusion is the behavior opposite to "downhill" diffusion (in the direction of the concentration gradient as shown in Fig. 7).

Early stages of spinodal decomposition are difficult to observe experimentally. The composition fluctuations are small, and thus the scattering of light is weak. The early stages also evolve very rapidly, leaving little time to study them carefully. Nevertheless, a great deal of work has gone into extrapolating experimental data back to early times. This makes it possible to determine the typical unstable length scale soon after a sample has been quenched. The quench depth dependence of the measured length scales is found to be in good agreement with theoretical predictions. Qualitatively, it is found that the deeper the quench the shorter the unstable wavelengths. In addition to characteristic length scales, such an analysis also yields a quench-depth–dependent diffusion coefficient associated with the speed at which the composition fluctuations grow. The diffusion coefficients, which are formally negative quantities since the diffusion of material is against the concentration gradient, are quantitatively not understood.

3.2 The Intermediate Regime

As can be seen in Fig. 13, the maximally unstable wavelength and thus the characteristic domain size grow as a function of time. Typical data for fluid mixtures are shown in Fig. 14. The inverse of the domain size obtained from light scattering and direct visualization techniques is plotted as a function of time in reduced units. The units used in the scaling procedure are the natural time, $[Dq_m^2(t = 0)]^{-1}$, and length, $q_m^{-1}(t = 0)$, of the early-stage linear analysis. For the experiments shown in this figure, the scaling procedure is appropriate and yields a universal representation of the data. In experiments on spinodally decomposing polymer blends, systematic deviations from such universality have been found (Kedrowski *et al.*, 1993). For the reasons described above, data at very early stages, where the wave vector q_m is independent of time, are not available.

A complete theoretical understanding is still not available. This is a complicated nonlinear problem that poses a serious challenge to analytical work. The solid curve through the data in Fig. 14 is based on an empirical formula given by Furukawa (Komura and Furukawa, 1988). A number of numerical simulations have been performed on two- and three-dimensional models of these systems, and typical results are shown in Fig. 15 (Bortz *et al.*, 1974; Huse, 1986; Amar *et al.*, 1988). These are extensive Monte-Carlo simulations of domain growth in the two-dimensional spin-exchange model of a system of spins that can point either up or down. Such a system has a high-temperature phase where the spins are disordered and a low-temperature phase where they want to be organized in domains. In this particular model, the total number of up and down spins is conserved. The lattice of spins is shown at three different times, and the coarsening of the domains is apparent. The average domain size in this simulation grows as the $\frac{1}{3}$ power of time, which interestingly is the same exponent as that predicted for the disconnected dilute droplets observed in the Lifshitz–Slyozov model. The same power-law

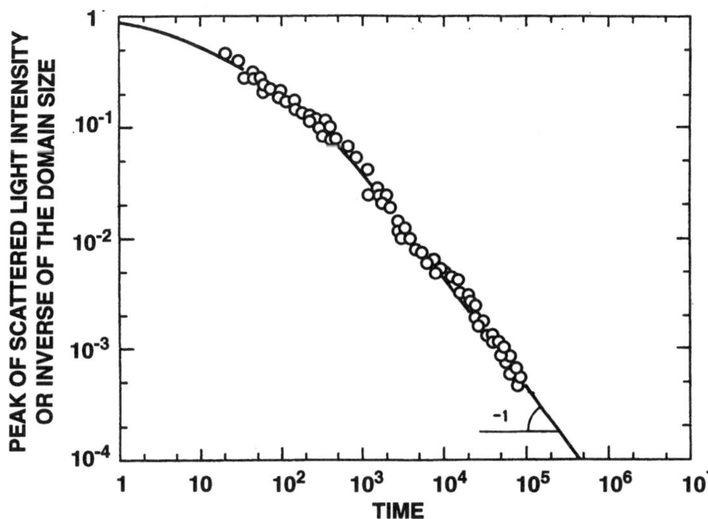

FIG. 14. The inverse domain size is plotted as a function of time for a variety of different liquid mixtures quenched to several temperatures in the unstable region. Both the domain sizes and the time have been scaled with appropriate scale factors. The asymptotic linear growth as a function of time is predicted for a growth of bicontinuous phases, where the growth mechanism happens via hydrodynamic flow. [Adapted from Beysens *et al.* (1988).]

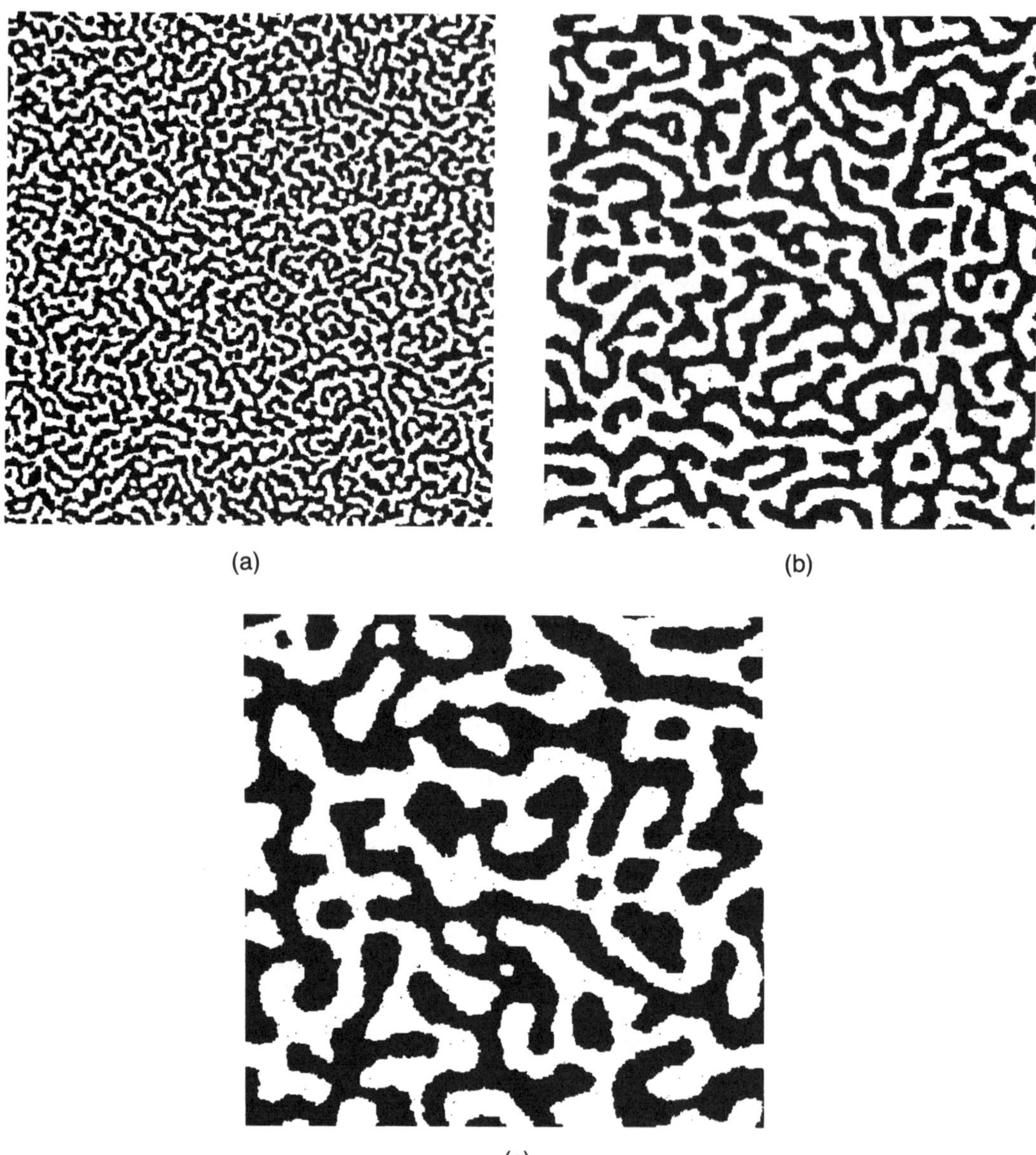

FIG. 15. Numerical computer simulations of a two-dimensional spin-exchange model displaying the qualitative features of phase separation in physical systems. Here a given site can be occupied by either a spin up (white) or a spin down (black). The total number of up and down spins is conserved. Parts (a)–(c) are at increasing times. [From Amar *et al.* (1988).]

behavior has also been observed in three-dimensional simulations. This growth law has proven remarkably universal and independent of details of the models.

3.3 The Late-Stage Regime

Once the composition within the phase-separated domains has reached the values given by the coexisting curves (see Fig. 11), the only driving force is reduction of the interfacial energy. Some of the mechanisms described above lead to the $\frac{1}{3}$-power law behavior of the growth. As can be seen in Fig. 14 (Beysens *et al.*, 1987), however, in fluid mixtures at late stages a linear growth law (slope of -1) of domain size as a function of time is observed. In a very insightful paper, Siggia (1979) described the basic physical mechanism leading to this growth law. It hinges on the fact that in phase-separating mixtures with both phases topologically fully connected, fluid flow can occur throughout these bicontinuous phases. Analysis of the underlying hydrodynamic equations leads to the prediction for the linear growth law.

It has been shown experimentally and in simulations that during the late stages of domain coarsening the structures are self similar. That is, properly rescaled, the three panels in Fig. 15 have the same global properties. It is impossible to tell which are at earlier times and which at later. The self similarity is evidenced by structure factors or pair-correlation functions, which have a generalized form that depends only on a scaled spatial variable. In Fig. 16(a), several

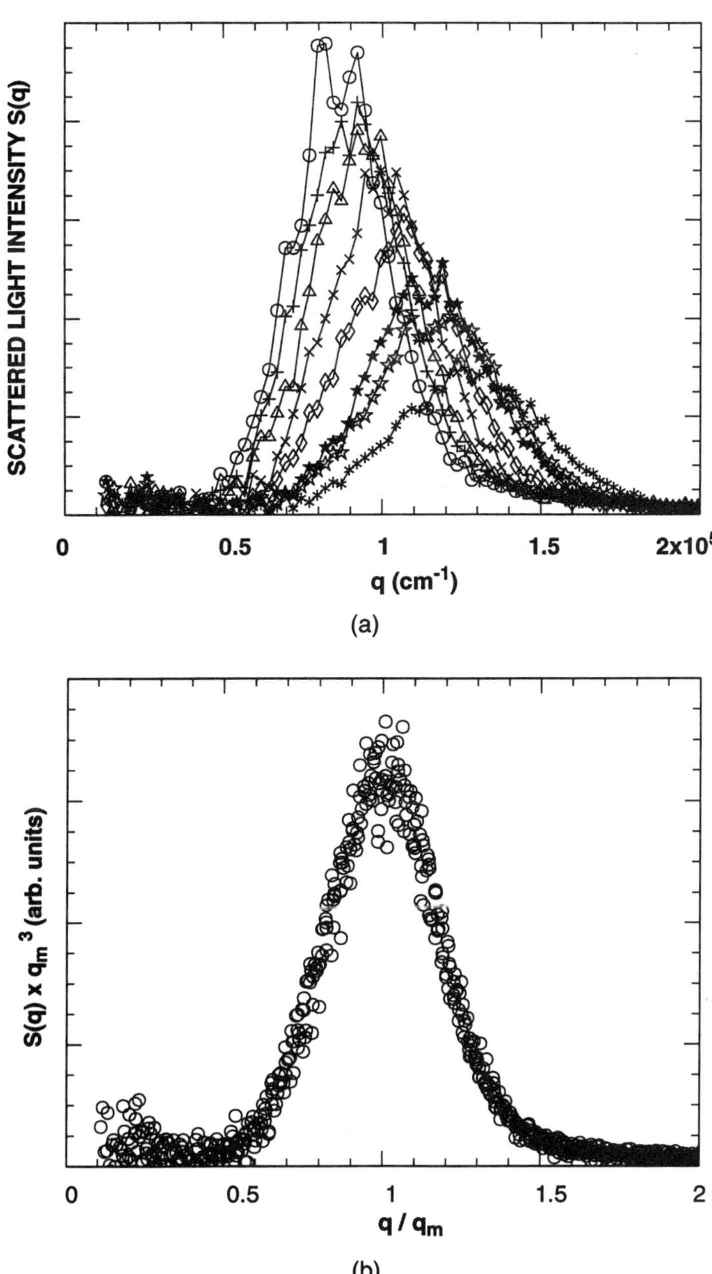

FIG. 16. (a) Light-scattering data for a polymer mixture undergoing spinodal decomposition for various times after the mixture has been quenched into the unstable region. The peak intensity is seen to increase with time (the earliest time is indicated by stars and the latest time by open circles) while its position decreases to smaller wave numbers. This indicates growth of spatial and compositional amplitude fluctuations. (b) The same data rescaled in a way appropriate to demonstrate self similarity of the time-dependent phase-separating structures. [Adapted from Kedrowski *et al.* (1993).]

such structure factors of phase-separating polymer blends are shown in their unscaled form. These are obtained from light-scattering measurements on samples similar to those shown in Fig. 13. In Fig. 16(b), the same structure factors are shown in a rescaled form. Superposition of the structure factors is in agreement with the concept of self similarity, which has been formalized by Furukawa in a scaling hypothesis.

4. CONCLUSIONS

Phase-separation phenomena can be broadly classified into two different categories depending on whether the processes occur in the thermodynamically unstable or metastable region of the phase diagram. Our insight both in nucleation and growth and in spinodal decomposition is based on work spanning well over a hundred years. Although the experimental and theoretical understanding is quite good, there are a number of issues that are not resolved. These include the quantitative understanding of nucleation barriers, the kinetics of phase separation, and finite-volume effects during late-stage coarsening. Most of the interest in the topic of phase separation has been driven by the challenging theoretical and experimental problems. There are some applications worth mentioning. Polymer-dispersed liquid crystals (PDLC) mentioned in Sec. 1.2 (Doane *et al.*, 1986; Smith, 1993) have emerged recently as switchable windows and even as candidates for flat-panel displays. PDLCs are made through a phase-separation process that results in a dispersion of micron-size droplets of liquid crystal trapped in a polymer matrix. An electric field can align the nematic textures inside the droplets, resulting in large changes in their scattering properties. It is possible to switch the optical appearance of a PDLC from transparent to milky white with reasonably small electric fields. The electro-optic properties of PDLCs depend largely on the morphology of the droplets formed during the phase-separation process. Predicting and controlling these morphologies is thus of great interest. In other applications, it can be desirable to tailor, e.g., the mechanical properties of a polymer blend by mixing various materials. Unfortunately, this results quite often in macroscopic phase separation of the components without the desired effect. A better understanding of the phase behavior and the separation processes could lead to better control.

GLOSSARY

Coarsening: Growth process of the minority-phase droplets during the nucleation and growth phase-separation process.

Coexistence Curve: Line separating the thermodynamically stable one-phase region of the phase diagram from the metastable two-phase region. Also known as binodal curve.

Critical Composition: Composition of a two-component mixture equal to the composition where the coexistence curve and the spinodal curve touch.

Critical Radius: Droplets of the minority phase smaller than the critical radius shrink and ultimately vanish from a phase-separating mixture. These droplets are also called subcritical.

Critical Temperature: Phase-separation temperature of a mixture prepared at the critical composition.

Depletion Layer: Region of lower concentration surrounding a droplet of the minority phase.

Liquid Crystal: Material with properties intermediate between those of liquids and crystals. Many different liquid-crystalline structures can be distinguished according to their degree of translational and/or orientational order.

Majority Phase: The phase in a two-component phase-separating mixture that occupies more volume than the minority phase.

Minority Phase: The phase in a two-component phase-separating mixture that occupies less volume than the majority phase.

Nucleation: Creation of a domain of the minority phase in a system undergoing phase separation in the metastable region of the phase diagram. The nucleation process can be heterogeneous, if it is induced by a third component, or homogeneous, if it is due to thermal fluctuations.

Self Similarity: Property of a system that changes, e.g., with time, but has qualitatively the same features if the dimensions are rescaled appropriately.

Spinodal Curve: Line separating the thermodynamically metastable two-phase region from the thermodynamically unstable two-phase region of the phase diagram.

Spinodal Decomposition: Phase-separation process in the unstable region of the phase diagram.

Supersaturation: Measure of the quench depth below the coexistence curve. See Eq. (5).

Works Cited

Amar, J. G., Sullivan, F. E., Mountain, R. D. (1988), *Phys. Rev. B* **37**, 196–208.

Bates, F. S., Wiltzius, P. (1989), *J. Chem. Phys.* **91**, 3258–3274.

Beysens, D., Guenoun, P., Perrot, F. (1987), *Phys. Rev. A* **38**, 4173–4185.

Binder, K. (1983), *J. Chem. Phys.* **79**, 6387–6409.

Bortz, A. B., Kalos, M. H., Lebowitz, J. L., Zendejas, M. A. (1974), *Phys. Rev. B* **10**, 535–541.

Cahn, J. W. (1965), *J. Chem. Phys.* **42**, 93–99.

Cahn, J. W., Charles, R. J. (1965), *Phys. Chem. Glasses* **6**, 181–191.

Cahn, J. W., Hilliard, J. E. (1958), *J. Chem. Phys.* **28**, 258–267; (1959), *J. Chem. Phys.* **31**, 688–699.

Cumming, A., Wiltzius, P., Bates, F. S., Rosedale, J. H. (1992), *Phys. Rev. A* **45**, 885–897.

de Gennes, P. G. (1979), *Scaling Concepts in Polymer Physics*, Ithaca: Cornell University Press.

Doane, J. W., Vaz, N. A., Wu, B.-G., Zumer, S. (1986), *Appl. Phys. Lett.* **48**, 269–271.

Flory, P. (1953), *Principles of Polymer Chemistry*, Ithaca: Cornell University Press.

Gibbs, J. W. (1948), *Collected Works*, Vol. 1, New Haven, CT: Yale University Press.

Goldburg, W. I. (1983), in: H. Z. Cummins, A. P. Levanyuk (Eds.), *Light Scattering Near Phase Transitions*, Amsterdam: North-Holland.

Gunton, J. D., San Miguel, M., Sahni, P. S. (1983), in: C. Domb, J. L. Lebowitz (Eds.), *Phase Transitions and Critical Phenomena*, Vol. 8, London: Academic Press.

Hardy, S. C., Voorhees, P. W. (1988), *Metall. Trans. A* **19**, 2713–2721.

Hashimoto, T. (1988), *Phase Transitions* **12**, 47–119.

Huse, D. A. (1986), *Phys. Rev. B* **34**, 7845–7850.

Kedrowski, C., Bates, F. S., Wiltzius, P. (1993), *Macromolecules* **26**, 3448–3454.

Komura, S., Furukawa, H. (1988), *Dynamics of Ordering Processes in Condensed Matter*, New York: Plenum Press.

Lifshitz, I. M., Slyozov, V. V. (1961), *J. Phys. Chem. Solids* **19**, 35–50.

Marder, M. (1987), *Phys. Rev. A* **36**, 858–874.

Safran, S. A., Clark, N. A. (1987), *Physics of Complex and Supermolecular Fluids*, New York: Wiley Interscience.

Siggia, E. D. (1979), *Phys. Rev. A* **20**, 595–605.

Smith, G. W. (1993), *Int. J. Mod. Phys. B* **7**, 4187–4213.

Stanley, H. E. (1971), *Introduction to Phase Transitions and Critical Phenomena*, New York: Oxford Science Publications.

Voorhees, P. W., Glicksman, M. E. (1984), *Metall. Trans. A* **15**, 1081–1088.

Voorhees, P. W. (1992), *Annu. Rev. Mater. Sci.* **22**, 197–215.

Further Reading

Binder, K. (1987), *Colloid Polym. Sci.* **265**, 273–288.

Cahn, J. W. (1968), *Trans. Metall. Soc. AIME* **242**, 166–180.

Gunton, J. D., San Miguel, M., Sahni, P. S. (1983), in: C. Domb, J. L. Lebowitz (Eds.), *Phase Transitions and Critical Phenomena*, Vol. 8, London: Academic Press.

Stanley, H. E. (1971), *Introduction to Phase Transitions and Critical Phenomena*, New York: Oxford Science Publications.

PHASE TRANSITIONS: RENORMALIZATION AND SCALING

Franz Schwabl, *Institut für Theoretische Physik, Physik-Department der Technischen Universität München, Garching, Germany*

Uwe Claus Täuber, *Lyman Laboratory of Physics, Harvard University, Cambridge, Massachusetts, U.S.A.*

	Introduction	343
1.	**Phase Transitions**	344
1.1	Phenomenology and Basic Concepts	344
1.2	Examples	347
1.3	Universality	348
1.4	Scaling Laws	349
2.	**Mathematical Representations**	350
2.1	Microscopic Hamiltonians	350
2.2	Ginzburg–Landau Functional	351
2.3	Landau–Ginzburg Approximation	352
2.3.1	Mean-Field Theory	352
2.3.2	Gaussian Fluctuations	353
3.	**Renormalization Group**	355
3.1	Introductory Remarks	355
3.2	Real-Space Renormalization Group	355
3.2.1	1D Ising Model	356
3.2.2	2D Ising Model	357
3.2.3	General Real-Space RG Transformations	360
3.2.4	Emergence of Scaling Laws	360
3.3	Momentum-Shell Renormalization Group	361
3.3.1	Wilson's RG Scheme	361
3.3.2	Gaussian Model	362
3.3.3	Perturbation Theory and ϵ Expansion	362
3.3.4	More Sophisticated Field-Theoretical Methods	365
4.	**Scaling Laws in Statistical Physics**	366
4.1	Phase Transitions and Properties of Phases	366
4.2	Disordered Systems	366
4.3	Polymers, Membranes	367
4.4	Critical Dynamics and Nonequilibrium Phenomena	368
	Glossary	369
	Works Cited	370
	Further Reading	371

INTRODUCTION

The fundamental laws of nature governing the properties of matter possess distinguished symmetry properties: They are invariant with respect to translation in space and time, to rotations, and to inversions. The states of matter that we observe, however, usually do not display the full symmetries of the underlying basic principles. Rather, the atoms or molecules of a specific substance can be found in certain condensed phases consisting of enormously large numbers (typically 10^{23} cm^{-3}) of constituents, and these phases may be classified according to some specific symmetry transformation under which the system remains invariant (a subgroup of the original symmetry group; see Group Theory). In passing, we note that even the "fundamental" laws themselves, like the electroweak interaction and quantum chromodynamics, may have originated in the transition to a broken-symmetry state of an even more fundamental and more symmetric basic law. In this article, we shall be solely concerned with matter governed by the Schrödinger equation and electromagnetic forces.

The macroscopic properties of the phases of condensed matter, whose stability regimes are determined by the internal interactions of the particles involved, as well as by the

3-527-28135-5/95/$5.00 + .50

external conditions (temperature, pressure, electric and magnetic fields, etc.), can be very different. Upon a change of the external control parameters, the system may undergo a phase transition to a new macroscopic phase, thus lowering its total free energy (see THERMODYNAMICS, EQUILIBRIUM). Frequently, the internal energy of the system as a function of the control variables is minimized by the change of its thermodynamic phase, but there are also important cases where the driving force toward a new equilibrium state is a gain of entropy.

The stability of a particular phase and the implications of phase transitions are of course of prominent importance for applications and technology. Moreover, these issues provide a considerable intellectual challenge for their understanding in the framework of statistical mechanics (see STATISTICAL MECHANICS, CLASSICAL; STATISTICAL MECHANICS, QUANTUM). Accordingly, methods need to be developed in order to deal with a strongly interacting system containing a large number of particles, and near its stability boundary.

Fortunately, it is not necessary to study all the microscopic degrees of freedom in detail; instead it suffices to investigate the statistical mechanics of a few macroscopic entities (which, however, usually still poses a formidable task), namely the order parameter characterizing the specific phase transition and distinguishing the phases involved and its correlation functions, as well as the dependence on the external fields serving as control parameters (see CATASTROPHE THEORY).

Furthermore, near a critical point in the phase diagram the fluctuations of the order parameter become increasingly strong and long-range correlations emerge in the system, although the interactions of the particles are genuinely short ranged. Therefore, any approximate approach that neglects these fluctuations entirely, or merely treats them in some perturbative manner, is at least questionable.

On the other hand, because of the almost entirely collective behavior in the vicinity of a critical point, the microscopic details of the system become irrelevant there, and its long-wavelength (and low-frequency) properties tend to display *universality* (*q.v.*), i.e., in some asymptotic regime they depend solely on the symmetry and number of components of the order parameter, on the space dimension, and possibly on the long-range nature of the interactions. Historically, this emergence of universal behavior has provided one of the essential keys for a mathematical description and profound theoretical understanding of critical phenomena, in the framework of *renormalization-group* (RG) theory. In turn, the RG approach has laid a solid foundation for the striking phenomena of scale invariance and universality near second-order phase transitions.

Originally, RG concepts were developed in the context of elementary-particle quantum field theory. Within the theory of phase transitions and critical phenomena, scaling ideas had been developed before, but their sound justification was only provided when Wilson formulated a version of the RG approach designed to focus on the underlying structure of the nonlinear field theories involved, a procedure that also allowed for systematic and detailed calculations (Wilson and Kogut, 1974; Wilson, 1975). This considerable breakthrough has not only led to an enormous increase of the amount of knowledge and to a deeper understanding of the properties of condensed matter (see, e.g., Chaikin and Lubensky, 1995), but has in turn also provided important insights for the formulation and treatment of quantum field theories, from where it had originated (see UNIFIED FIELD THEORIES). For a thorough account of the historical development of the field, see CRITICAL PHENOMENA.

Remarkably, RG techniques, along with the ensuing notions of fixed points governing universal physical properties, relevant and irrelevant perturbations, the emphasis on the long-wavelength and low-frequency collective behavior, and the emergence of scaling laws have matured into a most important paradigm and tool well beyond the original realm of second-order phase transitions, and have found fruitful applications in a large variety of different fields of intensive current research (see Sec. 4).

1. PHASE TRANSITIONS

1.1 Phenomenology and Basic Concepts

At a phase transition, the free energy F (or some other appropriate thermodynamic potential; see STATISTICAL MECHANICS, CLASSI-

CAL) is a nonanalytic function of a control parameter. According to a standard classification scheme introduced by Ehrenfest, a phase transition is called of nth order if at least one of the nth derivatives of F is discontinuous or singular at the transition, while all the $(n-1)$th derivatives are still continuous. In order to distinguish the different phases, a macroscopic *order parameter* ϕ is introduced, which is zero in one phase and finite in the other. If ϕ is given by the first derivative of the thermodynamic potential with respect to its conjugate external field h, its value will jump at a first-order transition; at a second-order transition, ϕ is to be a continuous function, but its derivatives will be singular or discontinuous.

To be specific, one may refer to a ferromagnetic material undergoing a second-order phase transition from the paramagnetic to the ferromagnetic phase at the critical (Curie) temperature T_c. The basic microscopic entities are the magnetic moments $\phi(\mathbf{x})$ at site $\mathbf{x}$, also called local order parameter (in this article, $\mathbf{x}$ denotes a d-dimensional vector referring to a point either in a discrete lattice or in a continuum). The macroscopic order parameter (the spontaneous magnetization) is then given by the thermal average $\phi = \langle\phi(\mathbf{x})\rangle$. In anisotropic crystals, the local order parameter may assume just the two values $\phi(\mathbf{x}) = \pm 1$ (referring to spin up and down, respectively). Accordingly, the (average) order parameter may point up or down. Whichever of the equivalent directions is chosen, the original discrete symmetry is spontaneously broken for $T < T_c$ [see Fig. 1(a)]. In cubic crystals, the order parameter may be a three-dimensional vector $\boldsymbol{\phi}(\mathbf{x})$, and point in an arbitrary direction. In this case, in the ordered phase a continuous (rotational) symmetry is broken.

The singularities in the vicinity of such continuous phase transitions generally assume the form of power laws, with certain *critical exponents* characterizing the transition. In case the temperature T serves as control parameter, with a second-order phase transition occurring at the critical temperature T_c, the most important examples of such power laws, and associated exponents, are the behavior of the order parameter near the critical point T_c [$\tau = (T - T_c)/T_c \ll 1$] [see Fig. 1(a)],

$$\phi \propto (-\tau)^{\beta} \quad \text{for } T < T_c, \tag{1}$$

the divergence of the isothermal susceptibility on approaching the critical point (from both sides of the transition),

$$\chi = \left(\frac{\partial \phi}{\partial h}\right)_T \propto |\tau|^{-\gamma}, \tag{2}$$

the power law for the critical isotherm,

$$h \propto \phi^{\delta} \quad \text{for } T = T_c, \tag{3}$$

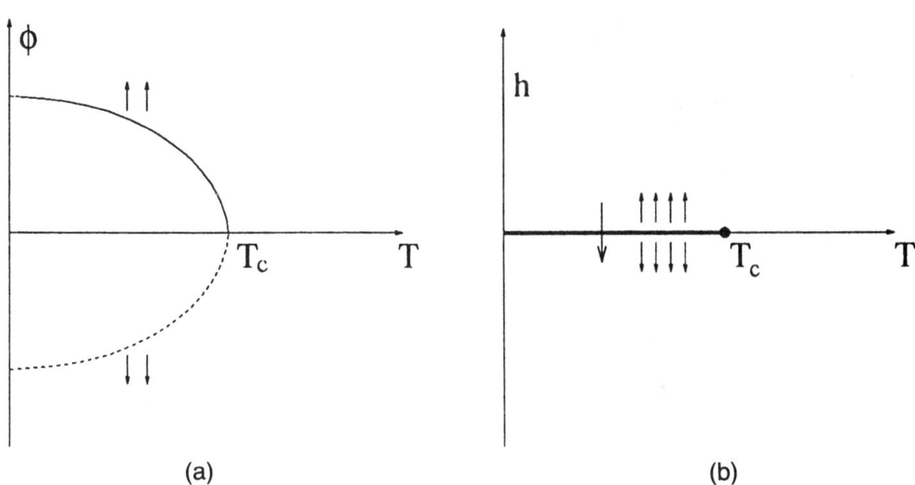

FIG. 1. (a) The order parameter ϕ of a ferromagnet for zero external field $h = 0$, as a function of the temperature T. ϕ sets in continuously at T_c and has at least two equivalent orientations. (b) Phase diagram (external field h vs temperature T) of a ferromagnet.

and the specific-heat singularity in the vicinity of T_c,

$$C_{h=0} \propto |\tau|^{-\alpha}. \tag{4}$$

Equation (2) leads to a most remarkable behavior of the order-parameter *correlation function*, defined as second cumulant (again, the brackets $\langle \ldots \rangle$ denote the thermal average)

$$G(\mathbf{x} - \mathbf{x}') = \langle \phi(\mathbf{x})\phi(\mathbf{x}') \rangle - \langle \phi \rangle^2 \tag{5}$$

(here translational symmetry has been assumed); namely from the fluctuation-response theorem in d dimensions,

$$\chi = \frac{1}{k_B T} \int d^d x G(\mathbf{x}), \tag{6}$$

it becomes apparent that a diverging susceptibility implies correlations of infinite range. This feature can be described by two additional critical exponents η and ν, for the power-law behavior of the correlation function precisely at the critical point,

$$G(\mathbf{x}) \propto \frac{1}{|\mathbf{x}|^{d-2+\eta}} \quad \text{for } T = T_c, \tag{7}$$

and the temperature dependence of the *correlation length* (measuring the spatial extent of the correlations) in the vicinity of T_c,

$$\xi \propto |\tau|^{-\nu}, \tag{8}$$

respectively. The growth of correlated regions upon approaching the critical point is illustrated in Fig. 2, where the spin-up/down configurations are plotted for a two-dimensional Ising model at different temperatures.

It turns out that for temperatures in the vicinity of T_c, the so-called *asymptotic region*, Eq. (7) may be generalized to the scaling form

$$G(\mathbf{x},\tau) = \frac{1}{|\mathbf{x}|^{d-2+\eta}} \hat{G}_\pm(|\mathbf{x}|/\xi), \tag{9}$$

or, by introducing the Fourier transform of the correlation function, $G(\mathbf{k}) = \int d^d x G(\mathbf{x}) e^{-i\mathbf{k}\cdot\mathbf{x}}$,

$$G(\mathbf{k},\tau) = \frac{1}{|\mathbf{k}|^{2-\eta}} \hat{g}_\pm(|\mathbf{k}|\xi). \tag{10}$$

(The suffix "$\pm$" distinguishes between the cases $T > T_c$ and $T < T_c$.) The *scaling functions* $\hat{G}_\pm(y)$ and $\hat{g}_\pm(p)$ approach constant values in the limits $y \rightarrow 0$ and $p \rightarrow \infty$, respectively (see Sec. 1.4).

Near a critical point, where the only relevant length scale is set by the correlation

(a)

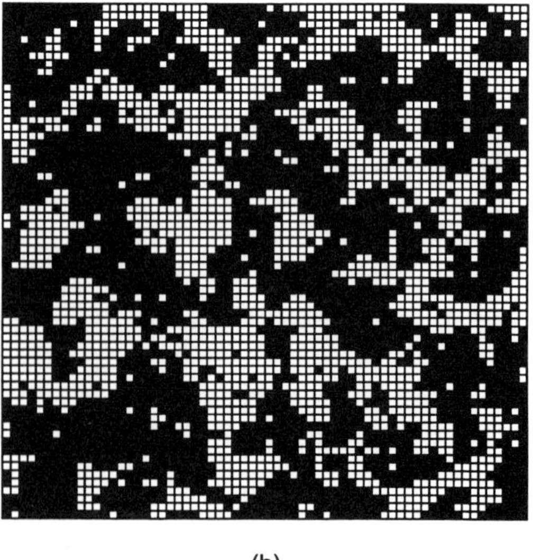

(b)

FIG. 2. Correlated clusters in the two-dimensional Ising model (the black and white points correspond to the two possible states at each site i in the lattice) for two different temperatures: (a) far away from T_c in the disordered phase, and (b) near the critical point.

length ξ, the increasing size of correlated regions also causes a *critical slowing down* of the dynamics of the system. In fact, the characteristic frequency ω_c of the system is expected to obey a scaling law similar to Eq. (10), namely

$$\omega_c(\mathbf{k},\tau) = |\mathbf{k}|^z \hat{\Omega}_{\pm}(|\mathbf{k}|\xi), \tag{11}$$

with $\hat{\Omega}_{\pm}(p) \rightarrow$ const for $p \rightarrow \infty$, such that at the critical temperature

$$\omega_c(\mathbf{k},0) \propto |\mathbf{k}|^z \quad \text{for } T = T_c, \tag{12}$$

with the *dynamic critical exponent* z (Hohenberg and Halperin, 1977). For example, in the case of purely relaxational dynamics ω_c may be identified with the inverse relaxation time t_0^{-1}; thus by using Eq. (11) and $\Omega_{\pm}(p) \rightarrow p^{-z}$ for $p \rightarrow 0$, the temperature dependence of the relaxation time is found to be $t_0 \propto \xi^z \propto |\tau|^{-z\nu}$.

In addition, certain universal features of the thermodynamic phases themselves can sometimes be characterized by power laws also. Specifically, the cost in free energy Σ of creating a domain wall in a phase with broken symmetry is assumed to scale with its linear dimension L according to

$$\Sigma = \sigma L^{\theta}, \tag{13}$$

where σ denotes the interfacial tension; a positive *stiffness exponent* θ means that long-range order is stable in the phase under consideration, whereas for $\theta < 0$ the system would spontaneously break up into domains (see, e.g., Fisher, 1993).

1.2 Examples

We have already referred to the phase transition from a paramagnetic to a ferromagnetic state, occurring at the Curie temperature T_c (in fact, the language used in this article is adopted from that particular physical realization). The order parameter is the macroscopic magnetization, χ is the magnetic susceptibility, and h the external magnetic field. A nonzero order parameter selects a particular direction; thus, depending on the Hamiltonian (see Sec. 2), the broken symmetry may be discrete (inversions) or continuous (rotations). At T_c, the order parameter sets in continuously [Fig. 1(a)], justifying the terminology "continuous transition." In anticipation of the comparison with the liquid–gas transition, we note, however, that reversing the external magnetic field for $T < T_c$ leads to a discontinuous change of sign of the magnetization [Fig. 1(b)]. The full phase diagram containing ϕ, h, and T is depicted in Fig. 3(a).

Probably the most familiar phase transitions are the solid–liquid and the liquid–gas

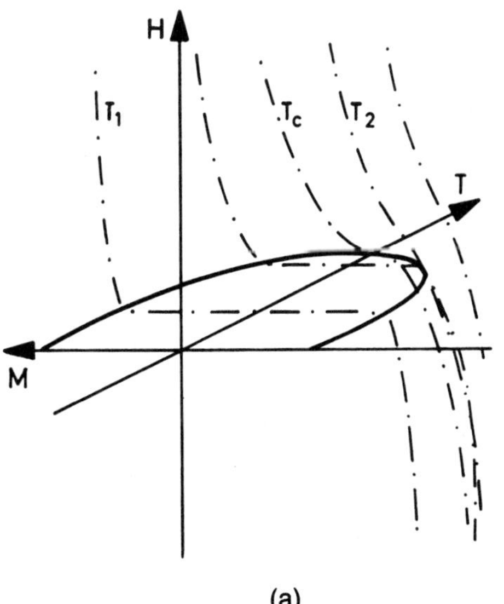

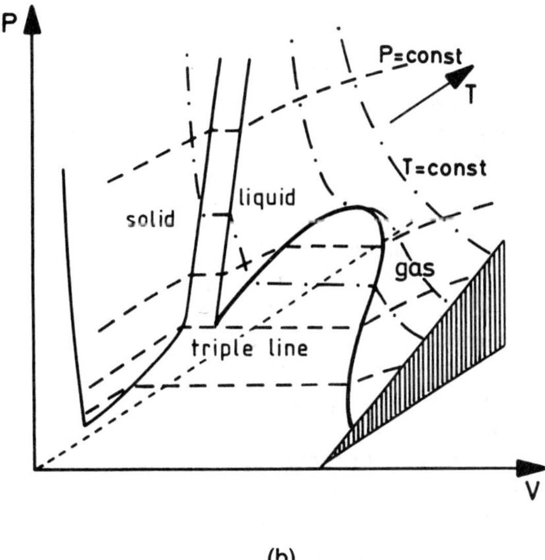

FIG. 3. Comparison of the phase diagrams (a) of a ferromagnet, and (b) of a one-component gas/liquid/solid.

transitions [Fig. 3(b)]. For the latter, the density at the coexistence line minus the critical density may serve as an order parameter, which thus has a low- and high-density, i.e., a gas and liquid, branch. The associated external field is the pressure P, the (isothermal) compressibility κ would have to be identified with the order-parameter susceptibility, and the surface tension of the liquid defines the stiffness. At the critical point, where the isotherm as a function of P and V has an inflection point, one may observe the critical power laws as stated in Eqs. (1)–(4), (7), and (8) above (e.g., the divergence of κ becomes apparent in the phenomenon of critical opalescence near T_c).

In everyday life, the liquid–gas transition is usually observed at constant pressure well below P_c. As the system is heated up, the density changes discontinuously as a function of the temperature. Consequently, the liquid–gas transition is frequently referred to as being of first order, and the critical point as the end point of the transition line where the distinction between liquid and gas disappears. The analogy between the liquid–gas and the ferromagnetic transitions may be exhibited vividly if one encloses a fluid of critical density ρ_c in a so-called Natterer tube. For $T > T_c$, there is a uniform fluid phase, which splits up into a liquid and a gas phase if the system is cooled down below T_c. These phases appear separated by a meniscus.

At the liquid–solid transition, the translational symmetry is broken, and the order parameter may be chosen as the Fourier component of the density at a reciprocal-lattice vector of the solid; this transition does not terminate at a critical point, however [Fig. 3(b)]. Somewhat more subtle and diverse are structural transformations between different solid phases, where suitable components either of the displacement field or of the strain tensor appear as order parameters (see PHASE TRANSITIONS, STRUCTURAL); and furthermore the various ordered phases of liquid crystals (see LIQUID CRYSTALS, STRUCTURE OF). In the former, the order parameter may be the polarization or staggered polarization at ferroelectric or antiferroelectric phase transitions, or some quantity characterizing the orientation of molecular groups. At elastic transitions, it is a component of the strain tensor, specifying the deformation involved.

It should be noted that the phase diagrams of systems as different as ferromagnets and fluids look remarkably similar near their respective critical points (Fig. 3). Again, there is a very rich variety of magnetically ordered systems (e.g., antiferromagnets, ferrimagnets, helical phases), the order parameters of which may be of a rather complicated structure (see MAGNETIC ORDERING IN SOLIDS). A powerful tool to study the various phases of solids is neutron scattering, which directly measures the Fourier transform of the correlation function [Eq. (5)]. Moreover, time-dependent phenomena can be explored by inelastic scattering of neutrons and light, which renders the critical slowing down [Eq. (12)] of the collective excitations (see COLLECTIVE PHENOMENA IN SOLIDS) directly observable.

Finally, phase transitions to macroscopic quantum states, namely superfluidity and superconductivity, should be mentioned. The order parameter in these cases is a complex field Φ, which may be identified with a macroscopic wave function, and the broken symmetry is a gauge invariance with respect to the phase of Φ (see SUPERFLUIDITY: LIQUID HELIUM SYSTEMS; SUPERCONDUCTIVITY, LOW TEMPERATURE). Nature also allows for most interesting interplays between several kinds of phase transitions, which can lead to so-called multicritical points, and fairly complicated phase diagrams (see, e.g., SUPERCONDUCTIVITY, HIGH TEMPERATURE).

1.3 Universality

Near critical points, the topologies of the phase diagrams of such diverse systems as a liquid–gas mixture and a ferromagnetic material look amazingly similar (Fig. 3). Moreover, both in computer simulations of simplified models and in real experiments the critical exponents for the corresponding phase transitions turn out to be identical for a broad variety of physical systems, and to depend only on the number of components and symmetry properties of the order parameter, the space dimension, and possibly the character of the particle interaction, namely if it is long range (e.g., Coulomb or dipolar forces). This surprising feature is usually termed *universality*. The microscopic details of the strongly interacting many-particle system merely enter the prefactors of the ensuing power laws, and even ratios of such amplitudes are in fact universal numbers.

A clue for this most remarkable fact may be found in the divergence of the correlation length [Eq. (8)] at a continuous phase transition. Upon approaching T_c, ξ will asymptotically become the only relevant length scale in the system, dominating over all microscopic scales. Even though the phase transition itself is caused by some (usually short range) interactions of the microscopic constituents, the ensuing large fluctuations will markedly reduce all dependences on the degrees of freedom acting on short length scales, like the underlying crystal structure, the interaction range (as long as it is short), etc. Therefore, the system's thermodynamic properties near T_c should be entirely determined by global features such as its overall symmetry and large-scale properties (this is why long-range interactions, though possibly weak, may become important). In this regime, the system will behave strongly collectively, and one expects universal laws to apply for its low-wavelength and low-frequency excitations.

1.4 Scaling Laws

Another striking experimental feature near a critical point is the apparent *scale invariance* of physical properties (Fig. 4 illustrates this spectacular effect for the specific heat of helium-4 near the normal–superfluid transition). Again, this may be related to the diverging correlation length. The appropriate mathematical description for this phenomenon of scale invariance is that the free energy and the correlation functions assume the form of generalized homogeneous functions. In the case that the only relevant control parameters are the temperature and the external field, and using the definitions of the critical exponents (also called critical indices), one is thus led to the formulation of the *scaling hypotheses* for the singular part of the free energy:

$$F_s = A_\pm |\tau|^{2-\alpha} \hat{f}_\pm (h/|\tau|^{\beta\delta}), \tag{14}$$

and the correlation function, Eq. (9) [or Eq. (10)] (Fisher, 1974). In Eq. (14), the scaling function $\hat{f}_\pm$ approaches a constant value in the limit $h \to 0$.

There are various ways to make Eq. (14) plausible. First, it can be considered as a generalization of the mean-field results [see Sec. 2.3.1] to general critical exponents. Second, one may start from the observation that singularities are present only at $\tau = 0$ and $h = 0$. To what extent these become apparent depends on the separation from the critical point, τ, and on the value of $h/|\tau|^{\beta\delta}$, i.e., the ratio of the external field to the field equivalent of τ, namely $h_\tau = \phi^\delta = |\tau|^{\beta\delta}$. As long as $h \ll h_\tau$, the system is essentially in zero field, and $F_s \approx A_\pm |\tau|^{2-\alpha} \hat{f}_\pm(0) \propto |\tau|^{2-\alpha}$. On the other hand, when τ gets so small that $|\tau| \leq h^{1/\beta\delta}$, the system is manifestly in a finite field. Any further decrease of τ will then not lead to any changes: F_s remains at the value it acquired for $|\tau| = h^{1/\beta\delta}$, i.e. $A_\pm h^{(2-\alpha)/\beta\delta} \hat{f}_\pm(1)$; that is to say, in the limit $\tau \to 0$ one must have $\hat{f}_\pm(y)$

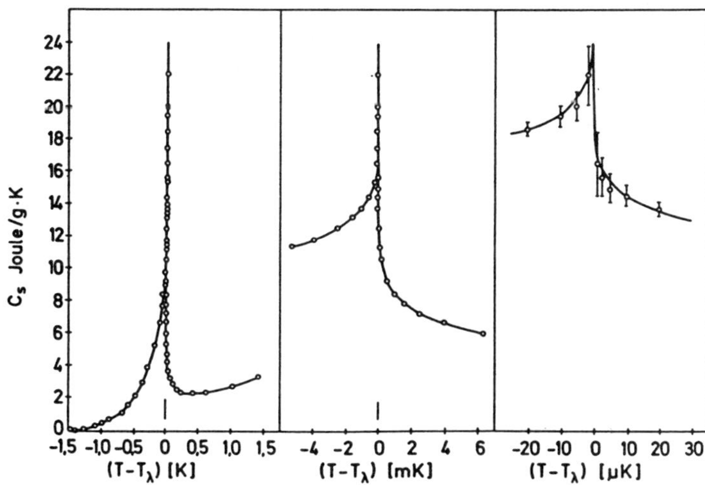

FIG. 4. Temperature dependence of the specific heat of helium-4 at the normal–superfluid phase transition, shown for successively reduced scales in the temperature variable (from Buckingham and Fairbank, 1961, p. 86).

$\rightarrow y^{(2-\alpha)/\beta\delta}$, in order to cancel the singular dependence of F_s on τ.

The scaling forms [Eqs. (14), (9), and (10)] incorporate the power laws [Eqs. (1)–(4), (7), and (8)]; moreover, by taking suitable derivatives with respect to τ and h, several *scaling relations* connecting the different critical indices may readily be obtained, namely

$$\alpha = 2 - \beta(\delta + 1), \tag{15}$$

$$\gamma = \beta(\delta - 1) = \nu(2 - \eta). \tag{16}$$

Relations of the type in Eqs. (14) and (9) are called *scaling laws*, because they are invariant under the following scale transformations:

$$\begin{gathered} x \rightarrow x/b, \quad \xi \rightarrow \xi/b, \\ \tau \rightarrow \tau b^{1/\nu}, \quad h \rightarrow h b^{\beta\delta/\nu}, \\ F_s \rightarrow F_s b^{(2-\alpha)/\nu}, \quad G \rightarrow G b^{(d-2+\eta)/\nu}. \end{gathered} \tag{17}$$

If, in addition, one assumes that underlying these scale transformations there is a microscopic elimination procedure mapping the original system with lattice constant a and N sites to a new one with the same lattice constant a, but a reduced number Nb^{-d} degrees of freedom, one has

$$\frac{F_s(\tau,h)}{N} = b^{-d}\,\frac{F_s(\tau b^{1/\nu}, h b^{\beta\delta/\nu})}{N b^{-d}}, \tag{18}$$

which in turn implies the *hyperscaling relation*

$$2 - \alpha = d\nu \tag{19}$$

containing the space dimension d. Equations (15), (16), and (19) leave only *two independent static critical exponents* characterizing the transition.

For dynamic critical phenomena, one arrives at a *dynamic scaling hypothesis* (Hohenberg and Halperin, 1977), originally proposed in the context of the λ transition of helium-4, by generalizing Eq. (10) to

$$G(\mathbf{k},\omega) = \frac{1}{|\mathbf{k}|^{2-\eta}}\,\hat{g}_{\pm}(|\mathbf{k}|\xi, \omega/\omega_c(\mathbf{k})) \tag{20}$$

for the frequency-dependent correlation function. Scaling laws of the form in Eqs. (14), (9), and (20) can be derived by the renormalization-group theory (see Sec. 3.2.4).

2. MATHEMATICAL REPRESENTATIONS

2.1 Microscopic Hamiltonians

In view of the expected universality of critical phenomena, simplified models containing the essential physics, namely the symmetry properties of the order parameter and the nature of the interactions, have provided a useful means for theoretical studies. For example, the *Ising* Hamiltonian describes a scalar order parameter, $\phi = \langle\sigma_i\rangle$, where the spin variables $\sigma_i = \pm 1$ are defined on a d-dimensional lattice, interacting with each other via a pair interaction J_{ij}, and subject to an external field with local strength h_i,

$$H[\sigma_i] = -\frac{1}{2}\sum_{i,j} J_{ij}\sigma_i\sigma_j - \sum_i h_i\sigma_i. \tag{21}$$

Positive exchange interaction J amounts to a ferromagnetic coupling, tending to align the spins, while an antiferromagnetic $J < 0$ favors alternating spin orientation. Its range is often restricted to nearest-neighbor interaction, but could also be taken as long-range. In the simplest case, the nonzero J_{ij} are chosen as constant, but in order to describe disordered systems they may as well be drawn from a certain random distribution (*random-bond* Ising model). Similarly, for most applications the external field is assumed to be uniform, $h_i = h$, but may in certain circumstances also be considered a random variable (*random-field* Ising model). Once the Hamiltonian is specified, in principle all thermodynamic properties of the model can of course be derived from the partition sum (see STATISTICAL MECHANICS, CLASSICAL)

$$Z = e^{-F/k_BT} = \mathrm{Tr}(e^{-H/k_BT}) \tag{22}$$

$$= \sum_{\{\sigma_i = \pm 1\}} e^{-H[\sigma_i]/k_BT}, \tag{23}$$

and the definition of thermodynamic averages

$$\langle A\rangle = \frac{1}{Z}\mathrm{Tr}(e^{-H/k_BT}A). \tag{24}$$

The short-range Ising model without disorder can be shown to display a second-order finite-temperature phase transition for all dimensions $d \geq 2$, whereupon the discrete symmetry $\sigma_i \leftrightarrow -\sigma_i$ is broken (see CRITICAL PHENOMENA). The statistical mechanics of lattice gases as well as binary liquid mixtures may be mapped onto Ising-type models; this once more illustrates the intimate analogies of phase transitions in very different systems.

Similarly, a three-component vector order parameter is incorporated in the anisotropic *Heisenberg* model,

$$H = -\frac{1}{2}\sum_{i,j}[J_{ij}^{\perp}(S_i^x S_j^x + S_i^y S_j^y) + J_{ij}^{\parallel} S_i^z S_j^z] - \sum_i h_i S_i^z \tag{25}$$

(we use the notation S_i^α, $\alpha = x, y, z$, for the three-component spin variables here; of course, this may be readily generalized to an n-component order parameter), where the symmetry-breaking field h_i is taken along the z direction. For nearest-neighbor interactions, in the cases $J^\parallel > J^\perp > 0$ and $J^\perp > J^\parallel > 0$ the Hamiltonian [Eq. (25)] describes a uniaxial ferromagnet or planar ferromagnet, respectively, while in the special situation $J_{ij}^\parallel = J_{ij}^\perp$ Eq. (25) yields the isotropic Heisenberg model, and for $J_{ij}^\parallel = 0$ the xy model, the latter being appropriate for the description of superfluid helium-4, for example (see Fisher, 1967; Ma, 1976a). As a consequence of the continuous rotational symmetry broken in the ordered phase, the Heisenberg and xy models may have long-range order only for $d > 2$ dimensions. The two-dimensional xy model displays another very interesting scenario, namely the Berezinskii–Kosterlitz–Thouless transition (Nelson, 1983): There appears a phase, although without long-range order, in which the correlation function decays as a power law. The system then disorders at higher temperatures via the creation and unbinding of pairs of topological defects (vortex-antivortex pairs of spin excitations). The relatively simple models introduced above have become of paradigmatic importance in the statistical mechanics of phase transitions. One should keep in mind, however, that in more complicated physical situations more elaborate models may be required.

2.2 Ginzburg–Landau Functional

Near a continuous phase transition, predominantly long-wavelength properties will be of significance, because of the divergence of the correlation length according to Eq. (8). Therefore, a continuum model for just the long-wavelength degrees of freedom should be sufficient and appropriate, at least for the description of the universal features. We remark that this concept of an *effective (field) theory* is of course defined on a certain energy scale, which in this context is set by the transition temperature. On different scales, the corresponding effective models may differ considerably, and one may in fact imagine them to be generated by a sequence of symmetry-breaking phase transitions occurring at successively lower energies.

Such an effective model for the long-wavelength degrees of freedom may either be derived from a microscopic Hamiltonian, such as Eq. (21) or (25), by mapping the partition function on a continuum problem and performing a gradient expansion; or merely be written down by carefully exploiting symmetry arguments. Let us denote the n-component vector order parameter by $\boldsymbol{\phi}$. For the case of a second-order phase transition breaking an O(n) symmetry, the effective free energy, also called Hamiltonian in the context of critical phenomena, which retains only the lowest orders in the gradients and nonlinear terms, is the *Ginzburg–Landau free energy functional* (in d dimensions)

$$\mathcal{F}[\phi] = \int d^d x \left[\frac{r}{2}\boldsymbol{\phi}^2 + \frac{u}{4}(\boldsymbol{\phi}^2)^2 + \frac{1}{2}(\nabla\boldsymbol{\phi})^2 - \mathbf{h}\cdot\boldsymbol{\phi}\right], \tag{26}$$

where $r = a(T - T_c)$, and a, u are positive coefficients, while $\mathbf{h}$ is the (n-component) external field. The coefficient of the gradient term in Eq. (26) was assumed to be positive, assuring that spatially uniform configurations be energetically preferred to inhomogeneous ones, and has then been absorbed in a redefinition of the field variable $\boldsymbol{\phi}$. For $n = 1$ the Hamiltonian [Eq. (26)] has the discrete symmetry of the Ising model [Eq. (21)] (i.e., for vanishing h), with a scalar order parameter, while for $n = 2$ and $n = 3$ the continuous symmetries of the xy model and the isotropic Heisenberg model are recovered.

Again, the thermodynamics can be retrieved from the partition function, which mathematically assumes the form of a functional integral of the probability distribution

$\mathcal{P} \propto \exp(-\mathcal{F}/k_BT)$ over the vector field $\boldsymbol{\phi}(x)$:

$$Z = Z_0(T) \int \mathcal{D}[\boldsymbol{\phi}(x)]e^{-\mathcal{F}[\boldsymbol{\phi}]/k_BT}. \tag{27}$$

Here the term Z_0 contains those contributions that are not explicitly stated in the Ginzburg–Landau functional [Eq. (26)] and that should be insignificant near the transition. Figure 5 depicts the probability distribution $\mathcal{P}$ in the case of a uniform one-component order parameter ϕ, containing the first two terms of the integrand in Eq. (26), for $r > 0$, and $r < 0$, respectively (the phase transition occurs at $r = 0$). We note that a very convenient way to calculate correlation functions uses derivatives with respect to the components of $\mathbf{h}$, e.g.,

$$G_{\alpha\beta}(x - x') = \frac{(k_BT)^2\partial^2 \ln Z[\mathbf{h}]}{\partial h_\alpha(x)\partial h_\beta(x')}. \tag{28}$$

2.3 Landau–Ginzburg Approximation

2.3.1 Mean-Field Theory Evaluating the functional integral [Eq. (27)] is a highly nontrivial task, and in general one is forced to resort to some approximation. Because of the exponential in the integrand, one might expect the probability distribution to be sharply peaked, and the thermodynamics of the system thus to be dominated by the most probable configuration, which is given by the solution to the stationarity equation

$$\begin{aligned} 0 &= \frac{\delta\mathcal{F}}{\delta\boldsymbol{\phi}(x)} \\ &= [r + u\boldsymbol{\phi}(x)^2 - \nabla^2]\boldsymbol{\phi}(x) - \mathbf{h}(x). \end{aligned} \tag{29}$$

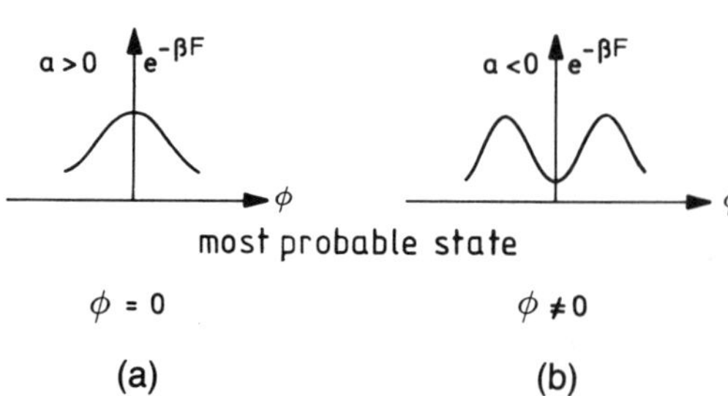

FIG. 5. Probability distribution $\mathcal{P} \propto \exp(-\mathcal{F}/k_BT)$ for the Landau part of the free energy $\mathcal{F}$, Eq. (26), as function of a single-component order parameter ϕ, for (a) $T > T_c$, and (b) $T < T_c$.

In order to determine the homogeneous order parameter $[\nabla\boldsymbol{\phi}(x) = 0]$ as a function of T, one must evaluate the Landau–Ginzburg equation [Eq. (29)] for $\mathbf{h} = h\mathbf{e}_n \to 0$ (note the order of limits here: *first* take the thermodynamic limit, and only *then* let the symmetry-breaking field vanish; with the reverse procedure there would be no nonzero order parameter for symmetry reasons, because one would have to average over the degenerate ground states). This yields

$$\boldsymbol{\phi} = \begin{cases} 0, & T \geq T_c, \\ \pm\sqrt{|r|/u}\mathbf{e}_n, & T < T_c; \end{cases} \tag{30}$$

and inserting Eq. (30) into Eq. (26) one arrives at the following approximate result for the singular part of the free energy F_s:

$$F_s[\boldsymbol{\phi}] = \begin{cases} 0, & T \geq T_c, \\ -r^2L^d/4u, & T < T_c \end{cases} \tag{31}$$

(here L^d is the volume of a cube in d-dimensional space). By comparison with Eqs. (1) and (4), one thus immediately finds the critical exponents $\beta = \frac{1}{2}$, and $\alpha = 0$; more precisely, the specific heat displays a discontinuity at the transition temperature T_c, the height of which is

$$\Delta C_{h=0} = T_c a^2 L^d/2u. \tag{32}$$

Similarly, for finite external field Eq. (29) leads to the equation of state in terms of the order-parameter component ϕ_n parallel to the symmetry-breaking field $\mathbf{h} = h\mathbf{e}_n$,

$$h = (r + u\phi_n^2)\phi_n, \tag{33}$$

from which the exponent for the critical isotherm is readily obtained to be $\delta = 3$. Finally, differentiating Eq. (33) with respect to h yields the isothermal susceptibility

$$\chi = \begin{cases} 1/r = C_+(T - T_c)^{-1}, & T > T_c, \\ 1/2|r| = C_-(T_c - T)^{-1}, & T < T_c, \end{cases} \tag{34}$$

and hence the critical exponent $\gamma = 1$ (on both sides of the transition), as well as the also universal amplitude ratio $C_+/C_- = 2$. The above results for the critical indices coincide with those of the mean-field approximation for the Ising and Heisenberg models, and with the exponents found for the van der Waals gas near its critical point. (In the mean-field approximation for the Ising model [Eq. (21)], for instance, the exchange field $\Sigma_j J_{ij}\sigma_j$ acting on spin σ_i is replaced by its average $\Sigma_j J_{ij}\langle\sigma_j\rangle$.)

We note that if u in Eq. (26) became negative, which may happen as some external parameter, e.g., the pressure, is varied, additional terms would have to be included in the Ginzburg–Landau functional for stability reasons. In the case of an underlying O(n) symmetry, the leading such nonlinearity would be $\propto (\phi^2)^3$, and the ensuing phase transition of first order. Its transition temperature $T_1 \neq T_c$ is to be defined by the condition that the free energies of the distinct coexisting phases be equal. In the special case that u vanishes, T_c becomes a tricritical point, displaying most interesting and complex scaling behavior.

2.3.2 Gaussian Fluctuations In keeping the most probable state only, fluctuations have been neglected entirely so far. However, the divergence of the susceptibility, Eq. (34), implies that fluctuations must indeed become very strong near the critical point. The next step is therefore to check whether the above results are altered by including fluctuations to lowest order. Thus, we introduce the deviations $\boldsymbol{\psi}(x)$ from the most probable state $\boldsymbol{\phi}(x)$, given by the solution to Eq. (29), by writing the order-parameter field as $\boldsymbol{\phi}(x) + \boldsymbol{\psi}(x)$, and expand the Ginzburg–Landau functional [Eq. (26)] with respect to $\boldsymbol{\psi}(x)$, keeping terms up to second order only. This is most conveniently done in Fourier space, where we define the Fourier transform of $\boldsymbol{\psi}(x)$ as

$$\boldsymbol{\psi}(x) = L^{-d/2} \sum_{k \in \mathrm{B}} \boldsymbol{\psi}_k e^{ikx}; \tag{35}$$

here B denotes the Brillouin zone for the allowed values of the k vectors. The partition function [Eq. (27)] then becomes

$$Z = Z_0(T) \int \prod_k d\boldsymbol{\psi}_k e^{-\mathcal{F}_G[\boldsymbol{\psi}]/k_B T}. \tag{36}$$

Beginning with the situation above T_c, for $h = 0$, the Fourier expansion renders the Ginzburg–Landau functional diagonal, $\mathcal{F}_G = \Sigma_k (r + k^2)\boldsymbol{\psi}_k \cdot \boldsymbol{\psi}_{-k}/2$. The functional integrals in the partition function thus reduce to simple Gaussian integrals, yielding for the free energy in *Gaussian approximation*

$$F_s(T,h = 0) = -k_B T \frac{n}{2} \sum_k \ln \frac{2\pi k_B T}{r + k^2}, \tag{37}$$

and by transforming the discrete sum over the Brillouin zone into an integral, the most singular part of the specific heat follows as

$$C_s = k_B L^d \frac{n}{2} (Ta)^2 \int \frac{d^d k}{(2\pi)^d} \frac{1}{(r + k^2)^2}. \tag{38}$$

The result of evaluating the integral is

$$C_s = \begin{cases} A_+(T - T_c)^{-(4-d)/2}, & d < 4, \\ \propto \ln(T - T_c), & d = 4, \\ B - A(T - T_c)^{(d-4)/2}, & d > 4; \end{cases} \tag{39}$$

hence while for $d > 4$ the specific heat shows a (nonanalytical) cusp at T_c, it diverges for $d \leq 4$, with the exponent $\alpha = (4 - d)/2$; here an exponent "0" is to be understood as a logarithmic singularity. Neglecting deviations from the most probable state therefore turns out to be entirely unjustifiable in the latter case, where the thermodynamics is obviously dominated by the order-parameter fluctuations.

Similarly, in the ordered phase one finds

$$C_s = TL^d a^2/2u + A_-(T_c - T)^{-(4-d)/2} \tag{40}$$

for $d < 4$; the critical exponent α is therefore the same above and below the transition. The universal (i.e., depending only on d and n) amplitude ratio for the specific heat is $A_+/A_- = n/2^{d/2}$, where the factor n stems from the fact that the $n - 1$ transverse fluctuations (with respect to the direction of the spontaneous order parameter) do not contribute to the specific heat in the ordered phase.

With the help of the result in Eq. (40), the *Ginzburg–Levanyuk criterion* (see Als-Nielsen and Birgeneau, 1977) may be formulated, stating that fluctuations are negligible as long as the first term ΔC dominates over the second contribution; introducing $\xi_0 = (aT_c)^{-1/2}$ this means that the Landau–Ginzburg approximation gives essentially correct answers, provided

$$|\tau|^{(4-d)/2} \gg \frac{\mathcal{N}}{(\xi_0 L)^d} \Delta C, \tag{41}$$

where $\mathcal{N}$ is a numerical factor. Therefore, for $d > d_c = 4$ fluctuations are not very important [$\alpha_{fl} < 0$, Eq. (39)], and mean-field types of approach yield reliable results. Below this *upper critical dimension*, however, such approximations will break down in the vicinity of T_c. Depending on the value of ξ_0, the nonclassical scaling behavior (with exponents different from the mean-field indices) will be important only very near to the critical point, or already noticeable even far away from it.

For example, for magnets typically $\xi_0 \approx 1$ Å, and fluctuations become important for $|\tau| < \tau_{\mathrm{GL}} \approx 0.01$. For the normal–superfluid transition of helium-4, $\xi_0 \approx 4$ Å, and $\tau_{\mathrm{GL}} \approx 0.3$, which renders this system a prominent candidate for the study of critical phenomena (also, by varying the pressure one can explore a line of critical points and thus examine universality issues in detail). On the other hand, for conventional (low-T_c) superconductors ξ_0 is of the order 1000 Å, and $\tau_{\mathrm{GL}} \approx 10^{-10}$ to 10^{-14}, which means that the Ginzburg–Landau theory is an excellent approximation there. This is in remarkable contrast with the high-T_c materials, where $\xi_0 \approx 1$ Å, and fluctuations play a dominant role; in fact, they account for an enormously rich phase diagram of the cuprate-based superconductors (see Blatter *et al.*, 1994).

For $|\tau| < \tau_{\mathrm{GL}}$, it does not at all suffice to add to the results of the Ginzburg–Landau approximation the Gaussian corrections, since in this temperature range higher-order nonlinear contributions will be even more infrared divergent and eventually exceed those from the harmonic term. This constitutes of course a serious breakdown of the standard perturbational procedure, and signals that an entirely new theoretical approach is required in the case $|\tau| < \tau_{\mathrm{GL}}$.

For finite external field, the partition function [Eq. (36)] becomes above T_c

$$Z[\mathbf{h}] = \exp\left[\frac{1}{2k_{\mathrm{B}}T}\sum_k \frac{\mathbf{h}_k\cdot\mathbf{h}_{-k}}{r+k^2}\right], \tag{42}$$

and the correlation function can be obtained by differentiating this generating functional [see Eq. (28)]. Above T_c, the result in k space is

$$\langle \psi_k^\alpha \psi_{-k'}^\beta \rangle = \frac{k_{\mathrm{B}}T}{r+k^2}\,\delta_{kk'}\delta^{\alpha\beta}; \tag{43}$$

consequently, for $r = 0$ one finds in real space

$$G(x) \propto 1/|x|^{d-2}, \tag{44}$$

and thus the critical exponent defined in Eq. (7) is $\eta = 0$ in the Landau–Ginzburg approximation. Also, the inverse of r defines the square of a characteristic length, which we identify with the correlation length, $\xi_+ = r^{-1/2}$, and hence $\nu = \frac{1}{2}$. In three dimensions, the Fourier transform of Eq. (43) is the *Ornstein–Zernicke correlation function*

$$G(x) = \frac{k_{\mathrm{B}}T}{4\pi|x|}\,e^{-|x|/\xi_+}. \tag{45}$$

In the ordered phase, one has to distinguish between the longitudinal and transverse correlation functions, with respect to the spontaneous order parameter, which are still diagonal in the indices α, β, and for the O(n) symmetric model [Eq. (26)] in Gaussian approximation read

$$G_\parallel(k) = k_{\mathrm{B}}T/(\xi_-^{-2} + k^2) \tag{46}$$

and

$$G_\perp(k) = k_{\mathrm{B}}T/k^2, \tag{47}$$

respectively. Hence, on this level $G_\parallel(k)$ is of the form in Eq. (43), with a different correlation length $\xi_- = (2|r|)^{-1/2}$, however. Note that the amplitude ratio for the correlation lengths above and below the transition is universal again, with the mean-field value $\sqrt{2}$. Remarkably, the transverse correlation function diverges for $k \to 0$, which means that the range for the transverse correlations is infinite in the entire low-temperature phase. This can be shown to be a rigorous consequence of the breaking of the continuous rotational symmetry. For $n \geq 2$ and $d \leq 4$, the large transverse fluctuations [Eq. (47)] via nonlinear effects imply a longitudinal correlation function surpassing the result obtained within the harmonic approximation [Eq. (46)], which for $k \to 0$ has to be replaced by

$$G_\parallel(k) \propto 1/k^{4-d}. \tag{48}$$

Remarkably, Eq. (48) is a universal power law characterizing not a phase transition singularity, but an *entire ordered phase*. Yet the longitudinal correlation length ξ_- still retains some physical meaning; namely, it provides the scale on which the crossover from the critical behavior [Eq. (7)] to the asymptotic form [Eq. (48)] occurs.

In the Ising case ($n = 1$), Eq. (46) is qualitatively correct outside the critical region, and ξ_- sets the characteristic length scale in the ordered phase. For instance, a domain-wall configuration $\phi_K(x)$ connecting the two degenerate ground states $\phi_\pm = \pm\sqrt{|r|/u}$, and centered at x_0, extends over the range ξ_-,

$$\phi_K(x) = \phi_\pm \tanh\frac{x - x_0}{2\xi_-}, \qquad (49)$$

as can be seen by solving the Landau–Ginzburg differential equation [Eq. (29)] for $h = 0$, allowing for a spatially varying order parameter with the appropriate boundary conditions.

3. RENORMALIZATION GROUP

3.1 Introductory Remarks

The term "renormalization" of a theory stands for a certain reparametrization, with the goal that the resulting renormalized theory be more tractable than the original one. Historically, renormalization was invented by Stückelberg and Feynman, in order to cure quantum field theories (e.g., quantum electrodynamics) from divergences (see UNIFIED FIELD THEORIES). Instead of the bare parameters (masses and coupling constants), the Lagrangian was expressed in terms of physical masses and coupling coefficients, such that ultraviolet-divergent (UV) virtual processes were systematically absorbed in the relationship between the bare and physical quantities, thus rendering the renormalized theory finite. The renormalization procedure is not unique; the renormalized quantities may for instance depend on a cutoff length scale up to which certain virtual processes are taken into account. The *renormalization-group* (RG) theory studies the dependence on this specific length scale, also called *flow parameter*. The name renormalization group derives from the fact that two successive RG transformations lead to a third such transformation.

In the context of critical phenomena, where one needs to explain the long-distance (or, in Fourier space, small–wave-number) regime, it is natural to implement the renormalization scheme by performing some kind of elimination of the short-wavelength fluctuations. Such a partial evaluation of the partition function will be much easier to achieve than a calculation of the complete partition sum, and also permits the systematic implementation of approximate methods. As a result of the elimination step, the remaining degrees of freedom will experience modified *effective* interactions.

Quite generally, various benefits can be expected from such a RG procedure:

1. The new coupling coefficients might be smaller. By repeating the renormalization process, one may thus arrive at an effectively free theory with no interactions.
2. The successively iterated coupling coefficients, also called the *parameter flow*, may have a finite *fixed point*, at which the system does not change any more under subsequent RG transformations. Since the elimination of some degrees of freedom is accompanied by a change of the underlying lattice spacing, or length scale, one may anticipate that the fixed points are under certain circumstances related to critical points. Furthermore, one may hope that the properties of the flow near these fixed points will yield information about the universal physical quantities in the vicinity of the critical point.

Below, we shall find scenario **1** in the case of the one-dimensional Ising model and **2** for the Ising model in two dimensions, as well as for the Ginzburg–Landau–Wilson effective field theory (ϕ^4 model).

3.2 Real-Space Renormalization Group

The idea of any RG approach in statistical mechanics is to exploit the scale invariance near a critical point in parameter space. In real-space RG transformations, which refers to procedures working in direct space (as opposed to Fourier space), one eliminates specific degrees of freedom that are defined on a lattice, thus performing a partial partition sum. Then the lattice constant of the ensuing system is readjusted, and also the internal variables are renormalized, in order to achieve a new Hamiltonian whose form resembles the original one. By comparison, effective couplings are defined, which thus become scale dependent. The fundamental paradigm is that certain fixed points in parameter space exist, at which these running couplings remain invariant. At these fixed points, the theory is manifestly scale invariant, and the critical exponents can be obtained by studying the linearized flow equations. As this concept of the RG may seem rather abstract at first sight, the method is explained by applying a simplified version to the ferromagnetic Ising model [Eq. (21)] in $d = 1$ and $d = 2$ dimensions (Niemeijer and van Leeuwen, 1976). Then a more general discussion, including an explanation of the emergence of scaling laws, is given. A continuous field-theoretical formulation for the RG in momentum space is presented in Sec. 3.3.

3.2.1 1D Ising Model For a one-dimensional Ising chain with constant ferromagnetic nearest-neighbor exchange coupling $J_{ij} = J > 0$ in zero external field, the Hamiltonian [Eq. (21)] becomes $H = -J\sum_i \sigma_i\sigma_{i+1}$, and introducing the abbreviation $K = J/k_BT$ the partition function [Eq. (23)] reads (for N spins, and periodic boundary conditions $\sigma_{N+1} = \sigma_1$)

$$Z_N = \sum_{\{\sigma_i=\pm 1\}} \exp\left[K\sum_i \sigma_i\sigma_{i+1}\right]. \tag{50}$$

A partial evaluation of Eq. (50) may be performed by summing over every other spin in the chain, marked by an "x" in Fig. 6. The sum over the possible orientations of spin i, for instance, yields

$$\sum_{\sigma_i=\pm 1} e^{K\sigma_i(\sigma_{i-1}+\sigma_{i+1})} = \exp[2g' + K'\sigma_{i-1}\sigma_{i+1}], \tag{51}$$

for the result can only depend on whether the adjacent spins σ_{i-1} and σ_{i+1} are oriented parallel or antiparallel to each other. By comparing the results for these two configurations, $\sigma_{i-1}\sigma_{i+1} = \pm 1$, one readily obtains the two recursion relations

$$K' = \tfrac{1}{2}\ln\cosh 2K, \tag{52}$$

$$g' = \tfrac{1}{2}(\ln 2 + K'). \tag{53}$$

Thus, via decimation of half of the spins in the chain one arrives at another Ising Hamiltonian, whose nearest-neighbor coupling is K'; in addition, a spin-independent contribution g' to the free energy is generated. If this transformation is repeated, the *exact* recursion relations [Eqs. (52) and (53)] quite generally describe how the effective couplings $K^{(k)}$ and $g^{(k)}$ in the kth step result from the corresponding values in the $(k-1)$th elimination step. In Fig. 7, the ensuing flow for the coupling K, starting with some initial value K_0, is depicted for several steps of the procedure. Clearly, the fixed-point equation $K^* = (\tfrac{1}{2})\ln\cosh 2K^*$ has two solutions: $K^* = 0$, which (for the original fixed J) corresponds to infinite temperature, or vanishing effective interaction; and $K^* = \infty$, the $T = 0$ fixed point. On starting at some initial value $K_0 > 0$, the parameter flow always drives the system into the $K^* = 0$ fixed point [by expanding Eq. (52) around $K^* = 0$, this fixed point is seen to be stable]. Physically, this means that there can be no ordered state at any finite temperature $T > 0$, because asymptotically the ratio of exchange coupling and tem-

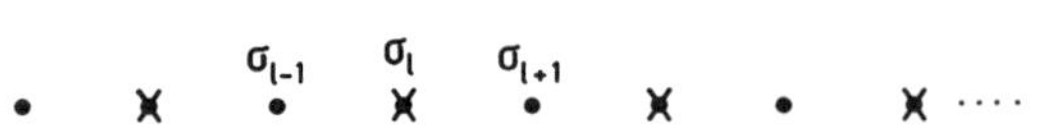

σ_{i-1} σ_i σ_{i+1}

• x • x • x • x ....

FIG. 6. Decimation transformation for the one-dimensional Ising chain with nearest-neighbor interaction.

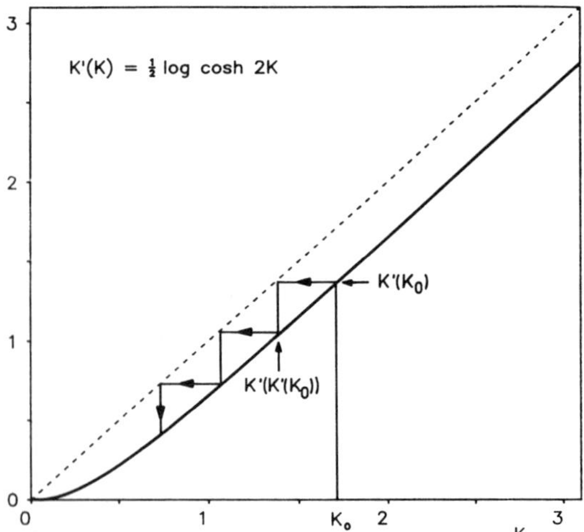

FIG. 7. Recursion relation for the nearest-neighbor coupling of the Ising chain (full; the dashed line marks the line $K' = K$). The iteration steps are indicated by the arrows.

perature vanishes, and the system becomes completely disordered.

At $T = 0$, however, the situation is rather special. Expanding Eq. (52) near $1/K^* = 0$ yields $1/K' = 1/K + \ln b/(2K^2)$, where $b = 2$ is just the factor rescaling all lengths. It now turns out to be useful to introduce $\delta = \ln b$, such that after l steps $b^l = e^{\delta l}$, and then take the limit $\delta \to 0$; thereby the discrete RG steps are mapped onto a continuous flow with the flow parameter l. The resulting differential flow equation for the now scale-dependent coupling $K(l)$ reads to lowest order

$$\frac{dK^{-1}(l)}{dl} = \tfrac{1}{2}\,[K^{-1}(l)]^2, \tag{54}$$

which is readily integrated to $K(1) - K(l) = l/2$. From this, the temperature dependence of the correlation length ξ can be determined by applying the ("matching") condition that $K(b = \xi)$ be a constant of order 1. This gives

$$\xi \propto \exp(2J/k_BT), \tag{55}$$

i.e., the correlation length diverges exponentially at the *marginal* critical point $T = 0$, a result that may also be readily obtained by more elementary methods.

Using the above recursion relations the partition function may be evaluated:

$$\begin{aligned} Z_N &= e^{N_g(K')} Z_{N/2}(K') = \cdots \\ &= \exp\left[N \sum_{k=1}^{n} \frac{g(K^{(k)})}{2^{k-1}} + \ln Z_{N/2^n}(K^{(n)})\right]. \end{aligned} \tag{56}$$

The factor 2 in Eq. (56) can be interpreted as stemming from a readjustment of the overall scale, as the number of degrees of freedom is being diminished by the elimination procedure. The reduced free energy (in units of k_BT) thus becomes $f = -(1/N)\ln Z_N(K)$. As is apparent from Fig. 7, $K^{(k)}$ rapidly becomes very small, and perturbative methods may be employed. Assuming $K^{(n)} \approx 0$, one arrives at the approximate result

$$f^{(n)}(K) = -\sum_{k=1}^{n} \frac{g(K^{(k)})}{2^{k-1}} - \frac{1}{2^n}\ln 2, \tag{57}$$

for in a noninteracting spin-1/2 system the free energy per spin is $-\ln 2$. Figure 8 shows the rapid approach toward the exact result,

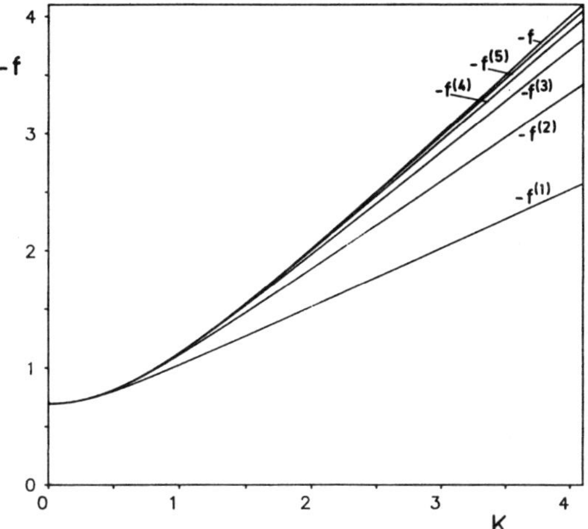

FIG. 8. Reduced free energy f for the nearest-neighbor Ising chain. The $f^{(k)}$ denote the approximation to the exact result f as obtained in the kth step of the RG procedure.

$f(K) = -\ln(2\cosh K)$, as the number of elimination steps is increased.

3.2.2 2D Ising Model More interesting is the case of the two-dimensional Ising model, which is known to have a phase transition at a finite temperature $T_c > 0$ (see CRITICAL PHENOMENA). Here a quadratic lattice with nearest-neighbor ferromagnetic interactions will be considered (Fig. 9). In order to find the recursion relations in this case, one may perform the partial partition sum over those spins

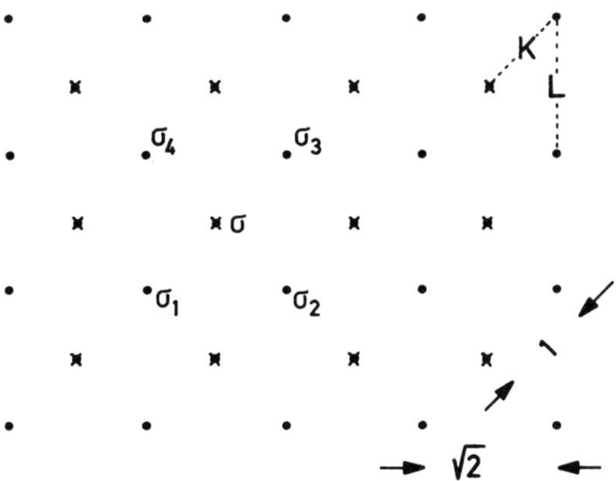

FIG. 9. Ising model on a quadratic lattice. Even if originally there were only nearest-neighbor interactions K present, by application of the decimation procedure next-nearest-neighbor couplings L are generated.

marked with an "x" in Fig. 9, which leads to a new square lattice with a lattice constant larger by a factor of $\sqrt{2}$ than that of the original one. A typical sum is

$$\sum_{\sigma=\pm 1} e^{K\sigma(\sigma_1+\sigma_2+\sigma_3+\sigma_4)} = \exp[A' + K'(\sigma_1\sigma_2 + \sigma_2\sigma_3 + \sigma_3\sigma_4 + \sigma_4\sigma_1)] \times \exp[L'(\sigma_1\sigma_3 + \sigma_2\sigma_4) + M'\sigma_1\sigma_2\sigma_3\sigma_4]; \tag{58}$$

thus some new interaction, namely a next-nearest-neighbor coupling L' and a four-spin interaction M', are *generated* by the transformation. In the vicinity of the actual critical value $K_c = J/\kappa_B T_c = 0.4406$, one finds $M' \ll L' \leq K'$. Therefore, M' shall henceforth be neglected here (a more thorough perturbational analysis is given in Niemeijer and van Leeuwen, 1976.) The terms A', K', and L' can readily be expressed by K. If these relations are expanded to second order in K, and if one realizes that a next-nearest-neighbor interaction L in the original Hamiltonian appears as a contribution to the nearest-neighbor interaction in the primed Hamiltonian, one obtains

$$K' = 2K^2 + L, \tag{59}$$

$$L' = K^2. \tag{60}$$

A simple physical picture behind these approximate recursion relations is that they just represent the numbers of paths between the remaining spins, weighted by the product of the bonds involved.

It should be clear, however, that with each step of the elimination new couplings appear, as opposed to the one-dimensional case. Of course, restricting the flow to the subspace (K,L) will not produce quantitatively satisfactory results, but still the structure and typical features of such recursion relations are preserved.

The fixed points of Eqs. (59) and (60) are

1. $K^* = L^* = 0$,
2. $K^* = L^* = \infty$, and
3. $K_c^* = \frac{1}{3}$, $L_c^* = \frac{1}{9}$ (Fig. 10).

As in the 1D Ising model, fixed points **1** and **2** refer to infinite temperature (disordered phase) and $T = 0$, respectively. Yet now the latter describes an entire (ordered) phase. Furthermore, there is an additional nontrivial finite fixed point, which governs the second-order transition separating these two phases. For at each step in the procedure, the effective lengths are reduced by a factor $b = \sqrt{2}$, which applies to the correlation length also, and therefore at a fixed point either $\xi = 0$ (which is true for the $T = 0$ and $T = \infty$ fixed points), or $\xi = \infty$ (for finite values of K^* and L^*), which implies scale invariance. Therefore, the fixed point (K_c^*,L_c^*) with diverging correlation length describes the critical state. Even more, actually all points lying on a trajectory leading into the fixed point represent critical behavior, because on these paths the correlation length is infinite as well.

In order to determine the critical behavior of the model, the recursion relations [Eqs. (59) and (60)] are linearized near the point (K_c^*,L_c^*) in parameter space; with $\delta K_l = K_l - K_c^*$ and $\delta L_l = L_l - L_c^*$ in the lth step of the

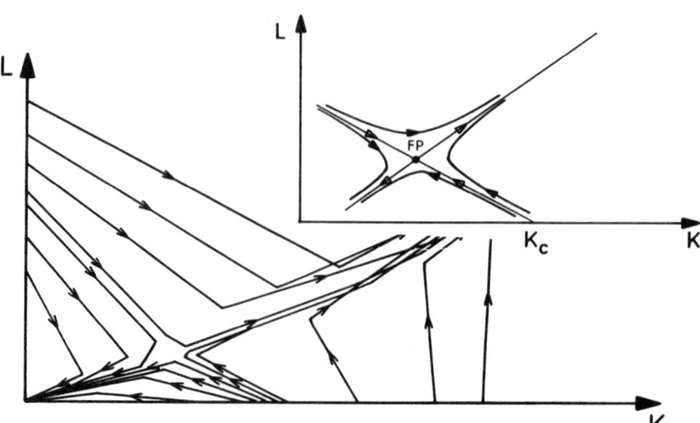

FIG. 10. Flow diagram for the two-dimensional Ising model on a square lattice near its critical point. (Only the results of every other recursion step are shown.) The inset shows the behavior in the vicinity of the critical fixed point, which is described by Eq. (63).

procedure, one finds

$$\begin{pmatrix} \delta K_l \\ \delta L_l \end{pmatrix} = \begin{pmatrix} \frac{4}{3} & 1 \\ \frac{2}{3} & 0 \end{pmatrix} \begin{pmatrix} \delta K_{l-1} \\ \delta L_{l-1} \end{pmatrix}. \tag{61}$$

The eigenvalues of the transformation matrix are $\lambda_{1,2} = (2 \pm \sqrt{10})/3$, and the corresponding eigenvectors $\mathbf{e}_{1,2} = (3, \pm\sqrt{10} - 2)$. Thus, if the original couplings are expanded in the basis of eigenvectors,

$$\begin{pmatrix} K_0 \\ L_0 \end{pmatrix} = \begin{pmatrix} K_c^* \\ L_c^* \end{pmatrix} + c_1\mathbf{e}_1 + c_2\mathbf{e}_2, \tag{62}$$

after l steps one arrives at

$$\begin{pmatrix} K_l \\ L_l \end{pmatrix} = \begin{pmatrix} K_c^* \\ L_c^* \end{pmatrix} + \lambda_1^l c_1\mathbf{e}_1 + \lambda_2^l c_2\mathbf{e}_2. \tag{63}$$

As is depicted in the flow diagram (Fig. 10), if $c_l \neq 0$, the flow will in general run away from the critical fixed point (K_c^*, L_c^*), which is thus unstable toward the *relevant perturbation* $c_1\mathbf{e}_1$, which is characterized by $|\lambda_1| > 1$. On the other hand, if $c_1 = 0$, the flow will always tend toward this fixed point, because $|\lambda_2| < 1$ ($\mathbf{e}_2$ is called the *irrelevant* direction; had we included more coupling parameters initially, additional irrelevant directions would have emerged). The original nearest-neighbor model had $L_0 = 0$. Inserting this in Eq. (62) for $c_1 = 0$, one finds the critical value of K (within the linear approximation), for which the decimation transformation will lead into the critical fixed point, to be

$$K_c = \tfrac{1}{3}\left(1 + \frac{1}{\sqrt{10} + 2}\right) \approx 0.3979, \tag{64}$$

which should be compared to the mean-field and exact values, $K_c = 0.25$ and $K_c = 0.4406$, respectively. Note that $K - K_c = -J(T - T_c)/k_BTT_c$, and hence $c_1 \neq 0$ whenever $T \neq T_c$. In this case, the flow will tend toward either the zero-temperature (for $T < T_c$) or the disordered fixed point (for $T > T_c$).

After a sufficient number of steps λ_2^l will be minute, and

$$(K - K_c)_l = b^{y_1 l}(K - K_c), \tag{65}$$

where $b = \sqrt{2}$, and $y_1 = 2\ln\lambda_1/\ln 2 \approx 1.566$. As also $\xi_l = b^{-l}\xi$, and the definition of the critical exponent ν for the correlation length translates to $\xi \propto |K - K_c|^{-\nu}$, one finds

$$\nu = 1/y_1 \approx 0.638, \tag{66}$$

which is already larger than the mean-field result $\nu = \frac{1}{2}$, but still far off the exact value $\nu = 1$.

Similarly, an external field h may be included in the scheme. The approximate recursion relation for h turns out to be

$$h' = (1 + K)h, \tag{67}$$

with the fixed point $h^* = 0$, and the eigenvalue for the linearized recursion relation is $\lambda_h = 1 + K_c^* = \frac{4}{3}$. Hence the external field constitutes a second relevant variable besides (essentially) $T - T_c$. Notice that the flows for K and h decouple. The two relevant fields τ and h precisely correspond to the two independent critical exponents of the model.

In addition, the universal properties of the phases, as the stiffness exponent θ in Eq. (13), can be determined in a similar way. For the stiffness, the most convenient way is to study the change in free energy of a finite system as the boundary conditions are changed, e.g., from periodic to antiperiodic (see Fisher, 1993). For an Ising system of linear dimension $L = b$ in d dimensions, the domain wall energy at $T = 0$ turns out to be

$$\Sigma \approx 2Jb^{d-1}, \tag{68}$$

and hence $\theta = d - 1$. Thus, a rescaling of the length by b is equivalent to a recursion relation for the temperature

$$T' = b^{-\theta}T \tag{69}$$

for the temperature, which shows that the stiffness exponent may be identified with the negative eigenvalue $\lambda_T = -\theta$ at the zero-temperature fixed point ($T^* = 0$). Equation (68) demonstrates that the ordered phase is stable for $d \geq 2$ in the Ising model. For Heisenberg models with continuous rotational symmetry, on the other hand, one finds $\theta = d - 2$, and no long-range order is possible in two dimensions. Note that mean-field theory fails to predict such a *lower critical dimension*, at and below which fluctuations destroy the formation of spatially homogeneous long-range order.

3.2.3 General Real-Space RG Transformations A general real-space RG transformation maps a certain spin system $\{\sigma\}$, defined on a lattice, and with Hamiltonian $H\{\sigma\}$, onto a new spin system $\{\sigma'\}$ with new Hamiltonian $H'\{\sigma'\}$ and by a factor of $N'/N = b^{-d}$ fewer degrees of freedom. It may be represented by a transformation $\tau\{\sigma',\sigma\}$, such that (k_BT has been set equal to unity here)

$$e^{-F-H'\{\sigma'\}} = \sum_{\{\sigma\}} \tau\{\sigma',\sigma\}e^{-H\{\sigma\}}, \qquad (70)$$

with the conditions

$$\sum_{\{\sigma'\}} H'\{\sigma'\} = 0 \qquad (71)$$

and

$$\sum_{\{\sigma'\}} \tau\{\sigma',\sigma\} = 1, \qquad (72)$$

which guarantee that $e^{-F}\,\mathrm{Tr}_{\{\sigma'\}}e^{-H'\{\sigma'\}} = \mathrm{Tr}_{\{\sigma\}}e^{-H\{\sigma\}}$.

Important examples for such transformations are decimation transformations, and linear as well as nonlinear block-spin transformations. Several types of approximations, like the cumulant expansion and cluster and finite-lattice approximations, have been applied to various kinds of lattices (quadratic, triangular, etc.), where a large number of interaction coefficients have been included. In two dimensions, critical exponents and thermodynamic properties as a whole have thereby been obtained to an impressive degree of accuracy (Niemeijer and van Leeuwen, 1976). A combination of numerical Monte-Carlo methods and real-space RG transformations leads to the *Monte-Carlo renormalization group* (Swendsen, 1984). Apart from that, however, the more versatile momentum-space methods (see Sec. 3.3 below) have been far more useful in higher dimensions ($d > 2$).

3.2.4 Emergence of Scaling Laws While the decimation procedure employing only a few parameters does not provide quantitatively satisfactory results, and also is inappropriate for the study of correlation functions, it still elucidates the general structure of RG transformations R, whereupon the original Hamiltonian H is mapped onto a new one, $H' = R(H)$ (this includes the rescaling of lengths in the problem, or for the degrees of freedom: $N' = Nb^{-d}$ in d dimensions). Near the fixed-point Hamiltonian $H^* = R(H^*)$, one may expand $R(H^* + \delta H) \approx H^* + L\delta H$, which yields the linearized recursion relation $\delta H' = L\delta H$. The eigenoperators $\delta H_1, \delta H_2 \ldots$ for this linear transformation are then determined by the eigenvalue equations

$$L\delta H_i = \lambda_i \delta H_i. \qquad (73)$$

A given Hamiltonian H that does not differ too much from the fixed-point Hamiltonian may be expanded according to

$$H = H^* + \tau\delta H_\tau + h\delta H_h + \sum_{j\geq 3} c_j\delta H_j, \qquad (74)$$

where δH_τ and δH_h denote the two relevant perturbations, $|\lambda_\tau| = b^{y_\tau} > 1$, $|\lambda_h| = b^{y_h} > 1$, connected with the temperature variable $T - T_c$ and the external field h, while $|\lambda_j| < 1$ (and hence $y_j < 0$) for $j \geq 3$ (in the Ising case, $\delta H_h = \Sigma_i\sigma_i$). The coefficients τ, h, and c_j are called *scaling fields*.

After l steps, the free energy per spin, $f(\{c_j\}) = F_N(\{c_j\})/N$, transforms according to

$$f(\tau,h,c_3,\ldots) = b^{-dl}f(\tau b^{y_\tau l},hb^{y_h l},c_3b^{y_3 l},\ldots) \qquad (75)$$

in the linear approximation. (Here we have omitted an additive term, which does not affect the following derivation of the scaling law, but would of course be important for the evaluation of the total free energy.)

The scale parameter l may now be chosen conveniently, for example $|\tau|b^{y_\tau l} = 1$, which sets the first argument of f equal to ± 1; this yields

$$f(\tau,h,\ldots) = |\tau|^{d/y_\tau}\hat{f}_\pm(h|\tau|^{-y_h/y_\tau},c_3|\tau|^{|y_3|/y_\tau},\ldots), \qquad (76)$$

where $\hat{f}_\pm(x,y,\ldots) = f(\pm 1,x,y,\ldots)$. Close to T_c, the dependence on the irrelevant fields $c_3, \ldots$ can be neglected, and Eq. (76) then precisely assumes the scaling form [Eq. (14)], with the exponents

$$\beta\delta = y_h/y_\tau \qquad (77)$$

and

$$2 - \alpha = d/y_\tau. \qquad (78)$$

Thus, the scaling law [Eq. (14)] has been derived within the RG theory for fixed points with precisely one relevant field, aside from the external magnetic field and the irrelevant operators. Furthermore, the dependence of the irrelevant fields $c_3, \ldots$ yields *corrections to scaling*, which may have to be taken into account in situations not yet close enough to T_c to be in the fully asymptotic regime.

In order to relate y_τ to the exponent ν, one has to recall that l iterations reduce the correlation length to $\xi' = b^{-l}\xi$, which implies $(\tau b^{y_\tau l})^{-\nu} = b^{-l}\tau^{-\nu}$, and thus

$$\nu = 1/y_\tau \tag{79}$$

[see Eq. (66) for the 2D Ising model]. Thus, with the basic assumption that there exists a fixed-point Hamiltonian with two relevant operators, the scaling form of the free energy could be derived, and moreover, a prescription for the calculation of the critical exponents is provided. A similar procedure may be followed for the correlation function [Eq. (9)]. At this point, the renormalization of the spin variables, $\sigma' = b^\zeta\sigma$, becomes important; it turns out that the choice

$$\zeta = (d - 2 + \eta)/2 \tag{80}$$

ensures the validity of Eq. (7) at the critical point.

We add some remarks on the generic structure of the flow diagram in the neighborhood of a critical fixed point (Fig. 11). In the multidimensional space of coupling coefficients, there will be one specific direction that leads away from the fixed point (the relevant direction). The other eigenvectors of the linearized RG transformation span the *critical hypersurface*. Further away from the fixed point, the hypersurface is no longer a plane but acquires some curvature. From any point on the critical hypersurface, the trajectories lead into the critical fixed point. If the starting point is close to but not at the critical hypersurface, the corresponding trajectory will first run parallel to the hypersurface until the relevant part becomes magnified sufficiently, so that finally the trajectory leaves the vicinity of the critical hypersurface, and turns off to either the low- or high-temperature fixed point. For a particular physical system (a ferromagnet, a binary mixture, etc.), the parameters will depend on the temperature (the long-dashed curve in Fig. 11). The temperature for which this curve intersects the critical hypersurface constitutes its transition temperature T_c.

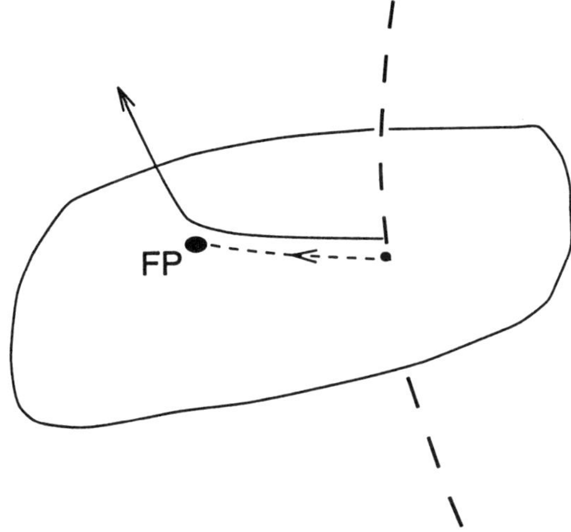

FIG. 11. Critical hypersurface. A trajectory in the critical hypersurface is shown dashed, and a trajectory starting close to the critical hypersurface is displayed as a full curve. The coupling coefficient of a particular physical system as a function of a control parameter is indicated by the long-dashed line.

From the above discussion, the universality property becomes obvious, too. All systems belonging to a particular part of parameter space, namely the region of attraction of a specific fixed point, are governed by the same scaling behavior and are described by identical power laws in the vicinity of the critical hypersurface of the fixed point.

3.3 Momentum-Shell Renormalization Group

3.3.1 Wilson's RG Scheme The RG method shall now be applied to the continuum model [Eq. (26)], called *Landau–Ginzburg–Wilson functional* in this context. A very intuitive approach was proposed by Wilson (Wilson and Kogut, 1974; Ma, 1976a). Basically, in momentum space the trace over the large-k degrees of freedom is performed, and thus recursion relations for the Ginzburg–Landau coefficients are obtained. Because the short-wavelength fluctuations are not expected to contribute essentially, the Brillouin zone may be approximated simply by a d-dimensional sphere with radius (cutoff)

Λ. The momentum-shell RG transformation then consists of the following steps:

1. Performing the trace over all Fourier components ϕ_k with $\Lambda/b < |k| < \lambda$ (Fig. 12) eliminates these short-wavelength modes.
2. By means of a scale transformation

$$k' = bk, \tag{81}$$

$$\phi' = b^{\zeta}\phi, \tag{82}$$

and therefore

$$\phi'_{k'} = b^{\zeta - d}\phi_k, \tag{83}$$

the ensuing effective Hamiltonian is brought to a form resembling the original model, by which then effective scale-dependent coupling parameters are defined.

By successive application of this RG transformation (which constitutes a semigroup, as there is no inverse element), the supposedly universal features of the long-wavelength regime are disclosed. The fixed points of the transformation will, as above, correspond to the distinct thermodynamic phases and the phase transitions between them. The eigenvalues of the linearized RG flow equations near the critical fixed point finally yield the critical exponents [see Eqs. (79), (77), and (78)].

Although a perturbational approach (with respect to u) is not at all justifiable in the critical region, it appears entirely sound in the region far from the critical point, where the fluctuations are negligible. The important observation now is that the RG flow connects these very different regimes, and thus results of the perturbation expansion obtained in the noncritical region can be transported to the vicinity of T_c, keeping track of the nonanalytic singularities on the way in a consistent, controlled, and reliable manner. Perturbative methods may also be employed in the elimination of the short-wavelength degrees of freedom (step **1**).

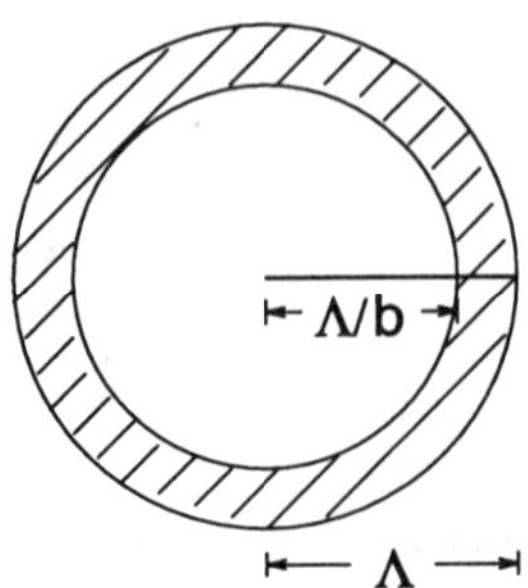

FIG. 12. Momentum-space RG: The partial trace is performed over Fourier components ϕ_k with momenta in the shell $\Lambda/b < |k| < \Lambda$.

3.3.2 Gaussian Model This concept shall now first be applied to the Gaussian model, where $u = 0$ (see Sec. 2.3.2),

$$\mathcal{F}_0[\phi_k] = \int_{|k|<\Lambda} \frac{r + k^2}{2} |\phi_k|^2 \tag{84}$$

[$\int_k = \int d^dk/(2\pi)^d$]. As Eq. (84) is diagonal in the Fourier modes, elimination of the large-k components simply produces constants; the form of the effective Hamiltonian is then unaltered, provided

$$\zeta = (d-2)/2, \quad \text{i.e.,}\ \eta = 0, \tag{85}$$

$$r' = b^2 r. \tag{86}$$

The fixed points of Eq. (86) are $r^* = \pm\infty$ corresponding to the high- and low-temperature phases, respectively, and the critical fixed point $r^* = 0$. The eigenvalue in the relevant temperature direction at this critical fixed point is obviously $y_\tau = 2$, and therefore the exponent $\nu = 1/2$ is recovered, as in mean-field theory.

3.3.3 Perturbation Theory and ϵ Expansion The nonlinear "interaction" term

$$\mathcal{F}_{\text{int}}[\phi_k] = \frac{u}{4} \int_{|k_i|<\Lambda} \phi_{k_1}\phi_{k_2}\phi_{k_3}\phi_{-k_1-k_2-k_3} \tag{87}$$

can now be treated perturbationally by expanding the exponential in Eq. (27) with respect to u. Separating the field variables into their parts in the outer and inner momentum shells, respectively, $\phi_k = \phi_{k_<} + \phi_{k_>}$, with $|k_<| < \Lambda/b$ and $\Lambda/b < |k_>| < \Lambda$, to first order in u one arrives at terms of the following (symbolically written) forms (henceforth, k_BT shall be set equal to 1):

1. $u \int \phi_<^4 e^{-\mathcal{F}_0}$ just needs to be reexponentiated, for these degrees of freedom are not eliminated;

2. any terms with odd numbers of $\phi_<$ or $\phi_>$ like $u\int \phi_<^3\phi_> e^{-\mathcal{F}_0}$ vanish;
3. $u\int \phi_>^4 e^{-\mathcal{F}_0}$ yields a constant, contributing to the free energy; and finally
4. $u\int \phi_<^2\phi_>^2 e^{-\mathcal{F}_0}$, for which the Gaussian integral over the $\phi_>$ may be performed by using Eq. (43) for the *propagator* $\langle \phi_{k_>}^\alpha \phi_{-k_>'}^\beta \rangle_0$, an average calculated with the statistical weight $e^{-\mathcal{F}_0}$.

Quite generally, *Wick's theorem* states that expressions of the form $\langle \Pi_i^m \phi_{k_{i>}} \rangle_0$ factorize into a sum of products of all possible pairs $\langle \phi_{k_>}\phi_{-k_>}\rangle_0$, if m is even, and vanish otherwise. Especially for treating higher orders in perturbation theory, *Feynman diagrams*, with lines symbolizing the propagators, and *interaction vertices* standing for the nonlinear coupling u, provide a very helpful representation for the large number of contributions to be summed in the perturbation series.

With these means, the two-point function $\langle \phi_{k_<}\phi_{-k_<}\rangle$ and the similarly defined four-point function can be evaluated. Using Eq. (80), the following recursions for r and u are then obtained to first nontrivial order ("*1-loop*", a notation stemming from the graphical representation):

$$r' = b^2[r + (n+2)A(r)u], \tag{88}$$

$$u' = b^{4-d}u[1 - (n+8)C(r)u], \tag{89}$$

where A and C denote the integrals

$$\begin{aligned} A(r) &= K_d \int_{\Lambda/b}^{\Lambda} (k^{d-1}/r + k^2)dk \\ &= K_d[\Lambda^{d-2}(1 - b^{2-d})/(d-2) \\ &\quad - r\Lambda^{d-4}(1 - b^{4-d})/(d-4)] + O(r^2), \end{aligned}$$

$$\begin{aligned} C(r) &= K_d \int_{\Lambda/b}^{\Lambda} [k^{d-1}/(r + k^2)^2]dk \\ &= K_d\Lambda^{d-4}(1 - b^{4-d})/(d-4) + O(r), \end{aligned}$$

with $K_d = 1/2^{d-1}\pi^{d/2}\Gamma(d/2)$, and the factors containing the number of components n of the order-parameter field originate from the combinatorics of counting the equivalent ways for "contracting" the fields $\phi_{k_>}$, i.e., performing the integrals over the large momenta. Note that again Eq. (85) is valid here.

By linearizing Eqs. (88) and (89) at the Gaussian fixed point $r^* = 0$, $u^* = 0$, one immediately finds the eigenvalues $y_r = 2$ and $y_u = 4 - d$. Thus for $d > d_c = 4$, the nonlinearity $\propto u$ turns out to be irrelevant, and the mean-field exponents are valid, as has already been anticipated in Sec. 2.3.2. However, for $d < 4$ the fluctuations become relevant, and any initial $u \neq 0$ grows under renormalization. In order to obtain the scaling behavior in this case, we thus have to search for a nontrivial, finite fixed point. This is most readily done by introducing the differential flow again, via $b^l = e^{\delta l}$, and $\delta \to 0$, whereby the number of RG steps effectively becomes a continuous variable, and studying the ensuing differential recursion relations

$$\begin{aligned} \frac{dr(l)}{dl} &= 2r(l) + (n+2)u(l)K_d\Lambda^{d-2} \\ &\quad - (n+2)r(l)u(l)K_d\Lambda^{d-4}, \end{aligned} \tag{90}$$

$$\frac{du(l)}{dl} = (4-d)u(l) - (n+8)u(l)^2K_d\Lambda^{d-4}. \tag{91}$$

Namely, a fixed point is defined by the condition that $dr/dl = 0 = du/dl$. In Fig. 13, the flow of $u(l)$ according to Eq. (91) is depicted; for any initial value $u_0 \neq 0$ one finds that asymptotically, i.e., for $l \to \infty$, the nontrivial fixed point

$$u_c^* K_d = \frac{\epsilon}{n+8}\Lambda^\epsilon, \quad \epsilon = 4 - d, \tag{92}$$

is approached, which therefore describes scale-invariant behavior and should govern the universal critical properties of the model. Here, too, the RG procedure generates new interactions—for instance, terms $\propto \phi^6$ and $\nabla^2\phi^4$, etc., which in successive steps couple back into the recursion relations for r and u. However, it turns out that up to order ϵ^3 these terms may be disregarded.

The original assumption that u be small, justifying a perturbation expansion, now means that the effective expansion parameter here is the deviation from the upper critical dimension ϵ. Inserting this in Eq. (90) and collecting the terms of $O(\epsilon)$, one finds

$$r_c^* = -\frac{n+2}{2}u_c^* K_d\Lambda^{d-2} = -\frac{(n+2)\epsilon}{2(n+8)}\Lambda^2; \tag{93}$$

the physical interpretation for this result is

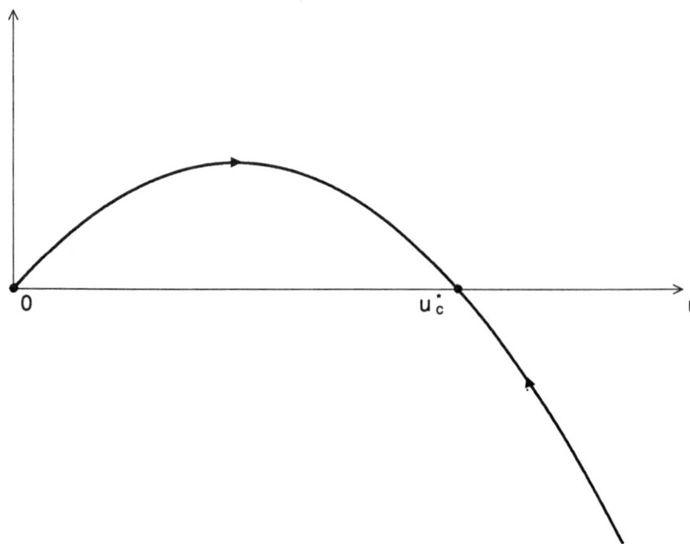

FIG. 13. Flow of the effective coupling $u(l)$, as determined from the right-hand side of Eq. (91). Both for initial values $u_0 > u_c^*$ and $0 < u_0 < u_c^*$, one has $u(l) \to u_c^*$ for $l \to \infty$.

that the fluctuations lead to a downward shift of the transition temperature. Finally, with $\tau = r - r_c^*$, the differential flow equation

$$\frac{d\tau(l)}{dl} = \tau(l)[2 - (n + 2)uK_d\Lambda^{d-4}] \tag{94}$$

yields the eigenvalue $y_\tau = 2 - (n + 2)\epsilon/(n + 8)$ near the critical fixed point [Eq. (92)]. To one-loop order, $O(\epsilon)$, one therefore finds the critical exponent

$$\nu = \tfrac{1}{2} + \frac{n + 2}{4(n + 8)}\epsilon + O(\epsilon^2). \tag{95}$$

Using $\eta = O(\epsilon^2)$ and the scaling relations (15), (16), and (19), one arrives at the following results [note the remarkable difference from the result of Eq. (39), obtained within the Gaussian approximation]:

$$\alpha = \frac{4 - n}{2(n + 8)}\epsilon + O(\epsilon^2), \tag{96}$$

$$\beta = \tfrac{1}{2} - \frac{3}{2(n + 8)}\epsilon + O(\epsilon^2), \tag{97}$$

$$\gamma = 1 + \frac{n + 2}{2(n + 8)}\epsilon + O(\epsilon^2), \tag{98}$$

$$\delta = 3 + \epsilon + O(\epsilon^2) \tag{99}$$

to first order in the expansion parameter $\epsilon = 4 - d$. The first nontrivial contribution to the exponent η turns out to appear at two-loop order,

$$\eta = \frac{n + 2}{2(n + 8)^2}\epsilon^2 + O(\epsilon^3). \tag{100}$$

The universality of these results becomes manifest by the fact that they depend solely on the space dimension d and the number of order-parameter components n, and not on the original "microscopic" Ginzburg–Landau parameters. (In passing, we note that for long-range power-law interactions, $\propto |x|^{-(d+\sigma)}$, the critical exponents acquire an additional dependence on the parameter σ.) At the upper critical dimension, $d_c = 4$, the solution of Eq. (91) becomes an inverse power law instead of an exponential, which leads to logarithmic corrections to the mean-field exponents.

Differential recursion relations of the form of Eqs. (90) and (91) also constitute the basis for the treatment of more subtle issues, like the calculation of the scaling functions, or even *crossover phenomena*, within the RG approach. For instance, an anisotropic perturbation in the n-component Heisenberg model favoring m directions leads to a crossover from the O(n) Heisenberg fixed point to an m-component fixed point (see Amit, 1984, Chap. 5-3). The instability of the former is described by a crossover exponent. For a small anisotropic perturbation, the RG flow trajectory still comes very close to this unstable fixed point, which implies that further away from T_c the system behaves like an n-component system, before it is finally dominated by the anisotropic critical behavior. The crossover from one RG fixed point to another may be illustrated (and measured) by introducing *effective exponents*, defined as logarithmic derivatives of appropriate physical quantities. Other important perturbations that have been treated in the framework of RG theory are so-called cubic terms reflecting the underlying crystal structure, and contributing a fourth power of the Cartesian components of ϕ, as well as the dipolar interaction that alters the harmonic part of the theory.

For systems characterized by certain spatial anisotropies, the upper critical dimension may be shifted from $d_c = 4$ to lower values as a result of restricted space for fluctuations. For example, for uniaxial dipolar ferromagnets $d_c = 3$, as well as for tricritical points, where $u = 0$ and a sixth-order term in ϕ is required. In the case of elastic structural phase transitions, for instance, one finds $d_c = \frac{5}{2}$ or $d_c = 3$, for a one- or two-dimensional soft sector, respectively (Schwabl, 1985).

3.3.4 More Sophisticated Field-Theoretical Methods Once one has to go beyond the first order in the perturbation expansion, Wilson's k-shell renormalization scheme, despite its very intuitively appealing features, is not the best choice for practical calculations. The technical reason is that the integrals in Fourier space involve nested momenta, and become rather tedious with a finite cutoff Λ. Thus, it appears more convenient to use a field-theoretical renormalization scheme, where $\Lambda \to \infty$ (Amit, 1984). Yet this leads to additional ultraviolet (UV) divergences in the integrals for $d \geq d_c$. Observing that at the critical dimension d_c both the ultraviolet and infrared (IR) singularities appear combined in logarithmic form [$\propto \ln(\Lambda^2/r)$], the idea is to treat the UV divergences with the methods originally developed in quantum field theory, and through this diversion arrive at the correct scaling equations for the opposite, IR limit.

Formally, this is done by taking advantage of the fact that the original unrenormalized theory does not depend on an arbitrarily chosen renormalization point; as a consequence one is led to the *Callan–Symanzik* or *RG equation*. This is a partial differential equation, which, when solved via the method of characteristics, leads to a set of ordinary differential equations that resemble the differential flow equations in Wilson's scheme. This procedure may be pursued provided the theory is renormalizable, which for the ϕ^4 theory is the case for $d \leq d_c$. However, this also means that irrelevant operators cannot, at least without considerable precautions, be included in the scheme. The field-theoretical treatments are therefore confined to the critical manifold in parameter space; yet, for the universal properties of the system under consideration, they yield fully equivalent results as compared with the momentum-shell approach.

ϵ expansions have been carried out up to seventh order; the resulting series is, however, only asymptotically convergent (quite obviously, the convergence radius of the perturbation series in u must be zero, as $u < 0$ corresponds to an unstable theory). Yet, by a combination of these fairly high-order results with the divergent asymptotic behavior and Borel-type resummation techniques, the critical exponents can be obtained with impressive accuracy; see Table 1 (Le Guillou and Zinn-Justin, 1985).

Near the lower critical dimension of the O(n) ($n \geq 2$) symmetric Heisenberg model, $d = 2$ (for the Ising case, $n = 1$, the lower critical dimension is $d = 1$; see Sec. 3.2.1), the ϵ expansion does not provide very accurate results, as might be expected. Near two dimensions, the investigation of the O(n)-symmetric nonlinear σ model by means of a $2 + \epsilon$ expansion, with a nontrivial fixed point of order ϵ corresponding to T_c, has proven to be more valuable (see Amit, 1984).

Besides the ϵ expansion, an expansion in terms of powers of $1/n$ is possible, where the limit $n \to \infty$ corresponds to the exactly solvable spherical model (Ma, 1976b). While this approach may help to clarify certain general issues, it is usually less accurate numerically, though, for the small values of n that are of practical interest.

Table 1. Best estimates for the static critical exponents γ, ν, β, and δ for the O(n)-symmetric ϕ^4 model in $d = 2$ and $d = 3$ dimensions, as obtained from large-order ϵ expansion in combination with Borel resummation techniques (from Le Guillou and Zinn-Justin, 1985). For comparison, the exact Onsager results for the 2D Ising model are also listed. The limit $n = 0$ describes the statistical mechanics of polymers.

		γ	ν	β	η
$d = 2$	$n = 0$	1.39 ± 0.04	0.76 ± 0.03	0.065 ± 0.015	0.21 ± 0.05
	$n = 1$	1.73 ± 0.06	0.99 ± 0.04	0.120 ± 0.015	0.26 ± 0.05
2D Ising	(exact)	1.75	1	0.125	0.25
$d = 3$	$n = 0$	1.160 ± 0.004	0.5885 ± 0.0025	0.3025 ± 0.0025	0.031 ± 0.003
	$n = 1$	1.239 ± 0.004	0.6305 ± 0.0025	0.3265 ± 0.0025	0.037 ± 0.003
	$n = 2$	1.315 ± 0.007	0.671 ± 0.005	0.3485 ± 0.0035	0.040 ± 0.003
	$n = 3$	1.390 ± 0.010	0.710 ± 0.007	0.368 ± 0.004	0.040 ± 0.003

4. SCALING LAWS IN STATISTICAL PHYSICS

4.1 Phase Transitions and Properties of Phases

Phenomenological Ginzburg–Landau expansions, scaling approaches, and the language and methods of RG methods have become fundamental tools in the study of phase transitions and of other universal properties of condensed-matter phases, in a large variety of systems. A list (incomplete) includes structural phase transitions of various kinds, namely distortive, ferroelectric, order-disorder, and elastic transitions (see PHASE TRANSITIONS, STRUCTURAL; COLLECTIVE PHENOMENA IN SOLIDS), liquid crystals (see LIQUID CRYSTALS, STRUCTURE OF), magnetic systems (see MAGNETIC ORDERING IN SOLIDS), magnetic impurities in a lattice (*Kondo problems*; see Wilson, 1975), superfluidity (see SUPERFLUIDITY: LIQUID HELIUM SYSTEMS), and superconductivity (see SUPERCONDUCTIVITY, LOW-TEMPERATURE); here specifically the numerous phases of type-II superconductors subject to an external magnetic field have aroused considerable interest lately (see SUPERCONDUCTIVITY, HIGH-TEMPERATURE, and Blatter *et al.*, 1994). Complex scaling and crossover behavior may ensue when either anisotropies or additional long-range interactions are included.

Finite-size scaling studies how the asymptotic power laws of the thermodynamic limit are approached as the system size grows, and is thus an important tool for numerical studies of continuous phase transitions (Privman, 1990). Furthermore, the fact that all real systems are finite leads to additional interesting phenomena. At surfaces, new critical exponents may come into play (Diehl, 1986). Also, interfacial structures may change drastically as external parameters are varied; an important example is the wetting transition of liquid drops on a substrate (Dietrich, 1988).

Scaling and RG concepts have also been applied to the theory of symmetry breaking in the fundamental field theories of matter (see UNIFIED FIELD THEORIES), which also have important implications for cosmology (see GALAXIES AND COSMOLOGY). (We note that in elementary-particle physics a Compton wavelength $\lambda_c = \hbar c/m$ may be assigned to a particle of rest mass m. For large momenta $\hbar k \gg mc$, the theory becomes scale invariant, and the propagator behaves as $k^{-2+\eta}$. The correlation length ξ thus corresponds to the Compton wavelength or inverse mass.)

4.2 Disordered Systems

In many physical systems, disorder plays an important role. The basic question therefore is, under what circumstances can quenched randomness affect the nature of a phase transition and its properties. Let the disorder induce a relevant perturbation $g(x)$ in the effective Hamiltonian, e.g., changing the harmonic part of the Landau–Ginzburg–Wilson functional (*random* T_c), the scaling dimension of which is $[g] \propto L^{\Delta-d}$, with $\Delta \leq d$. The *Harris criterion* then states that the character of the pure transition may be considerably affected by the disorder, provided $\Delta < d/2$ (Itzykson and Drouffe, 1989). If defects of this kind are correlated only over short distances, this criterion says that the critical behavior is changed if the specific-heat exponent of the pure system is positive ($\alpha > 0$).

On the other hand, in the case of *random fields*, both the upper and lower critical dimensions can be changed as compared with the pure case. The *Imry–Ma argument* (see Nattermann and Villain, 1988) compares the interfacial energy $\propto L^{\theta}$ [Eq. (13)] with the gain in bulk free energy $\propto L^{d/2}$ by splitting into domains of size L (the mean square order-parameter fluctuations due to the random field add up to a contribution $\propto L^{d}$ for each domain). Thus, for $\theta \le d/2$, the system will always break into domains and no long-range order can emerge. For Ising systems ($\theta = d - 1$), this will be true for $d \le 2$, while in the case of continuous symmetry ($\theta = d - 2$) the lower critical dimension becomes $d = 4$.

An example for such disorder-dominated systems is *spin glasses*, e.g., an Ising spin glass with bond disorder, where the couplings J_{ij} in Eq. (21) are drawn from some stochastic distribution. Similarly, disorder can be introduced in Heisenberg models [Eq. (25)]. Because spin glasses are model systems for phase transitions in highly disordered materials, a considerable amount of effort has been devoted to the examination of such spin glasses (Binder and Young, 1986). In addition to the thermal average, all physical quantities have to be averaged over all possible configurations of the quenched disorder. A frequently used method in analytical calculations is the *replica trick*, which replaces the computation of the mean free energy by the average of the nth power of the partition function, and then entails taking both the derivative with respect to n and the limit $n \to 0$:

$$\bar{F} = -\lim_{n\to 0} \frac{\partial}{\partial n} \overline{Z^{n}}. \qquad (101)$$

Extensions of these studies have led to the investigation of manifolds in random media, a direct application of which may be found in biological membranes, or flux lines in high-T_c superconductors, and the design of neural networks (see NEURAL NETWORKS).

Disorder may also play a prominent role in the quantum regime ($T \to 0$). While near a phase transition with finite T_c quantum fluctuations are of minor importance (although quantum effects may provide the fundamental mechanism for the transition itself, as in magnets, superfluids, and superconductors), at $T = 0$ the interplay between disorder and quantum mechanics can lead to very interesting novel phases with unusual phase transitions. Examples of such quantum critical phenomena (see Continentino, 1994) are electron localization (see ELECTRON STATES, LOCALIZED), metal-insulator transitions (see METAL-INSULATOR TRANSITIONS; Belitz and Kirkpatrick, 1994), helium-4 on disordered substrates (disordered bosons), and quantum antiferromagnets, which are of relevance in the microscopic theory of high-temperature superconducting materials and as one phase of solid helium-3. For a Fermi system, the interesting singularities and the universal behavior are expected in the vicinity of the Fermi surface, instead of the point $k = 0$ in Fourier space, and therefore the RG approach has to be modified accordingly (Shankar, 1994).

The mathematical theory of percolation also has immediate applications for the structure of certain disordered systems, and their transport properties (see FRACTAL GEOMETRY; Bunde and Havlin, 1991). The percolation threshold may be understood as a critical point, and the scaling behavior in its vicinity can be obtained by mapping onto the so-called Potts model (Wu, 1982), and with the help of certain continuum models of the same universality class.

4.3 Polymers, Membranes

An important ingredient to the physics of macromolecules like polymers is the interplay and competition of energetic and entropic optimization. De Gennes has demonstrated that the partition function of a polymer, or the statistics of a self-avoiding random walk, can be mapped to the $n \to 0$ limit of the n-component Heisenberg model (see Table 1). As a consequence, scaling and RG approaches to the static and dynamic properties of polymers have been successfully employed (see POLYMERS, STRUCTURE OF; De Gennes, 1979; Freed, 1987). The inverse separation from the critical point $|\tau|^{-1}$ formally corresponds to the degree of polymerization (chain length) N of a long flexible polymer, and, e.g., in a dilute solution in a good solvent the average end-to-end distance of the polymer scales as $R_e \propto N^{\nu}$, where $\nu \approx 0.588$ (in three dimensions) is the $n = 0$ result for the correlation length exponent of the O(n) model. Naturally, similar attempts have been used for biological systems like muscular fibers and membranes (see STATISTICAL PHYSICS OF MEMBRANES AND LAMELLAR SYSTEMS).

4.4 Critical Dynamics and Nonequilibrium Phenomena

In Sec. 1.4, the extension of the above scaling concepts to dynamical phenomena was mentioned. The order-parameter dynamics displays critical slowing down near a second-order phase transition, described by the critical exponent z [Eq. (11)]. One very powerful and common mathematical description of critical dynamics rests on *Langevin-type equations of motion* (see STOCHASTIC PROCESSES) for the order-parameter fluctuations and the conserved quantities, which constitute the *slow modes* of the system (Hohenberg and Halperin, 1977; Enz, 1979). For, as a consequence of the continuity equation related to a conservation law, the corresponding density shows diffusive behavior (inverse relaxation time $t_0^{-1} \propto k^2$). In a phase with broken continuous symmetry, in addition the fluctuations transverse to the order parameter have to be included, for their static susceptibility is also singular for $k \to 0$ [Eq. (47)], and thus the corresponding mode has frequency $\propto \chi^{-1/2}$, and damping coefficient $\propto \chi^{-1}$. Collecting all these slow modes in a multicomponent field $\mathbf{\Phi}(x,t)$, the equations of motion assume the general form

$$\frac{\partial \Phi^\alpha(x,t)}{\partial t} = K[\mathbf{\Phi}] - \Gamma^\alpha \frac{\delta \mathcal{F}[\mathbf{\Phi}]}{\delta \Phi^\alpha(x,t)} + \zeta^\alpha(x,t). \tag{102}$$

Here, $K[\mathbf{\Phi}]$ comprises the reversible forces, as obtained from the microscopic equations of motion. The second term provides a damping term, which after some small perturbation drives the system back into the equilibrium state, where $\delta\mathcal{F}/\delta\mathbf{\Phi} = 0$; the damping coefficients Γ^α are constants for the order-parameter fluctuations, if the latter is not conserved, and take the form $\lambda\nabla^2$ for each conserved quantity. Finally, all "fast" modes have been subsumed in the stochastic forces $\zeta^\alpha(x,t)$, whose mean is taken to be zero, $\langle\zeta^\alpha(x,t)\rangle = 0$, and whose second moments are, for simplicity, assumed to be drawn from a Gaussian distribution with a variance corresponding to "white noise":

$$\langle\zeta^\alpha(x,t)\zeta^\beta(x',t')\rangle = 2k_BT\Gamma^\alpha\delta^{\alpha\beta}\delta(x-x')\delta(t-t'). \tag{103}$$

The *fluctuation-dissipation theorem (Einstein relation)* [Eq. (103)] assures that for $t \to \infty$ the probability distribution is given by the equilibrium expression $\exp(-\mathcal{F}/k_BT)$.

The set of Eqs. (102) with (103) can then be very elegantly mapped onto a dynamic functional, and the ensuing field theory can be treated with precisely those perturbation expansions and RG methods explained above for the statics. In some cases, the underlying symmetries and/or conservation laws lead to so-called *Ward identities*, which yield a connection between the dynamic exponent z and already known static indices, prominently η. For example, for a model merely consisting of a conserved order-parameter field (model B according to the classification of Hohenberg and Halperin, 1977), the result is $z = 4 - \eta$; for the isotropic ferromagnet (model J) one finds $z = (d + 2 - \eta)/2$, and for planar ferromagnets as well as isotropic antiferromagnets (models E and G, respectively), $z = d/2$. In Table 2, a list of dynamical systems, with the corresponding values of z, and the physical quantities for which the critical slowing down can be observed in experiment, is presented. Note that although the range of the dynamical critical exponent is limited to $\frac{3}{2} \le z \le 4$ in three dimensions, the ensuing behavior of the transport coefficients can yet be as different as the diverging magnetization diffusion $D_M \propto |\tau|^{-1/3}$ for an isotropic antiferromagnet, and the vanishing order-parameter relaxation rate $\Gamma \propto |\tau|$ in the same system, or the vanishing magnetization diffusion constant $D_M \propto |\tau|^{1/3}$ for an isotropic Heisenberg ferromagnet.

An alternative route to finding these exponents, and probably superior for the issue of precisely calculating the scaling functions, is the *mode-coupling* approach, which at least in some cases may be represented as a self-consistent one-loop theory for the generalized (i.e., wave-number– and frequency-dependent) transport coefficients (Frey and Schwabl, 1995).

Another physical situation, where equations of motion of the type in Eq. (102) and a RG approach have been used, is the problem of quenching a system from above T_c into the low-temperature phase, and observing

Table 2. Dynamic critical exponents z in d dimensions, for various dynamical systems, along with a list of the physical quantities, and their temperature dependence in the vicinity of T_c, for which the ensuing critical slowing down may be observed experimentally (classification according to Hohenberg and Halperin, 1977). The approximate numerical results are for the three-dimensional case, using $\nu \approx \frac{2}{3}$, $\eta \approx 0$, and $\alpha \approx 0$.

System	Model	Dynamic exponent	Physical quantity
Anisotropic magnet	A	$z = 2 + c\eta,\ c = O(1)$	Relaxation rate $\Gamma \propto \lvert\tau\rvert^{z\nu} \approx \lvert\tau\rvert^{4/3}$
Uniaxial ferromagnet	B	$z = 4 - \eta$	Magnetization diffusion $D_M \propto \lvert\tau\rvert^{(z-2)\nu} \approx \lvert\tau\rvert^{4/3}$
Structural phase transitions	C	$z = 2 + \alpha/\nu$	Relaxation rate $\Gamma \propto \lvert\tau\rvert^{z\nu} \approx \lvert\tau\rvert^{4/3}$
Planar ferromagnet	E	$z = d/2$	Magnetization diffusion $D_M \propto \lvert\tau\rvert^{z\nu} \approx \lvert\tau\rvert$
Superfluid helium-4	F	$z = \frac{1}{2}(d + \tilde{\alpha}),\ \tilde{\alpha} = \max(\alpha,0)$	Thermal conductivity $\lambda_T \propto \lvert\tau\rvert^{(z-2)\nu} \approx \lvert\tau\rvert^{-1/3}$ (+ logarithmic corrections)
Isotropic antiferromagnet	G	$z = d/2$	Order-parameter relaxation $\Gamma_N \propto \lvert\tau\rvert^{z\nu} \approx \lvert\tau\rvert$ Magnetization diffusion $D_M \propto \lvert\tau\rvert^{(z-2)\nu} \approx \lvert\tau\rvert^{-1/3}$
Isotropic Heisenberg ferromagnet	J	$z = \frac{1}{2}(d + 2 - \eta)$	Magnetization diffusion $D_M \propto \lvert\tau\rvert^{(z-2)\nu} \approx \lvert\tau\rvert^{1/3}$

how the ordered domains grow with time (Bray, 1993).

Upon relaxing the condition of Eq. (103), the same methods may also be applied to phase transitions in nonequilibrium systems. A prototypical example might be the Burgers–Kardar–Parisi–Zhang equation, which is supposed to describe the kinetic roughening of growing surfaces (Family and Vicsek, 1991), but is also intimately connected to the problem of directed lines in random environment, and to the hydrodynamics of randomly stirred fluids. Related issues are the depinning transitions of lines and interfaces, and the dynamics of charge-density waves. The behavior of interfaces in disordered systems in general poses a number of most interesting issues and requires certain new conceptual approaches (see Forgacs *et al.*, 1991).

The properties of systems very far from equilibrium, well within the scope of scaling approaches and RG methods, turn out to be remarkably rich. On the one hand, there is the vast field of nonequilibrium pattern formation (Cross and Hohenberg, 1994). On the other hand, self-similar and self-affine structures emerge in many dynamical processes (see FRACTAL GEOMETRY), which are hence good candidates for the application of the above methods. The ubiquity of fractals and $1/f$ noise has led to the paradigm of *self-organized criticality (SOC)* (reviewed in Bak and Creutz, 1994). SOC refers to the tendency of large dissipative systems to drive themselves into a self-similar critical state without the special adjustment of parameters. SOC has been suggested for avalanches in sand piles, earthquake statistics, evolution, spreading of diseases, forest fires, and many others (Bak and Creutz, 1994). In the so-called forest-fire model, one finds SOC if a certain double separation of time scales holds (which represents still a large region in parameter space), while in other regions of the parameter space spiral structures emerge (Drossel and Schwabl, 1994).

At any rate, scaling ideas and the language of the renormalization group, including such concepts as effective Hamiltonians, relevant and irrelevant perturbations, fixed points, diverging correlation lengths, self-similarity and power laws, and universal long-wavelength and low-frequency behavior of collective modes, as well its powerful mathematical tools, have provided considerable new insight into the physics of many important research areas, and it seems likely that more interesting applications will be discovered in the future.

GLOSSARY

Asymptotic Regime: Region near the critical point where the relevant perturbations dominate, leading to pure power laws.

Corrections to Scaling: Contributions of irrelevant operators to thermodynamic quantities.

Correlation Length: Length scale for the spatial decay of correlations, usually a function of the external control parameters (temperature, magnetic field, etc.).

Critical Exponents: Exponents describing the power laws at continuous phase transitions.

Critical Slowing Down: Decrease of the order-parameter decay or oscillation rate ω_c, characterized by the dynamic exponent z; cf. Eq. (11).

Critical Surface: Hypersurface in parameter space with the property that all RG trajectories starting on it lead into the critical fixed point.

Crossover: Change in scaling behavior from power laws characterized by a certain unstable fixed point to different power laws governed by the asymptotically stable fixed point.

Effective Theory: Mathematical model describing the long-wavelength and low-frequency behavior of a physical system, at a certain energy scale.

Fixed Point: Point in the space of coupling coefficients that remains invariant under the RG transformation.

Lower Critical Dimension: Borderline dimension below which a certain model or physical system does not have a phase transition leading to long-range order.

Order Parameter: Certain macroscopic quantity that may be used to characterize different phases.

Relevant and Irrelevant Perturbations: Eigenoperators of the linearized RG transformation that grow and decrease, respectively, under successive RG steps.

Renormalization-Group (RG) Transformation: Operation by which a Hamiltonian is transformed into a new one, with a reduced number of degrees of freedom, accompanied by a subsequent scale transformation.

Scaling Law: Mathematical expression of the fact that at a critical point the free energy and correlation functions are scale invariant, i.e., are generalized homogeneous functions.

Scaling Relations: Equations that relate certain critical indices, thus reducing the number of independent exponents.

Universality: The fact that critical properties depend only on global properties such as the dimension and the symmetry and number of components of the order parameter.

Upper Critical Dimension: Borderline dimension above which the critical exponents of a certain model or physical system are correctly given by their mean-field values. Below the upper critical dimension, fluctuations are relevant and affect the scaling behavior.

Works Cited

Als-Nielsen, J., Birgeneau, R. J. (1977), *Am. J. Phys.* **45**, 554–560.

Amit, D. J. (1984), *Field Theory, the Renormalization Group, and Critical Phenomena*, 2nd ed., Singapore: World Scientific.

Bak, P., Creutz, M. (1994), in: Bunde, A., Havlin, S. (Eds.), *Fractals in Science*, Berlin: Springer-Verlag, pp. 26–47.

Belitz, D., Kirkpatrick, T. R. (1994), *Rev. Mod. Phys.* **66**, 261–380.

Binder, K., Young, A. P. (1986), *Rev. Mod. Phys.* **58**, 801–976.

Blatter, G., Feigel'man, M. V., Geshkenbein, V. B., Larkin, A. I., Vinokur, V. M. (1994), *Rev. Mod. Phys.* **66**, 1125–1388.

Bray, A. J. (1993), in: Riste, T., Sherrington, D. (Eds.), *Phase Transitions and Relaxation in Systems with Competing Energy Scales*, Dordrecht: Kluwer Academic Publishers, pp. 405–436.

Buckingham, M. J., Fairbank, W. M. (1961), in: Gorter, C. J. (Ed.), *Progress in Low Temperature Physics*, Vol. III, Amsterdam: North Holland Publishing Company, pp. 80–112.

Bunde, A., Havlin, S. (1991), in: Bunde, A., Havlin, S. (Eds.), *Fractals and Disordered Systems*, Berlin: Springer-Verlag, pp. 51–149.

Chaikin, P. M., Lubensky, T. C. (1995), *Principles of Condensed Matter Physics*, Cambridge, U.K.: Cambridge Univ. Press.

Continentino, M. A. (1994), *Phys. Rep.* **239**, 179–213.

Cross, M. C., Hohenberg, P. C. (1994), *Rev. Mod. Phys.* **65**, 851–1112.

de Gennes, P. G. (1979), *Scaling Concepts in Polymer Physics*, Ithaca: Cornell Univ. Press.

Diehl, H. W. (1986), in: Domb, C., Lebowitz, J. L. (Eds.), *Phase Transitions and Critical Phenomena*, Vol. 10, London: Academic Press, pp. 75–267.

Dietrich, S. (1988), in: Domb, C., Lebowitz, J. L. (Eds.), *Phase Transitions and Critical Phenomena*, Vol. 12, London: Academic Press, pp. 1–218.

Drossel, B., Schwabl, F. (1994), *Physica A* **204**, 212–229.

Enz, C. P. (Ed.) (1979), *Dynamic Critical Phenomena and Related Topics*, Lecture Notes in Physics Vol. 104, Berlin: Springer-Verlag.

Family, F., Vicsek, T. (Eds.) (1991), *Dynamics of Fractal Surfaces*, Singapore: World Scientific.

Fisher, D. S. (1993), in: Riste, T., Sherrington, D. (Eds.), *Phase Transitions and Relaxation in Systems with Competing Energy Scales*, Dordrecht: Kluwer Academic Publishers, pp. 1–25.

Fisher, M. E. (1967), *Rep. Prog. Phys.* **30**, 615–730.

Fisher, M. E. (1974), *Rev. Mod. Phys.* **46**, 597–616.

Forgacs, G., Lipowsky, R., Nieuwenhuizen, T. M. (1991), in: Domb, C., Lebowitz, J. L. (Eds.), *Phase Transitions and Critical Phenomena*, Vol. 14, London: Academic Press, pp. 135–363.

Freed, K. F. (1987), *Renormalization Group Theory of Macromolecules*, New York: Wiley.

Frey, E., Schwabl, F. (1995), *Adv. Phys.* (in press).

Hohenberg, P. C., Halperin, B. I. (1977), *Rev. Mod. Phys.* **49**, 435–479.

Itzykson, C., Drouffe, J.-M. (1989), *Statistical Field Theory*, Vols. 1, 2, Cambridge, U.K.: Cambridge Univ. Press.

Le Guillou, J. C., Zinn-Justin, J. (1985), *J. Phys. Lett.* **46**, L-137–L-141, and references therein.

Ma, S. (1976a), *Modern Theory of Critical Phenomena*, Reading, MA: Benjamin.

Ma, S. (1976b), in: Domb, C., Green, M. S. (Eds.), *Phase Transitions and Critical Phenomena*, Vol. 6, London: Academic Press, pp. 249–292.

Nattermann, T., Villain, J. (1988), *Phase Transitions* **11**, 5–51.

Nelson, D. R. (1983), in: Domb, C., Lebowitz, J. L. (Eds.), *Phase Transitions and Critical Phenomena*, Vol. 7, London: Academic Press, pp. 1–99.

Niemeijer, T., van Leeuwen, J. M. J. (1976), in: Domb, C., Green, M. S. (Eds.), *Phase Transitions and Critical Phenomena*, Vol. 6, London: Academic Press, pp. 425–505.

Privman, V. (Ed.) (1990), *Finite Size Scaling and Numerical Simulation of Statistical Systems*, Singapore: World Scientific.

Schwabl, F. (1985), *J. Stat. Phys.* **39**, 719–737.

Shankar, R. (1994), *Rev. Mod. Phys.* **66**, 129–192.

Swendsen, R. H. (1984), *J. Stat. Phys.* **34**, 963–973.

Wilson, K. G., Kogut, J. (1974), *Phys. Rep. C* **12**, 76–199.

Wilson, K. G. (1975), *Rev. Mod. Phys.* **47**, 773–840.

Wu, F. Y. (1982), *Rev. Mod. Phys.* **54**, 235–268.

Further Reading

General

Binney, J. J., Dowrick, N. J., Fisher, A. J., Newman, M. E. J. (1993), *The Theory of Critical Phenomena*, 2nd ed., New York: Oxford Univ. Press.

Brey, J., Jones, R. B. (Eds.) (1976), *Critical Phenomena*, Lecture Notes in Physics Vol. 54, Berlin: Springer-Verlag.

Domb, C., Green, M. S. (Eds.) (1972–1976), *Phase Transitions and Critical Phenomena*, Vols. 1–6, London: Academic Press.

Domb, C., Lebowitz, J. L. (Eds.) (1983–1992), *Phase Transitions and Critical Phenomena*, Vols. 7–15, London: Academic Press.

Goldenfeld, N. (1992), *Lectures on Phase Transitions and the Renormalization Group*, Reading, MA: Addison-Wesley.

Patashinskii, A. Z., Pokrovskii, V. L. (1979), *Fluctuation Theory of Phase Transitions*, New York: Pergamon Press.

Stanley, H. E. (1971), *Introduction to Phase Transitions and Critical Phenomena*, Oxford, U.K.: Clarendon Press.

Yeomans, J. M. (1991), *Statistical Mechanics of Phase Transitions*, Oxford, U.K.: Clarendon Press.

Field Theory

Parisi, G. (1988), *Statistical Field Theory*, Redwood, CA: Addison-Wesley.

Zinn-Justin, J. (1989), *Quantum Field Theory and Critical Phenomena*, Oxford, U.K.: Clarendon Press.

PHASE TRANSITIONS, STRUCTURAL

E. K. H. SALJE, *Department of Earth Sciences, University of Cambridge, Cambridge, United Kingdom*

	Introduction	373
1.	**Entropic Effects for Temperature-Driven SPTs**	374
1.1	Displacive Phase Transitions	375
1.2	"Almost" Displacive Transitions	376
1.3	The Order-Disorder Limit	376
2.	**The Order Parameter**	376
3.	**Ferroelasticity and Ferroelectricity**	379
4.	**Coupling between Order Parameters**	381
5.	**Influence of Defects**	382
6.	**Twins and Antiphase Boundaries as Fingerprints of SPTs: Surface Relaxations**	383
7.	**Kinetics of Site Ordering**	384
8.	**Conclusion**	386
	Glossary	387
	Works Cited	387

INTRODUCTION

When crystalline materials are subjected to changes of temperature T, pressure P, magnetic field H, etc., they will change their atomic arrangements. If such changes involve topological modifications of the crystal structure and/or changes of symmetry, the various structures represent phases with structural phase transitions (SPTs) between them.

A typical example for a phase transition between a hexagonal structure (at high temperatures) and a trigonal structure (at low temperatures) is shown for quartz (SiO_2) (Fig. 1). Small rotations of SiO_4 tetrahedra destroy the hexagonal symmetry and lead to a denser arrangement of tetrahedra at low temperatures. The amplitudes of the various polyhedra tilts increase with decreasing temperature, whereas no such tilts exist at temperatures above the transition point. This transition mechanism, with gradual and mostly very subtle atomic shifts, is termed "displacive." (A more precise definition of the term "displacive" is based on the appearance of soft phonons as fundamental excitation of such systems; a simple static picture is given later.)

Symmetry breaking can occur via many other mechanisms. A common mechanism involves the ordering of different sorts of atoms on various positions in a crystal structure. This process is often seen in alloys, which may form a random distribution of metal atoms such as Cu and Au at high temperatures but an ordered structure at low temperatures. Another example, more similar to that of quartz, is the feldspar structure. Feldspars are most common materials: ca. 60% of the outer crust of the planet Earth is made from feldspars and related materials. In Fig. 2, the framework of feldspar at high and low temperatures is depicted. In each case, the structure is made up from oxygen tetrahedra with a metal position inside. This metal atom can be Al or Si. At high temperatures, Al and Si are disordered; at low temperatures, they are ordered whereby the order starts at a well-defined transition temperature. The degree of order increases when cooled further. Note that the AlO_4 and SiO_4 tetrahedra have different sizes so that the ordering of Al and Si is accompanied by local strain in the structure, which accommodates for the appropriate atomic ordering. SPTs in which the essential mechanism is related to changes of atomic order are usually called "order-disorder" (or simply od).

Many other transition mechanisms exist, and it is not just temperature that acts as the relevant variable to generate SPTs but also pressure, variations of the chemical composition, etc. Furthermore, all these mecha-

3-527-28135-5/95/$5.00 + .50

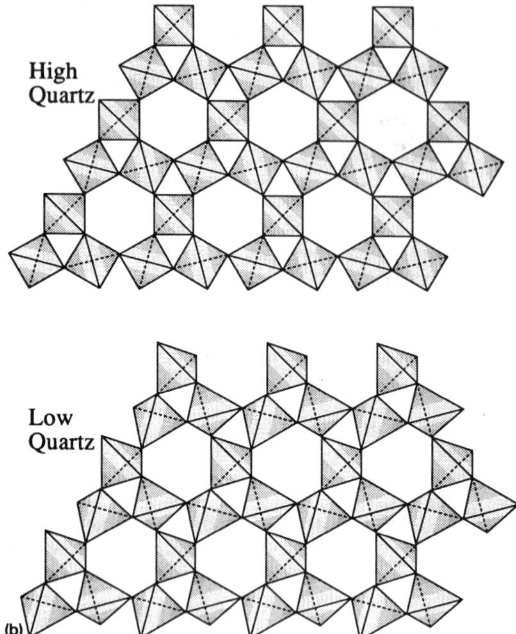

FIG. 1. Crystal structures of quartz at temperatures above and below the transition temperature. Rather rigid SiO_4 tetrahedra build up a framework that reduces its specific volume at low temperature via tilts of the tetrahedra. Courtesy A. Putnis, Cambridge.

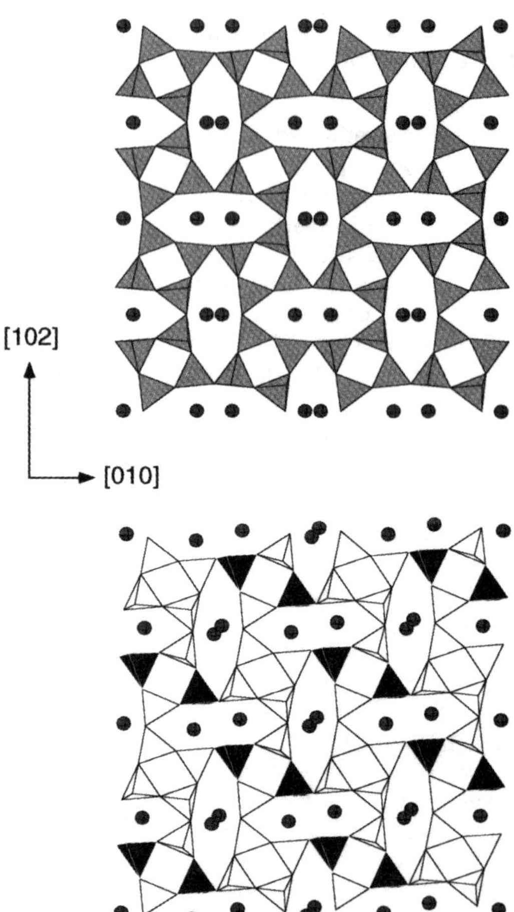

FIG. 2. Structural changes in alkali feldspars (Na,K)-$AlSi_3O_8$: the high-symmetry form is monoclinic with random distribution of Al and Si on tetrahedral sites. The alkali atom reflects the overall monoclinic symmetry (top). The triclinic, low-symmetry form shows Al ordered on the black tetrahedra whereas Si is inside the unshaded tetrahedra. The monoclinic symmetry is broken during the STP. The locations of alkali atoms are indicated by the circles. Courtesy D. Palmer, Cambridge.

nisms can interplay, so that materials often show several SPTs, each of which influences all others. It is an essential first step for research on SPTs to disentangle the various contributions (Landau and Lifshitz, 1980; Bruce and Cowley, 1981; Khatchaturyan, 1983; Levanyuk and Sigov, 1988; Salje, 1993b).

1. ENTROPIC EFFECTS FOR TEMPERATURE-DRIVEN SPTs

A SPT occurs when the enthalpic energy H is compensated by entropy S so that the Gibbs free energy $G = H - TS$ changes sign. A simple atomistic picture illustrates this mechanism (Salje *et al.*, 1991). Let us first consider local potentials that are anharmonic so that the midpoint of the potential is not necessarily at an energy minimum (Fig. 3). The second energy contribution is the coupling between the relevant atomic positions. In SPT, this communication is mainly achieved by elastic deformation of the structure (structural stress) or via electric dipolar forces. In both cases, the effective interaction decays as $1/R^3$ with increasing distance R. This long-ranging effective interaction together with its strong directional dependencies (in contrast to nearest-neighbor Ising models) allows us to summarize the effect of all interactions on each atom as a field ("mean field theory," MFT), and we may depict the interatomic coupling as a simple spring (Fig. 3). The effective Hamiltonian of this model is then the sum over the kinetic energy, the local potentials, and the coupling energy:

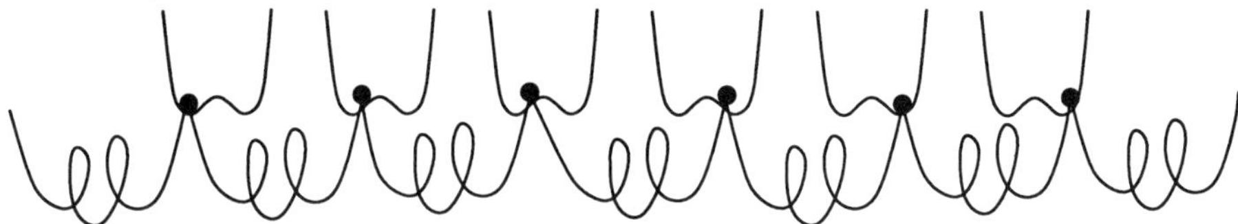

FIG. 3. Simple model representing a physical system with nonharmonic local potentials and harmonic intersite interactions.

$$H = \sum_l \left(\frac{1}{2M} P_l^2 + \frac{1}{2} M\Omega_{ol}^2 Q_l^2 + \frac{1}{2} uQ_l^4 + \cdots \right) - \frac{1}{2} \sum_{l \neq l'} v_{ll'} Q_l Q_{l'} . \quad (1)$$

Here M is the effective atomic mass, P is the momentum, Ω_{ol} is the local frequency for small oscillations around the local energy minimum at site l if it were purely parabolic in the absence of interatomic interactions, u gives the strength of the anharmonicity, and $v_{ll'}$ are the coupling constants that form the amplitudes Q_l on site l and $Q_{l'}$ on site l'. We can now define an order parameter Q as the time- and space-averaged value of the position coordinates Q_l on all sites l. This average value is denoted by angular brackets $Q = \langle Q_l \rangle$. If the average position is situated in the middle of the potential, we find $Q = 0$. If the average position is shifted to the right or the left, we find $Q > 0$ or $Q < 0$. One sees already in this simple model that $Q = 0$ represents a high-symmetry phase, whereas $Q > 0$ and $Q < 0$ represent a low-symmetry phase. If the low-symmetry phase contains macroscopic areas with $Q > 0$ and others with $Q < 0$, these areas are called "domains" with "domain walls" between them. The order parameter is a one-dimensional parameter in this simple model; higher dimensional order parameters are needed in 3D structures with high crystallographic symmetries. If the site ordering is not uniform, the order parameter is defined as the relevant Fourier component of $\langle Q_l \rangle$:

$$\langle Q_l \rangle = Q = Q_0 \exp(-i\mathbf{q} \cdot \mathbf{r}).$$

Thermal fluctuations occur around the order parameter, with

$$\sigma = \langle (Q_l - Q)^2 \rangle,$$

which depend on temperature via the dissipation-fluctuation theorem as

$$\sigma = \frac{\hbar}{2M\Omega} \coth \frac{\hbar\Omega}{2k_B T}. \quad (2)$$

The Gibbs free energy G can then be calculated in MFT at the wave vector q at which long-range order with the order parameter (OP) $Q = \langle Q_l \rangle$ occurs. The SPT occurs when G changes sign with $Q = 0$ in the high-symmetry phase and a gradual increase of $Q \neq 0$ in the low-symmetry phase.

Approximate solutions exist for three regimes, which are characterized by a parameter $\Delta = v/M\Omega_0^2 - 1$ that vanishes when the intersite coupling compensates the parabolic part of the on-site potential. We furthermore define two temperatures. Firstly, T_c defines the temperature below which the high-symmetry phase becomes unstable. Secondly, thermal fluctuations saturate at temperatures below $\theta_s = \hbar\Omega_0/2k_B$, whereas at $T \gg \theta_s$ fluctuations increase linearly with temperature [Eq. (2)].

The three solutions mostly used for the analysis of experimental observations are the following.

1.1 Displacive Phase Transitions

These have $0 \leqslant \Delta \ll 1$, and

$$G = \tfrac{1}{2} A\theta_s \left[\coth \frac{\theta_s}{T} - \coth \frac{\theta_s}{T_c} \right] Q^2 + \tfrac{1}{2} BQ^4 + \tfrac{1}{6} CQ^6, \quad (3)$$

which becomes for $\theta_s \ll T$ the standard Landau potential

$$G = \tfrac{1}{2} A(T - T_c)Q^2 + \tfrac{1}{4} BQ^4 + \tfrac{1}{6} CQ^4. \quad (4)$$

Here Q follows from the minimum condition

$\partial G/\partial Q = 0$ (see below). Note that the standard Landau potential is a high-temperature approximation and does not hold at temperatures at which the quantum-mechanical treatment of phonons becomes relevant.

1.2 "Almost" Displacive Transitions

These also have $\Delta \leqslant 1$, and

$$\sigma = \sigma_c \left(\frac{\eta \Omega_0}{\Omega}\right) \coth\left(\frac{\eta \Omega}{\Omega_0}\right), \tag{5}$$

$$\Omega^2 = \Omega_0^2[1 + \Delta(3 - 2\sigma/\sigma_c)], \tag{6}$$

where

$$\eta = \hbar/2M\Omega_0\sigma_c, \quad \sigma_c = (v - M\Omega^2)/3u,$$
$$Q^2 = 3(\sigma_c - \sigma). \qquad (7)\text{–}(9)$$

These "self-consistency equations" have to be solved numerically in order to obtain the value of Q for each temperature T. The resulting curves of Q vs T show smaller values of Q at low temperatures compared with the equivalent solution [Eq. (3)].

1.3 The Order-Disorder Limit

The order-disorder (od) limit is reasonably well described by the Bragg–Williams equation

$$Q = \tanh(QT_c/T), \quad \Delta \gg 1. \tag{10}$$

In Fig. 4, a sequence of measured OPs are compared with theoretical predictions for $\Delta < 1$. In all cases, Landau theory is a good approximation for $T \gg \theta_s$, and we list the analytical expressions for the temperature dependence of the OP for the Landau potential [Eq. (4)]:

1. a "second-order" transition:

$$Q = \left(\frac{T_c - T}{T_c}\right)^{1/2} \tag{11}$$

for $T < T_c$, $C = 0$, $B > 0$;

2. a "tricritical" transition:

$$Q = \left(\frac{T_c - T}{T_c}\right)^{1/4} \tag{12}$$

for $T < T_c$, $C > 0$, $B = 0$;

3. a "first-order" transition

$$Q^2 = -\frac{1}{2C}[B - \sqrt{B^2 - 4AC(T - T_c)}]$$

for

$$T < T_{\text{trans}} = T_c + B^2/4AC, \tag{13}$$

$$B < 0, C > 0.$$

Other Landau potentials including odd-order terms (when compatible with the symmetry of the crystal) or higher-order terms lead to similar expressions for $Q(T)$.

Experimental observations often follow the predicted T dependence of Q very well—in particular for displacive phase transitions (Fig. 4). In fact, the applicability of the generalized Landau expression in Eq. (3) is much wider than originally anticipated and includes thermal regimes with OPs close to unity (Salje, 1993b; Salje *et al.*, 1991). Furthermore, structural phase transitions for which the order parameter disappears continuously on heating through the transition point (cases **1** and **2**) do not necessarily show criticalities as in case of phase transitions with short-range interactions such as described in the nearest-neighbor Ising model. In most cases, critical fluctuations are suppressed by the long-ranging and strongly anisotropic elastic interactions in the crystal structure (Cowley, 1976; Folk *et al.*, 1979; Marais *et al.*, 1991).

2. THE ORDER PARAMETER

The order parameter, OP, was so far introduced as a measure for the deviation of the structural state in the low-symmetry phase ($Q \neq 0$) from that of the high-symmetry phase ($Q = 0$). Its meaning stems from thermodynamic requirements, and it is not always obvious how the order parameter is related to structural quantities that maintain the general structural topology of the high-symmetry phase. One finds that all symmetry elements of the low-symmetry structure are already present in the high-symmetry phase. Some symmetry elements of the high symmetry, on the other hand, disappear during the phase transition, i.e., specific symmetry elements are "broken." This requires that an

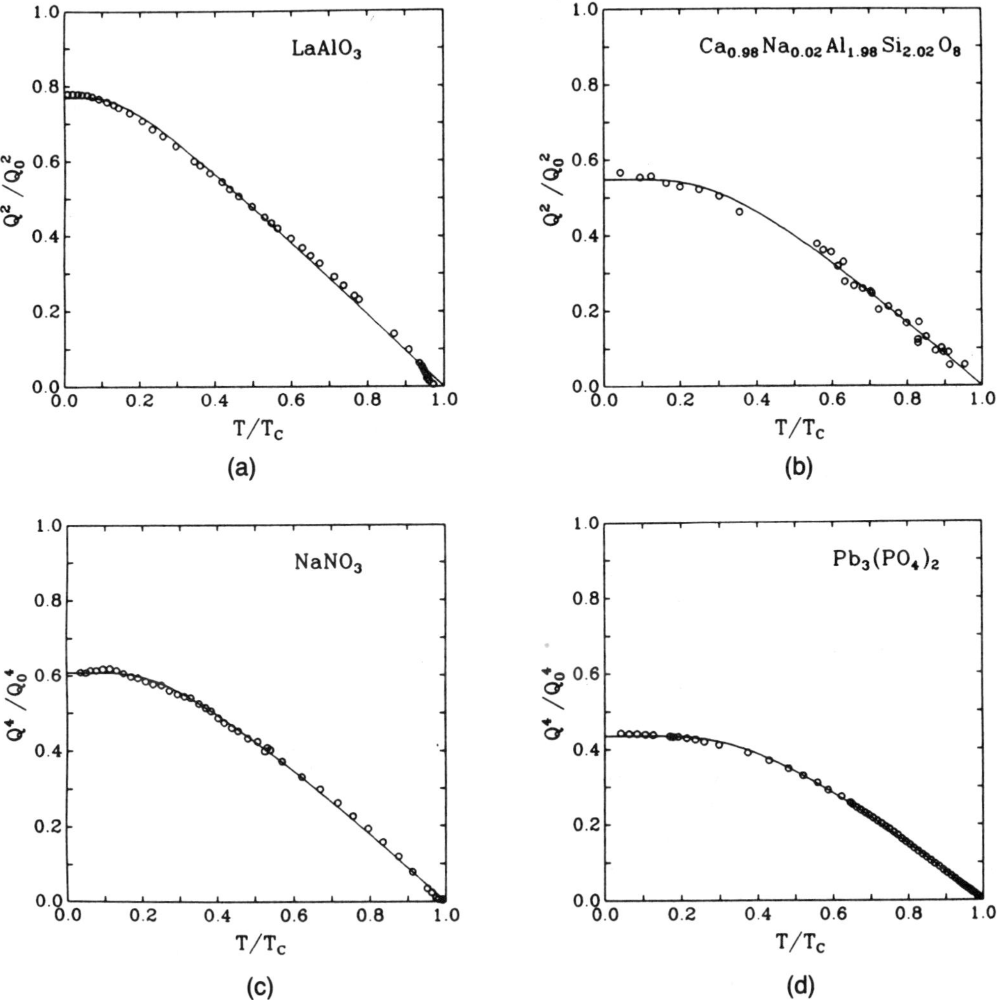

FIG. 4. Temperature evolution of the OP for four typical SPTs. In each case, the curves of a self-consistent solution [Eqs. (5)–(9)] are shown. The relevant parameters are, for (a) $LaAlO_3$, $T_c = 800\ k$, $\theta_s = 95\ k$, $\Delta = 0.116$; for (b) $Ca_{0.98}Na_{0.02}Al_{1.98}Si_{2.02}O_8$ (anorthite), $T_c = 513\ k$, $\theta_s = 226\ k$, $\Delta = 0.082$; for (c) $NaNO_3$, $T_c = 552\ k$, $\theta_s = 165\ k$, $\Delta = 0.174$; and for (d) $Pb_3(PO_4)_2$, $T_c = 455\ k$, $\theta_s = 259\ k$, $\Delta = 0.087$. In all cases $\Delta \ll 1$, so that the data points follow approximately the predictions of classic Landau theory.

additional parameter exists that specifies the thermodynamic state of the low-symmetry structure (which is already completely defined in the high-symmetry structure).

Let us illustrate this requirement using the example of quartz in Fig. 1. In the high-symmetry form, the space- and time-averaged orientation of the tetrahedra is fully determined by the symmetry of the crystal structure. In the low-symmetry structure, this is not the case because the symmetry is reduced so that small rotations of the tetrahedra are compatible with the symmetry constraints. In order to characterize the low-symmetry structure, we need to specify "how much" the tetrahedra rotated and "in which direction."

In a similar manner, the high-symmetry phase of the feldspar structure requires that Al and Si are randomly distributed over the relevant tetrahedra positions. In the low-symmetry structure, this distribution is no longer random with some structural sites being occupied preferentially by Al rather than Si. Again, one needs to quantify to what degree this site ordering has taken place.

As the basic process of breaking a symmetry constraint is always the same, it makes sense to introduce a thermodynamic quantity that describes the structural state of the low-symmetry phase in a universal manner. This thermodynamic parameter is called the order parameter Q. In order to connect Q with the thermodynamic properties of the material, we simply note that the Gibbs free energy G has to depend on $T, P, \ldots$ *and* Q in $G(T,P, \ldots, Q)$. With $Q = 0$ in the high-symmetry phase, we can then define $\Delta G = G - G(Q = 0)$ as the excess Gibbs free energy. Similarly, it is customary to define all other physical quantities as excess values with respect to the state defined by $Q = 0$. Once the dependence of ΔG on Q is either determined experimentally or anticipated by phenomenological models (such as in Sec. 1), the dependence of Q on $T, P \ldots$ simply follows from the equilibrium condition that G must have a minimum at the equilibrium value of Q, i.e., $dG/dQ = 0$, $d^2G/dQ^2 > 0$. All other physical parameters are then determined via their relationship with Q.

It is essential for the definition of an order parameter that a close structural relationship exists between the two phases. This allows us to describe a "difference structure" as the true crystal structure minus the high-symmetry structure as reference structure, i.e., we calculate the difference of atomic densities $\rho(\mathbf{r}) = \rho_{Q\neq 0}(\mathbf{r}) - \rho_{Q=0}(\mathbf{r})$. The pattern of this difference structure describes then directly the atomic displacements, changes in site occupancies, etc. The OP is then the amplitude of this reference structure and tells us "how much happens," whereas the pattern itself describes "what happens." The OP is usually normalized to unity for the maximum change and is zero when no change occurs, i.e., in the high-symmetry phase.

The required close structural relationship may not exist, however. If one crystal structure transforms into another without any topological relationship between them, the transition is called "reconstructive" and no order parameter besides a trivial normalized volume ratio of the two phases can be defined. If the two structures transform into each other with a well-defined three-dimensional topological relationship, the transition is called "topotactic." For two-dimensional relationships, the structural transition is called "epitactic." Several materials show layer-type structures. If the relative stacking of the layers changes during a phase transition, the transition is called "polytypic." The various structures that can be related by such polytypic phase transitions (hypothetical or real) are usually called "polytypes"; they are distinguished by the different stacking of identical (or almost so) slabs as the basic structural unit. The order parameter Q describes in this case the amount of material in a given stacking sequence.

Group-theoretical methods have been used to derive the symmetry properties of Q. The OP is a scalar quantity for phase transitions between structures with triclinic, monoclinic, or orthorhombic symmetry. For transitions from phases with higher symmetry, the order parameter can have up to three components $Q = \{Q_i, i \leq 3\}$ representing degeneracies of the symmetry constraints. Further degeneracies stem from the fact that the order parameter is a complex quantity $Q = Q_0 e^{-i\mathbf{q}\cdot\mathbf{r}}$, if the structural change is not uniform. In this case, the translational symmetry is broken with the excess wave vector $\mathbf{q}$ ($\mathbf{q} = 0$ if the translational symmetry is maintained). Degeneracies can occur if the relevant wave vector $\mathbf{q}$ describes several symmetry-equivalent transition mechanisms.

Similar symmetry constraints also exist for the terms in a polynomial expression of the excess Gibbs free energy. As this energy is invariant under all symmetry operations, only those terms Q^n or $Q_i^n Q_j^m$ are symmetry allowed that are invariant under all symmetry operations. The various possible combinations are listed in a great number of publications.

Let us consider a typical geometrical example. A structural unit such as an octahedron is embedded in a cubic crystal structure (Fig. 5). Now let us consider a movement of the structural unit such as a rotation, an elongation, etc. The movement is topologically identical if we interchange the three cubic axes, so that the crystal axes cannot depend on the choice of the Gibbs free energy. The OP must be threefold degenerate, therefore. The symmetry-allowed form of the Gibbs free energy includes then three components Q_1, Q_2, and Q_3 with

$$\begin{aligned} G = G_0 &+ \tfrac{1}{2}A'(Q_1^2 + Q_2^2 + Q_3^2) \\ &+ \tfrac{1}{4}B(Q_1^4 + Q_2^4 + Q_3^4) \\ &+ \tfrac{1}{4}B'(Q_1^2 Q_2^2 + Q_2^2 Q_3^2 + Q_1^2 Q_3^2) + \cdots . \end{aligned} \qquad (14)$$

For $B' = 0$, there is no correlation between

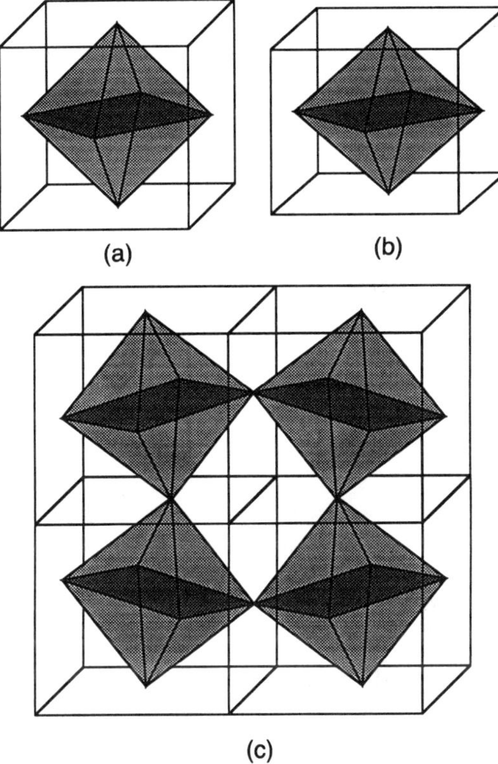

FIG. 5. Phase transitions generated by the distortion of (a) an octahedron in a cube-shaped cage. Two basic deformations can occur depending on the energy contribution of B'; (b) an elongation or shrinking along one of the cube axes (with three degenerate orientations) or along the cube diagonal (with four degenerate orientations). (c) If the translational symmetry is broken via rotation of the octahedra but not the cubes, tetragonal and trigonal structures are commonly formed.

Q_1, Q_2, and Q_3 with each component Q_i generating a phase transition. The term $Q_i^2Q_k^2$ couples the three components of the order parameter. With the condition $\delta G = 0$, $\delta^2 G > 0$ leads to only two types of solutions ($B' \neq 0$) (Slonczewski and Thomas, 1970):

$$\text{(a)}\ Q = (\pm Q,0,0), (0,\pm Q,0), (0,0,\pm Q), \qquad (15)$$

$$\text{(b)}\ Q = \frac{1}{\sqrt{3}}(\pm Q,\pm Q,\pm Q). \qquad (16)$$

The solution (a) represents a tetragonal phase with the fourfold axis along the cubic a, b, or c axis. The solution (b) describes a trigonal phase with the triad along the cubic [1 1 1], [$\bar{1}$ 1 1], [1 $\bar{1}$ 1], or [1 1 $\bar{1}$] axis. Such phase transitions between cubic and tetragonal or trigonal symmetries are often encountered, in particular in perovskite type crystal structures.

If the wave vector **q** of the order parameter Q is zero, the translational symmetry of both phases is equal. Such phase transitions are sometimes called "ferrodistortive." If **q** leads to a doubling of the translational unit, the transition is "antiferrodistortive." All simple multiplications of the size of the translational unit lead to a "commensurate" phase because the translational periodicity of the low-symmetry phase is commensurate with that of the high-symmetry phase. If this is not the case, the low-symmetry phase is called "incommensurate" (IC). Most IC phases exist only for small temperature intervals although notable exceptions occur (e.g., Na_2CO_3).

3. FERROELASTICITY AND FERROELECTRICITY

Atomic ordering may involve the reshuffling of atoms of different size or with different polyhedral environments. In such cases, the interchange of atoms will lead to a local distortion of the crystal structure. Local distortions lead to stresses that penetrate far into the structure and influence other atomic-ordering processes even when they happen at large distances from the original event. It is this long-ranging character of the elastic interactions that makes them so important. Experimental work has shown that in many SPTs a strong coupling between the OP and elastic degrees of freedom is observed (Salje, 1993b; Cowley, 1976; Folk *et al.*, 1979; Marais, 1991). As a consequence, the shape of the structural unit changes by typically 1%–6% with a smaller change of the specific volume. Even larger changes are seen in some metallic alloys. The excess deformation is called the "spontaneous strain." The generation of the spontaneous strain contributes significant elastic energy to the excess Gibbs free energy and leads to temperature dependences of elastic constants. In many crystals, a SPT was first discovered by the experimental observation of such anomalous thermal expansions using x-ray or neutron-diffraction methods. Materials with elastic anomalies are called coelastic. A specific case exists if the

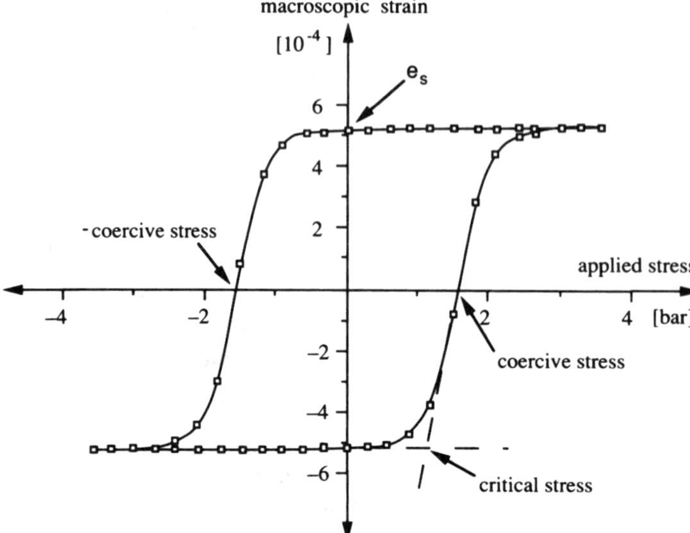

FIG. 6. Ferroelastic hysteresis for $Pb_3(PO_4)_2$.

spontaneous strain can be switched between symmetry-equivalent orientations (ferroelasticity). In Fig. 6, a ferroelastic hysteresis related to $Pb_3(PO_4)_3$ is shown for a macroscopic sample (Salje, 1993b). Mechanical shear stress of a monoclinic sample allows switching of the monoclinic axis by 120°, so that the macroscopic shape of the crystal changes accordingly. This switch is fully reversible.

Similar hysteresis phenomena exist for ferroelectric materials in which the reorientation of the order parameter involves the change of the spontaneous polarization (Lines and Glass, 1977). In Fig. 7, the spontaneous polarization in $BaTiO_3$ is shown as the dipole moment of the positively charged Ti and the negatively charged O with less contribution from Ba. The centers of gravity of the posi-

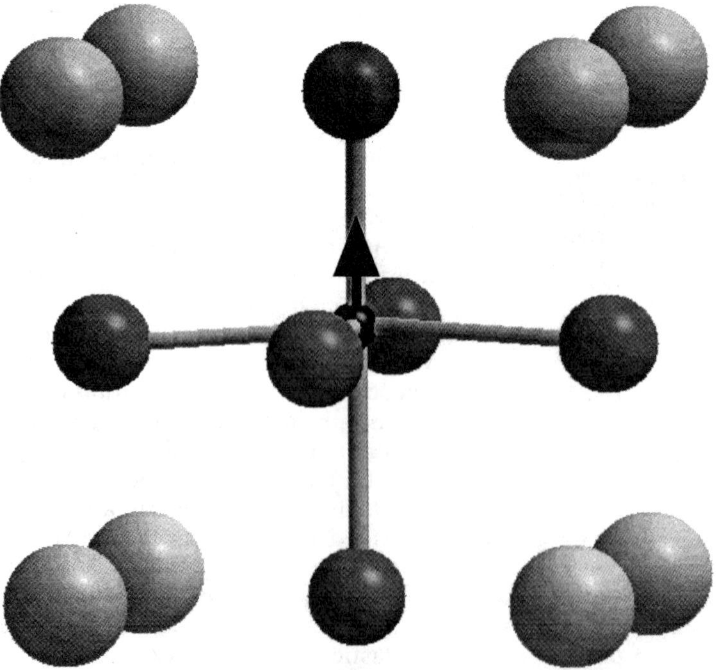

FIG. 7. Crystal structure of ferroelectric, tetragonal $BaTiO_3$. During the SPT from the paraelectric cubic phase, the black Ti cations are slightly shifted with respect to the darker gray oxygen anions. The charge separation generates a spontaneous dipole moment.

tive and negative changes are not identical, so that an electrical dipole moment builds up in the low-symmetry phase (i.e., the spontaneous polarization). Ferroelasticity and ferroelectricity sometimes occur simultaneously with comparable contributions of the electrical and mechanical energy to the excess Gibbs free energy (e.g., in KH_2PO_4).

4. COUPLING BETWEEN ORDER PARAMETERS

Several SPTs may occur in the same material if temperature or pressure is varied over large intervals. For example, almost all incommensurate transitions lead to a second, commensurate (lock-in) transition. Other examples include cascades of three or more transitions each of which breaks a different symmetry element of the same high-symmetry phase. If such elements are restored in a subsequent transition, the SPT is called "reentrant."

Even when the T and P conditions for the various SPTs are very different, the OPs may still interact. The total excess Gibbs free energy is then the sum of the individual contributions and the interaction energy G_{coupling}:

$$G = G_1 + G_2 + G_{\text{coupling}}.$$

Three types of coupling occur commonly whereby the dependence of the coupling energy on the OPs is different (Salje, 1993b):

1. If the coupled order parameters have the same symmetry (but, say, different transition points), $G_{\text{coupling}} = \lambda Q_1 Q_2$ is the coupling energy in lowest order of the OPs. The term λ is a measure for the coupling strength.
2. The coupling energy $\lambda Q_1^2 Q_2^2$ may occur under all symmetry conditions and is often considered as the "standard" coupling term.
3. Specific systems show $G_{\text{coupling}} = \lambda Q_1 Q_2^2$ or $\lambda Q_1^2 Q_2$, whereby this product of OPs must be invariant under the relevant symmetry operations.

The stability of phases has now to depend on both order parameters, which are commonly displayed in the order-parameter vector diagram (Fig. 8). Such diagrams contain all combinations of order parameters that represent thermodynamic equilibrium under the various T and P conditions. The origin is always $Q_1 = Q_2 = 0$, i.e., the high-symmetry phase. When T and P are changed, the structural states run along the trajectories in the diagram. Structural states outside the trajectories represent no minima of the Gibbs free energy and are, thus, not equilibrium states.

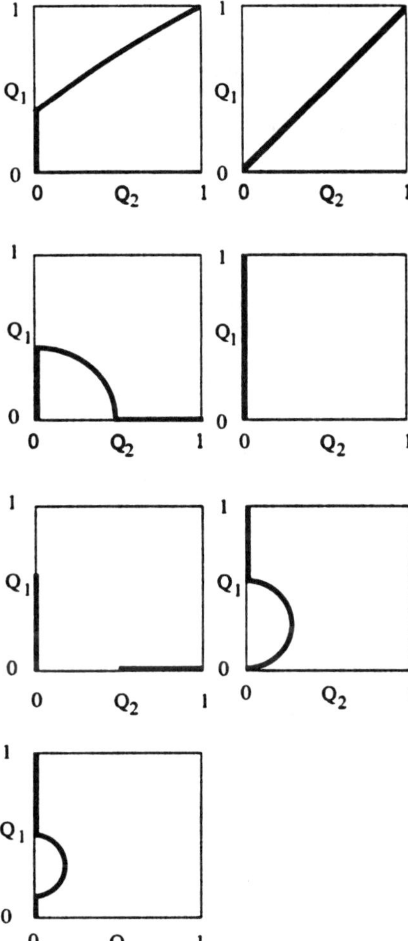

FIG. 8. The typical order parameter vector diagrams for two OPs coupled biquadratically ($\propto \lambda Q_1^2 Q_2^2$). An equilibrium structural state is always characterized by two OPs, which define a point in the diagram. All equilibrium points together form lines of equilibrium states. Changing T, P, etc. moves the equilibrium point along such equilibrium lines as shown in the figures. The high-symmetry phase is represented by the origin ($Q_1 = Q_2 = 0$). Additional phase transitions occur when one of the OPs vanishes [e.g., between the states ($Q_1 = 0$, $Q_2 \neq 0$) and ($Q_1 \neq 0$, $Q_2 \neq 0$)].

Bilinear coupling **1** leads to bending of the

equilibrium line, which is the locus of all stable configurations. Topological changes occur for nonlinear coupling with the creation of additional phase transitions that stem from the competition between the OPs and are *not* due to an independent, third, transition mechanism. All quantitative analyses of SPTs with several OPs have to account for these coupling effects, which greatly modify the T and P dependences of the OPs. A typical example of the experimentally observed T dependence of two coupled OPs is shown in Fig. 9.

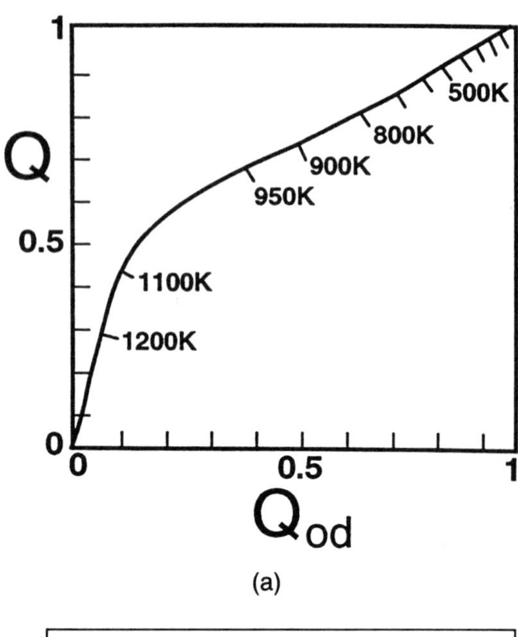

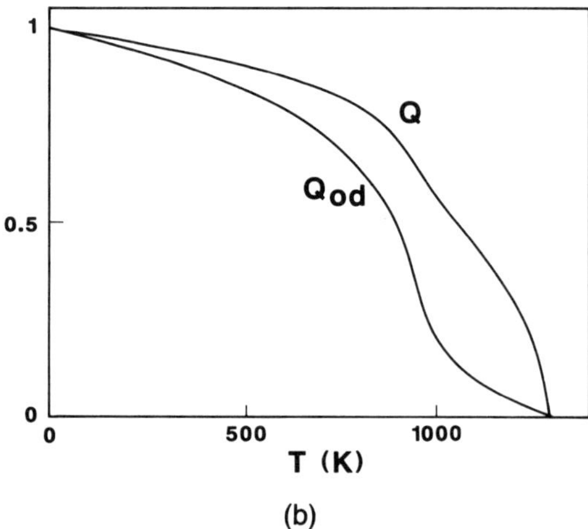

FIG. 9. Example for an experimentally determined order-parameter vector diagram of two coupled OPs (bilinear coupling $\propto \lambda Q_1 Q_2$) for Na feldspar in Fig. 2. Q_{od} represents the OP for the od process of ordering Si and Al on various lattice sites, Q represents an OP of a displacive SPT. (a) Vector diagram, (b) temperature evolution of both OPs [after Salje (1993b)].

5. INFLUENCE OF DEFECTS

SPTs are influenced by structural defects. Defects often couple with thermal fluctuations. Defects can generate anomalies such as excess heat capacities, softening of elastic constants, and parasitic short-range order in the high-symmetry phase. These anomalies tend to mimic closely the behavior of critical systems with short-range interactions (such as magnetic phase transitions). It makes sense, therefore, to describe these anomalies with the same effective exponents, α for the specific heat, β for the order parameter, and γ for the susceptibility. Note that the observed anomalies are not intrinsic to the SPT but they are generated by the lattice relaxations around the defects. Common exponents for α are $\frac{1}{2}$ for unpolarized defects and $\frac{3}{2}$ for polarized defects (in contrast to $\alpha = 0$ for the defect-free SPT in the MFT approximation and as observed in many ferroelastic crystals). Other values exist depending on the details of the relaxation (Levanyuk and Sigov, 1988).

At temperatures small compared with T_c, the OP increases (reduces) when the defects stabilize (destabilize) the low-symmetry phase. The extrapolated transition temperature is accordingly changed. These changes are linear for small defect concentrations ($\Delta G_{\text{defect}} \propto \zeta N Q^2$ changes T_c to $T_c + N\zeta$). For larger defect concentrations N, strongly nonlinear effects lead to a "plateau" of the extrapolated T_c vs N curves.

If the defects stabilize the low-symmetry phase, they may destroy the high-symmetry phase altogether. A typical example is that a point defect leads to a local lattice deformation that has a common component with the spontaneous strain. Both contributions generate the total spontaneous strain, which will not disappear (because of the defect contribution) at $T > T_c$ thereby preventing the high-symmetry phase from appearing under heating. A common example is that the defect acts as fields H_d conjugated to the OP with $\Delta G_{\text{defect}} \propto -\lambda H_d Q$. In this case, Q decreases near the defect-free transition point but then forms a tail at $T > T_c$ (Fig. 10). The structural phase with $Q = 0$ is no longer a minimum of

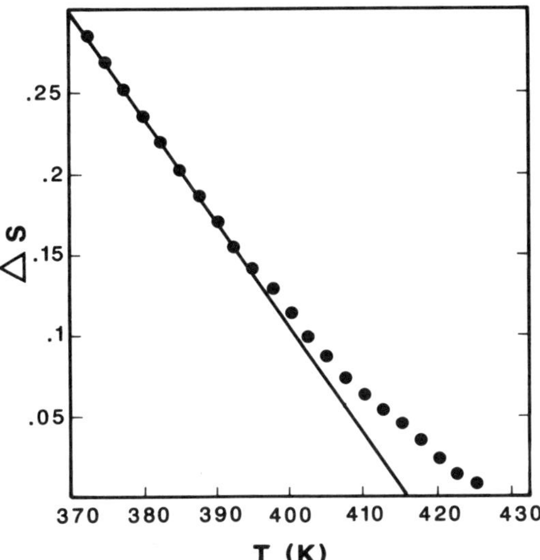

FIG. 10. Defect-related tail in displacive SPT in (Na,K) feldspar. The excess entropy ΔS is proportional to the square of the OP; the pure material has a SPT at 416 K. Defects generate a tail of the OP at higher temperatures.

the total Gibbs free energy (i.e., $\delta G = 0$ always leads to $Q \neq 0$). If the defect field is intrinsic to the crystal structure, this behavior is called a nonconvergent SPT, indicating that the high-symmetry form is asymptotically approached but never reached. Strictly speaking, a nonconvergent SPT is not a SPT because the actual transition does not take place, although the thermodynamic and structural features of nonconvergent behavior are appropriately described within the same theoretical framework as that of SPTs (including the intrinsic field term).

6. TWINS AND ANTIPHASE BOUNDARIES AS FINGERPRINTS OF SPTs; SURFACE RELAXATIONS

The generation of the OP may allow the crystal to become heterogeneous with different values of the OP in different domains. In ferrodistortive SPTs if two domains have OPs with opposite sign, the domains are called twins with a twin boundary between them. Such OPs usually carry large spontaneous strain e_{ik}. The locus of the boundary X_i is determined by the compatibility between the two twins S_1 and S_2, i.e.,

$$[e_{ik}(S_1) - e_{ik}(S_2)]X_iX_k = 0, \tag{17}$$

leading for traceless e_{ik} to two types of walls. These two types of walls would be strictly orthogonal in the metric of the high-symmetry phase. In the low-symmetry phase, the angle between the walls is $\pi/2 \pm \omega$ where the small angle ω is numerically identical to the relevant tensor component of the spontaneous strain ($\omega \lesssim 10^{-2}$). In many experimental observations, this angle ω is too small to be seen and it is customary to describe the wall pattern with the two types of wall as "right-angle domain walls." Walls that are fully localized by symmetry constraints (e.g., those along mirror planes of the high-symmetry phase) are called W walls. Those which have orientational degrees of freedom are called W' walls. W' walls occur, e.g., when the only symmetry constraint is that a twofold axis of the high-symmetry phase has to be contained in the wall. The wall can then rotate around the axis in order to minimize its energy. The rotation angle will then depend explicitly on the control parameter (e.g., T) (Salje, 1993b).

The internal structure of twin walls is usually derived from the Euler–Lagrange equation with $G = G(Q,T) + \frac{1}{2}g(\nabla Q)^2$:

$$\frac{dQ}{dt} \propto \frac{\partial G}{\partial Q} + g\frac{d^2Q}{dx^2}, \tag{18}$$

where x is the distance from the twin wall. The gradient energy describes the energy needed to modify the OP spatially. The solutions depend on the explicit form of $G(Q,T)$; in most cases, they are identical with or close to the solution for the simplest Landau potential with

$$Q = Q_0 \tanh(x/w), \tag{19}$$

where Q_0 is the equilibrium OP and w is the wall thickness. The term w increases close to the transition point as

$$w = \sqrt{\frac{2g}{A|T - T_c|}}, \tag{20}$$

where $\sqrt{2g/AT_c}$ is in the order of lattice repetition units.

The internal structure of twin walls is

modified when several order parameters couple (Houchmandzadeh *et al.*, 1991). Instead of having a rather smooth bend of the OP from one equilibrium value on one side of the wall to another equilibrium value on the other side of the wall, new structural states may occur inside the wall. These new states are described by the second order parameter. The true crystal structure inside the wall and close to the wall is then characterized by the space dependence of the two OPs. Within the spirit of order-parameter vector diagrams introduced in Fig. 8, we now depict solutions of the Euler–Lagrange equation for two OPs as a graph of one OP vs the second OP and obtain the "trajectory" of the domain wall in the order-parameter space. Two typical trajectories for walls with biquadratic coupling between two OPs are shown in Figs. 11(a) and 11(b). The first OP is Q_1 and shows a modified hyperbolic tangent profile. The second OP, Q_2, is zero in the uniform state [Fig. 12(a)] and is generated only inside the wall. In the spatial evolution of the order-parameter vector diagram [Fig. 12], the effective OP "rotates" in the wall, so that this wall is called chiral (similar to magnetic Bloch walls). Parts of walls with right-handed and left-handed chirality meet in lines of vanishing chirality. The second solution in Figs. 11(b) and 12(b) shows a right-handed chirality on one side of the wall and a left-handed chirality on the other side. Such walls are called bichiral.

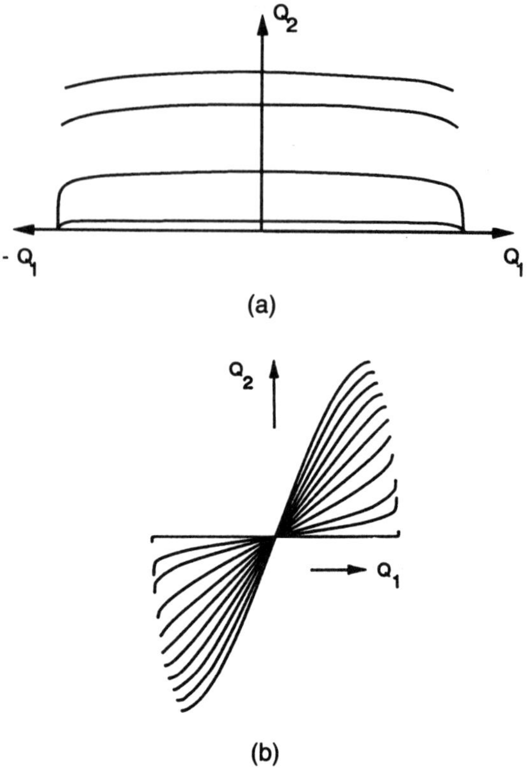

FIG. 11. Vector diagrams of biquadratically coupled OPs together with trajectories of the domain boundaries. Each line represents a trajectory with the stable OPs of the bulk at either end of the trajectory. In (a), only Q_1 changes sign in the boundary but not Q_2. Trajectories in (b) show Q_1 and Q_2 changing sign in the domain boundary.

Antiferrodistortive phase transitions can lead to the formation of antiphase boundaries (APBs) in which the translational symmetry is broken. In cases with weak strain coupling, APBs wiggle through the crystal without orientational preference. Increasing strain coupling leads to preferential orientations and lock-in phenomena along the optimum direction. The thickness of APBs is of the order of interatomic distances.

Structural relaxations close to surfaces and interfaces are enhanced by SPTs. In case of interfaces, such relaxations appear as ripples of the OP and decrease exponentially with distance from the interface. The characteristic length increases close to the transition point. Surfaces show additional relaxations in the high- and low-symmetry phases. Such relaxations act as precursor phenomena although the relaxational modes are not necessarily identical with the deformation pattern of the actual phase transitions. Theoretical work predicts a dominance of incommensurate relaxations and quasiperiodic pattern formation even when the phase transition is (anti)ferro-distortive.

7. KINETICS OF SITE ORDERING

A system out of equilibrium (i.e., $\partial G/\partial Q \neq 0$) will tend to lower its energy via kinetic processes. In displacive SPTs, equilibrium is reached via phonon processes so that the time scale is fast. Much slower processes occur in od systems, which involve the activation of site exchanges that is necessary to interchange atomic sites. Two principal processes dominate od SPTs. Firstly, local site ordering lowers the energy so that local state parameters equivalent to OPs can be used to describe the kinetic rate equation. If only two

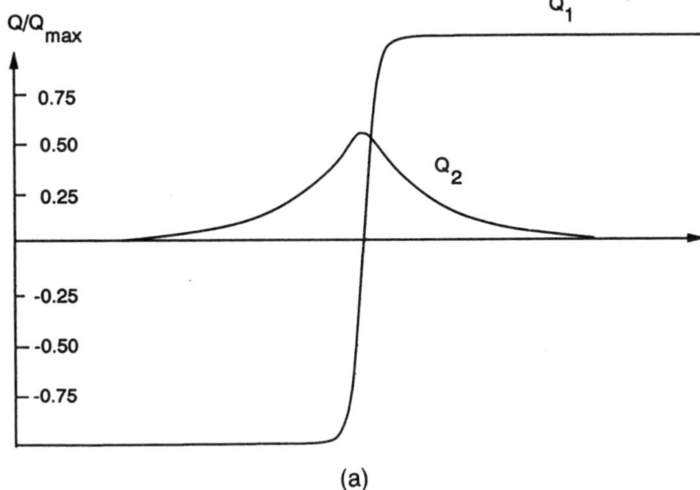

(a)

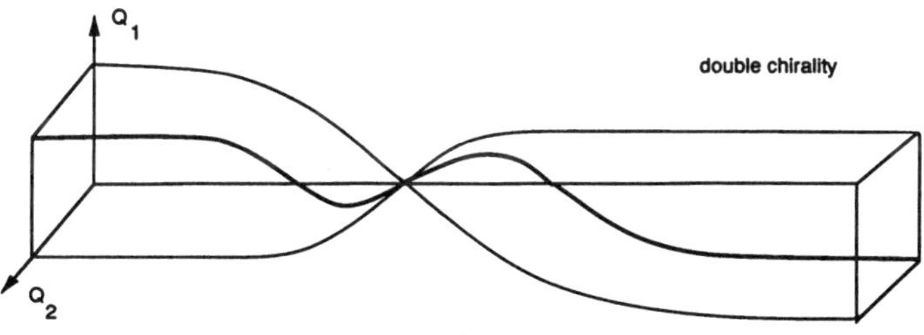

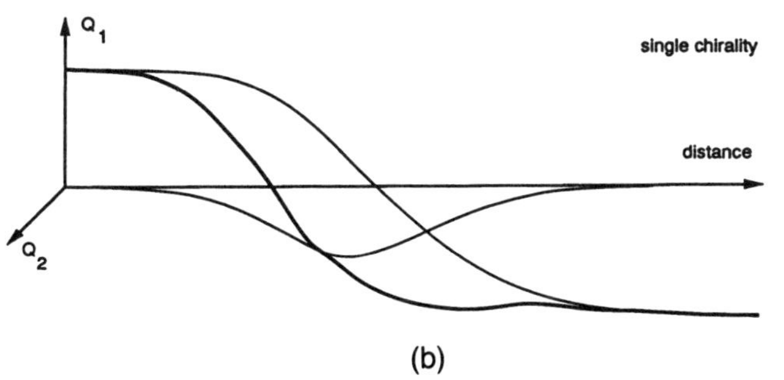

(b)

FIG. 12. Typical wall profiles for both types of trajectories shown in Fig. 11. (a) Q_1 changing sign but not Q_2 whereby the case $Q_2 = 0$ in the bulk is shown. Q_2 exists only in the wall indicating a SPT created in a domain boundary but not in the bulk. (b) The two types of wall profiles with both OPs changing sign (top) and only one changing sign (bottom). The latter is similar to a chiral Bloch wall in magnetism whereas the former self-compensates the chirality in opposite sign on either side of the wall. The abscissa is the distance from the center of the wall; typical wall thicknesses are measured in tens of ångstroms.

local states exist (e.g., like in spin systems), the rate is determined by the random choice of sites and subsequent state changes according to Boltzmann statistics [the so-called "Glauber" kinetics, often in the MF limit because of long-ranging correlations (Glauber, 1963)]. In more complex framework structures, many local exchange schemes exist and compete for the kinetic mechanism. Several structural states intermediate between the nonequilibrium starting configuration and the final (equilibrium) configurations have energy states between the initial and final states. These intermediate states can then serve as "stepping stones" for the system progressing kinetically toward equilibrium. If a sufficient number of such intermediate states exist, the kinetic rate law can be described in the continuum approximation (and in MFT due to long-ranging interactions) by the generalized time-dependent Ginzburg–Landau equation $dQ/dt \propto \partial G/\partial Q$, where G is the equilibrium Gibbs free energy.

The second process involves the transport

of atoms (e.g., via interstitial sites). This part of the ordering requires chemical neutrality and is "conserved" so that each local change of state is compensated by an equivalent change in the opposite direction somewhere else in the structure. This process dominates exsolution kinetics and follows for two-state systems the so-called "Kawasaki" rate law, i.e., now pairs of sites are chosen for an exchange of state variables according to Boltzmann statistics (Kawasaki, 1966; Langer, 1971). This kinetic mechanism is in contrast to Glauber statistics where each site is chosen individually so that it is only in the Kawasaki case that the global order parameter remains unchanged. In SPTs, both processes may exist simultaneously and a general rate equation for continuous state parameters can be formulated (Salje, 1993a):

$$\frac{dQ}{dt} \propto \left(1 - \frac{\xi_c^2}{\xi^2}\frac{\sinh(\xi\nabla)}{\xi\nabla}\right)\frac{dG}{dQ}, \tag{21}$$

where ξ is the Kawasaki length and ξ_c scales the amount of conservation in the individual rate process: With $\xi_c = 0$ we recover the time-dependent Ginzburg–Landau equation, and $\xi_c = 1$ defines in lowest order the so called "Cahn–Hilliard rate equation" (Langer, 1971; Amengual *et al.*, 1993; Salje, 1993a).

These rate laws have been used successfully to describe the kinetics of atomic ordering in framework structures and exsolution processes in alloys and other materials. The dependences of G and the kinetic operator $L = 1 - (\xi_c^2/\xi^2)\sinh(\xi\nabla)/\xi\nabla$ on the gradient operator ∇ lead to heterogeneous patterns for the OP. The exact form of these patterns and their time dependence (e.g., coarsening of domain structures) is still a matter of debate. Experimental observations often show that exsolution patterns ($\xi_c = 1$) show stripe patterns of (approximately) parallel domains with strong variations of the OP at the boundary between two domains. The internal structure of the boundaries appears to be similar to twin boundaries in ferroelastic crystals. SPTs involving atomic ordering show characteristic sequences of mesoscopic domain structures. The early stages contain tweed patterns consisting of patches of ordered material with different OPs (e.g., opposite signs in different patches) and wall-like boundaries between them. In the course of kinetic ripening of such microstructures, the tweed transforms into stripe patterns with well-developed domain walls. These stripe patterns then coarsen via an increase of the domain width (Plate 1).

Kinetic development of such mesoscopic microstructures with characteristic lengths of some 100 Å has been simulated by computer experiments (Parlinski *et al.*, 1993; Semenovskaya and Khatchaturyan, 1992). In most cases, the effective interaction between the sites is assumed to be elastic; the entropic part is determined by pseudospin variables at the sites, which describe the occupancy by one or the other atomic species. The standard Hamiltonian of such a system is (Marais *et al.*, 1991)

$$H_{\text{eff}} = \tfrac{1}{2}\mathbf{uAu} - \tfrac{1}{2}\mathbf{uFs}, \tag{22}$$

where $\mathbf{u}$ is the displacement vector, $\mathbf{A}$ is the dynamical matrix, $\mathbf{s}$ is the pseudospin variable, and $\mathbf{F}$ describes the spin-lattice coupling. This Hamiltonian can be renormalized to spin coordinates at sites R and R':

$$H_{\text{eff}} = -\sum_{\mathbf{RR}'} J(\mathbf{R} - \mathbf{R}')S(\mathbf{R})S(\mathbf{R}'). \tag{23}$$

The spin-spin coupling is now long-ranging with a strong directional dependence and nonanalytical behavior in the long-wavelength limit:

$$J(R) \propto \frac{1}{R^3}[A_2Y_{2m}(\theta,\varphi) + A_4Y_{4m}(\theta,\varphi)] + \frac{Z}{N} \tag{24}$$

with the constant term Z/N where N is the number of particles and Z is the Zehner–Eshelby constant describing the macroscopic relaxation of the sample. Using this Hamiltonian in computer simulation, the characteristic domain structures of tweed and stripes have been reproduced. Such mesoscopic structures are shown in Plate I for a simulation of 64×64×8 unit cells with thermodynamic conditions similar to those for the experimental results in Fig. 13.

8. CONCLUSION

Structural phase transitions are dominated by long-ranging interactions, which makes their description by mean field theories possible. Much work has been domi-

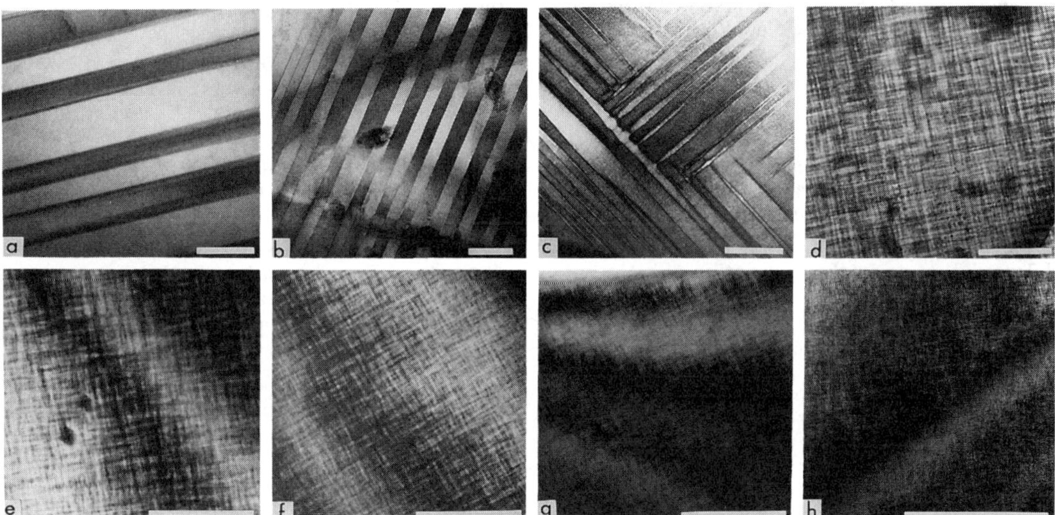

FIG. 13. Characteristic pattern for nonequilibrium SPTs. Quenched samples of Co-doped $YBa_2Cu_3O_7$ show the typical stripe (a)–(c) and tweed (d)–(h) pattern. Co-concentrations are (a) 0%, (b) 1%, (c) 2%, (d) 2.5%, (e) 2.8%, (f) 3%, (g) 5%, and (h) 7%. The scale bar is 0.1 μm. Note the formation of junctions between orthogonal stripes in (c). Courtesy A. Putnis, Cambridge.

nated by the application of Landau theory, including the coupling between several order parameters and their saturation at low temperatures. Future work is likely to concentrate on the time dependence (kinetics) of structural phase transitions.

GLOSSARY

Excess Gibbs Free Energy ΔG: Difference between the Gibbs free energies of the low-symmetry phase $G(Q)$ and the high-symmetry phase $G(Q = 0)$, also called the excess Gibbs free energy of the SPT.

Order Parameter (Q): Variable that describes the structural state, with $Q = 0$ for the high-symmetry phase and $Q \neq 0$ for the low-symmetry phase.

Phase Transition Temperature: (T_c for continuous phase transitions, T_{trans} for stepwise transition.) Temperature at which the Gibbs free energy of the high-symmetry form is identical to that of the low-symmetry form.

Saturation Temperature (θ_s): Temperature below which the temperature dependence of the order parameter diminishes rapidly.

Spontaneous Strain (e_{ik}): Strain of the crystal generated by the phase transition. The spontaneous strain is measured as change of lattice parameters with respect to the high-symmetry phase.

Works Cited

Amengual, A., Hernandez Garcia, E., San Miguel, M. (1993), "Ordering and Finite Size Effects in the Dynamics of One-Dimensional Transient Pattern," *Phys. Rev. E* **47**, 4151–4160.

Bruce, A. D., Cowley, R. A. (1981), *Structural Phase Transitions*, London: Taylor and Francis.

Cowley, R. A. (1976), "Acoustic Phonon Instabilities and Structural Phase Transitions," *Phys. Rev. B* **13**, 4877–4885.

Folk, R., Iro, H., Schwabl, F. (1979), "Critical dynamics of Elastic Phase Transitions," *Phys. Rev. B* **20**, 1229–1244.

Glauber, R. J. (1963), "Time-Dependent Statistics of the Ising Model," *J. Math. Phys.* **4**, 294–307.

Houchmandzadeh, B., Lajzerowicz, J., Salje, E. (1991), "Order Parameter Coupling and Chirality of Domain Walls," *J. Phys. Condensed Matter* **3**, 5163–5169.

Kawasaki, K. (1966), "Diffusion Constants Near Critical Points for Time Dependent Ising Models," *Phys. Rev.* **145**, 224–230.

Khatchaturyan, A. G. (1983), *Theory of Structural Transformations in Solids*, New York: Wiley.

Landau, L. D., Lifshitz, E. M. (1980), *Statistical Physics*, Oxford: Pergamon Press.

Langer, J. S. (1971), "Theory of Spinodal Decomposition in Alloys," *Ann. Phys. (New York)* **65**, 53–86.

Levanyuk, A. P., Sigov, A. S. (1988), *Defects and Structural Phase Transitions*. New York: Gordon and Breach.

Lines, M. E., Glass, A. M. (1977), *Principles and Applications of Ferroelectrics and Related Materials*, Oxford: Clarendon Press.

Marais, S., Heine, V., Nex, C., Salje, E. (1991) "Phenomena Due to Strain Coupling in Phase Transitions," *Phys. Rev. Lett.* **66**, 2480–2483.

Parlinski, K., Heine, V., Salje, E. (1993), "Origin of Tweed Texture in the Simulation of a Cuprate Superconductor," *J. Phys. Condensed Matter* **5**, 497–518.

Salje, E., Wruck, B., Thomas, H. (1991), "Order Parameter Saturation and a Low-Temperature Extension of Landau Theory," *Z. Phys. B Condensed Matter* **82**, 399–404.

Salje, E. (1993a), "On the Kinetics of Partially Conserved Order Parameters: A Possible Mechanism for Pattern Formation," *J. Phys. Condensed Matter* **5**, 4775–4784.

Salje, E. K. H. (1993b), *Phase Transitions in Ferroelastic and Co-elastic Crystals*, Cambridge: Cambridge Univ. Press.

Semenovskaya, S., Khatchaturyan, A. G. (1992), "Structural Transformations in Nonstoichiometric $YBa_2Cu_3O_{6+\delta}$," *Phys. Rev. B* **46**, 6511–6534.

Slonczewski, J. C., Thomas, H. (1970), "Interaction of Elastic Strain with Structural Transition of Strontium Titanite," *Phys. Rev. B* **1**, 3599–3608.

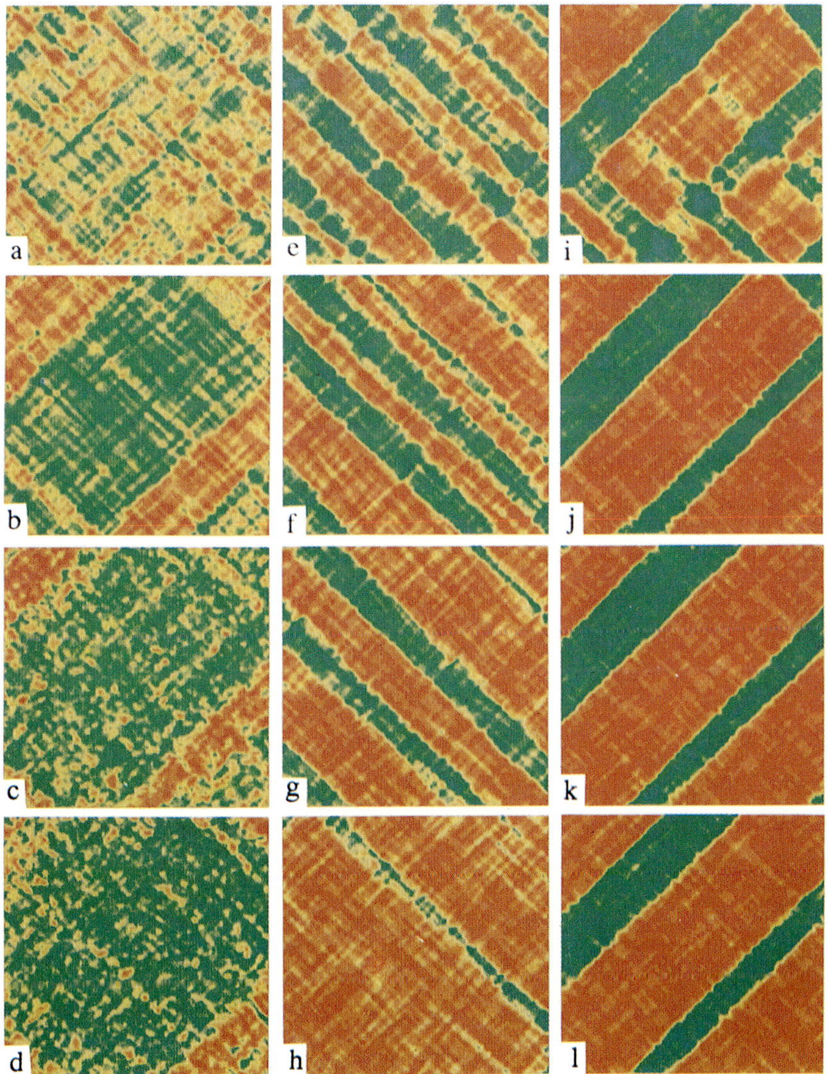

PLATE 1. Results of computer calculations of domain structures of a quench through a SPT. The system has been equilibrated at 1.75 T_c; the annealing temperatures are (a)–(d) 0.93 T_c, (e)–(h) 0.83 T_c, and (i)–(l) 0.73 T_c.

PHILOSOPHY OF PHYSICS

MAX JAMMER, *Bar-Ilan University, Ramat-Gan, Israel*

	Introduction	389
1.	**The Philosophy of Preclassical Physics**	390
1.1	The Early Development of the Philosophy of Physics	390
1.2	Philosophy and Experimentation in Preclassical Physics	391
2.	**The Philosophy of Classical Physics**	392
2.1	Philosophical Problems of Classical Mechanics	392
2.2	The Philosophy of Space and Time in Classical Physics	393
2.2.1	The Dimensionality Problem of Space	394
2.2.2	The Directionality Problem of Time	396
3.	**Antimechanistic Theories**	397
3.1	Dynamism	397
3.2	The Electromagnetic World Picture	398
3.3	Energetism	399
4.	**The Philosophy of Modern Physics**	400
4.1	The Philosophical Background of Modern Physics	400
4.2	Philosophical Problems of the Special Theory of Relativity	400
4.3	Philosophical Problems of the General Theory of Relativity	403
4.4	Philosophical Problems of Quantum Mechanics	407
	Works Cited	414

INTRODUCTION

Philosophy of physics, as currently understood, studies the conceptual foundations and philosophical implications of physics, and examines the interdependence between observation and theory as well as the structure and interrelation of physical theories. It also deals with issues traditionally classified as "metaphysical," such as the nature of space, time, matter, and energy. Last, but not least, it discusses the question of whether physical theories lead to, or are compatible with, a coherent and intellectually satisfying picture of the world.

Philosophy of physics differs from physics with respect to both the objectives aimed at and the techniques or tools employed. The objective of physics is the construction of a physical theory of nature by discovering new data, embedding them in a consistent scheme, and testing the consequences of the theory by experimental methods. The objective of the philosophy of physics is the study of the nature of a physical theory by examining its evidential and inferential foundations, its logical consistency, and its metaphysical assumptions and implications. The work of the philosopher of physics is therefore usually, in contrast to that of the physicist, *ex post facto*. The tools of the physicist are the laboratory and the computer, the tool of the philosopher of physics is the method of conceptual analysis.

This contradistinction does not, however, imply that a great physicist could not be also a philosopher of physics—a combination that, as we shall see, has been rather the rule than the exception, especially at periods of revolutionary innovations.

The historical development of physics can be divided into three major periods:

1. preclassical physics, which began in the sixth century B.C. with the natural philosophy of the school of Miletus (Thales and his disciples) and lasted until the scientific revolution of the 17th century;
2. classical or Newtonian physics, which came to an end at about 1900; and
3. modern physics, which began with the advent of the special theory of relativity and of quantum mechanics.

3-527-28135-5/95/$5.00 + .50

Because of the perpetual interaction between physics and its philosophy, it will prove useful to adhere to this historical division also in our discussion of the philosophy of physics.

1. THE PHILOSOPHY OF PRECLASSICAL PHYSICS

1.1 The Early Development of the Philosophy of Physics

The first to use the expression "philosophy of physics" as a *terminus technicus* was Aristotle. He used this term, in his treatise *Metaphysics* (Book 6, Chap. 1, 1026 a 18)*, in order to distinguish its subject matter from those of other branches of philosophy and science. In fact, at his time philosophy consisted already of several subdivisions, ranging from metaphysics to ethics and aesthetics. But prior to his time, when philosophy was born, when Thales and his school, the φυσικοι in Aristotle's terminology, emancipated themselves from mythological beliefs, rational reflexion was almost exclusively a philosophy of nature or φυσις, the Greek term for nature. These "physikoi" tried to explain the multifarious phenomena of nature by conceiving them merely as modifications of one and the same substratum or as manifestations of a single all-embracing principle. Their quest for a universally valid explanatory principle has ever since become one of the guidelines of physical research. The search for a unified theory of physics in our days is but a modern version of this quest.

In the fifth century B.C., the Pythagoreans claimed to have found this principle in the relation between numbers or in the structure of geometrical patterns. On the other hand, Empedocles, convinced that qualitative differences cannot be accounted for by only one principle, posited four principles, which he identified with the four elements earth, water, air, and fire; in addition, he introduced the two principles of Love and Strife, anticipating thereby the dynamics of attractive and repulsive forces in physics. Atomism, inaugurated by Leucippus and Democritus, conceived the universe to consist only of atoms and the void; devoid of qualitative properties except impermeability, atoms may differ in size, shape, arrangement, and position, and are in perpetual motion through the void. Although rejected by Aristotle and anathematized in the Middle Ages, atomism exerted a profound influence on the development of physics. True, the "atom" of Democritus is not that of Boyle, Dalton, or Rutherford, or the atom as conceived in the modern theory of elementary particles. But some of the fundamental tenets of ancient atomism persisted almost unchanged throughout the ages. Leucippus' dictum "Nothing comes into being at random, but everything comes from reason and by necessity" was the earliest enunciation of the principles of causality and determinism that dominated classical physics. Democritus' statement "Things appear to be colored, sweet or bitter, but in reality there are only atoms and the void" anticipated the distinction between primary and secondary qualities; the former, like solidity, extension, or motion, are properties inherent in bodies and inseparable from them, while the latter, like color, sound, or taste, are "nothing in the objects themselves" (Locke, 1690), but merely sensations in our consciousness. This distinction led Galileo, Descartes, Hobbes, and Locke to confine the task of physics to the study of only quantitative relations. It also raised, however, the philosophical question of whether macroscopic entities, like tables or trees, have an existence on their own, not only as aggregates of atoms—that is, the problem of whether the composite is more than the sum of its components and has a reality of its own.

An answer to this problem was given by Plato, who declared that the objects perceived by our senses are but imperfect copies or crude imitations of their ideas or "forms," which exist in a supersensible world beyond space and time and which are accessible to us only by pure thought. Influenced by Pythagoras, Plato contended that true knowledge of physical objects can be obtained only by mathematical studies of the ideas or forms of the objects. Consistent with his philosophy, Plato proposed in his influential dialogue *Timaeus* a purely geometrical theory of the structure of atoms as regular polyhedra.

Aristotle rejected the Platonic transcendentality of forms and argued that the form

*All references to works of Aristotle use what is known as the Bekker pagination (Bekker, 1831–1870).

or idea of an object inheres in its substance: Every object is a union of matter and form; matter endows it with its individuality, form makes it a member of a class of similar objects. On this metaphysical assumption, he based his inductive-deductive method, according to which physical research has to progress from sense experience and to derive by induction generalizations that serve as premises for the deduction and explanation of observed phenomena. To avoid an infinite regress of explanations, self-evident premises have to be posited without demonstration. In accordance with the rule that "demonstrations must be based on premises prior to and better known than the conclusions," as stated in his *Posterior Analytics* (Book 1, Chap. 3, 72 b 25), Aristotle laid the foundations of the axiomatic approach to the formulation of physical theories and developed a rather elaborate proof-theory, stating, e.g., that "that proof is the better which proceeds from the fewer postulates" (*ibid.*, Book 1, Chap. 25, 86 a 33). His invention of the axiomatic method had an incisive influence not only on mathematics, where it was first applied to geometry by Euclid in his *Elements*, but also on the development of physics. Archimedes, for example, adopted the axiomatic method in his presentation of the science of statics when he began his treatise *On the Equilibrium of Planes* with the postulate that "equal weights balance at equal distances," implying thereby that a symmetrically loaded lever is in equilibrium.

Already Aristotle himself axiomatized, though not yet in a syntactical formalism, his theory of enforced motion by postulating that "any object that is in motion must be moved by some agent" (256 a 3) and that the mover must be in direct contact with the moved. This statement requires some explanation. In his treatise *Τα Φυσικα* (the Greek for "nature," from which the discipline of "physics" derived its name), Aristotle defined "physics" as the science of things that move or change, and he distinguished enforced from natural motion. The motion of a vehicle, when drawn by a horse, or of a stone, when thrown by a hand, is an enforced motion, while the motion of an apple falling off a tree is a natural one. The Aristotelian theory of natural motion is intimately related to the doctrine of the natural places of the elements, which doctrine in turn is the basis of Aristotle's cosmological world picture. For he accepted the traditional theory of the four elements earth, water, air, and fire, the natural places of which determine the sublunary spheres with Earth at the center; the celestial spheres that enclose the finite universe consist of a fifth element, the incorruptible ether. Now, the natural motion of an object in the sublunary world is but the quest of the element of which it consists to return by the shortest way to its natural place; and the natural motion of celestial matter is the eternal and self-contained circular motion. This ontological and nomological differentiation between a physics of the heavens and a physics of the Earth complied later very well with the teachings of medieval theology and contributed, therefore, to the survival of Aristotelian physics throughout the Middle Ages.

1.2 Philosophy and Experimentation in Preclassical Physics

Ancient and medieval physicists based their science upon observation and speculation and not, as a rule, on experimentation. Sociologists explain this fact by pointing out that in those times the institution of slavery provided cheap labor and thus retarded the development of energy-saving technologies and hence also the construction of scientific instruments. Another no less important reason was undoubtedly the thesis of the ancient philosophy of physics that nature is endowed with a life of its own (hylozoism) and that things can behave naturally only within their natural habitat. An experiment, in contrast to passive observation, was regarded as an interference with the course of nature or even as an act of violence. In fact, simple devices, such as the lever, the wheel and axle, or the pulley, were not called "physical" instruments but rather "mechanical" contrivances, for the Greek term μεχανη, the root of the words "machine" and "mechanics," meant "trick" or "deceit"; by using these appliances, one supposedly bilked or cheated nature to behave contrary to itself. "When we have to produce an effect contrary to nature . . .," says the pseudo-Aristotelian treatise *Mechanical Problems*, "we call it mechanical" (*Mechanica*, 847 a 10). In short, since the performance of an experiment imposes artificial conditions, "contrary to nature," an experiment, it was argued, could never increase our knowledge about nature. This idea, though slightly modified, reappeared in modern physics, where according to certain interpretations of quantum mechanics the Heisenberg indeterminacies result from disturbances caused by the measuring process.

As a result of the invention of mechanical devices that, like the waterwheel, were widely applied to spare human effort, these arguments began to lose much of their cogency already in the early Middle Ages. During the Renaissance, the development of mechanical craftmanship in architecture, engineering, and the fine arts led gradually to an appreciation of experimentation, as illustrated by Leonardo da Vinci's experiments with small-scale models to solve architectural problems. Finally, though not yet for the purpose of discovering new physical effects or laws, Galileo Galilei established the method of controlled experimentation in order to test mathematically derived conclusions. Since, as Galileo (1623) put it, "The book of nature is written in mathematical language" and since an experiment *qua* quantitative test is likewise described in mathematical terms, the experiment was no longer regarded as an extraneous interference with nature.

With his experimental study of free fall, Galileo refuted the teleological conception of "natural motion" and replaced it by the purely descriptive notion of a uniformly accelerated motion satisfying the famous formula $s = \frac{1}{2}gt^2$, the earliest physical equation to contain "time" as the independent variable. Galileo also introduced the use of thought experiments, as, e.g., in his refutation of Aristotle's thesis that heavy bodies fall faster than light ones, by simply conceiving a heavy body as a combination of light components and showing that the Aristotelian thesis thus leads to a self-contradiction.

2. THE PHILOSOPHY OF CLASSICAL PHYSICS

2.1 Philosophical Problems of Classical Mechanics

The most devastating attack against Aristotelian physics was, of course, the announcement of the principle of inertia, the very antithesis of Aristotle's first axiom of motion. Historians differ on who was the first to announce the law of inertia, whether it was Galileo who applied it only to horizontal motions along the surface of the Earth, whether it was the Dutch physicist Isaac Beeckman who derived it from the properties of space, or whether it was Descartes who deduced it from the conservation of linear momentum (Jammer, 1990). In his *Philosophiae Naturalis Principia Mathematica*, which has been called "the most important work ever published in the physical sciences" (S. W. Hawking), Isaac Newton (1687) formulated the principle of inertia as the first law of motion as follows: "Every body continues in its state of rest, or of uniform motion in a right line, unless it is compelled to change that state by forces impressed upon it." Although Newton's first law has ever since been regarded as a cornerstone of classical physics, its philosophical status is still a matter of dispute. Opinions differ on whether it should be regarded as the result of experimental experience (Frank, 1946), whether it is an idealized construct, inaccessible to empirical realization, and hence merely a convention (Poincaré, 1902), or whether it should serve as a definition of the absence of forces or as a statement of the existence of inertial systems. As the Preface of the second edition of the *Principia* clearly indicates, for Newton the first law was firmly rooted in experience. True, Newton regarded physics, as the title of his work clearly shows, as a part of philosophy; but his philosophy was not a kind of speculative metaphysics but rather "experimental philosophy" in the sense of Francis Bacon's *philosophia experimentalis* (Bacon, 1620). Newtonians were convinced that the laws of motion express undeniable truths. In fact, according to Leonhard Euler (1750), these laws could even be used to reverse the traditional relation between physics and philosophy, for "knowledge of the truths of mechanics can serve as a guide in the thorny researches of philosophy . . . and it would be justified to reject all the reasons and all the ideas of metaphysics, however well founded they otherwise be, if they lead to conclusions contrary to these truths of mechanics." This statement proclaimed not only the independence of physics but even its epistemological priority over philosophy and was a protest against the traditional view that had been dominating physical thought in the past. Let us not forget in this context that philosophical, teleological, and theological thought had played an important role in the development of physics. Descartes, for example, derived the conservation laws of mass and of linear momentum from the immutability of God; for Newton, absolute space was an attribute of divine omnipresence; Maupertuis interpreted the principle of least action, from which can be derived the Lagrangian and Hamiltonian equations, the most frequently used equations in theoretical physics, as a teleological maxim, according to which nature strives to attain a maximum of parsimony. It took quite some time until it was understood that these extremum principles are merely integral formulations of differential equations and as such not more indicative of purposiveness in nature than any other law of physics.

In the 19th century, physics and philosophy were regarded as two independent and autonomous disciplines. Their alienation led, at the close of that century, to the program of a systematic elimination of all metaphysical ingredients from physics. The way to this positivistic enterprise had been prepared already by the sensationalistic school founded by George Berkeley, a contemporary of Newton. It claimed that physical objects in their *knowable* aspects are reducible to sensible properties (*"esse est percipi"*). But it did not completely exempt itself from metaphysics since the physical object continued to lurk behind the scenes as the source of its sensible appearances. The radical step of identifying the physical object with its sensible appearances was made by Ernst Mach. His positivistic replacement of metaphysical notions by logical constructions and operational definitions was instrumental for the rise of modern physics, as we shall see in Sec. 4.

2.2 The Philosophy of Space and Time in Classical Physics

Most of the philosophical problems of classical physics are conceptually related to Newton's laws of motion. It will have been noted, for example, that the first law, as stated by Newton, does not specify the reference system relative to which the state of rest or uniform motion is to be determined. The question of the identification and ontological status of this reference system became one of the major problems in the philosophy of space and time. Newton thought to have solved this problem with his theory of absolute space, absolute time, and absolute motion, the existence of which he claimed to have proved experimentally. A water-filled bucket, suspended by a twisted rope, was allowed to rotate with the unwinding of the rope; analyzing the different phases of this bucket experiment, Newton argued that the deformation of the rotating water surface could not be due to the motion of the water relative to the bucket but was rather the result of a centrifugal force and hence, according to the second law of motion, of an acceleration, which, since not relative to the bucket, must be relative to something else that he identified with absolute space.

Newton also maintained the absoluteness of time. Herein he followed his teacher Isaac Barrow (1670) who in his *Lectiones Geometricae* had already rejected the Aristotelian-scholastic conception of time as "the number (or measure) of motion" and argued that since God's creation of the world is itself a process that takes place in time, time must have existed prior to the act of creation. Influenced by Barrow and the Cambridge Platonists, like Henry More, whose notions of space and time were theologically biased, Newton declared in the famous Scholium of the second edition of the *Principia* that God, being eternal and infinite, and existing always and everywhere, "constitutes duration and space." In addition to these arguments, Newton also resorted—a point rarely mentioned—to purely epistemological considerations in support of his theory of absolute space and absolute time. Thus he wrote, for example, anticipating somewhat Immanuel Kant's critical approach in the "Transcendental Aesthetic" of the *Critique of Pure Reason* (Kant, 1781), "One can imagine that there is nothing in space, yet we cannot think that space does not exist, just as we cannot think that there is no duration, even though it would be possible to suppose that nothing whatever endures" (Hall and Hall, 1962).

Newton's theory of space and time was strongly opposed by G.W. Leibniz's relational theory according to which space and time are merely systems of relations between things or events. In a posthumously published essay, Leibniz (1715) presented a deductive construction of relational time and space that became the prototype of modern causal theories of space-time. Leibniz defined the temporal relations of simultaneity and priority in terms of causal relations and space as the order of simultaneous things or events. In a famous epistolary debate with Newton, who was represented by Samuel Clarke, Leibniz argued that absolute space offers no sufficient reason why physical objects have the particular position or orientation they happen to have and are not "placed the quite contrary way, for instance by changing East into West" (Clarke, 1717). Neither succeeded in convincing the other. Leibniz failed to counter Newton's dynamical arguments, while Newton, aware of the fact that mechanics obeys Galilean relativity so that only accelerations but not velocities can be absolute, failed to identify the privileged reference frame. He could only conjecture that "in the remote regions of the fixed stars, or perhaps beyond them, there may be some body absolute at rest," an idea taken up later by Carl Neumann (1869) in his theory of the "Alpha-Body" (Jammer, 1954).

Because of the spectacular success of Newtonian mechanics in its applications to terrestrial and astronomical problems, the absolutist theory dominated classical physics until the end of the 19th century. In the course of that century, the ancient idea of the ether, which had played an important role in the natural philosophy of Descartes, Huygens, Hooke, Euler, and others, gained renewed prominence as an explanatory principle for electrical, magnetical, and optical phenomena. With the general acceptance of Augustin Fresnel's theory of the ether as an all-pervasive stationary substratum, in which light propagates and through which the Earth is moving, the ether became the representative of the ultimate system of reference and assumed the role of absolute space. On this assumption, the time required for a light wave to pass between two points on the surface of the Earth would have to depend on whether the light wave is moving in the direction of the Earth's motion or opposite to this motion.

As is well known, the meticulously performed ether-drift experiments, like the famous Michelson–Morley interferometer experiment, showed with their "null results" the untenability of these ideas and led, with the rise of the special theory of relativity, to the abolishment of the ether and with it, presumably, of absolute space. This does not mean, however, that a non-null result of these experiments would have left Newtonian space-time as the only surviving alternative or that the issue "absolute vs relational space" has been definitely settled thereby. In fact, quite a few modern philosophers of space and time abstain from taking sides or even declare that "relativity theory, in either its special or general form, is more inimical to a relational conception ... than is classical physics" (Earman 1989; Jammer, 1993).

2.2.1 The Dimensionality Problem of Space One of the first to realize that the three-dimensionality of space, which had been usually taken for granted, poses a physical problem was Immanuel Kant. In the very first of his published writings, he claimed that "the threefold dimension of space is due to the law according to which the forces in substances act upon another" and "that the strength of their action holds inversely as the square of the distance" (Kant, 1747). Kant's reasoning can be explained most briefly by using the notation of modern vector analysis as follows. If **P** denotes the force produced by a substance of amount M which is enclosed by a surface S of radius r, then by the divergence theorem of Gauss M is equal (up to certain constants) to $\mathbf{F} \cdot d\mathbf{S}$; hence, if $\mathbf{F}$ varies like r^{-2}, S must be two-dimensional and space consequently three-dimensional.

What Kant had in mind was of course Newton's inverse-square law of gravitation. The discoveries, after 1760, of Tobias Mayer and of Charles Augustin Coulomb that also magnetic and electric forces obey the same inverse-square law seemed to support the objection, raised by Kant's critics, that the inverse-square dependence is a consequence of the three-dimensionality of space rather than vice versa as Kant maintained. It is remarkable that recent results in the study of the dimensionality problem of physical space, which Kant inaugurated in 1747, agree on this issue more with Kant than with his opponents. In order to understand this issue, let us divide the problem into three questions:

1. What, precisely, is meant by saying that physical space has a certain number of dimensions?
2. How do we know that it really has, as generally claimed, three dimensions (or, equivalently, that space-time has four dimensions)?

3. Is there a *raison d'être*, an inherent reason, for this dimensionality?

The first question is a problem of mathematics and has been dealt with by Poincaré, Brouwer, Menger, Urysohn, and other mathematicians. The second question has been answered by physicists who, like Paul Ehrenfest (1917), showed that the stability of planetary orbits, the stability of atoms and molecules, the unique properties of wave operators and many other phenomena are manifestations of the three-dimensionality of space (perhaps only after compactification if the recent Kaluza–Klein–type theories are admitted). Concerning the third question, which from the philosophical point of view is perhaps the most interesting one, it is generally agreed that there is no fully satisfactory answer. There are, however, some indications that, as Hermann Weyl (1949) phrased it, the "physical laws may cease to be indifferent to the number of dimensions on some deeper level that has hardly been touched by the physics of today. There is thus good reason to hope that our problem will one day find a cogent solution along such lines."

For a number of reasons, a promising candidate along these lines is perhaps the electromagnetic field. Referring to the fact that the Dirac equation of quantum electrodynamics, the best-established theory of elementary particles, makes sense only in a (3 + 1)-dimensional space-time, already Cornelius Lanczos (1962) declared that "we can hardly doubt that electricity has something to do with the specifically four-dimensional nature of the universe." The Maxwell equations, which can be established without "any appeal whatsoever to the notion of metric or length" (Misner and Wheeler, 1957), relate the electric field vector **E**, which as a polar vector has in an n-dimensional space n components, with the magnetic field vector **B**, which as an axial vector, or more precisely an antisymmetric second-rank tensor, has $\frac{1}{2}n(n-1)$ components. The necessary equality of the number of their components holds in an n-dimensional space only if $n = 3$, that is, only in a coordinate system in which the number of the coordinate axes equals the number of planes spanned by the axes. Only in an $(n + 1)$-dimensional space-time can Maxwell's theory be constructed upon an invariant integral form of the action and be conformally invariant. Only there do there exist two fields related as "duals," as Weyl once pointed out. Only in a (3 + 1)-dimensional space-time do Maxwell's equations determine the Maxwell field with the same "strength" as the Einstein equations determine the metric of the gravitational field (Einstein, 1956) or as the Weyl linear first-order matrix differential equation of the free neutrino determines the neutrino field (Penney, 1965). Furthermore, the Maxwell equations are fully compatible with quantum mechanics; if one calculates, for instance, the angular momentum in a multipole light-wave field, a quantity that must be an integral number of h, Maxwell's equations neatly avoid the impasse by yielding only ratios in which Planck's quantum of action h has cancelled out (Biedenharn, 1969). Finally, all information about the physical world reaches us only via the electromagnetic field. If then, in view of this truly astonishing all-pervasive reign of Maxwell's equations, we concede that the electromagnetic field is the prior datum of experience, then clearly the *raison d'être* of the (3 + 1)-dimensionality of space-time is the existence of the electromagnetic field and "that which constitutes the spatial character of reality is then simply the four-dimensionality of the field" (Einstein, 1953). But then also Kant's derivation of the three-dimensionality of space from the laws "according to which the forces in substances act upon another" is not so erroneous as his critics contend.

In contrast to the dimensionality of space, the dimensionality of time has scarcely ever been the subject of philosophical inquiry. It has played some role in certain scholastic debates about the Aristotelian doctrine of the unity and uniqueness of time. For if time, as defined by Aristotle, is the measure of motion, the obvious multiplicity of motion, it was argued, should imply a multiplicity of times ("*tot tempora quot motus*"). It was in this context that, e.g., Roger Bacon in his *Liber Primus Communium Naturalis* emphasized in 1245 that time "has only one dimension like length." A noteworthy exception is also a passage in Kant's *Inaugural Dissertation* in which time is represented as a one-dimensional line, intersected by surfaces (today we would say "hypersurfaces") of simultaneity, anticipating modern graphical representations of space-time (Kant, 1770). The controversial speculations of accounting for precognition or extrasensory perception, such as J. W. Dunne's (1927) theory of multidimensional time or C. D. Broad's thesis of two-dimensional time, had little relevance to physics. A serious attempt at rigorously deriving the one-dimensionality of time was made only within the context of the theory of relativity (Dorling, 1970).

2.2.2 The Directionality Problem of Time Just as Newton's first law, the law of inertia, by not specifying the reference system relative to which it holds, gave rise to problems in the philosophy of space (as explained in Sec. 2.2), so did Newton's second law give rise to problems in the philosophy of time. The second law, expressed in Euler's formulation, states that "force equals mass times acceleration" or $F = md^2x/dt^2$ and, being a second-order differential equation, is invariant under the time-reversal transformation $t \rightarrow -t$. Hence, if $f(t)$ is a solution of the equation of motion, $f(-t)$ is a solution too, so that, if any process occurs in accordance with the laws of mechanics, so does the reverse process. The same invariance holds for the Lagrangian or Hamiltonian equations of motion, and for the Maxwell equations of the electromagnetic field, as well as for the Lorentz force formula. Moreover, it also holds for the fundamental equations of modern physics, such as Einstein's field equations of the gravitational field in general relativity, the Schrödinger equation in quantum mechanics, and even the equations of quantum electrodynamics (Watanabe, 1951). And yet, experience shows that the time-reversed versions of processes such as the fall of an apple from the tree or of the emission of an electromagnetic wave or the heat flowing from a hotter to a colder body never occur in nature. The most elementary example of irreversibility is the case of two particles in a Newtonian universe. If initially they are approaching each other, they will eventually be moving apart; but if they are initially moving apart, they will never be approaching each other.

Already Aristotle mentioned in his *Physics* instances of such an irreversibility, and referring to the so-called "destructive power of time," he pointed out that "we are wont to say that time crumbles things" (221 a 32); but he emphasized that "after all, it is not really time itself that destroys things in this way, but the changes that do destroy take place in time" (222 b 26). Two thousand years later, physicists reaffirmed that the apparent asymmetry does not result from the properties of time itself but is rather due to an asymmetry inherent in the physical processes. They thus faced the question of how it is possible that, for instance, the changes or motions of a macroscopic body are not time reversible, whereas the laws of motion pertaining to its constitutive parts are.

The earliest attempt to answer this question was made in the context of thermodynamics and the statistical theory of entropy. In 1854, Rudolf Clausius announced what became known as the second law of thermodynamics, namely that the entropy of an isolated system never decreases. Ludwig Boltzmann, who had studied Maxwell's probability distribution of the velocities of molecules, made use of them in the construction of his famous H function, whose negative value behaves like entropy. By showing in his H theorem that on the molecular level a nonequilibrium state evolves toward equilibrium with an increase of entropy corresponding to an increase of randomness, he thought to have solved the problem of how individual time-reversible motions can account for collective time-irreversible motions.

However, his proof of the monotonic increase in time of the negative value of the H function contained a probabilistic postulate, the assumption of the molecular chaos (*Stoßzahlansatz*) according to which the velocity of a molecule is uncorrelated with its position. This probabilistic component of the proof was not only the ultimate reason of Josef Loschmidt's "reversibility paradox" and Ernst Zermelo's "recurrence paradox," it also led Boltzmann to admit that the entropy law itself has only probabilistic validity, so that there might exist in the vast expanses of space regions in which the entropy decreases and time "runs backward."

Although the conception of time "running backward" has sometimes been used in physics, as, e.g., in Richard P. Feynman's interpretation of an antiparticle as a particle moving "backward in time" for a fraction of a microsecond, in order to explain pair production and pair annihilation, there is little agreement among physicists and philosophers on what precisely it means that the direction of time is reversed throughout cosmic expanses envisaged by Boltzmann. Could intelligent beings, if they existed in such regions, regard their own time as "forward" and merely claim that the entropy decreases, as the philosopher Hans Reichenbach (1956) suggested? If it is true, as Norbert Wiener (1958) and F. Russell Stannard (1966) contended, that any exchange of information or any physical interactions between regions of opposing time directions are impossible, such considerations would always be doomed to be mere speculations.

Boltzmann's search for a reduction of thermodynamics to Newtonian mechanics was motivated primarily by his desire to save the mechanistical world picture and the unity of physics. His awareness of having ultimately failed in accomplishing this task may well have been one of the causes of his suicide. It is therefore noteworthy that more recently the incompetence of classical mechanics to describe irreversible processes has been challenged and it has been pointed out that there exist irreversible processes that do not involve any increase of entropy (Popper, 1956). In fact, time-irreversible solutions of Newton's equations actually have been constructed, albeit only by resorting to forces the existence of which is not vouchsafed by classical mechanics or by any other well-established theory in physics. Nonentropic, boundary-dependent, so-called *de facto* (in contrast to nomologically) irreversible processes have become a widely discussed topic in the philosophy of time (Grünbaum, 1963).

It has also been pointed out that it is a widespread error to invoke the second law of thermodynamics as an objective criterion of the direction of the "flow" of time from the past to the future, or what Arthur Stanley Eddington (1929) called "the arrow of time," to express the one-way property of time that has no analog in space. Since the entropy law has no factual content in the absence of an independent definition of the direction of increasing time, such a criterion would only debase this law to a mere tautology.

Alternative criteria for the definition of the direction of time, the so-called "nonentropic time arrows," have been proposed. The most important of these are the "electromagnetic time arrow," based on the exclusive acceptance of the retarded solutions of the electromagnetic wave equation, the "microscopic time arrow," based on the fact that a rare mode of the neutral K-meson decay can be explained only as a violation of the time-reversal invariance, the "cosmological time arrow," which derives the unidirectionality of time as a result of the expansion of the universe, the "quantum-mechanical time arrow," expressing the preparation-registration asymmetry of quantum states (Bohm *et al.*, 1994), and the "quantum-gravitational arrow of time" (Wald, 1980).

It has, however, also been maintained that the idea of a "time arrow" has no objective reality in nature nor that it is even required in theory, for "throughout the sciences of physics and chemistry, symmetrical or two-way time everywhere suffices ... we can no longer regard effect as subsequent to cause. If we think of the present as pushed into existence by the past, we must in precisely the same sense think of it as pulled into existence by the future" (Lewis, 1930). This theory of "symmetrical time" does not necessarily conceive of time as merely a psychological phenomenon, as has been done by numerous philosophers in the past; it only denies the existence of a preferential and structurally distinguished direction of time. It resembles, therefore, but is not identical with, the conception of spatialized time that characterizes the "block universe" interpretation of the space-time of relativity, which will be dealt with in the sequel.

3. ANTIMECHANISTIC THEORIES

3.1 Dynamism

The purely mechanistic world picture, most consistently proposed by Descartes according to whom only space (or extension, as he called it) and motion are the ultimate ingredients of physical reality, and subsequently modified in Boyle's kinetic theory of matter, suffered already a serious blow as soon as Newton ascribed to the notion of "force" physical reality. The "dynamization" of physics, which characterizes the renunciation of the mechanistic world picture in favor of a dynamical one, was a gradual and slow-paced process that began with Leibniz, who, in fact, introduced the term "dynamics" in his treatise *Specimen Dynamicum* (Leibniz, 1695). In opposition to Descartes, Leibniz declared that "agere est character substantiarum," "To act is the characteristic of substance; extension, on the other hand, is nothing but the continuation or diffusion of the substance already presupposed, which thrives, withstands and resists. . . ." This statement may well serve as the motto of "dynamism," the doctrine that "force" (in Greek δυναμις), which at that time denoted also what later was called "energy," has ontological supremacy in nature and is the ultimate element of physical reality. In contrast to the scholastic dictum "*operari sequitur esse*" ("action follows existence"), dynamism declares "*quod non agit, non existit*" ("what does not act, does not exist"). That in spite of such declarations Leibniz had to resort in his treatment of mechanical problems to the corpuscular theory of particle collisions had methodological reasons.

Physical arguments for the vindication of dynamism against the corpuscular-kinetic approach have been put forward especially by Roger Joseph Boscovich (1758) on the basis of the principle of continuity, which admits only continuous changes in nature in accordance with the proverbial "*natura non facit saltus.*" The ultimate particles of matter or atoms cannot be spatially extended, Boscovich argued, because this would imply a discontinuous change of density at their boundaries. Furthermore, in a collision of two particles their velocities would have to change abruptly when they rebound from each other. Hence, he concluded, atoms must be merely unextended or pointlike sources of force. To account for the impenetrability and gravitation of matter, he assumed that these forces are repulsive at small distances and become attractive in accordance with Newton's law of gravitation. Using modern terminology, one is tempted to say that the Boscovichian atom is a singularity of a field of forces.

Boscovich's ideas had a profound influence on many philosophers, amongst them Christian Wolff, Moses Mendelssohn, Immanuel Kant, Gustav Theodor Fechner, and Hermann Lotze, as well as on physicists like André Ampère, Augustin Cauchy, Siméon Poisson, and, most importantly, Michael Faraday. For Faraday, under the influence of Boscovich, was the first to propose experimental evidence for the physical reality of the notion of force. As a mathematical device that assigns to each point in a region of space some quantity, the value of which is a function of the coordinates of space and time, this notion had been used before by Lagrange, Laplace, and others. But if the criterion of the physical reality of a theoretical construct lies in the demonstration that this construct possesses detectable properties other than those for the sake of which it has been introduced, Faraday's famous experiments, described in his *Experimental Researches in Electricity and Magnetism* (1839), fully satisfy this criterion with respect to the notion of field. Although Faraday's philosophical interest focused primarily on the problem of whether electric and magnetic interactions propagate continuously through space or are actions at a distance, his work and its elaborations by James Clerk Maxwell and Heinrich Hertz led to a change in the conception of physical reality that Einstein (1934) called "the most profound and fruitful one that has come to physics since Newton." On another occasion, Einstein went even further when he wrote, "Under the influence of the ideas of Faraday and Maxwell the notion developed that the whole of physical reality could perhaps be represented as a field whose components depend on four space-time parameters. If the laws of this field are in general covariant, then the introduction of an independent (absolute) space is no longer necessary" (Einstein, 1953).

3.2 The Electromagnetic World Picture

Still after the introduction of the field conception the mechanistic world picture had a strong ally to keep it alive, the advantage of its psychological pictorability. The statement of William Thomson (Lord Kelvin) that we could really "understand a particular topic in physics" only "if we can make a mechanical model of it" characterizes this psychological need and explains why even a mathematically minded physicist, like Maxwell, described electromagnetic phenomena, and in particular the ether of optics, in terms of mechanical models. But Maxwell was fully aware that such pictorial representations should not be taken too seriously. "I wish merely to direct the mind of the reader to mechanical phenomena which will assist him in understanding the electrical ones. All such phrases in the present paper are to be considered as illustrative, not as explanatory," he wrote in his famous article "A Dynamical

Theory of the Electromagnetic Field" (Maxwell, 1865).

Only two years after his death, an electromagnetic field theory of dynamics made its debut with a study by Joseph John Thomson that envisaged the possibility of explaining mechanical inertia in terms of electromagnetism. A new research program intended to abolish the hegemony of mechanics by reducing it to electromagnetism rather than to reduce electromagnetism to mechanics. The central difficulty of explaining ordinary mass seemed to have been solved by showing that a moving charged particle has an effectively greater mass than its rest mass as a consequence of self-induction in accordance with Faraday's laws. Oliver Heaviside, George F. C. Searle, Wilhelm Wien, and most persistently Max Abraham, who declared in 1902 that "the mass of the electron is of purely electromagnetic nature," believed that ultimately all processes in nature would be reduced to convection currents and their radiations and that the "stuff" of the world would be stripped of its material substantiality. The metaphysical maxim underlying these ideas, namely that matter no longer does what it does because it is what it is, but is what it is because it does what it does, attracted much attention even on the part of more conservative physicists. Thus for example Boltzmann (1897), who had tried to reduce thermodynamics to mechanics, gave his blessings to this new program when he wrote, "... The advantage of deriving the whole science of mechanics from conceptions which anyhow are indispensable for the explanation of electromagnetism would be as important as if conversely electromagnetic phenomena were explained on the basis of mechanics. May the former succeed."

3.3 Energetism

In its attempt to renounce the supremacy of the mechanistic philosophy of nature, the electromagnetic doctrine had a serious competitor in the energetic monism or energetism, which claimed that the "stuff" of the physical universe is energy. Energetism referred to the empirical evidence that the various types of physical energy can be transformed one into another in accordance with the law of energy conservation. It interpreted these phenomena in terms of a single substratum that manifests itself as mechanical, thermal, chemical, electromagnetic, and possibly still other forms of energy. The metaphysical tenets of energetism can be traced back to the romantic *Naturphilosophie* of Friedrich W. J. Schelling, Lorenz Oken, and Wolfgang von Goethe, who criticized the mechanico-atomistic approach of Newtonian physics as incapable of giving a true account of nature, which they regarded as a vitalistic evolutionary system. Encouraged also by remarks made by Hans C. Oersted and Faraday, and searching for a scientific foundation of a nonmaterialistic world view, the energetists replaced the mechanistic-corpuscular conception of matter, motion, and the ether by the notion of energy.

One of the earliest energetists was the evolutionist Herbert Spencer, mostly known for his application of Darwinism to social developments. In his *First Principles*, using still "force" as synonymous with "energy," he wrote "We come down, then, finally to Force, as the ultimate of ultimates. Though Space, Time, Matter, and Motion, are apparently all necessary data of intelligence, yet a psychological analysis shows us that these are either built up of, or abstracted from, experience of Force," (Spencer, 1862). The most prominent advocate of energetism, however, was the chief founder of modern physical chemistry, Friedrich Wilhelm Ostwald. His energetism was based not only on his agreement with Ernst Mach that the atom is merely a convenient methodological device, but on much deeper metaphysical deliberations. What we sense, he claimed, is not the dubious and in itself entirely indefinite "matter" that classical physics thought to have brought within the compass of exact quantitative determination by its concept of inertial mass, it is rather a source of activities or energies that affect our senses. Ostwald claimed to have deduced Newton's second law of motion from the law of energy conservation, a deduction that would have decisively changed the course of modern physics. But as Boltzmann (1896) has shown, Ostwald's reasoning was mathematically flawed. Still, the idea of a "dissolution of matter into energy" became a topic of great interest. It was widely publicized, e.g., by Gustave Le Bon (1905) in a book published in the same year in which Einstein wrote his first paper on relativity and derived the mass-energy equivalence ($E = mc^2$) that prompted

Bertrand Russell (1948) to declare that "mass is only a form of energy . . . it is energy, not matter, that is fundamental in physics."

4. THE PHILOSOPHY OF MODERN PHYSICS

4.1 The Philosophical Background of Modern Physics

Twentieth-century physics extended the reach of scientific exploration to areas that previously had been inaccessible to research, such as high-speed phenomena or the structure of atoms. It soon became evident that, in order to cope with the newly discovered phenomena, not only the laws but even the conceptual foundations of traditional physics had to be subjected to a drastic revision. Questions concerning the precise meaning and the philosophical implications of the new conceptions confronted the physicist with the need of adopting a specific philosophical stance for his work. From the epistemological point of view, he could choose between *realism*, which maintains that the physical theory reflects, strictly or only structurally, the behavior of real objects, the existence of which was affirmed without doubt, and *positivism*, according to which the task of physics consists only in describing relations between measurable quantities or results of measurements, without specifying whether anything, yielding these results, does actually exist. From the metaphysical point of view, he has the option between *determinism*, according to which, like in classical physics, the final state of a system is unambiguously determined or predictable, provided the initial conditions and the laws are known, and *probabilism*, which admits chance as an unavoidable and inherent element at least in the realm of microscopic phenomena. Each of the four compatible combinations among these alternatives has a representative among the founders of modern physics.

4.2 Philosophical Problems of the Special Theory of Relativity

Special relativity was the first fully elaborated physical theory the conceptual foundations of which deviated fundamentally from those of classical physics. Traditionally, it is said to be based on two major principles; the principle of relativity, which for purely mechanico-kinematical processes had been announced already by Galileo (Galilean relativity), and the principle of the invariance of the velocity of light, which explained the null results of the well-known ether drift experiments, such as the Michelson–Morley experiment, performed at the end of the 19th century. Aware of the apparent contradiction between these two principles and puzzled by certain asymmetries in the explanations of electromagnetic phenomena, Einstein resolved these difficulties in 1905 by a drastic revision of the traditional conceptions of space and time. The critical step in these revisions was Einstein's operational definition of the simultaneity of spatially separated events. In fact, the only topic of the first section of his revolutionary paper "On the Electrodynamics of Moving Bodies" (Einstein, 1905) was precisely the concept of simultaneity from which, in the sequel of the paper, the relativistic effects, the relativity of simultaneity, the time dilatation, the length contraction, and even the Lorentz transformations are derived (in accordance with the two principles) as logical consequences. Taking into consideration that this procedure, as documentary evidence proves, was instrumental for the development also of quantum mechanics and modern physics as a whole, "It is hard to overestimate the impact of Einstein's definition of simultaneity" not only on the physics but also on the "philosophy of this century," as a contemporary philosopher of science expressed it (Nerlich, 1982).

Although the notion of simultaneity had been the subject of numerous discussions in the past (Jammer, 1986), Einstein realized that it had never been given a physically meaningful definition. For according to Einstein (1917a), "The concept does not exist for the physicist until he has the possibility of discovering whether or not the concept is applicable in a given case. We thus require a definition of simultaneity such that this definition supplies us with the method by means of which, in the present case, one can decide by experiment whether or not two events occur simultaneously." Einstein's approach, at that time, was obviously that of a staunch positivist, similar to that which had prompted Ernst Mach, 40 years earlier, to his well-known operational definition of the concept of mass (Jammer, 1961).

Einstein defined simultaneity in terms of clock synchronization as follows. Let e_A and e_B denote events that occur respectively at the points A and B of an inertial reference frame. A clock stationary at A, indicates "A time" and an identically constructed clock, stationary at B, indicates "B time." "Common time" can now be defined as follows. A light signal is emitted from A at "A time" t_A, is reflected from B at "B time" t_B, and returns to A at "A time" t'_A. If $t'_A - t_B = t_B - t_A$, then the two clocks are said to indicate "common time" and are called synchronized; and if they, thus synchronized, indicate the same time at the occurrences of e_A and e_B, these two events are by definition simultaneous.

Assuming that this procedure can be generalized for any number of points and that simultaneity is an equivalence relation, Einstein called the thus-generalized common time "the time of the reference system" or simply "coordinate time." In the sequel, he proved the relativity of simultaneity and of time by showing that two events simultaneous in one reference system are not so in another, and that the coordinate time of a given event differs in different reference systems.

Einstein's definition of simultaneity and the conclusions he drew from it gave rise to a number of logical, epistemological, and metaphysical problems. It was asked, for instance, whether Einstein's reasoning was not circular in view of the fact that his definition of time used the notion of an inertial system which, if defined as a system in which Newton's first law holds, obviously presupposes the notion of time. Another logical problem dealt with the question of whether the equivalence of simultaneity, rather than being postulated, can be obtained as a logical consequence of fundamental assumptions.

Interested in clarifying such problems and convinced that only an axiomatic formalism reveals the precise logical structure of a theory, Hans Reichenbach (1924) reformulated the theory of relativity as a rigorously deductive axiomatized system. Faced with the choice of what to posit as axioms and what as definitions, Reichenbach distinguished sharply between the empirical or factual components and the definitional or conventional components of the theory. Calling a definition, or more generally any theoretical construction, conventional if it can be replaced by an alternative that, though logically incompatible with the former, has the same observational consequences, he realized that Einstein's simultaneity definition is conventional. More precisely, he argued, the empirical facts of the theory of relativity allow replacing the Einsteinian condition $t_B = t_A + \frac{1}{2}(t'_A - t_A)$ by the more general alternative $t_B = t_A + \epsilon\,(t'_A - t_A)$, where (for causality reasons) $0 < \epsilon < 1$, without ever leading to conflict with experience.

The thesis of conventionality, namely that no facts of nature preclude the choice of an arbitrary value of ϵ between 0 and 1, gave rise to heated debates and is still today a controversial issue. The traditional method to establish simultaneity or, equivalently, to synchronize distant clocks, by sending a light signal—or any other signal for that matter—from one clock to the other and taking into account the time lapse due to the transmission velocity of the signal, was countered by Reichenbach by pointing out that "to determine the simultaneity of distant events we need to know a velocity, and to measure a velocity we require knowledge of the simultaneity of distant events. The occurrence of this circularity proves that simultaneity is not a matter of knowledge" but a matter of convention. In fact, if c denotes the two-way velocity of light that has been measured by Fizeau, Foucault, Michelson, and others with the help of light-reflecting mirrors, etc., that is, in the above-mentioned notation, twice the distance between A and B divided by $t'_A - t_A$, where only the clock in A has been used showing that c is convention free, then Reichenbach's alternative equation yields for the one-way velocity of light from A to B the value $c/2\epsilon$. Hence, knowledge of ϵ and knowledge of the one-way velocity imply each other as Reichenbach contended in his *The Philosophy of Space and Time* (Reichenbach, 1928), which has been called "the greatest work in the philosophy of physics in the 20th century" (Salmon, 1977).

The professional literature abounds in ingenious proposals for convention-free measurements of the one-way velocity of light and for other very sophisticated contrivances to disprove the conventionality thesis. A historical example is the "twin-rod experiment" of two rods of equal rest length, moving in opposite directions with equal speeds; the meetings of their extremities have been claimed to be simultaneous events. This argument, based on what is one of the most simple experiments conceivable, has misled, among others, even Wolfgang Pauli (1958), one of the shrewdest critics physics ever had, to declare that it "shows that the determination of the simultaneity of spatially separated events . . . can be carried out entirely with the help of measuring rods, without the use of clocks." The fallacy of this and other proposals to this effect follows also from the fact that John A. Winnie (1970) and others succeeded in foregoing deliberately any choice of ϵ and in deriving from premises not containing any hidden one-way velocity assumptions the equations of relativistic kinematics with ϵ as a free parameter, and yet in obtaining all experimentally testable or "factual" results of the theory. However, David Malament's (1977) challenge of Grünbaum's (1963) profound elaboration of the conventionality thesis within the context of the causal theory of time has been regarded by many, though not by all, experts in this field as a serious blow to the thesis.

The metaphysical import of this issue can be gauged by the following considerations. The concept of the "present" or "now," defined as the totality of simultaneous events, or more precisely according to Minkowski as the simultaneity hypersurface through the "here-now" of the observer, depends not only, as relativity implies, on the state of motion of the observer or on the inertial system in which he or she is at rest, but also on the arbitrary choice of ϵ. Moreover, if only the present is real and the past and future unreal, as, e.g., Arthur N. Prior (1970), one of the founders of modern tense logic, persuasively contended, then reality itself is not only relative but also a conventional conception, which sounds like a *contradictio in adjecto*.

The relativization of simultaneity and time was criticized by many philosophers, and most severely by Henri Bergson, on the following grounds. Relativization may be applicable, perhaps, to "spatialized time" as used by Minkowski in his geometrical representation of space-time, it was argued; but since real time, which exhibits itself in dynamical processes and in the unfolding of states of consciousness, loses its uniqueness if joined with, or substituted by, any spatial representation, the time of relativity is but a misleading fiction.

This dynamical point of view stands in sharp contrast to the static "block universe" conception of time, according to which "the objective world simply *is*, it does not *happen*. Only to the gaze of my consciousness, crawling upward along the lifeline of my body, does a section of this world come to life as a fleeting image in space that continuously changes in time" (Weyl, 1949). As has been objected, however, even were the block-universe conception successful in eliminating temporal succession in the physical world, it could not avoid acknowledging the progress of time at least in the movement of our consciousness to the future, and every attempt to exclude dynamical time introduces somewhere at least surreptitiously the dynamical reality of time.

Nevertheless, the idea of the static four-dimensional universe found support in certain chronogeometrical analogs to the scholastic logical arguments for predestination and fatalism, which can be traced back to Aristotle's treatment of determinism in his *De Interpretatione*. The logical problem was this: Since according to the law *tertium non datur* a proposition is either true or false, and is so at all times, one of two contradictory propositions, like "tomorrow there will be a sea-battle" and "tomorrow there will not be a sea-battle," must be true today; hence, future contingencies are determinate before they occur. The chronogeometrical analogs in question are Cornelius W. Rietdijk's (1966) "rigorous proof of determinism from the special theory of relativity" and Hilary Putnam's (1967) similar proof of his claim that future events are real already in the present. Both proofs are based on the fact that for any two events e_1 and e_2 in Minkowski space-time there always exists a third event e_3 in space-like separation from them and hence simultaneous with e_1 in one inertial frame and with e_2 in another. If in particular e_1 is my "here-now" and e_2 any event in my future, then e_3 is happening in my "now" and also in the "now" of e_2. But if what happens now is determinate and real, and if determinateness and reality are frame independent, then e_2 is just as determinate and real as e_1. It follows that the future is determinate and real already in the present and that the notions of an "open future" and "free will" are fictitious.

Both proofs have been rebutted by Howard Stein (1968) and Roberto Torretti (1983) on the grounds that both arguments resort surreptitiously to the prerelativistic temporal structure of using the tenseless present for the pronouncement of their ontological assertions.

The "twin-rod experiment," mentioned above, was originally designed by Einstein (1911) to decide between two interpretations of the relativistic length contraction, whether it was a purely metrological effect, resulting from the measurement procedure or the "perspective of observation," like the change of a rod when viewed from different angles, or whether it is an intrinsic phenomenon as real as "a change of the length of a rod, caused by the change of temperature" (Podlaha, 1975). That the ether-theoretic FitzGerald–Lorentz contraction, which had been proposed to account for the null result of the Michelson–Morley experiment, had to be regarded, within the context of the Lorentz theory of electrons, as a real phenomenon, has never been questioned since it was explained as a result of the variation of the intermolecular electromagnetic forces due to the motion through the ether. Nevertheless, it became the paradigm of an *ad hoc* hypothesis in the sense of not being testable by experiments that differ in principle from the Michelson–Morley experiment (Grünbaum, 1963). It thus initiated numerous discussions on the logical and epistemological status of *ad hoc* hypotheses in physics and science in general. The Lorentz–Larmor time dilatation, on the other hand, was hardly referred to in this context, probably because the relativistic time dilatation, in contrast to the length contraction, was generally regarded as empirically well established by the Ives–Stilwell transversal Doppler effect, the lifetime measurements of muon decay in cosmic rays, and the comparison of moving atomic clocks, even though the strength of this evidence is sometimes called into question (Huang, 1993).

4.3 Philosophical Problems of the General Theory of Relativity

A philosophical analysis of a physical theory has first of all to define the subject matter of the theory. That this is not a trivial issue is shown by the fact that the special theory of relativity has been variously classified as a geometry of light, as a theory of space and time, or as part of the electromagnetic theory of Maxwell. In fact, the German periodical *Fortschritte der Physik* used to list its abstracts of papers on relativity under the heading "Elektrizität und Magnetismus" until Max von Laue (1913) pointed out that the relativity theory deals with all branches of physics and their theories and serves "as a criterion for the admissibility of all of them." More recently, Eugene P. Wigner (1962) compared it with a "superlaw" that restricts the possible forms of laws to satisfy an invariance or symmetry principle.

In order to find out what the general theory of relativity is about, it should be recalled that Einstein's 1905 theory is named "special" or "restricted" because it deals with physical phenomena in inertial systems, a restriction that from the purely kinematical point of view seems unwarranted and philosophically unsatisfactory. The question of how to formulate the laws of physics for arbitrary reference systems motivated Einstein's search for a "general" theory, the "clearest logical characterization" of which is according to Einstein (1955) the statement that it is "the theory which avoids the introduction of the 'inertial system'." Clearly, for Einstein the general theory was an extension or "generalization" of the special theory. Studying at first the simplest noninertial motion, the uniformly accelerated motion of a freely falling body in a homogeneous gravitational field, and being aware of the experimentally well-established equality between inertial and gravitational mass, Einstein postulated the (strong) "principle of equivalence" according to which every freely falling, nonrotating reference system constitutes a local inertial frame. This principle, in turn, led him to the revolutionary discovery that gravitational phenomena are reducible to, and explainable by, the Riemannian geometry of curved spacetime. The fact that the ensuing theory of gravitation was generally convariant, i.e., valid in arbitrary reference systems, justified in Einstein's view calling it the "general theory of relativity."

That this term is misleading and confusing was argued by Vladimir Fock (1959); for already in Newtonian mechanics, he pointed out, "One deals with the generally covariant Lagrange equations of the second kind, so that one would have to say that Newtonian mechanics contains in itself 'general' relativity." Fock's proposal of calling Einstein's 1916 theory a relativistic "theory of gravitation" has been accepted by many physicists.

If one tries to resolve this controversy by investigating the asymptotic approximations of the theory, one will find that for vanishing gravitation the theory reduces to the special theory of relativity, justifying Einstein's point of view, but that for slow motions and weak static fields it reduces to the classical gravitation theory of Newton and Poisson, justifying Fock's point of view. But Newtonian gravitation, based on the notion of actions at a distance, and special relativity, excluding such interactions and even superluminal velocities, are irreconcilable. The general theory exhibits therefore the philosophically interesting feature, rarely encountered elsewhere, of having not only different but even incompatible asymptotic approximations. This poses a problem since quite frequently the meaning of a high-level scientific concept is given its intuitive clarification in terms of its application in an asymptotic approximation of the theory.

No less unsatisfactory than the restriction to inertial systems and "contrary to the spirit of science" was the causal asymmetry between matter and space: Space affects matter, as in the aforementioned bucket experiment performed by Newton, but matter does not affect space. In the general theory of relativity, space and matter, or more precisely mass-energy, become mutually interacting partners: Mass-energy determines the geometry of space-time, and the latter determines the motion of matter. The mathematical expression of this interdependence is provided by the famous covariant field equations, which have been called "one of the greatest achievements of the human mind." Reducing the whole mechanics of gravitation to the solution of a single system of covariant partial differential equations, these field equations proved very successful: Whenever their predictions deviated from those of classical physics, observation confirmed their predictions. The geometry defined by these equations changes constantly in accordance with the changing mass-energy distribution, while its curvature, i.e., the deviation of its space-time metric from the flatness of Euclidean geometry, accounts for the physical phenomena usually ascribed to a gravitational field. General relativity dispelled not only the previously prevailing doctrine of the Euclidicity of physical space—which Kant regarded as an *a priori* necessary condition of experience—but also the erroneous identification of abstract logically formalized geometry with the geometry of physical space, which latter is the product of empirical investigation. Referring to this differentiation, Einstein (1921) declared, "As far as the propositions of geometry refer to reality, they are not certain; and as far as they are certain, they do not refer to reality."

The question of whether geometry is a corollary of physics or physics is a corollary of geometry has been a subject of philosophical discussions long before the advent of geometrodynamics according to which empty curved space-time is the only "stuff" of physical reality (Wheeler, 1980a). Even the fundamental problem of how the geometrical structure of a region of space or space-time can be empirically determined by means of measuring devices that themselves are physical objects and hence subjected to the deforming influence of gravitation, a problem dealt with already by Poincaré, is still a controversial issue. The epistemological holism claims that geometry and physics form an inseparable whole; the validity or nonvalidity, say, of Euclidean geometry cannot be proved by empirical observation, because it is only the whole composed of physics and geometry that can be tested empirically. "We require this whole network of theoretical considerations to arrive at an experimental means of verifications" (Weyl, 1992). This argumentation has been criticized, e.g., on the ground that "it would then have to be acknowledged that there is a sense in which physical geometry itself does not provide a geometric characterization of physics reality" (Grünbaum, 1963).

To generalize the special to the general theory, Einstein (1916) formulated "the postulate of general covariance," demanding that "the general laws of nature are to be expressed by equations which hold good for all systems of coordinates, that is, are covariant with respect to any substitution [coordinate transformation] whatever." This postulate implies, he claimed, what he called "the postulate of general relativity" according to which "the laws of physics must be of such a nature that they apply to systems of reference in any kind of motion." To prove that the latter principle is a logical consequence of the former, he pointed out that "the sum *of all* substitutions [coordinate transformations] in any case includes those which correspond to all relative motions of three-dimensional systems of coordinates." Einstein's argument was soon criticized by Erich Kretschmann (1917) on the ground that all physical theories, independently of the physical content of their laws, are expressible in generally covariant form and hence would have to satisfy the general relativistic principle; this principle could therefore not lead to a new physics entailing factual consequences not encountered before. This controversy initiated extended discussions on the nature and status of the principles of relativity, covariance, invariance, and symmetry, and led to the distinctions between coordinate systems and reference frames, between laws and metalaws (Bunge, 1961), and between "dynamical objects" and "absolute objects" (Anderson, 1967), all designed to resolve the conceptual muddle.

Another principle, that has also played an important, though only heuristic, role in the attempt to extend the relativity principle to all frames of reference, and which has likewise intrigued philosophers of physics, is what Einstein has called "Mach's principle." Briefly stated, it claims that the inertial mass of a body results from its interaction with other bodies in the universe. Because of the vagueness of its formulation and its cosmic connotations, it has never lost its philosophical appeal. True, the field equations of the theory partly incorporate this principle. For according to them the metric tensor, i.e., the geometrical structure of the inertial or gravitational field, is determined by the energy-stress tensor, that is to say, by the distribution of matter and energy in the universe. Since, however, these differential field equations are of the second order, boundary conditions have to be imposed in order to obtain solutions. But the very idea of boundary conditions is contrary to the spirit of the Machian program, according to which the observed universe should self-referentially supply the terms of its own description. Some authors, like Grünbaum (1963), regard the need of boundary conditions as a substitute for absolute space, claiming that the general theory has not at all "exorcised the ghost of Newton's absolute space." Others, like Helmut Hönl (1953), interpret Mach's principle as a constraint to serve as a selection rule to separate admissible from inadmissible solutions of the field equations. Still other philosophers, following Lawrence Sklar (1974), dispose of the principle by simply accepting the existence of inertial and noninertial systems as "just a brute, inexplicable fact about the world."

The question of how to save the general theory from the need of imposing anti-Machian boundary conditions was precisely the issue that led to the establishment of modern cosmology. For it changed the cognitive status of traditional cosmology by shifting it from the realm of poetical, theological, and metaphysical speculations to that of a rigorously established physical science. The pivotal idea, first conceived by Paul Ehrenfest in 1916, but fully elaborated only by Einstein, was to eliminate boundary conditions by conceiving the four-dimensional universe as a spherically closed manifold, finite with respect to its spatial dimensions, and to let matter turn space back on itself—in analogy to the two-dimensional unbounded, but finite, surface of an ordinary three-dimensional sphere.

In his foundational paper "Cosmological Considerations in the General Theory of Relativity," Einstein (1917b) constructed on the basis of the field equations (modified by the addition of the so-called "cosmological constant") an isotropic, homogeneous, unbounded, spatially finite, and static model of the universe. Its material content was thought to exist of a pressureless fluid or "dust," uniformly smeared over all space-time. Surprisingly, not much later, Willem de Sitter found a non-flat solution of the same field equations devoid of all matter. Clearly, the existence of two disparate solutions of the same field equations was a fatal blow to Mach's principle. After strenuous efforts to save the principle, Einstein rejected it and its underlying positivist philosophy and declared that scientific theories are free creations of the human mind; they are not derivable from sense experience but have to be confirmed by it.

The subsequent development of relativistic cosmology, from Aleksandr A. Friedmann's construction of models of an expanding universe as solutions of the field equations (without the cosmological constant), their subsequent observational confirmation by Edwin P. Hubble, until, say, Kurt Gödel's solutions yielding an absolutely rotating universe, disproved not only the validity of Mach's principle, but also that of Einstein's cosmological model of 1917. But the same development also shows that sometimes it is not the answer but the question and method of tackling it that are of lasting importance.

The promotion of traditional cosmology to the status of a respectable physical science with the help of general relativity was greatly enhanced when in the recent decades relativity, in cooperation with particle physics, investigated not only the structure of the universe but also its physical history or what is traditionally called cosmogony. By developing, for example, the by now widely accepted Big Bang theory and studying baryosynthesis and nucleosynthesis, as well as the formation of galaxies and stars, cosmology could scientifically address the age-old problem of how the universe began and how it will end. Like other branches of physics, cosmology made specific quantitative predictions that subsequently have been borne out by observation as, e.g., in the case of the well-known 3-K microwave radiation.

However, the object of these studies is the universe as a whole and not, as in all other branches of physics, a spatially and temporally small-sized portion of it. This raises the question of whether this peculiarity of cosmology allows free use of the research methodology employed in ordinary physics. For if the universe is by its very definition necessarily unique in the sense of not being a member of an ensemble and *a fortiori* not being reproducible, then clearly inductive, statistical, and probabilistic methods, employed in ordinary physics, are inapplicable.

Some philosophers distinguish between essential or necessary and accidental or contingent uniqueness, the former being the case if any imaginable alternatives to the actual universe are necessarily impossible; the latter, if it is conceivable that different kinds of universes might have been formed. In this latter case, the additional question arises: Why is the actual universe just the one that has come into existence?

It is precisely this last question that has recently become a much-debated topic in the philosophy of physics. For modern physics has shown that the universe in which we live is "finely tuned" to the conditions of human life in the sense that, if the values of the fundamental physical constants, like the charge of the electron, the strength of the strong-nuclear force, or the gravitational constant, differed even only slightly from their actual values, life would have been impossible. One may, of course, answer that this state of affairs is simply a matter of pure coincidence, or the result of a purpose or design.

Eschewing teleological explanations, the proponents of what Brandon Carter (1974) called the "anthropic principle" (or "cognizability principle") have given a different answer, which departs, however, from the traditional method of scientific reasoning. It says "that what we can expect to observe must be restricted by the conditions necessary for our presence as observers." More precisely, Carter distinguished between the basic and less controversial *weak* version of the anthropic principle, according to which "our spatio-temporal location in the universe is necessarily privileged to the extent of being compatible with our existence as observers," and the philosophically much more audacious *strong* version, according to which the universe "must be such as to admit the creation of observers within it at some stage."

Historically viewed, both versions generalize Descartes' famous "Cogito, ergo sum" ("I think, therefore I exist") to "Cogito, ergo mundus talis est" ("I think, therefore the world is as it is"), and both versions qualify the so-called (but misnamed) "Copernican principle," which denies any privilege to man's location in the universe, by asserting man's epistemological centrality. Both versions maintain an ontological and nomological compatibility relation between man and the universe as it is; but the strong version affirms in addition that any counterfactual variation of the laws or initial conditions of the universe would exclude the possibility of man's existence. A still stronger version is John A. Wheeler's (1980b) "participatory anthropic principle," "Observers are necessary to bring the universe into being," implying that in a certain sense man "participates" in the creation of the universe. The epistemological basis of Wheeler's self-referential cosmogony, namely that "no elementary phenomenon is a phenomenon until it is observed or registered," is a logical consequence of his interpretation of quantum mechanics, which will be discussed in due course.

4.4 Philosophical Problems of Quantum Mechanics

Since its inception in 1925, modern quantum mechanics, established primarily by Werner Heisenberg, Erwin Schrödinger, Max Born, and Paul A. M. Dirac, has been stupendously successful in accounting for virtually all phenomena in atomic and molecular physics, in solid-state physics, and chemistry, as well as in its contributions to nuclear physics and to the physics of elementary particles. But in no other theory is there such an irreconcilable disagreement among physicists and philosophers as to what the theory tells us about physical reality, and are physics and philosophy so inextricably entangled, as in quantum mechanics. In spite of an unprecedented sharpening of the logical, mathematical, and semantical tools for the philosophical analysis of physical theories, the ontological and epistemological problems of quantum mechanics have not only not been resolved but have even been aggravated in the past few decades. These difficulties are partly the result of the fact that, unlike the situation in any other theory of physics, the formalism of quantum mechanics preceded its interpretation.

Opinions already disagree on the referent of the theory, that is, on what precisely the theory or its formalism is about. According to the standard version, the so-called Copenhagen interpretation, the mathematical formalism in terms of the state vector or Schrödinger wave function provides a description of the individual quantum system, whereas the ensemble interpretation refers the state vector to a statistical collection of systems.

Nor is there any consensus on whether the immediate referent is physical reality itself or merely knowledge of it. Those who agree with Einstein (1949) that "physics is an attempt conceptually to grasp reality as it is thought independently of its being observed," and that "the belief in an external world independent of the perceiving subject is the basis of all natural science," insist that quantum mechanics should be a realist theory. The opposing instrumentalist or subjectivist position is taken by those who agree with Bohr that "there is no quantum world; there is only an abstract quantum physical description; it is wrong to think that the task of physics is to find out how nature *is*, physics concerns what we can say about nature" (Petersen, 1963), or with Heisenberg (1958), that "the subject of research is no longer nature itself, but man's conception of nature." It should be noted that some positivistically inclined critics refuse to attach any philosophical significance to a controversy like that between Einstein and Bohr and regard it rather, as Ernest Nagel (1961) called it, as "a conflict over preferred modes of speech." But, as we shall see at the end of our discussion, the endorsement or rejection of a philosophical position may well be motivated by physical arguments rather than by the predilection for a certain mode of speech.

If the question concerning the referent of quantum mechanics is, indeed, a matter of physical considerations, it may be tempting to search for the answer by way of applying the method of asymptotic approximations, as it has been applied above in the case of general relativity. Although there exist, in fact, different and—surprisingly—nonequivalent approximation methods (Planck's method of $h \rightarrow 0$, Bohr's method of $n \rightarrow \infty$, where n is the principal quantum number, and Ehrenfest's method of taking the expectation values of the observables), a closer analysis (Liboff, 1984; Ajanopon, 1987) shows that this method cannot supply any answer.

The tenability of the ontology of classical physics and hence also the maintainability of its realism for microphysics have been questioned already by the "old quantum theory." This term denotes the microphysics of the period between 1900 and 1924, i.e., from Planck's explanation of the black-body radiation in terms of the quantum of action h, and Einstein's subsequent interpretation of the photoelectric effect, through Bohr's theory of the hydrogen atom until Louis de Broglie's thesis of the wavelike behavior of material particles. In spite of its name, the "old quantum theory" has never become a full-fledged scientific theory. Its principal fault was its indiscriminate conflation of classical and nonclassical conceptions, making a coherent interpretation impossible.

The development of a conceptually coherent and logically consistent quantum theory began in 1925 with the advent of Heisenberg's matrix mechanics and Schrödinger's wave mechanics, and reached its maturity about 1930 with John von Neumann's rigorous proof that both formalisms, in spite of their apparent disparities, are but different representations of the same operator calculus in Hilbert space. In view of the unvisualizability of matrix mechanics, wave mechanics with its presentation of particles in terms of wave packets gained greater popularity. But the "spread" or instability of these wave packets and other peculiarities caused serious difficulties. To resolve these difficulties, Max Born (1926) proposed the statistical interpretation of the wave function as a "probability wave," the amplitude of which specifies the probability of the particle's location. The wave function thus becomes a catalog of possible probability distributions, where the word "possible" refers to the choice of the observer as to what observable he wishes to measure, while the word "probability" refers to the observer-independent relative frequency of obtaining a specific value of the chosen observable. Born's introduction of the notion of probability into quantum mechanics is generally regarded as marking the demise of the determinism of classical physics. True, classical physics has made ample use of the probability concept, as for instance in the theory of errors in measurement procedures or in statistical mechanics; but classical probability was always thought to be due to human ignorance of a more refined specification of the physical data. In quantum mechanics, however, at least according to Born's interpretation, probability became an objectively existing, intrinsic, and irreducible component of physical ontology and the source of an uneliminable indeterminism of the theory.

Laws of nature no longer, as they do in classical physics, determine the occurrence of an event but only the probability of its occurrence. This raises the problem of the precise ontological status of this probability. Its temporal evolution and spatial propagation in accordance with the Schrödinger equation seem to indicate that it is not merely a mathematical construct. Since probability waves do not transmit energy or momentum and hence can hardly be regarded as physically "real," Heisenberg (1961) compared their ontological status with that of the Aristotelian notion of *potentia*.

Heisenberg's (1927) indeterminacy relations, which exclude the possibility of simultaneously specifying the values of two canonically conjugate observables, like position and momentum, to within arbitrary degrees of accuracy, have also been regarded as a confutation of determinism. For determinism states that *if* the initial state, e.g., position and momentum, of a system is exactly known, then the state of the system at any other time can be predicted. But Heisenberg's principle denies the possibility of ascertaining the initial state. The objection that the falsity of the premise of an implication does not necessarily entail the falsity of the implication was rebuffed by Heisenberg on the ground that any conception of a deterministic universe, hidden behind the statistical universe of perception, is a meaningless speculation since physics deals only with the relationships between perceptions.

The indeterminacy relations are intimately related to Bohr's (1927) complementarity principle and generally regarded as its quantitative aspect. Complementarity requires that the quantum object be described in two different, even contradictory, ways, which are complementary in the sense that both are necessary to exhaust the possibility of obtaining information about the object, even though the two ways are experimentally not simultaneously performable. Complementarity is in Bohr's view the only possibility of generalizing the classical mode of description without ignoring the quantum of action. His belief in the unity of knowledge led him to extend the epistemology of complementarity to other branches of science, like psychology or biology, and even to consider it as a means to reach greater mutual understanding among nations and cultures.

The best-known example of complementary descriptions is the wave-particle duality; and the easiest way to demonstrate it is the well-known two-slit experiment which, according to Richard Feynman (1951), "has in it the heart of quantum mechanics . . . and contains the *only* mystery" of this theory. In fact, this experiment, which, incidentally, has recently been actually performed with neutrons, implies a wave description because of the interference effect if both slits are uncovered, and a particle description if only one slit is uncovered. Moreover, since in any case, the interference case included, the photographic plate records pointlike and not spread-out events, the experiment may be regarded in its final stage also as a measurement of position and as illustrating thereby the so-called "collapse of the wave packet" or "reduction of the wave function," a characteristic feature of every measurement process in quantum mechanics.

The very notion of a quantum-mechanical measurement raises a number of philosophical problems not encountered in classical physics. Every measurement or observation requires an interaction between the measured or observed system and the measuring device or observer. In classical physics, in which the action of the latter on the former is assumed to be negligible or eliminable by correction, this interaction is a one-way process; but in quantum mechanics, because of the indivisibility of the quantum of action, it is a two-way process with the dire consequence that, in general, a measurement of an observation changes the quantum state of the system measured or observed. Furthermore, since such measurements necessarily involve macroscopic measurement devices to yield pointer readings, and since quantum mechanics should include classical physics as a special case, the ensuing logical circularity between quantum and classical concepts questions the self-consistency and conceptual autonomy of quantum mechanics.

Aware of this difficulty, Bohr regarded the measurement outcome as depending upon the entire experimental setup, combining object and apparatus, and conceived the object-apparatus interaction as epistemologically irreducible. In order to eliminate this irreducibility, it has been deemed necessary to include the measurement process into the formal account of quantum mechanics. How to do this is the task of the "measurement theory of quantum mechanics."

Such theories, most of them modifications of a proposal made by von Neumann (1932), face two difficulties, namely the questions of how and where the so-called "collapse of the wave packet" takes place, questions that are controversial issues still today. For if the measuring device itself is treated as a quantum system, it adds to each component of the superposition, which describes the initial state of the object, an additional term instead of reducing it to an eigenstate corresponding to the eigenvalue of the observable that is to be measured. In other words, object and measuring device form a combined system for the measurement of which a second measuring apparatus is required. This process obviously leads to an infinite regress.

Among the numerous suggestions to resolve these difficulties, the proposal made by Fritz London and Edmond Bauer (1939) attracted much attention because of its relevance to the mind-body problem. They suggested that it is the human "faculty of introspection" (*"connaissance immanente"*) that actualizes the reduction of the superposition in the consciousness of the sentient observer. The London-Bauer measurement theory has far-reaching philosophical implications. Since according to quantum mechanics every physical interaction between particles involves a superposition, a world without a sentient observer, who reduces the superposition, would be a limbo of mere potentialities that never actualize into factualities; it would be a world in which nothing whatever happens. It was precisely this consideration that led Wheeler to the formulation of his aforementioned "participatory anthropic principle."

Critics objected that this measurement theory, although free from logical inconsistencies, is hardly compatible with the intersubjective agreement among independent observers. Furthermore, they pointed out, the notion of a mental act should not serve as a primitive concept in physics, for too little is known about the mind-body problem or about the brain, which is supposedly the seat of consciousness and probably the most complex system in the whole universe. This observer-dependent measurement theory in their view simply explains a riddle by a mystery.

To avoid this impasse, alternative measurement theories have been proposed, some of them resorting to the irreversibility of macroscopic systems, others to a hypothetical micro-to-macro amplification process, and still others to a modification of the very structure of quantum mechanics into a nonlinear theory. But so far no satisfactory solution seems to have been obtained. The well-known "paradoxes" of "Schrödinger's cat" (1935a) and of "Wigner's friend" (1962) illustrate the bizarre conclusions of such measurement theories.

An attempt to dispense altogether with the reduction of the wave function has been made by Hugh Everett III (1957) and elaborated by Bryce S. DeWitt and Wheeler in the ontologically most daring and epistemologically most provoking theory ever made in physics. By regarding the quantum object, the measuring apparatus, and the observer as one single quantum system, described by a state vector that never collapses, their theory eliminates the need for an extraneous observer or consciousness and affirms that physical reality is rigorously deterministic. The theory implies a continual splitting of the universe into a multitude of noncommunicating worlds, each corresponding to a component of the superposition, and it is therefore called "the many-world interpretation." Since no experiment performed in one branch of the splitting universe can detect the outcome of a measurement obtained in another branch, appearances are saved. The theory has been criticized as violating Ockham's rule that entities should not be multiplied without necessity, as ambiguous in its determination of the branches of the splitting universe because a given state function can be expanded in different sets of functions, and as not providing an explanation of why the observer is just in that branch in which he happens to be.

As is well known, the orthodox interpretation of quantum mechanics has been strongly opposed by physicists, like Einstein and de Broglie, as well as by philosophers, like Karl Popper and Mario Bunge. Most noteworthy in this context is the long-standing debate between Einstein and Bohr (Bohr, 1949), in which Einstein challenged the probability interpretation and the indeterminacy relations by a number of thought experiments that Bohr successfully rebutted on the basis of his complementarity interpretation (Jammer, 1974). One of the highlights of this debate was the argument in which Einstein presented his thought experiment with a photon box in order to disprove the time-energy indeterminacy relation, and which Bohr refuted by referring Einstein to his own general theory of relativity (Bohr, 1949).

Although defeated by Bohr, Einstein did not lose interest in the photon box. To eliminate the troublesome gravitational effect, he regarded the box and the photon as two interacting systems that then move apart along a horizontal line. This led him to the physical scenario of the Einstein–Podolsky–Rosen (EPR) argument against the completeness of quantum mechanics. The published version (Einstein *et al.*, 1935) owes its central idea to Einstein, its mathematical elaboration to Nathan Rosen, and its final formulation to Boris Podolsky. Although Einstein, as we know from his correspondence with Schrödinger, did not fully agree with the printed text, it was, of course, the published version that affected the subsequent conceptual development of the theory.

The paper begins with a necessary condition for the completeness of a physical theory: "Every element of the physical reality must have a counterpart in the physical theory," and defines "physical reality" by the sufficient condition: "If, without in any way disturbing a system, we can predict with certainty . . . the value of a physical quantity, then there exists an element of physical reality corresponding to this physical quantity." Accordingly, physical reality is not to be defined by *a priori* philosophical considerations, "but must be found by an appeal to results of experiments and measurements." In addition to these two explicitly stated stipulations of completeness and reality, a third premise is assumed that is mentioned only *en passant*: If two systems no longer interact, "no real change can take place in the second system in consequence of anything that may be done to the first system." For Einstein this premise or "principle of separation," as he called it in a letter to Schrödinger, was an essential feature of his realist philosophy. "But on one supposition," he wrote (Einstein, 1949), "we should, in my opinion, absolutely hold fast: the real factual situation of the system S_2 is independent of what is done with the system S_1, which is spatially separated from the former." Actually, this supposition is a conflation of two different principles, the "separability principle" and the "locality principle." The former states that the spatio-temporal separation is a sufficient condition for the individuation of a system or its state, and the latter, that the state of a system at a given space-time location can be affected only by events within or on the past light cone of that location. The separability principle is a metaphysical maxim; the locality principle is a physical feature of the special theory of relativity and excludes, in particular, any simultaneous action at a distance.

EPR consider a system of two particles S_1 and S_2 that are widely separated in space after having interacted with each other. They are supposed to be in a state in which their total momentum $p_1 + p_2$ and their relative position $q_1 - q_2$ have definite known values. The existence of such eigenstates is assured by the fact that the operators for this sum and difference commute. From these assumptions, EPR infer that, if they measure the momentum of S_1, they can predict with certainty the momentum of S_2 without disturbing it; "if, instead of this, [they] had chosen" to measure the position of S_1, they could have predicted with certainty the position of S_2 without disturbing it. Hence, the reality criterion implies that the momentum as well as the position of S_2 has reality. Were quantum mechanics a complete theory, it should therefore, in accordance with the completeness criterion, ascribe definite values to the momentum and position of S_2. But quantum mechanics does not do this because the respective operators do not commute. Hence EPR conclude "The quantum mechanical description of physical reality given by the wave function is not complete." The possible objection that this conclusion would be valid only if the momentum and position of S_2 are simultaneously predictable is countered by pointing out that this condition implies simultaneity of the corresponding measurement processes on S_1; but then the reality of p_2 and q_2 "would depend upon the process of measurement carried out on the first system [S_1], which does not disturb the second system in any way. No reasonable definition of reality could be expected to permit this." Obviously, EPR resort to the principles of separability and locality to meet this objection and the related criticism of making use of the counterfactual definiteness, i.e., of the controversial assumption that unperformed experiments have definite results, an assumption ascribed to their above-quoted statement beginning with the words "if, instead . . ."

As might be expected, the EPR paper became the target of numerous criticisms, the most important of which was Bohr's. Faithful to his positivistic philosophy, Bohr (1935) rejected the argument by claiming that in order to speak validly of real properties of particles "it is necessary to take into account the experimental conditions that define the possible types of prediction about the system since these conditions are an inherent element in the description of any phenomenon to which the expression 'physical reality' can meaningfully be attached." In particular, this requirement illegalizes the use of counterfactual implications for defining physical reality. Bohr was particularly interested in refuting the EPR argument for the following reason. If Heisenberg's indeterminacy relations are interpreted not merely as a prohibition of the simultaneous measurability of conjugate observables but rather, in the spirit of the Copenhagen interpretation, as a denial of the very existence of sharp values of such observables, the EPR argument disproves even the consistency of quantum mechanics.

Schrödinger (1935b), in contrast to Einstein and Bohr, saw the importance of the EPR argument in its disclosure of the existence of the nonseparability, or entanglement ("*Verschränkung*") as he called it, of the subsystems or components S_1 and S_2 in a composite system S. This means that a measurement carried out on one of the components, whatever its distance from the other, interacts with S and affects thereby the other component; and it implies holistically that the state of S is not fully described in terms of the states of S_1 and S_2, a situation often epitomized by saying that "the whole is more than its parts." This entanglement, Schrödinger declared, is *the* characteristic feature of quantum mechanics, "the one that enforces its entire departure from classical lines of thought."

Schrödinger's notion of entanglement was eagerly seized by parapsychologists as scientific support for their theses of ESP (extrasensual perception), clairvoyance, psychokinesis, and other paranormal phenomena, as well as by interpreters of quantum mechanics as a revival of Eastern mysticism. Although some reputable physicists, like the Nobel laureate Brian Josephson or Richard Mattuck, expressed some sympathy with such ideas, the majority of physicists strongly reject them. Even Wheeler, whose name is often quoted by parapsychologists because of his participatory anthropic principle, vehemently opposed such ideas with the words ". . . Let no one use the EPR experiment to claim . . . any 'quantum interconnectedness' between separate consciousnesses," and he condemned those ideas as "baseless" and "moonshine."

Certain disadvantages of the published EPR

argument have been eliminated in its reformulation by David Bohm (1951) in terms of the states of a pair of spin-$\frac{1}{2}$ particles in the singlet state. Bohm's formulation became the standard presentation of the EPR argument in the philosophical literature on this issue.

The EPR incompleteness argument with its emphasis on "elements of physical reality" was soon regarded as an incentive for the construction of hidden-variable theories "to complete" quantum mechanics, although—as still unpublished documentary evidence shows—the argument was definitely not intended to serve as such an incentive. Attempts to construct such theories had been made already in the late 1920s for the purpose of reinstating classical determinism in quantum mechanics in quantum physics. But as a result of von Neumann's (1932) "proof" of the impossibility of hidden variables, interest in such theories had waned until it was revived by the publication of the EPR paper. Bohm (1952) formulated a logically consistent hidden-variable theory that ascribes to every particle definite values of position and momentum as well as a continuous trajectory by adding to the usual potential a "quantum mechanical potential." By averaging over the values of the hidden variables, he retrieved all empirically testable results of the standard theory. Since in Bohm's theory of measurement the hidden variables depend not only on the state of the observed object but also on that of the measuring device and thus differ from those considered by von Neumann, the latter's impossibility "proof" did not apply.

Still, the status of Bohm's theory and of other hidden-variable theories and their relation to von Neumann's argument were fully understood only when John Stewart Bell clarified the situation. His first contribution to these problems was his proof that von Neumann's reasoning, though mathematically impeccable, rests on a certain premise, the so-called additivity postulate, which need not hold for hidden variables (Bell, 1964a). Furthermore, he drew attention to the fact that in Bohm's theory the quantum potential exhibits the "curious feature of a grossly nonlocal character."

Studying the question of whether local deterministic hidden-variable theories, i.e., theories in which the hidden variables determine unambiguously the result of every single act of measurement and satisfy the locality principle, are possible, Bell made what Josephson called "the most important recent advance in physics" and Henry P. Stapp even "the most profound discovery of science" (Stapp, 1975). Using Bohm's EPR version, Bell showed that every deterministic hidden-variable theory that is local involves a certain inequality that became known as "Bell's inequality." This inequality—and this is the important point—is *not* satisfied by quantum physics. This result, "Bell's theorem," namely that no local deterministic hidden-variable theory can reproduce all the predictions of quantum mechanics, has been formulated by Bell as follows: "In a theory in which parameters are added to quantum mechanics to determine the results of individual measurements, without changing the statistical predictions, there must be a mechanism whereby the setting of one measuring device can influence the reading of another instrument, however remote" (Bell, 1964b).

Bell (1971) subsequently generalized his theorem for the case of stochastic hidden variables. Philippe H. Eberhard (1977) derived a Bell-type inequality without using hidden variables and without the assumption of determinism. Michael Redhead (1987) showed that on this assumption counterfactual conditionals are innocuous. Generalizations of Bell's inequality are still today a widely studied subject in the foundational research of quantum mechanics. But so also are experiments performed to test whether quantum mechanics satisfies these inequalities. Practically all these experiments, and among them the very sophisticated one performed by Alain Aspect (1981) and his team, confirm the predictions of quantum mechanics and violate Bell's inequality. These results imply that the classical world picture of local realism has to be given up irretrievably. For even should ever quantum mechanics turn out to be wrong, its successor must comply with these experimental facts. Moreover, these experiments can be regarded as inaugurating what has been called "experimental metaphysics," the existence of which strikingly disproves the aforementioned thesis that a disagreement between philosophical doctrines about the nature of physical reality is but "a conflict over preferred modes of speech."

As mentioned above, locality is a fundamental requirement of special relativity. The experimentally verified nonlocality of quantum mechanics may therefore be thought to imply the incompatibility of quantum mechanics with special relativity and, moreover, the possibility of sending messages with superluminal velocities. A detailed analysis (Jarrett, 1984) shows, however, that the locality principle used in the proof of Bell's theorem is a logical conjunction of two other conditions, one of which does not violate the principles of special relativity. Since quantum mechanics satisfies only this latter condition, it enjoys a peaceful coexistence with relativity.

The conceptual problems dealt with so far have been discussed in the language of the traditional framework of quantum mechanics. However, there exist other theoretical frameworks in which quantum mechanics, especially if axiomatized, is studied, each of them having its advantages and its deficiencies. Philosophically most remarkable is the so-called quantum logical approach according to which quantum mechanics should be formulated in a logic different from that applied in the description of classical physics.

Proponents of quantum logic (*q.v.*) justify their approach by comparing the relation between logic and quantum physics with the relation between geometry and physics in Poincaré's philosophy of geometry, referred to above in Sec. 4.3. Since always only an inseparable combination of two such components is empirically testable, one of them (e.g., logic) can be altered if also the other component (the formulation of the physical theory) is changed without affecting the factual situation; but the modified physical theory, it is claimed, may turn out to be conceptually more acceptable than its original unmodified version. Garrett Birkhoff and von Neumann (1936) were the first to propose a nonclassical quantum logic. In their logic, the distributivity law with respect to the logical connectives "and" ($\wedge$) and "or" ($\vee$) is not valid. In other words, if p, q, and r denote propositions, then the classical distributive law $p \wedge (q \vee r) = (p \wedge q) \vee (p \wedge r)$ does not always hold. If, e.g., it is known that the spin of a spin-$\frac{1}{2}$ particle is "up" in the sense that the proposition p, "$\sigma_z = +1$," in units of $\frac{1}{2}\hbar$, is true, and if q and r denote, respectively, the propositions "$\sigma_x = +1$" and "$\sigma_x = -1$," then it is easy to confirm that—at least in accordance with the realist interpretation of quantum mechanics—this distributivity law does not hold. A more technical exposition that does justice to the Birkhoff–von Neumann logic would have to specify it as a propositional calculus having the structure of an orthocomplemented nondistributive (modular) lattice, in contrast to the Boolean lattice of classical logic (see QUANTUM LOGIC).

Other versions of quantum logic soon followed. Reichenbach (1944) developed a three-valued logic that contains, in addition to the truth values "true" and "false" of classical logic, a third truth value "indeterminate" that applies to statements that were regarded as "meaningless" in the Copenhagen interpretation: but in Reichenbach's logic the question of whether a system has a certain value for a given observable has three answers: "yes," "no," and "maybe," where "maybe" represents a genuinely irreducible state of affairs. Carl Friedrich von Weizsäcker (1955) proposed a "many-valued complementarity logic" in which truth values correspond to probabilities.

Quantum logic has become an immensely rich and diverse field of new mathematical and philosophical ideas. It has stimulated the development of lattice theory and, more generally, of the theory of partially ordered sets and has even contributed to various mathematical generalizations of existing theories. Pavičić's (1992) bibliography on quantum logic, compiled for the *International Quantum Logics Association* and containing about 2000 entries, is indicative of the immensity of work done in this field.

Nevertheless, there are quite a few critics who challenge the status of quantum logic on the ground that it does not provide satisfactory answers to fundamental questions such as the following: Has quantum logic convincingly shown that the genuine and universally valid logic is non-Boolean and that the traditional logic is merely a kind of asymptotic approximation to it? Can logic be really empirical in the sense of "being read off" from empirical data, in contrast, e.g., to Einstein's famous remark to Heisenberg (1971) that "it is the theory which decides what we can observe"? How can quantum logic differ from classical logic if the mathematics used in quantum mechanics is necessarily based on classical logic? And, after all, does quantum logic truly resolve any of the problematic issues of quantum mechanics instead of explaining a riddle by mystery, as the observer-dependent measurement theory is said to have done?

In conclusion, it should be pointed out that this review of the philosophy of physics has said nothing about the philosophical implications of some physical theories that are at the forefront of current research. In fact, no philosophical comments have been made about quantum field theory or the *S*-matrix approach with its bootstrap conjecture, or about the standard model of elementary-particle physics, based on gauge theories of the electroweak and strong interactions, or about superstring theories designed to incorporate the standard model with general relativity in a logically consistent way and to unify all known fundamental forces in nature. Nor have any remarks been made on the prospects and philosophical significance of the so-called "Final Theory," which is to explain everything in nature from a minimum of basic assumptions.

These topics are widely publicized in popular magazine articles. But they are also rigorously discussed in professional writings of such competent physicists and philosophers of science as Cushing (1983), Redhead (1983), Stöckler (1984), Brown and Harré (1988), Kanitscheider (1989), Weinberg (1992), and Teller (1994). Since, however, research in these areas is at present making enormously rapid progress, we believe that it is still too early to grasp the full philosophical significance of these developments and that it is still premature to draw well-founded philosophical conclusions. We agree therefore with the statement, made by Teller, that "as yet we have no grip on what these theories are telling us about how to conceive the world."

Works Cited

Ajanopon, P. (1987), *Am. J. Phys.* **55**, 159–163.

Anderson, J. L. (1967), *Principles of Relativity Physics*, New York: Academic Press.

Aspect, A., Grangier, P., Roger, G. (1981), *Phys. Rev. Lett.* **47**, 460–463.

Bacon, F. (1620), *Novum Organum*, London: Billius.

Barrow, I. (1670), *Lectiones Geometricae*, London: Godbid.

Bekker, I. (Ed.) (1831–1870), *Aristotelis Opera*, Berlin: Reimer.

Bell, J. S. (1964a), *SLAC-PUB-44*; (1966), *Rev. Mod. Phys.* **38**, 447–452.

Bell, J. S. (1964b), *Physics (Long Island City)* **1**, 195–200.

Bell, J. S. (1971), "Introduction to the Hidden Variable Question," in: B. d'Espagnat (Ed.), *Foundations of Quantum Mechanics*, New York: Academic Press, pp. 171–181.

Biedenharn, L. C. (1969), Private communication.

Birkhoff, G., von Neumann, J. (1936), *Ann. Math.* **37**, 823–843.

Bohm, A., Antoniou, I., Kielanowski, P. (1994), *Phys. Lett. A* **189**, 442–448.

Bohm, D. (1951), *Quantum Theory*, Englewood Cliffs, NJ: Prentice-Hall.

Bohm, D. (1952), *Phys. Rev.* **85**, 166–193.

Bohr, N. (1928), *Nature Suppl.* **121**, 580–590.

Bohr, N. (1935), *Phys. Rev.* **48**, 696–702.

Bohr, N. (1949), "Discussion with Einstein," in: P. A. Schilpp (Ed.), *Albert Einstein: Philosopher-Scientist*, Evanston, IL: Library of Living Philosophers, pp. 199–241.

Boltzmann, L. (1896), *Ann. Phys. (Leipzig)* **57**, 39–71; **58**, 595–598.

Boltzmann, L. (1897), *Vorlesungen über die Prinzipien der Mechanik*, Leipzig: Barth.

Born, M. (1926), *Ann. Phys. (Leipzig)* **37**, 863–867; **38**, 803–827.

Boscovich, R. J. (1758), *Theoria Philosophiae Naturalis*, Vienna: Remondidiana.

Brown, R. H., Harré, R. (1988), *Philosophical Foundations of Quantum Field Theory*, Oxford: Clarendon Press.

Bunge, M. (1961), *Am. J. Phys.* **29**, 518–524.

Carter, B. (1974), "The Anthropic Principle and Large Number Coincidences," in: M. Longair (Ed.), *Confrontations of Cosmological Theories with Observation*, Dordrecht: Reidel.

Clarke, S. (1717), *A Collection of Papers which passed between the late Learned Mr. Leibniz and Dr. Clarke*, London: Knapton.

Cushing, J. T. (1983), *Philos. Sci. Assoc.* **2**, 31–56.

Dorling, J. (1970), *Am. J. Phys.* **38**, 539–540.

Dunne, J. W. (1927), *An Experiment with Time*, London: Black.

Earman, J. (1989), *World Enough and Space-Time*, Cambridge MA: MIT Press.

Eberhard, P. A. (1977), *Nuovo Cimento B* **38**, 75–80.

Eddington, A. S. (1929), *The Nature of the Physical World*, New York: Macmillan.

Ehrenfest, P. (1917), *Proc. K. Ned. Akad. Wet.* **20**, 200–209.

Einstein, A. (1905), *Ann. Phys. (Leipzig)* **17**, 891–921.

Einstein, A. (1911), *Phys. Z.* **12**, 509–510.

Einstein, A. (1916), *Ann. Phys. (Leipzig)* **49**, 769–822.

Einstein, A. (1917a), *Über die Spezielle und die Allgemeine Relativitätstheorie*, Braunschweig: Vieweg.

Einstein, A. (1917b), *Sitzungsber. K. Preuss. Akad. Wiss.*, 142–152.

Einstein, A. (1921), *Sitzungsber. K. Preuss. Akad. Wiss.*, 123–130.

Einstein, A. (1934), *Essays in Science*, New York: Philosophical Library.

Einstein, A., Podolsky, B., Rosen, N. (1935), *Phys. Rev.* **47**, 777–780.

Einstein, A. (1949), "Autobiographical Notes," in: P. A. Schilpp. (Ed.), *Albert Einstein: Philosopher-Scientist*, Evanston, IL: Library of Living Philosophers, pp. 1–95.

Einstein, A. (1953), "Preface," in: M. Jammer, *Concepts of Space*, Cambridge, MA: Harvard Univ. Press.

Einstein, A., Kaufman, B. (1955), *Ann. Math.* **62**, 128–138.

Einstein, A. (1956), *The Meaning of Relativity*, 5th ed., Princeton, NJ: Princeton Univ. Press.

Euler, L. (1750), *Hist. Acad. R. Sci. Belles Lett.* **4**, 324.

Everett, Hugh, III (1957), *Rev. Mod. Phys.* **29**, 454–462.

Faraday, M. (1839), *Experimental Researches in Electricity and Magnetism*, London: Quaritch.

Feynman, R. (1951), *The Feynman Lectures in Physics*, Vol. 2, Reading, MA: Addison-Wesley.

Fock, V. (1959), *The Theory of Space Time and Gravitation*, London: Pergamon Press.

Frank, P. (1946), *International Encyclopedia of Unified Science*, Vol. 1, Chicago, IL: Univ. of Chicago Press.

Galileo, G. (1623), *Il Saggiatore*, Rome: Mascardi.

Grünbaum, A. (1963), *Philosophical Problems of Space and Time*, New York: Knopf; (1973), 2nd ed., Dordrecht: Reidel.

Hall, A. R., Hall, M. B. (Eds.) (1962), *Unpublished Scientific Papers of Isaac Newton*, Cambridge: Cambridge Univ. Press.

Heisenberg, W. (1927), *Z. Phys.* **43**, 172–198.

Heisenberg, W. (1958), *The Physicist's Conception of Nature*, London: Hutchinson.

Heisenberg, W., Born, M., Schrödinger, E., Auger, P. (1961), *On Modern Physics*, New York: Collier Books.

Heisenberg, W. (1971), *Physics and Beyond*, London: George, Allen and Unwin.

Hönl, H. (1953), *Z. Naturforsch. A* **8**, 2–6.

Huang, Y. S. (1993), *Helv. Phys. Acta* **66**, 346–360.

Jammer, M. (1954), *Concepts of Space*, Cambridge, MA: Harvard Univ. Press; (1993), 3rd ed., New York: Dover Publications.

Jammer, M. (1961), *Concepts of Mass in Classical and Modern Physics*, Cambridge, MA: Harvard Univ. Press.

Jammer, M. (1974), *The Philosophy of Quantum Mechanics*, New York: Wiley.

Jammer, M. (1986), "The History of the Concept of Distant Simultaneity," *Rend. Accad. Naz. Sci. 40 Pt. 2: Mem. Sci. Fis. Nat.* **9**, 169–184.

Jammer, M. (1990), "The Program and Principles of Descartes' Physics," in: G. Belgioioso (Ed.), *Descartes—Il Metodo e i Saggi*, Rome: Istituto della Enciclopedia Italiana, pp. 303–334.

Jarrett, J. (1984), *Nous* **18**, 569–589.

Kanitscheider, B. (1989), "Cosmology and Particle Physics—The Philosophical Aspects of a New Scientific Symbiosis," in: P. Weingartner, G. Schurz (Eds.), *Philosophy of the Natural Sciences, Proceedings of the 13th International Wittgenstein Symposium, Kirchberg/Wechsel, Austria, 1988*, Vienna: Hölder, Pichler, Tempsky, pp. 60–70.

Kant, I. (1747), "Gedanken von der Wahren Schätzung der Lebendigen Kräfte," in: *Kant's Werke* (Akademie Ausgabe), Vol. 1 (1910), Berlin: Reimer, pp. 1–181.

Kant, I. (1770), "De Mundi Sensibilis atque Intelligibilis Forma et Principiis," in: *Kant's Werke*, Vol. 2 (1912), Berlin: Reimer, pp. 385–419.

Kant, I. (1781), *Kritik der Reinen Vernunft*, Riga: Hartknoch.

Kretschmann, E. (1917), *Ann. Phys. (Leipzig)* **83**, 575–614.

Lanczos, C. (1962), *Rev. Mod. Phys.* **34**, 379–389.

Le Bon, G. (1905), *L'Évolution de la Matière*, Paris: Flammarion.

Leibniz, G. W. (1695), "Specimen Dynamicum," in *Opera Omnia* (1768), Geneva: Dutens.

Leibniz, G. W. (1715), "Initium Rerum Mathematicarum Metaphysica," in: C. I. Gerhardt (Ed.) (1962), *Mathematische Schriften*, Hildesheim: Olms; Loemker, L. E. (Ed.) (1956), *Philosophical Papers and Letters*, Chicago, IL: Univ. of Chicago Press.

Lewis, G. N. (1930), *Science* **71**, 569–577.

Liboff, R. L. (1984), *Phys. Today* **37** (2), 30–55.

Locke, J. (1690), *An Essay on Human Understanding*, London: Bassett.

London, F., Bauer, E. (1939), *La Théorie de l'Observation en Mécanique Quantique*, Paris: Hermann.

Malament, D. (1977), *Nous* **11**, 293–300.

Maxwell, J. C. (1865), "A Dynamical Theory of the Electromagnetic Field," *Philos. Trans. R. Soc.* **155**, 459–508; reprinted (1990) in: W. D. Niven (Ed.), *The Scientific Papers*, Vol. 1, Cambridge, U.K.: Cambridge Univ. Press, pp. 526–597.

Misner, C. W., Wheeler, J. A. (1957), *Ann. Phys. (New York)* **2**, 525–603.

Nagel, E. (1961), *The Structure of Science*, New York: Harcourt.

Nerlich, G. (1982), "Simultaneity and Convention in Special Relativity," in: R. McLaughlin (Ed.), *What? Where? When? Why?*, Dordrecht: Reidel.

Neumann, C. (1869), *Über die Prinzipien der Galilei-Newton'schen Theorie*, Leipzig: Teubner.

Newton, I. (1687), *Philosophiae Naturalis Principia Mathematica*, London: Streater; (1947), *Mathematical Principles of Natural Philosophy*, Berkeley, CA: Univ. of California Press.

Pauli, W. (1958), *Theory of Relativity*, New York: Pergamon Press.

Pavičić, M. (1992), *Int. J. Theor. Phys.* **31**, 373–461.

Penney, R. (1965), *J. Math. Phys.* **6**, 1607–1611.

Petersen, A. (1963), *Bull. At. Sci.* **19**, 8–14.

Podlaha, M. F. (1975), *Indian J. Theor. Phys.* **23**, 69–75.

Poincaré, H. (1902), *La Science et l'Hypothèse*, Paris: Flammarion.

Popper, K. (1956), *Br. J. Philos. Sci.* **1**, 117–133, 173–195.

Prior, N. (1970), *Studium Generale* **23**, 245–248.

Putnam, H. (1967), *J. Philos.* **64**, 240–247.

Redhead, M. (1983), *Philos. Sci. Assoc.* **2**, 57–99.

Redhead, M. (1987), *Incompleteness, Nonlocality and Realism*, Oxford: Clarendon Press.

Reichenbach, H. (1924), *Axiomatik der Relativistischen Raum-Zeit-Lehre*, Braunschweig: Vieweg; *Axiomatization of the Theory of Relativity*, Berkeley, CA: Univ. of California Press.

Reichenbach, H. (1928), *Philosophie der Raum-Zeit-Lehre*, Berlin: Gruyter; (1958), *The Philosophy of Space and Time*, New York: Dover Publications.

Reichenbach, H. (1944), *Philosophical Foundations of Quantum Mechanics*, Berkeley, CA: Univ. of California Press.

Reichenbach, H. (1956), *The Direction of Time*, Berkeley, CA: Univ. of California Press.

Rietdijk, C. W. (1966), *Philos. Sci.* **33**, 341–344.

Russell, E. (1948), *Human Knowledge—Its Scope and Limits*, New York: Simon and Schuster.

Salmon, W. C. (1977), "Introduction," in: H. Reichenbach (Ed.), *Gesammelte Werke*, Vol. 1, Braunschweig: Vieweg.

Schrödinger E. (1935a), *Naturwissenschaften* **23**, 807–812, 824–828, 844–849.

Schrödinger E. (1935b), *Proc. Cambridge Philos. Soc.* **31**, 555–562.

Sklar, L. (1974), *Space, Time, and Spacetime*, Berkeley, CA: Univ. of California Press.

Spencer, H. (1962), *First Principles*, London: Williams and Norgate.

Stannard, F. R. (1966), *Nature* **211**, 693–695.

Stapp, H. P. (1975), *Nuovo Cimento B* **29**, 270–276.

Stein, H. (1968), *J. Philos.* **65**, 5–23.

Stöckler, M. (1984), *Philosophische Probleme der relativistischen Quantenmechanik*, Berlin: Duncker & Humblot.

Teller, P. (1983), *Philos. Sci. Assoc.* **2**, 100–111.

Teller, P. (1994), *An Interpretive Introduction to Quantum Field Theory*, Princeton, NJ: Princeton Univ. Press.

Torretti, R. (1983), *Relativity and Geometry*, New York: Pergamon Press.

von Laue, M. (1913), *Das Relativitätsprinzip*, Braunschweig: Vieweg.

von Neumann, J. (1932), *Mathematische Grundlagen der Quantenmechanik*, Berlin: Springer; (1958), *Mathematical Foundations of Quantum Mechanics*, Princeton, NJ: Princeton Univ. Press.

von Weizsäcker, C. F. (1955), *Naturwissenschaften* **42**, 521–529.

Wald, R. A. (1980), *Phys. Rev. D* **21**, 2742–2755.

Watanabe, S. (1951), *Phys. Rev.* **84**, 1008–1025.

Weinberg, S. (1992), *Dreams of a Final Theory*, New York: Pantheon.

Weyl, H. (1922), *Space-Time-Matter*, London: Methuen.

Weyl, H. (1949), *Philosophy of Mathematics and Natural Science*, Princeton, NJ: Princeton Univ. Press.

Wheeler, J. A. (1980a), *Epistemologia* **3**, 59–98.

Wheeler, J. A. (1980b), "Beyond the Black Hole," in: H. Woolf (Ed.), *Some Strangeness in the Proportion*, Reading, MA: Addison-Wesley.

Wiener, N. (1958), *Cybernetics*, Cambridge, MA: MIT Press.

Wigner, E. (1962), "Remarks on the Mind-Body Question," in: I. J. Good (Ed.), *The Scientist Speculates*, New York: Basic Books, pp. 284–302.

Wigner, E. (1964), *Phys. Today* **17** (3), 34–40.

Winnie, J. (1970), *Philos. Sci.* **37**, 81–89, 223–238.

PHILOSOPHY OF TECHNOLOGY

ERWIN MARQUIT, *School of Physics and Astronomy, University of Minnesota, Minneapolis, Minnesota, U.S.A.*

	Introduction	417
1.	**Technology and Societal Development**	419
1.1	Role of Technology in Human Evolution	419
1.2	Hunting and Foraging Cultures	420
1.3	Transition from Mobile to Sedentary Cultures: Crop Cultivation, Animal Domestication, and Social Differentiation	420
1.4	Occupational Specialization in the Bronze and Iron Ages	421
1.5	Reckoning Technology, Mathematics, and the Origins of Writing in Mesopotamia and the Americas	422
1.6	Technological Development and Blockage in Classical Greece	423
1.7	Technology and Society in China	423
1.8	The Middle Ages	424
1.9	Beginnings of the Technological Revolution in Feudal Europe	425
1.10	From Feudalism to the Industrial Revolution in Europe	426
1.11	From the Industrial Revolution to Modern Technology	428
2.	**Origins of Philosophy of Science and Philosophy of Technology**	429
3.	**Contemporary Views on Philosophy of Technology**	431
4.	**Unresolved Problems of Technological Development**	434
	Works Cited	436
	Further Reading	437

INTRODUCTION

Philosophy of technology, which only emerged in the nineteenth century as a field of philosophy, investigates interconnections among the human needs that invoke a technology for their satisfaction, the determinants of these needs, and how people come to know them.

The term *technology* is a combination of the Greek *technē*, meaning art, and *logos*, meaning word or discourse. In English, it first (1615) denoted a discourse on an art or arts (*Shorter Oxford English Dictionary*, 1933), later embracing the scientific study of the practical and industrial arts and their methods. In philosophical literature, the terms *technology*, *technic* (or *technics*), and *technique* are differentiated. Hood gives the traditional definition of technology originating with Aristotle as "a human arrangement of technics—tools, machines, instruments, materials, sciences, and personnel—to make possible and serve the attainment of human ends" (Hood, 1983, p. 347). *Technique* is usually used to convey the method of using *technics*, but because usage varies among different authors and languages, readers may find the terms used interchangeably. Martin Heidegger's concept is broader: "The manufacture and utilization of equipment, tools, and machines, the manufactured and used things themselves, and the needs and ends that they serve, all belong to what technology is." He adds that according to the current conception, technology is "a means and a human activity" (Heidegger, 1977, pp. 4–5).

Heidegger here incorporates two contemporary connotations of *technology*, the most common being the means and methods used to produce and deliver goods and services for industrial, commercial, and military use. An-

3-527-28135-5/95/$5.00 + .50

other, or anthropological, sense is "the body of knowledge available to a civilization that is of use in fashioning implements, practicing manual arts and skills, and extracting or collecting materials" (*American Heritage Dictionary*, 3rd edition, 1993). This second sense excludes aspects of modern technology, which depends more on the integration of science with technological applications and less on the manual arts and skills. Hidden in the differences among these definitions is a philosophical question: does technology deal with a body of knowledge, only its application, or both?

Technology refers in some usage to "spiritual," religious, and aesthetic needs also. In 1897, Alfred Espinas, noting that the Greeks applied *technē* to both the fine and the applied arts, argued that "we should give the applied arts the name techniques in order to distinguish them from the arts that tend to produce aesthetic emotion" (Espinas, 1897, p. 7). The case of a culture that regards a rain dance as necessary for agricultural production, however, shows that such a distinction is not always clear.

Ferré (1988, p. 26) suggests limiting the term *technology* to the "practical implementation of intelligence." He considers the purpose of "practical intelligence" to be "to survive or thrive," and that of "theoretical intelligence" as "to know or understand" (p. 33). This distinction, however, may obscure the social connections that form the essence of technology and constitute the principal stimulus for its development.

The following definition of *technology* is suggested for our discussion here: *Technology denotes the systematic, conscious application of knowledge by a society to meet its culturally conditioned real and perceived physical, biological, and spiritual needs and to guarantee its continued social existence. This knowledge is the socially accumulated understanding of how human labor and available material resources and artifacts are integrated by a society; the aspect of the society primarily involved here is its particular set of historically conditioned social relations of production.*

The association of technology here exclusively with human activity is deliberate to stress the central role that tool use and toolmaking play in human life, despite the fact that animals use and even make tools. The qualitative difference between these activities by animals on the one hand, and tool use and toolmaking in a deeply social context by humans on the other, is so sharp, however, that one scholar even suggests the term *utensil* be used for objects used or produced by animals, while the term *tool* be limited to objects produced by humans (Tobach, 1993). Only humans consciously create artifacts. Early examples are primitive tools made by combining or reshaping natural objects to be held in the hand for physical protection or for procuring or processing food. Shaped stone objects in the vicinity of fossil bones identify the remains as protohuman. Toolmaking has been a criterion to distinguish humans from other animals at least since the time of Benjamin Franklin, to whom James Boswell attributed the definition that man is "a toolmaking animal" (Boswell, 1934, p. 245). Thomas Carlyle's remark that "Man is a Tool-using Animal" has also been used to distinguish between humans and other animals (Carlyle, 1921, p. 37). Charles Darwin (1981) attributed to tool use a distinctive role in human evolution.

Since technology involves the conscious application of knowledge by a society, ontological and epistemological questions arise: What is this knowledge and how do we come to know it? This body of knowledge, in turn, involves three areas: material objects (frequently referred to in philosophical literature as "artifacts"), skill and knowledge in employing them, and the technological relations of production, that is, the way human labor is organized in the technological process.

Our species emerged in a process of biological evolution. The technology by which we provide our material needs was an integral part of this process. After the emergence and biological stabilization of our species, technological development continued. We begin this article with a discussion of the role of technology in human evolution, then go on to consider the interconnections between technological development and the evolution of human societies through their various stages. Readers are referred to PHYSICS AND TECHNOLOGY, HISTORY OF, for a detailed discussion of that history. Here, we focus our attention on stages of technological development that represent turning points in human history. We consider some theoretical and practical areas in which science and technology interact and conclude the article with a discussion of viewpoints being presented in contemporary philosophical literature on problems arising from technology today.

1. TECHNOLOGY AND SOCIETAL DEVELOPMENT

A principal concern of philosophy of technology is how technological development influences the course of societal interrelationships at different stages of societal organization, as well as how specific stages of societal organization and culture affect technological development. Among philosophical approaches, technological determinism sees technological development as a spontaneous evolutionary process requiring a given society to organize itself so as to make efficient use of the technologies becoming available. It also seizes upon individual technological innovations as immediate and direct causes for fundamental social change. A second approach views technological advance as a consequence of the development of human spirit and culture. The most generally held view is that technological change is the cause of profound societal change at certain points in human history, while at other times societal change stimulates technological development. Followers of this third approach attach widely varying weights to the two sides of this interaction.

The discussion in Sec. 1 of this interaction between technology and society is organized around the historical sequence of key stages of human societal development. The aim here is not to unfold a chronology of technology, but to examine the intricacies of the interrelationships between technology on the one hand, and the complex of science, culture, and socioeconomic organization on the other. (It will be noted that this section and PHYSICS AND TECHNOLOGY, HISTORY OF, are thus very different in content and focus.)

1.1 Role of Technology in Human Evolution

The interconnections between technology and society begin with the evolution of the human species. It is therefore appropriate to begin this section with the three-way interaction of technological, societal, and biological interactions that have led to the emergence of the modern human species *Homo sapiens*.

The unique character of human toolmaking and tool use along with the uniqueness of bipedality suggests these two factors as the primary stimulus for the evolution of the modern human being.

The evolutionary line of the modern human species, *Homo sapiens*, can be traced backward through at least two earlier species of the genus *Homo*, namely *Homo habilis* and *Homo erectus*, to the bipedal genus *Australopithecus*, originating in Africa some five million years ago, branching off from the family of the primates from which the modern chimpanzee descends (Hood, 1993). Bipedality frees the hands for many activities, including using tools. The first fossil evidence for tools appears later—some 2.5 million years ago—stone choppers, scrapers, and sharp flakes, probably used for stripping meat and skin from bones (Napier, 1993, pp. 43–44; Howells, 1993, p. 118; Isaac, 1976, p. 487), leading some scholars to suggest that the development of the brain was a more important factor in human evolution (Lovejoy, 1993). On the other hand, the tools that we observe among pongids like the chimpanzee today such as stripped branches for prodding termite nests or stone anvils for cracking open hard fruits (Candland, 1987; Goodall, 1968) are not likely to leave a fossil trail, so that it is not unreasonable to assume (even without evidence) that the early bipedal hominids not only did not lag behind the pongids in toolmaking, but also had begun to advance in their use of tools.

Apart from structural changes favoring bipedal locomotion, the fossil trail from the australopithecines to *Homo sapiens* is marked by significant increases in brain size (Falk, 1993, p. 64); changes in the bone structure of the hand (Napier, 1993); a lowering of the larynx so as to increase the space between the larynx and the back of the nasal cavity, thereby enhancing the possibility for articulate speech (Laitman, 1993); and a decrease in the size of the canine teeth (Skelton *et al.*, 1993). Toolmaking, development of the hand, socialization of food consumption, increasing differentiation in vocal communication, and mental development move forward in the evolutionary process as an integrative whole, a development in one area contributing to the development of the others.

1.2 Hunting and Foraging Cultures

The term *hunting and foraging* is generally used to characterize both the level of technological development and form of societal organization within which this technology is applied. The first evidence of hunting, probably by *Homo sapiens*, appears some 100 000 years ago in northwestern Germany, where a fossil elephant with a wooden spear nearly eight feet long between its ribs was found. Apart from the weapons themselves, the earliest symbolic representation of their use are the cave paintings with hunting scenes in northwestern Spain and southwestern France painted perhaps some 10 000 to 30 000 years ago.

To overcome the lack of information about the technological culture of ancient hunting and foraging societies, it is necessary to consider recent and contemporary hunting and foraging cultures, such as the Mbuti people in the Ituri rain forest in Zaire. These people lived until the 1980s much as they had thousands of years ago, with a cultural form that is essentially pre–Stone Age. The primary economic standard is one of adequacy, the mobile lifestyle mitigating against accumulation of private property. The adult men, women, and unmarried youth all participate in the hunt. Mbuti net hunters establish their symbiotic relationship with the forest ecology by a culture of minimal hunting, spending a daily average of only four to five hours away from camp (Turnbull, 1983, p. 138).

The close cooperation of women and men in the hunt and the division of labor between them are reinforced in their hunting songs, which are always sung in round form, utilizing a technique that recreates the cooperative patterns required by the hunt. The other major types of song, the foraging song and the death song, also reinforce the appropriate patterns for the corresponding activity. Except for the lullaby, none of the songs can be produced by a single singer. The ritualistic character of the singing suggests that this art form is viewed as a necessary component of the hunt itself.

The egalitarian character of Mbuti society is characteristic of most hunting and foraging societies, generally marked by the absence of hierarchical structures. Group members make decisions collectively or through councils of elders. The prestige of certain individuals is based on respect for their achievements as outstanding hunters or foragers. The individual function of the chief stands out; he gives leadership in dealing with conflict with other groups of people. A second individual function is that of the shaman, who is believed to have special powers in connection with the group's religious beliefs or mythology, and a third is one with skills in rituals practiced by the group, such as a dancer or singer (Coudert, 1991, p. 398). Although a certain degree of social division of labor, partly based on levels of skill, emerges in hunting and foraging societies, the society is essentially free of social conflicts; the disputes and tensions that do arise do so as conflicts among individuals and are frowned upon by the other members of the society. It is for this reason that many social philosophers reject the sociobiological views asserting that humans are genetically conditioned to be aggressive and acquisitive.

1.3 Transition from Mobile to Sedentary Cultures: Crop Cultivation, Animal Domestication, and Social Differentiation

The first major revolution in social organization associated with technological developments appears to have occurred toward the end of the Ice Age in the period 12 000 to 10 000 B.C. in the Near East. Sedentary settlements and the subsequent development of crop cultivation and domestication of animals were accompanied by relatively rapid technological changes in comparison with the preceding millennia, resulting not only in the increase in production above the subsistence level, but also in the emergence and deepening of social stratification within the population. This process, culminating in the formation of great city-states, occurred in Mesopotamia, Egypt, China, India, Mesoamerica, and the Andes.

Donald O. Henry, on the basis of suggestions from Binford (1980) and Gould (1982), sees the first signs of technological change as involving a transition from *simple foraging* to *complex foraging*. Henry (1989, p. 4) summarizes simple foraging as "a high mobility and movement to resources" and *complex foraging* as a "collecting strategy in which greater residential permanence is maintained through storage and logistical acquisition of resources."

The increased productivity that resulted from this transition led to both population increase and social differentiation. The new technology involved not only a new level of foraging activity, but also qualitative changes in the handling of the products as a result of the increased quantity of goods and people. Administrative problems seem to arise when more than two hundred people engage in coordinated activity. Structures that appear to have been communal storehouses suggest the need for administering and redistribution of the goods stored. Similarly, ceremonial structures suggest leaders (Henry, 1989, p. 212), and some central management may have been needed for housing the larger population.

A widely accepted model of social development through several stages of the social reorganization that began with the production of a food surplus has been put forward by Fried (1967). Fried suggests that the domestication of animals and plants initiated a series of changes in social organization from the egalitarian societies of hunters and foragers to the formation of what he calls rank societies, then into stratified societies, and finally to the formation of the state.

The rank society is characterized by the emergence of an individual, such as a chief, whose principal function is the redistribution of the products derived from the technological activity. According to Fried, in a stratified society, essential basic resources become the private property of a ranked segment of the community. Fried makes a distinction between what might be called the personal property of an individual or family, that is, objects of immediate use or consumption, and the basic resources that are needed to produce them (pp. 193–196).

The state emerges as an institution that stabilizes the stratification through control over the productive resources and products. Haas (1987, p. 2) defines the state society "as a type of society in which rulers have control over production or procurement of basic resources and as a result exercise coercive power over their respective populations."

Stone-walled dwellings and storage pits are evidence of sedentary settlements; stable storage of food was necessary beyond that immediately consumed (Henry, 1989, p. 19). The production of such a surplus is needed since life cycles of plants and, to a large extent, animals are tied to the seasons of the year. Stored supplies can be exchanged among different communities or even be produced specifically for exchange.

Once the technology to produce a surplus is available, the possibility for continual accumulation of a wealth of products arises, although this potential has not been realized uniformly everywhere, since subsistence-level economies persist to this day. Therefore, Fried's model of social evolution must be viewed as a tendency rather than a specific sequence of stages through which every coherent community of people must pass.

The change from mobile to sedentary cultures involves a division of labor in two distinct areas: the organization of production and the associated infrastructures for production and storage on the one hand, and the administration of the process of redistribution for consumption on the other. The managerial/administrative functions are often combined with the role of leader in ritualistic functions, since the rituals to a large extent are considered necessary components of the technology. The increase in the storable surplus also creates the conditions for the separation of the product from the producers through increasing bureaucratization of the administrative/managerial function. Those appropriating the product ultimately must establish state institutions based on a morality justifying the social stratification and also set in place means of physical force to enforce the new social morality.

1.4 Occupational Specialization in the Bronze and Iron Ages

Bronze metallurgy was introduced about 3300 B.C. in Mesopotamia and Egypt. This technological change was not a revolution in the sense of the speed with which it spread, but in the economic, political, and social consequences of the use of metal in weapons and tools. The new technologies significantly increased the level of agricultural production and thereby stimulated larger-scale irrigation projects. The agricultural surplus made possible an increase in the number of people who could separate themselves from agriculture, thus contributing to the growth of urban settlements. The rank societies turned into stratified state societies by the fourth century B.C., with the political and economic control centered in temple-cities ruled by priest-kings. Iron metallurgy appeared about 2200 B.C. and was in widespread use by 1100 B.C.

During the same period, the plow, the horse, and wheeled vehicles were introduced. The workmen involved in the mining, transport, and smelting of ores and in producing tools may have been the first specialists separated from agriculture who did not raise their own food and were not part of the ruling bureaucracy. It therefore appears that by the third millennium B.C. technological development in Mesopotamia and Egypt had reached a level that could sustain the principal types of social stratification we now find in contemporary populations except, perhaps, for a strata of intellectuals. The first intellectuals, the mathematicians of Babylon, may have also emerged in this period, but evidence for their presence does not appear until the second millennium B.C.

1.5 Reckoning Technology, Mathematics, and the Origins of Writing in Mesopotamia and the Americas

Along with first evidence of the cultivation of cereals in Syria and Iran, the harvesting or hoarding of grain, and the domestication of animals occur finds of small clay objects of various geometrical shapes—spheres, cones, disks, and cylinders. Each type of token served as a counter for a definite quantity of a particular food or other basic commodity stored or delivered. These tokens have been found throughout the Middle East, with the earliest dated about 8000 B.C. in the Zagros Mountains of Iran (Schmandt-Besserat, 1992, pp. 40, 168).

After the emergence of city-states in the fourth millennium B.C., accounting acquired a new importance. The tokens, originally sealed in small clay containers (envelopes) were now impressed on the clay envelopes into which they were placed so that the contents were visible on the envelopes themselves. Ultimately, the tokens became unnecessary, as their shapes and markings were reproduced on clay tablets. A conceptual leap toward abstract thinking then occurred. Quantitative and qualitative aspects of the goods became symbolically separated, so that a token of one shape represented a given commodity, while a second set of tokens represented the quantity. The wedge-shaped impressions developed into the system of writing known as cuneiform (Latin *cuneus*, a wedge).

In the Peruvian Andes in the eighth or ninth century, a period of major state formation, a system of knotted strings called *khipus* (Patterson, 1991, p. 43) was used to keep accounts. The *khipus* did not lead to writing, however. The Mayans later adopted a system of writing from other peoples and refined it with both ideographic and phonetic elements. They used it to record events and dates, deeds involving the nobles, the position of the stars and planets, religious rituals, and calendar reckoning, but there is no evidence of its use for accounting. The failure to develop written accounting and, except perhaps for calendars, the largely noneconomic use of writing were manifestations of different paths of technological development as compared, say, to Mesopotamia. The absence of large domesticated animals restricted the transport of goods, and the resulting fragmentation of the economies further limited technological development (Gille, 1986, p. 424). The Mayan, Inca, and Aztec rulers never developed the technological infrastructure that could tightly link the conquered peoples into vast, geographically expanding empires. The extraction of tribute and the continual pillaging of weaker neighbors by the stronger city-states in the Andes and Mesoamerica evoked resistance that could not be permanently repressed, ultimately allowing a small number of Spanish invaders, with their superior military technology, to form local alliances in order to destroy the Inca, Aztec, and Mayan states (Patterson, 1991).

The first sign of activity that can be called *scientific* (according to our current understanding of scientific knowledge) in antiquity is the development of mathematics in Babylon, known to us from tablets written between 1800 and 1500 B.C. The Babylonian tablets contain lists of conversions of measurement units, multiplication tables, and mathematical exercises. The calculations concerned practical problems, such as work quotas, food requirements, and wages for workers; materials for wells and building construction; weights and measures; and inheritance and compound interest (Hooke, 1954, pp. 786–791). Ancient Egyptian mathematics was more geometrically oriented than was Babylonian, its greater sophistication attributable to greater economic complexity and more diverse technological activity. The availability of mathematical methods coordinating elaborate economic projects made it possible to undertake construction projects of increasing complexity (pp. 790–792).

The tokens, *khipus*, and the types of calculations represented on the tablets seem to indicate that mathematics did not arise as an aesthetic intellectual pastime, but was an integral part of the application of technology to the economy. Subsequently, the Greek philosophers of the fourth to first centuries B.C. separated mathematics from technology and integrated it with logic, a process that culminated with Euclid's hypothetico-deductive system of geometry.

1.6 Technological Development and Blockage in Classical Greece

Despite the emergence of Greek science concurrent with the technological advances between the sixth and third centuries B.C., no revolution in production occurred on the scale that might be expected.

The usual explanations for this "technological blockage" are based on the economic and social structures of the Greek city-states. Most Athenians were peasants, and played no direct role in the political life. *Citizens* lived in the cities and neighboring villages, were generally landowners, and did not engage in manual labor or crafts. Their economic well-being was based on extensive domestic, craft, and field labor by slaves taken as spoils of war (Anderson, 1974, pp. 29–52; Lilley, 1965, pp. 32–33). It was simpler and cheaper to put slaves to heavy work than to construct machines to do it (Lilley, 1965, p. 32). The mechanical arts, increasingly being left to slave labor, could not be respected by the ruling elite. According to Aristotle, since artisans, historically, were slaves or foreigners, the "best form of state will not admit them to citizenship" (*Politics*, III.5.1278a.5–10). "Some men are by nature free, and others slaves, and that for these latter slavery is both expedient and right" (*Politics*, I.5.1255a.39–41). "For that some should rule and others be ruled is a thing not only necessary, but expedient; from the hour of their birth, some are marked out for subjection, others for rule" (*Politics*, I.4.1254a.20–25). "No man can practise virtue who is living the life of a mechanic or labourer" (*Politics*, III.5.1278a.20–25).

Free men, according to Aristotle, would best realize their nature through the sciences—above all, through politics, "truly the master art" (*Ethics*, I.2.1094a.25). Such ideological blockage could only have led to technological blockage. It has also been suggested that the Greek view that mechanics could be reduced to a rational, closed system based on the five simple machines (lever, pulley, screw, winch, and wedge) failed to leave room for the role of experiment in the development of mechanics, which in effect had been limited to statics (Vernant, 1957, p. 217).

The stagnation discussed here is relative to the far more rapid developments in Europe in the second millennium A.D. The gradual improvement in the technological system that characterized Greek technology after the third century B.C. was continued by the Romans, who acquired through their conquests the technological traditions of Greece, the Near East, and Egypt, absorbing whatever was useful for them from other peoples in their expanding empire. Individual Romans did gather information about technological practices in various parts of the empire, but little attempt to coordinate or codify it was made (Gille, 1986, pp. 320–326). Although the towns grew, the propertied classes maintained their traditional disdain for trade (Anderson, 1974, p. 81). While the Romans did not introduce revolutionary changes in the technological system, they did make important improvements in hand tools, agriculture, architecture, city planning, the use of water-powered grain mills on an industrial scale, and the building of aqueducts, cisterns, irrigation canals, and roads.

1.7 Technology and Society in China

The history of Chinese technology paralleled in time and in some fields even excelled the early technological developments of the classical civilizations of the Mediterranean and the Near East and, later, of Europe (Needham, 1954–1967). Chinese interest in philosophy and science unfolded at about the same time as in Greece. The Chinese had already learned to write by the fourteenth century B.C., well before the Greeks (Haudricourt and Needham, 1963, p. 162). Yet by the second half of the second millennium A.D. European technology moved decisively ahead of Chinese technology.

Needham discusses in detail what he sees as the reasons why China took an early lead in so many areas of technology and then failed to maintain it. The country's topography and agriculture required a vast network of waterworks for flood protection and irrigation, especially for wet rice cultivation, and a far-flung canal system whereby the tax grain could be brought to granary centers and to the capital. A highly centralized system emerged in which the emperor assumed the role of one great feudal lord governing and collecting taxes through a gigantic bureaucracy that became known as the "mandarinate." The mandarinate collected taxes and mobilized unpaid labor (similar to corvée labor of European feudalism in which a serf is required to perform a day of unpaid labor by a feudal lord) for the water projects. They enjoyed no hereditary succession, and were recruited afresh in each generation through state examinations based on literary and cultural subjects that rarely included what might be called scientific subjects (1969, pp. 195–197). Under this system, referred to by some Chinese scholars as "bureaucratic feudalism" (pp. 177–179), the economy was based on taxation in kind, so that money was not widely used. The mandarinate was not salaried by the state, but supported itself through its administrative powers over agricultural production and whatever wealth it could garner from its bureaucratic functions in the towns.

The value system that favored the development of merchant wealth in the feudal towns of Europe found no parallel in China. While capital accumulation in Chinese society was possible, "the application of [capital] in permanently productive industrial enterprises was constantly inhibited by the scholar-bureaucrats, as indeed any other social action which might threaten their supremacy" (Needham, p. 197). "There is nothing in Chinese history resembling the conception of a mayor or burgomaster, alderman, councillors, masters and journeymen of guilds, or any of those civic individuals who played such a large part in the development of city institutions in the west" (p. 185).

Since Chinese philosophy was not capable of allowing the mercantile mentality a leading place in the civilization, it was not capable of fusing together the techniques of the higher artisanate with the methods of mathematical and logical reasoning that the scholars had worked out, so that the passage from the Vincian to the Galilean stage in the development of modern natural science was probably not possible.

1.8 The Middle Ages

The scientific and technological heritage of the Egyptian, Greek, Roman, and Mesopotamian civilizations of antiquity was in part preserved and continued by the Byzantine and Arabic scholars. *Arabic* here includes Arab and other scholars of the Near and Middle East, Northern Africa, and Moslem Spain, where Arabic was the scientific language. From its beginnings, Islam encouraged the study of the sky and the earth to find proofs of one's faith (Arnaldez and Massignon, 1963, p. 385). "Arabian schools were organized on the pattern of the Greek, publishing commentaries, encyclopaedias, dictionaries and scientific manuals. The scholars versed in more than one science if not in all . . . were the rule rather than the exception" (p. 386). The best known of these encyclopedic minds were the Persian physician and philosopher Avicenna (980–1037 A.D.) and al-Biruni (973–1048 A.D., probably born in Uzbekistan). The Moslem scholars elaborated algebra and passed on Hindu–Arabic numerals from India (p. 407). Like the Romans, the Arabs absorbed the technological knowledge of the peoples they conquered and disseminated this knowledge over their empires, innovating, but not generating a technological revolution of their own.

In Europe proper, the period from the end of classical times following the collapse of Rome to beginning of the eleventh century is commonly referred to as the Early Middle Ages or the Dark Ages. The term *Dark Ages* suggests a darkening relative to Greek and Roman antiquity and in many ways is misleading. The so-called barbaric invaders had already achieved an iron-age culture and had access to the technology of the Roman empire. The principal regression relative to Greece and Rome was the collapse of an elaborate city-state culture. With the collapse of Greece and Rome, literacy in Europe outside of Byzantium retreated significantly.

The principal developments in technology during the early Middle Ages, between the collapse of Rome and the eleventh century, appear to have been the introduction of the heavy iron plow in northern and western Europe; the horse harness, which allowed the replacement of oxen by horses for plowing; and triennial crop rotation and fertilization with marl. These innovations caused a rapid increase in population and an agricultural surplus that permitted the growth of towns (Hilton and Sawyer, 1963; White, 1962; Anderson, 1974, pp. 182–183).

A technological determinist explanation for the specific direction of the development of feudalism in Europe was put forth by White, in his influential *Medieval Technology and Social Change* (1962). White asserts that the introduction of the foot-stirrup in the Frankish kingdom of the eighth century immediately changed the nature of warfare. A mounted knight could deliver a strong blow with a lance, and cavalry replaced foot soldiers as the main fighting force. Since "mounted warriors could only be maintained in large numbers by landed endowment . . . [the] estates of the Church were . . . seized and handed over to an enlarged body of followers on condition that they serve him on horseback. . . . Protofeudal and seignorial elements had, of course, saturated the very fluid Celtic, Germanic, late Roman, and Merovingian societies; but the need for cavalry . . . precipitated and crystallized these anticipations to form medieval feudalism" (pp. 4–5).

Hilton and Sawyer (1963) criticize White's technological determinism as being methodologically at fault, the conclusions based on it not standing up to the historical evidence. They show that the stirrup was not a new arrival in Europe but reached Europe well before the eighth century, and that a complete change in Frankish weapons did not take place at that time—despite the stirrup, mounted shock troops did not become the rule in the eighth century.

Cultural, economic, and environmental factors all interact with technology to affect the rate of historical change. The African Mbuti, for example, satisfied with their traditional hunting methods, rejected the use of guns for hunting despite clear advantages over nets and spears, while the differing circumstances of the Eskimos led them to accept guns.

1.9 Beginnings of the Technological Revolution in Feudal Europe

Toward the end of the ninth century, signs appeared that a new period of technological development was about to begin in Europe. The political disarray of constant invasions, marauding, and remapping of boundaries gave way to greater stability as central governments established authority. In 789, Charlemagne issued edicts to establish schools for the largely illiterate clergy. A ninth century mini-Renaissance produced educational advances, including adoption of a legible script and restoration of a good Latin, but with little or no scientific content (Beaujouan, 1963, p. 471).

Increasing artisan and merchant activity fueled rapid expansion of urban commercial centers in the tenth to thirteenth centuries. The growing use of metals—silver, gold, copper, lead, tin, and iron and steel—put mining, metallurgy, and metalworking in the center of industrial activity and trade. Grain mills powered by water wheels had already appeared throughout Europe, not all concentrated in the towns. They became a traditional source of revenue for the lords, who required that the peasants take their grain to be ground at the lord's mill. The Domesday Book, an English economic census completed in 1086, reported 5624 water mills in 3000 communities south of the Trent and Severn (cited by Forbes, 1956, p. 611). Water-powered cloth-fulling and grain mills and forges (many in monasteries) began in the mountainous regions of southern and central Europe, later spreading into France and northern Europe. By the fourteenth century, water-powered hammers, stamps, and bellows were fairly general in the entire Alpine region. "Such machines," writes Forbes, "entailed heavier capital expenditure than formerly, and the rise of western capitalism was largely tied up with the development of mining and metallurgy." The feudal lords promoted the rise of industrial–commercial towns (pp. 62–69). Manufacturing was at the heart of the urban economy, and initially a symbiotic relationship existed between the social environment of urban industry and trade and the feudal agrarian economy. Although more or less self-governing, the towns were still under the aegis of the feudal lords, who cornered markets or scooped off profits from long-distance trade and obtained their domestic or imported luxury goods from the towns. Production was regulated by the artisan guilds, which emerged in the tenth century.

The first great mathematician of the thirteenth century, and perhaps the most outstanding in the Middle Ages, Leonardo Fibonacci (1170–1250, also known as Leonardo Pisano or Leonardo of Pisa), came from the rising commercial social stratum (Smith, 1923, pp. 214–218). Italian merchants acquired computational methods from their North African counterparts. Double-entry bookkeeping first appeared in 1340 in Genoa (Beaujouan, 1963, p. 516). In parallel with the commercial interest in practical mathematics, the growing application of a variety of mechanical devices to manufacturing contributed to an interest in the representation of spatial relationships. The contributions of Leonardo da Vinci in this connection are well known.

A close relationship thereby developed between the artists and the artisans. This bond among those engaged in the arts and crafts continued to spread through Western Europe in the fifteenth and sixteenth centuries. Consisting of artist-engineers, painters, sculptors, architects, builders of locks and canals and fortifications, and toolmakers, this group of practitioners, who themselves did not study questions of a scientific character, stimulated others to do so. A related group included the instrument makers (who supplied navigators, geodesists, astronomers, and musicians with the instruments they needed), the clock-makers, the cartographers, and the military technicians. "Empirical knowledge did not have to be sought deliberately, but arose naturally from the pursuit of technical trades; the waiting was only for theoretical reflection, which, however, was helped by the fact that there is no single department of physics which calls more urgently for mathematical treatment and lends itself more naturally to it than mechanics. The first essential element of classical physics, the mathematical approach, thus came into its own as spontaneously as the second, the empirical foundation" (Dijksterhuis, 1969, p. 243).

The center of scholarship initially rested in the monasteries, later moving into the universities, themselves Church institutions. Interest in mathematics developed among Church scholars because the Church calculated the date of Easter by combining the lunar cycle with the solar year on the basis of the Ptolemaic geocentric model of the heavens. In the tenth century, the French monk Gerbert of Aurillac (who later became Pope Sylvester II) brought Hindu–Arabic numerals to Christian Europe from Spain. Gerbert also greatly simplified the methods for arithmetic computations. Within the Church, however, conflicts arose between the traditions of classical scholarship and the new approaches needed for the burgeoning crafts and trade.

The English monk Roger Bacon (1214?–1294) was among the first Europeans to recognize the need for empirical verification of theoretical knowledge: "Reasoning does not make a conclusion certain, unless the mind discover it by the method of experiment." He also recognized the need to deal with nature quantitatively: "Nature cannot be known without mathematics" (Beaujouan, 1963, p. 492).

Francis Bacon (1561–1626) may be considered the first to have seen a direct dependence of science on technology. He called for scholars to place technology on a scientific foundation by familiarizing themselves with empirical knowledge accumulated in the crafts (Dijksterjuis, 1969, p. 401). The encyclopedia Bacon proposed was accomplished by Diderot and d'Alembert. By including descriptions of all tools and the methods of their fabrication and utilization, it conferred the status of knowledge and social dignity on that which up to that time was a matter of utility, of spontaneous empiricism (Fontenay, 1982).

1.10 From Feudalism to the Industrial Revolution in Europe

In the historical shift that initiated the most rapid growth in material production in history, the interaction among factors that stimulate or are stimulated by technological development may be seen clearly. Small-scale manufacturing continued to grow throughout Europe and North America, but developments in Britain took on a distinctive character that led to what became known as the Industrial Revolution.

In the middle of the eighteenth century, Britain's manufacturing and commerce, especially the slave-based sea trade in colonial products such as tea, sugar, and tobacco, were prospering. Perhaps one-tenth of all fixed investments other than real estate were in the maritime trade. Britain had several times the number of ships as France, its nearest competitor. The rural economy of Britain developed into a substantial system of cash incomes and sales; tenant farmers employed large numbers of farm laborers, unlike the predominantly subsistence-farming peasant households that persisted on the continent (Hobsbawm, 1968, pp. 10–19). Manufacturing spread throughout rural areas, the typical workers being village artisans or smallholders in their cottages making specific products, such as cloth, hosiery, or metal goods. With rapid urbanization and wood being in short supply (Gille, 1986, p. 590), the growth in the production of coal, primarily for heating and cooking, formed a substantial preindustrial base for subsequent industrial expansion (Galloway, 1969, Chap. 4–5). Coal was the first industrial commodity to be produced at a rate of millions of tons a year (Hobsbawm, 1968, p. 31). Unlike on the continent, the large landowners, rather than the monarch, drew royalties from mining operations and had a direct interest in supporting village manufacturing, as well as canals and turnpikes to facilitate the sale of the goods produced. The British economy was thereby in a unique position to benefit from a major expansion in goods produced cheaply enough to be purchased by the middle classes, instead of the production of luxury goods for the wealthier strata as on the continent.

The main advantage of the preindustrial home market was its size and steadiness. "The domestic market may not have provided the spark, but it provided fuel and sufficient draught to keep it burning" (Hobsbawm, 1968, p. 32). The spark, according to Hobsbawm, was the export market. "Between 1700 and 1750 home industries increased their output by seven per cent, export industries by seventy-six percent; between 1750 and 1770 (which we may regard as the runway for the industrial 'take-off') by another seven per cent and eighty per cent respectively. Industrial production increased—but foreign demand multiplied."

Cotton manufacture was responsible for the bulk of the export trade. All raw material had to be imported, and by the end of the eighteenth century, about two-thirds of the industry's output was exported. Prior to the 1760s, the spinning wheels could not supply enough yarn to keep pace with looms now employing the flying shuttle, introduced in 1733. A breakthrough came with the spinning jenny in 1765. Further improvements in spinning followed, leading finally to the fully mechanized spinning mule in 1777 (Gille, 1986, p. 624). Subsequent improvements in the machines increased the yarn yield per worker 100-fold in relation to the spinning wheel. As the textile industry expanded, its machinery was increasingly driven by the steam engines made possible by James Watt's invention, although water power continued to be used when convenient. The widespread use of steam power made possible industrial activities in regions in which water power was not available. The entire technological basis of British manufacturing thereby shifted from water and wood to iron, coal, and steam.

The rapid advances of science and technology at this time were quickly absorbed by the various branches of manufacturing. The relatively simple technology required for cotton production, however, propelled the Industrial Revolution forward at an unprecedented rate without much interaction with science. The majority of the technological innovations were made by people closely involved with production—foremen, entrepreneurs, and skilled workers (Gille, 1986, p. 594). Scientific advance in eighteenth-century Britain came largely from societies established by craftsmen and industrialists, such as the Lunar Society of Birmingham and the Manchester Literary and Philosophical Society (Hill, 1967, p. 198). James Watt was an instrument maker at the University of Glasgow, where he learned about Joseph Black's discoveries of latent heat. Bernoulli's and Euler's work on ships no doubt influenced ship construction. It should not be forgotten, however, that Carnot's theoretical introduction of the cyclical engine came much later, in 1824, while the distinction between force and energy was not theoretically explained until the 1860s. In general, the sciences during the early years of the Industrial Revolution did not lead to direct results in design, but pinpointed quantitative relations that inventors and innovators could take into account in their empirical work.

The immediate result of the Industrial Revolution was great economic benefit for entrepreneurs and harm for workers. Mechanization of the textile industry threw large numbers of weavers out of work. The shift of the workplace into factories from rural cottages and homes destroyed this source of supplemental income. Wages were kept at subsistence levels. The shift to the cities was precipitous. In 1750, only London and Edinburgh had populations over 50 000. In 1851, there were twenty-nine cities with populations that big and nine with over 100 000. "And what cities! It was not merely that smoke hung over them and filth impregnated them, that the elementary public service—water supply, sanitation, street cleaning, open spaces, etc.—could not keep pace with the mass migration of men into the cities, thus producing, especially after 1830, epidemics of cholera, typhoid, and an appalling constant toll of the two great groups of urban killers—air pollution and water pollution, or respiratory and intestinal disease" (Hobsbawm, 1968, p. 67). As the industrial economy reached new heights quantitatively, qualitative changes in market relationships ushered in an era of cyclical economic crises of unprecedented depth. The first deep crises occurred in the 1810s, in response to which displaced textile workers directed their anger against the machines that replaced them. Crises occurred again in the 1830s, and again in the 1840s, leading to large-scale bankruptcies and high unemployment. The conditions of the laboring population were so harsh that advocates of slavery in North America argued that chattel slavery in the U.S. South was far more humane than wage slavery in Britain. In the 1850s, economists began to conclude that higher wages would put more money in circulation and thereby stimulate industrial production. Growing restiveness among the workers and their reliance on trade unions finally led to higher wages and reform in labor legislation.

Although the Industrial Revolution replaced many skilled workers with unskilled workers, a need arose for a literate segment in the work force able to read and follow somewhat complex instructions, read technical drawings, and perform calculations. Public primary education was therefore initiated in the first half of the nineteenth century in Britain, France, and the United States.

1.11 From the Industrial Revolution to Modern Technology

In the wake of the Industrial Revolution, manufacturing moved into a central position in the dynamics of the national economies of Western Europe, Canada, and the United States, spreading to Japan late in the nineteenth century, bringing with it an infrastructure of transportation, raw material sources, fuel, and finance. Raw materials were secured domestically, from other industrialized lands, or from colonial and semicolonial possessions. The role of the national governments in securing, maintaining, and regulating these infrastructures increased greatly. As the size of the enterprises increased, the pattern of ownership also changed from single owners and partnerships to the corporate form. Toward the end of the nineteenth century, investment banks formed, acquired, or merged with industrial firms. These developments took place simultaneously in the United States, Britain, France, Germany, and Japan; the United States overtook Britain in industrial potential around 1900. Competition for sources of raw materials and foreign markets grew intense. The economic, political, and military subjugation of almost all of Africa, Asia (outside of Russia and Japan), Central and South America, the Pacific, and the Caribbean was essentially completed by 1910 in an atmosphere of great political and military rivalry among the industrial powers, a rivalry that sharpened the interest of national governments in military technology.

Industrial managers near the end of the nineteenth century were not as closely connected personally with technological development as formerly. Research and development laboratories arose within the individual enterprises (see PHYSICS AND TECHNOLOGY, HISTORY OF). The universities remained centers of theoretical research and of the experimental research that necessarily preceded it.

The introduction of the assembly line at the Ford factory in 1913 based on the time and motion studies of Frederick W. Taylor and Frank and Lillian Gilbreth had a profound social and economic impact. The organization of work on the assembly line broke up complex manufacturing operations into simple jobs that could be performed by a worker with little or no training. By 1926, 79% of the Ford workers needed less than one to no more than eight days training (Gille, 1986, p. 1017). The economic success of this method of production led to its use wherever technologically feasible. The dehumanizing reduction of labor to essentially mindless tasks, already foreshadowed in the textile mills, reduced workers to appendages of machines, alienating their spirits from their hands and from the object of production. Assembly-line production continues to this day, but with increasing involvement of the workers in setting up and rotating tasks. Many big manufacturing plants are now experimenting with abandoning assembly lines altogether.

2. ORIGINS OF PHILOSOPHY OF SCIENCE AND PHILOSOPHY OF TECHNOLOGY

From the seventeenth century on, scientific research tended to concentrate in academic institutions, while technological research and development was conducted largely in private industry, but this has become less true since the mid-twentieth century. As the time between scientific discovery and technological innovation decreased, technology became increasingly dependent on science for its innovations. Toward the end of the twentieth century, the dominant view is that science is the primary stimulant to technological development and not the reverse.

Prior to the eighteenth century, no fundamental distinction was made between scientific knowledge and philosophy. The advances in physics, especially Newton's mechanics, stimulated the separation. The theoretical structure of Newton's mechanics is that of the hypothetico-deductive model employed by Euclid. The clearly empirical character of Euclid's postulates and axioms had not been understood, even in Newton's time. The simplicity of the axioms and their fewness of number continually stimulated the quest to find a logical basis for their origin, somewhat reminiscent of today's quest for a grand unification.

Kant (1966) proposed a compromise. Certain kinds of knowledge need not be demonstrated; logic, geometry, and time, for example, exist prior to and independently of material things. Knowledge about the material world is acquired through our senses. How does one identify *a priori* knowledge, however, since if it cannot be proved, it must be asserted? John Stuart Mill straddled the issue: "Axioms are but a class, the most universal class, of inductions from experience; the simplest and easiest cases of generalization from the facts furnished to us by our senses and by our internal consciousness" (Mill, 1904, p. 187).

Two logical problems immediately arise with what on the surface appears to be the experimental or empirical basis of Mill's view. One is the logical basis of facts furnished by our internal consciousness—this can only be a return to the Kantian *a priori*. The other is the source of the sensual stimuli that produce the "facts furnished to us by our senses."

Francis Bacon had already recognized the need to put technology on a scientific basis by incorporating technological experiences into scientific investigation. The early nineteenth-century philosopher Hegel (1969) saw (without pursuing its logical consequences in detail) that properties of space and time are not independent of, but originate with, matter. Riemann, confronted first with the non-Euclidian geometry of Lobachevsky and Bolyai, and then with his own, concluded that only experiment would establish the validity of a geometrical system.

Though known, perhaps, as the leading nineteenth-century proponent of philosophical idealism, Hegel uncharacteristically took a very earthy approach to technology by asserting the importance of means in relation to the ends: "The *means* is *superior* to the *finite* ends of *external* purposiveness: the *plough* is more honourable than are immediately the enjoyments procured by it and which are ends. The *tool* lasts, while the immediate enjoyments pass away and are forgotten. In his tools man possesses power over external nature, even though in respect of his ends he is, on the contrary, subject to it" (Hegel, 1976, p. 747). Thus, in Hegel's view, the ends are established through human reason, but this reason is transcended by the technology created to procure the ends and which can continue to be employed for other ends. Hegel here begins what might be called the praxis philosophy, according to which practical activity of humans serves as the ultimate source of their technology and science.

Karl Marx, the first to incorporate technology philosophically into a world outlook, never explicitly identifies the philosophy of technology in what came to be known as his materialist conception of history. In Marx's scheme, the primary motor of societal evolution is the development of the *productive forces*, by which he means natural resources, raw materials, implements and tools, and human labor. Implied, although not mentioned explicitly, is the technological and scientific knowledge associated with productive processes. Technological development, that is, the development of the productive forces, proceeds under more or less stable relations of production, which Marx equates with the dominant form of property relations associated with the use of the productive forces. He then projects that an incompatibility ultimately arises between the productive forces and the property relations so that property relations cannot be kept stable. Social revolution, to Marx, is simply the resolution of this incompatibility through a relatively rapid change in the dominant property relations. Thus, Marx (1987) explains the evolution of society from its early forms of communal property through slavery, feudalism, capitalism, and ultimately communism. A critical point of Marx's theoretical position is that "it is not the consciousness of men that determines their existence, but their social existence that determines their consciousness." The technological implications of this are profound, because production relations are not limited to property relations. People engaged in production also have technological relations of production to the tools and the object of production, and to one another when several people work together. It is therefore the productive activity of human beings that ultimately shapes their understanding of their relations to the material world around them and of the entire complex of social relations that they have to one another.

Such a view of the relationship between existence and consciousness leads Leonard Goldstein to attribute the changes in the way the Renaissance artists viewed space to the need to understand the spatial relationships for the design and use of large machines driven by natural power and for manufacturing processes involving the coordinated labor of a group of workers. Goldstein notes that linear perspective depicts the view grasped by the eye of an individual observer. He identifies the source of this one-point view as a reflection of the individualism that is necessarily shaped by vigorous activity of the ownership of a successful enterprise (Goldstein, 1988).

A propensity to mechanical innovation has sometimes been seen as a character trait common in some nations; Marx, however, would look rather to the conditions motivating people to engage in production as the stimulus for innovation. One source of motivation is production to meet one's own needs for direct consumption; another is the production of commodities for exchange for other commodities for direct consumption. Still another, and this is the principal characteristic of the capitalist economy, is the production of commodities for exchange at a profit on a continuing basis. Since the goal is accumulation, not consumption, the appetite for profit is unsatiable and technological improvements will help feed it. Marx regards technological development as intrinsically neutral, but leading to property relationships that provide the social dynamic. "The hand-mill gives you society with the feudal lord; the steam-mill society with the industrial capitalist" (Marx, 1935, p. 355). Fundamentally, Marx was not a critic of the application of technology or of the various types of social inequality that he associated with given levels of technological development. His radicalism was associated with his conviction that the level of the productive capacity under capitalism had reached a point where the market relations produced a conflict between distribution and consumption that manifested itself in periodic economic crises, that these crises would become deeper and deeper, and that public ownership, as distinct from private ownership, of capital resources was the next stage of the socioeconomic evolutionary process that began with the appearance of the surplus product. He is often incorrectly characterized as a technological determinist. If this were the case, he would not have had to assume the role of a political radical. He did so because he rejected Laplacian determinism in regard to the human spirit, so that he did not consider that the radical social consciousness necessary to bring about this next social transformation would emerge spontaneously, but required a political movement to bring it about, even if history was on its side.

In 1877, the term *philosophy of technology* appears first in the title of a German work, *Grundlinien einer Philosophie der Technik* by Ernst Kapp, who was strongly influenced by Hegel's concept of history being the realization of the Idea. He wrote that human beings using their first tools, their hands, in producing their first works became historical human beings progressing toward self-consciousness (Kapp, 1877, p. 39).

The goal of technology as a field of human endeavor is, like science, not the same for all its practitioners. For some, the "tinkerers," it can be an end in itself, just as mathematicians will sometimes say that they are interested only in the beauty of the logic and do not care if their work has any practical use. On the other hand, it is rare for successful tinkerers to neglect patenting the results of their tinkering.

The empiricist view of John Stuart Mill was further refined by the Austrian physicist Ernst Mach and, subsequently, further developed by a group of scientists and philosophers known as the Vienna Circle into what became the most influential school of philosophy of science in the first half of the twentieth century, *logical positivism*. According to this view, in its application to physics, "quantities such as the charge, temperature, mass, and length of a body are *defined* as the objective results of certain prescribed operations that can be carried out in the laboratory. ... This *operational viewpoint* is the basis of this philosophy." Further, "physical *laws* are relationships between operationally defined quantities that always occur when certain experiments are performed" (Halliday, 1955, pp. 4–5).

In the logical positivist view, the question of the source of our sense impressions or observations is meaningless, since there is no operational way of establishing existence of objects external to us. The attractiveness of logical positivism to the scientist is its apparent rejection of speculative activity. The practical task of scientific theory is to provide correlations between observational data.

3. CONTEMPORARY VIEWS ON PHILOSOPHY OF TECHNOLOGY

By the middle of the twentieth century, the logical problems connected with operational definitions began to lead to their demise. One objection, for example, was that an operational definition of a field would require an infinite set of measurements. Another objection was that no two instruments are identical and no two measurements by the same instrument of infinite precision would be identical; therefore each act of measurement would produce its own unique operational definition. It can be argued that operational definitions are means for standardizing scales of measurement of physical properties, the meaning of which arise from the laws that embrace them.

With the collapse of logical positivism, scientific realism rose to the fore. Realism itself is not one particular philosophical current but a category of approaches to what is often called the mind–body problem. Realism, in general, recognizes the relative independence of our ideas from that which exists outside our minds. At the same time, realism does not necessarily exclude the dependence of one on the other. Some realist philosophies may give primacy to the world of ideas in the sense that it is ideas that ultimately engender the world outside our minds, while others can argue that it is the world outside our minds that ultimately gives rise to our ideas, allowing for various degrees of this dependence.

According to Don Ihde, philosophy of science has been historically linked to the three prejudices that he associates with a *theory bias* that affects the attitude toward technology. The first goes back to Greek antiquity, according to which "the aim of highest humanity was, in effect to achieve the deepest knowledge and, once attained, meditate upon it." The second is more recent and holds that modern technology is both essentially different from all ancient or traditional technologies, and therefore in some fundamental sense, "better." The third prejudice is the belief that modern technology differs from all previous technology in that it is derived from modern science (Ihde, 1993, p. 20). Ihde cites the Platonic view, reflected by Aristotle, which holds that anything associated with the body and the material is inferior to the realm of ideas and forms, the latter being associated with the mind or soul.

Philosophical attention to technological concerns grew slowly during the first half of the twentieth century. Among the important commentators on the question were Friedrich Dessauer (1927), Ernst Jünger (1932), and José Ortega y Gasset (1941). Renewed interest developed after World War II, stimulated at first by the social, political, and economic concerns resulting from nuclear weapons. Simultaneous with the introduction of nuclear technology was the beginning of the computer age, large-scale application of electronic computers having begun during World War II in connection with coupling artillery (naval and land) to radar.

One work that attracted immediate attention was an essay by Martin Heidegger (1953), "The Question Concerning Technology." Heidegger begins his discussion of technology by pointing to the inadequacies today of the earlier instrumental and anthropological definitions of technology, the instrumental definition being the means to achieve particular ends, while the anthropological definition refers to the human activities employing these means. He thinks that it is necessary to go beyond these definitions to arrive at a deeper understanding of the relationship between humans and the technology they employ. This is because as humans attempt to control technology and bring it spiritually within their power, it continually slips through their fingers no matter how hard they try. It is necessary, above all, to distinguish modern technology from past technologies. What is fundamentally different in modern technology is that it views both nature and human labor as a continually available and extendible standing reserve whose possibilities confront humanity as a challenge for further disclosure or uncovering. Modern technology challenges and stockpiles the sun's warmth stored in coal "for heat, which in turn is ordered to deliver steam whose pressure turns the wheels that keep a factory running" (Heidegger, 1977, p. 15). The windmill of classical technology is of a different character. "Its sails do indeed turn in the wind; they are left entirely to the wind's blowing. But the windmill does not unlock energy from the air currents in order to store it" (1977, p. 14). "The gathering together of that setting-upon that sets upon man, i.e., challenges him forth, to reveal the real, in the mode of ordering, as standing reserve" is the way of revealing "which holds sway in the essence of modern technology and which is itself nothing technological" (p. 20).

Modern and classical technology are both ways of disclosure (or uncovering), but only in modern technology are humans challenged to this uncovering. This already occurred with the development of physics as an exact science, on which modern technology is based (p. 21). Heidegger, however, does not see science as the primary stimulant of technology. He points out that modern physics involves experiment and is therefore dependent on the development and application of technical apparatus. Since the essence of technology lies in the uncovering process that constitutes technological activity, technology must be seen as applying natural science, rather than being considered as applied natural science (1977, p. 14).

Heidegger sees the main danger from modern technology not so much in the threats of environmental disasters resulting from its applications but in human beings surrendering themselves to what is technologically revealed and being absorbed into the revealed technology, all the while believing that they are the masters and in full control of the earth. This is the illusion generated by looking at the technological world merely as a collection of objects and means to be set in motion by human activity, and it is this viewpoint that threatens to reduce humans themselves to a mere standing reserve. It is not enough to acquaint oneself with external objects and the relations among them, including their relationships to humans themselves. It is necessary to be conscious of how our own being is grounded in the complex of relationships among ourselves and the objects around us. An airplane is not merely an object of a given shape and construction. When revealed, it is a standing reserve for the transportation of people and goods, and to fulfill this role, it must be fully appropriate in structure and readied for flight. It needs airports from which to take off and land, and these airports must be planned for and constructed by persons who do not necessarily travel in the planes, etc. Heidegger argues that this type of understanding is necessary before we can do anything to dispel the illusion that humans are the masters of the world, and uncover the paths to reverse the domination that technology maintains.

By his repeated stress on the concept of uncovering, Heidegger in essence projects an *a priori* character onto that which can be uncovered. Moreover, he regards continual involvement in technological uncovering as human destiny, but not in a deterministic way. The danger of technology that he sees is therefore that people see themselves challenged to uncover technology in a one-sided way, a way that contributes to their virtual enslavement by technology and obliterates any free will. He sees human destiny to be the confrontation of the challenge for technological uncovering and thus has to reject the possibility of abandoning technological development. Rather, he calls for the implementation of human free will to proceed along a path of technological uncovering "that in no way confines us to a stultified compulsion to push on blindly with technology or, what comes to the same thing, to rebel hopelessly against it and curse it as the work of the devil" (1977, p. 26). Heidegger does not, however, see this enactment of human "free claim" as a purely human undertaking. That which is there to be uncovered was not put there by human beings. This last point underscores the essentially mystical character of Heidegger's philosophy, since the human response to an objectively existing situation will be indifferent to the process by which this objective situation arose if no knowledge of its supposed purpose, assuming there was one, is available.

Heidegger's *a priori* concept of technology was disputed by Herbert Marcuse in *One-Dimensional Man* (1964). "The technological *a priori* is a political *a priori* inasmuch as the transformation of nature involves that of man, and inasmuch as the 'man-made creations' issue from and re-enter a society ensemble" (p. 154). Marcuse saw the inherent rationality of modern technology as a source of human enslavement. "Scientific management and scientific division of labor vastly increased the productivity of the economic, political, and cultural enterprise. Result: the higher standard of living. At the same time and on the same ground, this rational enterprise produced a pattern of mind and behavior which justified and absolved even the most destructive and oppressive features of the enterprise. Scientific–technical rationality and manipulation are welded together into new forms of social control" (1964, p. 146). Marcuse criticizes the harm done by relying on positivist criteria in assessing technological performance. What is quantitatively measurable is profit and efficiency, not the toll in human suffering.

The views of another important writer on technology, Jacques Ellul, continue to be influential today. According to Ellul's summary of his own views (1983), first expressed in his book *The Technological Society* (1964), technology replaces nature by creating the specific milieu in which humans must exist. Technology thus becomes artificial, autonomous with respect to values, ideas, and the state, self-determining, growing in a process that is causal, but directed to ends. "It is incorrect to say that economics, politics, and the sphere of the cultural are influenced or modified *by* Technique; they are rather situated *in* it, a novel situation modifying all traditional social concepts. Politics, for example, is not modified by Technique as one factor among others which operate upon it; the political world is today defined through its relation to the technological society. Traditionally, politics formed a part of a larger social whole; at the present the converse is the case" (1983, p. 86). Ellul argues that technological progress is in itself neither good nor bad. "The fact is that, viewed objectively, technological progress produces values of unimpeachable merit, while simultaneously destroying values no less important. As a consequence it cannot be maintained that there is absolute progress or absolute regress" (p. 98). In summary, Ellul maintains that all technological progress exacts a price, raising more problems than it solves, pernicious and generally unforeseeable effects being inseparable from its favorable effects. The technological society cannot not be a humanist one. There is a contradiction between technological perfection and human development because technological perfection "is only to be achieved through quantitative development and necessarily aims exclusively at what is measurable. Human excellence, on the contrary, is of the domain of the qualitative and aims at what is not measurable" (p. 90). Ellul sees no escape from this situation. "There is no possibility of turning back, of annulling, or even arresting technical progress. What is done is done. It is our duty to find our place in our present situation and in no other" (p. 91).

The most widely read contemporary philosopher of technology is Landon Winner (1977, 1986), who also focuses on the theme of the autonomous character of technology, viewing it as self-moving and self-evolving according to its own laws. Megatechnical systems like the electric power system dominate humanity by exerting demands for their own maintenance. Humans can monitor and guard the system against destruction and deterioration, or introduce innovations to improve its efficiency, but are unable to replace it by another. "We indeed 'use' telephones, automobiles, electric lights, and computers in the conventional sense of picking them up and putting them down. But our world soon becomes one in which telephony, automobility, electric lighting, and computing are forms of life in the most powerful sense: life would scarcely be thinkable without them" (1986, p. 11).

In the 1950s and 1960s, the environmental hazards associated with the unrestricted development of nuclear and chemical technologies became evident. Problems also arose in connection with the disposal of nuclear and toxic chemical waste. The growing quantity of nonbiodegradable plastic waste created additional problems. The growing integration of electric power systems into national and even international networks generated the threat of a national and international catastrophe from a major breakdown in the electrical grid. With the development of supertankers, ecological disasters produced by oil spills became a reality. Standing above all other threats facing humanity was that of nuclear war. These developments stimulated a number of philosophers to focus their attention on the scientific and societal implications of contemporary technological development. Philosophy of technology, which hitherto was generally limited to sporadic study by philosophers primarily interested in other areas, emerged as a field of specialization within philosophy, with international organizations and publications.

One set of strategies put forward, according to Ihde (1993, pp. 119–27), to deal with environmental issues involves returning to some previous lower level of technology or to minimalist technologies. Ihde argues that such a strategy would produce social disruption as well as an immediate drastic worldwide decline in population comparable to times of vast warfare and plagues.

An *ameliorationist* set of strategies accepts high-technology developments, but turns them into regional and decentralized modes for such activities as food and energy production. Ihde believes that such strategies can only have limited success, since the rate of technological change outpaces the possibilities of such alternative strategies.

The third set of strategies Ihde calls *hi-tech reformist*. Solutions to the problems of continued technological development are to be provided by miniaturization, clean-up technologies, and alternative energy production. These strategies require a centralized administration and planning on an international scale and continual technological development of "clean" solutions. Efforts to deal with global problems must be approached with a multicultural perspective and, according to Ihde, a holistic view of the interconnections of technology, environment, and nature.

4. UNRESOLVED PROBLEMS OF TECHNOLOGICAL DEVELOPMENT

Modern technology is now capable of eradicating human life on earth through a number of separate paths. The most obvious has been through nuclear war. Although the principal sources of political tension that posed this threat were removed at the beginning of the 1990s, production and maintenance of nuclear weapons continues. New sources of political tension may emerge and pose the threat of nuclear war once again. Further development of chemical and biological weapons may bring them to the point that they pose the same scale of danger to human existence as nuclear weapons.

Scientists in the United States are divided over the question of nuclear power, although a majority of physicists probably believe that current technologies carry an unacceptable risk of nuclear disaster. The potential danger from commercial application of genetic engineering, which began on a significant scale in the 1980s, is twofold. Unanticipated effects of scale may occur—that is, harmful effects appearing only when artificially created genetic strains are available in much larger quantities and concentrations than in the research and development stage. The second danger is of accidental release of harmful organisms. Richard Leakey has pointed out how fortunate it is for humanity that the AIDS virus is not transmitted through sneezing. What if an accidental release occurred of some potentially lethal experimental microorganism capable of such rapid transmission?

The most commonly acknowledged threat to human existence at present is from industrial chemicals. International agreements designed to halt the release of chemical agents that destroy ozone in the upper atmosphere are now in effect. No international agency, however, has the authority to act independently to meet other dangers as they arise; each hazard requires a separate international agreement. Differences in the assessment of dangers arising among different countries, based on political and economic pressures rather than scientific considerations, may prevent averting the ultimate disaster.

Between 1700 and the end of World War II, industrialized countries pursued economic policies denying the export of technology. The breakup of the colonial systems, increasing resistance in the Third World to such imperialistic policies, and export of capital seeking lower wage scales are now bringing technological development to previously underdeveloped regions. Global resource and environmental crises loom, such as the destruction of the Brazilian rain forests. As the peoples of these countries aspire to the level of commodity consumption that they see in the industrialized countries, the potential for international conflicts is obvious.

An area of crisis limited to the industrialized countries is the growth of unemployment as a result of automation, robotization, and computerization of industrial and office work. The still-growing replacement of metal and wood by the less labor-intensive polymers in manufacturing and construction exacerbates this problem. In the early 1990s, over a period that is not usually characterized as a time of deep economic crisis, unemployment in the industrialized countries of Europe was often above 10%, while in the United States (with less strict statistical measures for unemployment), unemployment held at about 6%. While the financial consequences of this unemployment can be offset in part by social insurance and social welfare programs, the shifting of skilled labor into unskilled labor at lower pay as well as threatened future displacement for those not yet affected has spread insecurity among employees in all income levels.

The source of these problems lies outside of technology itself, in the driving force behind the application of technology. The application of technology to production is not stimulated directly by the need for products, but by the profit that can be made from the sale of products that satisfy real or imagined needs. The "unseen hand" that classical laissez-faire economics saw balancing economic life seems no longer to function. Various degrees of economic regulation and intervention by the state have been introduced into otherwise free-market economies. Marx suggested that such measures would not eliminate the business cycle with its growing swings of prosperity and depression and that the solution would lie in replacing production for profit by production for need, displacing private ownership of the capital resources by public ownership and planned utilization of these resources. The effort to establish such a system on a wide scale beginning in the U.S.S.R. in 1917, extending to Eastern Europe and China and some other countries in the 1940s and later, collapsed completely in the U.S.S.R. and Eastern Europe. It has survived in modified form in the less-industrialized countries of China, Vietnam, North Korea, and Cuba, where various mixtures of market and planned economies are being tried. Chinese, Vietnamese, and Cuban leaders attribute the European collapse to the extreme bureaucratization of political and economic life and excessive centralization of the economic planning mechanisms. Whether they can produce different results is still to be seen.

As long as economies function within a capitalist framework, technological development must proceed, for the inanimate corporation or its animate principal stockholders cannot consume all profits from a successful year of activity, but must continually reinvest them on an expanding scale. Global environmental and resource problems must ultimately be solved by limiting reinvestments in material production to levels that can be technologically tolerated. Expansion of economic activity in directions that minimally absorb material resources offers at present a partial solution to the problem. Commodities and services that take largely nonmaterial forms, such as computer software and artistic production, are of this character. Perhaps the welfare of humanity in the immediate future lies in the expansion of the range of such activities.

Works Cited

Anderson, P. (1974), *Passages from Antiquity to Feudalism*, London: Verso.

Aristotle (1947), *Politics* and *Ethics*, in: R. McKeon (Ed.), *Introduction to Aristotle*, New York: Modern Library.

Arnaldez, R., Massignon, L. (1963), in: R. Taton (Ed.), *Ancient and Medieval Science: From Prehistory to AD 1450*, London: Thames and Hudson, pp. 385–410.

Beaujouan, G. (1963), in: R. Taton (Ed.), *Ancient and Medieval Science: From Prehistory to AD 1450*, London: Thames and Hudson, pp. 468–532.

Binford, L. R. (1980), *American Antiquity* **45**, 4–20.

Boswell, J. (1934), *Boswell's Life of Johnson*, Vol. 3, Oxford, U.K.: Clarendon.

Candland, D. K. (1987), in: G. Mitchell, J. Erwin (Eds.), *Comparative Primate Biology*, Vol. 2B, *Behavior, Cognition, and Motivation*, New York: Alan R. Liss, pp. 85–103.

Carlyle, T. (1921), *Sartor Resartus*, New York: Harcourt.

Coudart, A. (1991), in: S. A. Gregg (Ed.), *Between Bands and States*, Occasional Paper No. 9, Center for Archaeological Investigations, Southern Illinois University at Carbondale, pp. 395–420.

Darwin, C. (1981), *The Descent of Man and Selection in Relation to Sex*, London: John Murry.

Dessauer, F. (1927), *Philosophie der Technik: Das Problem der Realisierung*, Bonn: Friedrich Cohen.

Dijksterhuis, E. J. (1969), *The Mechanization of the World Picture*, London: Oxford Univ. Press.

Ellul, J. (1964), *The Technological Society*, New York: Knopf.

Ellul, J. (1983), in: C. Mitcham, R. Mackey (Eds.), *Philosophy and Technology: Readings in the Philosophical Problems of Technology*, New York: Free Press, pp. 88–105.

Espinas, A. V. (1897), *Les Origines de la Technologie*, Paris: F. Alcon.

Falk, D. (1993), in: R. L. Ciochon, J. G. Fleagle (Eds.), *The Human Evolution Source Book*, Englewood Cliffs, NJ: Prentice-Hall, pp. 61–70.

Ferré, F. (1988), *Philosophy of Technology*, Englewood Cliffs, NJ: Prentice-Hall.

Fontenay, E. de (1982), *Diderot, Reason and Resonance*, New York: G. Braziller.

Forbes, R. J. (1956), in: C. Singer, E. J. Holmyard, A. R. Hall (Eds.), *A History of Technology*, Vol. 2, Oxford, U.K.: Clarendon, pp. 41–80, 589–628.

Fried, M. H. (1967), *The Evolution of Political Society*, New York: Random House.

Galloway, R. L. (1969), *A History of Coal Mining in Great Britain*, Newton Abbot, Devon, U.K.: David and Charles Reprints.

Gille, B. (1986), *The History of Techniques*, New York: Gordon and Breach, 2 vols.

Goldstein, L. (1988), *The Social and Cultural Roots of Linear Perspective*, Minneapolis: MEP Publications.

Goodall, J. (1968), *Anim. Behav. Monographs* **1**, 165–311.

Gould, R. (1982), in: N. M. Williams, E. S. Hienn (Eds.), *Resource Managers: North American and Australian Hunter-Gatherers*, AAAS Selected Symposium No. 67, Boulder, CO: Westview, pp. 69–92.

Haas, J. (1987), in: J. Haas, S. Pozorski, T. Pozorski (Eds.), *The Origins and Development of the Andean State*, Cambridge, U.K.: Cambridge Univ. Press, pp. 1–4.

Halliday, D. (1955), *Introductory Nuclear Physics*, 2nd ed., New York: Wiley.

Haudricourt, A., Needham, J. (1963), in: R. Taton (Ed.), *Ancient and Medieval Science: From Prehistory to AD 1450*, London: Thames and Hudson, pp. 161–77.

Hegel, G. W. F. (1969), *Hegel's Science of Logic*; translation by A. V. Miller, London: Allen & Unwin; reprint (1976), New York: Humanities.

Heidegger, M. (1977), *The Question Concerning Technology and Other Essays*, translation by W. Lovitt, New York: Harper & Row.

Henry, D. O. (1989), *From Foraging to Agriculture*, Philadelphia: Univ. of Pennsylvania Press.

Hill, C. (1967), *Reformation to Industrial Revolution: A Social and Economic History of Britain 1530–1780)*, London: Weidenfeld & Nicolson.

Hilton, R. H., Sawyer, P. H. (1963), *Past & Present*, No. 24.

Hobsbawm, E. J. (1968), *Industry and Empire: An Economic History of Britain since 1750*, London: Weidenfeld and Nicolson.

Hood, B. (1993), in: R. L. Ciochon, J. G. Fleagle (Eds.), *The Human Evolution Source Book*, Englewood Cliffs, NJ: Prentice-Hall, pp. 319–331.

Hood, W. F. (1983), in: C. Mitcham, R. Mackey (Eds.), *Philosophy and Technology: Readings in the Philosophical Problems of Technology*, New York: Free Press, pp. 347–363.

Hooke, S. H. (1954), in: C. Singer, E. J. Holmyard, A. R. Hall (Eds.), *A History of Technology*, Vol. 1, Oxford, U.K.: Clarendon, pp. xlviii–lv.

Howells, W. (1993), *Getting Here: The Story of Human Evolution*, Washington, DC: Compass.

Ihde, D. (1993), *Philosophy of Technology: An Introduction*, New York: Paragon House.

Isaac, G. L., McCown, E. R. (Eds.) (1976), *Hu-*

man Origins: Louis Leakey and the East African Evidence, Menlo Park, CA: Benjamin.

Jünger, E. (1932), *Der Arbeiter: Herrschaft und Gestalt*, Hamburg: Hanseatische Verlagsanstalt.

Kant, I. (1966), *Critique of Pure Reason*, Garden City, NY: Doubleday.

Kapp, E. (1877), *Grundlinien einer Philosophie der Technik: Zur Entstehungsgeschichte der Kultur*, Braunschweig: G. Westermann.

Laitman, J. (1993), in: R. L. Ciochon, J. G. Fleagle (Eds.), *The Human Evolution Source Book*, Englewood Cliffs, NJ: Prentice-Hall, pp. 56–60.

Lilley, S. (1965), *Men, Machines and History*, London: Lawrence & Wishart.

Lovejoy, C. O. (1993), in: R. L. Ciochon, J. G. Fleagle (Eds.), *The Human Evolution Source Book*, Englewood Cliffs, NJ: Prentice-Hall, pp. 230–243.

Marcuse, H. (1964), *One-Dimensional Man*, Boston: Beacon.

Marx, K. (1935), *The Poverty of Philosophy*, in: E. Burns (Ed.), *A Handbook of Marxism*, London: V. Gollancz.

Marx, K. (1987), Preface to *A Contribution to the Critique of Political Economy*, in: *Karl Marx, Frederick Engels: Collected Works*, Vol. 29, New York: International Publishers, pp. 257–265.

Mill, J. S. (1904), *A System of Logic: Ratiocinative and Inductive*, New York: Harper.

Mitcham, C., Mackey, R. (Eds.), *Philosophy and Technology: Readings in the Philosophical Problems of Technology*, New York: Free Press.

Napier, J. (1993), in: R. L. Ciochon, J. G. Fleagle (Eds.), *The Human Evolution Source Book*, Englewood Cliffs, NJ: Prentice-Hall, pp. 61–70.

Needham, J. (1954–1967), *Science and Civilisation in China*, Cambridge, U.K.: Cambridge Univ. Press, 7 vols.

Needham, J. (1969), *The Grand Titration: Science and Society in East and West*, Toronto: Univ. of Toronto Press.

Ortega y Gasset, J. (1941), *Toward a Philosophy of History*, New York: Norton.

Patterson, T. C. (1991), *The Inca Empire: The Formation and Disintegration of a Pre-Capitalist State*, Oxford, U.K.: Berg.

Schmandt-Bessarat, D. (1992), *From Counting to Writing*, Vol. 1, Austin: Univ. of Texas Press.

Skelton, R. R., McHenry, H. M., Drawhorn, G. M. (1993), in: R. L. Ciochon, J. G. Fleagle (Eds.), *The Human Evolution Source Book*, Englewood Cliffs, NJ: Prentice-Hall, pp. 112–134.

Smith, D. E. (1923), *History of Mathematics*, Vol. 1, Boston: Ginn.

Tobach, E. (1993), in: L. W. Martin, K. A. Nelson, E. Tobach (Eds.), *Social-Cultural Psychology: Theory and Practice of Knowing and Doing*, Cambridge, U.K.: Cambridge Univ. Press.

Turnbull, C. M. (1983), *The Mbuti Pygmies*, New York: Holt.

Vernant, J.-P. (1957), *Rev. Histoire Sci. Appl.* **10**, 205–225.

White, L., Jr. (1962), *Medieval Technology and Social Change*, Oxford, U.K.: Oxford Univ. Press.

Winner, L. (1977), *Autonomous Technology: Technics-Out-of-Control As a Theme in Political Thought*, Cambridge, MA: Massachusetts Institute of Technology Press.

Winner, L. (1986), *The Whale and the Reactor*, Chicago: Univ. of Chicago Press.

Further Reading

Anderson, P. (1974), *Passages from Antiquity to Feudalism*, London: Verso.

Ciochon, R. L., Fleagle, J. G. (Eds.) (1993), *The Human Evolution Source Book*, Englewood Cliffs, NJ: Prentice-Hall.

Dijksterhuis, E. J. (1969), *The Mechanization of the World Picture*, London: Oxford Univ. Press.

Durbin, P. (1994), "Philosophy of Science, Technology, and Medicine," in: C. Mitcham, W. Williams (Eds.), *The Reader's Adviser*, 14th ed., Vol. 5, New York: Bowker, pp. 59–104. Contains valuable bibliography.

Ellul, J. (1964), *The Technological Society*, New York: Knopf.

Gille, B. (1986), *The History of Techniques*, New York: Gordon and Breach, 2 vols.

Heidegger, M. (1977), *The Question Concerning Technology and Other Essays*, translation by W. Lovitt, New York: Harper & Row.

Henry, D. O. (1989), *From Foraging to Agriculture*, Philadelphia: Univ. of Pennsylvania Press.

Hobsbawm, E. J. (1968), *Industry and Empire: An Economic History of Britain since 1750*, London: Weidenfeld and Nicolson.

Ihde, D. (1993), *Philosophy of Technology: An Introduction*, New York: Paragon House.

Marcuse, H. (1964), *One-Dimensional Man*, Boston: Beacon.

Mitcham, C., Mackey, R. (Eds.) (1983), *Philosophy and Technology: Readings in the Philosophical Problems of Technology*, New York: Free Press.

Mumford, L. (1934), *Technics and Civilization*, New York: Harcourt and Brace.

Needham, J. (1954–1967), *Science and Civilisation in China*, Cambridge, U.K.: Cambridge Univ. Press, 7 vols.

White, L., Jr. (1962), *Medieval Technology and Social Change*, Oxford: Oxford Univ. Press.

Winner, L. (1977), *Autonomous Technology: Technics-Out-of-Control As a Theme in Political Thought*, Cambridge, MA: Massachusetts Institute of Technology Press.

PHONONS IN CRYSTAL LATTICES

A. G. Every, *Physics Department, University of the Witwatersrand, Johannesburg, South Africa*

	Introduction	439
1.	**Theoretical Background**	440
1.1	Quantization; the Phonon	442
1.2	Statistical Properties of Phonons	443
2.	**Phonon–Phonon Interactions**	444
3.	**Scattering Methods for Determining Phonon Dispersion Relations**	446
3.1	Inelastic X-Ray Scattering	446
3.2	Coherent Inelastic Neutron Scattering	447
3.3	Helium-Atom and Electron Scattering; Surface Phonons	448
4.	**Infrared Absorption and Light Scattering**	449
4.1	Raman Scattering	450
4.2	Brillouin Scattering	450
5.	**Electron–Phonon Interaction**	451
5.1	Metals	451
5.2	Metallic Superconductors	452
5.3	Semiconductors	453
5.4	Insulators	453
6.	**Phonon–Defect Interactions**	453
6.1	Local Modes	454
6.2	Electronic States	455
6.3	Spin-Lattice Relaxation	455
6.4	Mössbauer Effect	456
7.	**Closing Remarks**	456
	Works Cited	457
	Further Reading	458

INTRODUCTION

Crystal-lattice vibrations are quantized in much the same way that electromagnetic waves are, with the counterpart of the photon being the *phonon*. As one might expect, there are a number of parallels between the two classes of phenomena. For example, the phonon energy flux radiated into the interior of a cold crystal from a heated surface varies as the fourth power of the absolute temperature, just as for blackbody electromagnetic radiation. The blue of the sky has its counterpart in that small crystal inhomogeneities give rise to Rayleigh scattering of phonons, which varies with the fourth power of the frequency. It is common for an atom undergoing a transition between two quantum levels to conserve energy by absorbing or emitting a phonon rather than a photon.

There are, however, limits to how far the analogy can be taken. In contrast to electromagnetic waves in vacuum, lattice vibrations have a lower bound to their wavelength, equal to twice the lattice spacing in principal crystallographic directions, and associated with this is an upper cutoff in the spectrum of vibrational frequencies. For wavelengths approaching this limit, there is pronounced dispersion. Nonlinearity is important for lattice vibrations, leading in particular to phonon–phonon scattering.

The main emphasis of this article is on the interaction between phonons and other excitations or particles, on how these interactions influence the physical properties of crystals, and on how the scattering of x rays, neutrons, and other particles can be used to determine phonon dispersion relations. The concept of the phonon is introduced in Sec. 1, which also contains a synopsis of the main *lattice dynamics* results that are drawn on in this article. For a fuller discussion of lattice dynamics, the reader is referred to Lattice Vibrations, Statistics of (hereafter cited as LV). Subsequent sections deal with scattering and absorption processes involving phonons, γ rays, x rays, optical and infrared photons, neutrons, electrons, neutral atoms, and localized spin and other excitations. A major aspect of phonon physics that is touched on

3-527-28135-5/95/$5.00 + .50

only very briefly in the present article is *phonon transport*. This subject is treated extensively in LV and in HEAT PULSE PROPAGATION IN SOLIDS (HP).

1. THEORETICAL BACKGROUND

We start by recalling that in the harmonic approximation the Hamiltonian for the *vibrations of a crystal lattice* takes the form (LV)

$$H_L = \sum_{l\kappa\alpha} \frac{p_\alpha^2(l\kappa)}{2M_\kappa} + \tfrac{1}{2}\sum_{l\kappa\alpha}\sum_{l'\kappa'\beta} \Phi_{\alpha\beta}(l\kappa;\, l'\kappa')u_\alpha(l\kappa)u_\beta(l'\kappa'), \quad (1)$$

where $u_\alpha(l\kappa)$ are the displacements of the atoms from their equilibrium positions, with α denoting the Cartesian component, l the cell index, and κ ranging over the r atoms in the unit cell; $p_\alpha(l\kappa) = M_\kappa \partial u_\alpha(l\kappa)/\partial t$ are the momenta of the atoms, where M_κ are the atomic masses; and $\Phi_{\alpha\beta}(l\kappa;\, l'\kappa')$ are the *atomic force constants*, given by the derivatives

$$\Phi_{\alpha\beta}(l\kappa;\, l'\kappa') = \frac{\partial^2\Phi}{\partial u_\alpha(l\kappa)\partial u_\beta(l'\kappa')} \quad (2)$$

of the lattice potential energy Φ at equilibrium. The equations of motion derived from this Hamiltonian are

$$M_\kappa \frac{\partial^2 u_\alpha(l\kappa)}{\partial t^2} = -\sum_{l'\kappa'\beta} \Phi_{\alpha\beta}(l\kappa;\, l'\kappa')u_\beta(l'\kappa'). \quad (3)$$

Because of the *translational invariance* of the lattice, the normal modes can be taken as plane waves, and the displacements and momenta expressed in terms of *normal coordinates* $Q(\mathbf{k}j)$ and their *conjugate momenta* $P(\mathbf{k}j)$ thus:

$$u_\alpha(l\kappa) = \frac{1}{\sqrt{NM_\kappa}} \sum_{\mathbf{k}j} e_\alpha(\kappa|\mathbf{k}j)Q(\mathbf{k}j)e^{i\mathbf{k}\cdot\mathbf{x}(l)}, \quad (4)$$

$$p_\alpha(l\kappa) = \frac{1}{\sqrt{NM_\kappa}} \sum_{\mathbf{k}j} e_\alpha(\kappa|\mathbf{k}j)P(\mathbf{k}j)e^{i\mathbf{k}\cdot\mathbf{x}(l)}. \quad (5)$$

The factor N comes from invoking *cyclic boundary conditions* over a macrocell of $N = L \times L \times L$ unit cells. The sum extends over a regular grid of N values of the wave vector $\mathbf{k}$ within the first Brillouin zone and $3r$ values of the branch index j. The $e_\alpha(\kappa|\mathbf{k}j)$ for each value of $\mathbf{k}$ are the $3r$ normalized eigenvectors of the *dynamical matrix*

$$D_{\alpha\beta}(\kappa\kappa'|\mathbf{k}) = \frac{1}{\sqrt{M_\kappa M_{\kappa'}}} \sum_{l'} \Phi_{\alpha\beta}(l - l',\, \kappa;\, 0\kappa') \times e^{-i\mathbf{k}\cdot[\mathbf{x}(l)-\mathbf{x}(l')]}. \quad (6)$$

These eigenvectors satisfy the linear equations

$$\omega^2(\mathbf{k}j)e_\alpha(\kappa|\mathbf{k}j) = \sum_{\kappa'\beta} D_{\alpha\beta}(\kappa\kappa'|\mathbf{k})e_\beta(\kappa'|\mathbf{k}j), \quad (7)$$

with the normal-mode frequencies $\omega(\mathbf{k}j)$ being determined by the secular equation

$$|D_{\alpha\beta}(\kappa\kappa'|\mathbf{k}) - \omega^2\delta_{\alpha\beta}\delta_{\kappa\kappa'}| = 0. \quad (8)$$

Since $D^*_{\alpha\beta}(\kappa\kappa'|\mathbf{k}) = D_{\alpha\beta}(\kappa\kappa'|-\mathbf{k})$, we may choose the phases of the eigenvectors such that

$$e^*_\alpha(\kappa|\mathbf{k}j) = e_\alpha(\kappa|-\mathbf{k}j). \quad (9)$$

It follows from Eqs. (4), (5), and (9) and the requirement that $u_\alpha(l\kappa)$ and $p_\alpha(l\kappa)$ are real that

$$Q^*(\mathbf{k}j) = Q(-\mathbf{k}j), \quad (10)$$

$$P^*(\mathbf{k}j) = P(-\mathbf{k}j). \quad (11)$$

Three of the solutions $\omega(\mathbf{k}j)$ of Eq. (8) go to zero in the limit $\mathbf{k} \to 0$ and represent the *acoustic branches* of the *dispersion relation*. They pertain in the long-wavelength continuum limit to longitudinal and transverse acoustic waves. The remaining $3r - 3$ solutions $\omega(\mathbf{k}j)$ approach finite limiting values as $\mathbf{k} \to 0$ and represent the *optic branches*. In these latter modes near $\mathbf{k} = 0$ the sublattices vibrate against one another, and this motion can generally be detected by infrared absorption or Raman scattering.

In order to solve Eq. (8), the atomic-force constants are required. There are some successful *ab initio* theories, but most lattice dynamics calculations are based on a parametrized model of some kind (see Bilz and Kress, 1979). Over the years, many different models have been devised. The *valence force field* model is one of the simplest and is based on intuitive ideas that have a foundation in molecular vibrational studies. One pictures the atoms in the crystal as joined together by bonds and expresses the interatomic forces in terms of the forces and torques required to stretch, bend, and twist these bonds. This model has been applied with some success to covalent crystals such as diamond, silicon, and germanium. By and large though, most successful models incorporate long-range Coulomb forces in some form. This is obviously necessary for treating ionic crystals, but even in the case of covalent crystals the number of parameters required for a successful fit tends to be less when Coulomb forces are incorporated explicitly in the model in some way. The *rigid-ion model* is the simplest approach to ionic crystals, and there are a number of more sophisticated models that allow for deformation of the electron wave functions of the atoms as they move relative to one another. These include dipole models, various shell and deformation-shell models, and the bond-charge model. In metals, the interaction between the ions and the conduction electrons has to be taken into account. In the de Launay (1953) model for instance, the electron gas is attributed a finite bulk modulus and is assumed to adjust adiabatically to the motion of the ions, screening their long-range Coulomb interactions in the process. This coupling has a pronounced effect on the longitudinal lattice modes, while pure transverse modes are unaffected.

Whatever model is adopted for a particular application, it commonly features a number of parameters such as bond force constants, shell charges, atomic polarizabilities, or whatever. These parameters have to be determined from experiment. The phase velocities $v(j) = \omega(\mathbf{k}j)/k$ of long-wavelength acoustic modes are a function of the *bulk elastic constants* of the crystal, and this provides as many constraints on the parameters of the model as the crystal has independent elastic constants. Measured Raman frequencies, dielectric tensor, and optical absorption often provide further constraints on the parameters. If the model is simple enough, this information may be sufficient to determine all the parameters. The ultimate test of a model, though, is that it should fit the dispersion relations measured by neutron scattering or one of the other techniques discussed in Sec. 3. As an example, Fig. 1(a) shows dispersion curves for silicon in a number of principal crystallographic directions

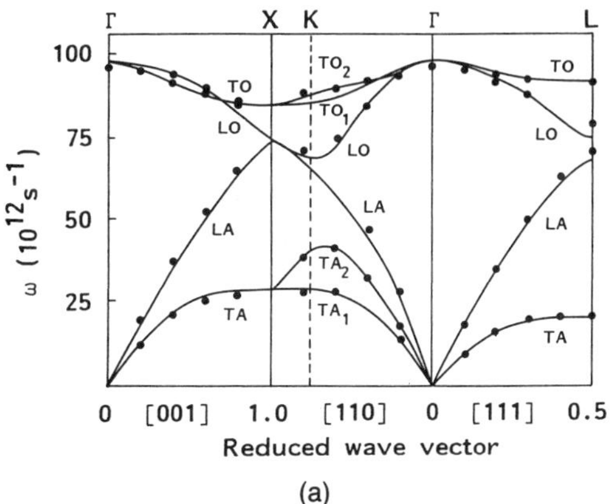

(a)

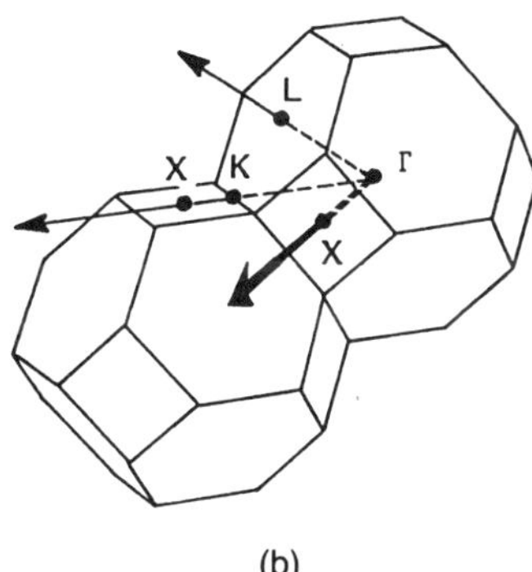

(b)

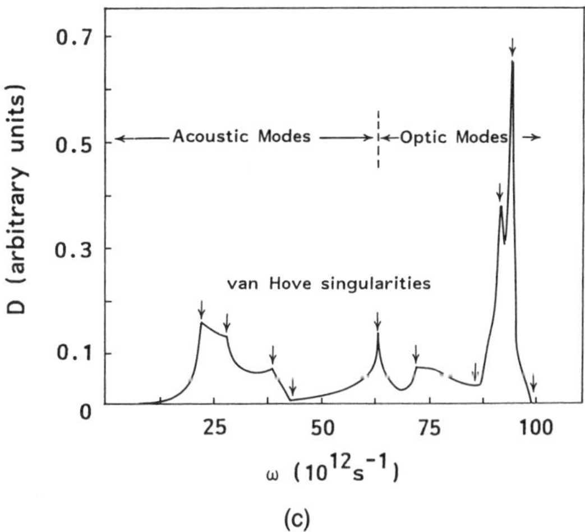

(c)

FIG. 1. (a) Phonon dispersion curves for silicon (adapted from Bilz and Kress, 1979). The directions are referred to the face-centered cubic (fcc) Brillouin zone shown in (b). The branches are distinguished as longitudinal acoustic (LA), longitudinal optic (LO), transverse acoustic (TA), and transverse optic (TO). The two TA and two TO branches are degenerate in the [001] and [111] directions because of symmetry. (c) Calculated frequency-distribution function $D(\omega)$ for Si.

[see Fig. 1(b)]. The points denote neutron-scattering data, and the lines represent calculations on the basis of a four-parameter *bond-charge model.*

Many thermodynamic quantities can be calculated from a knowledge of the *frequency distribution function* or *density of states* $D(\omega)$, defined as the number of normal-mode frequencies per unit frequency interval at ω. Figure 1(c) shows $D(\omega)$ calculated for Si. The region below about $20 \times 10^{12}\ \mathrm{s}^{-1}$ corresponds to long-wavelength acoustic modes. Here $D(\omega)$ is roughly proportional to ω^2, in accordance with the Debye model. At higher frequencies, $D(\omega)$ is punctuated by a number of sharp features known as *van Hove singularities* (Maradudin *et al.*, 1971). These correspond to critical points at the zone boundaries (and elsewhere) where the dispersion relation has an extremum. To one side of the critical frequency ω_0, $D(\omega) \sim \sqrt{|\omega - \omega_0|}$, giving a divergence in the derivative of $D(\omega)$ at ω_0, while on the other side there is no singular behavior. These singularities carry over to other spectral quantities such as defect-activated infrared absorption, and this provides scope for experimentally determining zone-boundary frequencies.

1.1 Quantization; the Phonon

Quantization of the lattice vibrations is effected by regarding the normal coordinates $Q(\mathbf{k}j)$ and their conjugate momenta $P(\mathbf{k}j)$ as operators satisfying the commutation relations (LV)

$$[P(\mathbf{k}'j'), Q(\mathbf{k}j)] = (\hbar/i)\, \delta_{jj'} \Delta(\mathbf{k} - \mathbf{k}'), \tag{12}$$

where $\hbar$ is Planck's constant, $\delta_{jj'}$ is the Kronecker delta, and

$$\Delta(\mathbf{k}) = \begin{cases} 1 & \text{if } \mathbf{k} = \mathbf{G}, \\ 0 & \text{otherwise,} \end{cases} \tag{13}$$

where $\mathbf{G}$ is a reciprocal lattice vector. The Hamiltonian is given in terms of these conjugate operators by

$$H_L = \tfrac{1}{2} \sum_{\mathbf{k}j} [P(\mathbf{k}j)P^*(\mathbf{k}j) + \omega_j^2(\mathbf{k})Q(\mathbf{k}j)Q^*(\mathbf{k}j)] \tag{14}$$

and is in the form of a sum of $3rN$ independent harmonic oscillator Hamiltonians. The eigenstates of H_L can thus be represented as products of eigenstates of the individual oscillators.

The usual approach to the harmonic oscillator, found in most elementary textbooks on quantum mechanics (see, e.g., Schiff, 1955), is to solve Schrödinger's equation in the coordinate or momentum representation and thereby obtain the eigenstates and eigenenergies. In the present context, of more interest than the form of eigenstates are the energy eigenvalues and the matrix representations of operators such as Q, Q^2, and P, etc., in the energy basis. This information is required for calculating expectation values and thermal averages, carrying out perturbation calculations, and many other purposes. The terms P and Q both appear quadratically in the Hamiltonian, and this symmetry allows the energy spectrum and required matrix representations to be obtained by purely algebraic means.

The trick, which goes by the name of *second quantization*, is to introduce operators $a^\dagger(\mathbf{k}j)$ and $a(\mathbf{k}j)$ for each mode $\mathbf{k}j$ by means of the equations

$$Q(\mathbf{k}j) = \sqrt{\frac{\hbar}{2\omega(\mathbf{k}j)}}\, [a^\dagger(-\mathbf{k}j) + a(\mathbf{k}j)], \tag{15}$$

$$P(\mathbf{k}j) = i \sqrt{\frac{\hbar\omega(\mathbf{k}j)}{2}}\, [a^\dagger(\mathbf{k}j) - a(-\mathbf{k}j)]. \tag{16}$$

It is easily verified (see, e.g., Ghatak and Kothari, 1972) that these operators satisfy the commutation relations

$$[a(\mathbf{k}'j'), a^\dagger(\mathbf{k}j)] = \delta_{jj'}\Delta(\mathbf{k} - \mathbf{k}') \tag{17}$$

and that the Hamiltonian is given in terms of them by

$$H_L = \sum_{\mathbf{k}j} \hbar\omega(\mathbf{k}j)\, [a^\dagger(\mathbf{k}j)a(\mathbf{k}j) + \tfrac{1}{2}]. \tag{18}$$

Considering an individual oscillator, it may be shown using the commutation relations that $a^\dagger(\mathbf{k}j)$ and $a(\mathbf{k}j)$ operating on an eigenstate of the Hamiltonian yield other eigenstates that are higher and lower in energy respectively by $\hbar\omega(\mathbf{k}j)$, that the complete set of eigenstates can thereby be generated, and that there is a ground state. The eigenstates can thus be denoted $|n(\mathbf{k}j)\rangle$, and the associated eigenvalues are given by

$$E(n(\mathbf{k}j)) = [n(\mathbf{k}j) + \tfrac{1}{2}]\hbar\omega(\mathbf{k}j), \tag{19}$$

where the quantum number $n(\mathbf{k}j)$ takes on all positive integer values and is zero for the ground state. When the oscillator undergoes a transition, energy is added or removed in multiples of $\hbar\omega(\mathbf{k}j)$. This quantum of energy is called the *phonon*, and $n(\mathbf{k}j)$ is known as the *phonon occupation number* for the state. The operators $a^\dagger(\mathbf{k}j)$ and $a(\mathbf{k}j)$ are called the *phonon creation* and *annihilation* (or *destruction*) *operators* respectively, and raise and lower the phonon occupation number of a mode in accordance with

$$a^\dagger(\mathbf{k}j)|n(\mathbf{k}j)\rangle = \sqrt{n(\mathbf{k}j) + 1}\,|n(\mathbf{k}j) + 1\rangle, \tag{20}$$

$$a(\mathbf{k}j)|n(\mathbf{k}j)\rangle = \sqrt{n(\mathbf{k}j)}\,|n(\mathbf{k}j) - 1\rangle. \tag{21}$$

The product $a^\dagger(\mathbf{k}j)a(\mathbf{k}j)$ is called the number operator because it extracts the phonon occupation number of a state:

$$a^\dagger(\mathbf{k}j)a(\mathbf{k}j)|n(\mathbf{k}j)\rangle = n(\mathbf{k}j)|n(\mathbf{k}j)\rangle. \tag{22}$$

Eigenstates of the combined system of all the oscillators can be represented as products of the individual oscillator states: $|\cdots n(\mathbf{k}j) \cdots n(\mathbf{k}'j') \cdots\rangle = |\cdots\rangle|n(\mathbf{k}j)\rangle|\cdots\rangle|n(\mathbf{k}j)\rangle \times |\cdots\rangle$, with the energy being the sum of the energies of the individual oscillators, i.e.,

$$E(\{n(\mathbf{k}j)\}) = \sum_{\mathbf{k}j} [n(\mathbf{k}j) + \tfrac{1}{2}]\hbar\omega(\mathbf{k}j). \tag{23}$$

The phonon can in many respects be regarded as a particle. If one thinks of it as occupying a wave packet centered on some value of $\mathbf{k}$, rather than an infinitely extended plane wave, then the phonon can be ascribed an instantaneous position in space. The more precisely that point is specified, the greater of course is the spread of $\mathbf{k}$'s in the wave packet, which is a manifestation of the Heisenberg uncertainty principle. The rate at which the position of the phonon changes with time is its group velocity, which is given by $\mathbf{V} = \nabla_{\mathbf{k}}\omega(\mathbf{k}j)$. The quantity $\hbar\mathbf{k}$ is called the *pseudomomentum* or *crystal momentum* of the phonon. Although not momentum in the true sense, it nevertheless plays a role very similar to that of physical momentum. For instance, when a photon of light is scattered inelastically from a crystal with the loss of energy being taken up in the creation of a phonon, the change in momentum of the photon is equal to the value of $-\hbar\mathbf{k}$ for the phonon, which seems to signify the conservation of total momentum. Momentum conservation is a consequence of translational invariance, and a crystal lattice is only invariant under a discrete set of translations, not all translations. The upshot of this is that in general a scattering event involving phonons takes place with conservation of real plus crystal momentum, but only to within $\hbar\mathbf{G}$, where $\mathbf{G}$ is a reciprocal-lattice vector. An inelastic neutron-scattering event, for instance, might involve a combination of Bragg scattering by $\mathbf{G}$ together with the creation of a phonon of wave vector $\mathbf{k}$, so that the momentum transfer to the lattice, $\hbar(\mathbf{G} + \mathbf{k})$, differs from the phonon momentum by $\hbar\mathbf{G}$.

1.2 Statistical Properties of Phonons

Collectively, phonons can be ascribed a phase-space distribution function $n(\mathbf{k}j;\,\mathbf{x})$, from which quantities such as energy density, heat current, and momentum flux can be derived. Phonon scattering can lead to a rearrangement of the distribution, the process being governed by the Boltzmann equation (LV). Phonons obey Bose statistics, with the chemical potential being zero. The average number of phonons in any mode in thermal equilibrium depends on the local temperature T and is given by the *Planck distribution function* (LV)

$$\bar{n}(\mathbf{k}j) = \{\exp[\hbar\omega(\mathbf{k}j)/k_{\mathrm{B}}T] - 1\}^{-1}, \tag{24}$$

where k_{B} is Boltzmann's constant. It follows that

$$\begin{aligned}\langle Q(\mathbf{k}j)Q^*(\mathbf{k}j)\rangle &= \frac{\hbar}{2\omega(\mathbf{k}j)}[2\bar{n}(\mathbf{k}j) + 1]\\ &= \frac{\hbar}{2\omega(\mathbf{k}j)}\coth[\hbar\omega(\mathbf{k}j)/2k_{\mathrm{B}}T].\end{aligned} \tag{25}$$

The evaluation of thermodynamic quantities such as the Helmholtz free energy, the internal energy, heat capacity, and entropy with the aid of the partition function is discussed in LV.

The thermal expectation values of operators of the form $\exp[isu_\alpha(l\kappa)]$, where s is a parameter, are met in many situations. In treating the scattering of x rays and neutrons from a vibrating crystal, for instance, the phase factor for each atom is multiplied by $\sim\exp[isu_\alpha(l\kappa)]$ to take account of the instantaneous displacement of that atom from equilibrium, and the result has to be averaged over the ensemble of vibrational states of the crystal.

The evaluation of this expectation value entails expanding the exponential in powers of $u_\alpha(l\kappa)$, calculating the diagonal matrix elements, and then averaging over all oscillator states with their Boltzmann weighting factors. The details of the calculation can be found in Maradudin *et al.* (1971), and the result is

$$\langle\exp[isu_\alpha(l\kappa)]\rangle = \exp[-\tfrac{1}{2}s^2\langle u_\alpha^2(l\kappa)\rangle]. \qquad (26)$$

This quantity exhibits a reduction with temperature as the mean square displacement of the atom increases and there is a growing degree of phase cancellation. Equation (26) is the basis of the *Debye–Waller factor* for the decrease in the intensity of x-ray Bragg peaks with temperature, related effects for neutron and other types of scattering, the temperature dependence of the recoil-free fraction in the Mössbauer effect, and zero-phonon line intensities in color center spectra. These phenomena are discussed individually in subsequent sections.

2. PHONON–PHONON INTERACTIONS

For any real crystal, the Hamiltonian considered thus far represents only part of the full picture. To arrive at Eq. (1), the lattice potential energy has been expanded in powers of the displacements of the atoms from equilibrium, and only the quadratic terms retained. Moreover, interactions between the atoms and other particles or excitations that might be present in the crystal have not been considered. These additional terms account for many interesting physical phenomena. They constitute the coupling to the probes such as light, neutrons, and x rays that are used to study lattice vibrations, they are the means by which thermal equilibrium is established, and they give rise to scattering processes that play a central role in phonon transport phenomena. In this and subsequent sections, we consider the forms that these interactions take, and some of their consequences.

Continuing the expansion of the lattice potential energy beyond second order, the terms of third order can be grouped together as $\Phi^{(3)}$, those of fourth order as $\Phi^{(4)}$, and so on. Expressed in terms of the previously defined normal coordinates, $\Phi^{(3)}$ takes the form (LV)

$$\Phi^{(3)} = \sum_{\mathbf{k}j}\sum_{\mathbf{k}'j'}\sum_{\mathbf{k}''j''} V^{(3)}(\mathbf{k}j, \mathbf{k}'j', \mathbf{k}''j'') \times Q(\mathbf{k}j)Q(\mathbf{k}'j')Q(\mathbf{k}''j''), \qquad (27)$$

where $V^{(3)}(\mathbf{k}j,\mathbf{k}'j',\mathbf{k}''j'')$ are a set of coefficients that preserve the translational invariance of the lattice. A number of effects can be handled by treating $\Phi^{(3)}$ by perturbation theory. In the first place, the vibrational states form a quasicontinuum, and Fermi's "golden rule" can be used to calculate transition rates. It follows from Eqs. (15) and (27) that a matrix element of $\Phi^{(3)}$ between eigenstates of the harmonic Hamiltonian will be zero unless precisely three phonon occupation numbers in the one state differ by +1 or −1 from the corresponding occupation numbers in the other state. The three occupation numbers cannot all increase or all decrease in a transition, since that would violate energy conservation. A transition in which, say, $n(\mathbf{k}j)$ decreases and $n(\mathbf{k}'j')$ and $n(\mathbf{k}''j'')$ both increase, i.e., phonons $\mathbf{k}'j'$ and $\mathbf{k}''j''$ are created whose combined energies equal that of the original phonon $\mathbf{k}j$, has transition probability

$$p = \frac{2\pi}{\hbar^2}|V^{(3)}(\mathbf{k}j,-\mathbf{k}'j',-\mathbf{k}''j'')|^2 \times \frac{\hbar^3}{8\omega(\mathbf{k}j)\omega(\mathbf{k}'j')\omega(\mathbf{k}''j'')} n(\mathbf{k}j)[n(\mathbf{k}'j') + 1] \times [n(\mathbf{k}''j'') + 1]\delta(\omega(\mathbf{k}j) - \omega(\mathbf{k}'j') - \omega(\mathbf{k}''j'')). \qquad (28)$$

Because of the translational symmetry of the lattice, $|V^{(3)}(\mathbf{k}j,-\mathbf{k}'j',-\mathbf{k}''j'')|^2$ is proportional to $\Delta(\mathbf{k} - \mathbf{k}' - \mathbf{k}'')$, and so there is conservation of crystal momentum to within a reciprocal-lattice vector times $\hbar$. The transition can be represented diagrammatically as shown in Fig. 2(a). Transitions represented by Fig. 2(b), in which two phonons are annihilated and one created, are calculated in a similar way. The total *decay rate*, or *inverse lifetime* of a phonon against two-phonon decay under thermal equilibrium conditions, is then obtained by summing p over all decay channels, using the thermal expectation values for the n's.

Phonon–phonon scattering processes play a central role in phonon transport. Transitions in which $\mathbf{k} - \mathbf{k}' - \mathbf{k}'' = \mathbf{G} \neq 0$ are termed *Umklapp* (U) processes and require that the participating phonons have wave vectors that are comparable with the zone-boundary values. Crystal momentum is not conserved in these processes, and they are the dominant source of thermal resistance in most insulating crystals at room temperature, where large–wave-vector phonons are present in appreciable numbers. Transitions in which $\mathbf{k} - \mathbf{k}' - \mathbf{k}'' = 0$ are termed *normal* (N) processes and conserve crystal momentum. They will rearrange a nonequilibrium distribution of phonons without changing the total momentum, and on their own do not represent a source of thermal resistance.

As a general observation, in thermal equilibrium at a temperature T well below the Debye temperature Θ_D, the average phonon lifetime against two-phonon decay is proportional to T^{-5} or $\bar{\omega}^{-5}$. The requirements of energy and momentum conservation, though, place constraints on the channels by which a specific phonon can decay, and there are considerable differences in lifetime depending on the particular phonon branch. The phonon dispersion curves for silicon shown in Fig. 1(a) are fairly representative of a large class of insulating crystals, in that the acoustic branches curve downward with increasing $|\mathbf{k}|$ (soft-mode materials discussed below provide exceptions to this rule). This being the case, a phonon cannot decay into two other phonons of the same branch with collinear momenta, since in order for momentum to be conserved the combined energies of the two phonons would exceed that of the original one. For an isotropic crystal, or in a slow direction in an anisotropic crystal, noncollinear decay would leave an even greater energy shortfall. Elaborating on this idea, Lax *et al.* (1981) have proved that a phonon cannot decay by anharmonic processes of any order into phonons each of whose phase velocities are higher than that of the original phonon. This is true for U as well as N processes. Slow transverse phonons can therefore, in the absence of other forms of scattering, be very long lived. This has important implications for *ballistic transport*.

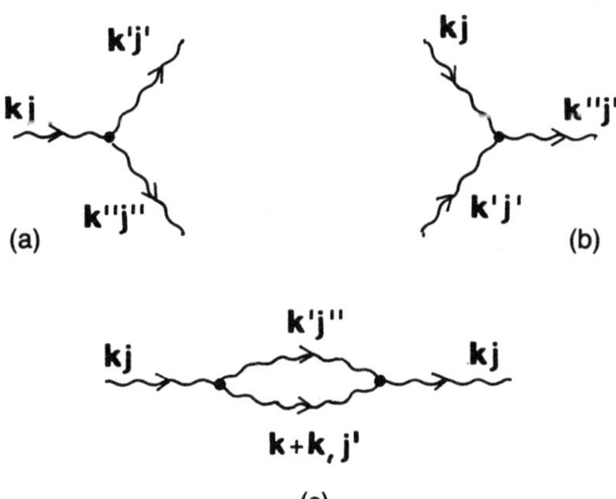

FIG. 2. Anharmonic processes involving $-\Phi^{(3)}$ in (a,b) first order and (c) second order.

Perturbation theory predicts a mixing of phonon states, and in second order a *renormalization of the phonon energies*, which can be represented by diagrams such as the one shown in Fig. 2(c). The phonon dispersion relations, phonon velocities, and thermodynamic quantities such as the elastic constants thereby become temperature dependent. In *quantum crystals* such as solid He, the effects of renormalization are pronounced, even at the absolute zero of temperature (Guyer, 1969). The anharmonic terms in the lattice potential account for thermal expansion, which in the Grüneisen approach is treated by considering the volume dependence of the phonon frequencies and minimizing the total free energy (Ashcroft and Mermin, 1981).

Phonon–phonon interactions play an important role in second-order *structural phase transitions* (see PHASE TRANSITIONS, STRUCTURAL). Many of these are accompanied by a *soft mode* in the high-temperature symmetrical phase, which shows up as a pronounced dip centered on a particular value $\mathbf{k}_0$ in the phonon dispersion relation. The soft-mode $Q(\mathbf{k}_0)$ has a negative harmonic restoring force constant $\omega_0^2(\mathbf{k}_0)$ but is dynamically stabilized through interactions with all the other modes.

The stabilization is most easily accounted for in terms of the quartic terms in the lattice potential, specifically those with $\mathbf{k} = \mathbf{k}' = \mathbf{k}_0$ and $\mathbf{k}'' = \mathbf{k}'''$. On taking a statistical average over the latter modes and summing, the mode $Q(\mathbf{k}_0)$ comes out to have a quasiharmonic frequency $\omega(\mathbf{k}_0)$ given by (Dorner and Comes, 1977)

$$\omega^2(\mathbf{k}_0) = \omega_0^2(\mathbf{k}_0) + \sum_{k''} V^{(4)}(\mathbf{k}_0,-\mathbf{k}_0,\mathbf{k}'',-\mathbf{k}'') \times \langle Q^*(\mathbf{k}'')Q(\mathbf{k}'')\rangle. \tag{29}$$

In the high-temperature phase, the second term in Eq. (29) is positive and exceeds the first in magnitude, and so $\omega(\mathbf{k}_0)$ is real. The soft-mode frequency is usually fairly small, though, and the mode heavily damped. As the temperature is lowered, $\omega(\mathbf{k}_0)$ approaches zero, the lattice becomes unstable, and a transition to a lower-symmetry structure takes place. In simple cases, the order parameter for the transition corresponds fairly closely to the soft-mode normal coordinate.

Varying the isotopic composition of a substance changes the phonon frequencies and zero-point motion amplitudes, and coupled with anharmonicity this results in changes to the lattice constant, elastic constants, and related physical quantities. The isotope effect in superconductivity is a classic example that is discussed later. Recently, the properties of isotopically enriched diamond have attracted much attention (see Hass *et al.*, 1992, and references contained therein).

3. SCATTERING METHODS FOR DETERMINING PHONON DISPERSION RELATIONS

Thermal diffuse scattering of x rays provided the earliest means of measuring the bulk phonon dispersion relations of crystals. Coherent inelastic neutron scattering arrived later on the scene but is a more powerful technique and has largely supplanted x-ray scattering in this task. Neutron scattering is also a valuable source of information on the displacement eigenvectors $e_\alpha(\kappa|\mathbf{k}j)$, phonon lifetimes, vibrations at defects, mode softening, and many other physical phenomena. Helium-atom and electron scattering are widely used in the study of lattice vibrations at surfaces.

The above-mentioned methods share certain common features. To permit measurement of the phonon dispersion relation over the entire Brillouin zone, the wavelength of the probe (de Broglie wavelength in the case of a particle) has to be comparable to the lattice spacing. The scattering is consequently dominated by elastic Bragg scattering from the crystal planes. Information about lattice vibrations is gleaned from the much weaker diffuse background scattering. This background is caused by inelastic scattering in which some of the energy of the probe is given over to the creation or annihilation of one, two, or more phonons. In the case of x rays, the limited wavelength resolution does not permit the energy shift to be measured directly, and the phonon frequencies are inferred from the temperature dependence of the diffuse intensity. For inelastic neutron and atom scattering, the energy shift is measured as a matter of course, and the phonon frequencies are obtained directly. As the temperature is raised, the vibrational amplitudes increase and the crystal appears less and less ordered to the probing particle. This results in a reduction in the intensities of the Bragg peaks and a compensatory increase in the inelastic scattering.

3.1 Inelastic X-Ray Scattering

In treating x-ray scattering from a vibrating crystal, the frequency shift is ignored. This change is of the order of the phonon frequency, $\sim 10^{12}$ Hz, which is small compared with the x-ray frequency, $\sim 10^{18}$ Hz. For simplicity, we will consider scattering from a monatomic Bravais lattice, which entails dropping the index κ. The scattering from any point within an atom is proportional to the local electron probability density, which is assumed to accompany the atom as it vibrates. The resulting scattering amplitude for an instantaneous configuration of all the atoms, taking into account the phase relationships between the scattering amplitudes for the individual atoms, is

$$\Psi \sim f(\mathbf{S}) \sum_{l} \exp\{i\mathbf{S}\cdot[\mathbf{x}(l) + \mathbf{u}(l)]\}, \tag{30}$$

where $f(\mathbf{S})$ is the *atomic structure factor*, $\mathbf{S} = \mathbf{q} - \mathbf{q}'$ is the scattering vector, and $\mathbf{q}$ and $\mathbf{q}'$ are the wave vectors of the incoming and outgoing radiation. The time-averaged *scattering intensity*

$$I = \langle \Psi^* \Psi \rangle = f^2 \sum_{ll'} \langle \exp\{i\mathbf{S} \cdot [\mathbf{x}(l) - \mathbf{x}(l') + \mathbf{u}(l) - \mathbf{u}(l')]\}\rangle \quad (31)$$

is obtained as an ensemble average over all vibrational states. Making use of Eq. (26) and expressing the atomic displacements in terms of the normal coordinates, and then expanding the exponential in powers of the averaged displacements, one obtains (see Maradudin *et al.*, 1971)

$$I = f^2 e^{-2W} [I_0 + I_1 + \ldots], \quad (32)$$

where

$$2W = \frac{2}{NM} \sum_{\mathbf{k}j} [\mathbf{S} \cdot \mathbf{e}(\mathbf{k}j)]^2 \langle Q^2(\mathbf{k}j)\rangle, \quad (33)$$

$$I_0 = \sum_{ll'} \exp\{i\mathbf{S} \cdot [\mathbf{x}(l) - \mathbf{x}(l')]\} = N^2 \Delta(\mathbf{S}), \quad (34)$$

$$I_1 = \frac{\hbar}{M} \sum_{j=1}^{3} [\mathbf{S} \cdot \mathbf{e}(\mathbf{K}j)]^2 \frac{\coth [\hbar\omega(\mathbf{K}j)/2k_B T]}{\omega(\mathbf{K}j)}, \quad (35)$$

with $e(\mathbf{k}j)$ standing for the mode eigenvector.

The scattering for the static lattice is $f^2 I_0$, which is zero except in the direction of the *Bragg peaks* where the *Laue condition* $\Delta(\mathbf{S}) = 1$ is satisfied. As a result of zero-point and thermal motion, the intensities of the Bragg peaks are reduced by the *Debye–Waller factor* e^{-2W}. The second term, I_1, describes the diffuse background due to one-phonon scattering, with $\mathbf{K}$ being the distance of $\mathbf{S}$ from the nearest reciprocal-lattice point. The term I_1 increases rapidly as $\mathbf{S}$ approaches the reciprocal-lattice point, and its behavior is determined by the three lattice modes of wave vector $\mathbf{K}$. For $\mathbf{K}$ along a symmetry axis, $\mathbf{S} \cdot \mathbf{e}(\mathbf{K}j)$ is nonzero except for the longitudinal mode, and so a measurement of the diffuse intensity as a function of temperature yields $\omega(\mathbf{K}j)$ for that mode. Measurements in other directions in reciprocal space yield data on the transverse branches. A drawback of this technique is that corrections have to be made for other sources of diffuse scattering. These include second-, third-, and higher-order processes, whose contributions to the diffuse background grow relative to the one-phonon scattering the further from the reciprocal-lattice point.

3.2 Coherent Inelastic Neutron Scattering

Coherent inelastic neutron scattering represents the most effective method for measuring phonon dispersion relations. The source is a collimated beam of monochromatic neutrons of wavelength comparable to the interatomic spacing. The crystal specimen is placed in the beam and gives rise to Bragg scattering, as well as inelastic scattering in which the change in energy and momentum of the neutron is taken up in the creation or annihilation of one or more phonons. Neutron scattering differs from x-ray scattering in the important respect that for given momentum transfer $\hbar\mathbf{k}$ the energy shift $\hbar\omega(\mathbf{k}j)$ is measured directly, usually with an accuracy of a few percent.

In neutron scattering, one measures the *differential cross section* $d^2\sigma/dEd\Omega$, which is the probability per unit solid angle and energy interval that an incoming neutron of wave vector $\mathbf{q}$ and energy $E = \hbar^2 q^2/2m$ is scattered into wave vector $\mathbf{q}'$ and energy $E' = \hbar^2 q'^2/2m$. For a single stationary nucleus, this is equal to $b^2\delta(E - E')$, where b is a quantity called the scattering length, which is independent of the neutron wavelength or angle of scattering.

For scattering from a vibrating monatomic crystal, in which there may be a change in the magnitude of the wave vector, the ensemble-averaged cross section taken over all possible initial vibrational states i of the crystal, with their Boltzmann probabilities p_i and final vibrational states f, is given by (Lovesey, 1977)

$$\frac{d^2\sigma}{dEd\Omega} = b^2 \frac{q'}{q} \sum_{if} p_i \left| \left\langle i \right| \sum_{l} \exp\{i\mathbf{S} \cdot [\mathbf{x}(l) + \mathbf{u}(l)]\} \left| f \right\rangle \right|^2 \times \delta(\hbar\omega + E_i - E_f), \quad (36)$$

where $\mathbf{S} = \mathbf{q} - \mathbf{q}'$ is the scattering vector and $\hbar\omega = E - E'$ is the energy transfer to the crystal.

The evaluation of Eq. (36) proceeds along similar lines to that of Eq. (31) for x-ray scattering. There are the Bragg diffraction peaks, whose intensities are reduced by the Debye–Waller factor e^{-2W}. The inelastic scattering is obtained by expanding the exponential in Eq. (36) in powers of the $\mathbf{u}(l)$ and expressing the $\mathbf{u}(l)$ in terms of the normal-mode phonon creation and annihilation operators. The linear terms yield energy- and momentum-conserving one-phonon creation and annihilation processes. The scattering cross section for these one-phonon processes is (Cowley, 1969)

$$\frac{d^2\sigma}{dE d\Omega} = b^2 \frac{q'}{q} \frac{N\hbar e^{-2W}}{2M\omega(\mathbf{k}j)} [\mathbf{S}\cdot\mathbf{e}(\mathbf{k}j)]^2[\bar{n}(\mathbf{k}j) + \tfrac{1}{2} \pm \tfrac{1}{2}] \times \delta(\hbar\omega \mp \hbar\omega(\mathbf{k}j))\Delta(\mathbf{S} - \mathbf{k} - \mathbf{G}), \tag{37}$$

with the upper sign corresponding to energy loss by the neutron through the creation of a phonon $\mathbf{k}j$, and the lower sign to energy gain through annihilation of that phonon.

The dispersion relations of a great many crystals have been measured with neutron scattering (see Bilz and Kress, 1979, for a survey of results). The measurements are fairly time consuming and are usually confined to principal crystallographic directions. Figure 1(a) shows the measured phonon dispersion relation of Si.

3.3 Helium-Atom and Electron Scattering; Surface Phonons

In recent years, a great deal has been learned about lattice vibrations at crystal surfaces using helium-atom scattering and *electron energy-loss spectroscopy* (EELS). The He atoms in these experiments are given energies of the order of 0.02 eV, which permits them to approach no closer than about 3–4 Å from the crystal surface, and they thus effectively sense only the motion of the top layer of atoms. In EELS, on the other hand, the electrons have energies of the order of 100 eV, which allows them to penetrate two to three atomic layers into the crystal and sense the motion of all of those atoms. The surfaces that are studied are usually parallel to principal crystallographic planes such as (001) and (111), and the scattered particles are generally measured in the plane of incidence, which is located in another principal plane (Toennies, 1991).

A bounded crystal retains the translational symmetry of the lattice parallel, but not perpendicular, to the boundary, and so it is only the component of the wave vector parallel to the surface, $\mathbf{k}_\parallel$, that serves for labeling the vibrational modes. There has to be another quantum number taking on a quasi-continuum of values, but it does not represent the perpendicular component of the wave vector. Surface-phonon dispersion relations thus have the general appearance of that of platinum (111) shown in Fig. 3. Corresponding to each value of $\mathbf{k}_\parallel$ there is a continuous band of modes, which far from the surface match up with the subset of bulk modes of the crystal whose wave vectors have $\mathbf{k}_\parallel$ as their in-plane component. There may be resonances in the continuum, for which the displacements of the associated modes become very large near the surface. In addition, there is a discrete Rayleigh branch, which is separated from the band modes from below. This corresponds to true surface modes for which the displacements fall off more or less exponentially with distance from the surface.

Both He atoms and electrons undergo elastic Bragg scattering in which the change in the in-plane component of the momentum of the projectile is equal to $\hbar\mathbf{G}$. The Bragg intensities are, as usual, reduced by a Debye–Waller factor. In addition, there is inelastic scattering accompanied by the creation or destruction of a surface phonon. The kinematic conditions are conservation of energy and of momentum in the surface to within $\hbar\mathbf{G}$.

The experimental data in Fig. 3 show res-

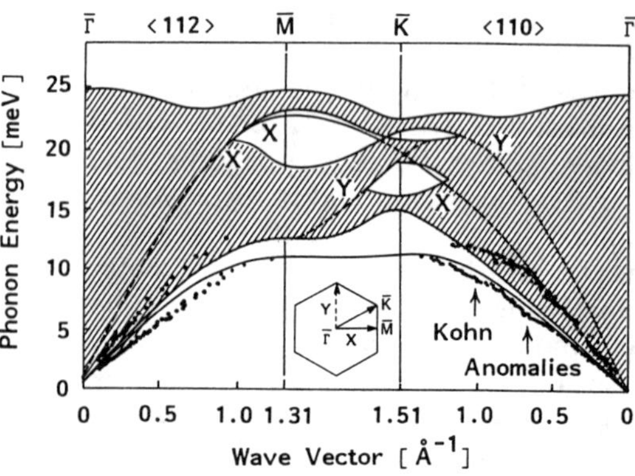

FIG. 3. Surface-phonon dispersion relation for Pt(111) (adapted from Toennies, 1991). Dots represent experimental data: curves and shaded area are theoretical. The letters X and Y indicate the polarization at the edge of the bulk bands.

onances that have been measured with He-atom scattering. These show a number of interesting features. In the ⟨110⟩ direction, there are two dips in the Rayleigh branch that have been ascribed to *Kohn anomalies* (see Sec. 5). In both the ⟨110⟩ and ⟨112⟩ directions, there is a resonance in the continuum due to a longitudinal mode concentrated near the surface by a weakening of the interatomic force constants there.

Helium-atom scattering has revealed a soft mode in tungsten (001) that precedes *surface reconstruction* as the temperature is lowered to 300 K. This is the first observation of a soft mode in a two-dimensional phase transition. Kress (1991) contains an extensive compilation of results obtained with He-atom scattering and EELS.

4. INFRARED ABSORPTION AND LIGHT SCATTERING

Phonons feature importantly in the infrared absorption and reflectivity of crystals, and in the Raman and Brillouin scattering of light.

The infrared properties of polar crystals are dominated by motion of the charged sublattices against each other. In the case of the alkali halides, for example, the near–zone-center longitudinal optical (LO) mode gives rise to an alternating electric field, which causes this branch to split away from the degenerate transverse optical (TO) branches, which are not accompanied by such a field. The well-known Lyddane–Sachs–Teller theory has these zone-center frequencies related by (Kittel, 1986)

$$\omega_L^2/\omega_T^2 = \epsilon(0)/\epsilon(\infty), \tag{38}$$

where $\epsilon(0)$ and $\epsilon(\infty)$ are the values of the dielectric constant at frequencies well below and well above ω_T, respectively. Electromagnetic waves, being transversely polarized, couple very strongly to the TO modes, but not to the LO mode. A coupled-mode polariton approach is required to describe wave propagation in the crystal, and this yields a band gap between ω_T and ω_L, where the crystal ideally should be perfectly reflecting to electromagnetic radiation. High values of reflectivity are indeed measured in this frequency range.

In nonpolar diamond-lattice–type crystals, one-phonon absorption is forbidden because there is a center of inversion midway between each pair of neighboring atoms, and there is therefore no electric dipole moment associated with the vibration of the two sublattices against each other. The presence of defects in the crystal breaks the symmetry, and all modes become infrared active, though not all to the same extent.

Two-phonon absorption is allowed in perfect crystals subject to the kinematic constraints of energy and momentum conservation. An infrared photon with the right energy to be absorbed will have a wavelength that is extremely large compared with the lattice spacing, and hence its wave vector $\mathbf{q}$ can be taken to be zero. The absorption is mediated mainly by the quadratic dependence of the electric dipole moment on the atomic displacements, which consists of momentum-conserving terms of the form $b(0)[a^\dagger(-\mathbf{k}j) + a(\mathbf{k}j)][a^\dagger(\mathbf{k}j') + a(-\mathbf{k}j')]$, where $b(0)$ is the photon-destruction operator. The absorption consists of a difference band, in which one phonon is created and another destroyed, and a summation band, in which two phonons are created. The absorptivity for the latter is given by (Reissland, 1973)

$$\alpha(\omega) = \sum_{\mathbf{k}jj'} \frac{A(\mathbf{k}jj')}{\omega(\mathbf{k}j)\omega(-\mathbf{k}j')}\{1 + n(\mathbf{k}j) + n(-\mathbf{k}j')\} \times \delta(\omega - \omega(\mathbf{k}j) - \omega(-\mathbf{k}j')), \tag{39}$$

where $A(\mathbf{k}jj')$ are not very easily determined coupling coefficients. As a crude first approximation, taking the combination of factors preceding the delta function as a constant, the absorption comes out to be proportional to the joint frequency distribution function $D_2(\omega)$. One can think of $D_2(\omega)$ as pertaining to a dispersion relation with $r(r + 1)/2$ branches representing $\omega(\mathbf{k}j) + \omega(-\mathbf{k}j')$ as a function of $\mathbf{k}$ for all combinations of j and j'. As with $D(\omega)$, $D_2(\omega)$ possesses critical points, but in much greater profusion, and it is a complex combinatorial exercise to extract the critical points of $D(\omega)$ from those of $D_2(\omega)$. Figure 4 shows the two-phonon infrared absorption spectrum of silicon and identifies some of the critical-point features.

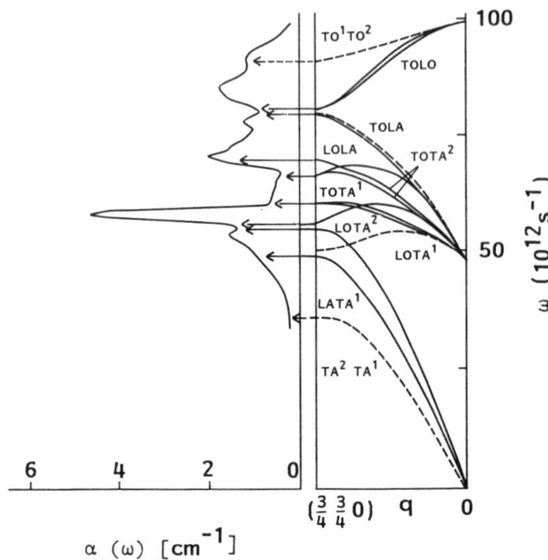

FIG. 4. Two-phonon infrared absorption $\alpha(\omega)$ in silicon at 20 K and the identification of some of the critical points in terms of the two-phonon dispersion relation in the [110] direction (adapted from Bilz, 1966).

4.1 Raman Scattering

In Raman scattering, a photon of light is scattered inelastically from a crystal, with the change in energy and momentum being taken up in the creation or destruction of one or more phonons. In the case where a single acoustic phonon is involved, the process is termed Brillouin scattering. Raman scattering comes about through the modulation of the frequency-dependent electronic polarizability and hence refractive index by the lattice modes. In first order, the interaction terms take the form $b(\mathbf{q}_i)b^\dagger(\mathbf{q}_f)[a^\dagger(-\mathbf{k}j) + a(\mathbf{k}j)]$. For all practical purposes, the wave vectors $\mathbf{q}_i \approx \mathbf{q}_f \approx 0$ for the incident and scattered photons, and so it is only the near–zone center optical modes, whose frequencies are independent of $\mathbf{k}$ to first order, that are affected.

The spectrum of the inelastically scattered light consists of two sets of lines with frequency shifts

$$\Delta\omega = \omega_i - \omega_f = \pm\omega(\mathbf{k}j), \quad \mathbf{k} \approx 0, \tag{40}$$

where ω_i and ω_f are the incident and scattered photon frequencies. The + sign denotes the *Stokes component*, which corresponds to phonon creation and has intensity proportional to $n(\mathbf{k}j) + 1$, and the − sign denotes the *anti-Stokes component*, which corresponds to phonon destruction and has intensity proportional to $n(\mathbf{k}j)$.

The quadratic dependence of the polarizability on the lattice displacements gives rise to second-order Raman scattering comprising Stokes and anti-Stokes summation and difference bands. In the Stokes summation band, for instance, the scattered intensity is given by (Hayes and Loudon, 1978)

$$\begin{aligned} I(\omega_f) = \sum_{\mathbf{k}jj'} & \frac{B(\mathbf{k}jj')}{\omega(\mathbf{k}j)\omega(-\mathbf{k}j')} \\ & \times \{1 + n(\mathbf{k}j)\}\{1 + n(-\mathbf{k}j')\} \\ & \times \delta(\omega_i - \omega_f - \omega(\mathbf{k}j) - \omega(-\mathbf{k}j')), \end{aligned} \tag{41}$$

where $B(kjj')$ are a set of coupling coefficients. Much the same comments apply here as earlier to two-phonon infrared absorption.

4.2 Brillouin Scattering

Brillouin scattering involves scattering by acoustic modes whose frequencies are proportional to k, and so the frequency shifts depend sensitively on the scattering geometry and incident photon frequency ω_i. Since these shifts are small compared with ω_i, it follows that $|\mathbf{q}_i| \approx |\mathbf{q}_f|$, and momentum conservation implies that $k = 2q_i \sin(\theta/2)$, where θ is the scattering angle (see Fig. 5). By changing θ, the direction of $\mathbf{k}$ can be varied. For each scattering geometry, there are in general three Stokes and three anti-Stokes components with frequency shifts (Reissland, 1973)

$$\Delta\omega = v(\mathbf{k}j)k = \frac{2v(\mathbf{k}j)\eta\omega_i}{c}\sin(\theta/2), \tag{42}$$

where η is the refractive index of the me-

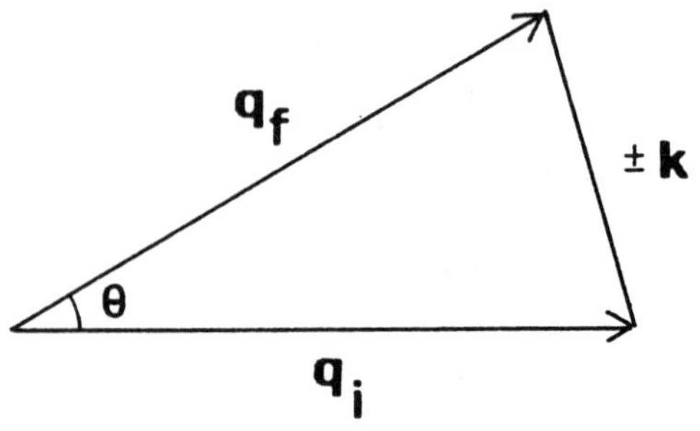

FIG. 5. Momentum-conservation condition for Brillouin scattering.

dium, c is the velocity of light in vacuum, and $v(\mathbf{k}j)$ are the three directionally dependent acoustic phase velocities. These velocities are functions of the bulk elastic constants, and this has led to Brillouin scattering being one of the more important methods for measuring the anisotropic elastic constants of crystals (see Every, 1994a). Such measurements, carried out under variable conditions of temperature and pressure, are a valuable source of information on phase transitions (see, e.g., Dil, 1982; Comins *et al.*, 1990).

There is considerable interest at present in *surface Brillouin scattering* (Nizzoli and Sandercock, 1990). The kinematic conditions are similar to those of He-atom scattering, except that because of the long wavelengths involved, the crystal behaves like an elastic continuum and there are only normal (N) processes to consider. One of the advantages of surface Brillouin scattering is that it can be applied to opaque materials. A drawback of the technique is that because the scattering volume is small, the scattered intensity is weak and sophisticated detection techniques are required.

5. ELECTRON–PHONON INTERACTION

Electrons are an inseparable part of any solid and lie at the very heart of lattice dynamics. They determine the crystal-lattice spacing and contribute in large part to the effective interatomic potential Φ. The problem of the coupled motion of the electrons and the nuclei is made manageable by the fact that the electrons are 2000 or more times less massive than the nuclei, move very fast in comparison, and are by and large able to adjust adiabatically to the instantaneous positions of the nuclei. These ideas are formalized in the *adiabatic* or *Born–Oppenheimer approximation* (see Maradudin, 1974). For many purposes, as we have seen earlier, the contribution of electrons to lattice vibrations can be treated phenomenologically by picturing them in bonds, or rigidly attached to atoms, or in shells that can be displaced relative to ion cores. Our concern in this section is with the more subtle effects that arise from the interaction between phonons and mobile electrons located within $\sim k_B T$ of the Fermi surface in metals and in semiconductors. These interactions come from small residual terms that are neglected in the Born–Oppenheimer approximation. This is a very extensive field, and we will merely touch on some of the many interesting effects that there are.

5.1 Metals

In a perfect metallic crystal, we may as a first approximation consider the electrons to be noninteracting and occupying Bloch states $\psi_{\mathbf{q}\sigma}$, which are characterized by crystal momentum $\hbar\mathbf{q}$ and a band index σ. The electrons are not scattered by the ions in their lattice positions, since the periodic potential of the lattice is already taken into account in constructing the Bloch states. Crystal-lattice vibrations represent departures from perfect periodicity and therefore do cause the electrons to be scattered, and this is the dominant source of temperature-dependent electrical resistance in most metals.

For small displacements of the ions from their equilibrium positions, the change in the potential the electrons experience at any point $\mathbf{x}$ is linear in the displacements. The electron-lattice interaction can thus be expressed as (Ziman, 1965)

$$H_{eL}(\mathbf{x}) = \sum_{\mathbf{k}j} U_{\mathbf{k}j}(\mathbf{x})(a^{\dagger}_{-\mathbf{k}j} + a_{\mathbf{k}j}), \tag{43}$$

where the form of the functions $U_{\mathbf{k}j}(\mathbf{x})$ depends on the specific model being considered, subject to preserving the symmetry of the lattice. In first order of perturbation theory, H_{eL} gives rise to the scattering processes depicted in Figs. 6(a) and 6(b), in which an

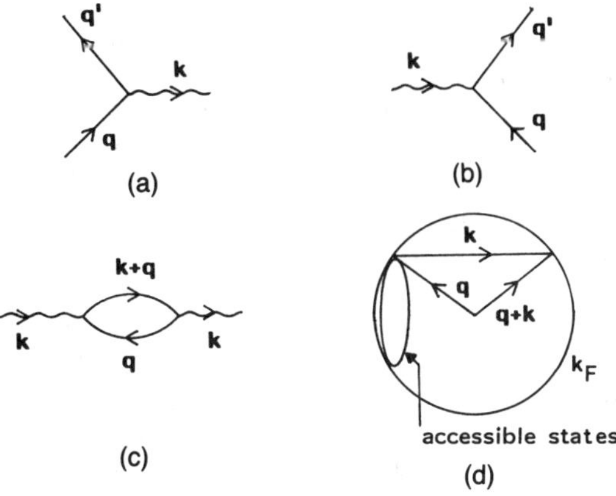

FIG. 6. The electron–phonon interaction in metals.

electron undergoes a transition from one Bloch state $\psi_{\mathbf{q}\sigma}$ to another $\psi_{\mathbf{q}'\sigma'}$, with the simultaneous creation or destruction of a phonon. Conservation of energy and momentum require that

$$E_{\mathbf{q}\sigma} - E_{\mathbf{q}'\sigma'} = \pm\hbar\omega(\mathbf{k}j), \tag{44}$$

$$\Delta(\mathbf{q} - \mathbf{q}' - \mathbf{k}) = 0. \tag{45}$$

The scattering probability is proportional to $n(\mathbf{k}j) + \frac{1}{2} \pm \frac{1}{2}$.

Electron–phonon scattering limits the electron mobility and is responsible for the *ideal electrical resistance* of metals, which at temperatures well below the Debye temperature is proportional to T^5 (Ziman, 1965). Likewise, phonons have greatly reduced mean free paths because of their coupling to the electrons, and so contribute minimally to thermal conduction in metals, and ballistic phonon transport is not observed. At low temperatures, where phonon U processes are rare, the displaced momentum distribution of the electrons associated with an electrical current is communicated to the phonons, so that they too acquire a skewed momentum distribution. This is known as the *phonon drag effect* and is observed as a heat flux in the direction of the current.

In second order of perturbation, H_{eL} leads to a renormalization of the phonon frequencies and electron effective masses. The change in phonon frequencies comes from the type of process depicted in Fig. 6(c), in which the intermediate state is that of the phonon **k** destroyed and an electron scattered from **q** to **q** + **k**. Figure 6(d) depicts these wave vectors in relation to a spherical Fermi surface, near which **q** and **q** + **k** must be located in order to make a significant contribution. Evidently, it is only when k is less than the diameter of the Fermi sphere that there is an effect. A singularity appears in the phonon dispersion relation as k crosses this threshold. The precise form and location of this singularity, which is known as a *Kohn anomaly*, depends on the geometry of the Fermi surface (Ziman, 1965). There may be more than one anomaly in a particular direction in the Brillouin zone, as is the case with the surface-phonon dispersion relation of Pt in Fig. 3.

5.2 Metallic Superconductors

The electron–lattice interaction is the origin of the attractive potential that is responsible for Cooper pair formation and superconductivity in metals. As the first of the two electrons in a *Cooper pair* moves through the lattice, it pulls the positive ions inward, and the second electron is then drawn toward the slightly compacted positively charged region. The net result is an attraction between the two electrons.

The effective interaction between the electrons is characterized by the matrix elements representing scattering of one electron from **q** to **q** − **k**, and a second electron from **q**′ to **q**′ + **k**. The scattering is mediated by the exchange of a *virtual phonon* of wave vector **k**. The process is second order in H_{eL} and involves intermediate states in which one of the electrons is scattered with the creation or destruction of a momentum-conserving phonon. The two processes, which have energy denominators $E_{\mathbf{q}} - E_{\mathbf{q}-\mathbf{k}} \mp \hbar\omega(\mathbf{k})$, are additive, and the resulting scattering matrix element therefore has the form (Ziman, 1965)

$$\langle \mathbf{q} - \mathbf{k}, \mathbf{q}' + \mathbf{k}|V|\mathbf{q}, \mathbf{q}'\rangle = \frac{|(U_{-\mathbf{k}j})_{\mathbf{q},\mathbf{q}-\mathbf{k}}|^2 2\hbar\omega(\mathbf{k})}{\{E_{\mathbf{q}} - E_{\mathbf{q}-\mathbf{k}}\}^2 - \{\hbar\omega(\mathbf{k})\}^2}. \tag{46}$$

For

$$|E_{\mathbf{q}} - E_{\mathbf{q}-\mathbf{k}}| < \hbar\omega(\mathbf{k}), \tag{47}$$

the effective interaction is negative, corresponding to an attractive force between the two electrons. The details of how this coupling leads to the formation of an energy gap and superconductivity lie beyond the scope of this article, but certain salient features of superconductivity are already evident in the form of Eq. (46). One is that high electrical resistance in the normal metal and a high superconducting transition temperature T_c go hand in hand, since they both depend on the electron lattice interaction H_{eL} being large. Another is that because of the factor $2\hbar\omega(\mathbf{k})$ in the numerator of Eq. (46), which is proportional to $M^{-1/2}$, where M is the atomic mass,

$$T_c \propto M^{-1/2}. \tag{48}$$

The observation of the dependence of T_c on

the isotopic composition has been one of the crucial tests of the Cooper pairing model of superconductivity in metals.

5.3 Semiconductors

The electron–phonon interaction plays an important role in the dynamics of electrons and holes in semiconductors. At low temperatures and in the absence of external excitation, it is the coupling to long-wavelength acoustic modes that is most important, and this is usually modeled by the *deformation potential* (see, e.g., Challis *et al.*, 1987). To illustrate the basic idea, let us consider electrons in a conduction band that has an energy minimum at $\mathbf{q} = 0$. For small lattice displacements, the band is assumed to be shifted vertically in proportion to the lattice dilatation $\Delta = \Sigma_\alpha \, \partial u_\alpha / \partial x_\alpha$, while the band curvature, which determines the electron effective mass m^*, is unaffected to first order. The electron–lattice interaction is thus given by

$$H_{eL}(\mathbf{x}) = \Xi\Delta = \Xi \sum_{\mathbf{k}j} i\mathbf{k}\cdot\mathbf{e}(\mathbf{k}j) \times \sqrt{\frac{\hbar}{2NM\omega(\mathbf{k}j)}}\, e^{i\mathbf{k}\cdot\mathbf{x}} \times [a^\dagger(-\mathbf{k}j) + a(\mathbf{k}j)], \quad (49)$$

where Ξ is the deformation-potential constant.

As in the case of metals, H_{eL} gives rise to electron scattering with the creation or annihilation of a phonon. This limits the carrier mobility and is a determining factor in the electrical resistance. There has been considerable work done at low temperatures on phonon emission and absorption and phonon drag by two-dimensional electron gases (2DEGs) in the inversion layer of Si MOSFETs, in GaAlAs heterostructures, and elsewhere (Challis *et al.*, 1987). The converse of phonon drag, known as the *phonon wind effect*, has been very beautifully demonstrated in germanium (Markiewicz *et al.*, 1980). Under intense optical excitation at low temperatures, a condensed state of electrons and holes known as electron-hole liquid is formed. Droplets of this liquid are propelled into the crystal by the capture of momentum from phonons generated in the region of excitation, and this motion is revealed by the spreading recombination luminescence (for a review, see Keldysh and Sibeldin, 1986).

Interband optical absorption in semiconductors involves predominantly direct or "vertical" transitions, in which the wave vector of the electron is unchanged. *Indirect band-gap transitions* take place with much lower probability, and do so because of the slight mixing of different electron–phonon product states of the same total momentum by H_{eL}. Conservation of momentum requires that a transition involving a change in electron momentum by $\hbar\mathbf{q}$ is accompanied by the creation or annihilation of a phonon of momentum $-\hbar\mathbf{q}$ or $\hbar\mathbf{q}$ respectively (Ziman, 1965). The optical absorptivity near an indirect band-gap threshold is thus much weaker than for direct band-gap transitions, and the lifetime of carriers against radiative recombination is much larger.

5.4 Insulators

In polar insulators such as the alkali halides, the effective mass of conduction electrons is increased by the electron–lattice interaction by as much as a factor of 2 or more (Kittel, 1986). An electron pulls the nearby positive ions toward itself while pushing the negative ions away, and this adds to the inertia of the electron, thereby increasing its effective mass. The combination of the electron and ion displacements is known as a *polaron*.

6. PHONON–DEFECT INTERACTIONS

Lattice imperfections such as vacancies, interstitial and substitutional impurities, isotopic differences, dislocations, and grain boundaries, etc., have a marked effect on the vibrational properties of crystals. They destroy the translational symmetry of the lattice, thereby giving rise to momentum-nonconserving phonon scattering, defect-activated infrared absorption, and various other phenomena. They modify the local dynamical response of the lattice, resulting under certain conditions in the appearance of in-band resonances or local vibrational modes. Local electronic and spin states couple to the vibrations of nearby atoms, allowing an exchange of energy between these states and the phonons. This gives rise to *energy relaxation* to the phonon bath and provides a mechanism for doing *phonon spectroscopy* on defects. When the coupling is strong, the local electronic and vibrational states need to be considered as a composite system of *vibronic states*, and optical transitions occur in which the electronic and vibrational states simultaneously change. An extensive review covering most of these topics is Maradudin (1966).

6.1 Local Modes

The vibrational properties of crystal lattices containing defects are usually treated by the *dynamical response-function* approach (Maradudin, 1966; Elliott, 1966). The method involves expressing the dynamical response of the defective crystal in terms of the response of the perfect crystal and the local modification to the dynamical matrix brought about by the presence of the defect, thus obviating the need to determine from scratch the eigenvalues and eigenvectors of $3rN$ coupled oscillators.

The method can be illustrated in a simple way with the example of an isotopic defect. Consider an otherwise perfect silicon crystal in which the atom at the origin is replaced by one of a different mass without altering any of the force constants. For the unperturbed crystal, the dynamical response of the silicon atom to a sinusoidal force $\exp(i\omega t)$ is given by

$$G_0(\omega) = \frac{1}{6NM}\sum_{\mathbf{k}j}\left\{\frac{1}{\omega^2(\mathbf{k}j) - \omega^2} + \frac{i\pi}{2\omega(\mathbf{k}j)}\delta(\omega - \omega(\mathbf{k}j))\right\}, \quad \omega > 0. \tag{50}$$

The terms in the sum represent the real and imaginary parts of the response function of an individual harmonic oscillator, and the factor $1/6\,NM$ comes from projecting the force acting on the atom in question onto each normal coordinate of the lattice, and then projecting the displacement response of that mode back onto that atomic coordinate. More generally, there would be a factor of the form $e_\alpha^*(\kappa|\mathbf{k}j)e_\beta(\kappa'|\mathbf{k}j)$ inside the sum, but this comes out to be equal to $\frac{1}{6}$ because of the symmetry of the Si lattice.

If the atom at the origin is now replaced by one of different mass $M + \delta M$, then the applied force has the additional inertia of δM to contend with, and it is readily shown that the perturbed lattice response function is given by (Elliott, 1966)

$$G(\omega) = \frac{G_0(\omega)}{1 - \delta M\omega^2 G_0(\omega)}. \tag{51}$$

Force-constant changes can be handled in a similar way. If the impurity is charged, it will give rise to one-phonon infrared absorption proportional to $\omega Im(G(\omega))$. For $\delta M = 0$, it follows from Eqs. (50) and (51) that the absorptivity is proportional to the lattice frequency distribution function $D(\omega)$. For $\delta M \neq 0$, the absorption still extends over the entire band of frequencies of the perfect lattice but is modified in intensity. In the case of a light impurity for which $\delta M < 0$, there may in addition be absorption outside the perfect-lattice spectrum at a discrete frequency ω satisfying (Elliott, 1966)

$$1 - \delta M\omega^2 G_0(\omega) = 0. \tag{52}$$

This corresponds to a local mode in which the displacement is mainly confined to the impurity atom and its first few shells of neighbors. Conversely, for a heavy impurity there may be an in-band resonance. These resonances feature in infrared absorption and also in the resonant scattering of bulk phonons. Figure 7 shows calculated infrared absorption spectra for various substitutional

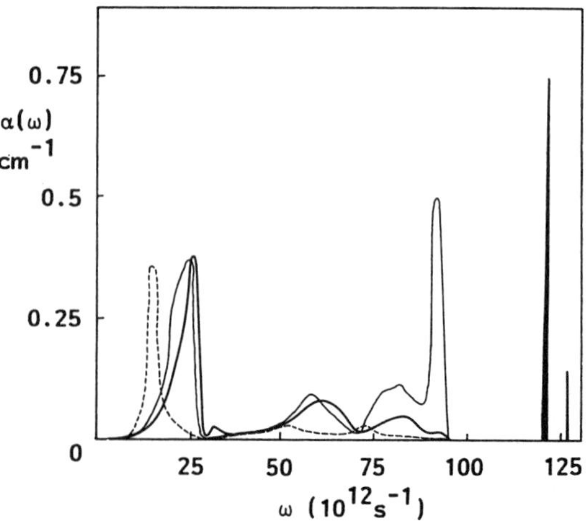

FIG. 7. Calculated infrared absorption spectra for substitutional boron and antimony in silicon (adapted from Elliott, 1966).

impurities in silicon. The full curves are for ^{10}B and ^{11}B in their natural abundances, which are lighter than the host lattice Si atom and give rise to the local modes above 10^{14} s^{-1}. The dashed curve is for Sb, which is heavier than Si and gives rise to the low-lying *in-band resonance* at $\sim 15 \times 10^{12}$ s^{-1}. Quantities such as correlation functions and mean square displacements, etc., are readily calculated once the modified response functions are known. For long-wavelength bulk acoustic modes well below any resonance, the scattering rate from point defects is proportional to $1/\lambda^4$, in accordance with the *Rayleigh scattering* law.

6.2 Electronic States

Electronic states localized at a defect are perturbed by the motion of neighboring atoms, and this provides a mechanism for the exchange of energy. In the case where the electronic energy level splitting lies within the band of vibrational frequencies, transitions can occur with the emission or absorption of a single phonon. The Cr^{3+} laser impurity center in ruby is a case in point that has a transition at 29 cm^{-1} between two excited states (see Renk, 1986; Kaplyanskii and Basun, 1986). By means of optical pumping into the upper of the two excited states and subsequent phonon emission, very high occupation numbers for 29-cm^{-1} phonons can be achieved, leading to *stimulated phonon emission*. The coupling also gives rise to the resonant scattering and *trapping* of 29-cm^{-1} phonons.

The effect of coupling between the localized electronic states and phonons is also seen in the optical absorption and fluorescence spectra of crystals. A grossly oversimplified model, which nevertheless conveys some of the essential features, is to assume that there is a single lattice mode of frequency ω_0 that is linearly coupled to the electronic states. The equilibrium positions about which the oscillations occur in the ground and excited electronic states are thus displaced relative to one another, by ΔQ, say. There are now a multitude of transitions possible, determined by the matrix of overlap integrals between the two sets of oscillator states, in which the optical and vibrational states simultaneously change. At $T = 0$, the probability of a transition occurring in which p phonons are created is proportional to (Fitchen, 1968)

$$W_p = e^{-S_0} S_0^p/p!, \tag{53}$$

where

$$S_0 = M\omega_0(\Delta Q)^2/2\hbar. \tag{54}$$

The $p = 0$ transition is called the *zero-phonon line*, and it remains sharp even when a band of coupled modes rather than a single mode is considered and the temperature is finite, whereas the $p > 0$ transitions are broadened. The temperature dependence of the zero-phonon line intensity is dominated by a Debye–Waller factor $\sim\exp(-S_0 \coth[\hbar\omega_0/2k_BT])$, known as the *Huang–Rhys factor*. Many such *vibronic spectra* have been observed in solids (see Fitchen, 1968; Maradudin, 1966).

In fluorescence, the mirror-image spectrum is observed, with p phonons being created, and the remaining energy being carried away by the photon. Figure 8 shows the absorption and emission spectra of the 4150-Å system in diamond, which is associated with a nitrogen impurity complex (Dyer and Mathews, 1957). *Nonradiative transitions* in which all the excitation energy is dumped as phonons occur at an increasing rate as the temperature is raised, thereby decreasing the luminescence efficiency.

6.3 Spin-Lattice Relaxation

Relaxation processes play an important role in paramagnetic resonance. A nuclear or unpaired electron spin belonging to an atom

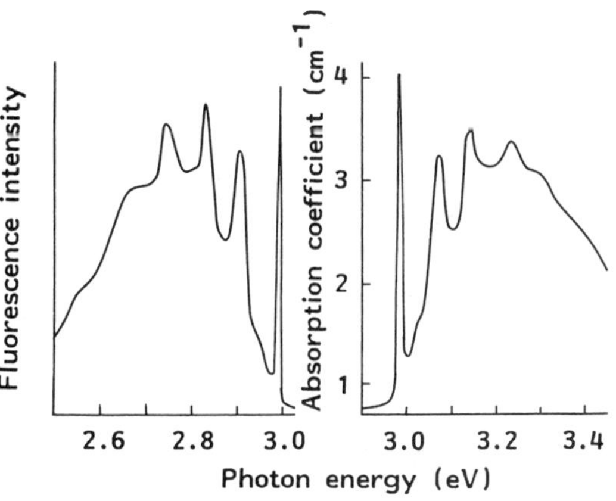

FIG. 8. The fluorescence and absorption spectra of the 4150-Å system in diamond (adapted from Dyer and Matthews, 1957).

in a crystal can exist in various states whose level splittings depend on the crystal field and the external magnetic field. A microwave or radio-frequency field is used to excite the spin system, and various relaxation processes then come into play to return the spin population to equilibrium. Of the many different relaxation processes that there are, we only discuss two that have a direct bearing on phonons.

An important relaxation process for isolated paramagnetic spins in the case where the level splitting is fairly large (which by and large excludes nuclear paramagnetism) is single-phonon emission into long-wavelength acoustic modes (Stevens, 1967). The parameters of the spin Hamiltonian depend on the precise location of the neighboring atoms around the paramagnetic atom and are thus functions of the strain field accompanying acoustic waves in the medium. The linear coupling to the strain yields a spin-lattice interaction Hamiltonian H_{SL} of the form

$$H_{SL} = \sum_{\mathbf{k}j} H_{\mathbf{k}j}(\mathbf{S})|\mathbf{k}| \sqrt{\frac{\hbar}{2NM\omega(\mathbf{k}j)}} \times [a^{\dagger}(-\mathbf{k}j) + a(\mathbf{k}j)], \quad (55)$$

where $H_{\mathbf{k}j}(\mathbf{S})$ is a spin operator. In first order of perturbation theory, the transition probability p between two spin levels separated in energy by $E = \hbar\omega_0$ is proportional to $D(\omega_0)\bar{n}(\omega_0)\ |\mathbf{k}|^2/\omega_0$. For $k_B T \gg \hbar\omega_0$, it follows that $p \propto \omega_0^2 T$.

If the rate at which phonons are pumped into the narrow band of frequencies around ω_0 exceeds the ability of the lattice to thermalize them, then there is a *phonon bottleneck* effect, in which repeatedly the phonons emitted by one paramagnetic spin are absorbed by another (Stevens, 1967).

For nuclear spins, where ω_0 tends to be relatively small, the number of lattice modes in that frequency region is very small, and the direct relaxation rate is extremely slow. Under these conditions, Raman processes, in which a phonon is absorbed and another of slightly higher frequency emitted, become important. The coupling for this process comes from the quadratic dependence of the spin Hamiltonian on the local strain. All phonons from $\omega = 0$ to $\omega \sim k_B T/\hbar$ contribute, and for $T \ll \Theta_D$ the relaxation rate is independent of the splitting and has a temperature dependence $p \propto T^7$.

6.4 Mössbauer Effect

The Mössbauer effect can be thought of as the nuclear analog of vibronic spectra, in which it is the nucleus rather than the electrons of the atom that undergoes a transition, and the photon that is absorbed or emitted is that of a γ ray rather than of visible light. Corresponding to the zero-phonon line there is a certain fraction of the γ rays that are emitted recoil-free, i.e., without any energy being absorbed by the vibrations of the lattice. The recoil momentum from these γ photons is taken up by the crystal as a whole, and because the crystal is so massive, the associated kinetic energy is totally negligible. The *recoil-free fraction* is given by the Debye–Waller factor (Maradudin, 1966)

$$f = \exp -\langle[\mathbf{K}\cdot\mathbf{u}(l\kappa)]^2\rangle, \quad (56)$$

where K is the wave vector of the γ ray and $\mathbf{u}(l\kappa)$ is the displacement coordinate of the resonant nucleus. With increasing temperature, as the mean square displacement of the atom increases, f decreases. At higher temperatures, anharmonicity plays an important role in this temperature dependence (Kolk, 1984). There has been considerable theoretical speculation concerning the single- and multi-phonon sidebands, which has greatly outstripped the technical means for measuring them. Part of the problem is that to measure directly a phonon energy shift of ~0.01 eV in a ~10 keV γ ray requires the velocity of the moving stage to be $\sim 10^{-6}c = 300$ m/s.

7. CLOSING REMARKS

This article has mainly concentrated on the interaction between phonons and other "-ons", the physical properties that these interactions impart to crystals, and how the scattering of x rays, neutrons, and other probes can be used to determine phonon dispersion relations. In an article of this nature, it is not possible to give complete coverage to the whole of phonon physics. There has in particular been very little discussion here of phonon transport, which is covered extensively in the articles LV and HP. The first of these two articles contains an extensive discussion of the interplay between boundary scattering, defect scattering, and anharmonic processes in limiting thermal conduction in dielectric crystals. The latter article contains a lot of recent material on *ballistic phonon transport* and its investigation by means of *phonon imaging*. This has shed light on various phenomena, including phonon transmission through surfaces, the scattering of phonons by dislocations, and electron–phonon scattering in semiconductors (for reviews on phonon imaging, see Wolfe, 1989; Every, 1994b). HP also discusses second sound, which is a collective mode of the phonon gas that comes about when N processes dominate over U and other momentum-nonconserving scattering processes.

A development that is attracting considerable attention at the present time is that of *phonon-mediated elementary-particle detection* (for a review, see Cabrera, 1990). This offers the possibility of much greater energy resolution than can be achieved by other available means, and the prospect of being able to detect the hypothetical particles that are invoked as an explanation of the missing dark matter in our galaxy. While much progress has been made, the technical difficulties are formidable, and the final word on this subject is some way off.

ACKNOWLEDGMENTS

This article was partly written while the author was on leave at the Institute of Applied Mechanics, National Taiwan University. The National Science Council of the Republic of China and the South African Foundation for Research Development are thanked for their financial support. J. Wesner is thanked for critical reading of the manuscript.

Works Cited

Ashcroft, N. W., Mermin, N. D. (1981), *Solid State Physics*, Philadelphia: Saunders.

Bilz, H. (1966), in: R. W. H. Stevenson (Ed.), *Phonons*, Edinburgh: Oliver & Boyd, pp. 208–234.

Bilz, H., Kress, W. (1979), *Phonon Dispersion Relations in Insulators*, Berlin: Springer.

Cabrera, B. (1990), in S. Hunklinger, W. Ludwig, G. Weiss (Eds.), *PHONONS 89*, Singapore: World Scientific, pp. 1373–1382.

Challis, L. J., Toombs, G. A., Sheard, F. W. (1987), in: T. Paszkiewicz (Ed.), *Physics of Phonons*, Berlin: Springer, pp. 348–374.

Comins, J. D., Ngoepe, P. E., Catlow, C. R. A. (1990), *J. Chem. Soc. Faraday Trans.* **86**, 1183–1192.

Cowley, R. A. (1969), in: R. H. Enns, R. R. Haering (Eds.), *Modern Solid State Physics*, Vol. 2, New York: Gordon and Breach, pp. 43–105.

de Launay, J. (1953), *J. Chem. Phys.* **21**, 1975–1986.

Dil, J. G. (1982), *Rep. Prog. Phys.* **45**, 285–334.

Dorner, B., Comes, R. (1977), in: S. W. Lovesey, T. Springer (Eds.), *Dynamics of Solids and Liquids by Neutron Scattering*, Berlin: Springer, pp. 127–196.

Dyer, H. B., Matthews, I. G. (1957), *Proc. R. Soc. A* **243**, 320–335.

Elliott, R. J. (1966), in: R. W. H. Stevenson (Ed.), *Phonons*, Edinburgh: Oliver & Boyd, pp. 377–402.

Every, A. G. (1994a), *NDT&E Int.*, **27**, 3–10.

Every, A. G. (1994b), in: T. Paszkiewicz, T. Rapcewicz (Eds.), *Die Kunst of Phonons*, New York: Plenum, pp. 55–72.

Fitchen, D. B. (1968), in: W. B. Fowler (Ed.), *Physics of Color Centers*, New York: Academic, pp. 293–350.

Ghatak, A. K., Kothari, L. S. (1972), *Lattice Dynamics*, London: Addison-Wesley.

Guyer, R. A. (1969), in: F. Seitz, D. Turnbull, H. Ehrenreich (Eds.), *Solid State Physics*, Vol. 23, New York: Academic, pp. 413–499.

Hass, K. C., Tamor, M. A., Anthony, T. R., Banholzer, W. F. (1992), *Phys. Rev. B* **45**, 7171–7182.

Hayes, W., Loudon, R. (1978), *Scattering of Light by Crystals*, New York: Wiley.

Kaplyanskii, A. A., Basun, S. A. (1986), in: W. Eisenmenger, A. A. Kaplyanskii (Eds.), *Nonequilibrium Phonons in Nonmetallic Crystals*, Amsterdam: North Holland, pp. 373–453.

Keldysh, L. V., Sibeldin, N. N. (1986), in: W. Eisenmenger, A. A. Kaplyanskii (Eds.), *Nonequilibrium Phonons in Nonmetallic Crystals*, Amsterdam: North Holland, pp. 455–686.

Kittel, C. (1986), *Introduction to Solid State Physics*, New York: Wiley.

Kolk, B. (1984), in: G. K. Horton, A. A. Maradudin (Eds.), *Dynamical Properties of Solids*, Vol. 5, Amsterdam: North-Holland, pp. 1–332.

Kress, W. (1991), in: W. Kress, F. W. de Wette (Eds.), *Surface Phonons*, Berlin: Springer, pp. 209–292.

Lax, M., Hu, P., Narayanamurti, V. (1981), *Phys. Rev. B* **23**, 3095–3097.

Lovesey, S. W. (1977), in: S. W. Lovesey, T. Springer (Eds.), *Dynamics of Solids and Liquids by Neutron Scattering*, Berlin: Springer, pp. 1–75.

Maradudin, A. A. (1966), in: F. Seitz, D. Turnbull (Eds.), *Solid State Physics*, Vol. 18, New York: Academic, pp. 273–420.

Maradudin, A. A., Montroll, E. W., Weiss, G. H.,

Ipatova, I. P. (1971), *Theory of Lattice Dynamics in the Harmonic Approximation*, in: F. Seitz, D. Turnbull (Eds.), *Solid State Physics*, Supplement 3, New York: Academic.

Maradudin, A. A. (1974), in : G. K. Horton, A. A. Maradudin (Eds.), *Dynamical Properties of Solids*, Vol. 1, Amsterdam: North-Holland, pp. 1–82.

Markiewicz, R. S., Greenstein, M., Wolfe, J. P. (1980), *Solid State Commun.* **35**, 339–344.

Nizzoli, F., Sandercock, J. R. (1990), in: G. K. Horton, A. A. Maradudin (Eds.), *Dynamical Properties of Solids*, Vol. 6, Amsterdam: North-Holland, pp. 281–335.

Reissland, J. A. (1973), *The Physics of Phonons*, New York: Wiley.

Renk, K. F. (1986), in: W. Eisenmenger, A. A. Kaplyanskii (Eds.), *Nonequilibrium Phonons in Nonmetallic Crystals*, Amsterdam: North-Holland, pp. 317–372.

Schiff, L. I. (1955), *Quantum Mechanics*, New York, McGraw-Hill.

Stevens, K. W. H. (1967), *Rep. Prog. Phys.* **30**, 189–226.

Toennies, J. P. (1991), in: W. Kress, F. W. de Wette (Eds.), *Surface Phonons*, Berlin: Springer, pp. 111–166.

Wolfe, J. P. (1989), *Festkörperprobleme* **29**, 75–105.

Ziman, J. M. (1965), *Principles of the Theory of Solids*, Cambridge, U.K. Cambridge Univ. Press.

Further Reading

Ghatak, A. K., Kothari, L. S. (1972), *Lattice Dynamics*, London: Addison-Wesley.

Maradudin, A. A., Montroll, E. W., Weiss, G. H., Ipatova, I. P. (1971), *Theory of Lattice Dynamics in the Harmonic Approximation*, in: F. Seitz, D. Turnbull (Eds.), *Solid State Physics*, Supplement 3, New York: Academic.

Reissland, J. A. (1973), *The Physics of Phonons*, New York: Wiley.

Srivastava, G. P. (1990), *The Physics of Phonons*, Bristol: Adam Hilger.

PHOTOACOUSTIC SPECTROSCOPY

See SPECTROSCOPY, PHOTOACOUSTIC

PHOTOCHEMISTRY

See REACTIONS INDUCED BY LIGHT

PHOTODETECTORS*

SUSAN R. SLOAN, *Hewlett-Packard Company, Lightwave Operation, Santa Rosa, California, U.S.A.*

	Introduction	459
1.	**Physics of Photodetectors**	459
2.	**Materials, Design, and Properties of Photodetectors**	460
2.1	Semiconductor Material Systems	460
2.2	Properties of Photodetectors	461
2.2.1	Dark Current	461
2.2.2	Photocurrent	462
2.2.3	Dynamic Resistance	462
2.2.4	Capacitance	462
2.2.5	Responsivity and Quantum Efficiency	463
2.2.6	Reflectivity	463
2.2.7	Response Time	464
2.2.8	Noise	465
3.	**Types of Photodetectors**	467
3.1	Photoconductors	467
3.2	*p-n* Junction Photodiodes	467
3.3	*p-i-n* Photodetectors	468
3.4	Metal–Semiconductor–Metal (Schottky Barrier) Photodetectors	469
3.5	Avalanche Photodetectors	470
3.6	Arrays	471
3.7	Waveguide Photodetectors	471
3.8	Quantum-Well Photodetectors	472
3.9	High-Speed Detectors	472
3.9.1	Device Types and Trends	472
4.	**Applications in Optical Receivers**	473

INTRODUCTION

An optical communications system requires both a source—laser or light-emitting diode (LED)—and a receiver—a photodetector. Historically, photomultiplier tubes and vacuum photodiodes were available as receivers, but smaller, less expensive, wide-bandwidth semiconductor devices are now widely used. The gas-filled tube devices are limited in bandwidth to around 10 kHz (Melchoir, 1972), and in the case of the vacuum photodiode to relatively high minimum detectable optical powers. The wavelength range over which a device functions is determined by its material structure. Choice of a photodetector is governed by the operating wavelengths of the system, the response to these wavelengths, the speed or frequency response requirements, and the dark current or noise performance of the device.

This article discusses the basic operation of a photodetector, the parameters commonly used for its characterization, semiconductor material systems, design and fabrication techniques for different types of photodetectors, and applications in optical receivers. This first section concentrates on basic photodetector physics. Section 2 covers semiconductor material choices and the properties of photodetectors that need to be characterized to understand their operation. Section 3 focuses on the operation of different types of photodetector devices, the semiconductor fabrication techniques for producing them, and high–speed-device trends. Finally, Sec. 4 examines applications for photodetectors in optical receivers.

1. PHYSICS OF PHOTODETECTORS

The function of a photodetector is to convert incoming modulated light energy into modulated electrical current. Light incident on a semiconductor photodetector is absorbed, and if the photon energy is greater than the band-gap energy, electron–hole pairs

*Material in this article is excerpted from Sloan, Susan R. (1994), "Photodetectors," in: Robert G. Huntsperger (Ed.), *Photonic Devices and Systems*, New York: Marcel Dekker, pp. 171–246.

3-527-28135-5/95/$5.00 + .50

are generated. These carriers, created in a region of high electric field within the depletion region of the device, are swept out by an applied reverse bias resulting in a photocurrent.

In photomultipliers, vacuum photodetectors, and phototubes, electrons are excited by photons over the energy barrier, or work function, of the material and become free electrons in a process called the external photoeffect (Melchoir, 1972). Semiconductor photodiodes operate on a principle known as the internal photoeffect, where the electrons and holes excited by the photon to higher energy levels remain within the material and contribute to photocurrent. The incident photon absorbed by the semiconductor generates a free electron, which is excited from the valence to the conduction band while a hole is generated in the valence band. The carriers created in the depletion region drift in opposite directions as they are separated by the electric field, and carriers created within a diffusion length of the depletion region diffuse to the edges of the region and are collected. Under reverse bias, the carriers' drift in the electric field induces a photocurrent in the load, R_L. The photocurrent is given by

$$I_P = \frac{P_0 q(1 - r')}{h\nu}(1 - e^{-\alpha d}), \quad (1)$$

where P_0 is the incident optical power, q is the electrical charge, r' is the Fresnel reflection at the semiconductor–air interface, $h\nu$ is the photon energy, α is the absorption coefficient of the material, and d is the width of the depletion layer.

Figure 1 illustrates a photon incident on a double-heterojunction *p-i-n* photodiode. The photon passes through the larger–band-gap cap material, shown here as *p* type, and is absorbed in the active layer, the *i* region, below, creating the electron–hole pair. As reverse bias is applied, the electron moves toward the *n* side of the device (cathode) and the hole toward the *p* side (anode), creating a photocurrent.

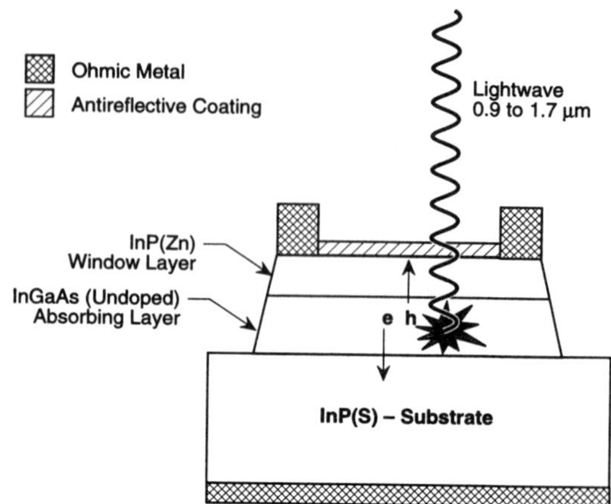

FIG. 1. A cross-sectional view of a *p-i-n* double-heterojunction photodiode illustrating the absorption of the photon, the creation of the electron–hole pair, and their respective motion under influence of the electric field (Sloan, 1989).

2. MATERIALS, DESIGN, AND PROPERTIES OF PHOTODETECTORS

2.1 Semiconductor Material Systems

The photodetector properties are strongly influenced by the choice of the semiconductor material system. For the device active layer, the material band gap is chosen to absorb photons in the wavelength range of interest. Figure 2 illustrates properties of various materials used for optical devices. The left vertical scale shows the energy gap or band gap E_g of materials, in electron volts (eV), the horizontal axis gives lattice constant values in ångströms, and the right vertical scale shows the usable wavelength range in microns. For example, the telecommunications market wavelengths of 1.3 and 1.55 μm can be detected by Ge or InGaAs active-region photodetectors. When InGaAs active regions are used, the III-V material InGaAs is lattice matched to InP substrate material by using 53% In and 47% Ga as the group III element, and As as the group V element, written as $In_{0.53}Ga_{0.47}As$. Germanium is also used for the telecommunications wavelengths, but its response to longer wavelengths, beyond 1500 nm, drops off sharply whereas the curve of response vs wavelength for InGaAs extends to 1700 nm.

The maximum wavelength for which a material will absorb is determined by its band-gap energy—the energy gap between the bottom of the conduction band and the top of the valence band. For wavelengths

$$\lambda_0 > hc/E_g, \quad (2)$$

the material will not absorb, but becomes transparent. The light absorption is depen-

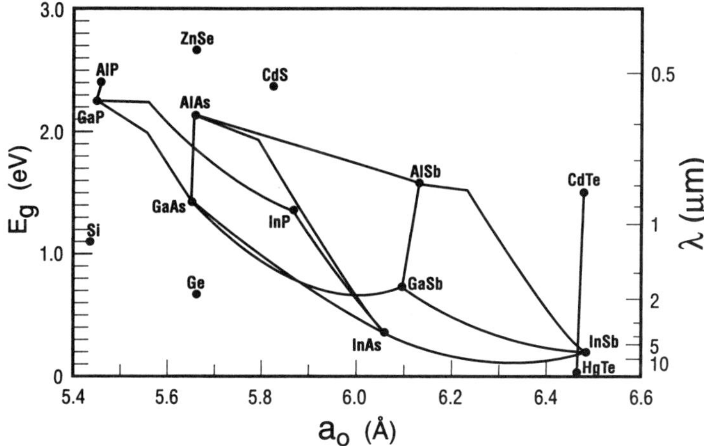

FIG. 2. Common optical-device materials showing their energy band gaps in eV along the left vertical scale, the lattice constants on the horizontal axis, and the usable wavelength range in microns along the right vertical axis. Compositions of the tertiary and quaternary III-V materials may be changed in order to change their detection range as shown by the curves joining the single-III/single-V element compounds (Manasreh, 1993).

dent upon the material absorption coefficient α. In direct–band-gap materials, transitions can occur between the conduction-band minimum and the valence-band maximum with the same wave vector. Indirect materials, where the conduction-band minimum is not at the same value of the wave vector as the valence-band maximum, require a phonon to conserve momentum in addition to the photon of energy $>E_g$. The direct transitions are more probable. The active layer must be thick enough, on the order of $1/\alpha$ (Lucovsky *et al.*, 1964), to absorb the light and create electron–hole pairs. The short-wavelength response cuts off because the absorption coefficient becomes very large at shorter wavelengths; thus absorption occurs near the surface and recombination time is short, and the carriers recombine before they are collected (Sze, 1981).

2.2 Properties of Photodetectors

This section describes important test parameters for photodetectors. A detector's operation is characterized by the wavelength range over which it operates, the response to these wavelengths, the speed of the response, and its noise performance. Low dark current for devices operated under reverse bias and high dynamic resistance for devices operated at zero bias are important for low noise in a receiver system. This section examines sources of noise in photodetectors. For the greatest signal-to-noise performance, responsivity or quantum efficiency should be maximized and reflections minimized. Capacitance, series resistance, and active-layer thicknesses are important considerations for determining the speed of response of the detector.

2.2.1 Dark Current Under reverse bias, the leakage current present in the device in the absence of illumination is referred to as dark current. Low dark current is important for low-noise performance of photodetectors in optical receivers. Dark current of the photodetector is given by the following equation:

$$I_d = I_{d\text{-}G\text{-}R} + I_{d\text{-diffusion}} + I_{d\text{-surface}}, \tag{3}$$

where $I_{d\text{-}G\text{-}R}$ is the dark current resulting from generation and recombination of carriers, $I_{d\text{-diffusion}}$ results from carrier diffusion, and $I_{d\text{-surface}}$ is dark current contributed by surface leakage (Muller, 1981). By testing devices of various sizes, the dominant dark-current mechanism can be determined for a particular device. A current proportional to the circumference of the device, $I_d \propto r$, where r is the radius of the device (assumed to be round), indicates that surface leakage is dominant; current proportional to the area, $I_d \propto r^2$, is indicative of bulk leakage mechanisms (Carey *et al.*, 1989).

Dark-current measurements are performed by probing or making other electrical connection to the device and applying bias voltage while keeping the device in the dark. Mesa-etched devices, where the active area is defined by removing the surrounding epitaxial material and the *p-n* junction is exposed (such as the InGaAs *p-i-n* structure shown in Fig. 11), typically are dominated by surface leakage. Dark currents in diffused-junction diodes are more likely due to bulk currents.

The following equations describe the dark-current terms in more detail (Forrest *et al.*, 1980; Sze, 1981). The generation-recombination current is given by

$$I_{d\text{-}G\text{-}R} = \frac{qn_iAd}{\tau_{\text{eff}}}\left[1 - \exp\left(\frac{-qV_r}{2kT}\right)\right], \quad (4)$$

where q is the charge, A is the photodetector area, d is the width of the depletion region, τ_{eff} is the effective generation-recombination lifetime, V_r is the applied reverse bias, k is Boltzmann's constant, T is the temperature in kelvins, and n_i, the intrinsic carrier concentration, is given by

$$n_i = 2\left(\frac{4\pi^2 m_e m_h k^2}{h^4}\right)^{3/4} T^{3/2} \exp\left[-\frac{E_g}{2kT}\right], \quad (5)$$

where m_e and m_h are the effective masses of the electron and holes, respectively, and h is Planck's constant. The diffusion current is given by

$$I_{d\text{-diffusion}} = \frac{qn_i^2(D_n/\tau_m)^{1/2}A}{N_a}\left[1 - \exp\left(\frac{-qV_r}{kt}\right)\right], \quad (6)$$

where the additional terms in this equation are D_n, the minority-carrier diffusion constant, τ_m, the minority-carrier lifetime, and N_a, the acceptor concentration. When the photodetector is operated under a reverse bias, the temperature dependence of the generation-recombination and diffusion currents is dominated by the n_i terms. That means that

$$I_{d\text{-}G\text{-}R} \propto \exp[-E_g/2kT] \quad (7)$$

and

$$I_{d\text{-diffusion}} \propto \exp[-E_g/kT]. \quad (8)$$

Plots of the logarithm of the dark current vs $1/T$ in kelvins show that the surface leakage-current behavior is due to generation currents believed to be due to traps near the mesa edge (Carey *et al.*, 1989; Forrest *et al.*, 1980).

2.2.2 Photocurrent A photodiode is generally operated under reverse bias in order to increase the width of the depletion region, thereby lowering the capacitance, and making a stronger electric field to increase the drift velocity of the electrons and holes and minimize the carrier transit time (Saleh and Teich, 1991). The photocurrent produced, I_p, is given by

$$I_p = \frac{P_0 q(1 - r')}{h\nu}(1 - e^{-\alpha d}) = \eta q \frac{P_0}{h\nu}, \quad (9)$$

where η is the quantum efficiency discussed in Sec. 2.2.5. The current–voltage (I-V) characteristic for a photodiode is given by

$$I = I_d\left[\exp\left(\frac{eV_0}{n_0kT}\right) - 1\right] - I_p, \quad (10)$$

including the photocurrent portion I_p, and the normal diode diffusion current discovered by Shockley (1949) where I_d is the dark current, n_0 is the diode ideality factor (Sze, 1981)

$$n_0 \equiv \frac{q}{kT}\frac{\partial V}{\partial(\ln I)}, \quad (11)$$

and V_0 is the junction voltage (Lee and Li, 1979).

2.2.3 Dynamic Resistance For photodetectors operated at 0 V bias, the parameter of interest becomes the dynamic resistance, sometimes called shunt resistance, instead of dark current. The zero bias allows for low-noise receiver designs because little or no dark current is present. Dynamic resistance R_d is measured by examining the I-V curve of a device and taking the slope of the curve ($\Delta V/\Delta I$) around the zero-volt region. R_d is controlled by the steepness of the forward voltage characteristic of the diode vs current and the reverse leakage current of the device. For the case of only thermal (Johnson) noise,

$$\langle i_t^2 \rangle = 4kTB/R_d, \quad (12)$$

where i_t is the thermal noise current and B is the bandwidth of the measurement. For low-noise operation, R_d should be maximized. Typical dynamic resistance values for 500-μm–active-area InGaAs *p-i-n* photodiodes are ~100 MΩ.

2.2.4 Capacitance Measuring a photodetector's capacitance vs voltage is a way to look at the degree to which the active region of the device is depleted of carriers. For a photodiode, the junction capacitance is given by

$$C = A\left[\frac{2}{q\epsilon}(\Phi + V_a)\left(\frac{1}{N_a} + \frac{1}{N_d}\right)\right]^{-1/2} = \frac{\epsilon A}{d} \tag{13}$$

(Lee and Li, 1979), where A is the junction area, q is the charge, ϵ is the dielectric constant of the i-layer material, Φ is the built-in voltage, V_a is the applied voltage, N_a and N_d are the acceptor and donor concentrations, respectively, and d is the width of the depletion region. Capacitance can be reduced by lowering the doping in the i layer or by increasing the applied voltage. Minimizing capacitance for a p-i-n photodetector means minimizing device active area and using a thick, low-doped i layer. The RC time constant of the device, series resistance times capacitance, is key in determining the frequency response of the device. This must be traded off against the increased carrier transit time due to a thicker i layer.

2.2.5 Responsivity and Quantum Efficiency Responsivity is a measure of photodetector sensitivity to incoming light energy. To be most efficient, the photodetector should turn as many of the incoming photons as possible into detectable photocurrent. Responsivity is measured by illuminating the device with a known optical power at a known wavelength, and measuring the magnitude of the resulting photocurrent at a particular bias value; the units are amperes/watt.

The thickness of the p-i-n device i layer determines the response. If the i layer is not thick enough to absorb the incoming photons, the response will be less than the maximum calculated value. The absorption coefficient is given as a function of wavelength. For InGaAs at the wavelengths of 1.06 and 1.3 μm, the absorption coefficients are $\alpha(1.06\ \mu\text{m}) = 1.8\ \mu\text{m}^{-1}$ and $\alpha(1.3\ \mu\text{m}) = 1.15\ \mu\text{m}^{-1}$. At a thickness of $1/\alpha$, the light intensity falls to $1/e$ of the original value. The fraction of power absorbed in a depth d is given by

$$P_{\text{abs}} = P_0[1 - \exp(-\alpha d)]. \tag{14}$$

The photon energy must be greater than the semiconductor band gap in order for absorption to occur and carriers contributing to photocurrent to be produced.

The number of electron–hole pairs created by a single incident photon is called the quantum efficiency,

$$\eta = (I_p/q)(h\nu/P_0), \tag{15}$$

where I_p is the photocurrent, q is the magnitude of the charge on a carrier, and P_0 is the optical power at the wavelength corresponding to the photon energy given by $h\nu$. Some authors and manufacturers quote responsivity values while others quote quantum efficiency as the measure of the device optical sensitivity. Quantum efficiency is related to responsivity R by

$$\eta = \frac{1}{q}\frac{hc}{\lambda}R = \frac{1.24}{\lambda}R. \tag{16}$$

By definition, the maximum value of η is 1. Maximum response values for a particular wavelength may be found by setting $\eta = 1$ and finding R.

Responsivity or quantum efficiency vs wavelength is measured using a spectrometer and large–wavelength-range sources. The power is monitored by using an optical coupler to sample a portion of the light energy, which is incident on a power meter. The photocurrent from the photodetector is measured and recorded at each wavelength. The resulting plot for two different InGaAs devices is shown in Fig. 3. One is a conventional p-i-n with an InP cap layer, and one has been modified to extend its range to shorter wavelengths. Here, the response was measured from 600 to 1700 nm. Figure 4 shows responses for a variety of material systems (Melchoir, 1972; Sze, 1981) allowing the receiver designer to select the material appropriate for her application. Maximizing response in a receiver system is a way of achieving high signal-to-noise performance.

2.2.6 Reflectivity If the surface of the photodetector is highly reflective, a large part of the incoming light will not be absorbed and not result in the creation of electron–hole pairs, or photocurrent. This will limit the device responsivity. The reflected light may also be coupled back into the light source, perturbing it. This is highly undesirable for laser sources. To prevent this, an antireflective

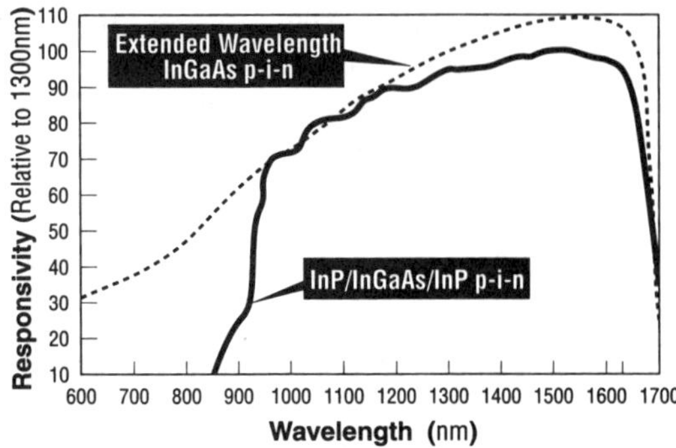

FIG. 3. Plot of dc responsivity vs wavelength for two InGaAs devices. The solid line illustrates a standard InP/InGaAs/InP response, and the dashed line shows an extended-wavelength InGaAs device. Both are used for the long telecommunications wavelengths (measurements courtesy of Mike McClendon and Norm Brewer, Hewlett-Packard Lightwave Operation Standards Lab).

coating is applied to the device surface. The goal is to minimize the percentage of optical power reflected or maximize the optical return loss (ORL). Reflection is measured as a percentage of incident light reflected or in optical receivers as the figure of merit optical return loss (ORL), which is in units of dB and is given by

$$\mathrm{ORL} = 10 \log(P_0/P_r), \tag{17}$$

where P_0 is the incident power and P_r is the reflected power. With no coating, a semiconductor surface reflects about 30% of the light that is incident upon it.

Antireflective coatings are usually made of dielectric materials deposited on the surface of the device. To first approximation, a quarter wavelength of dielectric material is deposited. A plot of the modeled power reflected vs wavelength is shown in Fig. 5 for a single-layer antireflection coating of silicon nitride on an InP/InGaAs/InP photodetector. If optimized for InP/InGaAs/InP devices, dielectric coatings can lower the reflected light to on the order of 0.5% at 1300 and 1550 nm (Braun, 1988). To improve the coating further, multiple layers are used. To prevent the reflected light from the detector from coupling back to the laser source, the device is usually tilted. A 6- to 8-degree angle of the device relative to the normal will prevent light from being coupled back into an optical single-mode fiber from the device.

2.2.7 Response Time Photodetector response time is governed by the transit time of carriers in the active region and the *RC* time constant of the device. The carrier transit time is the distance the carriers travel across the *i* layer divided by the velocity at which the carriers travel. Both electrons and holes are present, and each has a saturation velocity at which it travels:

$$t = d/v_{sat}. \tag{18}$$

To maximize the speed of response, the transit time needs to be minimized. This means making the active layer thin. By making the active region thin, the capacitance of the de-

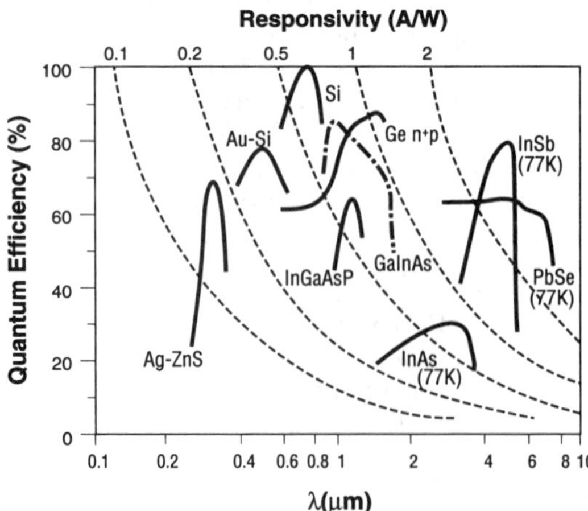

FIG. 4. Response curves for various detector materials (Sze, 1981).

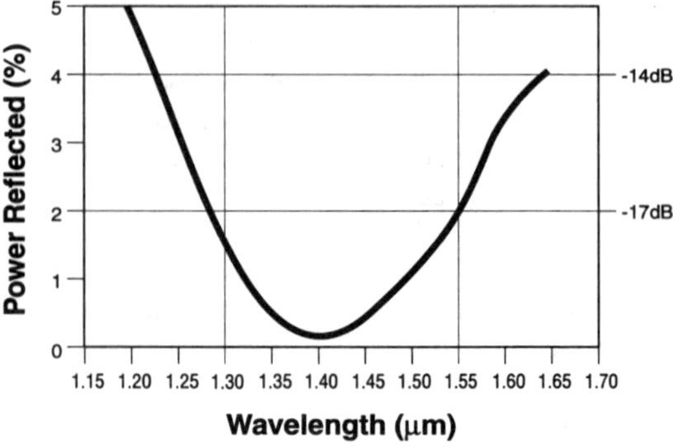

FIG. 5. Reflection vs wavelength characteristic for modeled single-layer antireflection coating for the telecommunications wavelength range (Beck, 1993).

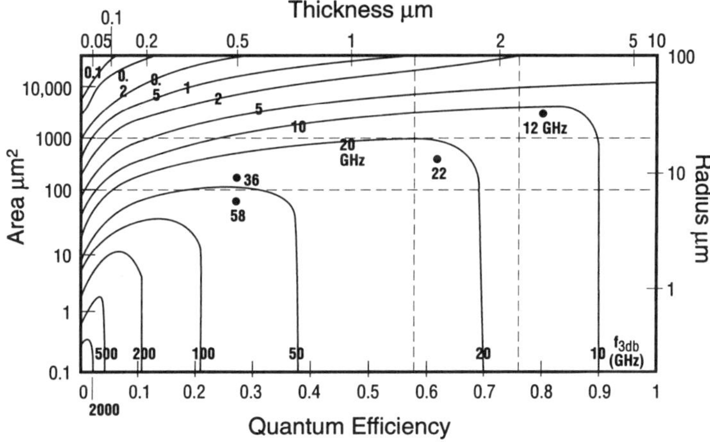

FIG. 6. Frequency-response contours vs active-area diameter, active area, and quantum efficiency (Bowers *et al.*, 1986a).

vice is increased and the quantum efficiency of the device is decreased. The increasing capacitance causes the *RC* time constant to rise and the speed is slowed. Tradeoffs must be made between transit time and *RC* considerations in order to optimize the photodetector speed of response. These tradeoffs have been modeled for InP/InGaAs/InP p-i-n devices and are plotted in Fig. 6 (Bowers *et al.*, 1986) where the device area, *i*-layer thickness, and active-area diameter are examined and frequency contours are established.

2.2.8 Noise Low-noise performance is important for photoreceivers where low power levels need to be detected. The minimum detectable optical signal is determined by the photodetector quantum efficiency and the noise sources present in the receiver. Sources of noise in photodetectors include shot noise and thermal or Johnson noise. Shot noise consists of noise due to dark current, surface leakage current, and quantum effects (Lee and Li, 1979). Assuming dark current and surface leakage are negligible by design, shot noise is due to the fluctuations in the photoexcitation of carriers. It is present whenever current is flowing, as a result of fluctuations in the arrival rate of electrons which in turn lead to random changes in the current flow (Wilson and Hawkes, 1989). This is the lowest noise achievable in the system. For devices with gain such as avalanche photodetectors (APDs), excess shot noise exists due to the multiplication process. Thermal noise from the load resistor in a receiver limits noise in a device without internal gain.

A simple model of a photodetector receiver is shown in Fig. 7. The photodiode is modeled as a current source with a small series resistance R_s due to the resistance of the p contacting layer and the metal–semiconductor resistance, a junction capacitance C_j, and a load resistance R_L. The amplifier capacitance and resistance are labeled as C_a and R_a respectively. When a time-varying optical power is incident, the resulting photocurrent is

$$i_{ph}(t) = \eta q p(t)/h\nu, \tag{19}$$

where η is the quantum efficiency, q the electrical charge, and $p(t)$ the modulated incident power, and $h\nu$ is Planck's constant times the frequency (Lee and Li, 1979). The term $i_{ph}(t)$ has a dc component

$$I_p = \eta q P_0/h\nu \tag{20}$$

(Sze, 1981), and a time-varying current, or signal component, i_p. For a device with gain,

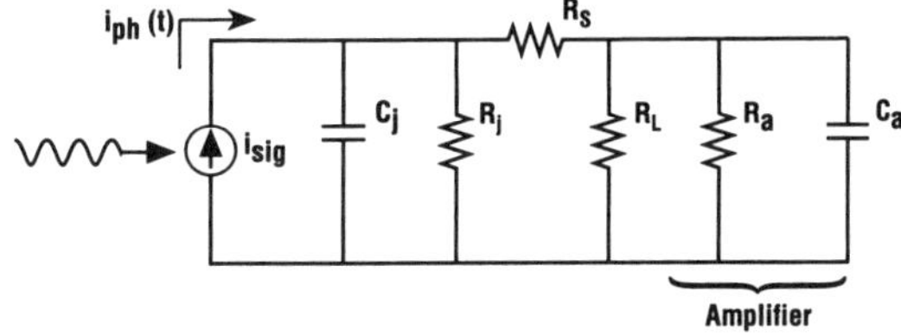

FIG. 7. Model of a photodetector as a current source with a junction resistance, junction capacitance, series resistance, resistance due to the load, R_L, and the resistance and capacitance due to the amplifier following the photodiode.

the signal current is multiplied by the internal avalanche gain. The average value of the signal current is then

$$\langle t_{sig}\rangle = i_p\langle m\rangle, \tag{21}$$

where $\langle m\rangle$ is an average of the varying multiplication factor. For devices without gain such as *p-i-n*, $\langle m\rangle = 1$.

The shot-noise current is determined by the dc component of the photocurrent, I_p, and is given by

$$\langle i_{sp}^2\rangle = 2qI_p\langle m^2\rangle B, \tag{22}$$

where $\langle m^2\rangle$ is the mean square multiplication factor and B is the bandwidth of the noise measurement. Noise-current terms are commonly written as mean squares ($\langle i^2\rangle$). The shot noise from photocurrent and dark current can be combined into a single expression. Dark currents can arise from different sources. Electrons and holes thermally generated in the *p-n* junction of a device contribute to bulk dark current (Keiser, 1983) and currents originating from surfaces due to mesa conditions, trapping, or surface recombination are referred to as surface leakage currents. The bulk currents I_{db} are subject to the avalanche gain mechanism, while the surface currents I_{ds} are not. The shot-noise expression becomes

$$\langle i_{spd}^2\rangle = 2q(I_p + I_{db})\langle m^2\rangle B + 2qI_{ds}B. \tag{23}$$

Random current fluctuations due to background radiation, stray light in the system striking the device active area and being collected by the photodetector, denoted here as I_b, can also be included in the shot-noise term (Sze, 1981):

$$\langle i_{spdb}^2\rangle = 2q(I_p + I_{db} + I_b)\langle m^2\rangle B + 2qI_{ds}B. \tag{24}$$

Thermal noise arises from all resistances present in the system. The series resistance is usually small compared with the other resistance values in the system and will be neglected. The equivalent resistance for the parallel combination is given by (Sze, 1981)

$$\frac{1}{R_{eq}} = \frac{1}{R_j} + \frac{1}{R_L} + \frac{1}{R_a}, \tag{25}$$

where R_j is the junction resistance. The mean square thermal, or Johnson, noise is

$$\langle i_T^2\rangle = 4kTB/R_{eq} \tag{26}$$

(Sze, 1981), where k is Boltzmann's constant, T is the temperature in kelvins, and B is the bandwidth of the measurement.

In a receiver, the signal-to-noise ratio determines its sensitivity and low-power detection capabilities. The signal-to-noise ratio is defined as the signal due to the photocurrent divided by the mean squared noise components from the photodetector and the amplifier. It is important to have high quantum efficiency in the detector and low noise in the amplifier in order to maximize receiver sensitivity. The signal-to-noise ratio is then

$$\frac{S}{N} = \frac{\langle i_p^2\rangle\langle m^2\rangle}{2q(I_p + I_{db} + I_b)\langle m^2\rangle B + 2qI_{ds}B + 4kTB/R_{eq} + \langle i_{amp}^2\rangle}, \tag{27}$$

where i_{amp} is the amplifier current. The minimum detectable power associated with a particular S/N ratio can be found [for $m = 1$, $I_{ds} + I_{db} = I_d$, i_{amp} negligible, and $\langle i_p^2\rangle = \frac{1}{2}(nqP_0/h\nu)^2$] (Sze, 1981) as

$$\frac{S}{N} \simeq \frac{\frac{1}{2}(q\eta P_0/h\nu)^2}{2q(I_p + I_d + I_b)B + 4kTB/R_{eq}} \tag{28}$$

and

$$P_{omin} = \frac{2h\nu B}{\eta}\left(\frac{S}{N}\right)\left[1 + \left(1 + \frac{I_{eq}}{qB(S/N)}\right)^{1/2}\right], \tag{29}$$

where

$$I_{eq} = I_d + I_b + 2kT/qR_{eq}. \tag{30}$$

If the receiver is in the shot-noise–limited regime, the dark-current shot noise sets the minimum detectable power limit. For $m = 1$, this is

$$P_{omin} = \frac{2qI_dB}{R} = \frac{2hcI_dB}{\eta\lambda} \tag{31}$$

in units of watts.

If the background radiation or thermal noise is the dominant noise source, a figure of merit for the receiver is the noise-equivalent power (NEP). The NEP is the rms optical power ($P_0/\sqrt{2}$). To calculate NEP, Eq. (29) is used where the bandwidth is assumed to be 1 Hz, $S/N = 1$, and $I_{eq}/qB(S/N)$ is assumed to be $\gg 1$:

$$\text{NEP} = (\sqrt{2}h\nu/\eta)(I_{eq}/q)^{1/2} \tag{32}$$

in units of $\text{W/cm}^2\ \text{Hz}^{1/2}$.

3. TYPES OF PHOTODETECTORS

The choice of a photodetector for a specific optical receiver application depends on the performance and cost tradeoffs for the design. The wavelength range of interest will determine the material used. More mature technologies with simple fabrication processes in Si, and Ge, and GaAs usually mean lower-cost devices with adequate performance. High-performance devices for the telecommunications wavelengths of 1300 and 1550 nm typically have more involved fabrication processes and more difficult-to-control dark currents. Speed of response, noise, quantum efficiency, and gain are parameters to consider in the choice of a photodetector.

3.1 Photoconductors

A photoconductor is fabricated by placing Ohmic contacts onto a semiconductor material. When the material is illuminated, electron–hole pairs are produced and the electrical conductivity of the material increases. The greater the photon flux, the greater the increase in conductivity (Saleh and Teich, 1991, Sze, 1981). When voltage is applied to the contacts, the electrons and holes move in opposite directions resulting in a photocurrent. The anode and cathode may be placed on opposite ends of the semiconductor or both on top in an interdigitated pattern as illustrated in Fig. 8.

Photoconductive detectors have photoconductive gain defined as the flow of electrons per second from the device divided by the rate of generation of electron–hole pairs within the device (Wilson and Hawkes, 1989). The device gain is expressed as

$$G = \tau_m/\tau_e, \tag{33}$$

where τ_m is the excess carrier recombination lifetime and τ_e is the electron transit time ($\tau_e = 1/\nu_e$) (Saleh and Teich, 1991). Gain may be larger than unity.

Photoconductive detectors have the advantages of photoconductive gain and ease of fabrication over *p-n* or *p-i-n* devices, but lack their speed capabilities. The transit time in a photoconductor is set by the peak velocity of the fastest carrier, as contrasted with the case of *p-i-n*'s where it is limited by the slowest carrier. In III-V materials where the velocity of electrons is greater than the velocity of holes, and low-field electron velocity is greater than the high-field peak saturation velocity, high gain–bandwidth products can be obtained, but the *p-i-n* (with no gain) has much higher bandwidth (Forrest, 1988).

3.2 *p-n* Junction Photodiodes

The *p-n* junction photodiode is formed by doping layers of semiconductor material *n* type and *p* type. When a photon is absorbed, it creates an electron–hole pair as in the photoconductive device. Here, the electric field is formed only in the depletion region of the

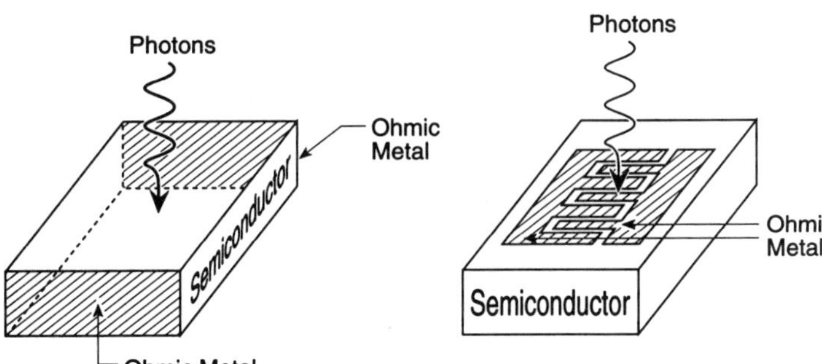

FIG. 8. Two designs for photoconductors showing Ohmic contacts on the ends of a semiconductor slab, and interdigitated contacts on the surface of a semiconductor.

device, and only carriers generated there, or near there (within a diffusion length), have the ability to be moved in a specific direction. The device is operated under reverse bias and the electrons are moved toward the n region while the holes drift to the p region under the influence of the electric field.

Response time of a p-n junction device is faster than a photoconductive detector as a result of the strong electric field in the depletion region giving carriers a larger velocity. The wider depletion region of the p-i-n photodetector makes it a better choice for high-speed devices.

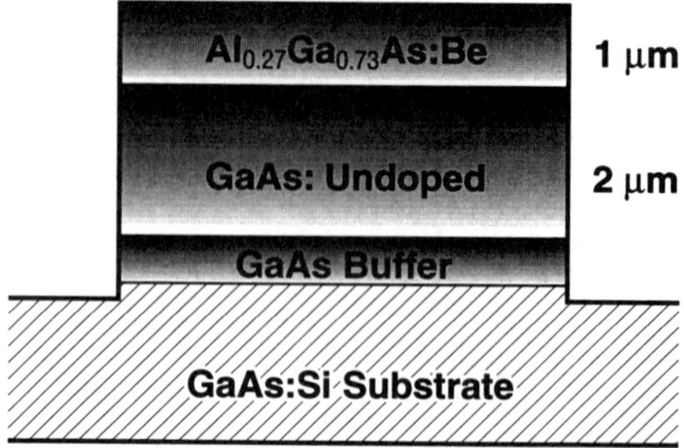

FIG. 9. The epitaxial layer structure for an AlGaAs/GaAs heterojunction p-i-n detector.

3.3 *p-i-n* Photodetectors

A p-i-n detector is typically formed by growing an intrinsic (lightly doped) layer on top of a highly doped n^+ substrate and growing a p^+ layer on top of the intrinsic layer or diffusing the p dopant into the top layer. The detector structure will also work in reverse, that is, growing an i layer and an n^+ layer on a p^+ substrate. The p-i-n has advantages over the p-n junction device in that it has a wider depletion region (thickness of i layer) and, therefore, a lower capacitance and lower RC time constant. The transit time, however, will increase (Saleh and Teich, 1991). High bandwidths, in excess of 100 GHz (Wey, 1993), are achievable with p-i-n photodetectors.

When the p-i-n device is formed out of a single semiconductor material, as in silicon p-i-n's, the device material is homogeneous and it is referred to as a homojunction device. A photodetector composed of two different semiconductor materials is referred to as a heterojunction device. An example of this structure is the $Al_{0.27}Ga_{0.73}As$/GaAs device for use from about 680 to 880 nm. This structure, which can have high bandwidths (>22 GHz) and high external quantum efficiency (73%) (Johnsen *et al.*, 1991), is shown in Fig. 9 and can be grown using molecular beam epitaxy (MBE). The higher–band-gap material sits on top where light incident at

$$\lambda_0 > hc/E_g \tag{2}$$

is not absorbed, but passes through to the intrinsic layer, which, under bias, is depleted of carriers as they are continuously swept out. This top transparent layer, the heterojunction window layer of higher–band-gap material, allows for the photons to pass through and be collected in the i layer below. A double heterojunction is formed when there are two junctions of differing semiconductor materials as in the InP/$In_{0.53}Ga_{0.47}As$/InP detector material system (Wang *et al.*, 1987).

p-n or p-i-n junction photodiodes can be formed using a planar diffusion process. These diffused devices can have high response and very low dark current or high dynamic resistance for low-noise receiver performance.

The fabrication process for InP/InGaAs/InP diffused heterojunction devices will be used to illustrate a typical planar device process (Beck, 1993). The epitaxial structure used is an InP buffer layer on the InP:S (InP sulfur doped) substrate, followed by the InGaAs active layer, which has a low (1×10^{15} cm^{-3}) background doping, and then the window layer of InP, which will be diffused with the p-type dopant zinc. To form the diffusion barrier, silicon nitride is deposited on the wafer, patterned using a photoresist mask, and removed from the areas where Zn is diffused into the semiconductor. The diffusion depth is targeted to be through the InP window layer and into the InGaAs layer beneath. An antireflection coating is used to provide low reflections from the top surface and also to act as passivation for the device. The p contact is formed by etching away a small region of the antireflection coating and depositing Ti/Pt/Au metallization by a liftoff process. A bonding pad can then be formed by gold plating. The wafer is lapped and polished to provide the desired chip thickness and polish on the back side. The n-side metal contact (AuGe alloy) is deposited on the back side of the wafer on the InP:S substrate. A cross sec-

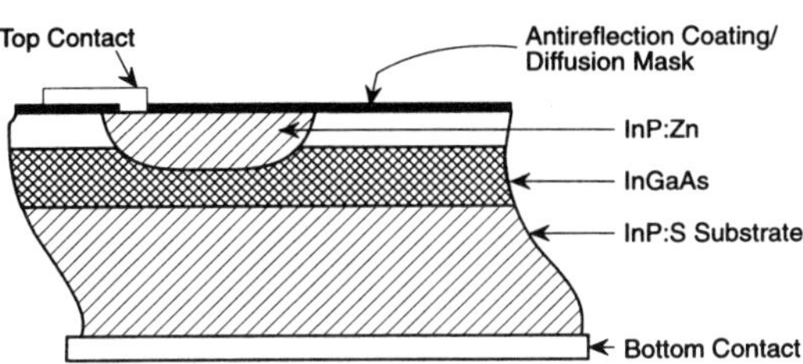

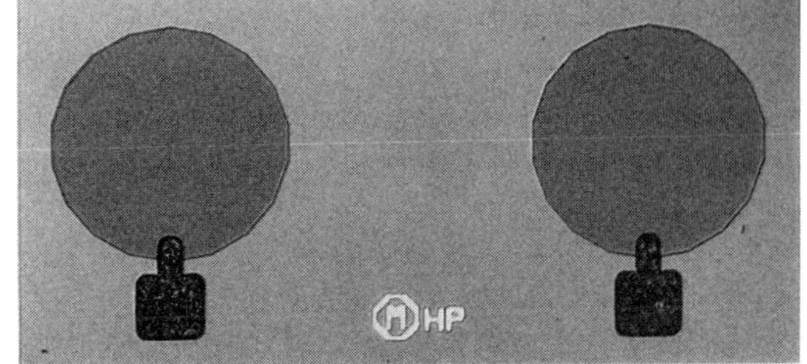

FIG. 10. (Left) Cross section of a diffused InGaAs detector showing the double-heterojunction material structure, Si_xN_y diffusion mask, AR coating, *p*-metal bondpad, and back-side *n* metals. (Right) Photograph of a pair of diffused detectors fabricated in this way (Sloan, 1989).

tion of the device and a photograph of a pair of diffused detectors are shown in Fig. 10.

Photodetector mesa structures provide a low-capacitance design for high-frequency performance. Mesa devices can be front-side or back-side illuminated. They can have grown-in doped layers or diffused layers (Ohnakea *et al.*, 1987) with mesas etched outside the diffused region. Mesa devices often have higher dark currents than diffused-junction devices because of the exposed junction surface.

A typical process for a *p-i-n* structure (Sloan, 1989) with a grown-in *p* layer involves deposition of the metal contact on the *p* layer by a lift-off process, deposition of an antireflection coating, etching the mesa with wet-chemical etches or by reactive-ion etching, deposition of the *n*-metal contact, and passivation of the mesa surface. A cross section and scanning electron micrograph of the device are shown in Fig. 11.

3.4 Metal–Semiconductor–Metal (Schottky Barrier) Photodetectors A thin metal layer deposited on a semiconductor can form a Schottky barrier (see SCHOTTKY BARRIERS). This metal–semiconductor contact can act as a *p*- or *n*-type layer forming a *p-n* junction photodetector. If made thin enough, the metal is transparent to the incident light, and a depletion region is formed at the semiconductor surface. An example of a Schottky or metal–semiconductor device fabricated in the GaAs material system is given as follows (Wang and Bloom, 1983). An n^+-GaAs layer and an *n*-GaAs layer are grown on a semi-insulating GaAs substrate. The area under the Schottky and Ohmic contacts is left active and conductive, while the surrounding area is isolated by proton bombardment, which destroys the ordered atomic lattice. This also minimizes the capacitance, which is necessary for high-speed performance. The Ohmic contact is placed on the n^+ layer after etch-

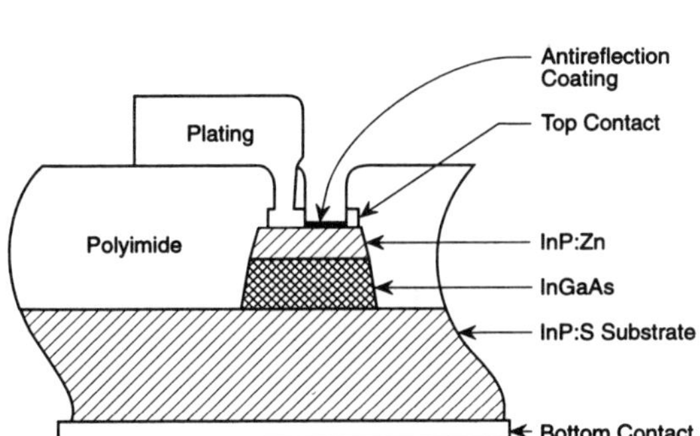

FIG. 11. (Left) Cross section of a mesa InGaAs detector showing the fabrication process steps—*p* metal, AR coating, wet-etched mesa, polyimide passivation, bond pad, and back-side metal. (Right) Photograph of a mesa *p-i-n* using this process (Sloan, 1989).

ing away the n layer of GaAs. A AuGe-type Ohmic contact to the n^+ layer is deposited using a liftoff process. The transparent contact to the n-GaAs layer is made by depositing a thin layer (100 Å) of platinum and patterning using photoresist and etching with an Ar ion mill. The light-absorbing area is covered with a dielectric passivation. For a small-area device, $5 \times 5\ \mu m^2$, a pulse response of 5.4 ps full-width-half-maximum (FWHM) or >100 GHz 3-dB frequency response has been demonstrated (Wang and Bloom, 1983).

Metal–semiconductor–metal (MSM) photodetectors commonly use interdigitated metal fingers to form the two electrical contacts for the device. These devices are attractive because of the low capacitance of the small active areas and the short transit time of the carriers between the electrodes. MSMs typically have lower responsivity or quantum efficiency than *p-i-n* devices because of the large amount of the incident light area covered by metal. Response can also be reduced by using thin *i* layers to try to minimize the transit-time effects of carriers needed to achieve high bandwidths. This can be a problem for *p-i-n* devices as well. The broad bandwidth possibilities and the ease and compatibility of processing with existing field-effect transistors (FETs) make MSMs a good choice for integrated optical devices. MSMs are being fabricated in GaAs and InGaAs material systems (Klingenstein *et al.*, 1992; Zirngibl *et al.*, 1989; Van Zeghbroeck *et al.*, 1988).

3.5 Avalanche Photodetectors

The avalanche photodetector (APD) operates under a strong reverse bias so that the electric field strength across the region between the p-type and n-type layers is large. When an incoming photon is absorbed, a cascading of carriers is created by a process known as impact ionization. This avalanche process of carriers produces carrier multiplication, or gain, in the device.

The process of impact ionization or multiplication that takes place in an APD is illustrated in Fig. 12 (Saleh and Teich, 1991). An incoming photon absorbed at point A creates an electron–hole pair, where the electron resides in the conduction band and the hole in the valence band. Under the strong electric field provided by the high reverse bias, the electrons are accelerated toward the n side of the device and the holes toward the p side. As the electron is accelerated, it gains energy with respect to the bottom of the conduction band and undergoes collisions with the semiconductor lattice, losing some of its energy. The electron reaches an average saturation velocity balancing the increase in velocity due to the acceleration in the electric field and decrease in velocity experienced in the random lattice collisions. An electron with energy greater than the band-gap energy E_g can generate another electron–hole pair as shown at point B in Fig. 12 by kicking an electron into the conduction band and leaving a hole in the valence band (Pearsall and Pollack, 1985), giving up energy in the process. The energy required is greater than E_g because of the energy used in the three-body collision (Capasso, 1985). This process can now continue with the two electrons each having the possibility of creating more electron–hole pairs and causing the cascading effect. The holes are likewise accelerated toward the p side of the device and can undergo impact ionization spawning more electron–hole pairs (see point C in Fig. 12).

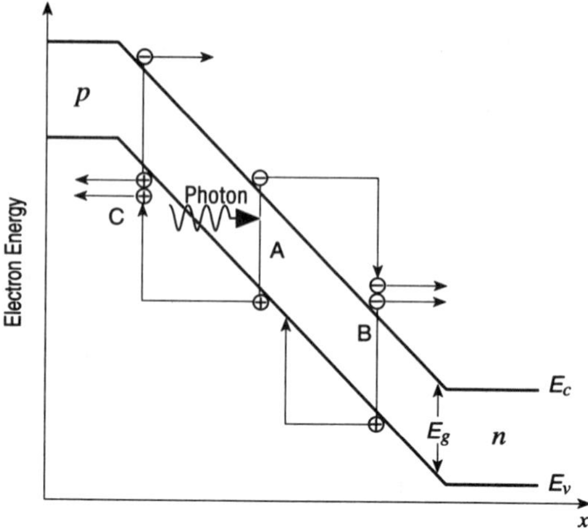

FIG. 12. The process of multiplication by impact ionization in an APD. A photon incident at point A creates an electron in the conduction band and a hole in the valence band. In a high electric field, the electron gains enough energy with respect to the bottom of the conduction band to generate another electron–hole pair at point B. Holes can likewise generate carrier pairs as shown at point C (Saleh and Teich, 1991).

Novel device designs are used to optimize APD performance. To increase quantum efficiency, the absorption layer of the APD needs to be thick enough to absorb the majority of the carriers incident, but to minimize the chance for the inclusion of an area of potential instability or area of uncontrolled avalanche (microplasmas), the multiplication region needs to be as small as possible. Various device structures have been designed in an attempt to reconcile these competing tendencies. The separate-absorption-multiplication (SAM) APD uses separate semiconductor layers to perform the necessary device functions and is able to optimize the features of each by maintaining spatially separate regions. Guard-ring structures are a way of lessening leakage currents, which can limit multiplication, by having a ring of low-doped, high-breakdown material around the active region of the device. Guard-ring structures are used in Si, Ge, and III-V devices.

For wavelengths from 0.8 to 1.0 μm, silicon is the material used for APDs. In the longer-wavelength range of 1.0–1.55 μm, germanium is used as well as III-V materials. In III-V semiconductors, because of the high leakage due to tunneling in homojunction APDs, heterojunction devices are used. Gain-bandwidth products of 100–130 GHz have been reported for various configurations.

Response times in APDs are limited by the same factors affecting *p-i-n*'s—transit times in the depletion region, *RC* time constants, diffusion times of carriers in the nearby undepleted regions—and in addition by avalanche buildup time.

In addition to the noise contributions listed in Sec. 2.2.8 for shot noise and thermal noise, another parameter of interest for APDs is multiplication noise described by the excess noise factor.

3.6 Arrays

Photodetector arrays are used in imaging and in spectroscopy. Detectors are placed close together, and light strikes many of them simultaneously. In spectroscopy, the light is dispersed into its various wavelengths before it is incident on the detector array. A current resulting from illumination of a particular array element indicates that light of that wavelength is present corresponding to a chemical signature of a substance under test. Such arrays have been fabricated in the InGaAs material system having over 500 elements sensitive to wavelengths from 600 to 1700 nm. A type of two-dimensional array called a charge-coupled device (CCD) is formed using a series of Schottky photodiodes and can be used to do visible or infrared imaging (see CHARGE-COUPLED DEVICES).

3.7 Waveguide Photodetectors

Detectors where the input light is coupled to the edge of the device are known as waveguide photodetectors. In conventional top-illuminated *p-i-n* detectors, a tradeoff exists between the speed of response and the quantum efficiency of the device. The *i* layer is made thin to minimize transit time and maximize frequency response, but thinning the *i* layer means there may not be enough absorption depth for a large fraction of the light to be absorbed and the quantum efficiency or responsivity is not optimized. In an edge-illuminated device, the current flow and the light absorption occur in perpendicular directions and both the speed and quantum efficiency can be high (Bowers and Burrus, 1986). Internal quantum efficiencies are determined by the wavelength of light and the absorption coefficient, the *i*-layer thickness, and the length of the chip. External quantum efficiency takes the coupling efficiency into account. Careful light coupling and packaging schemes are needed to get the light focused into the *i* layer of the device. Waveguide devices with 3-dB electrical bandwidths of 50 GHz and quantum efficiencies of 68% for antireflection-coated devices have been fabricated and tested (Kato *et al.*, 1992).

Fabrication processes have used both n^+ and semi-insulating substrates. The epitaxial structure is similar to a laser structure in that it contains waveguiding layers on either side of the active layer, though here the active layer absorbs rather than emits light. A typical epitaxial structure for the InP/InGaAs material system consists of a contact layer (either the n^+ substrate or an InP *n*-type grown layer), an InGaAsP *n*-type cladding layer to aid in the guiding of the mode, an InGaAs absorbing layer, an InGaAsP top guiding layer (p^+ or undoped), and *p* contacting layers (InP *p* type with an InGaAs and/or InGaAsP contacting layer on top to lower the contact resistance). A mesa or ridge is etched down through the active layer, and the mesa is passivated with polyimide. After a via hole is etched to contact the *p* layer, *p* contacts are formed on the top of the device by liftoff techniques. The wafers are thinned by lapping to facilitate cleaving. On the semi-insulating substrate, devices are etched down to the *n* layer and the *n* metal is deposited, and on the n^+-substrate devices the *n* metal is placed all over the back side of the device. Devices are cleaved and antireflection coated.

3.8 Quantum-Well Photodetectors

Quantum-well photodetectors take advantage of band-gap engineering in order to control their wavelength detection range and response. They depend on precisely controlled, often thin, alternating epitaxial layers grown by solid-source MBE, organometallic vapor-phase epitaxy, or gas-source MBE (Ritter *et al.*, 1991). Rapid switching between materials in order to achieve abrupt interfaces on an atomic scale (layers on the order of tens of ångströms) with well-characterized composition and doping profiles is necessary in order to grow the quantum wells (Manasreh and Brown, 1993). Electron confinement in these thin layers, caused by the conduction- or valence-band offset due to the difference in band-gap energies at the heterojunction interfaces, shows quantum effects where continuous energy levels become discrete (Dingle, 1987). A thin layer of small–band-gap material is placed between layers of higher–band-gap material forming a potential well where the electron or hole energy states are quantized (particle in a box). These discrete energy levels can be varied by adjusting the thickness and depth (band discontinuity) of the wells. More bound states with closer energy levels are found in wider quantum wells. Multiple quantum-well (MQW) devices are formed by growing stacked wells with barrier layers between them on the wafer. As the barrier layers become thinner than about 50 Å, the structure is referred to as a superlattice. Materials that are grown lattice mismatched to change the band structure of the device purposely are known as strained layers (Manasreh and Brown, 1993).

Detection at the long infrared wavelengths (3–14 μm) is achieved by using such quantum-well devices. Applications at these long wavelengths are for imaging with linear arrays, medical imaging, global and space surveillance, and astronomy. Information on other types of infrared detectors can be found in the article SENSORS, INFRARED.

3.9 High-Speed Detectors

As optical data rates increase in both the data communications and telecommunications markets, high-speed detectors will be necessary for the transportation of that information. 10-Gbit and 20-Gbit systems are now being designed in research environments. 2.488-Gbit SONET (Synchronous Optical Network) systems are already being implemented. Detectors designed for high speed that trade off thin *i* layers for short transit times and thick *i* layers for low capacitance will be needed. Designs that involve low capacitance, low series resistance, and short transit times and still maintain reasonable responsivities will be needed (Bowers and Burrus, 1987). *p-i-n* detectors with bandwidths of 100 GHz have been made in the GaAs and InGaAs material systems (Wang and Bloom, 1983; Wey, 1993). APDs with gain-bandwidth products of 120 GHz have also been demonstrated (Watanabe *et al.*, 1993).

3.9.1 Device Types and Trends As systems push to higher data rates, detector areas are shrunk in an effort to reduce junction capacitance, eventually making the devices difficult to fabricate and illuminate. Parasitic capacitances and inductances become very important to the speed of response of the devices, influencing the flatness and rolloff of the frequency response. Mesa *p-i-n* devices as small as 5 × 5 μm^2 have been fabricated for research applications (Wey, 1993), but would be difficult to reproduce repeatably in a device manufacturing environment. The small mesa size formed by wet etching means that the undercutting would need to be carefully monitored and traded off against the surface etches necessary to achieve low dark currents. Back-side–illuminated devices that can be attached by a solder bump technique known as flip-chip (Katsura *et al.*, 1990; Makiuchi *et al.*, 1993; see also ELECTRONIC COMPONENTS AND ASSEMBLIES: RELIABILITY AND TESTING) are a means of lowering device capacitance by removing the need for a front-side bondpad and lowering parasitics since bond wires are replaced with solder interconnects. In addition to the rf response increase, dc responsivity also increases for the back-side–illuminated flip-chip device over the front-side–illuminated device, where the benefit comes from using a double pass of the light through the *i* layer of the device, thus absorbing a larger fraction of the incident photons and converting them into usable photocurrent.

Optoelectronic integrated circuits (OEICs) are another way of forming interconnections between the devices by fabricating them on a single substrate, often using multiple material regrowths to achieve the layer structures necessary for the varied devices. OEICs using photodiodes as one of their elements have also been fabricated. Performance of the OEICs has not yet met the performance of the individual optical and electrical components, but improvements have been made and performance is beginning to approach that of the hybrid circuits, making OEICs useful for high-speed applications and lower-cost alternatives.

4. APPLICATIONS IN OPTICAL RECEIVERS

An optical receiver converts incoming optical information into an electrical signal that can then be analyzed by conventional electronic means. In long-haul communications systems, the optical signal incident on the photodetector may be weak and distorted, and so many receivers involve the photodetector, or optical-to-electrical converter (O/E), followed by an amplifier to boost the electrical signal to levels where it can be processed by the subsequent electronics. Some applications for receivers, however, involve looking at the source characteristics in the time domain on an oscilloscope where cleanly reproduced pulse response is a major concern. For these applications and those where the optical and electrical signals are at an adequate level, unamplified receivers are the choice (King *et al.*, 1993). Figure 13 shows a photograph of two such *p-i-n*–only photoreceivers attached to the test set of a high-speed oscilloscope. Desired characteristics of optical receivers are high sensitivity, low noise, wide dynamic range [ratio of maximum to minimum input signals (Senior, 1992)], and adequate bandwidth. These are necessary for long-haul systems where long repeater spacing and high capacity are important and, along with low cost, for local-area networks, local loops, cable TV, and data links (Williams, 1991).

In a digital system, a stream of data, or bits, is sent electrically as a series of ones and zeros. An electric pulse is applied to a laser or light-emitting diode (LED), and a series of optical pulses is transmitted down an optical fiber where it is attenuated and distorted somewhat before it impinges upon the pho-

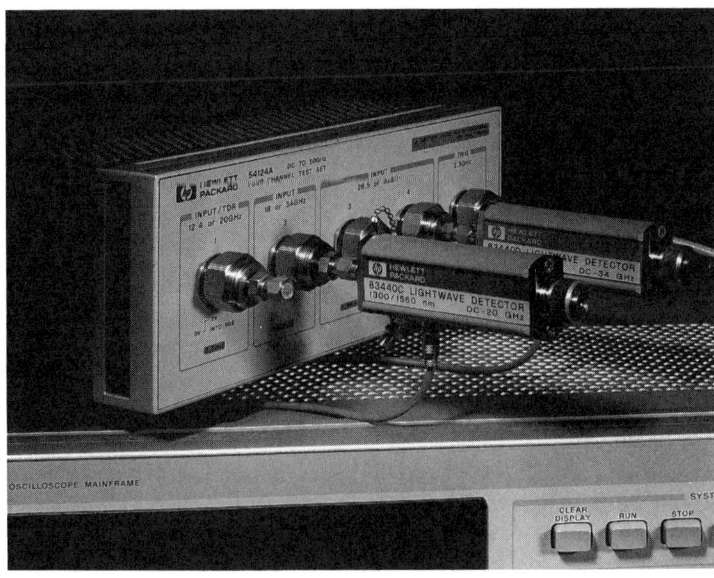

FIG. 13. Photograph of Hewlett-Packard 83440 series lightwave receivers.

todetector. An electrical wave form results that contains the fiber distortions plus any noise due to the photodetector. This is amplified and picks up any noise associated with the amplifier, and the pulses pass to a decision circuit where the ones and zeros must be discerned (Keiser, 1983). If the voltage is greater than some threshold voltage, then the conclusion is that a light pulse was present and a one is sensed. If the voltage falls below that threshold, than no light was present and a zero is detected. Detection can be done by asynchronous or synchronous methods. The bit error rate (BER) of the digital system is defined as the number of errors occurring in a given time interval N_e by the number of pulses transmitted during that time interval N_t (Keiser, 1983)

$$\mathrm{BER} = N_e/N_t. \tag{34}$$

Typical BERs for optical communications systems are 10^{-6}–10^{-10} (Keiser, 1983). Most systems now require BERs of 10^{-9} (Williams, 1991).

Noise, variable or undesirable components of the electrical signal, limits receiver performance. Noise can arise from several sources. Random fluctuation of currents in resistors is called Johnson or thermal noise, and in diodes it is called shot noise. Shot noise comes from photodiode dark current from both the semiconductor bulk and surface. APDs have additional noise due to the random nature of the gain process. Quantum noise is associated with the random process of converting light into electrical current. Interference of adjacent signals is noise referred to as cross talk or intersymbol interference. In a *p-n* or *p-i-n* photoreceiver, the thermal noise tends to dominate, and in an APD receiver, the random noise of the multiplication process adds additional shot noise (due to fluctuations in dark current and quantum noise) and this is the limiting factor (Senior, 1992).

Optical-receiver amplifier designs are usually of two types—high-impedance and transimpedance designs (Hullett and Muoi, 1976; Kasper, 1988). The high–input-impedance amplifier provides good sensitivity, but does not have wide bandwidth characteristics because of the effective integration of the detector output over a large time constant (Senior, 1992; Personick, 1973). The transimpedance amplifier uses a low-noise, high-impedance design where the high impedance is reduced by negative feedback provided by the feedback resistor (Senior, 1992). It has higher bandwidth performance than the high-impedance case. Transimpedance amplifiers are less sensitive than high-impedance designs (Williams, 1991). Transimpedance amplifier designs are being used in hybrid schemes and also in monolithic versions where an APD is integrated with an amplifier or a *p-i-n* diode is integrated with a FET to form a *p-i-n*-FET (Akahori *et al.*, 1991; Uchida *et al.*, 1991). *p-i-n*'s are generally used when speed is the major concern, and APDs can provide increased sensitivity as a result of their gain (Forrest, 1985).

Monolithic integration of photodetectors and amplifiers has been demonstrated using *p-i-n*'s and HBTs (heterojunction bipolar transistors) (Chandrasekhar *et al.*, 1990), MSMs and HEMTs (high–electron-mobility transistors) (Sasaki *et al.*, 1989; Yano *et al.*, 1990). These OEICs have been grown using InP/InGaAs heterostructures or InAIAs/InGaAs and fabricated on a single substrate. Photoreceivers that will respond to data rates of 1–4 Gbit/s have been made (Chandrasekhar *et al.*, 1990; Yano *et al.*, 1990).

As data rates increase, photodetector speeds will increase and OEICs will become necessary to overcome the parasitics associated with hybrid designs and interconnections. Photodetectors will continue to evolve as even lower-capacitance, lower–transit-time, high-response designs are in demand.

Works Cited

Akahori, Y., Ikeda, M., Uchida, N., Kohzen, A., Temmyo, J., Yoshida, J., Kokubun, T., Suto, K. (1991), *IEEE Photon. Technol. Lett.* **3**, 378–380.

Beck, P. A. (1993), *Hewlett-Packard J.* **44**, 49–51.

Bowers, J. E., Burrus, C. A. (1986), *Electron. Lett.* **22**, 905–906.

Bowers, J. E., Burrus, C. A. (1987), *J. Lightwave Technol.* **LT-5**, 1339–1350.

Bowers, J. E., Burrus, C. A., Mitschke, F. (1986), *Electron. Lett.* **22**, 633–635.

Braun, D. M. (1988), *Appl. Opt.* **27**, 2006–2011.

Capasso, F. (1985), "Physics of Avalanche Photodiodes," in: W. T. Tsang (Ed.), *Semiconductors and Semimetals*, Vol. 22, *Lightwave Communications Technology*, Pt. D, *Photodetectors*, Orlando, FL: Academic, pp. 2–172.

Carey, K. W., Wang, S. Y., Chang, J. S. C., Nauka, K. (1989), *J. Cryst. Growth* **98**, 90–97.

Chandrasekhar, S., Dentai, A. G., Joyner, C. H., Johnson, B. C., Gnauck, A. H., Qua, G. J. (1990), *Electron. Lett.* **26**, 1880–1882.

Dingle, R. (1987), "Applications of Multiquantum Wells, Selective Doping and Superlattices," in: R. K. Willardson, A. C. Beer (Eds.), *Semiconductor and Semimetals*, Vol. 24, San Diego: Academic Press.

Forrest, S. R., Leheny, R. F., Nahory, R. E., Pollack, M. A. (1980), *Appl. Phys. Lett.* **37**, 322–325.

Forrest, S. R. (1985), "Sensitivity of Avalanche Photodetector Receivers for High-Bit-Rate Long-Wavelength Optical Communication Systems," in: W. T. Tsang (Ed.), *Semiconductors and Semimetals*, Vol. 22, *Lightwave Communications Technology*, Pt. D, *Photodetectors*, Orlando, FL: Academic, pp. 329–387.

Forrest, S. R. (1988), "Optical Detectors for Lightwave Communication," in: S. E. Miller, I. P. Kaminow (Eds.), *Optical Fiber Telecommunications II*, San Diego: Academic, pp. 569–599.

Hullett, J. L., Muoi, T. V. (1976), *IEEE Trans. Commun.* **COM-24**, 1180–1185.

Johnsen, C., Sloan, S., Braun, D., Russell, J. L., Zurakowski, M., Lightner, M., Kellert, F., Patterson, G., Koo, R., Derrickson, D., Bowers, J. E. (1991), *IEEE Trans. Electron. Devices* **ED-38**, 1968–1970.

Kasper, B. L. (1988), "Receiver Design," in: S. E. Miller, I. P. Kaminow (Eds.), *Optical Fiber Telecommunications II*, New York: Academic, pp. 689–722.

Kato, K., Hata, S., Kawano, K., Yoshida, J., Kozen, A. (1992), *IEEE J. Quantum Electron.* **QE-28**, 2728–2735.

Katsura, K., Hayashi, T., Ohira, F., Hata, S., Iwashita, K. (1990), *J. Lightwave Technol.* **8**, 1323–1326.

Keiser, G. (1983), *Optical Fiber Communications*, New York: McGraw-Hill.

King, R., Braun, D. M., Hinch, S. W., Shubert, K. (1993), *Hewlett-Packard J.* **44**, 83–86.

Klingenstein, M., Kuhl, J., Notzel, R., Ploog, K., Rosenzweig, J., Moglestue, C., Hulsmann, A., Schneider, J. (1992), *Appl. Phys. Lett.* **60**, 627–629.

Lee, T. P., Li, T. (1979), "Photodetectors," in: S. E. Miller, A. G. Chynoweth (Eds.), *Optical Fiber Telecommunications*, New York: Academic.

Lucovsky, G., Schwarz, R. F., Emmons, R. B. (1964), *J. Appl. Phys.* **35**, 622.

Makiuchi, M., Norimatsu, M., Sakurai, T., Kondo, K., Yano, M. (1993), *IEEE Photon. Technol. Lett.* **5**, 518–520.

Manasreh, M. O., Brown, G. J. (1993), "Introduction to Long-Wavelength Infrared Quantum Detectors," in: M. O. Manasreh (Ed.), *Semiconductor Quantum Wells and Superlattices for Long-Wavelength Infrared Detectors*, Boston: Artech House.

Melchoir, H. (1972), "Demodulation and Photodetection Techniques," in: F. T. Arecchi, E. O. Schulz-Dubois (Eds.), *Laser Handbook*, Vol. 1, Amsterdam: North-Holland, pp. 703–835.

Muller, J. (1981), "Photodiodes for Optical Communication," in: L. Maron, C. Marton (Eds.), *Advances in Electronics and Electron Physics*, Vol. 55, New York: Academic.

Ohnaka, K., Kubo, M., Shibata, J. (1987), *IEEE Trans. Electron. Devices* **ED-34**, 199–204.

Pearsall, T. P., Pollack, M. A. (1985), "Compound Semiconductor Photodiodes," in: W. T. Tsang (Ed.), *Semiconductors and Semimetals*, Vol. 22, *Lightwave Communications Technology*, Pt. D, *Photodetectors*, Orlando, FL: Academic, pp. 174–245.

Personick, S. D. (1973), *Bell Syst. Tech. J.* **52 (6)**, 843–887.

Ritter, D., Hamm, R. A., Panish, M. B., Vandenber, J. M., Gershoni, D., Ganapala, S. D., Levine, B. F. (1991), *Appl. Phys. Lett.* **59**, 552–554.

Saleh, B. E., Teich, M. C. (1991), *Fundamentals of Photonics*, New York: Wiley.

Sasaki, G., Koike, K., Kuwata, N., Ono, K. (1989), *J. Lightwave Technol.* **7**, 1510–1514.

Senior, J. M. (1992), *Optical Fiber Communications: Principles and Practice*, 2nd ed., Englewood Cliffs, NJ: Prentice-Hall.

Shockley, W. (1949), *Bell Syst. Tech. J.* **28**, 435.

Sloan, S. R. (1989), *Hewlett-Packard J.* **40**, 69–75.

Sze, S. M. (1981), *Physics of Semiconductor Devices*, New York: Wiley.

Uchida, N., Akahori, Y., Ikeda, M., Kohzen, A., Yoshida, J., Kokubun, T., Suto, K. (1991), *IEEE Photon. Technol. Lett.* **3**, 540–542.

Van Zeghbroeck, B. J., Patrick, W., Halbout, J. M., Vettiger, P. (1988), *Electron. Device Lett.* **9**, 527–529.

Wang, S. Y., Bloom, D. M. (1983), *Electron. Lett.* **19**, 554–555.

Wang, S. Y., Carey, K. W., Kolner, B. H. (1987), *IEEE Trans. Electron. Devices* **ED-34**, 938–940.

Watanabe, I., Sugou, S., Ishikawa, H., Anan, T., Makita, K., Tusji, M., Taguchi, K. (1993), *IEEE Photon. Technol. Lett.* **5**, 675–677.

Wey, Y. G. (1993), *ECE Technical Report 93-04*, Department of Electrical and Computer Engineering, University of California at Santa Barbara.

Williams, G. F. (1991), "Lightwave Receivers," in: Tingye Li (Ed.), *Topics in Lightwave Transmission Systems*, New York: Academic, pp. 79–149.

Wilson, J., Hawkes, J. F. B. (1989), *Optoelectronics*, Englewood Cliffs, NJ: Prentice-Hall.

Yano, H., Aga, K., Kamei, H., Sasaki, G., Hayashi, H. (1990), *J. Lightwave Technol.* **8**, 1328–1333.

Zirngibl, M., Bischoff, J. C., Theron, D., Ilegems, M. (1989), *Electron. Device Lett.*, **10**, 336–338.

Further Reading

Lee, T. P., Li, T. (1979), *Photodetectors*, in: S. E. Miller, A. G. Chynoweth (Eds.), *Optical Fiber Telecommunications*, New York: Academic.

Melchoir, H. (1972), "Demodulation and Photodetection Techniques," in: F. T. Arecchi, E. O. Schulz-Dubois (Eds.), *Laser Handbook*, Vol. 1, Amsterdam: North-Holland, pp. 703–835.

Muller, J. (1981), "Photodiodes for Optical Communication," in: L. Maron, C. Marton (Eds.), *Advances in Electronics and Electron Physics*, Vol. 55, New York: Academic.

Pearsall, T. P., Pollack, M. A. (1985), "Compound Semiconductor Photodiodes," in: W. T. Tsang (Ed.), *Semiconductors and Semimetals*, Vol. 22, *Lightwave Communications Technology*, Pt. D, *Photodetectors*, Orlando, FL: Academic, pp. 174–245.

Saleh, B. E., Teich, M. C. (1991), *Fundamentals of Photonics*, New York: Wiley.

PHOTOEMISSION AND PHOTOELECTRON SPECTRA

F. J. HIMPSEL, *IBM T.J. Watson Research Center, Yorktown Heights, New York, U.S.A.*

I. LINDAU, *Department of Synchrotron Radiation Research, Lund University, Lund, Sweden*

1.	**Introduction**	477
1.1	History	477
1.2	Phenomena	478
1.3	Theory	479
2.	**Instrumentation**	480
2.1	Light Sources	480
2.2	Detectors	482
3.	**Photoelectric Yield and Absorption Spectroscopy**	482
3.1	Mechanism	482
3.2	Unoccupied Atomic and Molecular Orbitals	482
3.3	Solids	483
3.4	Magnetic Dichroism	483
3.5	Microspectroscopy	483
4.	**Core-Level Spectroscopy**	483
4.1	Origin of Core-Level Shifts	483
4.2	Atomic Case: Shakeup and Shakeoff Satellites	485
4.3	Adsorbates and Surfaces	485
4.4	Bulk Solids	486
5.	**Valence Spectroscopy**	487
5.1	Occupied Atomic and Molecular Orbitals	487
5.2	Surface States	489
5.3	Band Mapping	489
5.4	Time-Resolved, Pump-Probe Measurements	490
5.5	Inverse Photoemission	490
6.	**Structural Methods**	490
6.1	EXAFS and SEXAFS	490
6.2	Photoelectron Diffraction and Holography	491
7.	**Summary and Outlook**	492
	Glossary: Definitions, Acronyms, Jargon	492
	Databases	493
	Works Cited	493

1. INTRODUCTION

1.1 History

Photoemission, i.e., the emission of electrons by photons, was discovered as an unwanted side effect in the study of electromagnetic waves by Hertz (1887). At that time, neither the electron nor the photon was known as a particle. Einstein (1905) explained the photoelectric effect as a quantum phenomenon and derived the correlation

$$E_{\max} = h\nu - \Phi \tag{1}$$

between the maximum kinetic energy $E_{\max}$ of the photoelectrons, the photon energy $h\nu$, and the work function Φ of the emitting solid (compare Fig. 1). Many of the following decades were spent on finding clear experimental evidence for or against the Einstein formula (Millikan, 1916). The time was not ripe for photoelectron spectroscopy as we know it today, since photoemission turned out to be a rather surface-sensitive effect. The mean free path of photoelectrons in a solid limits the probing depth to a few atomic layers. The prerequisites for reproducible preparation of clean and well-ordered surfaces did not arrive until the 1960s, when the necessary ultrahigh vacuum could be produced and measured on a routine basis. At a residual gas pressure of 10^{-6} Torr (1.3×10^{-4} Pa), a surface atom is being hit by a residual gas molecule about once every second, thus requiring pressures in the 10^{-10}-Torr range for controlled photoemission studies. When surface characterization techniques, such as low-energy electron diffraction (LEED) and Auger-electron spectroscopy, became generally available, photoelectron spectroscopy took a big leap forward. Photoelectron spectroscopy

3-527-28135-5/95/$5.00 + .50

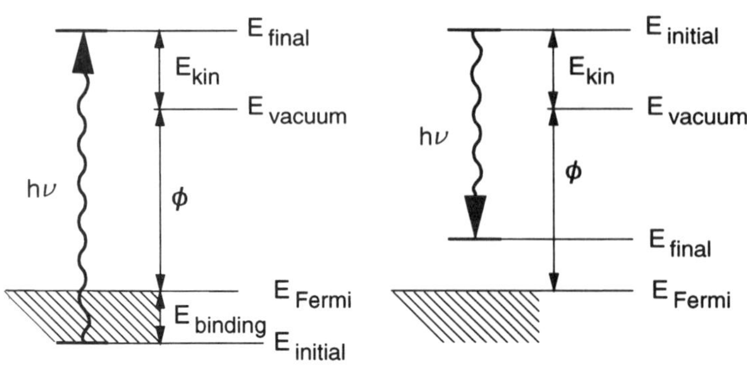

FIG. 1. Energy diagram for photoemission (photon in, electron out) and inverse photoemission (electron in, photon out). The energy of the lower state is determined by subtracting the photon energy $h\nu$ from the energy of the upper state. The latter corresponds to the kinetic energy of the emitted or incoming electron.

is now an indispensable technique for the determination of the electronic structure of solids, surfaces, and gas-phase species. Tunable and polarized synchrotron radiation became the light source of choice in the 1970s and gave rise to the construction of a series of storage rings dedicated to the production of light from the ultraviolet to the x-ray regime. In the 1980s, the measurement of a complete set of quantum numbers became feasible for electrons in solids, which involves measuring the momentum and spin polarization of photoelectrons as a function of the photon energy and photon polarization. Current developments point toward the use of temporal resolution for looking at dynamics down to the subpicosecond regime, and the use of spatial resolution for element-resolved microscopy down to the 10-nm level.

1.2 Phenomena

The basic energy diagram of the photoemission process is given in Fig. 1. An electron is lifted from an occupied initial state into an unoccupied final state. Energy conservation allows us to obtain the desired energy E_{initial} of the initial state from the measured energy E_{final} of the photoelectron in the final state by simply subtracting the photon energy $h\nu$. As reference energies, one can use either the vacuum level E_{vacuum}, corresponding to a photoelectron with zero velocity, or the Fermi level E_{Fermi}, corresponding to the highest occupied state. From there, one obtains energy differences, such as the kinetic energy E_{kin}, the work function Φ, and the binding energy E_{binding}. The Einstein relation for the maximum kinetic energy, $E_{\max} = h\nu - \Phi$, follows directly for photoemission from the highest occupied state at E_{Fermi}.

Figure 1 also shows the energy diagram for the time-reversed process, i.e., an incident electron dropping down into a lower unoccupied state by emission of a photon. This process is equivalent to the quantum description of bremsstrahlung, but in the low-energy regime of typically 10 eV kinetic energy it is known today as inverse photoemission. It characterizes unoccupied states in a fashion complementary to the information provided by photoemission about occupied states. The counterpart to the Einstein relation is the Duane–Hunt equation,

$$h\nu_{\max} = E_{\text{kin}} + \Phi, \tag{2}$$

for the highest photon energy emitted into bremsstrahlung by electrons with kinetic energy E_{kin}. Inverse photoemission as a surface technique got a late start compared with photoemission since its cross section is down by four to five orders of magnitude, requiring extremely sensitive photon detectors.

There are other electron- and photon-emission phenomena related to photoemission. As shown in Fig. 2, a core hole created in the photoemission process can decay by an Auger process (left) (see AUGER ELECTRON SPECTROSCOPY), which is most likely for shallow core levels, or by fluorescence (right), which is prevalent for deep core levels (Keski-Rahkonen and Krause, 1974) (see also CATHODOLUMINESCENCE). In addition, the photo-

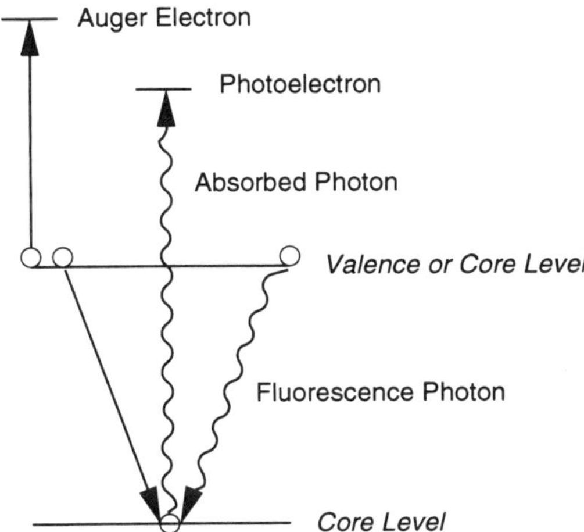

FIG. 2. Energy diagram for core-level photoemission, including the decay of the core hole by an Auger process (left) or by fluorescence (right).

electron can lose energy by creating electron-hole pairs or plasmons during its escape. These processes actually limit the escape depth. The end result is a spectrum of emitted photoelectrons that contains several secondary features in addition to the directly emitted photoelectrons (Fig. 3). For valence spectroscopy [often labeled ultraviolet photoelectron spectroscopy (UPS)], the main disturbance is a tail of secondary electrons at low kinetic energies, caused by energy losses to electron-hole pairs [Fig. 3(a)]. For core-level spectroscopy [often labeled x-ray photoelectron spectroscopy (XPS) or electron spectroscopy for chemical analysis (ESCA); Siegbahn *et al.*, 1967], there are additional Auger peaks and plasmon-loss satellites complicating the spectra [Fig. 3(b)]. With tunable synchrotron radiation, it is straightforward to separate out the Auger peaks, since they do not change their kinetic energy with increasing photon energy, in contrast to elastic core lines and their plasmon-loss satellites.

An important consideration in photoemission experiments is the probing depth, which is governed by the mean free path of the photoelectrons (Fig. 4). Electrons penetrate solids farther at high kinetic energies, but their mean free path increases again at very low energies, because they lose the ability to excite plasmons or excitons. The minimum of the mean free path in the 30–70-eV range is only a few atomic distances, making it possible to separate out surface or interface atoms. The tunability of synchrotron radiation provides the ability to produce photoelectrons with the desired kinetic energy, which has spawned a large field of surface core-level spectroscopy.

1.3 Theory

The most general theory of photoemission is given by an expression of the Fermi "golden rule" type, i.e., a differential cross section containing a matrix element between the initial and final states and a phase-space integration. For practical purposes, we have to consider mainly the various selection rules

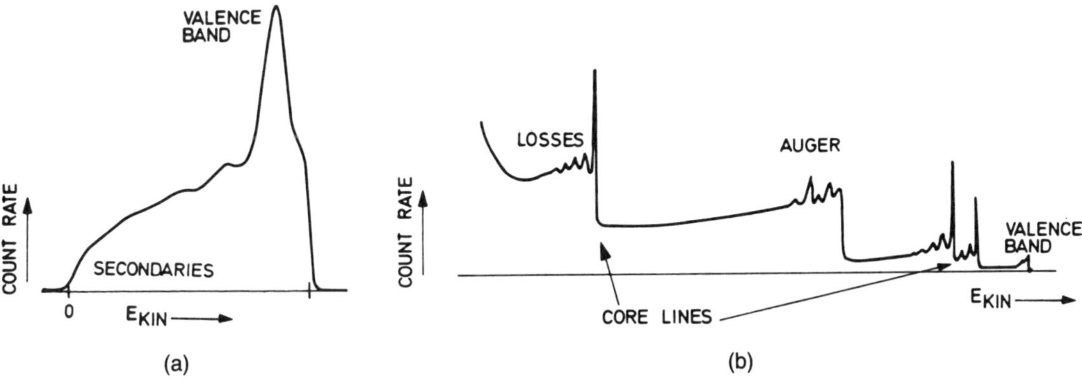

FIG. 3. (a) Typical valence-photoelectron spectrum, showing photoelectrons emitted directly from the valence band and secondary electrons emitted after electron–hole pair production. The two energy cutoffs are given by the vacuum level (zero kinetic energy E_{kin}) and the Fermi level (i.e., excitation of the highest occupied state). (b) Typical core-level photoelectron spectrum, showing core-level peaks, their plasmon-loss satellites, and Auger-electron features. From Feuerbacher *et al.* (1978).

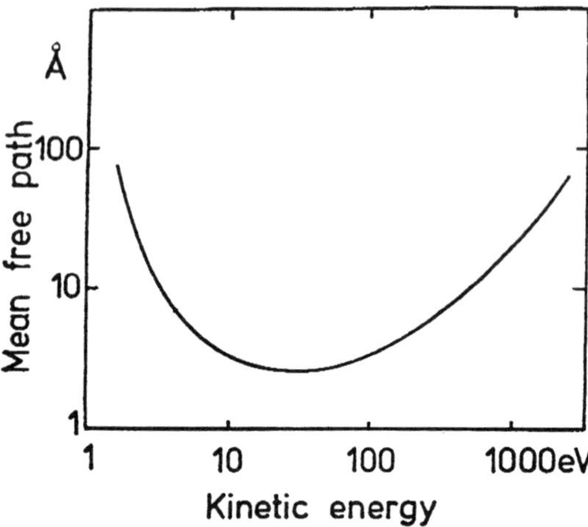

FIG. 4. Schematic dependence of the mean free path of photoelectrons on their kinetic energy. The minimum at 30–70 eV can be used to probe surface or interface layers. Compare Lindau and Spicer (1974).

inherent in this expression, and the wave functions of the electronic states involved.

An electron in a periodic crystal lattice is characterized by a set of quantum numbers, i.e., energy, momentum, angular symmetry, and spin. We mentioned already the energy-conservation law in Fig. 1. A similar conservation law holds for the momentum parallel to the surface, $\hbar k_{\parallel}$, where $k_{\parallel}$ is the corresponding wave vector. The wave vector of the photon is usually negligible compared with that of the electron for the energy range in question, since its wavelength is so much longer. The wavelength λ and wave vector $k = 2\pi/\lambda$ of the photon are given by the photon energy $h\nu$ via

$$\lambda/\text{Å} = 12\,399/(h\nu/\text{eV}) \qquad (3)$$

and the wave vector of the electron by its kinetic energy E_{kin} via

$$k_{\parallel}/\text{Å}^{-1} = 0.51(E_{\text{kin}}/\text{eV})^{1/2} \qquad (4)$$

(1 Å = 0.1 nm). The momentum component perpendicular to the surface is not conserved during the passage of the photoelectron across the surface-energy barrier, but it can be varied by tuning the photon energy. Angular symmetry can be inferred from polarization selection rules for the point group at a particular point in momentum space. Spin is conserved. The complete set of quantum numbers is usually displayed in the form of energy vs momentum band dispersion, labeled with the appropriate symmetry and spin symbols (see Sec. 5.1; Himpsel, 1983; Kevan, 1992).

While selection rules provide clear yes-no decisions, there are also more subtle effects of the matrix element that can be used to bring out specific electronic states. The atomic symmetry character determines the energy dependence of the cross section (Yeh and Lindau, 1985), allowing a selection of specific orbitals by varying the photon energy. For example, the (s,p) states in transition and noble metals dominate the spectra near the photoelectric threshold, while the d states turn on at 10 eV above threshold. It takes photon energies of 30 eV above threshold to make the f states in rare earths visible. Resonance effects at a threshold for a core-level excitation can also enhance particular orbitals. Conversely, the cross section for states with a radial node exhibits so-called Cooper minima, where they become almost invisible.

The wave functions of the electronic states in solids and at surfaces are usually approximated by the ground-state wave functions obtained from a variety of schemes, e.g., tight-binding, local-density approximation (LDA), or Hartree–Fock. Strictly speaking, one should use the excited-state wave functions and energies that represent the hole created in the photoemission process, or the extra electron added to the solid in the case of inverse photoemission. Such quasiparticle calculations have now become feasible and provide the most accurate band dispersions to date. Particularly in the case of semiconductors, the traditional ground-state methods are unable to determine the fundamental band gap from first principles, with Hartree–Fock overestimating it and LDA underestimating it, typically by a factor of 2.

2. INSTRUMENTATION

2.1 Light Sources

To be able to excite photoelectrons, a minimum photon energy of 5–10 eV is required. This is one of the reasons why conventional lasers have been of limited applicability in photoelectron spectroscopy. A detailed discussion of the possible sources for photoelectron excitation is given in Siegbahn and Karlsson (1982). In the ultraviolet-radiation region, the most common source is based on a capillary glow discharge. The most convenient line is the so-called He I, which provides monochromatic radiation with a narrow line width at 21.2 eV. This radiation originates from the 2*p*-to-1*s* transition in neutral He atoms. Emission can also be produced from the He ions, the primary emission line being He II at an energy of 40.8 eV. Similarly, emission lines of Ne I at 16.8 eV and Ne II at 26.9 eV have frequently been used in photoelectron spectroscopy.

When a target is bombarded with an electron beam of sufficient energy and intensity, x-ray emission will occur. In principle, a vast amount of x-ray-emitting materials should be available, but there are very few practical x-ray sources that fulfill the practical constraints, i.e., having high intensity, small energy width, and no satellite structures. Therefore, Mg $K\alpha$ radiation at 1253.6 eV and Al $K\alpha$ radiation at 1486.7 eV have been completely dominant. Since both magnesium and aluminum are easy to handle, and since their inherent line width is as small as 1 eV, most commercial x-ray anodes are based on Al and/or Mg as the target material. It is also possible to produce monochromatized Al $K\alpha$ and Mg $K\alpha$ radiation by using, for instance, a quartz-crystal monochromator. An excitation line width of 0.25 eV can be achieved but with considerable loss of intensity. These technical approaches of achieving higher resolution have found their way into commercial instruments.

During the last two decades, synchrotron radiation has emerged as both a powerful and a convenient excitation source in photoelectron spectroscopy (Winick and Doniach, 1980; Koch, 1983; Winick *et al.*, 1989). The term synchrotron radiation is usually associated with the radiation emitted by relativistic electrons accelerated in a circular orbit. The basic properties of synchrotron radiation were first presented in an elegant paper by Schwinger (1949), though they had been treated much earlier by Schott (1912). With E being the electron energy and R the radius of the circular orbit, the total synchrotron-radiation energy emitted per revolution by one electron is given by

$$\Delta E(R,E) = (4\pi/3)(e^2/R)(E/m_0c^2)^4. \quad (5)$$

The spectral radiation distribution is characterized by a critical energy ϵ_c, given by

$$\epsilon_c = (3hc/2R)(E/m_0c^2)^3. \quad (6)$$

The physical significance of this equation is that half of the total power is radiated above the critical energy and half below. The photon intensity decreases rapidly for photon energies above the critical energy ϵ_c. A universal curve for the synchrotron-radiation photon flux is shown in Fig. 5. Synchrotron radiation has a number of desirable properties. It provides

1. a continuous spectral distribution from the infrared region into the x-ray region;
2. high intensity;
3. a high degree of collimation;
4. a high degree of polarization (completely linearly polarized in the plane of the orbital and elliptically out of the plane); and
5. a pulsed time structure given by the orbital frequency of the circulating electron beam.

An important development occurred with the realization that magnetic structures inserted in straight sections of the storage ring, so-called undulators and wigglers, could drastically improve the radiation characteristics of conventional bending magnets. Presently, facilities are being built and commissioned that are optimized for these insertion devices, so-called third-generation sources.

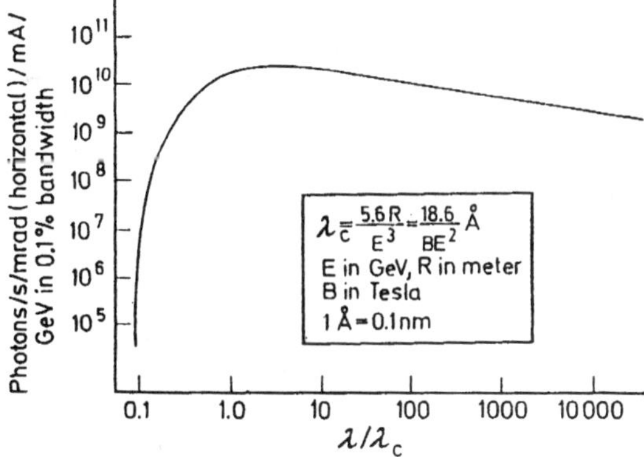

FIG. 5. The universal curve for the synchrotron-radiation flux, plotted with λ/λ_c on the horizontal axis. λ_c is the critical wavelength, given by the electron energy E, the radius R of the electron orbit, or the magnetic field of the bending magnet. After Koch (1983).

2.2 Detectors

The most common types of photoelectron spectrometers have been the cylindrical mirror analyzer (CMA) and the hemispherical analyzer, both of the electrostatic type. A key feature in efficient spectrometers is parallel detection. Energy multidetection in hemispherical analyzers has become a widely available option. There exist custom designs with more than one variable being detected simultaneously, such as energy and one angle in toroidal analyzers and two angles in display spectrometers. For detecting the spin of photoelectrons, one typically measures the right-left asymmetry in the electron-scattering cross section at heavy nuclei (Mott scattering), after accelerating the photoelectron to 100 keV. A less bulky detector uses scattering at the electron cloud surrounding a heavy nucleus, typically at 150 eV electron energy. It lacks the absolute calibration of a Mott detector, however. In general, spin detection costs three to four orders of magnitude in detection efficiency.

3. PHOTOELECTRIC YIELD AND ABSORPTION SPECTROSCOPY

3.1 Mechanism

This section starts with the simplest measurement in photoemission, which consists of determining the total photocurrent vs photon energy. Subsequent sections go into energy- and angle-resolved differential cross sections.

Before sophisticated measurements became routine, the determination of the quantum yield near the photoelectric threshold dominated the field. Such measurements were used to confirm the Einstein formula [Eq. (1)] (Millikan, 1916) and to obtain the work function Φ of various materials by finding the photon energy for the onset of photoemission. Accurate determinations involved various extrapolation schemes, predicting certain power laws for the photoelectric yield near threshold (Fowler, 1931; Cardona and Ley, 1978). The photoelectric yield increases roughly like the square of the excess energy, since photoemission is dominated near threshold by the narrowing of an "escape cone" in momentum space that contains the electronic states with sufficient momentum perpendicular to the surface to overcome the surface-potential step. Quantum yield measurements had technological consequences in the development of photocathodes for image tubes and photomultipliers.

The bulk of today's yield measurements is performed in a regime far above the photoelectric threshold, i.e., at core-level absorption edges. In this case, the photoelectric yield is closely related to the absorption coefficient, as long as the escape depth of the photoelectrons is short compared to the absorption length of the photons (Gudat and Kunz, 1972). Essentially, the number of photoelectrons is proportional to the number of core holes generated, which in turn is proportional to the absorption coefficient in the linear-absorption regime near the surface. In the following, we discuss applications of near-edge x-ray-absorption fine structure (NEXAFS) measurements. They involve transitions from a core level to the lowest unoccupied states, e.g., atomic and molecular orbitals and conduction-band states within the first tens of eV above threshold. Absorption measurements over a wider photon-energy range (hundreds of eV) are usually labeled EXAFS (extended x-ray-absorption fine structure). They provide structural information, which is discussed in Sec. 6.2.

3.2 Unoccupied Atomic and Molecular Orbitals

The unoccupied atomic and molecular orbitals that dominate the absorption fine structure near a core-level threshold can serve as a fingerprint of a molecule or polymer. The pattern of π^* and σ^* orbitals has been useful in the surface chemistry of small organic molecules during catalytic processes at surfaces. A large database of gas-phase spectra (Hitchcock and Mancini, 1994) helps in identifying surface species (Stöhr, 1992). In addition, there is the option of using the polarization dependence of the absorption from different orbitals to determine the orientation of adsorbed molecules. Similar applications exist with polymers. Figure 6 shows that two different polymers in a blend can easily be distinguished by their π^* orbitals (Ade *et al.*, 1992).

3.3 Solids

Absorption spectroscopy in solids provides information about the unoccupied states, like in the gas phase. In contrast to molecular systems, where the Coulomb interaction between the excited electron and the core hole significantly distorts the energy scale, there is enough screening in many solid-state systems to use the absorption spectrum directly as a measure of the density of unoccupied states. By using different core levels, it is possible to project the density of states onto a particular atomic species and to select states with a specific angular momentum l, using the l-to-($l \pm 1$) selection rules. A more thorough analysis has to include electron–hole interaction and atomic multiplet structure, particularly for localized $2p$-to-$3d$ and ($3d$, $4d$)-to-$4f$ transitions.

3.4 Magnetic Dichroism

By using circularly polarized light with magnetic samples, absorption spectroscopy can be extended to the detection of magnetic properties in bulk solids and at surfaces. Using atomic sum rules, it is possible to estimate both the spin and the angular magnetic moment from the dichroism, i.e., the change in absorption induced by flipping the helicity of the light relative to the magnetization of the sample. Thereby, the two pieces of independent information come from the dichroism of the two spin-orbit partner lines of a core level. In addition, one can play all the other selection rules, such as projecting the magnetic moment onto an atomic species or separating magnetism of d electrons from that of f and (s,p) electrons. Figure 7 gives an example where the magnetization of different constituents in a gadolinium iron garnet is determined (Rudolf *et al.*, 1992).

3.5 Microspectroscopy

The recent development of third-generation synchrotron light sources with high brilliance makes it possible to perform absorption measurements with a spatial resolution down to 0.1 μm. The two pictures of a polymer blend in Fig. 6 were taken at two different photon energies, where the two components absorb differently because of their different π^* orbital energies (Ade *et al.*, 1992).

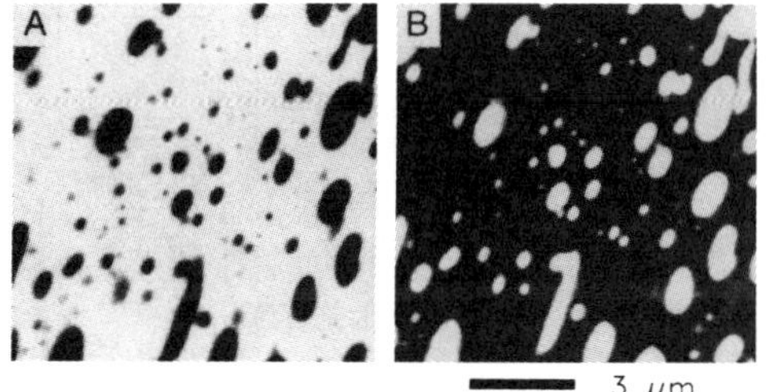

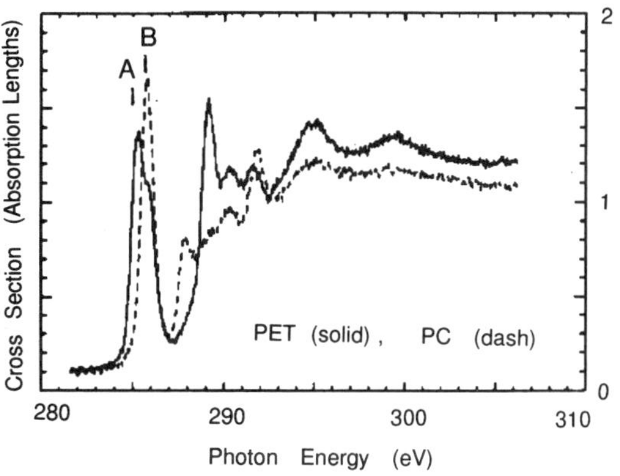

FIG. 6. Identification of different polymers in a blend by their π^* orbitals. The carbon 1s core-level absorption spectra (bottom) are used to identify the C 1s-to-π^* transition energies of two polymers. By tuning in to the respective π^* resonances, the two components can be distinguished in the micrograph (top). After Ade *et al.* (1992).

In this case, the absorption coefficient was measured directly in a transmission experiment, but such experiments can be performed with a variety of imaging photoelectron microscopes that are currently being commissioned. Microspectroscopy promises to become a versatile method that not only identifies the elemental species in a sample by the choice of the core-level absorption edge, but also their chemical state by selecting transitions into individual unoccupied valence states.

4. CORE-LEVEL SPECTROSCOPY

4.1 Origin of Core-Level Shifts

When an atom is irradiated with a monochromatic photon beam of sufficient energy to overcome the core-level binding energy, photoelectrons will be ejected from the different electron shells with a kinetic energy

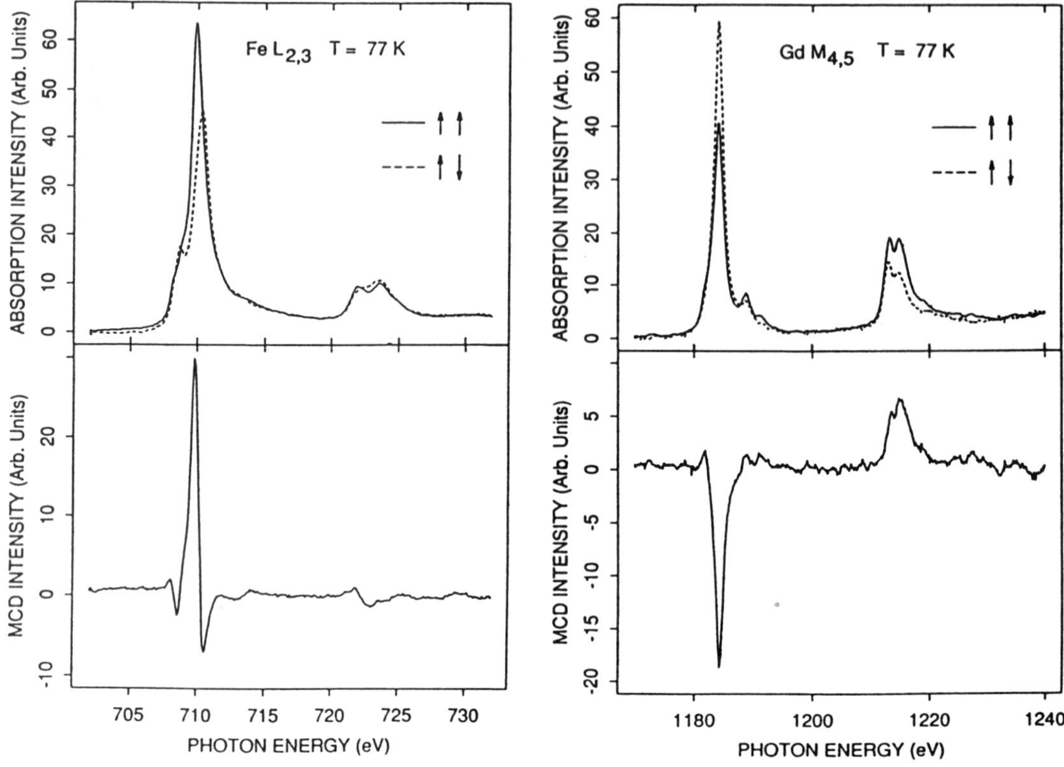

FIG. 7. Use of magnetic circular dichroism for determining the magnetization of the elemental species Fe (left) and Gd (right) in a garnet. The raw absorption spectra are shown on top for two different orientations of the magnetizing field relative to the helicity of the circularly polarized light. Differences are on the bottom. From Rudolf *et al.* (1992).

according to their binding energy [see Eq. (1) and Figs. 1, 2]. In a simplified picture, the electrons will thus appear at discrete energies characteristic for each element. X-ray photoelectron spectroscopy (XPS) is thus element specific. This technique is also known by the acronym ESCA, i.e., electron spectroscopy for chemical analysis. ESCA refers to the fact that the binding energy of a core level is slightly different depending on the chemical environment, hence the concept of chemical shift (Siegbahn *et al.*, 1967).

Theoretical interpretations of core-level chemical shifts have received much attention during the last 30 years and have been performed at different levels of sophistication. The major objective is to obtain chemically significant information about the initial-state electronic structure of the studied systems. However, in some instances, final-state effects will dominate the measured core-level spectra, which are caused by the removal of a photoelectron. We refer to the extensive literature for reviews of various theoretical approaches (Fadley, 1978). In the most simplified and naive picture, the magnitude of the chemical shift is correlated with the amount of charge transfer in the initial state. Such a simplified picture works, for instance, quite well for the different oxidation states of silicon, shown in Fig. 8 (Himpsel *et al.*, 1990). Oxygen pulls electrons away from silicon, creating a positive electrostatic potential at the Si core that lowers the Si core levels. All oxidation states of silicon are resolvable, with the three peaks between Si and SiO_2 originating from intermediate oxidation states, Si^{1+}, Si^{2+}, and Si^{3+}. The charge transfer is about half an electron per oxidation state. A similar trend is displayed for another electronegative ligand, F. An electropositive ligand gives up electrons to silicon, causing a chemical shift in the opposite direction, as shown in Fig. 8 for calcium. It is interesting to note that the chemical shift in SiO_2 increases slightly with increasing oxide thick-

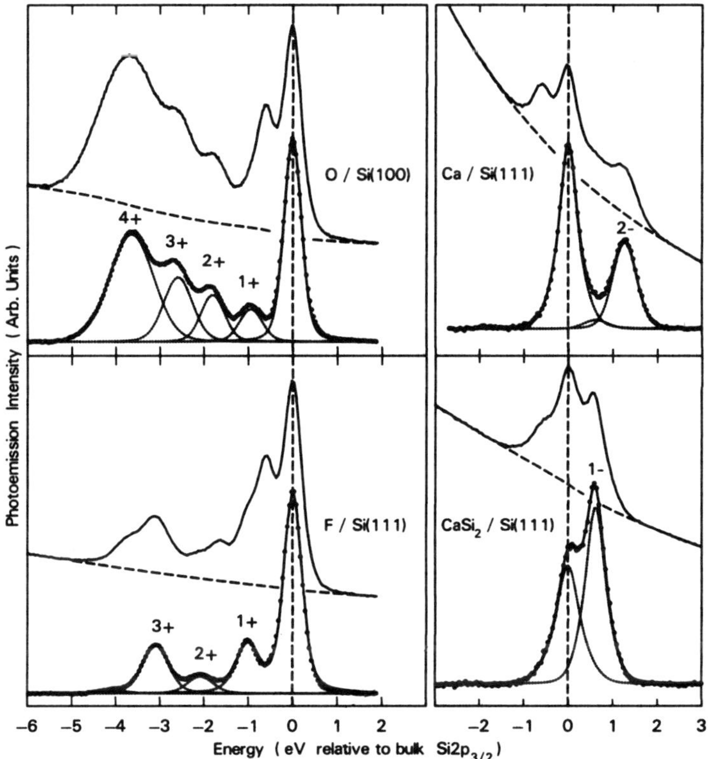

FIG. 8. Discrete chemical shifts, observed for different oxidation and reduction states of silicon. Electronegative ligands (O,F) give rise to chemical shifts toward higher binding energy and electropositive ligands (Ca) induce shifts to lower binding energy. Raw data are given on top of each panel, the Si $2p_{3/2}$ component on the bottom. From Himpsel *et al.* (1990).

ness (not shown), a manifestation of decreased screening in the final state. It should be further noted that the intensity of the different oxidation-state peaks in Fig. 8 can be used to estimate the amount of suboxides in the interfacial Si/SiO_2 region (Himpsel *et al.*, 1988).

4.2 Atomic Case: Shakeup and Shakeoff Satellites

The core-level spectra from a closed-shell noble-gas atom should in principle be very simple with discrete peaks corresponding to the different atomic subshells. However, final-state effects give rise to multielectron excitations, and atomic core-level spectra are characterized by a complex satellite structure, first discussed by Carlson (1967). The most likely multielectron process is a two-electron transition. But higher-order transitions and configuration interactions also contribute to the satellite spectra. It is common to distinguish between shakeup states and shakeoff states, where the second electron is excited to a higher bound state or an unbound continuum, respectively. The Ne 1*s* core-level spectrum and the magnified satellite structure of the one-electron 1*s* line are shown in Fig. 9. A detailed discussion and compilation of atomic core-level spectra and satellites can be found in Siegbahn and Karlsson (1982) and Sonntag and Zimmerer (1992).

4.3 Adsorbates and Surfaces

Chemical-shift studies of molecules on surfaces can provide information not only about the nature of chemisorption, dissociative or nondissociative, but also about structural arrangements. This is illustrated for the much-studied system of CO on Pt(111) (Björneholm *et al.*, 1993). Figure 10 shows core-level spectra for O 1*s*, C 1*s* (obtained with monochromatized Al *Ka* radiation) and Pt $4f_{5/2}$ (obtained with synchrotron radiation). The clean Pt $4f_{5/2}$ spectrum has two components where the low–binding-energy component can be attributed to the Pt atoms in the topmost atomic layer, a so-called surface chemical shift (Johansson and Mårtensson, 1980). CO forms a number of ordered phases on Pt(111): 4 × 4, *c*(4 × 2), and $c(5 \times \sqrt{3})$, dependent on the

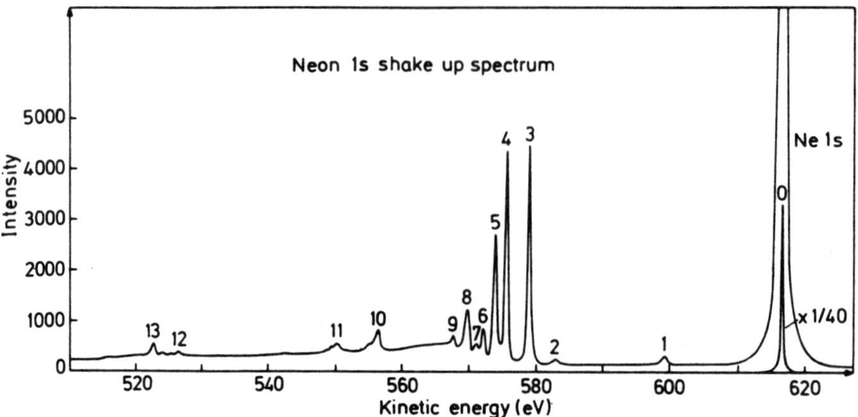

FIG. 9. The Ne 1s core-level photoelectron spectrum with accompanying satellites, labeled 1 through 13. From Gelius (1974).

coverage. A detailed correlation can be made between the structural arrangement of the CO molecule and the chemical shifts. For the 4 × 4 phase, there are single peaks of C 1s and O 1s, which is a manifestation of the fact that all CO molecules are absorbed at identical sites on top of Pt atoms. The surface component of clean Pt is reduced in intensity, and a new chemically shifted peak appears toward higher binding energy. For the $c(4 \times 2)$ and $c(5 \times \sqrt{3})$ phases, CO is occupying both on-top and bridge sites. This is reflected in double components for both the O 1s and C 1s peaks and an additional shifted Pt 4f peak. From the intensities of the chemically shifted 4f peaks, it is further possible to deduce the occupation of CO molecules on top and bridge sites. Chemically shifted peaks can thus provide information about both the bonding and structure of adsorbed systems (Björneholm *et al.*, 1993; Andersen *et al.*, 1991).

4.4 Bulk Solids

In Sec. 4.1, it was pointed out how core-level spectroscopy can be used for both qualitative (elemental and chemical specific) and

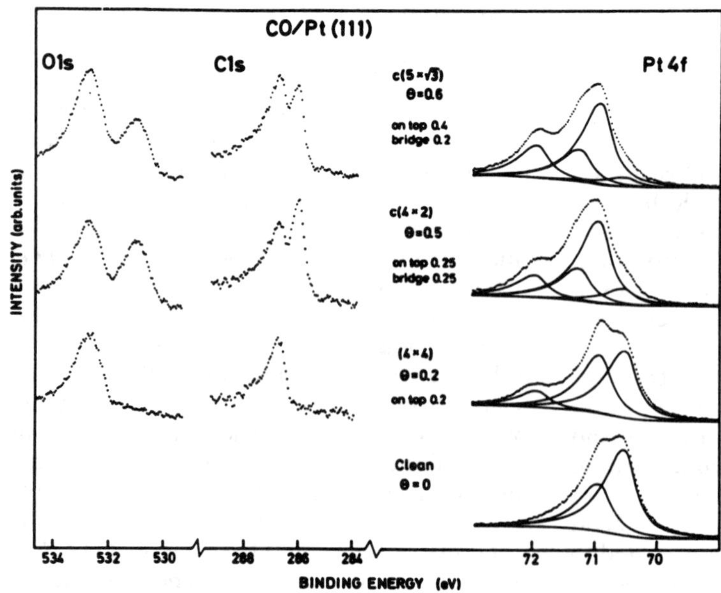

FIG. 10. The O 1s, C 1s, and Pt 4f core-level photoelectron spectra for various structures of CO chemisorbed on Pt(111). Adsorbate and substrate core-level shifts provide bonding-site information. From Björneholm *et al.* (1993).

quantitative analysis. Here, two other aspects of studies of solids are illustrated. The first example relates to the growth of thin films and their interfaces, a topic of particular importance for semiconductor structures (for an overview, see Weaver, 1988). The growth of a II–VI compound on a III–V semiconductor, viz. CdTe on InSb(100), is characterized by quantitative core-level spectroscopy using synchrotron radiation in Fig. 11 (Mackey *et al.*, 1986). Two different growth conditions are shown, i.e., (a) room temperature and (b) 500 K. The core-level spectra show the evolution of the intensity of the 4*d* levels of In and Sb (substrate) and Cd and Te (overlayers) with increasing CdTe deposition.

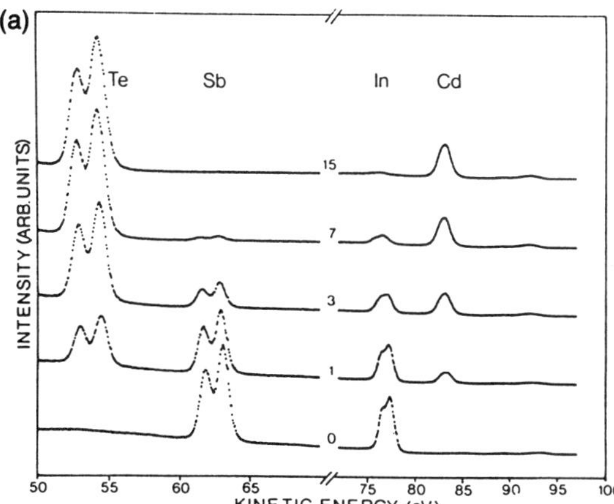

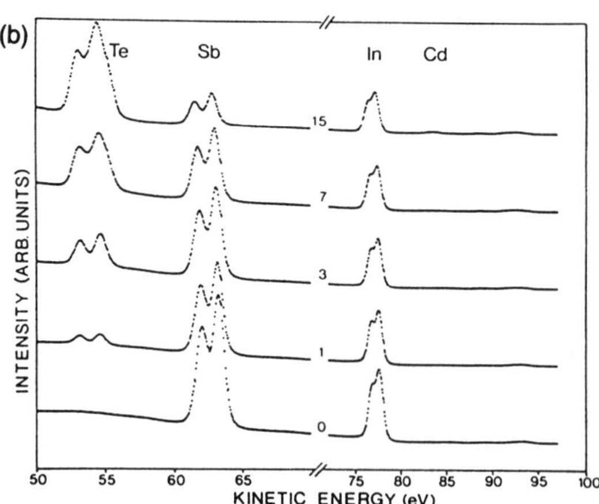

FIG. 11. Growth of CdTe on InSb(100), studied via the intensities of the 4*d* core levels (film thickness given in Å). (a) At room temperature the overlayer is stochiometric with a sharp interface. (b) At 500 K an interfacial layer of InTe is formed, together with segregated Sb. From Mackey *et al.* (1986).

At room temperature, the overlayer appears to be stoichiometric (no relative change in the Cd and Te core-level intensities), and the In/Sb 4*d* core levels are attenuated monotonically. The interface is obviously quite abrupt since the substrate core levels are covered up by 7–15 Å of CdTe. The situation is dramatically different for deposition at 500 K. The In 4*d* signal does not decrease as much as Sb 4*d*, the Te 4*d* is broadened, and the Cd signal is barely detectable. In this case, chemical reactions take place at the interface and an interfacial layer with indium telluride forms with segregated Sb. It can be concluded that the interface is neither abrupt nor stoichiometric on an atomic scale.

A second example deals with core-level spectra that are completely dominated by atomic final-state effects and very insensitive to the chemical environment. The most prominent representatives can be found among rare-earth compounds. The 4*f* core-level spectra for a few rare-earth antimonides are shown in Fig. 12 (Campagna *et al.*, 1976). The three spectra can all be described very well with final-state multiplets without any consideration of the effects from differences in the chemical environment. From the multiplet structure, one can directly read off the valence of the rare-earth atoms.

5. VALENCE SPECTROSCOPY

Valence states are more complex than core levels, since their wave functions extend out to neighbor atoms and interact with them. Consequently, the set of quantum numbers characterizing valence states is richer, containing the three momentum components as additional quantum numbers (in solids) and forming bonding or antibonding orbital combinations (in molecules). The extra quantum numbers are determined by using additional variables, such as the angular pattern of the photoelectron and the photon polarization. The optimum energy range for mapping valence states has an upper limit of about 50 eV, since the cross section of valence states decays rapidly at higher photon energies.

5.1 Occupied Atomic and Molecular Orbitals

The valence orbitals of free molecules can be resolved in great detail by ultraviolet photoelectron spectroscopy (Turner *et al.*, 1970),

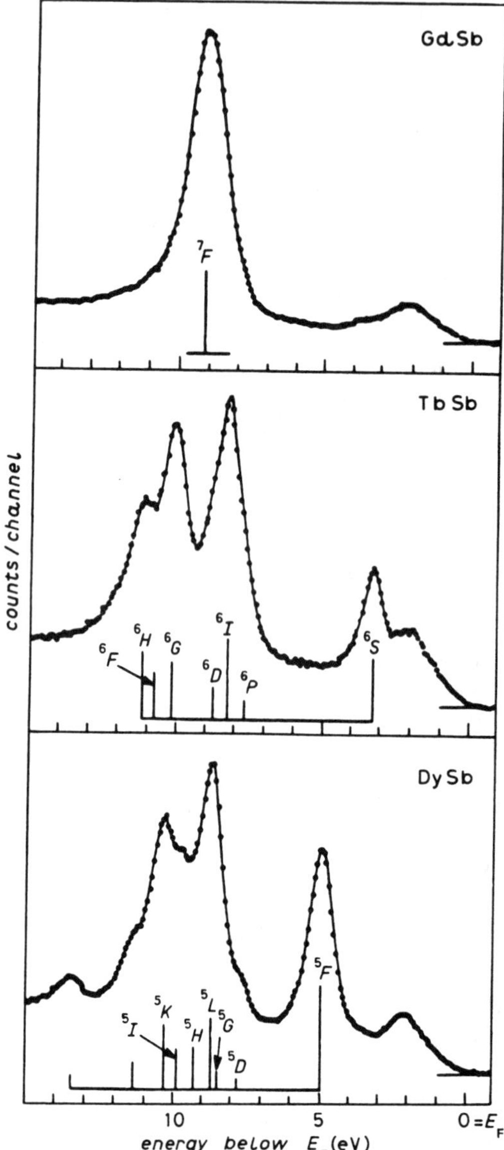

FIG. 12. Core-level spectra of the 4*f* levels for rare-earth antimonides. The agreement with calculated $4f^{n-1}$ multiplets shows that the multiplet structure reflects the number of 4*f* electrons in the final state. From Campagna *et al.* (1976).

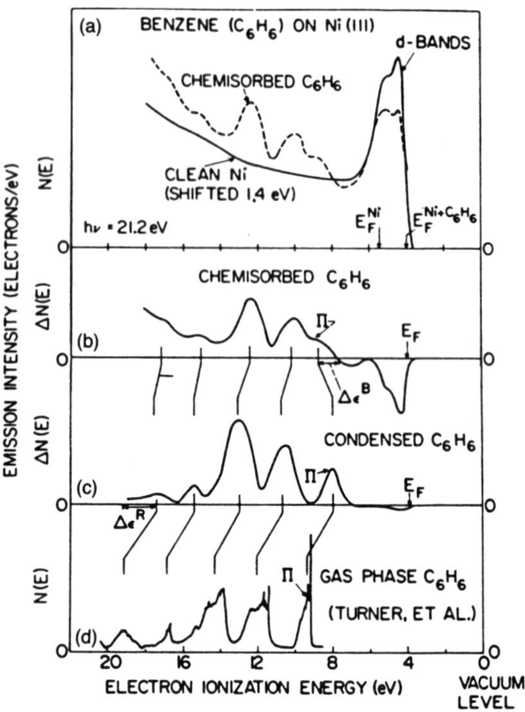

FIG. 13. (a)–(d) Fingerprinting of adsorbed molecules by photoelectron spectroscopy. Benzene in the gas phase exhibits a characteristic pattern of valence orbitals (Turner *et al.*, 1970), which is shifted rigidly in energy after condensing it at a surface. The stronger chemisorption bond on a nickel surface causes the uppermost bonding orbital to fall out of place. From Demuth and Eastman (1974).

including vibrational and rotational fine structure. During adsorption at a surface, e.g., in catalysis-oriented studies, the molecular orbitals can be used for fingerprinting the adsorbate [Fig. 13; Demuth and Eastman (1974)]. Orbitals not involved in the surface-to-adsorbate bond experience only a rigid, upward energy shift, which is due to dielectric screening at the surface. Therefore, the pattern of orbital energies at a surface is often sufficient to identify the adsorbate, to tell whether or not a molecule has decomposed, and to pinpoint the orbitals that are involved in the surface bond by their additional energy shift.

Angular effects are averaged out in randomly oriented molecules up to a single parameter β, which determines the angular distribution $d\sigma/d\Omega$ of photoelectrons in terms of the polar emission angle θ relative to the polarization of the light:

$$\frac{d\sigma}{d\Omega} = \left(\frac{\sigma}{4\pi}\right)[1 + \beta P_2(\cos\theta)]. \tag{7}$$

Here $P_2(x) = (3x^2 - 1)/2$ is the $l = 2$ Legendre polynomial. The asymmetry parameter β can take values between -1 and 2. At the magic angle $\theta = 54°44'$, the emission intensity is independent of θ, making this a useful geometry for quantitative measurements of the cross section σ of a given orbital.

Molecules adsorbed at surfaces can be oriented by a crystalline substrate. In such a case, it becomes possible to determine their orientation by changing the polarization of the light and using dipole selection rules. For example, a p_z orbital is excited with the electric field vector along the z direction, and a $p_{x,y}$ orbital with the component along the x, y direction.

5.2 Surface States

The surface disrupts the potential for electrons in a solid enough to give rise to separate surface states. These can have localized, broken-bond character, such as in semiconductors, or evanescent-wave character, such as delocalized (s,p) states in metals. The d states at metal surfaces lie somewhere in between. Surface states can be ideally probed with photoelectron spectroscopy when working at kinetic energies of 20–70 eV for the photoelectrons, where their escape depth goes through a minimum. Only a few atomic layers are seen in this case. There are several criteria that help in distinguishing surface states from bulk states. The "crud" test is based on the assumption that a surface state is quenched by an adsorbate, which is true in most cases, but not all. A cleaner method utilizes the two-dimensional nature of surface states, which implies that the energy of a surface state does not vary with the momentum perpendicular to the surface. This is tested by varying the photon energy at fixed parallel momentum. The most thorough test for a surface state consists of mapping all the bulk bands and ensuring that the state falls into a symmetry gap of the band structure. There are many surface resonances, however, that do not pass this last test but exhibit strong localization in the outermost layer according to calculation.

5.3 Band Mapping

The electronic structure of a crystalline solid is characterized by a set of quantum numbers. These are energy E, three-dimensional momentum $\mathbf{k}$, spin, and point-group symmetry, i.e., a symmetry equivalent to the angular momentum. This information is usually compounded into $E(\mathbf{k})$-band dispersion plots with the appropriate symmetry labels for each band (Fig. 14). The experimental techniques that can probe the complete set of quantum numbers are photoemission and inverse photoemission, with photoemission probing occupied states and inverse photoemission unoccupied states (see Fig. 1). There are just enough variables in these experiments to cover all the unknowns. Roughly speaking, E is determined from the kinetic energy of the electrons; the two momentum components parallel to the surface, $k_{\parallel}$, from the polar and azimuthal angles of the electrons; the perpendicular momentum from the photon energy; spin from the spin of the electrons; and the point-group symmetry from the photon polarization. A more thorough analysis, particularly concerning the perpendicular momentum, can be found in various re-

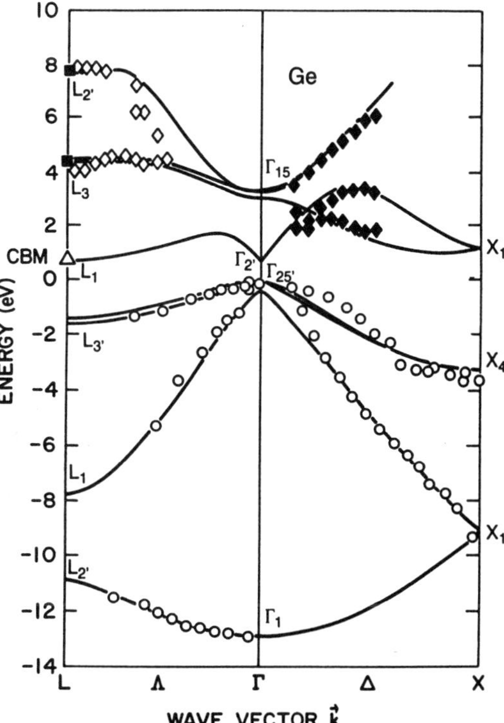

FIG. 14. Energy-vs-momentum band dispersions for germanium obtained from angle-resolved photoemission, inverse photoemission, and quasiparticle band calculations (Ortega and Himpsel, 1993; Hybertsen and Louie, 1986; Wachs *et al.*, 1985). Such band dispersions provide complete information about electrons in solids. The labels denote the point-group symmetry. The energy is given relative to the valence-band maximum.

views (Himpsel, 1983; Kevan, 1992). Figure 14 is characteristic for current band-mapping results, both theoretically and experimentally. Angle-resolved photoemission and inverse-photoemission results are compared to state-of-the-art quasiparticle calculations (Ortega and Himpsel, 1993; Hybertsen and Louie, 1986; Wachs *et al.*, 1985). Traditional local-density calculations would not give the correct band gap for germanium, and in fact predict it to be metallic. A substantial body of experimental and theoretical band dispersions has been accumulated and is being compiled as the basic set of electronic-structure data for solids (Goldmann and Koch, 1989). Current developments are directed toward high energy resolution, in order to map the electronic states at the Fermi level that determine transport phenomena and phase transitions, such as superconductivity and magnetism.

5.4 Time-Resolved, Pump-Probe Measurements

The lifetime of electronic excitations and their decay mechanisms are at the heart of technologies, such as optoelectronics. Time-resolved photoemission with pulsed lasers can go beyond optical experiments in not only determining the combined density of occupied and unoccupied states, but in addition sampling specific states with well-defined momentum. The experiments are typically performed in a pump-probe mode, where a low-energy photon pumps electrons from the valence band into a long-lived, low-lying conduction-band or gap state, and a higher-energy ultraviolet photon ejects them. For a review see Haight (1995).

5.5 Inverse Photoemission

Inverse photoemission may be viewed as the time-reversed photoemission process (see Fig. 1), with an incident electron and an emitted photon. It complements photoemission by probing unoccupied states, and in particular the unoccupied states between the Fermi level and the vacuum level that are not accessible to photoemission—neither as initial nor as final state. This gap contains many interesting states, e.g., minority-spin states in ferromagnets, broken-bond orbitals at semiconductor surfaces, and antibonding orbitals of adsorbed molecules. In order to visualize the territory covered by photoemission and inverse photoemission, one may look at the analogy between occupied and unoccupied states, bonding vs antibonding states, or donor vs acceptor states. Inverse photoemission poses experimental challenges, since its cross section is four to five orders of magnitude lower than for photoemission, as a result of a phase-space factor in the cross section. Typical elastic photoelectron yields are in the range of 10^{-3} electron per photon, while typical inverse photoemission yields are 10^{-8} photon per electron. Nevertheless, this technique has become a routine complement to photoemission work, as exemplified by the conduction-band data in Fig. 14. Results on semiconductors are reviewed by Himpsel (1990), results on metals and adsorbates by Dose (1985) and Smith (1988).

6. STRUCTURAL METHODS

6.1. EXAFS and SEXAFS

As mentioned in Sec. 3.1, EXAFS (extended x-ray-absorption fine structure) provides structural information. The ejected photoelectron from an absorbing atom can be described as an outgoing spherical wave. The outgoing photoelectron wave will be backscattered by neighboring atoms, thereby creating incoming waves, one from each atom. The interference between the outgoing and incoming waves gives rise to modulation of the absorption coefficient for photon energies above the threshold. The periodicity of this modulation contains information about the distance between the absorbing atom and its neighbors, and the amplitude of the modulation is a measure of the number of neighbors, i.e., the coordination number. The EXAFS technique is used extensively for structural determinations of solids, liquids, and biomolecules (Prins and Koningsberger, 1985; Teo, 1986; Teo and Jay, 1981). A surface-sensitive version of EXAFS, with the acronym SEXAFS, can be applied to determine the surface structure of monolayers of adsorbates on surfaces (Prins and Koningsberger, 1985). Experimentally, two parameters are measured: adsorbate–substrate bond lengths and absorption sites. Figure 15 shows SEXAFS results for the much-studied system (2 × 1)O/

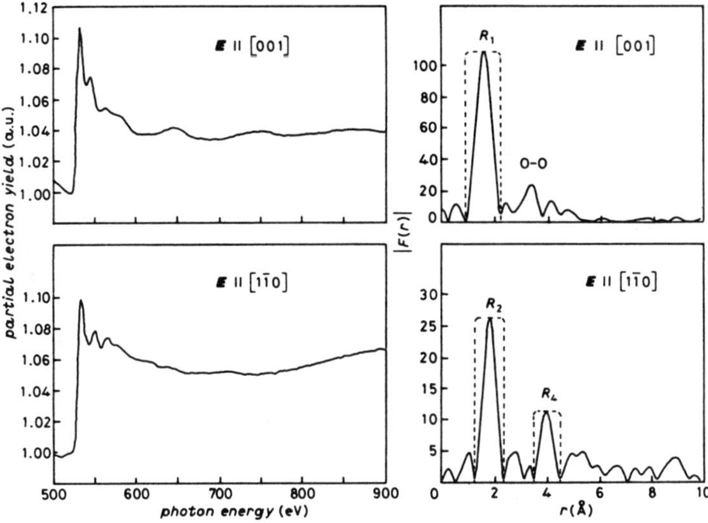

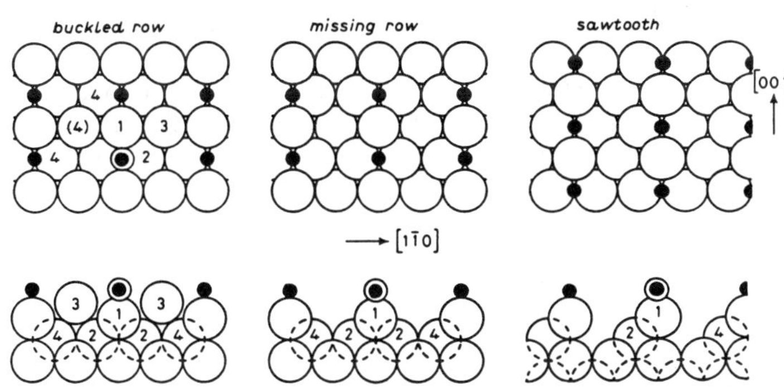

FIG. 15. (a) SEXAFS spectra using the oxygen 1*s* level of an ordered overlayer of oxygen on Cu(110), together with their Fourier transforms. R_1 and R_2 are the O–Cu bond lengths for the first- and second-nearest distances. R_4 corresponds to the bond length between O and Cu in the third layer, i.e., a fourth-nearest neighbor. O–O is a multiple scattering peak. (b) Different models for the reconstruction of the (2 × 1)O/Cu(110) surface in top and side views. After Bader *et al.* (1986).

Cu(110). The experimental spectra are shown for two different orientations of the polarization vector of the exciting synchrotron light. From a detailed analysis, it can be concluded that only the missing-row model is consistent with the SEXAFS data (Bader *et al.*, 1986).

6.2 Photoelectron Diffraction and Holography

While an EXAFS experiment integrates over all emission angles, there is additional information to be gained from the angular distribution of diffracted photoelectrons. The outgoing waves of photoelectrons emitted from localized core levels can scatter from nearby atoms and produce diffraction patterns. Considering the short scattering length of photoelectrons, it is intuitively clear that such diffraction patterns can potentially be used to determine surface structures (Fadley, 1993). Experimentally, the intensity of a core level is measured as a function of either the direction or the energy of the photoelectron, providing what is termed scanned-angle or scanned-energy data. Considerable theoretical effort has gone into establishing accurate models relating photoelectron-diffraction data to the surface structure (Fadley *et al.*, 1994). Compared with x-ray diffraction, the analysis is complicated by multiple scattering, which is inherent in the strong interaction of electrons with matter. That, by the same token, provides the surface sensitivity of the technique. We illustrate the method with two simplified cases. For photoelectrons with kinetic energies above 500 eV, it turns out that the scattering amplitude is highly peaked in the forward direction (forward focusing), which provides a direct determination of the bond directions for absorbed molecules and epitaxial relationships in thin films. A second simplified case is the possibility of determining the path-length difference between the source of the photoelectron and a scattering atom by using scanned-energy data. This requires access to synchrotron radiation so that the photoelectron energy can be tuned continuously. Figure 16 shows the results for an ordered overlayer of sulfur on chromium, $c(2 \times 2)$S/Cr(001), from Terminello *et al.* (1988). It was realized by Szöke (1986) and Barton (1988) that photoelectron-diffraction patterns can form the basis for a hologram. The unscattered photoelectron wave will serve as the reference wave of the hologram and the scattered waves as the object waves. It should thus be possible to get direct images in three dimensions. Preliminary results have been obtained (Fadley, 1993), and extensive research effort is presently underway to pursue this development further.

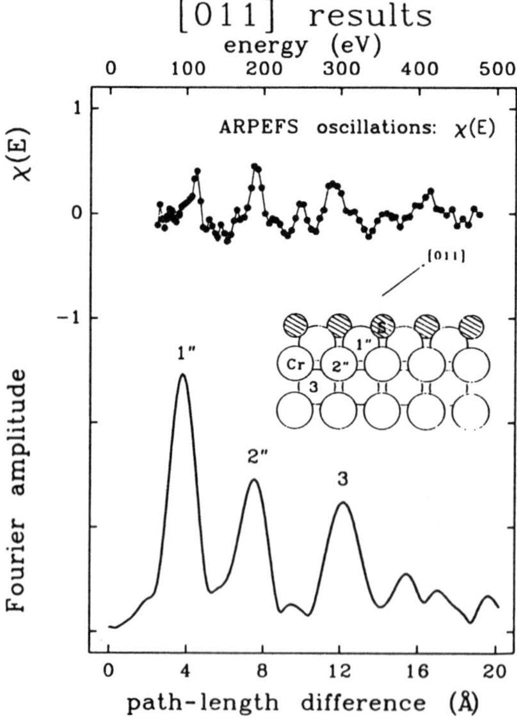

FIG. 16. Photoelectron diffraction from an ordered overlayer of sulfur on Cr(001). Each numbered atom in the side view of the adsorbate system is associated with a peak in the path-length-difference curve. From Terminello *et al.* (1988).

7. SUMMARY AND OUTLOOK

Photoelectron spectroscopy has made significant contributions to the basic understanding of the electronic structure of atoms, molecules, solids, and surfaces. The technique has also been adopted into commercial instrumentation for routine qualitative and quantitative chemical analysis and materials characterization. Presently, hundreds of research papers are published each year involving photoelectron spectroscopy in basic research or applications.

The development of the field of photoelectron spectroscopy, for both basic research and applications, has been closely tied to advancements in instrumentation, i.e., better light sources and detectors. This trend will certainly continue in the foreseeable future. The third-generation synchrotron-radiation sources are presently being implemented and will provide orders-of-magnitude improvements in brightness. It will be possible to study atoms, molecules, and solids with unprecedented spectral resolution. This will have a major impact for a better theoretical understanding of the complex photoelectron spectra of atoms and molecules. For solids and surfaces, major advances can be expected for studies of the electronic states around superconducting gaps in high-T_c materials, and of charge-density waves, surface transitions, and surface magnetism. Major breakthroughs should be anticipated in photoelectron microscopy where all the assets of photoelectron spectroscopy will be applied with a spatial resolution down to a few tens of nanometers. Therefore, these are all reasons to predict that the tremendous growth photoelectron spectroscopy has enjoyed during the last 30 years will continue well into the future.

GLOSSARY: DEFINITIONS, ACRONYMS, JARGON

Al *Kα*: X-ray line at 1.49 keV photon energy, emitted from the 1*s* level of aluminum. It is frequently used for core-level photoelectron spectroscopy (see Sec. 2.1).

ARPES: Angle-resolved photoelectron spectroscopy.

Auger Spectroscopy: Use of the Auger electrons emitted during the decay of core holes for elemental analysis at surfaces (see Fig. 2).

BIS: Bremsstrahlung isochromat spectroscopy, traditional label for inverse photoemission (see Sec. 5.6). Now used mainly for work in the x-ray regime.

ESCA: Electron spectroscopy for chemical analysis, mapping of the core-level energies and their chemical shifts, typically using the Al or Mg *Ka* lines at 1486.7 and 1253.6 eV, respectively. Synonym: XPS.

EXAFS: Extended X-ray-absorption fine structure, determination of nearest-neighbor distances by interference effects in the core-level absorption spectrum (see Sec. 6.2).

GW: Quasiparticle calculation method based on solving the Dyson equations with the full electron and photon propagators *G* and *W*, and a free-particle approximation for the vertex. It goes beyond the local-density approximation (LDA) in determining the electronic structure for the excited state, which is measured in photoemission and inverse photoemission (see Sec. 1.3).

He I: Resonance line of helium gas at 21.2 eV photon energy, frequently used for photoelectron spectroscopy of valence states (see Sec. 2.1).

LDA: Local-density approximation, representing one of the most common methods for determining electronic states in solids and at surfaces (see Sec. 1.3).

LEED: Low-energy electron diffraction, often used in conjunction with photoelectron spectroscopy to characterize the ordering at surfaces.

MCD: Magnetic circular dichroism, change of absorption between left- and right-handed circularly polarized light, used for the determination of the magnetization at specific atomic sites (see Sec. 3.4).

NEXAFS: Near-edge x-ray-absorption fine structure, mapping of unoccupied states via transitions from core levels (see Sec. 3). Synonyms: XANES, partial-yield spectroscopy.

PhD: Photoelectron diffraction, determination of the local atomic structure of surfaces (see Sec. 6.2).

SEXAFS: Surface extended x-ray–absorption fine structure, determination of nearest-neighbor distances at surfaces by interference effects in the core-level absorption spectrum (see Sec. 6.2).

Synchrotron Radiation: Versatile source of ultraviolet and x-ray light, emitted by high-energy electrons in a storage ring deflected by a magnetic field (see Sec. 2.1).

UPS: Ultraviolet photoelectron spectroscopy, mapping of the valence-band structure of solids and surfaces with photon energies in the 5–50-eV range (see Fig. 3).

XANES: X-ray–absorption near-edge structure, mapping of unoccupied states via transitions from core levels (see Sec. 3). Synonyms: NEXAFS, partial-yield spectroscopy.

XPS: X-ray photoelectron spectroscopy, mapping of the core-level energies and their chemical shifts, typically using the Al or Mg *Kα* lines at 1486.7 and 1253.6 eV, respectively. See also ESCA.

DATABASES

Core-level binding energies: Cardona and Ley (1978).

Atomic core-level spectra: Siegbahn and Karlsson (1982).

Absorption and core-level spectra of metal atoms: Sonntag and Zimmermann (1992).

Molecular core-level absorption: Hitchcock and Mancini (1994).

Molecular core-level energies: Bakke *et al.* (1980).

Molecular valence spectra: Turner *et al.* (1970).

Solid-state core and valence spectra: Goldmann and Koch (1989).

Cross sections: Yeh and Lindau (1985); Keski-Rahkonen and Krause (1974).

Various databases are published regularly in *Surface Science Spectra*, published by the American Institute of Physics for the American Vacuum Society.

Works Cited

Ade, H., Zhang, X., Cameron, S., Costello, C., Kirz, J., Williams, S. (1992), *Science* **258**, 972–975; the data shown in Fig. 6 were obtained from the *NSLS Newsletter*, November 1993.

Andersen, J. N., Qvarford, M., Nyholm, R., Sorensen, S. L., Wigren, C. (1991), *Phys. Rev. Lett.* **67**, 2822–2825.

Bader, M., Haase, J., Puschmann, A., Ocal, C. (1986), *Phys. Rev. Lett.* **57**, 3273–3276.

Bakke, A. A., Chen, H.-W., Jolly, W. (1980), *J. Electron Spectros.* **20**, 333–366.

Barton, J. (1988), *Phys. Rev. Lett.* **61**, 1356–1359.

Björneholm, O., Nilsson, A., Tillborg, H., Bennich, P., Sandell, A., Hernnäs, B., Puglia, C.,

Mårtensson, N. (1993), in: J. N. Andersen, L. S. O. Johansson, R. Nyholm (Eds.), *MAX-lab Activity Report 1992*, Lund, Sweden, p. 58.

Campagna, M., Wertheim, G. K., Bucher, E. (1976), *Struct. Bonding* **30**, 99–140.

Cardona, M., Ley, L. (1978), *Photoemission in Solids I*, Topics in Applied Physics, Vol. 26, Springer, Berlin.

Carlson, T. A. (1967), *Phys. Rev.* **156**, 142–149.

Demuth, J. E., Eastman, D. E. (1974), *Phys. Rev. Lett.* **32**, 1123–1127.

Dose, V. (1985), *Surf. Sci. Rep.* **5**, 337–378.

Einstein, A. (1905), *Ann. Phys.* **17**, 132–148.

Fadley, C. S. (1978), in: C. R. Brundle, A. D. Baker (Eds.), *Electron Spectroscopy: Theory, Techniques and Applications*, Vol. 2, New York: Academic Press, p. 1.

Fadley, C. S. (1993), in: R. Z. Bachrach (Ed.), *Synchrotron Radiation Research: Advances in Surface Science*, New York: Plenum.

Fadley, C. S. *et al.* (1994), *J. Electron Spectrosc. Relat. Phenom.* **68**, 19–47.

Feuerbacher, B., Fitton, B., Willis, R. F. (1978), *Photoemission and the Electronic Properties of Surfaces*, New York: Wiley.

Fowler, R. F. (1931), *Phys. Rev.* **38**, 45–56.

Gelius, U. (1974), *J. Electron Spectrosc. Relat. Phenom.* **5**, 985–1057.

Goldmann, A., Koch, E. (Eds.) (1989), *Landolt-Börnstein: Numerical Data and Functional Relationships in Science and Technology*, New Series, Group III, Vol. 23a, *Electronic Structure of Solids: Photoemission Spectra and Related Data*, Berlin: Springer; additional subvolumes in preparation.

Gudat, W., Kunz, C. (1972), *Phys. Rev. Lett.* **29**, 169–172.

Haight, R. (1995), *Surf. Sci. Rep.*, in press.

Hertz, H. (1887), *Ann. Phys.* **32**, 983.

Himpsel, F. J. (1983), *Adv. Phys.* **32**, 1–57.

Himpsel, F. J. (1990), *Surf. Sci. Rep.* **12**, 1–48.

Himpsel, F. J., McFeely, F. R., Taleb-Ibrahimi, A., Yarmoff, J. A., Hollinger, G. (1988), *Phys. Rev. B* **38**, 6084–6096.

Himpsel, F. J., Meyerson, B. S., McFeely, F. R., Morar, J. F., Taleb-Ibrahimi, A., Yarmoff, J. A. (1990), in: M. Campagna, R. Rosei (Eds.), *Photoemission and Absorption Spectroscopy of Solids and Interfaces with Synchrotron Radiation*, Amsterdam: North Holland, p. 203.

Hitchcock, A. P., Mancini, D. C. (1994), *J. Electron Spectrosc.* **67**, 1–132.

Hybertsen, M. S., Louie, S. G. (1986), *Phys. Rev. B* **34**, 5390–5413.

Johansson, B., Mårtensson, N. (1980), *Phys. Rev. B* **21**, 4427–4457.

Keski-Rahkonen, O., Krause, M. O. (1974), *Atomic Data Nucl. Data Tables* **14**, 139–146.

Kevan, S. D. (Ed.) (1992), *Angle-Resolved Photoemission*, Amsterdam: Elsevier North Holland.

Koch, E. E. (Ed.) (1983), *Handbook on Synchrotron Radiation*, Amsterdam: North Holland.

Lindau, I., Spicer, W. E. (1974), *J. Electron Spectrosc.* **3**, 409–413.

Mackey, K. J., Allen, P. M. G., Herrenden-Harker, W. G., Williams, R. H., Whitehouse, C. R., Williams, G. M. (1986), *Appl. Phys. Lett.* **49**, 354–356.

Millikan, R. A. (1916), *Phys. Rev.* **7**, 18–32 and 355–388.

Ortega, J. E., Himpsel, F. J. (1993), *Phys. Rev. B* **47**, 2130–2137.

Prins, R., Koningsberger, D. (Eds.) (1985), *X-Ray Absorption: Principles, Applications, Techniques of EXAFS, SEXAFS and XANES*, New York: Wiley.

Rudolf, P., Sette, F., Tjeng, L. H., Meigs, G., Chen, C. T. (1992), *J. Magn. Magn. Mater.* **109**, 109–112.

Schott, G. A. (1912), *Electromagnetic Radiation*, Cambridge, U.K.: Cambridge Univ. Press.

Schwinger, J. (1949), *Phys. Rev.* **75**, 1912–1925; see also (1946) *Phys. Rev.* **70**, 798–799 (A).

Siegbahn, K., Nordling, C., Fahlman, R., Nordbeck, R., Hamrin, K., Hedman, J., Johansson, G., Bengmark, T., Karlsson, S.-E., Lindgren, I., Lindberg, B. (1967), *ESCA: Atomic, Molecular, and Solid State Structure Studied by Means of Electron Spectroscopy*, Stockholm: Almquist and Wiksells.

Siegbahn, H., Karlsson, L. (1982), in: S. Flügge (Ed.), *Encyclopedia of Physics*, Vol. XXXI, New York: Springer, pp. 215–468.

Smith, N. V. (1988), *Rep. Prog. Phys.* **51**, 1227–1294.

Sonntag, B., Zimmermann, P. (1992), *Rep. Prog. Phys.* **55**, 911–987.

Stöhr, J. (1992), *NEXAFS Spectroscopy*, Springer Series in Surface Sciences, Vol. 25, Heidelberg: Springer.

Szöke, A. (1986), in: D. T. Attwood, J. Baker (Eds.), *Short Wavelength Coherent Radiation: Generation and Applications*, American Institute of Physics Conference Proceedings, No. 147, New York: AIP.

Teo, B. K., Jay, D. C. (Eds.) (1981), *EXAFS Spectroscopy: Techniques and Applications* New York: Plenum.

Teo, B. K. (1986), *EXAFS: Basic Principles and Data Analysis* New York: Springer.

Terminello, L. J., Zhang, X. S., Huang, Z. Q., Kim, S., Schach von Wittenau, A. E., Leung, K. T., Shirley, D. A. (1988), *Phys. Rev. B* **38**, 3879–3891.

Turner, D. W., Baker, C., Baker, A. D., Brundle, C. R. (1970), *Molecular Photoelectron Spectroscopy*, London: Wiley, Interscience.

Wachs, A. L., Miller, T., Hsieh, T. C., Shapire, A. P., Chiang, T. C. (1985), *Phys. Rev. B* **32**, 2326–2333.

Weaver, J. H. (1988), in: K. N. Tu, R. Rosenberg (Eds.), *Analytical Techniques for Thin Films*, Treatise on Materials Science and Technology, Vol. 27, New York: Academic, p. 15.

Winick, H., Doniach, S. (Eds.) (1980), *Synchrotron Radiation Research* New York: Plenum.

Winick, H., Xian, D., Ye, M.-H., Huang, T. (Eds.) (1989), *Applications of Synchrotron Radiation* New York: Gordon and Breach.

Yeh, J. J., Lindau, I. (1985), *Atomic Data Nucl. Data Tables* **32**, 1–155.

PHOTOGRAPHIC INSTRUMENTATION

See RECORDING, PHOTOGRAPHIC; TECHNOLOGY AND PHYSICS OF PHOTOGRAPHY

PHOTOLUMINESCENCE

CLAUDE BENOIT À LA GUILLAUME, *Groupe de Physique des Solides, Université Denis Diderot et Université Pierre et Marie Curie, Paris, France.*

	Introduction	497
1.	**Physical Principles**	498
1.1	General Scheme	498
1.2	Photoluminescence Quantum Efficiency	498
1.3	Main Types of Luminescence Materials	498
1.4	Specificity of Photoexcitation	499
2.	**Methods and Equipment**	499
2.1	Excitation Sources	499
2.2	Spectrometers and Detectors	499
2.3	Control of Sample Temperature	500
2.4	Photoluminescence Kinetics and Excitation Spectrum	500
2.5	Time Resolution	500
2.6	Spatial Resolution	501
2.7	External Perturbations	501
3.	**Applications**	502
3.1	Fundamental Applications	502
3.1.1	Organic Crystals	502
3.1.2	Luminescent Centers in Transparent Solids	502
3.1.3	Bulk Semiconductors	503
3.1.3.1	Intrinsic Luminescence	503
3.1.3.2	Effect of Impurities and Defects	504
3.1.4	Low-Dimensional Semiconductor Structures	506
3.1.5	Optical Pumping and Hot Luminescence	507
3.1.6	Exciton Transport	508
3.2	Excitation–Medium Interaction	508
3.2.1	Electron–Phonon Coupling, Polarons, Self-Trapping	508
3.2.2	Electron–Magnetic-Ion Exchange Interaction, Magnetic Polaron	508
3.3	High Excitation Effects	508
3.3.1	Saturation, Optical Gain, Laser	509
3.3.2	Many-Body Effects: Biexcitons, Electron–Hole Plasma or Liquid	509
3.3.3	Higher-Order Emission Processes	509
3.4	Practical Applications	509
3.4.1	Nondestructive Material Testing	509
3.4.2	Devices: Optically Pumped Solid-State Lasers and Amplifiers	510
	Glossary	510
	Works Cited	511
	Further Reading	511

INTRODUCTION

Luminescence is described as light emitted by a *cold* excited body, in contrast with the light emitted by incandescent bodies. There is a variety of excitation processes: In photoluminescence, the excitation results from light absorption. Because of the quantum nature of light and considering only low excitation conditions, one absorbed photon can induce at best one luminescence photon whose energy is necessarily smaller than (or equal to) that of the exciting photon (Stokes's law). Historically, luminescence was restricted to the visible part of the spectrum, but nothing prevents an extension toward the ultraviolet or infrared regions. By comparison with other excitation processes like electron beam (see CATHODOLUMINESCENCE), electric field (see ELECTROLUMINESCENCE), injection in *p-n* junctions (see LIGHT-EMITTING DIODES), chemiluminescence, and bioluminescence, photoluminescence allows the best control over the excitation conditions and for that reason is the most useful process for studying the fundamental properties of luminescent materials. Development of luminescent materials for fluorescent tubes and cathode-ray

3-527-28135-5/95/$5.00 + .50

tubes took place in the 1940s. Study of photoluminescence in organic and semiconductor materials started at the end of the 1950s and developed very fast with the advent of lasers. The tremendous improvements of semiconductor quality linked to microelectronic applications, in terms of purity and doping control, allowed the identification of well-established luminescence mechanisms. Conversely, photoluminescence has become a very common nondestructive test of semiconductor structures.

1. PHYSICAL PRINCIPLES

1.1 General Scheme

Photoluminescence can be described as a three-step process:

1. absorption of one photon of the excitation source creating in the material a primary excitation;
2. evolution of this primary excitation by energy relaxation or energy transfer toward secondary excitation at lower energy; and
3. coupling of this secondary excitation to light, giving rise to the spontaneous emission of one luminescence photon.

Steps **1** and **3** require, as in the field of optical properties (see OPTICAL PROPERTIES OF SOLIDS), the knowledge (a) of the electronic energy spectrum of the material and (b) of the strength of optical transition between any pair of energy levels. Traditionally, the energy spectrum of condensed matter is described in the framework of the *Born-Oppenheimer approximation*, which relies upon the fact that the movements of the nuclei are much slower than those of the electrons, because of the large mass ratio. Actually, electronic excitations and lattice vibrations are coupled with a strength depending on the material under consideration. Luminescence is linked mainly with electronic excitations; coupling between electronic and lattice excitations enters into play mainly in step **2** to allow for some energy relaxation. Let us note that Stokes's law is strictly valid only at temperature $T = 0$; at finite T, there is a finite probability to see the excitation jumping to a higher energy by phonon absorption: This leads to anti-Stokes emission.

1.2 Photoluminescence Quantum Efficiency

The photoluminescence quantum efficiency η is defined as the ratio of the number of photons emitted n_l to the number of excitation photons absorbed n_a: $\eta = n_l/n_a$. The fact that $\eta < 1$ is due to nonradiative processes that allow complete energy dissipation of the exciting photon. Among these are the multiphonon processes where an electronic excitation ΔE is dissipated into n lattice vibrations. Since the probability of such process decreases very fast with n, an energy gap large enough between the emitting state and the ground state is required in order to have a sizable quantum efficiency. This explains why metals, which exhibit a continuum of electronic excited states, are not luminescent materials.

1.3 Main Types of Luminescence Materials

To be more specific, we will consider three main types of luminescent materials:

1. Molecular organic materials (see SEMICONDUCTORS, ORGANIC—ELECTRONIC PROPERTIES). In that case, the building block is the isolated molecule (see MOLECULES), which is only weakly perturbed by its environment, e.g., a solvent or its inclusion in a molecular solid, tied by van der Waals interactions.
2. Luminescent centers in large-gap (hence transparent) insulators, like oxides, tungstates, silicates, or glasses (see ELECTRON STATES, LOCALIZED). The most studied are transition-metal or rare-earth ions; the isolated ions have usually a rich spectrum of excitations linked to their partially filled d or f shell, providing opportunities to obtain luminescence in the visible. Perturbation of these ions by the crystalline matrix is usually treated in the crystal-field approximation. Another example is the F center in alkali halides, where an electron is trapped in an anion vacancy.
3. Semiconductors. They present an energy gap E_g between a filled valence band and an empty conduction band. In that case, luminescence takes place at an energy close to the band gap, which spans a range of values from near ultraviolet to infrared, depending on the material. The building blocks are now free electrons in the conduction band and free holes in the valence

band, characterized by their effective masses (see ELECTRONIC STRUCTURES OF SOLIDS). When Coulomb interaction between these particles and with impurities is taken into account, weakly bound states are built, which can be described in the effective-mass approximation: free excitons (see EXCITONS) or bound excitons; they dominate the luminescence spectrum in pure semiconductors at low temperature.

1.4 Specificity of Photoexcitation

Through the use of monochromatic excitation, very specific energy levels can be excited; this can be used either to optimize the luminescence quantum efficiency or for fundamental research in order to get a better understanding of step **2**. In addition, one can take advantage of light polarization, which allows additional selectivity in anisotropic materials and is at the roots of a technique known as "optical pumping" (see Sec. 3.1.5).

Depending upon the energy difference $E_a - E_l$ between excitation and luminescence, one can distinguish three cases (applicable mainly to semiconductors):

1. Resonance luminescence, when $E_a = E_l$. This is sometimes called Rayleigh scattering.
2. Hot luminescence, when E_l spans a region between E_a and the main luminescence band.
3. Thermalized luminescence corresponding to the main luminescence band, close to the band gap.

The relative importance of these three contributions is governed by a system of rate equations describing such an out-of-equilibrium situation.

2. METHODS AND EQUIPMENT

A typical experimental photoluminescence setup is shown in Fig. 1. It is composed of an excitation source, optical elements to focus excitation on the sample and collect the luminescence light, a spectrometer (or filter) to analyze the luminescence, and a detector followed by a data acquisition system.

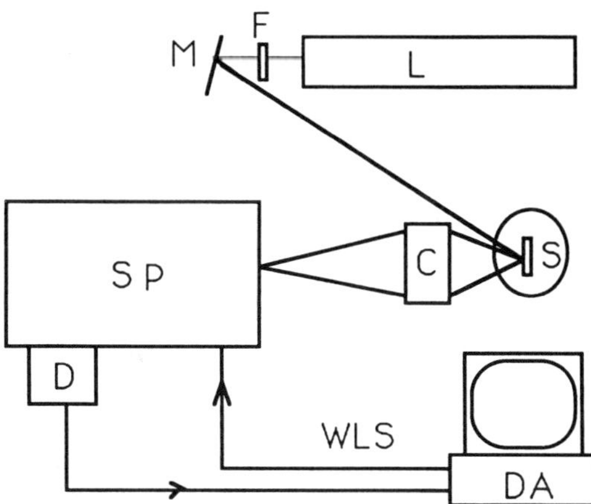

FIG. 1. Typical photoluminescence setup. *L*, laser; *F*, filter; *M*, mirror; *S*, sample; *C*, condenser; *SP*, spectrometer; *D*, detector; *DA*, data acquisition system; WLS, wavelength scanning.

2.1 Excitation Sources

Before the advent of lasers, the most useful sources were high-pressure mercury or xenon lamps. Both have a broad continuous spectrum extending from 300 nm to near infrared. Appropriate filters have to be inserted in order to cut source emission in the spectral region of sample luminescence. Lasers (see LASERS, GAS; LASERS, SOLID STATE) are excitation sources much more convenient for several reasons:

1. They are much more powerful.
2. Since they are well collimated, it is easy to use 100% of their power, as compared to 15% at best for conventional sources.
3. Their emission is essentially monochromatic.

Notice, however, that a filter may be required to suppress plasma emission lines in gas lasers. A wide range of wavelengths is now available from the infrared (CO_2 near 10 μm, CO near 5 μm, and Nd solid-state laser at 1.06 μm), through gas lasers in the visible (He-Ne, Ar, Kr), and recently gas lasers with intracavity second-harmonic generation for the 250-nm region.

2.2 Spectrometers and Detectors

If in some cases a filter could be enough to select a luminescence line, in most cases a spectrometer is needed to get the luminescence spectrum (see SPECTROMETERS, OPTICAL). The most widely used are grating spectrometers, either single or double. The light emitted by the sample should be focused on the entrance slit of the spectrometer by an objective or a combination of mirrors in the infrared to avoid absorption and dispersion.

In the ultraviolet, visible, and near infrared down to 1 μm, the most useful detectors are photomultipliers (see DETECTORS, RADIATION). Semiconductor detectors (photovoltaic, photoconductor, avalanche *p-n* junction) cover the visible and infrared range down to 10 μm (see DETECTORS, SEMICONDUCTOR). Chopping of the excitation at a frequency f followed by synchronous detection is common practice. Let us point out that this is mandatory only when using detectors exhibiting $1/f$ noise, since the detector contributes to the noise in a window δf around the frequency f. With photomultipliers, the best is to use a discriminator followed by a counter. Detector arrays, featuring up to 2048 (or even more) elements, provide a very efficient detection method, since wavelength scanning is suppressed. Fourier transform spectrometers (see INTERFEROMETERS AND INTERFEROMETRY; SPECTROMETERS, INFRARED) have also been used in the infrared; for a given spectral resolution, they can accommodate a luminescence beam of larger extension.

2.3 Control of Sample Temperature

Sample temperature is an important parameter in photoluminescence experiments, since, in general, line narrowing is expected on cooling. Moreover, new ground or excited states (see Sec. 3.2) can appear at low temperature, like some type of condensation (see Sec. 3.3.2). Optical cryostats (see CRYOGENICS) allow easy operation down to 1.5 K, the best cooling conditions being achieved by immersion of the sample in superfluid liquid helium. In peculiar cases, operation down to about 0.1 K has been obtained with a dilution refrigerator; in that case, excitation and luminescence light are transported through optical fibers, in order to minimize heat dissipation. Experiments between helium and room temperature are also easily performed—for example, using commercial closed-cycle refrigerators.

2.4 Photoluminescence Kinetics and Excitation Spectrum

It is often useful to study photoluminescence *kinetics*, i.e., how the luminescence intensity depends upon the excitation intensity, taking care of the fact that different luminescence lines may exhibit different kinetics. Another powerful way to obtain information about the photoluminescence process is the study of the dependence upon excitation wavelength of the luminescence spectrum. This is performed usually by recording the *photoluminescence excitation spectrum*, the excitation wavelength being scanned at fixed luminescence wavelength. A tunable source is required; this can be done either with a broad-band source and a monochromator or, more conveniently, with a tunable laser, like a dye laser (see ORGANIC DYE LASERS) or a solid-state tunable laser (titanium–sapphire or F-center). Often, the excitation spectrum is close to the absorption spectrum of the sample. Light *polarization* is an additional useful feature: by inserting a polarizer (either linear or circular) in the excitation and/or the luminescence beam, one can gain information about the symmetries of the primary excited state and the final emitting state, when the sample presents some kind of anisotropy, either natural or induced by external perturbations (see Sec. 2.7). Selection of the luminescence photon wave vector, realized by inserting a slit out of focus to limit the solid angle of the luminescence beam, can also be useful with anisotropic samples. Even with isotropic samples, a polarized excitation can induce some degree of polarization of the luminescence, a technique known as *optical pumping* (see Sec. 3.1.5).

2.5 Time Resolution

Time resolution adds new capabilities to directly measure characteristic times in the sample, like lifetime of states or transfer time between states. It requires first a pulsed excitation. For decay time longer than about 10 ns, modulation by electro-optical (see QUANTUM-OPTOELECTRONIC DEVICES) or acousto-optical (see MODULATORS AND DEMODULATORS, ACOUSTICAL) devices can be used. For shorter time scales down to 0.1 ns, mode-locked lasers can be used. Synchronously pumped dye (or titanium–sapphire) lasers can provide pulses reaching the subpicosecond range (see ULTRAFAST PULSE PHENOMENA), tunable in a substantial wavelength range.

Detection was done first using fast oscilloscopes or boxcar techniques, which require a scanning of the delay between excitation and gate defining the measurement window. A better method, in conjunction with photomultiplier detectors, is the so-called *single-photon counting* technique: the time delay between the excitation pulse and the first luminescence photon detected is measured with a time-to-amplitude converter, and the data are stored in the multichannel mode. This method is very efficient at low levels of excitation since essentially all luminescence photons are detected. With fast discriminators, time resolution below 50 ps is achievable. A second method relies upon *up-conversion* in a nonlinear medium (see OPTICS, NONLINEAR): Part of the excitation pulse is combined with the luminescence beam in a nonlinear crystal under phase-matching conditions; photons at an energy that is the sum of excitation and luminescence photon energies are detected. This method, less sensitive than the previous one, allows a time resolution down to the subpicosecond range; moreover, it can be used for infrared luminescence. A third method uses a *streak camera*: The luminescence light focused on a transparent photocathode creates electrons focused on a fluorescent screen after deflection by a sawtoothlike electric field synchronized with the excitation pulse. The bright trace on the screen, which is an image of the time-dependent luminescence, is read by an array of detectors. This method is fairly sensitive with a typical time resolution of about 10 ps and a spectral domain of wavelength below 850 nm. In all cases where the expected time resolution is below 10 ps, time broadening due to the passage on one grating has to be suppressed with the use of a double-grating spectrometer in the substractive mode.

Detection of *quantum beats* occurs when two nearby energy states separated by ΔE have been both excited by a short pulse; during their coherent decay, beats with a period $\Delta t = \hbar/\Delta E$ can be detected on the luminescence signal, under appropriate polarization conditions, thus allowing measurement in the time domain of small energy splitting.

2.6 Spatial Resolution

Using well-focused excitation in a spot of a few microns, any extension of the luminescence spot gives information about the diffusion of the excitations. This can be done by inserting a slit in the plane of an enlarged image of the emitting region. Such measurements can be performed in the time-resolved mode, allowing the measurement of a signal $S(\Delta x,\Delta\tau)$, where Δx is the distance to the exciting spot and $\Delta\tau$ is the delay after the exciting pulse. Under typical conditions, the analysis of $S(\Delta x,\Delta\tau)$ provides access to the lifetime and the diffusion coefficient of the excitations.

Another more powerful technique is provided by *four-wave mixing* experiments (see OPTICS, NONLINEAR), but this is outside the scope of this article.

2.7 External Perturbations

The purpose of external perturbations applied to the sample is to induce shift or splitting (see ELECTRON LEVEL SPLITTING) of the energy levels in order to help their theoretical identification. On solid samples, *hydrostatic pressure* up to 10 GPa can be applied by use of a diamond anvil cell: This induces shifts and in some cases phase transitions. *Uniaxial stress* of much lower intensity can be applied; it can induce splitting associated with polarization effects. It can also be applied as a modulation, using piezoelectric elements.

Magnetic fields up to about 10 T can be applied by use of split-coil superconducting magnets to allow light propagation perpendicular or parallel to the field. Different effects are induced, like Zeeman splitting, diamagnetic shift, or Landau level splitting; they can be detected on the photoluminescence spectrum, mainly at low temperature. Photoluminescence and electron paramagnetic resonance (see ELECTRON PARAMAGNETIC RESONANCE) can be combined into a technique known as *optical detection of magnetic resonance* (Cavenett, 1981): The change in spin populations at resonance is detected through modification of the intensity of some polarized components of photoluminescence lines. This method is quite sensitive and provides a correlation between different magnetic resonance components (i.e., different g factors) and different luminescence components. A related method, *level-crossing spectroscopy*, does not require microwaves: When two magnetic sublevels happen to cross, transfer between these two levels is enhanced and changes in polarized photoluminescence intensity can be recorded as bump or dip as a function of the field strength.

To conclude this section, let us point out that several extensions of photoluminescence as described in Secs. 2.3 to 2.7 can be combined in a given experiment, provided they are not mutually exclusive.

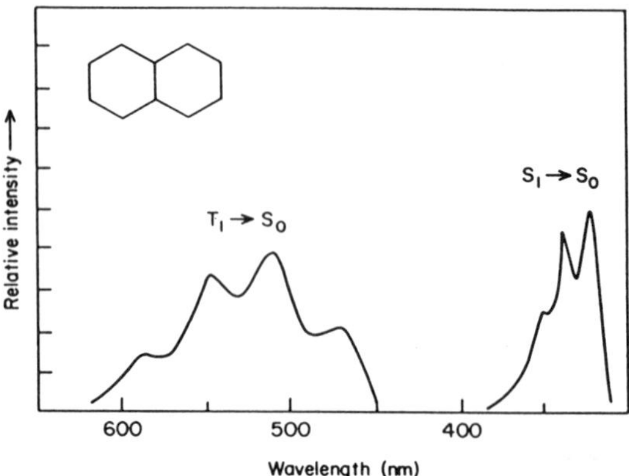

FIG. 2. Fluorescence ($S_1 \rightarrow S_0$) and phosphorescence ($T_1 \rightarrow S_0$) spectra of naphthalene at room temperature. The $T_1 \rightarrow S_0$ spectrum was obtained using a rigid polystyrene solution of naphthalene. The two spectra are plotted on different intensity scales.

3. APPLICATIONS

3.1 Fundamental Applications

In general, photoluminescence spectroscopy provides information about energy levels, as do other kinds of optical spectroscopy (see SPECTROSCOPY OF SOLIDS), but with special emphasis upon the lowest excited states. In addition, increasing the excitation level may induce new collective states (see Sec. 3.3.2). We give first examples of low-level photoluminescence in the three kinds of luminescence materials defined in Sec. 1.3. Particular emphasis is put on semiconductors.

3.1.1 Organic Crystals As an example, we will consider solid naphthalene or anthracene. Looking first into the lowest energy states of the isolated molecule, one finds a singlet ground state denoted S_0, and a first excited singlet state S_1 with a corresponding triplet state T_1 more than 0.5 eV below. These states are coupled to vibrational modes of the molecule. The optical transition $S_0 \rightarrow S_1$ is allowed whereas $S_0 \rightarrow T_1$ is forbidden. So the luminescence is dominated by the $S_1 \rightarrow S_0$ transition, accompanied on its low-energy side by several molecular vibration sidebands: This is called "prompt" fluorescence. The T_1 state is also populated from S_1 by intersystem crossing; it gives rise to slow phosphorescence (Fig. 2).

In the molecular crystal, where binding is due to van der Waals forces, the prompt fluorescence is not very different from that of the molecule; there is a moderate red shift due to the binding. The main difference is the possibility of rapid *energy transfer* to neighbors: At low temperature, this takes place through the formation of Frenkel excitons (see EXCITONS), which are delocalized excitations characterized by a wave vector **k**. When the temperature increases, the mean free path of the Frenkel exciton can decrease below the intermolecular distance. Then one has an incoherent hopping regime for these excitations. That rather fast energy transfer has several consequences: If the crystal is doped with a parent molecule, like naphthacene in anthracene, S_1 excitons can be trapped by the impurities; then the matrix prompt fluorescence is quenched at the benefit of naphthacene emission at lower energy. Another effect comes from collisions between two triplet excitons resulting in regeneration of a singlet S_1 (called also triplet fusion); this is the origin of "delayed" fluorescence.

Let us add that more complicated molecules, the organic dyes, are used in liquid solution to achieve tunable lasers (see ORGANIC DYE LASERS).

3.1.2 Luminescent Centers in Transparent Solids Most of the examples concern rare-earth or transition-metal ions in large-gap matrices (see SOLID, LIQUID, AND GASEOUS INSULATOR MATERIALS). Going from rare-earth elements to transition metals, as a result of the smaller diameter of f orbitals as compared with d orbitals, rare-earth ions are less perturbed by the crystal field and less coupled to lattice vibrations. Thus, emission and absorption lines of rare-earth ions are quite narrow. Some inhomogeneous broadening occurs in glasses (Er^{3+} in silica glass fiber, Nd^{3+} in glass). Strong coupling with lattice vibrations can lead to broad absorption and emission bands, like in the case of titanium-doped sapphire. Usually, ion concentration is in the few percent range so that samples of length in the centimeter range are required for laser emission. Concentration quenching is observed mainly for transition-metal ions (above 0.3% in ruby: Cr^{3+} in Al_2O_3). For rare-earth ions, stoichiometric material like $PrCl_3$ exhibits no luminescence quenching. Such material can give a micron-size optically pumped laser.

Another type of defect in ionic crystals has been widely studied: the *F* center, which corresponds to an electron trapped in an anion vacancy. Since the electron wave function extends toward the neighboring cations, strong electron–phonon coupling induces a large Stokes shift between absorption and emission (Fig. 3).

A nonlinear effect giving rise to "up-conversion" occurs in some transparent crystals doped with two types of luminescent ions, the first emitting at an energy E and the second at an energy E' smaller than but close to $2E$. If a lot of ions of the first type are excited, an energy transfer from two such ions to one of the second kind may occur, followed by the emission of a photon at energy E'. If E is in the near infrared, then an IR–visible conversion results.

3.1.3 Bulk Semiconductors

3.1.3.1 Intrinsic Luminescence. High-temperature photoluminescence in pure semiconductors is dominated by the band-to-band process: Free electrons recombine with free holes. This is a typical case of thermalized luminescence where electrons and holes are in thermal equilibrium with the lattice. However, since their densities n and p exceed the values of their densities n_0 and p_0 at thermal equilibrium in the unexcited semiconductor, their statistical distributions are described by separate *quasi-Fermi* levels, one for the electrons F_e and one for the holes F_h, with $F_e > F_h$. At vanishing excitation intensity, F_e and F_h both coincide with the Fermi level at equilibrium.

Another approach is given by the use of the principle of *detailed balance*, to get a relationship between light absorption and emission in a semiconductor at temperature T (Van Roosbroeck and Shockley, 1954). At thermodynamic equilibrium, at every photon energy, the number of photons emitted (through spontaneous and stimulated emission) is equal to the number of blackbody photons absorbed. Thus, a luminescence spectrum at temperature T can be deduced from the absorption spectrum, as well as a radiative lifetime (i.e., the lifetime if the only existing recombination process were the radiative one).

In fact, the effect of electron–hole Coulomb interaction is already present in the ex-

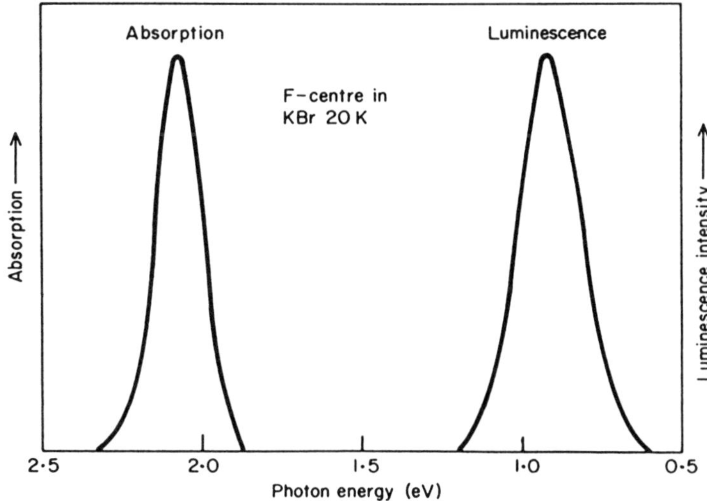

FIG. 3. Absorption and luminescence transitions between the ground state and the lowest electronic excited state of the *F* center in KBr.

perimental absorption spectrum and, thus, also in the luminescence. This interaction between an electron and a hole induces the formation of bound states, analogous to those of the hydrogen atom, called *Wannier excitons* (see EXCITONS). Their binding energy $\mathfrak{R}^*$ is scaled from the hydrogen Rydberg $\mathfrak{R}$ by the electron-hole reduced mass μ and the relative permittivity ϵ according to $\mathfrak{R}^* = (\mu/\epsilon^2)\mathfrak{R}$.

On the other hand, the optical band gap of a semiconductor can be either direct or indirect in reciprocal space. Since the photon wave vector is very small at the Brillouin-zone scale, first-order optical transitions take place vertically in $\mathbf{k}$ space. In direct-gap semiconductors, the extrema of conduction and valence bands are at the same point in the Brillouin zone; so first-order optical transitions are in general allowed (except in special cases). In indirect-gap semiconductors, band extrema are at different points $\mathbf{k}_v$ and $\mathbf{k}_c$ in the Brillouin zone; in that case, first-order optical transitions near the extrema are always forbidden. The next higher-order process, in pure semiconductors, has to involve the absorption or emission of a phonon of wave vector $\mathbf{k}_c - \mathbf{k}_v$. As a consequence, optical absorption near threshold for indirect-gap material is weak, and the radiative lifetime of the lowest exciton is large (in the millisecond range for Ge). In contrast, in direct-gap material, the optical absorption near threshold is strong; it exhibits well-resolved exciton lines, and the photoluminescence spectrum is in a region of strong optical absorption. This has several consequences. At high temperature, when electron–hole recombination dominates, once a photon has been emitted in the material, it has a good probability to be reabsorbed, creating again an electron–hole pair before escaping the material: Such a process, called *photon recycling*, enhances the diffusion of the electron–hole pairs and modifies the radiative lifetime, especially if the quantum efficiency is good. At lower temperature, when free excitons dominate the luminescence spectrum, a free exciton of small wave vector $\mathbf{k}$ will be absorbed, creating a photon with the same $\mathbf{k}$ that later on will create an exciton $\mathbf{k}$, and so on, unless it makes a collision with a phonon or a lattice defect. This occurs if the exciton–photon coupling is strong enough; the result of this coupling is a mixed exciton–photon excitation that has been called *polariton*. Under such conditions, a luminescence photon has a chance to be created when a polariton hits the sample surface. In fact, photoluminescence is not the best method to study polaritons; resonant Brillouin scattering is the best way to determine the polariton dispersion relation.

The tendency of low-temperature photoluminescence to accumulate the excitations in the lowest possible excited state is also valuable to detect weak transitions related to *triplet* states, which are only weakly coupled to light; this allows the measurement of electron–hole *exchange* interaction. As an example, consider the case of exciton–polaritons in CdS, a crystal of uniaxial symmetry; the exciton is built from a spin-$\frac{1}{2}$ electron and a spin-$\frac{3}{2}$ hole, quantized along the c axis of the crystal. So excitons of total angular momentum $J = 1$ or $J = 2$ are obtained. The first are strongly coupled to light of polarization $\perp c$ and show a strong polariton effect; the second, our triplet, are weakly coupled to light. By using polarization and wave-vector selection in a photoluminescence experiment on pure CdS platelets, it was possible to measure accurately the energy difference between the $J = 2$ excitons and the longitudinal $J = 1$ polariton (see Fig. 4).

3.1.3.2 Effect of Impurities and Defects (see ELECTRON STATES, LOCALIZED). At 2 K, low-level photoluminescence in the near–band-gap spectral region is usually dominated by the effect of *shallow impurities*, donors and acceptors that are well-known dopants. Under typical conditions, unintentionally doped semiconductors present some compensation; under weak photocreation of electron–hole pairs, ionized donors and acceptors capture respectively free electrons and free holes. The first recombination process comes from pairs D^0-A^0 of neutral donor and acceptor (Dean, 1973). At somewhat higher temperature, a fraction of electrons may remain free, thus allowing electron-to-neutral acceptor transitions, noted e-A^0. When all donors and acceptors are neutral, additional electron–hole pairs can form free excitons, and these excitons can form bound states with different impurities or defects. Among these are an exciton bound to a neutral donor, denoted D^0X, or to a neutral acceptor, denoted A^0X (Bebb and Williams, 1972). Depending on the electron-to-hole mass ratio, binding to an ionized impurity is possible, e.g., D^+X in CdS. These

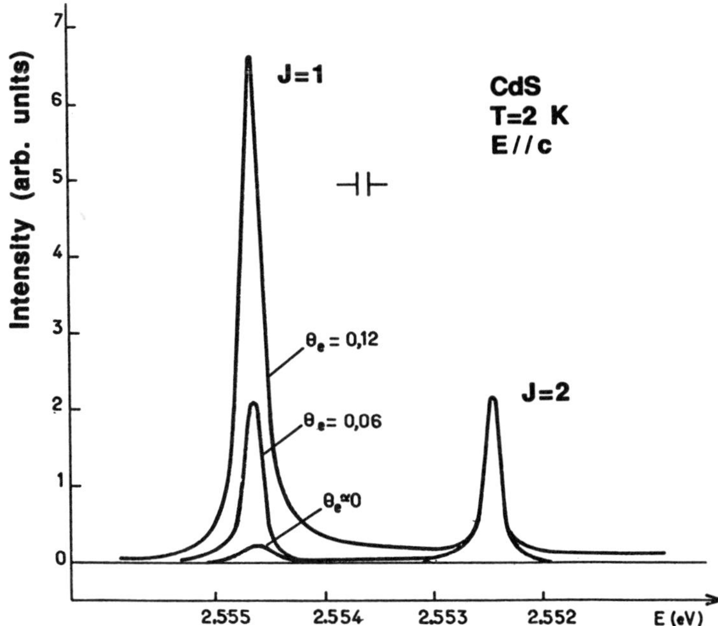

FIG. 4. Exchange splitting of free exciton in CdS. The photoluminescence spectrum is obtained at 2 K with polarization selection ($E \parallel c$) and photon wave-vector selection (**k** nearly normal to c), c being the axis of this hexagonal crystal: θ_e is the angle between **k** and a plane normal to c. For $\theta_e \rightarrow 0$, the $J = 1$ exciton becomes purely longitudinal, and the coupling with a photon $\parallel c$ vanishes.

complexes, in direct- as well as indirect-gap materials, give narrow luminescence lines that can be split in a magnetic field of a few tesla. If one considers the complex D^0X, its main decay process gives a photon and a neutral donor D^0 in its ground 1S state; however, there is less probable decay process, named *two-electron transition,* which leaves the neutral donor in its nS state (nP states are forbidden by symmetry considerations). So, from the energy difference between the two luminescence lines, the energy difference between the 1S and 2S states of the donor D^0 can be measured, which is very helpful for its identification. The same occurs for A^0X (Fig. 5).

When one atom of a semiconductor is replaced by an atom of the same column of the Mendeleev table, if they are not too different, an *alloy* can be obtained in the whole composition range. Examples of such behavior are $Ge_{1-x}Si_x$, $Ga_{1-x}Al_xAs$, and $GaAs_{1-x}P_x$. As a first approximation, they behave just like ordinary semiconductors whose parameters are now x dependent. However, specific features appear in the low-temperature photoluminescence spectra of these alloys. Since they are random alloys, statistical fluctuations of local concentration are unavoidable. They produce a tail of bound exciton states readily observable in the luminescence spectrum. In addition, time resolution shows the existence of energy transfer toward lower-energy states.

If the electronegativity of the impurity is too different from that of the host, in general, the solubility will be small. An impurity less electronegative can bind a hole; in contrast, it can bind an electron if it is more electronegative. Then, an additional carrier can neutralize the complex, which is called an *isoelectronic exciton*. This happens, for example, in GaP when Bi or N is substituted for P. Such complexes exhibit an electron–hole exchange splitting, readily observable

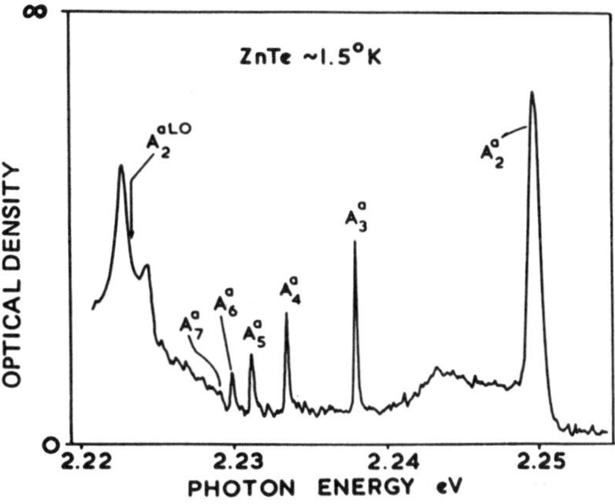

FIG. 5. Two-hole transitions in pure as-grown ZnTe at 1.5 K. The main line A_1^a, related to an exciton bound to a neutral acceptor, is at 2.375 eV; the lines A_n^a correspond to transitions starting from the ground state of A^0X, leaving A^0 in an nS excited state.

in low-temperature photoluminescence. Excitons can also be bound more strongly to pairs of isoelectronic impurities at different separation. Let us stress a strong apparent difference between GaAs and GaP. In GaAs at atmospheric pressure, it is impossible to detect any N-related luminescence signal. Under the action of a hydrostatic pressure larger than 2.2 GPa, photoluminescence lines related to N appear. This is because the energy levels of the N isoelectronic exciton are linked mainly to the conduction-band minima of X symmetry, at the Brillouin-zone edge, the lowest in GaP. In contrast, these minima are above the Γ minimum, at the Brillouin-zone center, in GaAs. The effect of pressure is to push up the Γ minimum and push down the X minima. With increasing pressure, GaAs becomes GaP-like, and the N-related lines appear.

Among the variety of *deep centers*, let us distinguish the *transition metals*. When they are incorporated in large-gap semiconductors, one recovers the situation encountered in Sec. 3.1.2. Mn in ZnS is such an example of a traditional luminescent material. In fact, Mn incorporated in most II-VI compounds with band gap larger than 2.2 eV produces an analogous yellow-orange luminescence band coming from the excitation of the ^5d shell of Mn^{2+}. Hybridization of transition-metal d orbitals with valence-band orbitals of the host crystal is somewhat larger in semiconductors. Another feature of interest is the magnetic moment carried by some of these transition-metal ions; consequences for the luminescence are examined in Sec. 3.2.2. Another class of deep centers may be called *lattice defects*, like vacancies, interstitials, antisites defects in compounds, and complex defects involving impurities and lattice defects. Photoluminescence is just a tool, among a lot of others, to help defect identification. Let us just mention the famous *EL*2 defect in GaAs; it is a native defect that exists in three charged states (0, +, and ++), giving energy levels near midgap and providing semi-insulating GaAs substrates. It involves likely an arsenic antisite As_{Ga} that might dissociate into a gallium vacancy V_{Ga} and an interstitial arsenic As_i in some of its possible charge states; this model is still under debate.

3.1.4 Low-Dimensional Semiconductor Structures With the advent in the seventies of refined growth methods [see CVD (CHEMICAL VAPOR DEPOSITION) and MOLECULAR BEAM EPITAXY], the growth of artificial layered structures became possible, with an accuracy approaching the atomic-layer limit. Other methods were developed later to obtain one-dimensional wires and zero-dimensional boxes (or dots) with somewhat poorer accuracy. Photoluminescence turned out to be a very efficient tool to study these structures since it requires no further preparation of the sample, in contrast with optical absorption measurements, which require substrate removal. A widely used substitute for absorption spectrum is the photoluminescence excitation spectrum. Figure 6 shows results for a GaAs quantum well; the photoluminescence excitation spectrum exhibits peaks that are related to the *quantization* of electron and hole movement along the growth axis (see ELECTRON STATES IN ZERO-, ONE-, AND TWO-DIMENSIONAL STRUCTURES). Excitonic effects are enhanced in two-dimensional structures, in agreement with theory. Another interesting feature is the Stokes shift between the lowest excitation peak and the photoluminescence peak. This shift, which depends on the well width and on the growth conditions, is related to the fact that it is impossible to grow perfect interfaces: There are always some steps or *interface roughness* and also some amount of interdiffusion. This specific kind of disorder induces a weak localization of the excitons, readily observed at 2 K.

Among the very large number of fundamental applications of photoluminescence in low-dimensional structures, let us single out the following, taken under extreme conditions: the study of a two-dimensional free-electron system in modulation-doped heterostructures in a transverse magnetic field. Through magnetotransport measurements, the integer quantum Hall effect and later the more unexpected fractional quantum Hall effect were discovered in such structures. Photoluminescence performed down to 0.1 K and in magnetic field up to 25 T did provide new insight into the nature of the ground state of this electron system: A plane of acceptors was introduced in GaAs at some distance from the interface; a weak photoexcitation neutralizes these acceptors, which can then recombine radiatively with the electrons, thus providing a precise measurement of the energy $E(H)$

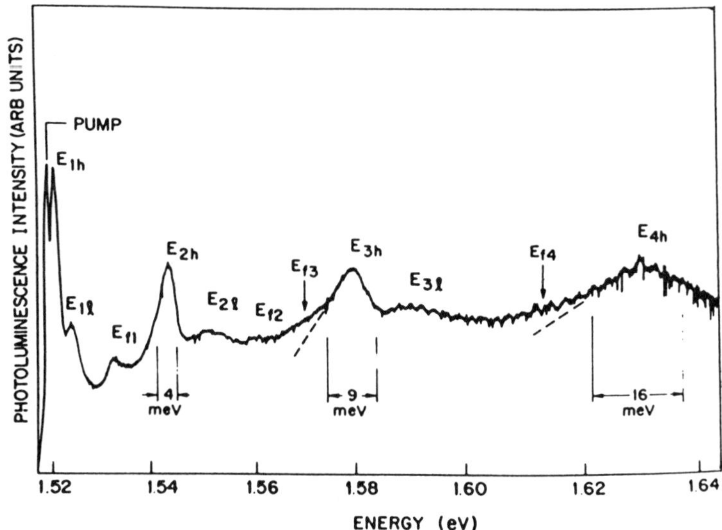

FIG. 6. Excitation spectrum of the exciton luminescence of a 26-nm GaAs quantum well in GaAlAs barriers. The strongest E_{nh} transitions (n = 1 to 4) are related to electron and heavy hole in their nth quantized states; transitions noted E_{nl} involve electron and light hole.

of the electron system as a function of the magnetic field H. Indeed, anomalies in the $E(H)$ curve were observed at the field values where the fractional quantum Hall effect is detected. This gives strong support to the theory, based on electron-electron Coulomb interaction, which predicts a new kind of condensation into an incompressible electron liquid for these particular fractional Landau level-filling factors. Experiments are still in progress to determine under what conditions the electron system could crystallize, as predicted by Wigner.

3.1.5 Optical Pumping and Hot Luminescence Absorption of circularly polarized light allows us to excite unequally sublevels in a degenerate set of levels. This *optical pumping* scheme, first developed on atoms, leads, through spin–orbit coupling, to *spin orientation*. The degree of spin orientation, $p_e = (n^+ - n^-)/(n^+ + n^-)$ where n^+ and n^- are the populations of the two spin states, can be monitored by measuring the polarization of the luminescence. This was done first in p-type GaSb (or GaAs). Because of spin–orbit effect near the top of the valence band, absorption of circular photons close to the band gap creates electrons with a degree of spin orientation $p_e = 0.5$; when they recombine with neutral acceptors, the degree of luminescence polarization, $p_L = (I^+ - I^-)/(I^+ + I^-)$ where I^+ and I^- are the intensities of plus or minus circular components of the luminescence, has a maximum value of 0.25. This value is obtained only if the electron lifetime τ is much shorter than the electron spin relaxation time τ_s. In general, one finds $p_L = 0.25/(1 + \tau/\tau_s)$. Then, using a magnetic field H perpendicular to the excitation beam, which is also the quantization axis for the spins, one can induce spin precession at the Larmor frequency $\Omega = g^*\mu_B H/\hbar$, where g^* is the effective gyromagnetic factor of the electron and μ_B is the Bohr magneton. This causes a depolarization of the luminescence known as the *Hanle effect*, the field-dependent polarization being given as

$$p_L(H) = p_L(0)/[1 + (\Omega T_s)^2]$$

where $1/T_s = 1/\tau + 1/\tau_s$. So, from these two measurements, the lifetime τ and the spin relaxation time τ_s can be determined, provided g^* is known (Parsons, 1969); in this way, fairly short times can be measured (down to the 10^{-11}-s range) with dc method; the trick was to use the Larmor frequency Ω as an internal clock.

Orientation of *excitons* is also possible: Under appropriate conditions, a polarized photon will be transmitted at the vacuum–crystal interface as an exciton (polariton) of the *same* polarization. In particular, one can create linearly polarized excitons; when submitted to a magnetic field parallel to their wave vector, the plane of their polarization will rotate, an effect that has been called resonant Faraday effect.

In a typical *hot photoluminescence* experiment, an excitation about 0.5 eV above the band gap is used; the very weak luminescence in the region between the excitation and the main near–band-gap luminescence was at the origin of a new effect: the *optical alignment of electron momenta*. For example, pumping between the heavy-hole band and the conduction band in GaAs with a linear photon of electric polarization along z creates preferentially electrons with wave vector perpendicular to z. Here again, alignment can be demonstrated by measuring the rate of polarization of hot luminescence; alignment can be relaxed by elastic collisions and destroyed by the action of a magnetic field, which induces a cyclotron motion in **k** space and introduces a new internal clock, the cyclotron frequency.

Let us add that analogous parameters can be measured in the time-resolved mode, but at the expense of more sophisticated equipment.

3.1.6 Exciton Transport Using a sharply focused pump beam, it is possible to detect the diffusion of the excitations by measuring the diameter of the luminescence spot, or even better using a time-resolved mode. For example, it was possible to show that excitons are indeed trapped at 2 K in type-II GaAs/AlAs superlattices, but become mobile at increasing temperature.

A new kind of exciton transport experiment occurs with the use of a nonuniform stress: Thus, a gradient of band gap is created so that an exciton experiences a force proportional to this gradient. With an excitation pulsed and localized, one can follow the movement of an exciton (or electron–hole) cloud by looking at the time- and space-resolved luminescence. The so-called "phonon wind" effect has been demonstrated (Wolfe, 1985).

3.2 Excitation–Medium Interaction

The excitation introduced into the semiconductor may interact with other (intrinsic) excitations of the medium; we consider two different cases.

3.2.1 Electron–Phonon Coupling, Polarons, Self-Trapping The excitation (electron–hole pair or exciton) can induce a lattice distortion, an effect described in terms of electron–phonon coupling. The strongest coupling occurs generally with longitudinal optical (LO) phonons in polar materials. Such an LO phonon carries with it an electric field to which any charge carrier couples. This coupling, known as Fröhlich interaction, increases in the series GaAs, ZnSe, CuBr. Such coupling has two effects on the luminescence: At moderate coupling strength, sidebands appear, separated from the no-phonon line by 1, 2, 3, . . . times the LO phonon energy; this occurs for bound states, like the D^0A^0 luminescence (e.g., the well-known "green edge" luminescence in CdS) as well as for free excitons. If the coupling is strong enough, self-trapping may occur: The exciton is localized in a strongly distorted region, which acts as a potential well for the exciton. This is observed in KI and NaI; in fact, free-running excitons and self-trapped excitons can coexist because these two states are separated by a potential barrier.

3.2.2 Electron–Magnetic-Ion Exchange Interaction, Magnetic Polaron Another type of interaction of the excitation with the medium is found in magnetic semiconductors: this is the carrier–ion exchange interaction. Starting from a paramagnetic medium, like a diluted magnetic semiconductor (see SEMICONDUCTORS, DILUTED MAGNETIC), a carrier (or an exciton) can lower its energy by the orientation of the ion spins with respect to the spin of the carrier. This can take place already in the ground state, e.g., for localized carriers like neutral donors or acceptors. For example, the increase of neutral acceptor binding energy due to exchange was determined by a shift of the D^0A^0 luminescence with temperature. Such shifts affect also the recombination of excitons weakly localized by disorder, either in bulk or in low-dimensional structures based on diluted magnetic semiconductors. Here again, as in Sec. 3.2.1, if the coupling between the exciton and the ions is strong enough, self-trapping (or free magnetic polaron) may occur. It should be favored in low-dimensional structures; experiments to detect true free magnetic polarons in low-dimensional structures are going on actively.

3.3 High-Excitation Effects

Increasing the excitation level induces two kinds of effects.

3.3.1 Saturation, Optical Gain, Laser We have already shown in Sec. 3.1.3.2 an example of luminescence saturation when all donors and acceptors are neutralized: the D^0A^0 luminescence is saturated. But another effect takes place when the population n_e of the excited state is larger than the population n_g of the ground state: Photons at the energy difference can be amplified rather than absorbed; absorption is proportional to $n_g - n_e$ and gain is just a negative absorption. With band-to-band transition, population inversion can be realized near the band gap because electrons and holes are fermions. When gain occurs, the luminescence is modified in a way that depends upon the geometry of the excited region and the existence of an optical cavity that provides feedback: Laser emission may result see (LASER PHYSICS).

3.3.2 Many-body Effects: Biexcitons, Electron–Hole Plasma or Liquid When the intensity of excitation increases, especially in semiconductors, the density of electrons, holes, and excitons may increase to a point where their mutual Coulomb interaction can shift the energy of the excited states, or even produce new kinds of excited states. At low temperature, with increasing exciton density in pure materials, two excitons can form a bound state, the *exciton molecule* (or *biexciton*), in close analogy with the hydrogen molecule. These have been detected as specific luminescence lines first in direct-gap material: CuCl and later CdS; then in indirect-gap materials like AgBr and Si. At a total density of electrons and holes exceeding the Mott density, the excited medium is described as an electron–hole plasma. In such a state, the mean energy of an electron–hole pair is modified, with respect to the band gap, by a positive term, the Fermi energy of the particles, and a negative one coming from the Coulomb interaction between the particles. As a consequence, the luminescence band due to an electron–hole plasma exhibits a broadening related to the Fermi energy and a shift related to the Coulomb term, both functions of the density. In direct-gap materials, a strong optical gain is also obtained; it may distort the luminescence band and lead to laser action. In indirect-gap materials, something special may occur: because one kind of carrier occupies several equivalent valleys, there is a reduction of the Fermi energy; then the mean energy of a pair exhibits, for some density n_0, a minimum that can stay below the free exciton energy. This phase at density n_0, which is the most stable one at $T = 0$, has been called *electron–hole liquid*; its existence was predicted by Keldysh and well studied mainly using photoluminescence in the 1970s, first in Ge and Si (Fig. 7), later in GaP and SiC. At finite temperature, thermal equilibrium between the electron–hole liquid and the free-exciton gas is nothing but a new example of liquid–gas coexistence. Another kind of condensation was considered in the case of free excitons, which are bosons, being made of two fermions: this is the Bose–Einstein condensation, which is likely occurring in Cu_2O.

3.3.3 Higher-Order Emission Processes A first kind may be called partial Auger process. For example, an exciton interacting with another particle (electron or exciton) can recombine with emission of a photon, some energy being transferred to the second particle. This was well studied in direct-gap materials like CdS. Such a process is likely playing a major role in the mechanism of blue laser emission in ZnSe-based heterostructures, a domain of intense activity.

In a second kind, two electron–hole pairs can recombine with emission of a single photon. For example, a weak green luminescence at about two times the band gap was detected from electron–hole liquid and biexcitons in Si.

3.4 Practical Applications

3.4.1 Nondestructive Material Testing Photoluminescence turns out to be an easy test of a class of materials since it requires

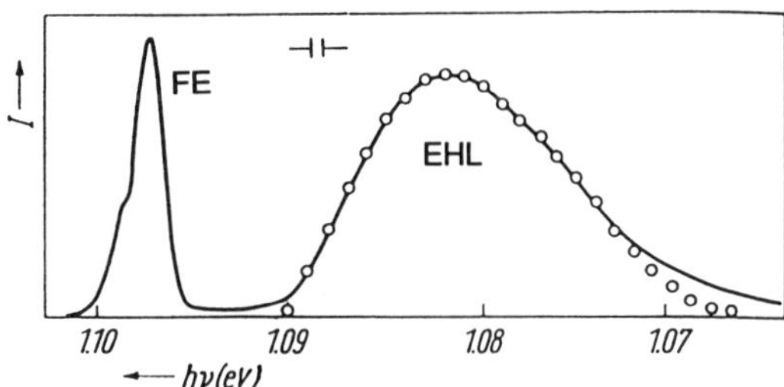

FIG. 7. Photoluminescence spectrum of pure Si at 4.2 K, showing the coexistence of free excitons (FE) and electron–hole liquid (EHL). Si being an indirect-gap material, both luminescence bands are phonon assisted with the same phonons. Open circles correspond to a theoretical fit of the EHL band.

generally no special treatment of the sample and is clearly nondestructive. It finds applications mainly with semiconductors.

In the case of bulk samples, it gives an indication about impurity content and crystalline perfection. It is especially useful for fairly pure samples, where some impurities can be identified from their specific narrow luminescence lines at low temperature. However, precise determination of impurity density is not possible, but comparison with calibrated samples may be useful. Indication about nonradiative centers may be obtained from luminescence decay.

Testing of low-dimensional structures is even more interesting. From the luminescence energy and the photoluminescence excitation spectrum, a precise determination of the energy spectrum of the structure is obtained, which is compared with a model using as inputs the nominal parameters of the growth process. Low-temperature line broadening and Stokes shift give also indications about the quality of the interfaces, to be correlated with the growth conditions.

3.4.2 Devices: Optically Pumped Solid-State Lasers and Amplifiers Whereas most practical luminescent sources use electrical excitation (e.g., semiconductor lamps or electroluminescent panels), optical excitation is used to produce optical gain in the medium, and this is applied to get light amplifiers and lasers. As an amplifier, one can cite Er^{3+}-doped silica fiber, which operates near 1.5 μm; the most efficient pump was found to be a GaAsP injection laser. Such an amplifier is now the best candidate for fiber-optics high-rate communications all around the world. Most solid-state lasers like Nd^{3+} in glass or crystals are optically pumped. Here again, the best overall efficiency is obtained using as a pump semiconductor injection lasers. In the case of a luminescent material like titanium-doped sapphire, the broad gain spectrum leads to a laser with wide tunability, from 680 to 1000 nm; in addition, since the broadening due to phonon coupling is homogeneous, very short laser pulses can be obtained, down to 100 fs.

GLOSSARY

Boxcar: Detection mode using a gate to select in a time window an electronic signal.

Brillouin Scattering: Light scattering by acoustical phonons in crystals.

Brillouin Zone: The polyhedron in reciprocal space over which the wave vector **k** is allowed to vary.

Effective Mass: In a crystal near a band extremum, the dispersion relation for an electron is a quadratic function of its wave vector **k**, written as $E = \hbar^2 k^2 / 2m^*$. The effective mass m^* is the quantity governing the response of the electron in the crystal to an external perturbation, in the same way as m_0, the free electron mass, for an electron in vacuum.

Exchange Interaction: Interaction between particles that satisfy the Pauli exclusion principle, electrons among them.

Frenkel Exciton: In a crystal, coherent propagation of an excitation primarily localized on a molecule or an ion.

Landau Level: Orbital quantized state of a free electron in a magnetic field.

Mode-Locked Laser: By forcing all the longitudinal modes of a laser to be in phase near an end mirror, a pulsed operation is obtained, with a delay between pulses equal to the round-trip time of a photon in the laser cavity.

Modulation-Doped Heterostructure: By introducing donors in the barrier material (e.g., GaAlAs) at some distance from the interface with the well (e.g., GaAs), part of the donor electrons transfer to the well, creating a high-mobility two-dimensional electron gas.

Mott Density: At increasing electron density, a transition from a collection of isolated impurities (or exciton gas) to a metallic state (or an electron–hole plasma) takes place at the Mott density.

Phase Matching: In a nonlinear process, conditions upon the propagation angles of the different beams such that, inside the crystal, wave-vector conservation is insured.

Phonon Wind: In semiconductors, a cloud of acoustical phonons created, e.g., at the focus of an optical beam can push electrons, holes or excitons, thus contributing to an anomalous carrier diffusion.

Quantum Hall Effect: In a two-dimensional electron gas, a plateau in the Hall effect and a dip in the longitudinal resistance are observed at integer (or fractional) Landau-level filling factors.

Rayleigh Scattering: Elastic scattering of light by fixed defects or surface roughness.

Reciprocal Space (k Space): A mathematical space, with dimension of inverse length, over which wave vectors are defined.

Synchronous Detection: Detection of an electronic signal involving the product with a sinusoidal reference signal at frequency *f*. The component of the signal at frequency *f* is extracted.

Subtractive Mode: In a double-grating spectrometer, geometry that insures constant optical path length for any ray.

Wannier Exciton: In crystalline semiconductors, excitation made of an electron–hole pair tied by Coulomb interaction, free to move in the crystal.

Works Cited

Bebb, H. Barry, Williams, E. W. (1972), "Photoluminescence I: Theory," in: R. K. Willardson, A. C. Beer, (Eds.), *Semiconductors and Semimetals*, Vol. 8, New York and London: Academic Press, pp. 181–320.

Cavenett B. C. (1981), "Optical Detection of Magnetic Resonance (O. D. M. R.) investigations of recombination processes in semiconductors," *Adv. Phys.* **30**, 475–538.

Dean P. J. (1973), in: J. O. McCaldin, G. Somorjai (Eds.), *Progress in Solid State Chemistry*, Vol. 8, Oxford: Pergamon, pp. 1–126.

Parsons, R. R. (1969), "Band-to-band optical pumping in solids and polarized photoluminescence," *Phys. Rev. Lett.* **23**, 1152–1154.

Van Roosbroeck, W., Shockley, W. (1954), "Photon-radiative recombination of electrons and holes in germanium," *Phys. Rev.* **94**, 1558–1560.

Wolfe J. P. (1985), "Transport of degenerate electron-hole plasma in Si and Ge," in: M. H. Pilkuhn (Ed.), *High Excitation and Short Pulse Phenomena*, Amsterdam: North-Holland, pp. 82–113.

Further Reading

Jeffries, C. D., Keldysh, L. V. (Eds.) (1983), *Electron-Hole Droplets in Semiconductors*, Amsterdam: North-Holland.

Lumb, M. D. (Ed.) (1978), *Luminescence Spectroscopy*, New York: Academic Press.

Meier, F., Zakharchenya, B. P. (Eds.) (1984), *Optical Orientation*, Amsterdam: North Holland.

Pankove, J. I. (1971), *Optical Processes in Semiconductors*, Englewood Cliffs, NJ: Prentice Hall.

Pope, M., Swenberg, C. E. (1982), *Electronic Processes in Organic Crystals*, New York: Oxford University Press.

Rashba, E. I., Sturge, M. D. (Eds.) (1982), *Excitons*, Amsterdam: North-Holland.

PHOTOSYNTHESIS

JOHN WHITMARSH, *Agricultural Research Service/USDA, Department of Plant Biology, University of Illinois, Urbana, Illinois, U.S.A.*

GOVINDJEE, *Department of Plant Biology, University of Illinois, Urbana, Illinois, U.S.A.*

	Introduction	513
1.	**Brief History**	514
2.	**Classification of Photosynthetic Organisms**	515
2.1	Oxygenic Photosynthetic Organisms	516
2.2	Anoxygenic Photosynthetic Organisms	516
3.	**General Principles of Photosynthetic Energy Transformation in Plants**	516
4.	**Oxygenic Photosynthesis in Plants**	518
4.1	Chloroplasts—Structure and Organization	518
4.2	Light Absorption—The Antenna System	519
4.3	Primary Photochemistry—Photosystem II and Photosystem I Reaction Centers	520
4.4	Electron Transport	522
4.5	Creation of a Proton Electrochemical Potential Across the Photosynthetic Membrane	523
4.6	Synthesis of ATP by the ATP-Synthase Enzyme	524
4.7	Synthesis of Carbohydrates from Atmospheric CO_2 by the Calvin Cycle Enzymes	524
4.8	Photosynthetic Quantum Yield and Energy Conversion Efficiency	525
5.	**Oxygenic Photosynthesis in Algae**	526
6.	**Oxygenic Photosynthesis in Bacteria**	526
6.1	Cyanobacteria	526
6.2	Prochlorophytes	527
7.	**Anoxygenic Photosynthesis in Bacteria**	527
7.1	Purple Bacteria	527
7.2	Green Sulfur Bacteria	528
7.3	Green Gliding Bacteria	528
7.4	Heliobacteria	529
8.	**Control of Intraprotein Electron Transfer**	529
9.	**Global Photosynthesis and the Atmosphere**	530
	Glossary	530
	Works Cited	532
	Further Reading	532

INTRODUCTION

Photosynthesis is the physicochemical process by which plants, algae, and photosynthetic bacteria use light energy to drive the synthesis of organic compounds. In plants, algae, and certain types of bacteria, the photosynthetic process results in the release of molecular oxygen and the removal from the atmosphere of carbon dioxide, which is used to synthesize carbohydrates. Other types of bacteria use light energy to create organic compounds but do not produce oxygen. Photosynthesis provides the energy and reduced carbon required for the survival of virtually all life on our planet, as well as the molecular oxygen necessary for the survival of oxygen-consuming organisms. In addition, the fossil fuels currently being burned to provide energy for human activity were produced by ancient photosynthetic organisms. Although photosynthesis occurs in cells or organelles that are typically only a few microns across, the process has a profound impact on the earth's atmosphere and climate. Each year more than 10% of the total atmospheric carbon dioxide is reduced to carbohydrate by photosynthetic organisms. Most, if not all, of

3-527-28135-5/95/$5.00 + .50

the reduced carbon is returned to the atmosphere as carbon dioxide by microbial, plant, and animal metabolism, and by biomass combustion. In turn, the performance of photosynthetic organisms depends on the earth's atmosphere and climate. Over the next century, the large increase in the amount of atmospheric carbon dioxide created by human activity is certain to have a profound impact on the performance and competition of photosynthetic organisms. Knowledge of the physicochemical process of photosynthesis is essential for understanding the relationship between living organisms and the atmosphere and the balance of life on earth.

1. BRIEF HISTORY

In the 1770s, Joseph Priestley, an English chemist and clergyman, performed experiments showing that plants release a type of air that allows combustion. He demonstrated this by burning a candle in a closed vessel until the flame went out. He placed a sprig of mint in the chamber and after several days showed that the candle could burn again. Although Priestley did not know about molecular oxygen, his work showed that plants release oxygen into the atmosphere. It is noteworthy that over 200 years later, investigating the mechanism by which plants produce oxygen is one of the most active areas of photosynthetic research. Building on the work of Priestley, Jan Ingenhousz, a Dutch physician, demonstrated that sunlight was necessary for photosynthesis and that only the green parts of plants could release oxygen. During this period, Jean Senebier, a Swiss botanist and naturalist, discovered that CO_2 is required for photosynthetic growth, and Nicolas-Théodore de Saussure, a Swiss chemist and plant physiologist, showed that water is required. It was not until 1845 that Julius Robert von Mayer, a German physician and physicist, proposed that photosynthetic organisms convert light energy into chemical free energy.

By the middle of the nineteenth century, the key features of plant photosynthesis were known, namely, that plants could use light energy to make carbohydrates from CO_2 and water. The empirical equation representing the net reaction of photosynthesis for oxygen-evolving organisms is

$$CO_2 + 2H_2O + \text{light energy} \rightarrow [CH_2O] + O_2 + H_2O, \quad (1)$$

where $[CH_2O]$ represents a carbohydrate (e.g., glucose, a six-carbon sugar). The synthesis of carbohydrate from carbon and water requires a large input of light energy. The standard free energy for the reduction of one mole of CO_2 to the level of glucose is +478 kJ/mol. Because glucose, a six-carbon sugar, is often an intermediate product of photosynthesis, the net equation of photosynthesis is frequently written as

$$6CO_2 + 12H_2O + \text{light energy} \rightarrow C_6H_{12}O_6 + 6O_2 + 6H_2O. \quad (2)$$

The standard free energy for the synthesis of glucose is +2870 kJ/mol.

Not surprisingly, early scientists studying photosynthesis concluded that the O_2 released by plants came from CO_2, which was thought to be split by light energy. In the 1930s, comparison of bacterial and plant photosynthesis lead Cornelis van Niel to propose the general equation of photosynthesis that applies to plants, algae, and photosynthetic bacteria (discussed by Wraight, 1982). Van Niel was aware that some photosynthetic bacteria could use hydrogen sulfide (H_2S) instead of water for photosynthesis and that these organisms released sulfur instead of oxygen. Van Niel, among others, concluded that photosynthesis depends on electron donation and acceptor reactions and that the O_2 released during photosynthesis comes from the oxidation of water. Van Niel's generalized equation is

$$CO_2 + 2H_2A + \text{light energy} \rightarrow [CH_2O] + 2A + H_2O. \quad (3)$$

In oxygenic photosynthesis, $2A$ is O_2, whereas in anoxygenic photosynthesis, which occurs in some photosynthetic bacteria, the electron donor can be an inorganic hydrogen donor, such as H_2S in which case A is elemental sulfur, or an organic hydrogen donor such as succinate in which case A is fumarate.

The biochemical conversion of CO_2 to carbohydrate is a reduction reaction that involves the rearrangement of covalent bonds between carbon, hydrogen, and oxygen. The energy for the reduction of carbon is provided by energy-rich molecules that are produced by the light-driven electron-transfer reactions. Carbon reduction can occur in the dark and involves a series of biochemical reactions that were elucidated by Melvin Calvin, Andrew Benson, and James Bassham in the late 1940s and 1950s. Using the radioisotope carbon-14, most of the intermediate steps that result in the production of carbohydrate were identified. Calvin was awarded the Nobel Prize for Chemistry in 1961 for this work.

In 1954, Daniel Arnon and co-workers discovered that plants, and A. Frenkel discovered that photosynthetic bacteria, use light energy to produce ATP, an organic molecule that serves as an energy source for many biochemical reactions. During the same period, L. N. M. Duysens showed that the primary photochemical reaction of photosynthesis is an oxidation/reduction reaction that occurs in a protein complex (the reaction center). Over the next few years, the work of several groups, including those of Robert Emerson, Bessel Kok, L. N. M. Duysens, Robert Hill, and Horst Witt, combined to prove that plant, algae, and cyanobacteria require two reaction centers, photosystem II and photosystem I, operating in series.

In 1961, Peter Mitchell suggested that cells can store energy by creating an electric field or a proton gradient across a membrane. Mitchell's proposal that energy is stored as an electrochemical gradient across a vesicular membrane opened the door for understanding energy transformation by membrane systems. He was awarded the Nobel Prize in Chemistry in 1978 for his theory of chemiosmotic energy transduction.

Most of the proteins required for the conversion of light energy and electron transfer reactions of photosynthesis are located in membranes. Despite decades of work, efforts to determine the structure of membrane-bound proteins had little success. This changed in the 1980s when Johann Deisenhofer, Hartmut Michel, and Robert Huber determined the structure of the reaction center of the purple bacterium *Rhodospeudomonas viridis*. They were awarded the Nobel Prize for Chemistry in 1988 for their work, which provided insight into the relationship between structure and function in membrane-bound proteins.

A key element in photosynthetic energy conversion is electron transfer within and between protein complexes and simple organic molecules. The electron-transfer reactions are rapid (as fast as a few picoseconds) and highly specific. Much of our current understanding of the physical principles that guide electron transfer is based on the pioneering work of Rudolph A. Marcus, who received the Nobel Prize in Chemistry in 1992 for his contributions to the theory of electron-transfer reaction in chemical systems.

The overall equation for photosynthesis is deceptively simple. In fact, a complex set of physical and chemical reactions must occur in a coordinated manner for the synthesis of carbohydrates. To produce a sugar molecule such as sucrose, plants require nearly 30 distinct proteins that work within a complicated membrane structure. Research into the mechanism of photosynthesis centers on understanding the structure of the photosynthetic components and the molecular processes that use radiant energy to drive carbohydrate synthesis. The research involves several disciplines, including physics, chemistry, structural biology, biochemistry, molecular biology, and physiology, and serves as an outstanding example of the success of multidisciplinary research. As such, photosynthesis presents a special challenge in understanding several interrelated molecular processes. From a physicist's viewpoint, the reactions that transform energy are of particular interest, and it is on these processes that this description will focus.

2. CLASSIFICATION OF PHOTOSYNTHETIC ORGANISMS

All life can be divided into three domains, Archaea, Bacteria, and Eucarya, which originated from a common ancestor (Woese *et al.*, 1990). Historically, the term photosynthesis has been applied to organisms that depend on chlorophyll (or bacteriochlorophyll) for the conversion of light energy into chemical free energy (Gest, 1993). These include organisms in the domains Bacteria (photosynthetic bacteria) and Eucarya (algae and higher plants). The most primitive domain, Archaea, includes organisms known as halobacteria, which convert light energy into chemical free energy. However, the mechanism by which halobacteria convert light is fundamentally different from that of higher organisms because there is no oxidation/reduction chemistry and halobacteria cannot use CO_2 as their carbon source. Consequently, some biologists do not consider halobacteria as photosynthetic (Gest, 1993). This article will follow the historical definition of photosynthesis and omit halobacteria.

2.1 Oxygenic Photosynthetic Organisms

The photosynthetic process in all plants and algae, as well as in certain types of photosynthetic bacteria, involves the reduction of CO_2 to carbohydrate and the removal of electrons from H_2O, which results in the release of O_2. In this process, known as oxygenic photosynthesis, water is oxidized by the photosystem II reaction center, a multisubunit protein located in the photosynthetic membrane. Years of research have shown that structure and function of photosystem II is similar in plants, algae, and certain bacteria, so that knowledge gained in one species can be applied to others. This homology is a common feature of proteins that perform the same reaction in different species. The importance of this homology at the molecular level is shown by the fact that there are an estimated 300 000–500 000 species of plants. If different species had evolved diverse mechanisms for oxidizing water, research aimed at a general understanding of photosynthetic water oxidation would be hopeless.

2.2 Anoxygenic Photosynthetic Organisms

Some photosynthetic bacteria can use light energy to extract electrons from molecules other than water. These organisms are of ancient origin, presumed to have evolved before oxygenic photosynthetic organisms. Anoxygenic photosynthetic organisms occur in the domain Bacteria and have representatives in four phyla: Purple Bacteria, Green Sulfur Bacteria, Green Gliding Bacteria, and Gram Positive Bacteria.

3. GENERAL PRINCIPLES OF PHOTOSYNTHETIC ENERGY TRANSFORMATION IN PLANTS

The energy that drives photosynthesis originates in the center of the sun, where mass is converted to heat by the fusion of hydrogen. Over time, the heat energy reaches the sun's surface, where some of it is converted to light by blackbody radiation that reaches the earth. A small fraction of the visible light incident on the earth is absorbed by plants. Through a series of energy-transducing reactions, plants are able to transform light energy into chemical free energy in a stable form that can last for hundreds of millions of years (e.g., fossil fuels). A simplified scheme describing how plants transform energy is presented in this section. The focus is on the structural and functional features essential for the energy-transforming reactions. For clarity, mechanistic and structural details are omitted. A more highly resolved description of oxygenic and anoxygenic photosynthesis is given in the remaining sections.

The photosynthetic process in plants and algae occurs in small organelles, known as chloroplasts, that are located inside cells. The photosynthetic reactions are traditionally divided into two types: the "light reactions," which consist of electron- and proton-transfer reactions, and the "dark reactions," which consist of the biosynthesis of carbohydrates from CO_2. The light reactions occur in a complex membrane system (the photosynthetic membrane) that is made up of protein complexes, electron carriers, and lipid molecules. The photosynthetic membrane is surrounded by water and can be thought of as a two-dimensional surface that defines a closed space, with an inner and an outer water phase. A molecule or ion must pass through the photosynthetic membrane to go from the inner space to the outer space. The protein complexes embedded in the photosynthetic membrane have a unique orientation with respect to the inner and outer phases. The asymmetrical arrangement of the protein complexes allows some of the energy released during electron transport to create an electrochemical gradient of protons across the photosynthetic membrane.

Photosynthetic electron transport consists of a series of individual electron-transfer steps from one electron carrier to another. The electron carriers are metal ion complexes and aromatic groups. The metal ion complexes and most of the aromatic groups are bound within proteins. Most of the proteins involved in photosynthetic electron transport are composed of numerous polypeptide chains that lace through the membrane, providing a scaffolding for metal ions and aromatic groups. An electron enters a protein complex at a specific site, is transferred within the protein from one carrier to another, and exits from the protein at a different site. The protein controls the pathway of electrons between the carriers by determining the location and environment of the metal ion complexes and aromatic groups. By setting the distance between electron carriers and controlling the electronic environment surrounding a metal ion complex or an aromatic group, the protein controls pairwise electron transfer reactions. Between proteins, electron transfer is controlled by distance and free energy, as for intraprotein transfer, and by the probability that the two proteins are in close contact. Protein association is controlled by a number of factors, including the structure of the two proteins, their surface electrical and chemical properties, and the probability that they collide with one another. Not all electron carriers are bound to proteins. The reduced forms of plastoquinone and nicotinamide adenine dinucleotide phosphate (NADPH) act as mobile electron carriers operating between protein complexes. For electron transfer to occur, these small molecules must bind to special pockets in the proteins known as binding sites. The binding sites are highly specific and are a critical factor in controlling the rate and pathway of electron transfer.

The light reactions convert energy into several forms (Fig. 1). The first step is the conversion of a photon to an excited electronic state of an antenna pigment molecule located in the antenna system. The antenna system consists of hundreds of pigment molecules (mainly chlorophyll and carotenoids) that are anchored to proteins within the photosynthetic membrane and serve a specialized protein complex known as a reaction center. The electronic excited state is transferred over the antenna molecules as an exciton. Some excitons are converted back into photons and emitted as fluorescence, some are converted to heat, and some are trapped by a reaction-center protein. Excitons trapped by a reaction center provide the energy for the primary photochemical reaction of photosynthesis—the transfer of an electron from a donor molecule to an acceptor molecule. Both the donor and acceptor molecules are attached to the reaction-center protein complex.

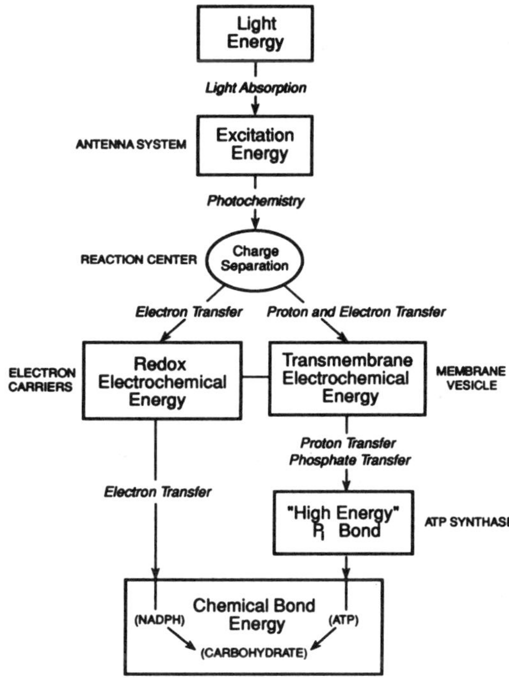

FIG. 1. Photosynthesis shown as a series of reactions that transform energy from one form to another. The different forms of energy are shown in boxes, and the direction of energy transformation is shown by the arrows. The energy-transforming reaction is shown by italics in the arrows. The site at which the energy is stored is shown in capital letters outside the boxes. The primary photochemical reaction, charge separation, is shown in the oval. Details of these reactions are given in the text.

Once primary charge separation occurs, the subsequent electron-transfer reactions are energetically downhill. In oxygenic photosynthetic organisms, two different reaction centers work in series: photosystem II and photosystem I. Electrons are transferred from photosystem II to the photosystem I reaction center by intermediate carriers. The net reaction is the transfer of electrons from a water molecule to $NADP^+$, producing the reduced form, NADPH. In the photosynthetic process, much of the energy initially provided by light energy is stored as redox free energy (a form of chemical free energy) in NADPH, to be used later in the reduction of carbon. In addition, the electron-transfer reactions concentrate protons inside the membrane vesicle and create an electric field across the photosynthetic membrane. In this process, the electron-transfer reactions convert redox free energy into an electrochemical potential of protons. The energy stored in the proton electrochemical potential is used by a membrane-bound protein complex (ATP synthase) to attach a phosphate group covalently to adenosine diphosphate (ADP), forming adenosine triphosphate (ATP). Protons pass through the ATP-synthase protein complex that transforms electrochemical free energy into a type of chemical free energy known as phosphate group-transfer potential (or a high-energy phosphate bond) (Klotz, 1967). The energy stored in ATP can be transferred to another molecule by transferring the phosphate group. The net effect of the light reactions is to convert radiant energy into redox free energy in the form of NADPH and phosphate group-transfer energy in the form of ATP. In the light reactions, the transfer of a single electron from water to $NADP^+$ involves about 30 metal ions and seven aromatic groups. The metal ions include 19 Fe, 5 Mg, 4 Mn, and 1 Cu. The aromatics include quinones, pheophytin, NADPH, tyrosine, and a flavoprotein.

The NADPH and ATP formed by the light reactions provide the energy for the dark reactions of photosynthesis, known as the Calvin cycle or the photosynthetic carbon reduction cycle. The reduction of atmospheric CO_2 to carbohydrate occurs in the aqueous phase of the chloroplast and involves a series of enzymatic reactions. The first step is catalyzed by the protein Rubisco (D-ribulose 1,5-bisphosphate carboxylase/oxygenase), which attaches CO_2 to a five-carbon compound. The reaction produces two molecules of a three-carbon compound. Subsequent biochemical reactions involve several enzymes that reduce carbon by hydrogen transfer and rearrange the carbon compounds to synthesize carbohydrates. The carbon reduction cycle involves the transfer and rearrangement of chemical bond energy.

4. OXYGENIC PHOTOSYNTHESIS IN PLANTS

4.1 Chloroplasts—Structure and Organization

Photosynthesis occurs inside chloroplasts, which are small organelles found in plant cells. Chloroplasts provide the energy and reduced carbon needed for plant growth and development, while the plant provides the chloroplast with CO_2, water, nitrogen, organic molecules, and minerals necessary for the chloroplast biogenesis. Most chloroplasts are located in specialized leaf cells, which often contain 50 or more chloroplasts per cell. Each chloroplast is defined by an inner and an outer envelope membrane and is shaped like a meniscus convex lens that is 5–10 μm in diameter (Fig. 2), although many different shapes and sizes can be found in plants. The inner envelope membrane acts as a barrier, controlling the flux of organic and charged molecules in and out of the chloroplast. Water passes freely through the envelope membranes, as do other small neutral molecules like CO_2 and O_2. There is convincing evidence that chloroplasts were once free-living bacteria that invaded a nonphotosynthetic cell long ago. They have retained some of the DNA necessary for their assembly, but much of the DNA necessary for their biosynthesis is located in the cell nucleus. This enables a cell to control the biosynthesis of chloroplasts within its domain.

Inside the chloroplast is a complicated membrane system, known as the photosynthetic membrane (or thylakoid membrane), that contains most of the proteins required for the light reactions. The proteins required for the fixation and reduction of CO_2 are located outside the photosynthetic membrane in the surrounding aqueous phase. The photosynthetic membrane is composed mainly of glycerol lipids and protein. The glycerol lipids are a family of molecules characterized by a polar head group that is hydrophilic and two fatty acid side chains that are hydrophobic. In membranes, the lipid mol-

FIG. 2. An electron micrograph of a plant chloroplast. Micrograph by A. D. Greenwood, courtesy of J. Barber.

ecules arrange themselves in a bilayer, with the polar head toward the water phase and the fatty acid chains aligned inside the membrane forming a hydrophobic core (Fig. 3). The photosynthetic membrane is vesicular, defining a closed space with an outer water space (stroma) and an inner water space (lumen). The organization of the photosynthetic membrane can be described as groups of stacked membranes (like stacks of pita bread, with the inner pocket representing the inner aqueous space), interconnected by nonstacked membranes that protrude from the edges of the stacks (Staehelin, 1986). Experiments indicate that the inner aqueous space of the photosynthetic membrane is likely continuous inside of the chloroplast. It is not known why the photosynthetic membrane forms such a convoluted structure. To understand the energetics of photosynthesis, the complicated structure can be ignored and the photosynthetic membrane can be viewed as a simple vesicle.

4.2 Light Absorption—The Antenna System

Plant photosynthesis is driven primarily by visible light (wavelengths from 400 to 700 nm) that is absorbed by pigment molecules (mainly chlorophyll *a* and *b* and carotenoids). Plants appear green because of chlorophyll, which is so plentiful that regions of the earth appear green from space. The absorption spectra of chlorophyll *a* and *b* and of a chloroplast are shown in the article EXCITONS. Light is collected by 200–300 pigment molecules that are bound to light-harvesting protein complexes located in the photosynthetic membrane. The three-dimensional structure of the light-harvesting complex (Kühlbrandt *et al.*, 1994) shows that the protein determines the position and orientation of the antenna pigments. The light-harvesting complexes surround the reaction centers and serve as antennae. Photosynthesis is initiated by the absorption of a photon by an antenna molecule, which occurs in

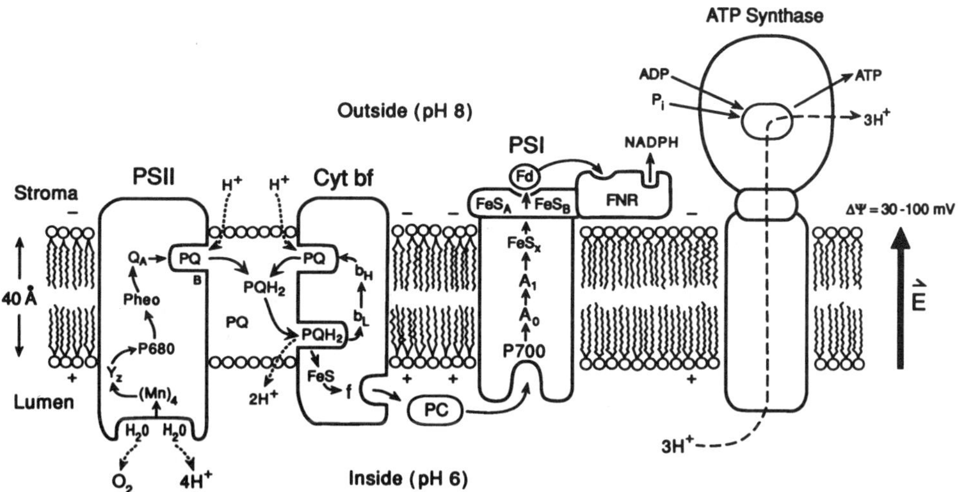

FIG. 3. Model of the photosynthetic membrane of plants showing the electron-transport components and the ATP-synthase enzyme (cross-sectional view). The complete membrane forms a vesicle. The pathways of electrons are shown by solid arrows. The membrane-bound electron-transport protein complexes involved in transferring electrons are the photosystem II and I reaction centers (PSII and PSI) and the cytochrome *bf* complex (Cyt *bf*). Abbreviations: Y_z, tyrosine; P680 and P700, the reaction-center chlorophyll of photosystem II and photosystem I, respectively; Pheo, pheophytin; Q_A, bound plastoquinone; PQ, free plastoquinone (oxidized form), PQH_2, free plastoquinone (reduced form); b_L and b_H, different forms of *b*-type cytochromes; FeS, iron-sulfur centers; *f*, cytochrome *f*; PC, plastocyanin; A_0, chlorophyll; A_1, phylloquinone; Fd, ferredoxin; FNR, ferredoxin/$NADP^+$ oxidoreductase; NADPH, nicotinomide adenine dinucleotide phosphate (reduced form); ADP, adenosine diphosphate; ATP, adenosine triphosphate; P_i, inorganic phosphate; H^+, protons; $\Delta\Psi$, the light-induced electrical potential across the membrane. The light-harvesting protein complexes are not shown. Details are given in the text.

10^{-15} s and causes a transition from the electronic ground state to an excited state. Within 10^{-13} s, the excited state decays by vibrational relaxation to the first excited singlet state. The fate of the excited-state energy is guided by the structure of the protein. Because of the proximity of other antenna molecules with the same or similar energy states, the excited-state energy has a high probability of being transferred by resonance energy transfer to a near neighbor. Exciton energy transfer between antenna molecules is due to the interaction of the transition dipole moment of the molecules. The probability of transfer is dependent on the distance between the transition dipoles of the donor and acceptor molecules ($1/R^6$), the relative orientation of the transition dipoles, and the overlap of the emission spectrum of the donor molecule with the absorption spectrum of the acceptor molecule (for a discussion of resonance energy transfer, see EXCITONS). Photosynthetic antenna systems are very efficient at this process. Under optimum conditions, over 90% of the absorbed quanta are transferred within a few hundred picoseconds from the antenna system to the reaction center, which acts as a trap for the excitons.

4.3 Primary Photochemistry—Photosystem II and Photosystem I Reaction Centers

Photosystem II uses light energy to drive two chemical reactions: the oxidation of water and the reduction of plastoquinone. The photosystem II complex is composed of more than 15 polypeptides, and at least nine different redox components (chlorophyll, pheophytin, plastoquinone, tyrosine, Mn, Fe, cytochrome b559, carotenoid, and histidine) have been shown to undergo light-induced electron transfer (Debus, 1992). However, only five of these redox components are known to be involved in transferring electrons from H_2O to the plastoquinone pool—the water-oxidizing manganese cluster $(Mn)_4$, the amino acid tyrosine, the reaction center chlorophyll (P680), pheophytin, and the plastoquinone molecules, Q_A and Q_B. Of these essential redox components, tyrosine, P680, pheophytin, Q_A, and Q_B have been shown to be bound to two key polypeptides that form the heterodimeric reaction center core of photosystem II (*D*1 and *D*2). Recent work indicates that the *D*1 and *D*2 polypeptides also provide ligands for the $(Mn)_4$ cluster. The three-dimensional structure of photosystem II is not known. Our knowledge of its structure is guided by the known structure of the reaction center in purple bacteria and biochemical and spectroscopic data. Figure 4 shows a schematic view of photosystem II that is consistent with current data.

Photochemistry in photosystem II is initiated by charge separation between P680 and pheophytin, creating $P680^+/Pheo^-$. Primary charge separation takes about 3 ps (Fig. 5). Subsequent electron-transfer steps have been designed through evolution to prevent the primary charge separation from recombining. This is accomplished by transferring the electron within 200 ps from pheophytin to a plastoquinone molecule (Q_A) that is permanently bound to photosystem II. Although plastoquinone normally acts as a two-electron acceptor, it works as a one-electron acceptor at the Q_A site. The electron on Q_A- is then transferred to another plastoquinone molecule that is loosely bound at the Q_B site. Plastoquinone at the Q_B site differs from Q_A in that it works as a two-electron acceptor, becoming fully reduced and protonated after two photochemical turnovers of the reaction center. The full reduction of plastoquinone requires the addition of two electrons and two

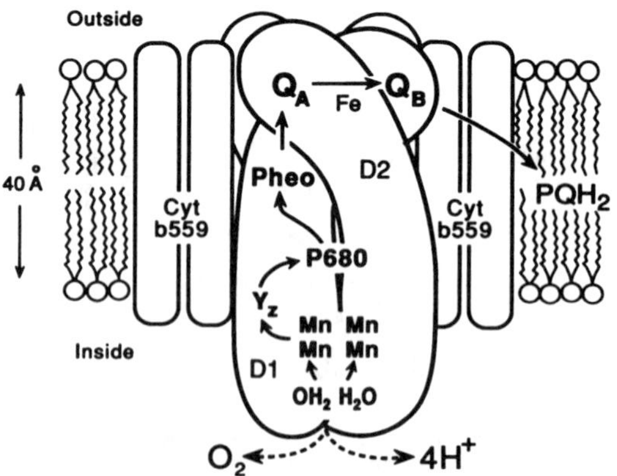

FIG. 4. Schematic drawing of photosystem II. Photosystem II is composed of numerous polypeptides, but only two of them, *D*1 and *D*2, bind the electron carriers involved in transferring electrons from Y_z to plastoquinone. Abbreviations: Y_z, tyrosine; P680, reaction-center chlorophyll (primary electron donor); Pheo, pheophytin; Q_A and Q_B, bound plastoquinone; PQH_2, reduced plastoquinone; Cyt *b*559, *b*-type cytochrome. Details are given in the text.

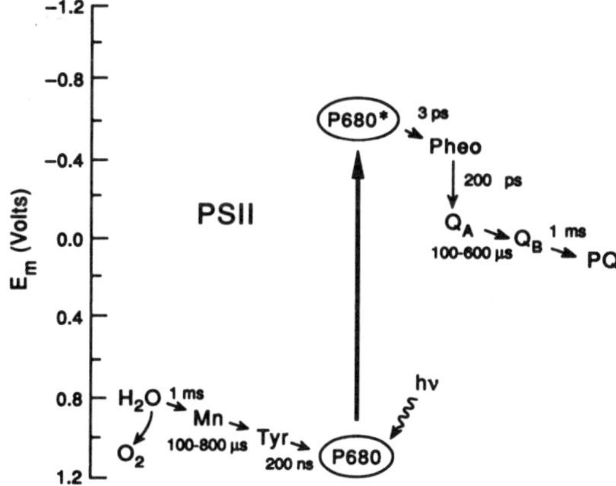

FIG. 5. Photosystem II electron-transport pathways and rates. The vertical axis shows the midpoint potential of the electron carriers. The heavy vertical arrow shows light absorption. P680* is the electronically excited state of P680. The abbreviations are given in the legend of Fig. 4.

protons, i.e., the addition of two hydrogen atoms (Fig. 6). The reduced plastoquinone then debinds from the reaction center and diffuses into the hydrophobic core of the membrane, then an oxidized plastoquinone molecule finds its way to the Q_B-binding site, and the process is repeated. Because the Q_B site is near the outer aqueous phase, the protons added to plastoquinone during its reduction are taken from the outside of the membrane.

Photosystem II is the only known protein complex that can oxidize water, resulting in the release of O_2 into the atmosphere. Despite years of research, little is known about the molecular events that lead to water oxidation. Energetically, water is a poor electron donor. The oxidation-reduction midpoint potential ($E_{m,7}$) of water is +0.82 V (*p*H 7). In photosystem II, this reaction is driven by the oxidized reaction center, $P680^+$ (the midpoint potential of $P680/P680^+$ is estimated to be +1.2 V at *p*H 7). How electrons are transferred from water to $P680^+$ remains a mystery (Govindjee and Coleman, 1990). It is known that $P680^+$ oxidizes a tyrosine on the *D*1 protein and that Mn plays a key role in water oxidation. Four Mn ions are present in the water-oxidizing complex. X-ray absorption spectroscopy shows that Mn undergoes light-induced oxidation. Water oxidation requires two molecules of water and involves four sequential turnovers of the reaction center. Each photochemical reaction creates an oxidant that removes one electron. The net reaction results in the release of one O_2 molecule, the deposition of four protons into the inner water phase, and the transfer of four electrons to the Q_B site (producing two reduced plastoquinone molecules).

The photosystem I complex catalyzes the oxidation of plastocyanin, a small soluble Cu protein, and the reduction of ferredoxin, a small FeS protein (Fig. 7). Photosystem I is

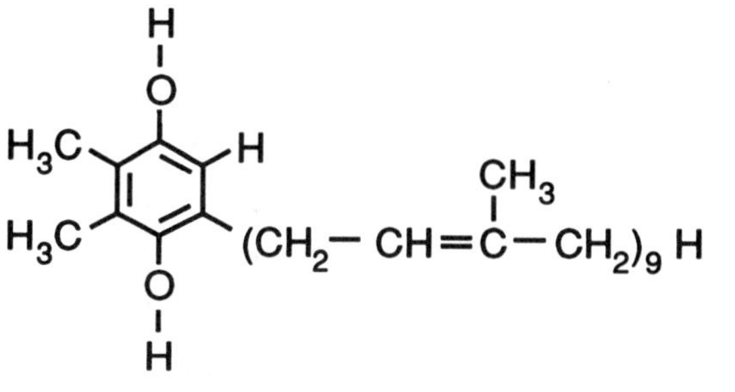

FIG. 6. Structure of plastoquinone (reduced form), an aromatic molecule that carries electrons and protons in photosynthetic electron transport.

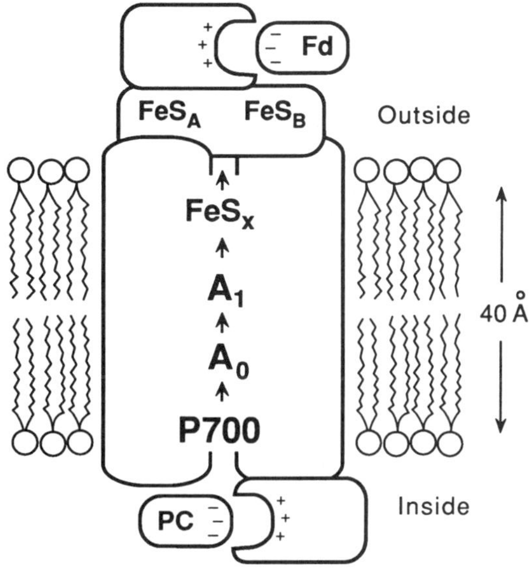

FIG. 7. Schematic drawing of photosystem I. Photosystem I is composed of numerous polypeptides, but only three of them bind the electron carriers. Abbreviations: PC, plastocyanin; P700, reaction-center chlorophyll (primary electron donor); A_0, chlorophyll; A_1, phylloquinone; FeS, FeS centers; Fd, ferredoxin. Details are given in the text.

composed of a heterodimer of proteins that act as ligands for most of the electron carriers (Krauss *et al.*, 1993). The reaction center is served by an antenna system that consists of about two hundred chlorophyll molecules (mainly chlorophyll *a*), and primary photochemistry is initiated by a chlorophyll *a* dimer, P700. In contrast to photosystem II, many of the antenna chlorophyll molecules in photosystem I are bound to the reaction-center proteins. Also, FeS centers serve as electron carriers in photosystem I, and so far as is known, photosystem I electron transfer is not coupled to proton translocation. Primary charge separation occurs between a primary donor, P700, a chlorophyll dimer, and a chlorophyll monomer (A_0). The subsequent electron transfer events and rates are shown in Fig. 8.

4.4 Electron Transport

Electron transport from water to $NADP^+$ requires three membrane-bound protein complexes operating in series: photosystem II, the cytochrome bf complex, and photosystem I (Fig. 3). Electrons are transferred between these large protein complexes by small mobile molecules. Because these small molecules carry electrons (or hydrogen atoms) over relatively long distances, they play a unique role in photosynthetic energy conversion. This is illustrated by plastoquinone (PQ), which serves two key functions. Plastoquinone transfers electrons from the photosystem II reaction center to the cytochrome *bf* complex and carries protons across the photosynthetic membrane. It does this by shuttling hydrogen atoms across the membrane from photosystem II to the cytochrome *bf* complex. Because plastoquinone is hydrophobic, its movement is restricted to the hydrophobic core of the photosynthetic membrane. Plastoquinone operates by diffusing through the membrane until, as a result of random collisions, it becomes bound to a specific site on the photosystem II complex. The photosystem II reaction center reduces plastoquinone at the Q_B site by adding two electrons and two protons, creating PQH_2. The reduced plastoqinone molecule debinds from photosystem II and diffuses randomly in the photosynthetic membrane until it encounters a specific binding site on the cytochrome *bf* complex. The cytochrome *bf* complex is a membrane-bound protein complex that contains four electron carriers, three cytochromes, and an FeS center. In a complicated reaction sequence that is not fully understood, the cytochrome *bf* complex removes the electrons from reduced plastoquinone and facilitates the release of the protons into the inner aqueous space. The electrons are eventually transferred to the photosystem I reaction center. The protons released into the inner aqueous space contribute to the proton chemical free energy across the membrane.

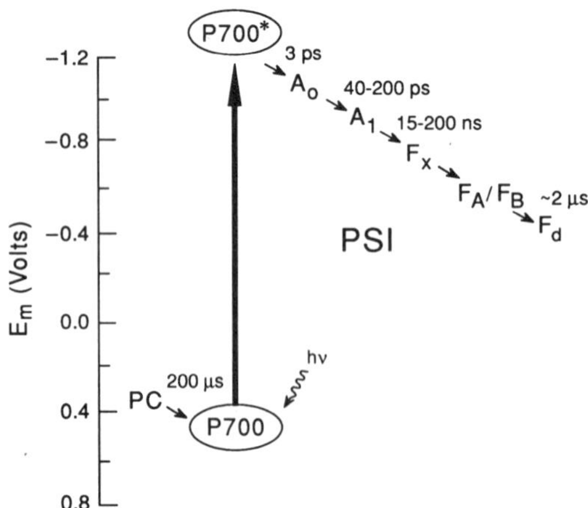

FIG. 8. Photosystem I electron transport pathways and rates. The vertical axis shows the midpoint potential of the electron carriers. Abbreviations are given in the legend of Fig. 7.

Electron transfer from the cytochrome *bf* complex to photosystem I is mediated by a small Cu protein, plastocyanin (PC). Plastocyanin is water soluble and operates in the inner water space of the photosynthetic membrane. Electron transfer from photosystem I to $NADP^+$ requires ferredoxin, a small FeS protein, and ferredoxin-NADP oxidoreductase, a peripheral flavoprotein that operates on the outer surface of the photosynthetic membrane. Ferredoxin and $NADP^+$ are water soluble and are found in the outer aqueous phase.

As discussed in Sec. 3, the pathway of electrons is largely determined by the energetics of the reaction and the distance between the carriers. The electron affinity of the carriers is represented in Fig. 9 by their midpoint potentials, which show the free energy available for electron-transfer reactions under equilibrium conditions. (It should be kept

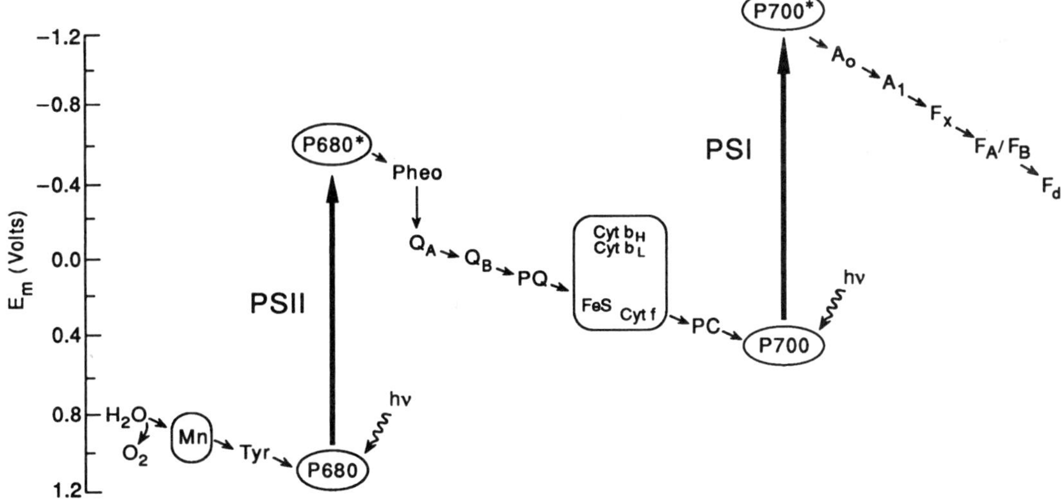

FIG. 9. The electron transport pathway of plants (oxygenic photosynthesis). Abbreviations are given the legend of Fig. 3. Details are given in the text.

in mind that reaction conditions during photosynthesis are not in equilibrium.) Subsequent to primary charge separation, electron transport is energetically downhill (from a lower to a higher redox potential). It is the downhill flow of electrons that provides free energy for the creation of a proton chemical gradient.

Photosynthetic membranes effectively limit electron transport to two dimensions. For mobile electron carriers, limiting diffusion to two dimensions increases the number of random encounters (Whitmarsh, 1986). Furthermore, because plastocyanin is mobile, any one cytochrome *bf* complex can interact with a number of photosystem I complexes. The same is true for plastoquinone, which commonly operates at a stoichiometry of about six molecules per photosystem II complex.

4.5 Creation of a Proton Electrochemical Potential

Electron transport creates the proton electrochemical potential of the photosynthetic membrane by two types of reactions.

1. The release of protons during the oxidation of water by photosystem II and the translocation of protons from the outer aqueous phase to the inner aqueous phase by the coupled reactions of photosystem II and the cytochrome *bf* complex in reducing and oxidizing plastoquinone on opposite sides of the membrane creates a concentration difference of protons across the membranes ($\Delta p\text{H} = p\text{H}_{\text{in}} - p\text{H}_{\text{out}}$).
2. Primary charge separation at a reaction center drives an electron across the photosynthetic membrane, which creates an electric potential across the membrane ($\Delta\Psi = \Psi_{\text{in}} - \Psi_{\text{out}}$).

Together, these two forms of energy make up the proton electrochemical potential across the photosynthetic membrane ($\Delta\mu_{\text{H}^+}$), which is related to the *p*H difference across the membrane and the electrical potential difference across the membrane by the following equation:

$$\Delta\mu_{\text{H}^+} = F\Delta\Psi - 2.3\text{R}T\,\Delta p\text{H},$$

where F is the Faraday constant, R is the gas constant, and T is the temperature in kelvins. Although the value of $\Delta\Psi$ across the photosynthetic membrane in chloroplasts can be as large as 100 mV, under normal conditions the proton gradient dominates. For example, during photosynthesis, the outer *p*H is typically near 8 and the inner *p*H is typically near 6, giving a *p*H difference of 2 across the membrane that is equivalent to 120 mV. Under these conditions, the free energy for proton transfer from the inner to the outer aqueous phase is −12 kJ/mol of protons.

4.6 Synthesis of ATP by the ATP-Synthase Enzyme

The conversion of proton electrochemical energy into chemical free energy is accomplished by a single protein complex known as ATP synthase. This enzyme catalyzes a phosphorylation reaction, which is the formation of ATP by the addition of inorganic phosphate (P_i) to ADP:

$$ADP^{-3} + P_i^{-2} + H^+ \rightarrow ATP^{-4} + H_2O.$$

The reaction is energetically uphill ($\Delta G = +32$ kJ/mol) and is driven by proton transfer through the ATP-synthase protein. The ATP-synthase complex is composed of two major subunits, CF_0 and CF_1 (Fig. 10). The CF_0 subunit spans the photosynthetic membrane and forms a proton channel through the membrane. The CF_1 subunit is attached to the top of the CF_0 on the outside of the membrane and is located in the aqueous space. The CF_1 subunit is composed of several different protein subunits, referred to as α, β, γ, δ, and ϵ. The top portion of the CF_1 subunit is composed of three $\alpha\beta$ dimers that contain the catalytic sites for ATP synthesis. The molecular processes that couple proton transfer through the protein to the chemical addition of phosphate to ADP are poorly understood. It is known that phosphorylation can be driven by a *p*H gradient, a transmembrane electric field, or a combination of the two. Experiments indicate that three protons must pass through the ATP-synthase complex for the synthesis of one molecule of ATP. However, the protons are not involved in the chemistry of adding phosphate to ADP. Paul Boyer and co-workers have proposed an alternating binding site mechanism for ATP synthesis (Boyer, 1993). One model based on their proposal is that there are three catalytic sites on each CF_1 that cycle among three different states. The states differ in their affinity for ADP, P_i, and ATP. At any one time, each site is in a different state. Initially, one catalytic site on CF_1 binds one ADP and one inorganic phosphate molecule relatively loosely. Through a conformational change of the protein, the site becomes a tight binding site, which stabilizes ATP. Next, proton transfer induces an alteration in protein conformation that causes the site to release the ATP molecule into the aqueous phase. In this model, the energy from the proton electrochemical gradient is used to lower the affinity of the site for ATP, allowing its release to the water phase. The three sites on CF_1 act cooperatively; i.e., the conformational states of the sites are linked. It has been proposed that protons affect the conformational change by driving the rotation of the top part (the three $\alpha\beta$ dimers) of CF_1. This revolving-site mechanism would require rates as high as 100 rev/s. It is worth noting that flagella that propel some bacteria are driven by a proton pump and can rotate at 60 rev/s.

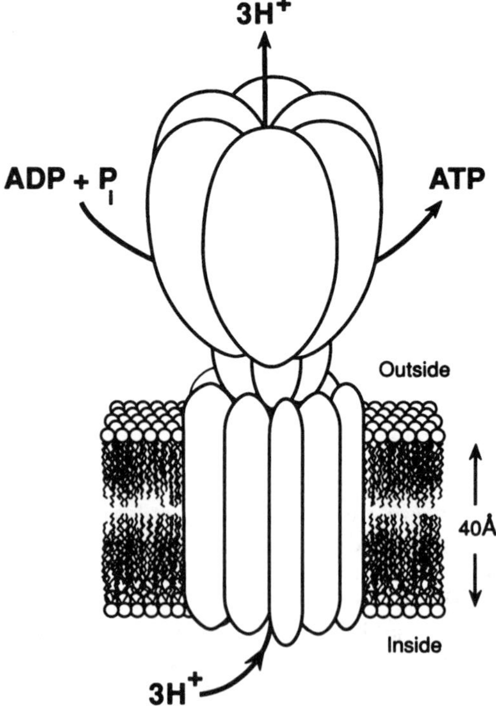

FIG. 10. Schematic drawing of the ATP-synthase enzyme embedded in the membrane. Proton transfer through the ATP synthase provides the energy for the creation of ATP from ADP and P_i. Abbreviations are given in the legend of Fig. 3. Details are given in the text.

4.7 Synthesis of Carbohydrates from Atmospheric CO_2 by the Calvin-Cycle Enzymes

All plants and algae remove CO_2 from the environment and reduce it to carbohydrate by the Calvin cycle. The process is a sequence of biochemical reactions that reduce carbon and rearrange bonds to produce carbohydrate from CO_2 molecules. The first step is the addition of CO_2 to a five-carbon compound (ribulose 1,5-bisphosphate) (Fig. 11). The six-carbon compound is split, giving two molecules of a three-carbon compound (3-phosphoglycerate). This key reaction is catalyzed by Rubisco, a large water-soluble protein complex. The carboxylation reaction is energetically downhill. The main energy input in the Calvin cycle is the phosphorylation by ATP and subsequent reduction by NADPH of the initial three-carbon compound forming a three-carbon sugar, triose phosphate. Some of the triose phosphate is exported from the chloroplast and provides the building block for synthesizing more complex molecules. In a process known as regeneration, the Calvin cycle uses some of the triose phosphate molecules to synthesize the energy-rich ribulose 1,5-bisphosphate needed for the initial carboxylation reaction. This reaction requires the input of energy in the form of one ATP. Overall, 13 enzymes are required to catalyze the reactions in the Calvin cycle. The energy-conversion efficiency of the Calvin cycle is approximately 90%. The reactions do not involve energy transduction but rather the rearrangement of chemical energy. Each molecule of CO_2 reduced to a sugar $[CH_2O]_n$ requires two molecules of NADPH and three molecules of ATP.

Rubisco is a bifunctional enzyme that, in addition to binding CO_2 to ribulose bisphosphate, can also bind O_2. This oxygenation reaction produces the 3-phosphoglycerate that is used in the Calvin cycle and a two-carbon compound (2-phosphoglycolate) that is not useful for the plant. In response, a complicated set of reactions (known as photorespiration) are initiated that serve to recover reduced carbon and to remove phosphoglycolate. The Rubisco oxygenation reaction appears to serve no useful purpose for the plant. Some plants have evolved specialized structures and biochemical pathways that concentrate CO_2 near Rubisco, which serves to decrease the fraction of oxygenation reactions.

4.8 Photosynthetic Quantum Yield and Energy Conversion Efficiency

The theoretical minimum quantum requirement for photosynthesis is eight quanta for each molecule of oxygen evolved (four quanta required by photosystem II and four by photosystem I). Measurements in leaves under optimal conditions (e.g., low light) give

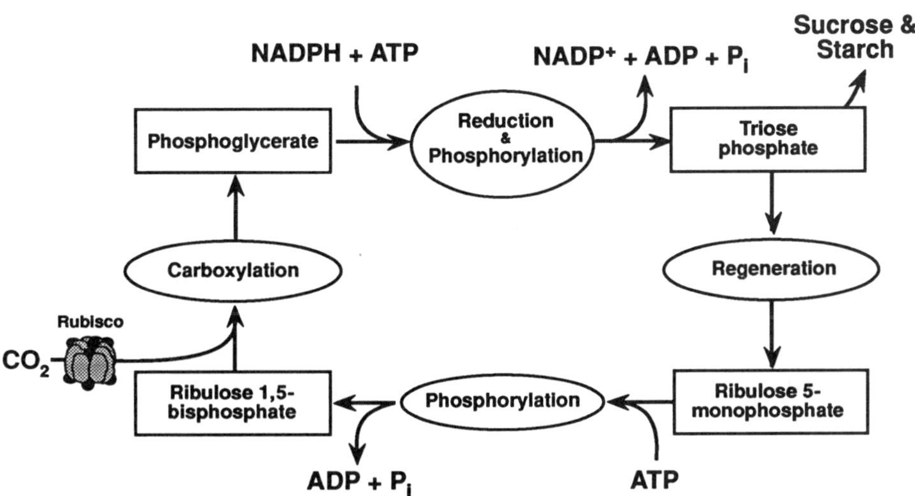

FIG. 11. An abbreviated scheme showing reduction of carbon dioxide by the Calvin cycle. The first step is carboxylation, in which ribulose 1,5-bisphosphate carboxylase/oxygenase (Rubisco) catalyzes the addition of CO_2 to the five-carbon compound ribulose 1,5-bisphosphate, which is subsequently split into two molecules of the three-carbon compound 3-phosphoglycerate. Next are reduction and phosphorylation reactions that form the carbohydrate triose phosphate. Some of the triose phosphate molecules are used to form the products of photosynthesis, sucrose and starch, while the rest is used to regenerate ribulose 1,5-bisphosphate needed for the continuation of the cycle. Details are given in the text.

quantum requirements of 8–10 photons per oxygen molecule released. These quantum yield measurements show that the quantum yields of photosystem II and photosystem I reaction centers under optimal conditions are near 100%. These values can be used to calculate the theoretical energy conversion efficiency of photosynthesis (free energy stored as carbohydrate/light energy absorbed). If eight red quanta are absorbed (8 mol of red photons are equivalent to 1400 kJ) for each CO_2 molecule reduced (480 kJ/mol), the theoretical maximum energy efficiency for carbon reduction is 34%. Under optimal conditions, plants can achieve energy conversion efficiencies within 90% of the theoretical maximum. However, under normal growing conditions, the actual performance of the plant is far below these theoretical values. The factors that conspire to lower the quantum yield of photosynthesis include limitations imposed by biochemical reactions in the plant and environmental conditions that limit photosynthetic performance. One of the most efficient crop plants is sugar cane, which has been shown to store up to 1% of the incident visible radiation over a period of 1 year. However, most crops are less productive. The annual conversion efficiency of corn, wheat, rice, potatoes, and soybeans typically ranges from 0.1% to 0.4% (Odum, 1971).

5. OXYGENIC PHOTOSYNTHESIS IN ALGAE

Algae are photosynthetic eukaryotic organisms that evolve O_2 and reduce CO_2. They represent a diverse group that include dinoflagellates, euglenoids, yellow-green algae, golden-brown algae, diatoms, red algae, brown algae, and green algae. The photosynthetic apparatus and biochemical pathways of carbon reduction of algae are similar to plants. Photosynthesis occurs in chloroplasts that contain photosystems II and I, the cytochrome bf complex, the Calvin-cycle enzymes, and pigment-protein complexes containing chlorophyll *a* and other antenna pigments (*e.g.*, chlorophyll *b* in green algae, chlorophyll *c* and fucoxanthol in brown algae and diatoms, and phycobilins in red algae). Green algae are thought to be the ancestral group from which land plants evolved. Algae are abundant and widespread on the earth, living mainly in fresh and sea water. Some algae live as single-celled organisms, while others form multicellular organisms, some of which can grow quite large, such as kelp and seaweed. Phytoplankton in the ocean is made up of algae and oxygenic photosynthetic bacteria. Most photosynthesis in the ocean is due to phytoplankton, which is an important source of food for marine life.

6. OXYGENIC PHOTOSYNTHESIS IN BACTERIA

6.1 Cyanobacteria

Cyanobacteria are photosynthetic prokaryotic organisms that evolve O_2. Fossil evidence indicates that cyanobacteria existed over 3 billion years ago, and it is thought that they were the first oxygen-evolving organisms on earth. Cyanobacteria are presumed to have evolved in water in an atmosphere that lacked O_2. Initially, the O_2 released by cyanobacteria reacted with ferrous iron in the oceans and was not released into the atmosphere. Geological evidence indicates that the ferrous Fe was depleted around 2 billion years ago, and earth's atmosphere became aerobic. The release of O_2 into the atmosphere by cyanobacteria has had a profound affect on how life evolved.

The photosynthetic apparatus of cyanobacteria is similar to that of chloroplasts. The main difference lies in the antenna system. Cyanobacteria depend on chlorophyll *a* and specialized protein complexes called phycobilisomes that gather light energy. They do not contain chlorophyll *b*. As in chloroplasts, the chlorophyll *a* is located in membrane-bound proteins. The phycobilisomes are bound to the outer side of the photosynthetic membrane and act to funnel exciton energy to the photosystem II reaction center. They are composed of phycobiliproteins, protein subunits that contain covalently attached open ring structures known as bilins that are the light-absorbing pigments. Primary photochemistry, electron transport, phosphorylation, and carbon reduction occur much as they do in chloroplasts. Cyanobacteria have a simpler genetic system than plants and algae, which enables them to be easily modified genetically. Because of this, cyanobacteria have been used as a model to understand photosynthesis in plants. By genetically altering photosynthetic proteins, researchers can investigate the relationship between molecular structure and function.

6.2 Prochlorophytes

Over the past three decades, several types of oxygenic bacteria known as prochlorophytes (or oxychlorobacteria) have been discovered that have light-harvesting protein complexes that contain chlorophyll *a* and *b* but do not contain phycobilisomes (Palenik and Haselkorn, 1992). Because prochlorophytes have chlorophyll *a/b* light-harvesting proteins like chloroplasts, they are being investigated as models for plant photosynthesis.

7. ANOXYGENIC PHOTOSYNTHESIS IN BACTERIA

Anoxygenic photosynthetic bacteria differ from oxygenic organisms in that each species has only one type of reaction center. In some photosynthetic bacteria, the reaction center is similar to photosystem II, and in others, it is similar to photosystem I. However, neither of these two types of bacterial reaction center is capable of extracting electrons from water, and so they do not evolve O_2. Many species can only survive in environments that have a low concentration of O_2. To provide electrons for the reduction of CO_2, anoxygenic photosynthetic bacteria must oxidize inorganic or organic molecules available in their environment. For example, the purple bacterium *Rhodobacter sphaeroides* can use succinate to reduce NAD^+ by a membrane-linked reverse electron transfer that is driven by a transmembrane electrochemical potential. Although many photosynthetic bacteria depend on Rubisco and the Calvin cycle for the reduction of CO_2, some are able to fix atmospheric CO_2 by other biochemical pathways.

Despite these differences, the general principles of energy transduction are the same in anoxygenic and oxygenic photosynthesis. Anoxygenic photosynthetic bacteria depend on bacteriochlorophyll, a family of molecules similar to chlorophyll that absorb strongly in the infrared between 700 and 1000 nm. The antenna system consists of bacteriochlorophyll and carotenoids that serve a reaction center where primary charge separation occurs. The electron carriers include quinones (e.g., ubiquinone, menaquinone) and the cytochrome *bc* complex, which is similar to the cytochrome *bf* complex of oxygenic photosynthetic apparatus. As in oxygenic photosynthesis, electron transfer is coupled to the generation of an electrochemical potential that drives phosphorylation by ATP synthase. The energy required for the reduction of CO_2 is provided by ATP and NADH, a molecule similar to NADPH.

7.1 Purple Bacteria

There are two divisions of photosynthetic purple bacteria: the nonsulfur purple bacteria (e.g., *Rhodobacter sphaeroides* and *Rhodospeudomonas viridis*) and the sulfur purple bacteria (e.g., *Chromatium vinosum*). Nonsulfur purple bacteria typically use an organic electron donor, such as succinate or malate, but they can also use hydrogen gas. The sulfur bacteria use an inorganic sulfur compound, such as hydrogen sulfide, as the electron donor. The only pathway for carbon fixation by purple bacteria is the Calvin cycle. Sulfur purple bacteria must fix CO_2 to live, whereas nonsulfur purple bacteria can grow aerobically in the dark by respiration on an organic carbon source.

The determination of the three-dimensional structures of the reaction center of the nonsulfur purple bacteria, *Rhodopseudomonas viridis* and *Rhodobacter sphaeroides,* provides an unprecedented opportunity to understand the structure and function of photosynthetic reaction centers (Deisenhofer *et al.*, 1984, 1985; Feher *et al.*, 1989). The positions of the electron-transfer components in the reaction center of *Rhodobacter sphaeroides* are shown in Fig. 12 (Norris and van Brakel, 1986). The reaction center contains four bacteriochlorophyll and two bacteriopheophytin molecules. Two of the bacteriochlorophyll molecules form the primary donor (P870). At present, there is controversy over whether a bacteriochlorophyll molecule is an intermediate in electron transfer from the P870 to bacteriopheophytin. However, there is agreement that the remaining steps involve two quinone molecules (Q_A and Q_B) and that two turnovers of the reaction center result in the release of reduced quinone (QH_2)

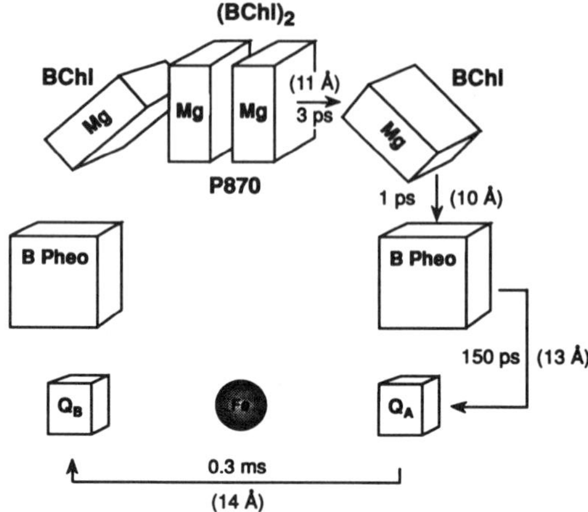

FIG. 12. Relative positions of the chromophores of the reaction center of *Rhodobacter sphaeroides* (from Norris and van Brakel, 1986). Abbreviations: P870, reaction center bacteriochlorophyll (primary electron donor); B Pheo, bacteriopheophytin; Q_A and Q_B, bound ubiquinone. Details are given in the text.

into the photosynthetic membrane. Although there is a nonheme Fe between the two quinone molecules, there is convincing evidence that this Fe is not involved directly in transferring an electron from Q_A to Q_B. Because the primary donor (P870), bacteriopheophytin, and quinone acceptors of the purple bacterial reaction center are similar to the photosystem II reaction center, the bacterial reaction center is used as a guide to understand the structure and function of photosystem II.

Light-driven electron transfer is cyclic in *Rhodobacter sphaeroides* and other purple bacteria (Fig. 13). The reaction center produces reduced quinone, which is oxidized by the cytochrome *bc* complex. Electrons from the cytochrome *bc* complex are transferred to a soluble electron carrier, cytochrome c_2, which reduces the oxidized primary donor $P870^+$. The product of the light-driven electron transfer reactions is ATP. The electrons for the reduction of carbon are extracted from an organic donor, such as succinate or malate, or from hydrogen gas, but not by the reaction center. The energy needed to reduce NAD^+ is provided by light-driven cyclic electron transport in the form of ATP. The energy transformation pathway is complicated. Succinate is oxidized by the membrane-bound enzyme succinate dehydrogenase that transfers the electrons to quinone—the source of electrons for the reduction of NAD^+. However, electron transfer from reduced quinone to NAD^+ is energetically uphill. By a mech-

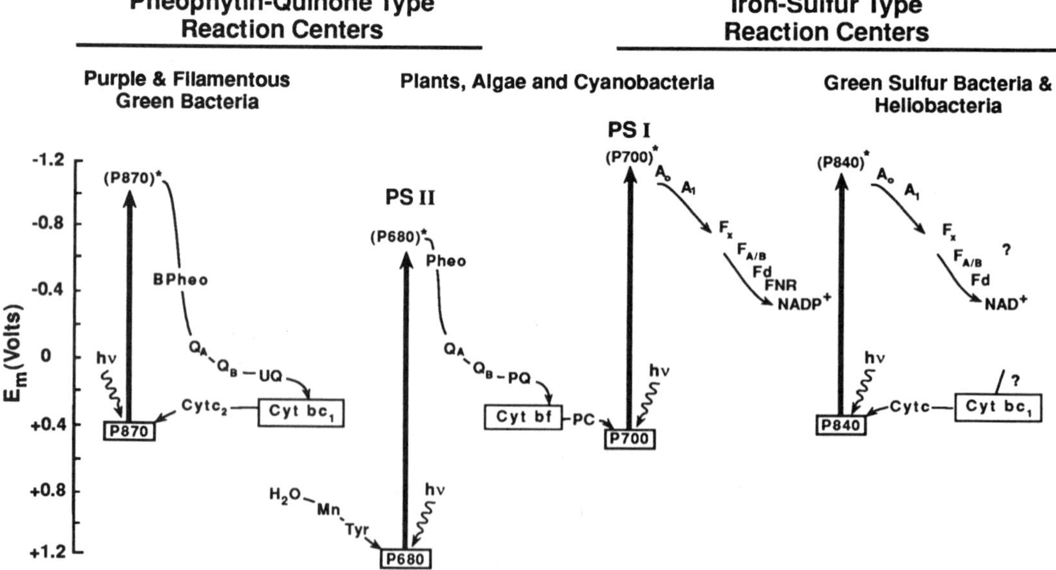

FIG. 13. Comparison of electron-transport pathways in oxygenic and anoxygenic organisms (from Blankenship, 1992). Abbreviations: Cyt bc_1, cytochrome *bc* complex; P840, reaction center bacteriochlorophyll; other abbreviations are given in the legends of Figs. 3 and 12.

anism that is poorly understood, a membrane-bound enzyme is able to use energy stored in the proton electrochemical potential to drive electrons from reduced quinone to NAD^+.

7.2 Green Sulfur Bacteria

Green sulfur bacteria (e.g., *Chlorobium thiosulfatophilum* and *Chlorobium vibrioforme)* can use sulfur compounds as the electron donor as well as organic hydrogen donors. As shown in Fig. 13, the reaction center of green sulfur bacteria is similar to the photosystem I reaction center of oxygenic organisms. The FeS centers in the reaction center can reduce NAD^+ (or $NADP^+$) by ferredoxin and the ferredoxin-$NAD(P)^+$ oxidoreductase enzyme; therefore green sulfur bacteria are not necessarily dependent on reverse electron flow for carbon reduction. The antenna system of the green sulfur bacteria is composed of bacteriochlorophyll and carotenoids and is contained in complexes known as chlorosomes that are attached to the surface of the photosynthetic membrane. This antenna arrangement is similar to the phycobilisomes of cyanobacteria. Green sulfur bacteria can fix CO_2 without Rubisco. It has been proposed that they accomplish this by using the respiratory chain that normally oxidizes carbon (known as the Krebs cycle), resulting in the release of CO_2. With the input of energy, this process can be run in the reverse direction, resulting in the uptake and reduction of CO_2.

7.3 Green Gliding Bacteria

Green gliding bacteria (e.g., *Chloroflexus aurantiacus*), also known as green filamentous bacteria, can grow photosynthetically under anaerobic conditions or in the dark by respiration under aerobic conditions. Like the green sulfur bacteria, green gliding bacteria harvest light using chlorosomes. The green gliding bacteria appear to have reaction centers similar to those of the purple bacteria (Fig. 13), but there are several notable differences. For example, instead of two monomer bacteriochlorophyll molecules, *C. aurantiacus* has one bacteriochlorophyll and one bacteriopheophytin, and the metal between the two quinones is Mn rather than Fe. *C. aurantiacus* appears to fix CO_2 by a scheme that does not involve the Calvin cycle or the reverse Krebs cycle (Ivanovsky *et al.*, 1993).

7.4 Heliobacteria

Heliobacteria (e.g., *Heliobacterium chlorum* and *Heliobacillus mobilis*) are in the phylum Gram Positive Bacteria that are strict anaerobes. Although the heliobacterial reaction center is similar to photosystem I in that it can reduce NAD^+ (or $NADP^+$), it contains a different type of chlorophyll known as bacteriochlorophyll *g*.

8. CONTROL OF INTRAPROTEIN ELECTRON TRANSFER

The three-dimensional structure of the reaction center of *Rhodopseudomonas viridis* and *Rhodobacter sphaeroides* reveals the distances between the electron donors and acceptors (Deisenhofer *et al.*, 1984,1985; Norris and van Brakel, 1986; Feher *et al.*, 1989) and has had an important influence on biophysical and molecular genetics studies designed to identify the factors that control the rate of electron transfer within proteins. There is currently a controversy concerning the importance of the amino acid composition of the protein on the rate of intraprotein electron transfer. In part, the disagreement centers on whether the protein between the donor and acceptor molecules can be treated as a uniform material or whether the specific amino acid composition of the protein significantly alters the rate. For example, it has been proposed that aromatic amino acids may provide a particular pathway that facilitates electron transfer between a donor and acceptor pair. This is the case in the photosystem II reaction center, where a tyrosine residue on one of the reaction center core proteins donates an electron to the primary donor chlorophyll, $P680^+$. However, in other cases, replacement of an aromatic residue by another nonaromatic residue results in relatively minor changes in the rate of electron transfer. L. Dutton and co-workers (Moser *et al.*, 1992) have analyzed electron-transfer reactions in biological and chemical systems in terms of electron tunneling theory developed by R. Marcus and others (DeVault, 1984). Dutton and co-workers argue that protein provides a uniform electronic barrier to electron tunneling and a uniform nuclear characteristic frequency. They suggest that the specific amino acid residues between an electron transfer pair are generally of less importance than the distance in determining the rate of pairwise electron transfer. In their view, protein controls the rate of electron transfer mainly through the distance between the donor and acceptor molecules, the free energy, and the reorganization energy of the reaction. The importance of distance is demonstrated by electron-transfer data from biological and synthetic systems showing that the dependence of the electron transport rate on the edge-to-edge distance is exponential over 12 orders of magnitude when the free energy is optimized (Moser *et al.*, 1992). Increasing the distance between two carriers by 1.7 Å slows the rate of electron transfer tenfold. The extent to which this view is generally applicable for intraprotein electron transfer remains to be established (Williams, 1992). One of the challenges in understanding pairwise electron-transfer rates from first principles is illustrated by the reaction center of *Rhodobacter sphaeroides*, in which the redox components are arranged along a twofold axis of symmetry that extends from the primary donor (P870) to the Fe. Despite the fact that the reaction center presents two spatially similar pathways for electron transfer from P870 to quinone, nearly all electrons are transferred down the right arm of the reaction center, as shown in Fig. 12. The same holds true for the reaction center of *Rhodopseudomonas viridis*, in which it is estimated that electron transfer down the left arm is less than 1:100 (Kellogg *et al.*, 1989). The challenge to theorists is to explain the surprisingly high probability of electron flow down the right arm. Since the distances are similar, it has been suggested that electron transfer down the left arm is less probable because of an endothermic free-energy change (Parson *et al.*, 1990) or to an unfavorable rearrangement energy for the reaction (Moser *et al.*, 1992).

9. GLOBAL PHOTOSYNTHESIS AND THE ATMOSPHERE

The amount of CO_2 removed from the atmosphere each year by oxygenic photosynthetic organisms is massive. It is estimated that photosynthetic organisms remove 100×10^{15} g/yr of carbon (Houghton and Woodwell, 1989). This is equivalent to 4×10^{18} kJ of free energy stored in reduced carbon, which is roughly 0.1% of the incident visible radiant energy incident on the earth per year. Each year the photosynthetically reduced carbon is oxidized, either by living organisms for their survival or by combustion. The result is that more CO_2 is released into the atmosphere from the biota than is taken up by photosynthesis. The amount of carbon released by the biota is estimated to be $(1\text{–}2) \times 10^{15}$ g/yr of carbon. Added to this amount is carbon released by the burning of fossil fuels, which amounts to 5×10^{15} g/yr of carbon. The oceans mitigate this increase by acting as a sink for atmospheric CO_2. It is estimated that the oceans remove about 2×10^{15} g/yr of carbon from the atmosphere. This carbon is eventually stored on the ocean floor. Although these estimates of sources and sinks are uncertain, the net global CO_2 concentration is increasing. Direct measurements show that each year the atmospheric carbon content is currently increasing by about 3×10^{15} grams. Over the past 200 years, CO_2 in the atmosphere has increased from about 280 parts per million (ppm) to its current level of 360 ppm. On the basis of predicted fossil fuel use and land management, it is estimated that the amount of CO_2 in the atmosphere will reach 700 ppm within the next century. The consequences of this rapid change in our atmosphere are unknown. Because CO_2 acts as a greenhouse gas, some climate models predict that the temperature of the earth's atmosphere may increase by 2–8 °C. Such a large temperature increase would lead to significant changes in rainfall patterns. Little is known about the impact of such drastic atmospheric and climatic changes on plant communities and crops. Current research is directed at understanding the interaction between global climate change and photosynthetic organisms.

GLOSSARY

ATP: Adenosine triphosphate, a small water-soluble molecule that acts as an energy currency in cells.

ATP Synthase: A membrane-bound protein complex that uses the energy stored across the photosynthetic membrane to add inorganic phosphate to ADP, thus creating ATP. (Also known as coupling factor.)

Calvin Cycle: The biochemical reactions, initiated by Rubisco, that result in the reduction of CO_2 to a carbohydrate (also known as the photosynthetic carbon reduction cycle).

Cytochrome: Heme-containing protein.

Cytochrome *bc* Complex: A membrane-bound electron-transfer protein complex, found in all anoxygenic photosynthetic organisms, that oxidizes reduced quinone and reduces a *c*-type cytochrome. The complex contains a *c*-type cytochrome, two *b*-type cytochromes, and an FeS center.

Cytochrome *bf* Complex: A membrane-bound electron-transfer protein complex, found in all oxygenic photosynthetic organisms, that oxidizes reduced plastoquinone and reduces plastocyanin (or cytochrome *c*). The complex contains a *c*-type cytochrome, two *b*-type cytochromes, and an FeS center.

Free Energy: The amount of energy in a reaction available to do work. Because most biochemical reactions occur at a constant temperature and pressure, the free energy is frequently the Gibbs energy.

Light-Harvesting Complex: A protein complex that harvests light energy and converts it to exciton energy that can migrate to a reaction center. The light is absorbed by pigment molecules (e.g., chlorophyll, bacteriochlorophyll, carotenoids, phycobilin) that are attached to the protein.

NADPH: Reduced form of nicotinamide adenine dinucleotide phosphate, a small water-soluble molecule that acts as a hydrogen carrier in biochemical reactions.

$NADP^+$: Oxidized form of nicotinamide adenine dinucleotide phosphate.

Oxidation: The removal of one or more electrons from an atom or molecule. In the case of a molecule, protons may be involved as well, resulting in hydrogen being removed.

Phosphorylation: The covalent attachment of a phosphate group to a molecule.

Photorespiration: The removal of O_2 from the atmosphere by Rubisco and the subsequent biochemical reactions that serve to recycle some of the reduced carbon.

Photosynthesis: The physical-chemical process by which certain chlorophyll- (or bacteriochlorophyll-) containing organisms use light energy for the biosynthesis of organic molecules.

Photosynthetic Membrane: A bilayer of lipid molecules in which are embedded proteins that transform light energy into chemical free energy. (Also known as the thylakoid membrane.)

Photosystem I: A protein complex located in the photosynthetic membrane. Photosystem I is one of two types of reaction centers found in higher plants, algae, and cyanobacteria. The photosystem I reaction center uses light energy to transfer an electron from a mobile electron-transfer protein (plastocyanin or cytochrome *c*) on one side of the photosynthetic membrane to a mobile electron-transfer protein (ferredoxin) on the opposite side of the photosynthetic membrane.

Photosystem II: A protein complex found in the photosynthetic membrane. Photosystem II is one of two types of reaction centers found in higher plants, algae, and cyanobacteria. The photosystem II reaction center uses light energy to transfer electrons from water to plastoquinone. Photosystem II is the source of the molecular oxygen in the atmosphere.

Plastoquinone: A small organic molecule involved in electron and proton transfer in photosynthesis.

Protein: A chemical structure composed of one or more polypeptides. In photosynthesis, proteins serve as the scaffolding that hold the cofactors that gather light energy, transfer electrons, and catalyze biochemical reactions.

Reaction Center: A protein complex that uses light energy to create a stable charge separation by transferring a single electron energetically uphill from a donor molecule to an acceptor molecule, both of which are located in the reaction center.

Reduction: The addition of one or more electrons to an atom or molecule. In the case of a molecule, protons may be involved as well, resulting in hydrogen being added.

Rubisco (D-ribulose 1,5-bisphosphate carboxylase/oxygenase): A water-soluble protein complex responsible for the removal of CO_2 from the atmosphere. The enzyme works by attaching CO_2 to a five-carbon compound (1,5 ribulose bisphosphate) that is split into two identical three-carbon compounds (phosphoglycerate). In addition to catalyzing the removal of CO_2 from the atmosphere, Rubisco also catalyzes the removal of O_2 from the atmosphere (less efficiently). The removal of O_2 is thought to be a consequence of poor design and leads to a complex set of compensatory reactions known as photorespiration.

Works Cited

Blankenship, R. (1992), *Photosynth. Res.* **33,** 91–111.

Boyer, P. D. (1993), *Biochim. Biophys. Acta* **1140,** 215–250.

Debus, R. (1992), *Biochim. Biophys. Acta* **1102,** 269–352.

Deisenhofer, J., Epp, O., Miki, K., Huber, R., Michel, H. (1984), *J. Mol. Biol.* **180,** 385–398.

Deisenhofer, J., Epp, O., Miki, K., Huber, R., Michel, H. (1985), *Nature* **318,** 618–624.

DeVault, D. (1984), *Quantum Mechanical Tunnelling in Biological Systems,* Cambridge, U.K.: Cambridge University Press.

Feher, G., Allen, J. P., Okamura, M. Y., Rees, D. C. (1989), *Nature* **339,** 111–116.

Gest, H. (1993), *FEMS Microbiol. Lett.* **112,** 1–6.

Govindjee and Coleman, W. J. (1990), *Sci. Am.* **262** (2), 50–58.

Houghton, R. A., Woodwell, G. M. (1989), *Sci. Am.* **260** (4), 36–44.

Ivanovsky, R. N., Krasilnikova, E. N, Fal, Y. I. (1993), *Arch. Microbiol.* **159,** 257–264.

Kellogg, E. C., Kolaczkowski, S., Wassielewski, M. R., Tiede, D. M. (1989), *Photosynth. Res.* **22,** 47–59.

Klotz, I. M. (1967), *Energy Changes in Biochemical Reactions,* New York: Academic Press, Chap. 6.

Krauss, N., Hinrichs, W., Witt, I., Fromme, P. Pritzkow, W., Dauter, Z. Betzel, C., Wilson, K. S., Witt, H. T., Saenger, W. (1993), *Nature* **361,** 326–331.

Kühlbrandt, W., Wang, D. N., Fujiyoshi, Y. (1994), *Nature* **367,** 614–621.

Moser, C. C., Keske, J. M., Warncke, K., Farid, R. S., Dutton, P. L. (1992), *Nature* **355,** 796–802.

Norris, J. F., van Brakel, G. (1986), in: Govindjee, J. Amesz, D. C. Fork (Eds.), *Light Emission by Plants and Bacteria,* Orlando: Academic Press, pp. 35–36.

Odum, E. P. (1971), *Fundamentals of Ecology,* Philadelphia: W.B. Saunders Co.

Palenik, B., Haselkorn, R. (1992), *Nature* **355,** 265–267.

Parson, W. W., Chu, Z. T., Warshel, A. (1990), *Biochim. Biophys. Acta* **1017,** 251–272.

Staehelin, L. A. (1986), *Encyclopedia Plant Physiol.* **19,** 1–84.

Whitmarsh, J. (1986), *Encyclopedia Plant Physiol.* **19,** 508–527.

Williams, R. J. P. (1992), *Nature* **355,** 770.

Woese, C. R., Kandler, O., Wheeler, M. L. (1990), *Proc. Natl. Acad. Sci. USA* **87,** 4576–4579.

Wraight, C. (1982), in: Govindjee (Ed.), *Photosynthesis,* New York: Academic Press, pp. 17–61

Further Reading

Cramer, W. A., Knaff, D. B. (1991), *Energy Transduction in Biological Membranes,* New York: Springer-Verlag.

Barber, J. (Ed.) (1992), *The Photosystems: Structure, Function and Molecular Biology,* New York: Elsevier.

Huzisige, H., Ke, B. (1993), "Dynamics of the History of Photosynthesis," in: *Photosynthesis Res.* **38,** 185–209.

Lawlor, D. W. (1993), *Photosynthesis: Molecular, Physiological and Environmental Processes,* Essex, U.K.: Longman Scientific & Technical.

Taiz, L., Zeiger, E. (1991), *Plant Physiology,* Redwood City, CA: Benjamin/Cummings Publishing Co.

Walker, D. (1992), *Energy, Plants and Man,* East Sussex, U.K.: Oxygraphics Limited.

PHOTOVOLTAIC DEVICES

JOSEPH J. LOFERSKI, *Division of Engineering, Brown University, Providence, Rhode Island, U.S.A.*

	Introduction	533
1.	**Physical Principles Underlying the PV Effect**	535
1.1	Light Absorption in Semiconductors	535
1.2	Charge Separation in PV Cells	536
1.3	Lifetime of Light-Generated Carriers	538
2.	**Equations Governing the PV Effect**	539
2.1	Diode Equation	539
2.2	Calculation of the Short-Circuit Current I_{SC}	541
2.3	The Reverse Saturation Current I_0 and the Junction Quality Factor A	543
2.4	Maximizing η_{max}	544
2.5	Dependence of η_{max} on E_G	545
2.6	Ultimate Efficiency Calculation	547
2.7	Higher Efficiencies through Spectrum Splitting	547
2.8	Dependence of η_{max} on Temperature	549
3.	**Semiconductor Materials Used in PV Cells: History and Current Status**	549
3.1	Silicon Monocrystalline and Polycrystalline Cells	549
3.2	Gallium Arsenide Cells	550
3.3	Cadmium Sulfide/Copper Sulfide	550
3.4	*p*-Copper Indium Selenide/*n*-CdS Cells	551
3.5	Cadmium Telluride Cells	551
3.6	Amorphous Silicon Solar Cells	551
3.7	Cells Based on Other Semiconductors	552
4.	**Effects of High-Energy Radiation on PV Cells**	552
5.	**Environmental Aspects of PV Cells**	553
6.	**Applications of PV Power Supplies**	553
6.1	Satellite Power Supplies	554
6.2	Remote Power	554
6.3	Other Applications	554
7.	**Economics of PV Power Plants for Large-Scale Electricity Generation**	554
8.	**Summary and Conclusions**	555
	Glossary	556
	Works Cited	556
	Further Reading	557

INTRODUCTION

The photovoltaic effect (PVE) is the generation of an emf as a result of absorption of ionizing radiation. The absorbing medium can be a gas, a liquid, or a solid. For the PVE to occur, an electrostatic potential-difference site must be present in the absorbing medium. The positive and negative charges resulting from the ionization must then be separated by this internal electrostatic field. The separated charges can pass through an external resistive load connected across the potential barrier, thus transforming a portion of absorbed ionizing energy directly into electricity.

The first paper describing the PVE was published by E. Becquerel (1839). He observed an emf when he directed light onto platinum or gold electrodes immersed in acidic or alkaline solutions. The first report of a PVE observed in a solid was published by W. G. Adams and R. E. Day in 1877. They observed the effect in a small selenium rod into whose ends they fused platinum electrodes. During the following 30 years, research and development resulted in an im-

3-527-28135-5/95/$5.00 + .50

proved selenium PV cell, which eventually found practical application as a light meter for photography and as a light sensor in optical instruments.

The next important step in the development of PV science and technology was the invention/discovery during the period 1930–1933 of the PV effect in copper/copper oxide devices by Lange (1930), Schottky (1930), and Grondahl (1933). Schottky's efforts to explain this PV effect made him a pioneer in the then fledgling science of semiconductors. A very important contribution by Schottky was his introduction of the concept of the metal–semiconductor electrostatic potential barrier, the Schottky barrier, to explain the PV effect in both the Cu/CuO_x and the metal/selenium cells. Because of simplicity of fabrication, the Cu/CuO_x cell became a competitor of the selenium cell in the gradually increasing commercial market for such devices.

However, a revolution in the application of the PV effect began with the invention of the transistor (Brattain and Bardeen, 1948; Shockley, 1949). The first transistors utilized as semiconductors the elements silicon and germanium. These new semiconductors could be grown as large single crystals whose electronic properties could be controlled by the addition of minute quantities of other elements. Furthermore, electric current in these new semiconductors could be carried by either positive or negative charges. The positive charges proved to be as mobile as the negative charges (known to be electrons). These mobile positive charged carriers are called "holes" because they result from holes in the electron bonding structures of these materials.

Germanium and silicon are elements from the fourth column of the periodic table; they have a valence of 4. Each atom shares its four valence electrons with its four nearest neighbors, thus forming a crystal consisting of two interlocking face-centered cubic lattices—the diamond structure. At absolute zero, all of the electrons are tied up in covalent bonds; there are no electrons free to move through the lattice if an electric field is applied across the semiconductor. As the temperature increases, some of the electrons are thermally excited and thus liberated from their covalent bonds, leaving behind "holes" in the bonding structure. The liberated electron and the hole can now move through the lattice; they are no longer bound to a specific atom. Generation of such free electrons and holes requires an external source of energy (heat, light, etc.). The minimum energy needed to generate a free electron-hole pair is the difference in energy between the highest valence band and the lowest conduction band in the semiconductor, i.e., the forbidden energy gap E_G.

In an intrinsic semiconductor, the concentrations of free holes and electrons are equal since a free hole is formed whenever a free electron is generated. If elements having a valence different from 4 are incorporated into the crystal, the concentrations of electrons and holes need not be equal any more. There can be an excess of electrons or an excess of holes, depending on the valence of the impurity atom. A semiconductor whose conductivity is controlled by such "extrinsic" elements is referred to as an extrinsic semiconductor. If the dominant impurity element has a valence of 5 (like phosphorus, arsenic, or antimony), then it can replace silicon or germanium atoms in the crystal lattice, forming four covalent bonds with its neighbors. The extra valence electron is not bound to the impurity atom as tightly as the electrons participating in the bonding with the column-four host element; it can move freely through the lattice at temperatures around room temperature. A semiconductor in which the dominant impurity contributes such free electrons to the conduction band is called an *n*-type semiconductor. If the dominant impurity element has a valence of 3 (like boron, aluminum, or indium), when it replaces a silicon or germanium atom in the lattice, it lacks one electron to satisfy the covalent bonding with its neighboring atoms, i.e., there is a "hole" in the bonds around the trivalent atom. This hole arises from the absence of an electron and, therefore, moves through the lattice as if it had a positive charge. A semiconductor in which the dominant impurity contributes free holes to the valence band is called a *p*-type semiconductor.

Shockley developed an elegant theory that explained what happens if a piece of semiconductor has a variable concentration of holes and electrons so that a transition from *p* type to *n* type occurs, thus forming a *p-n* junction. He showed that the *p-n* junction produced an internal electrostatic potential barrier whose properties were far more predictable and reproducible than the properties of Schottky barriers. Rectifiers based on such *p-n* junctions in germanium and silicon proved to be much better than those based on the metal-semiconductor junctions in metal-selenium and Cu/CuO_x rectifiers.

They also proved to be much better photovoltaic cells, as the scientific world was to learn from three brief papers published in journals of the American Physical Society in 1954 (Rappaport, 1954; Chapin *et al.*, 1954: Reynolds *et al.*, 1954). These papers, which marked the beginning of the modern photovoltaic age, described the electron voltaic and photovoltaic effects in *p-n* junction devices made from single crystals of germanium, silicon, or cadmium sulfide. The new PV cells, more specifically those based on silicon and cadmium sulfide, were able to convert the radiant energy of sunlight directly into electricity with efficiencies in the 4% to 6% range, about an order of magnitude higher than the highest solar-energy conversion efficiencies measured on Cu/CuO_x or metal-selenium devices of that time.

Subsequent developments in theory and in the experimental science and technology of PV cells are the subjects addressed in the balance of this article. These developments have led to the fabrication and study of PV cells made from many semiconductors besides silicon; to PV cells having solar-energy conversion efficiencies up to 35%; and to widespread application of PV cells as power supplies for earth satellites, pocket calculators, remote sites, etc. If continued—and it shows no signs of abating—this long-term trend toward improved performance accompanied by reduced cost will result in large-scale, environmentally benign PV power plants displacing conventional fossil fuel and nuclear plants as preferred sources of electricity.

1. PHYSICAL PRINCIPLES UNDERLYING THE PV EFFECT

To produce a PV effect, three events must occur. First, excess positive and negative carriers must be produced by the absorption of ionizing radiation. The ionizing radiation can be electromagnetic (infrared, visible, ultraviolet, x rays, γ rays) or charged-particle radiation (electrons, protons, α particles, β particles). The photons of the electromagnetic radiation must have an energy equal to or in excess of E_G, the forbidden energy gap of the semiconductor that is the internal ionization energy of the semiconductor. As energetic charged particles pass through the semiconductor, they produce internal ionization by liberating bonding electrons. The average energy expended by a charged particle to produce one hole-electron pair in silicon and germanium is about $3.5E_G$. When high-energy electrons are the ionizing radiation, the effect is called the electron-voltaic effect with similar nomenclature for other charged particles. All such effects are essentially the same and the photovoltaic effect can be viewed as a generic term describing them. Second, the excess charges of opposite signs must be separated by an internal electrostatic potential barrier like a *p-n* junction or a metal-semiconductor junction. Third, the mobile carriers generated by the ionizing radiation must "live" long enough to reach the charge-collecting or -separating barrier. For further details, see the following subsections and Loferski (1979).

1.1 Light Absorption in Semiconductors

Let a monochromatic beam of light containing photons whose energy exceeds the forbidden energy gap of a particular semiconductor be incident on that semiconductor. If the photon flux incident on the plane $x = 0$ is $N_{ph}(0)$, then the photon flux remaining after traversing a thickness x of the semiconductor, $N_{ph}(x)$, is given by the relation

$$N_{ph}(x) = N_{ph}(0)\exp(-\alpha x), \qquad (1)$$

where α is the absorption constant for that particular energy photon. The photon flux lost from the beam, $N_{ph}(0) - N_{ph}(x)$, is absorbed in the semiconductor and generates at least one hole-electron pair for each photon absorbed. The absorption constant α is a function of the wavelength λ or photon energy $h\nu$. The dependence of α on $h\nu$ is a fundamental characteristic of a semiconductor. Figure 1 shows curves of α vs $h\nu$ for a few semiconductors that have been used in PV cells. The threshold energy for the onset of absorption is the energy gap E_G of the semiconductor, which is a first-order parameter in determining the efficiency with which a PV cell can

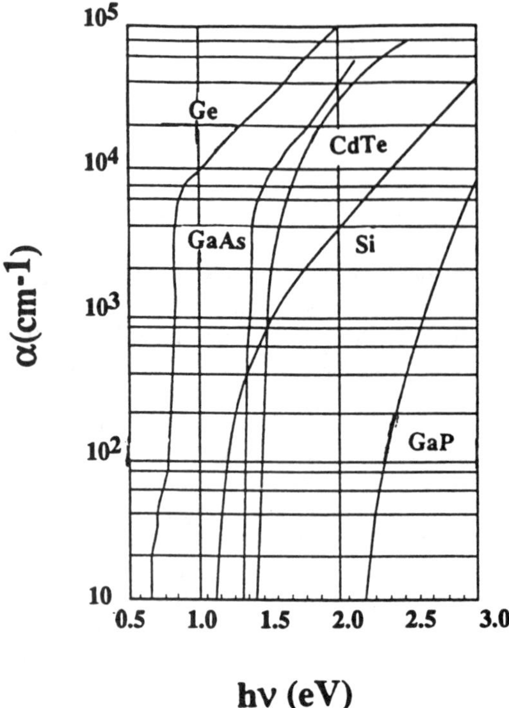

FIG. 1. Absorption constant α versus photon energy $h\nu$ for selected semiconductors of interest as photovoltaic cell materials.

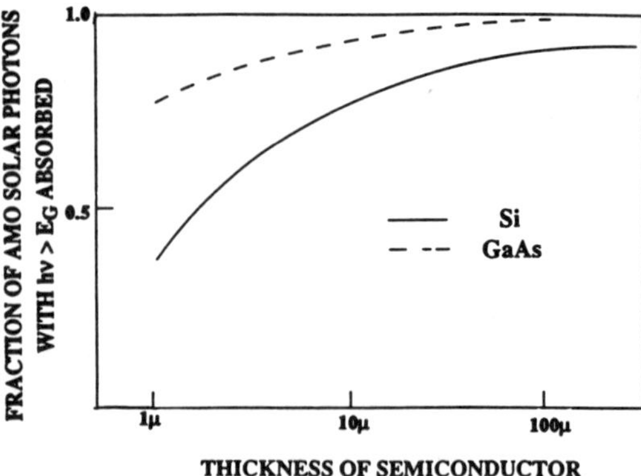

FIG. 2. Comparison of fraction of solar photons with energy greater than their energy gaps absorbed in a given thickness of the direct-gap semiconductor GaAs and the indirect-gap semiconductor silicon.

convert the energy of ionizing energy into electricity. Closer examination of these curves of α vs $h\nu$ leads to the conclusion that they can be divided into two groups:

1. those in which α rises very rapidly from very low values, say 10 cm^{-1}, to values in excess of 10^4 cm^{-1} (the GaAs and CdTe curves are in this category); and
2. those in which α rises more gradually (the Si, Ge, and GaP curves are in this category).

Semiconductors with rapidly rising curves of α vs $h\nu$ are called direct-gap semiconductors; those with gradually rising curves of α vs $h\nu$ are called indirect-gap semiconductors.

A practical consequence of this difference for PV cells used as solar-energy converters can be seen in Fig. 2. Here we compare the fraction of photons in the solar spectrum that have an energy greater than the forbidden energy gap E_G and are absorbed in a given thickness of the indirect-gap semiconductor silicon with the same fraction in the direct-gap semiconductor gallium arsenide (GaAs). For example, to absorb 90% of the photons that silicon is capable of absorbing, we need a thickness of approximately 100 μm, whereas for GaAs, the corresponding thickness is only 10 μm. This difference is very significant for large-scale conversion of sunlight into electricity, which would require covering tens of thousands of square kilometers with solar cells, because the amount of direct-gap semiconductor required is much smaller and, therefore, should ultimately be less expensive than the amount of an indirect-gap semiconductor required for the same purpose. Of course, the relative abundance of the materials of which the semiconductor is composed has a strong effect on the cost. For example, silicon is an extremely abundant element, whereas gallium is not. Nevertheless, the amount of gallium available is sufficient to make direct-gap semiconductors in which it is a constituent strong candidates as building blocks for large-scale PV applications.

1.2 Charge Separation in PV Cells

As has been noted, charge separation requires the presence of an electrostatic potential barrier inside the PV cell. Figure 3 shows the electron energy-band diagrams of the three basic kinds of charge-separation barriers. In this figure, E_C is the electron energy at the bottom of the conduction band, E_V is the electron energy at the top of the valence band, and E_F is the electron energy at the Fermi

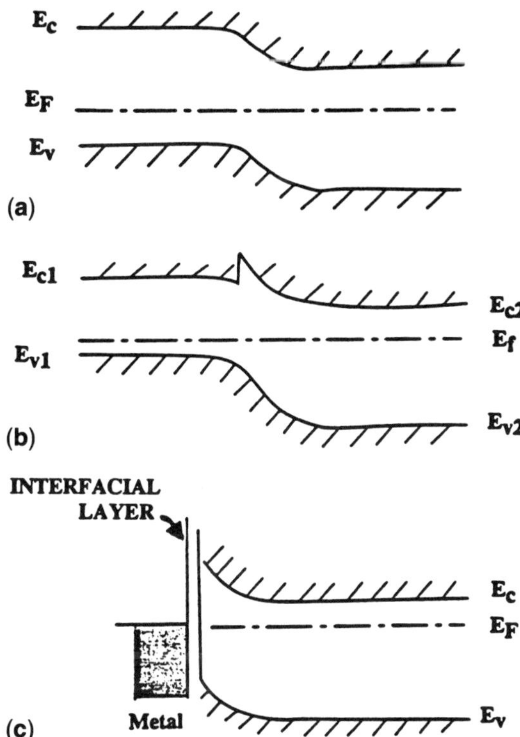

FIG. 3. Energy-band diagrams of (a) a *p-n* homojunction, (b) a *p-n* heterojunction, and (c) a metal-semiconductor (Schottky barrier) junction.

level. If the temperature across the barrier is uniform and there is no radiant energy input, i.e., the PV cells are in the dark, the Fermi energy is the same throughout the material including the ends of the junction. The energy differences $E_F - E_V$ and $E_C - E_F$ are determined by the temperature and the ionized-impurity (dopant) concentrations in the various regions of the semiconductor. The energy-band diagrams of Fig. 3 correspond to the case of uniform temperature and constant but different dopant concentrations in the *p* and *n* regions. When $E_F - E_V$ is less than $E_C - E_F$, the material is *p* type and the current carriers are predominantly holes. When $E_F - E_V$ is greater than $E_C - E_F$, the material is *n* type and electrons are the dominant charge carriers. So band diagrams (a) and (b) of Fig. 3 have *p*-type material on the left side of the barrier and *n*-type material on the right side of the barrier, i.e., they are the band diagrams of a *p-n* junction.

Figure 3(a) is the band diagram of a *p-n* homojunction because the *p-n* junction spans

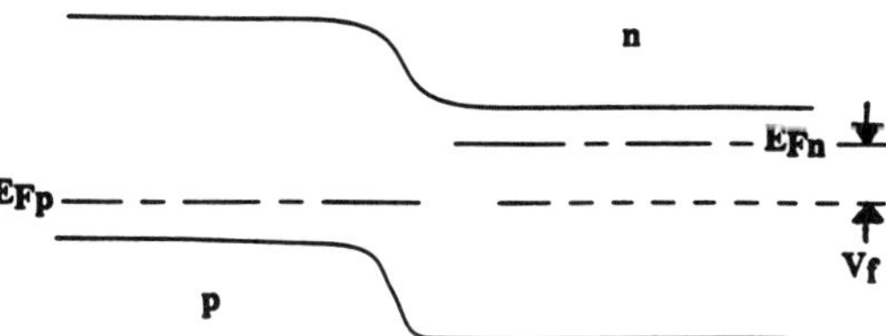

FIG. 4. Energy-band diagram of a *p-n* homojunction under forward bias.

a single semiconductor. Figure 3(b) is the band diagram of a *p-n* heterojunction because the *p* side (left side) consists of one semiconductor and the *n* side (right side) consists of another semiconductor. In Fig. 3(b), the energy gaps $E_C - E_V = E_G$ are substantially different on the *p* and *n* sides.

Figure 3(c) is the band diagram of a metal-semiconductor or Schottky barrier junction in which the work functions of metal and semiconductor bear such a relation to each other that the band edges of the *n*-type semiconductor bend upward at the junction. The sketch includes a thin interfacial layer (which may or may not be present) between the metal and the semiconductor. If this interfacial layer is thin enough, carriers can tunnel through it as they pass from the metal to the semiconductor or vice versa.

If a voltage *V* is applied across the *p-n* homojunction of Fig. 3(a), the band diagrams change to the forms shown in Figs. 4 and 5, depending on whether the *n* side is more negative than the *p* side (Fig. 4) or the *n* side is more positive than the *p* side (Fig. 5). The voltage bias of Fig. 4 is referred to as a forward bias, while the voltage bias of Fig. 5 is referred to as a reverse bias. When an open-circuited *p-n* junction absorbs ionizing radiation, it acquires a forward bias like that depicted in Fig. 4. This occurs because the elec-

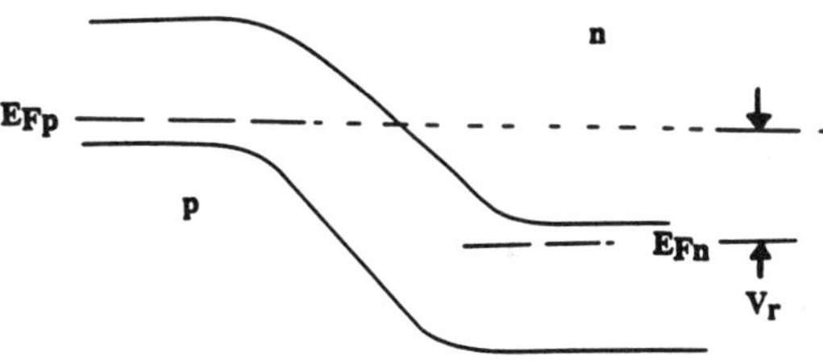

FIG. 5. Energy-band diagram of a *p-n* homojunction under reverse bias.

trostatic barrier associated with a *p-n* junction like that depicted in Fig. 3(a) is a "sink" for excess minority electrons generated on the *p* side and for excess minority holes generated on the *n* side. A flow of excess electrons from the *p* side to the *n* side and of excess holes from the *n* side to the *p* side would constitute an electric current flowing in the "reverse" direction. But if the junction is open circuited, by definition there can be no current flowing through it. So the junction reacts by biasing itself in the forward direction, i.e., it reduces the height of the electrostatic barrier just enough to produce a forward current equal in magnitude to the reverse current generated by the ionizing radiation. Thus, an illuminated *p-n* junction that is not open circuited acquires a forward bias and the current flows through it in the reverse direction, as shown in the *I–V* diagram of an illuminated solar cell of Fig. 6. The figure shows the open-circuit (forward) voltage (V_{oc}); the short-circuit (reverse) current I_{sc}; the maximum power point, P_{max}; the load voltage at maximum power, V_{mp}; and the load current at maximum power, I_{mp}.

1.3 Lifetime of Light-Generated Carriers

As noted above, the excess carriers generated by the absorbed ionizing radiation must remain mobile long enough to allow them to migrate to the charge-separating junction. Their lifetime is terminated when positive and negative carriers recombine with each other. Recombination can be intrinsic or extrinsic in origin; it can occur in the bulk or at the surface. The lifetime of the excess carriers is inversely proportional to the recombination rate.

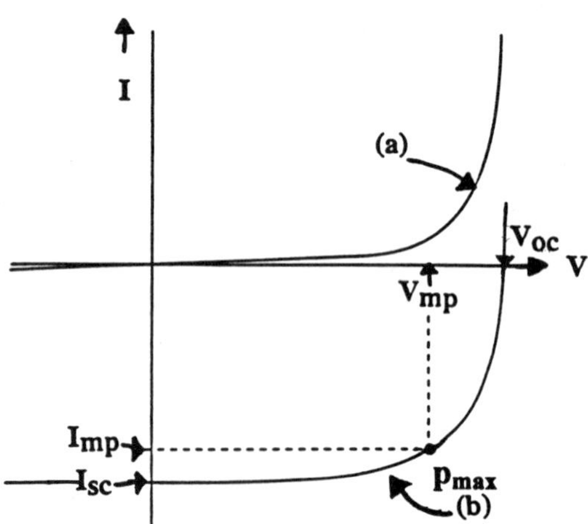

FIG. 6. Current–voltage characteristic of a solar cell (a) in the dark and (b) subjected to ionizing radiation, showing the open-circuit voltage V_{oc}; the short-circuit current I_{sc}, the maximum power point, P_{max}; the current at maximum power, I_{mp}; and the voltage at maximum power, V_{mp}.

For the case of bulk intrinsic recombination, the excess minority carriers generated in one of the bands by the ionizing radiation recombine with carriers of the opposite sign in the other band. Such band-to-band recombination sets an upper limit on the minority-carrier lifetime. Its value can be as high as 10^{-2} s in an indirect-gap semiconductor like Si and as low as 10^{-8} s in a direct-gap semiconductor like GaAs.

Indirect recombination requires the presence of recombination centers in the semiconductor. Recombination centers have energy levels that are closer to the middle of the forbidden energy gap than the energy levels of dopant impurities. In the case of silicon, typical energy levels E_r of recombination centers are located between 0.15 and 0.25 eV from one of the band edges. If the concentration of recombination centers N_r is high enough (10^{13} centers/cm^3 is high in silicon doped with 10^{16}/cm^3 donors or acceptors), the lifetime is determined by these centers. It is given by the expression

$$1/\tau_{ext} = N_r \sigma_r v f(E_r - E_F), \qquad (2)$$

where τ_{ext} is the extrinsic bulk minority-carrier lifetime; σ_r is the minority-carrier capture cross section of the recombination center (10^{-12} cm^2 $< \sigma_r <$ 10^{-16} cm^2); v is the thermal velocity of the carriers (about 10^7 cm/s at 300 K); and $f(E_r - E_F)$ is the probability that the recombination center is occupied by a majority carrier, i.e., it is the Fermi–Dirac statistics occupancy factor for the center. If $E_F - E_r > kT$, $f = 1$, and the range of values for $N_r \tau_{ext}$ is

$$10^5 \text{ s/cm}^3 \le N_r \tau_{ext} \le 10^9 \text{ s/cm}^3.$$

Therefore, with $N_r = 10^{13}$/cm^3 the range of values of extrinsic lifetime would be

$$10^{-8} \text{ s} \le \tau_{ext} \le 10^{-4} \text{ s}. \qquad (3)$$

The actual lifetime τ is related to τ_{ext} and the intrinsic lifetime τ_{int} by the relation

$$1/\tau = 1/\tau_{int} + 1/\tau_{ext}. \qquad (4)$$

In an indirect-gap semiconductor, like silicon, the extrinsic lifetime dominates. In direct-gap material like GaAs, the intrinsic lifetime (which is low anyway) can dominate unless the concentration of recombination centers is greater than about $10^{15}/cm^3$.

If a PV cell is to achieve its optimum performance, its minority-carrier lifetime must be as high as possible. This in turn means that the lifetime should be close to the intrinsic lifetime, i.e., the concentration of recombination centers N_r must be as low as possible. For an indirect-gap semiconductor like silicon in which τ_{int} is large (10^{-2} s), N_r must be of the order of $10^{11}/cm^3$. For a direct-gap semiconductor like GaAs, N_r can be as high as $10^{15}/cm^3$ without affecting τ.

2. EQUATIONS GOVERNING THE PV EFFECT

2.1 Diode Equation

The derivation of equations for the PVE begins with the ambipolar diffusion equation formulated by Shockley (1951). This equation governs the transport of excess carriers (like those produced by absorption of ionizing radiation) in a semiconductor. The appropriate one-dimensional form of this equation for the light-generated excess minority-carrier population Δn on the p side of a p-n junction is

$$D_e \frac{d^2\Delta n}{dx^2} + \alpha N_0 e^{-\alpha x} - \frac{\Delta n}{\tau} = 0. \qquad (5)$$

An analogous equation applies to the excess minority hole population on the n side. In this equation, D_e is the free-electron diffusion constant, α is the absorption constant for the wavelength of the incident light, N_0 is the photon flux of that wavelength incident on the surface of the planar PV cell, and τ is the minority-carrier lifetime.

These same equations with $N_0 = 0$ were used by Shockley to derive the I–V relations for a p-n junction diode under forward and reverse bias conditions. He showed that the current flowing through the junction I_j with a voltage V_j applied across it was

$$I_j = I_0[\exp(qV_j/kT) - 1]. \qquad (6)$$

The value of I_0 depends on semiconductor properties, junction parameters, and the boundary conditions appropriate to the particular cell. If the width w of each side of the junction is greater than the respective minority-carrier diffusion lengths (L_e and L_h), then

$$I_0 = I_{0n1} + I_{0p1}$$
$$= q[D_e n_p/L_e + D_h p_n/L_h], \qquad (7)$$

where I_{0n} and I_{0p} are the contributions to I_0 from the n and p sides, respectively; n_p and p_n are the minority electron concentration on the p side and minority hole concentration on the n side, respectively; and $L_e \equiv D_e\tau_p$ and $L_h = D_h\tau_n$. The term I_0 is proportional to n_i^2 because $n_p p_p = p_n n_n = n_i^2$, and therefore

$$I_0 = qn_i^2[D_e/N_A L_h + D_h/N_D L_e], \qquad (8)$$

because in extrinsic semiconductors, $N_A \sim p_p$ and $N_D \sim n_n$; N_A is the acceptor concentration on the p side and N_D the donor concentration on the n side of the junction. Since $n_i^2 \propto \exp(-E_G/kT)$ we can write

$$I_0 = K(T)\exp(-E_G/kT), \qquad (9)$$

where E_G is the energy gap and $K(T)$ is a weak function of absolute temperature T.

As we have already shown, when a PV cell is exposed to ionizing radiation, it acquires a forward bias (Fig. 6). The current generated by the ionizing radiation, I_{sc}, flows in the direction of a reverse current. From the principle of superposition, the equation for the current I_L through a load resistor R connected across the PV cell as shown in Fig. 7 is

$$I_L = I_j - I_{sc}, \qquad (10)$$

where I_j is the junction current and I_{sc} is the current that flows through the short-circuited PV cell and is of course the current generated by the ionizing radiation. Substitution of Eq. (6) into Eq. (10) leads to

$$I_L = I_0[\exp(\Lambda V_L - 1)] - I_{sc}, \qquad (11)$$

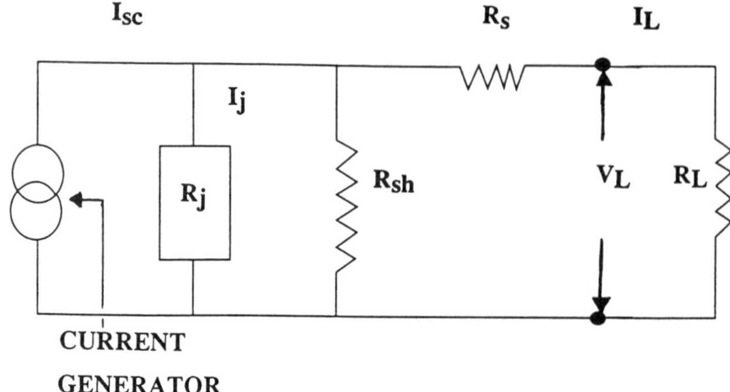

FIG. 7. Equivalent circuit of an illuminated solar cell. R_j is the nonlinear junction impedance, R_{sh} and R_s are the internal shunt and series resistances, respectively, and R_L is the load resistance.

where $\Lambda \equiv q/AkT$. The parameter A is equal to unity for the ideal Shockley diode [Eq. (6)]. However, in general $2 < A \leq 1$.

Figure 6(a) shows the I–V characteristic of a PV cell in the dark and Fig. 6(b) shows the I–V characteristic of the same cell under illumination. The dark I_j–V_j curve has a forward branch in the first quadrant where both voltage and current are positive and a reverse branch in the third quadrant where both the voltage and current are negative. Power (the product of current and voltage) is therefore consumed by the junction in either the forward or reverse directions. When the cell is illuminated, part of the characteristic drops into the fourth quadrant where the voltage is positive (a forward voltage) and the current is negative (a reverse current). Their product (the power) is negative, which means that the device is generating power. Figure 6(b) also shows the open-circuit voltage V_{oc}, the short-circuit current I_{sc}, the maximum power point P_{max}, the current at maximum power, I_{mp}, and voltage at maximum power, V_{mp}.

The expression for the open-circuit voltage V_{oc} is derived by setting $I_L = 0$ in Eq. (11). The result is

$$\exp(\Lambda V_{oc}) = (I_{sc} + I_0)/I_0. \tag{12}$$

The equivalent circuit of the PV cell (Fig. 7) must include an internal series (R_s) and an internal shunt (R_{sh}) resistance so that the complete equation for I_L assumes the form

$$I_L = I_0[\exp\Lambda(V_L - I_L R_s) - 1] - I_{sc} - (V_L - I_L R_s)/R_{sh}. \tag{13}$$

Observations on "good" PV cells lead to the conclusion that R_{sh} is usually large enough so that its effect can be neglected and R_s is small enough so that it can be neglected in a first-order calculation of the maximum power parameters. With these approximations ($R_s = 0$, $R_{sh} = \infty$) and applying the impedance-matching condition that must be satisfied for maximum power transfer from the PV cell to the load, i.e.,

$$R_{Lmp} = \left(\frac{\partial V}{\partial I}\right)_{mp}, \tag{14}$$

where R_{Lmp} is the load resistance for maximum power (mp) transfer and $(\partial V/\partial I)_{mp}$ is the dynamic impedance at the maximum power point, we can calculate the "mp" parameters. Algebraic manipulation leads to the following expressions for the load current at maximum power I_{Lmp} and the load voltage at maximum power V_{Lmp}:

$$(1 + \Lambda V_{Lmp}) \exp(\Lambda V_{Lmp}) = I_{sc} + I_0/I_{sc} = \exp(\Lambda V_{oc}), \tag{15}$$

$$I_{Lmp} = I_{sc}\Lambda V_{Lmp}/(1 + \Lambda V_{Lmp}). \tag{16}$$

These equations contain only three parameters that the PV cell fabricator can hope to control, i.e., I_{sc}, I_0, and the factor A (recall that $\Lambda \equiv q/AkT$). Their values are determined by the choice of semiconductor and PV cell parameters.

The maximum conversion efficiency η_{max}, i.e., the efficiency for matched load conditions, is given by

$$\eta_{\max} = \frac{\text{Maximum Power Out}}{\text{Power In}} = \frac{I_{Lmp}V_{Lmp}}{P_{in}} \quad (17)$$

$$= \frac{\Lambda V_{Lmp}}{(1 + \Lambda V_{Lmp})} \times \frac{\Lambda V_{Lmp}}{\Lambda V_{Lmp} + \ln(1 + \Lambda V_{Lmp})} \frac{I_{sc}V_{oc}}{P_{in}} \quad (18)$$

$$\equiv (CF)I_{sc}V_{oc}/P_{in}, \quad (19)$$

where CF is defined as the curve factor. The curve factor is the ratio of the area of the power rectangle I_{Lmp} V_{Lmp} to the area of the I_{sc} V_{oc} rectangle (see Fig. 6). Obviously it is always less than unity. In "good" cells, it can attain values close to 0.90.

2.2 Calculation of the Short-Circuit Current I_{sc}

The short-circuit current generated as a result of the absorption of ionizing radiation is calculated (Loferski and Wysocki, 1961) from the ambipolar diffusion equation [Eq. (5)] with two boundary conditions:

1. The junction is a sink for carriers, and therefore the excess carrier concentration at the edge of the space-charge region (SCR) is zero, i.e.,

$$\Delta n_{SCR} = 0. \quad (20)$$

2. At the outer surfaces of the PV cell,

$$(D_e \operatorname{grad}\Delta n)_{surf} = s n_{surf}, \quad (21)$$

where s is the surface recombination velocity.

The surface recombination velocity is a parameter that is inversely proportional to the recombination lifetime of minority carriers at the surface. If $s = 0$, the minority carriers do not recombine at the surface and are "reflected" back into the bulk region. When $s = 0$, the surface is a minority-carrier "mirror." The parameter s will be zero if there are no recombination centers at the surface or if there are surface charges that generate an electrostatic field of such a polarity that the minority carries are repelled from the surface. When the surface recombination velocity is infinite, all minority carriers arriving at the surface recombine with carriers of the opposite sign. For s to be infinite, the surface concentration of recombination centers must be infinite or else there are surface charges of such a polarity that they give rise to an electrostatic field that drives minority carriers toward the surface where they recombine at surface recombination centers. From Eq. (21), it is clear that when $s = 0$, gradΔn at the surface is zero. As s approaches infinity, n_{surf} approaches zero, and gradΔn at the surface remains finite. For intermediate values of s, both n_{surf} and gradΔn at the surface have finite values.

The solution of the ambipolar diffusion equation for an n "skin" on a p-base cell in which ionizing radiation is incident on the $x = 0$ surface, the p-n junction is located at $x = l$, and the cell thickness is b is divided into two parts. The first of these is the contribution to I_{sc} from the n region and has the form

$$I_p(l) = \frac{qN_0(1 - R)}{\alpha(1 - \nu_p^2/\alpha^2)} [\beta_p \nu_p \exp(\nu_p l) - \gamma_p \nu_p \exp(-\nu_p l)\alpha e^{-\alpha l}], \quad (22)$$

where N_0 is the flux of photons, characterized by having an absorption constant α, incident on the cell surface $x = 0$ and R is the fraction of these photons which are reflected, and the following definitions apply:

$$\beta_p \Delta_p = \exp(-\nu_p l)(h + \alpha) - e^{-\alpha l}(h + \nu_p); \quad (23)$$

$$\gamma_p \Delta_p = e^{-\alpha l}(h - \nu_p) - \exp(\nu_p l)(h + \alpha); \quad (24)$$

$$\Delta p = \exp(\nu_p l)(h + \nu_p) - \exp(-\nu_p l)(h - \nu_p); \quad (25)$$

$$h = s/D_p, \quad (26)$$

and

$$\nu_p = (D_p \tau_p)^{-1/2} = L_p^{-1}. \quad (27)$$

Similarly, the contribution to the current from the p base is

$$I_n(l) = \frac{qN_0(1 - R)}{\alpha(1 - \nu_n^2/\alpha^2)} [\beta_n \nu_n \exp(\nu_n l) - \gamma_n \exp(-\nu_n l) - \alpha e^{-\alpha l}], \quad (28)$$

$$\beta_n \Delta_n = \alpha \exp[-(\nu_n l + \alpha b)] - \nu_n \exp[-(\nu_n b + \alpha l)]; \quad (29)$$

$$\gamma_n \Delta_n = -\alpha \exp(\nu_n l - \alpha b) - \nu_n \exp(\nu_n b - \alpha l); \tag{30}$$

$$\Delta_n = 2\nu_n \cosh(\nu_n b - \nu_n l); \tag{31}$$

and

$$\nu_n = (D_n \tau_n)^{-1/2} = L_n^{-1}. \tag{32}$$

The total short-circuit current is

$$I_{sc} = I_p(l) + I_n(l). \tag{33}$$

It is convenient to define a collection efficiency for each value of α (i.e., for each wavelength of the incident light) as follows:

$$Q_{total}(\alpha) = \frac{I_p(l, \alpha)}{I_{max}(\alpha)} + \frac{I_n(l, \alpha)}{I_{max}(\alpha)} = Q_p(\alpha) + Q_n(\alpha), \tag{34}$$

where $I_{max}(\alpha)$ is the current that would have been produced if every photon with absorption constant α contributed one carrier to $I_{sc}(\alpha)$. Thus,

$$I_{max}(\alpha) = qN_0(\alpha)[I - R(\alpha)]. \tag{35}$$

Figures 8 and 9 show the results of calculations of $Q_p(\alpha)$ and $Q_n(\alpha)$ for various values of the cell parameters. The results of the calculations shown in Fig. 8 correspond to the case of a low surface recombination (about 100 cm/s) on the face $x = 0$. The parameter $\nu_p l = l/L_p$, where L_p is the hole diffusion length, is varied from 0.1, which corresponds to a cell in which the junction depth l is small compared to the minority-carrier diffusion length L, to $\nu_p l = 6.0$, a cell in which the inverse is true. If the calculation is conducted for higher values of $h = s/D$, i.e., higher values of surface recombination velocity, the values of Q_p are reduced for all values of α. Figure 9 shows the results of similar calculations for base collection efficiency Q_n as a function of α for various values of l/L_n with a fixed value for b (=0.5 cm). As expected, Q_n increases as the diffusion length in the base increases. Figure 10 shows $Q_{total} = Q_n + Q_p$ for a particular set of values of h, νl, and b.

Data like those presented in Figs. 8, 9, and 10 can be transformed into spectral response curves for specific semiconductors by transforming the abscissa to wavelength or photon energy with the help of curves of α vs λ like those shown in Fig. 1. If this is done for a Si cell having specific values of cell parameters, the result is a family of curves like those shown in Fig. 11, where νl in the diffused "skin" is varied. In the particular case chosen for Fig. 11, the surface recombination velocity $s = 0$ and so, for small values of l/L_p,

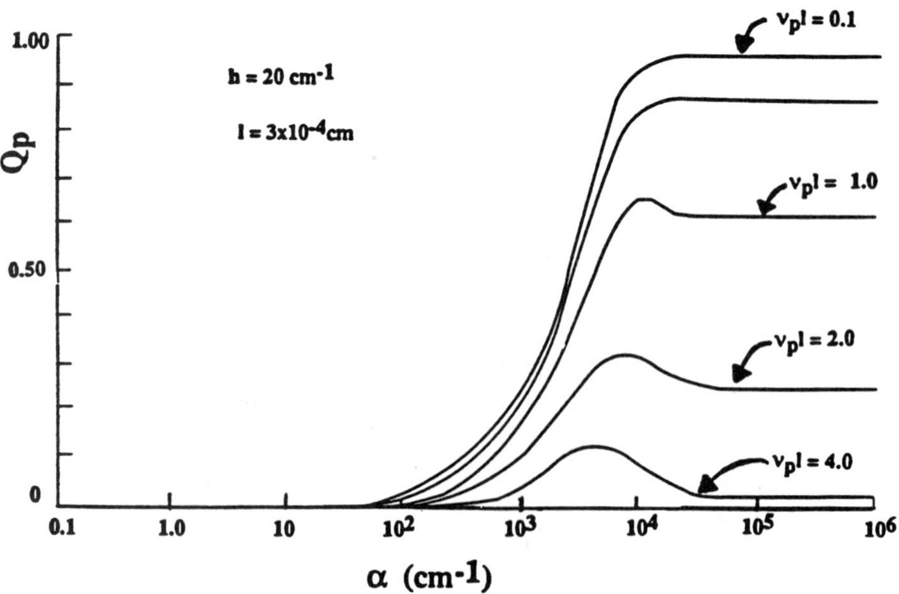

FIG. 8. Collection efficiency Q_p of the diffused p-type skin of a PV cell vs α for different values of νl ($n = L^{-1}$).

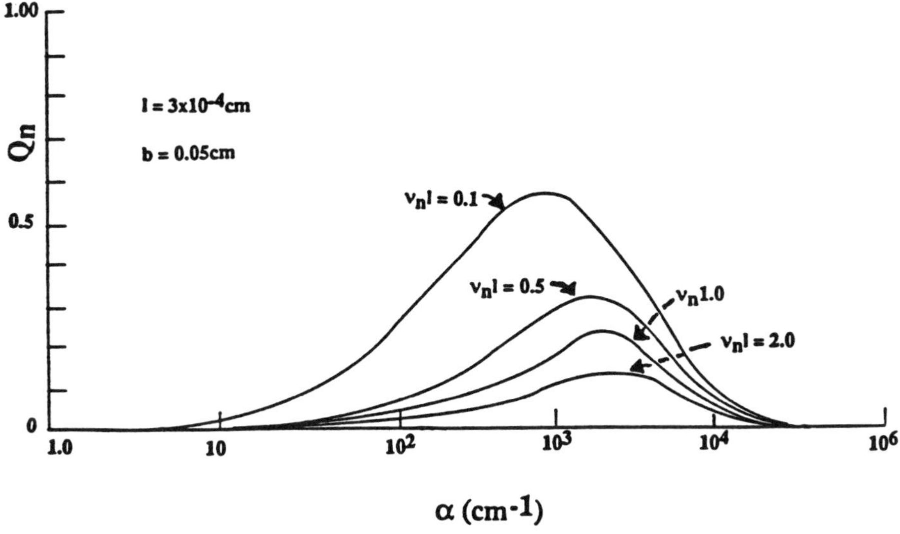

FIG. 9. Collection efficiency Q_n of the n-type base region of a PV cell vs α for different values of nl.

the total collection efficiency approaches unity value for short-wavelength radiation.

2.3 The Reverse Saturation Current I_0 and the Junction Quality Factor A

Most real junctions do not behave as predicted by the ideal Shockley diode equation, i.e.,

$$I_j = I_0[\exp(qV_j/kT) - 1], \tag{36}$$

where I_j is the current through the junction and V_j is the voltage across it. Rather their I_j–V_j characteristic can be approximated by the following expression:

$$I_j = \Sigma_i I_{0i}[\exp(\Lambda_i V_j) - 1], \tag{37}$$

where

$$\Lambda_i = q/A_i kT, \tag{38}$$

and A_i is the diode quality factor (Loferski, 1979). For the ideal diode ($i = 1$), $A_1 = 1$, and I_0 is given by Eq. (8).

A fundamental assumption underlying the derivation of the ideal diode equation and of the corresponding expression for I_0 is that there is no loss of carriers as they traverse the space-charge region. If this restriction is

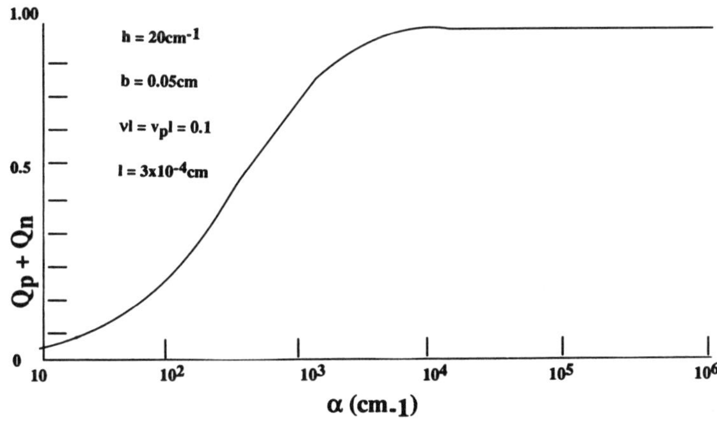

FIG. 10. Total collection efficiency $Q_p + Q_n$ vs α for $\nu l = 0.1$ in both the skin and base regions and with a low surface recombination velocity on the illuminated surface of the p-type skin.

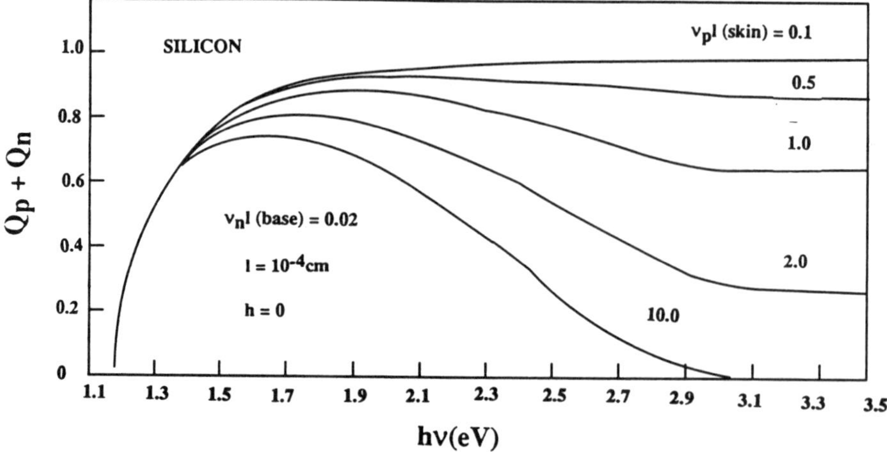

FIG. 11. Family of silicon PV-cell spectral response curves ($Q_p + Q_n$) vs photon energy for different values of $l/L_p = \nu_p l$ in the skin and a large diffusion length in the base.

removed and recombination and generation of carriers is assumed to occur in the space-charge region at recombination centers having their energy levels in the middle of the forbidden energy gap, then the dependence of I_0 on n_i can be shown to have the form

$$I_{02} = K_2(T)n_i. \tag{39}$$

Therefore,

$$I_{02} \propto \exp(-E_G/2kT) \tag{40}$$

and

$$I_{j2} = I_{02}[\exp(qV/2kT) - 1]. \tag{41}$$

More generally,

$$I_{0i} = K_i(T)\exp(-E_G/B_i kT) \tag{42}$$

and

$$I_{ji} = I_{0i}[\exp(qV/A_i kT) - 1]. \tag{43}$$

In the ideal diode case $A_1 = B_1 = 1$, while in the recombination–generation case $A_2 = B_2 = 2$. In real diodes, $2 \leq A \leq 1$ and $2 \leq B \leq 1$.

It is easy to show that a PV cell that obeys the ideal diode equation ($A = B = 1$) should have a higher value of η_{max} than a PV cell with $A = B = 2$. The short-circuit current I_{sc} is not affected by the values of A and B and, therefore, will be the same in either cell. There is, however, a difference in V_{oc} values for the two cases, namely

$$V_{oc2} - V_{oc1} = (kT/q)\ln(I_{01}/I_{02}). \tag{44}$$

Since I_{02} is always greater than I_{01}, V_{oc1} is always greater than V_{oc2}. Consequently, according to Eq. (19), $\eta_{max1} > \eta_{max2}$. It is, therefore, a goal of PV cell technology to produce junctions whose I_j–V_j characteristics are governed by the ideal diode equation.

There is evidence that ideal diode behavior is more likely to be observed in *p-n* junctions made from smaller-bandgap semiconductors like germanium (0.7 eV) than in larger-bandgap semiconductors like GaAs (1.35 eV). The bandgap of silicon (1.1 eV) places it in a transition region between low- and high-E_G materials; consequently, it has been difficult but not impossible to produce silicon cells having ideal diode behavior and, therefore, exhibiting the highest η_{max} of which silicon is capable.

2.4 Maximizing η_{max}

The equations governing the PV effect provide guidance on extracting the highest efficiency from a PV cell made from any semiconductor. Most important of all, they show that both surface and bulk recombination rates must be minimized. Minimizing surface losses means passivation of the surface. In the case of silicon, passivation is achieved by covering the outer surfaces with a silicon oxide that has a low interface state density and/or that produces a band bending at the surface in such a direction that light-generated minority carriers are repelled from the surface region.

A more generally applicable method of reducing front-surface recombination losses is to employ the wide-bandgap window concept. The idea is to cover the photovoltaically active semiconductor (PVAS), i.e., the absorber of the ionizing radiation, with a different semiconductor whose bandgap is so large that none of the photons in the light source are absorbed in it. This can result in a heterojunction device like that shown in Fig. 12. There is an added small benefit to efficiency in that the contribution from the wide-E_G semiconductor to I_0 will be orders of magnitude lower than the contribution to I_0 from the PV-active material. Lower I_0, of course, leads to somewhat higher V_{oc} and η_{max}.

Minimizing bulk recombination means minimizing the concentration of bulk recombination centers throughout the light-absorbing volume of the PV cell. In the case of an indirect-gap material like silicon, the concentration of elements like Cu and Fe and defects like vacancies and interstitial atoms, all of which are effective recombination centers, must be minimized. Concentration of these defects must be held to values in the range 10^{12}–$10^{13}/\text{cm}^3$. Such defects can be introduced during crystal growth, during growth of thin films, or during preparation of the cell. As noted above, highest efficiency requires A values of unity. This in turn requires minimizing recombination-center concentration in the space-charge region.

If the surfaces that serve as boundaries of the PV-active region of the cell have zero surface recombination velocity, the value of I_0 calculated from the one-dimensional ambipolar diffusion equations for a cell in which $w \ll L$ has the value (Loferski, 1988)

$$I_0 = I_\infty \tanh(w/L), \tag{45}$$

where w is the distance between the edge of the space-charge region and the edge of the cell and I_∞ is the value of reverse saturation current for a cell in which $w \gg L$.

Since $\tanh x \to x$ as $x \to 0$,

$$I_0 \to I_\infty(w/L). \tag{46}$$

Thus, I_0 decreases as the thickness w decreases. But as w decreases, the photon-absorbing volume decreases and so does I_{sc}. This situation can be remedied by resorting to light trapping, i.e., by making the internal surfaces of the cell optically reflecting. Figure 13 shows how the cell efficiency depends on w for $s = 0$ and $s = \infty$ with 100% efficient optical reflection $R = 1$ and with zero internal reflection $R = 0$. Clearly the goal of the designer is to fabricate cells that approach the case $s = 0$ and $R = 1$, and indeed the highest-performance solar cells approach such values for s and R.

2.5 Dependence of η_{max} on E_G

Up to this point, the discussion has been focused on the performance of PV cells made from a given semiconductor. Clearly, the forbidden energy gap E_G has a first-order effect on the maximum efficiency that can be extracted from a semiconductor (Loferski, 1956). As E_G increases, I_{sc}, which is given by the

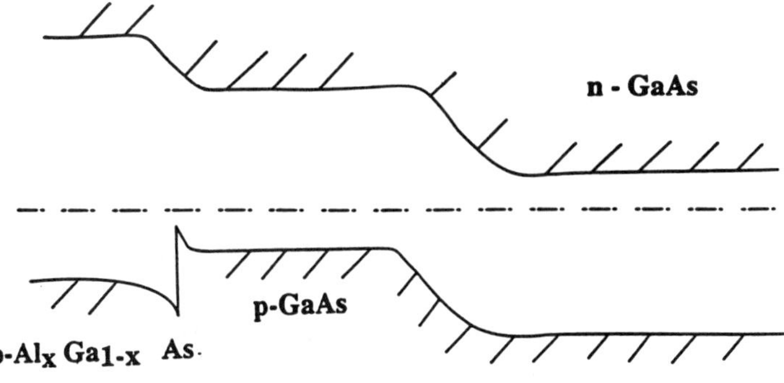

FIG. 12. Energy-band diagram of the p-$Al_xGa_{(1-x)}As$, p-GaAs/n-GaAs solar cell.

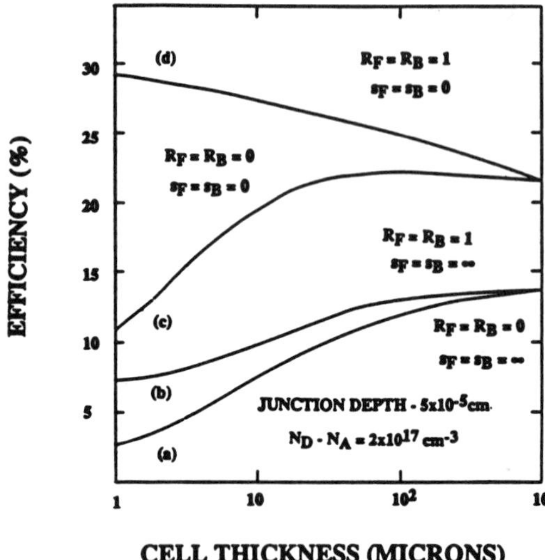

FIG. 13. Solar-energy conversion efficiency vs cell thickness calculated for a silicon solar cell for the four combinations of two extreme values of surface recombination velocity ($s = 0$ and $s = \infty$) and of internal reflectance ($R = 0$ and $R = 1$).

equation

$$I_{sc} = \int_{E_G}^{\infty} Q(h\nu)n_{ph}(h\nu)d(h\nu), \tag{47}$$

decreases. For the solar spectrum, it turns out that

$$I_{sc} = I_{sc0}\exp(-K_E E_G). \tag{48}$$

The value of I_{sc0} is the value of I_{sc} if every photon in the incident solar spectrum contributed a carrier to I_{sc}. The term K_E has a value between 1.5 and 2, depending on the distribution of photons in the spectrum.

The value of I_0 also decreases with E_G because

$$I_0 = K_i(T)\exp[-(E_G/B_i kT)]. \tag{49}$$

At $T = 300$ K and with $B = 1$,

$$I_0 \sim K_1(300\text{ K})\exp(-40E_G). \tag{50}$$

Now, Eq. (12) can be written in the form

$$V_{oc} = \frac{A_i kT}{q}\ln\left(\frac{I_{sc}}{I_0}\right). \tag{51}$$

For $A = 1$ and 300 K, and with $K_E = 2$,

$$V_{oc}(E_G) \sim 0.025[38E_G + \ln(I_{sc0}/K_1)]. \tag{52}$$

So V_{oc} increases as E_G increases. Therefore, since

$$\eta_{max} = I_{sc}V_{oc}(CF)/P_{in}, \tag{53}$$

η_{max} passes through a maximum.

Figure 14 shows η_{max} vs E_G for the Air Mass Zero (AM0) solar spectrum for two cases: $A = B = 1$, i.e., the ideal diode case, and $A = B = 2$, i.e., the case of recombination-generation in the SCR. (The AM0 solar spectrum is the spectrum above the earth's atmosphere.) In generating these curves, the value

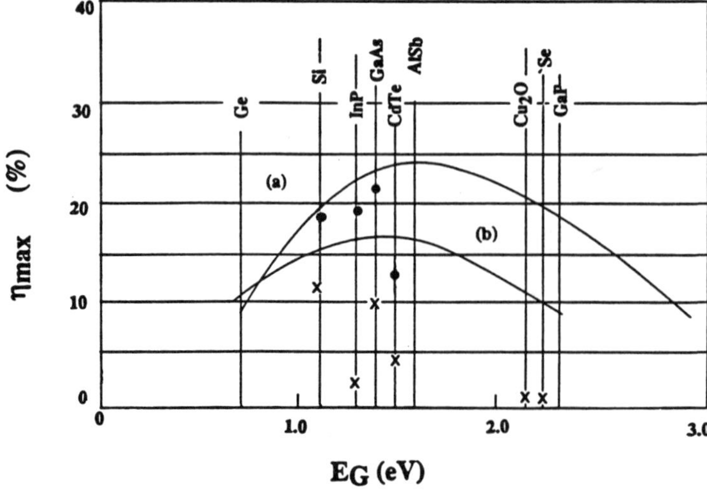

FIG. 14. Efficiency η_{max} vs energy gap E_G for solar cells exposed to AM0 solar spectrum for (a) an ideal p-n junction ($A = B = 1$) and (b) a p-n junction including recombination generation in the space-charge region ($A = B = 2$). The crosses indicate the best experimental values of η_{max} when this theoretical curve was constructed (Loferski, 1956) and the solid dots show the best experimental values in 1994.

of I_0 was calculated for Si for these two cases and it was assumed that the values of K_i are the same as in Si for all other semiconductors. This of course is not true; a separate calculation of K_i should be made for each semiconductor. However, the values of K_i do not vary by large amounts from semiconductor to semiconductor and so the curves of η_{max} vs E_G based on the indicated assumptions correctly predict the gross dependence of η_{max} on E_G.

The major conclusions of this exercise are as follows:

1. The maximum efficiency for solar energy conversion for a single-junction solar cell is around 26%.
2. The best semiconductors for PV solar cells have bandgaps between 1.0 and 2.2 eV with the highest η_{max} occurring around 1.5 eV for AM0 illumination.
3. Cells in which $A = B = 1$ have much higher potential efficiencies than cells with $A = B = 2$.
4. Although the data of Fig. 14 do not show it, the shape of the curves of η_{max} vs E_G changes as the solar spectrum changes.

2.6 Ultimate Efficiency Calculation

The η_{max} vs E_G calculations presented above are based on empirical values of the parameters of semiconductors and of the *p-n* junctions made from them. An important result of these calculations is that qV_{oc} can never be equal to E_G and the curve factor is less than unity. If, however, one assumes that $qV_{oc} = E_G$, which is the maximum imaginable value qV_{oc} can attain, that $CF = 1$, and that every absorbed photon contributes one carrier to I_{sc}, then the efficiency for a cell made from a semiconductor with energy gap E_G would be

$$\eta(E_G) = E_G N_G / E_{\text{total inc}}, \tag{54}$$

where

$$N_G = \int_{h\nu_G = E_G}^{\infty} n(h\nu)d(h\nu); \tag{55}$$

i.e., N_G is the total number of photons having energy equal to or greater than E_G and $n(h\nu)d(h\nu)$ is the photon distribution in the spectrum. The total incident energy is given by

$$E_{\text{total inc}} = \int_0^{\infty} h\nu n(h\nu)d(h\nu). \tag{56}$$

If the solar spectrum is approximated by the spectrum of a 6000-K blackbody, the peak in the curve of η vs E_G calculated from Eq. (54) lies at $E_G \sim 1.1$ eV with $\eta(1.1 \text{ eV}) = 44\%$ (Trivich and Flinn, 1955).

2.7 Higher Efficiencies through Spectrum Splitting

The maximum imaginable energy that a light-generated carrier can deliver to a load in a PV cell is the energy gap E_G. The energy in excess of E_G delivered to the cell by photons with $h\nu > E_G$ is converted into heat inside the cell. This loss of photon energy can be reduced by splitting the spectrum among a stack of cells made from semiconductors having different energy gaps. For example, suppose the smallest bandgap of an absorber semiconductor in a PV cell stack is E_{G1}, and suppose further that the solar spectrum incident on this cell produces a current I_{sc1}. Construct a stack of N cells whose energy gaps are chosen so that each cell in the stack absorbs such a number of photons that the value of $I_{scN} = I_{sc1}/N$. Then the conversion efficiency of the stack will be higher than the conversion efficiency of any one cell in the stack acting alone because the fraction of energy deposited by absorbed photons that is converted into heat (i.e., all the energy of the photon in excess of E_G) is less in the stack than in a single cell. For a three-cell stack with $E_G = 0.94$, 1.34, and 1.91 eV and the AM0 solar spectrum, the theoretical efficiency is about 37%, a value considerably higher than the 26% calculated for a single cell made from the semiconductor having the "optimum" bandgap.

The upper limit on the conversion efficiency of such a stack of PV cells has been the subject of much discussion in the literature. Calculations of the efficiency of a stack of cells with absorber bandgaps varying from 0 to ∞ has shown that for the case of maximum possible concentration of sunlight with lenses or mirrors, namely a concentration ratio of 46 747, the limiting efficiency is $\eta_L \sim 86\%$. The concentration ratio C is the ratio of radiant power per unit area delivered to the PV cell by the concentrator (a lens or a mirror) to the radiant power per unit area that the PV cell receives from the sun if no concentrator is used, in which case $C = 1$. The concentrator focuses all of the sun's rays that it intercepts into an image of the sun. If the cell is a disk and it is placed in the focal plane and the image of the sun just covers the cell, then C is equal to the ratio of the geometrical area of the concentrator or aperture to the area of the solar image it produces, which, in this example, is equal to the area of the PV cell. The maximum attainable value of C is the ratio of the maximum solid angle aperture of a concentrating mirror that is focusing the solar image onto a cell of the same dimension and shape as the image (this solid angle is π steradians) to the solid angle subtended by the solar disc [$\pi \sin^2(\frac{1}{2}\theta)$ steradians; $\frac{1}{2}\theta = 15.9$ minutes]. In this case, $C_{max} = 46\,747$. For the case of no concentration, $\eta_L \sim 68\%$ (Landsberg *et al.*, 1991). The efficiency is higher as the concentration is increased because the value of I_{sc} increases with light intensity while the voltage increases as the logarithm of the light intensity. Figures 15 and 16 show the results of empirical calculations of the efficiency of a finite number of cells

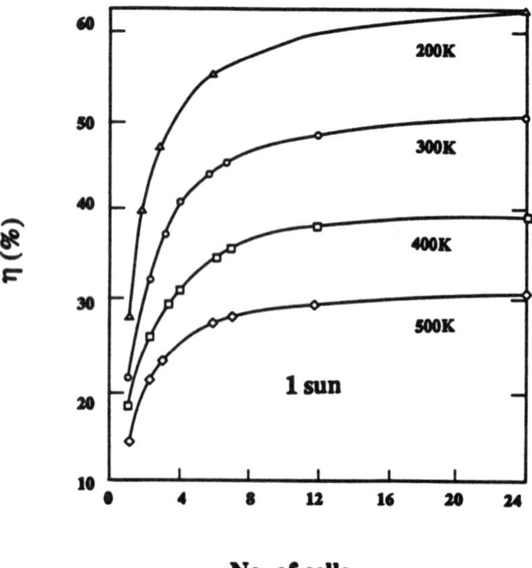

FIG. 15. Calculated optimized efficiencies at various temperatures for tandem PV cell stacks consisting of N solar cells with bandgaps of PV active semiconductors selected so that $I_{sc1} = I_{sc2} = I_{scN}$. AM0 spectrum, no concentration, i.e., $C = 1$.

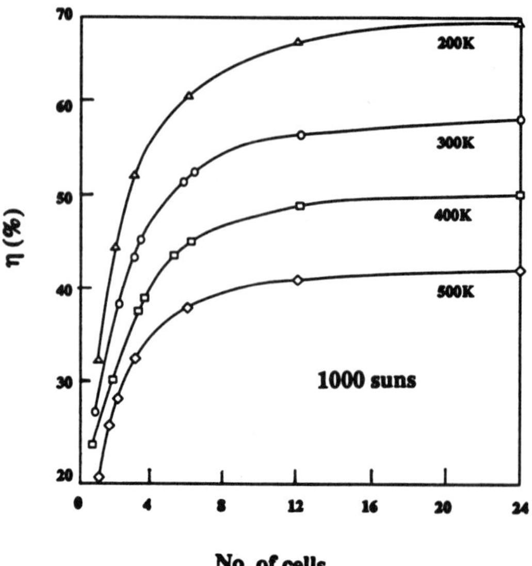

FIG. 16. Same as Fig. 15 except that concentration $C = 1000$.

based on the model of PV cells used to calculate the curves of η_{max} vs E_G of Fig. 14 for the case $A = B = 1$ (Gocsen and Loferski, 1979). For this calculation, the bandgaps of the cells were selected so that each of them would absorb the same number of photons and, therefore, generate the same I_{sc}. Figure 15 shows the results for unconcentrated AM0 sunlight and Fig. 16 for a concentration C of 1000. For example, the efficiency of a 26-cell optimized stack at 300 K is about 48% for $C = 1$ and about 59% for $C = 1000$. Note that as T increases, η decreases.

These calculations do not take into account losses, which are unavoidable in tandem cell systems or for that matter in any PV cell. For example, there will always be an internal resistive loss ρ, a grid transmissivity loss γ, and a reflection loss β. In a three-cell stack, if we assume that each of these loss factors is only 2%, a very optimistic value, then the accumulated fractional loss F_L for a three-cell stack would be (Bennett and Olsen, 1978)

$$F_L = \prod_1^3 (1 - \rho)(1 - \gamma)(1 - \beta) = 0.83, \qquad (57)$$

i.e., the efficiency of the third cell in the stack would be 83% of the lossless value. For more realistic loss factors of 0.05, $F_L = 0.63$. Such losses would limit the number of cells in a stack to perhaps four cells.

2.8 Dependence of η_{max} on Temperature

The open-circuit voltage V_{oc} is a strong function of temperature and, therefore, so is η_{max} (Wysocki and Rappaport, 1960). From Eqs. (12), (42), and (48),

$$V_{oc} = \frac{AkT}{q}\left[\frac{E_G}{BkT} - E_G K_E + \ln(I_{so0}/K_i)\right] \quad (58)$$

$$\approx \frac{AE_G}{Bq} - \frac{AkT}{q}[K_E E_G + \ln(I_{so0}/K_i)]. \quad (59)$$

It turns out that I_{sc0} is always larger than K_i and, therefore, V_{oc} always decreases as T increases. Figure 17 shows a family of curves of η_{max} vs E_G calculated for the AM0 solar spectrum using the same assumptions as those used to establish the curves of η_{max} vs E_G of Fig. 14 with $A = B = 1$. The curves of Fig. 17 show that η_{max} of all cells decreases as T increases and that the peak in the curves of η_{max} vs E_G shifts toward larger E_G as T increases.

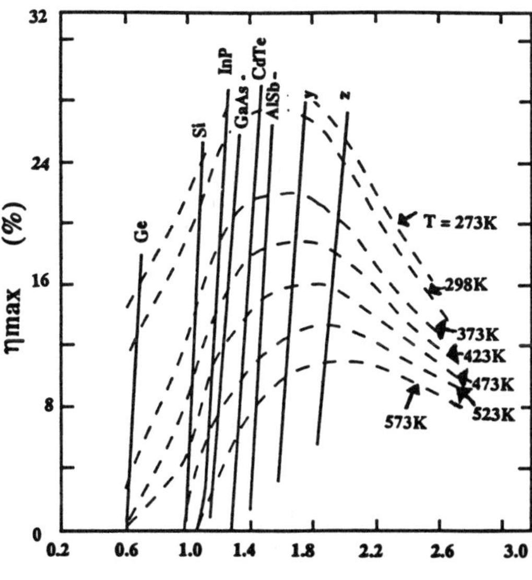

FIG. 17. Efficiency h_{max} vs energy gap E_G as a function of temperature.

3. SEMICONDUCTOR MATERIALS USED IN PV CELLS: HISTORY AND CURRENT STATUS

The calculations of η_{max} vs E_G show that the best semiconductors for PV cells have energy gaps in the range 1.0 eV $\leq E_G \leq$ 2.2 eV and that the maximum efficiency to be expected is around 26%. To attain the highest efficiencies, the minority-carrier diffusion lengths in the photovoltaically active volume (PVAV) of the cell must be as large as possible. Furthermore, the surface recombination velocity must be vanishingly small on the surfaces of the PVAV.

Photovoltaic cells have been fabricated from many materials having bandgaps in the interesting range. The most successful of these materials have crystal lattice structures related to the diamond lattice, which consists of two interpenetrating face-centered cubic lattices displaced with respect to each other by one-fourth of a lattice constant in each of the three mutually perpendicular directions. These materials include silicon; the III-V semiconductor compounds [gallium arsenide (GaAs), indium phosphide (InP)]; III-V semiconductor alloys thereof like $Ga_xAl_{1-x}As$, $GaAs_yP_{1-y}$, etc.; certain II-VI semiconductor compounds [cadmium telluride (CdTe); cadmium sulfide (CdS); zinc telluride (ZnTe)] and their alloys; and the chalcopyrite semiconductors [copper indium diselenide (CISe), copper gallium diselenide (CGSe), copper indium disulphide (CISu)] and their alloys. Most of the cited materials have served as the constituent of the PVAV, but some of them, notably CdS, have served mainly as wide-bandgap windows or have other functions to perform in the cells. Other materials having different crystal lattices, like ZnP_2, WSe_2, CuO, and CuO_2, have also been explored, but none of them has performed as well as the best of the diamond-lattice materials. There is another important class of PVAV materials that holds promise, namely certain amorphous semiconductors, notably hydrogenated amorphous silicon and its "alloys" with germanium and carbon. The next few paragraphs are devoted to brief commentaries on solar cells made from these materials.

3.1 Silicon Monocrystalline and Polycrystalline Cells

The silicon cell has been the workhorse of the modern PV age. The most successful, most utilized, and oldest of the modern PV cells is the single-crystal silicon cell. In its original form (Chapin *et al.*, 1954), it consisted of a diffused boron-doped *p*-type "skin" region on a phosphorus-doped, *n*-type single-crystal base wafer. The *p-n* junction was located about 1 μm below the surface. As shown in Fig. 18, a gridded metallic Ohmic contact was deposited over the diffused region and the back surface was covered with a metallic Ohmic contact. The area of individual cells was (and continues to be) controlled by the diameter of silicon single-crystal ingots, which had a maximum value of about 3 cm in 1954 and about 15 cm and even 20 cm in 1994. The solar-energy conversion efficiency of the 1954 cell was around 10% for terrestrial sunlight. In 1993, the highest efficiency of crystalline cells having an area of 4 cm^2 was about 23% (Green *et al.*, 1987; Green and Emery, 1993). The cells that produce such efficiencies are of a design radically different from that of the simple 1953 cell. These high-performance cells incorporate surface passivation to reduce surface losses; light trapping to improve utilization of the incident photons; and high minority-carrier lifetime silicon material and cell fabrication techniques that preserve the lifetime. The PVAV has a relatively high resistivity; the heavily doped *p* and *n* regions are both located on the side opposite the one on which light is incident.

Special-design, high-efficiency single-crystal silicon cells have also been fabricated for concentrator applications. The highest efficiency reported for such cells is about 28% with $C \sim 200$. These cells had a small area (3×5 mm^2); they had a high-resistivity PVAV and "point-contact" p^+ and n^+ collector regions on the back of the cell (Verlinden *et al.*, 1993).

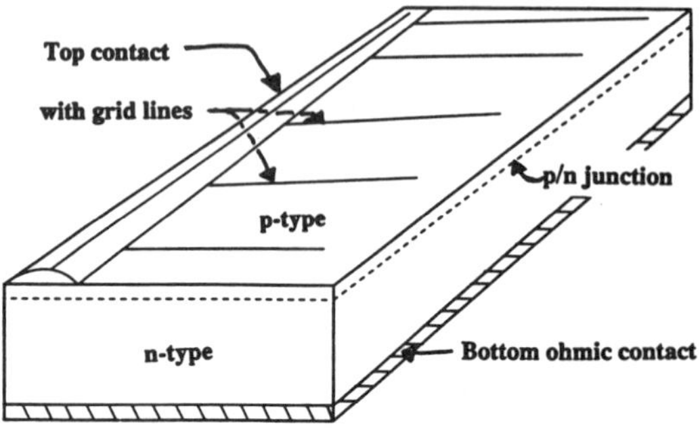

FIG. 18. Schematic representation of standard planar solar cell.

These concepts have been applied to polycrystalline silicon having grain volumes of at least several cubic millimeters. The best of the cells made from such polycrystalline silicon have exhibited solar-energy conversion efficiencies in excess of 17% in cells having an area of 4 cm^2 (Green and Emery, 1993).

3.2 Gallium Arsenide Cells

The first GaAs solar cells were reported in 1956 (Jenny *et al.*, 1956). They were prepared by diffusing cadmium into *n*-type single-crystal wafers. Their area was small (between 0.1 and 6.0 mm^2) because only small single crystals were available at that time, and the highest efficiency was 6.5%. By 1961, the size of GaAs single-crystal ingots had increased and processing had improved so that 10% efficient cells having areas up to 2 cm^2 were being fabricated.

A significant breakthrough in efficiency was achieved in 1972 by the invention of the wide-bandgap window concept and its application to GaAs (Woodall and Hovel, 1972). Analyses of extant GaAs cells showed that, in this direct-gap material, surface recombination losses exacted a high toll on excess carriers generated in the diffused skin. Incorporation of a wide-bandgap p-$Ga_xAl_{1-x}As$ window (see Fig. 12) reduced surface losses dramatically and pushed the efficiency up to 15%. Further refinement of this type of cell has resulted in GaAs cells with efficiencies up to 25% in 1993.

Another approach to reducing surface losses is to make the diffused surface layer so thin that a negligible fraction of photons are absorbed in it (Gale *et al.*, 1984). This also has resulted in 23% efficient cells by 1993 (Green and Emery, 1993).

3.3 Cadmium Sulfide/Copper Sulfide

The 1954 CdS cell was made by depositing a thin layer of copper onto a cadmium sulfide single crystal; its actual solar-energy conversion efficiency was only about 1.5% (Reynolds *et al.*, 1954). By 1956, the efficiency of such cells had increased to about 5%. The first thin-film CdS cell was reported in 1954 and the attention of researchers was thereafter directed to the thin-film cell (Nadjakov *et al.*, 1954). At first it was thought that the cell was based on a *p-n* junction in CdS since it was observed that incorporation of copper into n-type CdS eventually resulted in a p-type material. However, by the mid-1960s it was clear that this cell was based on a heterojunction between n-CdS having a bandgap of 2.4 eV and a p-type copper sulfide having a bandgap around 1.0 eV. The copper sulfide was the photovoltaically active material; the CdS was serving as a wide-bandgap window. The Cu-S binary system includes many compounds having various crystal structures. Only the chalcocite form, whose chemical composition is close to Cu_2S, could produce good PV cells. However, the compound is not stable. Efficiencies up to 10% were achieved in n-CdS/p-Cu_xS cells, but the efficiency was not stable. Consequently, interest in this type of cell waned and the cell was practically abandoned.

3.4 *p*-Copper Indium Selenide/*n*-CdS Cells

CdS continues to play an important role in PV cell science and technology because of its role as the wide-bandgap window in the p-$CuInSe_2$ (CISe)/n-CdS cell. The first of these cells was made in 1974 by depositing a thin layer of CdS onto a small-area CISe crystal; its efficiency was 12% and it was stable (Shay *et al.*, 1975). By 1976, thin-film CISe/CdS cells with efficiencies up to 6% were made by vacuum deposition of the constituents (Kazmerski *et al.*, 1976). In 1981, a 1-cm^2-area, 10% efficient, stable CISe/CdS cell was reported (Mickelsen and Chen, 1980). It was made by simultaneous controlled-rate evaporation of Cu, In, and Se onto a substrate heated to a high enough temperature (about 500 °C) to react the elements so that CISe is formed. These cells have been further improved by using alloys of CISe with $CuGaSe_2$ and alloys of CdS with ZnS. In 1993, the efficiency of 1-cm^2-area cells of this type had reached 16.9% active area, 14.8% total area (Hedstrom *et al.*, 1993, 1994). In 1994, a total-area efficiency of 16.4% was reported (Gabor *et al.*, 1994). At the time of this writing, this is the highest confirmed efficiency for a thin-film polycrystalline cell measured under standard conditions.

CISe cells have been produced from electrodeposited films, by reactively sputtered films, and by sputtering from a previously synthesized CISe target. These deposition methods are less expensive than the simultaneous evaporation technique, but the cells produced by these methods have not as yet been as efficient as cells made by three-source evaporation.

3.5 Cadmium Telluride Cells

Cadmium telluride has a bandgap of about 1.45 eV, which is close to the peak of the curve of η_{max} vs E_G. It can be made n or p type so that p-n junction CdTe cells are possible. The first of these cells was made in 1954 but its efficiency was only around 2%. In 1963, a thin-film p-Cu_xTe/n-CdTe cell having $\eta \sim 6\%$ was reported (Cusano, 1963). Subsequent research showed that the most successful CdTe cells were heterojunction devices with wide-bandgap window semiconductors including CdS, ZnO, ZnS, and alloys of these materials. Values of η_{max} up to 15.8% (total area) have been achieved (Chu and Chu, 1993).

The films for CdTe cells have been made by close spaced sublimation ($\eta \sim$ 15.8%), electrodeposition (12%), screen printing (11.3%), vacuum evaporation (11%), and chemical spray pyrolysis (12.7%). These low-cost production methods make CdTe cells among the most promising candidates for large-scale PV power plants.

3.6 Amorphous Silicon Solar Cells

The invention of the amorphous Si solar cell at RCA Laboratories in 1974 unleashed a worldwide research effort on this promising device. The first 1974 cells had $\eta \sim 0.2\%$, but by 1976 devices using a-Si had reached $\eta \sim 5.5\%$ (Carlson and Wronski, 1976).

The amorphous silicon layers were prepared by electrodeless rf discharge in silane (SiH_4) so that the resulting layers were actually "alloys" between Si and H_2. The best solar cells were made from hydrogenated a-Si (a-Si:H) films containing between 10% and 20% H_2. The H_2 passivates the dangling bonds in the Si atoms with the result that the "forbidden energy gap" has a much lower density of states than a pure amorphous Si layer contains and, therefore, the material can be doped substitutionally with elements from column 3 or 5 of the periodic table. To make p-type layers, some borane (B_2H_6) is added to the SiH_4; to make n-type layers, some phosphene (PH_3) is added to the SiH_4. Consequently it is possible to make p-n junctions in a-Si:H. The best solar cells utilize a p-i-n structure.

The optical bandgap E_o of a-Si:H varies with H_2 content between 1.5 and 1.8 eV. The absorption constant rises rapidly from E_o to values around 10^5 cm^{-1} so that the layer thicknesses of the order of 1 μm absorb all the photons with energies in excess of E_o.

The optical bandgap can be increased by incorporating carbon and decreased by incorporating germanium. Therefore, tandem cell structures can be made from a-Si:H. By 1993, the best a-Si:H single-junction cells had initial η values of about 11%. Multijunction cells reached initial η values of about 12% (Green and Emery, 1993).

Unfortunately, a-Si:H cells have exhibited degradation during illumination. In the best cases, the decay of efficiency from its initial value is about 10%. However, in 1994, it was reported that after several thousand hours of operation under constant illumination by simulated sunlight, the efficiency of certain a-Si:H modules stabilized at a value slightly in excess of 10%.

3.7 Cells Based on Other Semiconductors

Many semiconductors have been explored as the PVA material in PV cells. They include InP, WSe_2, ZnSe, Zn_3P_2, $GaAs_xP_{(1-x)}$, organic semiconductors, etc.

The most notable in this group is indium phosphide (InP), which has a bandgap around 1.25 eV. The first InP cells were made in 1956; their efficiency was only about 2%. For the next few decades, InP was neglected because GaAs with its larger energy gap was a more promising material. Research on InP was revived in the late 1980s and cells with efficiencies up to 22% were prepared by epitaxial growth techniques in 1990. These InP cells proved to be extremely radiation-resistant high-efficiency cells, and, therefore, they may play a role in power supplies of future satellites.

There may be other materials which could produce high-performance solar cells, but the research effort that must be expended to realize the potential of each material is formidable. The dominance of Si, GaAs, and InP results, to a considerable degree, from the large effort expended on these materials for more commercially lucrative applications like integrated circuits and optoelectronics devices. The amount of research expended on materials whose only application appears to be in PV cells (like CISe and a-Si:H) is very modest and this impedes progress in the field.

4. EFFECTS OF HIGH-ENERGY RADIATION ON PV CELLS

From Eqs. (23) through (33), it is clear that the short-circuit current I_{sc} of an illuminated PV cell is a strong function of the minority-carrier diffusion lengths L_n and L_p in the n and p sides of the junction. The diffusion lengths in turn are functions of the respective minority-carrier lifetimes τ through relations of the type

$$L^2 = D\tau, \tag{60}$$

where D is the respective minority-carrier diffusion constant. From Eq. (2), τ is inversely proportional to the product of the concentration of recombination centers of interest, N_r; the capture cross section σ_r of the centers; and $f(E_r - E_F)$, their Fermi–Dirac occupancy factor, where E_r is the energy level in the forbidden energy gap associated with the particular recombination centers under consideration.

When a solid is exposed to irradiation by high-energy particles (electrons, protons, α particles, neutrons, γ rays, β rays, etc.), collisions between the high-energy particles and lattice atoms can lead to dislodging of lattice atoms resulting in the generation of vacancy-interstitial pairs (Frenkel defects), provided that the energy transferred to the lattice atom during the collision exceeds the threshold energy for the displacement, E_{th}. For silicon and germanium, E_{th} has a value of about 15 V (Loferski and Rappaport, 1958). This amount of energy can be delivered to a Si atom by an electron having an energy of about 130 keV or a proton having an energy of about 500 eV. The corresponding quantities for Ge, which has a higher mass than Si, are about 350 keV for electrons and about 1100 eV for protons.

In a semiconductor, radiation-generated vacancies and interstials act as recombination centers having energy levels lying between about 0.10 and about 0.25 eV from the band edges. Sometimes the vacancies and interstials form complexes with impurity atoms already present in the semiconductor. The resulting complexes have E_r values in the same 0.10- to 0.25-eV range, and therefore they too can serve as recombination centers. In either case, τ decreases as the concentration of recombination centers introduced by bombardment N_{rB} increases. If the lifetime before irradiation is τ_0, then after bombardment by N_{rB} particles

$$1/\tau = 1/\tau_0 + N_{rB}\sigma_{cB}f_B V. \tag{61}$$

The reason why these irradiation effects are important for PV solar cells is that the cells are used as primary sources of power on artificial earth satellites. In this application, they are exposed to irradiation by electrons and protons trapped in the radiation belts surrounding the earth. Indeed, the discovery of the so-called Van Allen radiation belts by the first satellite launched by the U.S. in 1960 raised questions about the longevity and, therefore, the usefulness of solar cells in space. The cells used on that first satellite were silicon *p* skin on *n*-base cells. Experiments with electron beams from accelerators showed that these cells would lose about 25% of their power output after about 1 year of exposure to the electrons in the Van Allen belts of low Earth orbits. When these irradiation experiments were repeated on silicon *n* skin on *p*-base cells, it was found that they were more tolerant of electron irradiation by almost an order of magnitude. Consequently, *n*-on-*p*-base cells became the silicon cell of choice for use in satellite power supplies. The search for even more radiation-resistant cells eventually led to the discoveries that GaAs and InP cells are substantially more tolerant of radiation than any of the cells made from silicon. They are also more efficient converters of AM0 radiation than silicon, and therefore they are strong candidates for eventually replacing silicon in space applications.

5. ENVIRONMENTAL ASPECTS OF PV CELLS

PV solar cells are completely silent, nonpolluting generators of electricity. They produce no smoke, no carbon dioxide, and no radioactive wastes and are very attractive from an environmental point of view.

One objection to their large-scale use is, however, that generation of large amounts of electricity would require covering large areas with solar panels because sunlight, the primary source of power, is so diffuse. At high noon at the equator, the density of sunlight normally incident on a receiver is about 1 kW/m^2. Over the course of 1 year, such a square meter of surface area would receive about 2 MWh of solar energy. If this solar energy is converted into electricity by 15% efficient solar PV panels, each square meter covered by the panels would generate about 0.3 MWh of electrical energy per annum. For comparison, a 1000-MW(e) fossil fuel or nuclear power plant generates about 8.7×10^6 MWh(e) per annum. A solar power plant based on 15% efficient PV panels would need to cover about 30 km^2 to equal the capacity of one such conventional electrical power plant. This is a large area; however, the solar power plant need not be concentrated at a single site. In fact, it may be preferable to distribute the PV panels over a wide geographical area so that the power would be produced close to where it will be consumed.

There are other potential causes of environmental concern. For example, the production of solar cells requires the use of toxic substances, like silane, phosphene, arsene, etc. Furthermore, the major constituents of some solar cells can be hazardous in themselves, elements like cadmium, selenium, and tellurium. However, studies of these potential problems have lead to the conclusion that they will not be serious, provided that standard safe industrial procedures are followed (Goetz *et al.*, 1994; Patterson *et al.*, 1994).

6. APPLICATIONS OF PV POWER SUPPLIES

The current (1994) annual production of PV cells is such that if all the cells were gathered in some one place and illuminated by solar radiation having an intensity of 1 kW/m^2, their total power output would be about 40 MW. This amount of solar cells is referred to as 40 MW peak or 40 MW$_p$. By 1994, the world production had been at this level for about 15 years so that the accumulated production of PV cells was about 600 MW$_p$ in 1994. About 70% of these cells have been single-crystal or polycrystalline silicon cells; most of the balance have been thin-film amorphous silicon cells. These cells have been used in satellite power supplies, for remote applications, and in electronic products.

6.1 Satellite Power Supplies

Satellite PV power supplies have always utilized the highest-performance solar cells available because the penalty for failure of the power supply is unacceptably high; it is difficult or impossible to repair or replace cells in satellites. The cells have therefore always been single-crystal cells, overwhelmingly silicon, although gallium arsenide cells have begun to be used on some satellites, partly because GaAs cells are more radiation resistant than silicon and even more importantly because they have a higher conversion efficiency. The cost of solar cells is not as important in space applications as in most terrestrial applications because the cost of the power supply for the satellite is a small fraction of the combined cost of the satellite and of launching it. Space applications account for only a few hundred kilowatts of the annual solar-cell production.

6.2 Remote Power

When there is a need for up to a few kilowatts at a remote site in a developed country, it is often more economical to install a PV power supply than to extend the electric grid. Such sites include power for microwave repeater stations, for cathodic protection of oil and gas pipelines, for railroad crossing signals, etc.

In developing countries, where the electric power grid covers only small fractions of the country, PV power is the cost-effective way to provide a few kilowatts for villages far from the grid. The power can be used for educational television receivers, night lighting, refrigerators for medical supplies, etc. The cells used in such remote applications are based either on single-crystal or on polycrystalline silicon.

6.3 Other Applications

Most hand-held calculators use a small solar-cell power supply. These power supplies usually utilize amorphous silicon cells because they cost less and are more suited to indoor artificial lighting where the calculators are commonly used. Amorphous silicon cells have also been used in toys, clocks, battery trickle chargers, etc.

7. ECONOMICS OF PV POWER PLANTS FOR LARGE-SCALE ELECTRICITY GENERATION

The principal impediment to the utilization of PV cells to produce electricity on a large scale has been their cost, even though the cost has decreased dramatically over the past 40 years. In 1955, the price of single-crystal silicon PV cells was about $2/cm² (in 1955 dollars) for 10% efficient cells. In 1994, the price of 12% efficient single-crystal cells has dropped to about $0.04/cm² (in 1994 dollars).

Economic studies of large-scale PV power generation lead to the conclusion that the price must still drop by a factor of about 5, if electricity produced by PV cells is to be competitive with electricity produced by fossil-fuel and nuclear power plants (Wolf, 1981). Furthermore, these studies conclude that there is a minimum efficiency that these low-cost solar PV cells and panels made from them must exhibit if the PV plant is to produce electricity at a competitive price. This is because the solar power plant is not built from PV cells alone. The other components of the system can be classified into two groups:

1. components whose cost is proportional to the area occupied by the PV plant, which in turn is inversely proportional to the PV panel efficiency η (e.g., these components include the site and its preparation, panel support structures, wiring, etc.), and
2. components whose cost is proportional to the power generated but not to the area (e.g., dc-to-ac converters, etc.).

The cost C_T of a PV plant per unit power produced will therefore be given by

$$C_T = A(C_A + C_{PV}) + C_{PW},$$

where A is the area of the PV plant required to produce unit power when the incident solar power is I; C_A and C_{PV} are the costs per unit power produced per unit area of the area-proportional components and of the PV panels, respectively, and C_{PW} is the cost per unit power of the power-proportional components. The total power produced by the PV panels is given by

$$P_0 = A_0 \eta I,$$

where η is the panel efficiency. From the definition of A, $P_0 = 1$ unit when $A_0 = A$. Therefore, $A = 1/\eta I$ per unit power produced. Consequently,

$$C_T = C_{PV}/\eta I + C_A/\eta I + C_{PW}.$$

The intended price of unit power at the output of the plant sets the value of C_T. If PV panels are free, i.e., $C_{PV} = 0$, then for given values of C_A and C_{PW}, there exists a minimum efficiency η that the panels must have if the target value of C_T is to be achieved. Figure 19 shows the results of one such calculation for a solar power plant with specific (reasonable) costs for C_A and C_{PW}. The calculation includes a "planning target" for the price of electricity at the output busbar of the solar PV power plant; it was set at 6 cents (U.S.) per kWh. The figure shows that if the solar-energy conversion efficiency of the PV modules used in the plant is 10%, then the allowable module cost (C_{PV}) will be about \$10/m^2. Since 10% cells produce about 100 W/m^2 when illuminated by 1 kW/m^2 sunlight, this translates to \$0.10/peak watt. For 15% modules, the numbers are \$45/m^2 and \$0.30/peak watt; for 20% modules, \$75/m^2 and \$0.37/peak watt; and for 25% modules, \$120/m^2 and \$0.48/peak watt. PV modules with efficiencies of 15% have been fabricated from silicon cells. Individual tandem cells having efficiencies in excess of 30% and an area of 1 cm^2 have also been made. Thus, cells with the efficiencies required for large-scale power plants have been made but their price remains too high.

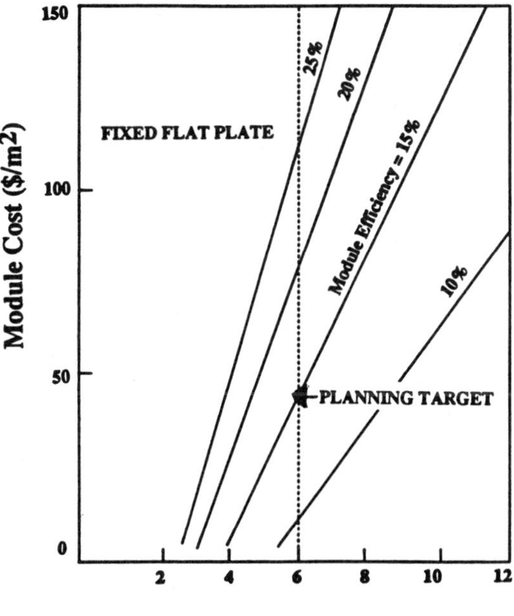

FIG. 19. Allowable cost/unit area of solar-cell panels vs busbar cost of electrical energy produced by a PV power plant.

Note also that if the "planning target" cost of electricity is increased to say 12 cents (U.S.)/kWh, then the allowable module cost for 10% efficient modules rises to about \$70/m^2 and \$0.70/peak watt. For 15% modules, the corresponding numbers are \$160/m^2 and \$1.10 peak watt. For orientation purposes, note that the cost of 10% efficient modules in 1994 is about \$4/peak watt so that reaching the allowable cost goals for 12 cents (U.S.)/kWh electricity is in sight, but the requirements for 6 cents (U.S.)/kWh module cost goals are formidable indeed. It can be expected that the price of PV modules will continue to decrease and that PV power plants will eventually become economically competitive.

8. SUMMARY AND CONCLUSIONS

1. Although the PV effect was first reported in 1839, its first limited applications occurred around 1910 and its first significant applications occurred only after 1953, which date marks the beginning of the modern PV age.
2. By 1955, theoretical analyses of the new PV cells led to the conclusion that solar-energy conversion efficiencies of about 26% were possible in single-junction PV cells. Multijunction PV cell systems can have efficiencies in excess of 30%, according to theory.
3. By 1955, single-crystal silicon cells with efficiencies of about 10% were being fabricated. By 1985, the gap between the 1955 theory and experimentally observed Si efficiencies was closed. The same was true

for cells made from single crystal of GaAs. Multicell efficiencies in excess of 30% have also been reported by the mid-1980s.

4. Thin-film cells have attracted attention since the earliest times because of the anticipated lower cost of such cells. By 1994, stable polycrystalline thin-film cells having efficiencies up to 16.4% were reported.
5. Amorphous silicon-hydride PV cells were invented in 1974 and continue to attract attention because of ease of fabrication and the promise of low cost. Their highest stabilized efficiencies in 1994 were around 10%.
6. Applications of PV cells include satellite power supplies, remote power, hand-held calculators, etc. The technology for large-scale generation of electricity from PV power plants exists, but its application awaits reduction in cost of high-performance PV cells.

GLOSSARY

Air-Mass Zero (AM0) Solar Spectrum: The solar spectrum just outside the Earth's atmosphere, which has an intensity of about 1.4 kW/m^2 and a specific, never changing spectral composition.

Air-Mass One Solar Spectrum: The solar spectrum at the surface of the Earth when the sun is at the zenith. This is necessarily an idealized spectrum since water vapor and other impurities in the atmosphere can change the intensity and spectral content of the sunlight. The AM1 intensity is about 1.0 kW/m^2, and the spectral composition is a best estimate of what it would be if water vapor and other impurity content were negligible. It can be approximated by the spectrum of a 5800-K blackbody.

***p-n* Junction:** An interface between an *n*-type and a *p*-type region in a semiconductor structure. It is a homojunction if the *n* and *p* sides are differently doped regions of the same semiconductor. It is a heterojunction if the *n*-type semiconductor is a different material than the *p*-type semiconductor.

Peak Watt Definition: An amount of solar cells that deliver 1 watt of electrical power to a matched load when the cells are illuminated by sunlight having a specified intensity and spectrum at normal incidence. The appropriate intensity for terrestrial solar cells is 1 kW/m^2; for space cells, 1.4 kW/m^2. Thus, one peak watt of 10% efficient solar cells would have an area of 100 cm^2, whereas one peak watt of 20% efficient solar cells would have an area of 50 cm^2.

Photovoltaically Active Volume (PVAV): That part of a PV cell in which absorption of light leads to the generation of excess carriers that contribute to the PV effect. Wide-bandgap windows are not part of the PVAV.

PV Solar-Energy Efficiency: The ratio of electrical power output to the solar power input. If comparisons of efficiency are to be made, the efficiency must be measured under comparable conditions. These conditions have been standardized so that the spectral composition and intensity of the solar radiation is fixed (usually AM1.5 for terrestrial PV cells) and the temperature of the cell is kept at some specified value (usually room temperature). The power delivered to a load whose resistance is equal to the resistive part of the impedance of the PV cell is measured. This is the maximum power that can be transferred to a load and the efficiency is called the maximum efficiency η_{max}.

Semiconductor: A material whose electrical conductivity is higher than that of an insulator but lower than that of a conductor. The room-temperature conductivities of semiconductors lie between about 100 S/cm and 10^{-5} S/cm. The conductivity of semiconductors increases with temperature, whereas the conductivity of metals decreases with increasing temperature.

Works Cited

Becquerel, E. (1839), *Comptes Rendues* **9**, 561–565.

Benett, A., Olsen, C. (1978), *Conf. Rec. 13th IEEE PVSC.*, New York; IEEE Press, 968–975.

Brattain, W., Bardeen, J. (1948), *Phys. Rev.* **74**, 230–233.

Carlson, D. E., Wronski, C. R. (1976), *Appl. Phys. Lett.* **28**, 671–673.

Chapin, D. M., Fuller, C. S., Pearson, G. L. (1954), *Phys. Rev.* **96**, 533.

Chu, T. L., Chu, S. S. (1993), *Prog. Photovoltaics*, **1**, 31–42.

Cusano, D. A. (1963), *Solid State Electron.* **6**, 217–232.

Gabor, A., Tuttle, J., Contreras, M., Alban, D., Tennant, A., Qu, Y., Du, D., Carapella, J., Noufi, R. (1994), *Proc. 12th European PV Solar Energy Con-*

ference, Amsterdam, 1994, Dordrecht, The Netherlands: Kluwer Science Publishers, pp. 939–943.

Gale, R. F., Fan, J. C. C., Turner, G. W., Chapman R. L. (1984), *Conf. Rec 17th IEEE PV Spec. Conf.*, New York: IEEE Press.

Gocsen, N., Loferski, J. J. (1979), *Solar Energy Mater.* **1**, 271–286.

Goetz, M., Vaucher, S., Tscharner, R., Shah, A. (1994), *Proc. 12th European PV Solar Energy Conference, Amsterdam, 1994*, Dordrecht, The Netherlands: Kluwer Science Publishers, pp. 1229–1231.

Green, M. A., Blakers, A. W., Wenham, S. R., Jianhua, Z., Chong, C. M., Taouk, M., Narayanan, S., Willison, M. R. (1987), *Conf. Rec. 19th IEEE PVSC, May 1987, New Orleans, LA*, New York: IEEE Press, pp. 49–52.

Green, M. A., Emery, K. (1993), *Prog. Photovoltaics* **1**, 225–227.

Grondahl, L. O. (1933), *Rev. Mod. Phys.* **5**, 141–160.

Hedstrom, J., Ohlsen, H., Bodegard, M., Kylner, A., Stolt, L., Hariskos, D., Ruckh, M. Schock, H.-W. (1993), *Conf. Rec. 23rd IEEE PVSC May 1993, Louisville, KY*, New York: IEEE Press, pp. 364–371.

Hedstrom, J., *et al.* (1994), *Proc. 12th European PV Solar Energy Conference, Amsterdam, 1994*, Dordrecht, The Netherlands: Kluwer Science Publishers, pp. 1743–1746.

Jenny, D. A., Loferski, J. J., Rappaport, P. (1956), *Phys. Rev.* **101**, 1208–1209.

Kazmerski, L. L., White, F. R., Morgan, G. K. (1976), *Appl. Phys. Lett.* **29**, 268–270.

Landsberg, P. T., DeVos, A., Baruch, P. (1991), *J. Condens. Matter* (U.K.) **3**, 6415–6421.

Lange, B. (1930), *Z. Phys.* **31**, 139–145.

Loferski, J. J. (1956), *J. Appl. Phys.* **27**, 777–784.

Loferski, J. J., Rappaport, P. (1958), *RCA Rev.* **19**, 536–554.

Loferski, J. J., Wysocki, J. J. (1961), *RCA Rev.* **22**, 38–68.

Loferski, J. J. (1979), *Materials Science in Energy Technology*, New York: Academic Press, 219–253.

Loferski, J. J. (1988), *Proc. of Int. Elec. Dev. and Mat. Symp., Taiwan, ROC*, pp. 221–226.

Mickelsen, R. A., Chen, W. S. (1980), *Appl. Phys. Lett.* **36**, 371–373.

Nadjakov, G., Andreitchine, R., Borissov, M. (1954), *Izv. Bulg. Aka. Nauk.* **4**, 10–15.

Patterson, M. H., Turner, A. K., Sadeghi, M., Marshall, R. I., (1994), *Proc. 12th European PV Solar Energy Conference, Amsterdam, 1994*, Dordrecht, The Netherlands: Kluwer Science Publishers, pp. 951–953.

Rappaport, P. (1954), *Phys. Rev.* **93**, 246–247.

Reynolds, D. C., Leis, G., Antes, L. L., Marburger, R. E. (1954), *Phys. Rev.* **96**, 533–534.

Schottky, W. (1930), *Z. Phys.* **31**, 913–920.

Shay, J. L., Wagner, S., Kasper, M. (1975), *Appl. Phys. Lett.* **27**, 89–90.

Shockley, W. (1949), *Bell Syst. Tech. J.* **28**, 435–450.

Shockley, W. (1951), *Electrons and Holes in Semiconductors*, New York: Van Nostrand.

Trivich D., Flinn, P. A. (1955), in F. Daniels, J. Duffie (Eds.), *Solar Energy Research*, Madison, WI: Univ. of Wisconsin Press, pp. 143–145.

Verlinden, P. J., Swanson, R. M., Sinton, R. A., Crane, R. A., Tilford, C., Perkins, J., Garrison, K. (1993), *Conf. Rec. 23rd IEEE PVSC, 1993, Louisville, KY*, New York: IEEE Press, pp. 58–64.

Wolf, M. (1981), *Solar Cells* **3**, 327–336.

Woodall, J. M., Hovel, J. J. (1972), *Appl. Phys. Lett.* **21**, 379–381.

Wysocki, J. J., Rappaport, P. (1960), *J. Appl. Phys.* **32**, 571–578.

Further Reading

A detailed history and current status of PV science and technology are to be found in the conference records and proceedings of the three sets of international photovoltaic conferences. The oldest continuous set is the Institute of Electrical and Electronic Engineers Photovoltaic Specialists Conferences, the first of which was held in 1956 and the twenty-third in 1993. The conference proceedings are published by the IEEE Press, New York. The second set is the European Community Photovoltaic Solar Energy Conferences, the first of which was held in 1977 and the 12th in 1994. Their proceedings are published by Kluwer Academic Publishers, Dordrecht, Germany. The third set of conferences is held in the countries of the Eastern Pacific and is entitled the Photovoltaic Science and Engineering Conferences, the first of which was held in 1980 and the seventh in 1993. Their proceedings are published as separate books. These three sets of conferences are so synchronized that one of them takes place every 6 months.

There are two archival journals devoted to PV, namely *Solar Cells and Solar Energy Materials*, Elsevier, which appears monthly, and *Progress in Photovoltaics*, John Wiley & Sons, which appears quarterly.

Fahrenbruch, A., Bube, R. H. (1983), *Fundamentals of Solar Cells*, New York: Academic.

Green, Martin A. (1992), *Solar Cells*, Englewood Cliffs, NJ: Prentice-Hall.

PHYSICAL GEOLOGY

See STRUCTURAL AND PHYSICAL GEOLOGY

PHYSICS AND ENGINEERING ORGANIZATIONS

H. WILLIAM KOCH, *Department of Physics, University of Denver, Denver, Colorado, U.S.A.*

F. KARL WILLENBROCK, *Washington, DC, U.S.A.*

	Introduction	560
1.	**Physics Organizations**	560
1.1	Science Disciplines in Natural and Engineering Sciences	560
1.2	Chronology and Operational Tenets of Physics Organizations	561
2.	**Precursers and Physics Societies**	562
2.1	National Academies of Science	563
2.1.1	ICSU and IUPAP	563
2.1.2	National Academy of Sciences, Washington	563
2.1.3	The Royal Society, London	564
2.2	Associations for the Advancement of Science	564
2.2.1	American Association for the Advancement of Science	564
2.2.2	British Association for the Advancement of Science	564
2.3	Physics Societies	565
2.3.1	Overview of Physics Societies in Three Regions of the World	565
2.3.2	Details of Selected Societies	568
2.3.2.1	The American Physical Society	568
2.3.2.2	American Institute of Physics	569
2.3.2.3	Other U.S. Physics and Related-Science Societies	571
2.3.2.4	American Center for Physics	572
2.3.2.5	The Institute of Physics, London	572
2.3.2.6	German Physical Society	572
2.3.3	Physics Societies in the Rest of the World	573
3.	**Engineering Organizations**	573
3.1	General	573
3.1.1	Organization of This Section	574
3.1.2	International Counterparts	574
3.2	Inherent Characteristics of Engineering	575
3.2.1	Integration of Knowledge	575
3.2.2	Relation to Mathematics and Science	575
3.2.3	Societal Impact	576
3.3	Technically Oriented Organizations	576
3.3.1	Services of Professional Societies	576
3.3.2	Founder Societies	577
3.3.3	Organizational Characteristics	577
3.3.4	Unique Character of the IEEE	578
3.3.5	Nontechnical Activities	578
3.4	Educationally Oriented Organizations	578
3.4.1	Studies of Engineering Education	578
3.4.2	Accreditation of Academic Programs	579
3.5	Industrially Oriented Organizations	580
3.5.1	Standards-Writing Organizations	580
3.5.1.1	International Standardization	581
3.5.2	Trade Associations	581
3.6	Public Policy Oriented Organizations	582
3.6.1	National Academy of Engineering	582
3.6.2	National Society of Professional Engineers	582
3.6.3	Umbrella Organizations	582
3.7	Challenges for Engineering Organizations	583
	Appendix A. Engineering's "Founder Societies"	584
	Appendix B. Comparison of Two Leading U.S. Computer Societies	586
	Works Cited	587
	Further Reading	587

3-527-28135-5/95/$5.00 + .50

INTRODUCTION

The foundation of physical sciences is physics, which emphasizes fundamental understanding of physical processes in nature. Physics is based on theoretical concepts supported by experimental data. In contrast to the physical sciences, engineering is concerned with the man-made world. It applies the understanding gained from all the sciences and also generates new knowledge in support of the creation of products and services useful to mankind. The principal engineering disciplines have been civil, mechanical, electrical, and chemical engineering. These names alone suggest that the central sciences applied by engineering are the physical sciences.

The professional institutions that serve physicists and engineers are physics and engineering organizations, respectively. This article summarizes the national and international histories and roles of these organizations, primarily from a U.S. perspective.

1. PHYSICS ORGANIZATIONS

Physics organizations provide the institutional framework that enables physicists and those in related fields to develop and distribute their knowledge; thus, they advance science and technology to meet the needs of government, of industry, and of society at large. They do so by arranging scientific meetings, publishing peer-reviewed research and education journals, and doing other profession-related services.

From the earliest days of the study of physics, the body of fundamental knowledge grew slowly, nourished by physicists working almost exclusively in academia. That slow growth changed dramatically in the twentieth century. With the exigencies of wars, scientific advances brought the world into the nuclear and the space ages.

Critical contributions by physicists in meeting national needs led to recognition of physics worldwide as a science to be encouraged by governments to obtain technological and economic advancements. The result was a 25-year period beginning in 1946 when physics and physicists prospered and basic physics received an increasing share of national funds in many industrialized countries. Then in the 1970s, with worsening economic conditions, due in part to the world oil shortage and increasing national debts, abundant financial support for basic physics research slackened. The emphasis for the next two decades or so was increasingly on applications of physics.

As physics expanded in the twentieth century, so, too, was there a need for expansion within existing physics organizations and for different types of organizations. What are those physics organizations? How do they serve the science of physics and related disciplines, the nation, and physicists themselves? This section and Sec. 2 aim to answer those questions.

1.1 Science Disciplines in Natural and Engineering Sciences

The characteristics of physics organizations are determined by the science that requires them. An understanding of that science can be obtained from the diagram in Fig. 1 that schematically relates the "things" in our natural universe and the science disciplines in the natural and engineering sciences that study them.

The shaded, outer ring of Fig. 1 identifies the four categories of things in our natural universe. They are nonliving materials, living materials, man-made systems, and natural systems.

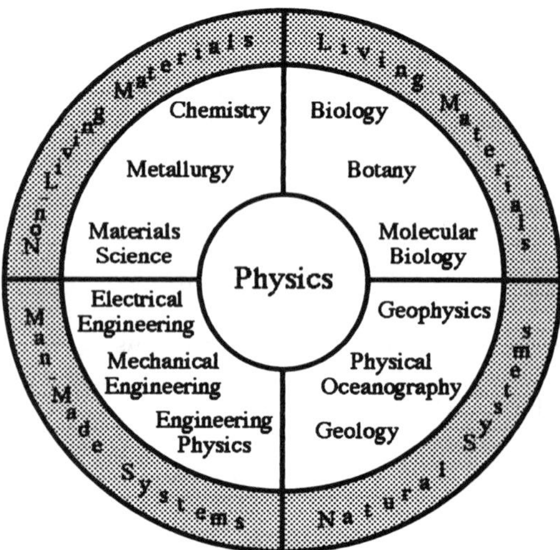

FIG. 1. "Things" in our natural world (shaded areas) and their relation to science disciplines in natural and engineering sciences.

Within the ring are four quadrants containing examples of the names of four groups of science disciplines. The sciences listed in the quadrant study "things" named in the part of the outer ring closest to the quadrant. For example, chemistry, metallurgy, and materials science study nonliving materials, while biology, botany, and molecular biology study living materials. Similarly for the parts of the ring and quadrants that deal with systems. Engineering physics and other sciences fundamental to electrical, mechanical, civil, or chemical engineering, for example, are usually involved with man-made systems, while geophysics, physical oceanography, and geology, for example, study natural systems.

The sciences in the four groupings are largely observational in character. Their scientists observe materials and systems as found in our natural world.

Physics is a science that is in its own grouping in the central small circle of Fig. 1. It is separate from, but coupled to, the other four. Partly observational in character, physics usually confines its observations to basic materials and systems that are simple and sometimes idealized. Physicists also use observational data obtained by others in other science disciplines, and provide the analysis and interpretation that clarify the underlying simplifications. Thus, physics is primarily regarded as a conceptual science because its involvement is principally with concepts of matter, motion, and energy that are fundamental to all of the natural and engineering sciences. Theoretical concepts form a significant part of all work done in physics. Consider, for example, the roughly 30% of all U.S. doctoral theses in physics that are devoted to theoretical topics.

Because of its coupling to all the other sciences, physics is very interdisciplinary. Evidence of this characteristic can be found in the interdisciplinary nature of subdisciplines, organizations, conferences, and journals in which physicists routinely participate. Examples of such subdisciplines are chemical physics, biophysics, geophysics, and physical oceanography. Other evidence to be discussed later is the unusual interdisciplinary mix of primary science disciplines represented in the members of physics membership societies in the U.S.

The diagram of Fig. 1 is not to be taken too literally, even though it does give some useful understanding of the way sciences relate to one another. As an example of such understanding, the relationship among the sciences is perhaps confirmed from the number of graduate doctorate degrees awarded each year by the physics graduate schools in the U.S.A. Data on these degrees have been collected by the U.S. National Academy of Sciences (NAS) since 1920. For each year since that time, the ratio of the number of degrees awarded in physics to the total in natural and engineering sciences has been remarkably constant at 11% to within one percentage point. "Rough statistics from Canada and the United Kingdom, with rather different educational systems, show this same fraction and constancy" (Physics Survey Committee, 1972). Thus, the science enterprises in the U.S., Canada, and the United Kingdom (at least) invest a fixed fraction (11%) of their respective resources each year in the development of concepts in physics to maintain the "health" of their natural science communities.

1.2 Chronology and Operational Tenets of Physics Organizations

By definition, the science of physics has benefited from physics organizations, ever since physics was established as an important branch of science in the mid 1800s. The precursors of physics organizations were science organizations, in which a physics component developed.

Thus, the history of physics organizations really began with the formation of an all-science, honorific society in 1660, the Royal Society of London. The Berlin Academy of Science followed in about 1720 (Evan, 1981). Membership in these organizations and counterparts elsewhere was limited because of the selection criteria, and they were followed by all-science, open-membership societies called associations for the advancement of the sciences. The British Association for the Advancement of Science (BAAS) was founded in 1831, while the American Association (AAAS) followed in 1848. These organizations satisfied the need for meetings and journals of scientific record, even in specialty sciences such as physics, until the German Physical Society (GPS) was founded in 1845, the Physical Society in London (PS) in 1873, and The American Physical Society (APS) in

1899, to cite only a few examples. The chronological development of physics organizations in Europe and the U.S. was consistent with the increases in the numbers of trained physicists.

As the numbers increased, so did the understanding of the basic tenets under which most physics organizations believe they must operate. The tenets are based on the demonstration that physics, with its heavy emphasis on pure physics and original ideas, requires for its healthy development strong interdisciplinary coupling plus free, open, voluntary interchanges of scientific information among all participating scientists. Thus, the important operational tenets are these:

1. All physics publishers should subject proposed contributions for archival journals to peer review.
2. A single language—English—should be accepted internationally for communicating physics information.
3. National policies should allow free, open, voluntary interchanges of scientific information among scientists.
4. National policies should honor intellectual property rights by use of patents and copyrights.

Examples of the application of these tenets during the last 50 years are the following:

1. "Rochester" conferences, named after the first high-energy-physics conference at Rochester University, Rochester, New York, have been held annually since 1952 in different parts of the world (including Moscow). Soviet scientists were able to convince their authorities that open communication with physicists in the U.S. and Europe was essential for their participation in high-energy research. Permission by the communist regimes in the USSR (before its dissolution) was given to Soviet high-energy physicists to participate openly in international conferences.
2. An important requirement for an international conference to be endorsed by the International Union of Pure and Applied Physics (IUPAP) is that participation in the conference must be open to physicists independent of their home country's politics.
3. English-language journals of American physics societies were sent most articles on pure physics of high quality by Japanese, Chinese, and Russian physicists during the last two decades. The selection of American journals is motivated by the desire to reach as large an audience of readers doing similar research as possible. Thereby, these foreign physicists attempt to establish their scientific credentials in the western world and to communicate more effectively with their American peers.
4. In about 1976, a decision was made by the German Physical Society (GPS) and the German government to encourage journal articles and abstracting services published by GPS to be published in the English language, because German physicists were reluctant to publish in journals read by limited audiences. One indirect result of that decision led to the creation of an English-language abstracting service called PHYSICS BRIEFS that replaced a service in the German language called PHYSIKALISCHE BERICHTE, which had been started in 1885.

2. PRECURSORS AND PHYSICS SOCIETIES

The physics organizations and their precursors to be treated in this article are those that now serve effectively the international community of physicists. They are

1. national academies of sciences,
2. the International Union of Pure and Applied Physics (IUPAP) of the International Council of Scientific Unions (ICSU),
3. associations for the advancement of science, and
4. physics societies.

A general discussion of each type in the order listed follows, with details for several outstanding examples. Many examples are from the United States of America (U.S.) and England, because here the information was readily available and because societies in the U.S. and England are large with interesting evolutionary characteristics. The information for both the U.S. and England will show the interrelationships of the four types of organizations, including their roles in the development of strong national physics programs.

2.1 National Academies of Sciences

National academies of sciences are defined as honorary organizations that afford membership and preferential treatment to those who have made outstanding contributions to the advancement of science. An example is the Japan Academy, which was established in 1877 and which now has two sections. Section 1 in the Humanities and Social Sciences has 59 members, none of whom is a physicist; Section 2 in the Pure Sciences and their Applications has 68 members, of whom 12 are physicists (Anon., 1994). Similarly, the Royal Societies in the separate territories within Australia and New Zealand are also honorary only.

Some national academies not only have an honorific role but also officially provide scientific advice to their federal governments. The U.S. National Academy of Sciences (NAS) is an example.

In addition to providing honor and advice, some academies are also the chief coordinating bodies of scientific research in their countries. Two important examples are in the Soviet Union (before its breakup) and in the People's Republic of China.

2.1.1 ICSU and IUPAP An important international role of many national academies of sciences is their adherence to the International Council of Scientific Unions (ICSU). ICSU was founded in 1919 "largely through the representatives of the National Academy of Sciences, Washington, and of the Royal Society, London, to coordinate international efforts in the different branches of sciences, under whose aegis international associations or unions in different branches of science could be formed" (International Union of Pure and Applied Physics Secretariat, 1992). Ninety-five countries are now members of ICSU, headquartered in Paris.

According to ICSU principles, physicists at the General Assembly in 1922 decided that an International Union of Pure and Applied Physics (IUPAP) should be formed. National physics committees or groups of physicists from 43 countries adhered to IUPAP in 1990. IUPAP "promotes international cooperation in physics and the use of international symbols, units, nomenclature, and standards; provides assistance in organizing committees and meetings; encourages research and publication of papers and tables; and fosters free circulation of scientists" (Irvin, 1993). General assemblies are held triennially.

The list of 18 IUPAP scientific and technical commissions in Table 1 (International Union of Pure and Applied Physics Secretariat, 1992) shows the major subdisciplines regarded by IUPAP to be active in the field of physics.

2.1.2 National Academy of Sciences, Washington A Congressional charter approved by President Abraham Lincoln created the National Academy of Sciences (NAS) in 1863, not only to obtain scientific advice regarding the execution of the Civil War, but also as an honorific, private, nonprofit, self-governing membership body responsible for advising the federal government, upon request and without fee, on questions of science and technology. Under this charter, the National Academy of Engineering (NAE) was established in 1964 and the Institute of Medicine (IOM) in 1970.

Specific examples of three of the NRC committees are the Space Science Board that advises on the activities of the National Aeronautics and Space Administration (NASA), the Defense Science Board that advises the Department of Defense (DOD), and the Visiting Committees to the National Institute of Standards and Technology (NIST) in the U.S. Department of Commerce.

Surveys of Physics in 1964 by a committee chaired by George Pake, in 1972 chaired by D. Allan Bromley (Physics Survey Committee, 1972), and in 1984 chaired by William Brinkman are examples of *ad hoc* assignments to the NRC.

Table 1. IUPAP scientific and technical commissions.

1. (Administrative)	12. Nuclear Physics
2. Symbols, Units, and Nomenclature	13. Physics for Development
3. Thermodynamics and Statistical Mechanics	14. Physics Education
4. Cosmic Rays	15. Atomic and Molecular Physics and Spectroscopy
5. Very Low Temperature	16. Plasma Physics
6. Biological Physics	17. Quantum Electronics
7. Acoustics	18. Mathematical Physics
8. Semiconductors	19. Astrophysics
9. Magnetism	
10. Condensed Matter	
11. Particles and Fields	

2.1.3 The Royal Society, London The Royal Society of London (RS) for Improving Natural Knowledge was founded as The Royal Society in 1660. RS was granted a Royal Charter under the full title in 1663 (Anon., 1976).

Most academies provide honor and advice and/or coordinate scientific research largely for their own nations, and they tend to pursue their scholarly activities in like manner. The single exception is the Royal Society, London. The Russian physicist, Peter Kapitza, has referred to "Newton's active role in the affairs of the Royal Society, observing that the Royal Society 'in contrast to all academies, has still the character of a non-state society. . . . Since Newton's time, the Royal Society has maintained a lively contact with scientific societies the world over'" (Evan, 1981).

The Royal Society has 763 Fellows and 74 Foreign members. Fellows are elected, on the basis of distinction in science, at each annual meeting of the Society. The Society has a long tradition in publishing with *Philosophical Transactions* (Series A and B) and *Proceedings* (Series A and B).

2.2. Associations for the Advancement of Science

Because academies are highly selective in choosing members, scientists in most countries require organizations that reach out also to the rest of the scientific community. Associations for the advancement of science are one such organizational type.

These associations attempt to be broadly inclusive of all sciences, interested individuals, other societies, companies, and audiences. Examples are the Australian and New Zealand Association for the Advancement of Science (founded in 1886 and now with 1000 members), the China Association for Science and Technology (founded in 1958), the British Association for the Advancement of Science (founded in 1831 and now with 2000 members), the French-Canadian Association for the Advancement of Science (founded in 1923 and now with 23 societies as members), and the American Association for the Advancement of Science (AAAS). The founding, history, and current status are treated next.

2.2.1 American Association for the Advancement of Science The American Association for the Advancement of Science (AAAS) was founded in 1848 with about 400 members. Then, physics was still not fully established as a separate discipline, but as part of natural philosophy or science. It was not until 1876 that the first Professor of physics was appointed at a U.S. university. Henry A. Rowland of the Johns Hopkins University had that distinction (Phillips, 1990). Physics received further recognition in the AAAS Constitution of 1874 that established physics in the name of a new Section: Section A—Mathematics, Physics, and Chemistry; Section B was Natural History. Nine Sections date from 1882, and they include a Section B, Physics. By 1895, there were 12 Sections, and the number has now expanded to 21. Section B in 1899 decided that their growing numbers required them to have a separate society. Thus was formed The American Physical Society, with Professor Rowland as the first president. Currently, the AAAS has 140 000 individual members and 300 affiliated societies and regional and state academies.

The AAAS holds large national meetings, publishes a weekly news magazine—*Science*—, publishes books, and manages many advisory committees. Some of the committees serve important sociopolitical roles. They address matters that influence science and technology generally, such as secrecy in government, civil rights of scientists, and scientific information exchange.

As the membership of AAAS grew, particularly after 1950, management by a very large governing council became unwieldy. At that time, AAAS was an association in which representatives of member societies on a large Council (then 500 members) had direct involvement in governance and which was advised by a Board of Directors (then 13 members). In 1970, it changed to an association in which the council was the group advising the Board (Wolfle, 1989).

2.2.2 British Association for the Advancement of Science The British Association (BAAS) was established at York in 1831 by Royal Charter. Membership was and still is open to all interested in science without proposal or election. 2000 members are active.

The annual meeting, held by invitation in a different city each year, is the main activity to achieve the ends of advancing science. It is generally held in provincial cities of the United Kingdom and the Republic of Ireland and in the past has been held in certain cities of the Commonwealth. It lasts for a week, and the program of the meeting is arranged by scientists from the major disciplines of modern science (Anon., 1976).

2.3. Physics Societies

Science societies were founded by small groups of scientists who wanted to share their ideas and scientific discoveries. They originated "in Renaissance Italy in the sixteenth century, spreading across Europe in the seventeenth, and over the American continents in the eighteenth and nineteenth century, and now in the twentieth century, rapidly invading the cities of Asia . . ." (Bates, 1965).

Thus, the concept of a science society was well established before the concept of a science of physics was established in the mid nineteenth century. Once the German physicists founded a German Physical Society in 1845, others followed in a relatively short time in France in 1873, England in 1873, Switzerland in 1887, Hungary in 1891, Netherlands in 1921, and Poland in 1921. The spread to the Pacific Rim began first with astronomical societies in China and Japan, both in 1908, and then later with physical societies in China in 1932 and Japan in 1946. In the case of the Americas, the Canadians and the Mexicans founded their astronomical societies in 1890 and 1878, respectively, followed in the mid 1900s by physical societies. In the U.S., physicists and astronomers founded both of their independent societies, The American Physical Society and the American Astronomical Society, respectively, in 1899 (Anon., 1994). Incidentally, in this article, physical and astronomical societies are both regarded as physics societies, because of the close relationship the subjects of physics and astronomy have in academic departments and because the American Astronomical Society (AAS) elected to be affiliated with other U.S. physics societies in 1966, as is discussed later.

Physics societies are now organized in many industrialized nations of the world. They currently establish the standards of their profession. They do so by arranging open scientific meetings, specialized review conferences, and peer-reviewed scientific journals. These societies usually are open to anyone interested in joining who meets certain minimum qualifications. Other societies can sometimes join if permitted by the bylaws of the parent societies. Physics societies are outstanding examples of organizations that accomplish their purposes of advancing and distributing knowledge in physics.

They may elect to focus only on science-related matters, profession-related matters (such as employment and career information), or education, or any combination of these. They may function in a national sphere only, or they can be international as well. They may also be business or commercially oriented. If they limit themselves to cultural matters, such as education, science, literature, philosophy, and charity, the societies have little in the way of resources with which to operate. To help such societies, their federal governments often provide direct or indirect financial subsidies. In some countries, the subsidies consist of free housing; in the U.S., the subsidies consist usually of freedom from the payment of certain federal and state taxes. Physics societies vary enormously in size. They range from the 20 or so physicists in a school's physics club to the 40 702 members of the world's largest physics society, The American Physical Society (APS).

For the purposes of this article, the highlights of physics societies will be covered by

1. an overview of the principal physics societies in three key regions of the world for which information can be found in the published literature;
2. a description of The American Physical Society headquartered near Washington, DC, in College Park, Maryland;
3. a description of a membership corporation of U.S. physics societies that includes The American Physical Society, called the American Institute of Physics;
4. comments on other U.S. physics and related-science societies;
5. a report on a new American Center for Physics at the APS address; and
6. comments on the Institute of Physics, London, on the German Physical Society (GPS), and on a few other societies in other regions of the world.

2.3.1 Overview of Physics Societies in Three Regions of the World The data in Tables 2 through 4 give an overview of some

Table 2. Principal physics societies in Europe (including Nation Group, Society, Year Founded, and Total Current Members). The 29 societies listed are incorporated as members in a membership society, the European Physical Society. EPS, founded in 1968, currently has 3700 individual members.

Nation group	Society	Year founded[a]	Members
Austria	Austrian Physical Society	1950	779
Belgium	Belgian Physical Society	NA	496
Bulgaria	Bulgarian Academy of Science	1971	105
Czechoslovakia	Physics Section, Mathematics/Physics Union	NA	1 000
Denmark	Danish Physical Society	NA	274
England	The Institute of Physics	1873/1970	14 500
Finland	Finnish Physical Society	NA	915
France	French Physical Society	1873	2 405
Germany	German Physical Society	1845	18 400
Greece	Hellenic Physical Society	NA	55
Hungary	Eotvos Lorand Physical Society	1891	628
Iceland	Icelandic Physical Society	NA	45
Ireland	Royal Irish Academy	NA	60
Israel	Israeli Physical Society	1954	250
Italy	Italian Physical Society	1897	1 100
Netherlands	The Netherlands Physical Society	1921	2 653
Norway	Norwegian Physical Society	NA	709
Poland	Polish Physical Society	1921	1 968
Portugal	Portuguese Physical Society	NA	200
Romania	Romanian Physical Society	NA	235
Spain	Spanish Royal Society of Physics	1903	300
Sweden	Swedish Physical Society	1920	1 043
Switzerland	Swiss Physical Society	1887	1 370
Turkey	Turkish Physical Society	NA	302
USSR	Department of General Physics & Astronomy	NA	2 000
Yugoslavia	Assoc. of Mathematics, Physics, & Astronomy	1950	200
Zagreb	Ruder Boskovic Institute	NA	95

[a]NA = Not available.

Table 3. Principal physics societies in Pacific Rim (including Nation, Society, Year Founded, and Total Current Members).

Nation	Society	Year	Members
Japan	Physical Society of Japan	1946	17 022
	Japan Society of Applied Physics	1932	9 000
	Crystallography Society of Japan	1950	10 007
	Acoustical Society of Japan	1936	3 600
	Astronomical Society of Japan	1908	2 640
China	Chinese Physical Society with 9 subsocieties in:	1932	27 095
	1. High Energy Physics		
	2. Nuclear Physics		
	3. Liquid Crystals		
	4. Electron Microscopy		
	5. Mass Spectroscopy		
	6. Luminescence		
	7. Particle Accelerators		
	8. Gravitational/Relativistic Astrophysics		
	9. Phase Diagrams		
	Chinese Optical Society	1979	9 018
	Acoustical Society of China	1988	2 879
	Chinese Astronomical Society	1908	2 400
Australia	Australian Institute of Physics	1963	2 400
	Australian Acoustical Society	1971	390
	Astronomical Society of Australia	1966	250

Table 4. Principal physics societies in North America (including Nation, Society, Year Founded, and Total Current Members). The 10 U.S.A. societies listed are incorporated as members in a membership society, the American Institute of Physics. The dates of affiliation with AIP are given in parentheses.

Nation	Society	Year[a]	Members
U.S.A.	The American Physical Society (1931)	1899	40 702
	Optical Society of America (1931)	1916	11 634
	Acoustical Society of America (1931)	1929	6 900
	The Society of Rheology (1931)	1929	1 269
	American Assoc. of Physics Teachers (1932)	1930	11 178
	American Crystallographic Association (1966)	1949	2 100
	American Astronomical Society (1966)	1899	6 119
	Amer. Assoc. of Physicists in Medicine (1973)	1958	3 678
	American Vacuum Society (1976)	1953	5 455
	American Geophysical Union (1986)	1919	29 766
Canada	Canadian Association of Physicists	1945	2 000
	Royal Astronomical Society of Canada	1890	3 500
	Spectroscopy Society of Canada	1967	500
Mexico	Mexican Physical Society	NA	NA
	Mexican Astronomical Society	1902	500

[a]NA = Not available.

of the principal physics societies in the world today (Anon., 1994; Burek, 1993; Irvin, 1993).

In Europe in 1968, 29 nation groups organized their physics societies (but not including their astronomical societies) in the European Physical Society (EPS) (see Table 2). These societies have grown in the 1990s to a total of more than 50 000 members (European Physical Society, 1992). Although EPS is a membership corporation of societies similar to the American Institute of Physics (AIP), the EPS selected at its formation a name that was analogous to The American Physical Society (rather than to the AIP). The reason is presumed to be that APS is the U.S. physics organization that European physicists wanted the EPS most to emulate.

Before the founding of the EPS, a major problem for European physicists had been the many individual societies in Europe. These societies tended to arrange local or regional meetings and local or regional journals. Very few Pan-European physics meetings had been available. The result was that European physicists had to come to meetings in the U.S. arranged by the APS, and others, to participate in free, large, and multidisciplinary exchanges of useful scientific information. Besides the meaningful choice of name, the EPS has identified appropriate operations for itself such as *Europhysics Journals*, and for individual nation groups who are qualified to contract to do publishing on behalf of the EPS community. Also, the EPS has begun holding triennial congresses, the most recent of which was held in 1993. It is evident that the European Physical Society represents a big step forward in meeting the organizational needs of European physicists.

In the Pacific Rim, three major sets of physics societies are to be found in the nations of Japan, Australia, and China (see Table 3).

Of these three, the nation that has had considerable investments in technology, but small investments before 1980 in basic physics, is Japan. The result has been that Japan has not generally developed a strong tradition for publishing journal articles in pure physics, although their journal *Progress in Theoretical Physics* has been an important exception since the 1950s. They have emphasized applications of physics. It is to be expected that the investments in the 1990s will be more directed to pure, long-range research in physics, in order to improve Japan's scientific and technological competitiveness. The general physics society is the Physical Society of Japan (PSJ), founded in Tokyo in 1946. PSJ has 15 859 members and publishes *Butsuri* (monthly, in Japanese), *Journal* (monthly), *Progress in Theoretical Physics* (monthly), and, jointly with the Japan Society of Applied Physics (JSAP), *Japanese Journal of Applied Physics* (monthly). JSAP was founded in 1932 and presently has 9000 members. The Crystallographic Society of Japan (founded 1950—now 10 007 members), the Acoustical Society of Japan (founded 1936—now 2600 members), and the Astronomical Society of Japan (founded 1908—now 2400 members) are the other Japanese physics societies.

Australians have made significant contributions to physics, in spite of having had limited resources of funds and personnel. Their general physics society is the Australian Institute of Physics. Their Institute was founded in Victoria in 1963, has 2400 members, and publishes *The Australian Physicist* (monthly). Two other Australian physics societies are the Australian Acoustical Society (founded 1971—now 390 members) and the Astronomical Society of Australia (founded 1966—now 250 members).

The final nation in this grouping is China. China has been slow to develop technologically; nevertheless, there is tremendous potential that will eventually be tapped when the Chinese accept the importance of free and open exchange of scientific and technological information. Their general physics society is the Chinese Physical Society (CPS) founded in Beijing in 1932. The CPS has 25 000 members and publishes various journals, mostly in the Chinese language, such as *Advances in Physics, Chinese Physics Letters, Wuli (Physics), University Physics*, and *Physics Education*. The CPS has nine subsocieties: the Chinese High Energy Physics Society, the Chinese Nuclear Physics Society, the Chinese Liquid Crystal Society, the Chinese Electron Microscopy Society, the Chinese Mass Spectrometry Society, the Chinese Luminescence Society, the Chinese Particle Accelerator Society, the Chinese Gravitation and Relativity Astrophysics Society, and a Special Society on Phase Diagrams (Anon., 1990). The Chinese Optical Society (founded 1979—now 9019 members), the Acoustical Society of China (founded 1985—now 2879 members), and the Chinese Astronomical Society (founded 1908—now 2400 members) are three other Chinese physics societies.

Of the three nations that make up North America, Canada has two physics societies, Mexico has two, and the U.S.A. has ten member societies of the American Institute of Physics (AIP) (see Table 4). Representatives of the Canadian Association of Physicists, the Mexican Physical Society (Sociedad Mexicana de Fisica), and The American Physical Society have met to identify feasible steps to strengthen physics in North America. Joint meetings, long-term exchange agreements, and reciprocal memberships are some of those steps, some of which have been carried out.

2.3.2 Details of Selected Societies

2.3.2.1 The American Physical Society. The American Physical Society (APS) is the oldest and largest of the U.S. physics societies (Phillips, 1990). The APS was formed in 1899, the same year the American Astronomical Society (AAS) was formed. The two organizations went their independent ways until 1966.

The original intent of the APS was to focus on meetings. They did so by organizing joint meetings with AAAS's Section B until 1940. Journals joined meetings in 1903 as important communication vehicles. At that time, the APS began "cooperation" in the publication of *The Physical Review* with Cornell University editors who had initiated the journal in 1893. *The Physical Review* was subsequently published by the APS beginning in 1913, published for the APS by the AIP from 1932, and, as of 1975, is now published by the APS through the AIP.

The meetings and journals arranged by the APS displayed that pure physics, very narrowly defined, was to be the province of the APS. "APS was elitist from the beginning. It was expedient, and undoubtedly wise, to persuade the most prestigious physicists in the country to take leadership positions in the new society. . . . Members might raise questions of applications and of pedagogy, but the decisions of the APS Council did not reflect these concerns. . . . Applied physics and even fundamental physics related to applications suffered much the same neglect" (Phillips and Wilson, 1989).

Obviously, the most prestigious physicists who were leaders of the APS in the early 1900s recognized their responsibilities to the profession. They were defining the very nature of physics by controlling the scientific content of APS meetings and journals. As some unknown seer has observed, "physics is what physicists do." Therefore, the contents of the APS meetings and journals defined what physicists did and so defined physics.

The role of APS leaders in establishing strict standards for defining physics during the formative years of the APS was appropriate. Now, a century later, when the membership has grown to more than 40 000, the society leadership is no longer autocratic. The APS has become a democratically run society that is much influenced by the current needs and interests of its membership. The Governing Council of the APS now consists of present and past officers, 23 Divisional Councillors, 12 Councillors-at-large, and two committee chairs. The total is 44. The Council meets twice a year.

Any 200 or more members who wish to advance the knowledge of a subfield of physics may petition the Council to establish a Topical Group. If a Topical Group attains a size specified in the By-laws for two consecutive years, it can become a division. Continuation of a division depends on continuation of membership interests in that division according to stated criteria. If at least 200 members wish to advance knowledge regarding the interrelation of physics with matters not exclusively in physics, they may petition the council to establish a forum.

The names of divisions, forums, and topical groups listed in Table 5 show the wide breadth of coverage of what is today considered physics appropriate for the meetings and journals of the APS. These organizational units plus 19 bylaw committees, three constitutional committees, and 31 prize committees provide the membership input to two annual general meetings, plus numerous subunit meetings and a journal program in which some 20 000 articles are published annually.

Table 5. Divisions, forums, and topical groups of the APS.

APS Divisions	
1. Astrophysics	7. Fluid Dynamics
2. Atomic, Molecular, & Optical Physics	8. High Polymer Physics
3. Biological Physics	9. History of Physics
4. Chemical Physics	10. Materials Physics
5. Computational Physics	11. Nuclear Physics
6. Condensed Matter Physics	12. Particles & Fields
	13. Physics of Beams
	14. Plasma Physics

APS Forums
1. Education
2. International Physics
3. Physics and Society

APS Topical Groups
1. Few-Body Systems
2. Fundamental Constants
3. Instrument and Measurement Science
4. Laser Science
5. Shock Compression of Compressed Matter

Several measures of the success of The American Physical Society in achieving its purpose of "advancement and diffusion of the knowledge of physics" are available. One measure is the continued, considerable growth in the total membership of the society as plotted as a function of time in Fig. 2: from 30 000 in 1980 to 40 702 in 1993 (American Institute of Physics, 1993). Another measure is the continued growth in attendance at meetings and in the total number of pages published annually in APS journals. A final measure is the many interdisciplinary activities sponsored by the APS.

2.3.2.2 American Institute of Physics. During the first 25 years of the 1900s, in which the Council of the APS exerted strict management and editorial control, other scientific and educational interests became so important to some physicists as to threaten the formation of several new societies and the fragmentation of the APS. For example, the Optical Society of America (OSA) and the Acoustical Society of America (ASA) were founded in 1916 and 1929, respectively, as societies independent of the APS. This dispersal of physicists was of great concern to the APS. However, "unlike the American Chemical Society, which embraced all aspects of chemistry, the APS continued to take a very narrow view of its role" (Phillips and Wilson, 1989).

An APS Committee on Applied Physics in 1930 took steps to stop the organizational disintegration as well as to overcome the financial distress in their and other societies. They did so by proposing a federation of societies to be called the American Institute of Physics (Barton, 1956). The societies supporting the proposal were the APS, OSA, and ASA. Invited to join also was the Society of Rheology (SOR); and when the American Association of Physics Teachers (AAPT) was incorporated in 1932, it was invited also. Table 4 gives the name, year founded, year of affiliation with the AIP, and total current membership for each of the five founder societies together with the same data for the five societies that have joined the AIP in the more than 50 years since.

The original assignments to the AIP were the complete responsibility, including journal ownership, for the publishing programs

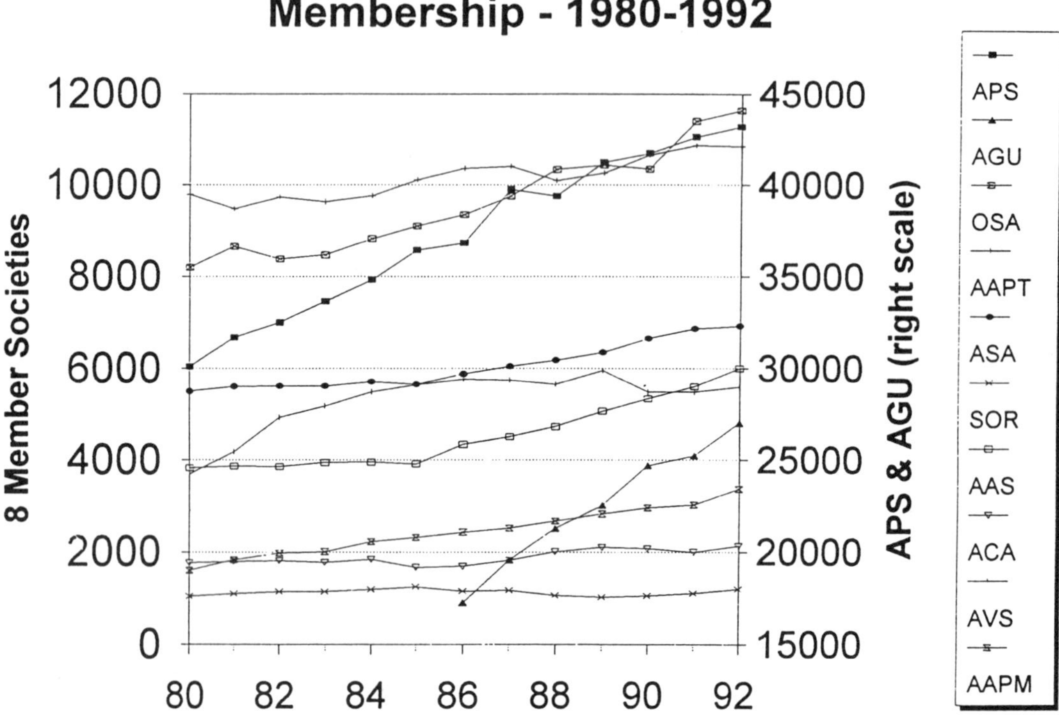

FIG. 2. AIP member society growth chart: the logarithms of total society membership plotted as a function of time.

of the societies and also some limited scientific responsibility. More specifically, the AIP was not to be a mere joint publishing department, but was "to be the joint instrument of the societies for the gaining of any of their collective ends in which success would be gained by combining" (Barton, 1956). One area of scientific responsibility was the applications of physics not covered by any other member society. Thus, several applications journals were begun by the AIP and continue to be the sole responsibility of the AIP.

Although the AIP was formed to be a "federation," in which some autonomy of the member society was to be transferred to the AIP, the development over the years has been to increase the independence of each of the AIP societies from each other and from the AIP. For this reason, the change of the publishing arrangement between the APS and the AIP cited above was made in 1975. For this reason, also, the AIP is referred to as a membership corporation and not a federation, since it is now clear that the members of the AIP are principally the societies and not their individual members.

In retrospect, the organizational relationship of the AIP and its member societies has been of great mutual benefit. The societies, who have relied on volunteer help from their memberships, have been able to focus on scientific matters, while assigning some business matters, including the development of new facilities as required, to the permanent professional staff at the AIP. Also, the early fixation on "pure" physics has made it possible for the APS to spawn new societies in close liaison with the rest of the physics community.

The first 25 years in the history of the APS must have been difficult years for their members interested in "pure" physics as well as for physicists about to form new societies focusing on physics applications and education. The difficulties were little indication of the flourishing success that has resulted in the 1990s both for the APS and for the other U.S. physics societies that have been formed separate from the APS.

Several measures of success in the development of all U.S. physics societies affiliated in the AIP are available. The most obvious is

Table 6. Type of employer for each society, 1990.

	AAPT	ACA	AAS	APS	AGU	ASA	SOR	AAPM	OSA	AVS
Academic	89	53	51	45	38	37	35	29	26	17
Industry/self	7	24	16	30	26	40	59	16	56	71
Federal	3	16	28	23	31	16	4	5	15	12
Nonprofit	1	7	5	2	5	7	2	50	3	0

the healthy growth in the memberships of all of the member societies as shown in the graph of Fig. 2. The total number of individuals represented by the 10 member societies is about 110 000 (American Institute of Physics, 1993).

Another measure is the mix of the types of employers represented among the physicist members of the member societies as shown by the percentages in Table 6 (American Institute of Physics, 1990). Note particularly that the AAPT has the largest percentage in academic institutions and that the AVS has the largest percentage in the industry/self category.

A similar measure is the unusual interdisciplinary mix of professions represented by the memberships of AIP member societies as shown by the data in Table 7 (American Institute of Physics, 1990). The mixtures of professions and of employers should lead to excellent exposures of ideas, methods, and applications in meetings and journals operated by the various member societies.

A final measure of success of the AIP created by its societies is the variety of programs operated by the AIP for the benefit of the member societies and their members that are exemplified by the listing in Table 8. Further details on publishing by U.S. physics societies can be found in the related article PHYSICS LITERATURE AND PUBLICATIONS.

2.3.2.3 Other U.S. Physics and Related-Science Societies. An additional class of membership in the AIP is the Affiliated Society that is any local, regional, or national organization interested in physics. Eighteen societies have so affiliated. They include such societies as the American Nuclear Society, the American Meteorological Society, the Geological Society of America, and the Materials Research Society. The total membership in the Member and Affiliated Societies is more than 200 000 and can be assumed to be most of the U.S. physics organizations (except for the NAS, the U.S. Liasion Committee to IUPAP, and the AAAS) of interest in this article.

Each of the Member and Affiliated Societies successfully serves important and well-defined constituencies. Several of them have organizational and historical characteristics that are particularly noteworthy in understanding U.S. physics organizations, as follows:

1. The American Astronomical Society has a history equivalent to that of the APS in that they were both founded in 1899. The AAS presently includes most U.S. astronomers though its membership is only 5900. Although physicists and astronomers are closely affiliated in U.S. academic departments, the AAS was not organizationally affiliated with other physics societies until 1966, when it became a Member Society of the AIP.
2. The American Geophysical Union, now a

Table 7. Professional self-identification for each society (1990).

Prof	AAPM	APS	OSA	AAPT	AVS	ASA	AGU	AAS	ACA	SOR
Physicist	86	73	48	44	31	21	14	12	9	4
Engineer	2	11	36	5	41	42	12	4	4	68
Chemist	1	8	4	4	15	...	4	...	57	22
Geologist	...	...	...	1	...	...	26	1	2	...
Astronomer	...	12	1	1	...	...	4	72	1	...
Geophysicist	...	...	...	...	...	1	21	1	1	...
Computer scientist	1	2	1	1	...	1	1	2	1	...
Other scientist	3	3	6	3	8	22	16	3	20	4
Other	7	3	4	42	6	13	5	6	7	3

Table 8. AIP publishing and professional programs for member societies.

Publishing	Professional
Publishing of research/ educ. journals	Studies of physics history
Monthly magazine: *Physics Today*	Press and public information programs
Computerized publishing facility	Projects on physics education
PINET electronic network for physicists	Corporate associates programs
Cooperation on *Phys. Abstracts* database	Job placement and career studies
	Society of Physics Students Management

very large society with 29 766 members, is like AAS in that it developed independently of the APS. Although the AGU was founded in 1919, it did not elect to join other Member Societies in the AIP until 1986.

3. The American Vacuum Society is a society that splintered off from the APS in 1953 and that elected to become an AIP Member Society in 1976.
4. The Materials Research Society is a relatively new society (founded in 1973) that has mostly physicists or chemists as members. The society considered affiliation with other Member Societies of the AIP as a Member in 1986. Instead, it elected to become an Affiliated Society. Thereby, it has developed strong ties to the other Societies in the AIP without being characterized as either a physics or chemistry society, because it indeed is both.

2.3.2.4 American Center for Physics. A common and highly visible home for the physics community in the U.S. called the American Center for Physics has been built jointly by The American Physical Society, the American Association of Physics Teachers, and the American Institute of Physics. The Center is built on a 24-acre tract of land in College Park, Maryland, 12 miles from the nation's capital in Washington, DC. The building contains 120 000 square feet of floor space, enough initially to house the headquarters of the three founder societies and at least one other member society of the AIP, the American Association of Physicists in Medicine. Most of the AIP professional programs itemized in Table 8 have been relocated to the Center. A featured element in the new building is the Niels Bohr Library, used for studies of the history and philosophy of physics.

2.3.2.5 The Institute of Physics, London. The Physical Society (founded in 1873 and since having developed into a society primarily for university physicists) and the original Institute of Physics (founded in 1918 and since having developed into a society primarily for industrial physicists) were amalgamated in 1970 to form the current Institute of Physics (IOP). Both organizations were based in London, and, upon their amalgamation, they received a Royal Charter.

The object of the Institute is to promote the advancement and dissemination of a knowledge of and education in the science of physics, pure and applied. The total membership is over 16 000, including 2400 Fellows of the Institute and 820 Fellows of the former Physical Society.

The Institute makes a great effort to be a "democratic organization which needs contributions from physicists with wide experiences if it is to be fully representive. By working through the Boards, Committees, Branches, Divisions, and Groups, members can actively participate in the work of the Institute influencing its activities and policies and representing its views to official bodies" (Institute of Physics Secretariat, 1993).

There are currently five divisions of the Institute (see Table 9) that act as forums for reviewing research and development in their respective fields of physics. The divisions collaborate with 32 subject and professional groups in arranging meetings and conferences.

IOP Publishing is a wholly owned subsidiary company of the Institute and one of the largest publishers of physics information in the world. The publishing activity has its origins in the *Proceedings of the Physical Society*, first published over a century ago. Today, it publishes over 70 new books and proceedings a year and over 30 journals.

2.3.2.6 German Physical Society. Germany has probably the longest tradition for educating physicists of any nation in the world. The universities in Germany were the schools of choice for most of the nineteenth century for many students who wanted to enter careers in physics and for many U.S. physicists. Also, the German Physical Society

Table 9. Divisions and groups of IOP.

IOP Divisions	
1. Applied Optics	4. Engineering Physics
2. Atomic, Molecular, and Optical Physics	5. Nuclear and Particle Physics
3. Condensed Matter and Materials Physics	
Subject and Professional Groups	
1. Atomic Collisions in Solids	17. Neutron Scattering
2. Carbon	18. Nuclear Physics
3. Combustion Physics	19. Optical
4. Computational Physics	20. Physical Acoustics
5. Consultancy	21. Physical Crystallography
6. Education	22. Plasma Physics
7. Electronics	23. Polymer Physics
8. Electron Microscopy and Analysis	24. Printing, Packaging, and Papermaking
9. Environmental Physics	25. Quantum Electronics
10. High Energy Particle Physics	26. Semiconductor Physics
11. History of Physics	27. Spectroscopy
12. Instrument Science and Technology	28. Static Electrification
13. Low Temperature	29. Stress Analysis
14. Magnetism	30. Thin Films and Surfaces
15. Materials and Testing	31. Tribology
16. Mathematical Physics	32. Vacuum

(GPS) (Deutsche Physikalische Gesellschaft) founded in 1845 represents the oldest physics society in the world as well as the largest in Europe with 18 400 members.

Thus, it is useful to compare this tradition and history with the relatively small size of the publishing program of GPS. In 1992, they published only two primary publications, called *Physikalische Blätter* (monthly) and *Verhandlungen der DPG* (eight issues per year). They also had an abstracting service called *Physikalische Berichte* (every two weeks). After a study of the large number of members and the small number of authors in their publications, they concluded that their solution would be to encourage authors in their publications to switch to the use of the English language, not only for their primary journals but also for their abstracting service. The English-language abstracting service is now called *Physics Briefs* and is highly successful because it is available worldwide on a computer network called STN International (Scientific and Technical Network International).

2.3.3 Other Physics Societies in the Rest of the World Many industrialized nations may not have physics societies, though they may have chemical, geology, mineralogy, meteorology, or other natural-science societies. The small number of physicists that are educated each year compared with the other natural-science professionals in all countries is the reason. Therefore, the occurrence of physics societies may be spotty in any given region of the world.

A few of the outstanding physics societies not covered by this article are the following:

1. The Indian Physics Association (IPA) was founded in Bombay in 1970. IPA has 30 chapters in different cities of India and has a total membership of about 2500.
2. The Malaysian Institute of Physics was founded in Malaysia in 1973 and has 60 members.
3. The Institute of Physics, Singapore, was founded in 1972. It has 237 members.
4. The African Union of Physics was founded in Oyo, Nigeria, in 1982.

3. ENGINEERING ORGANIZATIONS

3.1 General

Engineering has strongly influenced the development of human society. As the capability of engineers has increased, many technologies have been created that have extended and are continuing to extend human abilities. Examples are the ability to travel more widely and rapidly, to communicate globally and almost instantaneously, to control large amounts of energy, to make many new types of materials, and to modify environmental conditions. Engineers have assisted in effecting these changes and, in turn, have been affected by them. A primary means by which engineers interact with society is through the many organizations the profession has developed. An examination of these engineering organizations can provide insight on the functioning of the engineering profession and how it interacts with society.

In the U.S.A., engineers have created a number of nongovernmental and not-for-profit organizations during the last 150 years during which the modern profession emerged. These organizations range from those concerned with advancing the knowledge base of the profession and educating future engineers to those concerned with industrial practice and public policy on engineering-related issues. They reflect the common interests of engineers as a community and represent these interests to other elements of society. They operate in local, regional, national, and international contexts.

3.1.1 Organization of This Section Engineering-related organizations may be considered in four categories. The first includes the technically oriented organizations whose major concern is the advance of engineering knowledge and know-how—the research and development that make advances possible. The major group in this category, the engineering professional societies, have developed along both disciplinary and industrial sector lines.

In a second category are the education-oriented societies. Although the professional societies and some other engineering-related organizations carry on educational activities, there are several organizations primarily devoted to education-related activities. One such is the Accreditation Board for Engineering and Technology, which accredits engineering and engineering technology postsecondary educational programs.

The third category includes industrially oriented organizations that carry on engineering-related activities. In many industrial sectors, trade associations have been organized that support the engineering interests of their member companies. Another type of organization in this category comprises the standards-writing organizations, which develop the standardized procedures and processes used in many commercial transactions.

A fourth category includes organizations with a major focus on public policy related to engineering. Principal among these are the National Academy of Engineering, the National Society of Professional Engineers, the American Association of Engineering Societies, and the American Association for the Advancement of Science.

These four categories are in no sense mutually exclusive. Most engineering organizations have activities in several categories. Each organization is described here in the category that represents its major activity and the one in which the organization is most widely recognized.

3.1.2 International Counterparts There are counterpart organizations to these U.S.-based organizations in other nations. In the highly industrialized nations, there are usually well-developed organizations with a broad range of activities similar to those in the U.S.A. The national professional societies are typically organized along disciplinary lines, with the holding of meetings, publication of technical periodicals, and recognition of outstanding technical achievements as their principal functions.

However, there are also significant differences among the developed countries. In the United Kingdom, for example, the certification of professional competence, or registration, of engineers is linked to the membership category of the relevant engineering professional society. Thus, "chartered engineer" is a title conferred by a professional society and is used to designate someone recognized as qualified for the public practice of a particular branch of engineering. In the U.S.A., registration of engineers is a responsibility of the state governments, which they carry out with the assistance of the various professional societies. Such national differences occur in many of the industrialized countries.

Strong professional organizations are not the norm in most developing countries. Professional engineers in these countries often lack the benefits of good technical periodicals and regular meetings and conferences. As a consequence, they are not as able to develop effective technical networks with fellow specialists. Having recognized this weakness, some developing countries are in the process of organizing a professional society infrastructure to compensate.

It is beyond the scope of this article to survey the numerous engineering organizations that have been established throughout the world. But it is noteworthy that, in general, engineers everywhere have found that professional societies provide them with services essential to the well-being and health of their professional communities.

There are a number of international organizations in which U.S. engineering societies are members; among these are the World Federation of Engineering Organizations and the Union of International Technical Associations. Typically, their purpose is to facilitate collaboration among engineers throughout the world. While these two organizations operate on a global basis, others operate on a regional basis in a major geographical area, such as Latin America or Southeast Asia.

Some exceptions to the norm of engineering professional societies being organized on a national basis can be found. In the U.S., the most notable is the Institute of Electrical and Electronics Engineers (IEEE), which has a transnational organizational structure. It is described in detail in Sec. 3.3.4. Also, the United Kingdom system of "chartered engineers" is recognized in many other countries that are or have been part of the British Commonwealth.

3.2 Inherent Characteristics of Engineering

3.2.1 Integration of Knowledge Engineers both generate and apply knowledge. Thus, the activities of engineers range from research and development, in which the creation of new knowledge is the principal objective, to manufacturing, construction, and other operations in which knowledge is applied to deliver products and services. To encompass this wide range of functions, engineers draw on and integrate a variety of sources of information and knowledge. The process of integration, drawing on knowledge and experience from many fields, is an inherent characteristic of engineering. The services and products that are the output of engineering must meet the requirements of customers and competition in the marketplace.

3.2.2 Relation to Mathematics and Science Mathematics and the natural sciences provide much of the knowledge needed by engineers. As these disciplines advance, they give engineers the ability to improve on existing products and services and to create new ones. However, as the complexity of the products and services has increased, engineers have had to undertake fundamental research to improve the performance of the systems they have created and to understand their limitations. For example, as communication systems became more powerful and complex, a deeper understanding of their characteristics was needed. This need was met by the development of information theory at the Bell Telephone Laboratories in the 1940s. It enabled design engineers to determine the information-carrying–capacity limitations of existing systems and to improve their performance. While engineers, particularly in academic environments, carry on fundamental research driven by individual initiative as well as research in strategic areas, engineering research in industrial laboratories is primarily driven by needs of a strategic nature. The traditional engineering disciplines are linked closely to the most relevant science: chemical engineers to chemistry, electrical and mechanical engineers to physics, and civil and mining engineers to geology.

As the complexity of engineering-designed systems has increased, these traditional knowledge sources have been complemented by others. New analytical techniques have been created to handle the greater complexity. The field of systems engineering has gradually evolved to enable highly complex systems to be designed, produced, and operated. To optimize the performance of systems that involve both complex instrumentation and human operators requires improved analysis, modeling, and simulation techniques. Engineers who specialize in the design and production of such complex systems are no longer closely linked to one of the natural sciences.

The characteristics of engineering research and an identification of some research fields in which there are fruitful opportunities are discussed in a report written by a panel of the National Research Council Engineering Research Board (National Research Council, 1987). Figure 3, from this report, illustrates the relationship between

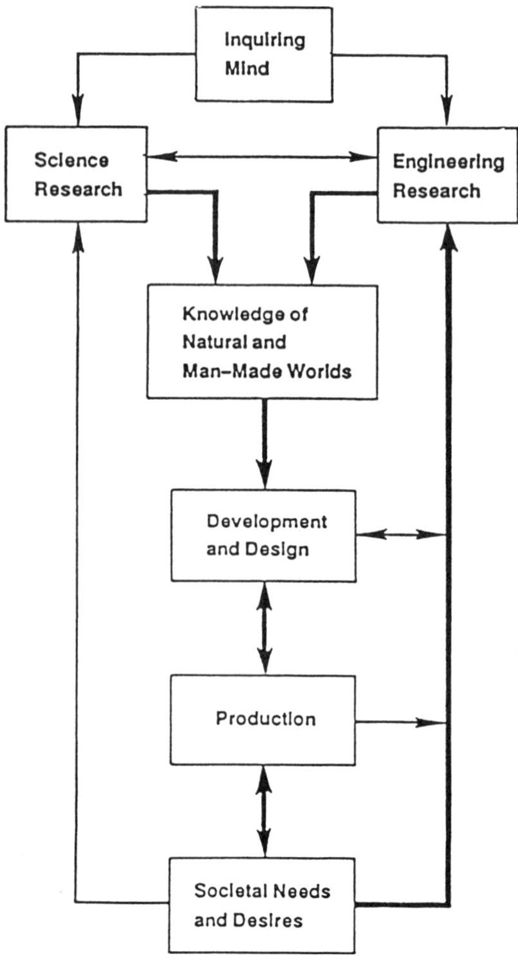

FIG. 3. Relationship between engineering and scientific research (from National Research Council, 1987).

engineering and scientific research in the perception of this panel. The broad distinction is made that scientific research seeks to gain more knowledge about the natural universe, while engineering research seeks more knowledge of the man-made world. Thus, technological advances can open up new branches of engineering, which become areas in which new research efforts are launched. One such branch of engineering, which is still in an early stage of development, is software engineering (*q.v.*). Its fundamental concepts are still in the process of being identified (Gibbs, 1991).

3.2.3 Societal Impact As technologies of increasing power have been developed, engineering's effect on human society has become more far-reaching, and the knowledge sources that engineers must draw on have widened. As medical technology has advanced, engineers have worked more closely with physicians and biologists. The need to control pollution and to protect the environment has required engineers to work closely with behavioral scientists as well as government officials. To meet competition in the global marketplace, engineers must work with economists and international experts. To meet changing conditions, engineers must integrate not only quantitative knowledge in their design and production of products and services but also the nonquantitative knowledge that characterizes many nonscientific disciplines.

The existing engineering societies have responded to these new demands by creating new publications and jointly sponsoring meetings. In some cases, new societies have been formed to facilitate the development of new fields of knowledge. The new modes of handling information have resulted in computer experts and other engineers working with specialists in medicine, education, organizational behavior, and governmental affairs as well as the natural sciences. Expanding frontiers have put new requirements on both individual engineers and the organizations that serve their interests.

3.3 Technically Oriented Organizations

3.3.1 Services of Professional Societies The engineering professional societies, organized to provide services needed by the developing engineering professions, have, in turn, influenced the actual development of engineering. Their primary service is to provide an efficient and cost-effective way for their members to communicate with other specialists in their fields. Through meetings and publications, the societies bring geographically dispersed engineers together intellectually, to learn from each other both the theory-based and the practice-based knowledge needed to advance the engineering state of the art. New research results as well as information about new instrumentation and techniques are rapidly disseminated to those working in the field or those interested in new applications. In addition, they provide the crucially important archival function of recording and preserving the evolving knowledge base in their discipline.

Second, they provide their members an opportunity to address issues of common concern, whether related to education of future engineering practitioners, to industrial practice, or to governmental policies. In a number of these arenas, professional societies provide the most effective means of developing, presenting, and implementing positions that reflect the judgment of the engineering community.

Third, they give their members recognizable credentials. By requiring those seeking society membership to meet specified professional criteria as determined by peers, the societies give their members professional recognition. By identifying the outstanding work of individual engineers, they encourage and reward above-average professional achievements. Through a system of membership grades and awards, individual accomplishments are publicized to peer groups and to external communities.

3.3.2 Founder Societies In the U.S.A., the first engineering professional societies were organized in the latter half of the nineteenth and early twentieth centuries. These first societies later became known as the Founder Societies.

The first, the American Society of Civil Engineers founded in 1852 (Wiseley, 1974), had the task of establishing the concept of civil as distinct from military engineering. Prior to the 1850s, the Military Academy at West Point was the major source of U.S. engineering graduates; there was little recognition of engineering as a civilian activity crucial to the construction of bridges, railroads, canals, and water supply and sewage disposal systems. Within the next few decades, the American Institute of Mining Engineers (1871), the American Society of Mechanical Engineers (1880), and the American Institute of Electrical Engineers (1884) were formed. As a result of the frequent explosions of poorly designed boilers during the late 1800s and the consequent danger to the public, the new mechanical engineering society was immediately involved with the development of safety standards for boilers, an activity that persists to the present day. Also during this period, there were major new technical developments in electrical power generation, lighting, and telegraphic and telephonic communications. The new electrical engineering society had a major influence in these fields by initiating the publication of technical papers on topics in which secrecy to protect commercial advantages had previously prevailed.

In the early 1900s, chemical engineering developed as a discipline distinct from both chemistry and mechanical engineering. The American Institute of Chemical Engineers (Reynolds, 1983) played a major role in defining the field and designing appropriate undergraduate curricula. Also at this time, the invention and use of wireless communications resulted in the formation of a new but rapidly growing society, the Institute of Radio Engineers.

These Founder Societies combined their efforts to build facilities in New York City to house their headquarters. A gift of more than $1 million in 1903 from Andrew Carnegie, an engineer who became a successful businessman, provided impetus. Appendix A includes detailed information about the organizational structure, budget, membership, publications and meetings, and other activities of the Founder Societies.

3.3.3 Organizational Characteristics As the various engineering disciplines matured, the societies developed internal structures with both geographical and technological components. The geographical units typically feature local and regional meetings and support student activities at local schools and universities; the technological units focus on national conferences and technical publications. Both entities provide useful membership services. The volunteer officers, elected by the members, typically reflect both of these structural components.

The membership requirements of the Founder Societies as well as the more recently formed engineering societies are usually similar. The typical requirement is a baccalaureate degree for the entering member grade, while senior grades are given to more experienced engineers. Student grades are used to encourage those preparing for engineering careers to develop university-based engineering activities and to facilitate their joining the appropriate society after graduation. The top membership grade is frequently restricted to a small percentage of the members elected in recognition of outstanding accomplishments.

3.3.4 Unique Character of the IEEE Many of the U.S. engineering societies have the term "American" in their title and consider themselves national organizations. Although they accept overseas members, they do not directly compete with their counterpart societies in other countries. A major exception is the Institute of Electrical and Electronics Engineers (IEEE), which was formed by the merger of the American Institute of Electrical Engineers and the Institute of Radio Engineers in 1962 (McMahon, 1984).

Although headquartered in the United Engineering Center in New York with the other Founder Societies, the IEEE organized and structured its geographical and technological activities on a global basis. It has attracted approximately 50 000 overseas members, one-sixth of its total membership of more than 300 000. Many meetings and conferences are held outside the U.S.A. These meetings and the Institute's publications are widely considered the most prestigious in their fields.

The IEEE has developed into a federation of more than 30 societies in the various specialties into which the electrical, electronics, and computer fields have evolved. The Institute's internal structure readily permits the organization of additional specialty groups as new technological fields emerge (Ryder and Fink, 1984). For example, the explosive growth of the computer field after World War II was accompanied by the rapid expansion of the newly organized Computer Society of the IEEE. It now has a membership of more than 70 000 and publishes 17 technical periodicals.

The IEEE Computer Society growth has been paralleled by the growth of another professional society in the computer field, the Association for Computing Machinery (ACM). In their early development, these societies were primarily focused on hardware and software, respectively. But as the computer technology advanced, this separation was no longer valid; both societies currently have publications in software engineering. The emergence of this latter field of engineering is being studied by joint committees of the two societies. With strongly overlapping memberships, the two societies have both flourished and have cooperated. Appendix B gives information about these societies' membership and activities and illustrates some of their complementarity.

3.3.5 Nontechnical Activities While the primary activity of the engineering professional societies has focused on the advance of the discipline, there has also been concern about engineering careers. For example, are enough students being attracted to the field? What are the salary ranges and fringe benefits available to practicing engineers? What are the legal obligations of engineers in public practice? The societies have responded by undertaking professional—in contrast to technical—activities; many standing committees and boards have been established to address such issues.

Currently, almost all major U.S. engineering societies have offices or representatives in Washington, DC, and devote both staff and volunteer time to relations with the federal government. On one hand, they serve as a valuable resource to federal agencies and to the Congress in supplying specialized technical information. On the other, they seek to avoid legislation or regulations that are considered to be inimical to the engineering profession. In this arena, the societies operate differently; most are still exploring how to respond to changes in the environment in which engineering is practiced.

3.4 Educationally Oriented Organizations

3.4.1 Studies of Engineering Education The engineering professional societies have had a strong interest in academic programs in engineering and have had considerable effect on their development. In addition to the educational activities of the disciplinary societies, another society, the American Society for Engineering Education (ASEE) devoted solely to engineering education, was founded in 1893. It has served as the sponsor for a number of major reports on engineering education that have influenced the philosophy and content of engineering education at the undergraduate and graduate levels. Its membership includes primarily engineering faculty members with strong interest in undergraduate and continuing education and teaching methods (Reynolds and Seely, 1993). In addition to engineering faculty members, it also includes engineering technology faculty members who teach in two- and four-year colleges offering degrees in engineering technology. Most ASEE members also belong to at least one disciplinary society.

There have been a number of major ASEE studies of engineering education spanning more than 70 years. Although none of them has had an effect as great as the Flexner report on medical education in the 1920s, they have served to highlight the problems that engineering educators have faced and to point to new directions. In the post–World War II era, one report written in 1952–1955 was notable for involving a large segment of the academic engineering community in gathering data and in stimulating discussions. The wartime experience at laboratories such as the MIT Radiation Laboratory had demonstrated that engineers educated before the War did not have the academic background needed to contribute to such new engineering developments as radar and microwave communications. The fact that the authors of the 25-volume series of books published by the MIT Radiation Laboratory in the late 1940s were primarily physicists and mathematicians had a profound effect on engineering educators. As a result, much of the handbook-based material in undergraduate curricula was replaced with more advanced courses in calculus, physics, and chemistry. The "Report on the Evaluation of Engineering Education" by an ASEE committee chaired by L. E. Grinter (American Society for Engineering Education, 1955) reflected this trend. It urged increased emphasis on what were labeled the engineering sciences and less emphasis on current engineering practice. Thus, postwar engineering curricula differed from the prewar in requiring engineering students to gain much greater depth of understanding in mathematics and the physical sciences.

Another major ASEE study resulted in the publication of a report entitled "The Goals of Engineering Education" (American Society for Engineering Education, 1968). By drawing an analogy between engineering and medical education, this report sought to replace the four-year undergraduate programs as preparation for an engineering career and require a master's degree as the first "professional" degree. Such proposals, although frequently discussed, have never gained the endorsement of either the academic community or industrial companies, the major employers of baccalaureate-degree engineers. Thus, engineering remains one of the few professions for which the baccalaureate is considered to be a professional degree. In some engineering specialties, such as sanitary engineering, the first professional degree is at the master's level. However, students undertaking such programs frequently complete a four-year civil engineering program first and so already have a professional degree.

A 1987 ASEE report, "A National Action Agenda for Engineering Education" (American Society for Engineering Education, 1987), chaired by Edward E. David, Jr., endorsed the four-year baccalaureate programs as the appropriate "launching platform" for an engineering career. The report also recommended a practice-oriented master's degree as a follow-on for students interested in technical specialties where greater academic depth is needed. This degree would complement the research-oriented Ph.D. degree. Much current thinking in engineering education is reflected in this report.

3.4.2 Accreditation of Academic Programs The disciplinary societies are also active in various aspects of engineering education. They frequently support programs concerned with the continuing education of their members as well as for students majoring in their respective specific disciplines. They exert significant influence on the content of undergraduate curricula through their participation in the accreditation process.

The Accreditation Board for Engineering and Technology (Accreditation Board for Engineering and Technology, 1993) is a creation of the engineering professional societies. ABET accredits programs in the various engineering disciplines in schools of engineering and of engineering technology. Both federal and state agencies recognize ABET as the accepted accrediting body for these fields. The ABET board of directors includes representatives of the professional societies that have curricular responsibilities, as well as some other engineering-related organizations. The professional societies develop the criteria used by accreditation teams that visit engineering schools and inspect their academic programs at three- or six-year intervals. Team members, selected from lists of engineers submitted by the disciplinary societies, typically include engineers employed by industry as well as by universities.

Essentially, the accreditation process sets a floor for the academic quality of the approved programs. While the accreditation process keeps pressure on all the more than 300 U.S. engineering schools and 120 engineering technology schools to keep their courses and laboratories up-to-date, many academicians feel that the accreditation process stultifies innovation. Currently, a number of innovative undergraduate programs are being developed as a result of the availability of curriculum development funding from the National Science Foundation. These programs are testing the ability of ABET to respond to nontraditional programs.

ABET has a powerful position with respect to the engineering schools because of its link to the legal process of registration. Registration, which is required for engineers in the public practice of engineering, is controlled in the U.S.A. by state governments. The state registration boards recognize graduation from an ABET-accredited school as completing a major requirement for registration. Since schools want their graduates to qualify for registration, they must retain ABET-accredited programs. Some schools offer both accredited and nonaccredited programs to allow some students greater flexibility in course selection. The National Council of Engineering Examiners, the organization of the state engineering examiners, is also represented on the ABET board of directors.

3.5 Industrially Oriented Organizations

In this category are two major types of engineering-related organizations: those that develop standards and the industrial trade associations. While these latter organizations typically engage in many nontechnical activities, some of their programs have significant engineering content.

3.5.1 Standards-Writing Organizations Much engineering practice depends directly on the use of standards, that is, concepts and procedures systematically developed and approved by experts in a technical specialty. Standards development is primarily an industry-based activity since industrial products and processes must usually meet standardized requirements set by customers, regulatory authorities, and other external bodies. The term *standards* is used to include methods of test, technical specifications, definitions and terminology, and recommended practices.

In the U.S.A., in contrast to most other industrialized countries, the writing of voluntary standards other than those related to national defense is largely a private-sector activity (Toth, 1991). Committees of engineers are convened from a variety of companies, testing laboratories, public agencies, and universities to draft voluntary standards. These documents are then subjected to wide review and comment before being accepted and put into general use. Since a consensus is sought among divergent interests, the time needed to develop a standard can vary from about six months to several years. Standards were first widely used in the specification of commercially important materials such as metals and plastics and hardware devices but have now also become an important means of specifying processes and procedures in many industrial sectors such as telecommunications and computers.

The U.S. government is involved in the standards-writing process through technical support provided by federal agencies such as the National Institute for Standards and Technology (NIST), formerly the National Bureau of Standards, and through the participation of federal engineers on standards-writing committees. Of great importance in both domestic and international commerce, standards can serve to facilitate or impede trade. Some privately developed voluntary standards adopted by public agencies attain the force of law. A prime example is building construction: Most mandatory building codes cite many voluntary standards that have been developed in the private sector.

The U.S. standards-writing community currently maintains some 94 000 standards. While about a third of these relate to the Department of Defense, more than a half were developed and maintained by the private sector. Among the leading standards-writing professional societies are the Society of Automotive Engineers, the American Society of Mechanical Engineers, and the Institute of Electrical and Electronics Engineers. Among the leading trade associations active in standards-writing are the Aerospace Industries

Association, the Association of American Railroads, and US Pharmacopeia.

There are also several other organizations whose primary function is standards development. Principal among these are the American Society for Testing and Materials, the National Fire Protection Association, and Underwriters Laboratories, which also provides certification services. The private-sector American National Standards Institute (ANSI) serves as a coordinator for domestic standards and represents the U.S.A. internationally in standards-writing activities.

3.5.1.1 International Standardization. Complementing the domestic U.S. activity are the international standards-writing organizations and national standards-writing societies in other countries. In most countries, standards-writing is considered both a private- and a public-sector activity, and public funding supplements private-sector resources. Such countries are better organized than the U.S.A. to participate systematically in international activities since they are not as totally dependent on private company personnel. However, when the industrial sector is well organized and U.S. companies are in a technological leadership position, the U.S.A. is very effective in international standards activities.

There are several hundred regional and international organizations conducting standards-related activities (Breitenberg, 1989); the U.S.A. participates in almost 200. Some organizations have primarily scientific concerns in fields such as physics, chemistry, and geology and devote their efforts to developing consistent terminology and agreed-on procedures for data handling. There are also engineering-related groups that cover the entire range of technical subjects with which U.S. domestic groups are concerned.

The major umbrella-type international organizations for engineering-related issues, the International Organization for Standardization (ISO) and the International Electrotechnical Commission (IEC), establish standards that are increasingly accepted internationally. The absence of agreed-on international standards can be an effective nontariff barrier to trade. The U.S. member in these organizations is ANSI. The major federal agency concerned with standards, the Department of Commerce, has no direct role in representing U.S. interests in ISO or IEC. There is general agreement in both the private and public sectors that this arrangement is far from optimum.

3.5.2 Trade Associations The 1993 issue of the Encyclopedia of Associations lists more than 23 000 national associations in the U.S.A. Among these are trade associations that represent the interests of their industrial sectors in political and public relations as well as in technical arenas. These arenas are not always separable in issues such as environmental protection, where there are strong political forces as well as scientific and engineering factors involved.

Trade associations in some major industrial sectors such as chemical processing, semiconductors, gas technology, and electric power have undertaken research programs to solve problems generic to their technology. Most such efforts have been funded from private sources; but increasingly, public funding is being provided to support generic technology development. Since changes in antitrust legislation have encouraged industrial companies to carry on cooperative research programs, it is probable that even more industrial associations will undertake engineering projects in the future.

Trade associations are also active in many other industrialized countries. In the United Kingdom, for example, there are trade associations in nearly every industrial sector. Their activities include not only presenting the point of view of their industry to the government but also the dissemination of information about issues such as health and safety, environmental protection, and other topics of general concern. They also recommend "best practices" for use by their member companies. In other European countries, such associations are also active. Currently, these associations are developing links throughout the European Community to bring their influence to bear throughout the Community.

In Japan, trade associations are very important actors in both the technical and the public policy arenas (Bloom, 1990). Associations can be licensed by the Ministry of International Trade and Industry (MITI) and in some cases receive modest funding. They have the significant function of facilitating communication between public policy makers and the various industrial sectors and providing a forum where consensus positions are developed. On the technical side, they are sometimes organized on a temporary basis to carry out a specific research program. On a continuing basis, many associations provide services for their members including collecting and disseminating information from both domestic and international sources, development of standards, and educational programs.

3.6 Public Policy Oriented Organizations

3.6.1 National Academy of Engineering In varying degrees, all the engineering-related organizations discussed are concerned with public policy issues. Most of them operate at the federal level, although some have significant operations at the state and local levels. Their ability to affect legislative action or executive branch program decisions varies. Among the most influential is the National Academy of Engineering (NAE). The Academy was organized in 1964 under the congressional charter of the National Academy of Sciences, granted in 1863, which charges the Academy to respond to requests from federal entities for advice and recommendations. As part of the National Academy complex, the NAE participates in the governance and operation of the National Research Council, which produces about 300 reports a year in response to requests from executive and legislative federal agencies. These reports, which cover a wide variety of scientific and technological topics, involve some 10 000 scientists and engineers serving as volunteers. The NAE has initiated a number of its own studies on broad issues such as sustainable technology and national strategies in an era of global technologies. As an honorary society of approximately 1700 leading engineers from industry, academia, and government, the NAE has become increasingly effective in its relatively brief span of existence (Edson, 1989).

3.6.2 National Society of Professional Engineers The National Society of Professional Engineers (NSPE) is a society that restricts its membership to registered engineers (Robbins, 1984). It was established more than 50 years ago to consolidate the activities of a number of engineering associations operating at the state level. The state societies are primarily concerned with the registration process by which engineers are certified as competent to practice engineering publicly. These state societies frequently bring considerable influence to bear on engineering-related issues with which state governments are involved. These issues range from policies of state agencies concerned with engineering projects to budgetary allocations for engineering schools in state universities. NSPE's headquarters in Washington is primarily concerned with federal legislation and monitors federal programs that involve the interests of registered engineers. The society is active in promoting the recruitment of young engineers, in surveying salaries of practicing engineers, and in analyzing ethical and legal issues confronting engineers.

3.6.3 Umbrella Organizations To increase its effectiveness in influencing public policy, the engineering community has made several attempts since the 1920s to form a single organization that speaks for the entire engineering profession. The current umbrella society is the American Association of Engineering Societies (AAES), which includes in its membership the engineering professional societies and associations listed in Table 10. The combined membership of its constituent societies is some 700 000 U.S. engineers. The positions it develops on public policy issues must be compatible with those of the major member societies, which are generally unwilling to abdicate power on issues vital to their members. Since the member societies possess technical expertise in their disciplinary fields, they usually prefer to speak to such issues directly before congressional or executive branch committees.

However, there are issues on which most engineers can agree. Examples are more portable pension plans, improving engineering's public image, and forging better links with international engineering organizations. In such arenas, the AAES serves as the voice and surrogate for U.S. engineering societies. It is the U.S. member in international engineering organizations such as the World Federation of Engineering Organizations (WFEO/FMOI) and also in regional groups such as the Panamerican Federation of Engineering Societies (UPADI), a federation of engineering societies in Latin and North America. In general, these groups have not been as effective as the International Council

Table 10. List of member, associate, and regional societies in the AAES as of 1993 1994.

Member Societies
American Academy of Environmental Engineers
American Indian Science and Engineering Society
American Institute of Chemical Engineers
American Institute of Mining, Metallurgical and Petroleum Engineers
American Institute of Plant Engineers
American Nuclear Society
American Society for Engineering Education
American Society of Agricultural Engineers
American Society of Civil Engineers
American Society of Mechanical Engineers
Institute of Electrical and Electronics Engineers
National Institute of Ceramic Engineers
National Society of Professional Engineers
Optical Society of America
Society of Fire Protection Engineers
Society of Women Engineers
Associate Societies
American Council of Independent Laboratories
Association for International Practical Training
Federation of Materials Societies
National Action Council for Minorities in Engineering
National Association of Corrosion Engineers
Operations Research Society of America
Tau Beta Pi Association
Regional Societies
Engineering Society of Detroit
Engineers' Society of Western Pennsylvania
Washington Society of Engineers
Western Society of Engineers

of Scientific Unions (ICSU) in launching programs that involve technical specialists and financial resources from many nations.

Within the U.S.A., the American Association for the Advancement of Science (AAAS) (see Sec. 2.2.1) has developed activities that affect both science and technology policy. While most engineers regard the AAAS as primarily a spokesman for the sciences, many engineers have been prominent in its activities. The major engineering professional societies are affiliates of the AAAS. They are active participants in the AAAS Congressional Fellows program, which has supplied the Congress with fellows drawn from the ranks of midcareer scientists and engineers to serve as members of congressional staffs. The AAAS annual analysis of the proposed federal research and development budget has involved several of the engineering societies. The editorial board of the AAAS's weekly publication *Science* usually includes engineers.

Although engineers constitute one of the largest professional groups in the U.S.A., they have not been a powerful influence on public policies. This lack of effectiveness may be traced to the reluctance of individual engineers to participate actively in politics. Engineers who work in industrial companies may also feel obligated to defer to their employer's views on public policy issues. Few engineering organizations have developed the skilled staffs or devoted the necessary resources to become forceful advocates of engineering interests in public policy arenas.

3.7 Challenges for Engineering Organizations

Engineering organizations are challenged to respond effectively to both rapid technological changes and worldwide changes in the industrial, political, and social environment. Engineers who are at the forefront of technological advances must be able to adapt to both types of change and still retain their operational effectiveness. Cutting-edge engineers need appropriate means to communicate with colleagues throughout the world. Those professional societies that provide such domestic and global communication services efficiently and cost-effectively will continue to grow. Others may well disappear.

Engineers will also require educational services beyond those currently available. Education is being increasingly recognized as a career-long activity for practicing engineers. The content of such educational programs should be determined cooperatively by active engineers in industry, government, and universities. The engineering organizations can contribute by providing educational materials and also the means for delivering them. While the problem of designing and delivering career-long educational programs is largely unsolved, the need is growing and engineering organizations have many assets to offer.

As the marketplace for engineering-intensive products and services becomes more global, engineering operations are being carried on in many cultural contexts. Many multinational corporations currently operate with relative ease in many different nations with little concern about national boundaries. Both individual engineers and the organizations they have created must learn how to operate effectively in this changing environment.

As new technologies affect more and more the way industry, government, and academia operate, engineers must not only add to their technical competence but must also be concerned with the nontechnical implications of what they create. The broader issues can no longer be considered someone else's responsibility. If engineering organizations help impart such broader perspectives, they will continue to serve as well as influence the future development of the profession.

APPENDIX A. ENGINEERING'S "FOUNDER SOCIETIES"

The first organized of the major engineering professional societies are known as the Founder Societies. These societies have played a crucial role in the development of the United Engineering Center in New York City, where many engineering societies maintain their headquarters. This Appendix presents data about each of these five societies: the number of members, their publications, and their stated objectives. The Founder Societies are listed in the order of their formation: American Society of Civil Engineers (ASCE) in 1852; American Institute of Mining, Metallurgical and Petroleum Engineers (AIME) in 1871; American Society of Mechanical Engineers (ASME) in 1880; Institute of Electrical and Electronics Engineers (IEEE) in 1884; and American Institute of Chemical Engineers (AIChE) in 1908. The AIME has undergone several organizational changes since its founding and is now a federation of four societies, with the Society for Mining, Metallurgy and Exploration (SME) as the continuing society.

The data about these societies have been excerpted from the *International Directory of Engineering Societies and Related Organizations 1993*, published by the American Association of Engineering Societies, and are used with permission.

- **American Society of Civil Engineers—ASCE**
  345 East 47th Street, New York, NY 10017, U.S.A.

  MEMBERSHIP:

  Individual 56 736
  Student 14 177
  Honorary 26 574

OBJECTIVE: The advancement of the science and profession of engineering to advance the welfare of mankind.

PUBLICATIONS: Publications catalog is available on request. The following is a sampling of periodicals published by the Society: *ASCE News*, monthly; *Civil Engineering*, monthly; *ASCE Publications Information*, bimonthly; journal *Hydraulic Engineering*, monthly; journal *Environmental Engineering*, bimonthly; journal *Structural Engineering*, monthly; journal *Engineering Mechanics*, monthly; journal *Irrigation and Drainage Engineering*, quarterly; journal *Geotechnical Engineering*, quarterly; journal *Surveying Engineering*, quarterly; *Transactions*, annual; journal *Aerospace Engineering*, quarterly; *Combined Index*, annual; *Cumulative Index*, quinquennial. Approximately 70 books are published annually, in addition to periodicals.

- **American Institute of Mining, Metallurgical and Petroleum Engineers, Inc.—AIME**
  345 East 47th Street, New York, NY 10017, U.S.A.

  MEMBERSHIP:

  Societies 4
  AIME is a separate corporation controlled jointly and severally by its four Member Societies: Society of Mining Engineers, The Metallurgical Society, Iron and Steel Society, and Society of Petroleum Engineers. Individual nonvoting membership consists of AIME Honorary Members and Legion of Honor Fifty-Year Members.

OBJECTIVES: The advancement of knowledge of the arts and sciences involved in the production and use of useful minerals, metals, energy sources, and materials and to record and disseminate significant developments in these areas of technology for the benefit of mankind.

- **Society for Mining, Metallurgy and Exploration, Inc.—SME**

  MEMBERSHIP:

  Individual 18 702
  Student 1 095

OBJECTIVES: To furnish a common ground to members of AIME interested in the fields of mining engineering, minerals exploration, geological engineering, rock mechanics, and minerals benefication; to promote the advancement of the profession through the exchange of information, the preparation and presentation of technical papers, their discussion and circulation; to advance education in the fields covered above and to promote high standards among the members of the profession.

PUBLICATIONS: *Mining Engineering*, monthly; *Minerals and Metallurgical Processing*; various proceedings and textbooks.

- **The American Society of Mechanical Engineers—ASME**
  345 East 47th Street, New York, NY 10017, U.S.A.
  MEMBERSHIP:

Individual	96 100
Student	20 000
Honorary	100
Emeritus	1 800

OBJECTIVES: To promote the arts and sciences connected with engineering and mechanical construction for scientific purposes, and to that end to meet and associate together to read and discuss professional papers, and to circulate by means of publication among its members the information thus obtained, and for the purpose of maintaining a library. Also, in 1971, the following Goal was adopted: to move vigorously from what is now a Society with essentially technical concerns to a Society that, while serving the technical interests of its members ever better, is increasingly professional in its outlook, sensitive to the engineer's responsibility to the public interest, and dedicated to a leadership role in making technology a true servant of man.

PUBLICATIONS: *Mechanical Engineering*, monthly; *ASME News*, monthly; *Transactions Journals* (18), quarterly; *Applied Mechanics Reviews*, monthly; Technical Preprints, approximately 1200/yr; Symposia and Proceedings, approximately 165/yr; *Codes and Standards*, approximately 270/yr.

- **The Institute of Electrical and Electronics Engineers—IEEE**
  345 East 47th Street, New York, NY 10017, U.S.A.
  MEMBERSHIP:

Individual	270 800
Student	49 500

OBJECTIVES: Its purposes are: (1) Scientific and educational, directed toward the advancement of the theory and practice of electrical engineering, electronics, radio, and the allied branches of engineering and the related arts and sciences. Means to these ends include, but are not limited to, the holding of meetings for the reading and discussion of professional papers and the publication and circulation of works of literature, science, and art pertaining thereto. (2) Professional, directed toward the advancement of the standing of the members of the professions it serves. Means to this end include, but are not limited to, the conduct and publication of surveys and reports on matters of professional concern to the members of such professions, collaboration with public bodies and with other societies for the benefit of the engineering profession as a whole, and the establishment of standards of qualification and ethical conduct. The IEEE shall not engage in collective bargaining on such matters as salaries, wages, benefits, and working conditions customarily dealt with by labor unions.

PUBLICATIONS: Sixty-seven periodicals, including *IEEE Proceedings*, *Spectrum*, *Transactions & Journals*, plus more than 50 000 pages annually in nonperiodicals (conference records, reports, standards, etc.).

MEMBER SOCIETIES: Aerospace and Electronic Systems Society (AES), Antennas and Propagation Society (AP), Broadcast Technology Society (BT), Circuits and Systems Society (CAS), Communications Society (COM), Components Hybrids and Manufacturing Technology Society (CHMT), Computer Society (C), Consumer Electronics Society (CE), Control Systems Society (CS), Dielectrics and Electrical Insulation Society (DEI), Education Society (Ed), Electromagnetic Compatibility Society (EMC), Electron Devices Society (ED), Engineering Management Society (EM), Engineering in Medicine and Biology (EMB), Geoscience and Remote Sensing (GRS), Industrial Electronics Society (EI), Industry Applications Society (IA), Information Theory (IT), Instrumentation and Measurement Society (IM), Lasers and Electro-Optics Society (LEO), Magnetics Society (Mag), Microwave Theory and Techniques Society (MTT), Nuclear and Plasma Sciences Society (NPS), Oceanic Engineering Society (OE), Power Electronics Society (PEL), Power Engineering Society (PE), Professional Communication Society (PC), Reliability Society (R), Robotics and Automation Society (RA), Signal Processing Society (SP), Social Implications of Technology Society (SIT), Systems Man and Cybernetics Society (SMC), Ultrasonics Ferroelectrics and Frequency Control Society (UFFC), Vehicular Technology Society (VT), Neural Networks Council (NN), and Solid-State Circuits Council (SSC).

- **American Institute of Chemical Engineers—AIChE**
  345 East 47th Street, New York, NY 10017, U.S.A.
  MEMBERSHIP:
  Individual 40 800
  Student 6 500
  Honorary 800
  Emeritus 3 900

OBJECTIVES: The objectives of the Institute are the advancement of chemical engineering in theory and in practice, the maintenance of a high professional standard among its members, and to serve society, particularly where chemical engineering can contribute to the public interest.

PUBLICATIONS: *Chemical Engineering Progress*, monthly; *AIChExtra*, monthly; *Chapter One*, 3/yr; *AIChE Journal*, bimonthly; *International Chemical Engineering*, quarterly; *Energy Progress*, quarterly; *Environmental Progress*, quarterly; *Plant/Operations Progress*, quarterly; *Biotechnology Progress*, bimonthly; *Symposium Series*, bimonthly; *The Directory of Chemical Engineering Consultants*, annual; *Chemical Engineering Faculties*, annual. Also, process safety procedures, education series, symposium proceedings.

APPENDIX B. COMPARISON OF TWO LEADING U.S. COMPUTER SOCIETIES (FROM *INTERNATIONAL DIRECTORY OF ENGINEERING SOCIETIES, 1993*).

- **Association for Computing Machinery Inc.—ACM**
  1515 Broadway, 17th Floor, New York, NY 10036, U.S.A.
  MEMBERSHIP:
  Individual 65 000
  Student 16 000
  Corporate 800

OBJECTIVES: To advance the sciences and arts of information processing including, but not restricted to, the study, design, development, construction, and application of modern machinery, computing techniques and appropriate languages for general information processing, for scientific computation, for the recognition, storage retrieval, and processing of data of all kinds, and for the automatic control and simulation of processes; to promote the free interchange of information about the sciences and arts of information processing both among specialists and among the public in the best scientific and professional tradition; to develop and maintain the integrity and competence of individuals engaged in the practices of the sciences and arts of information processing.

PUBLICATIONS: *Communications of the ACM (CACM)*, monthly; *Computing Reviews*, monthly; *Journal of the ACM (JACM)*, quarterly; *Computing Surveys*, quarterly; *CALGO*, quarterly; *Transactions on Database Systems*, quarterly; *Transactions on Mathematical Software*, quarterly; *Transactions on Programming Language and Systems (TOPLAS)*, quarterly; *Transactions on Office Information Systems (TOOIS)*, quarterly; *Transactions on Computer Systems (TOCS)*, quarterly; *Transactions on Graphics (TOG)*, quarterly; *Transactions on Software Engineering and Methodology (TOSEM)*, quarterly; *ACM Guide to Computing Literature*, annual.

- **Computer Society of the IEEE**
  IEEE Computer Society, 1730 Massachusetts Avenue, Washington, DC 20036, U.S.A.
  MEMBERSHIP:
  Individual 69 756
  Student 9 995

OBJECTIVES: To advance the theory, practice, and application of computer and information-processing science and technology and to maintain a high professional standing among its members. The society shall promote cooperation and exchange of technical information among its members and to this end shall hold meetings for the presentation and discussion of technical papers, shall publish technical journals, and shall through its organization and other appropriate means provide for the needs of its members. The scope of the Society shall encompass all aspects of theory, design, practice, and application relating to computer and information-processing science and technology.

PUBLICATIONS: *Annals of the History of Computing*, quarterly; *Computer Magazine*, monthly; *Transactions on Knowledge and Data Engineering*, bimonthly; *Transactions on Parallel and Distributed Systems*, bimonthly; *Transactions on Pattern Analysis and Machine Intelligence*, monthly; *Transactions on Software Engineering*, monthly; *Computer Graphics and Applications Magazine*, bimonthly; *Design and Test of Computers Magazine*, quarterly; *Expert Magazine*, bimonthly; *Micro Magazine*, bimonthly; *Software Magazine*, bimonthly; *Transactions on Computers*, monthly.

Works Cited

Accreditation Board for Engineering and Technology (1993), *ABET Accreditation Yearbook*, New York: ABET.

American Institute of Physics (1990), *Society Membership Profile: Employment Mobility and Career Change*, New York: American Institute of Physics.

American Institute of Physics (1993), *AIP Handbook for Member Society Officers, 1993–1994*, New York: American Institute of Physics.

American Society for Engineering Education (1955), "Report on Evaluation of Engineering Education (1952–1955)," *J. Eng. Educ.* **46**, 25–60.

American Society for Engineering Education (1968), "Final Report: Goals of Engineering Education," *J. Eng. Educ.* **58**, 367–446.

American Society for Engineering Education (1987), *A National Action Agenda for Engineering Education*, Washington, DC: ASEE.

Anon. (Ed.) (1994), *World of Learning*, London: Europa Publications.

Barton, H. A. (1956), *Phys. Today* **9** (1), 93–97.

Bates, Ralph S. (1965), *Scientific Societies in the United States*, Cambridge, MA: MIT Press, p. 1.

Bloom, J. L. (1990), *Japan as a Scientific and Technical Superpower*, NTIS PB 90-234923, Springfield, VA: National Technical Information Service.

Breitenberg, M. (Ed.) (1989), *Directory of International and Regional Organizations Conducting Standards-Writing Activities*, NIST Special Publication No. 767, Washington, DC: U.S. G.P.O.

Burek, Deborah M. (Ed.) (1993), *Encyclopedia of Associations: National Organizations of the U.S., 1993*, Detroit: Gale Research.

Edson, L. (1989), *The Making of the NAE, the First 25 Years*, Washington, DC: National Academy of Engineering.

European Physical Society (1992), *Europhys. News* **23**, 121 (Directory issue).

Evan, William M. (Ed.) (1981), *Knowledge and Power in a Global Society*, Beverly Hills, CA: Sage Publications.

Gibbs, N. E. (1991), "Software Engineering and Computer Science: the Impending Split," in: *Education and Computing*, Amsterdam: North-Holland, pp. 111–117.

Institute of Physics Secretariat (1993), *IOP Diary, Membership Brochure*, London: Institute of Physics.

International Union of Pure and Applied Physics Secretariat (1992), *IUPAP 1922–1992, Anniversary Brochure*, Geneva: IUPAP.

Irvin, Linda (Ed.) (1993), *Encyclopedia of Associations: International Organizations, 1993*, Detroit: Gale Research.

McMahon, A. M. (1984), *The Making of a Profession: A Century of Electrical Engineering in America*, New York: IEEE Press.

National Research Council (1987), *Directions in Engineering Research: An Assessment of Opportunities and Needs*, Washington, DC: National Academy of Sciences.

Phillips, Melba (1990), *Am. J. Phys.* **58**, 219–227.

Phillips, Melba, Wilson, Jack M. (1989), *AAPT Announcer Suppl.*, Sept.

Physics Survey Committee (1972), *Physics in Perspective*, NAS–NRC Report, Washington, DC: National Academy of Sciences.

Reynolds, T. S. (1983), *75 Years of Progress—A History of the American Institute of Chemical Engineers, 1908–1983*, New York: AIChE.

Reynolds, T. S., Seely, B. E. (1993), "Striving for Balance: A Hundred Years of the American Society for Engineering Education," *J. Eng. Educ.* **82**, 136–151.

Robbins, P. H. (1984), *Building for Professional Growth*, Washington, DC: National Society of Professional Engineers.

Ryder, J. D., Fink, D. G. (1984), *Engineers and Electrons, a Century of Electrical Progress*, New York: IEEE Press.

Tell, Norman, Chadwick, John (Eds.) (1976), *Professional Organizations in the Commonwealth*, London: Hutchinson.

Toth, R. B. (Ed.) (1991), *Standards Activities of Organizations in the United States*, NIST Special Publication No. 806, Washington, DC: U.S. G.P.O.

Wang Yulong (Ed.) (1990), *Directory of Chinese Learned Societies*, Berlin: World Publishing.

Wisely, W. H. (1974), *The American Civil Engineer, 1852–1974*, New York: American Society of Civil Engineers.

Wolfle, Dael (1989), *Renewing a Scientific Society*, Washington, DC: American Association for the Advancement of Science.

Further Reading

There is not a well-defined body of literature on the various types of engineering organizations discussed in this article. A description of the historical context in which the engineering societies emerged is given by Haber, Samuel (1991), *The Quest for Authority and Honor in the American Professions, 1750–1900*, Chicago: Univ. of Chicago Press. He devotes a chapter to "The Engineers: A Profession in a Business Bureaucracy."

One of the series of reports issued by the Carnegie Commission on Science, Technology, and Government (1993), *Facing Toward Governments, Nongovernmental Organizations and Scientific and Technical Advice*, Washington, DC: Carnegie Commission, views the activities of the many types of nongovernmental organizations (NGOs) as policy advisers to government. The engineering organizations discussed in this article fit under the broad NGO rubric.

Most of the engineering professional societies, typically on a special anniversary of their founding, have commissioned the writing of a history of the organization. Many of these are long on organizational aspects and short on technical development of their field.

There are a large number of directories that provide detailed information about organizations. Some contain lists of engineering-related organizations. There are two such directories that give detailed information about the engineering-related organizations discussed in this article. One is the *International Directory of Engineering Societies and Related Organizations 1993*, 14th ed., Washington, DC: American Association of Engineering Societies. This edition includes updated data about 950 organizations in 70 countries. The second is a more general directory of some 7000 entries including trade associations and labor unions as well as professional and scientific societies: *National Trade and Professional Associations of the United States*, Washington, DC: Columbia Books, published annually. By scanning such directories, a reader can gain a feeling for the number, diversity, and scope of currently existing organizations related to engineering.

CONTENTS OF PREVIOUS VOLUMES

Volume 1

03-A. Accelerators, Linear 1
03-A. Accelerators, Potential-Drop .. 27
11-D. Acoustic Properties of Liquids 43
02-A. Acoustical Instrumentation ... 63
07-C. Acoustical Tomography 89
07-E. Acoustics, Architectural 109
07-E. Acoustics, Linear 137
07-E. Acoustics, Nonlinear 183
07-E. Acoustics, Physiological 213
07-E. Acoustics, Psychological 261
07-D. Acoustics, Underwater 293
09-E. Aerodynamics 329
20-E. Aeronautics 379
17-B. Aerosols 415
19-E. Aerospace Medical Physics 443
18-C. Air-Pollution: Components, Causes, and Cures 489
01-C. Algebraic Methods 515
16-B. Aluminum 539
10-B. Amorphous Materials, Structure of 559
05-A. Amplifiers, Electronic 593
05-A. Amplifiers, Optical 607
01-C. Analytic Methods 623

Volume 2

01-E. Artificial Intelligence 1
20-E. Astronautics 25
20-A. Astronomical Telescopes 65
20-E. Astronomy 99
20-E. Astrophysics 117
07-D. Atmospheric Acoustics 145
20-E. Atmospheric Physics 181
20-D. Atmospheric Structure 201
04-C. Atomic Cooling and Trapping 225
04-C. Atomic Spectroscopy 245
04-D. Atoms 285
17-C. Auger Spectroscopy 297
18-A. Automotive Propulsion Systems 323
03-A. Betatrons 347
19-D. Biological Effects of Electromagnetic and Particle Radiation 365
19-D. Biological Effects of Sound and Ultrasound 403
19-D. Biomagnetism and Geomagnetic Field Detection by Organisms 421
19-C. Biomagnetism, Medical Aspects 453
18-C. Biomass Energy 473
19-E. Biomechanics of Living Organisms 489
19-E. Biomedical Engineering 513
19-C. Biomedical Uses of Radiation 549
19-E. Biophysics 571
19-D. Biorheology 609
02-A. Bus Systems and Computer Interfacing 639
07-E. Erratum: Acoustics, Physiological 653

Volume 3

02-C. Calibration and Maintenance of Test and Measuring Equipment 1
16-B. Carbon Materials 21
20-E. Cartography 41
17-D. Catalysis 67
01-E. Catastrophe Theory 85
14-D. Cathodoluminescence 121
19-E. Cellular Biomechanics 141
16-B. Ceramics, Material Properties of................ 169
08-D. Chaotic Phenomena 189
16-C. Characterization and Analysis of Materials........ 215
05-A. Charge-Coupled Devices..... 241
05-E. Charged-Particle Optics 273
17-C. Chemical Analysis 307
17-D. Chemical Kinetics 345
17-D. Chemical Reactions......... 377
17-D. Chemiluminescence 413
17-C. Chromatography............ 429
05-A. Circuit Elements............ 455
05-A. Circuits, Analog 479
05-A. Circuits, Digital 507
18-B. Coal....................... 527
12-D. Cohesive Energy 559
12-D. Collective Phenomena in Solids 579

Volume 4

17-D. Combustion 1
16-B. Composite Materials 17
09-D. Compressible Flows.......... 43
01-C. Computer-Aided Design in Electronics 71
01-C. Computer-Aided Manufacturing............... 89
01-C. Computer Databases 103
01-C. Computer Graphics 127
01-A. Computer Hardware 145
01-C. Computer Programming Languages.................. 163
01-A. Computers 189
17-C. Conformational Analysis 229
02-D. Constants, Fundamental 243
16-B. Copper 267
08-D. Critical Phenomena 277
02-A. Cryogenics 311
16-D. Crystal Growth 335
10-B. Crystalline State............ 365
10-C. Crystallography............. 385
16-C. CVD (Chemical Vapor Deposition) 409
03-A. Cyclotrons.................. 427
02-C. Data Acquisition............ 463
02-A. Detectors, Particle, Calorimetric 485
02-A. Detectors, Scintillation...... 503
02-A. Detectors, Semiconductor 513
17-C. Dialysis 533
13-D. Diamagnetism 557
16-B. Erratum: Carbon Materials 573

Volume 5

16-B. Diamond and Diamondlike Carbon 1
14-D. Dielectric Properties of Insulators 25
08-A. Diesel Engines 47
11-D. Diffusion and Ionic Conduction in Liquids 61
11-D. Diffusion in Thin Films 75
17-B. Disperse Systems 87
02-A. Display Technology 101
20-B. Earth, Interior Structure of . 127
18-E. Electric Power Engineering . 149
19-D. Electrical Conduction and Dielectric Behavior in Biological Systems 177
02-A. Electrical Instrumentation .. 201
17-E. Electrochemistry 223
05-E. Electrodynamics, Classical .. 259
14-D. Electroluminescence 295
02-A. Electromagnetic Quantities, Basic, Measurement of 327
05-D. Electromagnetic Radiation .. 345
05-D. Electromagnetic Wave Propagation 379
10-C. Electron Diffraction 405
12-D. Electron Level Splitting 431
10-C. Electron Microscopy 453
04-D. Electron Paramagnetic Resonance 475
04-D. Electron Scattering by Atoms and Molecules 499
15-D. Electron States in Zero-, One-, and Two-Dimensional Structures 531
12-D. Electron States: Localized .. 549
12-D. Electron Structure of Liquids 571
12-D. Electron Structure of Solids 595

Volume 6

05-A. Electronic Circuits 1
05-C. Electronic Components and Assemblies: Reliability and Testing 21
04-D. Electronic Structure of Atoms and Molecules 45
15-D. Electronic Structure of Surfaces 99
12-D. Electronic Transport in Solids 121
10-D. Electrorheology 145
05-A. Electrostatic Capacitative Energy Storage 155
05-E. Electrostatics 177
02-A. Ellipsometers 191
19-C. Energetics of Biological Processes 207
18-C. Energy Conversion 229
07-E. Engineering Acoustics 245
18-E. Environmental Health and Safety 271
11-D. Equations of State 291
12-D. Excitons 311
17-B. Explosives 327
06-B. Fiber Optics 359
12-D. Field Emission and Field Ionization 379
10-C. Field Ion Microscopy 393
09-E. Fluid Dynamics, Equilibrium 409
09-E. Fluid Dynamics, Non-equilibrium 437
01-C. Fourier & Other Mathematical Transforms ... 481
01-D. Fractal Geometry 501
16-B. Fullerenes 515
09-C. Fusion, Inertial Confinement 545
09-C. Fusion, Magnetic Confinement 575

Volume 7

09-A Fusion Technologies 1
20-E Galaxies and Cosmology 21
16-B Gallium Arsenide 43
01-C Game Theory 69
20-E Geodesy 89
20-D Geoelectromagnetism 109
01-C Geometrical Methods 125
20-E Geophysics 149
18-C Geothermal Energy 197
16-B Germanium 219
20-D Glaciers and Ice Sheets 239
16-B Glasses 251
15-D Grain Boundaries 271
16-B Graphite 289
03-D Gravitation and General Relativity 303
01-C Green's Functions 341
01-C Group Theory 365
08-A Heat Engines and Refrigerators 383
08-D Heat Transfer 405
11-D Heat-Pulse Propagation in Solids 417
13-D Heavy-Fermion Phenomena . 435
15-D Heterostructures and Superlattices, Metallic 465
15-D Heterostructures and Superlattices, Semiconductor 477
02-C High-Pressure Techniques ... 495
07-C Holography, Acoustical 511
06-C Holography, Electron 531
06-C Holography, Optical 541
20-E Hydrology 563
19-C Imaging Techniques, Biomedical 581

Volume 8

05-A Inductors 1
01-E Information Science 21
12-D Insulators and Semiconductors, Conductivity in 41
10-B Insulators, Structure of 85
09-D Intense Particle Beams 103
14-D Interaction of Solids with Particles and Radiation 117
15-B Intercalation Compounds ... 133
02-A Interferometers and Interferometry 157
16-C Ion-Beam Modification of Materials 173
11-D Ionic Conduction and Diffusion in Solids 193
20-D Ionosphere 223
02-A Laboratory Instrumentation 247
06-D Laser Photochemistry 283
06-D Laser Physics 299
06-C Laser Technology 321
06-A Lasers, Dye, Technology and Engineering 331
06-A Laser, Free-Electron 353
06-A Lasers, Gas 371
06-C Lasers, Industrial Uses of ... 397
06-A Lasers, Semiconductor 425
06-A Lasers, Solid State 443
11-D Lattice Vibrations, Statistics of 465
06-A Light-Emitting Diodes 485
10-B Liquid Crystals, Structure of 515
10-B Liquids, Simple, Structure of 533
05-A Machines, Electrical 561
13-A Magnetic Devices 587

Volume 9

16-B Magnetic Materials 1
16-D Magnetic Ordering in Solids . 15
04-D Magnetic Resonance and Quadrupole Resonance, Nuclear 31
19-C Magnetic Resonance Imaging 47
02-A Magnetic Resonance Spectrometers 71
13-A Magnetic Storage of Information 95
09-D Magnetohydrodynamics 111
13-A Magneto-optical Devices 157
20-B Magnetospheres of the Earth and Planets 187
05-E Magnetostatics 207
13-A Magnetostrictive Materials and Devices 229
05-A Magnets 245
02-C Maintenance and Calibration of Laboratory Instrumentation 261
01-E Manufacturing Engineering . 279
20-E Marine Geophysics 305
02-A Mass and Density, Measurement of 335
16-D Materials Degradation 349
16-C Materials Preparation: Solids 365
16-C Materials Treatment 399
01-C Mathematical Modeling 417
09-A Measurement and Instrumentation of Flow 445
02-A Measurement of Magnetic Properties and Quantities ... 463
07-A Measurement, Acoustical 491
10-D Mechanical and Elastic Waves in Solids 515
10-D Mechanical Properties of Liquids 531
10-D Mechanical Properties of Solids 545
07-D Mechanical Vibration and Damping 561
03-D. Erratum: Gravitation and General Relativity 591
02-A. Erratum: Laboratory Instrumentation 591
06-A. Lasers, Dye, Technology and Engineering 591
15-B. Erratum: Intercalation Compounds 601

Volume 10

01-E Mechanics, Classical 1
19-C Medical Use of Lasers 33
01-A Memories 61
16-D Mesoscopic Systems 81
12-D Metal-Insulator Transitions . 113
10-B Metallic Glasses 129
16-E Metallurgy, Physical 141
12-D Metals and Alloys, Conductivity in 163
10-B Metals and Alloys, Structure of 199
20-E Meteorology and Climatology 215
02-E Metrology 239
05-A Microelectronics 253
15-C Microlithography 281
01-A Microprocessors 297
02-A Microsensors 321
05-E Microwave Circuits 349
01-A Modulators and Demodulators, Electrical 379
01-A Modulators and Demodulators, Optical 393
04-D Molecular and Atomic Clusters 411
04-D Molecular and Atomic Collison Processes with Ion Beams 437
15-C Molecular Beam Epitaxy 471
04-C Molecular Spectroscopy 491
04-D Molecules 525
06-A Monochromators 553
01-C Monte-Carlo Methods 567

Volume 11

13-D	Mössbauer Effect	1
02-C	Muon Spin Rotation/ Relaxation/Resonance	23
04-D	Muonic, Mesonic, and Baryonic Atoms and Molecules	55
04-B	Muonium and Positronium	79
07-E	Music, Electronic	97
07-A	Musical Instruments	129
16-B	Nanophase Materials	173
18-B	Natural Gas	201
05-C	Network Theory, Electrical	221
01-C	Networks, Computer	263
01-C	Neural Networks	275
19-E	Neurobiophysics	297
04-D	Neutral Atomic and Molecular Collision Processes	323
10-C	Neutron Diffraction	339
10-C	Neutron Scattering	353
09-D	Nonhomogeneous Flows	379
01-D	Nonlinear Systems	417
18-C	Nuclear Energy, Fission	429
18-B	Nuclear Fuels and Isotopes	479
19-E	Nuclear Medicine	513
03-D	Nuclear Reactions	543
03-D	Nuclear Structure	571

Volume 12

18-C	Nuclear Waste Management	1
19-B	Nucleic Acids	19
20-E	Oceanography	47
20-B	Oil Shales and Tar Sands	77
01-C	Operations Research	101
06-C	Optical Communications	119
06-A	Optical Components and Systems	157
01-A	Optical Computing	177
06-A	Optical Filters	195
02-A	Optical Instrumentation	215
06-A	Optical Interconnections	243
06-C	Optical Microscopy	259
14-D	Optical Properties of Solids	285
06-A	Optical Scanners	337
01-A	Optical Storage	369
19-C	Optical Tomography	395
06-D	Optics, Atmospheric	405
06-E	Optics, Geometrical	435
06-E	Optics, Linear	451
06-E	Optics, Nonlinear	487
19-E	Optics, Physiological	541
06-D	Optics, Underwater	571

LIST OF RECOMMENDED UNITS AND SYMBOLS

RECOMMENDED UNITS AND CONVERSION FACTORS

The SI system provides six basic units: meter m, kilogram kg, second s, ampere A, kelvin K, candela cd, and mole mol.

Some important derived units are also allowed and bear special names, e.g.:

1 N (newton) = 1 kg m s^{-2}
1 J (joule) = 1 N m = 1 kg m^2 s^{-2}
1 W (watt) = 1 J s^{-1} = 1 kg m^2 s^{-3}
1 Pa (pascal) = 1 N m^{-2} = 1 kg m^{-1} s^{-2}

For mass, the gram g, or the metric ton t which equals 1000 kg, may be used instead of kilogram kg.

The so-called "long ton" (UK) and "short ton" (US) have been abandoned and will not be used in the *Encyclopedia of Applied Physics*.

For pressure, the bar (name and symbol alike) may be used, and for temperature, the degree celsius °C.

From all of these, decimal multiples or fractions can be derived:

Power of ten	Prefix	Symbol	Power of ten	Prefix	Symbol
10	deca	da	10^{-1}	deci	d
10^2	hecto	h	10^{-2}	centi	c
10^3	kilo	k	10^{-3}	milli	m
10^6	mega	M	10^{-6}	micro	μ
10^9	giga	G	10^{-9}	nano	n
10^{12}	tera	T	10^{-12}	pico	p
10^{15}	peta	P	10^{-15}	femto	f
10^{18}	exa	E	10^{-18}	atto	a

For mass, multiples or fractions are derived from g, not from kg (since the latter already contains a prefix), e.g., mg.

Units with the prefixes are considered as one entity and can, therefore, be raised to any power, e.g., cm^3.

The liter is now considered as synonymous with dm^3 (which does not hold in the older literature!). Please use the capital letter L as unit symbol (following a recent IUPAC recommendation). The use of the Ångström unit is discouraged; it should be replaced by fractional meters (1 Å = 100 pm = 0.1 nm).

SELECTED QUANTITIES, UNITS, AND SYMBOLS

Name of quantity[a]	Symbol[b]	SI unit[c]	Name of unit	Other units
Space and time				
length*	l	m	meter[d]	
breadth, width	b	m		
height	h	m		
radius	r	m		
thickness	d	m		
area	A,S	m^2		a (are) h (hectare)
volume	V	m^3		L, l(liter)[e]
plane angle	$\alpha,\beta,\gamma,$ ϑ,φ	1, rad	radian	° (degree) ′ (minute) ″ (second)
solid angle	ω,Ω	1, sr	steradian	
wavelength	λ	m		
wave number	σ,ν	m^{-1}		
time*	t	s	second	min (minute) h (hour) d (day)
frequency	ν, f	s^{-1}		Hz (hertz)[f]
relaxation time	τ	s		
velocity	u,v	$m\ s^{-1}$		km/h
acceleration	a	$m\ s^{-2}$		
Mechanics				
mass*	m	kg	kilogram	g (gram) t (tonne)[g]
(mass) density	ρ	$kg\ m^{-3}$		g/cm^3
momentum	$\mathbf{p}$	$kg\ m\ s^{-1}$		
angular momentum	$\mathbf{L}$	$kg\ m^2\ s^{-1}$		
force	$\mathbf{F}$	N	newton[f]	
moment of force	$\mathbf{M}$	N m		
weight	G, W	N		
pressure	p	Pa	pascal	bar (bar)[h]
energy	E, W	J	joule	W h (watt hour)[i] eV (electron volt)[j]
work	W,A	J		
power	P	W	watt	J/s, V A[k]
Molecular Physics and Thermodynamics				
thermodynamic temperature*	T	K	kelvin	°C (degrees Celsius)
Celsius temperature	ϑ,t			°C
number of entities	N			
Avogadro constant[l]	N_A,L	mol^{-1}	(particles) per mole	
Boltzmann constant[m]	k,k_B	$J\ K^{-1}$		
Planck constant[n]	h	J s		
(molar) gas constant[o]	R	$J\ mol^{-1}\ K^{-1}$		
(quantity of) heat	Q	J		
entropy[p]	S	$J\ K^{-1}$		
internal energy[p]	U	J		
Helmholtz function,[p] (Helmholtz) free energy, Helmholtz energy	F,A	J		
enthalpy[p]	H	J		
Gibbs function,[p] (Gibbs) free energy, Gibbs energy	G	J		
heat capacity[p]	C_p,C_V	$J\ K^{-1}$		

Name of quantity[a]	Symbol[b]	SI unit[c]	Name of unit	Other units
Chemical Physics				
amount of substance*	n	mol	mole	
relative atomic mass	A_r	1		
relative molecular mass	M_r	1		
atomic mass constant	m_u	kg		u (atomic mass unit)[q]
mass of a portion (of substance B)	$m_B, m(B)$	kg		g (gram)
molar mass (of substance B)	$M_B, M(B)$	kg mol^{-1}		
concentration (of substance B)	$c_B, c(B)$	mol m^{-3}		mol/L
mole fraction[r] (of substance B)	$\kappa_B, \kappa(B)$	1		
mass fraction (of substance B)	$\omega_B, \omega(B)$	1		%,‰,ppm,ppb
volume fraction (of substance B)	$\varphi_B, \varphi(B)$ $\phi_B, \phi(B)$	1		%,‰,ppm,ppb
mass concentration	ρ	kg m^{-3}		g/L
molality		mol kg^{-1}		mmol/kg
volume concentration[s]	σ	1		
molar volume	V_m	m^3 mol^{-1}		L/mol
molar heat capacity	C_m	J mol^{-1} K^{-1}		
molar conductivity	Λ_m	S m^2 mol^{-1}	(S: siemens)	
Faraday constant[t]	F	C mol^{-1}		
Electricity and Magnetism				
quantity of electricity[u]	Q	C	coulomb	
charge density	ρ	C m^{-3}		
electric potential	ϕ, V	V	volt	
electric potential difference, voltage	$U, \Delta\phi, \Delta V$	V		
electric dipole moment	$\mathbf{p}, \mathbf{p}_c$	C m		
electric current*	I	A	ampere	
electric current density	j	A m^{-2}		
electric field strength	$\mathbf{E}$	V m^{-1}		
electric displacement	$\mathbf{D}$	C m^{-2}		
capacitance	C	F	farad	
permittivity	ε	F m^{-1}		
relative permittivity	ε	1		
dielectric polarization	$\mathbf{P}$	C m^{-2}		
electric susceptibility	χ_c	1		
polarization (of a particle)	α	m^2 C V^{-1}		
magnetic flux	Φ	Wb	weber	
magnetic flux density	$\mathbf{B}$	T	tesla	
magnetic field strength	$\mathbf{H}$	A m^{-1}		
permeability	μ	H m^{-1}, N A^{-2}	(H: henry)	
relative permeability	μ_r	1		
magnetization	M	A m^{-1}		
magnetic susceptibility	χ	1		
molar magnetic susceptibility	χ_m	m^3 mol^{-1}		
(electrical) resistance	R	Ω	ohm	
(electrical) conductance	G	S	siemens	
(electrical) resistivity	ρ	Ω m		
(electrical) conductivity	κ,o	S m^{-1}		
self-inductance	L	H	henry	
Radiation				
radiant energy	Q, W, Q_c	J		
luminous intensity*	I	cd	candela	
radiant intensity	I_c	W sr^{-1}, W		
emissivity, emittance	ε	1		

Name of quantity[a]	Symbol[b]	SI unit[c]	Name of unit	Other units
absorptance	α	1		
reflectance	ρ, R	1		
transmittance	τ	1		
absorption coefficient:				
linear (decadic)	a	m^{-1}		
molar (decadic)	ε	$m^2\ mol^{-1}$		
refractive index	n	1		
molar refraction	R_m	$m^3\ mol^{-1}$		
angle of optical rotation	α	1, rad		
Transport Properties				
flux of quantity X	J_X, J	(varies)		
mass flow rate	q_m, $\dot{m}$	$kg\ s^{-1}$		
volume flow rate	q_V, $\dot{V}$	$m^3\ s^{-1}$		
heat flow rate	Φ	W		
thermal conductivity	κ, k, λ	$W\ m^{-1}\ K^{-1}$		
coefficient of heat transfer	h	$W\ m^{-2}\ K^{-1}$		
thermal diffusivity	a	$m^2\ s^{-1}$		
diffusion coefficient	D	$m^2\ s^{-1}$		
thermal diffusion coefficient	D_T	$m^2\ s^{-1}$		
viscosity	η, μ	Pa s		
kinematic viscosity	ν	$m^2\ s^{-1}$		

[a]SI base quantities are marked by asterisks (*)

[b]Recommended by IUPAC

[c]SI base units as well as derived and supplementary units are listed; all are to be used with prefixes as needed

[d]do not use "metre"

[e]do not use "litre"; $1\ L = 10^{-3}\ m^3$

[f]$1\ Hz = 1\ s^{-1}$; $1\ N = 1\ kg\ m\ s^{-2}$

[g]Formerly metric ton; $1\ t = 10^3\ kg$

[h]$1\ bar = 10^5\ Pa$

[i]$1\ W\ h = 3.6 \times 10^3\ J$

[j]$1\ eV = 1.602\ 189 \times 10^{-19}\ J$

[k]$1\ W = 1\ J/s = 1\ VA$

[l]$N_A = 6.022\ 136\ 7 \times 10^{23}\ mol^{-1}$

[m]$k = 1.380\ 658 \times 10^{-23}\ J\ K^{-1}$

[n]$h = 6.626\ 075\ 5 \times 10^{-34}\ J\ s$

[o]$R = 8.314\ 510\ J\ mol^{-1}\ K^{-1}$

[p]Molar quantities can be distinguished from the quantity of a system by adding the subscript m; e.g., molar internal energy U_m, in $J\ mol^{-1}$

[q]$1\ u = 1.660\ 565\ 5 \times 10^{-27}\ kg$

[r]A more accurate, but rather uncommon, name is "amount-of-substance fraction"

[s]σ refers to the total volume of a mixture, whereas the volume fraction φ relates the volume of a substance to the volume of several components before mixing

[t]$F = 9.648\ 530\ 9 \times 10^4\ C\ mol^{-1}$

[u]Also called electric charge